Essentials of Genetics

SEVENTH EDITION

William S. Klug
The College of New Jersey

Michael R. Cummings
Illinois Institute of Technology

Charlotte A. Spencer
University of Alberta

Michael A. Palladino
Monmouth University

with contributions by
Sarah M. Ward, Colorado State University

Benjamin Cummings
Boston Columbus Indianapolis New York San Francisco Upper Saddle River
Amsterdam Cape Town Dubai London Madrid Milan Munich Paris Montréal Toronto
Delhi Mexico City São Paulo Sydney Hong Kong Seoul Singapore Taipei Tokyo

Editor-in-Chief: Beth Wilbur
Executive Editor: Gary Carlson
Executive Director of Development: Deborah Gale
Project Editor: Dusty Friedman
Editorial Assistant: Lindsay White
Managing Editor: Michael Early
Production Supervisor: Lori Newman
Senior Media Producer: Laura Tommasi
Design Manager: Marilyn Perry
Production Management and Composition: Prepare Inc.
Interior Designer: Hespenheide Design
Cover Designer: Seventeenth Street Studios
Illustrators: Imagineering Media Services, Inc.
Photo Researcher: Yvonne Gerin
Image Rights and Permissions Manager: Zina Arabia
Image Permissions Coordinator: Debbie Hewitson
Manufacturing Buyer: Michael Penne
Director of Marketing: Christy Lawrence
Executive Marketing Manager: Lauren Harp

Credits and acknowledgments borrowed from other sources and reproduced, with permission, in this textbook appear on p. C-1.

ISBN 10: 0-321-66999-1
ISBN 13: 978-0-321-66999-5

1 2 3 4 5 6 7 8 9 10—CRK—13 12 11 10 09
Manufactured in the United States of America.

Benjamin Cummings
is an imprint of
PEARSON

About the Authors

William S. Klug is currently Professor of Biology at the College of New Jersey (formerly Trenton State College) in Ewing, New Jersey. He served as Chair of the Biology Department for 17 years, a position to which he was first elected in 1974. He received his B.A. degree in Biology from Wabash College in Crawfordsville, Indiana, and his Ph.D. from Northwestern University in Evanston, Illinois. Prior to coming to the College of New Jersey, he was on the faculty of Wabash College as an Assistant Professor. His research interests have involved ultrastructural and molecular genetic studies of oogenesis in *Drosophila.* He has taught the genetics course as well as the senior capstone seminar course in human and molecular genetics to undergraduate biology majors for each of the last 35 years. In 2002, he was the recipient of the initial teaching award given at the College of New Jersey granted to the faculty member who most challenges students to achieve high standards. He also received the 2004 Outstanding Professor Award from the Sigma Pi International, and in the same year, he was nominated as the Educator of the Year, an award given by the Research and Development Council of New Jersey.

Michael R. Cummings is currently Research Professor in the Department of Biological, Chemical and Physical Sciences at Illinois Institute of Technology, Chicago, Illinois. For more than 25 years, he was a faculty member in the Department of Biological Sciences and in the Department of Molecular Genetics at the University of Illinois at Chicago. He has also served on the faculties of Northwestern University and Florida State University. He received his B.A. from St. Mary's College in Winona, Minnesota, and his M.S. and Ph.D. from Northwestern University in Evanston, Illinois. In addition to ***Essentials of Genetics*** and its companion volumes, he has also authored several texts in human genetics and general biology for nonmajors. His research interests center on the molecular organization and physical mapping of the heterochromatic regions of human acrocentric chromosomes. At the undergraduate level, he teaches courses in Mendelian and molecular genetics, human genetics, and general biology, and has received numerous awards for teaching excellence given by university faculty, student organizations, and graduating seniors.

Charlotte A. Spencer is currently Adjunct Associate Professor in the Department of Oncology at the University of Alberta in Edmonton, Alberta, Canada. She has also served as a faculty member in the Department of Biochemistry at the University of Alberta. She received her B.Sc. in Microbiology from the University of British Columbia and her Ph.D. in Genetics from the University of Alberta, followed by postdoctoral training at the Fred Hutchinson Cancer Research Center in Seattle, Washington. Her research interests involve the regulation of RNA polymerase II transcription in cancer cells, cells infected with DNA viruses, and cells traversing the mitotic phase of the cell cycle. She has authored booklets in the Prentice Hall Exploring Biology series.

Michael A. Palladino is Dean of the School of Science and Associate Professor of Biology at Monmouth University in West Long Branch, New Jersey. He received his B.S. degree in Biology from Trenton State College (now known as The College of New Jersey) and his Ph.D. in Anatomy and Cell Biology from the University of Virginia. He directs an active laboratory of undergraduate student researchers studying molecular mechanisms involved in innate immunity of mammalian male reproductive organs and genes involved in oxygen homeostasis and ischemic injury of the testis. He has taught a wide range of courses for both majors and nonmajors and currently teaches genetics, biotechnology, endocrinology, and laboratory in cell and molecular biology. He has received several awards for research and teaching, including the 2009 Young Investigator Award from the American Society of Andrology, the 2005 Distinguished Teacher Award from Monmouth University, and the 2005 Caring Heart Award from the New Jersey Association for Biomedical Research. In addition to *Essentials of Genetics* and its companion volumes, he is coauthor of the undergraduate textbook *Introduction to Biotechnology,* Series Editor for the Benjamin Cummings Special Topics in Biology booklet series, and author of the first booklet in the series, *Understanding the Human Genome Project.*

Brief Contents

Contents

Preface

Essentials of Genetics is written for courses requiring a text that is shorter and more basic than its more comprehensive companion, *Concepts of Genetics*. While coverage is thorough and modern, *Essentials* is written to be more accessible to biology majors early in their undergraduate careers, as well as to students majoring in a number of other disciplines, including agriculture, animal husbandry, chemistry, engineering, forestry, psychology, and wildlife management. Because the text is shorter than many other books, *Essentials of Genetics* is also more manageable in one-quarter and one-semester courses.

Goals

In this edition of *Essentials of Genetics,* one of the most important goals has been to make the text more affordable to students by reducing its length by about 30 pages. We subsequently expect the text to be priced 25% lower than its predecessor. Our other overarching goals remain the same as for previous editions. Specifically, we seek to

- Emphasize concepts rather than excessive detail.
- Write clearly and directly to students in order to provide understandable explanations of complex analytical topics.
- Maintain careful organization within and between chapters.
- Propagate the rich history of genetics that so beautifully elucidates how information is acquired as the discipline develops and grows.
- Emphasize problem solving, thereby guiding students to think analytically and to apply and extend their knowledge of genetics.
- Provide the most modern and up-to-date coverage of this exciting field.
- Create inviting, engaging, and pedagogically useful full-color figures enhanced by meaningful photographs to support concept development.
- Provide outstanding Online Media Tutorials to guide students in understanding important concepts through animations, tutorial exercises, and self-assessment tools.

These goals serve as the cornerstone of *Essentials of Genetics*. This pedagogic foundation allows the book to accommodate courses with many different approaches and lecture formats. While the book presents a coherent table of contents that represents one approach to offering a course in genetics, chapters are nevertheless written to be independent of one another, allowing instructors to utilize them in various sequences.

New to This Edition

The Seventh Edition has several new and enhanced features as well as updated content, making this edition even more pedagogically useful for students of genetics. Collectively, these improvements create a platform that seeks to challenge students to think more deeply about, and thus understand more comprehensively, the information he or she has just finished studying.

- **Exploring Genomics** New to this edition, *Exploring Genomics* appears in 12 chapters. Its presence is in keeping with the impact of genomics on most aspects of genetics. Each entry asks students to access one or more genomics-related Web sites that collectively are among the best publicly available resources and databases. Students work through interactive exercises that ensure their familiarity with the type of genomic or proteomic information available. Exercises instruct students on how to explore specific topics and how to access significant data. Questions guide student exploration and challenge them to further explore the sites on their own. Importantly, *Exploring Genomics* integrates genomics information throughout the text, as this emerging field is linked to chapter content. This feature provides the basis for individual or group assignments in or out of the classroom.

- **Case Study** A new feature called *Case Study* appears in each chapter and provides the basis for enhanced classroom interactions. In each entry, a short scenario related to one of the chapter topics is presented, followed by several questions. These ask students to apply their newly acquired knowledge to real-life issues that may be explored in small-group discussions or serve as individual assignments.

- **Essential Points** Within each chapter, at the conclusion of the coverage of all major topics, we have created a new feature called *Essential Points.* Found numerous times throughout each chapter, these provide important summaries that succinctly review the central focus of the previous coverage. By integrating these within the text of each chapter, they serve as "signposts" for students as they read or review each chapter.

- **Genetics, Technology, and Society Essays** We continue to offer these highly relevant essays, but in a new "interactive" format. After each essay, a new section entitled "Your Turn" appears in which questions are posed to students along with various resources to help answer them. This innovation provides yet another format to enhance classroom interactions. While all previous topics have been revised and updated, two new essays have been added:

"Down Syndrome, Prenatal Testing, and the New Eugenics" (Chapter 6) and "Personalized Genome Projects and the Race for the $1000 Genome" (Chapter 19).

- **Modernization of Topics** As with every new edition, our major efforts have been directed toward updating coverage of all aspects of the text. Special attention was paid to genomics and proteomics (Chapter 18), as well as to all areas of molecular genetics, including gene regulation (Chapter 15) and cancer genetics (Chapter 16). Sarah Ward again invigorated our coverage of conservation genetics (Chapter 24).
- **New Chapter** In keeping with our intent to present in an accessible fashion highly relevant, emerging areas of genetics, we have created a new chapter "Genetics and Behavior" (Chapter 21). The chapter reflects our growing knowledge of the role genes play in all aspects of the behavior of organisms. This topic will be as interesting to students because the findings surrounding it intersect our knowledge of our own species.
- **Updated Instructor and Student Media** The instructor and student media have been updated to reflect the changes made in this new edition. Support for lecture presentations and other teaching responsibilities includes electronic access to text photos and tables and a great variety of PowerPoint offerings on the book's Instructor Resource DVD. Media found on the revamped Companion Website reflect the growing awareness that today's students must use their limited study time as wisely as possible. Also available on the student Web site will be questions that direct students to research articles to summarize, compare, or evaluate.

Emphasis on Concepts and Problem Solving

Essentials of Genetics focuses on conceptual issues in genetics and uses problem solving to develop a deep understanding of them. Our experience shows that students whose primary focus is on concepts more easily comprehend and take with them to succeeding courses the most important ideas in genetics. Most importantly, linking conceptual issues to problem solving provides an enhanced analytic view of biology. To aid students in identifying conceptual aspects of a major topic, each chapter begins with a section called "Chapter Concepts," which outlines the most important ideas about to be presented.

Another valued pedagogic feature, first introduced in the Sixth Edition, appears in each chapter: *How Do We Know?* Entries ask the student to identify the experimental basis of the most important concepts presented in the chapter. As an extension of the learning approach in biology called "Science as a Way of Knowing," the use of this feature enhances students' understanding of many key topics covered in each chapter.

The feature called *Now Solve This*, found numerous times throughout each chapter, asks students to link conceptual understanding in a more immediate way to problem-solving. Each one directs the student to a problem found at the end of the chapter that is closely related to the current text discussion. Pedagogic hints are provided to aid in solving the problem. This feature more closely links text discussions to the problems.

Beyond the above features, all chapters conclude with what has become a popular and successful section called "Insights and Solutions." This section provides sample problems and solutions that demonstrate approaches useful in genetic analysis. These help students develop analytical thinking and experimental reasoning skills. Digesting the information in "Insights and Solutions" primes students as they move on to the lengthier "Problems and Discussion Questions" section that concludes each chapter. Here, we present questions that review topics in the chapter and problems that ask students to think in an analytical and applied way about genetic concepts. Problems are of graduated difficulty, with the most demanding near the end of each section.

Acknowledgments

Reviewers

All comprehensive texts are dependent on the valuable input provided by many colleagues. The following individuals provided valuable advice, constructive criticism, and/or suggestions regarding the content of the Seventh Edition:

Althea K. Alton, *Western Illinois University*
Thomas H. Alton, *Western Illinois University*
Brian Ashburner, *University of Toledo*
Mark Brick, *Colorado State University*
Susan Capasso, *St. Vincent's College*
Aaron Cassill, *University of Texas – San Antonio*
Steve Denison, *Eckerd College*
John P. Doucet, *Nicholls State University*
Kurt Elliott, *Northwest Vista College*
Lehman L. Ellis, *Our Lady of Holy Cross College*
Victor Fet, *Marshall University*
Clarence E. Fouche, *Virginia Intermont College*
Gail Fraizer, *Kent State University*
Alexandros Georgakilas, *East Carolina University*
Edward F. Golenberg, *Wayne State University*
John Gray, *University of Toledo*
Danielle Hamill, *Ohio Wesleyan University*
David Kass, *Eastern Michigan University*
Mary Kimble, *Northeastern Illinois University*
Joan Kuh, *University of Hawaii*

Jospeh Kulkosky, *Chestnut Hill College*

Alan C. Leonard, *Florida Institute of Technology*

Janet Lewis, *Michigan State University*

Jeannette M. Loutsch, *University of Science and Arts of Oklahoma*

Roy B. Mason, *Mt. San Jacinto College*

Shawn Meagher, *Western Illinois University*

Philip McClean, *North Dakota State University*

Sudhir Nayak, *The College of New Jersey*

Harry Nickla, *Creighton University*

Phillip A. Ortiz, *Empire State College*

Terrence Puryear, *Northeastern Illinois University*

Thomas F. Savage, *Oregon State University*

Brian W. Schwartz, *Columbus State University*

Thomas Smith, *Southern Arkansas University*

Allan Showalter, *Ohio University*

Tatiana Tatum, *Saint Xavier University*

Paul Wilson, *Trent University*

While we take full responsibility for any errors in this book, we gratefully acknowledge the help and input provided by the above individuals.

Contributors

We offer a special acknowledgment to those who have provided direct input during many revisions of this text. We thank: Sarah Ward at Colorado State University for her contribution of the conservation genetics chapter; Sudhir Nayak from The College of New Jersey for his input into all things genomic; David Kass at Eastern Michigan University, Katherine Uyhazi at Yale Medical School, and Tamara Mans at North Hennepin Community College for their numerous contributions to the Genetics, Technology, and Society essays; Elliott Goldstein from Arizona State University for his perpetual input into molecular genetics coverage, and Mike Guidry of LightCone Interactive and Karen Hughes of the University of Tennessee for their original contributions to the media program. We also offer special thanks to Harry Nickla, recently retired from Creighton University. In his role as author of the *Student Handbook* and the *Instructor's Manual,* he has written many new problems and authored the Selected Answers section that appear in Appendix A. We are grateful to all of the above contributors not only for sharing their genetic expertise, but for their dedication to this project as well as the pleasant interactions they provided.

Editorial and Production Input

At Benjamin-Cummings, we express appreciation and praise for Gary Carlson's editorial guidance. We also appreciate the production efforts of those at Preparé Inc., whose quest for perfection is reflected throughout the text, and the preparation and beauty of the art work, which is the product of Imagineering of Toronto.

Proofreading a manuscript of a 600-page text deserves more thanks than words can offer. Our utmost appreciation is extended to the four individuals who confronted this task with patience and diligence, including Michael Rossa, Susan Baglino, Cheryl Ingram-Smith, and Thomas H. Alton. In addition, Betty Pessagno successfully confronted the task of copyediting the entire manuscript. Her contribution to the overall quality of the text is very much appreciated.

Coordinating all of the above efforts was not a task for mere mortals, and thus we were fortunate to have had the talented oversight of Dusty Friedman, our Project Editor. Not only did she make everything work, but she always managed to maintain a much appreciated pleasant, unruffled demeanor.

As these many acknowledgments make clear, a text such as this is a collective enterprise. All of the above individuals deserve to share in any success this text enjoys. We want them to know that our gratitude is equaled only by the extreme dedication evident in their efforts.

For the Student

Companion Website—www.geneticsplace.com

The Companion Website, designed to enable users of the Seventh Edition to focus on those chapter sections and topics where they need review or further explanation, contains numerous in-depth tutorials that allow students to interactively comprehend genetic concepts. The Chapter Guide pages provide students with a focused, section-by-section access to the e-book, and questions that offer hints and answer feedback. Interactive Chapter Quizzes give students a way to test their compehension of chapter topics before exams. The grading function offers precise feedback and allows instructors to track their students' scores. The Website also contains RSS feeds to breaking news in genetics and links to related websites. The site also provides access via Pearson's **Research Navigator™** database to EBSCO, the world's leading online journal library, containing scholarly articles from over 79,000 publications. Online writing-focused **Research Navigator™** assignments allow students to evaluate and synthesize information from selected readings, then submit their work online directly to their instructor. Finally, *Exploring Genomics* sections from the text have been excerpted and placed on the Companion Website.

Online Web Tutorials

The most sophisticated learning and tutorial package available for students of genetics, these online tutorials address the concepts that students find most difficult. Each of the tutorials comprises animations and interactive exercises, along with a post-tutorial quiz of self-grading questions to reinforce important concepts. Found on the student Companion Website, the tutorials offer timely support for students. Look for this icon in the margin of this page to identify media tutorials that support concepts presented in the textbook.

WEB TUTORIAL

Study Guide and Solutions Manual

Harry Nickla, Creighton University ***(0-32-161870-X)***

This valuable handbook provides a detailed step-by-step solution or discussion for every problem in the text. It also features additional study aids, including extra study problems, chapter outlines, vocabulary exercises, and an overview of how to study genetics.

For the Instructor

Instructor Resource DVD

(0-32-161874-2)

The Instructor Resource DVD for the Seventh Edition offers adopters convenient access to a comprehensive and innovative set of teaching tools. Developed to meet the needs of veteran and newer instructors alike, these resources include:

- The JPEG files of all text line drawings with labels enhanced for optimal projection results (as well as unlabeled versions)
- Pedagogically important textbook photos and all textbook tables as JPEG files
- The JPEG files of line drawings, photos, and tables preloaded into comprehensive PowerPoint® presentations for each chapter
- PowerPoint® presentations containing a comprehensive set of in-class Classroom Response System (CRS) questions for each chapter.
- Over 100 Instructor Animations dynamically illustrating the most important concepts in the course preloaded into PowerPoint® presentations (PC and Macintosh versions).

Instructor Manual and Test Bank

Harry Nickla, Creighton University ***(0-32-161872-6)***

This manual and testbank contains over 1000 questions and problems for use in preparing exams. The manual also provides optional course sequences, a guide to audiovisual supplements, and a section on searching the Web. The testbank portion of the manual is also available in electronic format.

TestGen EQ Computerized Testing Software

(0-32-161873-4)

In addition to the printed volume, the test questions are also available as part of the TestGen EQ Testing Software, a text-specific testing program that is networkable for administering tests. It also allows instructors to view and edit questions, export the questions as tests, and print them out in a variety of formats.

Transparency Acetates

(0-32-161871-8)

Electronic versions of transparencies, in JPEG format, are available on the Instructor Resource DVD and instructors can copy them onto their desktop.

Newer model organisms in genetics include the roundworm C. elegans, *the plant* A. thaliana, *and the zebrafish,* D. rerio.

1

Introduction to Genetics

Chapter Concepts

- Transmission genetics is the general process by which traits controlled by factors (genes) are transmitted through gametes from generation to generation. Its fundamental principles were first put forward by Gregor Mendel in the mid-nineteenth century. Later work by others showed that genes are on chromosomes and that mutant strains can be used to map genes on chromosomes.
- The recognition that DNA encodes genetic information, the discovery of DNA's structure, and elucidation of the mechanism of gene expression form the foundation of molecular genetics.
- Recombinant DNA technology, which allows scientists to prepare large quantities of specific DNA sequences, has revolutionized genetics, laying the foundation for new fields—and for endeavors such as the Human Genome Project—that combine genetics with information technology.
- Biotechnology includes the use of genetically modified organisms and their products in a wide range of activities involving agriculture, medicine, and industry.
- Some of the model organisms used in genetics research since the early part of the twentieth century are now used in combination with recombinant DNA technology and genomics to study human diseases.
- Genetic technology is developing faster than the policies, laws, and conventions that govern its use.

In December 1998, following months of heated debate, the Icelandic Parliament passed a law granting deCODE Genetics, a biotechnology company with headquarters in Iceland, a license to create and operate a database containing detailed information drawn from medical records of all of Iceland's 270,000 residents. The records in this Icelandic Health Sector Database (or HSD) were encoded to ensure anonymity. The new law also allowed deCODE Genetics to cross-reference medical information from the HSD with a comprehensive genealogical database from the National Archives. In addition, deCODE Genetics would be able to correlate information in these two databases with results of deoxyribonucleic acid (DNA) profiles collected from Icelandic donors. This combination of medical, genealogical, and genetic information would be a powerful resource available exclusively to deCODE Genetics for marketing to researchers and companies for a period of 12 years, beginning in 2000.

This is not a science fiction scenario from a movie such as *Gattaca* but a real example of the increasingly complex interaction of genetics and society at the beginning of the twenty-first century. The development and use of these databases in Iceland have generated similar projects in other countries as well. The largest is the "UK Biobank" effort launched in Great Britain in 2003. There, a huge database containing the genetic information of 500,000 Britons will be compiled from an initial group of 1.2 million residents. The database will be used to search for susceptibility genes that control complex traits. Other projects have since been announced in Estonia, Latvia, Sweden, Singapore, and the Kingdom of Tonga, while in the United States, smaller-scale programs, involving tens of thousands of individuals, are underway at the Marshfield Clinic in Marshfield, Wisconsin; Northwestern University in Chicago, Illinois; and Howard University in Washington, D.C.

deCODE Genetics selected Iceland for this unprecedented project because the people of Iceland have a level of genetic uniformity seldom seen or accessible to scientific investigation. This high degree of genetic relatedness derives from the founding of Iceland about 1000 years ago by a small population drawn mainly from Scandinavian and Celtic sources. Subsequent periodic population reductions by disease and natural disasters further reduced genetic diversity there, and until the last few decades, few immigrants arrived to bring new genes into the population. Moreover, because Iceland's health-care system is state-supported, medical records for all residents go back as far as the early 1900s. Genealogical information is available in the National Archives and church records for almost every resident and for more than 500,000 of the estimated 750,000 individuals who have ever lived in Iceland. For all these reasons, the Icelandic data are a tremendous asset for geneticists in search of genes that control complex disorders. The project already has a number of successes to its credit. Scientists at deCODE Genetics have isolated genes associated with 12 common diseases including asthma, heart disease, stroke, and osteoporosis.

On the flip side of these successes are questions of privacy, consent, and commercialization—issues at the heart of many controversies arising from the applications of genetic technology. Scientists and nonscientists alike are debating the fate and control of genetic information and the role of law, the individual, and society in decisions about how and when genetic technology is used. For example, how will knowledge of the complete nucleotide sequence of the human genome be used? Will disclosure of genetic information about individuals lead to discrimination in jobs or insurance? Should genetic technology such as prenatal diagnosis or gene therapy be available to all, regardless of ability to pay? More than at any other time in the history of science, addressing the ethical questions surrounding an emerging technology is as important as the information gained from that technology.

This introductory chapter provides an overview of genetics in which we survey some of the high points of its history and give preliminary descriptions of its central principles and emerging developments. All the topics discussed in this chapter will be explored in far greater detail elsewhere in the book. Later chapters will also revisit the controversies alluded to above and discuss many other issues that are current sources of debate. There has never been a more exciting time to be part of the science of inherited traits, but never has the need for caution and awareness of social consequences been more apparent. This text will enable you to achieve a thorough understanding of modern-day genetics and its underlying principles. Along the way, enjoy your studies, but take your responsibilities as a novice geneticist very seriously.

1.1 Genetics Has a Rich and Interesting History

We don't know when people first recognized the existence of heredity, but archeological evidence (e.g., primitive art, preserved bones and skulls, and dried seeds) documents the successful domestication of animals and cultivation of plants thousands of years ago by artificial selection of genetic variants within populations. Between 8000 and 1000 B.C. horses, camels, oxen, and various breeds of dogs (derived from the wolf family) had been domesticated, and selective breeding soon followed. Cultivation of many plants, including maize, wheat, rice, and the date palm, began around 5000 B.C. Remains of maize dating to this period have been recovered in caves in the Tehuacan Valley of Mexico. Such evidence documents our ancestors' successful attempts to manipulate the genetic composition of species.

While few, if any, significant ideas were put forward to explain heredity during prehistoric times, during the Golden Age of Greek culture, philosophers wrote about this subject as it relates to humans. This is evident in the writings of the Hippocratic School of Medicine (500–400 B.C.), and of the philosopher and naturalist Aristotle (384–322 B.C.). The Hippocratic treatise *On the Seed* argued that active "humors" in various parts of the body served as the bearers of hereditary traits. Drawn from various parts of the male body to the semen and passed on to offspring, these humors could be healthy or diseased, the diseased condition accounting for the appearance of newborns with congenital disorders or deformities. It was also believed that these humors could be altered in individuals before they were passed on to offspring, explaining how newborns could "inherit" traits that their parents had "acquired" because of their environment.

Aristotle, who studied under Plato for some 20 years, extended Hippocrates' thinking and proposed that the generative power of male semen resided in a "vital heat" contained within it that had the capacity to produce offspring of the same "form" (i.e., basic structure and capacities) as the parent. Aristotle believed that this heat cooked and shaped the menstrual blood produced by the female, which was the "physical substance" that gave rise to an offspring. The embryo developed not because it already contained the parts in miniature (as some Hippocratics had thought) but because of the shaping power of the vital heat. Although the ideas of Hippocrates and Aristotle sound primitive and naive today, we should recall that prior to the 1800s neither sperm nor eggs had been observed in mammals.

1600–1850: The Dawn of Modern Biology

During the ensuing 1900 years (from 300 B.C. to A.D. 1600), our understanding of genetics was not extended by any new and significant ideas. However, between 1600 and 1850, major strides provided insight into the biological basis of life, setting the scene for the revolutionary work and principles presented by Charles Darwin and Gregor Mendel. In the 1600s, the English anatomist William Harvey (1578–1657) wrote a treatise on reproduction and development patterned after Aristotle's work. He is credited with the earliest statement of the theory of **epigenesis;** which posits that an organism is derived from substances present in the egg that differentiate into adult structures during embryonic development. Epigenesis holds that structures such as body organs are not initially present in the early embryo but instead are formed *de novo* (anew).

The theory of epigenesis directly conflicted with the theory of **preformation,** first proposed in the seventeenth century, which stated that sex cells contain a complete miniature adult, perfect in every form, called a **homunculus** (Figure 1–1). Although this popular theory was later discounted, other significant chemical and biological discoveries made during this same period affected future scientific thinking. Around 1830, Matthias Schleiden and Theodor Schwann proposed the **cell theory,** stating that all organisms are composed of basic units called cells, which are derived from similar preexisting structures. The idea of **spontaneous generation,** the creation of living organisms from nonliving components, was disproved by Louis Pasteur later in the century, and living organisms were considered to be derived from preexisting organisms and to consist of cells.

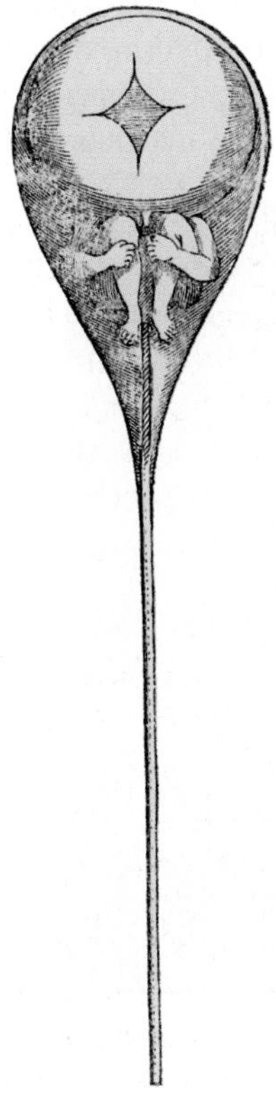

FIGURE 1–1 Depiction of the "homunculus," a sperm containing a miniature adult, perfect in proportion and fully formed. (Hartsoeker, N. Essay de dioptrique Paris, 1694, p. 230. National Library of Medicine)

Another influential notion prevalent in the nineteenth century was the **fixity of species.** According to this doctrine, animal and plant groups have remained unchanged in form since the moment of their appearance on Earth. This doctrine was particularly embraced by those who believed in special creation, including the Swedish physician and plant taxonomist Carolus Linnaeus (1707–1778), who is better known for devising the binomial system of species classification.

Charles Darwin and Evolution

With this background, we turn to a brief discussion of the work of Charles Darwin, who published the book-length statement of his evolutionary theory, *The Origin of Species*, in 1859. Darwin's geological, geographical, and biological observations convinced him that existing species arose by descent with modification from other ancestral species. Greatly influenced by his voyage on the HMS *Beagle* (1831–1836), Darwin's thinking culminated in his formulation of the theory of **natural selection,** which presented an explanation of the causes of evolutionary change. Formulated and proposed independently by Alfred Russel Wallace, natural selection was based on the observation that populations tend to consist of more offspring than the environment can support, leading to a struggle for survival among them. Those organisms with heritable traits that allow them to adapt to their environment are better able to survive and reproduce than those with less adaptive traits. Over a long period of time, slight but advantageous variations will accumulate. If a population bearing these inherited variations becomes reproductively isolated, a new species may result.

Darwin, however, lacked an understanding of the genetic basis of variation and inheritance, a gap that left his theory open to reasonable criticism well into the twentieth century. As Darwin's work ensued, Gregor Johann Mendel conducted experiments between 1856 and 1863. He published a paper in 1866 showing a number of statistical patterns underlying inheritance; he developed a theory involving hereditary factors in the germ cells to explain these patterns. His research was virtually ignored until it was partially duplicated and then cited by Karl Correns, Hugo de Vries, and Erich Tschermak around 1900, and subsequently championed by William Bateson.

By the early part of the twentieth century, support for the epigenetic interpretation of development had grown considerably. It gradually became clear that heredity and development were dependent on genetic information residing in genes contained in chromosomes, which were then contributed to each individual by gametes—the so-called **chromosomal theory of inheritance.** The gap in Darwin's theory had narrowed considerably, and Mendel's research has continued to serve as the foundation of genetics.

1.2 Genetics Progressed from Mendel to DNA in Less Than a Century

Because genetic processes are fundamental to life itself, the science of genetics unifies biology and serves as its core. The starting point for this exciting branch of science was a monastery garden in central Europe in the 1860s.

Mendel's Work on Transmission of Traits

Gregor Mendel, an Augustinian monk, conducted a decade-long series of experiments using pea plants. He applied quantitative data analysis to his results and showed that traits are passed from parents to offspring in predictable ways. He further concluded that each trait in the plant is controlled by a pair of genes and that during gamete formation (the formation of egg cells and sperm), members of a gene pair separate from each other. His work was published in 1866 but was largely unknown until it was partially duplicated and cited in papers published by others around 1900. Once confirmed, Mendel's findings became recognized as explaining the transmission of traits in pea plants and all other higher organisms. His work forms the foundation for **genetics,** which is defined as the branch of biology concerned with the study of heredity and variation. The story of Gregor Mendel and the beginning of genetics is told in an engaging book, *The Monk in the Garden: The Lost and Found Genius of Gregor Mendel, the Father of Genetics*, by Robin M. Henig. Mendelian genetics will be discussed in Chapters 3 and 4.

ESSENTIAL POINT

Mendel's work on pea plants established the principles of gene transmission from parent to offspring that serve as the foundation for the science of genetics.

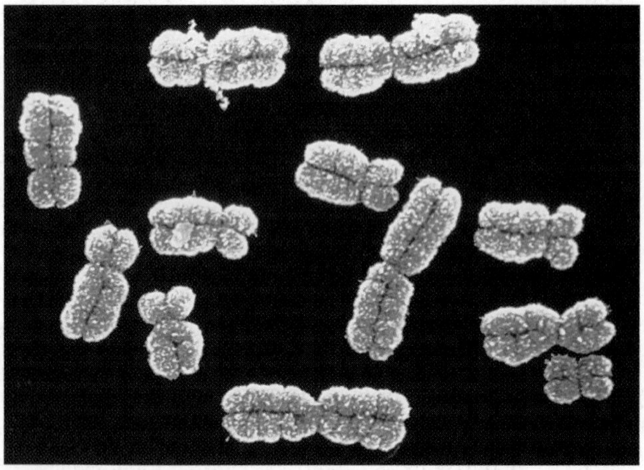

FIGURE 1–2 Colorized image of human chromosomes that have duplicated in preparation for cell division, as visualized under the scanning electron microscope.

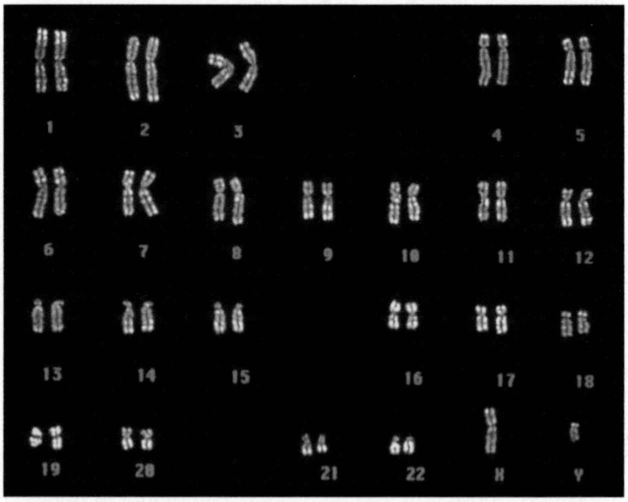

FIGURE 1–3 A colorized image of the human male chromosome set. Arranged in this way, the set is called a karyotype.

The Chromosome Theory of Inheritance: Uniting Mendel and Meiosis

Mendel did his experiments before the structure and role of chromosomes were known. About 20 years after his work was published, advances in microscopy allowed researchers to identify chromosomes (Figure 1–2) and establish that, in most eukaryotes, members of each species have a characteristic number of chromosomes called the **diploid number (2*n*)** in most of its cells. For example, humans have a diploid number of 46 (Figure 1–3). Chromosomes in diploid cells exist in pairs, called **homologous chromosomes.** Members of a pair are identical in size and location of the centromere, a structure to which spindle fibers attach during cell division.

Researchers in the last decades of the nineteenth century also described chromosome behavior during two forms of cell division, **mitosis** and **meiosis.** In mitosis (Figure 1–4), chromosomes are copied and distributed so that each daughter cell

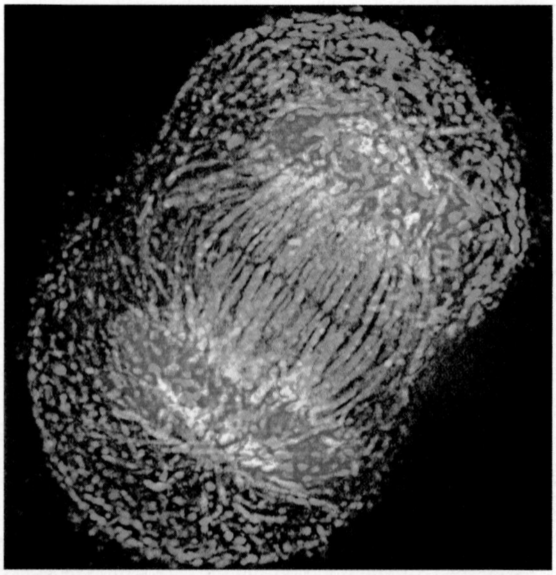

FIGURE 1–4 A stage in mitosis when the chromosomes (stained blue) move apart.

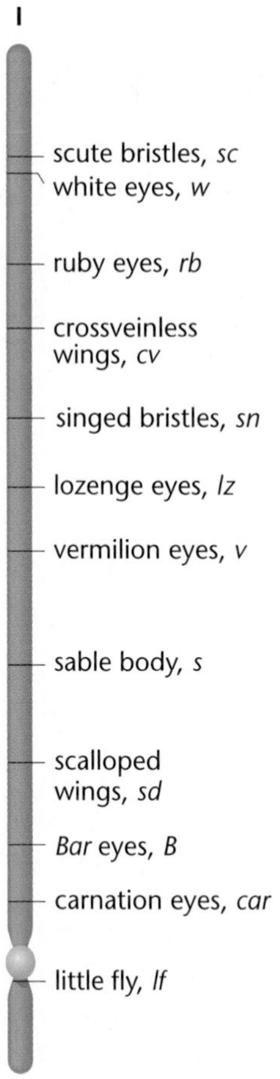

FIGURE 1–5 A drawing of chromosome I (the X chromosome, meaning one of the sex-determining chromosomes) of *D. melanogaster*, showing the locations of various genes. Chromosomes can contain hundreds of genes.

receives a diploid set of chromosomes. Meiosis is associated with gamete formation. Cells produced by meiosis receive only one chromosome from each chromosome pair, and the resulting number of chromosomes is called the **haploid (*n*) number.** This reduction in chromosome number is essential if the offspring arising from the union of two parental gametes are to maintain, over the generations, a constant number of chromosomes characteristic of their parents and other members of their species.

Early in the twentieth century, Walter Sutton and Theodor Boveri independently noted that genes, as hypothesized by Mendel, and chromosomes, as observed under the microscope, have several properties in common and that the behavior of chromosomes during meiosis is identical to the behavior of genes during gamete formation described by Mendel. For example, genes and chromosomes exist in pairs, and members of a gene pair and members of a chromosome pair separate from each other during gamete formation. Based on these parallels, Sutton and Boveri each proposed that genes are carried on chromosomes (Figure 1–5). This proposal is the basis of the **chromosome theory of inheritance,** which states that inherited traits are controlled by genes residing on chromosomes faithfully transmitted through gametes, maintaining genetic continuity from generation to generation.

ESSENTIAL POINT

The chromosome theory of inheritance explains how genetic information is transmitted from generation to generation.

Genetic Variation

At about the same time as the chromosome theory of inheritance was proposed, scientists began studying the inheritance of traits in the fruit fly, *Drosophila melanogaster*. A white-eyed fly (Figure 1–6) was discovered in a bottle containing normal (wild-type) red-eyed flies. This variation was produced by a **mutation** in one of the genes controlling eye color. Mutations are defined as any heritable change and are the source of all genetic variation.

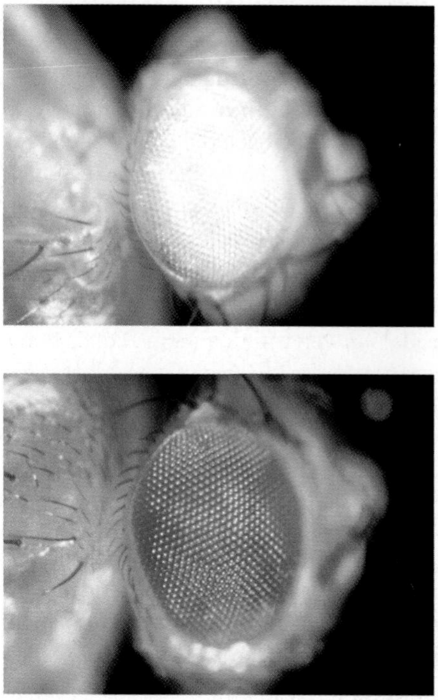

FIGURE 1–6 The normal red eye color in *D. melanogaster* (bottom) and the white-eyed mutant (top).

The variant eye color gene discovered in *Drosophila* is an **allele** of a gene controlling eye color. Alleles are defined as alternative forms of a gene. Different alleles may produce differences in the observable features, or **phenotype,** of an organism. The set of alleles for a given trait carried by an organism is called the **genotype.** Using mutant genes as markers, geneticists were able to map the location of genes on chromosomes (Figure 1–5).

The Search for the Chemical Nature of Genes: DNA or Protein?

Work on white-eyed *Drosophila* showed that the mutant trait could be traced to a single chromosome, confirming the idea that genes are carried on chromosomes. Once this relationship was established, investigators turned their attention to identifying which chemical component of chromosomes carried genetic information. By the 1920s, scientists were aware that proteins and DNA were the major chemical components of chromosomes. Proteins are the most abundant component in cells. There are a large number of different proteins, and because of their universal distribution in the nucleus and cytoplasm, many researchers thought proteins would be shown to be the carriers of genetic information.

In 1944, Oswald Avery, Colin MacLeod, and Maclyn McCarty, researchers at the Rockefeller Institute in New York, published experiments showing that DNA was the carrier of genetic information in bacteria. This evidence, though clear-cut, failed to convince many influential scientists. Additional evidence for the role of DNA as a carrier of genetic information came from other researchers who worked with viruses that infect and kill cells of the bacterium *Escherichia coli* (Figure 1–7). Viruses that infect bacteria are called **bacteriophages,** or **phages** for short, and like all viruses, consist of a protein coat surrounding a DNA core. Experiments showed that during infection the protein coat of the virus remains outside the bacterial cell, while the viral DNA enters the cell and directs the synthesis and assembly of more phage. This evidence that DNA carries genetic information, along with other research over the next few years, provided solid proof that DNA, not protein, is the genetic material, setting the stage for work to establish the structure of DNA.

1.3 Discovery of the Double Helix Launched the Era of Molecular Genetics

Once it was accepted that DNA carries genetic information, efforts were focused on deciphering the structure of the DNA molecule and the mechanism by which information stored in it is expressed to produce an observable trait, called the phenotype. In the years after this was accomplished, researchers learned how to isolate and make copies of specific regions of DNA molecules, opening the way for the era of recombinant DNA technology.

The Structure of DNA and RNA

DNA is a long, ladder-like macromolecule that twists to form a double helix (Figure 1–8). Each strand of the helix is a linear polymer made up of subunits called **nucleotides.** In DNA, there are four different nucleotides. Each DNA nucleotide contains one of four nitrogenous bases, abbreviated A (adenine), G (guanine), T (thymine), or C (cytosine). These four bases, in various sequence combinations, ultimately specify the amino acid sequences of proteins. One of the great discoveries of the twentieth century was made in 1953 by James Watson and Francis Crick, who established that the two strands of DNA are exact complements of one another, so that the rungs of the ladder in the double helix always consist of

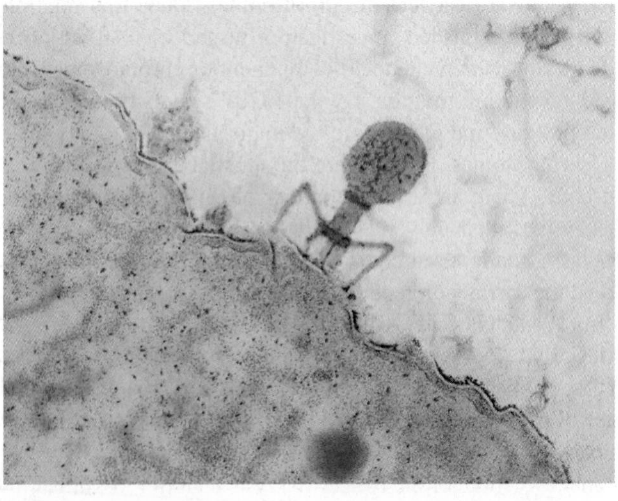

FIGURE 1–7 An electron micrograph showing T phage infecting a cell of the bacterium *E. coli.*

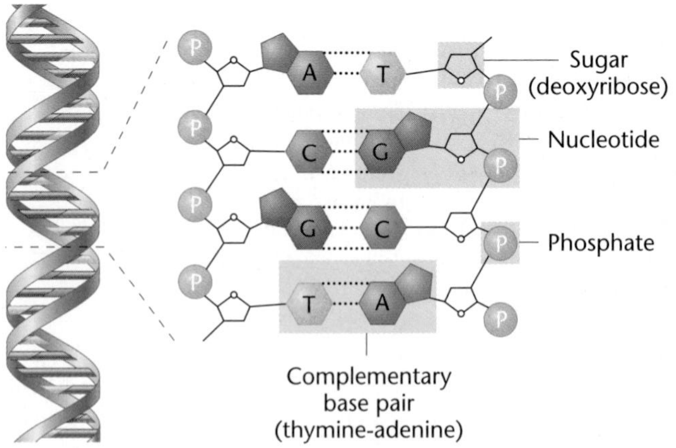

FIGURE 1–8 Summary of the structure of DNA, illustrating the arrangement of the double helix (on the left) and the chemical components making up each strand (on the right).

A $=$ T and G $\equiv$ C base pairs. Along with Maurice Wilkins, Watson and Crick were awarded a Nobel Prize in 1962 for their work on the structure of DNA. A first-hand account of the race to discover the structure of DNA is told in the book *The Double Helix,* by James Watson. We will discuss the structure of DNA in Chapter 9.

RNA, another nucleic acid, is chemically similar to DNA but contains a different sugar (ribose rather than deoxyribose) in its nucleotides and contains the nitrogenous base uracil in place of thymine. In addition, in contrast to the double helix structure of DNA, RNA is generally single stranded. Importantly, RNA can form complementary structures with a strand of DNA.

Gene Expression: From DNA to Phenotype

As noted earlier, nucleotide complementarity is the basis for gene expression, the chain of events that causes a gene to produce a phenotype. This process begins in the nucleus with **transcription,** in which the nucleotide sequence in one strand of DNA is used to construct a complementary RNA sequence (top part of Figure 1–9). Once an RNA molecule is produced, it moves to the cytoplasm. In protein synthesis, the RNA—called **messenger RNA,** or **mRNA** for short—binds to a **ribosome.** The synthesis of proteins under the direction of mRNA is called **translation** (center part of Figure 1–9). Proteins, the end product of many genes, are polymers made up of amino acid monomers. There are 20 different amino acids commonly found in proteins.

How can information contained in mRNA direct the addition of specific amino acids onto protein chains as they are synthesized? The information encoded in mRNA and called the **genetic code** consists of linear series of nucleotide triplets. Each triplet, called a **codon,** is complementary to the information stored in DNA and specifies the insertion of a specific amino acid into a protein. Protein assembly is accomplished with the aid of adapter molecules called **transfer RNA (tRNA).** Within the ribosome, tRNAs recognize the information encoded in the mRNA codons and carry the proper amino acids for construction of the protein during translation.

As the preceding discussion shows, DNA makes RNA, which most often makes protein. This sequence of events,

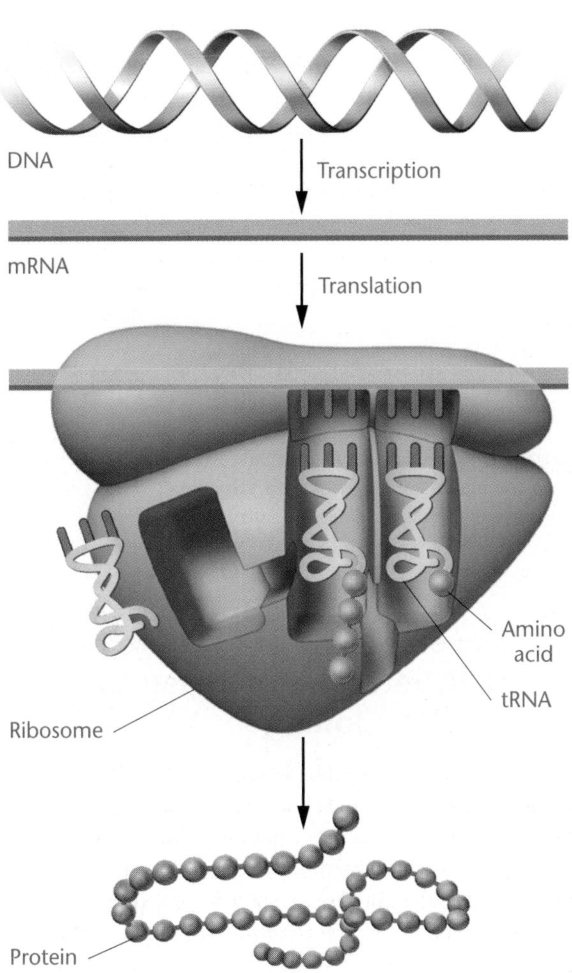

FIGURE 1–9 Gene expression consists of transcription of DNA into mRNA (top) and the translation (center) of mRNA (with the help of a ribosome) into a protein (bottom).

known as the **central dogma** of genetics, occurs with great specificity. Using an alphabet of only four letters (A, T, C, and G), genes direct the synthesis of highly specific proteins that collectively serve as the basis for all biological function.

Proteins and Biological Function

As we have mentioned, proteins are the end products of gene expression. These molecules are responsible for imparting the properties of living systems. The diversity of proteins and of the biological functions they can perform—the diversity of life itself—arises from the fact that proteins are made from combinations of 20 different amino acids. Consider that a protein chain containing 100 amino acids can have at each position any one of 20 amino acids; the number of possible different 100 amino acid proteins, each with a unique sequence, is therefore equal to

$$20^{100}$$

Because 20^{100} exceeds 5 trillion, imagine how large a number 20^{100} is! The tremendous number of possible amino acid sequences in proteins leads to enormous variation in their possible three-dimensional conformations (Figure 1–10). Obviously, proteins are molecules with the potential for enormous structural diversity and serve as the mainstay of biological systems.

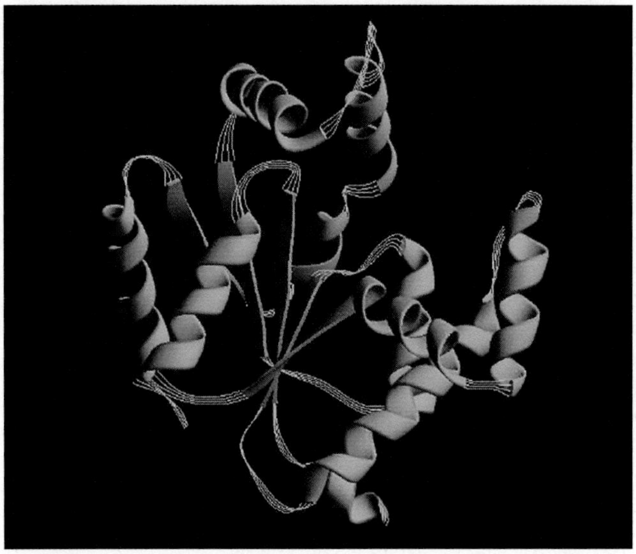

FIGURE 1–10 A three-dimensional conformation of a protein. The protein shown here is an enzyme.

The **enzymes** form the largest category of proteins. These molecules serve as biological catalysts, essentially causing biochemical reactions to proceed at rates necessary for sustaining life. By lowering the energy of activation in reactions, enzymes enable cellular metabolism to proceed at body temperatures, when otherwise those reactions would require intense heat or pressure in order to occur.

Proteins other than enzymes are critical components of cells and organisms. These include hemoglobin, the oxygen-binding pigment in red blood cells; insulin, the pancreatic hormone; collagen, the connective tissue molecule; keratin, the structural molecule in hair; histones, proteins integral to chromosome structure in eukaryotes (that is, organisms whose cells have nuclei); actin and myosin, the contractile muscle proteins; and immunoglobulins, the antibody molecules of the immune system. A protein's shape and chemical behavior are determined by its linear sequence of amino acids, which is dictated by the stored information in the DNA of a gene that is transferred to RNA, which then directs the protein's synthesis. To repeat, DNA makes RNA, which then makes protein.

Linking Genotype to Phenotype: Sickle-Cell Anemia

Once a protein is constructed, its biochemical or structural behavior in a cell plays a role in producing a phenotype. When mutation alters a gene, it may modify or even eliminate the encoded protein's usual function and cause an altered phenotype. To trace the chain of events leading from the synthesis of a given protein to the presence of a certain phenotype, we will examine sickle-cell anemia, a human genetic disorder.

Sickle-cell anemia is caused by a mutant form of hemoglobin, the protein that transports oxygen from the lungs to cells in the body. Hemoglobin is a composite molecule made up of two different proteins, α-globin and β-globin, each encoded

NORMAL β-GLOBIN

DNA.............TGA	GGA	CTC	CTC.......	
mRNA..........ACU	CCU	GAG	GAG........	
Amino acid....... Thr	Pro	Glu	Glu	
4	5	6	7	

MUTANT β-GLOBIN

DNA.............TGA	GGA	CAC	CTC.......	
mRNA..........ACU	CCU	GUG	GAG........	
Amino acid....... Thr	Pro	Val	Glu	
4	5	6	7	

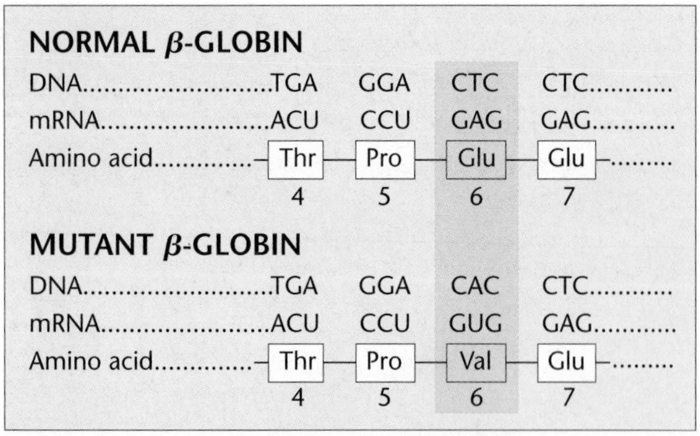

FIGURE 1–11 A single nucleotide change in the DNA encoding β-globin (CTC → CAC) leads to an altered mRNA codon (GAG → GUG) and the insertion of a different amino acid (Glu → Val), producing the altered version of the β-globin protein that is responsible for sickle-cell anemia.

by a different gene. Each functional hemoglobin molecule contains two α-globin and two β-globin proteins. In sickle-cell anemia, a mutation in the gene encoding β-globin causes an amino acid substitution in 1 of the 146 amino acids in the protein. Figure 1–11 shows part of the DNA sequence, and the corresponding mRNA codons and amino acid sequence, for the normal and mutant forms of β-globin. Notice that the mutation in sickle-cell anemia consists of a change in one DNA nucleotide, which leads to a change in codon 6 in mRNA from GAG to GUG, which in turn changes amino acid number 6 in β-globin from glutamic acid to valine. The other 145 amino acids in the protein are not changed by this mutation.

Individuals with two mutant copies of the β-globin gene have sickle-cell anemia. Their mutant β-globin proteins cause hemoglobin molecules in red blood cells to polymerize when the blood's oxygen concentration is low, forming long chains of hemoglobin that distort the shape of red blood cells (Figure 1–12). The deformed cells are fragile and break eas-

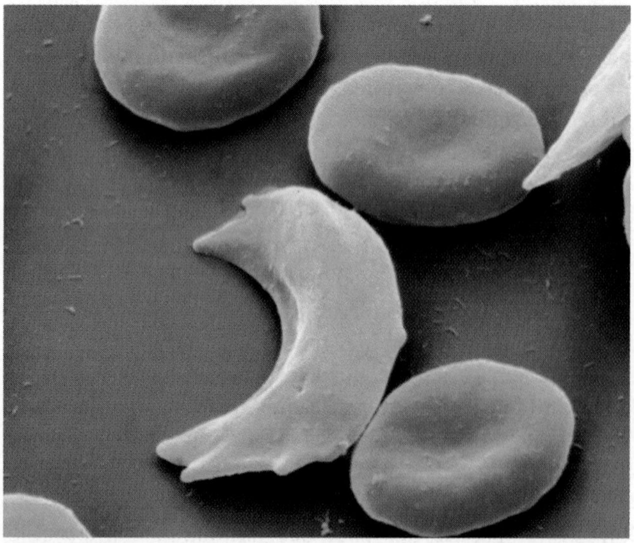

FIGURE 1–12 Normal red blood cells (round) and sickled red blood cells. The sickled cells block capillaries and small blood vessels.

ily, so that the number of red blood cells in circulation is reduced (anemia is an insufficiency of red blood cells). Moreover, when blood cells are sickle shaped, they block blood flow in capillaries and small blood vessels, causing severe pain and damage to the heart, brain, muscles, and kidneys. Sickle-cell anemia can cause heart attacks and stroke, and can be fatal if left untreated. All the symptoms of this disorder are caused by a change in a single nucleotide in a gene that changes one amino acid out of 146 in the β-globin molecule, demonstrating the close relationship between genotype and phenotype.

ESSENTIAL POINT ▪ ▪ ▪

The central dogma of molecular biology—that DNA is a template for making RNA, which in turn directs the synthesis of proteins—explains how genes control phenotypes.

1.4 Development of Recombinant DNA Technology Began the Era of DNA Cloning

The era of recombinant DNA began in the early 1970s, when researchers discovered that bacteria protect themselves from viral infection by producing enzymes that cut viral DNA at specific sites. When cut by these enzymes, the viral DNA cannot direct the synthesis of phage particles. Scientists quickly realized that such enzymes, called **restriction enzymes,** could be used to cut any organism's DNA at specific nucleotide sequences, producing a reproducible set of fragments. This set the stage for the development of DNA cloning, or making large numbers of copies of DNA sequences.

Soon after researchers discovered that restriction enzymes produce specific DNA fragments, methods were developed to insert these fragments into carrier DNA molecules called **vectors** to make **recombinant DNA** molecules and transfer them into bacterial cells. As the bacterial cells reproduce, thousands of copies, or **clones,** of the combined vector and DNA fragments are produced (Figure 1–13). These cloned copies can be recovered from the bacterial cells, and large amounts of the cloned DNA fragment can be isolated. Once large quantities of specific DNA fragments became available by cloning, they were used in many different ways: to isolate genes, to study their organization and expression, and to study their nucleotide sequence and evolution.

As techniques became more refined, it became possible to clone larger and larger DNA fragments, paving the way to establish collections of clones that represented an organism's **genome,** which is the complete haploid DNA content of a specific organism. Collections of clones that contain an entire genome are called genomic libraries. Genomic libraries are now available for hundreds of organisms.

Recombinant DNA technology has not only greatly accelerated the pace of research but has also given rise to the biotechnology industry, which has grown over the last 25 years to become a major contributor to the U.S. economy.

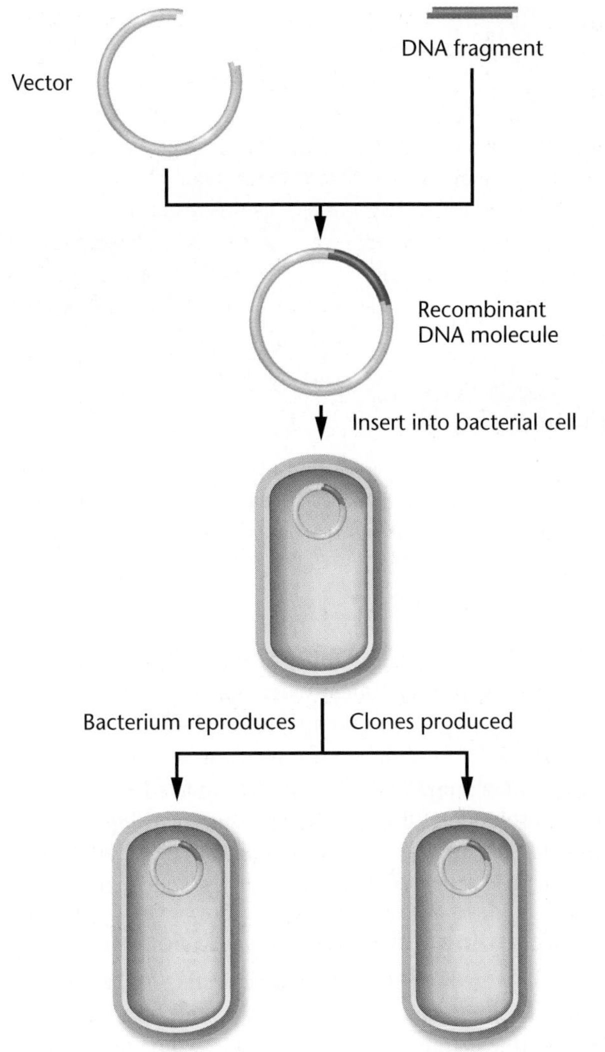

FIGURE 1–13 In cloning, a vector and a DNA fragment produced by cutting with a restriction enzyme are joined to produce a recombinant DNA molecule. The recombinant DNA is transferred into a bacterial cell, where it is cloned into many copies by replication of the recombinant molecule and by division of the bacterial cell.

1.5 The Impact of Biotechnology Is Continually Expanding

Quietly and without arousing much notice in the United States, biotechnology has revolutionized many aspects of everyday life. Humans have used microorganisms, plants, and animals for thousands of years, but the development of recombinant DNA technology and associated techniques allows us to genetically modify these organisms in new ways and use them or their products to enhance our lives. **Biotechnology** is the use of these modified organisms or their products. It is now in evidence at the supermarket; in doctors' offices; at drug stores, department stores, hospitals, and clinics; on farms and in orchards; in law enforcement and court-ordered child support; and even in industrial chemicals. There is a detailed discussion of biotechnology in Chapter 19, but for now, let's look at biotechnology's impact on just a small sampling of everyday examples.

TABLE 1.1 Some Genetically Altered Traits in Crop Plants

Herbicide Resistance
Corn, soybeans, rice, cotton, sugarbeets, canola
Insect Resistance
Corn, cotton, potato
Virus Resistance
Potato, yellow squash, papaya
Nutritional Enhancement
Golden rice
Altered Oil Content
Soybeans, canola
Delayed Ripening
Tomato

Plants, Animals, and the Food Supply

The genetic modification of crop plants is one of the most rapidly expanding areas of biotechnology. Efforts have been focused on traits such as resistance to herbicides, insects, and viruses; enhancement of oil content; and delay of ripening (Table 1.1). Currently, over a dozen genetically modified crop plants have been approved for commercial use in the United States, with over 75 more being tested in field trials. Herbicide-resistant corn and soybeans were first planted in the mid-1990s, and now about 45 percent of the U.S. corn crop and 85 percent of the U.S. soybean crop are genetically modified. In addition, more than 50 percent of the canola crop and 75 percent of the cotton crop are grown from genetically modified strains. It is estimated that more than 60 percent of the processed food in the United States contains ingredients from genetically modified crop plants.

This agricultural transformation is a source of controversy. Critics are concerned that the use of herbicide-resistant crop plants will lead to dependence on chemical weed management and may eventually result in the emergence of herbicide-resistant weeds. They also worry that traits in genetically engineered crops could be transferred to wild plants in a way that leads to irreversible changes in the ecosystem.

Biotechnology is also being used to enhance the nutritional value of crop plants. More than one-third of the world's population uses rice as a dietary staple, but most varieties of rice contain little or no vitamin A. Vitamin A deficiency causes more than 500,000 cases of blindness in children each year. A genetically engineered strain, called golden rice, has high levels of two compounds that the body converts to vitamin A. Golden rice should reduce this burden of disease. Other crops, including wheat, corn, beans, and cassava, are also being modified to enhance nutritional value by increasing their vitamin and mineral content.

Livestock such as sheep and cattle have been commercially cloned for more than 25 years, mainly by a method called embryo splitting. In 1996, Dolly the sheep (Figure 1–14, p. 10) was cloned by nuclear transfer, a method in which the nucleus of a differentiated adult cell (meaning a cell recognizable as belonging to some type of tissue) is transferred into an egg that has had its nucleus removed. This nuclear transfer

FIGURE 1–14 Dolly, a Finn Dorset sheep cloned from the genetic material of an adult mammary cell, shown next to her first-born lamb, Bonnie.

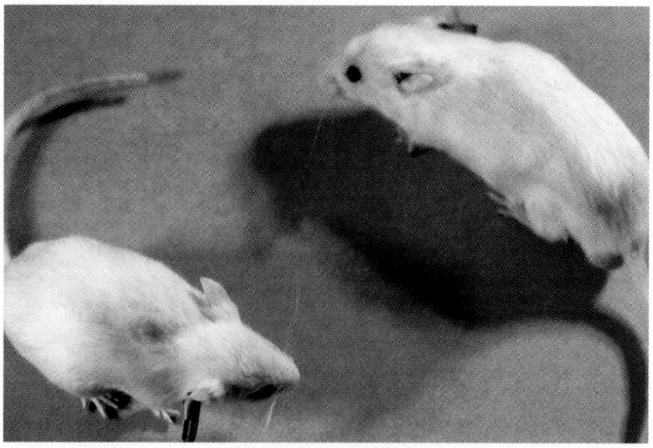

FIGURE 1–15 The first genetically altered organism to be patented, the *onc* strain of mouse, genetically engineered to be susceptible to many forms of cancer. These mice were designed for studying cancer development and the design of new anticancer drugs.

method makes it possible to produce dozens or hundreds of offspring with desirable traits. Cloning by nuclear transfer has many applications in agriculture, sports, and medicine. Some desirable traits, such as high milk production in cows, or speed in race horses, do not appear until adulthood; rather than mating two adults and waiting to see if their offspring inherit the desired characteristics, animals that are known to have these traits can now be produced by cloning differentiated cells from an adult with a desirable trait. For medical applications, researchers have transferred human genes into animals—so-called transgenic animals—so that as adults, they produce human proteins in their milk. By selecting and cloning animals with high levels of human protein production, biopharmaceutical companies can produce a herd with uniformly high rates of protein production. Human proteins from transgenic animals are now being tested as drug treatments for diseases such as emphysema. If successful, these proteins will soon be commercially available.

Who Owns Transgenic Organisms?

Once produced, can a transgenic plant or animal be patented? The answer is yes. In 1980 the United States Supreme Court ruled that living organisms and individual genes can be patented, and in 1988 an organism modified by recombinant DNA technology was patented for the first time (Figure 1–15). Since then, dozens of plants and animals have been patented. The ethics of patenting living organisms is a contentious issue. Supporters of patenting argue that without the ability to patent the products of research to recover their costs, biotechnology companies will not invest in large-scale research and development. They further argue that patents represent an incentive to develop new products because companies will reap the benefits of taking risks to bring new products to market. Critics argue that patents for organisms such as crop plants will concentrate ownership of food production in the hands of a small number of biotechnology companies, making farmers economically dependent on seeds and pesticides produced by these companies, and reducing the genetic diversity of crop plants as farmers discard local crops that might harbor important genes for resistance to pests and disease. Resolution of these and other issues raised by biotechnology and its uses will require public awareness and education, enlightened social policy, and carefully written legislation.

Biotechnology in Genetics and Medicine

Biotechnology in the form of genetic testing and gene therapy, already an important part of medicine, will be a leading force deciding the nature of medical practice in the twenty-first century. More than 10 million children or adults in the United States suffer from some form of genetic disorder, and every childbearing couple stands an approximately 3 percent risk of having a child with some form of genetic anomaly. The molecular basis for hundreds of genetic disorders is now known (Figure 1–16). Genes for sickle-cell anemia, cystic fibrosis, hemophilia, muscular dystrophy, phenylketonuria, and many other metabolic disorders have been cloned and are used for the prenatal detection of affected fetuses. In addition, tests are now available to inform parents of their status as "carriers" of a large number of inherited disorders. The combination of genetic testing and genetic counseling gives couples objective information on which they can base decisions about childbearing. At present, genetic testing is available for several hundred inherited disorders, and this number will grow as more genes are identified, isolated, and cloned. The use of genetic testing and other technologies, including gene therapy, raises ethical concerns that have yet to be resolved.

Instead of testing one gene at a time to discover whether someone carries a mutant gene that can produce a disorder in his or her offspring, a new technology is being developed that will allow screening of an entire genome to determine an individual's risk of developing a genetic disorder or of having a child with a genetic disorder. This technology uses devices called **DNA microarrays,** or **DNA chips** (Figure 1–17). Each

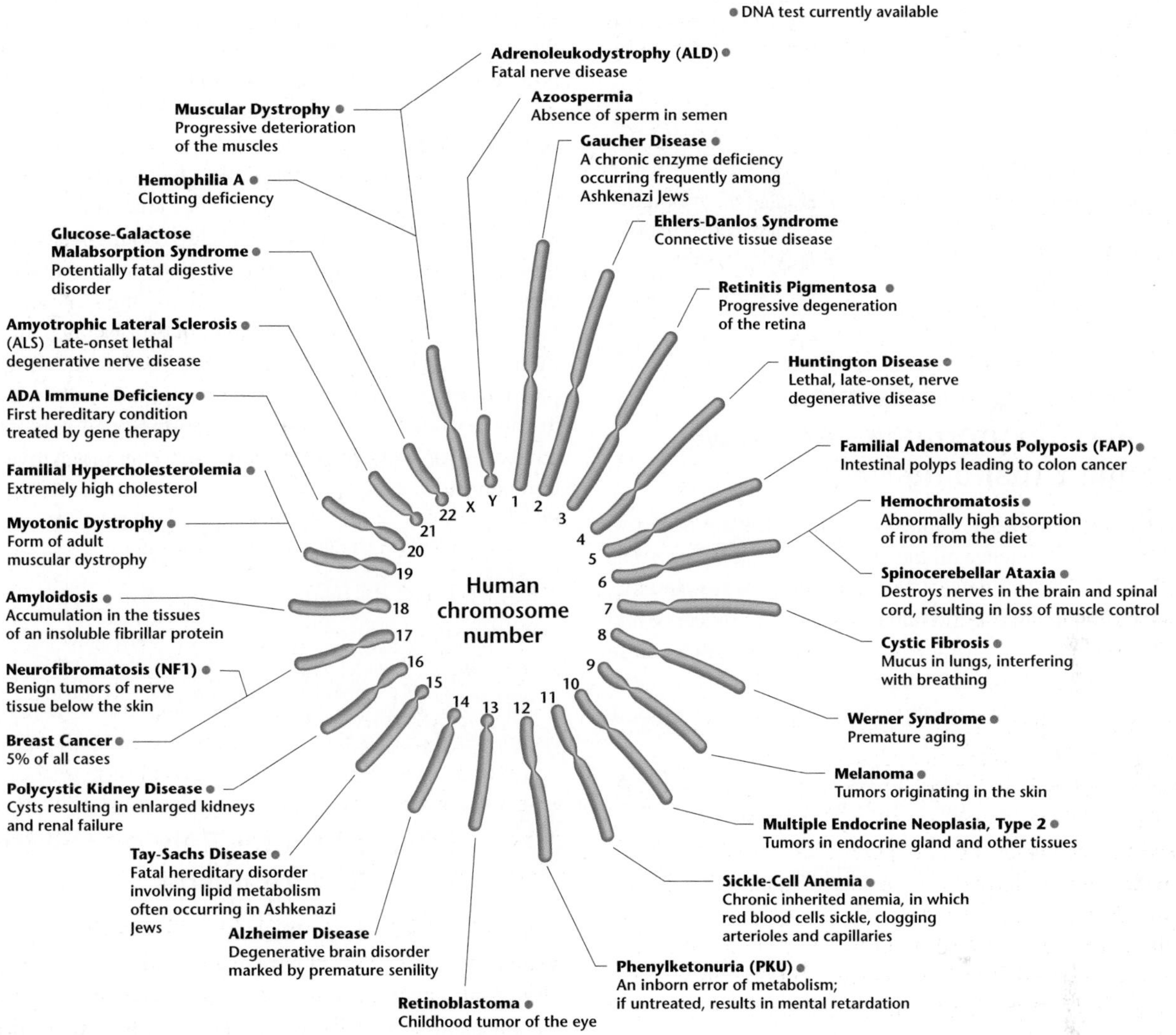

FIGURE 1–16 Diagram of the human chromosome set, showing the location of some genes whose mutant forms cause hereditary diseases. Conditions that can be diagnosed using DNA analysis are indicated by a red dot.

microarray can carry thousands of genes. In fact, microarrays carrying the entire human genome are now commercially available and are being used to test for gene expression in cancer cells as a step in developing therapies tailored to specific forms of cancer. As the technology develops further, it will be possible to scan an individual's genome in one step to identify risks for genetic and environmental factors that may trigger disease.

In **gene therapy,** clinicians transfer normal genes into individuals affected with genetic disorders. Unfortunately, although many attempts at gene therapy appeared initially to be successful, therapeutic failures and patient deaths have slowed the development of this technology. New methods of gene transfer are expected to reduce these risks, however, so it seems certain that gene therapy will become an important

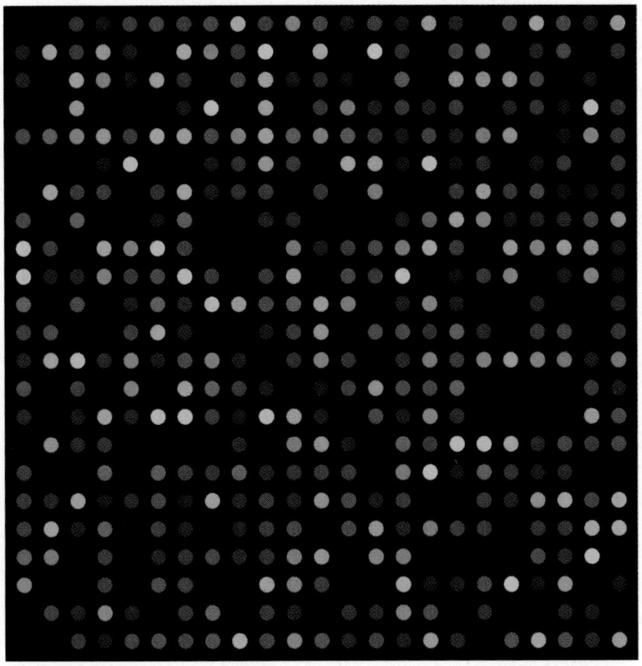

FIGURE 1–17 A portion of a DNA microarray. These arrays contain thousands of fields (the circles) to which DNA molecules are attached. Mounted on a microarray, DNA from an individual can be tested to detect mutant copies of genes.

tool in treating inherited disorders and that, as more is learned about the molecular basis of human diseases, more such therapies will be developed.

ESSENTIAL POINT

Biotechnology has revolutionized agriculture and the pharmaceutical industry, while genetic testing and gene therapy have had a profound impact on the treatment of genetic diseases.

1.6 Genomics, Proteomics, and Bioinformatics Are New and Expanding Fields

Once genomic libraries became available, scientists began to consider ways to sequence all the clones in such a library so as to spell out the nucleotide sequence of an organism's genome. Laboratories around the world initiated projects to sequence and analyze the genomes of different organisms, including those that cause human diseases. To date, the genomes of over 850 organisms have been sequenced, with over 3000 other genome projects underway.

The Human Genome Project began in 1990 as an international, government-sponsored effort to sequence the human genome and the genomes of five of the model organisms used in genetics research (the importance of model organisms is discussed below). At about the same time, various industry-sponsored genome projects also got underway. The first sequenced genome from a free-living organism, a bacterium, was published in 1995 by scientists at a biotechnology company.

In 2001, the publicly funded Human Genome Project and a private genome project sponsored by Celera Corporation reported the first draft of the human genome sequence, covering about 96 percent of the gene-containing portion of the genome. In 2003, the remaining portion of the gene-coding sequence was completed and published. Efforts are now focused on sequencing the remaining noncoding regions of the genome. The five model organisms whose genomes were also sequenced by the Human Genome Project are *Escherichia coli* (a bacterium), *Saccharomyces cerevisiae* (a yeast), *Caenorhabditis elegans* (a roundworm), *Drosophila melanogaster* (the fruit fly), and *Mus musculus* (the mouse).

As genome projects multiplied and more and more genome sequences were acquired, several new biological disciplines arose. One, called **genomics** (the study of genomes), sequences genomes and studies the structure, function, and evolution of genes and genomes. A second field, **proteomics,** is an outgrowth of genomics. Proteomics identifies the set of proteins present in a cell under a given set of conditions and additionally studies the post-translational modification of these proteins, their location within cells, and the protein–protein interactions occurring in the cell. To store, retrieve, and analyze the massive amount of data generated by genomics and proteomics, a specialized subfield of information technology called **bioinformatics** was created to develop hardware and software for processing nucleotide and protein data. Consider that the human genome contains over 3 billion nucleotides, representing some 25,000 genes encoding tens of thousands of proteins, and you can appreciate the need for databases to store this information.

These new fields are drastically changing biology from a laboratory-based science to a science that combines lab experiments with information technology. Geneticists and other biologists now use information in databases containing nucleic acid sequences, protein sequences, and gene-interaction networks to answer experimental questions in a matter of minutes instead of months and years. A feature called "Exploring Genomics," located at the end of many of the chapters in this textbook, gives you the opportunity to explore these databases for yourself while completing an interactive genetics exercise.

ESSENTIAL POINT

Recombinant DNA technology gave rise to several new fields, including genomics, proteomics, and bioinformatics, which allow scientists to explore the structure and evolution of genomes and the proteins they encode.

1.7 Genetic Studies Rely on the Use of Model Organisms

After the rediscovery of Mendel's work in 1900, genetic research on a wide range of organisms confirmed that the principles of inheritance he described were of universal significance among plants and animals. Although work continued on the genetics of many different organisms, geneticists gradually came to focus particular attention on a small number of organisms including the fruit fly (*Drosophila melanogaster*) and the mouse (*Mus musculus*) (Figure 1–18). This trend developed for two main reasons: first, it was clear that genetic mechanisms were the same in most organisms, and second, the preferred species had several characteristics that made them especially

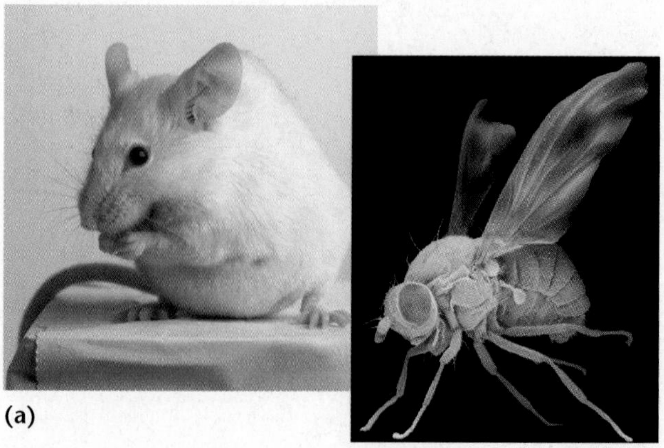
(a) (b)

FIGURE 1–18 The first generation of model organisms in genetic analysis included (a) the mouse and (b) the fruit fly.

suitable for genetic research. They were easy to grow, had relatively short life cycles, produced many offspring, and their genetic analysis was fairly straightforward. Over time, researchers created a large catalog of mutant strains for these species, and the mutations were carefully studied, characterized, and mapped. Because of their well-characterized genetics, these species became **model organisms,** defined as organisms used for the study of basic biological processes. In later chapters, we will see how discoveries in model organisms are shedding light on many aspects of biology, including aging, cancer, the immune system, and behavior.

The Modern Set of Genetic Model Organisms

Gradually, geneticists added other species to their collection of model organisms: viruses (such as the T phages and lambda phage) and microorganisms (the bacterium *Escherichia coli* and the yeast *Saccharomyces cerevisiae*) (Figure 1–19).

More recently, three additional species have been developed as model organisms, shown in the chapter opening photograph on page 1. Each was chosen to study some aspect of embryonic development. To study the nervous system and its role in behavior, the nematode *Caenorhabditis elegans* was chosen as a model system because it has a nervous system with only a few hundred cells. *Arabidopsis thaliana*, a small plant with a short life cycle, has become a model organism for the study of many aspects of plant biology. The zebrafish, *Danio rerio,* is used to study vertebrate development: it is small, it reproduces rapidly, and its egg, embryo, and larvae are all transparent.

Model Organisms and Human Diseases

The development of recombinant DNA technology and the results of genome sequencing have confirmed that all life has a common origin. Because of this common origin, genes with similar functions in different organisms tend to be similar or identical in structure and nucleotide sequence. Much of what scientists learn by studying the genetics of other species can therefore be applied to humans and serve as the basis for understanding and treating human diseases. In addition, the ability to transfer genes between species has enabled scientists to develop models of human diseases in organisms ranging from bacteria to fungi, plants, and animals (Table 1.2).

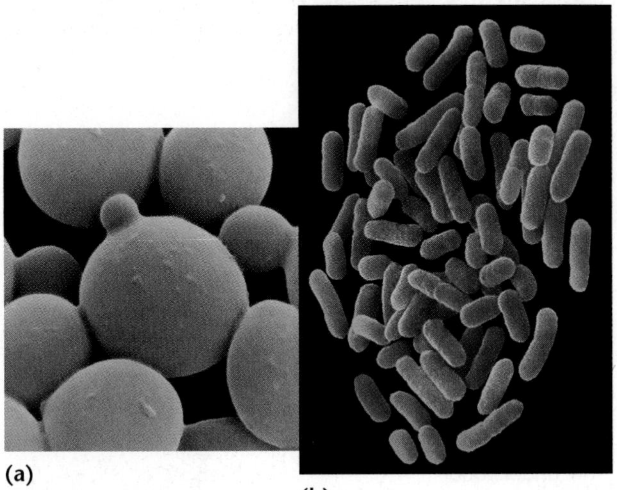

(a)

(b)

FIGURE 1–19 Microbes that have become model organisms for genetic studies include (a) the yeast *S. cerevisiae* and (b) the bacterium *E. coli.*

TABLE 1.2 Model Organisms Used to Study Some Human Diseases

Organism	Human Diseases
E. coli	Colon cancer and other cancers
S. cerevisiae	Cancer, Werner syndrome
D. melanogaster	Disorders of the nervous system, cancer
C. elegans	Diabetes
D. rerio	Cardiovascular disease
M. musculus	Lesch–Nyhan disease, cystic fibrosis, fragile-X syndrome, and many other diseases

The idea of studying a human disease such as colon cancer by using *E. coli* may strike you as strange, but the basic steps of DNA repair (a process that is defective in some forms of colon cancer) are the same in both organisms, and the gene involved (*mutL* in *E. coli* and *MLH1* in humans) is found in both organisms. More importantly, *E. coli* has the advantage of being easier to grow (the cells divide every 20 minutes), so that researchers can easily create and study new mutations in the bacterial *mutL* gene in order to figure out how it works. This knowledge may eventually lead to the development of drugs and other therapies to treat colon cancer in humans.

The fruit fly, *D. melanogaster,* is also being used to study specific human diseases. Mutant genes have been identified in *D. melanogaster* that produce phenotypes with abnormalities of the nervous system, including abnormalities of brain structure, adult-onset degeneration of the nervous system, and visual defects such as retinal degeneration. The information from genome-sequencing projects indicates that almost all these genes have human counterparts. As an example, genes involved in a complex human disease of the retina called retinitis pigmentosa are identical to *Drosophila* genes involved in retinal degeneration. Study of these mutations in *Drosophila* is helping to dissect this complex disease and identify the function of the genes involved.

Another approach to using *Drosophila* for studying diseases of the human nervous system is to transfer human disease genes into the flies by means of recombinant DNA technology. The transgenic flies are then used for studying the mutant human genes themselves, the genes affecting the expression of the human disease genes, and the effects of therapeutic drugs on the action of those genes, all studies that are difficult or impossible to perform in humans. This gene-transfer approach is being used to study almost a dozen human neurodegenerative disorders, including Huntington disease, Machado–Joseph disease, myotonic dystrophy, and Alzheimer's disease.

As you read this textbook, you will encounter these model organisms again and again. Remember that each time you meet them, they not only have a rich history in basic genetics research but are also at the forefront in the study of human genetic disorders and infectious diseases. As discussed in the next section, however, we have yet to reach a consensus on how and when this technology is determined to be safe and ethically acceptable.

ESSENTIAL POINT

The study of model organisms for understanding human health and disease is one of many ways genetics and biotechnology are rapidly changing everyday life.

1.8 We Live in the Age of Genetics

Mendel described his decade-long project on inheritance in pea plants in an 1865 paper presented at a meeting of the Natural History Society of Brünn in Moravia. Just 100 years later, the 1965 Nobel Prize was awarded to François Jacob, André Lwoff, and Jacques Monod for their work on the molecular basis of gene regulation in bacteria. This time span encompassed the years leading up to the acceptance of Mendel's work, the discovery that genes are on chromosomes, the experiments that proved DNA encodes genetic information, and the elucidation of the molecular basis for DNA replication. The rapid development of genetics from Mendel's monastery garden to the Human Genome Project and beyond is summarized in a timeline in Figure 1–20.

The Nobel Prize and Genetics

Although other scientific disciplines have also expanded in recent years, none has paralleled the explosion of information and excitement generated by the discoveries in genetics. Nowhere is this impact more apparent than in the list of Nobel Prizes related to genetics, beginning with those awarded in the early and mid-twentieth century and continuing into the present (see inside front cover). Nobel Prizes in Medicine or Physiology and Chemistry have been consistently awarded for work in genetics and associated fields. The first Nobel Prize awarded for such work was given to Thomas Morgan in 1933 for his research on the chromosome theory of inheritance. That award was followed by many others, including prizes for the discovery of genetic recombination, the relationship between genes and proteins, the structure of DNA, and the genetic code. In this century, geneticists continue to be recognized for their impact on biology in the current millennium. The 2002 Nobel Prize for Medicine or Physiology was awarded to Sydney Brenner, H. Robert Horvitz, and John E. Sulston for their work on the genetic regulation of organ development and programmed cell death. In 2006, prizes went to Andrew Fire and Craig Mello for their discovery that RNA molecules play an important role in regulating gene expression and to Roger Kornberg for his work on the molecular basis of eukaryotic transcription. The 2007 Nobel Prize went to M. R. Capecchi, O. Smithies, and M. J. Evans for the development of gene-targeting technology essential to the creation of knockout mice serving as animal models of human disease.

Genetics and Society

Just as there has never been a more exciting time to study genetics, the impact of this discipline on society has never been more profound. Genetics and its applications in biotechnology are developing much faster than the social conventions, public policies, and laws required to regulate their use (see this chapter's essay on "Genetics, Technology, and Society"). As a society, we are grappling with a host of sensitive genetics-related issues, including concerns about prenatal testing, insurance coverage, genetic discrimination, ownership of genes, access to and safety of gene therapy, and genetic privacy. By the time you finish this course, you will have seen more than enough evidence to convince you that the present is the Age of Genetics, and you will understand the need to think about and become a participant in the dialogue concerning genetic science and its use.

ESSENTIAL POINT

Genetic technology is having a profound effect on society, but policies and legislation governing its use are lagging behind the resulting innovations.

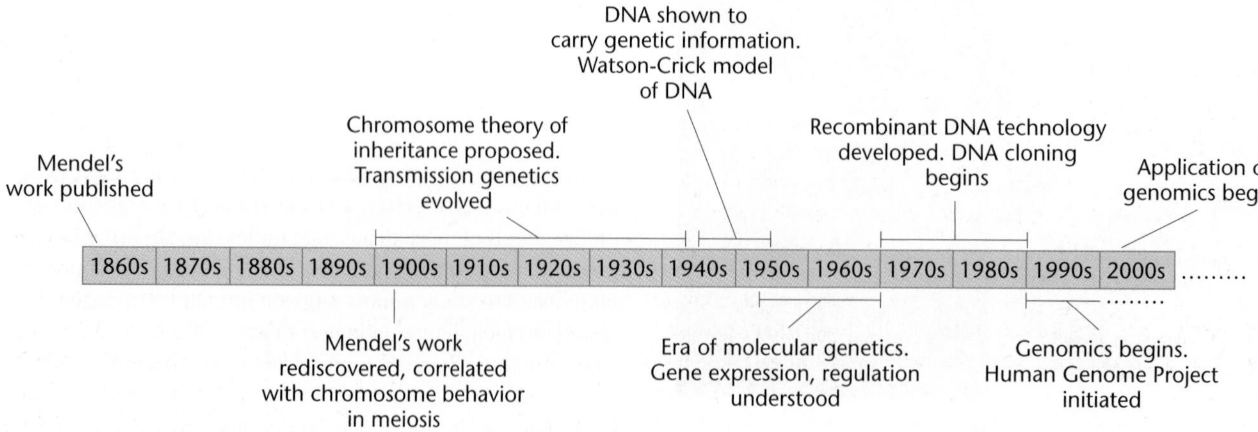

FIGURE 1–20 A timeline showing the development of genetics from Gregor Mendel's work on pea plants to the current era of genomics and its many applications in research, medicine, and society. Having a sense of the history of discovery in genetics should provide you with a useful framework as you proceed through this textbook.

GENETICS, TECHNOLOGY, AND SOCIETY

Genetics and Society: The Application and Impact of Science and Technology

One of the special features of this text is the series of essays on "Genetics, Technology, and Society" that you will find at the conclusion of many chapters. These essays explore genetics-related topics that have an impact on the lives of each of us and thus on society in general. Today, genetics touches all aspects of modern life, bringing rapid changes in medicine, agriculture, law, the pharmaceutical industry, and biotechnology. Physicians now use hundreds of genetic tests to diagnose and predict the course of disease and to detect genetic defects *in utero*. DNA-based methods allow scientists to trace the path of evolution taken by many species, including our own. Farmers grow disease-resistant and drought-resistant crops, and raise more productive farm animals, created by gene-transfer techniques. DNA profiling methods are applied to paternity testing and murder investigations. Biotechnologies resulting from genomics research have had dramatic effects on industry in general. Meanwhile, the biotechnology industry itself generates over 700,000 jobs and $50 billion in revenue each year and doubles in size every decade.

Along with these rapidly changing gene-based technologies comes a challenging array of ethical dilemmas. Who owns and controls genetic information? Are gene-enhanced agricultural plants and animals safe for humans and the environment? Do we have the right to patent organisms and profit from their commercialization? How can we ensure that genomic technologies will be available to all and not just to the wealthy? What are the likely social consequences of the new reproductive technologies? It is a time when everyone needs to understand genetics in order to make complex personal and societal decisions.

The "Genetics, Technology, and Society" essays explore the interface of society and genetic technology. It is our hope that these essays will act as entry points for your exploration of the myriad applications of modern genetics and their social implications. Even should your genetics course not cover certain chapters, we hope that you will find the essays in those chapters to be of interest. Good reading!

EXPLORING GENOMICS

Internet Resources for Learning about Genomics, Bioinformatics, and Proteomics

Genomics is one of the most rapidly changing disciplines in genetics. New information in this field is accumulating at an astounding rate. Keeping up with current developments in genomics, proteomics, bioinformatics, and other examples of the "omics" era of modern genetics is a challenging task indeed! As a result, geneticists, molecular biologists, and other scientists are relying on online databases to share and compare new information.

The purpose of the Exploring Genomics feature, which appears at the end of many chapters, is to introduce you to a range of Internet databases that scientists around the world depend on for sharing, analyzing, organizing, comparing, and storing data from studies in genomics, proteomics, and related fields. We will explore this incredible pool of new information—comprising some of the best publicly available resources in the world—and show you how to use bioinformatics approaches to analyze the sequence and structural data to be found there. Each set of Exploring Genomics exercises will provide a basic introduction to one or more especially relevant or useful databases or programs and then guide you through exercises that use the databases to expand on or reinforce important concepts discussed in the chapter. The exercises are designed to help you learn to navigate the databases, but your explorations need not be limited to these experiences. Part of the fun of learning about genomics is exploring these outstanding databases on your own, so that you can get the latest information on any topic that interests you. Enjoy your explorations!

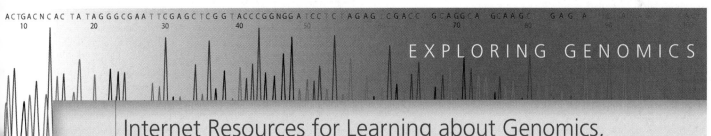

CASE STUDIES Extending essential ideas of genetics beyond the classroom

To make genetics relevant to situations that you may encounter outside the classroom, beginning with Chapter 2, each chapter will contain a short case study accompanied by questions. Each case is designed to draw attention to one of the basic concepts covered in the chapter. The case will use examples and scenarios to illustrate how this concept can be applied to genetic issues in everyday life. These will include personal decisions such as genetic testing and diagnosis, reproductive options, as well as broader issues such as public health, the use and limits of genetic technology, as well as examples illustrating the unity and diversity of organisms.

Many of the case studies and their accompanying questions can serve as the basis of classroom discussions, group projects, or as starting points for papers and presentations. Use these resources to enhance your learning experience and to bring genetics out of the book and into your life.

PROBLEMS AND DISCUSSION QUESTIONS

1. Describe Mendel's conclusions about how traits are passed from generation to generation.
2. What is the chromosome theory of inheritance, and how is it related to Mendel's findings?
3. Define genotype and phenotype, and describe how they are related.
4. What are alleles? Is it possible for more than two alleles of a gene to exist?
5. Given the state of knowledge at the time of the Avery, MacLeod, and McCarty experiment, why was it difficult for some scientists to accept that DNA is the carrier of genetic information?
6. Contrast chromosomes and genes.
7. How is genetic information encoded in a DNA molecule?
8. Describe the central dogma of molecular genetics and how it serves as the basis of modern genetics.
9. How many different proteins, each with a unique amino acid sequence, can be constructed with a length of five amino acids?
10. Outline the roles played by restriction enzymes and vectors in cloning DNA.
11. What are some of the impacts of biotechnology on crop plants in the United States?
12. Summarize the arguments for and against patenting genetically modified organisms.
13. We all carry about 20,000 genes in our genome. So far, patents have been issued for more than 6000 of these genes. Do you think that companies or individuals should be able to patent human genes? Why or why not?
14. How has the use of model organisms advanced our knowledge of the genes that control human diseases?
15. If you knew that a devastating late-onset inherited disease runs in your family (in other words, a disease that does not appear until later in life) and you could be tested for it at the age of 20, would you want to know whether you are a carrier? Would your answer be likely to change when you reach age 40?
16. The "Age of Genetics" has been brought on by remarkable advances in the applications of biotechnology to manipulate plant and animal genomes. Given that the world population has topped 6 billion and is expected to double in the next 50 years, some scientists have proposed that only the worldwide introduction of genetically modified (GM) foods will make it possible for future nutritional demands to be met. Pest resistance, herbicide, cold, drought, and salinity tolerance, along with increased nutrition, are seen as positive attributes of GM foods. However, after more than ten years of use, some caution that unintended harm to other organisms, reduced effectiveness of pesticides, gene transfer to nontarget species, allergenicity, and as yet, unknown effects to human health are potential concerns regarding GM foods. If you were in a position to control the introduction of a GM primary food product (rice, for example), what criteria would you establish before allowing such introduction?
17. The BIO (Biotechnology Industry Organization) meeting held in Philadelphia (June 2005) brought together worldwide leaders from the biotechnology and pharmaceutical industries. Concurrently, BioDemocracy 2005, a group composed of people seeking to highlight hazards from widespread applications of biotechnology, met in Philadelphia. The benefits of biotechnology are outlined in your text. Predict some of the risks that were no doubt discussed at the BioDemocracy meeting.

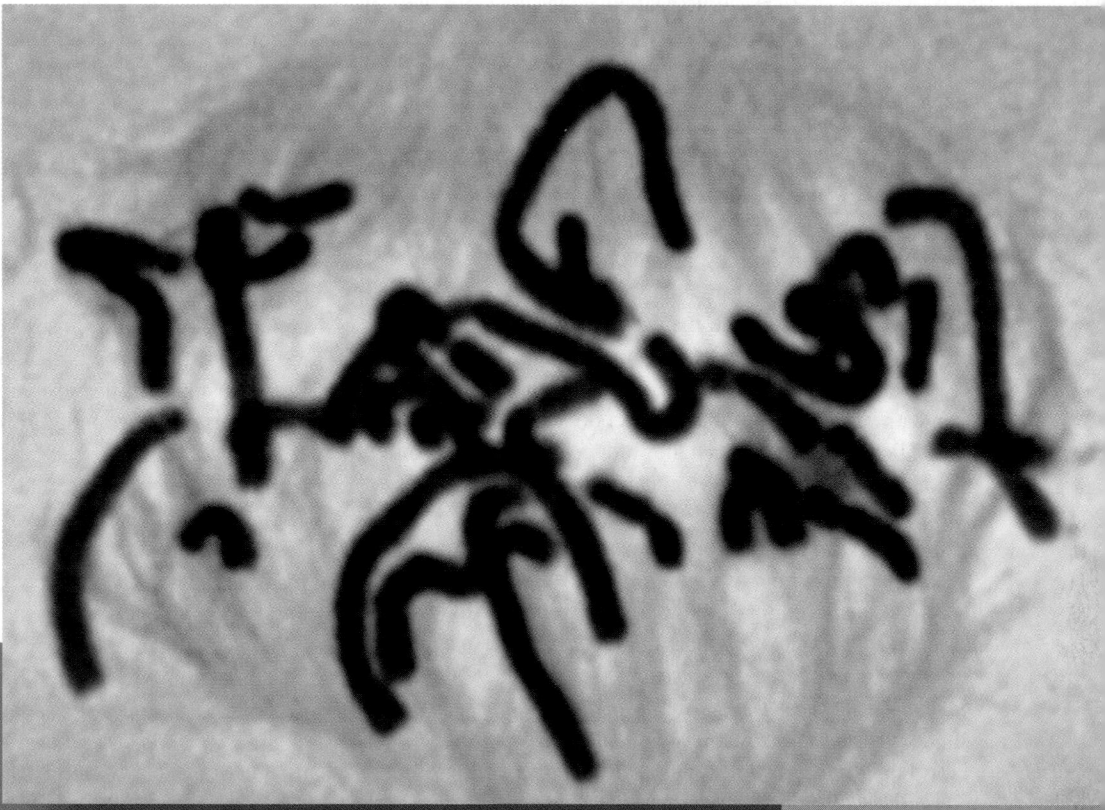

Chromosomes in the prometaphase stage of mitosis, from a cell in the flower of Haemanthus

2

Mitosis and Meiosis

Chapter Concepts

- Genetic continuity between generations of cells and between generations of sexually reproducing organisms is maintained through the processes of mitosis and meiosis, respectively.
- Diploid eukaryotic cells contain their genetic information in pairs of homologous chromosomes, with one member of each pair being derived from the maternal parent and one from the paternal parent.
- Mitosis provides a mechanism to distribute chromosomes that have duplicated into progeny cells during cell reproduction.
- Mitosis converts a diploid cell into two diploid daughter cells.
- The process of meiosis distributes one member of each homologous pair of chromosomes into each gamete or spore, thus reducing the diploid chromosome number to the haploid chromosome number.
- Meiosis generates genetic variability by distributing various combinations of maternal and paternal members of each homologous pair of chromosomes into gametes or spores.
- It is during the stages of mitosis and meiosis that the genetic material has been condensed into discrete structures called chromosomes.

In every living thing, there exists a substance referred to as the genetic material. Except in certain viruses, this material is composed of the nucleic acid, DNA. A molecule of DNA is organized into units called genes, the products of which direct the metabolic activities of cells. DNA, with its array of genes, is organized into structures called chromosomes, which serve as vehicles for transmitting genetic information. The manner in which chromosomes are transmitted from one generation of cells to the next and from organisms to their descendants must be exceedingly precise. In this chapter, we consider exactly how genetic continuity is maintained between cells and organisms.

Two major processes are involved in eukaryotes: **mitosis** and **meiosis.** Although the mechanisms of the two processes are similar in many ways, the outcomes are quite different. On one hand, mitosis leads to the production of two cells, each with the same number of chromosomes as the parent cell. Meiosis, on the other hand, reduces the genetic content and the number of chromosomes to precisely half. This reduction is essential if sexual reproduction is to occur without doubling the amount of genetic material at each generation. Strictly speaking, mitosis is that portion of the cell cycle during which the hereditary components are equally divided into daughter cells. Meiosis is part of a special type of cell division that leads to the production of sex cells: **gametes** or **spores.** This process is an essential step in the transmission of genetic information from an organism to its offspring.

Normally, chromosomes are visible only during mitosis and meiosis. When cells are not undergoing division, the genetic material making up chromosomes unfolds and uncoils into a diffuse network within the nucleus, generally referred to as chromatin. Before describing mitosis and meiosis, we will briefly review the structure of cells, emphasizing components that are of particular significance to genetic function. We will also compare the structural differences between the prokaryotic (nonnucleated) cells of bacteria and the eukaryotic cells of higher organisms. We then devote the remainder of the chapter to the behavior of chromosomes during cell division.

How Do We Know?

In this chapter, we will focus on how chromosomes are distributed during cell division, both in dividing somatic cells (mitosis) and in gamete- and spore-forming cells (meiosis). We will find many opportunities to consider the methods and reasoning by which much of this information was acquired. From the explanations given in the chapter, you should answer the following fundamental questions:

1. How do we know that chromosomes exist in homologous pairs?
2. How do we know that DNA replication occurs during interphase, and not early in mitosis?
3. How do we know that mitotic chromosomes are derived from chromatin?

2.1 Cell Structure Is Closely Tied to Genetic Function

Before 1940, our knowledge of cell structure was limited to what we could see with the light microscope. Around 1940, the transmission electron microscope was in its early stages of development, and by 1960, many details of cell ultrastructure were emerging. Under the electron microscope, cells were seen as highly organized, precise structures. A new world of whorled membranes, organelles, microtubules, granules, and filaments was revealed. These discoveries revolutionized thinking in the entire field of biology, but we will be concerned only with those aspects of cell structure that relate to genetic study. The typical animal cell shown in Figure 2–1 illustrates most of the structures we will discuss.

All cells are surrounded by a **plasma membrane,** an outer covering that defines the cell boundary and delimits the cell from its immediate external environment. This membrane is not passive but instead actively controls the movement of materials into and out of the cell. In addition to this membrane, plant cells have an outer covering called the **cell wall** whose major component is a polysaccharide called cellulose.

Many, if not most, animal cells have a covering over the plasma membrane, referred to as the **glycocalyx.** Consisting of glycoproteins and polysaccharides, this covering coat has a chemical composition that differs from comparable structures in either plants or bacteria. The glycocalyx, among other functions, provides biochemical identity at the surface of cells, and these forms of cellular identity are under genetic control. For example, various antigenic determinants such as the AB and MN antigens are found on the surface of red blood cells. In other cells, histocompatibility antigens, which elicit an immune response during tissue and organ transplants, are present. A variety of receptor molecules are also important components at the surface of cells. These molecules, synthesized under the direction of specific genes, are recognition sites that transfer specific chemical signals across the cell membrane into the cell.

Living organisms are categorized into two major groups depending on whether or not their cells contain a nucleus. The presence of a nucleus and other membranous organelles characterizes **eukaryotic cells.** The **nucleus** houses the genetic material, DNA, which is complexed with an array of acidic and basic proteins into thin fibers. During nondivisional phases of the cell cycle, these fibers are uncoiled and dispersed into **chromatin.** During mitosis and meiosis, chromatin fibers coil and condense into structures called **chromosomes.** Also present in the nucleus is the **nucleolus,** an amorphous component where ribosomal RNA (rRNA) is synthesized and where the initial stages of ribosomal assembly occur. The areas of DNA that encode rRNA are collectively referred to as the **nucleolus organizer region,** or the **NOR.**

Prokaryotic cells lack a nuclear envelope and membranous organelles. In many bacteria such as *Escherichia coli,* the genetic material is present as a long, circular DNA molecule compacted into the **nucleoid** area. Part of the DNA may be attached to the cell membrane, but in general the nucleoid constitutes a large area throughout the cell. Although the DNA is

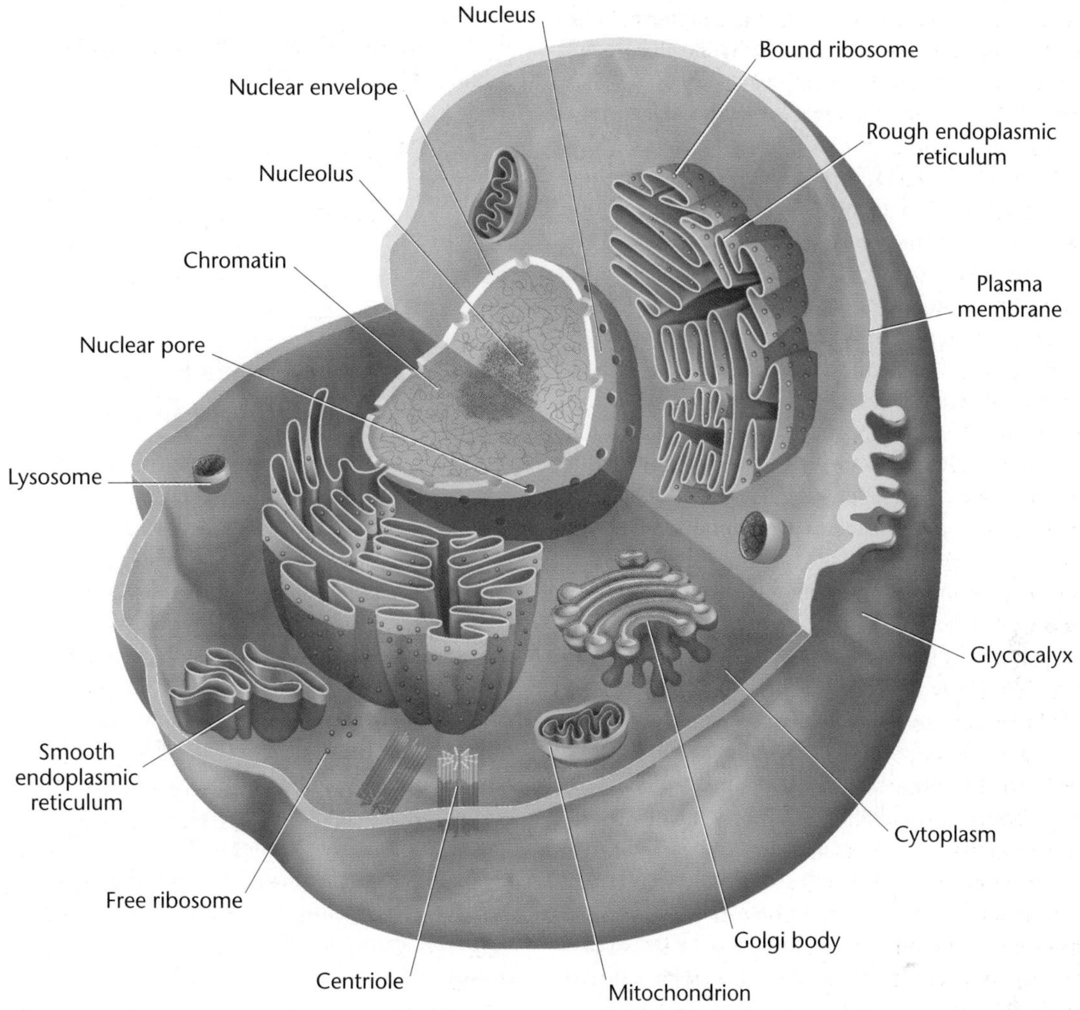

FIGURE 2–1 A generalized animal cell. The cellular components discussed in the text are emphasized here.

compacted, it does not undergo the extensive coiling characteristic of the stages of mitosis where, in eukaryotes, chromosomes become visible. Nor is the DNA in these organisms associated as extensively with proteins as is eukaryotic DNA. Figure 2–2, in which two bacteria are forming during cell division, illustrates the nucleoid regions that house the bacterial chromosome. Prokaryotic cells do not have a distinct nucleolus, but they do contain genes that specify rRNA molecules.

The remainder of the eukaryotic cell enclosed by the plasma membrane, excluding the nucleus, is composed of **cytoplasm** and all associated cellular organelles. Cytoplasm is a nonparticulate, colloidal material referred to as the cytosol, which surrounds and encompasses the cellular organelles. Beyond these components, an extensive system of tubules and filaments comprising the cytoskeleton provides a lattice of support structures within the cytoplasm. Consisting primarily of tubulin-derived microtubules and actin-derived microfilaments, this structural framework maintains cell shape, facilitates cell mobility, and anchors the various organelles.

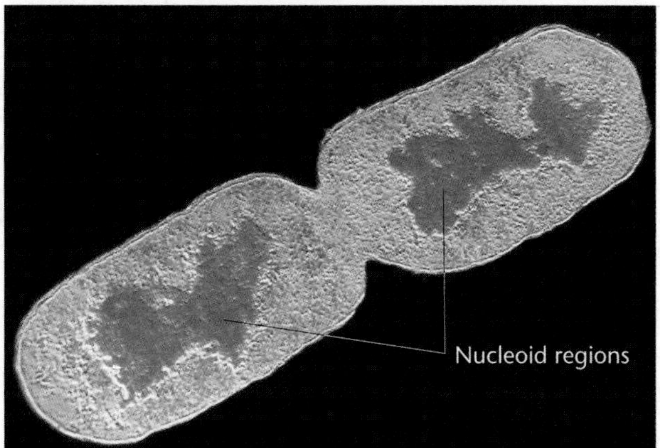

FIGURE 2–2 Color-enhanced electron micrograph of *E. coli* undergoing cell division. Particularly prominent are the two chromosomal areas (shown in red), called nucleoids, that have been partitioned into the daughter cells.

One such organelle, the membranous **endoplasmic reticulum (ER),** compartmentalizes the cytoplasm, greatly increasing the surface area available for biochemical synthesis. The ER may appear smooth, in which case it serves as the site for synthesizing fatty acids and phospholipids, or it may appear rough because it is studded with ribosomes. Ribosomes serve as sites where genetic information contained in messenger RNA (mRNA) is translated into proteins.

Three other cytoplasmic structures are very important in the eukaryotic cell's activities: mitochondria, chloroplasts, and centrioles. **Mitochondria** are found in both animal and plant cells and are the sites of the oxidative phases of cell respiration. These chemical reactions generate large amounts of adenosine triphosphate (ATP), an energy-rich molecule. **Chloroplasts** are found in plants, algae, and some protozoans. This organelle is associated with photosynthesis, the major energy-trapping process on Earth. Both mitochondria and chloroplasts contain a type of DNA distinct from that found in the nucleus. Furthermore, these organelles can duplicate themselves and transcribe and translate their genetic information. It is interesting to note that the genetic machinery of mitochondria and chloroplasts closely resembles that of prokaryotic cells. This and other observations have led to the proposal that these organelles were once primitive free-living organisms that established a symbiotic relationship with a primitive eukaryotic cell. This theory, which describes the evolutionary origin of these organelles, is called the **endosymbiotic hypothesis.**

Animal cells and some plant cells also contain a pair of complex structures called the **centrioles.** These cytoplasmic bodies, located in a specialized region called the centrosome, are associated with the organization of spindle fibers that function in mitosis and meiosis. In some organisms, the centriole is derived from another structure, the basal body, which is associated with the formation of cilia and flagella. Over the years, many reports have suggested that centrioles and basal bodies contain DNA, which could be involved in the replication of these structures. Currently, this is thought not to be the case.

The organization of **spindle fibers** by the centrioles occurs during the early phases of mitosis and meiosis. These fibers play an important role in the movement of chromosomes as they separate during cell division. They are composed of arrays of microtubules consisting of polymers of polypeptide subunits of the protein tubulin.

ESSENTIAL POINT

Most components of cells are involved directly or indirectly with genetic processes.

2.2 Chromosomes Exist in Homologous Pairs in Diploid Organisms

As we discuss the processes of mitosis and meiosis, it is important that you understand the concept of homologous chromosomes. Such an understanding will also be of critical importance in our future discussions of Mendelian genetics. Chromosomes are most easily visualized during mitosis. When they are examined carefully, distinctive lengths and shapes are apparent. Each chromosome contains a condensed or constricted region called the **centromere,** which establishes the general appearance of each chromosome. Figure 2–3 shows chromosomes with centromere placements at different points along their lengths. Extending from either side of the centromere are the arms of the chromosome. Depending on the position of the centromere, different arm ratios are produced. As Figure 2–3 illustrates, chromosomes are classified as **metacentric, submetacentric, acrocentric,** or **telocentric** on the basis of the centromere location. The shorter arm, by convention, is shown above the centromere and is called the **p arm** (p, for "petite"). The longer arm is shown below the centromere and is called the **q arm** (because q is the next letter in the alphabet).

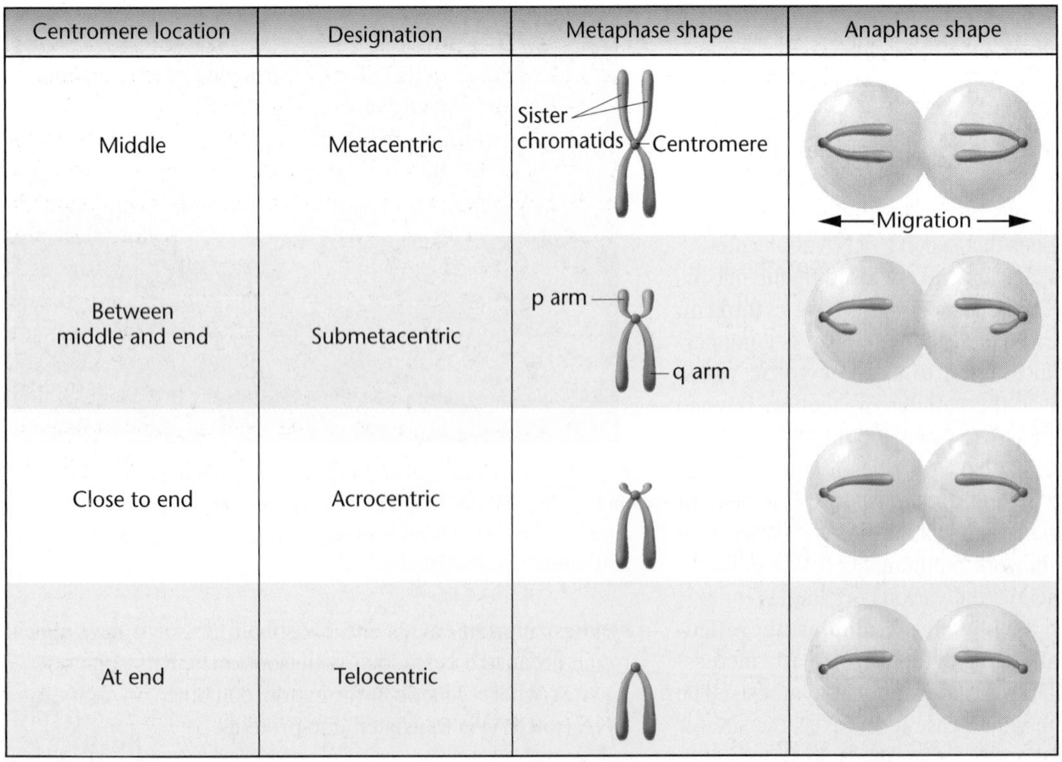
Centromere location
Designation
Metaphase shape
Anaphase shape
Middle
Metacentric
Sister chromatids
Centromere
Migration
Between middle and end
Submetacentric
p arm
q arm
Close to end
Acrocentric
At end
Telocentric

FIGURE 2–3 Centromere locations and designations of chromosomes based on centromere location. Note that the shape of the chromosome during anaphase is determined by the position of the centromere.

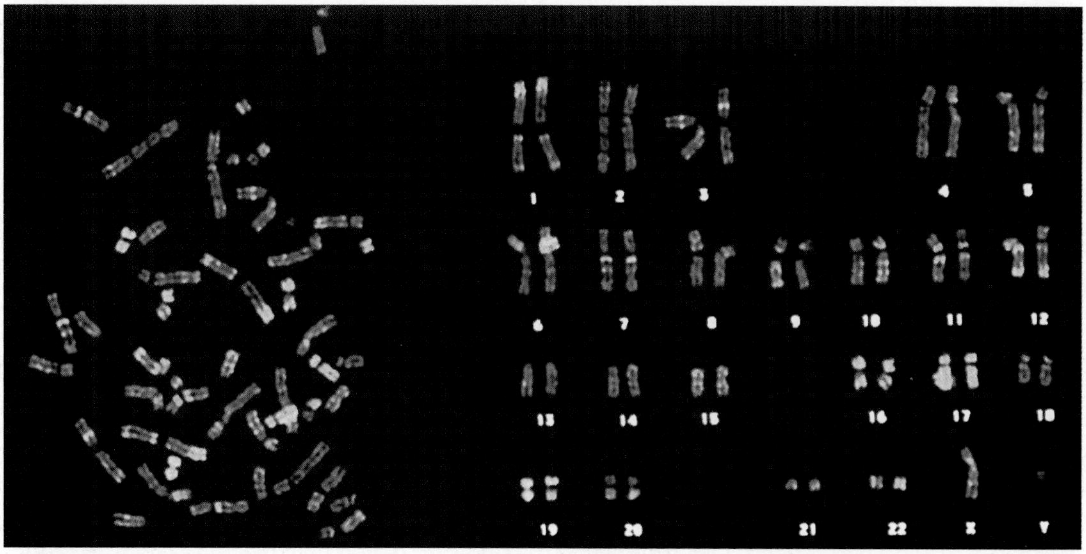

FIGURE 2–4 A metaphase preparation of chromosomes derived from a dividing cell of a human male (left), and the karyotype derived from the metaphase preparation (right). All but the X and Y chromosomes are present in homologous pairs. Each chromosome is clearly a double structure, constituting a pair of sister chromatids joined by a common centromere.

When studying mitosis, several observations are of particular relevance. First, all somatic cells derived from members of the same species contain an identical number of chromosomes. In most cases, this represents the **diploid number (2*n*).** When the lengths and centromere placements of the chromosomes are examined, a second general feature is apparent. Nearly all chromosomes exist in pairs with regard to these two criteria, and the members of each pair are called **homologous chromosomes.** So for each chromosome exhibiting a specific length and centromere placement, another exists with identical features.

There are exceptions to this rule. Most bacteria and viruses have only one chromosome, and organisms such as yeasts and molds and certain plants such as bryophytes (mosses) spend the predominant phase of the life cycle in the haploid stage. That is, they contain only one member of each homologous pair of chromosomes during most of their lives.

Figure 2–4 illustrates the physical appearance of different pairs of homologous chromosomes. There, the human mitotic chromosomes have been photographed, cut out of the print, and matched up, creating a **karyotype.** As you can see, humans have a 2*n* number of 46, which on close examination exhibit a diversity of sizes and centromere placements. Note also that each of the 46 chromosomes is clearly a double structure consisting of two parallel **sister chromatids** connected by a common centromere. Had these chromosomes been allowed to continue dividing, the sister chromatids, which are replicas of one another, would have separated into two new cells as division continued.

The **haploid number (*n*)** of chromosomes is equal to one-half the diploid number. Collectively, the genetic information contained in a haploid set of chromosomes constitutes the **genome** of the species. This, of course, includes copies of all genes as well as a large amount of noncoding DNA. The examples listed in Table 2.1 demonstrate the wide range of *n* values found in plants and animals.

Homologous pairs of chromosomes have important genetic similarities. They contain identical gene sites along their lengths, each called a **locus** (pl., loci). Thus, they are identical in their genetic potential. In sexually reproducing organisms, one member of each pair is derived from the maternal parent (through the ovum), and one is derived from the paternal parent (through the sperm). Therefore, each diploid organism contains two copies of each gene as a consequence of **biparental inheritance.** As we shall see in the following chapters on transmission genetics, the members of each pair of genes, while influencing the same characteristic or trait, need not be identical. In a population of members of the same species, many different alternative forms of the same gene, called **alleles,** can exist.

The concepts of haploid number, diploid number, and homologous chromosomes are important in understanding the process of meiosis. During the formation of gametes or spores, meiosis converts the diploid number of chromosomes to the haploid number. As a result, haploid gametes or spores contain precisely one member of each homologous pair of chromosomes—that is, one complete haploid set. Following fusion of two gametes in fertilization, the diploid number is reestablished; that is, the zygote contains two complete sets of haploid chromosomes, one set from each parent. The constancy of genetic material is thus maintained from generation to generation.

There is one important exception to the concept of homologous pairs of chromosomes. In many species, one pair, the **sex-determining chromosomes,** is often not homologous in size, centromere placement, arm ratio, or genetic content. For example, in humans, while females carry two homologous X chromosomes, males carry one Y chromosome and one X chromosome (Figure 2–4). These X and Y chromosomes are not strictly homologous. The Y is considerably smaller and lacks most of the loci contained on the X. Nevertheless, they contain homologous regions and behave as homologs in meiosis so that gametes produced by males receive either one X or one Y chromosome.

ESSENTIAL POINT

In diploid organisms, chromosomes exist in homologous pairs, where each member is identical in size, centromere placement, and gene sites. One member of each pair is derived from the maternal parent, and one is derived from the paternal parent.

TABLE 2.1 The Haploid Number of Chromosomes for a Variety of Organisms

Common Name	Scientific Name	Haploid Number
Black bread mold	*Aspergillus nidulans*	8
Broad bean	*Vicia faba*	6
Cat	*Felis domesticus*	19
Cattle	*Bos taurus*	30
Chicken	*Gallus domesticus*	39
Chimpanzee	*Pan troglodytes*	24
Corn	*Zea mays*	10
Cotton	*Gossypium hirsutum*	26
Dog	*Canis familiaris*	39
Evening primrose	*Oenothera biennis*	7
Frog	*Rana pipiens*	13
Fruit fly	*Drosophila melanogaster*	4
Garden onion	*Allium cepa*	8
Garden pea	*Pisum sativum*	7
Grasshopper	*Melanoplus differentialis*	12
Green alga	*Chlamydomonas reinhardii*	18
Horse	*Equus caballus*	32
House fly	*Musca domestica*	6
House mouse	*Mus musculus*	20
Human	*Homo sapiens*	23
Jimson weed	*Datura stramonium*	12
Mosquito	*Culex pipiens*	3
Mustard plant	*Arabidopsis thaliana*	5
Pink bread mold	*Neurospora crassa*	7
Potato	*Solanum tuberosum*	24
Rhesus monkey	*Macaca mulatta*	21
Roundworm	*Caenorhabditis elegans*	6
Silkworm	*Bombyx mori*	28
Slime mold	*Dictyostelium discoidium*	7
Snapdragon	*Antirrhinum majus*	8
Tobacco	*Nicotiana tabacum*	24
Tomato	*Lycopersicon esculentum*	12
Water fly	*Nymphaea alba*	80
Wheat	*Triticum aestivum*	21
Yeast	*Saccharomyces cerevisiae*	16
Zebrafish	*Danio rerio*	25

2.3 Mitosis Partitions Chromosomes into Dividing Cells

The process of mitosis is critical to all eukaryotic organisms. In some single-celled organisms, such as protozoans and some fungi and algae, mitosis (as a part of cell division) provides the basis for asexual reproduction. Multicellular diploid organisms begin life as single-celled fertilized eggs called **zygotes.** The mitotic activity of the zygote and the subsequent daughter cells is the foundation for the development and growth of the organism. In adult organisms, mitotic activity is prominent in wound healing and other forms of cell replacement in certain tissues. For example, the epidermal skin cells of humans are continuously sloughed off and replaced. Cell division also results in the continuous production of reticulocytes (immature red blood cells) that eventually shed their nuclei and replenish the supply of red blood cells in vertebrates. In abnormal situations, somatic cells may lose control of cell division and form a tumor.

The genetic material is partitioned into daughter cells during nuclear division or **karyokinesis.** This process is quite complex and requires great precision. The chromosomes must first be exactly replicated and then accurately partitioned. The end result is the production of two daughter nuclei, each with a chromosome composition identical to that of the parent cell.

Karyokinesis is followed by cytoplasmic division, or **cytokinesis.** The less complex division of the cytoplasm requires a mechanism that partitions the volume into two parts, then encloses both new cells within a distinct plasma membrane. Cytoplasmic organelles either replicate themselves, arise from existing membrane structures, or are synthesized *de novo* (anew) in each cell. The subsequent proliferation of these structures is a reasonable and adequate mechanism for reconstituting the cytoplasm in daughter cells.

Following cell division, the initial size of each new daughter cell is approximately one-half the size of the parent cell. However, the nucleus of each new cell is not appreciably smaller than the nucleus of the original cell. Quantitative measurements of DNA confirm that there is an amount of genetic material in the daughter nuclei equivalent to that in the parent cell.

Interphase and the Cell Cycle

Many cells undergo a continuous alternation between division and nondivision. The events that occur from the completion of one division until the beginning of the next division constitute the **cell cycle** (Figure 2–5). We will consider the initial **interphase** stage of the cycle as the interval between divisions. It was once thought that the biochemical activity during interphase was devoted solely to the cell's growth and its normal function. However, we now know

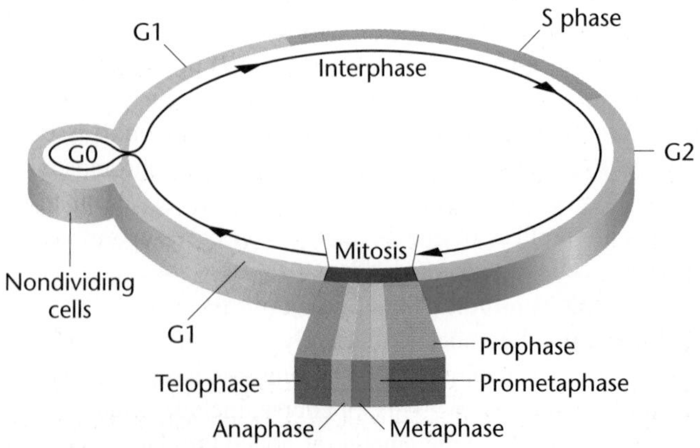

FIGURE 2–5 The intervals comprising an arbitrary cell cycle. Following mitosis, cells enter the G1 stage of interphase, initiating a new cycle. Cells may become nondividing (G0) or continue through G1, where they become committed to begin DNA synthesis (S) and complete the cycle (G2 and mitosis). Following mitosis, two daughter cells are produced and the cycle begins anew for both cells.

that another biochemical step critical to the ensuing mitosis occurs during interphase: *replication of the DNA of each chromosome*. This period during which DNA is synthesized occurs before the cell enters mitosis and is called the **S phase.** The initiation and completion of DNA synthesis can be detected by monitoring the incorporation of radioactive precursors into DNA.

Investigations of this nature show two periods during interphase when no DNA synthesis occurs, one before and one after S phase. These are designated **G1 (gap I)** and **G2 (gap II),** respectively. During both of these intervals, as well as during S, intensive metabolic activity, cell growth, and cell differentiation occur. By the end of G2, the volume of the cell has roughly doubled, DNA has been replicated, and mitosis (M) is initiated. Following mitosis, continuously dividing cells then repeat this cycle (G1, S, G2, M) over and over, as shown in Figure 2–5.

Much is known about the cell cycle based on *in vitro* (in glass) studies. When grown in culture, many cell types in different organisms traverse the complete cycle in about 16 hours. The actual process of mitosis occupies only a small part of the overall cycle, often less than an hour. The lengths of the S and G2 phases of interphase are fairly consistent among different cell types. Most variation is seen in the length of time spent in the G1 stage. Figure 2–6 shows the length of these intervals in a typical cell.

G1 is of great interest in the study of cell proliferation and its control. At a point late in G1, all cells follow one of two paths. They either withdraw from the cycle, become quiescent, and enter the **G0 stage** (see Figure 2–5), or they become committed to initiating DNA synthesis and completing the cycle. Cells that enter G0 remain viable and metabolically active but are nonproliferative. Cancer cells apparently avoid entering G0 or pass through it very quickly. Other cells enter G0 and never reenter the cell cycle. Still others remain in G0, but they can be stimulated to return to G1 and thereby reenter the cycle.

Cytologically, interphase is characterized by the absence of visible chromosomes. Instead, the nucleus is filled with chromatin fibers that are formed as the chromosomes are uncoiled and dispersed after the previous mitosis [Figure 2–7(a), p. 24]. Once G1, S, and G2 are completed, mitosis is initiated. Mitosis is a dynamic period of vigorous and continual activity. For discussion purposes, the entire process is subdivided into discrete stages, and specific events are assigned to each one. These stages, in order of occurrence, are prophase, prometaphase, metaphase, anaphase, and telophase. They are diagrammed in Figure 2–7 along with photomicrographs of each stage.

Prophase

Often, over half of mitosis is spent in **prophase** [Figure 2–7(b)], a stage characterized by several significant activities. One of the early events in prophase of all animal cells involves the migration of two pairs of centrioles to opposite ends of the cell. These structures are found just outside the nuclear envelope in an area of differentiated cytoplasm called the **centrosome.** It is believed that each pair of centrioles consists of one mature unit and a smaller, newly formed centriole.

Interphase			Mitosis
G1	S	G2	M
5	7	3	1

Hours

Pro	Met	Ana	Tel
36	3	3	18

Minutes

FIGURE 2–6 The time spent in each phase of one complete cell cycle of a human cell in culture. Times vary according to cell types and conditions.

The centrioles migrate to establish poles at opposite ends of the cell. After migrating, the centrioles are responsible for organizing cytoplasmic microtubules into the spindle fibers that run between these poles, creating an axis along which chromosomal separation occurs. Interestingly, the cells of most plants (there are a few exceptions), fungi, and certain algae seem to lack centrioles. Spindle fibers are nevertheless apparent during mitosis. Therefore, centrioles are not universally responsible for the organization of spindle fibers.

As the centrioles migrate, the nuclear envelope begins to break down and gradually disappears. In a similar fashion, the nucleolus disintegrates within the nucleus. While these events are taking place, the diffuse chromatin fibers condense, continuing until distinct threadlike structures, the chromosomes, become visible. It becomes apparent near the end of prophase that each chromosome is actually a double structure split longitudinally except at the centromere (review Figure 2–4). The two parts of each chromosome are called **chromatids.** Because the DNA contained in them represents the duplication of a single chromosome, they are genetically identical. Therefore, they are called *sister chromatids*. In humans, with a diploid number of 46, a cytological preparation of late prophase reveals 46 chromosomes randomly distributed in the area formerly occupied by the nucleus.

Prometaphase and Metaphase

The distinguishing event of the ensuing stages is the migration of each chromosome, led by the centromeric region, to the equatorial plane. The equatorial plane, or **metaphase plate,** is the midline region of the cell, a plane that lies perpendicular to the axis established by the spindle fibers. In some descriptions, the term **prometaphase** refers to the period of chromosome movement [Figure 2–7(c)], and **metaphase** is applied strictly to the chromosome configuration after migration.

Migration is made possible by the binding of spindle fibers to a structure associated with the centromere of each chromosome called the **kinetochore.** This structure, consisting of multilayered plates of proteins, forms on opposite sides of each centromere, intimately associating with the two sister chromatids of each chromosome. Once attached to microtubules

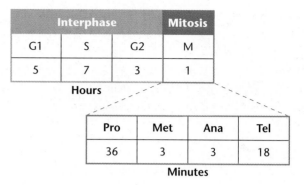

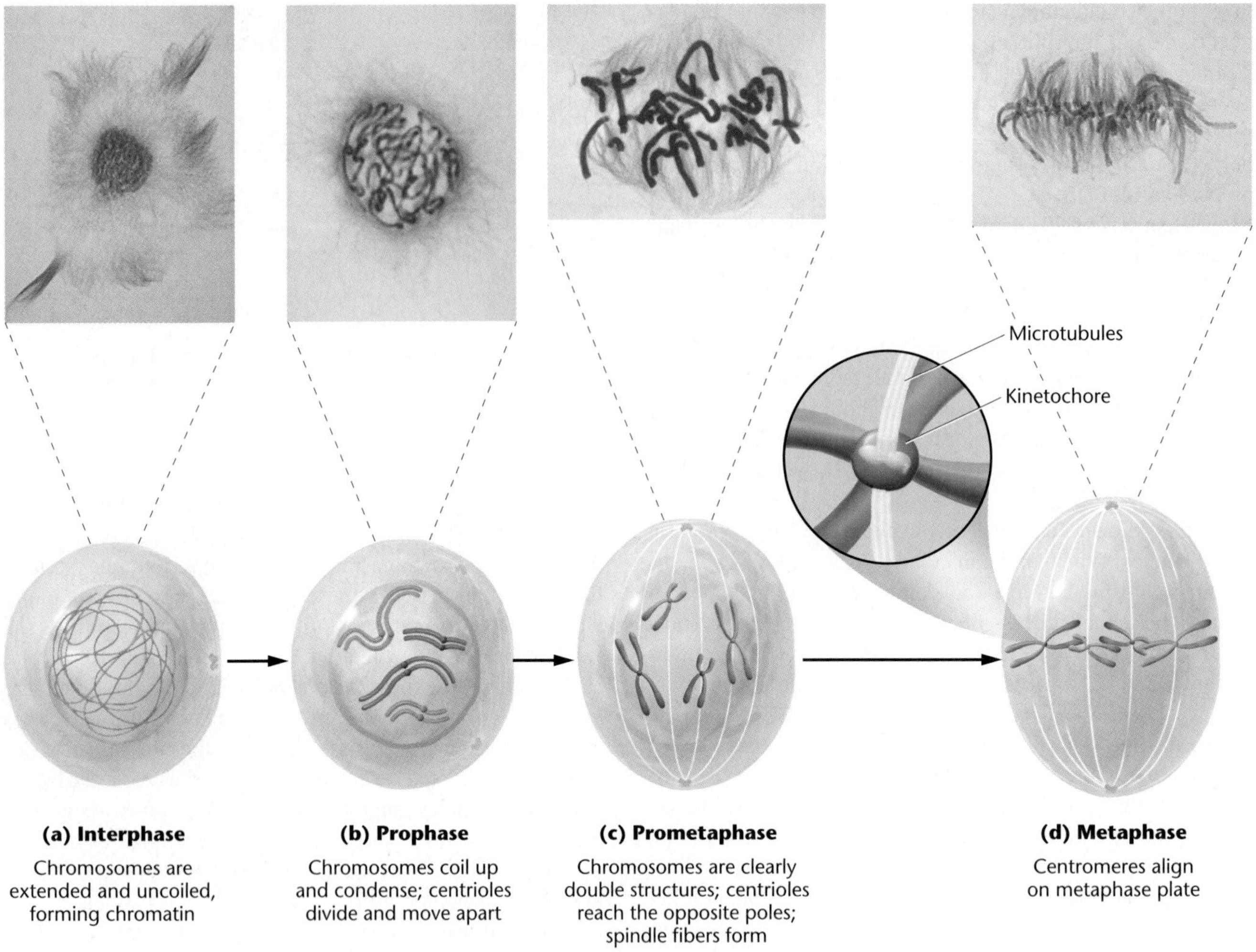

FIGURE 2–7 Drawings depicting mitosis in an animal cell with a diploid number of 4. The events occurring in each stage are described in the text. Of the two homologous pairs of chromosomes, one contains longer, metacentric members and the other shorter, submetacentric members. The maternal chromosome and the paternal chromosome of each pair are shown in different colors. In (f), the late telophase stage in a plant cell illustrates the formation of the cell plate and lack of centrioles. The cells shown in light micrographs came from the flower of *Haemanthus,* a plant that has a diploid number of 8.

making up the spindle fibers, the sister chromatids are now ready to be pulled to opposite poles during the ensuing anaphase stage.

At the completion of metaphase, each centromere is aligned at the metaphase plate, with the chromosome arms extending outward in a random array. This configuration is shown in Figure 2–7(d).

Anaphase

Events critical to chromosome distribution during mitosis occur during **anaphase,** the shortest stage of mitosis. During this phase, sister chromatids of each chromosome *disjoin* (separate) from each other and migrate to opposite ends of the cell. For complete disjunction to occur, each centromeric region must be split in two. This event signals the initiation of anaphase. Once it occurs, each chromatid is referred to as a **daughter chromosome.**

Movement of daughter chromosomes to the opposite poles of the cell is dependent on the centromere–spindle fiber attachment. Recent investigations reveal that chromosome migration results from the activity of a series of specific proteins, generally called motor proteins. These proteins use the energy generated by the hydrolysis of ATP, and their activity is said to constitute **molecular motors** in the cell. These motors act at several positions within the dividing cell, but all of them are involved in the activity of microtubules and ultimately serve to propel the chromosomes to opposite ends of the cell. The centromeres of each chromosome *appear* to lead the way during migration, with the chromosome arms trailing behind. The location of the centromere determines the shape of the chromosome during separation, as you saw in Figure 2–3.

The steps that occur during anaphase are critical in providing each subsequent daughter cell with an identical set of chromosomes. In human cells, there would now be 46 chromosomes at each pole, one from each original sister pair. Figure 2–7(e) shows anaphase just prior to its completion.

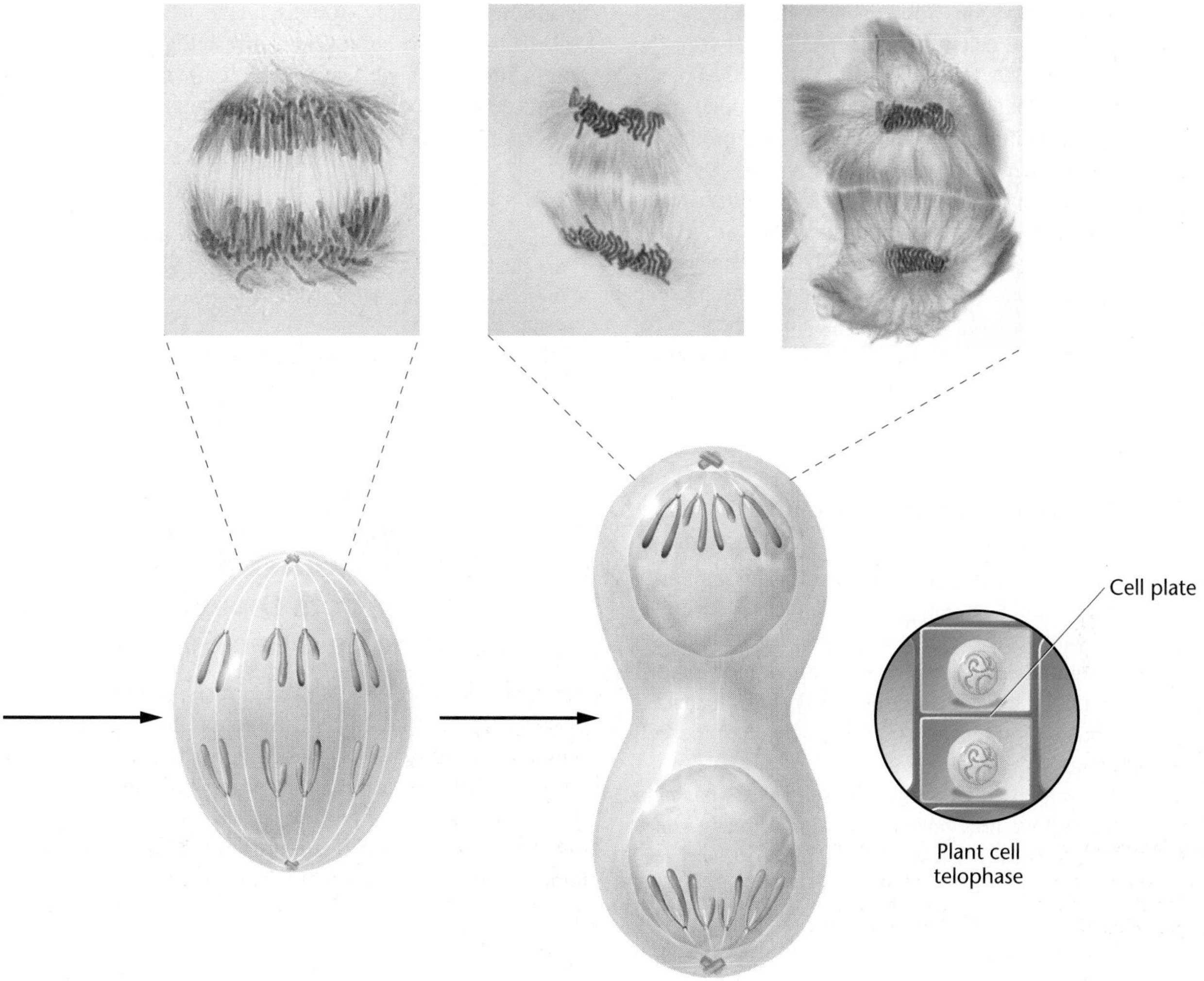

(e) Anaphase
Centromeres split and daughter chromosomes migrate to opposite poles

(f) Telophase
Daughter chromosomes arrive at the poles; cytokinesis commences

FIGURE 2–7 (Continued)

NOW SOLVE THIS

With the initial appearance of the feature we call "Now Solve This," a short introduction is in order. Occurring several times in this and all ensuing chapters, each entry identifies a problem from the "Problems and Discussion Questions" section at the end of each chapter related to the discussion just presented. A comment is made about the problem, and then a Hint is offered providing you with an analytical insight that will be useful as you solve the problem. Here is the first one:

Problem 5 on p. 36 involves an understanding of what happens to each pair of homologous chromosomes during mitosis, asking you to extrapolate your understanding to chromosome behavior in an organism with a diploid number of 16.

Hint: The major issue in solving this problem is to understand that, throughout mitosis, members of each homologous pair do not pair up but instead behave individually.

Telophase

Telophase is the final stage of mitosis and is depicted in Figure 2–7(f). At its beginning, there are two complete sets of chromosomes, one set at each pole. The most significant event is **cytokinesis,** the division or partitioning of the cytoplasm. Cytokinesis is essential if two new cells are to be produced from one cell. The mechanism differs greatly in plant and animal cells. In plant cells, a **cell plate** is synthesized and laid down across the region of the metaphase plate. Animal cells, however, undergo a constriction of the cytoplasm in the same way a loop of string might be tightened around the middle of a balloon. The end result is the same: Two distinct cells are formed.

It is not surprising that the process of cytokinesis varies among cells of different organisms. Plant cells, which are more regularly shaped and are structurally rigid, require a mechanism for depositing new cell wall material around the plasma membrane. The cell plate, laid down during telophase, becomes the **middle lamella.** Subsequently, the primary and secondary layers of the cell wall are deposited between the cell membrane and middle lamella on both sides of the boundary between the two daughter cells. In animals, complete constriction of the cell membrane produces a **cell furrow** characteristic of newly divided cells.

Other events necessary for the transition from mitosis to interphase are initiated during late telophase. They are generally a reversal of the events that occurred during prophase. In each new cell, the chromosomes begin to uncoil and become diffuse chromatin once again while the nuclear envelope reforms around them. The spindle fibers disappear, and the nucleolus gradually reforms and becomes visible in the nucleus during early interphase. At the completion of telophase, the cell enters interphase.

Cell-Cycle Regulation and Checkpoints

The cell cycle, culminating in mitosis (Figure 2–5), is fundamentally the same in all eukaryotic organisms. This similarity in many diverse organisms suggests that the cell cycle is governed by a genetically regulated program that has been conserved throughout evolution. Because disruption of this regulation may underlie the uncontrolled cell division characterizing malignancy, interest in how genes regulate the cell cycle is particularly strong.

A mammoth research effort over the past 15 years has paid high dividends, and we now have knowledge of many genes involved in the control of the cell cycle. This work was recognized by the awarding of the 2001 Nobel Prize in Medicine or Physiology to Lee Hartwell, Paul Nurse, and Tim Hunt. As with other studies of genetic control over essential biological processes, investigation has focused on the discovery of mutations that interrupt the cell cycle and on the effects of those mutations.

Many mutations are now known that exert an effect at one or another stage of the cell cycle. First discovered in yeast, but now evident in all organisms, including humans, such mutations were originally designated as ***cell division cycle (cdc)* mutations.** The normal products of many of the mutated genes are enzymes called **kinases** that can add phosphates to other proteins. They serve as "master control" molecules functioning in conjunction with proteins called **cyclins.** Cyclins bind to these kinases, activating them at appropriate times during the cell cycle. Activated kinases then phosphorylate other target proteins that regulate the progress of the cell cycle.

The study of *cdc* mutations has established that the cell cycle contains at least three major **checkpoints,** when the processes culminating in normal mitosis are monitored, or "checked," by these master control molecules before the next stage of the cycle commences. The importance of cell-cycle control and these checkpoints, discussed in detail in Chapter 16, is illustrated by considering what happens when this regulatory system is impaired. Let's assume, for example, that the DNA of a cell has incurred damage leading to one or more mutations impairing cell-cycle control. If allowed to proceed through the cell cycle as one of the population of dividing cells, this genetically altered cell would divide uncontrollably—precisely the definition of a tumor cell. If instead the cell cycle is arrested at one of the checkpoints, the cell may effectively be removed from the population of dividing cells, preventing its potential malignancy.

ESSENTIAL POINT

Mitosis is subdivided into discrete stages that initially depict the condensation of chromatin into the diploid number of chromosomes, each of which is initially a double structure, each composed of a pair of sister chromatids. During mitosis, sister chromatids are pulled apart and directed toward opposite poles, after which cytoplasmic division creates two new cells with identical genetic information.

2.4 Meiosis Reduces the Chromosome Number from Diploid to Haploid in Germ Cells and Spores

The process of meiosis, unlike mitosis, reduces the amount of genetic material to half. Whereas in diploids, mitosis produces daughter cells with a full diploid complement, meiosis produces gametes or spores with only one haploid set of chromosomes. During sexual reproduction, gametes then combine at fertilization to reconstitute the diploid complement found in parental cells. Figure 2–8 compares the two processes by following two pairs of homologous chromosomes.

Meiosis must be highly specific since, by definition, haploid gametes or spores contain precisely one member of each homologous pair of chromosomes. When successfully completed, meiosis ensures genetic continuity from generation to generation.

The process of sexual reproduction also ensures genetic variety among members of a species. As you study meiosis, you will see that this process results in gametes with many unique combinations of maternally and paternally derived chromosomes among the haploid complement. With such a tremendous genetic variation among the gametes, a large number of chromosome combinations are possible at fertilization. Furthermore, the meiotic event referred to as **crossing over** results in genetic exchange between members of each homologous pair of chromosomes. This creates intact chromosomes that are mosaics of the maternal and paternal homologs from which they arise, further enhancing the potential genetic variation in gametes and the offspring derived from them. Sexual reproduction therefore reshuffles the genetic material, producing offspring that often differ greatly from either parent. Thus, meiosis is the major form of genetic recombination within species.

An Overview of Meiosis

In the preceding discussion, we established what might be considered the goals of meiosis. Before we consider the phases of this process systematically, we will briefly examine how diploid cells give rise to haploid gametes or spores. You

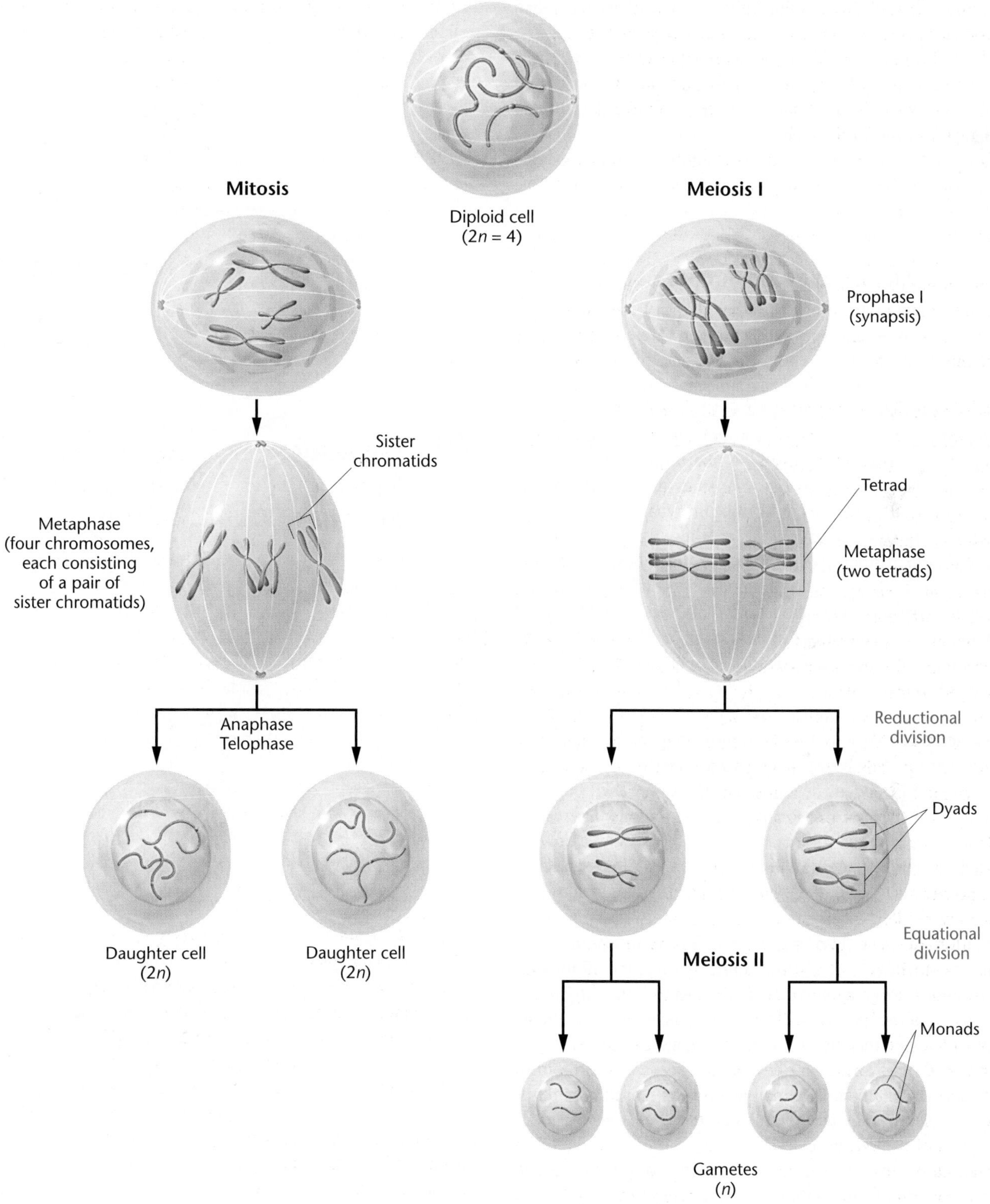

FIGURE 2–8 Overview of the major events and outcomes of mitosis and meiosis. As in Figure 2–7, two pairs of homologous chromosomes are followed.

should refer to the right side (Meiosis I) of Figure 2–8 during the following discussion.

You have seen that in mitosis each paternally and maternally derived member of any homologous pair of chromosomes behaves autonomously during division. By contrast, early in meiosis, homologous chromosomes form pairs; that is, they synapse. Each synapsed structure is initially called a **bivalent,** which eventually gives rise to a unit, the **tetrad,** consisting of four chromatids. The presence of four chromatids demonstrates that *both* homologs (making up the bivalent) have, in fact, duplicated. Therefore, in order to achieve haploidy, two divisions are necessary.

The first division occurs in meiosis I and is described as a **reductional division** (because the number of centromeres, each representing one chromosome, is *reduced* by one-half). Components of each tetrad—representing the two homologs—separate, yielding two **dyads.** Each dyad is composed of two sister chromatids joined at a common centromere. The second division occurs during meiosis II and is described as an **equational division** (because the number of centromeres remains *equal*). Here each dyad splits into two **monads** of one chromosome each. Thus, the two divisions potentially produce four haploid cells.

The First Meiotic Division: Prophase I

We turn now to a more detailed account of meiosis. As in mitosis, meiosis is a continuous process. We name the parts of each stage of division only to facilitate discussion. From a genetic standpoint, three events characterize the initial stage, **prophase I** (Figure 2–9). First, as in mitosis, the chromatin present in interphase condenses and coils into visible chromosomes. Second, unlike mitosis, members of each homologous pair of chromosomes undergo **synapsis**. Third, crossing over occurs between synapsed homologs. Because of the complexity of these genetic events, prophase I is further divided into five substages: leptonema, zygonema, pachynema, diplonema,* and diakinesis. As we discuss these substages, be aware that even though it is not immediately apparent in the earliest phases of meiosis, the DNA of chromosomes has already been replicated during the prior interphase.

Leptonema During the **leptotene stage,** the interphase chromatin material begins to condense, and the chromosomes, though still extended, become visible. Along each chromosome are **chromomeres,** localized condensations that resemble beads on a string. Recent evidence suggests that a process called homology search, which precedes and is essential to the initial pairing of homologs, begins during leptonema.

Zygonema The chromosomes continue to shorten and thicken during the **zygotene stage**. During the **homology search,** homologous chromosomes undergo initial alignment with one another. This so-called rough pairing is complete by the end of zygonema. In yeast, homologs are separated by about 300 nm, and near the end of zygonema, structures called lateral elements are visible between paired homologs. As meiosis proceeds, the overall length of the lateral elements along the chromosome increases, and a more extensive ultrastructural component, the **synaptonemal complex,** begins to form between the homologs.

At the completion of zygonema, the paired homologs are referred to as bivalents. Although both members of each bivalent have already replicated their DNA, it is not yet visually apparent that each member is a double structure. The number of bivalents in each species is equal to the haploid (n) number.

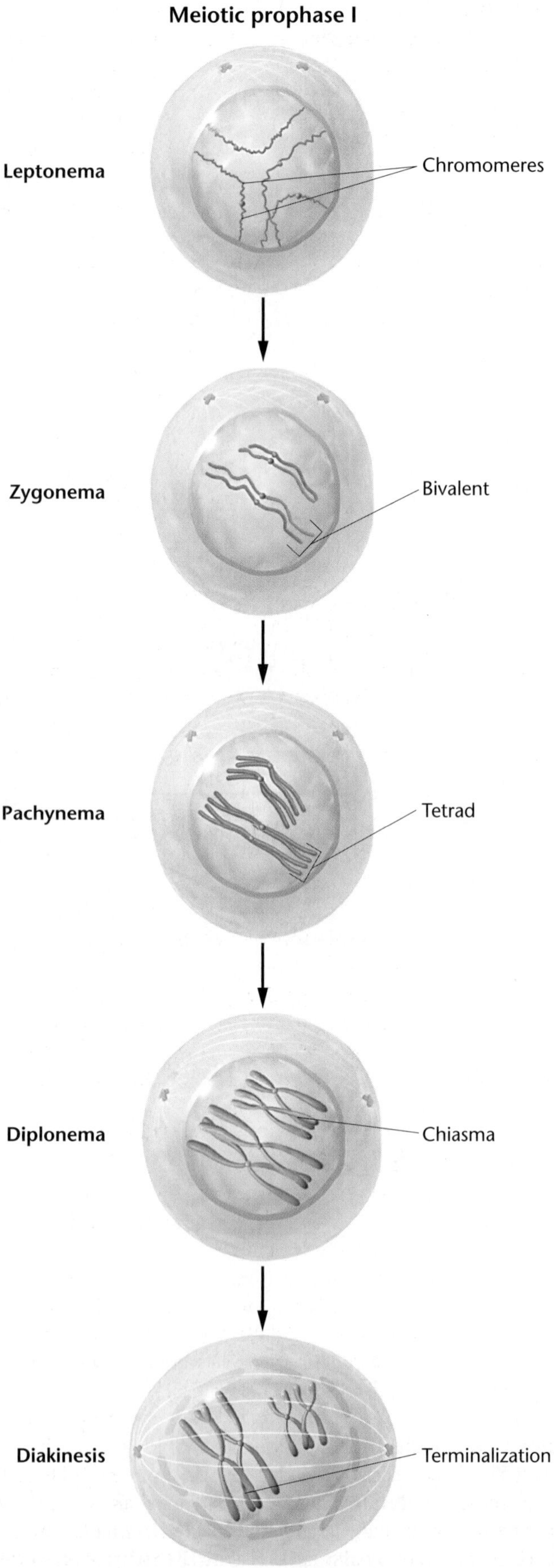

FIGURE 2–9 The substages of meiotic prophase I for the chromosomes depicted in Figure 2–8.

* These are the noun forms of these substages. The adjective forms (leptotene, zygotene, pachytene, and diplotene) are also used.

Pachynema In the transition from the zygotene to the **pachytene stage,** the chromosomes continue to coil and shorten, and further development of the synaptonemal complex occurs between the two members of each bivalent. This leads to synapsis, a more intimate pairing. Compared to the rough-pairing characteristic of yeast zygonema, the homologs are now separated by only 100 nm.

During pachynema, each homolog is first evident as a double structure, providing visual evidence of the earlier replication of the DNA of each chromosome. Thus, each bivalent contains four chromatids. As in mitosis, replicates are called sister chromatids, while chromatids from maternal and paternal members of a homologous pair are called nonsister chromatids. The four-membered structure is a tetrad, and each tetrad contains two pairs of sister chromatids.

Diplonema During the ensuing **diplotene stage,** it is even more apparent that each tetrad consists of two pairs of sister chromatids. Within each tetrad, each pair of sister chromatids begins to separate. However, one or more areas remain in contact where chromatids are intertwined. Each such area, called a **chiasma** (pl., chiasmata), is thought to represent a point where nonsister chromatids have undergone genetic exchange through the process of crossing over. Although the physical exchange between chromosome areas occurred during the previous pachytene stage, the result of crossing over is visible only when the duplicated chromosomes begin to separate. Crossing over is an important source of genetic variability, and new combinations of genetic material are formed during this process.

Diakinesis The final stage of prophase I is **diakinesis.** The chromosomes pull farther apart, but the nonsister chromatids remain loosely associated via the chiasmata. As separation proceeds, the chiasmata move toward the ends of the tetrad. This process of **terminalization** begins in late diplonema and is completed during diakinesis. During this final substage period, the nucleolus and nuclear envelope break down, and the two centromeres of each tetrad attach to the recently formed spindle fibers. By the completion of prophase I, the centromeres of each tetrad structure are present on the metaphase plate of the cell.

Metaphase, Anaphase, and Telophase I

The remainder of the meiotic process is depicted in Figure 2–10, p. 30. After meiotic prophase I, steps similar to those of mitosis occur. In the first division, **metaphase I,** the chromosomes have maximally shortened and thickened. The terminal chiasmata of each tetrad are visible and appear to be the only factor holding the nonsister chromatids together. Each tetrad interacts with spindle fibers, facilitating movement to the metaphase plate. The alignment of each tetrad prior to the first anaphase is random. Half of each tetrad is pulled randomly to one or the other pole, and the other half then moves to the opposite pole.

During the stages of meiosis I, a single centromere holds each pair of sister chromatids together. It does *not* divide. At **anaphase I,** one-half of each tetrad (the dyad) is pulled toward each pole of the dividing cell. This separation process is the physical basis of **disjunction,** the separation of chromosomes from one another. Occasionally, errors in meiosis occur and separation is not achieved. The term **nondisjunction** describes such an error. At the completion of a normal anaphase I, a series of dyads equal to the haploid number is present at each pole.

If crossing over had not occurred in the first meiotic prophase, each dyad at each pole would consist solely of either paternal or maternal chromatids. However, the exchanges produced by crossing over create mosaic chromatids of paternal and maternal origin.

In many organisms, **telophase I** reveals a nuclear membrane forming around the dyads. Next, the nucleus enters into a short interphase period. If interphase occurs, the chromosomes do not replicate since they already consist of two chromatids. In other organisms, the cells go directly from anaphase I to meiosis II. In general, meiotic telophase is much shorter than the corresponding stage in mitosis.

The Second Meiotic Division

A second division, **meiosis II,** is essential if each gamete or spore is to receive only one chromatid from each original tetrad. The stages characterizing meiosis II are shown in the right half of Figure 2–10. During **prophase II,** each dyad is composed of one pair of sister chromatids attached by a common centromere. During **metaphase II,** the centromeres are positioned on the metaphase plate. When they divide, **anaphase II** is initiated, and the sister chromatids of each dyad are pulled to opposite poles. Because the number of dyads is equal to the haploid number, **telophase II** reveals one member of each pair of homologous chromosomes at each pole. Each chromosome is now a monad. Following cytokinesis in telophase II, four haploid gametes may result from a single meiotic event. At the conclusion of meiosis II, not only has the haploid state been achieved, but if crossing over has occurred, each monad is also a combination of maternal and paternal genetic information. As a result, the offspring produced by any gamete receives a mixture of genetic information originally present in his or her grandparents. Meiosis thus significantly increases the level of genetic variation in each ensuing generation.

ESSENTIAL POINT

Meiosis converts a diploid cell into a haploid gamete or spore, making sexual reproduction possible. As a result of chromosome duplication and two subsequent meiotic divisions, each haploid cell receives one member of each homologous pair of chromosomes.

NOW SOLVE THIS

Problem 14 on p. 36 involves an understanding of what happens to the maternal and paternal members of each pair of homologous chromosomes during meiosis, asking you to extrapolate your understanding to chromosome behavior in an organism with a diploid number of 16.

Hint: The major issue in solving this problem is to understand that maternal and paternal homologs synapse during meiosis. Once it is evident that each chromatid has duplicated, creating a tetrad in the early phases of meiosis, each original pair behaves as a unit and leads to two dyads during anaphase I.

MEIOSIS WEB TUTORIAL

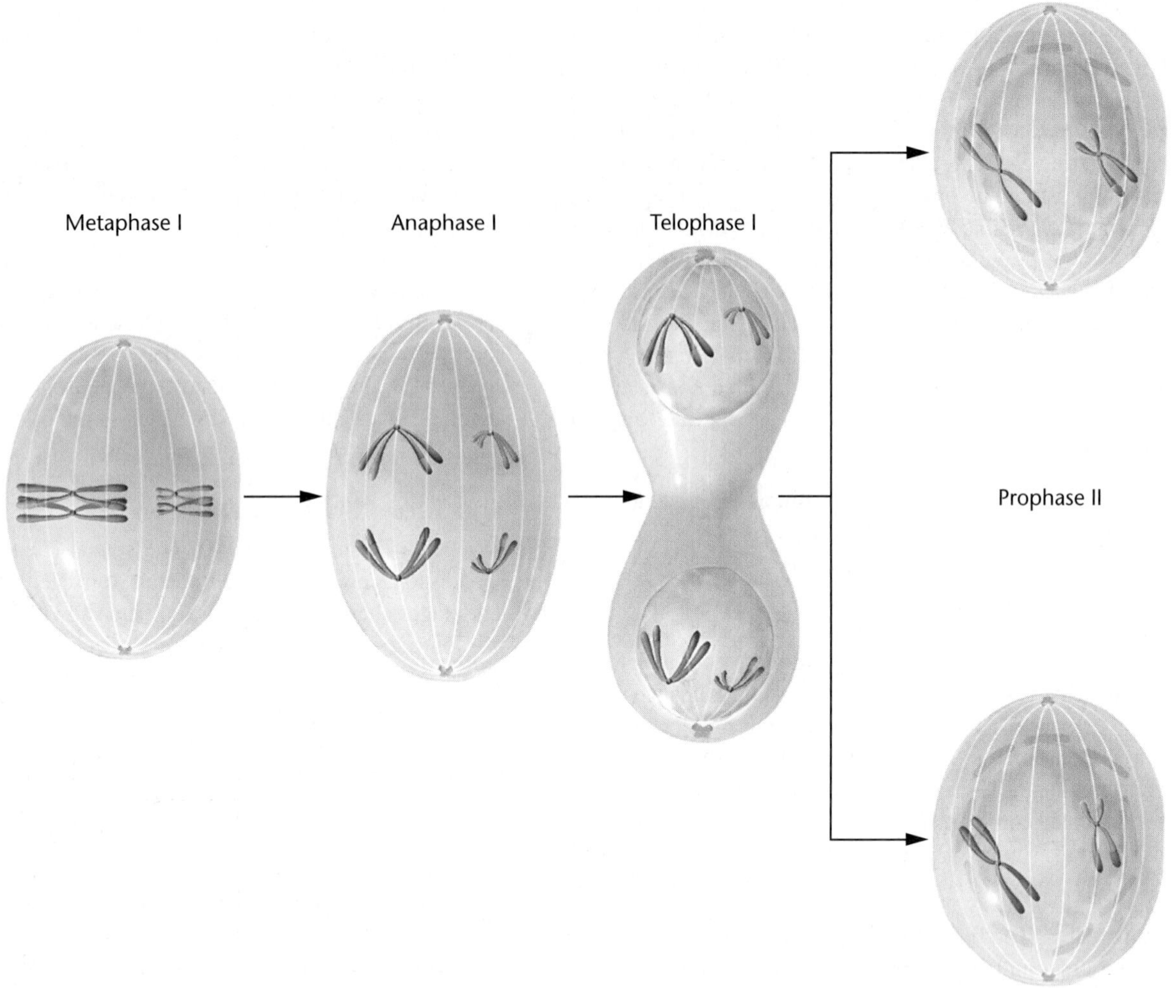

FIGURE 2–10 The major events in meiosis in an animal cell with a diploid number of 4, beginning with metaphase I. Note that the combination of chromosomes in the cells produced following telophase II is dependent on the random alignment of each tetrad and dyad on the equatorial plate during metaphase I and metaphase II. Several other combinations, which are not shown, can also be formed. The events depicted here are described in the text.

2.5 The Development of Gametes Varies during Spermatogenesis and Oogenesis

Although events that occur during the meiotic divisions are similar in all cells that participate in gametogenesis in most animal species, there are certain differences between the production of a male gamete (spermatogenesis) and a female gamete (oogenesis). Figure 2–11, p. 32, summarizes these processes.

Spermatogenesis takes place in the testes, the male reproductive organs. The process begins with the expanded growth of an undifferentiated diploid germ cell called a **spermatogonium.** This cell enlarges to become a **primary spermatocyte,** which undergoes the first meiotic division. The products of this division, called **secondary spermatocytes,** contain a haploid number of dyads. The secondary spermatocytes then undergo meiosis II, and each of these cells produces two haploid **spermatids.** Spermatids go through a series of developmental changes, **spermiogenesis,** and become highly specialized, motile **spermatozoa** or **sperm.** All sperm cells produced during spermatogenesis contain the haploid number of chromosomes and equal amounts of cytoplasm.

Spermatogenesis may be continuous or may occur periodically in mature male animals; its onset is determined by the species' reproductive cycle. Animals that reproduce year-round produce sperm continuously, whereas those whose breeding period is confined to a particular season produce sperm only during that time.

In animal **oogenesis,** the formation of **ova** (sing., ovum), or eggs, takes place in the ovaries, the female reproductive organs. The daughter cells resulting from the two meiotic divisions receive equal amounts of genetic material, but they do *not* receive equal amounts of cytoplasm. Instead, during each

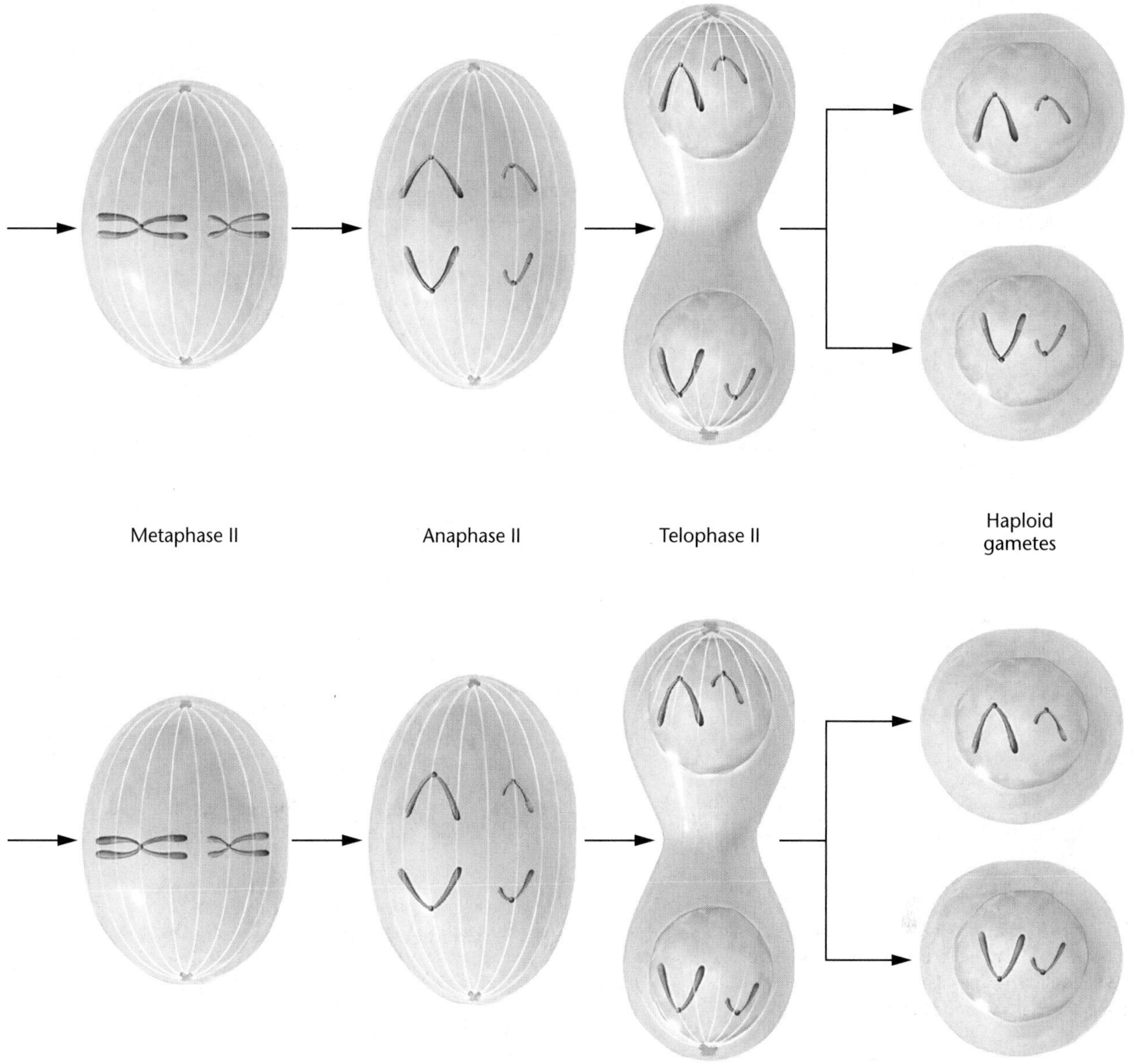

FIGURE 2–10 (Continued)

division almost all the cytoplasm of the **primary oocyte,** which is derived from the **oogonium,** is concentrated in one of the two daughter cells. This concentration of cytoplasm is necessary because a major function of the mature ovum is to nourish the developing embryo after fertilization.

During anaphase I in oogenesis, the tetrads of the primary oocyte separate, and the dyads move toward opposite poles. During telophase I, the dyads at one pole are pinched off with very little surrounding cytoplasm to form the **first polar body.** The first polar body may or may not divide again to produce two small haploid cells. The other daughter cell produced by this first meiotic division contains most of the cytoplasm and is called the **secondary oocyte.** The mature ovum will be produced from the secondary oocyte during the second meiotic division. During this division, the cytoplasm of the secondary oocyte again divides unequally, producing an **ootid** and a **second polar body.** The ootid then differentiates into the mature ovum.

Unlike the divisions of spermatogenesis, the two meiotic divisions of oogenesis may not be continuous. In some animal species, the two divisions may directly follow each other. In others, including humans, the first division of all oocytes begins in the embryonic ovary but arrests in prophase I. Many years later, meiosis resumes in each oocyte just prior to its ovulation. The second division is completed only after fertilization.

ESSENTIAL POINT

There is a major difference between meiosis in males and in females. On the one hand, spermatogenesis partitions the cytoplasmic volume equally and produces four haploid sperm cells. Oogenesis, on the other hand, collects the bulk of cytoplasm in one egg cell and reduces the other haploid products to polar bodies. The extra cytoplasm in the egg contributes to zygote development following fertilization.

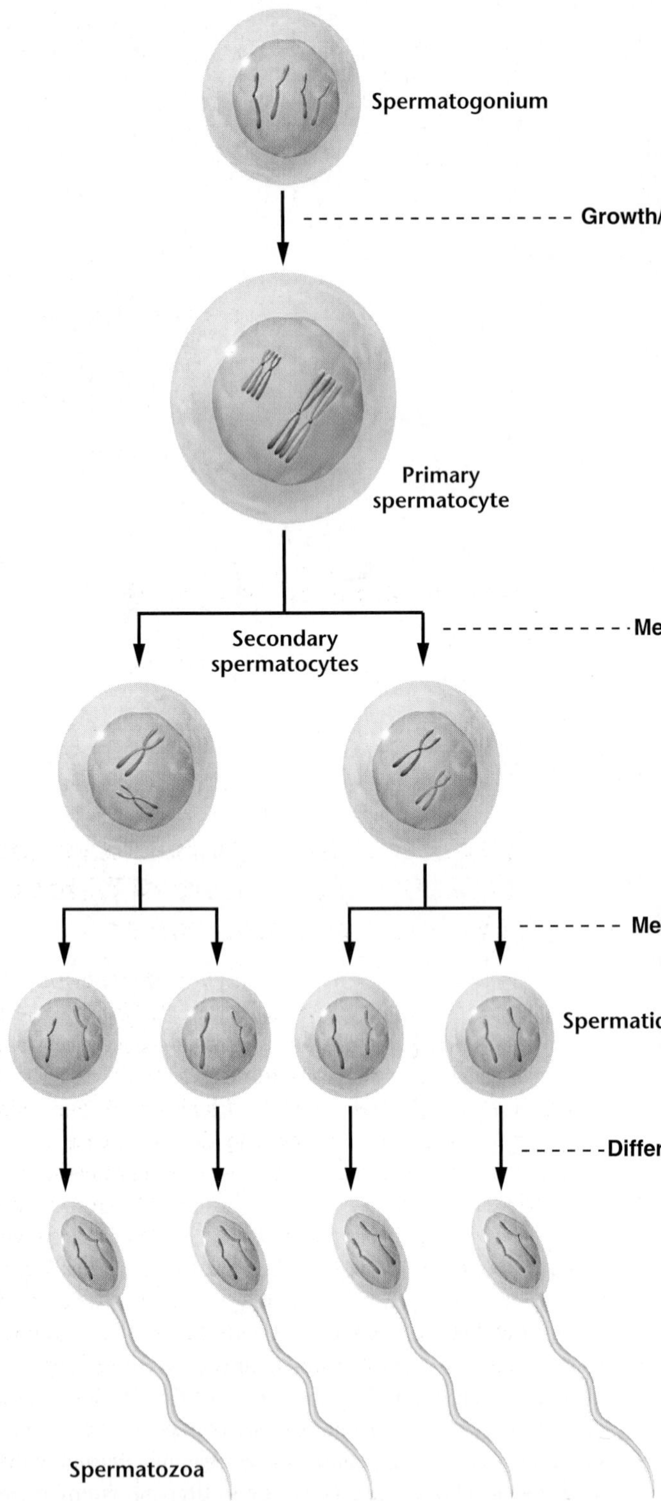

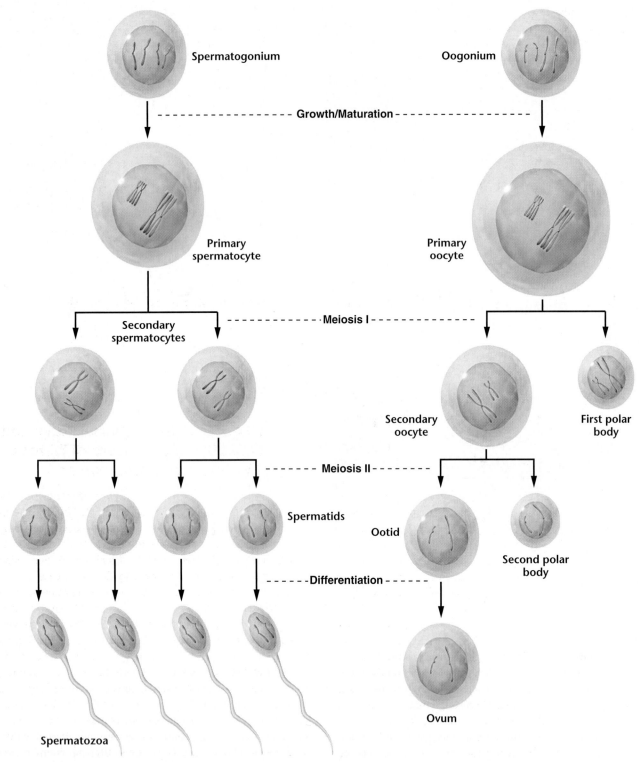

FIGURE 2–11 Spermatogenesis and oogenesis in animal cells.

NOW SOLVE THIS

Problem 9 on p. 36 involves an understanding of meiosis during oogenesis, asking you to demonstrate your knowledge of polar bodies.

Hint: To answer this question, you must take into account that crossing over occurred during meiosis I between each pair of homologs.

2.6 Meiosis Is Critical to the Sexual Reproduction Cycle of All Diploid Organisms

The process of meiosis is critical to the successful sexual reproduction of all diploid organisms. It is the mechanism by which the diploid amount of genetic information is reduced to the haploid amount. In animals, meiosis leads to the formation of

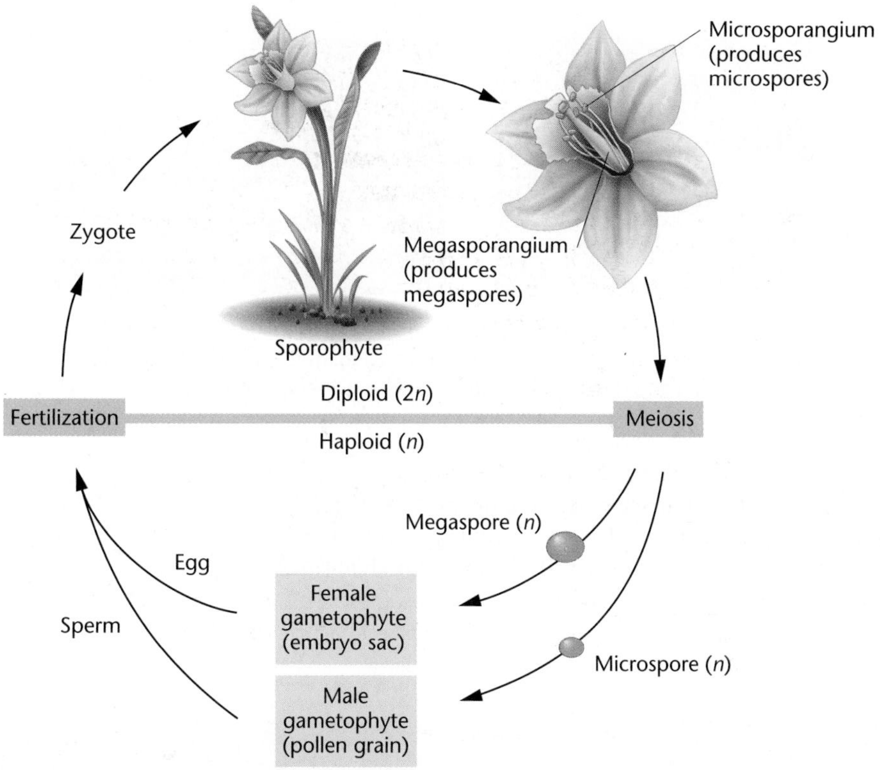

FIGURE 2–12 Alternation of generations between the diploid sporophyte ($2n$) and the haploid gametophyte (n) in a multicellular plant. The processes of meiosis and fertilization bridge the two phases of the life cycle. This is an angiosperm, where the sporophyte stage is the predominant phase.

gametes, whereas in plants, haploid spores are produced, which in turn leads to the formation of haploid gametes.

Each diploid organism contains its genetic information in the form of homologous pairs of chromosomes. Each pair consists of one member derived from the maternal parent and one from the paternal parent. Following meiosis, haploid cells potentially contain either the paternal or maternal representative of each homologous pair of chromosomes. However, the process of crossing over, which occurs in the first meiotic prophase, reshuffles alleles between the maternal and paternal members of each homologous pair, which then segregate and assort independently into gametes. This results in the great amounts of genetic variation in gametes.

It is important to touch briefly on the significant role that meiosis plays in the life cycles of fungi and plants. In many fungi, the predominant stage of the life cycle consists of haploid vegetative cells. They arise through meiosis and proliferate by mitotic cell division. In multicellular plants, the life cycle alternates between the diploid **sporophyte stage** and the haploid **gametophyte stage** (Figure 2–12). While one or the other predominates in different plant groups during this "alternation of generations," the processes of meiosis and fertilization constitute the "bridges" between the sporophyte and gametophyte stages. Therefore, meiosis is an essential component of the life cycle of plants.

ESSENTIAL POINT

Meiosis results in extensive genetic variation by virtue of the exchange during crossing over between maternal and paternal chromatids and their random segregation into gametes. In addition, meiosis plays an important role in the life cycles of fungi and plants, serving as the bridge between alternating generations.

2.7 Electron Microscopy Has Revealed the Cytological Nature of Mitotic and Meiotic Chromosomes

Thus far in this chapter, we have focused on mitotic and meiotic chromosomes, emphasizing their behavior during cell division and gamete formation. An interesting question is why chromosomes are invisible during interphase but visible during the various stages of mitosis and meiosis. Studies using electron microscopy clearly show why this is the case.

Recall that during interphase, only dispersed chromatin fibers are present in the nucleus [Figure 2–13(a), p. 34]. Once mitosis begins, however, the fibers coil and fold, condensing into typical metaphase chromosomes, as shown in the scanning electron micrograph in Figure 2–13(b). At the periphery of the chromosome, individual fibers appear similarly to those seen in interphase chromatin. Individual fibers always seem to loop back into the interior where they are twisted and coiled around one another, forming the regular pattern of the mitotic chromosome. Starting in late telophase of mitosis and continuing during G1 of interphase, the process is reversed and chromosomes unwind to form the long fibers characteristic of chromatin. It is in this physical arrangement that DNA can most efficiently function during transcription and replication.

Electron microscopic observations of metaphase chromosomes in varying states of coiling led Ernest DuPraw to postulate the **folded-fiber model** shown in Figure 2–13(c). During metaphase, each chromosome consists of two sister chromatids joined at the centromeric region. Each arm of the chromatid appears to be a single fiber wound much like a skein of yarn. An orderly coiling–twisting–condensing process appears to be involved in the transition of the interphase chromatin to the more

FIGURE 2–13 Comparison of (a) the chromatin fibers characteristic of the interphase nucleus with (b) and (c) a metaphase chromosome that was derived from chromatin during mitosis. Part (c) diagrams the mitotic chromosome and its various components, showing how chromatin is condensed into it. Part (a) is a transmission electron micrograph, while part (b) is a scanning electron micrograph.

condensed, mitotic chromosomes. Geneticists believe that during the transition from interphase to prophase, a 5000-fold compaction occurs in the length of DNA within the chromatin fiber! This process must be extremely precise given the highly ordered and consistent appearance of mitotic chromosomes in all eukaryotes. We will return to this topic in Chapter 11 when we consider chromosome structure in further detail.

ESSENTIAL POINT

Mitotic chromosomes are produced as a result of the coiling and condensation of chromatin fibers characteristic of interphase and are thus visible only during cell division.

EXPLORING GENOMICS

PubMed: Exploring and Retrieving Biomedical Literature

In this era of rapidly expanding information on genomics and the biomedical sciences, scientists must be conversant in the use of multiple online databases. These resources provide access to DNA and protein sequences, genomic data, chromosome maps, microarray gene-expression data and molecular structures, as well as to the bioinformatics tools necessary for data manipulation. **PubMed** is a key tool for conducting literature searches and accessing biomedical publications.

PubMed is an Internet-based search system developed by the National Center of Biotechnology Information (NCBI) at the National Library of Medicine. Using PubMed, one can access over 15 million articles in over 4600 biomedical journals. The full text of many of the journals can be obtained electronically through college or university libraries, and some journals provide free public access to articles within certain time frames.

In this exercise, we will explore PubMed to answer questions about relationships between tubulin, human cancers, and cancer therapies, as well as the genetics of spermatogenesis.

■ Exercise I: Tubulin, Cancer, and Mitosis

In this chapter we were introduced to tubulin and the dynamic behavior of microtubules during the cell cycle. Cancer cells are characterized by continuous and uncontrolled mitotic divisions.

Is it possible that tubulin and microtubules contribute to the development of cancer? Could these important structures be targets for cancer therapies?

1. To begin your search for the answers, access the PubMed site at http://www.ncbi.nlm.nih.gov/pubmed/.
2. In the SEARCH box, type "tubulin cancer" and then select the "Go" button to perform the search.
3. Select several research papers and read the abstracts.

To answer the question about tubulin's association with cancer, you may want to limit your search to fewer papers, perhaps those that are review articles. To do this:

1. Select the "Limits" tab near the top of the page.
2. Scroll down the page and select "Review" in the "Type of Article" list.
3. Select "Go" to perform the search.

Explore some of the articles, as abstracts or as full text, if access is available through your library, by personal subscription, or by free public access. Prepare a brief report or verbally share your experiences with your class. Describe two of the most important things you learned during your exploration and identify the information sources you encountered during the search.

■ Exercise II: Human Disorders of Spermatogenesis

Using the methods described in Exercise I, identify some human disorders associated with defective spermatogenesis. Which human genes are involved in spermatogenesis? How do defects in these genes result in fertility disorders? Prepare a brief written or verbal report on what you have learned and what sources you used to acquire your information.

Case Study Timing is everything

A man in his early 20s received chemotherapy and radiotherapy as treatment every 60 days for Hodgkin's disease. After unsuccessful attempts to have children, he had his sperm examined at a fertility clinic, upon which multiple chromosomal irregularities were discovered. When examined within 5 days of a treatment, extra chromosomes were often present or one or more chromosomes were completely absent. However, such irregularities were not observed 38 days and later after a treatment.

1. How might a geneticist explain the time-related differences in chromosomal irregularities?
2. Do you think that exposure to chemotherapy and radiotherapy of a spermatogonium would cause more problems than exposure to a secondary spermatocyte?
3. What is the obvious advice that the man received regarding fertility while he remained under treatment?

Insights and Solutions

This initial appearance of "Insights and Solutions" begins a feature that will have great value to you as a student. From this point on, "Insights and Solutions" precedes the "Problems and Discussion Questions" at each chapter's end to provide sample problems and solutions that demonstrate approaches you will find useful in genetic analysis. The insights you gain by working through the sample problems will improve your ability to solve the ensuing problems in each chapter.

1. In an organism with a diploid number of $2n = 6$, how many individual chromosomal structures will align on the metaphase plate during (a) mitosis, (b) meiosis I, and (c) meiosis II? Describe each configuration.

Solution:

(a) In mitosis, where homologous chromosomes do not synapse, there will be six double structures, each consisting of a pair of sister chromatids. The number of structures is equivalent to the diploid number.

(b) In meiosis I, the homologs have synapsed, reducing the number of structures to three. Each is a tetrad and consists of two pairs of sister chromatids.

(c) In meiosis II, the same number of structures exist (three), but in this case they are dyads. Each dyad is a pair of sister chromatids. When crossing over has occurred, each chromatid may contain parts of one of its nonsister chromatids obtained during exchange in prophase I.

2. Disregarding crossing over, draw all possible alignment configurations that can occur during metaphase I for the chromosomes shown in Figure 2–10.

Solution: As shown in the diagram below, four cases are possible when $n = 2$.

3. For the chromosomes in the previous problem, assume one gene is present on both of the larger chromosomes with two alleles, *A* and *a*, as shown. Also assume a second gene with two alleles (*B, b*) is present on the smaller chromosomes. Calculate the probability of generating each gene combination (*AB, Ab, aB, ab*) following meiosis I.

Solution: As shown in the diagram below:

Case I	*AB* and *ab*	
Case II	*Ab* and *aB*	
Case III	*aB* and *Ab*	
Case IV	*ab* and *AB*	
Total:	$AB = 2$	$(p = 1/4)$
	$Ab = 2$	$(p = 1/4)$
	$aB = 2$	$(p = 1/4)$
	$ab = 2$	$(p = 1/4)$

4. Describe the composition of a meiotic tetrad as it exists during prophase I, assuming no crossover event has occurred. What impact would a single crossover event have on this structure?

Solution: Such a tetrad contains four chromatids, existing as two pairs. Members of each pair are sister chromatids. They are held together by a common centromere. Members of one pair are maternally derived, whereas members of the other are paternally derived. Maternal and paternal members are nonsister chromatids. A single crossover event has the effect of exchanging a portion of a maternal *and* a paternal chromatid, leading to a chiasma, where the two chromatids overlap physically in the tetrad.

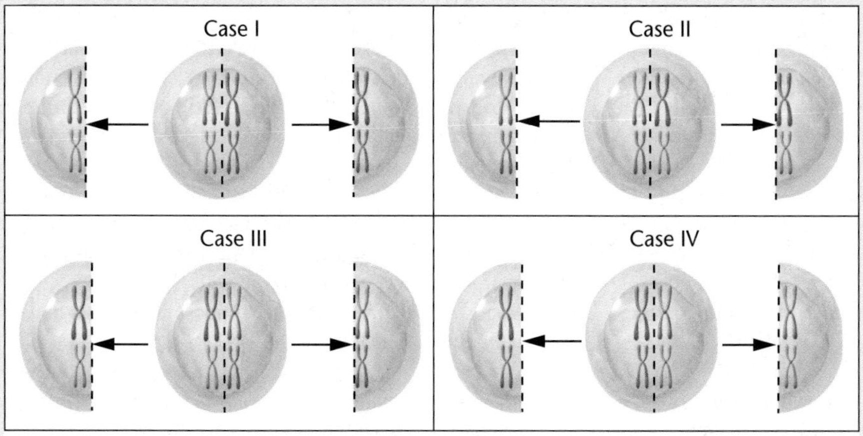

Solution for #2

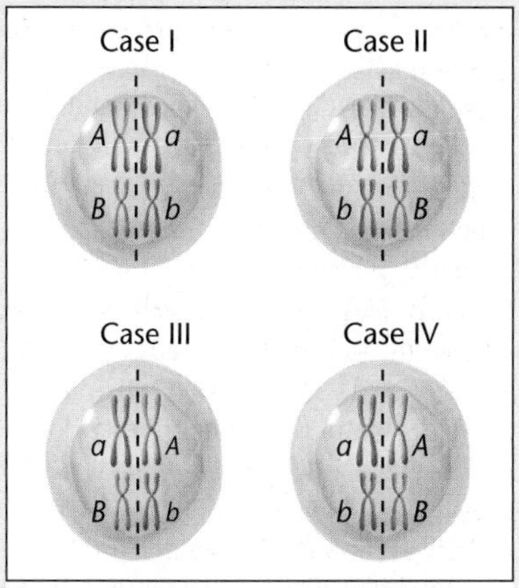

Solution for #3

Problems and Discussion Questions

1. What role do the following cellular components play in the storage, expression, or transmission of genetic information: (a) chromatin, (b) nucleolus, (c) ribosome, (d) mitochondrion, (e) centriole, (f) centromere?
2. Discuss the concepts of homologous chromosomes, diploidy, and haploidy. What characteristics are shared between two homologous chromosomes?
3. If two chromosomes of a species are the same length and have similar centromere placements, yet are not homologous, what is different about them?
4. Describe the events that characterize each stage of mitosis.
5. If an organism has a diploid number of 16, how many chromatids are visible at the end of mitotic prophase? How many chromosomes are moving to each pole during anaphase of mitosis? See **Now Solve This** on page 25.
6. How are chromosomes named on the basis of their centromere placement?
7. Contrast telophase in plant and animal mitosis.
8. Describe the phases of the cell cycle and the events that characterize each phase.
9. Examine Figure 2–11, which shows oogenesis in animal cells. Will the genotype of the second polar body (derived from meiosis II) always be identical to that of the ootid? Why or why not? See **Now Solve This** on page 32.
10. Contrast the end results of meiosis with those of mitosis.
11. Define and discuss these terms: (a) synapsis, (b) bivalent, (c) chiasmata, (d) crossing over, (e) chromomere, (f) sister chromatids, (g) tetrad, (h) dyad, (i) monad.
12. Contrast the genetic content and the origin of sister versus nonsister chromatids during their earliest appearance in prophase I of meiosis. How might the genetic content of these change by the time tetrads have aligned at the metaphase plate during metaphase I?
13. Given the end results of the two types of division, why is it necessary for homologs to pair during meiosis and not desirable for them to pair during mitosis?
14. An organism has a diploid number of 16 in a primary oocyte. (a) How many tetrads are present in prophase I? (b) How many dyads are present in prophase II? (c) How many monads migrate to each pole during anaphase II? See **Now Solve This** on page 29.
15. Contrast spermatogenesis and oogenesis. What is the significance of the formation of polar bodies?
16. Explain why meiosis leads to significant genetic variation while mitosis does not.
17. A diploid cell contains three pairs of homologous chromosomes designated C1 and C2, M1 and M2, and S1 and S2; no crossing over occurs. What possible combinations of chromosomes will be present in (a) daughter cells following mitosis, (b) the first meiotic metaphase, and (c) haploid cells following both divisions of meiosis?
18. Considering the preceding problem, predict the number of different haploid cells that will occur if a fourth chromosome pair (W1 and W2) is added.
19. During oogenesis in an animal species with a haploid number $n = 6$, one dyad undergoes nondisjunction during meiosis II. After the second meiotic division, the dyad ends up intact in the ovum. How many chromosomes are present in (a) the mature ovum and (b) the second polar body? (c) Following fertilization by a normal sperm, what chromosome condition is created?
20. What is the probability that, in an organism with a haploid number of 10, a sperm will be formed that contains all 10 chromosomes whose centromeres were derived from maternal homologs?
21. During the first meiotic prophase, (a) when does crossing over occur; (b) when does synapsis occur; (c) during which stage are the chromosomes least condensed; and (d) when are chiasmata first visible?
22. Describe the role of meiosis in the life cycle of a plant.
23. Contrast the chromatin fiber with the mitotic chromosome. How are the two structures related?
24. Describe the folded-fiber model of the mitotic chromosome.
25. You are given a metaphase chromosome preparation (a slide) from an unknown organism that contains 12 chromosomes. Two are clearly smaller than the rest, appearing identical in length and centromere placement. Describe all that you can about these two chromosomes.

For Problems 26–31, consider a diploid cell that contains three pairs of chromosomes designated AA, BB, and CC. Each pair contains a maternal and a paternal member (e.g., A^m and A^p, etc.). Using these designations, demonstrate your understanding of mitosis and meiosis by drawing chromatid combinations as requested. Be sure to indicate when chromatids are paired as a result of replication and/or synapsis. You may wish to use a large piece of brown manila wrapping paper or a large cut-up paper bag and work with another student as you deal with these problems. Such cooperative learning may be a useful approach as you solve problems throughout the text.

26. In mitosis, what chromatid combination(s) will be present during metaphase? What combination(s) will be present at each pole at the completion of anaphase?
27. During meiosis I, assuming no crossing over, what chromatid combination(s) will be present at the completion of prophase? Draw all possible alignments of chromatids as migration begins during early anaphase.
28. Are any possible combinations present during prophase of meiosis II other than those you drew in Problem 27? If so, draw them.
29. Draw all possible combinations of chromatids during anaphase in meiosis II.
30. Assume that during meiosis I, none of the *C* chromosomes disjoin at metaphase, but they separate into dyads (instead of monads) during meiosis II. How would this change the alignments that you constructed during the anaphase stages in meiosis I and II? Draw them.
31. Assume that each resultant gamete (Problem 30) participated in fertilization with a normal haploid gamete. What combinations will result? What percentage of zygotes will be diploid, containing one paternal and one maternal member of each chromosome pair?
32. A species of cereal rye (*Secale cereale*) has a chromosome number of 14, while a species of Canadian wild rye (*Elymus canadensis*) has a chromosome number of 28. Sterile hybrids can be produced by crossing *Secale* with *Elymus*.
    (a) What would be the expected chromosome number in the somatic cells of the hybrids?
    (b) If we assume that none of the chromosomes pair at meiosis I in the sterile hybrid, speculate on the anaphase I separation patterns of these chromosomes.

Gregor Johann Mendel, who in 1866 put forward the major postulates of transmission genetics as a result of experiments with the garden pea.

3

Mendelian Genetics

Chapter Concepts

- Inheritance is governed by information stored in discrete factors called genes.
- Genes are transmitted from generation to generation on vehicles called chromosomes.
- Chromosomes, which exist in pairs, provide the basis of biparental inheritance.
- During gamete formation, chromosomes are distributed according to postulates first described by Gregor Mendel, based on his nineteenth-century research with the garden pea.
- Mendelian postulates prescribe that homologous chromosomes segregate from one another and assort independently with other segregating homologs during gamete formation.
- Genetic ratios, expressed as probabilities, are subject to chance deviation and may be evaluated statistically.
- The analysis of pedigrees allows predictions involving the genetic nature of human traits.

Although inheritance of biological traits has been recognized for thousands of years, the first significant insights into the mechanisms involved occurred about 140 years ago. In 1866, Gregor Johann Mendel published the results of a series of experiments that would lay the foundation for the formal discipline of genetics. Mendel's work went largely unnoticed until the turn of the century, but in the ensuing years the concept of the gene as a distinct hereditary unit was established. The ways in which genes, as members of chromosomes, are transmitted to offspring and control traits were clarified. Research has continued unabated throughout the twentieth century and into the present century—indeed, studies in genetics, most recently at the molecular level, have remained continually at the forefront of biological research since the early 1900s.

When Mendel began his studies of inheritance using *Pisum sativum*, the garden pea, chromosomes and the role and mechanism of meiosis were totally unknown. Nevertheless, he determined that discrete *units of inheritance* exist and predicted their behavior during the formation of gametes. Subsequent investigators, with access to cytological data, were able to relate their observations of chromosome behavior during meiosis to Mendel's principles of inheritance. Once this correlation was made, Mendel's postulates were accepted as the basis for the study of what is known as **transmission genetics.**

3.1 Mendel Used a Model Experimental Approach to Study Patterns of Inheritance

Johann Mendel was born in 1822 to a peasant family in the central European village of Heinzendorf. An excellent student in high school, he studied philosophy for several years afterward, and in 1843 he was admitted to the Augustinian Monastery of St. Thomas in Brno, now part of the Czech Republic, taking the name Gregor. In 1849, he was relieved of pastoral duties and accepted a teaching appointment that lasted several years. From 1851 to 1853, he attended the University of Vienna, where he studied physics and botany. He returned to Brno in 1854, where he taught physics and natural science for the next 16 years. Mendel received support from the monastery for his studies and research throughout his life.

In 1856, Mendel performed his first set of hybridization experiments with the garden pea. The research phase of his career lasted until 1868, when he was elected abbot of the monastery. Although he retained his interest in genetics, his new responsibilities demanded most of his time. In 1884, Mendel died of a kidney disorder. The local newspaper paid him the following tribute:

> "His death deprives the poor of a benefactor, and mankind at large of a man of the noblest character, one who was a warm friend, a promoter of the natural sciences, and an exemplary priest."

Mendel first reported the results of some simple genetic crosses between certain strains of the garden pea in 1865. Although his was not the first attempt to provide experimental evidence pertaining to inheritance, Mendel's success where others failed can be attributed, at least in part, to his elegant model of experimental design and analysis.

Mendel showed remarkable insight into the methodology necessary for good experimental biology. He chose an organism that is easy to grow and hybridize artificially. The pea plant is self-fertilizing in nature but is easy to crossbreed experimentally. It reproduces well and grows to maturity in a single season. Mendel followed seven visible features (unit characters), each represented by two contrasting forms, or **traits** (Figure 3–1). For the character stem height, for example, he experimented with the traits *tall* and *dwarf*. He selected six other visibly contrasting pairs of traits involving seed shape and color, pod shape and color, and pod and flower arrangement. From local seed merchants, Mendel obtained true-breeding strains—those in which each trait appeared unchanged generation after generation in self-fertilizing plants.

There were several reasons for Mendel's success. In addition to his choice of a suitable organism, he restricted his examination to one or very few pairs of contrasting traits in each experiment. He also kept accurate quantitative records, a necessity in genetic experiments. From the analysis of his data, Mendel derived certain postulates that became principles of transmission genetics.

The results of Mendel's experiments were unappreciated until the turn of the century, well after his death. However, once Mendel's publications were rediscovered by geneticists investigating the function and behavior of chromosomes, the implications of his postulates were immediately apparent. He had discovered the basis for the transmission of hereditary traits!

How Do We Know?

In this chapter, we focus on how Mendel was able to derive the essential postulates that explain inheritance. We also consider some of the methods and reasoning by which these ideas, concepts, and techniques were developed. As you study this topic, you should try to answer several fundamental questions:

1. How was Mendel able to derive postulates concerning the behavior of "unit factors" during gamete formation, when he could not directly observe them?
2. How can we know whether an organism expressing a dominant trait is homozygous or heterozygous?
3. In analyzing genetic data, how do we know whether deviation from the expected ratio is due to chance rather than to another, independent factor?
4. Since experimental crosses are not performed in humans, how do we know how traits are inherited?

Character	Contrasting traits		F_1 results	F_2 results	F_2 ratio
Seed shape	round/wrinkled		all round	5474 round 1850 wrinkled	2.96:1
Seed color	yellow/green		all yellow	6022 yellow 2001 green	3.01:1
Pod shape	full/constricted		all full	882 full 299 constricted	2.95:1
Pod color	green/yellow		all green	428 green 152 yellow	2.82:1
Flower color	violet/white		all violet	705 violet 224 white	3.15:1
Flower position	axial/terminal		all axial	651 axial 207 terminal	3.14:1
Stem height	tall/dwarf		all tall	787 tall 277 dwarf	2.84:1

FIGURE 3–1 Seven pairs of contrasting traits and the results of Mendel's seven monohybrid crosses of the garden pea *(Pisum sativum)*. In each case, pollen derived from plants exhibiting one trait was used to fertilize the ova of plants exhibiting the other trait. In the F_1 generation, one of the two traits was exhibited by all plants. The contrasting trait reappeared in approximately 1/4 of the F_2 plants.

3.2 The Monohybrid Cross Reveals How One Trait Is Transmitted from Generation to Generation

Mendel's simplest crosses involved only one pair of contrasting traits. Each such experiment is a **monohybrid cross,** which is made by mating true-breeding individuals from two parent strains, each exhibiting one of the two contrasting forms of the character under study. Initially, we examine the first generation of offspring of such a cross, and then we consider the results of **selfing,** the offspring of self-fertilizing individuals from this first generation. The original parents constitute the $\mathbf{P_1}$, or **parental generation,** their offspring are the $\mathbf{F_1}$, or **first filial generation,** and the individuals resulting from the selfed F_1 generation are the $\mathbf{F_2}$, or **second filial generation.** We can continue to follow subsequent generations.

The cross between true-breeding pea plants with tall stems and dwarf stems is representative of Mendel's monohybrid crosses. *Tall* and *dwarf* are contrasting traits of the character of stem height. Unless tall or dwarf plants are crossed together or with another strain, they will undergo self-fertilization and breed true, producing their respective traits generation after generation. However, when Mendel crossed tall plants with dwarf plants, the resulting F_1 generation consisted only of tall plants. When members of the F_1 generation were selfed, Mendel observed that 787 of 1064 F_2 plants were tall, while the remaining 277 were dwarf. Note that in this cross (Figure 3–1) the dwarf trait disappears in the F_1, only to reappear in the F_2 generation.

Genetic data are usually expressed and analyzed as ratios. In this particular example, many identical P_1 crosses were made, and many F_1 plants—all tall—were produced. As noted, of the 1064 F_2 offspring, 787 were tall and 277 were dwarf—a ratio of 2.84:1.0, or about 3:1.

Mendel made similar crosses between pea plants exhibiting other pairs of contrasting traits; the results of these crosses are shown in Figure 3–1. In every case, the outcome was similar to the tall/dwarf cross just described. All F_1 offspring were identical to one of the parents, but in the F_2 offspring, an approximate ratio of 3:1 was obtained. That is, three-fourths looked like the F_1 plants, while one-fourth exhibited the contrasting trait, which had disappeared in the F_1 generation.

We will point out one further aspect of Mendel's monohybrid crosses. In each cross, the F_1 and F_2 patterns of inheritance were similar regardless of which P_1 plant served as the source of pollen (sperm) and which served as the source of the ovum (egg). The crosses could be made either way—pollination of dwarf plants by tall plants or vice versa. These are called **reciprocal crosses.** Therefore, the results of Mendel's monohybrid crosses were not sex-dependent.

To explain these results, Mendel proposed the existence of particulate unit factors for each trait. He suggested that these factors serve as the basic units of heredity and are passed unchanged from generation to generation, determining the various traits expressed by each individual plant. Using these general ideas, Mendel proceeded to hypothesize precisely how unit factors could account for the results of the monohybrid crosses.

Mendel's First Three Postulates

Using the consistent pattern of results in the monohybrid crosses, Mendel derived the following three postulates or principles of inheritance.

1. Unit Factors in Pairs

Genetic characters are controlled by unit factors that exist in pairs in individual organisms.

In the monohybrid cross involving tall and dwarf stems, a specific **unit factor** exists for each trait. Because the factors occur in pairs, three combinations are possible: two factors for tallness, two factors for dwarfness, or one factor for each trait. Every individual contains one of these three combinations, which determines stem height.

2. Dominance/Recessiveness

When two unlike unit factors responsible for a single character are present in a single individual, one unit factor is dominant to the other, which is said to be recessive.

In each monohybrid cross, the trait expressed in the F_1 generation is controlled by the **dominant** unit factor. The trait not expressed is controlled by the **recessive** unit factor. Note that this dominance/recessiveness relationship pertains only when unlike unit factors are present in pairs. The terms *dominant* and *recessive* are also used to designate traits. In this case, tall stems are said to be dominant over the recessive dwarf stems.

3. Segregation

During the formation of gametes, the paired unit factors separate or segregate randomly so that each gamete receives one or the other with equal likelihood.

If an individual contains a pair of like unit factors (e.g., both specific for tall), then all gametes receive one tall unit factor. If an individual contains unlike unit factors (e.g., one for tall and one for dwarf), then each gamete has a 50 percent probability of receiving either the tall or the dwarf unit factor.

These postulates provide a suitable explanation for the results of the monohybrid crosses. Let's use the tall/dwarf cross to illustrate. Mendel reasoned that P_1 tall plants contain identical paired unit factors, as do the P_1 dwarf plants. The gametes of tall plants all receive one tall unit factor as a result of **segregation.** Similarly, the gametes of dwarf plants all receive one dwarf unit factor. Following fertilization, all F_1 plants receive one unit factor from each parent: a tall factor from one and a dwarf factor from the other, reestablishing the paired relationship—but because tall is dominant to dwarf, all F_1 plants are tall.

When F_1 plants form gametes, the postulate of segregation demands that each gamete randomly receives either the tall or the dwarf unit factor. Following random fertilization events during F_1 selfing, four F_2 combinations result in equal frequency:

1. tall/tall
2. tall/dwarf
3. dwarf/tall
4. dwarf/dwarf

Combinations (1) and (4) result in tall and dwarf plants, respectively. According to the postulate of dominance/recessiveness, combinations (2) and (3) both yield tall plants. Therefore, the F_2 is predicted to consist of 3/4 tall and 1/4 dwarf, or a ratio of 3:1. This is approximately what Mendel observed in the cross between tall and dwarf plants. A similar pattern was observed in each of the other monohybrid crosses (see Figure 3–1).

ESSENTIAL POINT

Mendel's postulates help describe the basis for the inheritance of phenotypic traits. He hypothesized that unit factors exist in pairs and exhibit a dominant/recessive relationship in determining the expression of traits. He further postulated that unit factors segregate during gamete formation, such that each gamete receives one or the other factor, with equal probability.

Modern Genetic Terminology

To analyze the monohybrid cross and Mendel's first three postulates, we must first introduce several new terms as well as a symbol convention for the unit factors.

Traits such as tall or dwarf are visible expressions of the information contained in unit factors. The physical appearance of a trait is the **phenotype** of the individual. Mendel's unit factors represent units of inheritance called **genes** by modern geneticists. For any given character, such as plant height, the phenotype is determined by alternative forms of a single gene called **alleles.** For example, the unit factors representing tall and dwarf are alleles determining the height of the pea plant.

Geneticists have several different systems for using symbols to represent genes. In Chapter 4, we will review a number of these conventions, but for now, we will adopt one to use consistently throughout this chapter. According to this convention, the first letter of the recessive trait symbolizes the character in question; in lowercase italic, it designates the allele for the recessive trait, and in uppercase italic, it designates the allele for the dominant trait. Thus for Mendel's pea plants, we use *d* for the *d*warf allele and *D* for the tall allele. When alleles are written in pairs to represent the two unit factors present in any individual (*DD, Dd,* or *dd*), the resulting symbol is called the **genotype.** This term reflects the genetic makeup of an individual, whether it is haploid or diploid. By reading the genotype, we know the phenotype of the individual: *DD* and *Dd* are tall, and *dd* is dwarf. When both alleles are the same (*DD* or *dd*), the individual is **homozygous** or a **homozygote;** when the alleles are different (*Dd*), we use the term **heterozygous** or a **heterozygote.** These symbols and terms are used in Figure 3–2 to illustrate the monohybrid cross.

Because he operated without the hindsight that modern geneticists enjoy, Mendel's analytical reasoning must be considered a truly outstanding scientific achievement. On the basis of rather simple but precisely executed breeding experiments, he not only proposed that discrete particulate units of heredity exist, he also explained how they are transmitted from one generation to the next.

NOW SOLVE THIS

Problem 5 on page 57 describes a set of crosses in pigeons and asks you to analyze the data produced and to determine the mode of inheritance and the genotypes of the parents and offspring in a number of instances.

Hint: The first step is to determine whether or not this is a monohybrid cross. To do so, convert the data to ratios that are characteristic of Mendelian crosses. In the case of this problem, ask first whether any of the F_2 ratios match Mendel's 3:1 monohybrid ratio. If so, the second step is to determine which trait is dominant and which is recessive.

Punnett Squares

The genotypes and phenotypes resulting from the recombination of gametes during fertilization can be easily visualized by constructing a **Punnett square,** named after the person who first devised this approach, Reginald C. Punnett. Figure 3–3 demonstrates this method of analysis for our $F_1 \times F_1$ monohybrid cross. Each of the possible gametes is assigned a column or a row; the vertical columns represent those of the female parent, and the horizontal rows represent those of the male parent. After assigning the gametes to the rows and columns, we predict the new generation by entering the male and female gametic information into each box and thus producing every possible resulting genotype. By filling out the Punnett square, we are listing all possible random fertilization events. The genotypes and phenotypes of all potential offspring are ascertained by reading the combinations in the boxes.

The Punnett square method is particularly useful when you are first learning about genetics and how to solve problems. Note the ease with which the 3:1 phenotypic ratio and the 1:2:1 genotypic ratio is derived in the F_2 generation in Figure 3–3.

WEB TUTORIAL
MONOHYBRID CROSS AND PUNNETT SQUARES

The Testcross: One Character

Tall plants produced in the F_2 generation are predicted to be either the *DD* or *Dd* genotype. You might ask if there is a way to distinguish the genotype. Mendel devised a rather simple method that is still used today in breeding plants and animals: the **testcross.** The organism expressing the dominant phenotype, but of unknown genotype, is crossed to a known *homozygous recessive individual.* For example, as shown in Figure 3–4(a), p. 42, if a tall plant of genotype *DD* is testcrossed to a dwarf plant, which must have the *dd* genotype, all offspring will be tall phenotypically and *Dd* genotypically. However, as shown in Figure 3–4(b), if a tall plant is *Dd* and it is crossed to a dwarf plant (*dd*), then one-half of the offspring will be tall (*Dd*) and the other half will be dwarf (*dd*). Therefore, a 1:1 tall/dwarf ratio demonstrates the heterozygous nature of the tall plant of unknown genotype. The testcross reinforced Mendel's conclusion that separate unit factors control traits.

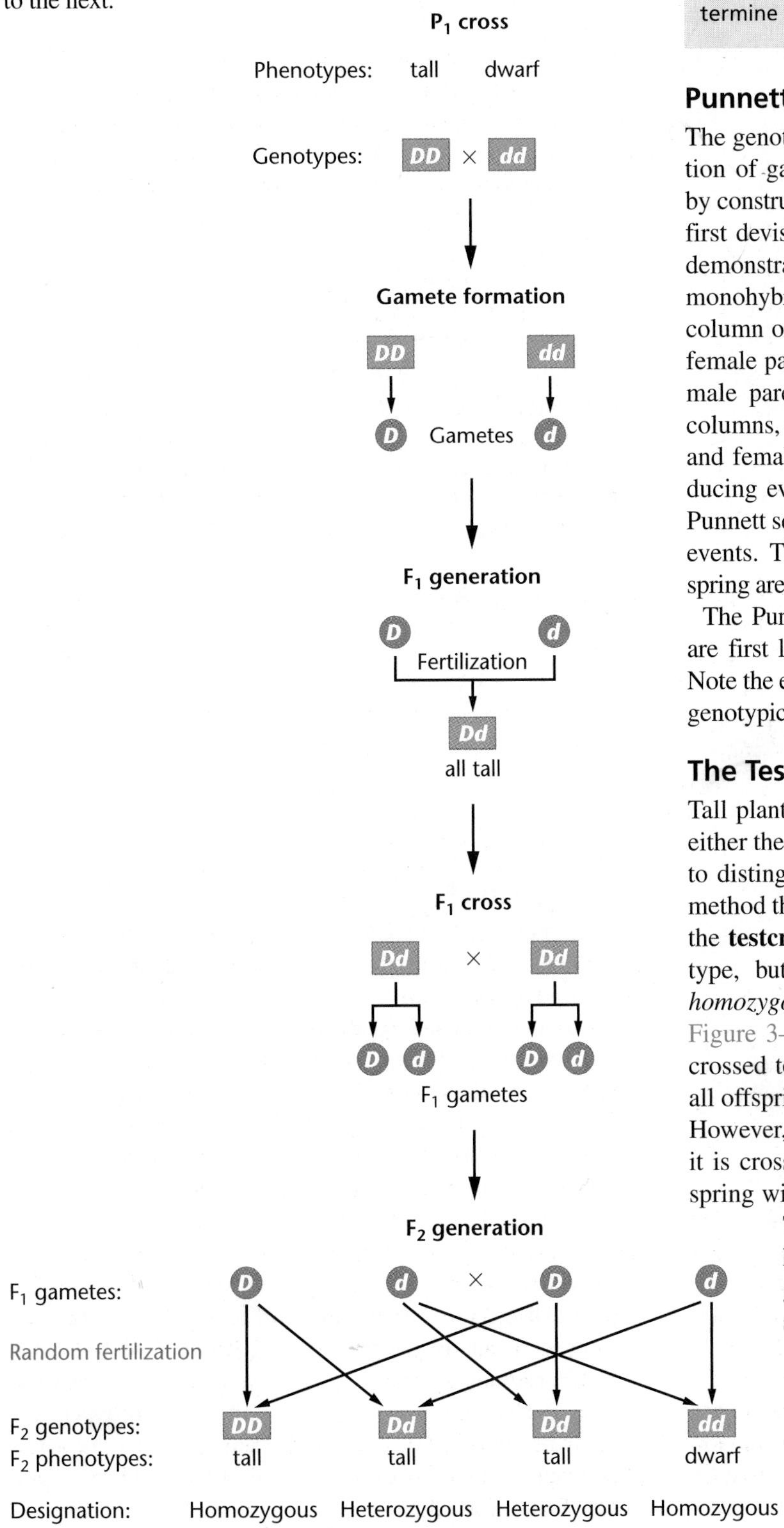

FIGURE 3–2 The monohybrid cross between tall (*DD*) and dwarf (*dd*) pea plants. Individuals are shown in rectangles, and gametes in circles.

INDEPENDENT ASSORTMENT
WEB TUTORIAL

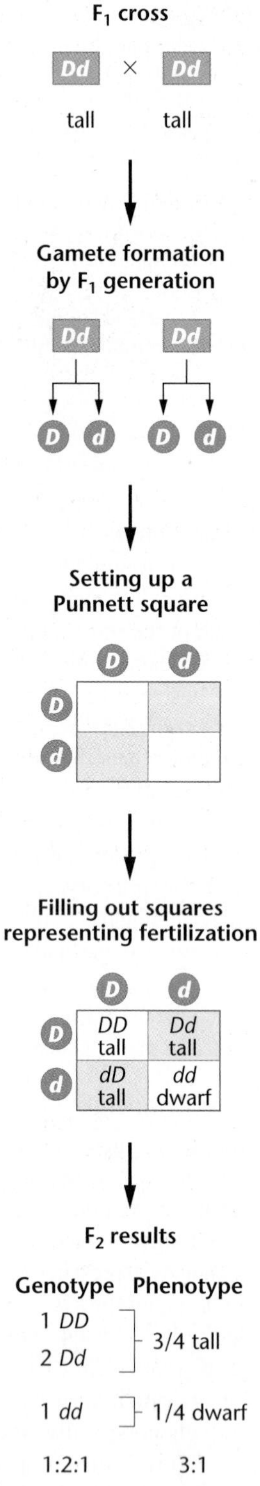

FIGURE 3–3 A Punnett square generating the F_2 ratio of the $F_1 \times F_1$ cross shown in Figure 3–2.

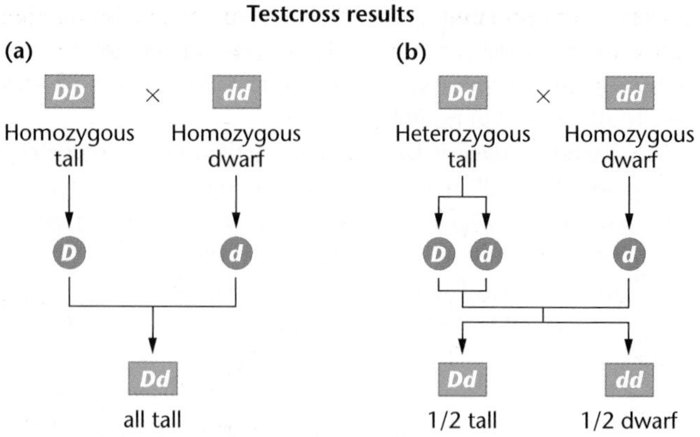

FIGURE 3–4 Testcross of a single character. In (a), the tall parent is homozygous, but in (b), the tall parent is heterozygous. The genotype of each tall P_1 plant can be determined by examining the offspring when each is crossed to a homozygous recessive dwarf plant.

3.3 Mendel's Dihybrid Cross Generated a Unique F_2 Ratio

As a natural extension of the monohybrid cross, Mendel also designed experiments in which he examined two characters simultaneously. Such a cross, involving two pairs of contrasting traits, is a **dihybrid cross,** or *two-factor cross.* For example, if pea plants having yellow seeds that are round are bred with those having green seeds that are wrinkled, the results shown in Figure 3–5 will occur: the F_1 offspring will all be yellow and round. It is therefore apparent that yellow is dominant to green and that round is dominant to wrinkled. When the F_1 individuals are selfed, approximately 9/16 of the F_2 plants express yellow and round, 3/16 express yellow and wrinkled, 3/16 express green and round, and 1/16 express green and wrinkled.

A variation of this cross is also shown in Figure 3–5. Instead of crossing one P_1 parent with both dominant traits (yellow, round) and one with both recessive traits (green, wrinkled), plants with yellow, wrinkled seeds are crossed with those with green, round seeds. In spite of the change in the P_1 phenotypes, both the F_1 and F_2 results remain unchanged. It will become clear in the next section why this is so.

Mendel's Fourth Postulate: Independent Assortment

We can most easily understand the results of a dihybrid cross if we consider it theoretically as consisting of two monohybrid crosses conducted separately. Think of the two sets of traits as inherited independently of each other; that is, the chance of any plant having yellow or green seeds is not at all influenced by the chance that this plant will have round or wrinkled seeds. Thus, because yellow is dominant to green, all F_1 plants in the first theoretical cross would have yellow seeds. In the second theoretical cross, all F_1 plants would have round seeds because round is dominant to wrinkled. When Mendel examined the F_1 plants of the dihybrid cross, all were yellow and round, as we just predicted.

The predicted F_2 results of the first cross are 3/4 yellow and 1/4 green. Similarly, the second cross would yield 3/4 round and 1/4 wrinkled. Figure 3–5 shows that in the dihybrid cross, 12/16 F_2 plants are yellow while 4/16 are green, exhibiting the expected 3:1 (3/4:1/4) ratio. Similarly, 12/16 F_2 plants have round seeds while 4/16 have wrinkled seeds, again revealing the 3:1 ratio.

These numbers demonstrate that the two pairs of contrasting traits are inherited independently, so we can predict the frequencies of all possible F_2 phenotypes by applying the **product law** of probabilities: *When two independent events occur simultaneously, the combined probability of the two outcomes is equal to the product of their individual probabilities of occurrence.* For example, the probability of an F_2 plant having yellow *and* round seeds is (3/4)(3/4), or 9/16, because 3/4 of all F_2 plants should be yellow and (3/4) of all F_2 plants should be round.

In a like manner, the probabilities of the other three F_2 phenotypes can be calculated: yellow (3/4) and wrinkled (1/4) are predicted to be present together 3/16 of the time; green (1/4) and round (3/4) are predicted 3/16 of the time; and green (1/4) and wrinkled (1/4) are predicted 1/16 of the time. These calculations are shown in Figure 3–6, p. 44.

It is now apparent why the F_1 and F_2 results are identical whether the initial cross is yellow, round plants bred with green, wrinkled plants, or if yellow, wrinkled plants are bred with green, round plants. In both crosses, the F_1 genotype of all plants is identical. Each plant is heterozygous for both gene pairs. As a result, the F_2 generation is also identical in both crosses.

On the basis of similar results in numerous dihybrid crosses, Mendel proposed a fourth postulate called **independent assortment:** *During gamete formation, segregating pairs of unit factors assort independently of each other.*

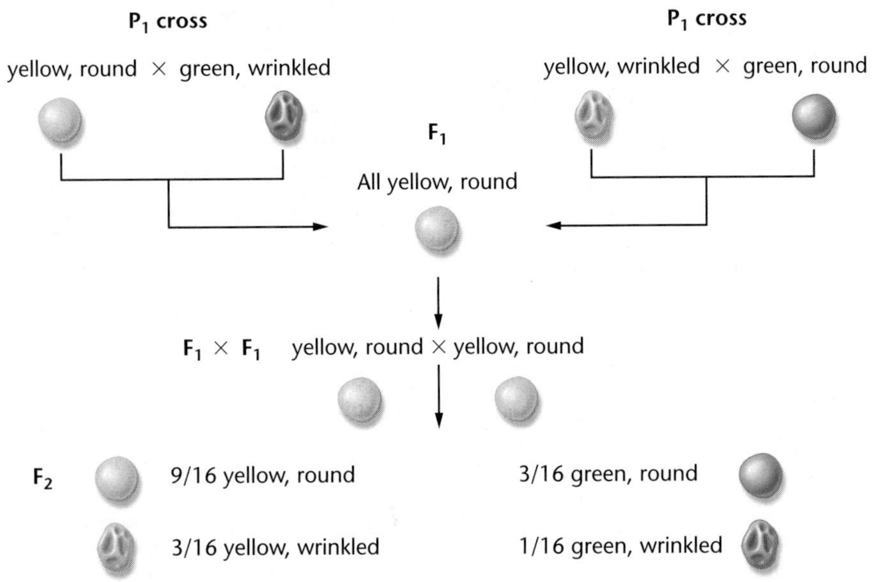

FIGURE 3–5 F_1 and F_2 results of Mendel's dihybrid crosses, where the plants on the top left with yellow, round seeds are crossed with plants having green, wrinkled seeds, and the plants on the top right with yellow, wrinkled seeds are crossed with plants having green, round seeds.

How Mendel's Peas Become Wrinkled: A Molecular Explanation

Only recently, well over a hundred years after Mendel used wrinkled peas in his groundbreaking hybridization experiments, have we come to find out how the *wrinkled* gene makes peas wrinkled. The wild-type allele of the gene encodes a protein called **starch-branching enzyme (SBEI).** This enzyme catalyzes the formation of highly branched starch molecules as the seed matures.

Wrinkled peas, which result from the homozygous presence of the mutant form of the gene, lack the activity of this enzyme. The production of branch points is inhibited during the synthesis of starch within the seed, which in turn leads to the accumulation of more sucrose and a higher water content while the seed develops. Osmotic pressure inside rises, causing the seed to lose water internally, and ultimately results in the wrinkled appearance of the seed at maturation. In contrast, developing seeds that bear at least one copy of the normal gene (being either homozygous or heterozygous for the dominant allele) synthesize starch and reach an osmotic balance that minimizes the loss of water. The end result is a smooth-textured outer coat.

The *SBEI* gene has been cloned and analyzed, providing greater insight into the relationship between genotypes and phenotypes. Interestingly, the mutant gene contains a foreign sequence of some 800 base pairs that disrupts the normal coding region. This foreign segment closely resembles other such units, called **transposable elements.** These elements have the ability to move from place to place in the genome of organisms. Transposable elements have been found in maize (corn), parsley, snapdragons, fruit flies, and humans, among many other organisms.

Wrinkled and round garden peas, the phenotypic traits in one of Mendel's monohybrid crosses.

This postulate stipulates that segregation of any pair of unit factors occurs independently of all others. As a result of random segregation, each gamete receives one member of every pair of unit factors. For one pair, whichever unit factor is received does not influence the outcome of segregation of any other pair. Thus, according to the postulate of independent assortment, all possible combinations of gametes are formed in equal frequency.

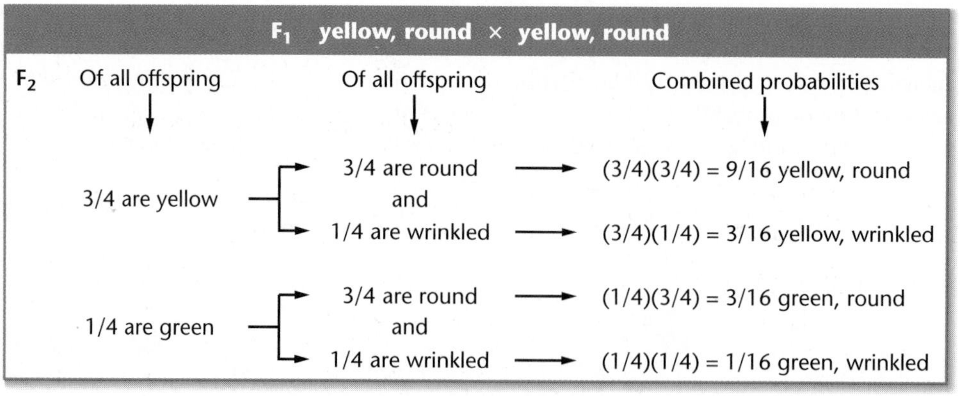

FIGURE 3–6 Computation of the combined probabilities of each F_2 phenotype for two independently inherited characters. The probability of each plant's seeds being yellow or green is independent of the probability of its seeds being round or wrinkled.

The Punnett square in Figure 3–7 shows how independent assortment works in the formation of the F_2 generation. Examine the formation of gametes by the F_1 plants; segregation prescribes that every gamete receives either a *G* or *g* allele and a *W* or *w* allele. Independent assortment stipulates that all four combinations (*GW, Gw, gW,* and *gw*) will be formed with equal probabilities.

In every $F_1 \times F_1$ fertilization event, each zygote has an equal probability of receiving one of the four combinations from each parent. If many offspring are produced, 9/16 have yellow, round seeds, 3/16 have yellow, wrinkled seeds, 3/16 have green, round seeds, and 1/16 have green, wrinkled seeds, yielding what is designated as **Mendel's 9:3:3:1 dihybrid ratio.** This is an ideal ratio based on probability events involving segregation, independent assortment, and random fertilization. Because of deviation due strictly to chance, particularly if small numbers of offspring are produced, actual results are highly unlikely to match the ideal ratio.

ESSENTIAL POINT

Mendel's postulate of independent assortment states that each pair of unit factors segregates independently of other such pairs. As a result, all possible combinations of gametes are formed with equal probability.

The Testcross: Two Characters

The testcross can also be applied to individuals that express two dominant traits but whose genotypes are unknown. For example, the expression of the yellow, round seed phenotype in the F_2 generation just described may result from the *GGWW, GGWw, GgWW,* and *GgWw* genotypes. If an F_2 yellow, round plant is crossed with a homozygous recessive green, wrinkled plant (*ggww*), analysis of the offspring will indicate the actual genotype of that yellow, round plant. Each of the above genotypes results in a different set of gametes, and in a testcross, a different set of phenotypes in the resulting offspring. You should work out the results of each of these four crosses to be sure you understand this concept.

NOW SOLVE THIS

Problem 7 on page 57 involves a series of Mendelian dihybrid crosses where you are asked to determine the genotypes of the parents in a number of instances.

Hint: In each case, write down everything that you know for certain. This reduces the problem to its bare essentials, clarifying what you need to determine. For example, the wrinkled, yellow plant in case (b) must be homozygous for the recessive wrinkled alleles and bear at least one dominant allele for the yellow trait. Having established this, you need only determine the remaining allele for cotyledon color.

3.4 The Trihybrid Cross Demonstrates That Mendel's Principles Apply to Inheritance of Multiple Traits

Thus far, we have considered inheritance by individuals of up to two pairs of contrasting traits. Mendel demonstrated that the identical processes of segregation and independent assortment apply to three pairs of contrasting traits in what is called a **trihybrid cross,** or *three-factor cross*.

Although a trihybrid cross is somewhat more complex than a dihybrid cross, its results are easily calculated if the principles of segregation and independent assortment are followed. For example, consider the cross shown in Figure 3–8, p. 46, where the gene pairs of theoretical contrasting traits are represented by the symbols *A, a, B, b, C,* and *c*. In the cross between *AABBCC* and *aabbcc* individuals, all F_1 individuals are heterozygous for all three gene pairs. Their genotype, *AaBbCc,* results in the phenotypic expression of the dominant *A, B,* and *C* traits. When F_1 individuals serve as parents, each produces eight different gametes in equal frequencies. At this point, we could construct a Punnett square with 64 separate boxes and read out the phenotypes—but such a method is cumbersome in a cross involving so many factors. Therefore another method has been devised to calculate the predicted ratio.

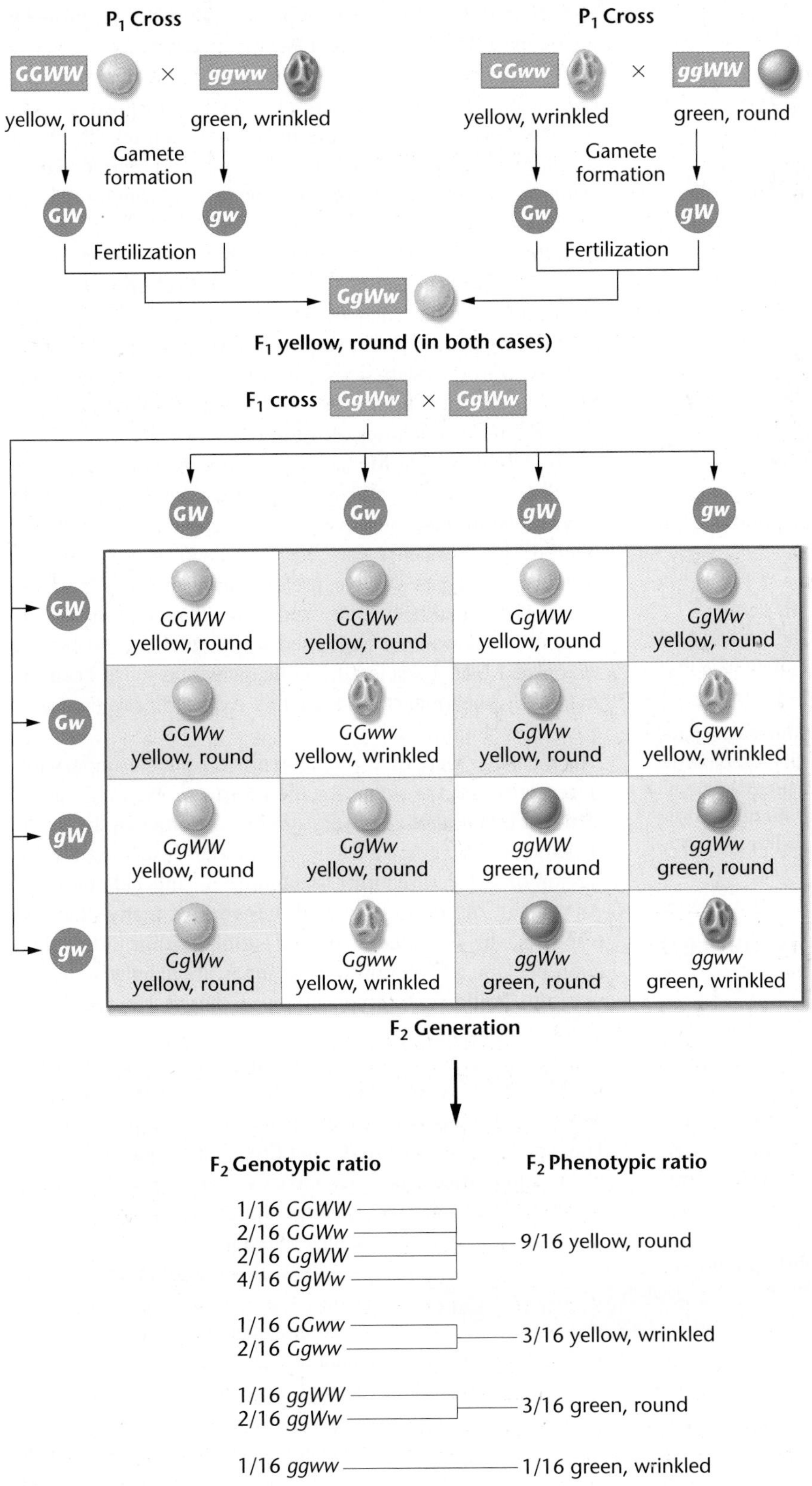

FIGURE 3–7 Analysis of the dihybrid crosses shown in Figure 3–5. The F_1 heterozygous plants are self-fertilized to produce an F_2 generation, which is computed using a Punnett square. Both the phenotypic and genotypic F_2 ratios are shown.

Each gene pair is assumed to behave independently during gamete formation.

When the monohybrid cross $AA \times aa$ is made, we know that:

1. All F_1 individuals have the genotype *Aa* and express the phenotype represented by the *A* allele, which is called the *A* phenotype in the following discussion.
2. The F_2 generation consists of individuals with either the *A* phenotype or the *a* phenotype in the ratio of 3:1.

The same generalizations can be made for the $BB \times bb$ and $CC \times cc$ crosses. Thus, in the F_2 generation, 3/4 of all organisms express phenotype *A*, 3/4 express *B*, and 3/4 express *C*. Similarly, 1/4 of all organisms express phenotype *a*, 1/4 express *b*, and 1/4 express *c*. The proportions of organisms that express each phenotypic combination can be predicted by assuming that fertilization, following the independent assortment of these three gene pairs during gamete formation, is a random process—we simply apply the product law of probabilities once again. Figure 3–9, p. 46, uses the forked-line method to calculate the phenotypic proportions of the F_2 generation. They fall into the trihybrid ratio of 27:9:9:9:3:3:3:1. The same method can be used to solve crosses involving any number of gene pairs, *provided that all gene pairs assort independently of each other*. We shall see later that this is not always the case. However, it appeared to be true for all of Mendel's characters.

The Forked-Line Method

It is much less difficult to consider each contrasting pair of traits separately and then to combine these results by using the **forked-line method,** first shown in Figure 3–6. This method, also called a **branch diagram,** relies on the simple application of the laws of probability established for the dihybrid cross.

ESSENTIAL POINT

The forked-line method is less complex than, but just as accurate as, the Punnett square in predicting the probabilities of phenotypes or genotypes from crosses involving two or more gene pairs.

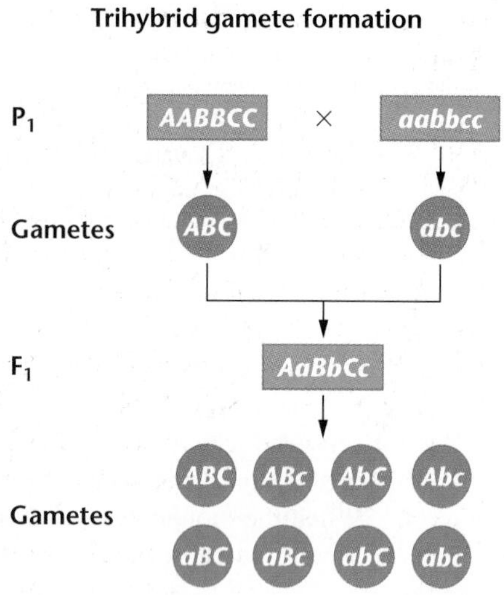

FIGURE 3–8 Formation of P_1 and F_1 gametes in a trihybrid cross.

NOW SOLVE THIS

Problem 15 on page 58 asks you to use the forked-line method to determine the outcome of a number of trihybrid crosses.

Hint: In using the forked-line method, consider each gene pair separately. For example, in this problem, first predict the outcome of each cross for *A*/*a* genes, then for the *B*/*b* genes, and finally, for the *C*/*c* genes. Then you are prepared to pursue the outcome of each cross using the forked-line method.

3.5 Mendel's Work Was Rediscovered in the Early Twentieth Century

Mendel initiated his work in 1856, presented it to the Brünn Society of Natural Science in 1865, and published it the following year. Although his findings were often cited and discussed, their significance went unappreciated for about 35 years. Many explanations have been proposed for this delay.

First, Mendel's adherence to mathematical analysis of probability events was an unusual approach in those days for biological studies. Perhaps his approach seemed foreign to his contemporaries. More important, his conclusions did not fit well with existing theories on the cause of variation among organisms. The source of natural variation intrigued students of evolutionary theory. These individuals, stimulated by the proposal developed by Charles Darwin and Alfred Russel Wallace, believed in **continuous variation,** which held that offspring were a *blend* of their parents' phenotypes. As we mentioned earlier, Mendel theorized that variation was due to discrete or particulate units, resulting in **discontinuous variation.** For example, Mendel proposed that the F_2 offspring of a dihybrid cross are expressing traits produced by new combinations of previously existing unit factors. As a result, Mendel's theories did not fit well with the evolutionists' preconceptions about causes of variation.

It is also likely that Mendel's contemporaries failed to realize that Mendel's postulates explained *how* variation was transmitted to offspring. Instead, they may have attempted to interpret his work in a way that addressed the issue of *why* certain phenotypes survive preferentially. It was this latter question that had been addressed in the theory of natural selection, but it was not addressed by Mendel. The collective vision of Mendel's scientific colleagues may have been obscured by the impact of Darwin's extraordinary theory of organic evolution.

In the latter part of the nineteenth century, a remarkable observation set the scene for the rebirth of Mendel's work: Walter Flemming's discovery of chromosomes in the nuclei of salamander cells. In 1879, Flemming described the behavior of these threadlike structures during cell division. As a result of his findings and the work of many other cytologists, the presence of discrete units within the nucleus soon became an integral part of ideas about inheritance. It was this mind-set that prompted scientists to reexamine Mendel's findings.

In the early twentieth century, research led to renewed interest in Mendel's work. Hybridization experiments similar to Mendel's were performed independently by three botanists: Hugo de Vries, Karl Correns, and Erich Tschermak. De Vries's work demonstrated the principle of segregation in his experiments with several plant species. Apparently, he searched the existing literature and found that Mendel's work anticipated his own conclusions! Correns and Tschermak also reached conclusions similar to those of Mendel.

In 1902, two cytologists, Walter Sutton and Theodor Boveri, independently published papers linking their discoveries of the behavior of chromosomes during meiosis to the Mendelian principles of segregation and independent assortment. They pointed out that the separation of chromosomes during meiosis could serve as the cytological basis of these two postulates. Although they thought Mendel's unit factors were probably chromosomes rather than

Generation of F_2 trihybrid phenotypes

A or *a*	*B* or *b*	*C* or *c*	Combined proportion
3/4 *A*	3/4 *B*	3/4 *C*	→ (3/4)(3/4)(3/4) *ABC* = 27/64 *ABC*
		1/4 *c*	→ (3/4)(3/4)(1/4) *ABc* = 9/64 *ABc*
	1/4 *b*	3/4 *C*	→ (3/4)(1/4)(3/4) *AbC* = 9/64 *AbC*
		1/4 *c*	→ (3/4)(1/4)(1/4) *Abc* = 3/64 *Abc*
1/4 *a*	3/4 *B*	3/4 *C*	→ (1/4)(3/4)(3/4) *aBC* = 9/64 *aBC*
		1/4 *c*	→ (1/4)(3/4)(1/4) *aBc* = 3/64 *aBc*
	1/4 *b*	3/4 *C*	→ (1/4)(1/4)(3/4) *abC* = 3/64 *abC*
		1/4 *c*	→ (1/4)(1/4)(1/4) *abc* = 1/64 *abc*

FIGURE 3–9 Generation of the F_2 trihybrid phenotypic ratio using the forked-line method. This method is based on the expected probability of occurrence of each phenotype.

genes on chromosomes, their findings also reestablished the importance of Mendel's work, which became the basis of ensuing genetic investigations. Sutton and Boveri are credited with initiating the **chromosomal theory of inheritance,** which was developed during the next two decades.

ESSENTIAL POINT

The discovery of chromosomes in the late 1800s, along with subsequent studies of their behavior during meiosis, led to the rebirth of Mendel's work, linking the behavior of his unit factors to that of chromosomes during meiosis.

Unit Factors, Genes, and Homologous Chromosomes

Because the correlation between Sutton's and Boveri's observations and Mendelian principles serves as the foundation for the modern interpretation of transmission genetics, we will examine this correlation in some depth before moving on to other topics.

As we know, each species possesses a specific number of chromosomes in each somatic cell nucleus (except in gametes). For diploid organisms, this number is called the **diploid number (2*n*)** and is characteristic of that species. During the formation of gametes, this number is precisely halved (*n*), and when two gametes combine during fertilization, the diploid number is reestablished. During meiosis, however, the chromosome number is not reduced in a random manner. It was apparent to early cytologists that the diploid number of chromosomes is composed of homologous pairs identifiable by their morphological appearance and behavior. The gametes contain one member of each pair—thus the chromosome complement of a gamete is quite specific, and the number of chromosomes in each gamete is equal to the haploid number.

With this basic information, we can see the correlation between the behavior of unit factors and chromosomes and genes. Figure 3–10 shows three of Mendel's postulates and the chromosomal explanation of each. Unit factors are really genes located on homologous pairs of chromosomes [Figure 3–10(a)]. Members of each pair of homologs separate, or segregate, during gamete formation [Figure 3–10(b)]. Two different alignments are possible, both of which are shown.

To illustrate the principle of independent assortment, we must distinguish between members of any given homologous pair of chromosomes. One member of each pair comes from the **maternal parent,** while the other member comes from the **paternal parent.** (We represent the different parental origins with different colors.) As shown in Figure 3–10(c), following independent segregation of each

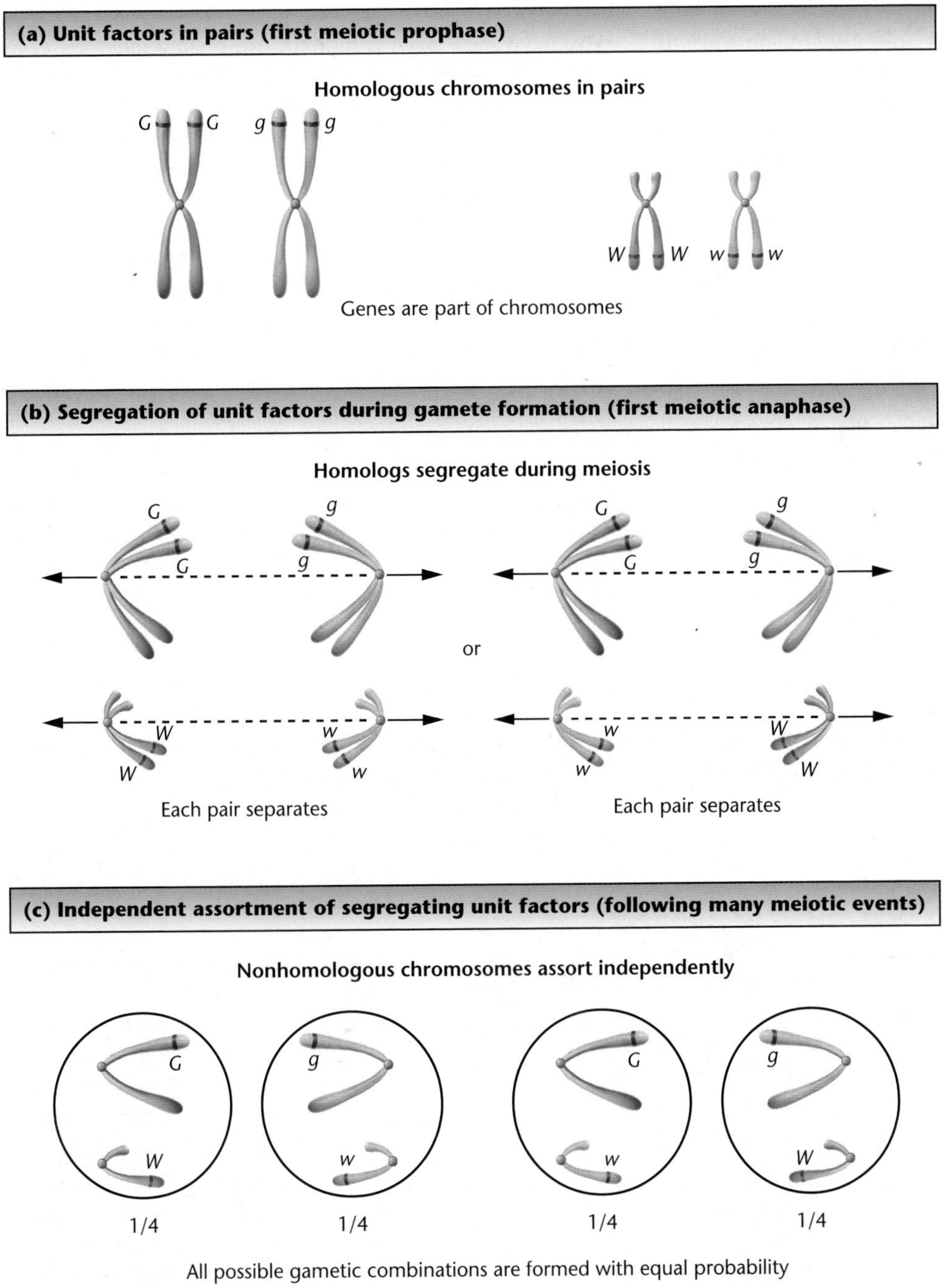

FIGURE 3–10 Illustrated correlation between the Mendelian postulates of (a) unit factors in pairs, (b) segregation, and (c) independent assortment, showing the presence of genes located on homologous chromosomes and their behavior during meiosis.

pair of homologs, each gamete receives one member from each pair of chromosomes. All possible combinations are formed with equal probability. If we add the symbols used in Mendel's dihybrid cross (*G*, *g* and *W*, *w*) to the diagram, we can see why equal numbers of the four types of gametes are formed. The independent behavior of Mendel's pairs of unit factors (*G* and *W* in this example) is due to their presence on separate pairs of homologous chromosomes.

Observations of the phenotypic diversity of living organisms make it logical to assume that there are many more genes than chromosomes. Therefore, each homolog must carry genetic information for more than one trait. The currently accepted concept is that a chromosome is composed of a large number of linearly ordered, information-containing *genes*. Mendel's unit factors (which determine tall or dwarf stems, for example) actually constitute a pair of genes located on one pair of homologous chromosomes. The location on a given chromosome where any particular gene occurs is called its **locus** (pl., loci). The different forms taken by a given gene, called *alleles* (*G* or *g*), contain slightly different genetic information that determines the same character (seed color in this case). Although we have examined only genes with two alternative alleles, most genes have more than two allelic forms. We conclude this section by reviewing the criteria necessary to classify two chromosomes as a homologous pair:

1. During mitosis and meiosis, when chromosomes are visible as distinct figures, both members of a homologous pair are the same size and exhibit identical centromere locations. The sex chromosomes are an exception.
2. During early stages of meiosis, homologous chromosomes form pairs, or synapse.
3. Although not generally microscopically visible, homologs contain identical, linearly ordered gene loci.

3.6 Independent Assortment Leads to Extensive Genetic Variation

One major consequence of independent assortment is the production by an individual of genetically dissimilar gametes. Genetic variation results because the two members of any homologous pair of chromosomes are rarely, if ever, genetically identical. As the maternal and paternal members of all pairs are distributed to gametes through independent assortment, all possible chromosome combinations are produced, leading to extensive genetic diversity.

We have seen that the number of possible gametes, each with different chromosome compositions, is 2^n, where n equals the haploid number. Thus, if a species has a haploid number of $n = 4$, then $2^4 = 16$ different gamete combinations can be formed as a result of independent assortment. Although this number is not high, consider the human species, where $n = 23$. If 2^{23} is calculated, we find that in excess of 8×10^6, or over 8 million, different types of gametes are represented. Because fertilization represents an event involving only one of approximately 8×10^6 possible gametes from each of two parents, each offspring represents only one of $(8 \times 10^6)^2$, or one of only 64×10^{12} potential genetic combinations! No wonder that, except for identical twins, each member of the human species demonstrates a distinctive appearance and individuality—this number of combinations is far greater than the number of humans who have ever lived on Earth! Genetic variation resulting from independent assortment has been extremely important to the process of evolution in all sexually reproducing organisms.

Tay–Sachs Disease: The Molecular Basis of a Recessive Disorder in Humans

An interesting question involving Mendelian traits centers around how mutant genes result in mutant phenotypes. Insights are gained by considering a modern explanation of the gene that causes **Tay–Sachs disease** (TSD), a devastating inherited recessive disorder involving unalterable destruction of the central nervous system. Infants with TSD are unaffected at birth and appear to develop normally until they are about six months old. Then, a progressive loss of mental and physical abilities occurs. Afflicted infants eventually become blind, deaf, mentally retarded, and paralyzed, often within only a year or two, seldom living beyond age five. Typical of rare autosomal recessive disorders, two unaffected heterozygous parents, who most often have no immediate family history of the disorder, have a probability of one in four of having a Tay–Sachs child.

We know that proteins are the end products of the expression of nearly all genes. The protein product involved in TSD has been identified, and we now have a clear understanding of the underlying molecular basis of the disorder. TSD results from the loss of activity of a single enzyme **hexosaminidase A (Hex-A)**. Hex-A, normally found in lysosomes within cells, is needed to break down the ganglioside GM2, a lipid component of nerve cell membranes. Without functional Hex-A, gangliosides accumulate within neurons in the brain and cause deterioration of the nervous system. Heterozygous carriers of TSD with one normal copy of the gene produce only about 50 percent of the normal amount of Hex-A, but they show no symptoms of the disorder. The observation that the activity of only one gene (one wild-type allele) is sufficient for the normal development and function of the nervous system explains and illustrates the molecular basis of recessive mutations. Only when both genes are disrupted by mutation is the mutant phenotype evident. The responsible gene is located on chromosome 15 and codes for the alpha subunit of the Hex-A enzyme. More than 50 different mutations within the gene have been identified that lead to TSD phenotypes.

3.7 Laws of Probability Help to Explain Genetic Events

Recall that genetic ratios are expressed as probabilities—for example, 3/4 tall: 1/4 dwarf. These values predict the outcome of each fertilization event, such that the probability of each zygote having the genetic potential for becoming tall is 3/4, while the potential for becoming dwarf is 1/4. Probabilities range from 0.0, when an event is *certain not to occur*, to 1.0, when an event is *certain to occur*. When two or more events occur independently but at the same time, we can calculate the probability of possible outcomes when they occur together. This is accomplished by applying the *product law*—the probability of two or more events occurring simultaneously is equal to the product of their individual probabilities. Two or more events are independent of one another if the outcome of each one does not affect the outcome of any of the others under consideration.

To illustrate the product law, consider the possible results if you toss a penny (*P*) and a nickel (*N*) at the same time and examine all combinations of heads (*H*) and tails (*T*) that can occur. There are four possible outcomes:

$$(P_H{:}N_H) = (1/2)\,(1/2) = 1/4$$
$$(P_T{:}N_H) = (1/2)\,(1/2) = 1/4$$
$$(P_H{:}N_T) = (1/2)\,(1/2) = 1/4$$
$$(P_T{:}N_T) = (1/2)\,(1/2) = 1/4$$

The probability of obtaining a head or a tail in the toss of either coin is 1/2 and is unrelated to the outcome of the toss of the other coin. Thus, all four possible combinations are predicted to occur with equal probability.

If we want to calculate the probability where the possible outcomes of two events are independent of one another but can be accomplished in more than one way, we apply the **sum law.** For example, what is the probability of tossing our penny and nickel and obtaining one head and one tail? In such a case, we do not care whether it is the penny or the nickel that comes up heads, provided the other coin has the alternative outcome. As we saw above, there are two ways in which the desired outcome can be accomplished, each with a probability of 1/4. Thus, according to the sum law, the overall probability is equal to

$$(1/4) + (1/4) = 1/2$$

One-half of all coin tosses are predicted to yield the desired outcome.

These simple probability laws will be useful throughout our discussions of transmission genetics and for solving genetics problems. In fact, we already applied the product law when we used the forked-line method to calculate the phenotypic results of Mendel's dihybrid and trihybrid crosses. When we wish to know the results of a cross, we need only calculate the probability of each possible outcome. The results of this calculation then allow us to predict the proportion of offspring expressing each phenotype or each genotype.

An important point to remember when you deal with probability is that predictions of possible outcomes are based on large sample sizes. If we predict that 9/16 of the offspring of a dihybrid cross will express both dominant traits, it is very unlikely that, in a small sample, exactly 9 of every 16 offspring will express this phenotype. Instead, our prediction is that, of a large number of offspring, approximately 9/16 of them will do so. The deviation from the predicted ratio in smaller sample sizes is attributed to chance, a subject we examine in our discussion of statistics in the next section. As you shall see, the impact of deviation due strictly to chance diminishes as the sample size increases.

ESSENTIAL POINT

Since genetic ratios are expressed as probabilities, deriving outcomes of genetic crosses requires an understanding of the laws of probability.

3.8 Chi-Square Analysis Evaluates the Influence of Chance on Genetic Data

Mendel's 3:1 monohybrid and 9:3:3:1 dihybrid ratios are hypothetical predictions based on the following assumptions: (1) Each allele is dominant or recessive; (2) segregation is operative; (3) independent assortment occurs; and (4) fertilization is random. The final two assumptions are influenced by chance events and are therefore subject to random fluctuation. This concept of **chance deviation** is most easily illustrated by tossing a single coin numerous times and recording the number of heads and tails observed. In each toss, there is a probability of 1/2 that a head will occur and a probability of 1/2 that a tail will occur. Therefore, the expected ratio of many tosses is 1:1. If a coin is tossed 1000 times, usually *about* 500 heads and 500 tails will be observed. Any reasonable fluctuation from this hypothetical ratio (e.g., 486 heads and 514 tails) is attributed to chance.

As the total number of tosses is reduced, the impact of chance deviation increases. For example, if a coin is tossed only four times, you would not be too surprised if all four tosses result in only heads or only tails. For 1000 tosses, however, 1000 heads or 1000 tails would be most unexpected. In fact, you might believe that such a result would be impossible. Actually, all heads or all tails in 1000 tosses can be predicted to occur with a probability of $(1/2)^{1000}$. Since $(1/2)^{20}$ is equivalent to less than one in a million times, an event occurring with a probability of $(1/2)^{1000}$ is virtually impossible. Two major points are significant before we consider *chi-square analysis*:

PROBABILITY ▶ WEB TUTORIAL

1. The outcomes of independent assortment and fertilization, like coin tossing, are subject to random fluctuations from their predicted occurrences as a result of chance deviation.
2. As the sample size increases, the average deviation from the expected results decreases. Therefore, a larger sample size diminishes the impact of chance deviation on the final outcome.

Chi-Square Calculations and the Null Hypothesis

In genetics, the ability to evaluate observed deviation is a crucial skill. When we assume that data will fit a given ratio such as 1:1, 3:1, or 9:3:3:1, we establish what is called the **null hypothesis (H_0)**. It is so named because the hypothesis assumes that *no real difference* exists between *measured values* (or ratio) and *predicted values* (or ratio). The apparent difference can be attributed purely to chance. The null hypothesis is evaluated using statistical analysis. On this basis, the null hypothesis may either (1) be rejected or (2) fail to be rejected. If it is rejected, the observed deviation from the expected result is not attributed to chance alone. The null hypothesis and the underlying assumptions leading to it must be reexamined. If the null hypothesis fails to be rejected, any observed deviations are attributed to chance.

One of the simplest statistical tests devised to assess the null hypothesis is **chi-square (χ^2) analysis.** This test takes into account the observed deviation in each component of an expected ratio as well as the sample size and reduces them to a single numerical value. The value for χ^2 is then used to estimate how frequently the observed deviation can be expected to occur strictly as a result of chance. The formula for chi-square analysis is

$$\chi^2 = \sum\frac{(o - e)^2}{e}$$

where o is the observed value for a given category, e is the expected value for that category, and $\sum$ (the Greek letter sigma) represents the sum of the calculated values for each category of the ratio. Because $(o - e)$ is the deviation (d) in each case, the equation reduces to

$$\chi^2 = \sum\frac{d^2}{e}$$

Table 3.1(a) shows a χ^2 calculation for the F_2 results of a hypothetical monohybrid cross. To analyze these data, you work from left to right, calculating and entering the appropriate numbers in each column. Regardless of whether the deviation d is positive or negative, d^2 always becomes positive after the number is squared. In Table 3.1(b) the F_2 results of a hypothetical dihybrid cross are analyzed. Be sure that you understand how each number was calculated in the dihybrid example.

The final step in chi-square analysis is to interpret the χ^2 value. To do so, you must initially determine the value of the **degrees of freedom (df),** which is equal to $n - 1$, where n is the number of different categories into which each datum point may fall. For the 3:1 ratio, $n = 2$, so df $= 2 - 1 = 1$. For the 9:3:3:1 ratio, $n = 4$ and df $= 3$. Degrees of freedom must be taken into account because the greater the number of categories, the more deviation is expected as a result of chance.

Once you have determined the degrees of freedom, we can interpret the χ^2 value in terms of a corresponding **probability value (p).** Since this calculation is complex, we usually take the p value from a standard table or graph. Figure 3–11 shows a wide range of χ^2 and p values for various degrees of freedom in both a graph and a table. Let's use the graph to determine the p value. The caption for Figure 3–11(b) explains how to use the table.

To determine p, execute the following steps:

1. Locate the χ^2 value on the abscissa (the horizontal or x-axis).
2. Draw a vertical line from this point up to the angled line on the graph representing the appropriate df.
3. Extend a horizontal line from this point to the left until it intersects the ordinate (the vertical or y-axis).
4. Estimate, by interpolation, the corresponding p value.

We used these steps for the monohybrid cross in Table 3.1(a) to estimate the p value of 0.48 shown in Figure 3–11(a). For the dihybrid cross, try this method to see if you can determine the p value. Since the χ^2 value is 4.16 and df $= 3$, an

TABLE 3.1 Chi-Square Analysis

(a) Monohybrid Cross

Expected Ratio	Observed (o)	Expected (e)	Deviation (o − e)	Deviation² (d²)	d²/e
3/4	740	3/4(1000) = 750	740 − 750 = −10	(−10)² = 100	100/750 = 0.13
1/4	260	1/4(1000) = 250	260 − 250 = +10	(+10)² = 100	100/250 = 0.40
	Total = 1000				χ^2 = 0.53
					p = 0.48

(b) Dihybrid Cross

Expected Ratio	(o)	(e)	(o − e)	(d²)	d²/e
9/16	587	567	+20	400	0.71
3/16	197	189	+8	64	0.34
3/16	168	189	−21	441	2.33
1/16	56	63	−7	49	0.78
	Total = 1008				χ^2 = 4.16
					p = 0.26

(a)

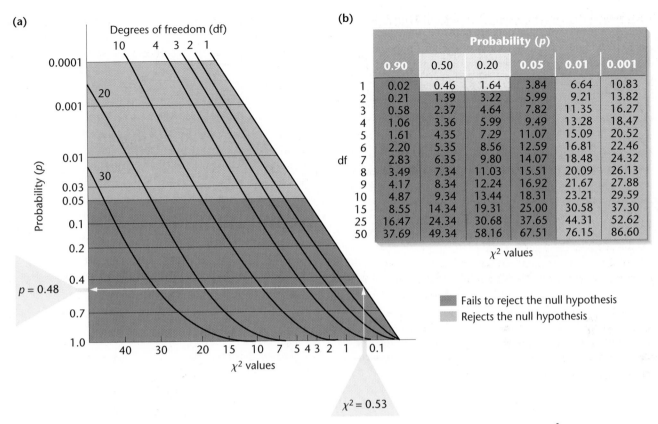

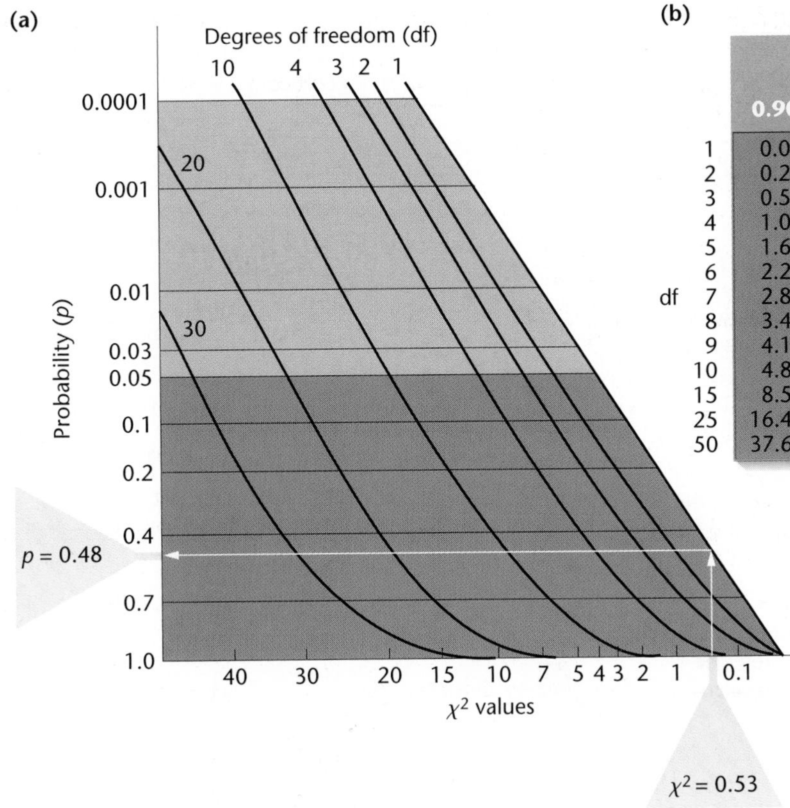

(b)

df	Probability (p) 0.90	0.50	0.20	0.05	0.01	0.001
1	0.02	0.46	1.64	3.84	6.64	10.83
2	0.21	1.39	3.22	5.99	9.21	13.82
3	0.58	2.37	4.64	7.82	11.35	16.27
4	1.06	3.36	5.99	9.49	13.28	18.47
5	1.61	4.35	7.29	11.07	15.09	20.52
6	2.20	5.35	8.56	12.59	16.81	22.46
7	2.83	6.35	9.80	14.07	18.48	24.32
8	3.49	7.34	11.03	15.51	20.09	26.13
9	4.17	8.34	12.24	16.92	21.67	27.88
10	4.87	9.34	13.44	18.31	23.21	29.59
15	8.55	14.34	19.31	25.00	30.58	37.30
25	16.47	24.34	30.68	37.65	44.31	52.62
50	37.69	49.34	58.16	67.51	76.15	86.60

χ^2 values

Fails to reject the null hypothesis
Rejects the null hypothesis

FIGURE 3–11 (a) Graph for converting χ^2 values to p values. (b) Table of χ^2 values for selected values of df and p. χ^2 values that lead to a p value of 0.05 or greater (darker blue areas) justify failure to reject the null hypothesis. Values leading to a p value of less than 0.05 (lighter blue areas) justify rejecting the null hypothesis. For example, using the table in part (b), where $\chi^2 = 0.53$ for 1 degree of freedom, the corresponding p value is between 0.20 and 0.50. The graph in (a) gives a more precise p value of 0.48 by interpolation. Thus, we fail to reject the null hypothesis.

approximate p value is 0.26. Checking this result in the table confirms that p values for both the monohybrid and dihybrid crosses are between 0.20 and 0.50.

Interpreting Probability Values

So far, we have been concerned with calculating χ^2 values and determining the corresponding p values. The most important aspect of chi-square analysis is understanding the meaning of the p value. Let's use the example of the dihybrid cross in Table 3.1(b) ($p = 0.26$). In these discussions, it is simplest to think of the p value as a percentage (e.g., $0.26 = 26\%$). In our example, the p value indicates that if we repeat the same experiment many times, 26 percent of the trials would be expected to exhibit chance deviation as great as or greater than that seen in the initial trial. Conversely, 74 percent of the trials would show less deviation than initially observed as a result of chance. Thus, the p value reveals that a hypothesis (the 9:3:3:1 ratio in this case) is never proved or disproved absolutely. Instead, a relative standard is set that enables us to either *reject* or *fail to reject* the null hypothesis—this standard is most often a p value of 0.05. When applied to chi-square analysis, a p value less than 0.05 means that the observed deviation in the set of results will be obtained by chance alone less than 5 percent of the time. Such a p value indicates that the difference between the observed and predicted results is substantial and thus enables us to reject the null hypothesis.

On the other hand, p values of 0.05 or greater (0.05 to 1.0) indicate that the observed deviation will be obtained by chance alone 5 percent or more of the time. The conclusion is not to reject the null hypothesis. Thus, for the p value of 0.26, assessing the hypothesis that independent assortment accounts for the results fails to be rejected. Therefore, the observed deviation can be reasonably attributed to chance.

A final note is relevant here for the case where the null hypothesis is rejected, that is, where $p \leq 0.05$. Suppose we had tested a data set to assess a possible 9:3:3:1 ratio, as in Table 3.1(b), but we rejected the null hypothesis based on our χ^2 calculation. What are alternative interpretations of the data? Researchers will reassess the assumptions that underlie the null hypothesis. In our example, we assumed that segregation operates faithfully for both gene pairs. We also assumed that fertilization is random and that the viability of all gametes is equal regardless of genotype—that is, that all gametes are equally likely to participate in fertilization. Finally, we assumed that, following fertilization, all preadult stages and adult offspring are equally viable, regardless of their genotype. If any of these assumptions is incorrect, the original hypothesis is not necessarily invalid.

An example will clarify this. Suppose our null hypothesis is that a dihybrid cross between fruit flies will result in 3/16 mutant wingless fly zygotes. However, not as many of the mutant embryos may survive their preadult development or as young adults, compared to flies whose genotype gives rise to wings. As a result, when the data are gathered, there are fewer than 3/16 wingless flies. Rejection of the null hypothesis is not in itself cause for us to reject the validity of the postulates of segregation and independent assortment, because other factors we are unaware of may also be affecting the outcome.

ESSENTIAL POINT

Chi-square analysis allows us to assess the null hypothesis, which states that there is no real difference between the expected and observed values. As such, it tests the probability of whether observed variations can be attributed to chance deviation.

NOW SOLVE THIS

Problem 18 on page 58 asks you to apply χ^2 analysis to a set of data and to determine whether those data fit any of several ratios.

Hint: In calculating χ^2, first determine the expected outcomes using the predicted ratios. Then follow a stepwise approach, determining the deviation in each case, and calculating d^2/e for each category.

3.9 Pedigrees Reveal Patterns of Inheritance of Human Traits

We now explore how to determine the mode of inheritance of phenotypes in humans, where designed crosses are not possible and where relatively few offspring are available for study. The traditional way to study inheritance has been to construct a family tree, indicating the presence or absence of the trait in question for each member of each generation. Such a family tree is called a **pedigree.** By analyzing a pedigree, we may be able to predict how the trait under study is inherited—for example, is it due to a dominant or recessive allele? When many pedigrees for the same trait are studied, we can often ascertain the mode of inheritance.

Pedigree Conventions

Figure 3–12 illustrates a number of conventions geneticists follow in constructing pedigrees. Circles represent females and squares designate males. Parents are connected by a single horizontal line, and vertical lines lead to their offspring. If the parents are related—that is, they are **consanguineous,** such as first cousins—they are connected by a double line. Offspring are called **sibs** (short for **siblings**) and are connected by a horizontal **sibship line.** Sibs are placed from left to right according to birth order, and are labeled with Arabic numerals. Each generation is indicated by a Roman numeral. If the sex of an individual is unknown, a diamond is used. When a pedigree traces only a single trait, the circles, squares, and diamonds are shaded if the phenotype being considered is expressed and unshaded if not. In some pedigrees, those individuals that fail to express a recessive trait, but are known with certainty to be a heterozygous carrier, have a shaded dot

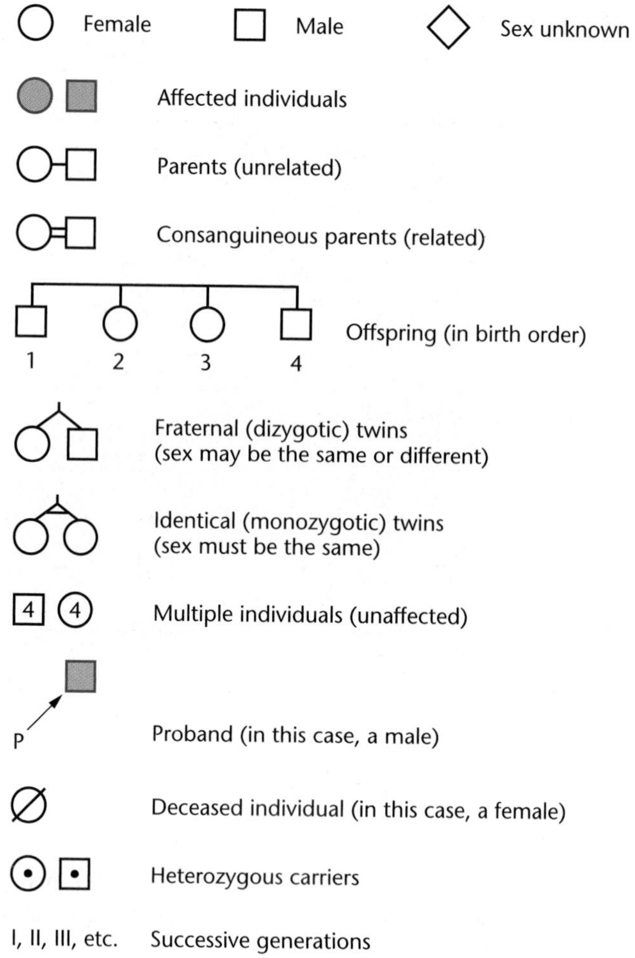

FIGURE 3–12 Conventions commonly encountered in human pedigrees.

within their unshaded circle or square. If an individual is deceased and the phenotype is unknown, a diagonal line is placed over the circle or square.

Twins are indicated by diagonal lines stemming from a vertical line connected to the sibship line. For **identical** (or **monozygotic**) **twins,** the diagonal lines are linked by a horizontal line. **Fraternal** (or **dizygotic**) **twins** lack this connecting line. A number within one of the symbols represents numerous sibs of the same or unknown phenotypes. The individual whose phenotype first brought attention to the investigation and construction of the pedigree is called the **proband** and is indicated by an arrow connected to the designation **p.** This term applies to either a male or a female.

Pedigree Analysis

In Figure 3–13, two pedigrees are shown. The first illustrates a representative pedigree for a trait that demonstrates autosomal recessive inheritance, such as **albinism.** The male parent of the first generation (I-1) is affected. Characteristic of a rare recessive trait with an affected parent, the trait "disappears" in the offspring of the next generation. Assuming recessiveness, we might predict that the unaffected female parent (I-2) is a homozygous normal individual because none of the offspring show the disorder. Had she been heterozygous, one

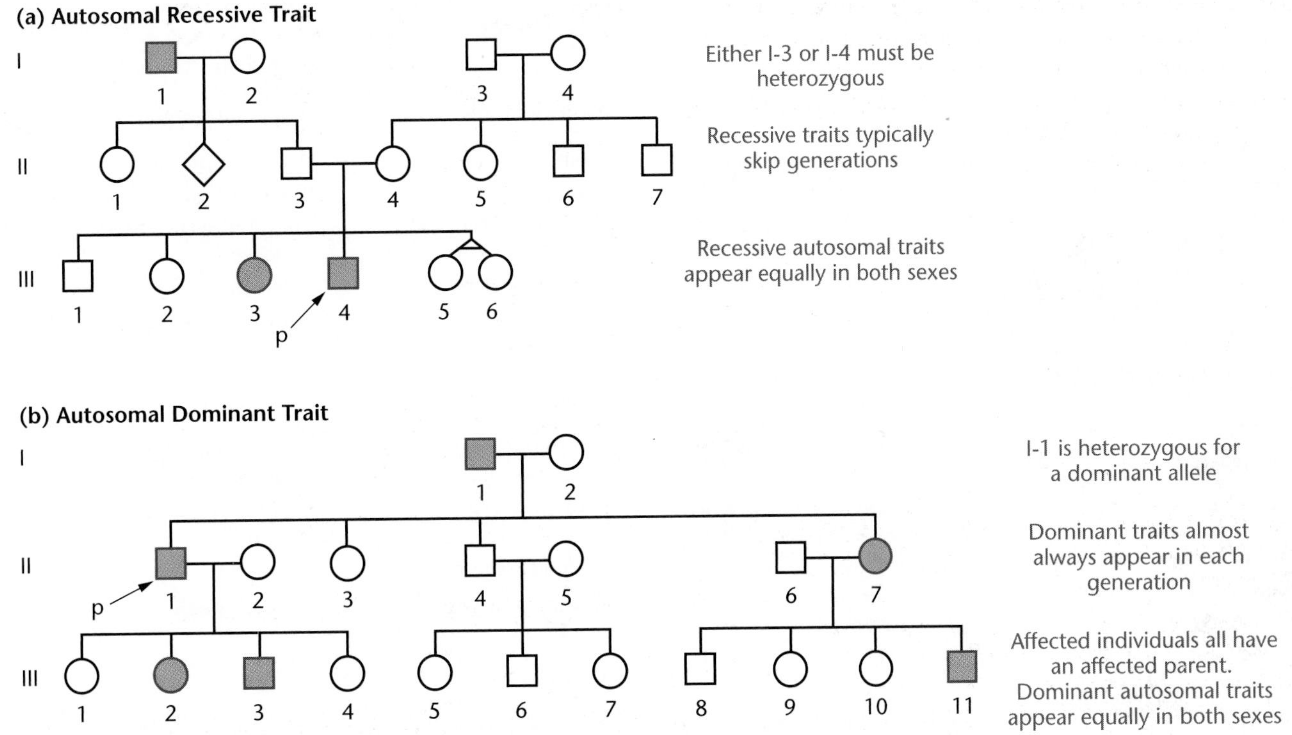
(a) Autosomal Recessive Trait
Either I-3 or I-4 must be heterozygous
Recessive traits typically skip generations
Recessive autosomal traits appear equally in both sexes
(b) Autosomal Dominant Trait
I-1 is heterozygous for a dominant allele
Dominant traits almost always appear in each generation
Affected individuals all have an affected parent.
Dominant autosomal traits appear equally in both sexes

FIGURE 3–13 Representative pedigrees for two characteristics, each followed through three generations.

half of the offspring would be expected to exhibit albinism, but none do. However, such a small sample (three offspring) prevents us from knowing for certain.

Further evidence supports the prediction of a recessive trait. If albinism were inherited as a dominant trait, individual II-3 would have to express the disorder in order to pass it to his offspring (III-3 and III-4), but he does not. Inspection of the offspring constituting the third generation (row III) provides still further support for the hypothesis that albinism is a recessive trait. If it is, parents II-3 and II-4 are both heterozygous, and approximately one-fourth of their offspring should be affected. Two of the six offspring do show albinism. This deviation from the expected ratio is not unexpected in crosses with few offspring. Once we are confident that albinism is inherited as an autosomal recessive trait, we could portray the II-3 and II-4 individuals with a shaded dot within their larger square and circle. Finally, we can note that, characteristic of pedigrees for autosomal traits, both males and females are affected with equal probability. In Chapter 4, we will examine a pedigree representing a gene located on the sex-determining X chromosome. We will see certain limitations imposed on the transmission of X-linked traits, such as that these traits are more prevalent in male offspring and are never passed from affected fathers to their sons.

The second pedigree illustrates the pattern of inheritance for a trait such as Huntington disease, which is caused by an autosomal dominant allele. The key to identifying such a pedigree that reflects a dominant trait is that all affected offspring will have a parent that also expresses the trait. It is also possible, by chance, that none of the offspring will inherit the dominant allele. If so, the trait will cease to exist in future generations. Like recessive traits, provided that the gene is autosomal, both males and females are equally affected.

When autosomal dominant diseases are rare within the population, and most are, then it is highly unlikely that affected individuals will inherit a copy of the mutant gene from both parents. Therefore, in most cases, affected individuals are heterozygous for the dominant allele. As a result, approximately one-half of the offspring inherit it. This is borne out in the second pedigree in Figure 3–13. Furthermore, if a mutation is dominant, and a single copy is sufficient to produce a mutant phenotype, homozygotes are likely to be even more severely affected, perhaps even failing to survive. An illustration of this is the dominant gene for **familial hypercholesterolemia.** Heterozygotes display a defect in their receptors for low-density lipoproteins, the so-called LDLs. As a result, too little cholesterol is taken up by cells from the blood, and elevated plasma levels of LDLs result. Such heterozygous individuals have heart attacks during the fourth decade of their life, or before. While heterozygotes have LDL levels about double that of a normal individual, rare homozygotes have been detected. They lack LDL receptors altogether and have LDL levels nearly ten times above the normal range. They are likely to have a heart attack very early in life, even before age 5, and almost inevitably before they reach the age of 20.

Pedigree analysis of many traits has historically been an extremely valuable research technique in human genetic studies. However, the approach does not usually provide the certainty in drawing conclusions afforded by designed crosses yielding large numbers of offspring. Nevertheless, when many independent pedigrees of the same trait or disorder are

WEB TUTORIAL ANALYSIS OF HUMAN PEDIGREES

analyzed, consistent conclusions can often be drawn. Table 3.2 lists numerous human traits and classifies them according to their recessive or dominant expression.

ESSENTIAL POINT

Pedigree analysis is a method for studying the inheritance pattern of human traits over several generations, providing the basis for predicting the mode of inheritance of characteristics and disorders in the absence of extensive genetic crossing and large numbers of offspring.

NOW SOLVE THIS

Problem 26 on page 58 asks you to examine a pedigree for myopia and predict whether the trait is dominant or recessive.

Hint: One of the first things to look for are individuals who express the trait, but neither of whose parents also expresses the trait. Such an observation makes it highly unlikely that the trait is dominant.

TABLE 3.2 Representative Recessive and Dominant Human Traits

Recessive Traits	Dominant Traits
Albinism	Achondroplasia
Alkaptonuria	Brachydactyly
Ataxia telangiectasia	Congenital stationary night blindness
Color blindness	Ehler–Danlos syndrome
Cystic fibrosis	Hypotrichosis
Duchenne muscular dystrophy	Huntington disease
Galactosemia	Hypercholesterolemia
Hemophilia	Marfan syndrome
Lesch–Nyhan syndrome	Neurofibromatosis
Phenylketonuria	Phenylthiocarbamide (PTC) tasting
Sickle-cell anemia	Porphyria
Tay–Sachs disease	Widow's peak

EXPLORING GENOMICS

Online Mendelian Inheritance in Man

The **Online Mendelian Inheritance in Man (OMIM)** database is a catalog of human genes and human genetic disorders that are inherited in a Mendelian manner. Genetic disorders that arise from major chromosomal aberrations, such as monosomy or trisomy (the loss of a chromosome or the presence of a superfluous chromosome, respectively), are not included. The OMIM database is a daily updated version of the book *Mendelian Inheritance in Man*, originally edited by Dr. Victor McKusick of Johns Hopkins University. Scientists use OMIM as an important information source to accompany the sequence data generated by the **Human Genome Project.**

The OMIM entries will give you links to a wealth of information, including DNA and protein sequences, chromosomal maps, disease descriptions, and relevant scientific publications. In this exercise, you will explore OMIM to answer questions about the recessive human disease sickle-cell anemia and other Mendelian inherited disorders.

■ Exercise I: Sickle-cell Anemia

In this chapter, you were introduced to sickle-cell anemia as an example of a single-gene recessive disease. You will now discover more about sickle-cell anemia by exploring the OMIM database.

1. To begin the search, access the OMIM site at: **www.ncbi.nlm.nih.gov/entrez/query.fcgi?=OMIM&itool=toolbar**.
2. In the "SEARCH" box, type "sickle-cell anemia" and click on the "Go" button to perform the search.
3. Select the first entry (#603903).
4. Examine the list of subject headings in the left-hand column and read some of the information about sickle-cell anemia.
5. Select one or two references at the bottom of the page and follow them to their abstracts in PubMed.
6. Using the information in this entry, answer the following questions:
   a. Which gene is mutated in individuals with sickle-cell anemia?
   b. What are the major symptoms of this disorder?
   c. What was the first published scientific description of sickle-cell anemia?
   d. Describe two other features of this disorder that you learned from the OMIM database and state where in the database you found this information.

■ Exercise II: Other Recessive or Dominant Disorders

Select another human disorder that is inherited as either a dominant or recessive trait and investigate its features, following the general procedure presented above. Follow links from OMIM to other databases if you choose.

Describe several interesting pieces of information you acquired during your exploration and cite the information sources you encountered during the search.

CASE STUDY To test or not to test

Thomas first discovered a potentially devastating piece of family history when he learned the medical diagnosis for his brother's increasing dementia, muscular rigidity, and frequency of seizures. His brother, at age 49, was diagnosed with Huntington disease (HD), a dominantly inherited condition that typically begins with such symptoms around the age of 45 and leads to death in one's early 60s. As depressing as the news was to Thomas, it helped explain his father's suicide. Thomas, 38, now wonders what are his chances of carrying the gene for HD, and he and his wife have discussed the pros and cons of him getting tested. Thomas and his wife have two teenaged children, a boy and a girl.

1. What role might a genetic counselor play in this real-life scenario?
2. How might the preparation and analysis of a pedigree help explain the dilemma facing Thomas and his family?
3. If Thomas decides to go ahead with the genetic test, what should be the role of the health insurance industry in such cases?
4. If Thomas tests positive for HD, and you were one of his children, would you want to be tested?

INSIGHTS AND SOLUTIONS

As a student, you will be asked to demonstrate your knowledge of transmission genetics by solving genetics problems. Success at this task represents not only comprehension of theory but its application to more practical genetic situations. Most students find problem solving in genetics to be challenging and rewarding. This section will provide you with basic insights into the reasoning essential to this process.

Genetics problems are in many ways similar to word problems in algebra. The approach to solving them is identical: (1) Analyze the problem carefully; (2) translate words into symbols, first defining each one; and (3) choose and apply a specific technique to solve the problem. The first two steps are critical. The third step is largely mechanical.

The simplest problems state all necessary information about the P_1 generation and ask you to find the expected ratios of the F_1 and F_2 genotypes and/or phenotypes. Always follow these steps when you encounter this type of problem:

(a) Determine insofar as possible the genotypes of the individuals in the P_1 generation.

(b) Determine what gametes may be formed by the P_1 parents.

(c) Recombine gametes by the Punnett square or the forked-line methods, or if the situation is very simple, by inspection. Read the F_1 phenotypes.

(d) Repeat the process to obtain information about the F_2 generation.

Determining the genotypes from the given information requires that you understand the basic theory of transmission genetics. Consider this problem: *A recessive mutant allele, black, causes a very dark body in Drosophila (a fruit fly) when homozygous. The wild-type (normal) color is gray. What F_1 phenotypic ratio is predicted when a black female is crossed with a gray male whose father was black?*

To work out this problem, you must understand dominance and recessiveness, as well as the principle of segregation. Furthermore, you must use the information about the male parent's father. Here is one way to solve this problem:

(a) The female parent is black, so she must be homozygous for the mutant allele (*bb*).

(b) The male parent is gray; therefore, he must have at least one dominant allele (*B*). His father was black (*bb*), and he received one of the chromosomes bearing these alleles, so the male parent must be heterozygous (*Bb*).

With this information, the problem is simple:

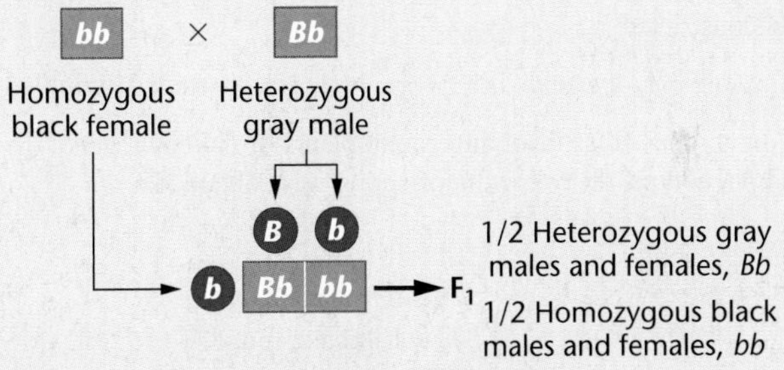

Apply this approach to the following problems.

1. Mendel found that full pods are dominant over constricted pods, while round seeds are dominant over wrinkled seeds. One of his crosses was between full, round plants and constricted, wrinkled plants. From this cross, he obtained an F_1 generation that was all full and round. In the F_2 generation, Mendel obtained his classic 9:3:3:1 ratio. Using this information, determine the expected F_1 and F_2 results of a cross between homozygous constricted, round plants and full, wrinkled plants.

Solution: Define gene symbols for each pair of contrasting traits. Use the lowercase first letter of the recessive traits to designate those phenotypes and the uppercase first letter to designate the dominant traits. Thus, *C* and *c* indicate full and constricted, and *W* and *w* indicate round and wrinkled phenotypes, respectively.

Determine the genotypes of the P_1 generation, form the gametes, reconstitute the F_1 generation, and read off the phenotype(s):

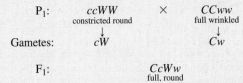

P_1: *ccWW* (constricted round) × *CCww* (full wrinkled)

↓ ↓

Gametes: *cW* *Cw*

F_1: *CcWw* (full, round)

You can immediately see that the F_1 generation expresses both dominant phenotypes and is heterozygous for both gene pairs. Thus, you expect that the F_2 generation will yield the classic Mendelian ratio of 9:3:3:1. Let's work it out anyway just to confirm this, using the forked-line method. Both gene pairs are heterozygous and can be expected to assort independently, so we can predict the outcomes from each gene pair separately and then proceed with the forked-line method.

Every F_2 offspring is subject to the following probabilities:

$Cc \times Cc$		$Ww \times Ww$	
↓		↓	
CC, Cc, cC	full	WW, Ww, wW	round
cc	constricted	ww	wrinkled

The forked-line method then confirms the 9:3:3:1 phenotypic ratio. Remember that this represents proportions of 9/16:3/16:3/16:1/16. Note that we are applying the product law as we compute the final probabilities:

3/4 full
- 3/4 round $\xrightarrow{(3/4)(3/4)}$ 9/16 full, round
- 1/4 wrinkled $\xrightarrow{(3/4)(1/4)}$ 3/16 full, wrinkled

1/4 constricted
- 3/4 round $\xrightarrow{(1/4)(3/4)}$ 3/16 constricted, round
- 1/4 wrinkled $\xrightarrow{(1/4)(1/4)}$ 1/16 constricted, wrinkled

2. In another cross involving parent plants of unknown genotype and phenotype, the following offspring were obtained.

F_1: 3/8 full, round
3/8 full, wrinkled
1/8 constricted, round
1/8 constricted, wrinkled

Determine the genotypes and phenotypes of the parents.

Solution: This problem is more difficult and requires keener insight because you must work backward. The best approach is to consider the outcomes of pod shape separately from those of seed texture.

Of all the plants, $3/8 + 3/8 = 3/4$ are full and $1/8 + 1/8 = 1/4$ are constricted. Of the various genotypic combinations that can serve as parents, which combination will give rise to a ratio of 3/4:1/4? This ratio is identical to Mendel's monohybrid F_2 results, and we can propose that both unknown parents share the same genetic characteristic as the monohybrid F_1 parents; they must both be heterozygous for the genes controlling pod shape and thus are *Cc*.

Before we accept this hypothesis, let's consider the possible genotypic combinations that control seed texture. If we consider this characteristic alone, we see that the traits are expressed in a ratio of $3/8 + 1/8 = 1/2$ round: $3/8 + 1/8 = 1/2$ wrinkled. To generate such a ratio, the parents cannot both be heterozygous, or their offspring would yield a 3/4:1/4 phenotypic ratio. They cannot both be homozygous, or all of their offspring would express a single phenotype. Thus, we are left with testing the hypothesis that one parent is homozygous and one is heterozygous for the alleles controlling texture. The potential case of $WW \times Ww$ does not work, since it yields only a single phenotype. This leaves us with the potential case of $Ww \times ww$. Offspring in such a mating will yield 1/2 *Ww* (round):1/2 *ww* (wrinkled), exactly the outcome we are seeking.

Now, let's combine the hypotheses and predict the outcome of the cross. In our solution, we use a dash (–) to indicate that the second allele may be either dominant or recessive, since we are only predicting phenotypes.

3/4 *C*–
- 1/2 *Ww* → 3/8 *C–Ww* full, round
- 1/2 *ww* → 3/8 *C–ww* full, wrinkled

1/4 *cc*
- 1/2 *Ww* → 1/8 *ccWw* constricted, round
- 1/2 *ww* → 1/8 *ccww* constricted, wrinkled

As you can see, this cross produces offspring according to our initial information, and we have solved the problem. Note that in this solution, we used genotypes in the forked-line method, in contrast to the use of phenotypes in the earlier solution.

3. Determine the probability that a plant of genotype *CcWw* will be produced from parental plants with the genotypes *CcWw* and *Ccww*.

Solution: The two gene pairs demonstrate straightforward dominance and recessiveness and assort independently during gamete formation. We need only calculate the individual probabilities of obtaining the two separate outcomes (*Cc* and *Ww*) and apply the product law to calculate the final probability:

$$Cc \times Cc \longrightarrow 1/4\ CC{:}1/2\ Cc{:}1/4\ cc$$

$$Ww \times ww \longrightarrow 1/2\ Ww{:}1/2\ ww$$

$$p = (1/2\ Cc)(1/2\ Ww) = 1/4\ CcWw$$

4. In the laboratory, a genetics student crossed flies that had normal, long wings with flies expressing the *dumpy* mutation (truncated wings), which she believed was a recessive trait. In the F_1 generation, all flies had long wings. The following results were obtained in the F_2 generation:

792 long-winged flies
208 dumpy-winged flies

The student tested the hypothesis that the dumpy wing is inherited as a recessive trait, using chi-square analysis of the F_2 data.

(a) What ratio was hypothesized?

(b) Did the analysis support the hypothesis?

(c) What do the data suggest about the *dumpy* mutation?

Solution:

(a) The student hypothesized that the F_2 data (792:208) fit Mendel's 3:1 monohybrid ratio for recessive genes.

(b) The initial step in χ^2 analysis is to calculate the expected results (e) if the ratio is 3:1. Then we can compute deviation $o - e$ (d) and the remaining numbers.

Ratio	o	e	d	d^2	d^2/e
3/4	792	750	42	1764	2.35
1/4	208	250	−42	1764	7.06
	Total = 1000				

$$\chi^2 = \sum \frac{d^2}{e}$$

$$= 2.35 + 7.06$$

$$= 9.41$$

We consult Figure 3–11 to determine the probability (p) and determine whether the deviations can be attributed to chance. There are two possible outcomes (n), so the degrees of freedom (df) $= n - 1$ or 1. The table in Figure 3–11(b) shows that p is a value between 0.01 and 0.001; the graph in Figure 3–11(a) gives an estimate of about 0.001. Since $p < 0.05$, we reject the null hypothesis. The data do not fit a 3:1 ratio.

(c) When we accepted Mendel's 3:1 ratio as a valid expression of the monohybrid cross, numerous assumptions were made. Examining our underlying assumptions may explain why the null hypothesis was rejected. We assumed that all genotypes are equally viable—that genotypes yielding long wings are equally likely to survive from fertilization through adulthood as the genotype yielding dumpy wings. Further study may reveal that dumpy-winged flies are somewhat less viable than normal flies. As a result, we would expect less than 1/4 of the total offspring to express dumpy wings. This observation is borne out in the data, although we have not proven that this is true.

Problems and Discussion Questions

When working out genetics problems in this and succeeding chapters, always assume that members of the P_1 generation are homozygous, unless the information given, or the data, indicates otherwise.

1. In a cross between a black and a white guinea pig, all members of the F_1 generation are black. The F_2 generation is made up of approximately 3/4 black and 1/4 white guinea pigs. Diagram this cross, and show the genotypes and phenotypes.
2. Albinism in humans is inherited as a simple recessive trait. Determine the genotypes of the parents and offspring for the following families. When two alternative genotypes are possible, list both. (a) Two nonalbino (normal) parents have five children, four normal and one albino. (b) A normal male and an albino female have six children, all normal.
3. In a problem involving albinism (see Problem 2), which of Mendel's postulates are demonstrated?
4. Why was the garden pea a good choice as an experimental organism in Mendel's work?
5. Pigeons exhibit a checkered or plain feather pattern. In a series of controlled matings, the following data were obtained:

	F_1 Progeny	
P_1 Cross	**Checkered**	**Plain**
(a) checkered × checkered	36	0
(b) checkered × plain	38	0
(c) plain × plain	0	35

Then F_1 offspring were selectively mated with the following results. (The P_1 cross giving rise to each F_1 pigeon is indicated in parentheses.) See **Now Solve This** on page 41.

	F_2 Progeny	
$F_1 \times F_1$ Crosses	**Checkered**	**Plain**
(d) checkered (a) × plain (c)	34	0
(e) checkered (b) × plain (c)	17	14
(f) checkered (b) × checkered (b)	28	9
(g) checkered (a) × checkered (b)	39	0

How are the checkered and plain patterns inherited? Select and assign symbols for the genes involved, and determine the genotypes of the parents and offspring in each cross.

6. Mendel crossed peas having round seeds and yellow cotyledons with peas having wrinkled seeds and green cotyledons. All the F_1 plants had round seeds with yellow cotyledons. Diagram this cross through the F_2 generation, using both the Punnett square and forked-line methods.
7. Determine the genotypes of the parental plants by analyzing the phenotypes of the offspring from the following crosses: See **Now Solve This** on page 44.

Parental Plants	Offspring
(a) round, yellow × round, yellow	3/4 round, yellow 1/4 wrinkled, yellow
(b) round, yellow × wrinkled, yellow	6/16 wrinkled, yellow 2/16 wrinkled, green 6/16 round, yellow 2/16 round, green
(c) round, yellow × wrinkled, green	1/4 round, yellow 1/4 round, green 1/4 wrinkled, yellow 1/4 wrinkled, green

8. Are any of the crosses in Problem 7 testcrosses? If so, which one(s)?
9. Which of Mendel's postulates can be demonstrated in the crosses of Problem 7 but not in those in Problems 1 and 5? Define this postulate.
10. Correlate Mendel's four postulates with what is now known about homologous chromosomes, genes, alleles, and the process of meiosis.

11. What is the basis for homology among chromosomes?
12. Distinguish between homozygosity and heterozygosity.
13. In *Drosophila,* gray body color is dominant over ebony body color, while long wings are dominant over vestigial wings. Work the following crosses through the F_2 generation, and determine the genotypic and phenotypic ratios for each generation. Assume that the P_1 individuals are homozygous:
    (a) gray, long × ebony, vestigial
    (b) gray, vestigial × ebony, long
    (c) gray, long × gray, vestigial
14. How many different types of gametes can be formed by individuals of the following genotypes? What are they in each case? (a) *AaBb,* (b) *AaBB,* (c) *AaBbCc,* (d) *AaBBcc,* (e) *AaBbcc,* and (f) *AaBbCcDdEe*?
15. Using the forked-line method, determine the genotypic and phenotypic ratios of these trihybrid crosses:
    (a) *AaBbCc* × *AaBBCC*
    (b) *AaBBCc* × *aaBBCc*
    (c) *AaBbCc* × *AaBbCc* See **Now Solve This** on page 46.
16. Mendel crossed peas with round, green seeds with peas having wrinkled, yellow seeds. All F_1 plants had seeds that were round and yellow. Predict the results of testcrossing these F_1 plants.
17. Shown are F_2 results of two of Mendel's monohybrid crosses. State a null hypothesis that you will test using chi-square analysis. Calculate the χ^2 value and determine the *p* value for both crosses, then interpret the *p* values. Which cross shows a greater amount of deviation?

(a)	Full pods	882
	Constricted pods	299
(b)	Violet flowers	705
	White flowers	224

18. In one of Mendel's dihybrid crosses, he observed 315 round, yellow; 108 round, green; 101 wrinkled, yellow; and 32 wrinkled, green F_2 plants. Analyze these data using chi-square analysis to see whether (a) they fit a 9:3:3:1 ratio; (b) the round, wrinkled traits fit a 3:1 ratio; or (c) the yellow, green traits fit a 3:1 ratio. See **Now Solve This** on page 52.
19. A geneticist, in assessing data that fell into two phenotypic classes, observed values of 250:150. He decided to perform chi-square analysis using two different null hypotheses: (a) The data fit a 3:1 ratio; and (b) the data fit a 1:1 ratio. Calculate the χ^2 values for each hypothesis. What can you conclude about each hypothesis?
20. The basis for rejecting any null hypothesis is arbitrary. The researcher can set more or less stringent standards by deciding to raise or lower the critical *p* value. Would the use of a standard of $p = 0.10$ be more or less stringent in failing to reject the null hypothesis? Explain.
21. Consider three independently assorting gene pairs, *A*/*a*, *B*/*b*, and *C*/*c*, where each demonstrates typical dominance (*A*–, *B*–, *C*–) and recessiveness (*aa*, *bb*, *cc*). What is the probability of obtaining an offspring that is *AABbCc* from parents that are *AaBbCC* and *AABbCc*?
22. What is the probability of obtaining a triply recessive individual from the parents shown in Problem 21?
23. Of all offspring of the parents in Problem 21, what proportion will express all three dominant traits?
24. For the following pedigree, predict the mode of inheritance and the resulting genotypes of each individual. Assume that the alleles *A* and *a* control the expression of the trait.

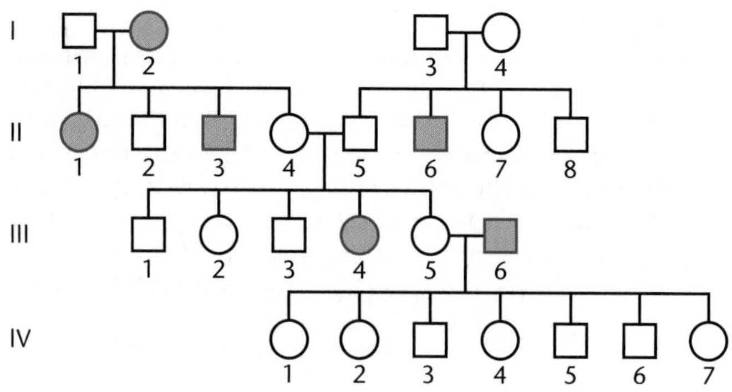

25. Which of Mendel's postulates are demonstrated by the pedigree in Problem 24? List and define these postulates.
26. The following pedigree follows the inheritance of myopia (nearsightedness) in humans. Predict whether the disorder is inherited as a dominant or a recessive trait. Based on your prediction, indicate the most probable genotype for each individual. See **Now Solve This** on page 54.

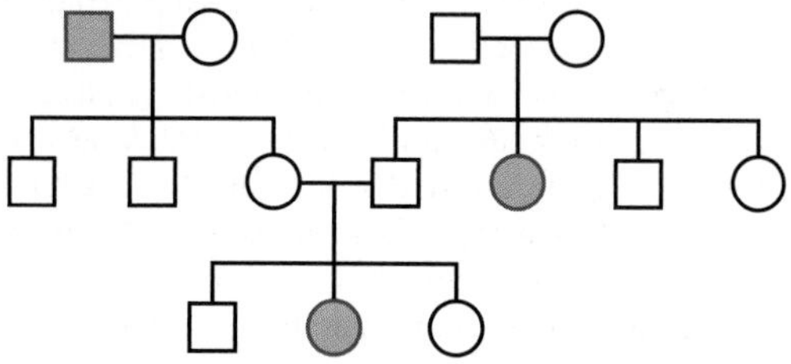

27. Draw all possible conclusions concerning the mode of inheritance of the trait expressed in each of the following limited pedigrees. (Each case is based on a different trait.)

(a)

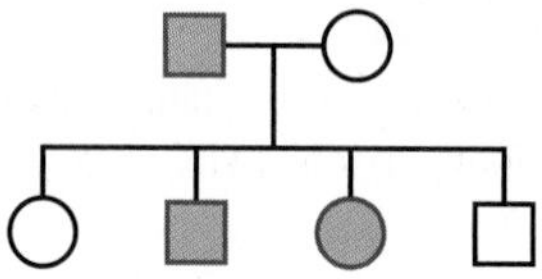

(b)

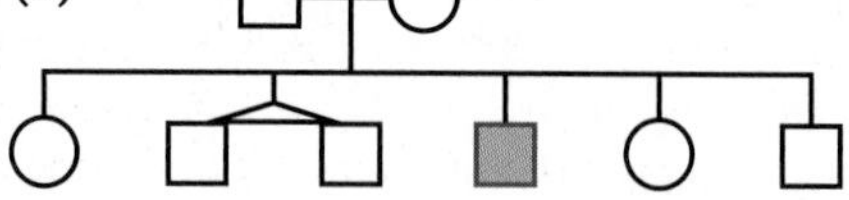

(c)

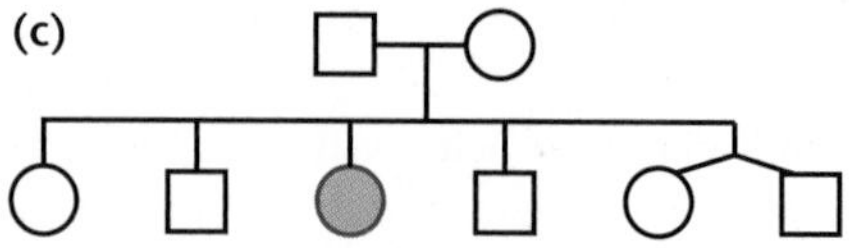

(d)

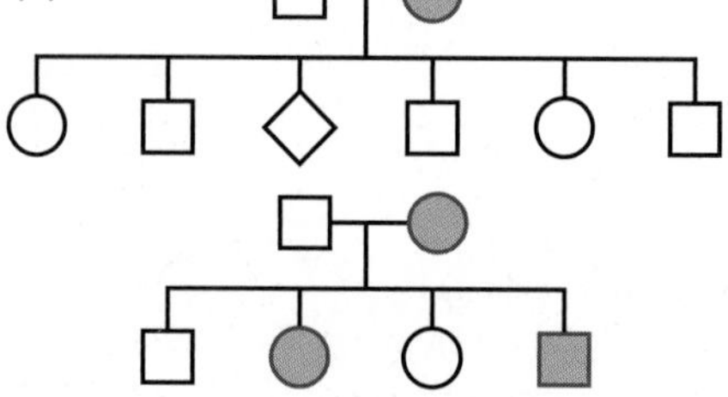

28. Two true-breeding pea plants are crossed. One parent is round, terminal, violet, constricted, while the other expresses the contrasting phenotypes of wrinkled, axial, white, full. The four pairs of contrasting traits are controlled by four genes, each located on a separate chromosome. In the F_1 generation, only round, axial, violet, and full are expressed. In the F_2 generation, all possible combinations of these traits are expressed in ratios consistent with Mendelian inheritance.
    (a) What conclusion can you draw about the inheritance of these traits based on the F_1 results?
    (b) Which phenotype appears most frequently in the F_2 results? Write a mathematical expression that predicts the frequency of occurrence of this phenotype.
    (c) Which F_2 phenotype is expected to occur least frequently? Write a mathematical expression that predicts this frequency.
    (d) How often is either P_1 phenotype likely to occur in the F_2 generation?
    (e) If the F_1 plant is testcrossed, how many different phenotypes will be produced, and how does this number compare to the number of different phenotypes in the F_2 generation discussed in part (b)?
29. Tay–Sachs disease (TSD) is an inborn error of metabolism that results in death, usually before the age of five. You are a genetic counselor, and you interview a phenotypically normal couple who consult you because the man had a female first cousin (on his father's side) who died from TSD and the woman had a maternal uncle with TSD. There are no other known cases in either family, and none of the matings were/are between related individuals. Assume that this trait is rare in this population.
    (a) Using standard pedigree symbols, draw a pedigree of these individuals' families, showing the relevant individuals.
    (b) The couple asks you to calculate the probability that they both are heterozygous for the TSD allele.
    (c) They also want to know the probability that neither of them is heterozygous.
    (d) They also ask you for the probability that one of them is heterozygous but the other is not.
    [*Hint:* The answers to (b), (c), and (d) should add up to 1.0.]
30. The wild-type (normal) fruit fly, *Drosophila melanogaster,* has straight wings and long bristles. Mutant strains have been isolated with either curled wings or short bristles. The genes representing these two mutant traits are located on separate chromosomes. Carefully examine the data from the five crosses below. (a) For each mutation, determine whether it is dominant or recessive. In each case, identify which crosses support your answer; and (b) define gene symbols, and determine the genotypes of the parents for each cross.

	Number of Progeny			
Cross	**straight wings, long bristles**	**straight wings, short bristles**	**curled wings, long bristles**	**curled wings, short bristles**
1 straight, short × straight, short	30	90	10	30
2 straight, long × straight, long	120	0	40	0
3 curled, long × straight, short	40	40	40	40
4 straight, short × straight, short	40	120	0	0
5 curled, short × straight, short	20	60	20	60

31. To assess Mendel's law of segregation using tomatoes, a true-breeding tall variety (*SS*) is crossed with a true-breeding short variety (*ss*). The heterozygous tall plants (*Ss*) were crossed to produce the two sets of F_2 data as follows:

Set I	Set II
30 tall	300 tall
5 short	50 short

    (a) Using chi-square analysis, analyze the results for both data sets. Calculate χ^2 values, and estimate the *p* values in both cases.
    (b) From the analysis in part (a), what can you conclude about the importance of generating large data sets in experimental settings?
32. *Datura stramonium* (the Jimsonweed) expresses flower colors of purple and white and pod textures of smooth and spiny. The results of two crosses in which the parents were not necessarily true-breeding were observed to be

    white spiny × white spiny → 3/4 white spiny:1/4 white smooth

    purple smooth × purple smooth → 3/4 purple smooth
    1/4 white smooth

    (a) Based on these results, put forward a hypothesis for the inheritance of the purple/white and smooth/spiny traits.
    (b) Assuming that true-breeding strains of all combinations of traits are available, what single cross could you execute and carry to an F_2 generation that will prove or disprove your hypothesis? Assuming your hypothesis is correct, what results of this cross will support it?

Labrador retriever puppies, which may display golden, black, or brown coats.

4

Modification of Mendelian Ratios

CHAPTER CONCEPTS

- While alleles are transmitted from parent to offspring according to Mendelian principles, they sometimes fail to display the clear-cut dominant/recessive relationship observed by Mendel.
- In many cases, in contrast to Mendelian genetics, two or more genes are known to influence the phenotype of a single characteristic.
- Still another exception to Mendelian inheritance is the presence of genes on sex chromosomes, whereby one of the sexes contains only a single member of that chromosome.
- Phenotypes are often the combined result of both genetics and the environment within which genes are expressed.
- The result of the various exceptions to Mendelian principles is the occurrence of phenotypic ratios that differ from those resulting from standard monohybrid, dihybrid, and trihybrid crosses.
- Extranuclear inheritance, resulting from the expression of genes present in the DNA found in mitochondria and chloroplasts, modifies Mendelian inheritance patterns. Such genes are most often transmitted through the female gamete.

In Chapter 3, we discussed the fundamental principles of transmission genetics. We saw that genes are present on homologous chromosomes and that these chromosomes segregate from each other and assort independently with other segregating chromosomes during gamete formation. These two postulates are the basic principles of gene transmission from parent to offspring. However, when gene expression does not adhere to a simple dominant/recessive mode or when more than one pair of genes influences the expression of a single character, the classic 3:1 and 9:3:3:1 ratios are usually modified. In this and the next several chapters, we consider more complex modes of inheritance. In spite of the greater complexity of these situations, the fundamental principles set down by Mendel still hold.

In this chapter, we restrict our initial discussion to the inheritance of traits controlled by only one set of genes. In diploid organisms, which have homologous pairs of chromosomes, two copies of each gene influence such traits. The copies need not be identical because alternative forms of genes (alleles) occur within populations. How alleles influence phenotypes is our primary focus. We will then consider **gene interaction,** a situation in which a single phenotype is affected by more than one set of genes. Numerous examples will be presented to illustrate a variety of heritable patterns observed in such situations.

Thus far, we have restricted our discussion to chromosomes other than the X and Y pair. By examining cases where genes are present on the X chromosome, illustrating **X-linkage,** we will see yet another modification of Mendelian ratios. Our discussion of modified ratios also includes the consideration of sex-limited and sex-influenced inheritance, cases where the sex of the individual, but not necessarily genes on the X chromosome, influences the phenotype. We will also consider how a given phenotype often varies depending on the overall environment in which a gene, a cell, or an organism finds itself. This discussion points out that phenotypic expression depends on more than just the genotype of an organism. Finally, we conclude with a discussion of extranuclear inheritance, cases where DNA within organelles influences an organism's phenotype.

How Do We Know?

In this chapter, we focus on many extensions and modifications of Mendelian principles and ratios. In the process, we will encounter many opportunities to consider how this information was acquired. As you study this topic, you should try to answer the following fundamental questions:

1. How were early geneticists able to ascertain inheritance patterns that did not fit typical Mendelian ratios?
2. How did geneticists determine that inheritance of some phenotypic characteristics involves the interactions of two or more gene pairs? How were they able to determine how many gene pairs were involved?
3. How do we know that specific genes are located on the sex-determining chromosomes rather than on autosomes?
4. For genes whose expression seems to be tied to the sex of individuals, how do we know whether a gene is X-linked in contrast to exhibiting sex-limited or sex-influenced inheritance?
5. How was extranuclear inheritance discovered?

4.1 Alleles Alter Phenotypes in Different Ways

After Mendel's work was rediscovered in the early 1900s, researchers focused on the many ways in which genes influence an individual's phenotype. Each type of inheritance was more thoroughly investigated when observations of genetic data did not conform precisely to the expected Mendelian ratios, and hypotheses that modified and extended the Mendelian principles were proposed and tested with specifically designed crosses. The explanations were in accord with the principle that a phenotype is under the control of one or more genes located at specific loci on one or more pairs of homologous chromosomes.

To understand the various modes of inheritance, we must first examine the potential function of alleles. Alleles are alternative forms of the same gene. The allele that occurs most frequently in a population, the one that we arbitrarily designate as normal, is called the **wild-type allele.** This is often, but not always, dominant. Wild-type alleles are responsible for the corresponding wild-type phenotype and are the standards against which all other mutations occurring at a particular locus are compared.

A mutant allele contains modified genetic information and often specifies an altered gene product. For example, in human populations, there are many known alleles of the gene that encodes the β chain of human hemoglobin. All such alleles store information necessary for the synthesis of the β-chain polypeptide, but each allele specifies a slightly different form of the same molecule. Once the allele's product has been manufactured, the function of the product may or may not be altered.

The process of mutation is the source of alleles. For a new allele to be recognized when observing an organism, it must cause a change in the phenotype. A new phenotype results from a change in functional activity of the cellular product specified by that gene. Often, the mutation causes the diminution or the loss of the specific wild-type function. For example, if a gene is responsible for the synthesis of a specific enzyme, a mutation in that gene may ultimately change the conformation of this enzyme and reduce or eliminate its affinity for the substrate. Such a case is designated as a **loss-of-function mutation.** If the loss is complete, the mutation has resulted in what is called a **null allele.**

Conversely, other mutations may enhance the function of the wild-type product. Most often when this occurs, it is the result of increasing the quantity of the gene product. In such cases, the mutation may be affecting the regulation of transcription of the gene under consideration. Such cases are designated **gain-of-function mutations,** which generally result in dominant alleles since one copy in a diploid organism is sufficient to alter the normal phenotype. Examples of gain-in-function mutations include the genetic conversion of *proto-oncogenes,* which regulate the cell cycle, to *oncogenes,* where regulation is overridden by excess gene product. The result is the creation of a cancerous cell.

Having introduced the concept of gain- or loss-of-function mutations, it is important to note the possibility that a mutation will create an allele where no change in function can be detected. In this case, the mutation would not be immediately apparent since no phenotypic variation would be evident. However, such a mutation could be detected if the DNA sequence of the gene was examined directly. These are sometimes referred to as **neutral mutations** because the gene product presents no change to either the phenotype or the evolutionary fitness of the organism.

Finally, we note here that while a phenotypic trait may be affected by a single mutation in one gene, traits are often influenced by more than one gene. For example, enzymatic reactions are most often part of complex metabolic pathways leading to the synthesis of an end product, such as an amino acid. Mutations in any of the various reactions have a common effect—the failure to synthesize the end product. Therefore, phenotypic traits related to the end product are often influenced by more than one gene.

In each of the many crosses discussed in the next few chapters, only one or a few gene pairs are involved. Keep in mind that in each cross discussed, all genes that are not under consideration are assumed to have no effect on the inheritance patterns described.

4.2 Geneticists Use a Variety of Symbols for Alleles

In Chapter 3, we learned a standard convention used to symbolize alleles for very simple Mendelian traits. The initial letter of the name of a recessive trait, lowercased and italicized, denotes the recessive allele, and the same letter in uppercase refers to the dominant allele. Thus, in the case of *tall* and *dwarf,* where *dwarf* is recessive, *D* and *d* represent the alleles responsible for these respective traits. Mendel used upper- and lowercase letters such as these to symbolize his unit factors.

Another useful system was developed in genetic studies of the fruit fly *Drosophila melanogaster* to discriminate between wild-type and mutant traits. This system uses the initial letter, or a combination of two or three letters, of the name of the mutant trait. If the trait is recessive, lowercase is used; if it is dominant, uppercase is used. The contrasting wild-type trait is denoted by the same letter, but with a superscript +. For example, *ebony* is a recessive body color mutation in *Drosophila.* The normal wild-type body color is gray. Using this system, we denote *ebony* by the symbol *e*, and we denote gray by e^+. The responsible locus may be occupied by either the wild-type allele (e^+) or the mutant allele (*e*). A diploid fly may thus exhibit one of three possible genotypes:

e^+/e^+	gray homozygote (wild type)
e^+/e	gray heterozygote (wild type)
e/e	ebony homozygote (mutant)

The slash between the letters indicates that the two allele designations represent the same locus on two homologous chromosomes. If we instead consider a dominant wing mutation such as *Wrinkled* (*Wr*) wing in *Drosophila*, the three possible designations are Wr^+/Wr^+, Wr^+/Wr, and Wr/Wr. The latter two genotypes express the wrinkled-wing phenotype.

One advantage of this system is that further abbreviation can be used when convenient: The wild-type allele may simply be denoted by the + symbol. With *ebony* as an example, the designations of the three possible genotypes become

$+/+$	gray homozygote (wild type)
$+/e$	gray heterozygote (wild type)
e/e	ebony homozygote (mutant)

Another variation is utilized when no dominance exists between alleles. We simply use uppercase italic letters and superscripts to denote alternative alleles (e.g., R^1 and R^2, L^M and L^N, I^A and I^B). Their use will become apparent later in this chapter.

Many diverse systems of genetic nomenclature are used to identify genes in various organisms. Usually, the symbol selected reflects the function of the gene or even a disorder caused by a mutant gene. For example, the yeast *cdk* is the abbreviation for the *c*yclin *d*ependent *k*inase gene, whose product is involved in cell-cycle regulation. In bacteria, leu^- refers to a mutation that interrupts the biosynthesis of the amino acid leucine, where the wild-type gene is designated leu^+. The symbol *dnaA* represents a bacterial gene involved in DNA replication (and DnaA is the protein made by that gene). In humans, capital letters are used to name genes: *BRCA*1 represents the first gene associated with susceptibility to *br*east *ca*ncer. Although these different systems may seem complex, they are useful ways to symbolize genes.

4.3 Neither Allele Is Dominant in Incomplete, or Partial, Dominance

A cross between parents with contrasting traits may generate offspring with an intermediate phenotype. For example, if plants such as four-o'clocks or snapdragons with red flowers are crossed with white-flowered plants, the offspring have pink flowers. Some red pigment is produced in the F_1 intermediate pink-colored flowers. Therefore, neither red nor white flower color is dominant. This situation is known as **incomplete,** or **partial, dominance.**

If this phenotype is under the control of a single gene and two alleles where neither is dominant, the results of the F_1 (pink) × F_1 (pink) cross can be predicted. The resulting F_2 generation shown in Figure 4–1 confirms the hypothesis that only one pair of alleles determines these phenotypes. The genotypic ratio (1:2:1) of the F_2 generation is identical to that of Mendel's monohybrid cross. However, because neither allele is dominant, the phenotypic ratio is identical to the genotypic ratio. Note that because neither allele is recessive, we have chosen not to use upper- and lowercase letters as symbols. Instead, we denoted the red and white alleles as R^1 and R^2, respectively. We could have used W^1 and W^2 or still other designations such as C^W and C^R, where C indicates "color" and the W and R superscripts indicate white and red.

Clear-cut cases of incomplete dominance, which result in intermediate expression of the overt phenotype, are relatively rare. However, even when complete dominance seems apparent, careful examination of the gene product, rather than the phenotype, often reveals an intermediate level of gene expression. An example is the human biochemical disorder **Tay–Sachs disease,** in which homozygous recessive individuals are severely affected with a fatal lipid storage disorder (see the text box in the previous chapter on page 48). There is almost no activity of the enzyme **hexosaminidase** in afflicted individuals. Heterozygotes, with only a single copy of the mutant gene, are phenotypically normal but express only about 50 percent of the enzyme activity found in homozygous normal individuals. Fortunately, this level of enzyme activity is adequate to achieve normal biochemical function—a situation not uncommon in enzyme disorders.

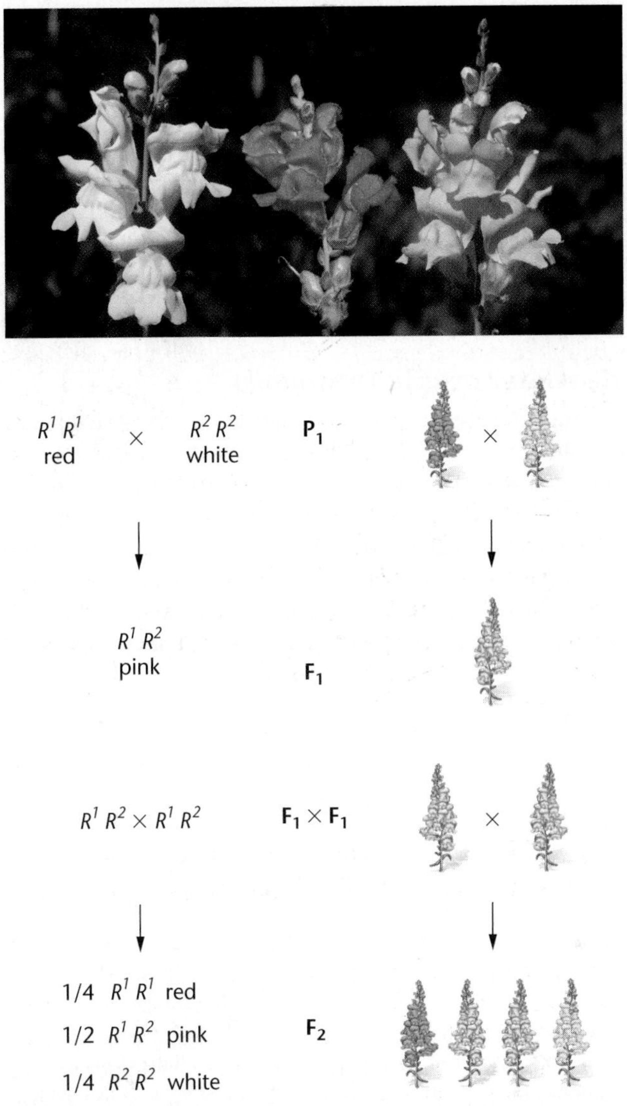

FIGURE 4–1 Incomplete dominance shown in the flower color of snapdragons.

4.4 In Codominance, the Influence of Both Alleles in a Heterozygote Is Clearly Evident

If two alleles of a single gene are responsible for producing two distinct, detectable gene products, a situation different from incomplete dominance or dominance/recessiveness arises. In this case, *the joint expression of both alleles in a heterozygote* is called **codominance.** The **MN blood group** in humans illustrates this phenomenon and is characterized by an antigen called a glycoprotein, found on the surface of red blood cells. In the human population, two forms of this glycoprotein exist, designated M and N; an individual may exhibit either one or both of them.

The MN system is under the control of an autosomal locus found on chromosome 4 and two alleles designated L^M and L^N. Humans are diploid, so three combinations are possible, each resulting in a distinct blood type:

Genotype	Phenotype
$L^M L^M$	M
$L^M L^N$	MN
$L^N L^N$	N

As predicted, a mating between two heterozygous MN parents may produce children of all three blood types, as follows:

$$L^M L^N \times L^M L^N$$
$$\downarrow$$
$$1/4\ L^M L^M$$
$$1/2\ L^M L^N$$
$$1/4\ L^N L^N$$

Once again the genotypic ratio, 1:2:1, is upheld.

Codominant inheritance is characterized by *distinct expression of the gene products of both alleles*. This characteristic distinguishes it from incomplete dominance, where heterozygotes express an intermediate, blended phenotype. We shall see another example of codominance when we examine the ABO blood-type system in the following section.

4.5 Multiple Alleles of a Gene May Exist in a Population

The information stored in any gene is extensive, and mutations can modify this information in many ways. Each change produces a different allele. Therefore, for any specific gene, the number of alleles within members of a population need not be restricted to two. When three or more alleles of the same gene are found, **multiple alleles** are present that create a unique mode of inheritance. It is important to realize that *multiple alleles can be studied only in populations*. An individual diploid organism has, at most, two homologous gene loci that may be occupied by different alleles of the same gene. However, among many members of a species, numerous alternative forms of the same gene can exist.

The ABO Blood Group

The simplest case of multiple alleles is that in which three alternative alleles of one gene exist. This situation is illustrated by the **ABO blood group** in humans, discovered by Karl Landsteiner in the early 1900s. The ABO system, like the MN blood group, is characterized by the presence of antigens on the surface of red blood cells. The A and B antigens are distinct from MN antigens and are under the control of a different gene, located on chromosome 9. As in the MN system, one combination of alleles in the ABO system exhibits a codominant mode of inheritance.

When individuals are tested using antisera that contain antibodies against the A or B antigen, four phenotypes are revealed. Each individual has either the A antigen (A phenotype), the B antigen (B phenotype), the A and B antigens (AB phenotype), or neither antigen (O phenotype). In 1924, it was hypothesized that these phenotypes were inherited as the result of three alleles of a single gene. This hypothesis was based on studies of the blood types of many different families.

Although different designations can be used, we use the symbols I^A, I^B, and I^O to distinguish these three alleles; the *I* designation stands for *isoagglutinogen*, another term for antigen. If we assume that the I^A and I^B alleles are responsible for the production of their respective A and B antigens and that I^O is an allele that does not produce any detectable A or B antigens, we can list the various genotypic possibilities and assign the appropriate phenotype to each:

Genotype	Antigen	Phenotype
I^AI^A	A	A
I^AI^O	A	
I^BI^B	B	B
I^BI^O	B	
I^AI^B	A, B	AB
I^OI^O	Neither	O

In these assignments the I^A and I^B alleles are dominant to the I^O allele, but are codominant to each other. Our knowledge of human blood types has several practical applications, the most important of which are compatible blood transfusions and organ transplantations.

The Bombay Phenotype

The biochemical basis of the ABO blood-type system has been carefully worked out. The A and B antigens are actually carbohydrate groups (sugars) that are bound to lipid molecules (fatty acids) protruding from the membrane of the red blood cell. The specificity of the A and B antigens is based on the terminal sugar of the carbohydrate group. Both the A and B antigens are derived from a precursor molecule called the **H substance,** to which one or two terminal sugars are added.

In extremely rare instances, first recognized in a woman in Bombay in 1952, the H substance is incompletely formed. As a result, it is an inadequate substrate for the enzyme that normally adds the terminal sugar. This condition results in the expression of blood type O and is called the **Bombay phenotype.** Research has revealed that this condition is due to a rare recessive mutation at a locus separate from that controlling the A and B antigens. The gene is now designated *FUT1* (encoding an enzyme, fucosyl transferase), and individuals that are homozygous for the mutation cannot synthesize the complete H substance. Thus, even though they may have the I^A and/or I^B alleles, neither the A nor B antigen can be added to the cell surface. This information explains why the woman in Bombay expressed blood type O, even though one of her parents was type AB (thus she should not have been type O), and why she was able to pass the I^B allele to her children (Figure 4–2).

The *white* Locus in *Drosophila*

Many other phenotypes in plants and animals are known to be controlled by multiple allelic inheritance. In *Drosophila*, many alleles are known at practically every locus. The recessive mutation that causes white eyes, discovered by Thomas H. Morgan and Calvin Bridges in 1912, is one of over 100 alleles that can occupy this locus. In this allelic series, eye colors range from complete absence of pigment in the *white* allele to deep ruby in the *white-satsuma* allele, to orange in

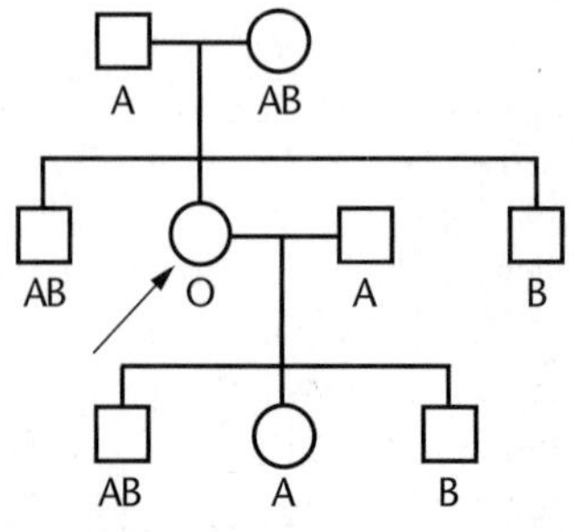

FIGURE 4–2 A partial pedigree of a woman with the Bombay phenotype. Functionally, her ABO blood group behaves as type O. Genetically, she is type B.

the *white-apricot* allele, to a buff color in the *white-buff* allele. These alleles are designated w, w^{sat}, w^a, and w^{bf}, respectively. In each case, the total amount of pigment in these mutant eyes is reduced to less than 20 percent of that found in the brick-red, wild-type eye. Table 4.1 lists these and other *white* alleles and their color phenotypes.

TABLE 4.1 Alleles of the *white* Locus of *Drosophila melanogaster* and Their Phenotype

Allele	Name	Eye Color
w	*white*	pure white
w^a	*white-apricot*	yellowish orange
w^{bf}	*white-buff*	light buff
w^{bl}	*white-blood*	yellowish ruby
w^{cf}	*white-coffee*	deep ruby
w^e	*white-eosin*	yellowish pink
w^{mo}	*white-mottled orange*	light mottled orange
w^{sat}	*white-satsuma*	deep ruby
w^{sp}	*white-spotted*	fine grain, yellow mottling
w^t	*white-tinged*	light pink

NOW SOLVE THIS

Problem 27 on page 89 involves a series of multiple alleles controlling coat color in rabbits. You are asked to determine the parental and F_1 genotypes and to predict the results of various crosses between F_1 individuals.

Hint: Note particularly the hierarchy of dominance of the various alleles. Remember also that even though there can be more than two alleles in a population, an individual can have at most two of these. Thus, the allelic distribution into gametes adheres to the principle of segregation.

4.6 Lethal Alleles Represent Essential Genes

Many gene products are essential to an organism's survival. Mutations resulting in the synthesis of a gene product that is nonfunctional can often be tolerated in the heterozygous state; that is, one wild-type allele may be sufficient to produce enough of the essential product to allow survival. However, such a mutation behaves as a *recessive lethal allele,* and homozygous recessive individuals will not survive. The time of death will depend on when the product is essential. In mammals, for example, this might occur during development, early childhood, or even adulthood.

In some cases, the allele responsible for a lethal effect when homozygous may also result in a distinctive mutant phenotype when present heterozygously. It is behaving as a recessive lethal allele but is dominant with respect to the phenotype. For example, a mutation that causes a yellow coat in mice was discovered in the early part of this century. The yellow coat varies from the normal agouti (wild-type) coat phenotype, as shown in Figure 4–3, p. 66. Crosses between the various combinations of the two strains yield unusual results:

Crosses				
(A) agouti	×	agouti	⟶	all agouti
(B) yellow	×	yellow	⟶	2/3 yellow: 1/3 agouti
(C) agouti	×	yellow	⟶	1/2 yellow: 1/2 agouti

These results are explained on the basis of a single pair of alleles. With regard to coat color, the mutant *yellow* allele (A^Y) is dominant to the wild-type *agouti* allele (A), so heterozygous mice will have yellow coats. However, the *yellow* allele is also a homozygous recessive lethal. When present in two copies, the mice die before birth. Thus, there are no homozygous yellow mice. The genetic basis for these three crosses is shown in Figure 4–3.

In other cases, a mutation may behave as a *dominant lethal allele.* In such cases, the presence of just one copy of the allele results in the death of the individual. In humans, a disorder called **Huntington disease** (previously referred to as Huntington's chorea) is due to a dominant autosomal allele H, where the onset of the disease in heterozygotes (Hh) is delayed, usually well into adulthood. Affected individuals then undergo gradual nervous and motor degeneration until they die. This lethal disorder is particularly tragic because it has such a late onset, typically at about age 40. By that time, the affected individual may have produced a family, and each of the children has a 50 percent probability of inheriting the lethal allele, transmitting the allele to his or her offspring, and eventually developing the disorder. The American folk singer and composer Woody Guthrie (father of modern-day folk singer Arlo Guthrie) died from this disease at age 39.

Dominant lethal alleles are rarely observed. For these alleles to exist in a population, the affected individuals must reproduce before the lethal allele is expressed, as can occur in Huntington disease. If all affected individuals die before reaching reproductive age, the mutant gene will not be passed to future generations, and the mutation will disappear from the population unless it arises again as a result of a new mutation.

ESSENTIAL POINT

Since Mendel's work was rediscovered, transmission genetics has been expanded to include many alternative modes of inheritance, including the study of incomplete dominance, codominance, multiple alleles, and lethal alleles.

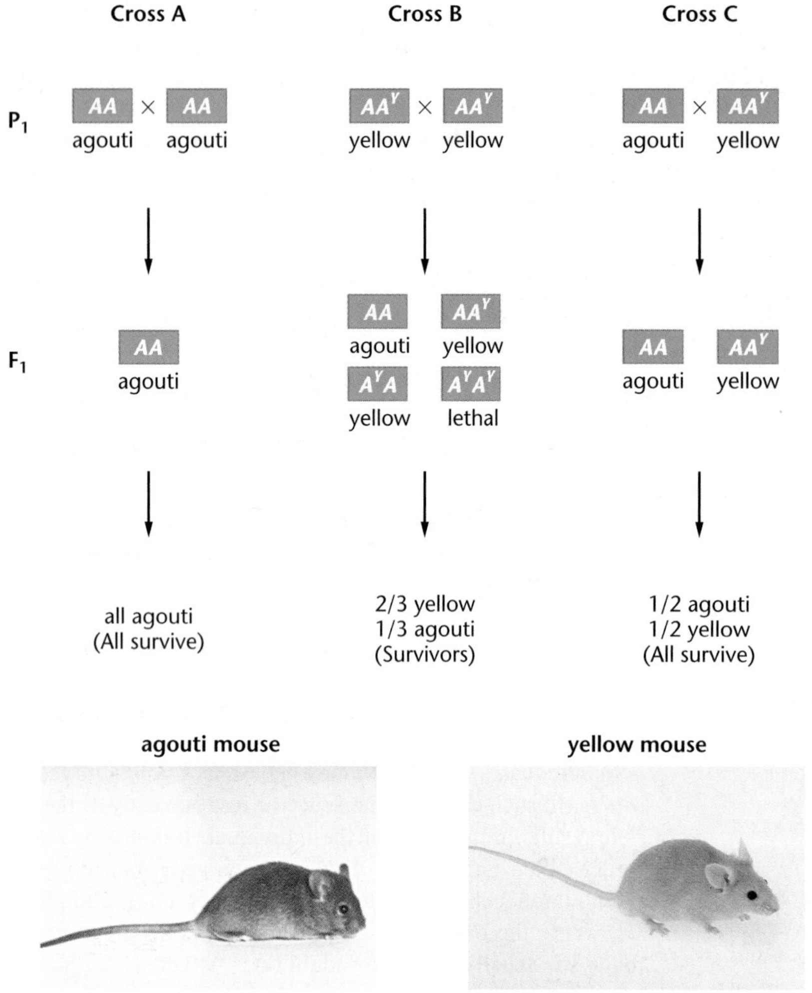

FIGURE 4–3 Inheritance patterns in three crosses involving the normal wild-type *agouti* allele (A) and the mutant *yellow* allele (A^Y) in the mouse. Note that the mutant allele behaves dominantly to the normal allele in controlling coat color, but it also behaves as a homozygous recessive lethal allele. The genotype A^YA^Y mice do not survive.

The Molecular Basis of Dominance and Recessiveness: The Agouti Gene

Molecular analysis of the gene resulting in agouti and yellow mice has provided insight into how a mutation can be both dominant for one phenotypic effect (hair color) and recessive for another (embryonic development). The A^Y allele is a classic example of a gain-of-function mutation. Animals homozygous for the wild-type A allele have yellow pigment deposited as a band on the otherwise black hair shaft, resulting in the agouti phenotype (see Figure 4–3). Heterozygotes deposit yellow pigment along the entire length of hair shafts as a result of the deletion of the regulatory region preceding the DNA coding region of the A^Y allele. Without any means to regulate its expression, one copy of the A^Y allele is always turned on in heterozygotes, resulting in the gain of function leading to the dominant effect.

The homozygous lethal effect has also been explained by molecular analysis of the mutant gene. The extensive deletion of genetic material that produced the A^Y allele actually extends into the coding region of an adjacent gene (*Merc*), rendering it nonfunctional. It is this gene that is critical to embryonic development, and the loss of its function in A^Y/A^Y homozygotes is what causes lethality. Heterozygotes exceed the threshold level of the wild-type *Merc* gene product and thus survive.

4.7 Combinations of Two Gene Pairs with Two Modes of Inheritance Modify the 9:3:3:1 Ratio

Each example discussed so far modifies Mendel's 3:1 F_2 monohybrid ratio. Therefore, combining any two of these modes of inheritance in a dihybrid cross will likewise modify the classical 9:3:3:1 ratio. Having established the foundation for the modes of inheritance of incomplete dominance, codominance, multiple alleles, and lethal alleles, we can now deal with the situation of two modes of inheritance occurring simultaneously. Mendel's principle of independent assortment applies to these situations, provided that the genes controlling each character are not linked on the same chromosome—in other words, that they do not demonstrate what is called *genetic linkage*.

Consider, for example, a mating that occurs between two humans who are both heterozygous for the autosomal recessive gene that causes albinism and who are both of blood type AB. What is the probability of a particular phenotypic combination

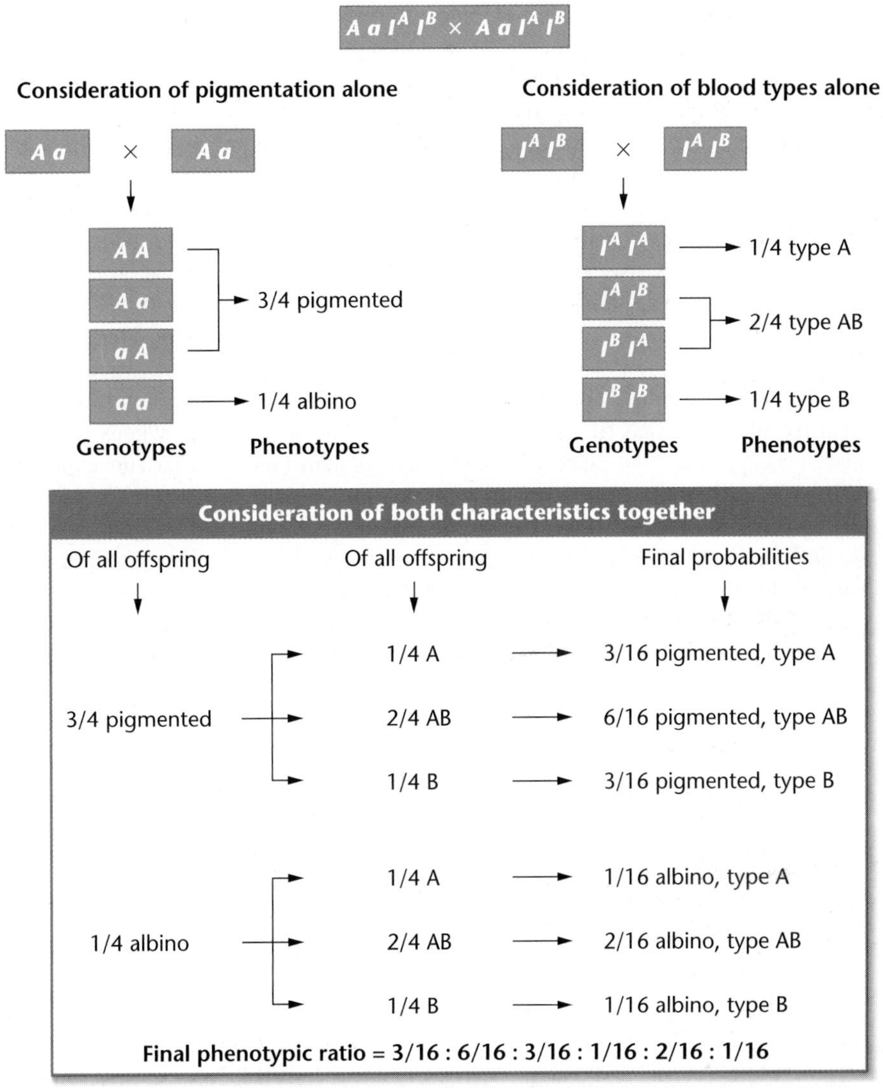

FIGURE 4–4 Calculation of the mating probabilities involving the ABO blood type and albinism in humans, using the forked-line method.

occurring in each of their children? Albinism is inherited in the simple Mendelian fashion, and the blood types are determined by the series of three multiple alleles, I^A, I^B, and I^O. The solution to this problem is diagrammed in Figure 4–4, using the forked-line method. This dihybrid cross does not yield the classical four phenotypes in a 9:3:3:1 ratio. Instead, six phenotypes occur in a 3:6:3:1:2:1 ratio, establishing the expected probability for each phenotype. This is just one of the many variants of modified ratios that are possible when different modes of inheritance are combined.

4.8 Phenotypes Are Often Affected by More Than One Gene

Soon after Mendel's work was rediscovered, experimentation revealed that individual characteristics displaying discrete phenotypes are often under the control of more than one gene. This was a significant discovery because it revealed that genetic influence on the phenotype is often much more complex than Mendel had envisioned. Instead of single genes controlling the development of individual parts of the plant or animal body, it soon became clear that phenotypic characters can be influenced by the interactions of many different genes and their products.

The term **gene interaction** is often used to describe the idea that several genes influence a particular characteristic. This does not mean, however, that two or more genes, or their products, necessarily interact directly with one another to influence a particular phenotype. Rather, the cellular function of numerous gene products contributes to the development of a common phenotype. For example, the development of an organ such as the compound eye of an insect is exceedingly complex and leads to a structure with multiple phenotypic manifestations—such as specific size, shape, texture, and color. The development of the eye is a complex cascade of developmental events leading to its formation. This process exemplifies the developmental concept of **epigenesis,** whereby each step of development increases the complexity of this sensory organ and is under the control and influence of one or more genes.

An enlightening example of *epigenesis* and *multiple gene interaction* involves the formation of the inner ear in mammals. The inner ear consists of distinctive anatomical features to capture, funnel, and transmit external sound waves and to convert them into nerve impulses. During the formation of the ear, a cascade of intricate developmental events occur, influenced by many genes. Mutations that interrupt many of the steps of ear development lead to a common phenotype: **hereditary deafness.** In a sense, these many genes "interact" to produce a common phenotype. In such situations, the mutant phenotype is described as a **heterogeneous trait,** reflecting the many genes involved. In humans, while a few common alleles are responsible for the vast majority of cases of hereditary deafness, over 50 genes are involved in development of the ability to discern sound.

Epistasis

Some of the best examples of gene interaction are those that reveal the phenomenon of **epistasis** (Greek for "stoppage"). Epistasis occurs when the expression of one gene or gene pair masks or modifies the expression of another gene or gene pair. Sometimes the genes involved control the expression of the same general phenotypic characteristic in an antagonistic manner, as when masking occurs. In other cases, however, the genes involved exert their influence on one another in a complementary, or cooperative, fashion.

For example, the homozygous presence of a recessive allele prevents or overrides the expression of other alleles at a second locus (or several other loci). In this case, the alleles at the first locus are said to be *epistatic* to those at the second locus, and the alleles at the second locus are *hypostatic* to those at the first locus. In another example, a single dominant allele at the first locus influences the expression of the alleles at a second gene locus. In a third example, two gene pairs **complement** one another such that at least one dominant allele at each locus is required to express a particular phenotype.

The Bombay phenotype discussed earlier is an example of the homozygous recessive condition at one locus masking the expression of a second locus. There, we established that the homozygous presence of the mutant form of the *FUT1* gene masks the expression of the I^A and I^B alleles. Only individuals containing at least one wild-type *FUT1* allele can form the A or B antigen. As a result, individuals whose genotypes include the I^A or I^B allele and who lack a wild-type allele are of the type O phenotype, regardless of their potential to make either antigen. An example of the outcome of matings between individuals heterozygous at both loci is illustrated in Figure 4–5. If many

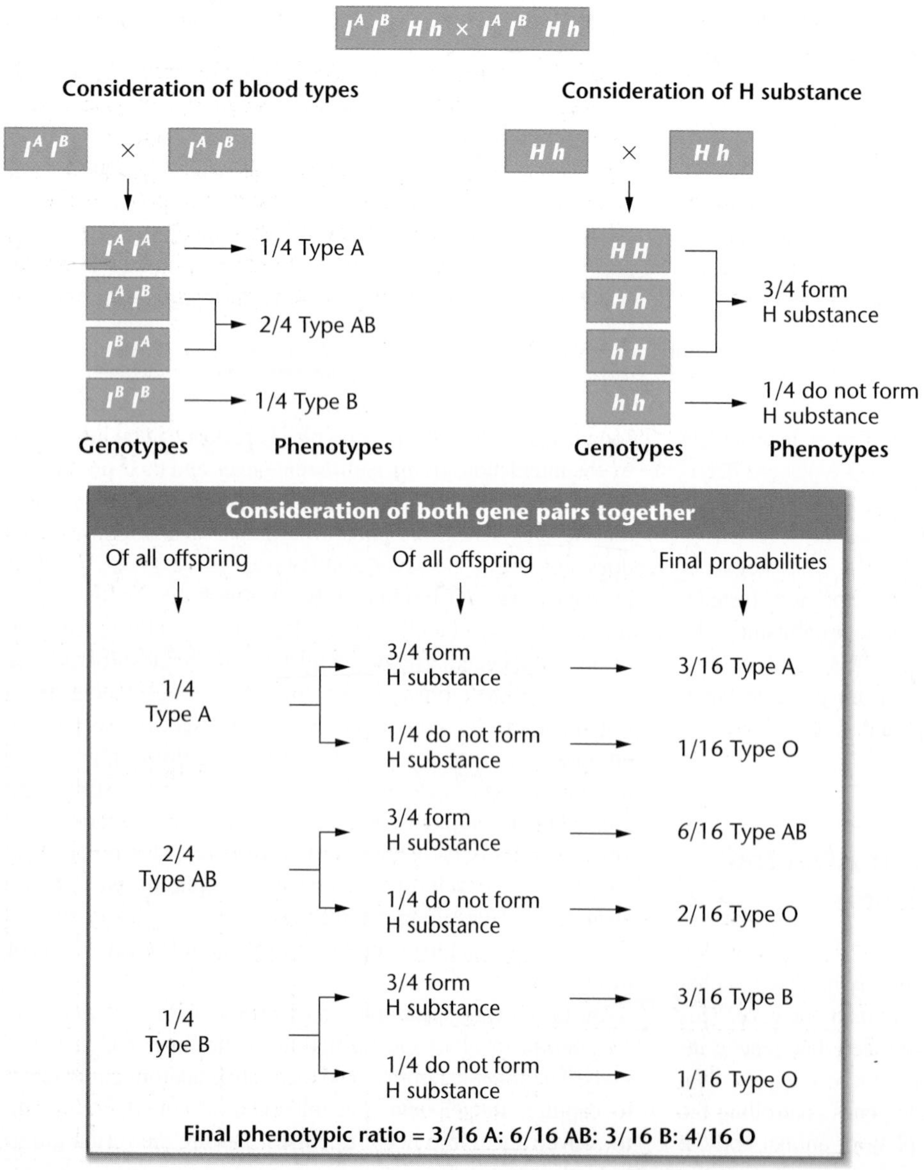

FIGURE 4–5 The outcome of a mating between individuals who are heterozygous at two genes determining their ABO blood type. Final phenotypes are calculated by considering both genes separately and then combining the results using the forked-line method.

such individuals have children, the phenotypic ratio of 3 A: 6 AB: 3 B: 4 O is expected in their offspring.

It is important to note the following points when examining this cross and the predicted phenotypic ratio:

1. A key distinction exists in this cross compared to the modified dihybrid cross shown in Figure 4–4: *only one characteristic—blood type—is being followed.* In the modified dihybrid cross of Figure 4–4, blood type *and* skin pigmentation are followed as separate phenotypic characteristics.
2. Even though only a single character was followed, the phenotypic ratio is expressed in sixteenths. If we knew nothing about the H substance and the genes controlling it, we could still be confident that a second gene pair, other than that controlling the A and B antigens, is involved in the phenotypic expression. *When studying a single character, a ratio that is expressed in 16 parts (e.g., 3:6:3:4) suggests that two gene pairs are "interacting" during the expression of the phenotype under consideration.*

The study of gene interaction reveals inheritance patterns that modify the classical Mendelian dihybrid F_2 ratio (9:3:3:1) in other ways as well. In these examples, epistasis combines one or more of the four phenotypic categories in various ways. The generation of these four groups is reviewed in Figure 4–6, along with several modified ratios.

As we discuss these and other examples, we will make several assumptions and adopt certain conventions:

1. In each case, distinct phenotypic classes are produced, each clearly discernible from all others. Such traits illustrate discontinuous variation, where phenotypic categories are discrete and qualitatively different from one another.
2. The genes considered in each cross are not linked and therefore assort independently of one another during gamete formation. To allow you to easily compare the results of different crosses, we designated alleles as *A*, *a* and *B*, *b* in each case.
3. When we assume that complete dominance exists between the alleles of any gene pair, such that *AA* and *Aa* or *BB* and *Bb* are equivalent in their genetic effects, we use the designations *A–* or *B–* for both combinations, where the dash (–) indicates that either allele may be present, without consequence to the phenotype.
4. All P_1 crosses involve homozygous individuals (e.g., *AABB* × *aabb*, *AAbb* × *aaBB*, or *aaBB* × *AAbb*). Therefore, each F_1 generation consists of only heterozygotes of genotype *AaBb*.
5. In each example, the F_2 generation produced from these heterozygous parents is our main focus of analysis. When two genes are involved (as in Figure 4–6), the F_2 genotypes fall into four categories: 9/16 *A–B–*, 3/16 *A–bb*, 3/16 *aaB–*, and 1/6 *aabb*. Because of dominance, all genotypes in each category have an equivalent effect on the phenotype.

Case 1 is the inheritance of coat color in mice (Figure 4–7, p. 70). Normal wild-type coat color is agouti, a grayish pattern formed by alternating bands of pigment on each hair. Agouti is dominant to black (non-agouti) hair, which is caused by a recessive mutation, *a*. Thus, *A–* results in agouti, while *aa* yields black coat color. When it is homozygous, a recessive mutation, *b*, at a separate locus, eliminates pigmentation altogether, yielding albino mice (*bb*), regardless of the genotype

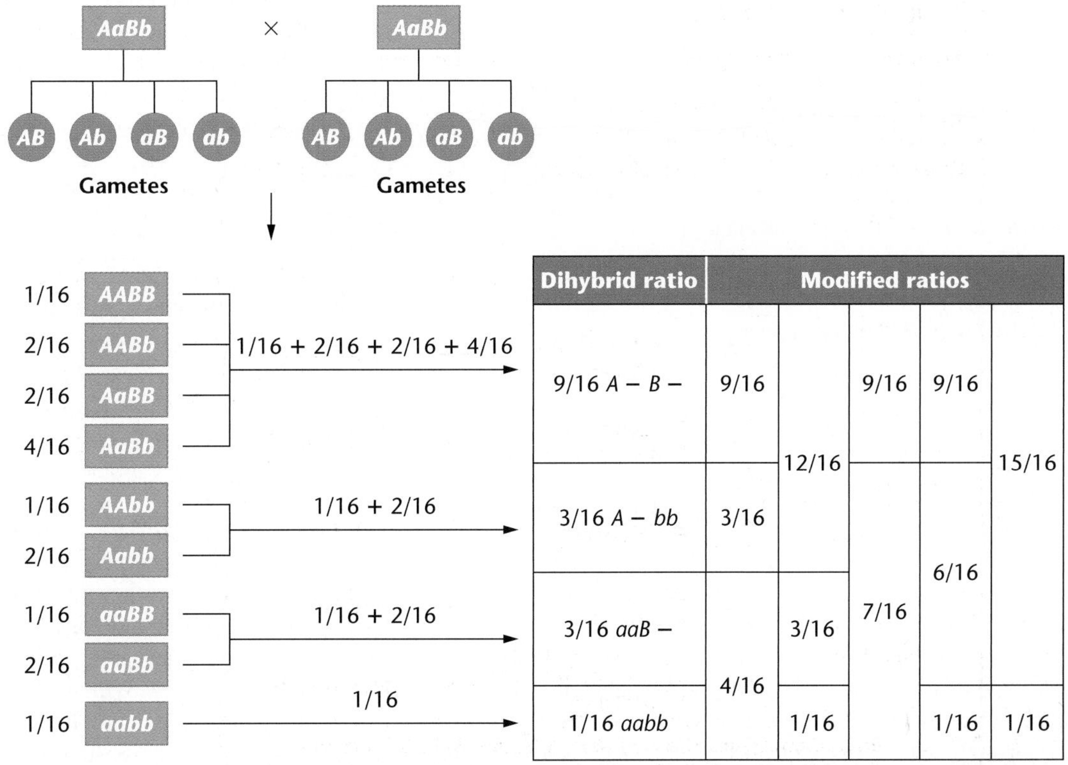

FIGURE 4–6 Generation of the various modified dihybrid ratios from the nine unique genotypes produced in a cross between individuals who are heterozygous at two genes.

Case	Organism	Character	F_2 Phenotypes 9/16	3/16	3/16	1/16	Modified ratio
1	Mouse	Coat color	agouti	albino	black	albino	9:3:4
2	Squash	Color	white		yellow	green	12:3:1
3	Pea	Flower color	purple	white			9:7
4	Squash	Fruit shape	disc	sphere		long	9:6:1
5	Chicken	Color	white		colored	white	13:3
6	Mouse	Color	white-spotted	white	colored	white-spotted	10:3:3
7	Shepherd's purse	Seed capsule	triangular			ovoid	15:1
8	Flour beetle	Color	6/16 sooty and 3/16 red	black	jet	black	6:3:3:4

FIGURE 4–7 The basis of modified dihybrid F_2 phenotypic ratios, resulting from crosses between doubly heterozygous F_1 individuals. The four groupings of the F_2 genotypes shown in Figure 4–6 and across the top of this figure are combined in various ways to produce these ratios.

at the other locus. The presence of at least one *B* allele allows pigmentation to occur in much the same way that the *H* allele in humans allows the expression of the ABO blood types. In a cross between agouti (*AABB*) and albino (*aabb*), members of the F_1 are all *AaBb* and have agouti coat color. In the F_2 progeny of a cross between two F_1 heterozygotes, the following genotypes and phenotypes are observed:

$$F_1: AaBb \times AaBb$$
$$\downarrow$$

F_2 Ratio	Genotype	Phenotype	Final Phenotypic Ratio
9/16	*A–B–*	agouti	9/16 agouti
3/16	*A–bb*	albino	3/16 black
3/16	*aa B–*	black	4/16 albino
1/16	*aa bb*	albino	

We can envision gene interaction yielding the observed 9:3:4 F_2 ratio as a two-step process:

	Gene *B*		Gene *A*	
Precursor Molecule (colorless)	↓ ⟶ *B–*	Black Pigment	↓ ⟶ *A–*	Agouti Pattern

In the presence of a *B* allele, black pigment can be made from a colorless substance. In the presence of an *A* allele, the black pigment is deposited during the development of hair in a pattern that produces the agouti phenotype. If the *aa* genotype occurs, all of the hair remains black. If the *bb* genotype occurs, no black pigment is produced, regardless of the presence of the *A* or *a* alleles, and the mouse is albino. Therefore, the *bb* genotype masks or suppresses the expression of the *A* gene. As a result, this is referred to as *recessive epistasis*.

A second type of epistasis, called *dominant epistasis* occurs when a dominant allele at one genetic locus masks the expression of the alleles at a second locus. For instance, Case 2 of Figure 4–7 deals with the inheritance of fruit color in summer squash. Here, the dominant allele *A* results in white fruit color regardless of the genotype at a second locus, *B*. In the absence of the dominant *A* allele (the *aa* genotype), *BB* or *Bb* results in yellow color, while *bb* results in green color. Therefore, if two white-colored double heterozygotes (*AaBb*) are crossed, this type of epistasis generates an interesting phenotypic ratio:

$$F_1: AaBb \times AaBb$$
$$\downarrow$$

F_2 Ratio	Genotype	Phenotype	Final Phenotypic Ratio
9/16	*A–B–*	white	12/16 white
3/16	*A–bb*	white	3/16 yellow
3/16	*aaB–*	yellow	1/16 green
1/16	*aabb*	green	

Of the offspring, 9/16 are $A{-}B{-}$ and are thus white. The 3/16 bearing the genotypes $A{-}bb$ are also white. Finally, 3/16 are yellow ($aaB{-}$), while 1/16 are green (*aabb*); and we obtain the modified ratio of 12:3:1.

Our third type of gene interaction (Case 3 of Figure 4–7) was first discovered by William Bateson and Reginald Punnett (of Punnett square fame). It is demonstrated in a cross between two true-breeding strains of white-flowered sweet peas. Unexpectedly, the results of this cross yield all purple F_1 plants, and the F_2 plants occur in a ratio of 9/16 purple to 7/16 white. The proposed explanation suggests that the presence of at least one dominant allele of each of two gene pairs is essential for flowers to be purple. Thus, this cross represents a case of *complementary gene interaction.* All other genotype combinations yield white flowers because the homozygous condition of *either* recessive allele masks the expression of the dominant allele at the other locus. The cross is shown as follows:

$$P_1 : AAbb \times aaBB$$

white white

↓

$$F_1 : \text{All } AaBb \text{ (purple)}$$

↓

F_2 Ratio	Genotype	Phenotype	Final Phenotypic Ratio
9/16	*A–B–*	purple	
3/16	*A–bb*	white	9/16 purple
3/16	*aaB–*	white	7/16 white
1/16	*aabb*	white	

We can now see how two gene pairs might yield such results:

	Gene *A*		Gene *B*	
Precursor	↓	Intermediate	↓	Final
Substance	⟶	Product	⟶	Product
(colorless)	*A–*	(colorless)	*B–*	(purple)

At least one dominant allele from each pair of genes is necessary to ensure both biochemical conversions to the final product, yielding purple flowers. In our cross, this will occur in 9/16 of the F_2 offspring. All other plants (7/16) have flowers that remain white.

The preceding examples illustrate how the products of two genes "interact" to influence the development of a common phenotype. In other instances, more than two genes and their products are involved in controlling phenotypic expression.

Novel Phenotypes

Other cases of gene interaction yield novel, or new, phenotypes in the F_2 generation, in addition to producing modified dihybrid ratios. Case 4 in Figure 4–7 depicts the inheritance of fruit shape in the summer squash *Cucurbita pepo*. When plants with disc-shaped fruit (*AABB*) are crossed to plants with long fruit (*aabb*), the F_1 generation all have disc fruit. However, in the F_2 progeny, fruit with a novel shape—sphere—appear, along with fruit exhibiting the parental phenotypes. A variety of fruit shapes are shown in Figure 4–8.

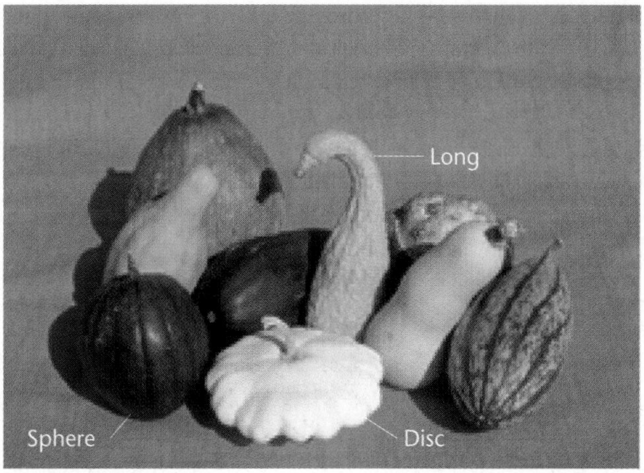

FIGURE 4–8 Summer squash exhibiting the fruit-shape phenotypes disc (white), long (orange gooseneck), and sphere (bottom left).

The F_2 generation, with a modified 9:6:1 ratio, is generated as follows:

$$F_1 : AaBb \times AaBb$$

disc ↓ disc

F_2 Ratio	Genotype	Phenotype	Final Phenotypic Ratio
9/16	*A–B–*	disc	9/16 disc
3/16	*A–bb*	sphere	6/16 sphere
3/16	*aaB–*	sphere	1/16 long
1/16	*aabb*	long	

In this example of gene interaction, both gene pairs influence fruit shape equally. A dominant allele at either locus ensures a sphere-shaped fruit. In the absence of dominant alleles, the fruit is long. However, if both dominant alleles (*A* and *B*) are present, the fruit displays a flattened, disc shape.

NOW SOLVE THIS

Problem 7 on page 87 describes a plant in which flower color, a single characteristic, can take on one of three variations. You are asked to determine how many genes are involved in the inheritance of this characteristic and what genotypes are responsible for what phenotypes.

Hint: The most important information is the data provided. You must analyze the raw data and convert the numbers to a meaningful ratio. This will guide you in determining how many gene pairs are involved. Then you can categorize the genotypic ratio in a way to match the phenotypic ratio.

Other Modified Dihybrid Ratios

The remaining cases (5–8) in Figure 4–7 show additional modifications of the dihybrid ratio and provide still other examples of gene interactions. However, all eight cases have two things in common. First, we have not violated the principles of segregation and independent assortment to explain the inheritance pattern of each case. Therefore, the added complexity of inheritance in these examples does not detract from the validity of Mendel's conclusions. Second, the F_2 phenotypic ratio in each example has been expressed in sixteenths. When similar observations are made in crosses where the inheritance pattern is unknown, it suggests to geneticists that two gene pairs are controlling the observed phenotypes. You should make the same inference in your analysis of genetics problems.

ESSENTIAL POINT

Mendel's classic F_2 ratio is often modified in instances when gene interaction controls phenotypic variation. Such instances can be identified when the final ratio is divided into eighths or sixteenths.

4.9 Complementation Analysis Can Determine If Two Mutations Causing a Similar Phenotype Are Alleles of the Same Gene

An interesting situation arises when two mutations, both of which produce a similar phenotype, are isolated independently. Suppose that two investigators independently isolate and establish a true-breeding strain of wingless *Drosophila* and demonstrate that each mutant phenotype is due to a recessive mutation. We might assume that both strains contain mutations in the same gene. However, since we know that many genes are involved in the formation of wings, mutations in any one of them might inhibit wing formation during development. The experimental approach called **complementation analysis** allows us to determine whether two such mutations are in the same gene—that is, whether they are alleles of the same gene or whether they represent mutations in separate genes.

To repeat, our analysis seeks to answer this simple question: *Are two mutations that yield similar phenotypes present in the same gene or in two different genes?* To find the answer, we cross the two mutant strains and analyze the F_1 generation. Two alternative outcomes and interpretations of this cross are shown in Figure 4–9. We discuss both cases, using the designations m^a for one of the mutations and m^b for the other one. Now we will determine experimentally whether or not m^a and m^b are alleles of the same gene.

Case 1. *All offspring develop normal wings.*

Interpretation: The two recessive mutations are in separate genes and are not alleles of one another. Following the cross, all F_1 flies are heterozygous for both genes. *Complementation* is said to occur. Since each mutation is in a separate gene and each F_1 fly is heterozygous at both loci, the normal products of both genes are produced (by the one normal copy of each gene), and wings develop.

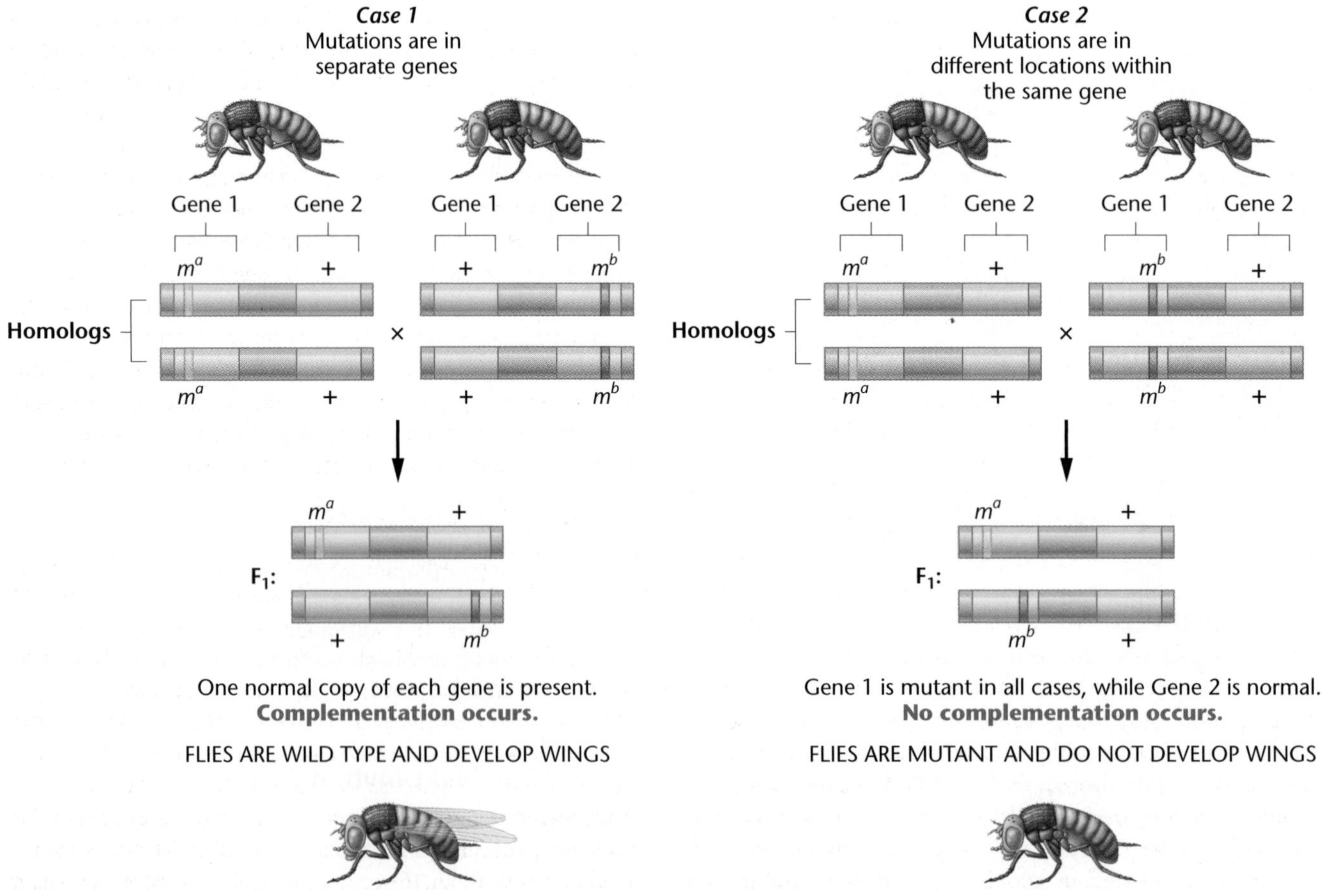

FIGURE 4–9 Complementation analysis of alternative outcomes of two wingless mutations in *Drosophila* (m^a and m^b). In Case 1, the mutations are not alleles of the same gene, whereas in Case 2, the mutations are alleles of the same gene.

Case 2. *All offspring fail to develop wings.*

Interpretation: The two mutations affect the same gene and are alleles of one another. Complementation does *not* occur. Since the two mutations affect the same gene, the F_1 flies are homozygous for the two mutant alleles (the m^a allele and the m^b allele). No normal product of the gene is produced, and in the absence of this essential product, wings do not form.

Complementation analysis, as originally devised by the *Drosophila* geneticist Edward B. Lewis, may be used to screen any number of individual mutations that result in the same phenotype. Such an analysis may reveal that only a single gene is involved or that two or more genes are involved. All mutations determined to be present in any single gene are said to fall into the same **complementation group,** and they will complement mutations in all other groups. When large numbers of mutations affecting the same trait are available and studied using complementation analysis, it is possible to predict the total number of genes involved in the determination of that trait.

ESSENTIAL POINT

Complementation analysis determines whether independently isolated mutations producing similar phenotypes are alleles of one another or whether they represent separate genes.

4.10 Expression of a Single Gene May Have Multiple Effects

While the previous sections have focused on the effects of two or more genes on a single characteristic, the converse situation, where expression of a single gene has multiple phenotypic effects, is also quite common. This phenomenon, which often becomes apparent when phenotypes are examined carefully, is referred to as **pleiotropy.** We will review two such cases involving human genetic disorders to illustrate this point.

Marfan syndrome is a human malady resulting from an autosomal dominant mutation in the gene encoding the connective tissue protein fibrillin. Because this protein is widespread in many tissues in the body, one would expect multiple effects of such a defect. In fact, fibrillin is important to the structural integrity of the lens of the eye, to the lining of vessels such as the aorta, and to bones, among other tissues. As a result, the phenotype associated with Marfan syndrome includes lens dislocation, increased risk of aortic aneurysm, and lengthened long bones in limbs. This disorder is of historical interest in that speculation abounds that Abraham Lincoln was afflicted.

Our second example involves another human autosomal dominant disorder, **porphyria variegata.** Afflicted individuals cannot adequately metabolize the porphyrin component of hemoglobin when this respiratory pigment is broken down as red blood cells are replaced. The accumulation of excess porphyrins is immediately evident in the urine, which takes on a deep red color. The severe features of the disorder are due to the toxicity of the buildup of porphyrins in the body, particularly in the brain. Complete phenotypic characterization includes abdominal pain, muscular weakness, fever, a racing pulse, insomnia, headaches, vision problems (that can lead to blindness), delirium, and ultimately convulsions. As you can see, deciding which phenotypic trait best characterizes the disorder is impossible.

Like Marfan syndrome, porphyria variegata is also of historical significance. George III, king of England during the American Revolution, is believed to have suffered from episodes involving all of the above symptoms. He ultimately became blind and senile prior to his death. We could cite many other examples to illustrate pleiotropy, but suffice it to say that if one looks carefully, most mutations display more than a single manifestation when expressed.

ESSENTIAL POINT

Pleiotropy refers to multiple phenotypic effects caused by a single mutation.

4.11 X-Linkage Describes Genes on the X Chromosome

In many animal and some plant species, one of the sexes contains a pair of unlike chromosomes that are involved in sex determination. In many cases, these are designated as the X and Y. For example, in both *Drosophila* and humans, males contain an X and a Y chromosome, whereas females contain two X chromosomes. While the Y chromosome must contain a region of pairing homology with the X chromosome if the two are to synapse and segregate during meiosis, much of the remainder of the Y chromosome in humans and other species is considered to be relatively inert genetically. Thus, it lacks most genes that are present on the X chromosome. As a result, genes present on the X chromosome exhibit unique patterns of inheritance in comparison with autosomal genes. The term **X-linkage** is used to describe these situations.

In the following discussion, we will focus on inheritance patterns resulting from genes present on the X but absent from the Y chromosome. This situation results in a modification of Mendelian ratios, the central theme of this chapter.

X-Linkage in *Drosophila*

One of the first cases of X-linkage was documented by Thomas H. Morgan around 1920 during his studies of the *white* mutation in the eyes of *Drosophila* (Figure 4–10, p. 74). The normal wild-type red eye color is dominant to white. We will use this case to illustrate X-linkage.

Morgan's work established that the inheritance pattern of the white-eye trait is clearly related to the sex of the parent carrying the mutant allele. Unlike the outcome of the typical monohybrid cross, reciprocal crosses between white- and red-eyed flies did not yield identical results. In contrast, in all of Mendel's monohybrid crosses, F_1 and F_2 data were similar regardless of which P_1 parent exhibited the recessive mutant trait. Morgan's analysis led to the conclusion that the *white* locus is present on the

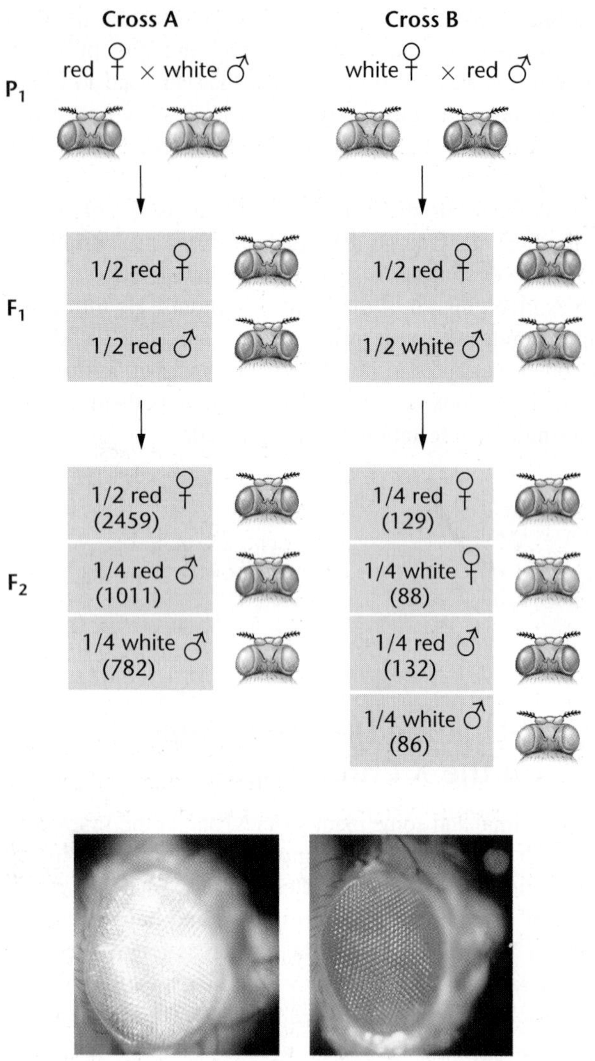

FIGURE 4–10 The F_1 and F_2 results of T. H. Morgan's reciprocal crosses involving the X-linked *white* mutation in *Drosophila melanogaster*. The actual F_2 data are shown in parentheses. The photographs show white eyes and the brick-red wild-type eye color.

X chromosome rather than on one of the autosomes. As such, both the gene and the trait are said to be X-linked.

Results of reciprocal crosses between white-eyed and red-eyed flies are shown in Figure 4–10. The obvious differences in phenotypic ratios in both the F_1 and F_2 generations are dependent on whether or not the P_1 white-eyed parent was male or female.

Morgan was able to correlate these observations with the difference found in the sex-chromosome composition between male and female *Drosophila*. He hypothesized that the recessive allele for white eyes is found on the X chromosome, but its corresponding locus is absent from the Y chromosome. Females thus have two available gene sites, one on each X chromosome, whereas males have only one available gene site on their single X chromosome.

Morgan's interpretation of X-linked inheritance, shown in Figure 4–11, provides a suitable theoretical explanation for his results. Since the Y chromosome lacks homology with most genes on the X chromosome, whatever alleles are present on the X chromosome of the males will be expressed directly in their phenotype. Males cannot be homozygous or heterozygous for X-linked genes, and this condition is referred to as being **hemizygous.**

One result of X-linkage is the **crisscross pattern of inheritance,** whereby phenotypic traits controlled by recessive X-linked genes are passed from homozygous mothers to all sons. This pattern occurs because females exhibiting a recessive trait carry the mutant allele on both X chromosomes. Because male offspring receive one of their mother's two X chromosomes and are hemizygous for all alleles present on that X, all sons will express the same recessive X-linked traits as their mother.

Morgan's work has taken on great historical significance. By 1910, the correlation between Mendel's work and the behavior of chromosomes during meiosis had provided the basis for the **chromosome theory of inheritance.** Work involving the X chromosome around 1920 is considered to be the first solid experimental evidence in support of this theory. In the ensuing two decades, these findings inspired further research, which provided indisputable evidence in support of this theory.

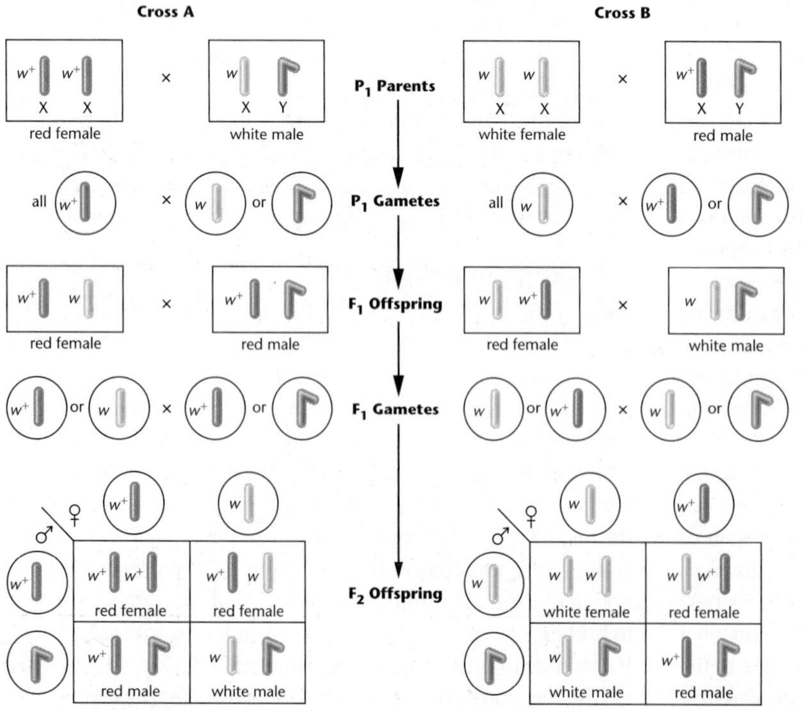

FIGURE 4–11 The chromosomal explanation of the results of the X-linked crosses shown in Figure 4–10.

TABLE 4.2 Human X-Linked Traits

Condition	Characteristics
Color blindness, deutan type	Insensitivity to green light.
Color blindness, protan type	Insensitivity to red light.
Fabry disease	Deficiency of galactosidase A; heart and kidney defects, early death.
G-6-PD deficiency	Deficiency of glucose-6-phosphate dehydrogenase; severe anemic reaction following intake of primaquines in drugs and certain foods, including fava beans.
Hemophilia A	Classical form of clotting deficiency; absence of clotting factor VIII.
Hemophilia B	Christmas disease; absence of clotting factor IX.
Hunter syndrome	Mucopolysaccharide storage disease resulting from iduronate sulfatase enzyme deficiency; short stature, clawlike fingers, coarse facial features, slow mental deterioration, and deafness.
Ichthyosis	Deficiency of steroid sulfatase enzyme; scaly dry skin, particularly on extremities.
Lesch–Nyhan syndrome	Deficiency of hypoxanthine-guanine phosphoribosyl transferase enzyme (HGPRT) leading to motor and mental retardation, self-mutilation, and early death.
Duchenne muscular dystrophy	Progressive, life-shortening disorder characterized by muscle degeneration and weakness; sometimes associated with mental retardation; absence of the protein dystrophin.

X-Linkage in Humans

In humans, many genes and the respective traits they control are recognized as being linked to the X chromosome (see Table 4.2). These X-linked traits can be easily identified in a pedigree because of the crisscross pattern of inheritance. A pedigree for one form of human **color blindness** is shown in Figure 4–12. The mother in generation I passes the trait to all her sons but to none of her daughters. If the offspring in generation II marry normal individuals, the color-blind sons will produce all normal male and female offspring (III-1, 2, and 3); the normal-visioned daughters will produce normal-visioned female offspring (III-4, 6, and 7), as well as color-blind (III-8) and normal-visioned (III-5) male offspring.

The way in which X-linked genes are transmitted causes unusual circumstances associated with recessive X-linked disorders, in comparison to recessive autosomal disorders. For example, if an X-linked disorder debilitates or is lethal to the affected individual prior to reproductive maturation, the disorder occurs exclusively in males. This is so because the only sources of the lethal allele in the population are in heterozygous females who are "carriers" and do not express the disorder. They pass the allele to one-half of their sons, who develop the disorder because they are hemizygous but rarely, if ever, reproduce. Heterozygous females also pass the allele to one-half of their daughters, who become carriers but do not develop the disorder. An example of such an X-linked disorder is Duchenne muscular dystrophy. The disease has an onset prior to age 6 and is often lethal around age 20. It normally occurs only in males.

ESSENTIAL POINT

Genes located on the X chromosome result in a characteristic mode of genetic transmission referred to as X-linkage, displaying so-called crisscross inheritance, whereby affected mothers pass X-linked traits to all of their sons.

NOW SOLVE THIS

Problem 23 on page 89 asks you to analyze three pedigrees and then to determine if each is consistent with X-linkage.

Hint: In X-linkage, because of hemizygosity, the genotype of males is immediately evident. Therefore, the key to solving this type of problem is to consider the possible genotypes of females that do not express the trait.

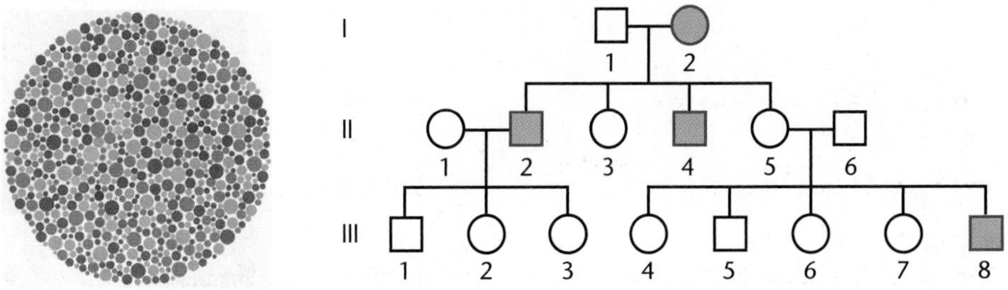

FIGURE 4–12 A human pedigree of the X-linked color-blindness trait. The photograph is of an Ishihara color-blindness chart. Red-green color-blind individuals see the number 3 rather than the number 8 visualized by those with normal color vision.

4.12 In Sex-Limited and Sex-Influenced Inheritance, an Individual's Sex Influences the Phenotype

In contrast to X-linked inheritance, patterns of gene expression may be affected by the sex of an individual even when the genes are not on the X chromosome. In numerous examples in different organisms, the sex of the individual plays a determining role in the expression of certain phenotypes. In some cases, the expression of a specific phenotype is absolutely limited to one sex; in others, the sex of an individual influences the expression of a phenotype that is not limited to one sex or the other. This distinction differentiates **sex-limited inheritance** from **sex-influenced inheritance.**

In both types of inheritance, autosomal genes are responsible for the existence of contrasting phenotypes, but the expression of these genes is dependent on the hormone constitution of the individual. Thus, the heterozygous genotype may exhibit one phenotype in males and the contrasting one in females. In domestic fowl, for example, tail and neck plumage is often distinctly different in males and females (Figure 4–13), demonstrating *sex-limited inheritance*. Cock feathering is longer, more curved, and pointed, whereas hen feathering is shorter and less curved. Inheritance of these feather phenotypes is controlled by a single pair of autosomal alleles whose expression is modified by the individual's sex hormones.

As shown in the following chart, hen feathering is due to a dominant allele, *H*, but regardless of the homozygous presence of the recessive *h* allele, all females remain hen-feathered. Only in males does the *hh* genotype result in cock feathering.

Genotype	Phenotype	
	Females	Males
HH	Hen-feathered	Hen-feathered
Hh	Hen-feathered	Hen-feathered
hh	Hen-feathered	Cock-feathered

FIGURE 4–13 Hen feathering (left) and cock feathering (right) in domestic fowl. Note that the hen's feathers are shorter and less curved.

In certain breeds of fowl, the hen-feathering or cock-feathering allele has become fixed in the population. In the Leghorn breed, all individuals are of the *hh* genotype; as a result, males always differ from females in their plumage. Sebright bantams are all *HH*, resulting in no sexual distinction in feathering phenotypes.

Another example of sex-limited inheritance involves the autosomal genes responsible for milk yield in dairy cattle. Regardless of the overall genotype that influences the quantity of milk production, those genes are obviously expressed only in females.

Cases of *sex-influenced inheritance* include pattern baldness in humans, horn formation in certain breeds of sheep (e.g., Dorset Horn sheep), and certain coat-color patterns in cattle. In such cases, autosomal genes are responsible for the contrasting phenotypes displayed by both males and females, but the expression of these genes is dependent on the hormonal constitution of the individual. Thus, the heterozygous genotype exhibits one phenotype in one sex and the contrasting one in the other. For example, **pattern baldness** in humans, where the hair is very thin on the top of the head (Figure 4–14), is inherited in this way:

Genotype	Phenotype	
	Females	Males
BB	Bald	Bald
Bb	Not bald	Bald
bb	Not bald	Not bald

Females can display pattern baldness, but this phenotype is much more prevalent in males. When females do inherit the *BB* genotype, the phenotype is less pronounced than in males and is expressed later in life.

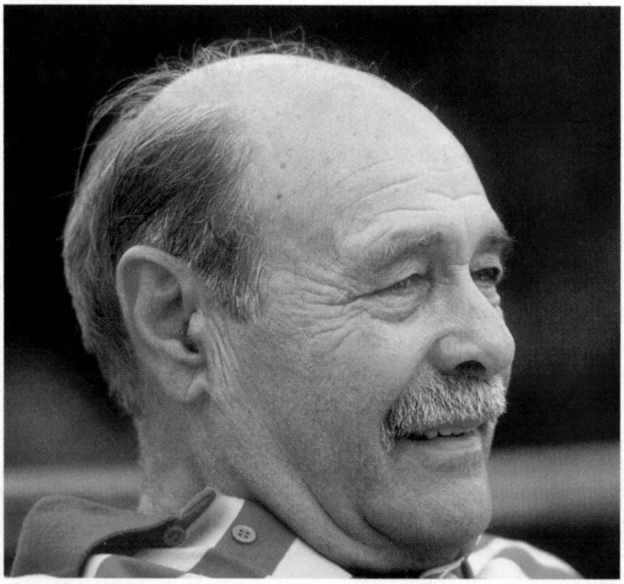

FIGURE 4–14 Pattern baldness, a sex-influenced autosomal trait in humans.

ESSENTIAL POINT

Sex-limited and sex-influenced inheritance occurs when the sex of the organism affects the phenotype controlled by a gene located on an autosome.

NOW SOLVE THIS

Problem 12 on page 88 involves the inheritance of the beard in goats and asks you to analyze the F_1 and F_2 ratios to determine the mode of inheritance.

Hint: Note particularly that the data are differentiated into male and female offspring and that the ratios in the F_2 vary according to sex (i.e., 3/8 of the males are bearded, whereas only 1/8 of the females are bearded, etc.). This should immediately alert you to consider the possible influences of sex differences on the outcome of crosses. In this case, you should consider whether X-linkage, sex-limited inheritance, or sex-influenced inheritance might be involved.

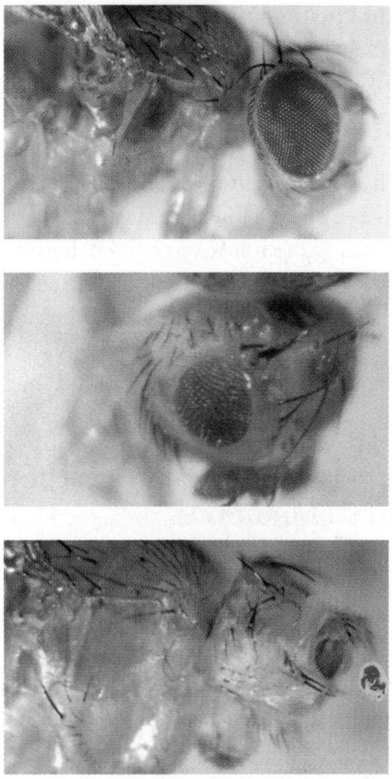

FIGURE 4–15 Variable expressivity, as shown in flies homozygous for the *eyeless* mutation in *Drosophila*. Gradations in phenotype range from wild type to partial reduction to eyeless.

4.13 Genetic Background and the Environment Affect Phenotypic Expression

We now focus on **phenotypic expression.** In previous discussions, we assumed that the genotype of an organism is always directly expressed in its phenotype. For example, pea plants homozygous for the recessive *d* allele (*dd*) will always be dwarf. We discussed gene expression as though the genes operate in a closed system in which the presence or absence of functional products directly determines the collective phenotype of an individual. The situation is actually much more complex. Most gene products function within the internal milieu of the cell, and cells interact with one another in various ways. Furthermore, the organism exists under diverse environmental influences. Thus, gene expression and the resultant phenotype are often modified through the interaction between an individual's particular genotype and the external environment. Here, we deal with several important variables that are known to modify gene expression.

Penetrance and Expressivity

Some mutant genotypes are always expressed as a distinct phenotype, whereas others produce a proportion of individuals whose phenotypes cannot be distinguished from normal (wild type). The degree of expression of a particular trait can be studied quantitatively by determining the *penetrance* and *expressivity* of the genotype under investigation. The percentage of individuals that show at least some degree of expression of a mutant genotype defines the **penetrance** of the mutation. For example, the phenotypic expression of many mutant alleles in *Drosophila* can overlap with wild type. If 15 percent of mutant flies show the wild-type appearance, the mutant gene is said to have a penetrance of 85 percent.

By contrast, **expressivity** reflects the *range of expression* of the mutant genotype. Flies homozygous for the recessive mutant *eyeless* gene yield phenotypes that range from the presence of normal eyes to a partial reduction in size to the complete absence of one or both eyes (Figure 4–15). Although the average reduction of eye size is one-fourth to one-half, expressivity ranges from complete loss of both eyes to completely normal eyes.

Examples such as the expression of the *eyeless* gene provide the basis for experiments to determine the causes of phenotypic variation. If, on one hand, a laboratory environment is held constant and extensive phenotypic variation is still observed, other genes may be influencing or modifying the *eyeless* phenotype. On the other hand, if the genetic background is not the cause of the phenotypic variation, environmental factors such as temperature, humidity, and nutrition may be involved. In the case of the *eyeless* phenotype, experiments have shown that both genetic background and environmental factors influence its expression.

Genetic Background: Position Effects

Although it is difficult to assess the specific effect of the **genetic background** and the expression of a gene responsible for determining a potential phenotype, one effect of genetic background has been well characterized, the **position effect.** In such instances, the physical location of a gene in relation to other genetic material may influence its expression. For example, if a region of a chromosome is relocated or rearranged (called a translocation or an inversion event), normal expression of genes in that chromosomal region may be modified.

This is particularly true if the gene is relocated to or near certain areas of the chromosome that are prematurely condensed and genetically inert, referred to as **heterochromatin.** An example of a position effect involves female *Drosophila* heterozygous for the X-linked recessive eye color mutant *white* (w). The w^+/w genotype normally results in a wild-type brick red eye color. However, if the region of the X chromosome containing the wild-type w^+ allele is translocated so that it is close to a heterochromatic region, expression of the w^+ allele is modified. Instead of having a red color, the eyes are variegated, or mottled with red and white patches. Apparently, heterochromatic regions inhibit the expression of adjacent genes.

Temperature Effects—An Introduction to Conditional Mutations

Chemical activity depends on the kinetic energy of the reacting substances, which in turn depends on the surrounding temperature. We can thus expect temperature to influence phenotypes. An example is seen in the evening primrose, which produces red flowers when grown at 23°C and white flowers when grown at 18°C. An even more striking example is seen in Siamese cats and Himalayan rabbits, which exhibit dark fur in certain regions where their body temperature is slightly cooler, particularly the nose, ears, and paws (Figure 4–16). In these cases, it appears that the enzyme normally responsible for pigment production is functional only at the lower temperatures present in the extremities, but it loses its catalytic function at the slightly higher temperatures found throughout the rest of the body.

Mutations whose expression is affected by temperature, called **temperature-sensitive mutations,** are examples of **conditional mutations,** whereby phenotypic expression is determined by environmental conditions. Examples of temperature-sensitive mutations are known in viruses and a variety of organisms, including bacteria, fungi, and *Drosophila.* In extreme cases, an organism carrying a mutant allele may express a mutant phenotype when grown at one temperature but express the wild-type phenotype when reared at another temperature. This type of temperature effect is useful in studying mutations that interrupt essential processes during development and are thus normally detrimental or lethal. For example, if bacterial viruses are cultured under *permissive conditions* of 25°C, the mutant gene product is functional, infection proceeds normally, and new viruses are produced and can be studied. However, if bacterial viruses carrying temperature-sensitive mutations infect bacteria cultured at 42°C—the *restrictive condition*—infection progresses up to the point where the essential gene product is required (e.g., for viral assembly) and then arrests. Temperature-sensitive mutations are easily induced and isolated in viruses, and have added immensely to the study of viral genetics.

Onset of Genetic Expression

Not all genetic traits become apparent at the same time during an organism's life span. In most cases, the age at which a mutant gene exerts a noticeable phenotype depends on events during the normal sequence of growth and development. In humans, the prenatal, infant, preadult, and adult phases require different genetic information. As a result, many severe inherited disorders are not manifested until after birth. For example, as we saw in Chapter 3, **Tay–Sachs disease,** inherited as an autosomal recessive, is a lethal lipid metabolism disease involving an abnormal enzyme, hexosaminidase A. Newborns appear to be phenotypically normal for the first few months. Then developmental retardation, paralysis, and blindness ensue, and most affected children die around the age of 3.

The **Lesch–Nyhan syndrome,** inherited as an X-linked recessive disease, is characterized by abnormal nucleic acid metabolism (biochemical salvage of nitrogenous purine bases), leading to the accumulation of uric acid in blood and tissues, mental retardation, palsy, and self-mutilation of the lips and fingers. The disorder is due to a mutation in the gene encoding hypoxanthine-guanine phosphoribosyl transferase (HGPRT). Newborns are normal for six to eight months prior to the onset of the first symptoms.

Still another example involves **Duchenne muscular dystrophy (DMD),** an X-linked recessive disorder associated with progressive muscular wasting. It is not usually diagnosed until the child is 3 to 5 years old. Even with modern medical intervention, the disease is often fatal in the early 20s.

Perhaps the most age-variable of all inherited human disorders is **Huntington disease.** Inherited as an autosomal dominant, Huntington disease affects the frontal lobes of the cerebral cortex, where progressive cell death occurs over a period of more than a decade. Brain deterioration is accompanied by spastic uncontrolled movements, intellectual and emotional deterioration, and ultimately death. Onset of this disease has been reported at all ages, but it most

(a) (b)

FIGURE 4–16 (a) A Himalayan rabbit. (b) A Siamese cat. Both species show dark fur color on the snout, ears, and paws. The patches are due to the temperature-sensitive allele responsible for pigment production.

frequently occurs between ages 30 and 50, with a mean onset age of 38 years.

These examples support the concept that the critical expression of genes varies throughout the life cycle of all organisms, including humans. Gene products may play more essential roles at certain life stages, and it is likely that the internal physiological environment of an organism changes with age.

Genetic Anticipation

Interest in studying the genetic onset of phenotypic expression has intensified with the discovery of heritable disorders that exhibit a progressively earlier age of onset and an increased severity of the disorder in each successive generation. This phenomenon is called **genetic anticipation.**

Myotonic dystrophy (DM), the most common type of adult muscular dystrophy, clearly illustrates genetic anticipation. Individuals afflicted with this autosomal dominant disorder exhibit extreme variation in the severity of symptoms. Mildly affected individuals develop cataracts as adults but have little or no muscular weakness. Severely affected individuals demonstrate more extensive myopathy and may be mentally retarded. In its most extreme form, the disease is fatal just after birth. In 1989, C. J. Howeler and colleagues confirmed the correlation of increased severity and earlier onset with successive generations. They studied 61 parent-child pairs, and in 60 cases, age of onset was earlier and more severe in the child than in his or her affected parent.

In 1992, an explanation was put forward for the molecular cause of the mutation responsible for DM, as well as the basis of genetic anticipation. A particular region of the DM gene—a short **trinucleotide DNA sequence—**is repeated a variable number of times and is unstable. Normal individuals average about five copies of this region; minimally affected individuals have about 50 copies; and severely affected individuals have over 1000 copies. The most remarkable observation was that in successive generations, the size of the repeated segment increases. Although it is not yet clear how this expansion in size affects onset and phenotypic expression, the correlation is extremely strong. Several other inherited human disorders, including the fragile-X syndrome, Kennedy disease, and Huntington disease, also reveal an association between the size of specific regions of the responsible gene and disease severity.

Genomic (Parental) Imprinting

Our final example of modification of the laws of Mendelian inheritance involves the variation of phenotypic expression that results during early development after one or the other member of a gene pair has been silenced, depending on the parental origin of the chromosome on which a particular allele is located. This phenomenon is called **genomic,** or **parental, imprinting.** In some species, certain chromosomal regions and the genes contained within them are somehow "imprinted," depending on their parental origin, in a way that determines whether specific genes will be expressed or remain genetically silent. Such "silencing," for example, leads to the direct phenotypic expression of the allele that is not being silenced.

The imprinting step, the critical issue in understanding this phenomenon, is thought to occur before or during gamete formation, leading to differentially marked genes (or chromosome regions) in sperm-forming versus egg-forming tissues. The process differs from mutation because the imprint is eventually erased and can be reversed in succeeding generations as genes pass from mother to son to granddaughter, and so on.

The first example of genomic imprinting was discovered in 1991, when three specific mouse genes were shown to undergo imprinting. One is the gene encoding insulinlike growth factor II (*Igf2*). A mouse that carries two normal wild-type alleles of this gene is normal in size, whereas a mouse that carries two mutant alleles lacks a growth factor and is dwarf. The size of a heterozygous mouse (one allele normal and one mutant) depends on the parental origin of the wild-type allele (Figure 4–17). The mouse is normal in size if the normal allele came from the father, but it will be dwarf if the normal allele came from the mother. From this, we can deduce that the normal *Igf2* gene is imprinted during egg production in a way that causes it to be silenced, but it functions normally when it has passed through sperm-producing tissue in males. Imprinting in the next generation depends on whether the gene now passes through sperm-producing or egg-forming tissue.

In humans, two distinct genetic disorders are thought to be caused by differential imprinting of the same region of chromosome 15. In both cases, the disorders appear to be the result of a deletion of an identical region in one member of the chromosome-15 pair. The first disorder, **Prader–Willi syndrome (PWS),** results when only an undeleted maternal

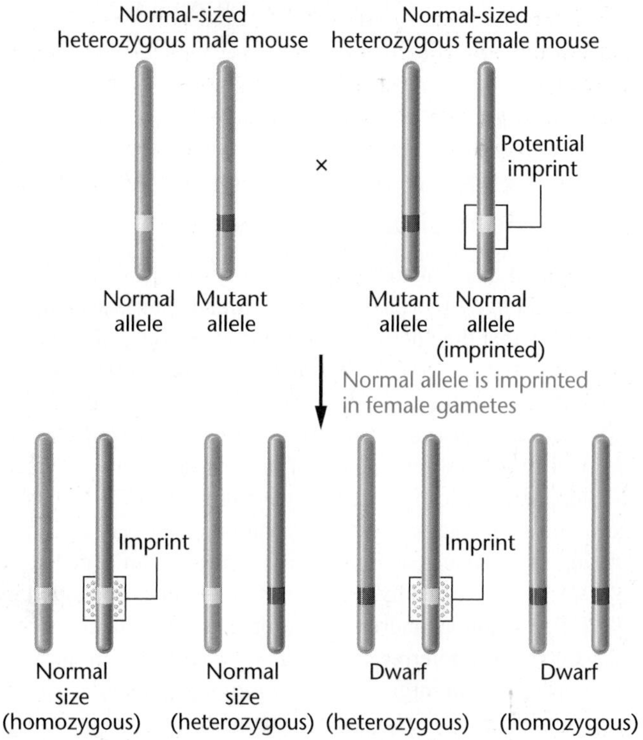

FIGURE 4–17 The effect of imprinting on the mouse *Igf2* gene, which produces dwarf mice in the homozygous condition. Heterozygotes that receive an imprinted normal allele from their mother are dwarf.

chromosome remains. If only an undeleted paternal chromosome remains, an entirely different disorder, **Angelman syndrome (AS),** results.

These two conditions exhibit different phenotypes. PWS entails mental retardation, a severe eating disorder marked by uncontrollable appetite, obesity, and diabetes. AS also exhibits mental retardation, but involuntary muscle contractions (chorea) and seizures accompany the disorder. We can conclude that the involved region of chromosome 15 is imprinted differently in male and female gametes and that an undeleted maternal and paternal region is required for normal development.

Although numerous questions remain unanswered regarding genomic imprinting, it is now clear that many genes are subject to this process. More than 50 have been identified in mammals thus far. How a region of a chromosome may be imprinted is the main focus of investigation. Current thinking is that chemical methylation of certain dinucleotide bases found in DNA (called CpG islands) are involved in imprinting and gene silencing. A high level of methylation can inhibit gene activity, while active genes (or their regulatory sequences) are often undermethylated.

ESSENTIAL POINT

Phenotypic expression is not always the direct reflection of the genotype. A percentage of organisms may not express the expected phenotype at all, the basis of the "penetrance" of a mutant gene. In addition, the phenotype can be modified by genetic background, temperature, and nutrition. The onset of expression of a gene may vary during the lifetime of an organism, and it may even be "imprinted" so that it is expressed differently depending on parental origin.

4.14 Extranuclear Inheritance Modifies Mendelian Patterns

Throughout the history of genetics, occasional reports have challenged the basic tenet of Mendelian transmission genetics—that the phenotype is determined solely by nuclear genes located on the chromosomes of both parents. In this final section of the chapter, we consider several examples of inheritance patterns that vary from those predicted by the traditional biparental inheritance of nuclear genes, upon which Mendelian genetics is based. In the following cases, we will focus on two situations where transmission of genetic information is extranuclear. In the first category, an organism's phenotype is affected by the expression of genes contained in the DNA of mitochondria or chloroplasts rather than the nucleus, generally referred to as *organelle heredity*. In the second category, an organism's phenotype is determined by genetic information expressed in the gamete of one parent—usually the mother—such that, following fertilization, the zygote is influenced by gene products from only one of the parents during development. As a result, the phenotype is influenced not by the individual's genotype, but by the genotype of the mother. This general phenomenon is referred to as a **maternal effect.**

Initially, such observations met with skepticism. However, increasing knowledge of molecular genetics and the discovery of DNA in mitochondria and chloroplasts caused the phenomenon of **extranuclear inheritance** to be recognized as an important aspect of genetics.

Organelle Heredity: DNA in Chloroplasts and Mitochondria

Let us examine examples of inheritance patterns related to chloroplast and mitochondrial function. Before DNA was discovered in these organelles, the exact mechanism of transmission of the traits was not clear, except that their inheritance appeared to be linked to something in the cytoplasm rather than to genes in the nucleus. Furthermore, transmission was most often from the maternal parent through the ooplasm, causing the results of reciprocal crosses to vary. Such an extranuclear pattern of inheritance is now appropriately called **organelle heredity.**

Analysis of the inheritance patterns resulting from mutant alleles in chloroplasts and mitochondria has been difficult for two reasons. First, the function of these organelles is dependent on gene products from both nuclear and organelle DNA, making the discovery of the genetic origin of mutations affecting organelle function difficult. Second, many mitochondria and chloroplasts are contributed to each progeny. Thus, if only one or a few of the organelles contain a mutant gene in a cell among a population of mostly normal mitochondria, the corresponding mutant phenotype may not be revealed. This condition, referred to as **heteroplasmy,** may lead to normal cells since the organelles lacking the mutation provide the basis of wild-type function. Analysis is therefore much more complex than for Mendelian characters.

Chloroplasts: Variegation in Four-o'clock Plants

In 1908, Karl Correns (one of the rediscoverers of Mendel's work) provided the earliest example of inheritance linked to chloroplast transmission. Correns discovered a variant of the four-o'clock plant, *Mirabilis jalapa*, that had branches with either white, green, or variegated white-and-green leaves. The white areas in variegated leaves and the completely white leaves lack chlorophyll that provides the green color to normal leaves. Chlorophyll is the light-absorbing pigment made within chloroplasts.

Correns was curious about how inheritance of this phenotypic trait occurred. As shown in Figure 4–18, inheritance in all possible combinations of crosses is strictly determined by the phenotype of the ovule source. For example, if the seeds (representing the progeny) were derived from ovules on branches with green leaves, all progeny plants bore only green leaves, regardless of the phenotype of the source of pollen. Correns concluded that inheritance was transmitted through the cytoplasm of the maternal parent because the pollen, which contributes little or no cytoplasm to the zygote, had no apparent influence on the progeny phenotypes.

Since leaf coloration is related to the chloroplast, genetic information contained either in that organelle or somehow present in the cytoplasm and influencing the chloroplast must be responsible for the inheritance pattern. It now seems certain that the genetic "defect" that eliminates the green chlorophyll in the white patches on leaves is a mutation in the DNA housed in the chloroplast.

Source of Pollen	*Location of Ovule*		
	White branch	Green branch	Variegated branch
White branch	White	Green	White, green, or variegated
Green branch	White	Green	White, green, or variegated
Variegated branch	White	Green	White, green, or variegated

FIGURE 4–18 Offspring from crosses between flowers from various branches of four-o'clock plants. The photograph illustrates variegation, seen in the pink flowers.

Mitochondrial Mutations: *poky* in *Neurospora* and *petite* in *Saccharomyces*

Mutations affecting mitochondrial function have been discovered and studied, revealing that they too contain a distinctive genetic system. As with chloroplasts, mitochondrial mutations are transmitted through the cytoplasm. In our current discussion, we will emphasize the link between mitochondrial mutations and the resultant extranuclear inheritance patterns.

In 1952, Mary B. Mitchell and Hershel K. Mitchell studied the bread mold *Neurospora crassa*. They discovered a slow-growing mutant strain and named it *poky*. Slow growth is associated with impaired mitochondrial function, specifically in relation to certain cytochromes essential for electron transport. Results of genetic crosses between wild-type and *poky* strains suggest that *poky* is an extranuclear trait inherited through the cytoplasm. If one mating type is *poky* and the other is wild type, all progeny colonies are *poky*. The reciprocal cross, where *poky* is transmitted by the other mating type, produces normal wild-type colonies.

Another extensive study of mitochondrial mutations has been performed with the yeast *Saccharomyces cerevisiae*. The first such mutation, described by Boris Ephrussi and his coworkers in 1956, was named *petite* because of the small size of the yeast colonies (Figure 4–19). Many independent *petite* mutations have since been discovered and studied, and all have a common characteristic—a deficiency in cellular respiration involving abnormal electron transport. This organism is a facultative anaerobe and can grow by fermenting glucose through glycolysis; thus, it may survive the loss of mitochondrial function by generating energy anaerobically.

The complex genetics of *petite* mutations has revealed that a small proportion are the result of nuclear mutations. They exhibit Mendelian inheritance and illustrate that mitochondria function depends on both nuclear and organellar gene products. The majority of them demonstrate cytoplasmic transmission, indicating mutations in the DNA of the mitochondria.

Mitochondrial Mutations: Human Genetic Disorders

Our knowledge of genetics of mitochondria has now greatly expanded. The DNA found in human mitochondria has been completely sequenced and contains 16,569 base pairs. Mitochondrial gene products have now been identified and include the following:

13 proteins, required for aerobic cellular respiration
22 transfer RNAs (tRNAs), required for translation
2 ribosomal RNAs (rRNAs), required for translation

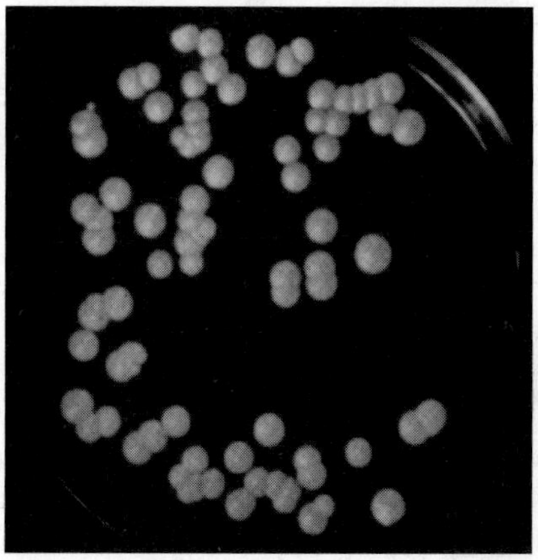

Normal colonies

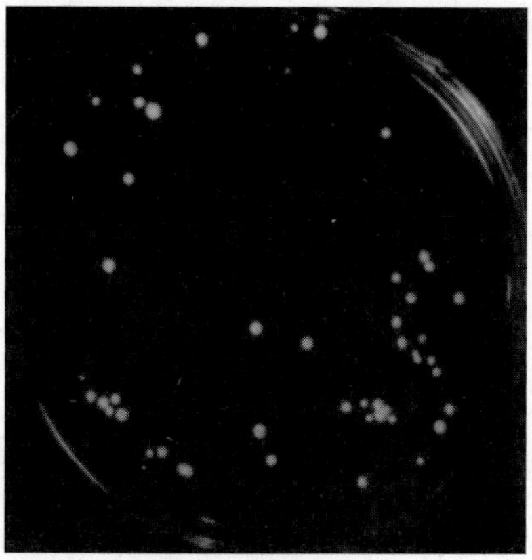

Petite colonies

FIGURE 4–19 Photos comparing normal versus petite colonies of the yeast *Saccharomyces cerevisiae*.

Because a cell's energy supply is largely dependent on aerobic cellular respiration, disruption of any mitochondrial gene by mutation may potentially have a severe impact on that organism. We have seen this in our previous discussion of the *petite* mutation in yeast, which would be lethal were it not for this organism's ability to respire anaerobically. In fact, mtDNA is particularly vulnerable to mutations, for two possible reasons. First, the ability to repair mtDNA damage does not appear to be equivalent to that of nuclear DNA. Second, the concentration of highly mutagenic free radicals generated by cell respiration that accumulate in such a confined space very likely raises the mutation rate in mtDNA.

Fortunately, a zygote receives a large number of organelles through the egg, so if only one organelle or a few of them contain a mutation (an illustration of *heteroplasmy*), the impact is greatly diluted by the many mitochondria that lack the mutation and function normally. If a deleterious mutation arises or is present in the initial population of organelles, adults will have cells with a variable mixture of both normal and abnormal organelles. From a genetic standpoint, this condition of heteroplasmy makes analysis quite difficult.

Several criteria must be met in order for a human disorder to be attributable to genetically altered mitochondria:

1. Inheritance must exhibit a maternal rather than Mendelian pattern.
2. The disorder must reflect a deficiency in the bioenergetic function of the organelle.
3. There must be a specific genetic mutation in one or more of the mitochondrial genes.

Several disorders in humans are known to demonstrate these characteristics. For example, **myoclonic epilepsy and ragged-red fiber disease (MERRF)** demonstrates a pattern of inheritance consistent with maternal transmission. Only the offspring of affected mothers inherit this disorder, while the offspring of affected fathers are normal. Individuals with this rare disorder express ataxia (lack of muscular coordination), deafness, dementia, and epileptic seizures. The disease is named for the presence of "ragged-red" skeletal-muscle fibers that exhibit blotchy red patches resulting from the proliferation of aberrant mitochondria (Figure 4–20). Brain function, which has a high energy demand, is also affected in this disorder, leading to the neurological symptoms described above.

The mutation that causes MERRF has now been identified and is in a mitochondrial gene whose altered product interferes with the capacity for translation of proteins within the organelle. This, in turn, leads to the various manifestations of the disorder.

The cells of MERRF individuals exhibit heteroplasmy, containing a mixture of normal and abnormal mitochondria. Different patients display different proportions of the two, and even different cells from the same patient exhibit various levels of abnormal mitochondria. Were it not for heteroplasmy, the mutation would very likely be lethal, testifying to the essential nature of mitochondrial function and its reliance on the genes encoded by DNA within the organelle.

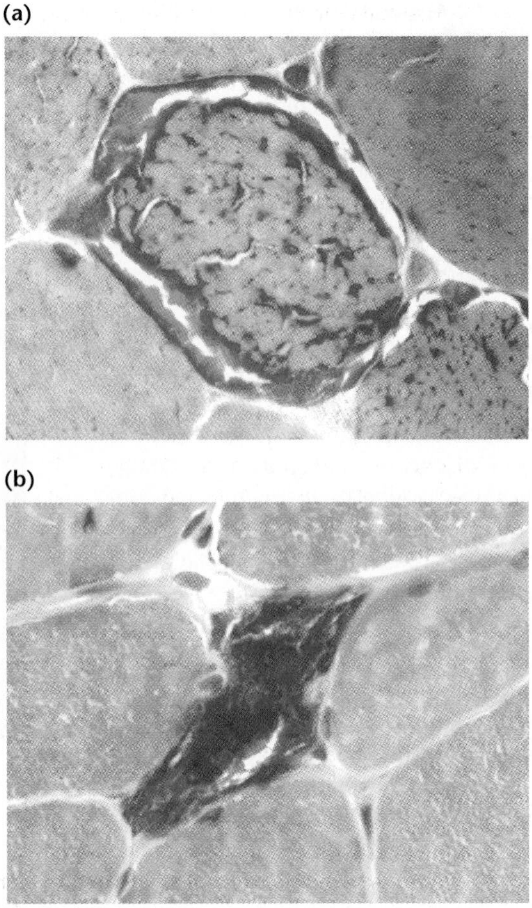

FIGURE 4–20 Ragged-red fibers in skeletal muscle cells from patients with MERRF. (a) The muscle fiber has mild proliferation (see red rim and speckled cytoplasm). (b) Marked proliferation where mitochondria have replaced most cellular structures.

A second human disorder, **Leber's hereditary optic neuropathy (LHON)**, also exhibits maternal inheritance as well as mitochondrial DNA lesions. The disease is characterized by sudden bilateral blindness. The average age of vision loss is 27, but onset is quite variable. Four mutations have been identified, all of which disrupt normal oxidative phosphorylation. Over 50 percent of cases are due to a mutation at a specific position in the mitochondrial gene encoding a subunit of NADH dehydrogenase, an enzyme essential to cell respiration. This mutation is transmitted maternally through the mitochondria to all offspring. It is interesting to note that in many instances of LHON, there is no family history; a significant number of cases are "sporadic," resulting from newly arisen mutations.

Maternal Effect

In **maternal effect,** also referred to as *maternal influence*, an offspring's phenotype for a particular trait is under the control of nuclear gene products present in the egg. This is in contrast to biparental inheritance, where both parents transmit information on genes in the nucleus that determines the offspring's phenotype. In cases of maternal effect, the nuclear genes of the female gamete are transcribed, and the genetic products

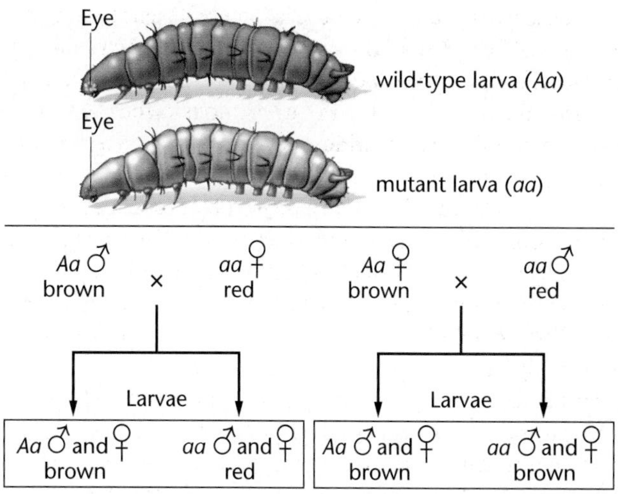

FIGURE 4–21 Maternal influence in the inheritance of eye pigment in the meal moth *Ephestia kuehniella*.

(either proteins or yet untranslated mRNAs) accumulate in the egg cytoplasm. After fertilization, these products are distributed among newly formed cells and influence the patterns or traits established during early development. Two examples will illustrate such an influence of the maternal genome on particular traits.

Ephestia Pigmentation

A straightforward illustration of a maternal effect is seen in the Mediterranean meal moth *Ephestia kuehniella*. The wild-type larva of this moth has a pigmented skin and brown eyes as a result of the dominant gene *A*. The pigment is derived from a precursor molecule, kynurenine, which is in turn a derivative of the amino acid tryptophan. A mutation, *a*, interrupts the synthesis of kynurenine and, when homozygous, may result in red eyes and little pigmentation in larvae. However, as illustrated in Figure 4-21, results of the cross *Aa* × *aa* depend on which parent carries the dominant gene. When the male is the heterozygous parent, a 1:1 brown- to red-eyed ratio is observed in larvae, as predicted by Mendelian segregation. When the female is heterozygous for the *A* gene, however, all larvae are pigmented and have brown eyes, in spite of half of them being *aa*. As these larvae develop into adults, one-half of them gradually develop red eyes, reestablishing the 1:1 ratio.

One explanation for these results is that the *Aa* oocytes synthesize kynurenine or an enzyme necessary for its synthesis and accumulate it in the ooplasm prior to the completion of meiosis. Even in *aa* progeny, if the mothers were *Aa*, this pigment is distributed in the cytoplasm of the cells of the developing larvae; thus, they develop pigmentation and brown eyes. In these progeny, however, the pigment is eventually diluted among many cells and depleted, resulting in the conversion to red eyes as adults. The *Ephestia* example demonstrates the maternal effect in which a cytoplasmically stored nuclear gene product influences the larval phenotype and, at least temporarily, overrides the genotype of the progeny.

Embryonic Development in *Drosophila*

A more recently documented example of maternal effect involves various genes that control embryonic development in *Drosophila melanogaster*. The genetic control of embryonic development in *Drosophila*, discussed in greater detail in Chapter 20, is a fascinating story. The protein products of the maternal-effect genes function to activate other genes, which may in turn activate still other genes. This cascade of gene activity leads to a normal embryo whose subsequent development yields a normal adult fly. The extensive work by Edward B. Lewis, Christiane Nüsslein-Volhard, and Eric Wieschaus (who shared the 1995 Nobel Prize for Physiology or Medicine for their findings) has clarified how these and other genes function. Genes that illustrate maternal effect have products that are synthesized by the developing egg and stored in the oocyte prior to fertilization. Following fertilization, these products specify molecular gradients that determine spatial organization as development proceeds.

For example, the gene *bicoid* (*bcd*) plays an important role in specifying the development of the anterior portion of the fly. Embryos derived from mothers who are homozygous for this mutation (bcd^-/bcd^-) fail to develop anterior areas that normally give rise to the head and thorax of the adult fly. Embryos whose mothers contain at least one wild-type allele (bcd^+) develop normally, even if the genotype of the embryo is homozygous for the mutation. Consistent with the concept of maternal effect, the *genotype of the female parent*, not the *genotype of the embryo*, determines the phenotype of the offspring.

When we return to our discussion of this general topic in Chapter 20, we will see examples of other genes illustrating maternal effect, as well as many "zygotic" genes whose expression occurs during early development and that behave genetically in the conventional Mendelian fashion.

ESSENTIAL POINT

When patterns of inheritance vary from that expected due to biparental transmission of nuclear genes, phenotypes are often found to be under the control of DNA present in mitochondria or chloroplasts, or are influenced during development by the expression of the maternal genotype in the egg.

GENETICS, TECHNOLOGY, AND SOCIETY

Improving the Genetic Fate of Purebred Dogs

For dog lovers, nothing is quite so heartbreaking as watching a dog slowly go blind, struggling to adapt to a life of perpetual darkness. That's what happens in progressive retinal atrophy (PRA), a group of inherited disorders first described in Gordon setters in 1909. Since then, PRA has been detected in more than 100 other breeds of dogs, including Irish setters, border collies, Norwegian elkhounds, toy poodles, miniature schnauzers, cocker spaniels, and Siberian huskies.

The products of many genes are required for the development and maintenance of healthy retinas, and a defect in any one of these genes may cause retinal dysfunction. Decades of research have led to the identification of five such genes (*PDE6A, PDE6B, PRCD, rhodopsin*, and *PRGR*), and more may be discovered. Different mutant alleles are present in different breeds, and each allele is associated with a different form of PRA that varies slightly in its clinical symptoms and rate of progression. Mutations of *PDE6A, PDE6B*, and *PRCD* genes are inherited in a recessive pattern, mutations of the *rhodopsin* gene (such as those found in Mastiffs) are dominant, and *PRGR* mutations (in Siberian huskies and samoyeds) are X-linked.

PRA is almost ten times more common in certain purebred dogs than in mixed breeds. The development of distinct breeds of dogs has involved intensive selection for desirable attributes, such as a particular size, shape, color, or behavior. Many desired characteristics are determined by recessive alleles. The fastest way to increase the homozygosity of these alleles is to mate close relatives, which are likely to carry the same alleles. For example, dogs may be mated to a cousin or a grandparent. Some breeders, in an attempt to profit from impressive pedigrees, also produce hundreds of offspring from individual dogs that have won major prizes at dog shows. This "popular sire effect," as it has been termed, further increases the homozygosity of alleles in purebred dogs.

Unfortunately, the generations of inbreeding that have established favorable characteristics in purebreeds have also increased the homozygosity of certain harmful recessive alleles, resulting in a high incidence of inherited diseases. More than 300 genetic diseases have been characterized in purebred dogs, and many breeds have a predisposition to more than 20 of them. According to researchers at Cornell University, purebred dogs suffer the highest incidence of inherited disease of any animal: 25 percent of the 20 million purebred dogs in America are affected with one genetic ailment or another.

Fortunately, advances in canine genetics are beginning to provide new tools to increase the health of purebred dogs. Genetic tests are available to detect 30 different retinal diseases in dogs. Tests for PRA are now being used to identify heterozygous carriers of *PRCD* mutations. These carriers show no symptoms of PRA but, if mated with other carriers, pass the trait on to about 25 percent of their offspring. Eliminating PRA carriers from breeding programs has almost eradicated this condition from Portuguese water dogs and has greatly reduced its prevalence in other breeds.

Scientists will be able to identify more genes underlying canine inherited diseases thanks to the completion of the Dog Genome Project in 2005. In addition, new therapies that correct gene-based defects will emerge.

The Dog Genome Project may have benefits for humans beyond the reduction of disease in their canine companions. Eighty-five percent of the genes in the dog genome have equivalents in humans, and over 300 diseases affecting dogs also affect humans, including heart disease, epilepsy, allergies, and cancer. Identification of a disease-causing gene in dogs can be a shortcut to isolation of the corresponding gene in humans. By contributing to the cure of human diseases, dogs may prove to be "man's best friend" in an entirely new way.

Your Turn

Take time, individually or in groups, to answer the following questions. Investigate the references and links, to help you understand some of the issues surrounding the genetics of purebred dogs.

1. What are some of the limitations of genetic tests, especially as they apply to purebred dog genetic diseases?

This topic is discussed on the OptiGen Web site (**http://www.optigen.com**). *OptiGen is a company that offers gene tests for all known forms of PRA in dogs and is developing tests for other inherited disorders. From their* TESTS *list, select prcd-PRA, and visit the link "Benefits and Limitations of All Genetic Tests."*

2. Which human disease is similar to PRA in the Siberian husky?

To learn more about these genes and diseases, visit the "Inherited Diseases in Dogs" database (**http://server.vet.cam.ac.uk**) *and search it for Progressive Retinal Atrophy in the Siberian husky. Refer to the OMIM link to learn about the human version of PRA in the Siberian husky.*

3. Recently, commercial laboratories have cloned dogs for research purposes and for people who want their beloved pet to return. Do you approve of cloning pet dogs? Why or why not? Do you think that a cloned dog would be identical to the original dog?

To learn about a recent pet dog cloning, read a Manchester Guardian *article entitled "Pet Cloning Service Bears Five Baby Boogers."* (**http://www.guardian.co.uk/science/2008/aug/05/genetics.korea**)

CASE STUDY A twin difference

Recent news of a professional baseball player's "mysterious cell disease" being traced to a mitochondrial disorder (Rocco Baldelli of the Tampa Bay Rays) has brought not only personal clarification to one set of identical twin brothers, but also national awareness of an array of similar disorders that affect thousands of children and adults regardless of age, gender, race, or position in society. For 14 years the brothers competed on equal terms until one began to suffer loss of visual acuity and muscularity. Diagnosis of a mitochondrial disorder explained the cause and forecasted supportive treatment, but also raised some important questions.

1. The parents and other relatives of the twins are apparently healthy, so how could this genetic disorder have arisen?
2. How could one identical twin have an inherited mitochondrial disorder and not the other?
3. Will the unaffected brother be guaranteed continued good health?

INSIGHTS AND SOLUTIONS

Genetic problems take on added complexity if they involve two independent characters and multiple alleles, incomplete dominance, or epistasis. The most difficult types of problems are those that pioneering geneticists faced during laboratory or field studies. They had to determine the mode of inheritance by working backward from the observations of offspring to parents of unknown genotype.

1. Consider the problem of comb-shape inheritance in chickens, where walnut, pea, rose, and single are the observed distinct phenotypes (see the photographs below). How is comb shape inherited, and what are the genotypes of the P_1 generation of each cross? Use the following data:

Cross 1:	single × single	⟶	all single
Cross 2:	walnut × walnut	⟶	all walnut
Cross 3:	rose × pea	⟶	all walnut
Cross 4:	$F_1 \times F_1$ of cross 3		
	walnut × walnut	⟶	93 walnut
			28 rose
			32 pea
			10 single

Walnut

Pea

Rose

Single

Solution: At first glance, this problem appears quite difficult. However, applying a systematic approach and breaking the analysis into steps usually simplifies it. Our approach involves two steps. First, analyze the data carefully for any useful information. Then, once you identify something that is clearly helpful, follow an empirical approach—that is, formulate a hypothesis and, in a sense, test it against the given data. Look for a pattern of inheritance that is consistent with all cases.

This problem gives two immediately useful facts. First, in cross 1, P_1 singles breed true. Second, while P_1 walnuts breed true in cross 2, a walnut phenotype is also produced in cross 3 between rose and pea. When these F_1 walnuts are crossed in cross 4, all four comb shapes are produced in a ratio that approximates 9:3:3:1. This observation immediately suggests a cross involving two gene pairs, because the resulting data display the same ratio as in Mendel's dihybrid crosses. Since only one trait is involved (comb shape), epistasis may be occurring. This could serve as your working hypothesis, and you must now propose how the two gene pairs "interact" to produce each phenotype.

If you call the allele pairs *A*, *a* and *B*, *b*, you can predict that because walnut represents 9/16 in cross 4, $A{-}B{-}$ will produce walnut. You might also hypothesize that in cross 2, the genotypes are *AABB* × *AABB*, where walnut bred true. (Recall that $A{-}$ and $B{-}$ mean *AA* or *Aa* and *BB* or *Bb*, respectively.)

The phenotype representing 1/16 of the offspring of cross 4 is single; therefore, you could predict that this phenotype is the result of the *aabb* genotype. This is consistent with cross 1.

Now you have only to determine the genotypes for rose and pea. The most logical prediction is that at least one dominant *A* or *B* allele combined with the double recessive condition of the other allele pair accounts for these phenotypes. For example,

$$A{-}bb \longrightarrow \text{rose}$$

$$aaB{-} \longrightarrow \text{pea}$$

If *AAbb* (rose) is crossed with *aaBB* (pea) in cross 3, all offspring will be *AaBb* (walnut). This is consistent with the data, and you must now look at only cross 4. We predict these walnut genotypes to be *AaBb* (as above), and from the cross *AaBb* (walnut) × *AaBb* (walnut) we expect

9/16	*A– B–*	(walnut)
3/16	*A– bb*	(rose)
3/16	*aaB–*	(pea)
1/16	*aabb*	(single)

Our prediction is consistent with the information given. The initial hypothesis of the epistatic interaction of two gene pairs proves consistent throughout, and the problem is solved.

This problem demonstrates the need for a basic theoretical knowledge of transmission genetics. Then, you can search for appropriate clues that will enable you to proceed in a stepwise fashion toward a solution. Mastering problem solving requires practice but can give you a great deal of satisfaction. Apply this general approach to the following problems.

2. In radishes, flower color may be red, purple, or white. The edible portion of the radish may be long or oval. When only flower color is studied, no dominance is evident, and red × white crosses yield all purple. If these F_1 purples are interbred, the F_2 generation consists of 1/4 red: 1/2 purple: 1/4 white. Regarding radish shape, long is dominant to oval in a normal Mendelian fashion.

(a) Determine the F_1 and F_2 phenotypes from a cross between a true-breeding red, long radish and one that is white and oval. Be sure to define all gene symbols initially.

Solution: This is a modified dihybrid cross in which the gene pair controlling color exhibits incomplete dominance. Shape is controlled conventionally. First, establish gene symbols:

RR = red *O*− = long

Rr = purple *oo* = oval

rr = white

P_1: *RROO* × *rroo*

(red long) (white oval)

F_1: all *RrOo* (purple long)

$F_1 \times F_1$: *RrOo* × *RrOo*

F_2:	1/4 *RR*	3/4 *O*−	3/16 *RRO*−	red long
		1/4 *oo*	1/16 *RRoo*	red oval
	2/4 *Rr*	3/4 *O*−	6/16 *RrO*−	purple long
		1/4 *oo*	2/16 *Rroo*	purple oval
	1/4 *rr*	3/4 *O*−	3/16 *rrO*−	white long
		1/4 *oo*	1/16 *rroo*	white oval

Note that to generate the F_2 results, we have used the forked-line method. First, we consider the outcome of crossing F_1 parents for the color genes (*Rr* × *Rr*). Then the outcome of shape is considered (*Oo* × *Oo*).

(b) A red oval plant was crossed with a plant of unknown genotype and phenotype, yielding the following offspring:

103 red long: 101 red oval

98 purple long: 100 purple oval

Determine the genotype and phenotype of the unknown plant.

Solution: The two characters appear to be inherited independently, so consider them separately. The data indicate a 1/4: 1/4: 1/4: 1/4 proportion. First, consider color:

P_1: red × ??? (unknown)

F_1: 204 red (1/2)

198 purple (1/2)

Because the red parent must be *RR*, the unknown must have a genotype of *Rr* to produce these results. Thus it is purple. Now, consider shape:

P_1: oval × ??? (unknown)

F_1: 201 long (1/2)

201 oval (1/2)

Since the oval plant must be *oo*, the unknown plant must have a genotype of *Oo* to produce these results. Thus it is long. The unknown plant is

RrOo purple long

3. In humans, red-green color blindness is inherited as an X-linked recessive trait. A woman with normal vision whose father is color blind marries a male who has normal vision. Predict the color vision of their male and female offspring.

Solution: The female is heterozygous since she inherited an X chromosome with the mutant allele from her father. Her husband is normal. Therefore, the parental genotypes are

Cc × C↾ (↾ represents the *Y* chromosome)

All female offspring are normal (*CC* or *Cc*). One-half of the male children will be color blind (*c*↾), and the other half will have normal vision (*C*↾).

4. Consider the two very limited unrelated pedigrees shown here. Of the four combinations of X-linked recessive, X-linked dominant, autosomal recessive, and autosomal dominant, which modes of inheritance can be absolutely ruled out in each case?

(a)

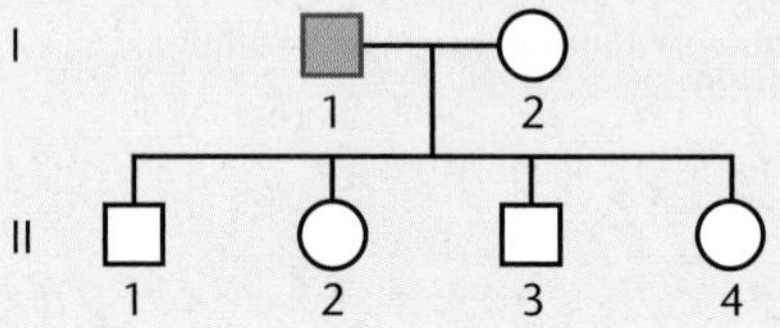

(b)

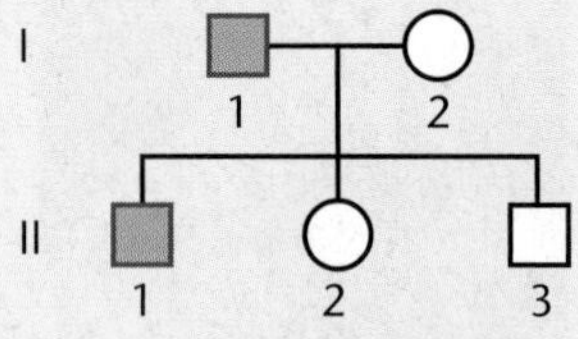

Solution: For both pedigrees, X-linked recessive and autosomal recessive remain possible, provided that the maternal parent is heterozygous in pedigree (b). At first glance autosomal dominance seems unlikely in pedigree (a), since at least half of the offspring should express a dominant trait expressed by one of their parents. However, while it is true that if the affected parent carries an autosomal dominant gene heterozygously, each offspring has a 50 percent chance of inheriting and expressing the mutant gene, the sample size of four offspring is too small to rule out this possibility. In pedigree (b), autosomal dominance is clearly possible. In both cases, one can rule out X-linked dominance because the female offspring would inherit and express the dominant allele, and they do not express the trait in either pedigree.

Problems and Discussion Questions

1. In Shorthorn cattle, coat color may be red, white, or roan. Roan is an intermediate phenotype expressed as a mixture of red and white hairs. The following data are obtained from various crosses:

red	×	red	⟶	all red
white	×	white	⟶	all white
red	×	white	⟶	all roan
roan	×	roan	⟶	1/4 red: 1/2 roan: 1/4 white

How is coat color inherited? What are the genotypes of parents and offspring for each cross?

2. Contrast incomplete dominance and codominance.
3. With regard to the ABO blood types in humans, determine the genotypes of the male parent and female parent:

Male parent: blood type B whose mother was type O

Female parent: blood type A whose father was type B

Predict the blood types of the offspring that this couple may have and the expected ratio of each.

4. In foxes, two alleles of a single gene, *P* and *p*, may result in lethality *(PP)*, platinum coat (*Pp*), or silver coat (*pp*). What ratio is obtained when platinum foxes are interbred? Is the *P* allele behaving dominantly or recessively in causing (a) lethality; (b) platinum coat color?
5. Three gene pairs located on separate autosomes determine flower color and shape as well as plant height. The first pair exhibits incomplete dominance, where color can be red, pink (the heterozygote), or white. The second pair leads to the dominant personate or recessive peloric flower shape, while the third gene pair produces either the dominant tall trait or the recessive dwarf trait. Homozygous plants that are red, personate, and tall are crossed with those that are white, peloric, and dwarf. Determine the F_1 genotype(s) and phenotype(s). If the F_1 plants are interbred, what proportion of the offspring will exhibit the same phenotype as the F_1 plants?

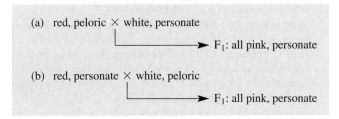

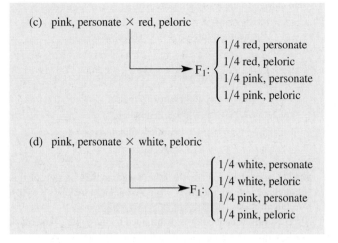

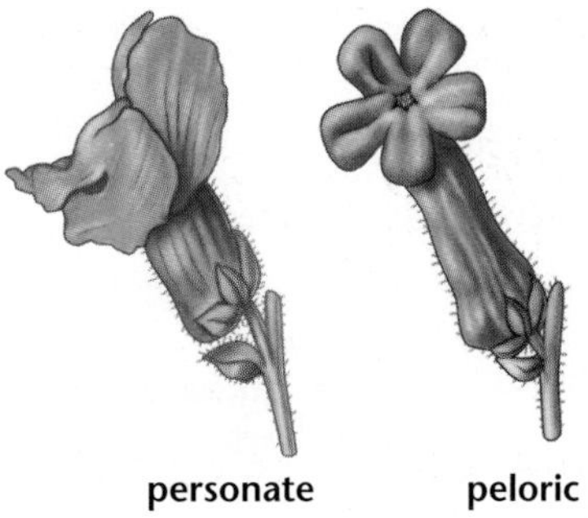

6. As in the plants of Problem 5, color may be red, white, or pink; and flower shape may be personate or peloric. Determine the P_1 and F_1 genotypes for the following crosses:

(a) red, peloric × white, personate
⟶ F_1: all pink, personate

(b) red, personate × white, peloric
⟶ F_1: all pink, personate

(c) pink, personate × red, peloric
⟶ F_1:
- 1/4 red, personate
- 1/4 red, peloric
- 1/4 pink, personate
- 1/4 pink, peloric

(d) pink, personate × white, peloric
⟶ F_1:
- 1/4 white, personate
- 1/4 white, peloric
- 1/4 pink, personate
- 1/4 pink, peloric

What phenotype ratios would result from crossing the F_1 of (a) with the F_1 of (b)?

7. In some plants, a red pigment, cyanidin, is synthesized from a colorless precursor. The addition of a hydroxyl group (—OH) to the cyanidin molecule causes it to become purple. In a cross between two randomly selected purple plants, the following results are obtained:

94 purple:31 red:43 colorless

How many genes are involved in determining these flower colors? Which genotypic combinations produce which phenotypes? Diagram the purple × purple cross.

See **Now Solve This** on page 71.

8. The following genotypes of two independently assorting autosomal genes determine coat color in rats:

A–B–(gray); *A–bb* (yellow); *aaB–*(black); *aabb* (cream)

A third gene pair on a separate autosome determines whether any color will be produced. The *CC* and *Cc* genotypes allow color according to the expression of the *A* and *B* alleles. However, the *cc* genotype results in albino rats regardless of the *A* and *B* alleles present. Determine the F_1 phenotypic ratio of the following crosses: (a) *AAbbCC* × *aaBBcc*; (b) *AaBBCC* × *AABbcc*; (c) *AaBbCc* × *AaBbcc*.

9. Given the inheritance pattern of coat color in rats described in Problem 8, predict the genotype and phenotype of the parents that produced the following F_1 offspring: (a) 9/16 gray: 3/16 yellow: 3/16 black: 1/16 cream; (b) 9/16 gray: 3/16 yellow: 4/16 albino; (c) 27/64 gray: 16/64 albino: 9/64 yellow: 9/64 black: 3/64 cream.
10. A husband and wife have normal vision, although both of their fathers are red-green color-blind, inherited as an X-linked recessive condition. What is the probability that their first child will be (a) a normal son, (b) a normal daughter, (c) a color-blind son, (d) a color-blind daughter?
11. In humans, the ABO blood type is under the control of autosomal multiple alleles. Red-green color blindness is a recessive X-linked trait. If two parents who are both type A and have normal vision produce a son who is color-blind and type O, what is the probability that their next child will be a female who has normal vision and is type O?

12. In goats, development of the beard is due to a recessive gene. The following cross involving true-breeding goats was made and carried to the F_2 generation:

P_1: bearded female × beardless male

↓

F_1: all bearded males and beardless females

$F_1 \times F_1 \rightarrow$ { 1/8 beardless males, 3/8 bearded males, 3/8 beardless females, 1/8 bearded females }

Offer an explanation for the inheritance and expression of this trait, diagramming the cross. Propose one or more crosses to test your hypothesis. See **Now Solve This** on page 77.

13. In cats, yellow coat color is determined by the *b* allele, and black coat color is determined by the *B* allele. The heterozygous condition results in a coat pattern known as tortoiseshell. These genes are X-linked. What kinds of offspring would be expected from a cross of a black male and a tortoiseshell female? What are the chances of getting a tortoiseshell male?
14. In *Drosophila*, an X-linked recessive mutation, *scalloped* (*sd*), causes irregular wing margins. Diagram the F_1 and F_2 results if
(a) A scalloped female is crossed with a normal male.
(b) A scalloped male is crossed with a normal female.
Compare these results to those that would be obtained if the *scalloped* gene were autosomal.
15. Another recessive mutation in *Drosophila*, *ebony* (*e*), is on an autosome (chromosome 3) and causes darkening of the body compared with wild-type flies. What phenotypic F_1 and F_2 male and female ratios will result if a scalloped-winged female with normal body color is crossed with a normal-winged ebony male? Work this problem by both the Punnett square method and the forked-line method.
16. While *vermilion* is X-linked in *Drosophila* and causes eye color to be bright red, *brown* is an autosomal recessive mutation that causes the eye to be brown. Flies carrying both mutations lose all pigmentation and are white-eyed. Predict the F_1 and F_2 results of the following crosses:
(a) vermilion females × brown males
(b) brown females × vermilion males
(c) white females × wild males
17. In pigs, coat color may be sandy, red, or white. A geneticist spent several years mating true-breeding pigs of all different color combinations, even obtaining true-breeding lines from different parts of the country. For crosses 1 and 4 in the following table, she encountered a major problem: her computer crashed and she lost the F_2 data. She nevertheless persevered and, using the limited data shown here, was able to predict the mode of inheritance and the number of genes involved, as well as to assign genotypes to each coat color. Attempt to duplicate her analysis, based on the available data generated from the crosses shown.

Cross	P_1	F_1	F_2
1	sandy × sandy	All red	Data lost
2	red × sandy	All red	3/4 red:1/4 sandy
3	sandy × white	All sandy	3/4 sandy:1/4 white
4	white × red	All red	Data lost

When you have formulated a hypothesis to explain the mode of inheritance and assigned genotypes to the respective coat colors, predict the outcomes of the F_2 generations where the data were lost.

18. A geneticist from an alien planet that prohibits genetic research brought with him two true-breeding lines of frogs. One frog line croaks by *uttering* "rib-it rib-it" and has purple eyes. The other frog line croaks by *muttering* "knee-deep knee-deep" and has green eyes. He mated the two frog lines, producing F_1 frogs that were all utterers with blue eyes. A large F_2 generation then yielded the following ratios:

27/64 blue, utterer
12/64 green, utterer
9/64 blue, mutterer
9/64 purple, utterer
4/64 green, mutterer
3/64 purple, mutterer

(a) How many total gene pairs are involved in the inheritance of both eye color and croaking?
(b) Of these, how many control eye color, and how many control croaking?
(c) Assign gene symbols for all phenotypes, and indicate the genotypes of the P_1, F_1, and F_2 frogs.
(d) After many years, the frog geneticist isolated true-breeding lines of all six F_2 phenotypes. Indicate the F_1 and F_2 phenotypic ratios of a cross between a blue, mutterer and a purple, utterer.

19. In another cross, the frog geneticist from Problem 18 mated two purple, utterers with the results shown here. What were the genotypes of the parents?

9/16 purple, utterers
3/16 purple, mutterers
3/16 green, utterers
1/16 green, mutterers

20. In cattle, coats may be solid white, solid black, or black-and-white spotted. When true-breeding solid whites are mated with true-breeding solid blacks, the F_1 generation consists of all solid white individuals. After many $F_1 \times F_1$ matings, the following ratio was observed in the F_2 generation:

12/16 solid white
3/16 black-and-white spotted
1/16 solid black

Explain the mode of inheritance governing coat color by determining how many gene pairs are involved and which genotypes yield which phenotypes. Is it possible to isolate a true-breeding strain of black-and-white spotted cattle? If so, what genotype would they have? If not, explain why not.

21. In the guinea pig, a locus controlling coat color may be occupied by any of four alleles: *C* (full color), c^k (sepia), c^d (cream), or c^a (albino). A progressive order of dominance exists among these alleles when they are present heterozygously: $C > c^k > c^d > c^a$. Determine the genotype of each individual, and predict the phenotypic ratios of the offspring in the following crosses:
(a) sepia × cream, where both had an albino parent

(b) sepia × cream, where the sepia individual had an albino parent and the cream individual had two sepia parents

(c) sepia × cream, where the sepia individual had two full-color parents and the cream individual had two sepia parents

(d) sepia × cream, where the sepia individual had two full-color parents and the cream individual had two full-color parents

22. In parakeets, two autosomal genes (located on different chromosomes) control the production of feather pigment. Gene *B* controls the production of blue pigment, and gene *Y* controls the production of yellow pigment. Known recessive mutations in each gene result in the loss of pigment synthesis. Two green parakeets are mated and produce green, blue, yellow, and albino progeny.
    (a) Based on this information, explain the pattern of inheritance. Be sure to include the genotypes of the green parents and all four phenotypic classes, as well as the fraction of total progeny that each phenotypic class represents in your answer.
    (b) The parental (green) parakeets are the progeny of a cross between two true-breeding strains. What two types of crosses between true-breeding strains could have produced the green parents? Indicate the genotypes and phenotypes for each cross.

23. Below are three pedigrees. For each, consider whether it could or could not be consistent with an X-linked recessive trait. In a sentence or two, indicate why or why not.

See **Now Solve This** on page 75.

(a)

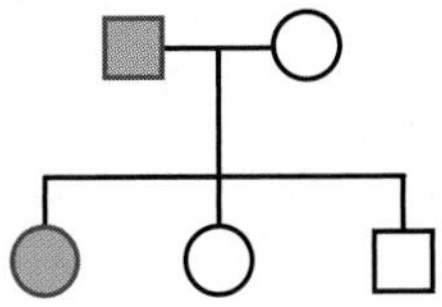

(b)

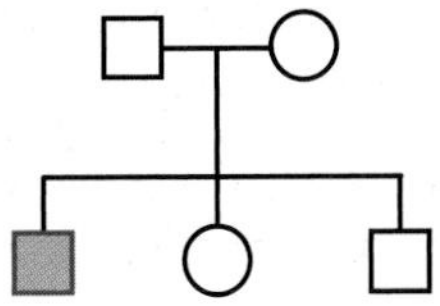

(c)

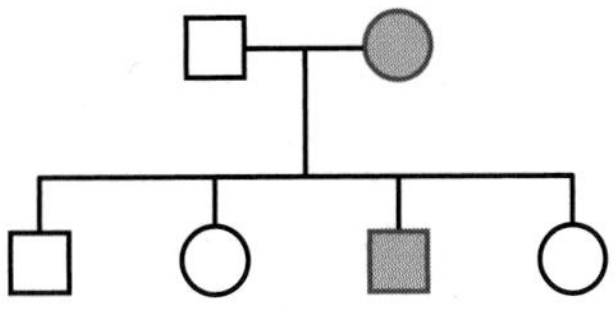

24. Consider the following three pedigrees, all involving the same human trait:

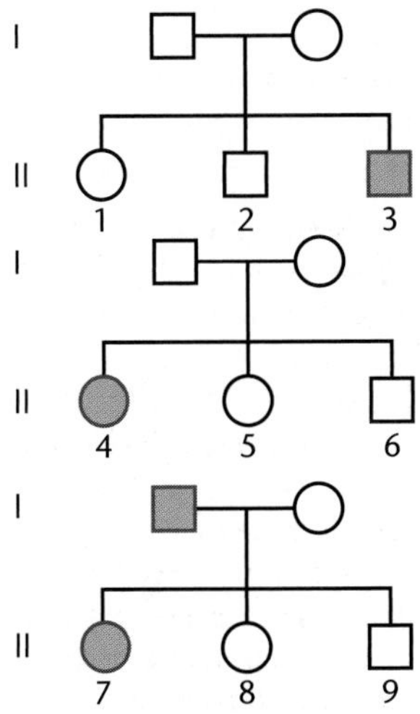

(a) Which sets of conditions, if any, can be excluded?
    dominant and X-linked
    dominant and autosomal
    recessive and X-linked
    recessive and autosomal

(b) For any set of conditions that you excluded, indicate the *single individual* in generation II (1–9) that was instrumental in your decision to exclude that condition. If none were excluded, answer "none apply."

(c) Given your conclusions in parts (a) and (b), indicate the *genotype* of individuals II-1, II-6, and II-9. If more than one possibility applies, list all possibilities. Use the symbols *A* and *a* for the genotypes.

25. Three autosomal recessive mutations in *Drosophila*, all with tan eye color (*r1*, *r2*, and *r3*), are independently isolated and subjected to complementation analysis. Of the results shown here, which, if any, are alleles of one another? Predict the results of the cross that is not shown—that is, *r2* × *r3*.

Cross 1: *r1* × *r2* ⟶	F_1: all wild-type eyes
Cross 2: *r1* × *r3* ⟶	F_1: all tan eyes

26. Labrador retrievers may be black, brown, or golden in color (see the chapter opening photograph on page 60). Although each color may breed true, many different outcomes occur if numerous litters are examined from a variety of matings, where the parents are not necessarily true-breeding. The following results show some of the possibilities. Propose a mode of inheritance that is consistent with these data, and indicate the corresponding genotypes of the parents in each mating. Indicate as well the genotypes of dogs that breed true for each color.

(a) black	×	brown	⟶	all black
(b) black	×	brown	⟶	1/2 black 1/2 brown
(c) black	×	brown	⟶	3/4 black 1/4 golden
(d) black	×	golden	⟶	all black
(e) black	×	golden	⟶	4/8 golden 3/8 black 1/8 brown
(f) black	×	golden	⟶	2/4 golden 1/4 black 1/4 brown
(g) brown	×	brown	⟶	3/4 brown 1/4 golden
(h) black	×	black	⟶	9/16 black 4/16 golden 3/16 brown

27. In rabbits, a series of multiple alleles controls coat color in the following way: *C* is dominant to all other alleles and causes full color. The chinchilla phenotype is due to the c^{ch} allele, which is dominant to all alleles other than *C*. The c^h allele, dominant only to c^a (albino), results in the Himalayan coat color. Thus, the order of dominance is $C > c^{ch} > c^h > c^a$. For each of the following three cases, the phenotypes of the P_1 generations of two crosses are shown, as well as the phenotype of one member of the F_1 generation.

	P₁ Phenotypes		F₁ Phenotypes
(a)	Himalayan × Himalayan	⟶	albino
			× ⟶ ??
	full color × albino	⟶	chinchilla
(b)	albino × chinchilla	⟶	albino
			× ⟶ ??
	full color × albino	⟶	full color
(c)	chinchilla × albino	⟶	Himalayan
			× ⟶ ??
	full color × albino	⟶	Himalayan

Determine the genotypes of the P_1 generation and the F_1 offspring for each case, and predict the results of making each cross between F_1 individuals as shown.

See **Now Solve This** on page 65.

Full color Chinchilla

Himalayan Albino

28. Horses can be cremello (a light cream color), chestnut (a reddish brown color), or palomino (a golden color with white in the horse's tail and mane).

Chestnut

Palomino

Cremello

Of these phenotypes, only palominos never breed true. The following results have been observed:

cremello × palomino	⟶	1/2 cremello 1/2 palomino
chestnut × palomino	⟶	1/2 chestnut 1/2 palomino
palomino × palomino	⟶	1/4 chestnut 1/2 palomino 1/4 cremello

(a) From these results, determine the mode of inheritance by assigning gene symbols and indicating which genotypes yield which phenotypes.
(b) Predict the F_1 and F_2 results of many initial matings between cremello and chestnut horses.

29. Pigment in the mouse is produced only when the *C* allele is present. Individuals of the *cc* genotype have no color. If color is present, it may be determined by the *A* and *a* alleles. *AA* or *Aa* results in agouti color, whereas *aa* results in black coats.
(a) What F_1 and F_2 genotypic and phenotypic ratios are obtained from a cross between *AACC* and *aacc* mice?
(b) In the three crosses shown here between agouti females whose genotypes were unknown and males of the *aacc* genotype, what are the genotypes of the female parents for each of the following phenotypic ratios?

(1)	(2)	(3)
8 agouti 8 colorless	9 agouti 10 black	4 agouti 5 black 10 colorless

30. Five human matings numbered 1–5 are shown in the following table. Included are both maternal and paternal phenotypes for ABO and MN blood-group antigen status.

Parental Phenotypes	Offspring
(1) A, M × A, N	(a) A, N
(2) B, M × B, M	(b) O, N
(3) O, N × B, N	(c) O, MN
(4) AB, M × O, N	(d) B, M
(5) AB, MN × AB, MN	(e) B, MN

Each mating resulted in one of the five offspring shown to the right (a–e). Match each offspring with one correct set of parents, using each parental set only once. Is there more than one set of correct answers?

31. In Dexter and Kerry cattle, animals may be polled (hornless) or horned. The Dexter animals have short legs, whereas the Kerry animals have long legs. When many offspring were obtained from matings between polled Kerrys and horned Dexters, half were found to be polled Dexters and half polled Kerrys. When these two types of F_1 cattle were mated to one another, the following F_2 data were obtained:

3/8 polled Dexters	3/8 polled Kerrys
1/8 horned Dexters	1/8 horned Kerrys

A geneticist was puzzled by these data and interviewed farmers who had bred these cattle for decades. She learned that Kerrys were true-breeding. Dexters, on the other hand, were not true-breeding and never produced as many offspring as Kerrys. Provide a genetic explanation for these observations.

Kerry cow

Dexter bull

32. What genetic criteria distinguish a case of extranuclear inheritance from (a) a case of Mendelian autosomal inheritance; (b) a case of X-linked inheritance?
33. The specification of the anterior–posterior axis in *Drosophila* embryos is initially controlled by various gene products that are synthesized and stored in the mature egg following oogenesis. Mutations in these genes result in abnormalities of the axis during embryogenesis, illustrating maternal effect. How do such mutations vary from those involved in organelle heredity that illustrate extranuclear inheritance? Devise a set of parallel crosses and expected outcomes involving mutant genes that contrast maternal effect and organelle heredity.
34. The maternal-effect mutation *bicoid* (*bcd*) is recessive. In the absence of the bicoid protein product, embryogenesis is not completed. Consider a cross between a female heterozygous for the bicoid mutation (bcd^+/bcd^-) and a homozygous male (bcd^-/bcd^-).
    (a) How is it possible for a male homozygous for the mutation to exist?
    (b) Predict the outcome (normal vs. failed embryogenesis) in the F_1 and F_2 generations of the cross described.
35. *Chlamydomonas*, a eukaryotic green alga, is sensitive to the antibiotic erythromycin, which inhibits protein synthesis in prokaryotes. There are two mating types in this alga, mt^+ and mt^-. If an mt^+ cell sensitive to the antibiotic is crossed with an mt^- cell that is resistant, all progeny cells are sensitive. The reciprocal cross (mt^+ resistant and mt^- sensitive) yields all resistant progeny cells. Assuming that the mutation for resistance is in the chloroplast DNA, what can you conclude from the results of these crosses?
36. Below is a partial pedigree of hemophilia in the British Royal Family descended from Queen Victoria, who is believed to be the original "carrier" in this pedigree. Analyze the pedigree and indicate which females are also certain to be carriers. What is the probability that Princess Irene is a carrier?

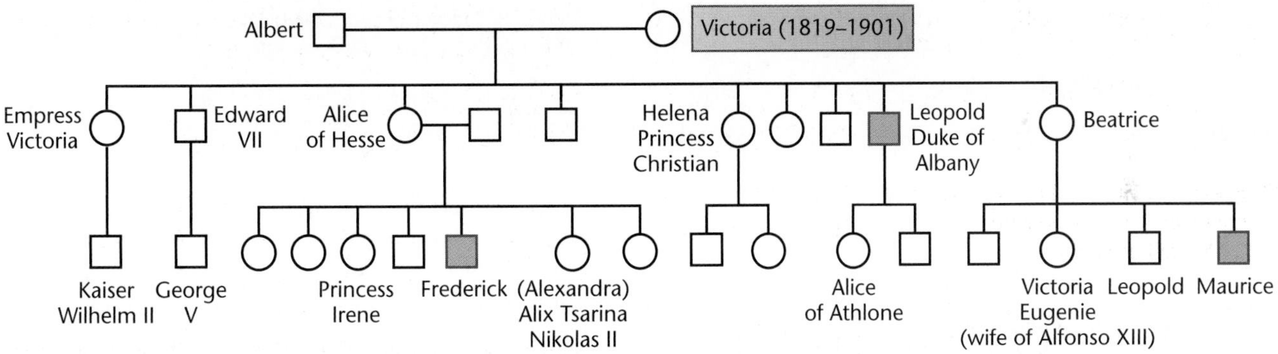

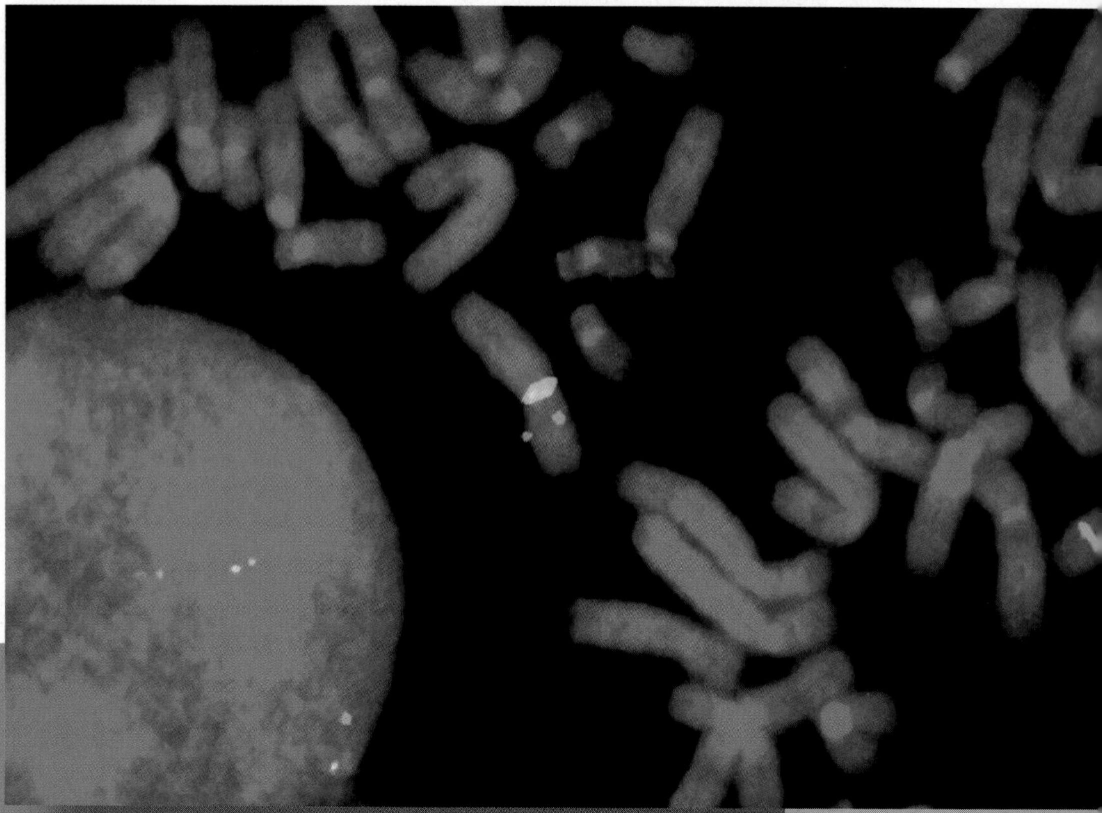

Human X chromosomes highlighted using fluorescence in situ *hybridization (FISH), a method in which specific probes bind to specific sequences of DNA. The probe producing green fluorescence binds to DNA at the centromere of X chromosomes. The probe producing red fluorescence binds to the DNA sequence of the X-linked Duchenne muscular dystrophy (DMD) gene.*

5

Sex Determination and Sex Chromosomes

Chapter Concepts

- Sexual reproduction, which greatly enhances genetic variation within species, requires mechanisms that result in sexual differentiation.
- A wide variety of genetic mechanisms lead to sexual dimorphism.
- Often, specific genes, usually on a single chromosome, cause maleness or femaleness during development.
- In humans, the presence of extra X or Y chromosomes beyond the diploid number may be tolerated but often leads to syndromes demonstrating distinctive phenotypes.
- While segregation of sex-determining chromosomes should theoretically lead to a one-to-one sex ratio of males to females, in humans the actual ratio greatly favors males at conception.
- In mammals, females inherit two X chromosomes compared to one in males, but the extra genetic information in females is compensated for by random inactivation of one of the X chromosomes early in development.
- In some reptilian species, temperature during incubation of eggs determines the sex of offspring.

In the biological world, a wide range of reproductive modes and life cycles are observed. Some organisms are entirely asexual, displaying no evidence of sexual reproduction. Some organisms alternate between short periods of sexual reproduction and prolonged periods of asexual reproduction. In most diploid eukaryotes, however, sexual reproduction is the only natural mechanism for producing new members of the species. Orderly transmission of genetic material from parents to offspring, and the resultant phenotypic variability, rely on the processes of segregation and independent assortment that occur during meiosis. Meiosis produces haploid gametes so that, following fertilization, the resulting offspring maintain the diploid number of chromosomes characteristic of their kind. Thus, meiosis ensures genetic constancy within members of the same species.

These events, seen in the perpetuation of all sexually reproducing organisms, depend ultimately on an efficient union of gametes during fertilization. In turn, successful fertilization depends on some form of sexual differentiation in the reproductive organisms. Even though it is not overtly evident, this differentiation occurs in organisms as low on the evolutionary scale as bacteria and single-celled eukaryotic algae. In more complex forms of life, the differentiation of the sexes is more evident as phenotypic dimorphism of males and females. The ancient symbol for iron and for Mars, depicting a shield and spear (♂), and the ancient symbol for copper and for Venus, depicting a mirror (♀), have also come to symbolize maleness and femaleness, respectively.

Dissimilar, or **heteromorphic, chromosomes,** such as the XY pair in mammals, characterize one sex or the other in a wide range of species, resulting in their label as **sex chromosomes.** Nevertheless, it is genes, rather than chromosomes, that ultimately serve as the underlying basis of **sex determination.** As we will see, some of these genes are present on sex chromosomes, but others are autosomal. Extensive investigation has revealed a wide variation in sex-chromosome systems—even in closely related organisms—suggesting that mechanisms controlling sex determination have undergone rapid evolution many times in the history of life.

In this chapter, we review some representative modes of sexual differentiation by examining the life cycles of three model organisms often studied in genetics: the green alga *Chlamydomonas*; the maize plant, *Zea mays*; and the nematode (roundworm), *Caenorhabditis elegans*. These organisms contrast the different roles that sexual differentiation plays in the lives of diverse organisms. Then, we delve more deeply into what is known about the genetic basis for the determination of sexual differences, with a particular emphasis on two organisms: our own species, representative of mammals; and *Drosophila*, on which pioneering sex-determining studies were performed.

How Do We Know?

In this chapter, we will focus on sex differentiation, sex chromosomes, and genetic mechanisms involved in sex determination. At the same time, we will find many opportunities to consider the methods and reasoning by which much of this information was acquired. From the explanations given in the chapter, you should answer the following fundamental questions:

1. How do we know that in humans the X chromosomes play no role in sex determination, while the Y chromosome causes maleness and its absence causes femaleness?
2. How did we learn that, although the sex ratio at birth in humans favors males slightly, the sex ratio at conception favors them much more?
3. How do we know that X chromosomal inactivation of either the paternal or maternal homolog is a random event during early development in mammalian females?
4. How do we know that *Drosophila* utilizes a different sex determination mechanism than mammals, even though it has the same sex chromosome compositions in males and females?

5.1 Life Cycles Depend on Sexual Differentiation

In describing sexual dimorphism (differences between males and females) in multicellular animals, biologists distinguish between **primary sexual differentiation,** which involves only the gonads, where gametes are produced, and **secondary sexual differentiation,** which involves the overall appearance of the organism, including clear differences in such organs as mammary glands and external genitalia as well as in nonreproductive organs. In plants and animals, the terms **unisexual, dioecious,** and **gonochoric** are equivalent; they all refer to an individual containing only male *or* only female reproductive organs. Conversely, the terms **bisexual, monoecious,** and **hermaphroditic** refer to individuals containing both male *and* female reproductive organs, a common occurrence in both the plant and animal kingdoms. These organisms can produce both eggs and sperm. The term **intersex** is usually reserved for individuals of an intermediate sexual condition; these are most often sterile.

Chlamydomonas

The life cycle of the green alga *Chlamydomonas*, shown in Figure 5–1, p. 94, is representative of organisms exhibiting only infrequent periods of sexual reproduction. Such organisms spend most of their life cycle in a haploid phase, asexually producing daughter cells by mitotic divisions. However, under unfavorable nutrient conditions, such as nitrogen depletion, certain daughter cells function as gametes, joining together in fertilization. Following fertilization, a diploid zygote, which may withstand the unfavorable environment, is formed. When conditions change for the better, meiosis ensues and haploid vegetative cells are again produced.

In such species, there is little visible difference between the haploid vegetative cells that reproduce asexually and the haploid gametes that are involved in sexual reproduction.

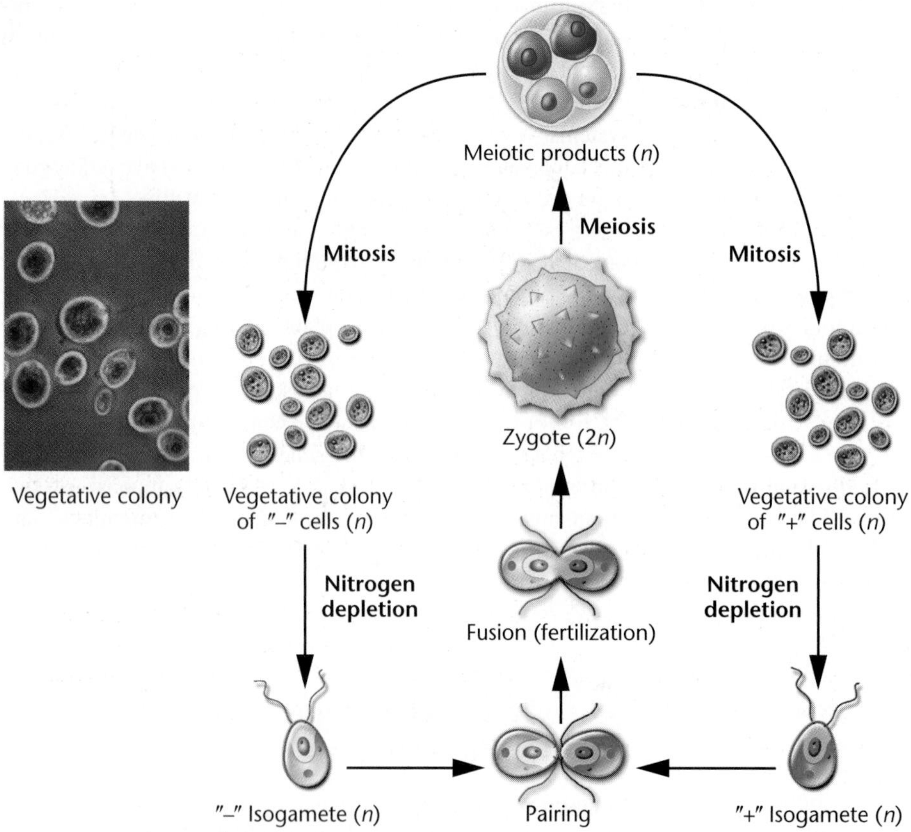

FIGURE 5–1 The life cycle of *Chlamydomonas*. Unfavorable conditions stimulate the formation of isogametes of opposite mating types that may fuse in fertilization. The resulting zygote undergoes meiosis, producing two haploid cells of each mating type. The photograph shows vegetative cells of this green alga.

Moreover, the two gametes that fuse during mating are not usually morphologically distinguishable from one another, which is why they are called **isogametes** (*iso*- means equal, or uniform). Species producing them are said to be **isogamous.**

In 1954, Ruth Sager and Sam Granik demonstrated that gametes in *Chlamydomonas* could be subdivided into two **mating types.** Working with clones derived from single haploid cells, they showed that cells from a given clone mate with cells from some but not all other clones. When they tested the mating abilities of large numbers of clones, all could be placed into one of two mating categories, either mt^+ or mt^- cells. "Plus" cells mate only with "minus" cells, and vice versa. Following fertilization and meiosis, the four haploid cells, or **zoospores,** produced were found to consist of two plus types and two minus types.

Further experimentation established that plus and minus cells differ chemically. When extracts are prepared from cloned *Chlamydomonas* cells (or their flagella) of one type and then added to cells of the opposite mating type, clumping, or agglutination, occurs. No such agglutination occurs if the extracts are added to cells of the mating type from which they were derived. These observations suggest that despite the morphological similarities between isogametes, they are differentiated chemically. Therefore, in this alga, a primitive means of sex differentiation exists, even though there is no morphological indication that such differentiation has occurred. Further research has pinpointed the *mt* locus to *Chlamydomonas* chromosome VI and has identified the gene that mediates the expression of the mating type, which is essential for cell fusion in response to nitrogen depletion.

Zea mays

The life cycles of many plants alternate between the haploid gametophyte stage and the diploid sporophyte stage (see Figure 2–12). The processes of meiosis and fertilization link the two phases during the life cycle. The relative amount of time spent in the two phases varies between the major plant groups. In some nonseed plants, such as mosses, the haploid gametophyte phase and the morphological structures representing this stage predominate. The reverse is true in seed plants.

Maize (*Zea mays*), familiar to you as corn, exemplifies a monoecious seed plant in which the sporophyte phase and the morphological structures representing that phase predominate during the life cycle. Both male and female structures are present on the adult plant. Thus, sex determination occurs differently in different tissues of the same organism, as shown in the life cycle of this plant (Figure 5–2). The stamens, which collectively constitute the tassel, produce diploid microspore mother cells, each of which undergoes meiosis and gives rise to four haploid microspores. Each haploid microspore in turn develops into a mature male microgametophyte—the pollen grain—which contains two haploid sperm nuclei.

Equivalent female diploid cells, known as megaspore mother cells, are produced in the pistil of the sporophyte. Following meiosis, only one of the four haploid megaspores survives. It usually divides mitotically three times, producing a total of eight haploid nuclei enclosed in the embryo sac. Two

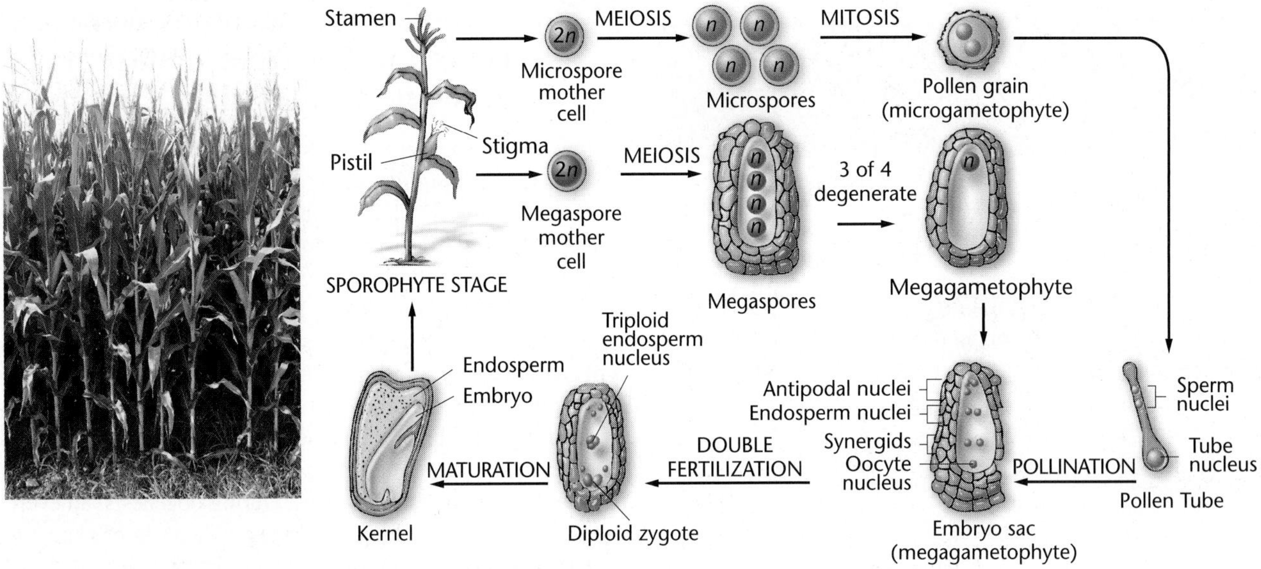

FIGURE 5–2 The life cycle of maize (*Zea mays*). The diploid sporophyte bears stamens and pistils that give rise to haploid microspores and megaspores, which develop into the pollen grain and the embryo sac that ultimately house the sperm and oocyte, respectively. Following fertilization, the embryo develops within the kernel and is nourished by the endosperm. Germination of the kernel gives rise to a new sporophyte (the mature corn plant), and the cycle repeats itself.

of these nuclei unite near the center of the embryo sac, becoming the endosperm nuclei. At the micropyle end of the sac, where the sperm enters, sit three other nuclei: the oocyte nucleus and two synergids. The remaining three, antipodal nuclei, cluster at the opposite end of the embryo sac.

Pollination occurs when pollen grains make contact with the silks (or stigma) of the pistil and develop long pollen tubes that grow toward the embryo sac. When contact is made at the micropyle, the two sperm nuclei enter the embryo sac. One sperm nucleus unites with the haploid oocyte nucleus, and the other sperm nucleus unites with two endosperm nuclei. This process, known as *double fertilization*, results in the diploid zygote nucleus and the triploid endosperm nucleus, respectively. Each ear of corn can contain as many as 1000 of these structures, each of which develops into a single kernel. Each kernel, if allowed to germinate, gives rise to a new plant, the *sporophyte*.

The mechanism of sex determination and differentiation in a monoecious plant such as *Zea mays*, where the tissues that form the male and female gametes have the same genetic constitution, was difficult to unravel at first. However, the discovery of a large number of mutant genes that disrupt normal tassel and pistil formation supports the concept that normal products of these genes play an important role in sex determination by affecting the differentiation of male or female tissue in several ways.

For example, mutant genes that cause sex reversal provide valuable information. When homozygous, all mutations classified as *tassel seed (ts)* interfere with tassel production and induce the formation of female structures instead. Thus, a single gene can cause a normally monoecious plant to become exclusively female. On the other hand, the recessive mutations *silkless (sk)* and *barren stalk (ba)* interfere with the development of the pistil, resulting in plants with only male-functioning reproductive organs.

Data gathered from studies of these and other mutants suggest that the products of many wild-type alleles of these genes interact in controlling sex determination. During development, certain cells are "directed" to become male or female structures. Following sexual differentiation into either male or female structures, male or female gametes are produced.

Caenorhabditis elegans

The nematode worm *Caenorhabditis elegans* [*C. elegans*; Figure 5–3(a), p. 96] has become a popular organism in genetic studies, particularly for investigating the genetic control of development. Its usefulness is based on the fact that adults consist of approximately 1000 cells, the precise lineage of which can be traced back to specific embryonic origins. There are two sexual phenotypes in these worms: males, which have only testes, and hermaphrodites, which contain both testes and ovaries. During larval development of hermaphrodites, testes form that produce sperm, which is stored. Ovaries are also produced, but oogenesis does not occur until the adult stage is reached several days later. The eggs that are produced are fertilized by the stored sperm in a process of self-fertilization.

The outcome of this process is quite interesting [Figure 5–3(b)]. The vast majority of organisms that result are hermaphrodites, like the parental worm; less than 1 percent of the offspring are males. As adults, males can mate with hermaphrodites, producing about half male and half hermaphrodite offspring.

The genetic signal that determines maleness in contrast to hermaphroditic development is provided by genes located on

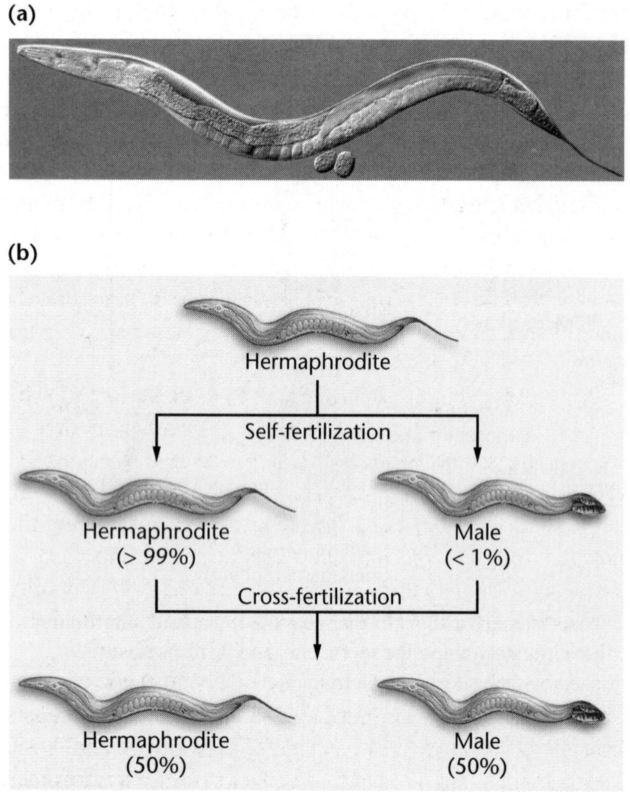

FIGURE 5–3 (a) Photomicrograph of a hermaphroditic nematode, *C. elegans*; (b) the outcomes of self-fertilization in a hermaphrodite, and a mating of a hermaphrodite and a male worm.

both the X chromosome and autosomes. *C. elegans* lacks a Y chromosome altogether—hermaphrodites have two X chromosomes, while males have only one X chromosome. It is believed that the ratio of X chromosomes to the number of sets of autosomes ultimately determines the sex of these worms. A ratio of 1.0 (two X chromosomes and two copies of each autosome) results in hermaphrodites, and a ratio of 0.5 results in males. The absence of a heteromorphic Y chromosome is not uncommon in organisms.

ESSENTIAL POINT

Sexual reproduction depends on the differentiation of male and female structures responsible for the production of male and female gametes, which in turn is controlled by specific genes, most often housed on specific sex chromosomes.

NOW SOLVE THIS

Problem 23 on page 109 involves a case in which environmental factors control sex determination in a marine worm. You are asked to devise an experiment to elucidate the original findings.

Hint: An obvious approach would be to attempt to isolate the unknown factor affecting sex determination and devise experiments making use of it. You might also use a genetic approach involving mutations.

5.2 X and Y Chromosomes Were First Linked to Sex Determination Early in the Twentieth Century

How sex is determined has long intrigued geneticists. In 1891, H. Henking identified a nuclear structure in the sperm of certain insects, which he labeled the X-body. Several years later, Clarence McClung showed that some of the sperm in grasshoppers contain an unusual genetic structure, which he called a *heterochromosome*, but the remainder of the sperm lack such a structure. He mistakenly associated the presence of the heterochromosome with the production of male progeny. In 1906, Edmund B. Wilson clarified Henking and McClung's findings when he demonstrated that female somatic cells in the butterfly *Protenor* contain 14 chromosomes, including two X chromosomes. During oogenesis, an even reduction occurs, producing gametes with seven chromosomes, including one X chromosome. Male somatic cells, on the other hand, contain only 13 chromosomes, including one X chromosome. During spermatogenesis, gametes are produced containing either six chromosomes, without an X, or seven chromosomes, one of which is an X. Fertilization by X-bearing sperm results in female offspring, and fertilization by X-deficient sperm results in male offspring [Figure 5–4(a)].

The presence or absence of the X chromosome in male gametes provides an efficient mechanism for sex determination in this species and also produces a 1:1 sex ratio in the resulting offspring. This mechanism, now called the **XX/XO,** or ***Protenor*, mode of sex determination,** depends on the random

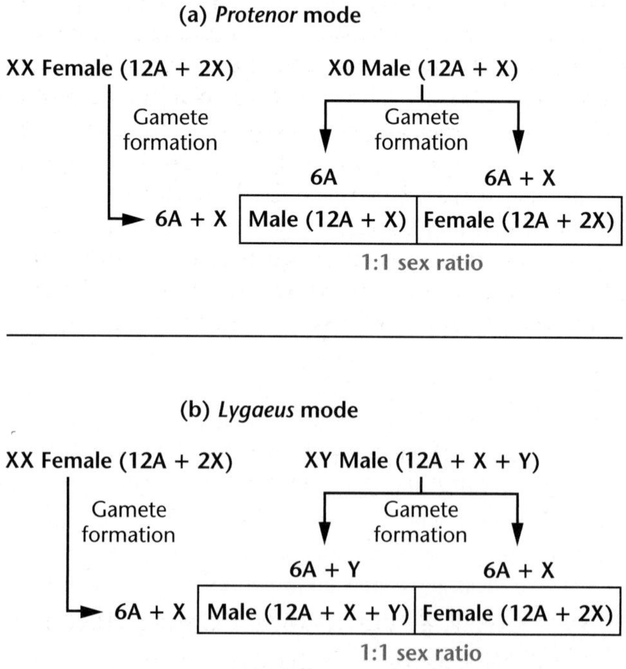

FIGURE 5–4 (a) The *Protenor* mode of sex determination where the heterogametic sex (the male in this example) is X0 and produces gametes with or without the X chromosome; (b) the *Lygaeus* mode of sex determination, where the heterogametic sex (again, the male in this example) is XY and produces gametes with either an X or a Y chromosome.

distribution of the X chromosome into one-half of the male gametes during segregation. As we saw earlier, *C. elegans* exhibits this system of sex determination.

Wilson also experimented with the milkweed bug *Lygaeus turcicus*, in which both sexes have 14 chromosomes. Twelve of these are autosomes (A). In addition, the females have two X chromosomes, while the males have only a single X and a smaller heterochromosome labeled the **Y chromosome.** Females in this species produce only gametes of the (6A + X) constitution, but males produce two types of gametes in equal proportions, (6A + X) and (6A + Y). Therefore, following random fertilization, equal numbers of male and female progeny will be produced with distinct chromosome complements. This mode of sex determination is called the ***Lygaeus,*** or **XX/XY, mode of sex determination** [Figure 5–4(b)].

In *Protenor* and *Lygaeus* insects, males produce unlike gametes. As a result, they are described as the **heterogametic sex,** and in effect, their gametes ultimately determine the sex of the progeny in those species. In such cases, the female, who has like sex chromosomes, is the **homogametic sex,** producing uniform gametes with regard to chromosome numbers and types.

The male is not always the heterogametic sex. In some organisms, the female produces unlike gametes, exhibiting either the *Protenor* XX/XO or *Lygaeus* XX/XY mode of sex determination. Examples include certain moths and butterflies; most birds; some fish; reptiles; amphibians; and at least one species of plants (*Fragaria orientalis*). To immediately distinguish situations in which the female is the heterogametic sex, some geneticists use the notation **ZZ/ZW,** where ZW is the heterogamous female, instead of the XX/XY notation.

Geneticists' experience with fowl (chickens) illustrates the difficulty of establishing which sex is heterogametic and whether the *Protenor* or *Lygaeus* mode is operable. Although genetic evidence supported the hypothesis that the female is the heterogametic sex, the cytological identification of the sex chromosome was not accomplished until 1961, because of the large number of chromosomes (78) characteristic of chickens. When the sex chromosomes were finally identified, the female was shown to contain an unlike chromosome pair including a heteromorphic chromosome (the W chromosome). Thus, in fowl, the female is indeed heterogametic and is characterized by the *Lygaeus* type of sex determination.

ESSENTIAL POINT

Specific sex chromosomes contain genetic information that controls sex determination and sexual differentiation.

5.3 The Y Chromosome Determines Maleness in Humans

The first attempt to understand sex determination in our own species occurred almost 100 years ago and involved the visual examination of chromosomes in dividing cells. Efforts were made to accurately determine the diploid chromosome number of humans, but because of the relatively large number of chromosomes, this proved to be quite difficult. In 1912, H. von Winiwarter counted 47 chromosomes in a spermatogonial metaphase preparation. It was believed that the sex-determining mechanism in humans was based on the presence of an extra chromosome in females, who were thought to have 48 chromosomes. However, in the 1920s, Theophilus Painter counted between 45 and 48 chromosomes in cells of testicular tissue and also discovered the small Y chromosome, which is now known to occur only in males. In his original paper, Painter favored 46 as the diploid number in humans, but he later concluded incorrectly that 48 was the chromosome number in both males and females.

For 30 years, this number was accepted. Then, in 1956, Joe Hin Tjio and Albert Levan discovered a better way to prepare chromosomes for viewing. This improved technique led to a strikingly clear demonstration of metaphase stages showing that 46 was indeed the human diploid number. Later that same year, C. E. Ford and John L. Hamerton, also working with testicular tissue, confirmed this finding. The familiar karyotypes of a human male (Figure 2–4) illustrate the difference in size between the human X and Y chromosomes.

Of the normal 23 pairs of human chromosomes, one pair was shown to vary in configuration in males and females. These two chromosomes were designated the X and Y sex chromosomes. The human female has two X chromosomes, and the human male has one X and one Y chromosome.

We might believe that this observation is sufficient to conclude that the Y chromosome determines maleness. However, several other interpretations are possible. The Y could play no role in sex determination; the presence of two X chromosomes could cause femaleness; or maleness could result from the lack of a second X chromosome. The evidence that clarified which explanation was correct came from the study of the effects of human sex-chromosome variations, described in the following section. As such investigations revealed, the Y chromosome does indeed determine maleness in humans.

Klinefelter and Turner Syndromes

Around 1940, scientists identified two human abnormalities characterized by aberrant sexual development, **Klinefelter syndrome (47,XXY)** and **Turner syndrome (45,X).*** Individuals with Klinefelter syndrome are generally tall and have long arms and legs and large hands and feet. They usually have genitalia and internal ducts that are male, but their testes are rudimentary and fail to produce sperm. At the same time, feminine sexual development is not entirely suppressed. Slight enlargement of the breasts (gynecomastia) is common, and the hips are often rounded. This ambiguous sexual development, referred to as intersexuality, can lead to abnormal social development. Intelligence is often below the normal range as well.

In Turner syndrome, the affected individual has female external genitalia and internal ducts, but the ovaries are rudimentary. Other characteristic abnormalities include short stature (usually

*Although the possessive form of the names of eponymous syndromes is sometimes used (e.g., Klinefelter's) the current preference is to use the non-possessive form.

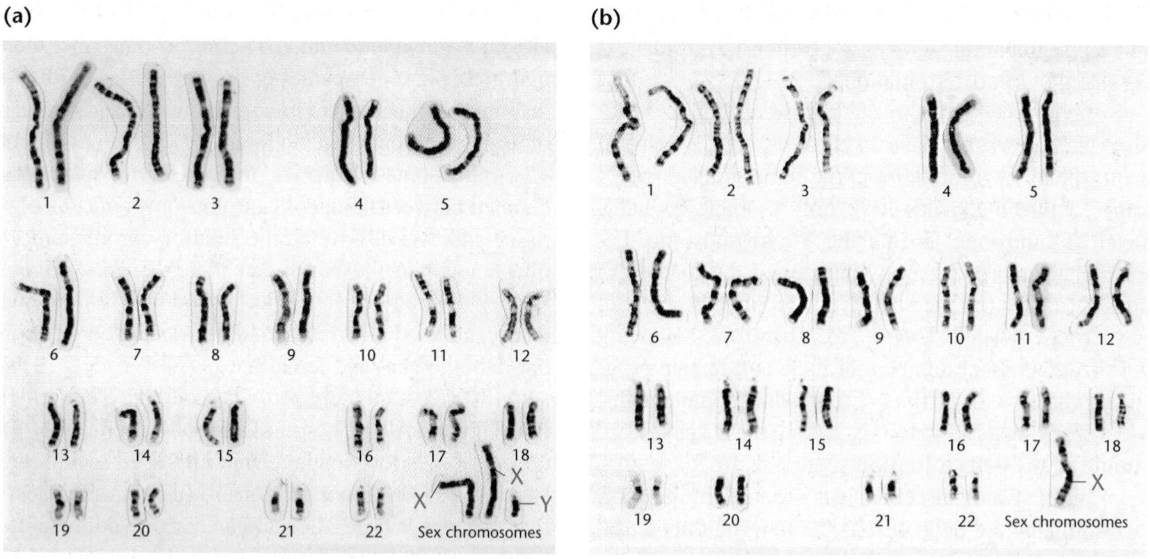

FIGURE 5–5 The karyotypes of individuals with (a) Klinefelter syndrome (47,XXY) and (b) Turner syndrome (45,X).

under 5 feet), skin flaps on the back of the neck, and underdeveloped breasts. A broad, shieldlike chest is sometimes noted. Intelligence is usually normal.

In 1959, the karyotypes of individuals with these syndromes were determined to be abnormal with respect to the sex chromosomes. Individuals with Klinefelter syndrome have more than one X chromosome. Most often they have an XXY complement in addition to 44 autosomes [Figure 5–5(a)], which is why people with this karyotype are designated 47,XXY. Individuals with Turner syndrome most often have only 45 chromosomes, including just a single X chromosome; thus, they are designated 45,X [Figure 5–5(b)]. Note the convention used in designating these chromosome compositions. The number states the total number of chromosomes present, and the information after the comma indicates the deviation from the normal diploid content. Both conditions result from **nondisjunction,** the failure of the X chromosomes to segregate properly during meiosis (nondisjunction is described in Chapter 6 and illustrated in Figure 6–1).

These Klinefelter and Turner karyotypes and their corresponding sexual phenotypes led scientists to conclude that the Y chromosome determines maleness in humans. In its absence, the person's sex is female, even if only a single X chromosome is present. The presence of the Y chromosome in the individual with Klinefelter syndrome is sufficient to determine maleness, even though male development is not complete. Similarly, in the absence of a Y chromosome, as in the case of individuals with Turner syndrome, no masculinization occurs. Note that we cannot conclude anything regarding sex determination under circumstances where a Y chromosome is present without an X because Y-containing human embryos lacking an X chromosome (designated 45,Y) do not survive.

Klinefelter syndrome occurs in about 1 of every 660 male births. The karyotypes **48,XXXY, 48,XXYY, 49,XXXXY,** and **49,XXXYY** are similar phenotypically to 47,XXY, but manifestations are often more severe in individuals with a greater number of X chromosomes.

Turner syndrome can also result from karyotypes other than 45,X, including individuals called **mosaics,** whose somatic cells display two different genetic cell lines, each exhibiting a different karyotype. Such cell lines result from a mitotic error during early development, the most common chromosome combinations being **45,X/46,XY** and **45,X/46,XX.** Thus, an embryo that began life with a normal karyotype can give rise to an individual whose cells show a mixture of karyotypes and who exhibits varying aspects of this syndrome.

Turner syndrome is observed in about 1 in 2000 female births, a frequency much lower than that for Klinefelter syndrome. One explanation for this difference is the observation that a substantial majority of 45,X fetuses die *in utero* and are aborted spontaneously. Thus, a similar frequency of the two syndromes may occur at conception.

47,XXX Syndrome

The abnormal presence of three X chromosomes along with a normal set of autosomes (**47,XXX**) results in female differentiation. The highly variable syndrome that accompanies this genotype, often called **triplo-X,** occurs in about 1 of 1000 female births. Frequently, 47,XXX women are perfectly normal and may remain unaware of their abnormality in chromosome number unless a karyotype is done. In other cases, underdeveloped secondary sex characteristics, sterility, delayed development of language and motor skills, and mental retardation may occur. In rare instances, **48,XXXX** (tetra-X) and **49,XXXXX** (penta-X) karyotypes have been reported. The syndromes associated with these karyotypes are similar to but more pronounced than the 47,XXX syndrome. Thus, in many cases, the presence of additional X chromosomes appears to disrupt the delicate balance of genetic information essential to normal female development.

47,XYY Condition

Another human condition involving the sex chromosomes is **47,XYY.** Studies of this condition, in which the only deviation from diploidy is the presence of an additional Y chromosome in an otherwise normal male karyotype, were initiated in 1965 by Patricia Jacobs. She discovered that 9 of 315 males in a Scottish maximum security prison had the 47,XYY karyotype. These males were significantly above average in height and had been incarcerated as a result of antisocial (nonviolent) criminal acts. Of the nine males studied, seven were of subnormal intelligence, and all suffered personality disorders. Several other studies produced similar findings.

The possible correlation between this chromosome composition and criminal behavior piqued considerable interest, and extensive investigation of the phenotype and frequency of the 47,XYY condition in both criminal and noncriminal populations ensued. Above-average height (usually over 6 feet) and subnormal intelligence have been generally substantiated, and the frequency of males displaying this karyotype is indeed higher in penal and mental institutions compared with unincarcerated populations (see Table 5.1). A particularly relevant question involves the characteristics displayed by XYY males who are not incarcerated. The only nearly constant association is that such individuals are over 6 feet tall.

A study addressing this issue was initiated to identify 47,XYY individuals at birth and to follow their behavioral patterns during preadult and adult development. By 1974, the two investigators, Stanley Walzer and Park Gerald, had identified about 20 XYY newborns in 15,000 births at Boston Hospital for Women. However, they soon came under great pressure to abandon their research. Those opposed to the study argued that the investigation could not be justified and might cause great harm to individuals who displayed this karyotype. The opponents argued that (1) no association between the additional Y chromosome and abnormal behavior had been previously established in the population at large, and (2) "labeling" the individuals in the study might create a self-fulfilling prophecy. That is, as a result of participation in the study, parents, relatives, and friends might treat individuals identified as 47,XYY differently, ultimately producing the expected antisocial behavior. Despite the support of a government funding agency and the faculty at Harvard Medical School, Walzer and Gerald abandoned the investigation in 1975.

Since Walzer and Gerald's work, it has become clear that there are many XYY males present in the population who do not exhibit antisocial behavior and who lead normal lives. Therefore, we must conclude that there is a high, but not constant, correlation between the extra Y chromosome and the predisposition of these males to exhibit behavioral problems.

Sexual Differentiation in Humans

Once researchers had established that, in humans, it is the Y chromosome that houses genetic information necessary for maleness, they attempted to pinpoint a specific gene or genes capable of providing the "signal" responsible for sex determination. Before we delve into this topic, it is useful to consider how sexual differentiation occurs in order to better comprehend how humans develop into sexually dimorphic males and females. During early development, every human embryo undergoes a period when it is potentially hermaphroditic. By the fifth week of gestation, gonadal primordia (the tissues that will form the gonad) arise as a pair of **gonadal (genital) ridges** associated with each embryonic kidney. The embryo is potentially hermaphroditic because at this stage its gonadal phenotype is sexually indifferent—male or female reproductive structures cannot be distinguished, and the gonadal ridge tissue can develop to form male or female gonads. As development progresses, primordial germ cells migrate to these ridges, where an outer cortex and inner medulla form (*cortex* and *medulla* are the outer and inner tissues of an organ, respectively). The cortex is capable of developing into an ovary, while the medulla may develop into a testis. In addition, two sets of undifferentiated ducts called the Wolffian and Müllerian ducts exist in each embryo. Wolffian ducts differentiate into other organs of the male reproductive tract, while Müllerian ducts differentiate into structures of the female reproductive tract.

Because gonadal ridges can form either ovaries or testes, they are commonly referred to as **bipotential gonads.** What switch triggers gonadal ridge development into testes or ovaries? The presence or absence of a Y chromosome is the key. If cells of the ridge have an XY constitution, development of the medulla into a testis is initiated around the seventh week. However, in the absence of the Y chromosome, no male development occurs, the cortex of the ridge subsequently forms ovarian tissue, and the Müllerian duct forms oviducts (Fallopian tubes), uterus, cervix, and portions of the vagina. Depending on which pathway is initiated, parallel development of the appropriate male or female duct system then

TABLE 5.1 Frequency of XYY Individuals in Various Settings

Setting	Restriction	Number Studied	Number XYY	Frequency XYY
Control population	Newborns	28,366	29	0.10%
Mental–penal	No height restriction	4,239	82	1.93
Penal	No height restriction	5,805	26	0.44
Mental	No height restriction	2,562	8	0.31
Mental–penal	Height restriction	1,048	48	4.61
Penal	Height restriction	1,683	31	1.84
Mental	Height restriction	649	9	1.38

Source: Compiled from data presented in Hook, 1973, Tables 1–8. Copyright 1973 by the American Association for the Advancement of Science.

occurs, and the other duct system degenerates. If testes differentiation is initiated, the embryonic testicular tissue secretes hormones that are essential for continued male sexual differentiation. As we will discuss in the next section, the presence of a Y chromosome and the development of the testes also inhibit formation of female reproductive organs.

In females, as the twelfth week of fetal development approaches, the oogonia within the ovaries begin meiosis, and primary oocytes can be detected. By the twenty-fifth week of gestation, all oocytes become arrested in meiosis and remain dormant until puberty is reached some 10 to 15 years later. In males, on the other hand, primary spermatocytes are not produced until puberty is reached (see Figure 2–11).

The Y Chromosome and Male Development

The human Y chromosome, unlike the X, was long thought to be mostly blank genetically. It is now known that this is not true, even though the Y chromosome contains far fewer genes than does the X. Data from the Human Genome Project indicate that the Y chromosome has at least 75 genes, compared to 900–1400 genes on the X. Current analysis of these genes and regions with potential genetic function reveals that some have homologous counterparts on the X chromosome and others do not. For example, present on both ends of the Y chromosome are so-called **pseudoautosomal regions (PARs)** that share homology with regions on the X chromosome and synapse and recombine with it during meiosis. The presence of such a pairing region is critical to segregation of the X and Y chromosomes during male gametogenesis. The remainder of the chromosome, about 95 percent of it, does not synapse or recombine with the X chromosome. As a result, it was originally referred to as the *nonrecombining region of the Y* (*NRY*). More recently, researchers have designated this region as the **male-specific region of the Y (MSY).** As you will see, some portions of the MSY share homology with genes on the X chromosome, and others do not.

The human Y chromosome is diagrammed in Figure 5–6. The MSY is divided about equally between *euchromatic* regions, containing functional genes, and *heterochromatic* regions, lacking genes. Within euchromatin, adjacent to the PAR of the short arm of the Y chromosome, is a critical gene that controls male sexual development, called the ***sex-determining region Y (SRY).*** In humans, the absence of a Y chromosome almost always leads to female development; thus, this gene is absent from the X chromosome. At six to eight weeks of development, the *SRY* gene becomes active in XY embryos. *SRY* encodes a protein that causes the undifferentiated gonadal tissue of the embryo to form testes. This protein is called the **testis-determining factor (TDF).** *SRY* (or a closely related version) is present in all mammals thus far examined, indicative of its essential function throughout this diverse group of animals.

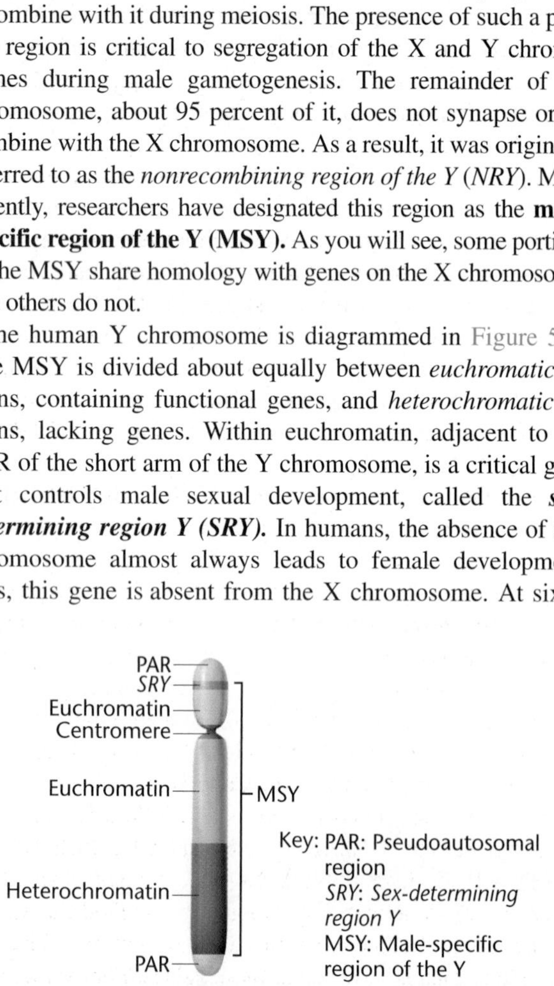

FIGURE 5–6 The regions of the human Y chromosome.

Our ability to identify the presence or absence of DNA sequences in rare individuals whose sex-chromosome composition does not correspond to their sexual phenotype has provided evidence that *SRY* is the gene responsible for male sex determination. For example, there are human males who have two X and no Y chromosomes. Often, attached to one of their X chromosomes is the region of the Y that contains *SRY*. There are also females who have one X and one Y chromosome. Their Y is almost always missing the *SRY* gene. These observations argue strongly in favor of the role of *SRY* in providing the primary signal for male development.

Further support of this conclusion involves an experiment using **transgenic mice.** These animals are produced from fertilized eggs injected with foreign DNA that is subsequently incorporated into the genetic composition of the developing embryo. In normal mice, a chromosome region designated *Sry* has been identified that is comparable to *SRY* in humans. When mouse DNA containing *Sry* is injected into normal XX mouse eggs, most of the offspring develop into males.

The question of how the product of this gene triggers development of embryonic gonadal tissue into testes rather than ovaries has been under investigation for a number of years. TDF is now believed to function as a *transcription factor*, a DNA-binding protein that interacts directly with the regulatory sequences of other genes to stimulate their expression. Thus, while TDF behaves as a master switch that controls other genes downstream in the process of sexual differentiation, identifying TDF target genes has been elusive. To date, no targets for TDF have been identified. However, one potential target for activation by TDF that has been extensively studied is the gene for **Müllerian inhibiting substance** (**MIS,** also called Müllerian inhibiting hormone, MIH, or anti-Müllerian hormone). Cells of the developing testes secrete MIS. As its name suggests, MIS protein causes regression (atrophy) of cells in the Müllerian duct. Degeneration of the duct prevents formation of the female reproductive tract.

Other autosomal genes, as studied in humans, are believed to be part of a cascade of genetic expression initiated by *SRY*. Examples include the *SOX9* gene and the *WT1* gene (on chromosome 11), originally identified as an oncogene associated with Wilms tumor, which affects the kidney and gonads. Another gene, *SF1*, is involved in the regulation of enzymes affecting steroid metabolism. In mice, this gene is initially active in both the male and female bisexual genital ridge, persisting until the point in development when testis formation is apparent. At that time, its expression persists in males but is extinguished in females. Establishment of the link between these various genes and sex determination has brought us closer to a complete understanding of how males and females arise in humans, but much work remains to be done.

Results of the Human Genome Project and findings by David Page and his many colleagues have now provided a reasonably complete picture of the MSY region of the human Y chromosome. Page has spearheaded the detailed study of the Y chromosome for the past several decades. The MSY consists of about 23 million base pairs (23 Mb) and can be divided into three regions. The first region is the *X-transposed region*. It comprises about 15 percent of the MSY and was originally derived from the X chromosome in the course of human evolution (about 3 to 4 million years ago). The X-transposed region is 99 percent identical to region Xq21 of the modern human X chromosome. Two genes, both with X chromosome homologs, are present in this region.

The second area is designated the *X-degenerative region*. Comprising about 20 percent of the MSY, this region contains DNA sequences that are even more distantly related to those present on the X chromosome. The X-degenerative region contains 27 single-copy genes and a number of *pseudogenes* (genes whose sequences have degenerated sufficiently during evolution to render them nonfunctional). As with the genes present in the X-transposed region, all share some homology with counterparts on the X chromosome. These 27 genetic units include 14 that are capable of being transcribed, and each is present as a single copy. One of these is the *SRY* gene, discussed earlier. Other X-degenerative genes that encode protein products are expressed ubiquitously in all tissues in the body, but *SRY* is expressed only in the testes.

The third area, the *ampliconic region*, contains about 30 percent of the MSY, including most of the genes closely associated with testes development. These genes lack counterparts on the X chromosome, and their expression is limited to the testes. There are 60 transcription units (genes that yield a product) divided among nine gene families in this region, most represented by multiple copies. Members of each family have nearly identical ($>$98 percent) DNA sequences. Each repeat unit is an **amplicon** and is contained within seven segments scattered across the euchromatic regions of both the short and long arms of the Y chromosome. Genes in the ampliconic region encode proteins specific to the development and function of the testes, and the products of many of these genes are directly related to fertility in males. It is currently believed that a great deal of male sterility in our population can be linked to mutations in these genes. Research by David Page and others has also revealed that sequences called **palindromes**—sequences of base pairs that read the same but in the opposite direction on complementary strands—are present throughout the MSY. Recombination between palindromes on sister chromatids of the Y during replication is a mechanism used to repair mutations in the Y. This discovery has fascinating implications concerning how the Y chromosome may maintain its size and structure, given that homologous recombination between the X and Y occurs primarily in PARs.

This recent work has greatly expanded our picture of the genetic information carried by this unique chromosome. It clearly refutes the so-called *wasteland theory*, prevalent only 20 years ago, that depicted the human Y chromosome as almost devoid of genetic information other than a few genes that cause maleness. The knowledge we have gained provides the basis for a much clearer picture of how maleness is determined. In addition, it provides important clues to the origin of the Y chromosome during human evolution.

ESSENTIAL POINT

The presence or absence of a Y chromosome that contains an intact *SRY* gene is responsible for causing maleness in humans.

NOW SOLVE THIS

Problem 29 on page 109 concerns the autosomal gene *SOX9*, which when mutated appears to inhibit normal human male development. You are asked to analyze a sequence of observations involving individuals with campomelic dysplasia (CMD1) and to draw appropriate conclusions.

Hint: Some genes are activated and produce their normal product as a result of expression of products of other genes found on different chromosomes—in this case, perhaps one that is on the Y chromosome.

5.4 The Ratio of Males to Females in Humans Is Not 1.0

The presence of heteromorphic sex chromosomes in one sex of a species but not the other provides a potential mechanism for producing equal proportions of male and female offspring. This potential depends on the segregation of the X and Y (or Z and W) chromosomes during meiosis, such that half of the gametes of the heterogametic sex receive one of the chromosomes and half receive the other one. As we learned in the previous section, small pseudoautosomal regions of pairing homology do exist at both ends of the human X and Y chromosomes, suggesting that the X and Y chromosomes do synapse and then segregate into different gametes. Provided that both types of gametes are equally successful in fertilization and that the two sexes are equally viable during development, a 1:1 ratio of male and female offspring should result.

The actual proportion of male to female offspring, referred to as the **sex ratio,** has been assessed in two ways. The **primary sex ratio** reflects the proportion of males to females conceived in a population. The **secondary sex ratio** reflects the proportion of each sex that is born. The secondary sex ratio is much easier to determine but has the disadvantage of not accounting for any disproportionate embryonic or fetal mortality.

When the secondary sex ratio in the human population was determined in 1969 by using worldwide census data, it did not equal 1.0. For example, in the Caucasian population in the United States, the secondary ratio was a little less than 1.06, indicating that about 106 males were born for each 100 females. (In 1995, this ratio dropped to slightly less than 1.05.) In the African-American population in the United States, the ratio was 1.025. In other countries the excess of male births is even greater than is reflected in these values. For example, in Korea, the secondary sex ratio was 1.15.

Despite these ratios, it is possible that the primary sex ratio is 1.0 and is later altered between conception and birth. For the secondary ratio to exceed 1.0, then, prenatal female mortality would have to be greater than prenatal male mortality. However, this hypothesis has been examined and shown to be false. In fact, just the opposite occurs. In a Carnegie Institute study, reported in 1948, the sex of approximately 6000 embryos and fetuses recovered from miscarriages and abortions was determined, and fetal mortality was actually higher in males. On the basis of the data derived from that study, the primary sex ratio in U.S. Caucasians was estimated to be 1.079. It is now believed that the figure is much higher—between 1.20 and 1.60, suggesting that many more males than females are conceived in the human population.

It is not clear why such a radical departure from the expected primary sex ratio of 1.0 occurs. To come up with a suitable explanation, researchers must examine the assumptions on which the theoretical ratio is based:

1. Because of segregation, males produce equal numbers of X- and Y-bearing sperm.
2. Each type of sperm has equivalent viability and motility in the female reproductive tract.
3. The egg surface is equally receptive to both X- and Y-bearing sperm.

No direct experimental evidence contradicts any of these assumptions; however, the human Y chromosome is smaller than the X chromosome and therefore of less mass. Thus, it has been speculated that Y-bearing sperm are more motile than X-bearing sperm. If this is true, then the probability of a fertilization event leading to a male zygote is increased, providing one possible explanation for the observed primary ratio.

ESSENTIAL POINT

In humans, while many more males than females are conceived, and although fewer male than female embryos and fetuses survive *in utero*, there are nevertheless more males than females born.

5.5 Dosage Compensation Prevents Excessive Expression of X-Linked Genes in Humans and Other Mammals

The presence of two X chromosomes in normal human females and only one X in normal human males is unique compared with the equal numbers of autosomes present in the cells of both sexes. On theoretical grounds alone, it is possible to speculate that this disparity should create a "genetic dosage" difference between males and females, with attendant problems, for all X-linked genes. There is the potential for females to produce twice as much of each product of all X-linked genes. The additional X chromosomes in both males and females exhibiting the various syndromes discussed earlier in this chapter are thought to compound this dosage problem. Embryonic development depends on proper timing and precisely regulated levels of gene expression. Otherwise, disease phenotypes or embryonic lethality can occur. In this section, we will describe research findings regarding X-linked gene expression that demonstrate a genetic mechanism of **dosage compensation** that balances the dose of X chromosome gene expression in females and males.

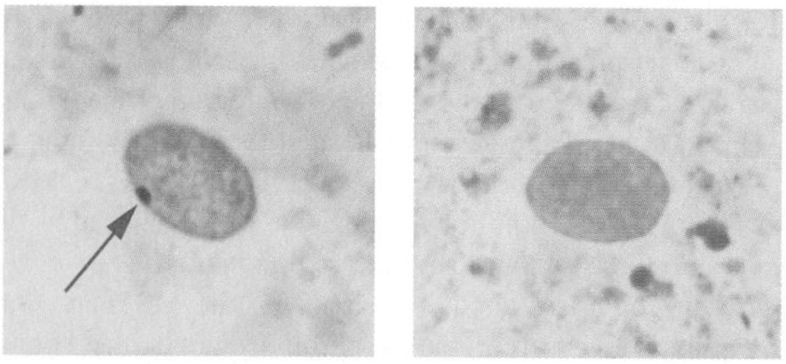

FIGURE 5–7 Photomicrographs comparing cheek epithelial cell nuclei from a male that fails to reveal Barr bodies (right) with a nucleus from a female that demonstrates a Barr body (indicated by the arrow in the left image). This structure, also called a sex chromatin body, represents an inactivated X chromosome.

Barr Bodies

Murray L. Barr and Ewart G. Bertram's experiments with female cats, as well as Keith Moore and Barr's subsequent study in humans, demonstrate a genetic mechanism in mammals that compensates for X chromosome dosage disparities. Barr and Bertram observed a darkly staining body in the interphase nerve cells of female cats that was absent in similar cells of males. In humans, this body can be easily demonstrated in female cells derived from the buccal mucosa (cheek cells) or in fibroblasts (undifferentiated connective tissue cells), but not in similar male cells (Figure 5–7). This highly condensed structure, about 1 μm in diameter, lies against the nuclear envelope of interphase cells, and it stains positively for a number of different DNA-binding dyes.

This chromosome structure, called a **sex chromatin body,** or simply a **Barr body,** is an inactivated X chromosome. Susumo Ohno was the first to suggest that the Barr body arises from one of the two X chromosomes. This hypothesis is attractive because it provides a possible mechanism for dosage compensation. If one of the two X chromosomes is inactive in the cells of females, the dosage of genetic information that can be expressed in males and females will be equivalent. Convincing, though indirect, evidence for this hypothesis comes from the study of the sex-chromosome syndromes described earlier in this chapter. Regardless of how many X chromosomes a somatic cell possesses, all but one of them appear to be inactivated and can be seen as Barr bodies. For example, no Barr body is seen in the somatic cells of Turner 45,X females; one is seen in Klinefelter 47,XXY males; two in 47,XXX females; three in 48,XXXX females; and so on (Figure 5–8). Therefore, the number of Barr bodies follows an $N - 1$ rule, where N is the total number of X chromosomes present.

Although this apparent inactivation of all but one X chromosome increases our understanding of dosage compensation, it further complicates our perception of other matters.

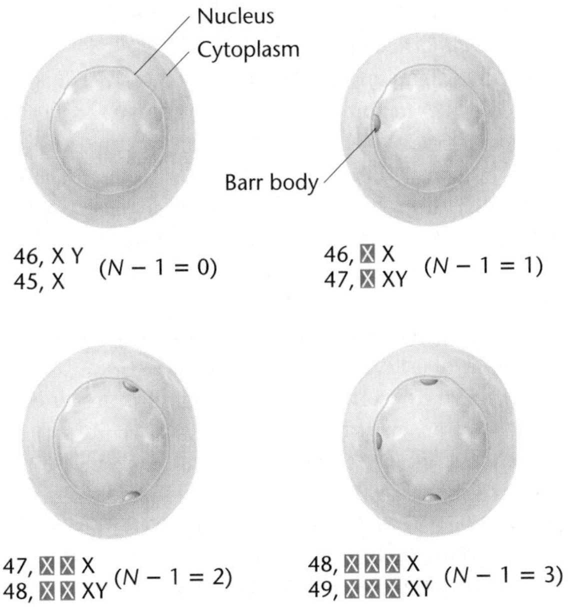

FIGURE 5–8 Occurrence of Barr bodies in various human karyotypes, where all X chromosomes except one ($N - 1$) are inactivated.

For example, because one of the two X chromosomes is inactivated in normal human females, why then is the Turner 45,X individual not entirely normal? Why aren't females with the triplo-X and tetra-X karyotypes (47,XXX and 48,XXXX) completely unaffected by the additional X chromosome? Furthermore, in Klinefelter syndrome (47,XXY), X chromosome inactivation effectively renders the person 46,XY. Why aren't these males unaffected by the extra X chromosome in their nuclei?

One possible explanation is that chromosome inactivation does not normally occur in the very early stages of development of those cells destined to form gonadal tissues. Another possible explanation is that not all genes on each X chromosome forming a Barr body are inactivated. Recent studies have indeed demonstrated that as many as 15 percent of the human X chromosomal genes actually escape inactivation. Clearly, then, not every gene on the X requires inactivation. In either case, excessive expression of certain X-linked genes might still occur at critical times during development despite apparent inactivation of superfluous X chromosomes.

The Lyon Hypothesis

In mammalian females, one X chromosome is of maternal origin, and the other is of paternal origin. Which one is inactivated? Is the inactivation random? Is the same chromosome inactive in all somatic cells? In 1961, Mary Lyon and Liane Russell independently proposed a hypothesis that answers these questions. They postulated that the inactivation of X chromosomes occurs randomly in somatic cells at a point early in embryonic development, most likely sometime during the blastocyst stage of development. Once inactivation has occurred, all descendant cells have the same X chromosome inactivated as their initial progenitor cell.

This explanation, which has come to be called the **Lyon hypothesis,** was initially based on observations of female mice heterozygous for X-linked coat-color genes. The pigmentation of these heterozygous females was mottled, with large patches expressing the color allele on one X and other patches expressing the allele on the other X. This is the phenotypic pattern that would be expected if different X chromosomes were inactive in adjacent patches of cells. Similar mosaic patterns occur in the black and yellow-orange patches of female tortoiseshell and calico cats (Figure 5–9). Such X-linked coat-color patterns do not occur in male cats because all their cells contain the single maternal X chromosome and are therefore hemizygous for only one X-linked coat-color allele.

The most direct evidence in support of the Lyon hypothesis comes from studies of gene expression in clones of human fibroblast cells. Individual cells are isolated following biopsy and cultured *in vitro*. A culture of cells derived from a single cell is called a **clone.** The synthesis of the enzyme glucose-6-phosphate dehydrogenase (G6PD) is controlled by an X-linked gene. Numerous mutant alleles of this gene have been detected, and their gene products can be differentiated from the wild-type enzyme by their migration pattern in an electrophoretic field.

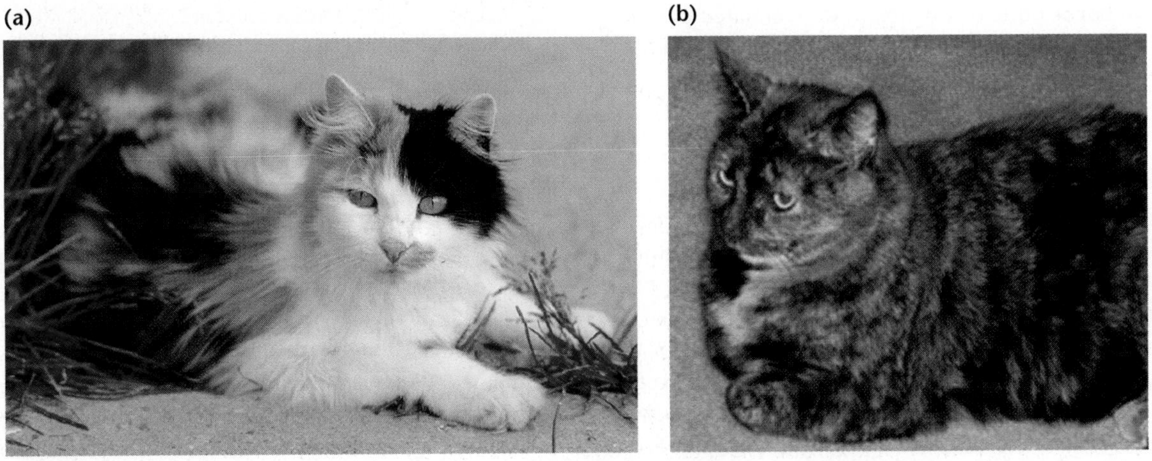

FIGURE 5–9 (a) The random distribution of orange and black patches in a calico cat illustrates the Lyon hypothesis. The white patches are due to another gene, distinguishing calico cats from tortoiseshell cats (b), which lack the white patches.

Fibroblasts have been taken from females heterozygous for different allelic forms of *G6PD* and studied. The Lyon hypothesis predicts that if inactivation of an X chromosome occurs randomly early in development, and thereafter all progeny cells have the same X chromosome inactivated as their progenitor, such a female should show two types of clones, each containing only one electrophoretic form of *G6PD*, in approximately equal proportions.

In 1963, Ronald Davidson and colleagues performed an experiment involving 14 clones from a single heterozygous female. Seven showed only one form of the enzyme, and 7 showed only the other form. Most important was the finding that none of the 14 clones showed both forms of the enzyme. Studies of *G6PD* mutants thus provide strong support for the random permanent inactivation of either the maternal or paternal X chromosome.

The Lyon hypothesis is generally accepted as valid; in fact, the inactivation of an X chromosome into a Barr body is sometimes referred to as **lyonization.** One extension of the hypothesis is that mammalian females are mosaics for all heterozygous X-linked alleles—some areas of the body express only the maternally derived alleles, and others express only the paternally derived alleles. Two especially interesting examples involve **red-green color blindness** and **anhidrotic ectodermal dysplasia,** both X-linked recessive disorders. In the former case, hemizygous males are fully color-blind in all retinal cells. However, heterozygous females display mosaic retinas, with patches of defective color perception and surrounding areas with normal color perception. Males hemizygous for anhidrotic ectodermal dysplasia show absence of teeth, sparse hair growth, and lack of sweat glands. The skin of females heterozygous for this disorder reveals random patterns of tissue with and without sweat glands. In both examples, random inactivation of one or the other X chromosome early in the development of heterozygous females has led to these occurrences.

NOW SOLVE THIS

Problem 32 on page 110 describes Carbon Copy, the first cat created by cloning, who was derived from a somatic nucleus of a calico cat. You are asked to comment on the likelihood that CC will appear identical to her genetic donor.

Hint: The donor nucleus was from a differentiated ovarian cell of an adult female cat, which itself had inactivated one of its X chromosomes.

The Mechanism of Inactivation

The least understood aspect of the Lyon hypothesis is the mechanism of chromosome inactivation in mammals. Somehow, either the DNA or the attached histone proteins, or both, of one (or more) of the mammalian X chromosomes of females is modified, silencing most genes that are part of that chromosome. Whatever the modification, a memory is created that keeps the same chromosome inactivated in the following chromosome replications and cell divisions. Such a process, whereby genetic expression of one homolog, but not the other, is affected, is referred to as **imprinting.** This term also applies to a number of similar processes in which genetic information is modified epigenetically.

Recent investigations are beginning to clarify this issue. A region of the mammalian X chromosome is the major control unit. This region, located on the proximal end of the p arm in humans (the end toward the centromere), is called the **X inactivation center** ***(Xic),*** and its genetic expression occurs only on the X chromosome that is inactivated. The *Xic* is about 1 Mb (10^6 base pairs) in length and is known to contain several putative regulatory units and four genes. One of these genes, ***X-inactive specific transcript (XIST),*** is now known to play a critical role in X-inactivation.

Two interesting observations have been made regarding the RNA that is transcribed from the *XIST* gene, in part based on experiments that used the equivalent gene in the mouse (*Xist*). First, the RNA product is quite large and lacks what is called an extended **open reading frame (ORF).** An ORF is comprised of the information necessary for translating the RNA product into a protein. Thus, in this case, the RNA is transcribed but is not translated. It appears to serve a structural role in the nucleus, presumably in the mechanism of chromosome inactivation. This finding has led to the belief that the RNA products of *Xist* spread over and coat the X chromosome bearing the gene that produced them, creating some sort of molecular "cage" that entraps and inactivates the chromosome. Inactivation is therefore said to be *cis*-acting. Two other noncoding genes at the *Xic* locus, *Tsix* (an antisense partner of *Xist*) and *Xite*, are also believed to play important roles in X-inactivation.

Second, transcription of *Xist* initially occurs at low levels on all X chromosomes. As the inactivation process begins, however, transcription continues, and is enhanced, only on the X chromosome that becomes inactivated. In 1996, a research group led by Graeme Penny provided convincing evidence that transcription of *Xist* is the critical event in chromosome inactivation. These researchers were able to introduce a targeted deletion (7 kb) into this gene. As a result, the chromosome bearing the mutation lost its ability to become inactivated.

ESSENTIAL POINT

In mammals, female somatic cells randomly inactivate one of two X chromosomes during early embryonic development, a process important for balancing the expression of X-chromosome linked genes in males and females.

5.6 The Ratio of X Chromosomes to Sets of Autosomes Determines Sex in *Drosophila*

Because males and females in *Drosophila melanogaster* (and other *Drosophila* species) have the same general sex-chromosome composition as humans (males are XY and females are XX), we might assume that the Y chromosome also causes maleness in these flies. However, the elegant work of Calvin Bridges in 1916 showed this not to be true. His studies of flies with quite varied chromosome compositions led him to the conclusion that the Y chromosome is not involved in sex determination in this organism. Instead, Bridges proposed that the X chromosomes and autosomes together play a critical role in sex determination. Recall that in the nematode *C. elegans*, which lacks a Y chromosome, the sex chromosomes and autosomes are also both critical to sex determination.

Bridges' work can be divided into two phases: (1) A study of offspring resulting from nondisjunction of the X chromosomes during meiosis in females and (2) subsequent work with progeny of females containing three copies of each chromosome, called triploid ($3n$) females. As we have seen previously in this chapter (and as you will see in Figure 6–1), nondisjunction is the failure of paired chromosomes to segregate or separate during the anaphase stage of the first or second meiotic divisions. The result is the production of two types of abnormal gametes, one of which contains an extra chromosome ($n + 1$) and the other of which lacks a chromosome ($n - 1$). Fertilization of such gametes with a haploid gamete produces $2n + 1$ or $2n - 1$ zygotes. As in humans, if nondisjunction involves the X chromosome, in addition to the normal complement of autosomes, both an XXY and an X0 sex-chromosome composition may result. (The "0" signifies that neither a second X nor a Y chromosome is present, as occurs in X0 genotypes of individuals with Turner syndrome.)

Contrary to what was later discovered in humans, Bridges found that the XXY flies were normal females and the X0 flies were sterile males. The presence of the Y chromosome in the XXY flies did not cause maleness, and its absence in the X0 flies did not produce femaleness. From these data, Bridges concluded that the Y chromosome in *Drosophila* lacks male-determining factors, but since the X0 males were sterile, it does contain genetic information essential to male fertility.

Bridges was able to clarify the mode of sex determination in *Drosophila* by studying the progeny of triploid females ($3n$), which have three copies each of the haploid complement of chromosomes. *Drosophila* has a haploid number of 4, thereby possessing three pairs of autosomes in addition to its pair of sex chromosomes. Triploid females apparently originate from rare diploid eggs fertilized by normal haploid sperm. Triploid females have heavy-set bodies, coarse bristles, and coarse eyes, and they may be fertile. Because of the odd number of each chromosome (3), during meiosis, a variety of different chromosome complements are distributed into gametes that give rise to offspring with a variety of abnormal chromosome constitutions. Correlations between the sexual morphology and chromosome composition, along with Bridges' interpretation, are shown in Figure 5–10.

Bridges realized that the critical factor in determining sex is the ratio of X chromosomes to the number of haploid sets of autosomes (A) present. Normal (2X:2A) and triploid (3X:3A) females each have a ratio equal to 1.0, and both are fertile. As the ratio exceeds unity (3X:2A, or 1.5, for example), what was once called a *superfemale* is produced. Because such females are most often inviable, they are now more appropriately called **metafemales.**

Normal (XY:2A) and sterile (X0:2A) males each have a ratio of 1:2, or 0.5. When the ratio decreases to 1:3, or 0.33, as in the case of an XY:3A male, infertile **metamales** result. Other flies recovered by Bridges in these studies had an X:A ratio intermediate between 0.5 and 1.0. These flies were generally larger, and they exhibited a variety of morphological abnormalities and rudimentary bisexual gonads and genitalia. They were invariably sterile and expressed both male and female morphology, thus being designated as **intersexes.**

Bridges' results indicate that in *Drosophila*, factors that cause a fly to develop into a male are not located on the sex chromosomes but are instead found on the autosomes. Some female-determining factors, however, are located on the X chromosomes. Thus, with respect to primary sex determination, male gametes containing one of each autosome plus a Y chromosome result in male offspring not because of the presence of the Y but because they fail to contribute an X chromosome. This mode of sex determination is explained by the **genic balance theory.** Bridges proposed that a threshold for maleness is reached when the X:A ratio is 1:2 (X:2A), but that the presence of an additional X (XX:2A) alters the balance and results in female differentiation.

Numerous genes involved in sex determination in *Drosophila* have been identified. Even though the Y chromosome does not cause maleness, recent work has demonstrated that genes on the Y may regulate the activity of hundreds of genes on autosomes as well as genes on the X chromosome. The recessive autosomal gene *transformer* (*tra*), discovered over 50 years ago by Alfred H. Sturtevant, clearly demonstrated that a single autosomal gene could have a profound impact on sex determination. Females homozygous for *tra* are transformed into sterile males, but homozygous males are unaffected.

More recently, another gene, *Sex-lethal* (*Sxl*), has been shown to play a critical role, serving as a "master switch" in sex determination. Activation of the X-linked *Sxl* gene, which

Normal diploid male

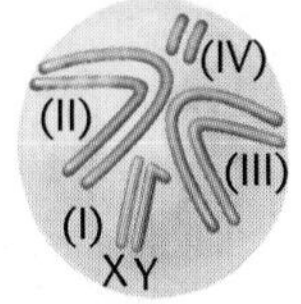

2 sets of autosomes
+
X Y

Chromosome formulation	Ratio of X chromosomes to autosome sets	Sexual morphology
3X/2A	1.5	Metafemale
3X/3A	1.0	Female
2X/2A	1.0	Female
3X/4A	0.75	Intersex
2X/3A	0.67	Intersex
X/2A	0.50	Male
XY/2A	0.50	Male
XY/3A	0.33	Metamale

FIGURE 5–10 The ratios of X chromosomes to sets of autosomes and the resultant sexual morphology seen in *Drosophila melanogaster*.

relies on a ratio of X chromosomes to sets of autosomes that equals 1.0, is essential to female development. In the absence of activation—as when, for example, the X:A ratio is 0.5—male development occurs. It is interesting to note that mutations that inactivate the *Sxl* gene, as originally studied in 1960 by Hermann J. Muller, kill female embryos but have no effect on male embryos, consistent with the role of the gene. Although it is not yet exactly clear how this ratio influences the *Sxl* locus, we do have some insights into the question. The *Sxl* locus is part of a hierarchy of gene expression and exerts control over other genes, including *tra* (discussed in the previous paragraph) and *dsx* (*doublesex*). The wild-type allele of *tra* is activated by the product of *Sxl* only in females and in turn influences the expression of *dsx*. Depending on how the initial RNA transcript of *dsx* is processed (spliced, as explained below), the resultant dsx protein activates either male- or female-specific genes required for sexual differentiation. Each step in this regulatory cascade requires a form of processing called **RNA splicing,** in which portions of the RNA are removed and the remaining fragments are "spliced" back together prior to translation into a protein. In the case of the *Sxl* gene, the RNA transcript may be spliced in different ways, a phenomenon called **alternative splicing.** A different RNA transcript is produced in females than in males. In potential females, the transcript is active and initiates a cascade of regulatory gene expression, ultimately leading to female differentiation. In potential males, the transcript is inactive, leading to a different pattern of gene activity, whereby male differentiation occurs. We will return to this topic in Chapter 15, where alternative splicing is again addressed as one of the mechanisms involved in the regulation of genetic expression in eukaryotes.

5.7 Temperature Variation Controls Sex Determination in Reptiles

We conclude this chapter by discussing several cases involving reptiles, in which the environment—specifically temperature—has a profound influence on sex determination. In contrast to **chromosomal,** or **genotypic, sex determination (CSD** or **GSD**), in which sex is determined genetically (as is true of all examples thus far presented in the chapter), the cases that we will now discuss are categorized as **temperature-dependent sex determination (TSD).** As we shall see, the investigations leading to this information may well have come closer to revealing the true nature of the underlying basis of sex determination than any findings previously discussed.

In many species of reptiles, sex is predetermined at conception by sex-chromosome composition, as is the case in many organisms already considered in this chapter. For example, in many snakes, including vipers, a ZZ/ZW mode is in effect, in which the female is the heterogamous sex (ZW). However, in boas and pythons, it is impossible to distinguish one sex chromosome from the other in either sex. In many lizards, both the XX/XY and ZZ/ZW systems are found, depending on the species. In still other reptilian species, including all crocodiles, most turtles, and some lizards, sex determination is achieved according to the incubation temperature of eggs during a critical period of embryonic development.

Three distinct patterns of TSD emerge (Cases I–III in Figure 5–11). In Case I, low temperatures yield 100 percent females, and high temperatures yield 100 percent males. Just the opposite occurs in Case II. In Case III, low *and* high temperatures yield 100 percent females, while intermediate temperatures yield various proportions of males. The third pattern is seen in various species of crocodiles, turtles, and lizards, although other members of these groups are known to exhibit the other patterns.

Two observations are noteworthy. First, in all three patterns, certain temperatures result in both male and female offspring; second, this pivotal temperature (T_p) range is fairly narrow, usually spanning less than 5°C, and sometimes only 1°C. The central question raised by these observations is: What are the metabolic or physiological parameters affected by temperature that lead to the differentiation of one sex or the other?

The answer is thought to involve steroids (mainly estrogens) and the enzymes involved in their synthesis. Studies clearly demonstrate that the effects of temperature on estrogens, androgens, and inhibitors of the enzymes controlling their synthesis are involved in the sexual differentiation of ovaries and testes. One enzyme in particular, **aromatase,** converts andro-

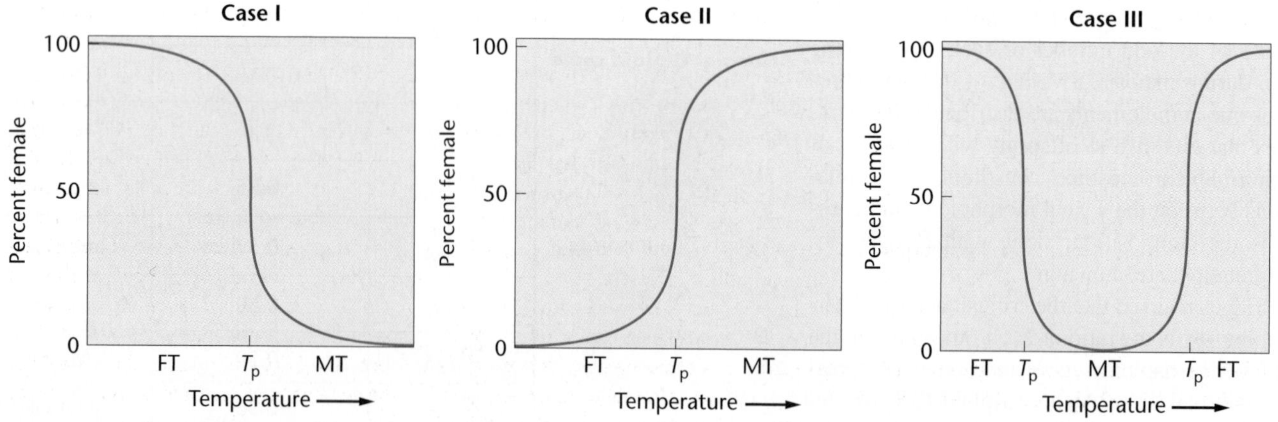

FIGURE 5–11 Three different patterns of temperature-dependent sex determination (TSD) in reptiles, as described in the text. The relative pivotal temperature T_p is crucial to sex determination during a critical point during embryonic development. FT = Female-determining temperature; MT = male-determining temperature.

gens (male hormones such as testosterone) to estrogens (female hormones such as estradiol). The activity of this enzyme is correlated with the pathway of reactions that occurs during gonadal differentiation activity and is high in developing ovaries and low in developing testes. Researchers in this field, including Claude Pieau and colleagues, have proposed that a thermosensitive factor mediates the transcription of the reptilian aromatase gene, leading to temperature-dependent sex determination. Several other genes are likely to be involved in this mediation.

The involvement of sex steroids in gonadal differentiation has also been documented in birds, fishes, and amphibians. Thus, sex-determining mechanisms involving estrogens seem to be characteristic of nonmammalian vertebrates. The regulation of such systems, while temperature-dependent in many reptiles, appears to be controlled by sex chromosomes (XX/XY or ZZ/ZW) in many of these other organisms. A final intriguing thought on this matter is that the product of *SRY*, a key component in mammalian sex determination, has been shown to bind *in vitro* to a regulatory portion of the aromatase gene, suggesting a mechanism whereby it could act as a repressor of ovarian development.

ESSENTIAL POINT

Many reptiles show temperature-dependent effects on sex determination. Although specific sex chromosomes determine genotypic sex in many reptiles, temperature effects on genes involved in sexual determination affect whether an embryo develops a male or female phenotype.

GENETICS, TECHNOLOGY, AND SOCIETY

A Question of Gender: Sex Selection in Humans

Throughout history, people have attempted to influence the gender of their unborn offspring by following varied and sometimes bizarre procedures. In medieval Europe, prospective parents would place a hammer under the bed to help them conceive a boy, or a pair of scissors to conceive a girl. Other practices were based on the ancient belief that semen from the right testicle created male offspring and that from the left testicle created females. As late as the eighteenth century, European men might tie off or remove their left testicle to increase the chances of getting a male heir.

In some cultures, efforts to control the sex of offspring has had a darker outcome—female infanticide. In ancient Greece, the murder of female infants was so common that the male:female ratio in some areas approached 4 : 1. In some parts of rural India, hundreds of families admitted to female infanticide as late as the 1990s. In 1997, the World Health Organization reported population data showing that about 50 million women were "missing" in China, likely because of selective abortion of female fetuses and institutionalized neglect of female children.

In recent times, sex-specific abortion has replaced much of the traditional female infanticide. For a fee, some companies offer amniocentesis and ultrasound tests for prenatal sex determination. Studies in India estimate that hundreds of thousands of fetuses are aborted each year because they are female. As a result of sex-selective abortion, the female:male ratio in India was 927 : 1000 in 1991. In some northern states, the ratio was as low as 600 : 1000.

In Western industrial countries, new genetics and reproductive technologies offer parents ways to select their children's gender prior to implantation of the embryo in the uterus—called *preimplantation gender selection (PGS)*. Following *in vitro* fertilization, embryos are biopsied and assessed for gender. Only sex-selected embryos are then implanted. The simplest method involves separating X and Y chromosome-bearing spermatozoa based on their DNA content. Because of the difference in size of the X and Y chromosomes, X-bearing sperm contain 2.8 to 3.0 percent more DNA than Y-bearing sperm. Sperm samples are treated with a fluorescent DNA stain, then passed through a laser beam in a Fluorescence-Activated Cell Sorter machine that separates the sperm into two fractions based on the intensity of their DNA-fluorescence. The sorted sperm are then used for standard intrauterine insemination.

The emerging PGS methods raise a number of legal and ethical issues. Some people feel that prospective parents have the legal right to use sex-selection techniques as part of their fundamental procreative liberty. Proponents state that PGS will reduce the suffering of many families. For example, people at risk for transmitting X-linked diseases such as hemophilia or Duchenne muscular dystrophy can now enhance their chance of conceiving a female child, who will not express the disease.

The majority of people who undertake PGS, however, do so for nonmedical reasons—to "balance" their families. A possible argument in favor of this use is that the ability to intentionally select the sex of an offspring may reduce overpopulation and economic burdens for families who would repeatedly reproduce to get the desired gender. By the same token, PGS may reduce the number of abortions. It is also possible that PGS may increase the happiness of both parents and children, as the children would be more "wanted."

On the other hand, some argue that PGS serves neither the individual nor the common good. They argue that PGS is inherently sexist, having its basis in the idea that one sex is superior to the other, and leads to an increase in linking a child's worth to gender. Other critics fear that social approval of PGS will open the door to other genetic manipulations of children's characteristics. It is difficult to predict the full effects that PGS will bring to the world. But the gender-selection genie is now out of the bottle and is unwilling to return.

(*Cont. on the next page*)

Your Turn

Take time, individually or in groups, to answer the following questions. Investigate the references and links to help you understand some of the issues that surround the topic of gender selection.

1. What do you think are valid arguments for and against the use of PGS?

2. A generally accepted moral and legal concept is that of reproductive autonomy—the freedom to make individual reproductive decisions without external interference. Are there circumstances under which reproductive autonomy should be restricted?

The above questions, and others, are explored in a series of articles in the American Journal of Bioethics, Volume 1 (2001). *See the article by J. A. Robertson on pages 2–9, for a summary of the moral and legal issues surrounding PGS.*

3. What do you think are the reasons that some societies practice female infanticide and prefer the birth of male children?

For a discussion of this topic, visit the "Gendercide Watch" Web site: **http://www.gendercide.org.**

4. If safe and efficient methods of PGS were available to you, do you think that you would use them to help you with family planning? Under what circumstances might you use them?

The Genetics and IVF Institute (Fairfax, Virginia) is presently using PGS techniques based on sperm sorting, in an FDA-approved clinical trial. As of 2008, over 1000 human pregnancies have resulted, with an approximately 80 percent success rate. Read about these methods on their Web site: **http://www.microsort.net.**

CASE STUDY Doggone it!

A dog breeder discovers one of her male puppies has abnormal genitalia. After a visit to the veterinary clinic at a nearby university, the breeder learns that the dog's karyotype lacks a Y chromosome, but instead has an XX chromosome pair, with one of the X chromosomes slightly larger than usual (being mammals, male dogs are normally XY and females are XX). The veterinarian tells her that in other breeds, some females display an XY chromosome pair, with the Y chromosome being slightly shorter than normal. These observations raise several interesting questions:

1. Can you offer a chromosomal explanation of these two cases?
2. How could such cases be used to locate the gene(s) responsible for maleness?
3. Suppose you discover a female dog with a normal-sized XY chromosome pair. What kind of mutation might be involved in this case?
4. Suppose you discover a female dog with only a single X chromosome. What predictions might you make about the sex organs and reproductive capacity of this dog?

INSIGHTS AND SOLUTIONS

1. In *Drosophila*, the X chromosomes may become attached to one another ($\widehat{XX}$) such that they always segregate together. Some flies thus contain a set of attached X chromosomes plus a Y chromosome.

(a) What sex would such a fly be? Explain why this is so.

(b) Given the answer to part (a), predict the sex of the offspring that would occur in a cross between this fly and a normal one of the opposite sex.

(c) If the offspring described in part (b) are allowed to interbreed, what will be the outcome?

Solution:

(a) The fly will be a female. The ratio of X chromosomes to sets of autosomes—which determines sex in *Drosophila*—will be 1.0, leading to normal female development. The Y chromosome has no influence on sex determination in *Drosophila*.

(b) All progeny flies will have two sets of autosomes along with one of the following sex-chromosome compositions:

(1) $\widehat{XX}X \rightarrow$ a metafemale with 3 X's (called a trisomic)

(2) $\widehat{XX}Y \rightarrow$ a female like her mother

(3) $XY \rightarrow$ a normal male

(4) $YY \rightarrow$ no development occurs

(c) A stock will be created that maintains attached-X females generation after generation.

2. The Xg cell-surface antigen is coded for by a gene located on the X chromosome. No equivalent gene exists on the Y chromosome. Two codominant alleles of this gene have been identified: *Xg1* and *Xg2*. A woman of genotype *Xg2/Xg2* bears children with a man of genotype *Xg1/Y*, and they produce a son with Klinefelter syndrome of genotype *Xg1/Xg2Y*. Using proper genetic terminology, briefly explain how this individual was generated. In which parent and in which meiotic division did the mistake occur?

Solution: Because the son with Klinefelter syndrome is *Xg1/Xg2Y*, he must have received both the *Xg1* allele and the Y chromosome from his father. Therefore, nondisjunction must have occurred during meiosis I in the father.

Problems and Discussion Questions

1. As related to sex determination, what is meant by
   (a) homomorphic and heteromorphic chromosomes and
   (b) isogamous and heterogamous organisms?
2. Contrast the life cycle of a plant such as *Zea mays* with an animal such as *C. elegans*.
3. Discuss the role of sexual differentiation in the life cycles of *Chlamydomonas*, *Zea mays*, and *C. elegans*.
4. Distinguish between the concepts of sexual differentiation and sex determination.
5. Contrast the *Protenor* and *Lygaeus* modes of sex determination.
6. Describe the major difference between sex determination in *Drosophila* and in humans.
7. How do mammals, including humans, solve the "dosage problem" caused by the presence of an X and Y chromosome in one sex and two X chromosomes in the other sex?
8. What specific observations (evidence) support the conclusions about sex determination in *Drosophila* and humans?
9. Describe how nondisjunction in human female gametes can give rise to Klinefelter and Turner syndrome offspring following fertilization by a normal male gamete.
10. An insect species is discovered in which the heterogametic sex is unknown. An X-linked recessive mutation for *reduced wing* (*rw*) is discovered. Contrast the F_1 and F_2 generations from a cross between a female with reduced wings and a male with normal-sized wings when
    (a) the female is the heterogametic sex and
    (b) the male is the heterogametic sex.
11. Given your answers to Problem 10, is it possible to distinguish between the *Protenor* and *Lygaeus* mode of sex determination based on the outcome of these crosses?
12. When cows have twin calves of unlike sex (fraternal twins), the female twin is usually sterile and has masculinized reproductive organs. This calf is referred to as a freemartin. In cows, twins may share a common placenta and thus fetal circulation. Predict why a freemartin develops.
13. An attached-X female fly, $\widehat{XX}Y$ (see the "Insights and Solutions" box), expresses the recessive X-linked *white*-eye phenotype. It is crossed to a male fly that expresses the X-linked recessive miniature wing phenotype. Determine the outcome of this cross in terms of sex, eye color, and wing size of the offspring.
14. Assume that on rare occasions the attached X chromosomes in female gametes become unattached. Based on the parental phenotypes in Problem 13, what outcomes in the F_1 generation would indicate that this has occurred during female meiosis?
15. It has been suggested that any male-determining genes contained on the Y chromosome in humans cannot be located in the limited region that synapses with the X chromosome during meiosis. What might be the outcome if such genes were located in this region?
16. What is a Barr body, and where is it found in a cell?
17. Indicate the expected number of Barr bodies in interphase cells of individuals with Klinefelter syndrome; Turner syndrome; and karyotypes 47,XYY, 47,XXX, and 48,XXXX.
18. Define the Lyon hypothesis.
19. Can the Lyon hypothesis be tested in a human female who is homozygous for one allele of the X-linked *G6PD* gene? Why, or why not?
20. Predict the potential effect of the Lyon hypothesis on the retina of a human female heterozygous for the X-linked red-green color-blindness trait.
21. Cat breeders are aware that kittens expressing the X-linked calico coat pattern and tortoiseshell pattern are almost invariably females. Why?
22. What does the apparent need for dosage compensation mechanisms suggest about the expression of genetic information in normal diploid individuals?
23. The marine echiurid worm *Bonellia viridis* is an extreme example of environmental influence on sex determination. Undifferentiated larvae either remain free-swimming and differentiate into females, or they settle on the proboscis of an adult female and become males. If larvae that have been on a female proboscis for a short period are removed and placed in seawater, they develop as intersexes. If larvae are forced to develop in an aquarium where pieces of proboscises have been placed, they develop into males. Contrast this mode of sexual differentiation with that of mammals. Suggest further experimentation to elucidate the mechanism of sex determination in *B. viridis*. See **Now Solve This** on page 96.
24. What type of evidence supports the conclusion that the primary sex ratio in humans is as high as 1.20 to 1.60?
25. Devise as many hypotheses as you can that might explain why so many more human male conceptions than human female conceptions occur.
26. In mice, the *Sry* gene (see Section 5.3) is located on the Y chromosome very close to one of the pseudoautosomal regions that pairs with the X chromosome during male meiosis. Given this information, propose a model to explain the generation of unusual males who have two X chromosomes (with an *Sry*-containing piece of the Y chromosome attached to one X chromosome).
27. The genes encoding the red- and green-color-detecting proteins of the human eye are located next to one another on the X chromosome and probably evolved from a common ancestral pigment gene. The two proteins demonstrate 76 percent homology in their amino acid sequences. A normal-visioned woman with both genes on each of her two X chromosomes has a red-color-blind son who was shown to have one copy of the green-detecting gene and no copies of the red-detecting gene. Devise an explanation for these observations at the chromosomal level (involving meiosis).
28. In mice, the X-linked dominant mutation *Testicular feminization* (*Tfm*) eliminates the normal response to the testicular hormone testosterone during sexual differentiation. An XY mouse bearing the *Tfm* allele on the X chromosome develops testes, but no further male differentiation occurs—the external genitalia of such an animal are female. From this information, what might you conclude about the role of the *Tfm* gene product and the X and Y chromosomes in sex determination and sexual differentiation in mammals? Can you devise an experiment, assuming you can "genetically engineer" the chromosomes of mice, to test and confirm your explanation?
29. Campomelic dysplasia (CMD1) is a congenital human syndrome featuring malformation of bone and cartilage. It is caused by an autosomal dominant mutation of a gene located on chromosome 17. Consider the following observations in sequence, and in each case, draw whatever appropriate conclusions are warranted.

(a) Of those with the syndrome who are karyotypically 46,XY, approximately 75 percent are sex reversed, exhibiting a wide range of female characteristics.
(b) The nonmutant form of the gene, called *SOX9*, is expressed in the developing gonad of the XY male, but not the XX female.
(c) The *SOX9* gene shares 71 percent amino acid coding sequence homology with the Y-linked *SRY* gene.
(d) CMD1 patients who exhibit a 46,XX karyotype develop as females, with no gonadal abnormalities.

See **Now Solve This** on page 101.

30. In the wasp *Bracon hebetor*, a form of parthenogenesis (the development of unfertilized eggs into progeny) resulting in haploid organisms is not uncommon. All haploids are males. When offspring arise from fertilization, females almost invariably result. P. W. Whiting has shown that an X-linked gene with nine multiple alleles (X_a, X_b, etc.) controls sex determination. Any homozygous or hemizygous condition results in males, and any heterozygous condition results in females. If an X_a/X_b female mates with an X_a male and lays 50 percent fertilized and 50 percent unfertilized eggs, what proportion of male and female offspring will result?
31. Shown here is a graph that plots the percentage of fertilized eggs containing males against the atmospheric temperature during early development in (a) snapping turtles and (b) most lizards. Interpret these data as they relate to the effect of temperature on sex determination.

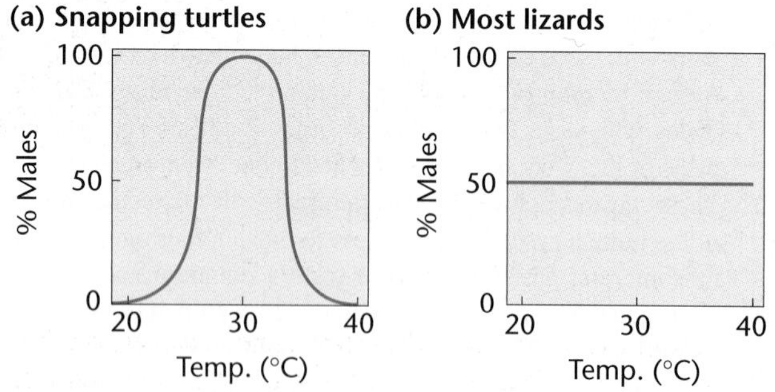

32. CC (Carbon Copy), the first cat produced from a clone, was created from an ovarian cell taken from her genetic donor, Rainbow. The diploid nucleus from the cell was extracted and then injected into an enucleated egg. The resulting zygote was then allowed to develop in a petri dish, and the cloned embryo was implanted in the uterus of a surrogate mother cat, who gave birth to CC. Rainbow is a calico cat. CC's surrogate mother is a tabby. Geneticists were very interested in the outcome of cloning a calico cat, because they were not certain if the cat would have patches of orange and black, just orange, or just black. Taking into account the Lyon hypothesis, explain the basis of the uncertainty. See **Now Solve This** on page 104.

Carbon Copy with her surrogate mother.

33. Let's assume hypothetically that Carbon Copy (see Problem 32) is indeed a calico with black and orange patches, along with the patches of white characterizing a calico cat. Would you expect CC to appear identical to Rainbow? Explain why or why not.
34. When Carbon Copy was born (see Problem 32), she had black patches and white patches, but completely lacked any orange patches. The knowledgeable students of genetics were not surprised at this outcome. Starting with the somatic ovarian cell used as the source of the nucleus in the cloning process, explain how this outcome occurred.

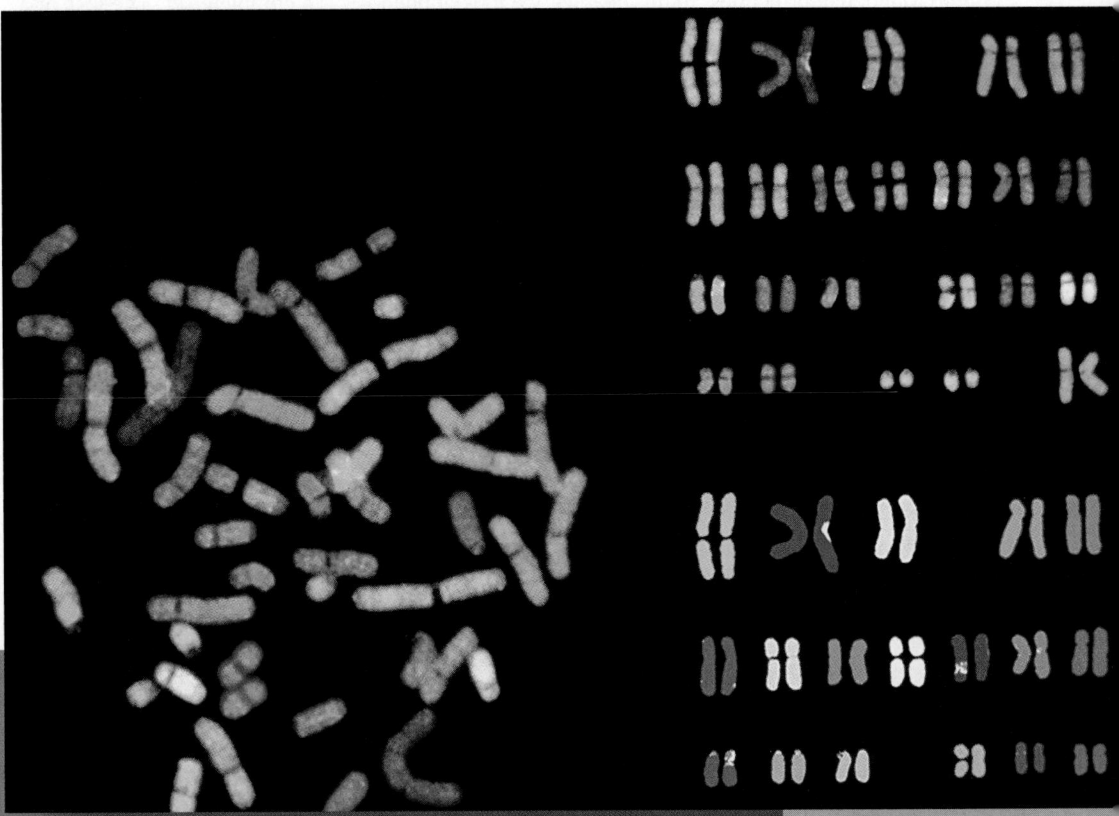

Spectral karyotyping of human chromosomes utilizing differentially labeled "painting" probes.

6

Chromosome Mutations: Variation in Number and Arrangement

Chapter Concepts

- The failure of chromosomes to properly separate during meiosis results in variation in the chromosome content of gametes and subsequently in offspring arising from such gametes.
- Plants often tolerate an abnormal genetic content, but, as a result, they often manifest unique phenotypes. Such genetic variation has been an important factor in the evolution of plants.
- In animals, genetic information is in a delicate equilibrium whereby the gain or loss of a chromosome, or part of a chromosome, in an otherwise diploid organism often leads to lethality or to an abnormal phenotype.
- The rearrangement of genetic information within the genome of a diploid organism may be tolerated by that organism but may affect the viability of gametes and the phenotypes of organisms arising from those gametes.
- Chromosomes in humans contain fragile sites—regions susceptible to breakage, which leads to abnormal phenotypes.

In previous chapters, we have emphasized how mutations and the resulting alleles affect an organism's phenotype and how traits are passed from parents to offspring according to Mendelian principles. In this chapter, we look at phenotypic variation that results from more substantial changes than alterations of individual genes—modifications at the level of the chromosome.

Although most members of diploid species normally contain precisely two haploid chromosome sets, many known cases vary from this pattern. Modifications include a change in the total number of chromosomes, the deletion or duplication of genes or segments of a chromosome, and rearrangements of the genetic material either within or among chromosomes. Taken together, such changes are called **chromosome mutations** or **chromosome aberrations,** to distinguish them from gene mutations. Because the chromosome is the unit of genetic transmission, according to Mendelian laws, chromosome aberrations are passed to offspring in a predictable manner, resulting in many unique genetic outcomes.

Because the genetic component of an organism is delicately balanced, even minor alterations of either content or location of genetic information within the genome can result in some form of phenotypic variation. More substantial changes may be lethal, particularly in animals. Throughout this chapter, we consider many types of chromosomal aberrations, the phenotypic consequences for the organism that harbors an aberration, and the impact of the aberration on the offspring of an affected individual. We will also discuss the role of chromosome aberrations in the evolutionary process.

How Do We Know?

In this chapter, we will focus on chromosomal mutations resulting from a change in number or arrangement of chromosomes. As you study this topic, you should try to answer several fundamental questions:

1. How do we know that the extra chromosome causing Down syndrome is usually maternal in origin?
2. How do we know that human aneuploidy for each of the 22 autosomes occurs at conception, even though most often human aneuploids do not survive embryonic or fetal development and thus are never observed at birth?
3. How do we know that specific mutant phenotypes are due to changes in chromosome number or structure?
4. How do we know that the mutant *Bar*-eye phenotype in *Drosophila* is due to a duplicated gene region rather than to a change in the nucleotide sequence of a gene?

CHROMOSOME SEPARATION IN: DISJUNCTION
WEB TUTORIAL

6.1 Variation in Chromosome Number: Terminology and Origin

Variation in chromosome number ranges from the addition or loss of one or more chromosomes to the addition of one or more haploid sets of chromosomes. Before we embark on our discussion, it is useful to clarify the terminology that describes such changes. In the general condition known as **aneuploidy,** an organism gains or loses one or more chromosomes but not a complete set. The loss of a single chromosome from an otherwise diploid genome is called *monosomy.* The gain of one chromosome results in *trisomy.* These changes are contrasted with the condition of **euploidy,** where complete haploid sets of chromosomes are present. If more than two sets are present, the term **polyploidy** applies. Organisms with three sets are specifically *triploid,* those with four sets are *tetraploid,* and so on. Table 6.1 provides an organizational framework for you to follow as we discuss each of these categories of aneuploid and euploid variation and the subsets within them.

As we consider cases that include the gain or loss of chromosomes, it is useful to examine how such aberrations originate. For instance, how do the syndromes arise where the number of sex-determining chromosomes in humans is altered, as described in Chapter 5? As you may recall, the gain (47,XXY) or loss (45,X) of an X chromosome from an otherwise diploid genome affects the phenotype, resulting in **Klinefelter syndrome** or **Turner syndrome,** respectively (see Figure 5–5). Human females may contain extra X chromosomes (e.g., 47,XXX, 48,XXXX), and some males contain an extra Y chromosome (47,XYY).

Such chromosomal variation originates as a random error during the production of gametes, a phenomenon referred to as **nondisjunction,** whereby paired homologs fail to disjoin during segregation. This process disrupts the normal distribution of chromosomes into gametes. The results of nondisjunction during meiosis I and meiosis II for a single chromosome of a diploid organism are shown in Figure 6–1. As you can see, abnormal gametes can form containing either two members of the affected chromosome or none at all. Fertilizing these with a normal haploid gamete produces a zygote with either three members (trisomy) or only one member (monosomy) of this chromosome. Nondisjunction leads to a variety of aneuploid conditions in humans and other organisms.

TABLE 6.1 Terminology for Variation in Chromosome Numbers

Term	Explanation
Aneuploidy	$2n \pm x$ chromosomes
Monosomy	$2n - 1$
Disomy	$2n$
Trisomy	$2n + 1$
Tetrasomy, pentasomy, etc.	$2n + 2$, $2n + 3$, etc.
Euploidy	Multiples of n
Diploidy	$2n$
Polyploidy	$3n$, $4n$, $5n$, ...
Triploidy	$3n$
Tetraploidy, pentaploidy, etc.	$4n$, $5n$, etc.
Autopolyploidy	Multiples of the same genome
Allopolyploidy (amphidiploidy)	Multiples of closely related genomes

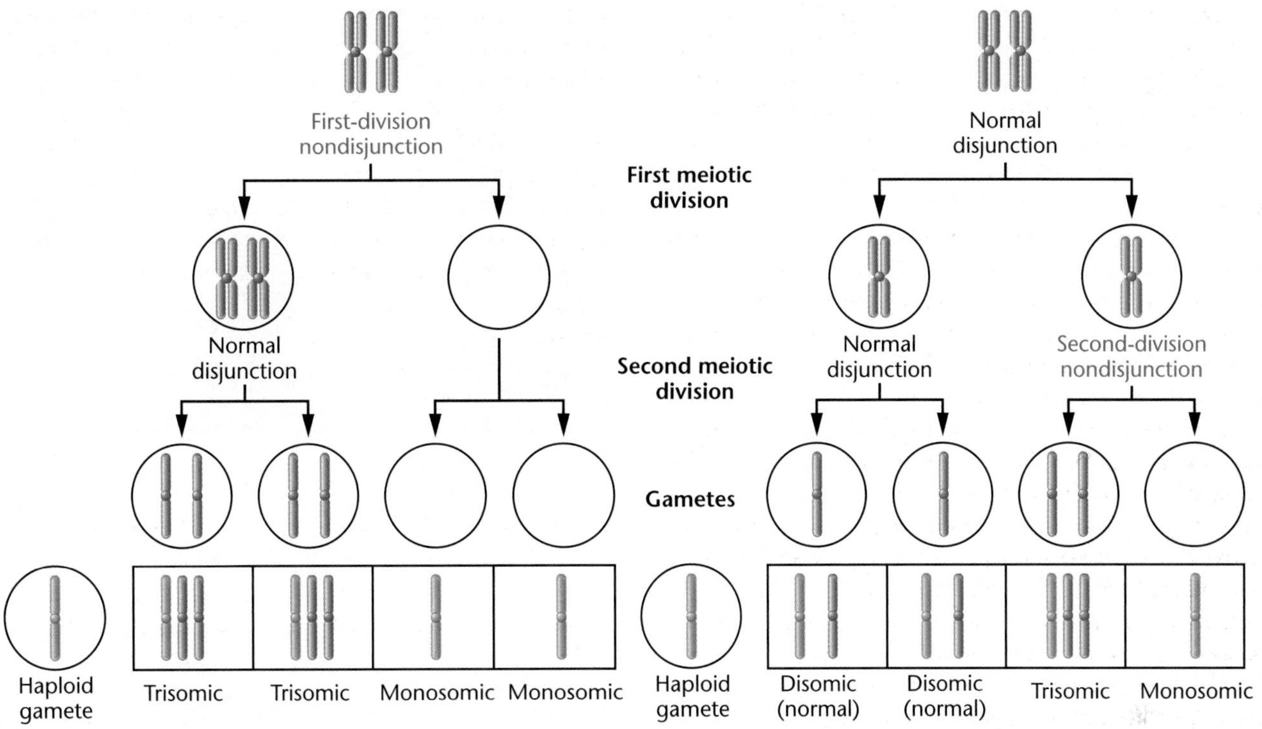

FIGURE 6–1 Nondisjunction during the first and second meiotic divisions. In both cases, some of the gametes that are formed either contain two members of a specific chromosome or lack that chromosome. After fertilization by a gamete with a normal haploid content, monosomic, disomic (normal), or trisomic zygotes are produced.

ESSENTIAL POINT

Alterations of the precise diploid content of chromosomes are referred to as chromosomal aberrations or chromosomal mutations.

NOW SOLVE THIS

Problem 14 on page 130 considers a female with Turner syndrome who expresses hemophilia, as did her father. You are asked which of her parents was responsible for the nondisjunction event leading to her syndrome.

Hint: The parent who contributed a gamete with an X chromosome underwent normal meiosis.

6.2 Monosomy and Trisomy Result in a Variety of Phenotypic Effects

We turn now to a consideration of variations in the number of autosomes and the genetic consequence of such changes. The most common examples of *aneuploidy,* where an organism has a chromosome number other than an exact multiple of the haploid set, are cases in which a single chromosome is either added to, or lost from, a normal diploid set.

Monosomy

The loss of one chromosome produces a $2n - 1$ complement called **monosomy.** Although monosomy for the X chromosome occurs in humans, as we have seen in 45,X Turner syndrome, monosomy for any of the autosomes is not usually tolerated in humans or other animals. In *Drosophila,* flies that are monosomic for the very small chromosome IV (containing less than 5 percent of the organism's genes) develop more slowly, exhibit reduced body size, and have impaired viability. Monosomy for the larger chromosomes II and III is apparently lethal because such flies have never been recovered.

The failure of monosomic individuals to survive is at first quite puzzling, since at least a single copy of every gene is present in the remaining homolog. However, if just one of those genes is represented by a lethal allele, the unpaired chromosome condition leads to the death of the organism. This occurs because monosomy unmasks recessive lethals that are tolerated in heterozygotes carrying the corresponding wild-type alleles. In other cases, a single copy of a recessive gene may be insufficient to provide adequate function for the organism, a phenomenon called **haploinsufficiency.**

Aneuploidy is better tolerated in the plant kingdom. Monosomy for autosomal chromosomes has been observed in maize, tobacco, the evening primrose (*Oenothera*), and the jimson weed (*Datura*), among many other plants. Nevertheless, such monosomic plants are usually less viable than their diploid derivatives. Haploid pollen grains, which undergo extensive development before participating in fertilization, are particularly sensitive to the lack of one chromosome and are seldom viable.

Trisomy

In general, the effects of **trisomy** ($2n + 1$) parallel those of monosomy. However, the addition of an extra chromosome produces somewhat more viable individuals in both animal

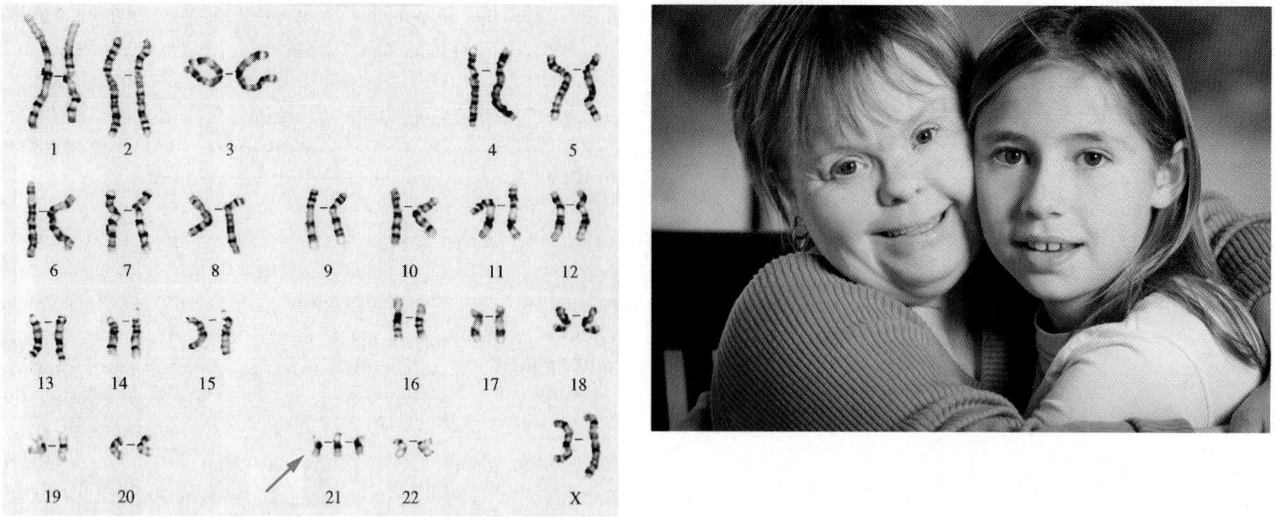

FIGURE 6–2 The karyotype and a photograph of a child with Down syndrome (hugging her unaffected sister on the right). In the karyotype, three members of the G-group chromosome 21 are present, creating the 47,21+ condition.

and plant species than does the loss of a chromosome. In animals, this is often true, provided that the chromosome involved is relatively small. However, the addition of a large autosome to the diploid complement in both *Drosophila* and humans has severe effects and is usually lethal during development.

In plants, trisomic individuals are viable, but their phenotype may be altered. A classical example involves the jimson weed, *Datura,* whose diploid number is 24. Twelve primary trisomic conditions are possible, and examples of each one have been recovered. Each trisomy alters the phenotype of the plant's capsule sufficiently to produce a unique phenotype. These capsule phenotypes were first thought to be caused by mutations in one or more genes.

Still another example is seen in the rice plant (*Oryza sativa*), which has a haploid number of 12. Trisomic strains for each chromosome have been isolated and studied—the plants of 11 strains can be distinguished from one another and from wild-type plants. Trisomics for the longer chromosomes are the most distinctive, and the plants grow more slowly. This is in keeping with the belief that larger chromosomes cause greater genetic imbalance than smaller ones. Leaf structure, foliage, stems, grain morphology, and plant height also vary among the various trisomies.

Down Syndrome: Trisomy 21

The only human autosomal trisomy in which a significant number of individuals survive longer than a year past birth was discovered in 1866 by Langdon Down. The condition is now known to result from trisomy of chromosome 21, one of the G group* (Figure 6–2), and is called **Down syndrome** or simply **trisomy 21** (designated **47,21+**). This trisomy is found in approximately 1 infant in every 800 live births. While this might seem to be a rare, improbable event, there are approximately 4000–5000 such births annually in the United States, and there are currently over 250,000 individuals with Down syndrome.

Typical of other conditions classified as syndromes, many phenotypic characteristics *may* be present in trisomy 21, but any single affected individual usually exhibits only a subset of these. In the case of Down syndrome, there are 12 to 14 such characteristics, with each individual, on average, expressing 6 to 8 of them. Nevertheless, the outward appearance of these individuals is very similar, and they bear a striking resemblance to one another. This is, for the most part, due to a prominent epicanthic fold in each eye* and the typically flat face and round head. People with Down syndrome are also characteristically short and may have a protruding, furrowed tongue (which causes the mouth to remain partially open) and short, broad hands with characteristic palm and fingerprint patterns. Physical, psychomotor, and mental development are retarded, and poor muscle tone is characteristic. While life expectancy is shortened to an average of about 50 years, individuals are known to survive into their 60s.

Children afflicted with Down syndrome are prone to respiratory disease and heart malformations, and they show an incidence of leukemia approximately 20 times higher than that of the normal population. However, careful medical scrutiny and treatment throughout their lives can extend their survival significantly. A striking observation is that death in older Down syndrome adults is frequently due to Alzheimer disease. The onset of this disease occurs at a much earlier age than in the normal population.

* On the basis of size and centromere placement, human autosomal chromosomes are divided into seven groups: A (1–3), B (4–5), C (6–12), D (13–15), E (16–18), F (19–20), and G (21–22).

* The epicanthic fold, or epicanthus, is a skin fold of the upper eyelid, extending from the nose to the inner side of the eyebrow. It covers and appears to lower the inner corner of the eye, giving the eye a slanted, or almond-shaped, appearance. The epicanthus is a prominent normal component of the eyes in many Asian groups.

Because Down syndrome is common in our population, a comprehensive understanding of the underlying genetic basis has long been a research goal. Investigations have given rise to the idea that a critical region of chromosome 21 contains the genes that are dosage sensitive in this trisomy and responsible for the many phenotypes associated with the syndrome. This hypothetical portion of the chromosome has been called the **Down syndrome critical region (DSCR).** A mouse model was created in 2004 that is trisomic for the DSCR, but some mice did not exhibit the characteristics of the syndrome. Nevertheless, this remains an important investigative approach.

The Origin of the Extra 21st Chromosome in Down Syndrome

Most frequently, this trisomic condition occurs through nondisjunction of chromosome 21 during meiosis. Failure of paired homologs to disjoin during either anaphase I or II may lead to gametes with the $n + 1$ chromosome composition. About 75 percent of these errors leading to Down syndrome are attributed to nondisjunction during meiosis I. Subsequent fertilization with a normal gamete creates the trisomic condition.

Chromosome analysis has shown that, while the additional chromosome may be derived from either the mother or father, the ovum is the source in about 95 percent of 47,21+ trisomy cases. Before the development of techniques using polymorphic markers to distinguish paternal from maternal homologs, this conclusion was supported by the more indirect evidence derived from studies of the age of mothers giving birth to infants afflicted with Down syndrome. Figure 6–3 shows the relationship between the incidence of Down syndrome births and maternal age, illustrating the dramatic increase as the age of the mother increases. While the frequency is about 1 in 1000 at maternal age 30, a tenfold increase to a frequency of 1 in 100 is noted at age 40. The frequency increases still further to about 1 in 30 at age 45. A very alarming statistic is that as the age of childbearing women exceeds 45, the probability of a Down syndrome birth continues to increase substantially. In spite of this high probability, substantially more than half of Down syndrome births occur to women younger than 35 years, because the overwhelming proportion of pregnancies in the general population involve women under 35.

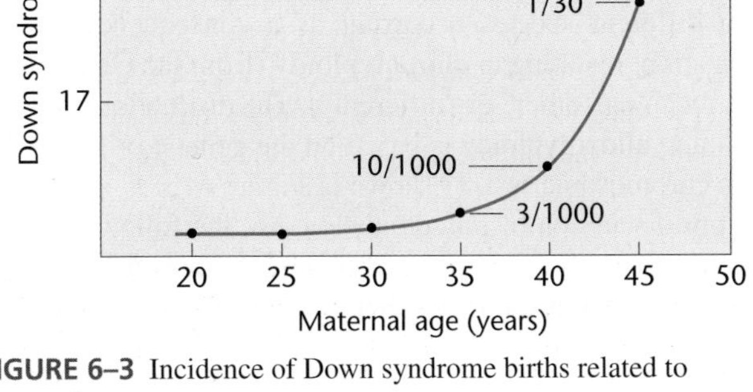

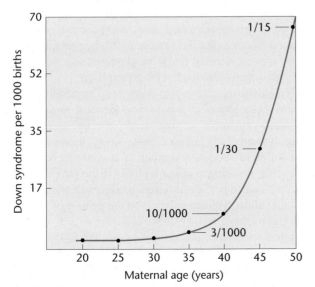

FIGURE 6–3 Incidence of Down syndrome births related to maternal age.

Although the nondisjunctional event that produces Down syndrome seems more likely to occur during oogenesis in women over the age of 35, we do not know with certainty why this is so. However, one observation may be relevant. Meiosis is initiated in all the eggs of a human female when she is still a fetus, until the point where the homologs synapse and recombination has begun. Then oocyte development is arrested in meiosis I. Thus, all primary oocytes have been formed by birth. When ovulation begins at puberty, meiosis is reinitiated in one egg during each ovulatory cycle and continues into meiosis II. The process is once again arrested after ovulation and is not completed unless fertilization occurs.

The end result of this progression is that each ovum that is released has been arrested in meiosis I for about a month longer than the one released during the preceding cycle. As a consequence, women 30 or 40 years old produce ova that are significantly older and that have been arrested longer than those they ovulated 10 or 20 years previously. However, no direct evidence proves that ovum age is the cause of the increased incidence of nondisjunction leading to Down syndrome.

These statistics obviously pose a serious problem for the woman who becomes pregnant late in her reproductive years. Genetic counseling early in such pregnancies is highly recommended. Counseling informs prospective parents about the probability that their child will be affected and educates them about Down syndrome. Although some individuals with Down syndrome must be institutionalized, others benefit greatly from special education programs and may be cared for at home. (Down syndrome children in general are noted for their affectionate, loving nature.) A genetic counselor may also recommend a prenatal diagnostic technique in which fetal cells are isolated and cultured. In **amniocentesis** and **chorionic villus sampling (CVS),** the two most familiar approaches, fetal cells are obtained from the amniotic fluid or the chorion of the placenta, respectively. In a newer approach, fetal cells are derived directly from the maternal circulation. Once perfected, this approach will be preferable because it is noninvasive, posing no risk to the fetus. After fetal cells are obtained, the karyotype can be determined by cytogenetic analysis. If the fetus is diagnosed as having Down syndrome, a therapeutic abortion is one option currently available to parents. Obviously, this is a difficult decision involving a number of religious and ethical issues.

Since Down syndrome is caused by a random error—nondisjunction of chromosome 21 during maternal or paternal meiosis—the occurrence of the disorder is *not* expected to be inherited. Nevertheless, Down syndrome occasionally runs in families. These instances, referred to as *familial Down syndrome,* involve a translocation of chromosome 21, another type of chromosomal aberration, which we will discuss later in the chapter.

Viability in Human Aneuploid Conditions

The reduced viability of individuals with recognized monosomic and trisomic conditions is evident. Only two other trisomies in humans survive to term. Both **Patau** and **Edwards syndromes** (**47,13+** and **47,18+**, respectively) result in severe malformations and early lethality. Figure 6–4 illustrates the abnormal karyotype and the many defects characterizing Patau infants.

Such observations lead us to believe that many other aneuploid conditions arise but that the affected fetuses do not survive to term. This observation has been confirmed by karyotypic analysis of spontaneously aborted fetuses. These studies reveal some rather striking statistics. At least 15 to 20 percent of all conceptions terminate in spontaneous abortion (some estimates are considerably higher). About 30 percent of all spontaneously aborted fetuses demonstrate some form of chromosomal anomaly, and approximately 90 percent of all chromosomal anomalies are terminated prior to birth as a result of spontaneous abortion.

A large percentage of fetuses demonstrating chromosomal abnormalities are aneuploids. The aneuploid with highest incidence among abortuses is the 45,X condition, which produces an infant with Turner syndrome if the fetus survives to term.

An extensive review of this subject by David H. Carr also reveals that a significant percentage of aborted fetuses are trisomic for one of the chromosome groups. Trisomies for every human chromosome have been recovered. Monosomies are seldom found, however, even though nondisjunction should produce $n - 1$ gametes with a frequency equal to $n + 1$ gametes. This finding suggests that gametes lacking a single chromosome are functionally impaired to a serious degree or that the embryo dies so early in its development that recovery occurs infrequently. Various forms of polyploidy and other miscellaneous chromosomal anomalies were also found in Carr's study.

These observations support the hypothesis that normal embryonic development requires a precise diploid complement of chromosomes to maintain the delicate equilibrium in the expression of genetic information. The prenatal mortality of most aneuploids provides a barrier against the introduction of these genetic anomalies into the human population.

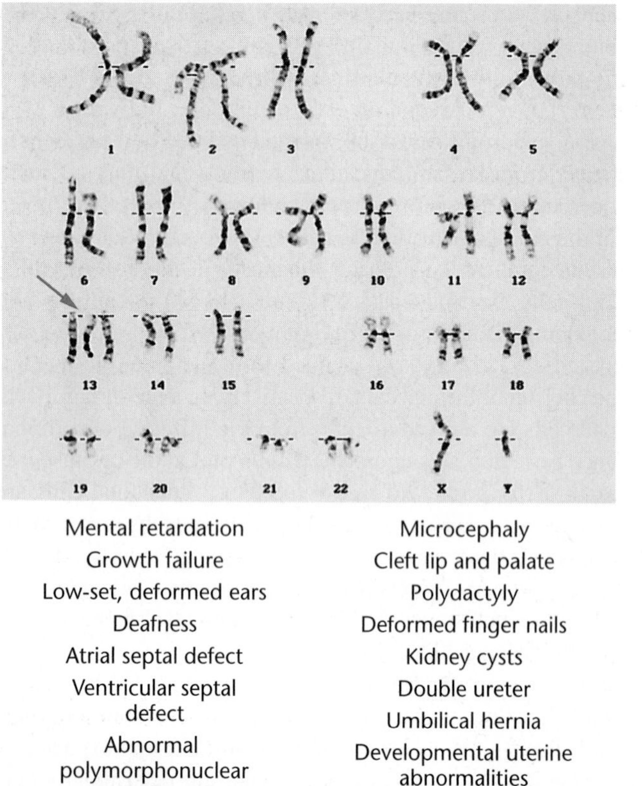

Mental retardation
Growth failure
Low-set, deformed ears
Deafness
Atrial septal defect
Ventricular septal defect
Abnormal polymorphonuclear granulocytes

Microcephaly
Cleft lip and palate
Polydactyly
Deformed finger nails
Kidney cysts
Double ureter
Umbilical hernia
Developmental uterine abnormalities
Cryptorchidism

FIGURE 6–4 The karyotype and phenotypic description of an infant with Patau syndrome, where three members of the D-group chromosome 13 are present, creating the 47,13+ condition.

ESSENTIAL POINT

Studies of monosomic and trisomic disorders are increasing our understanding of the delicate genetic balance that is essential for normal development.

6.3 Polyploidy, in Which More Than Two Haploid Sets of Chromosomes Are Present, Is Prevalent in Plants

POLYPLOIDY WEB TUTORIAL

The term *polyploidy* describes instances in which more than two multiples of the haploid chromosome set are found. The naming of polyploids is based on the number of sets of chromosomes found: A triploid has $3n$ chromosomes; a tetraploid has $4n$; a pentaploid, $5n$; and so forth (Table 6.1). Several general statements can be made about polyploidy. This condition is relatively infrequent in many animal species but is well known in lizards, amphibians, and fish, and is much more common in plant species. Usually, odd numbers of chromosome sets are not reliably maintained from generation to generation because a polyploid organism with an uneven number of homologs often does not produce genetically balanced gametes. For this reason, triploids, pentaploids, and so on, are not usually found in plant species that depend solely on sexual reproduction for propagation.

Polyploidy originates in two ways: (1) The addition of one or more extra sets of chromosomes, identical to the normal haploid complement of the same species, resulting in **autopolyploidy**; or (2) the combination of chromosome sets from different species occurring as a consequence of hybridization, resulting in **allopolyploidy** (from the Greek word *allo*, meaning "other" or "different"). The distinction between auto- and allopolyploidy is based on the genetic origin of the extra chromosome sets, as shown in Figure 6–5.

In our discussion of polyploidy, we use the following symbols to clarify the origin of additional chromosome sets. For example, if A represents the haploid set of chromosomes of any organism, then

$$A = a_1 + a_2 + a_3 + a_4 + \cdots + a_n$$

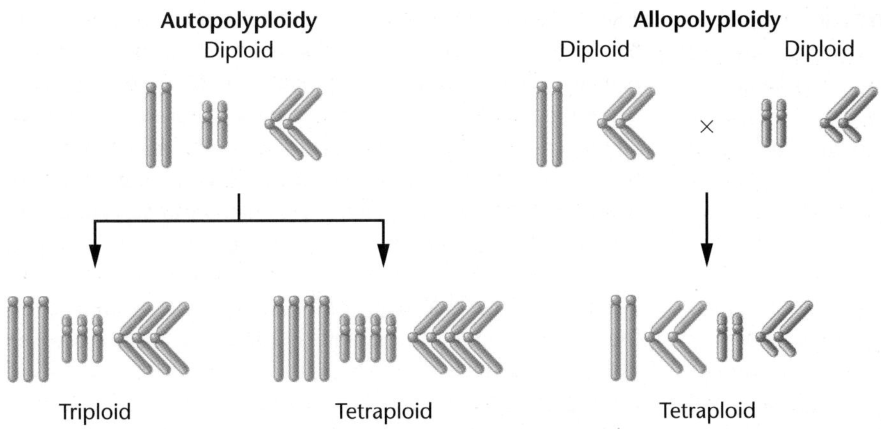

FIGURE 6–5 Contrasting chromosome origins of an autopolyploid versus an allopolyploid karyotype.

where a_1, a_2, and so on, are individual chromosomes and n is the haploid number. A normal diploid organism is represented simply as *AA*.

Autopolyploidy

In autopolyploidy, each additional set of chromosomes is identical to the parent species. Therefore, triploids are represented as *AAA,* tetraploids are *AAAA,* and so forth.

Autotriploids arise in several ways. A failure of all chromosomes to segregate during meiotic divisions can produce a diploid gamete. If such a gamete is fertilized by a haploid gamete, a zygote with three sets of chromosomes is produced. Or, rarely, two sperm may fertilize an ovum, resulting in a triploid zygote. Triploids are also produced under experimental conditions by crossing diploids with tetraploids. Diploid organisms produce gametes with n chromosomes, while tetraploids produce $2n$ gametes. Upon fertilization, the desired triploid is produced.

Because they have an even number of chromosomes, **autotetraploids** ($4n$) are theoretically more likely to be found in nature than are autotriploids. Unlike triploids, which often produce genetically unbalanced gametes with odd numbers of chromosomes, tetraploids are more likely to produce balanced gametes when involved in sexual reproduction.

How polyploidy arises naturally is of great interest to geneticists. In theory, if chromosomes have replicated, but the parent cell never divides and instead reenters interphase, the chromosome number will be doubled. That this very likely occurs is supported by the observation that tetraploid cells can be produced experimentally from diploid cells. This is accomplished by applying cold or heat shock to meiotic cells or by applying colchicine to somatic cells undergoing mitosis. **Colchicine,** an alkaloid derived from the autumn crocus, interferes with spindle formation, and thus replicated chromosomes cannot separate at anaphase and do not migrate to the poles. When colchicine is removed, the cell can reenter interphase. When the paired sister chromatids separate and uncoil, the nucleus contains twice the diploid number of chromosomes and is therefore $4n$. This process is shown in Figure 6–6.

In general, autopolyploids are larger than their diploid relatives. This increase seems to be due to larger cell size rather than greater cell number. Although autopolyploids do not contain new or unique information compared with their diploid relatives, the flower and fruit of plants are often increased in size, making such varieties of greater horticultural or commercial value. Economically important triploid plants include several potato species of the genus *Solanum,* Winesap apples, commercial bananas, seedless watermelons, and the cultivated tiger lily *Lilium tigrinum.* These plants are propagated asexually. Diploid bananas contain hard seeds, but the commercial, triploid, "seedless" variety has edible seeds. Tetraploid alfalfa, coffee, peanuts, and McIntosh apples are also of economic value because they are either larger or grow more vigorously than do their diploid or triploid counterparts. Many of the most popular varieties of hosta plant are tetraploid. In each case, leaves are thicker and larger, the foliage is more vivid, and the plant grows more vigorously. The commercial strawberry is an octoploid.

We have long been curious about how cells with increased ploidy values, where no new genes are present, express different phenotypes from their diploid counterparts. Our current ability to examine gene expression using modern biotechnology has provided some interesting insights. For example, Gerald Fink and his colleagues have been able to create strains of the yeast *Saccharomyces cerevisiae* with one, two, three, or four copies of the genome. Thus, each strain contains identical genes (they are said to be isogenic) but different ploidy values. These scientists then examined the expression levels of all genes during the entire cell cycle of the organism. Using the rather stringent standards of a tenfold increase or decrease of gene expression, Fink

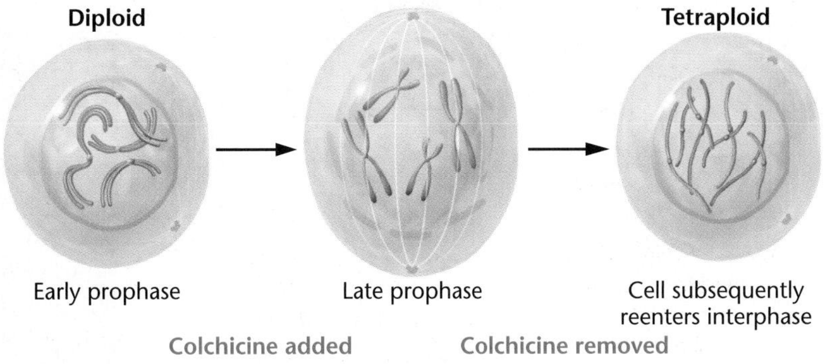

FIGURE 6–6 The potential involvement of colchicine in doubling the chromosome number. Two pairs of homologous chromosomes are shown. While each chromosome had replicated its DNA earlier during interphase, the chromosomes do not appear as double structures until late prophase. When anaphase fails to occur normally, the chromosome number doubles if the cell reenters interphase.

and coworkers proceeded to identify ten cases where, as ploidy increased, gene expression was increased at least tenfold and seven cases where it was reduced by a similar level.

One of these genes provides insights into how polyploid cells become larger than their haploid or diploid counterparts. In polyploid yeast, two **G1 cyclins,** Cln1, and Pcl1, are repressed when ploidy increases, while the size of the yeast cells increases. This is explained by the observation that G1 cyclins facilitate the cell's movement through G1, which is delayed when expression of these genes is repressed. The polyploid cell stays in the G1 phase longer and, on average, grows to a larger size before it moves beyond the G1 stage of the cell cycle. Yeast cells also show different morphology as ploidy increases. Several of the other genes, repressed as ploidy increases, have been linked to cytoskeletal dynamics that account for the morphological changes.

Allopolyploidy

Polyploidy can also result from hybridizing two closely related species. If a haploid ovum from a species with chromosome sets *AA* is fertilized by sperm from a species with sets *BB*, the resulting hybrid is *AB*, where $A = a_1, a_2, a_3, \ldots a_n$ and $B = b_1, b_2, b_3, \ldots b_n$. The hybrid plant may be sterile because of its inability to produce viable gametes. Most often, this occurs when some or all of the *a* and *b* chromosomes are not homologous and therefore cannot synapse in meiosis. As a result, unbalanced genetic conditions result. If, however, the new *AB* genetic combination undergoes a natural or an induced chromosomal doubling, two copies of all *a* chromosomes and two copies of all *b* chromosomes will be present, and they will pair during meiosis. As a result, a fertile *AABB* tetraploid is produced. These events are shown in Figure 6–7. Since this polyploid contains the equivalent of four haploid genomes derived from separate species, such an organism is called an **allotetraploid.** When both original species are known, an equivalent term, **amphidiploid,** is preferred in describing the allotetraploid.

Amphidiploid plants are often found in nature. Their reproductive success is based on their potential for forming balanced gametes. Since two homologs of each specific chromosome are present, meiosis occurs normally (Figure 6–7) and fertilization successfully propagates the plant sexually. This discussion assumes the simplest situation, where none of the chromosomes in set *A* are homologous to those in set *B*. In amphidiploids, formed from closely related species, some homology between *a* and *b* chromosomes is likely. Allopolyploids are rare in most animals because mating behavior is most often species-specific, and thus the initial step in hybridization is unlikely to occur.

A classical example of amphidiploidy in plants is the cultivated species of American cotton, *Gossypium* (Figure 6–8). This species has 26 pairs of chromosomes: 13 are large and 13 are much smaller. When it was discovered that Old World cotton had only 13 pairs of large chromosomes, allopolyploidy was suspected. After an examination of wild American cotton revealed 13 pairs of small chromosomes, this speculation was strengthened. J. O. Beasley reconstructed the origin of cultivated cotton experimentally by crossing the Old World strain with the wild American strain and then treating the hybrid with colchicine to double the chromosome number. The result of these treatments was a fertile amphidiploid variety of

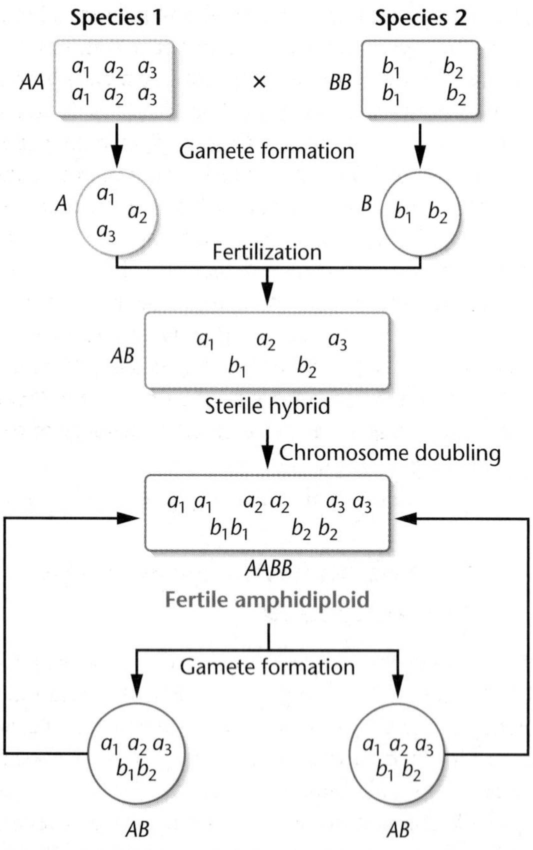

FIGURE 6–7 The origin and propagation of an amphidiploid. Species 1 contains genome *A* consisting of three distinct chromosomes, a_1, a_2, and a_3. Species 2 contains genome *B* consisting of two distinct chromosomes, b_1 and b_2. Following fertilization between members of the two species and chromosome doubling, a fertile amphidiploid containing two complete diploid genomes (*AABB*) is formed.

FIGURE 6–8 The pods of the amphidiploid form of *Gossypium,* the cultivated American cotton plant.

cotton. It contained 26 pairs of chromosomes as well as characteristics similar to the cultivated variety.

Amphidiploids often exhibit traits of both parental species. An interesting example, but one with no practical economic importance, is that of the hybrid formed between the radish *Raphanus sativus* and the cabbage *Brassica oleracea.* Both species have a haploid number $n = 9$. The initial hybrid consists of nine *Raphanus* and nine *Brassica* chromosomes (9R + 9B). Although hybrids are almost always sterile, some fertile amphidiploids (18R + 18B) have been produced. Unfortunately, the root of this plant is more like the cabbage and its shoot more like the radish; had the converse occurred, the hybrid might have been of economic importance.

A much more successful commercial hybridization uses the grasses wheat and rye. Wheat (genus *Triticum*) has a basic haploid genome of seven chromosomes. In addition to normal diploids ($2n = 14$), cultivated autopolyploids exist, including tetraploid ($4n = 28$) and hexaploid ($6n = 42$) species. Rye (genus *Secale*) also has a genome consisting of seven chromosomes. The only cultivated species is the diploid plant ($2n = 14$).

Using the technique outlined in Figure 6–7, geneticists have produced various hybrids. When tetraploid wheat is crossed with diploid rye and the F_1 plants are treated with colchicine, a hexaploid variety ($6n = 42$) is obtained; the hybrid, designated *Triticale,* represents a new genus. Fertile hybrid varieties derived from various wheat and rye species can be crossed or backcrossed. These crosses have created many variations of the genus *Triticale.* The hybrid plants demonstrate characteristics of both wheat and rye. For example, certain hybrids combine the high-protein content of wheat with rye's high content of the amino acid lysine. (The lysine content is low in wheat and thus is a limiting nutritional factor.) Wheat is considered to be a high-yielding grain, whereas rye is noted for its versatility of growth in unfavorable environments. *Triticale* species, which combine both traits, have the potential of significantly increasing grain production. Programs designed to improve crops through hybridization have long been under way in several developing countries.

ESSENTIAL POINT

When complete sets of chromosomes are added to the diploid genome, these sets can have an identical or diverse genetic origin, creating either autopolyploidy or allopolyploidy, respectively.

NOW SOLVE THIS

In Problem 7 on page 130, you are asked to consider the sterility of a hybrid plant derived from two different species that is more ornate than either of its parents.

Hint: Allopolyploid plants are often sterile when they contain an odd number of each chromosome, resulting in unbalanced gametes during meiosis.

6.4 Variation Occurs in the Composition and Arrangement of Chromosomes

WEB TUTORIAL: ABERRATIONS

The second general class of chromosome aberrations includes changes that delete, add, or rearrange substantial portions of one or more chromosomes. Included in this broad category are deletions and duplications of genes or part of a chromosome and rearrangements of genetic material in which a chromosome segment is inverted, exchanged with a segment of a nonhomologous chromosome, or merely transferred to another chromosome. Exchanges and transfers are called translocations, in which the locations of genes are altered within the genome. These types of chromosome alterations are illustrated in Figure 6–9, p. 120.

In most instances, these structural changes are due to one or more breaks along the axis of a chromosome, followed by either the loss or rearrangement of genetic material. Chromosomes can break spontaneously, but the rate of breakage may increase in cells exposed to chemicals or radiation. Although the actual ends of chromosomes, known as telomeres, do not readily fuse with newly created ends of "broken" chromosomes or with other telomeres, the ends produced at points of breakage are "sticky" and can rejoin other broken ends. If breakage and rejoining do not reestablish the original relationship and if the alteration occurs in germ plasm, the gametes will contain the structural rearrangement, which is heritable.

If the aberration is found in one homolog, but not the other, the individual is said to be *heterozygous for the aberration.* In such cases, unusual but characteristic pairing configurations are formed during meiotic synapsis. These patterns are useful in identifying the type of change that has occurred. If no loss or gain of genetic material occurs, individuals bearing the aberration "heterozygously" are likely to be unaffected phenotypically. However, the unusual pairing arrangements often lead to gametes that are duplicated or deficient for some chromosomal regions. When this occurs, the offspring of "carriers" of certain aberrations have an increased probability of demonstrating phenotypic changes.

6.5 A Deletion Is a Missing Region of a Chromosome

When a chromosome breaks in one or more places and a portion of it is lost, the missing piece is called a **deletion** (or a **deficiency**). The deletion can occur either near one end or within the interior of the chromosome. These are **terminal** and **intercalary deletions,** respectively [Figure 6–10(a) and (b)], p. 120. The portion of the chromosome that retains the centromere region is usually maintained when the cell divides, whereas the segment without the centromere is eventually lost in progeny cells following mitosis or meiosis. For synapsis to occur between a chromosome with a large intercalary deletion and a normal homolog, the unpaired region of the normal homolog must "buckle out" into a **deletion,** or **compensation, loop** [Figure 6–10(c)], p. 120.

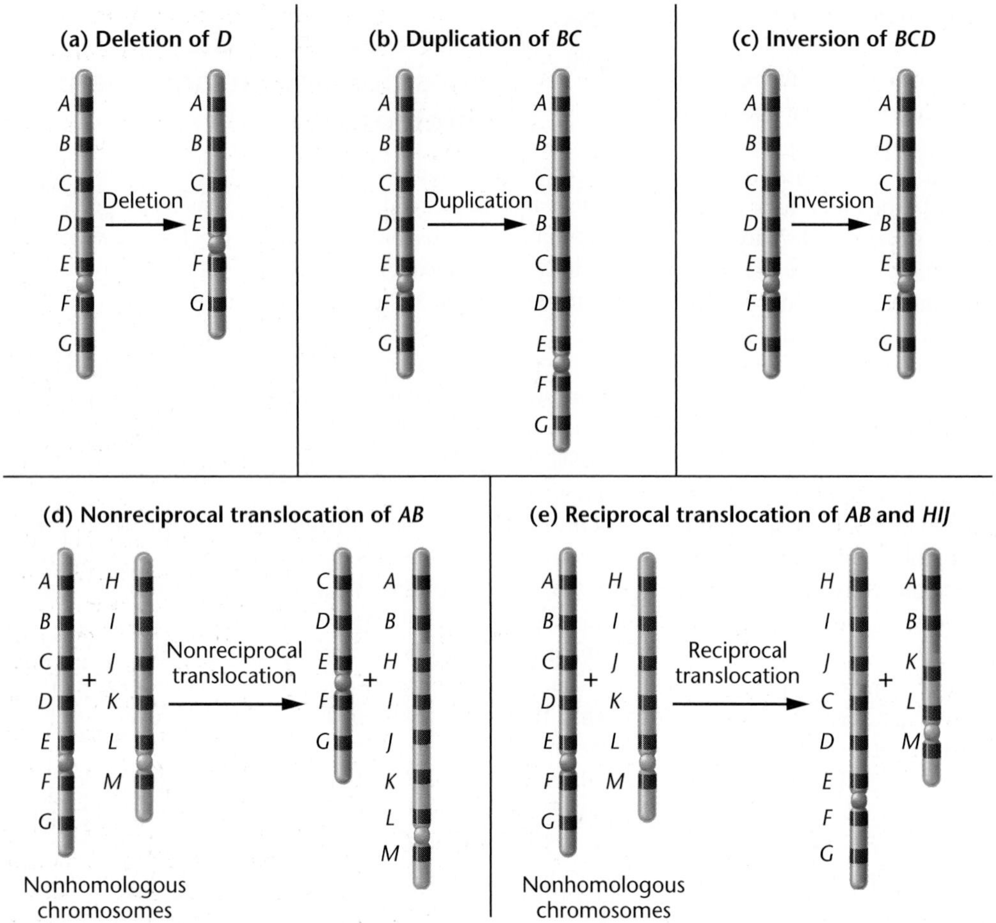

FIGURE 6–9 An overview of the five different types of gain, loss, or rearrangement of chromosome segments.

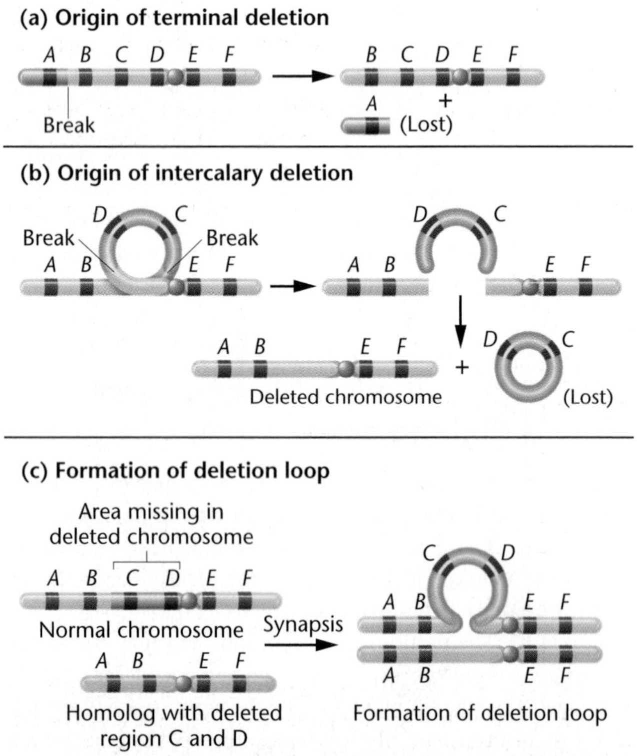

FIGURE 6–10 Origins of (a) a terminal and (b) an intercalary deletion. In (c), pairing occurs between a normal chromosome and one with an intercalary deletion by looping out the undeleted portion to form a deletion (or compensation) loop.

If only a small part of a chromosome is deleted, the organism might survive. However, a deletion of a portion of a chromosome need not be very great before the effects become severe. We see an example of this in the following discussion of the cri du chat syndrome in humans. If even more genetic information is lost as a result of a deletion, the aberration is often lethal, in which case the chromosome mutation never becomes available for study.

Cri du Chat Syndrome in Humans

In humans, the **cri du chat syndrome** results from the deletion of a small terminal portion of chromosome 5. It might be considered a case of *partial monosomy,* but since the region that is missing is so small, it is better referred to as a **segmental deletion.** This syndrome was first reported by Jérôme LeJeune in 1963, when he described the clinical symptoms, including an eerie cry similar to the meowing of a cat, after which the syndrome is named. This syndrome is associated with the loss of a small, variable part of the short arm of chromosome 5 (Figure 6-11). Thus, the genetic constitution may be designated as **46, 5p−,** meaning that the individual has all 46 chromosomes but that some or all of the p arm (the petite, or short, arm) of one member of the chromosome 5 pair is missing.

Infants with this syndrome may exhibit anatomic malformations, including gastrointestinal and cardiac complications, and they are often mentally retarded. Abnormal development

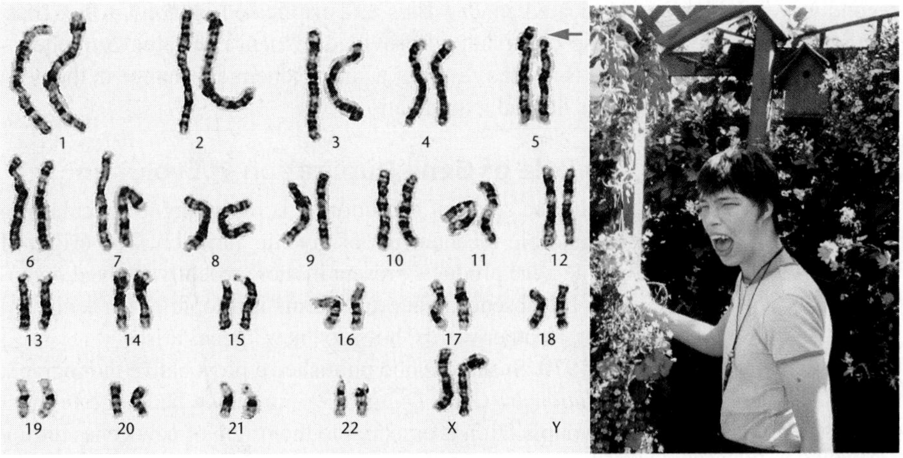

FIGURE 6–11 A representative karyotype and a photograph of a child exhibiting cri du chat syndrome (46,5p–). In the karyotype, the arrow identifies the absence of a small piece of the short arm of one member of the chromosome 5 homologs.

of the glottis and larynx (leading to the characteristic cry) is typical of this syndrome.

Since 1963, hundreds of cases of cri du chat syndrome have been reported worldwide. An incidence of 1 in 25,000–50,000 live births has been estimated. Most often, the condition is not inherited but instead results from the sporadic loss of chromosomal material in gametes. The length of the short arm that is deleted varies somewhat; longer deletions appear to have a greater impact on the physical, psychomotor, and mental skill levels of those children who survive. Although the effects of the syndrome are severe, most individuals achieve motor and language skills and may be home-cared. In 2004, it was reported that the portion of the chromosome that is missing contains the *TERT* gene, which encodes telomerase reverse transcriptase, an enzyme essential for the maintenance of telomeres during DNA replication. Whether the absence of this gene on one homolog is related to the multiple phenotypes of cri du chat infants is still unknown.

6.6 A Duplication Is a Repeated Segment of a Chromosome

When any part of the genetic material—a single locus or a large piece of a chromosome—is present more than once in the genome, it is called a **duplication.** As in deletions, pairing in heterozygotes can produce a compensation loop. Duplications may arise as the result of unequal crossing over between synapsed chromosomes during meiosis (Figure 6–12) or through a replication error prior to meiosis. In the former case, both a duplication and a deletion are produced.

We consider three interesting aspects of duplications. First, they may result in gene redundancy. Second, as with deletions, duplications may produce phenotypic variation. Third, according to one convincing theory, duplications have also been an important source of genetic variability during evolution.

Gene Redundancy and Amplification—Ribosomal RNA Genes

Although many gene products are not needed in every cell of an organism, other gene products are known to be essential components of all cells. For example, ribosomal RNA must be present in abundance to support protein synthesis. The more metabolically active a cell is, the higher the demand for this molecule. We might hypothesize that a single copy of the gene encoding rRNA is inadequate in many cells. Studies using the technique of molecular hybridization, which enables us to determine the percentage of the genome that codes for specific RNA sequences, show that our hypothesis is correct. Indeed, multiple copies of genes code for rRNA. Such DNA is called **rDNA,** and the general phenomenon is referred to as *gene redundancy.* For example, in the common intestinal bacterium *Escherichia coli* (*E. coli*), about 0.7 percent of the haploid genome consists of rDNA—the equivalent of seven copies of the gene. In *Drosophila melanogaster,* 0.3 percent of the haploid genome, equivalent to 130 gene copies, consists of rDNA. Although the presence of multiple copies of the same gene is not restricted to those coding for rRNA, we will focus on them in this section.

In some cells, particularly oocytes, even the normal redundancy of rDNA is insufficient to provide adequate amounts of rRNA needed to construct ribosomes. Oocytes store abundant nutrients, including huge quantities of ribosomes, for use by the embryo during early development. More ribosomes are included in oocytes than in any other cell type. By considering how the amphibian *Xenopus laevis* acquires this

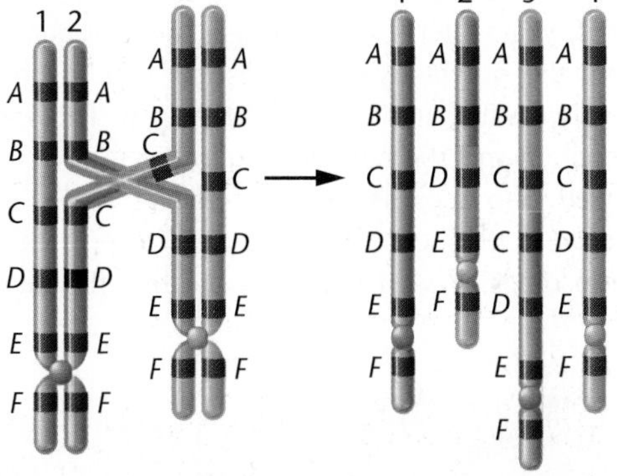

FIGURE 6–12 The origin of duplicated and deficient regions of chromosomes as a result of unequal crossing over. The tetrad on the left is mispaired during synapsis. A single crossover between chromatids 2 and 3 results in the deficient (chromosome 2) and duplicated (chromosome 3) chromosomal regions shown on the right. The two chromosomes uninvolved in the crossover event remain normal in gene sequence and content.

abundance of ribosomes, we shall see a second way in which the amount of rRNA is increased. This phenomenon is called **gene amplification.**

The genes that code for rRNA are located in an area of the chromosome known as the **nucleolar organizer region (NOR).** The NOR is intimately associated with the nucleolus, which is a processing center for ribosome production. Molecular hybridization analysis has shown that each NOR in the frog *Xenopus* contains the equivalent of 400 redundant gene copies coding for rRNA. Even this number of genes is apparently inadequate to synthesize the vast amount of ribosomes that must accumulate in the amphibian oocyte to support development following fertilization.

To further amplify the number of rRNA genes, the rDNA is selectively replicated, and each new set of genes is released from its template. Because each new copy is equivalent to NOR, multiple small nucleoli are formed around each NOR in the oocyte. As many as 1500 of these "micronucleoli" have been observed in a single oocyte. If we multiply the number of micronucleoli (1500) by the number of gene copies in each NOR (400), we see that amplification in *Xenopus* oocytes can result in over half a million gene copies! If each copy is transcribed only 20 times during the maturation of the oocyte, in theory, sufficient copies of rRNA are produced to result in well over 12 million ribosomes.

The *Bar* Mutation in *Drosophila*

Duplications can cause phenotypic variation that might at first appear to be caused by a simple gene mutation. The *Bar*-eye phenotype in *Drosophila* (Figure 6–13) is a classic example. Instead of the normal oval-eye shape, *Bar*-eyed flies have narrow, slitlike eyes. This phenotype is inherited in the same way as a dominant X-linked mutation.

In the early 1920s, Alfred H. Sturtevant and Thomas H. Morgan discovered and investigated this "mutation." Normal wild-type females (B^+/B^+) have about 800 facets in each eye. Heterozygous females (B/B^+) have about 350 facets, while homozygous females (B/B) average only about 70 facets. Females were occasionally recovered with even fewer facets and were designated as *double Bar* (B^D/B^+).

About 10 years later, Calvin Bridges and Herman J. Muller compared the polytene X chromosome banding pattern of the *Bar* fly with that of the wild-type fly. These chromosomes contain specific banding patterns that have been well categorized into regions. Their studies revealed that one copy of the region designated as 16A is present on both X-chromosomes of wild-type flies but that this region was duplicated in *Bar* flies and triplicated in *double Bar* flies. These observations provided evidence that the *Bar* phenotype is not the result of a simple chemical change in the gene but is instead a duplication.

The Role of Gene Duplication in Evolution

During the study of evolution, it is intriguing to speculate on the possible mechanisms of genetic variation. The origin of unique gene products present in more recently evolved organisms but absent in ancestral forms is a topic of particular interest. In other words, how do "new" genes arise?

In 1970, Susumo Ohno published a provocative monograph, *Evolution by Gene Duplication,* in which he suggested that gene duplication is essential to the origin of new genes during evolution. Ohno's thesis is based on the supposition that the gene products of many genes, present as only a single copy in the genome, are indispensable to the survival of members of any species during evolution. Therefore, unique genes are not free to accumulate mutations sufficient to alter their primary function and give rise to new genes.

However, if an essential gene is duplicated in the germ line, major mutational changes in this extra copy will be tolerated in future generations because the original gene provides the genetic information for its essential function. The duplicated copy will be free to acquire many mutational changes over extended periods of time. Over short intervals, the new genetic information may be of no practical advantage. However, over long evolutionary periods, the duplicated gene may change sufficiently so that its product assumes a divergent role in the cell. The new function may impart an "adaptive" advantage to organisms, enhancing their fitness. Ohno has outlined a mechanism through which sustained genetic variability may have originated.

Ohno's thesis is supported by the discovery of genes that have a substantial amount of their DNA sequence in common, but whose gene products are distinct. For example, trypsin and chymotrypsin fit this description, as do myoglobin and hemoglobin. The DNA sequence is so similar (homologous) in each case that we may conclude that members of each pair of genes arose from a common ancestral gene through duplication. During evolution, the related genes diverged sufficiently that their products became unique.

Other support includes the presence of **gene families**—groups of contiguous genes whose products perform the same function. Again, members of a family show DNA sequence homology sufficient to conclude that they share a common origin. Examples are the various types of human hemoglobin

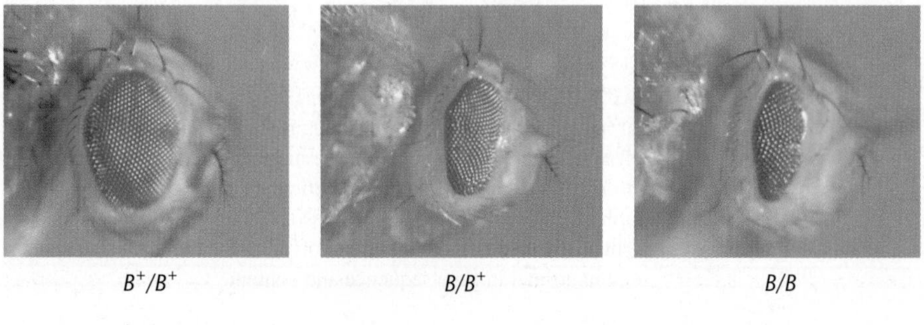

FIGURE 6–13 *Bar*-eye phenotypes in contrast to the wild-type eye in *Drosophila*.

polypeptide chains, as well as the immunologically important T-cell receptors and antigens encoded by the major histocompatibility complex.

Many recent findings derived from our ability to sequence entire genomes lend support to the idea that gene duplication has been a common feature of evolutionary progression. For example, Jurg Spring has compared a large number of genes in *Drosophila* and their counterparts in humans. In 50 genes studied, the fruit fly has only one copy of each, while there are multiple copies present in the human genome. Many other investigations have provided similar findings. For example, in the mustard plant *Arabidopsis thaliana,* approximately 70 percent of the genome is duplicated. In humans, there are 1077 duplicated blocks of genes, with 781 of them containing five or more copies. Chromosomes 18 and 20 contain large duplicated areas accounting for almost half of each chromosome.

A new debate has begun concerning a second aspect of Ohno's thesis—that major evolutionary jumps, such as the transition from invertebrates to vertebrates, may have involved the duplication of entire genomes. Ohno has suggested that this might have occurred several times during the course of evolution. Although it seems clear that genes, and even segments of chromosomes, have been duplicated, there is not yet any compelling evidence to convince evolutionary biologists that genome duplication has been responsible. However, from the standpoint of gene expression, duplicating all genes proportionately seems more likely to be tolerated than duplicating just a portion of them. Whatever may be the case, it is interesting to examine evidence based on new technology that tests a hypothesis that was proposed almost 40 years ago, long before that technology was available.

ESSENTIAL POINT

Deletions or duplications of segments of a gene or a chromosome may be the source of mutant phenotypes such as cri du chat syndrome in humans and *Bar* eyes in *Drosophila,* while duplications can be particularly important as a source of redundant or new genes.

6.7 Inversions Rearrange the Linear Gene Sequence

The **inversion,** another class of structural variation, is a type of chromosomal aberration in which a segment of a chromosome is turned around 180 degrees within a chromosome. An inversion does not involve a loss of genetic information but simply rearranges the linear gene sequence. An inversion requires breaks at two points along the length of the chromosome and subsequent reinsertion of the inverted segment. Figure 6–14, p. 124, illustrates how an inversion might arise. By forming a chromosomal loop prior to breakage, the newly created "sticky ends" are brought close together and rejoined.

The inverted segment may be short or quite long and may or may not include the centromere. If the centromere is not part of the rearranged chromosome segment, it is a **paracentric inversion.** If the centromere is part of the inverted segment, it is described as a **pericentric inversion,** which is the type shown in Figure 6–14.

Although inversions appear to have a minimal impact on the individuals bearing them, their consequences are of great interest to geneticists. Organisms that are heterozygous for

Copy Number Variants (CNVs)—Duplications and Deletions at the Molecular Level

Genomic investigations that focus on the DNA sequences in humans are providing new insights into our understanding of duplications and deletions. These variations, often involving thousands or even millions of base pairs, and even entire genes, may prove to play crucial roles in many of our individual attributes, such as sensitivity to drugs and susceptibility to disease. These differences, because they represent large DNA sequences, are termed **copy number variants (CNVs),** and are found in both coding and noncoding regions of the genome. Mutant CNVs show an increase or a decrease in copy number in comparison to a reference genome from a normal individual.

In 2004, two research groups independently described the presence of CNVs in the genomes of healthy individuals with no known genetic disorders. CNVs were defined as regions of DNA at least 1 kb in length (1000 base pairs) that display at least 90 percent sequence identity. This initial study revealed 50 loci consisting of CNVs, and in 2005, several other groups began sifting through the genome in search of CNVs, defining almost 300 additional sites. The current number of CNV sites has now risen to include approximately 12 percent of the human genome.

Current CNV studies have focused on finding associations with human diseases. CNVs appear to have both positive and negative associations with many diseases in which the genetic basis is not yet fully understood. For example, an association has been reported between CNVs and autism, a neurodevelopmental disorder that impairs communication, behavior, and social interaction. Interestingly, a mutant CNV site has been found to appear *de novo* (anew) in 10 percent of so-called sporadic cases of autism, where unaffected parents lack the CNV mutation. This is in contrast to only 2 percent of affected individuals where the disease appears to be familial (run in the family). Similarly, a higher than average copy number of the gene *CCL3L1* imparts an HIV-suppressive effect during viral infection, diminishing the progression to AIDS. Another research group has associated specific mutant CNV sites with certain subset populations of individuals with lung cancer—the greater number of copies of the *EGFR* (*Epidermal Growth Factor Receptor*) gene, the more responsive are patients with non–small-cell lung cancer to treatment. Finally, the greater the reduction in copy number of the gene designated *DEFB,* the greater the risk of developing Crohn's disease, a condition affecting the colon. Relevant to this chapter, these findings reveal that duplications and deletions are no longer restricted to textbook examples of these chromosomal mutations.

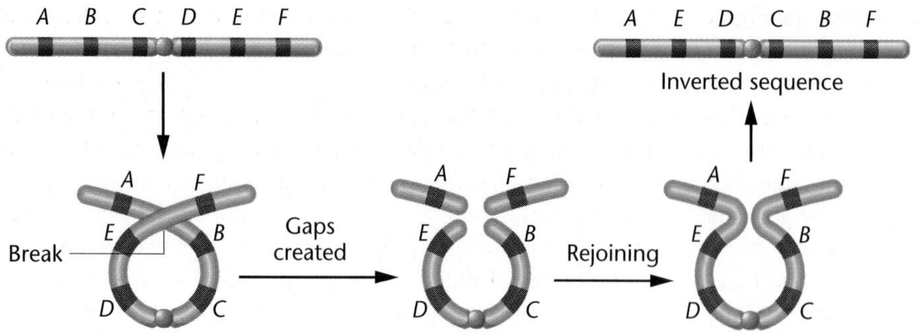

FIGURE 6–14 One possible origin of a pericentric inversion.

inversions may produce aberrant gametes that have a major impact on their offspring.

Consequences of Inversions during Gamete Formation

If only one member of a homologous pair of chromosomes has an inverted segment, normal linear synapsis during meiosis is not possible. Organisms with one inverted chromosome and one noninverted homolog are called **inversion heterozygotes.** Pairing between two such chromosomes in meiosis is accomplished only if they form an **inversion loop** (Figure 6–15).

If crossing over does not occur within the inverted segment of the inversion loop, the homologs will segregate, which results in two normal and two inverted chromatids that are distributed into gametes. However, if crossing over does occur within the inversion loop, abnormal chromatids are produced. The effect of a single crossover (SCO) event within a paracentric inversion is diagrammed in Figure 6–15(a).

In any meiotic tetrad, a single crossover between nonsister chromatids produces two parental chromatids and two recombinant chromatids. When the crossover occurs within a paracentric inversion, however, one recombinant **dicentric chromatid** (two centromeres) and one recombinant **acentric chromatid** (lacking a centromere) are produced. Both contain duplications and deletions of chromosome segments as well. During anaphase, an acentric chromatid moves randomly to one pole or the other or may be lost, while a dicentric chromatid is pulled in two directions. This polarized movement produces *dicentric bridges* that are cytologically recognizable. A dicentric chromatid usually breaks at some point so that part of the chromatid goes into one gamete and part into another gamete during the reduction divisions. Therefore, gametes containing either recombinant chromatid are deficient in genetic material. When such a gamete participates in fertilization, the zygote most often develops abnormally, if at all.

A similar chromosomal imbalance is produced as a result of a crossover event between a chromatid bearing a pericentric

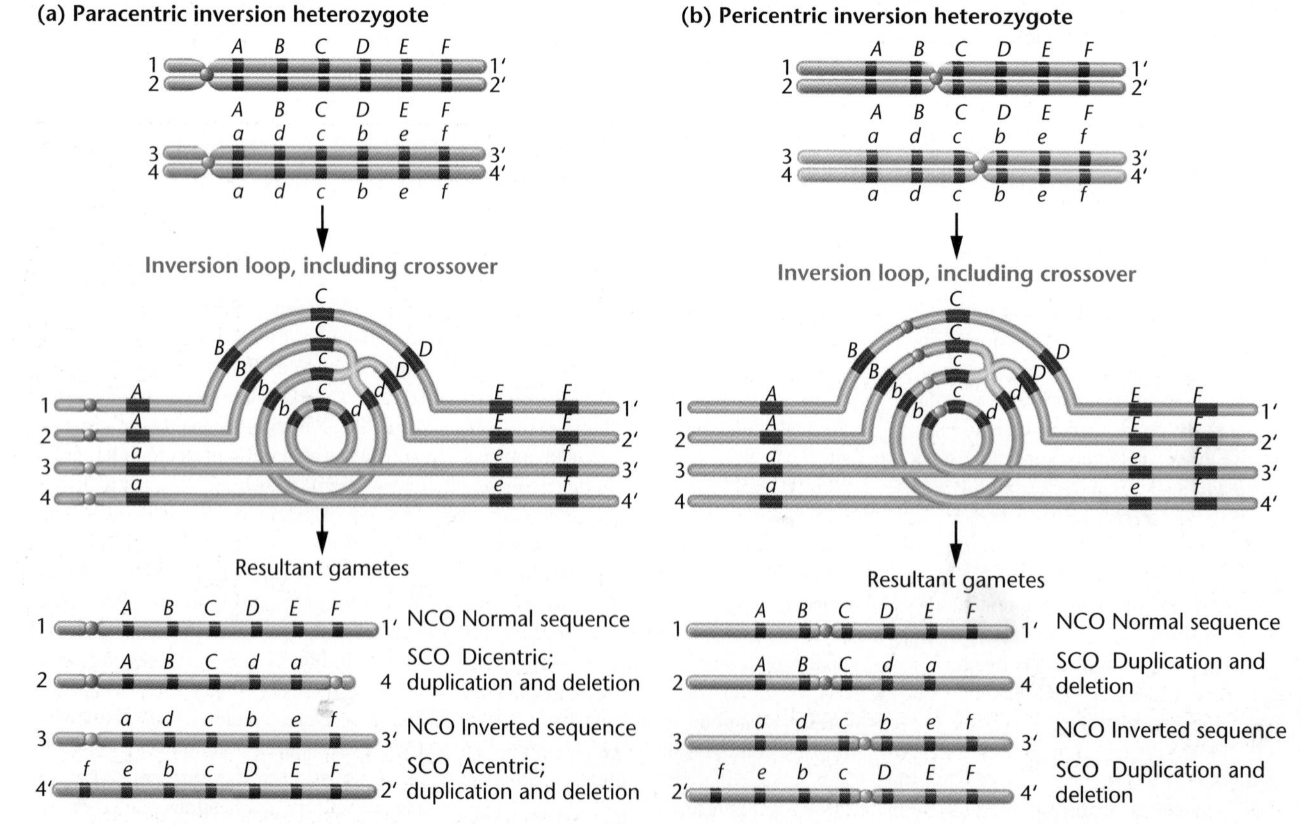

FIGURE 6–15 (a) The effects of a single crossover (SCO) within an inversion loop in a paracentric inversion heterozygote, where two altered chromosomes are produced, one acentric and one dicentric. Both chromosomes also contain duplicated and deficient regions. (b) The effects of a crossover in a pericentric inversion heterozygote, where two altered chromosomes are produced, both with duplicated and deficient regions.

inversion and its noninverted homolog, as shown in Figure 6–15(b). The recombinant chromatids that are directly involved in the exchange have duplications and deletions. In plants, gametes receiving such aberrant chromatids fail to develop normally, leading to aborted pollen or ovules. Thus, lethality occurs prior to fertilization, and inviable seeds result. In animals, the gametes have developed prior to the meiotic error, so fertilization is more likely to occur in spite of the chromosome error. However, the end result is the production of inviable embryos following fertilization. In both cases, viability is reduced.

Because offspring bearing crossover gametes are inviable and not recovered, it *appears* as if the inversion suppresses crossing over. Actually, in inversion heterozygotes, the inversion has the effect of *suppressing the recovery of crossover products* when chromosome exchange occurs within the inverted region. If crossing over always occurred within a paracentric or pericentric inversion, 50 percent of the gametes would be ineffective. The viability of the resulting zygotes is therefore greatly diminished. Furthermore, up to one-half of the viable gametes have the inverted chromosome, and the inversion will be perpetuated within the species. The cycle will be repeated continuously during meiosis in future generations.

Evolutionary Advantages of Inversions

One major effect of an inversion is the preservation of a set of specific alleles at a series of adjacent loci, provided that they are contained within the inversion. Because the recovery of crossover products is suppressed in inversion heterozygotes, a particular combination of alleles is preserved intact in the viable gametes. If the alleles of the involved genes confer a survival advantage on organisms maintaining them, the inversion is beneficial to the evolutionary survival of the species. For example, say that the set of alleles *ABcDef* is more adaptive than the sets *AbCdeF* or *abcdEF*. Effective gametes will contain this favorable set of genes, undisrupted by crossing over, if it is located within a heterozygous inversion.

NOW SOLVE THIS

Problem 23 on page 131 considers a prospective male parent with a family history of stillbirths and malformed babies. He was shown to contain an inversion covering 70 percent of one member of chromosome 1. You are asked to explain the cause of the stillbirths.

Hint: Human chromosome 1 is metacentric; therefore, an inversion covering 70 percent of the chromosome is no doubt a pericentric inversion.

6.8 Translocations Alter the Location of Chromosomal Segments in the Genome

Translocation, as the name implies, is the movement of a chromosomal segment to a new location in the genome. Reciprocal translocation, for example, involves the exchange of segments between two nonhomologous chromosomes. The least complex way for this event to occur is for two nonhomologous chromosome arms to come close to each other so that an exchange is facilitated. Figure 6–16(a) shows a simple reciprocal translocation in which only two breaks are required. If the exchange includes internal chromosome segments, four breaks are required, two on each chromosome.

The genetic consequences of reciprocal translocations are, in several instances, similar to those of inversions. For example, genetic information is not lost or gained. Rather, there is only a rearrangement of genetic material. The presence of a translocation does not, therefore, directly alter the viability of individuals bearing it.

Homologs that are heterozygous for a reciprocal translocation undergo unorthodox synapsis during meiosis. As shown in Figure 6–16(b), pairing results in a crosslike configuration.

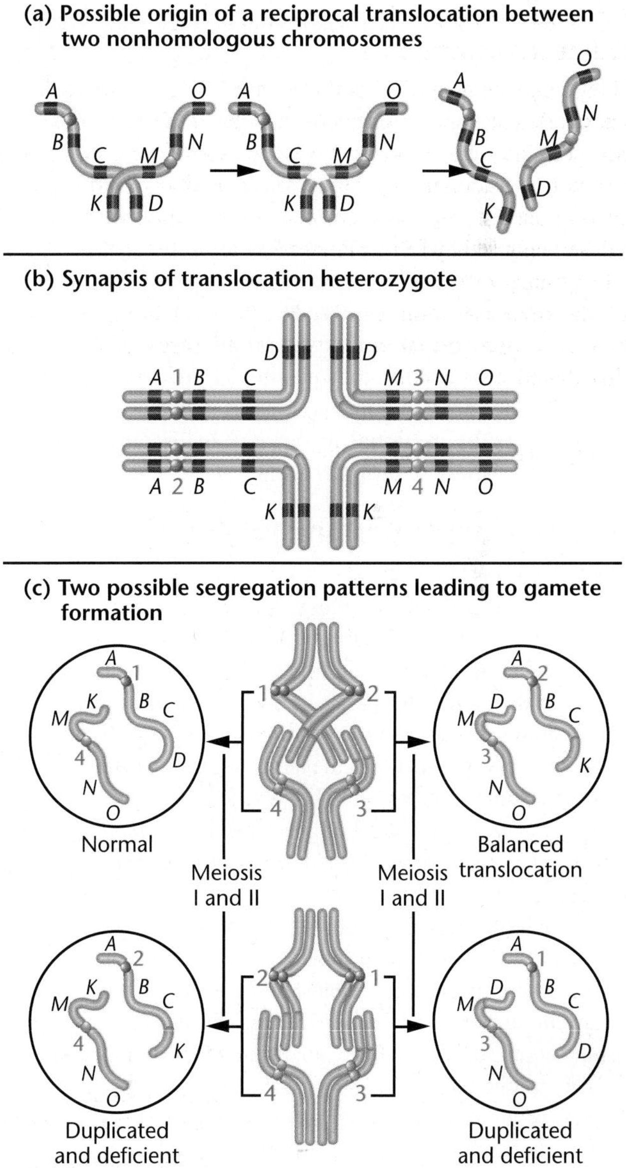

FIGURE 6–16 (a) Possible origin of a reciprocal translocation. (b) Synaptic configuration formed during meiosis in an individual that is heterozygous for the translocation. (c) Two possible segregation patterns, one of which leads to a normal and a balanced gamete (called alternate segregation) and one that leads to gametes containing duplications and deficiencies (called adjacent segregation).

As with inversions, genetically unbalanced gametes are also produced as a result of this unusual alignment during meiosis. In the case of translocations, however, aberrant gametes are not necessarily the result of crossing over. To see how unbalanced gametes are produced, focus on the homologous centromeres in Figure 6–16(b) and Figure 6–16(c), p. 125. According to the principle of independent assortment, the chromosome containing centromere 1 migrates randomly toward one pole of the spindle during the first meiotic anaphase; it travels along with *either* the chromosome having centromere 3 *or* the chromosome having centromere 4. The chromosome with centromere 2 moves to the other pole along with *either* the chromosome containing centromere 3 *or* centromere 4. This results in four potential meiotic products. The 1,4 combination contains chromosomes that are not involved in the translocation. The 2,3 combination, however, contains translocated chromosomes. These contain a complete complement of genetic information and are balanced. The other two potential products, the 1,3 and 2,4 combinations, contain chromosomes displaying duplicated and deleted segments. To simplify matters, crossover exchanges are ignored here.

When incorporated into gametes, the resultant meiotic products are genetically unbalanced. If they participate in fertilization, lethality often results. As few as 50 percent of the progeny of parents that are heterozygous for a reciprocal translocation survive. This condition, called *semisterility,* has an impact on the reproductive fitness of organisms, thus playing a role in evolution. Furthermore, in humans, such an unbalanced condition results in partial monosomy or trisomy, leading to a variety of birth defects.

Translocations in Humans: Familial Down Syndrome

Research conducted since 1959 has revealed numerous translocations in members of the human population. One common type of translocation involves breaks at the extreme ends of the short arms of two nonhomologous acrocentric chromosomes. These small segments are lost, and the larger segments fuse at their centromeric region. This type of translocation produces a new, large submetacentric or metacentric chromosome, often called a **Robertsonian translocation.**

One such translocation accounts for cases in which Down syndrome is familial (inherited). Earlier in this chapter, we pointed out that most instances of Down syndrome are due to trisomy 21. This chromosome composition results from nondisjunction during meiosis in one parent. Trisomy accounts for over 95 percent of all cases of Down syndrome. In such instances, the chance of the same parents producing a second affected child is extremely low. However, in the remaining families with a Down child, the syndrome occurs in a much higher frequency over several generations.

Cytogenetic studies of the parents and their offspring from these unusual cases explain the cause of **familial Down syndrome.** Analysis reveals that one of the parents contains a 14/21, D/G translocation (Figure 6–17). That is, one parent has the majority of the G-group chromosome 21 translocated to one end of the D-group chromosome 14. This individual is phenotypically normal, even though he or she has only 45 chromosomes. During meiosis, one-fourth of the individual's gametes have two copies of chromosome 21: a normal chromosome and a second copy translocated to chromosome 14. When such a gamete is fertilized by a standard haploid gamete, the resulting zygote has 46 chromosomes but three copies of chromosome 21. These individuals exhibit Down syndrome. Other potential surviving offspring contain either the standard diploid genome (without a translocation) or the balanced translocation like the parent. Both cases result in normal individuals. Knowledge of translocations has allowed geneticists to resolve the seeming paradox of an inherited trisomic phenotype in an individual with an apparent diploid number of chromosomes.

It is interesting to note that the "carrier," who has 45 chromosomes and exhibits a normal phenotype, does not contain the *complete* diploid amount of genetic material. A small region is lost from both chromosomes 14 and 21 during the translocation event. This occurs because the ends of both

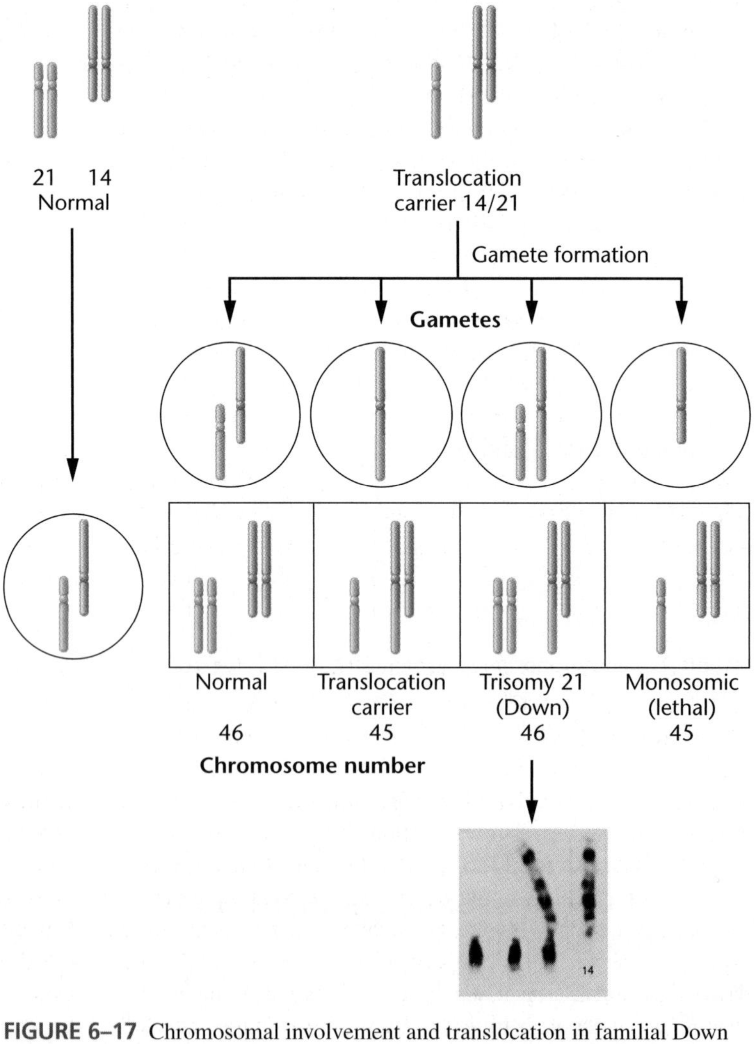

FIGURE 6–17 Chromosomal involvement and translocation in familial Down syndrome. The photograph shows the relevant chromosomes from a trisomy 21 offspring produced by a translocation carrier parent.

chromosomes have broken off prior to their fusion. These specific regions are known to be two of many chromosomal locations housing multiple copies of the genes encoding rRNA, the major component of ribosomes. Despite the loss of up to 20 percent of these genes, the carrier is unaffected.

ESSENTIAL POINT

Inversions and translocations may initially cause little or no loss of genetic information or deleterious effects. However, heterozygous combinations of the involved chromosome segments may result in genetically abnormal gametes following meiosis, with lethality or inviability often ensuing.

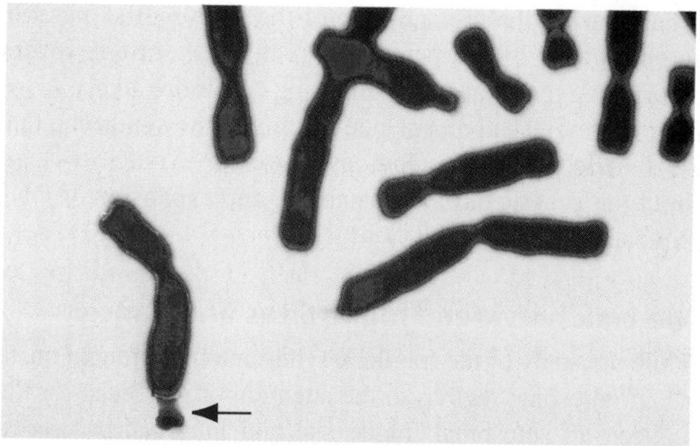

FIGURE 6–18 A fragile human X chromosome (bottom left). The "gap" region (near the bottom of the chromosome) is associated with the fragile X syndrome.

6.9 Fragile Sites in Human Chromosomes Are Susceptible to Breakage

We conclude this chapter with a brief discussion of the results of an intriguing discovery made around 1970 during observations of metaphase chromosomes prepared following human cell culture. In cells derived from certain individuals, a specific area along one of the chromosomes failed to stain, giving the appearance of a gap. In other individuals whose chromosomes displayed such morphology, the gaps appeared at other positions within the set of chromosomes. Such areas eventually became known as **fragile sites,** since they appeared to be susceptible to chromosome breakage when cultured in the absence of certain chemicals such as folic acid, which is normally present in the culture medium. Fragile sites were at first considered curiosities, until a strong association was subsequently shown to exist between one of the sites and a form of mental retardation.

The cause of the fragility at these sites is unknown. Because they represent points along the chromosome that are susceptible to breakage, these sites may indicate regions where the chromatin is not tightly coiled. Note that even though almost all studies of fragile sites have been carried out *in vitro* using mitotically dividing cells, clear associations have been established between several of these sites and the corresponding altered phenotype, including mental retardation and cancer.

Fragile X Syndrome (Martin-Bell Syndrome)

Most fragile sites do not appear to be associated with any clinical syndrome. However, individuals bearing a folate-sensitive site on the X chromosome (Figure 6–18) exhibit the **fragile X syndrome** (or **Martin-Bell syndrome**), the most common form of inherited mental retardation. This syndrome affects about 1 in 4000 males and 1 in 8000 females. Females carrying only one fragile X chromosome can be mentally retarded because it is a dominant trait. Fortunately, the trait is not fully expressed, as only about 30 percent of fragile X females are retarded, whereas about 80 percent of fragile X males are mentally retarded. In addition to mental retardation, affected males have characteristic long, narrow faces with protruding chins, enlarged ears, and increased testicular size.

A gene that spans the fragile site may be responsible for this syndrome. This gene, known as *FMR-1,* is one of a growing number of genes that have been discovered in which a sequence of three nucleotides is repeated many times, expanding the size of the gene. This phenomenon, called **trinucleotide repeats,** is also recognized in other human disorders, including Huntington disease. In *FMR-1,* the trinucleotide sequence CGG is repeated in an untranslated area adjacent to the coding sequence of the gene (called the "upstream" region). The number of repeats varies immensely within the human population, and a high number correlates directly with expression of fragile X syndrome. Normal individuals have between 6 and 54 repeats, whereas those with 55 to 230 repeats are considered "carriers" of the disorder. More than 230 repeats lead to expression of the syndrome.

It is thought that when the number of repeats reaches this level, the CGG regions of the gene become chemically modified so that the bases within and around the repeat are methylated, causing inactivation of the gene. The normal product of the gene is an RNA-binding protein, FMRP, known to be expressed in the brain. Evidence is now accumulating that directly links the absence of the protein with the cognitive defects associated with the syndrome.

From a genetic standpoint, perhaps the most interesting aspect of fragile X syndrome is the instability of the CGG repeats. An individual with 6 to 54 repeats transmits a gene containing the same number to his or her offspring. However, those with 55 to 230 repeats, though not at risk to develop the syndrome, may transmit to their offspring a gene with an increased number of repeats. The number of repeats continues to increase in future generations, demonstrating the phenomenon known as **genetic anticipation.** Once the threshold of 230 repeats is exceeded, expression of the malady becomes more severe in each successive generation as the number of trinucleotide repeats increases.

Although the mechanism that leads to the trinucleotide expansion has not yet been established, several factors are known that influence the instability. Most significant is the observation that expansion from the carrier status (55 to 230 repeats) to the syndrome status (over 230 repeats) occurs during the transmission of the gene by the maternal parent but not by the paternal parent. Furthermore, several reports suggest that male offspring are more likely to receive the increased repeat size leading to the syndrome than are female offspring. Obviously, we have much to learn about the genetic basis of instability and expansion of DNA sequences.

The Link between Fragile Sites and Cancer

While the study of the fragile X syndrome first brought unstable chromosome regions to the attention of geneticists, a link between an autosomal fragile site and lung cancer was reported in 1996 by Carlo Croce, Kay Huebner, and their colleagues. They have subsequently postulated that the defect is associated with the formation of a variety of different tumor types. Croce and Huebner first showed that the *FHIT* gene (standing for *fr*agile *hi*stidine *t*riad), located within the well-defined fragile site designated as *FRA3B* on the p arm of chromosome 3, is often altered or missing in cells taken from tumors of individuals with lung cancer. More extensive studies have now revealed that the normal protein product of this gene is absent in cells of many other cancers, including those of the esophagus, breast, cervix, liver, kidney, pancreas, colon, and stomach. Genes such as *FHIT* that are located within fragile regions undoubtedly have an increased susceptibility to mutations and deletions.

More recently, Muller Fabbri and Kay Huebner, working with others in Croce's lab, have identified and studied another fragile site, with most interesting results. Found within the *FRA16D* site on chromosome 16 is the *WWOX* gene. Like the *FHIT* gene, it has been implicated in a range of human cancers. In particular, like *FHIT,* it has been found to be either lost or genetically silenced in the large majority of lung tumors, as well as in cancer tissue of the breast, ovary, prostate, bladder, esophagus, and pancreas. When the gene is present but silent, its DNA is thought to be heavily methylated, rendering it inactive. Furthermore, the active gene is also thought to behave as a tumor suppressor, providing a surveillance function by recognizing cancer cells and inducing apoptosis, effectively eliminating them before malignant tumors can be initiated.

ESSENTIAL POINT

Fragile sites in human mitotic chromosomes have sparked research interest because one such site on the X chromosome is associated with the most common form of inherited mental retardation, while other autosomal sites have been linked to various forms of cancer.

GENETICS, TECHNOLOGY, AND SOCIETY

Down Syndrome, Prenatal Testing, and the New Eugenics

Down syndrome is the most common chromosomal abnormality seen in newborn babies. Each year, about 4000–5000 babies are born with Down syndrome in the United States. Prenatal diagnostic tests for Down syndrome have been available for decades, especially to older pregnant women who have an increased risk of bearing a Down syndrome child. Scientists estimate that there is an abortion rate of about 30 percent for fetuses that test positive for Down syndrome in the United States, and rates of up to 85 percent in other parts of the world, such as Taiwan and France. The reasons for aborting a Down syndrome fetus are varied but include the wish to have a normal healthy child and the desire to avoid the financial, physical, and emotional suffering that may occur in families that must raise these children.

In addition to Down syndrome testing, prenatal genetic tests for other chromosomal and gene-specific defects are entering medical practice. Over 1000 gene mutation tests are now offered to pregnant women. In addition, ultrasound and enzyme tests are standard procedures in obstetrical care. As new genomic methods gain prevalence, we will soon see inexpensive genetic screens for hundreds of genetic defects and gene variants.

Some prenatal tests have had a marked effect on infant mortality and disease rates. For example, in Taiwan, the genetic screening program for thalassemia mutations (resulting in severe anemia) has reduced the newborn disease rate from 5.6 per 100,000 births to 1.2. In some countries, infant mortality rates have declined by one-half, due to prenatal screening for congenital defects.

Many people agree that it is morally acceptable to prevent the birth of a genetically abnormal fetus. However, many others argue that prenatal genetic testing and the elimination of congenital disorders is unethical. These arguments often center around their views about abortion and self-determination. In addition, some people argue that prenatal genetic tests followed by selective abortion is eugenic.

These eugenics arguments are highly emotionally charged, and some people contend they are inappropriate arguments to use when discussing screening for genetic diseases. It is safe to say that most people on both sides of the eugenics debate are unclear about the definition of eugenics and even less clear about how it may relate to prenatal genetic testing.

So, what is eugenics? And how does eugenics apply, if at all, to screening for Down syndrome and other human genetic defects? The term *eugenics* was first defined by Francis Galton in 1883 as "the science which deals with all influences that improve the inborn qualities of a race; also with those that develop them to the utmost advantage." Galton believed that human traits such as intelligence and personality were hereditary and that humans could selectively mate with each other to create gifted groups of people—analogous to the creation of purebred dogs with specific traits. Galton did not propose coercion but thought that people would voluntarily enter matings for beneficial outcomes.

In the early to mid-twentieth century, countries throughout the world adopted eugenic policies with the aim of enhancing desirable human traits and eliminating undesirable ones. Two distinct approaches to eugenics developed: *positive eugenics,* in which those with desirable mental and physical traits were encouraged to reproduce, and *negative eugenics,* in which those with undesirable traits were discouraged or prevented from reproducing. Many countries, including Britain, Canada, and the United States, enacted compulsory sterilization programs for the "feebleminded," mentally ill, and criminals. The eugenic policies of Nazi Germany were particularly infamous, resulting in forced human genetic experimentation and the killing of tens of thousands of disabled people. The eugenics movement was discredited after World War II, and the evils perpetuated in its name have tainted the term *eugenics* ever since.

Given the history of the eugenics movement, is it fair to use the term *eugenics* when we speak about genetic testing for Down syndrome and other genetic disorders? Some people argue that it is not eugenic to select for healthy children because there is no coercion, the state is not involved, and the goal is the elimination of suffering. Others point out that such voluntary actions still constitute eugenics, since they involve a form of bioengineering for better human beings.

Now that we are entering an era of unprecedented knowledge about our genomes and our predisposition to genetic disorders, we must make decisions about whether our attempts to control or improve human genomes is ethical, and what limits we should place on these efforts. In theory, current and future genetic technologies may make possible a "new eugenics" with a scope and power that Francis Galton could not have imagined. The story of the eugenics movement provides us with a powerful cautionary tale about the potential misuses of genetic information.

Your Turn

Take time, individually or in groups, to answer the following questions. Investigate the references and links to help you discuss some of the issues surrounding genetic testing and eugenics.

1. Do you think that modern prenatal and preimplantation genetic testing followed by selective abortion is eugenic? Why or why not?

For background on these questions, see an annotated discussion of eugenics and its history on the Wikipedia Web site: **http://en.wikipedia.org/wiki/Eugenics.** *Another useful discussion can be found in* Scully, J. L., 2008. Disability and genetics in the era of genomic medicine. *Nature Reviews Genetics* 9: 797–802.

2. Is it desirable, or even possible, for genetic counseling to be nondirective?

One of the arguments that genetic tests followed by selective abortion are eugenic is that there are strong pressures on pregnant women to produce "normal" children. Some of these pressures may come from genetic counselors. For a discussion of this topic, see: King, D. S., 1999. Preimplantation genetic diagnosis and the "new" eugenics. *J. Med. Ethics* 25: 176–182.

3. If genetic technologies were more advanced than today, and you could choose the traits of your children, would you take advantage of that option? Which traits would you choose—height, weight, intellectual abilities, athleticism, artistic talents? If so, would this be eugenic?

To read about similar questions answered by groups of Swiss law and medical students, read: Elger, B. and Harding, T., 2003. Huntington's disease: Do future physicians and lawyers think eugenically? *Clin. Genet.* 64: 327–338.

Case Study Fish tales

Aquatic vegetation overgrowth, usually controlled by dredging or herbicides, represents a significant issue in maintaining private and public waterways. In 1963, diploid grass carp were introduced in Arkansas to consume vegetation, but they reproduced prodigiously and spread to eventually become a hazard to aquatic ecosystems in 35 states. In the 1980s, many states adopted triploid grass carp as an alternative because of their high, but not absolute, sterility and their longevity of seven to ten years. Today, most states require permits for vegetation control by triploid carp, requiring their containment in the body of water to which they are introduced. Genetic modifications of organisms to achieve specific outcomes will certainly become more common in the future and raise several interesting questions.

1. Taking triploid carp as an example, what controversies may emerge as similar modified species become available for widespread use?
2. If you were a state employee in charge of a specific waterway, what questions would you ask before you approved the introduction of a laboratory-produced, polyploid species into your waterway?
3. Why would the creation and use of a tetraploid carp species be less desirable in the above situation?

Insights and Solutions

1. In a cross using maize that involves three genes, *a, b,* and *c,* a heterozygote ($abc/+++$) is testcrossed to abc/abc. Even though the three genes are separated along the chromosome, thus predicting that crossover gametes and the resultant phenotypes should be observed, only two phenotypes are recovered: *abc* and $+++$. In addition, the cross produced significantly fewer viable plants than expected. Can you propose why no other phenotypes were recovered and why the viability was reduced?

Solution: One of the two chromosomes may contain an inversion that overlaps all three genes, effectively precluding the recovery of any "crossover" offspring. If this is a paracentric inversion and the genes are clearly separated (assuring that a significant number of crossovers occurs between them), then numerous acentric and dicentric chromosomes will form, resulting in the observed reduction in viability.

(Cont. on the next page)

2. A male *Drosophila* from a wild-type stock is discovered to have only seven chromosomes, whereas normally $2n = 8$. Close examination reveals that one member of chromosome IV (the smallest chromosome) is attached to (translocated to) the distal end of chromosome II and is missing its centromere, thus accounting for the reduction in chromosome number. (a) Diagram all members of chromosomes II and IV during synapsis in meiosis I.

Solution:

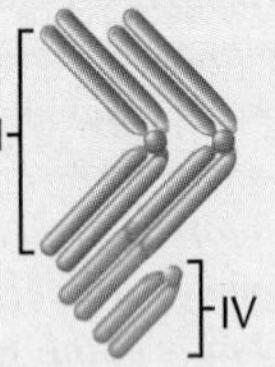

(b) If this male mates with a female with a normal chromosome composition who is homozygous for the recessive chromosome IV mutation *eyeless* (*ey*), what chromosome compositions will occur in the offspring regarding chromosomes II and IV?

Solution:

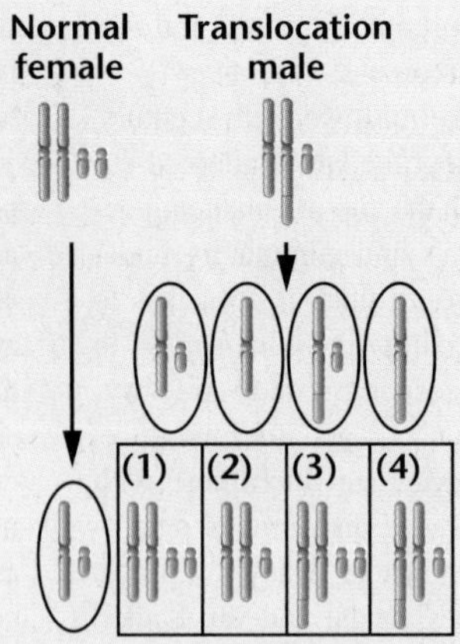

(c) Referring to the diagram in the solution to part (b), what phenotypic ratio will result regarding the presence of eyes, assuming all abnormal chromosome compositions survive?

Solution:

(1) normal (heterozygous)

(2) eyeless (monosomic, contains chromosome IV from mother)

(3) normal (heterozygous; trisomic and may die)

(4) normal (heterozygous; balanced translocation)

The final ratio is 3/4 normal: 1/4 eyeless.

Problems and Discussion Questions

1. For a species with a diploid number of 18, indicate how many chromosomes will be present in the somatic nuclei of individuals that are haploid, triploid, tetraploid, trisomic, and monosomic.
2. Define these pairs of terms, and distinguish between them.
   aneuploidy/euploidy
   monosomy/trisomy
   Patau syndrome/Edwards syndrome
   autopolyploidy/allopolyploidy
   autotetraploid/amphidiploid
   paracentric inversion/pericentric inversion
3. Contrast the relative survival times of individuals with Down, Patau, and Edwards syndromes. Speculate as to why such differences exist.
4. What evidence suggests that Down syndrome is more often the result of nondisjunction during oogenesis rather than during spermatogenesis?
5. What evidence indicates that humans with aneuploid karyotypes occur at conception but are usually inviable?
6. Contrast the fertility of an allotetraploid with an autotriploid and an autotetraploid.
7. When two plants belonging to the same genus but different species are crossed, the F_1 hybrid is more viable and has more ornate flowers. Unfortunately, this hybrid is sterile and can only be propagated by vegetative cuttings. Explain the sterility of the hybrid. How might a horticulturist attempt to reverse its sterility? See **Now Solve This** on page 119.
8. Describe the origin of cultivated American cotton.
9. Predict how the synaptic configurations of homologous pairs of chromosomes might appear when one member is normal and the other member has sustained a deletion or duplication.
10. Inversions are said to "suppress crossing over." Is this terminology technically correct? If not, restate the description accurately.
11. Contrast the genetic composition of gametes derived from tetrads of inversion heterozygotes where crossing over occurs within a paracentric versus a pericentric inversion.
12. Discuss Ohno's hypothesis on the role of gene duplication in the process of evolution.
13. What roles have inversions and translocations played in the evolutionary process?
14. A human female with Turner syndrome also expresses the X-linked trait hemophilia, as did her father. Which of her parents underwent nondisjunction during meiosis, giving rise to the gamete responsible for the syndrome? See **Now Solve This** on page 113.
15. The primrose, *Primula kewensis,* has 36 chromosomes that are similar in appearance to the chromosomes in two related species, *P. floribunda* ($2n = 18$) and *P. verticillata* ($2n = 18$). How could *P. kewensis* arise from these species? How would you describe *P. kewensis* in genetic terms?
16. Certain varieties of chrysanthemums contain 18, 36, 54, 72, and 90 chromosomes; all are multiples of a basic set of nine chromosomes. How would you describe these varieties genetically? What feature do the karyotypes of each variety share? A variety with 27 chromosomes has been discovered, but it is sterile. Why?
17. *Drosophila* may be monosomic for chromosome 4, yet remain fertile. Contrast the F_1 and F_2 results of the following crosses involving the recessive chromosome 4 trait, *bent* bristles: (a) monosomic IV, bent bristles × diploid, normal bristles; (b) monosomic IV, normal bristles × diploid, bent bristles.

18. Mendelian ratios are modified in crosses involving autotetraploids. Assume that one plant expresses the dominant trait green seeds and is homozygous (*WWWW*). This plant is crossed to one with white seeds that is also homozygous (*wwww*). If only one dominant allele is sufficient to produce green seeds, predict the F_1 and F_2 results of such a cross. Assume that synapsis between chromosome pairs is random during meiosis.
19. Having correctly established the F_2 ratio in Problem 18, predict the F_2 ratio of a "dihybrid" cross involving two independently assorting characteristics (e.g., $P_1 = WWWWAAAA \times wwwwaaaa$).
20. In a cross between two varieties of corn, $gl_1 gl_1 Ws_3 Ws_3$ (egg parent) $\times$ $Gl_1 Gl_1 ws_3 ws_3$ (pollen parent), a triploid offspring was produced with the genetic constitution $Gl_1 Gl_1 gl_1 Ws_3 ws_3 ws_3$. From which parent, egg or pollen, did the $2n$ gamete originate? Is another explanation possible? Explain.
21. What is the effect of a rare double crossover (a) within a pericentric inversion present heterozygously, (b) within a paracentric inversion present heterozygously?
22. The outcome of a single crossover between nonsister chromatids in the inversion loop of an inversion heterozygote varies depending on whether the inversion is of the paracentric or pericentric type. What differences are expected?
23. A couple planning their family are aware that through the past three generations on the husband's side a substantial number of stillbirths have occurred and several malformed babies were born who died early in childhood. The wife has studied genetics and urges her husband to visit a genetic counseling clinic, where a complete karyotype-banding analysis is performed. Although the tests show that he has a normal complement of 46 chromosomes, banding analysis reveals that one member of the chromosome 1 pair (in group A) contains an inversion covering 70 percent of its length. The homolog of chromosome 1 and all other chromosomes show the normal banding sequence. (a) How would you explain the high incidence of past stillbirths? (b) What can you predict about the probability of abnormality/normality of their future children? (c) Would you advise the woman that she will have to bring each pregnancy to term to determine whether the fetus is normal? If not, what else can you suggest? See Now Solve This on page 125.
24. Briefly compare the fate of gametes in plants and animals carrying aberrant chromatids or unbalanced chromosomal complements.
25. A woman who sought genetic counseling is found to be heterozygous for a chromosomal rearrangement between the second and third chromosomes. Her chromosomes, compared to those in a normal karyotype, are diagrammed here:

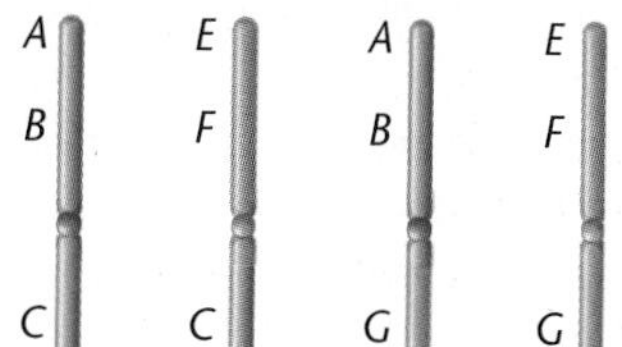

(a) What kind of chromosomal aberration is shown?
(b) Using a drawing, demonstrate how these chromosomes would pair during meiosis. Be sure to label the different segments of the chromosomes.
(c) This woman is phenotypically normal. Does this surprise you? Why or why not? Under what circumstances might you expect a phenotypic effect of such a rearrangement?

26. The woman in Problem 25 has had two miscarriages. She has come to you, an established genetic counselor, with these questions: (a) Is there a genetic explanation of her frequent miscarriages? (b) Should she abandon her attempts to have a child of her own? (c) If not, what is the chance that she could have a normal child? Provide an informed response to her concerns.
27. In a recent cytogenetic study on 1021 cases of Down syndrome, 46 were the result of translocations, the most frequent of which was symbolized as t(14;21). What does this symbol represent, and how many chromosomes would you expect to be present in t(14;21) Down syndrome individuals?
28. A boy with Klinefelter syndrome is born to a mother who is phenotypically normal and a father who has the X-linked skin condition called anhidrotic ectodermal dysplasia. The mother's skin is completely normal with no signs of the skin abnormality. In contrast, her son has patches of normal skin and patches of abnormal skin. (a) Which parent contributed the abnormal gamete? (b) Using the appropriate genetic terminology, describe the meiotic mistake that occurred. Be sure to indicate in which division the mistake occurred. (c) Using the appropriate genetic terminology, explain the son's skin phenotype.
29. To investigate the origin of nondisjunction, 200 human oocytes that had failed to be fertilized during *in vitro* fertilization procedures were subsequently examined (Angel, R. 1997. *Am. J. Hum. Genet.* 61: 23–32). These oocytes had completed meiosis I and were arrested in metaphase II (MII). The majority (67 percent) had a normal MII-metaphase complement, showing 23 chromosomes, each consisting of two sister chromatids joined at a common centromere. The remaining oocytes all had abnormal chromosome compositions. Surprisingly, when trisomy was considered, none of the abnormal oocytes had 24 chromosomes. (a) Interpret these results in regard to the origin of trisomy, as it relates to nondisjunction, and when it occurs. Why are the results surprising? (b) A large number of the abnormal oocytes contained 22 1/2 chromosomes; that is, 22 chromosomes plus a single chromatid representing the 1/2 chromosome. What chromosome compositions will result in the zygote if such oocytes proceed through meiosis and are fertilized by normal sperm? (c) How could the complement of 22 1/2 chromosomes arise? Provide a drawing that includes several pairs of MII chromosomes. (d) Do your answers support or dispute the generally accepted theory regarding nondisjunction and trisomy, as outlined in Figure 6–1?
30. In a human genetic study, a family with five phenotypically normal children was investigated. Two were "homozygous" for a Robertsonian translocation between chromosomes 19 and 20 (they contained two identical copies of the fused chromosome). These children have only 44 chromosomes but a complete genetic complement. Three of the children were "heterozygous" for the translocation and contained 45 chromosomes, with one translocated chromosome plus a normal copy of both chromosomes 19 and 20. Two other pregnancies resulted in stillbirths. It was later discovered that the parents were first cousins. Based on this information, determine the chromosome compositions of the parents. What led to the stillbirths? Why was the discovery that the parents were first cousins a key piece of information in understanding the genetics of this family?
31. A 3-year-old child exhibited some early indication of Turner syndrome, which results from a 45,X chromosome composition. Karyotypic analysis demonstrated two cell types: 46,XX (normal) and 45,X. Propose a mechanism for the origin of this mosaicism.
32. A normal female is discovered with 45 chromosomes, one of which exhibits a Robertsonian translocation containing most of chromosomes 18 and 21. Discuss the possible outcomes in her offspring when her husband contains a normal karyotype.

Chiasmata between synapsed homologs during the first meiotic prophase.

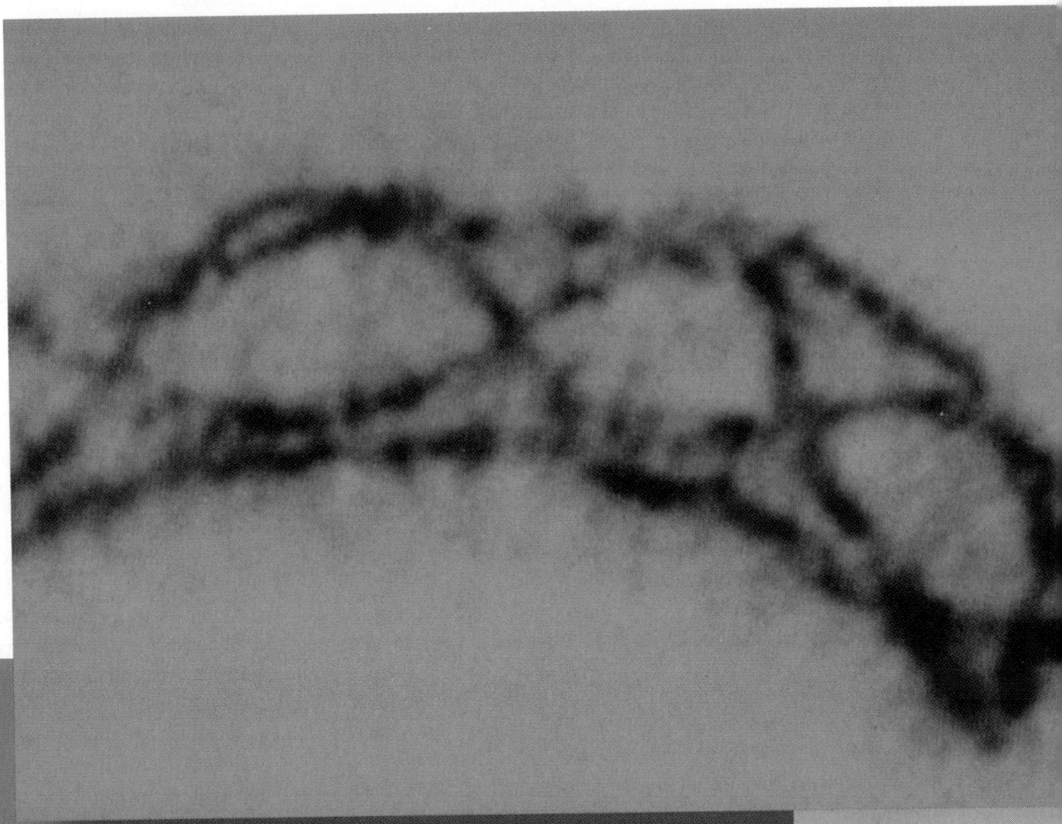

7

Linkage and Chromosome Mapping in Eukaryotes

Chapter Concepts

- Chromosomes in eukaryotes contain many genes whose locations are fixed along the length of the chromosomes.
- Unless separated by crossing over, alleles present on a chromosome segregate as a unit during gamete formation.
- Crossing over between homologs during meiosis creates recombinant gametes with different combinations of alleles that enhance genetic variation.
- Crossing over between homologs serves as the basis for the construction of chromosome maps.
- Genetic maps depict the relative locations of genes on chromosomes in a species.

Walter Sutton, along with Theodor Boveri, was instrumental in uniting the fields of cytology and genetics. As early as 1903, Sutton pointed out the likelihood that there must be many more "unit factors" than chromosomes in most organisms. Soon thereafter, genetics investigations revealed that certain genes segregate as if they were somehow joined or linked together. Further investigations showed that such genes are part of the same chromosome and may indeed be transmitted as a single unit. We now know that most chromosomes contain a very large number of genes. Those that are part of the same chromosome are said to be *linked* and to demonstrate **linkage** in genetic crosses.

Because the chromosome, not the gene, is the unit of transmission during meiosis, linked genes are not free to undergo independent assortment. Instead, the alleles at all loci of one chromosome should, in theory, be transmitted as a unit during gamete formation. However, in many instances this does not occur. During the first meiotic prophase, when homologs are paired or synapsed, a reciprocal exchange of chromosome segments can take place. This **crossing over** results in the reshuffling, or **recombination,** of the alleles between homologs and always occurs during the tetrad stage.

The frequency of crossing over between any two loci on a single chromosome is proportional to the distance between them. Therefore, depending on which loci are being studied, the percentage of recombinant gametes varies. This correlation allows us to construct chromosome maps, which give the relative locations of genes on chromosomes.

In this chapter, we will discuss linkage, crossing over, and chromosome mapping in more detail. We will conclude by entertaining the rather intriguing question of why Mendel, who studied seven genes in an organism with seven chromosomes, did not encounter linkage. Or did he?

How Do We Know?

In this chapter, we will focus on genes located on the same eukaryotic chromosome, a concept called linkage. We are particularly interested in how such linked genes are transmitted to offspring and how the results of such transmission enable us to map them. As you study this topic, you should try to answer several fundamental questions:

1. How do we know that specific genes are linked on a single chromosome, in contrast to being located on separate chromosomes?
2. How was it established experimentally that the frequency of recombination (crossing over) between two genes is related to the distance between them along the chromosome?
3. How do we know that crossing over results from a physical exchange between chromatids?
4. When designed matings cannot be conducted in an organism (for example, in humans), how do we learn that genes are linked, and how do we map them?

7.1 Genes Linked on the Same Chromosome Segregate Together

A simplified overview of the major theme of this chapter is given in Figure 7–1, p. 134, which contrasts the meiotic consequences of (a) independent assortment, (b) linkage *without* crossing over, and (c) linkage *with* crossing over. In Figure 7–1(a), we see the results of independent assortment of two pairs of chromosomes, each containing one heterozygous gene pair. No linkage is exhibited. When a large number of meiotic events are observed, four genetically different gametes are formed in equal proportions, and each contains a different combination of alleles of the two genes.

Now let's compare these results with what occurs if the same genes are linked on the same chromosome. If no crossing over occurs between the two genes [Figure 7–1(b)], only two genetically different gametes are formed. Each gamete receives the alleles present on one homolog or the other, which is transmitted intact as the result of segregation. This case demonstrates *complete linkage,* which produces only **parental** or **noncrossover gametes.** The two parental gametes are formed in equal proportions. Though complete linkage between two genes seldom occurs, it is useful to consider the theoretical consequences of this concept.

Figure 7–1(c) shows the results of crossing over between two linked genes. As you can see, this crossover involves only two nonsister chromatids of the four chromatids present in the tetrad. This exchange generates two new allele combinations, called **recombinant** or **crossover gametes.** The two chromatids not involved in the exchange result in noncrossover gametes, like those in Figure 7–1(b). The frequency with which crossing over occurs between any two linked genes is generally proportional to the distance separating the respective loci along the chromosome. In theory, two randomly selected genes can be so close to each other that crossover events are too infrequent to be easily detected. As shown in Figure 7–1(b), this complete linkage produces only parental gametes. On the other hand, if a small but distinct distance separates two genes, few recombinant and many parental gametes will be formed. As the distance between the two genes increases, the proportion of recombinant gametes increases and that of the parental gametes decreases.

As we will discuss later in this chapter, when the loci of two linked genes are far apart, the number of recombinant gametes approaches, but does not exceed, 50 percent. If 50 percent recombinants occur, the result is a 1:1:1:1 ratio of the four types (two parental and two recombinant gametes). In this case, transmission of two linked genes is indistinguishable from that of two unlinked, independently assorting genes. That is, the proportion of the four possible genotypes is identical, as shown in Figures 7–1(a) and (c).

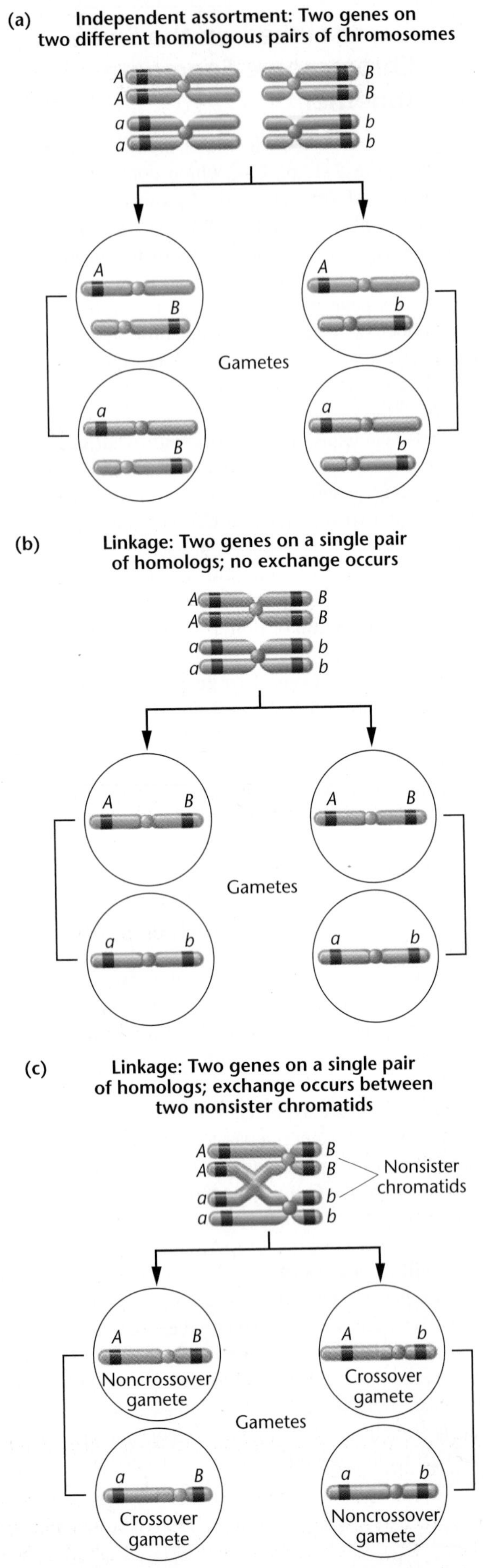

NOW SOLVE THIS

Problem 9 on page 155 asks you to contrast the results of a testcross when two genes are unlinked versus linked, and when they are linked, if they are very far apart or relatively close together.

Hint: The results are indistinguishable when two genes are unlinked compared to the case where they are linked but so far apart that crossing over always intervenes between them during meiosis.

The Linkage Ratio

If complete linkage exists between two genes because of their close proximity, and organisms heterozygous at both loci are mated, a unique F_2 phenotypic ratio results, which we designate the **linkage ratio.** To illustrate this ratio, let's consider a cross involving the closely linked, recessive, mutant genes *brown* eye (*bw*) and *heavy* wing vein (*hv*) in *Drosophila melanogaster* (Figure 7–2). The normal, wild-type alleles bw^+ and hv^+ are both dominant and result in red eyes and thin wing veins, respectively.

In this cross, flies with mutant brown eyes and normal thin wing veins are mated to flies with normal red eyes and mutant heavy wing veins. In more concise terms, brown-eyed flies are crossed with heavy-veined flies. If we extend the system of genetic symbols established in Chapter 4, linked genes are represented by placing their allele designations above and below a single or double horizontal line. Those above the line are located at loci on one homolog, and those below are located at the homologous loci on the other homolog. Thus, we represent the P_1 generation as follows:

$$P_1: \frac{bw\ hv^+}{bw\ hv^+} \times \frac{bw^+\ hv}{bw^+\ hv}$$

brown, thin   red, heavy

These genes are located on an autosome, so no distinction between males and females is necessary.

In the F_1 generation, each fly receives one chromosome of each pair from each parent. All flies are heterozygous for both gene pairs and exhibit the dominant traits of red eyes and thin wing veins:

$$F_1: \frac{bw\ \ hv^+}{bw^+\ hv}$$

red, thin

As shown in Figure 7–2(a), when the F_1 generation is interbred, each F_1 individual forms only parental gametes because of complete linkage. After fertilization, the F_2 generation is produced in a 1:2:1 phenotypic and genotypic ratio. One-fourth of this generation shows brown eyes and thin wing

FIGURE 7–1 Results of gamete formation where two heterozygous genes are (a) on two different pairs of chromosomes; (b) on the same pair of homologs, but where no exchange occurs between them; and (c) on the same pair of homologs, where an exchange occurs between two nonsister chromatids.

LINKAGE AND RECOMBINATION

WEB TUTORIAL

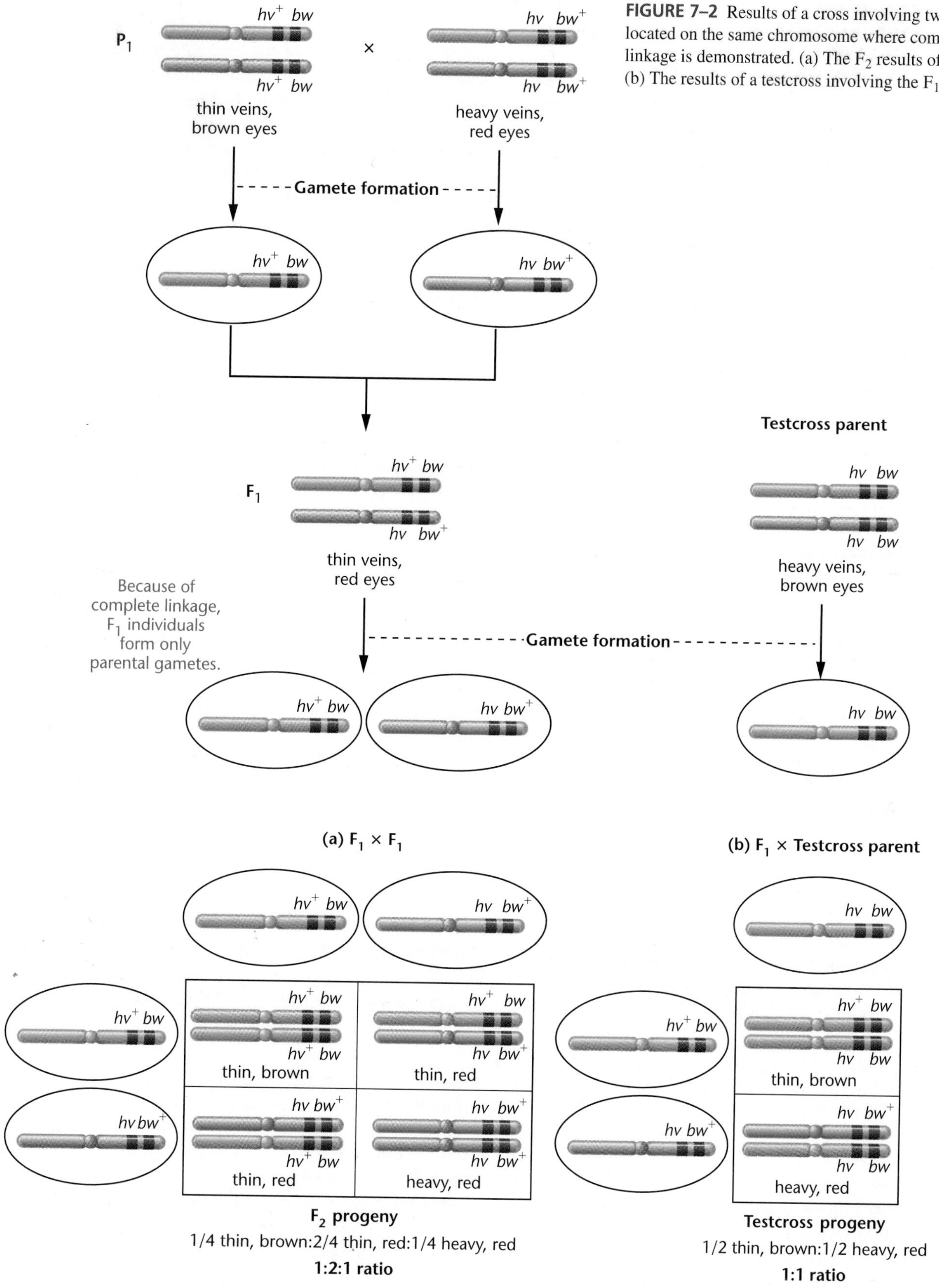

FIGURE 7–2 Results of a cross involving two genes located on the same chromosome where complete linkage is demonstrated. (a) The F_2 results of the cross. (b) The results of a testcross involving the F_1 progeny.

veins; one-half shows both wild-type traits, namely, red eyes and thin wing veins; and one-fourth shows red eyes and heavy wing veins. In more concise terms, the ratio is 1 brown: 2 wild:1 heavy. Such a 1:2:1 ratio is characteristic of complete linkage. Complete linkage is usually observed only when genes are very close together and the number of progeny is relatively small.

Figure 7–2(b) also gives the results of a testcross with the F_1 flies. Such a cross produces a 1:1 ratio of brown, thin and red, heavy flies. Had the genes controlling these traits been incompletely linked or located on separate autosomes, the testcross would have produced four phenotypes rather than two.

When large numbers of mutant genes present in any given species are investigated, genes located on the same chromosome show evidence of linkage to one another. As a result, **linkage groups** can be established, one for each chromosome. In theory, the number of linkage groups should correspond to the haploid number of chromosomes. In diploid organisms in which large numbers of mutant genes are available for genetic study, this correlation has been confirmed.

ESSENTIAL POINT

Alleles of linked genes located close together on the same homolog are usually transmitted together during gamete formation.

7.2 Crossing Over Serves as the Basis of Determining the Distance between Genes during Mapping

It is highly improbable that two randomly selected genes linked on the same chromosome will be so close to one another along the chromosome that they demonstrate complete linkage. Instead, crosses involving two such genes almost always produce a percentage of offspring resulting from recombinant gametes. This percentage is variable and depends on the distance between the two genes along the chromosome. This phenomenon was first explained around 1910 by two *Drosophila* geneticists, Thomas H. Morgan and his undergraduate student, Alfred H. Sturtevant.

Morgan, Sturtevant, and Crossing Over

In his studies, Morgan investigated numerous *Drosophila* mutations located on the X chromosome. When he analyzed crosses involving only one trait, he deduced the mode of X-linked inheritance. However, when he made crosses involving two X-linked genes, his results were initially puzzling. For example, female flies expressing the mutant *yellow* body (y) and *white* eyes (w) alleles were crossed with wild-type males (gray bodies and red eyes). The F_1 females were wild type, while the F_1 males expressed both mutant traits. In the F_2, the vast majority of the offspring showed the expected parental phenotypes—either yellow-bodied, white-eyed flies or wild-type flies (gray-bodied, red-eyed). However, the remaining flies, less than 1.0 percent, were either yellow-bodied with red eyes or gray-bodied with white eyes. It was as if the two mutant alleles had somehow separated from each other on the homolog during gamete formation in the F_1 female flies. This cross is illustrated in cross A of Figure 7–3, using data later compiled by Sturtevant.

When Morgan studied other X-linked genes, the same basic pattern was observed, but the proportion of the unexpected F_2 phenotypes differed. For example, in a cross involving the mutant *white* eye (w), *miniature* wing (m) alleles, the majority of the F_2 again showed the parental phenotypes, but a much higher proportion of the offspring appeared as if the mutant genes had separated during gamete formation. This is illustrated in cross B of Figure 7–3, again using data subsequently compiled by Sturtevant.

Morgan was faced with two questions: (1) What was the source of gene separation, and (2) why did the frequency of the apparent separation vary depending on the genes being studied? The answer he proposed for the first question was based on his knowledge of earlier cytological observations made by F. A. Janssens and others. Janssens had observed that synapsed homologous chromosomes in meiosis wrapped around each other, creating **chiasmata** (sing., *chiasma*) where points of overlap are evident (see the photo on p. 132). Morgan proposed that chiasmata could represent points of genetic exchange.

In the crosses shown in Figure 7–3, Morgan postulated that if an exchange occurs during gamete formation between the mutant genes on the two X chromosomes of the F_1 females, the unique phenotypes will occur. He suggested that such exchanges led to **recombinant gametes** in both the *yellow–white* cross and the *white–miniature* cross, in contrast to the **parental gametes** that have undergone no exchange. On the basis of this and other experiments, Morgan concluded that linked genes exist in a linear order along the chromosome and that a variable amount of exchange occurs between any two genes during gamete formation.

In answer to the second question, Morgan proposed that two genes located relatively close to each other along a chromosome are less likely to have a chiasma form between them than if the two genes are farther apart on the chromosome. Therefore, the closer two genes are, the less likely a genetic exchange will occur between them. Morgan was the first to propose the term *crossing over* to describe the physical exchange leading to recombination.

Sturtevant and Mapping

Morgan's student, Alfred H. Sturtevant, was the first to realize that his mentor's proposal could be used to map the sequence of linked genes. According to Sturtevant,

> *In a conversation with Morgan ... I suddenly realized that the variations in strength of linkage, already attributed by Morgan to differences in the spatial separation of the genes, offered the possibility of determining sequences in the linear dimension of a chromosome. I went home and spent most of the night (to the neglect of my undergraduate homework) in producing the first chromosomal map.*

Sturtevant compiled data from numerous crosses made by Morgan and other geneticists involving recombination between the genes represented by the *yellow, white,* and *miniature* mutants. These data are shown in Figure 7–3. The following recombination between each pair of these

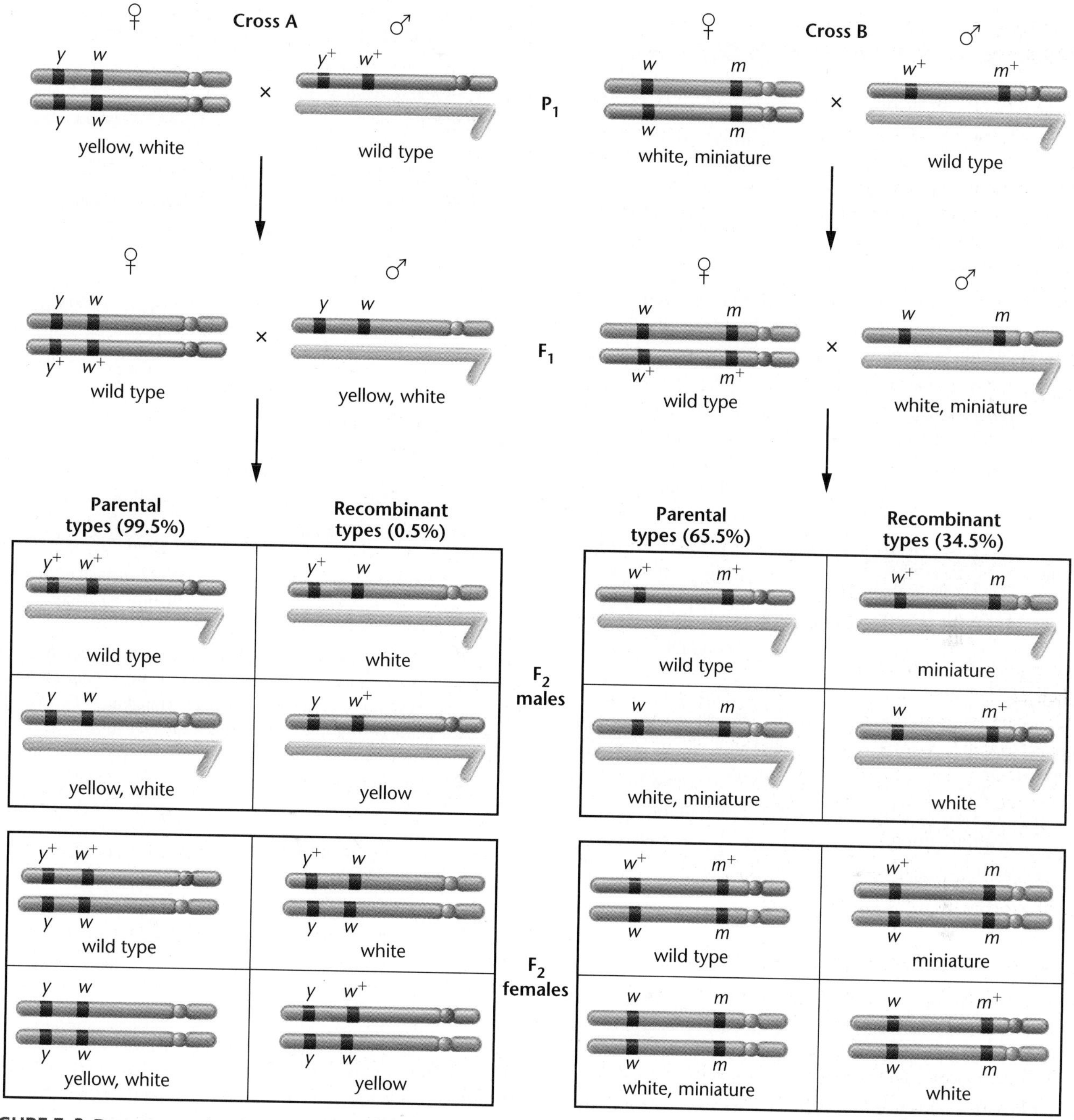

FIGURE 7–3 Data representing the F_1 and F_2 results of crosses involving the *yellow* body (y), *white* eye (w) mutations (cross A), and the *white* eye, *miniature* wing (m) mutations (cross B), as compiled by Sturtevant. In cross A, 0.5 percent of the F_2 flies (males and females) demonstrate recombinant phenotypes, which express either *white* or *yellow.* In cross B, 34.5 percent of the F_2 flies (males and females) demonstrate recombinant phenotypes, which express either *miniature* or *white.*

three genes, published in Sturtevant's paper in 1913, is as follows:

(1) *yellow–white*	0.5%
(2) *white–miniature*	34.5%
(3) *yellow–miniature*	35.4%

Because the sum of (1) and (2) approximately equals (3), Sturtevant suggested that the recombination frequencies between linked genes are additive. On this basis, he predicted that the order of the genes on the X chromosome is *yellow–white–miniature.* In arriving at this conclusion, he reasoned as follows: The *yellow* and *white* genes are apparently close to each other because the recombination frequency is low. However, both of these genes are much farther apart from *miniature* genes because the *white–miniature* and *yellow–miniature* combinations show larger recombination frequencies. Because *miniature* shows more recombination with *yellow* than with *white* (35.4 percent versus 34.5 percent), it follows that *white* is located between the other two genes, not outside of them.

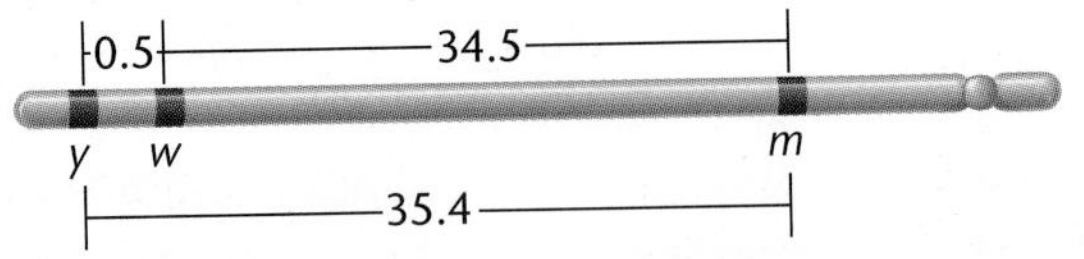

FIGURE 7–4 A map of the *yellow* body (*y*), *white* eye (*w*), and *miniature* wing (*m*) genes on the X chromosome of *Drosophila melanogaster.* Each number represents the percentage of recombinant offspring produced in one of three crosses, each involving two different genes.

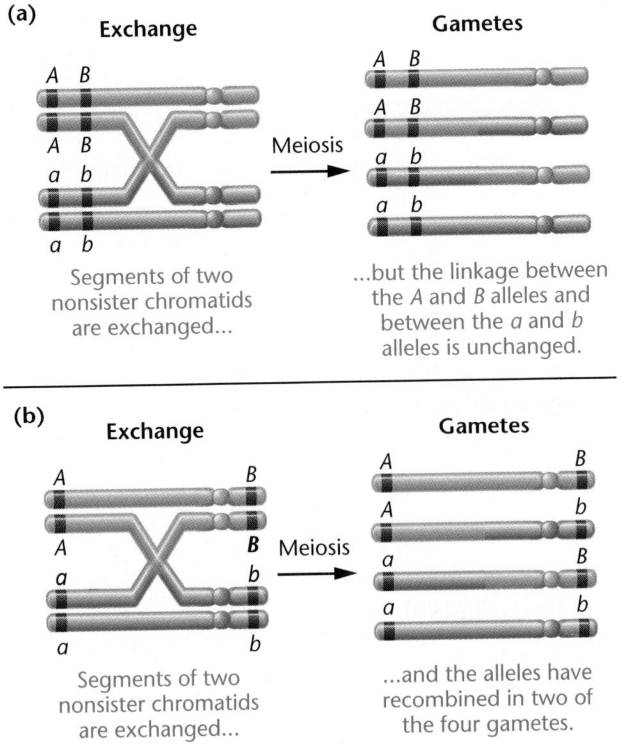

FIGURE 7–5 Two examples of a single crossover between two nonsister chromatids and the gametes subsequently produced. In (a), the exchange does not alter the linkage arrangement between the alleles of the two genes, only parental gametes form, and the exchange goes undetected. In (b), the exchange separates the alleles and results in recombinant gametes, which are detectable.

Sturtevant knew from Morgan's work that the frequency of exchange could be used as an estimate of the distance between two genes or loci along the chromosome. He constructed a **chromosome map** of the three genes on the X chromosome, setting 1 map unit (mu) equal to 1 percent recombination between two genes.* In the preceding example, the distance between *yellow* and *white* is thus 0.5 mu, and the distance between *yellow* and *miniature* is 35.4 mu. It follows that the distance between *white* and *miniature* should be $35.4 - 0.5 = 34.9$ mu. This estimate is close to the actual frequency of recombination between *white* and *miniature* (34.5 percent). The map for these three genes is shown in Figure 7–4. The fact that these do not add up perfectly is due to the imprecision of independently conducted mapping experiments.

In addition to these three genes, Sturtevant considered two other genes on the X chromosome and produced a more extensive map that included all five genes. He and a colleague, Calvin Bridges, soon began a search for autosomal linkage in *Drosophila.* By 1923, they had clearly shown that linkage and crossing over are not restricted to X-linked genes but can also be demonstrated with autosomes. During this work, they made another interesting observation. Crossing over in *Drosophila* was shown to occur only in females. The fact that no crossing over occurs in males made genetic mapping much less complex to analyze in *Drosophila.* However, crossing over does occur in both sexes in most other organisms.

Although many refinements in chromosome mapping have been developed since Sturtevant's initial work, his basic principles are considered to be correct. These principles are used to produce detailed chromosome maps of organisms for which large numbers of linked mutant genes are known. Sturtevant's findings are also historically significant to the broader field of genetics. In 1910, the **chromosomal theory of inheritance** was still widely disputed—even Morgan was skeptical of this theory before he conducted his experiments. Research has now firmly established that chromosomes contain genes in a linear order and that these genes are the equivalent of Mendel's unit factors.

Single Crossovers

Why should the relative distance between two loci influence the amount of recombination and crossing over observed between them? During meiosis, a limited number of crossover events occur in each tetrad. These recombinant events occur randomly along the length of the tetrad. Therefore, the closer two loci reside along the axis of the chromosome, the less likely any single crossover event will occur between them. The same reasoning suggests that the farther apart two linked loci, the more likely a random crossover event will occur between them.

In Figure 7–5(a), a single crossover occurs between two nonsister chromatids but not between the two loci; therefore, the crossover is not detected because no recombinant gametes are produced. In Figure 7–5(b), where two loci are quite far apart, a crossover does occur between them, yielding gametes in which the traits of interest are recombined.

When a single crossover occurs between two nonsister chromatids, the other two chromatids of the tetrad are not involved in this exchange and enter the gamete unchanged. Even if a single crossover occurs 100 percent of the time between two linked genes, recombination is subsequently observed in only 50 percent of the potential gametes formed. This concept is diagrammed in Figure 7–6. Theoretically, if we consider only single exchanges and observe 20 percent recombinant gametes, crossing over actually occurred in 40 percent of the tetrads. Under these conditions, the general rule is that the percentage of tetrads involved in an exchange between two genes is twice the percentage of recombinant gametes produced. Therefore, the theoretical limit of observed recombination due to crossing over is 50 percent.

* In honor of Morgan's work, map units are often referred to as centiMorgans (cM).

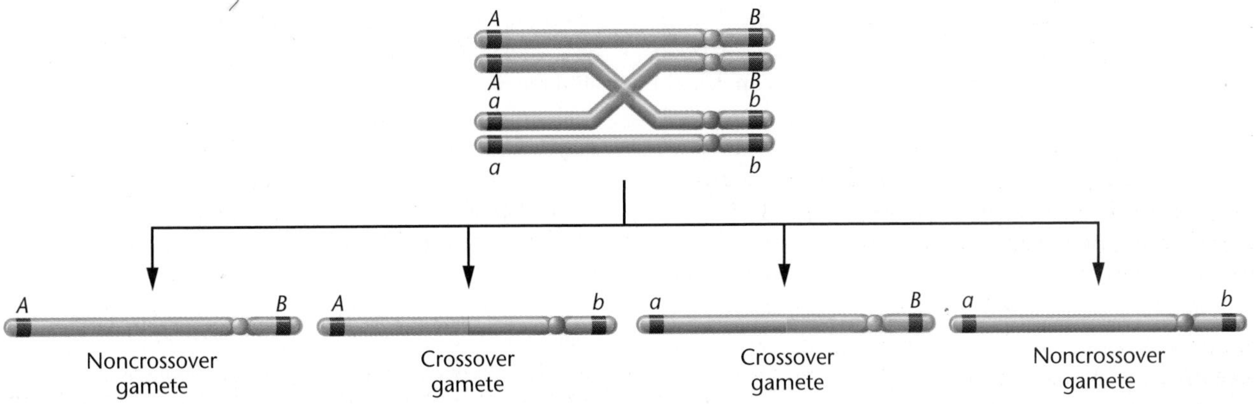

FIGURE 7–6 The consequences of a single exchange between two nonsister chromatids occurring in the tetrad stage. Two noncrossover (parental) and two crossover (recombinant) gametes are produced.

When two linked genes are more than 50 mu apart, a crossover can theoretically be expected to occur between them in 100 percent of the tetrads. If this prediction were achieved, each tetrad would yield equal proportions of the four gametes shown in Figure 7–6, just as if the genes were on different chromosomes and assorting independently. However, this theoretical limit is seldom achieved.

ESSENTIAL POINT

Crossover frequency between linked genes during gamete formation is proportional to the distance between genes, providing the experimental basis for mapping the location of genes relative to one another along the chromosome.

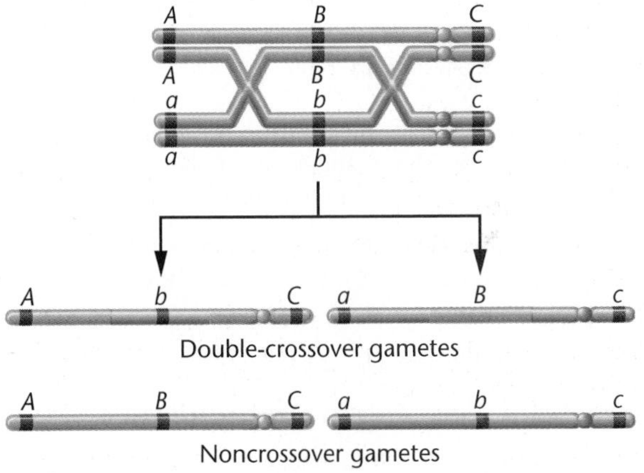

FIGURE 7–7 Consequences of a double exchange between two nonsister chromatids. Because the exchanges involve only two chromatids, two noncrossover gametes and two double-crossover gametes are produced. See the chapter opening photograph on page 132 that shows multiple chiasmata between synapsed homologs.

7.3 Determining the Gene Sequence during Mapping Requires the Analysis of Multiple Crossovers

The study of single crossovers between two linked genes provides the basis of determining the *distance* between them. However, when many linked genes are studied, their *sequence* along the chromosome is more difficult to determine. Fortunately, the discovery that multiple exchanges occur between the chromatids of a tetrad has facilitated the process of producing more extensive chromosome maps. As we shall see next, when three or more linked genes are investigated simultaneously, it is possible to determine first the sequence of the genes and then the distances between them.

Multiple Crossovers

It is possible that in a single tetrad, two, three, or more exchanges will occur between nonsister chromatids as a result of several crossover events. Double exchanges of genetic material result from **double crossovers (DCOs),** as shown in Figure 7–7. For a double exchange to be studied, three gene pairs must be investigated, each heterozygous for two alleles. Before we determine the frequency of recombination among all three loci, let's review some simple probability calculations.

As we have seen, the probability of a single exchange occurring between the *A* and *B* or the *B* and *C* genes relates directly to the distance between the respective loci. The closer *A* is to *B* and *B* is to *C*, the less likely a single exchange will occur between either of the two sets of loci. In the case of a double crossover, two separate and independent events or exchanges must occur simultaneously. The mathematical probability of two independent events occurring simultaneously is equal to the product of the individual probabilities (the **product law**).

Suppose that crossover gametes resulting from single exchanges are recovered 20 percent of the time ($p = 0.20$) between *A* and *B*, and 30 percent of the time ($p = 0.30$) between *B* and *C*. The probability of recovering a double-crossover gamete arising from two exchanges (between *A* and *B*, and between *B* and *C*) is predicted to be $(0.20)(0.30) = 0.06$, or 6 percent. It is apparent from this calculation that the frequency of double-crossover gametes is always expected to be much lower than that of either single-crossover class of gametes.

If three genes are relatively close together along one chromosome, the expected frequency of double-crossover gametes is extremely low. For example, suppose the *A–B* distance in

Figure 7–7 is 3 mu and the *B–C* distance is 2 mu. The expected double-crossover frequency is $(0.03)(0.02) = 0.0006$, or 0.06 percent. This translates to only 6 events in 10,000. Thus, in a mapping experiment where closely linked genes are involved, very large numbers of offspring are required to detect double-crossover events. In this example, it is unlikely that a double crossover will be observed even if 1000 offspring are examined. Thus, it is evident that if four or five genes are being mapped, even fewer triple and quadruple crossovers can be expected to occur.

ESSENTIAL POINT

Determining the sequence of genes in a three-point mapping experiment requires the analysis of the double-crossover gametes, as reflected in the phenotype of the offspring receiving those gametes.

NOW SOLVE THIS

Problem 14 on page 156 asks you to contrast the results of crossing over when the arrangement of alleles along the homologs differs in an organism heterozygous for three genes.

Hint: Homologs enter noncrossover gametes unchanged, and all crossover results must be derived from the noncrossover sequence of alleles.

Three-Point Mapping in *Drosophila*

The information in the preceding section enables us to map three or more linked genes in a single cross. To illustrate the mapping process in its entirety, we examine two situations involving three linked genes in two quite different organisms.

To execute a successful mapping cross, three criteria must be met:

1. The genotype of the organism producing the crossover gametes must be heterozygous at all loci under consideration.
2. The cross must be constructed so that genotypes of all gametes can be determined accurately by observing the phenotypes of the resulting offspring. This is necessary because the gametes and their genotypes can never be observed directly. To overcome this problem, each phenotypic class must reflect the genotype of the gametes of the parents producing it.
3. A sufficient number of offspring must be produced in the mapping experiment to recover a representative sample of all crossover classes.

These criteria are met in the three-point mapping cross from *Drosophila* shown in Figure 7–8. In this cross, three X-linked recessive mutant genes—*yellow* body color (*y*), *white* eye color (*w*), and *echinus* eye shape (*ec*)—are considered. To diagram the cross, *we must assume some theoretical sequence, even though we do not yet know if it is correct.* In Figure 7–8, we initially assume the sequence of the three genes to be *y–w–ec*. If this assumption is incorrect, our analysis will demonstrate this and reveal the correct sequence.

In the P_1 generation, males hemizygous for all three wild-type alleles are crossed to females that are homozygous for all three recessive mutant alleles. Therefore, the P_1 males are wild type with respect to body color, eye color, and eye shape. They are said to have a *wild-type phenotype.* The females, on the other hand, exhibit the three mutant traits—yellow body color, white eyes, and echinus eye shape.

This cross produces an F_1 generation consisting of females that are heterozygous at all three loci and males that, because of the Y chromosome, are hemizygous for the three mutant alleles. Phenotypically, all F_1 females are wild type, while all F_1 males are yellow, white, and echinus. The genotype of the F_1 females fulfills the first criterion for mapping; that is, it is heterozygous at the three loci and can serve as the source of recombinant gametes generated by crossing over. Note that because of the genotypes of the P_1 parents, all three mutant alleles in the F_1 female are on one homolog and all three wild-type alleles are on the other homolog. With other females, *other arrangements are possible that could produce a heterozygous genotype.* For example, a heterozygous female could have the *y* and *ec* mutant alleles on one homolog and the *w* allele on the other. This would occur if, in the P_1 cross, one parent was yellow, echinus and the other parent was white.

In our cross, the second criterion is met by virtue of the gametes formed by the F_1 males. Every gamete contains either an X chromosome bearing the three mutant alleles or a Y chromosome, which is genetically inert for the three loci being considered. Whichever type participates in fertilization, the genotype of the gamete produced by the F_1 female will be expressed phenotypically in the F_2 male and female offspring derived from it. Thus, all F_1 noncrossover and crossover gametes can be detected by observing the F_2 phenotypes.

With these two criteria met, we can now construct a chromosome map from the crosses shown in Figure 7–8. First, we determine which F_2 phenotypes correspond to the various noncrossover and crossover categories. To determine the noncrossover F_2 phenotypes, we must identify individuals derived from the parental gametes formed by the F_1 female. Each such gamete contains an X chromosome *unaffected by crossing over.* As a result of segregation, approximately equal proportions of the two types of gametes and, subsequently, the F_2 phenotypes, are produced. Because they derive from a heterozygote, the genotypes of the two parental gametes and the resultant F_2 phenotypes complement one another. For example, if one is wild type, the other is completely mutant. This is the case in the cross being considered. In other situations, if one chromosome shows one mutant allele, the second chromosome shows the other two mutant alleles, and so on. They are therefore called **reciprocal classes** of gametes and phenotypes.

The two noncrossover phenotypes are most easily recognized because *they exist in the greatest proportion.* Figure 7–8 shows that gametes 1 and 2 are present in the greatest numbers. Therefore, flies that express yellow, white, and echinus phenotypes and flies that are normal (or wild type) for all three characters constitute the noncrossover category and represent 94.44 percent of the F_2 offspring.

The second category that can be easily detected is represented by the double-crossover phenotypes. Because of their

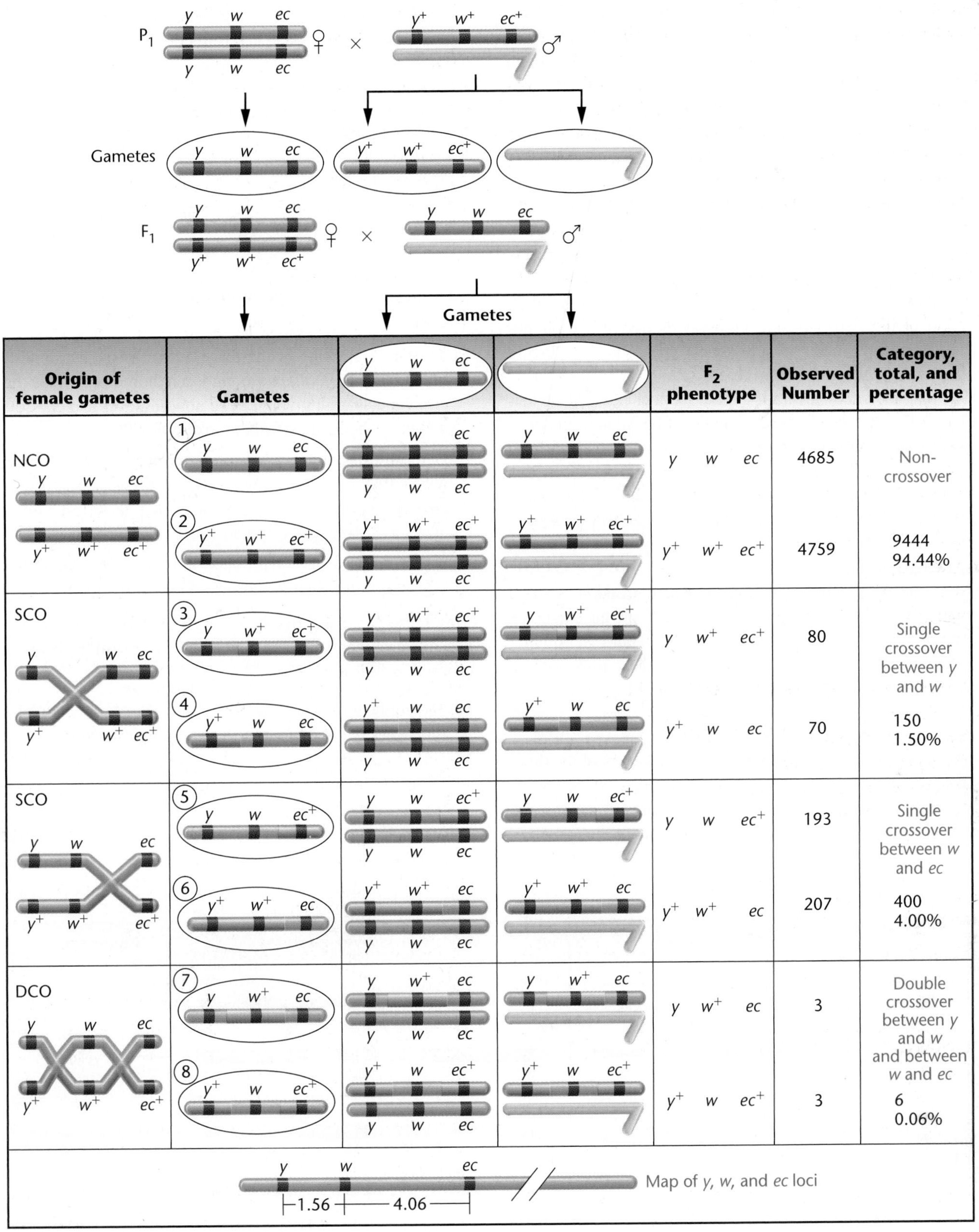

FIGURE 7–8 A three-point mapping cross involving the *yellow* (y or y^+), *white* (w or w^+), and *echinus* (ec or ec^+) genes in *D. melanogaster.* NCO, SCO, and DCO refer to noncrossover, single-crossover, and double-crossover groups, respectively. Centromeres are not included on the chromosomes, and only the two nonsister chromatids are shown initially.

low probability of occurrence, *they must be present in the least numbers*. Remember that this group represents two independent but simultaneous single-crossover events. Two reciprocal phenotypes can be identified: gamete 7, which shows the mutant traits yellow, echinus but normal eye color; and gamete 8, which shows the mutant trait white but normal body color and eye shape. Together these double-crossover phenotypes constitute only 0.06 percent of the F_2 offspring.

The remaining four phenotypic classes represent two categories resulting from single crossovers. Gametes 3 and 4, reciprocal phenotypes produced by single-crossover events occurring between the *yellow* and *white* loci, are equal to 1.50 percent of the F_2 offspring; gametes 5 and 6, constituting 4.00 percent of the F_2 offspring, represent the reciprocal phenotypes resulting from single-crossover events occurring between the *white* and *echinus* loci.

The map distances separating the three loci can now be calculated. The distance between *y* and *w* or between *w* and *ec* is equal to the percentage of all detectable exchanges occurring between them. For any two genes under consideration, this includes all appropriate single crossovers as well as all double crossovers. *The latter are included because they represent two simultaneous single crossovers.* For the *y* and *w* genes, this includes gametes 3, 4, 7, and 8, totaling 1.50% + 0.06%, or 1.56 mu. Similarly, the distance between *w* and *ec* is equal to the percentage of offspring resulting from an exchange between these two loci: gametes 5, 6, 7, and 8, totaling 4.00% + 0.06%, or 4.06 mu. The map of these three loci on the X chromosome is shown at the bottom of Figure 7–8.

Determining the Gene Sequence

In the preceding example, the sequence (or order) of the three genes along the chromosome was assumed to be *y–w–ec*. Our analysis shows this sequence to be consistent with the data. However, in most mapping experiments the gene sequence is not known, and this constitutes another variable in the analysis. In our example, had the gene sequence been unknown, it could have been determined using a straightforward method.

This method is based on the fact that there are only three possible arrangements, each containing one of the three genes between the other two:

(I)	*w–y–ec*	(*y* in the middle)
(II)	*y–ec–w*	(*ec* in the middle)
(III)	*y–w–ec*	(*w* in the middle)

Use the following steps during your analysis to determine the gene order:

1. Assuming any one of the three orders, first determine the *arrangement of alleles* along each homolog of the heterozygous parent giving rise to noncrossover and crossover gametes (the F_1 female in our example).
2. Determine whether a double-crossover event occurring within that arrangement will produce the *observed double-crossover phenotypes*. Remember that these phenotypes occur least frequently and are easily identified.
3. If this order does not produce the predicted phenotypes, try each of the other two orders. One must work!

In Figure 7–9, the above steps are applied to each of the three possible arrangements (I, II, and III above). A full analysis can proceed as follows:

1. Assuming that *y* is between *w* and *ec*, arrangement I of alleles along the homologs of the F_1 heterozygote is

$$\frac{w \quad y \quad ec}{w^+ \quad y^+ \quad ec^+}$$

We know this because of the way in which the P_1 generation was crossed: The P_1 female contributes an X chromosome

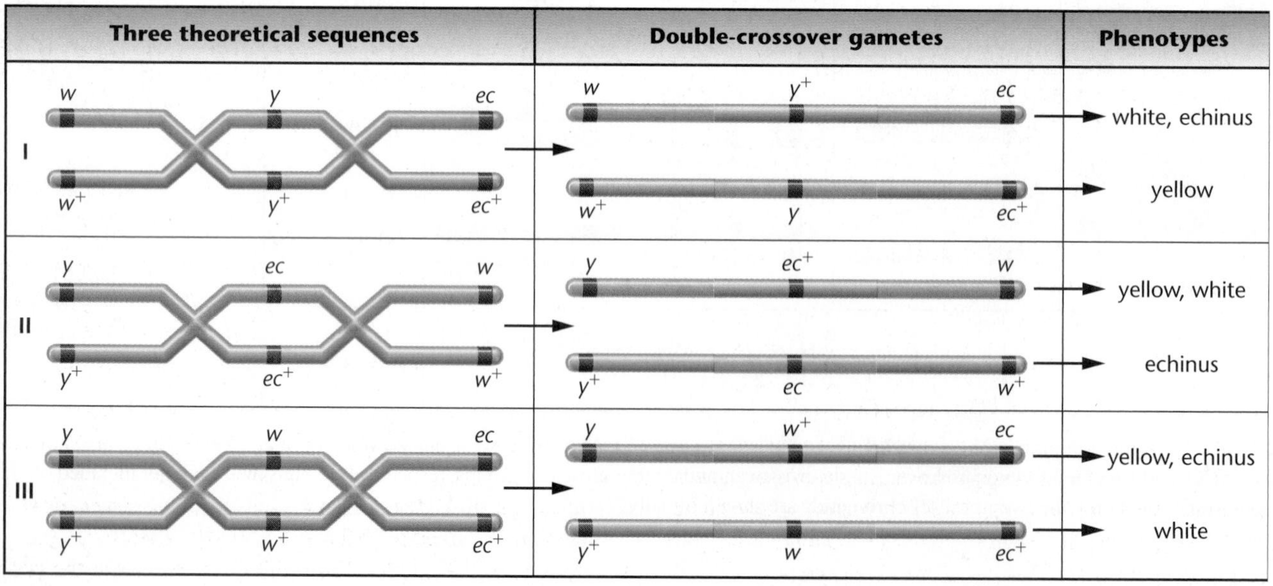

FIGURE 7–9 The three possible sequences of the *white, yellow,* and *echinus* genes, the results of a double crossover in each case, and the resulting phenotypes produced in a testcross. For simplicity, the two noncrossover chromatids of each tetrad are omitted.

bearing the *w*, *y*, and *ec* alleles, while the P_1 male contributes an X chromosome bearing the w^+, y^+, and ec^+ alleles.

2. A double crossover within that arrangement yields the following gametes:

$$\underline{w\quad y^+\quad ec}\quad \text{and}\quad \underline{w^+\quad y\quad ec^+}$$

Following fertilization, if *y* is in the middle, the F_2 double-crossover phenotypes will correspond to these gametic genotypes, yielding offspring that express the white, echinus phenotype and offspring that express the yellow phenotype. Instead, however, determination of the actual double-crossover phenotypes reveals them to be yellow, echinus flies and white flies. *Therefore, our assumed order is incorrect.*

3. If we consider arrangement II with the ec/ec^+ alleles in the middle or arrangement III with the w/w^+ alleles in the middle:

$$\text{(II)}\ \frac{y\quad ec\quad w}{y^+\quad ec^+\quad w^+}\quad \text{or}\quad \text{(III)}\ \frac{y\quad w\quad ec}{y^+\quad w^+\quad ec^+}$$

we see that arrangement II again provides *predicted* double-crossover phenotypes that *do not* correspond to the *actual* (observed) double-crossover phenotypes. The predicted phenotypes are yellow, white flies and echinus flies in the F_2 generation. *Therefore, this order is also incorrect.* However, arrangement III produces the observed phenotypes—yellow, echinus flies and white flies. *Therefore, this arrangement, with the w gene in the middle, is correct.*

To summarize, first determine the arrangement of alleles on the homologs of the heterozygote yielding the crossover gametes by locating the reciprocal noncrossover phenotypes. Then, test each of three possible orders to determine which yields the observed double-crossover phenotypes—*the one that does so represents the correct order.*

A Mapping Problem in Maize

Having established the basic principles of chromosome mapping, we will now consider a related problem in maize (corn). This analysis differs from the preceding example in two ways. First, the previous mapping cross involved X-linked genes. Here, we consider autosomal genes. Second, in the discussion of this cross we have changed our use of symbols, as first suggested in Chapter 4. Instead of using the gene symbols and superscripts (e.g., bm^+, v^+, and pr^+), we simply use + to denote each wild-type allele. This system is easier to manipulate but requires a better understanding of mapping procedures.

When we look at three autosomally linked genes in maize, the experimental cross must still meet the same three criteria we established for the X-linked genes in *Drosophila:* (1) One parent must be heterozygous for all traits under consideration; (2) the gametic genotypes produced by the heterozygote must be apparent from observing the phenotypes of the offspring; and (3) a sufficient sample size must be available for complete analysis.

In maize, the recessive mutant genes *brown* midrib (*bm*), *virescent* seedling (*v*), and *purple* aleurone (*pr*) are linked on chromosome 5. Assume that a female plant is known to be heterozygous for all three traits, but we do not know (1) the arrangement of the mutant alleles on the maternal and paternal homologs of this heterozygote, (2) the sequence of genes, or (3) the map distances between the genes. What genotype must the male plant have to allow successful mapping? To meet the second criterion, the male must be homozygous for all three recessive mutant alleles. Otherwise, offspring of this cross showing a given phenotype might represent more than one genotype, making accurate mapping impossible.

Figure 7–10, p. 144, diagrams this cross. As shown, we know neither the arrangement of alleles nor the sequence of loci in the heterozygous female. Several possibilities are shown, but we have yet to determine which is correct. We don't know the sequence in the testcross male parent either, and so we must designate it randomly. Note that we have initially placed *v* in the middle. *This may or may not be correct.*

The offspring are arranged in groups of two for each pair of reciprocal phenotypic classes. The two members of each reciprocal class are derived from no crossing over (NCO), one of two possible single-crossover events (SCO), or a double crossover (DCO).

To solve this problem, refer to Figures 7–10 and 7–11 as you consider the following questions.

1. *What is the correct heterozygous arrangement of alleles in the female parent?* Determine the two noncrossover classes, those that occur with the highest frequency. In this case, they are $\underline{+\ v\ bm}$ and $\underline{pr\ +\ +}$. Therefore, the alleles on the homologs of the female parent must be arranged as shown in Figure 7–11(a), p. 145. These homologs segregate into gametes, unaffected by any recombination event. Any other arrangement of alleles will not yield the observed noncrossover classes. (Remember that $\underline{+\ v\ bm}$ is equivalent to $\underline{pr^+\ v\ bm}$, and that $\underline{pr\ +\ +}$ is equivalent to $\underline{pr\ v^+\ bm^+}$.)

2. *What is the correct sequence of genes?* We know that the arrangement of alleles is

$$\frac{+\quad v\quad bm}{pr\quad +\quad +}$$

But is the gene sequence correct? That is, will a double-crossover event yield the observed double-crossover phenotypes after fertilization? *Observation shows that it will not* [Figure 7–11(b)]. Now try the other two orders [Figures 7–11(c) and (d)] *maintaining the same arrangement:*

$$\frac{+\quad bm\quad v}{pr\quad +\quad +}\quad \text{or}\quad \frac{v\quad +\quad bm}{+\quad pr\quad +}$$

Only the order on the right yields the observed double-crossover gametes [Figure 7–11(d)]. Therefore, the *pr* gene is in the middle. From this point on, work the problem using this arrangement and sequence, with the *pr* locus in the middle.

(a) Some possible allele arrangements and gene sequences in a heterozygous female

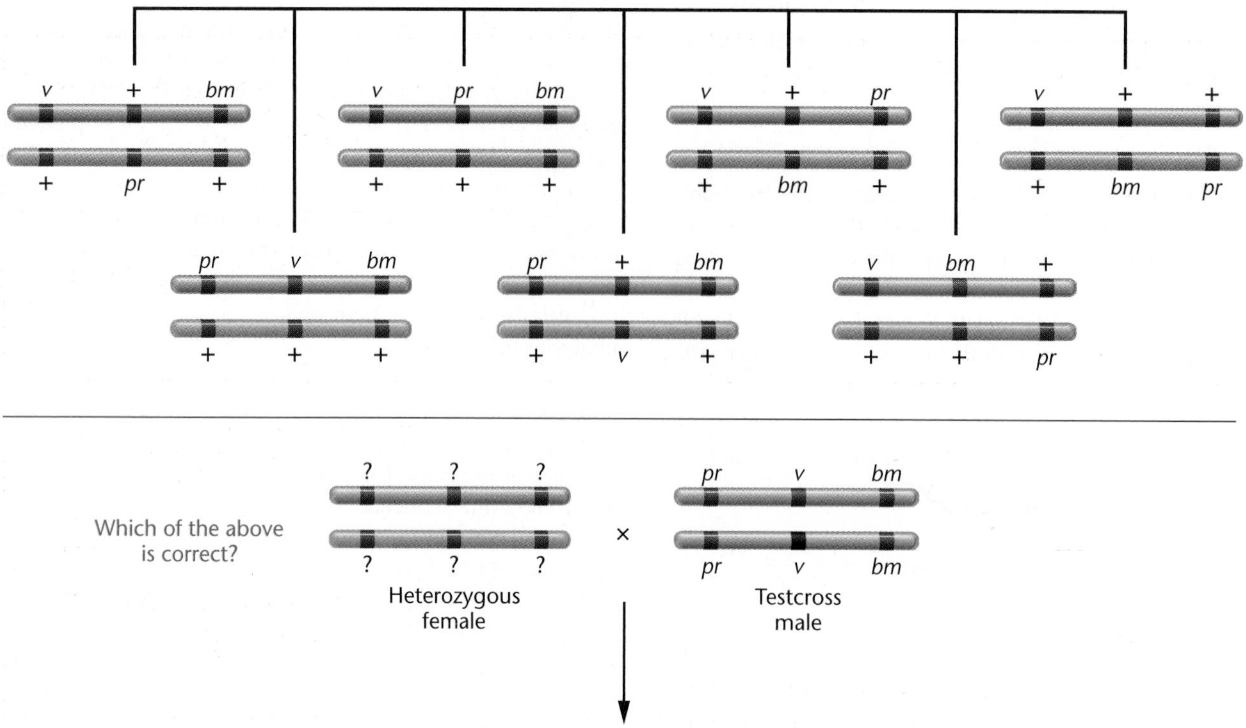

(b) Actual results of mapping cross*

Phenotypes of offspring	Number	Total and percentage	Exchange classification
+ *v* *bm*	230	467 42.1%	Noncrossover (NCO)
pr + +	237		
+ + *bm*	82	161 14.5%	Single crossover (SCO)
pr *v* +	79		
+ *v* +	200	395 35.6%	Single crossover (SCO)
pr + *bm*	195		
pr *v* *bm*	44	86 7.8%	Double crossover (DCO)
+ + +	42		

* The sequence *pr* – *v* – *bm* may or may not be correct.

FIGURE 7–10 (a) Some possible allele arrangements and gene sequences in a heterozygous female. The data from a three-point mapping cross depicted in (b), where the female is testcrossed, determine which combination of arrangement and sequence is correct [see Figure 7–11(d)].

3. *What is the distance between each pair of genes?* Having established the sequence of loci as *v–pr–bm*, we can determine the distance between *v* and *pr* and between *pr* and *bm*. Remember that the map distance between two genes is calculated on the basis of all detectable recombination events occurring between them. This includes both single- and double-crossover events.

Figure 7–11(e) shows that the phenotypes $\underline{v\ pr\ +}$ and $\underline{+\ +\ bm}$ result from single crossovers between the *v* and *pr* loci, accounting for 14.5 percent of the offspring [according to data in Figure 7–10(b)]. By adding the percentage of double crossovers (7.8 percent) to the number obtained for single crossovers, the total distance between the *v* and *pr* loci is calculated to be 22.3 mu.

Figure 7–11(f) shows that the phenotypes $\underline{v\ +\ +}$ and $\underline{+\ pr\ bm}$ result from single crossovers between the *pr* and *bm* loci, totaling 35.6 percent. Added to the double crossovers (7.8 percent), the distance between *pr* and *bm* is calculated to be 43.4 mu. The final map for all three genes in this example is shown in Figure 7–11(g).

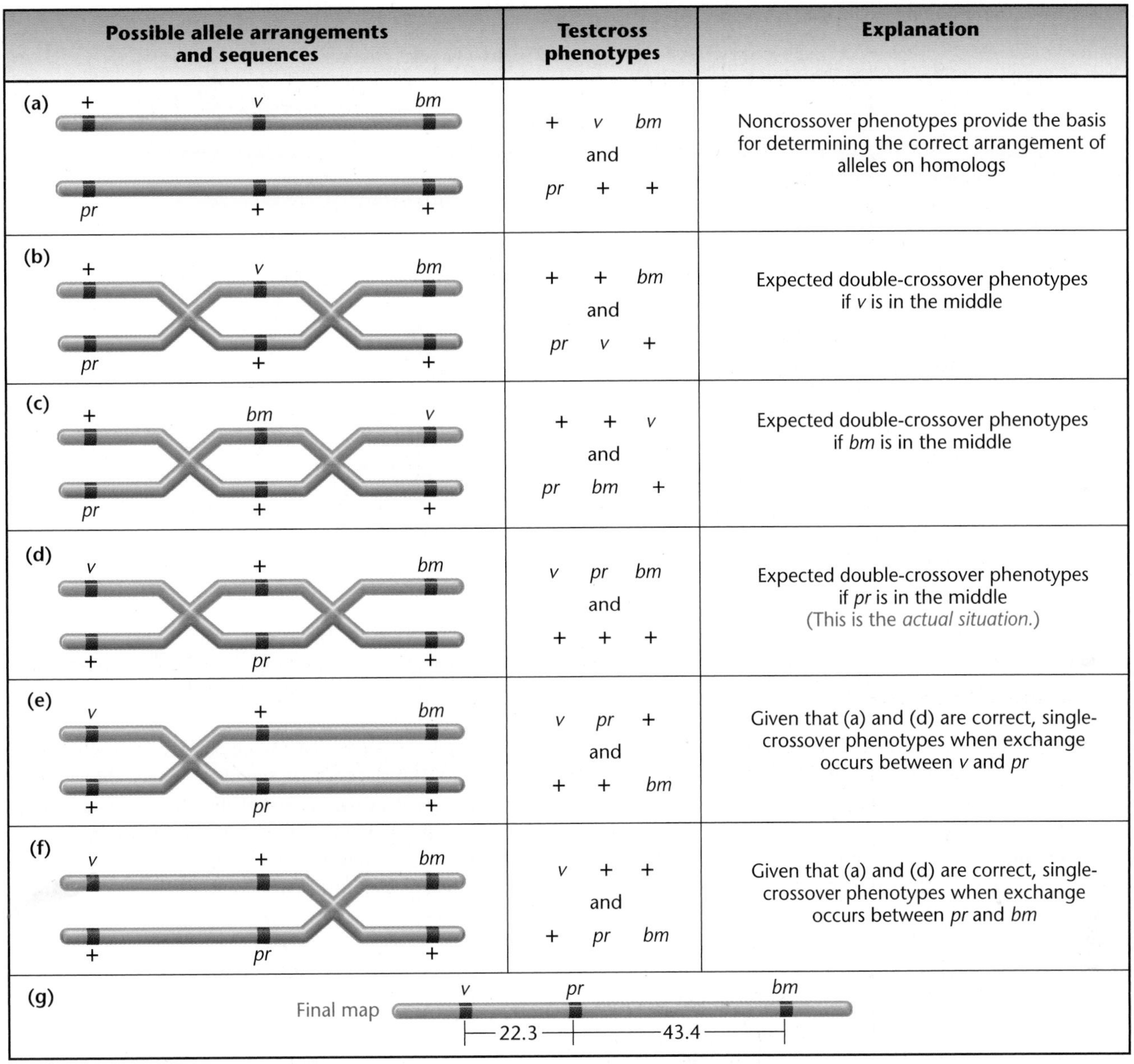

FIGURE 7–11 Steps utilized in the creation of a map of the three genes in the testcross of Figure 7–10, where neither the arrangement of alleles nor the sequence of genes in the heterozygous female parent is known.

NOW SOLVE THIS

Problem 20 on page 157 asks you to solve a three-point mapping problem where only six phenotypic categories are observed, when eight categories are typical of such a cross.

Hint: If the distances between each pair of genes is relatively small, reciprocal pairs of single crossovers will be evident, but the sample size may be too small to recover the predicted number of double crossovers, excluding their appearance. You should write the missing gametes down, assign them as double crossovers, and record zeroes for their frequency of appearance.

7.4 As the Distance between Two Genes Increases, Mapping Estimates Become More Inaccurate

So far, we have assumed that crossover frequencies are directly proportional to the distance between any two loci along the chromosome. However, it is not always possible to detect all crossover events. A case in point is a double exchange that occurs between the two loci in question. As shown in Figure 7–12(a), p. 146, if a double exchange occurs, the original arrangement of alleles on each nonsister homolog is

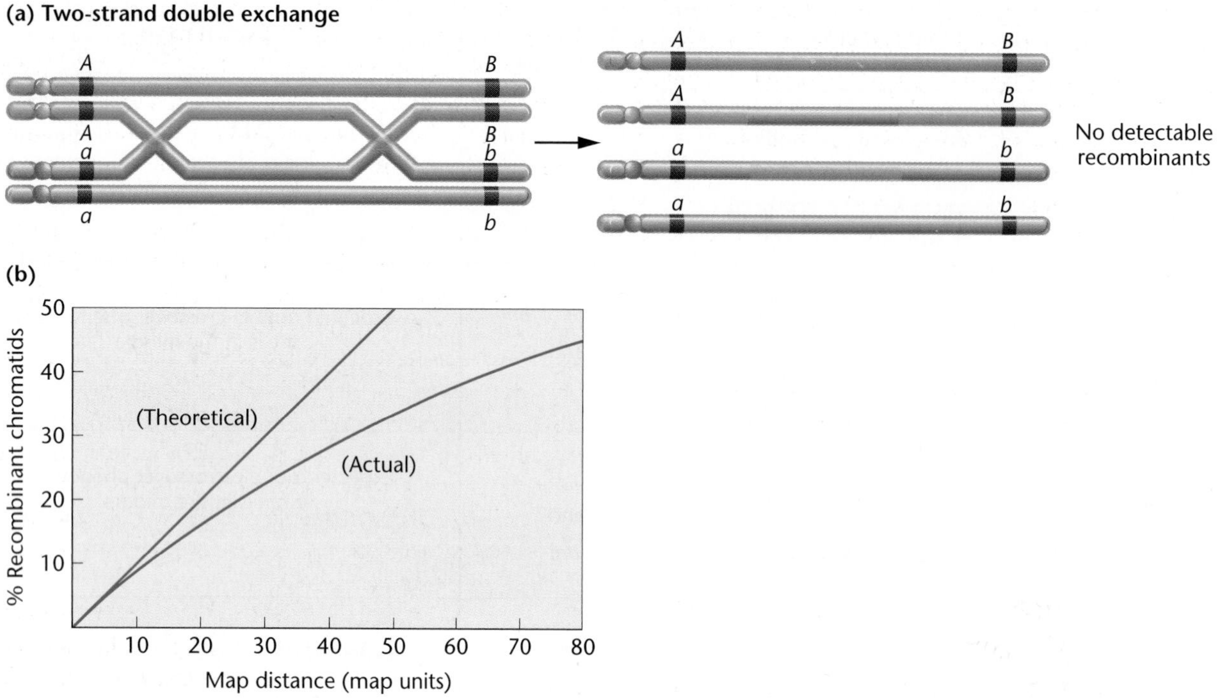

FIGURE 7–12 (a) A double crossover is undetected because no rearrangement of alleles occurs. (b) The theoretical and actual percentage of recombinant chromatids versus map distance. The straight line shows the theoretical relationship if a direct correlation between recombination and map distance exists. The curved line is the actual relationship derived from studies of *Drosophila, Neurospora,* and *Zea mays.*

recovered. Therefore, even though crossing over has occurred, it is impossible to detect. This phenomenon is true for all even-numbered exchanges between two loci.

Furthermore, as a result of complications posed by **multiple-strand exchanges,** mapping determinations usually underestimate the actual distance between two genes. The farther apart two genes are, the greater the probability that undetected crossovers will occur. While the discrepancy is minimal for two genes relatively close together, the degree of inaccuracy increases as the distance increases, as shown in the graph of recombination frequency versus map distance in Figure 7–12(b). There, the theoretical frequency where a direct correlation between recombination and map distance exists is contrasted with the actual frequency observed as the distance between two genes increases. The most accurate maps are constructed from experiments where genes are relatively close together.

Interference and the Coefficient of Coincidence

As shown in our maize example, we can predict the expected frequency of multiple exchanges, such as double crossovers, once the distance between genes is established. For example, in the maize cross, the distance between *v* and *pr* is 22.3 mu, and the distance between *pr* and *bm* is 43.4 mu. If the two single crossovers that make up a double crossover occur independently of one another, we can calculate the expected frequency of double crossovers (DCO_{exp}):

$$DCO_{exp} = (0.223) \times (0.434) = 0.097 = 9.7\%$$

Often in mapping experiments, the observed DCO frequency is less than the expected number of DCOs. In the maize cross, for example, only 7.8 percent DCOs are observed when 9.7 percent are expected. **Interference (*I*),** the phenomenon through which a crossover event in one region of the chromosome inhibits a second event in nearby regions, causes this reduction.

To quantify the disparities that result from interference, we calculate the **coefficient of coincidence (*C*):**

$$C = \frac{\text{Observed DCO}}{\text{Expected DCO}}$$

In the maize cross, we have

$$C = \frac{0.078}{0.097} = 0.804$$

Once we have found *C*, we can quantify interference using the simple equation

$$I = 1 - C$$

In the maize cross, we have

$$I = 1.000 - 0.804 = 0.196$$

If interference is complete and no double crossovers occur, then $I = 1.0$. If fewer DCOs than expected occur, *I* is a positive number and positive interference has occurred. If more DCOs than expected occur, *I* is a negative number and negative interference has occurred. In the maize example, *I* is a positive number (0.196), indicating that 19.6 percent fewer double crossovers occurred than expected.

Positive interference is most often the rule in eukaryotic systems. In general, the closer genes are to one another along the chromosome, the more positive interference occurs. In fact, interference in *Drosophila* is often complete within a distance of 10 mu, and no multiple crossovers are recovered. This observation suggests that physical constraints preventing the formation of closely aligned chiasmata contribute to interference. This interpretation is consistent with the finding that interference decreases as the genes in question are located farther apart. In the maize cross in Figures 7–10 and 7–11, the three genes are relatively far apart and 80 percent of the expected double crossovers are observed.

ESSENTIAL POINT

Interference describes the extent to which a crossover in one region of a chromosome influences the occurrence of a crossover in an adjacent region of the chromosome and is quantified by calculating the coefficient of coincidence (C).

7.5 Lod Score Analysis and Somatic Cell Hybridization Were Historically Important in Creating Human Chromosome Maps

In humans, genetic experiments involving carefully planned crosses and large numbers of offspring are neither ethical nor feasible, so the earliest linkage studies were based on pedigree analysis. These studies attempted to establish whether certain traits were X-linked or autosomal. As we showed in Chapter 4, traits determined by genes located on the X chromosome result in characteristic pedigrees; thus, such genes were easier to identify. For autosomal traits, geneticists tried to distinguish clearly whether pairs of traits demonstrated linkage or independent assortment. When extensive pedigrees are available, it is possible to conclude that two genes under consideration are closely linked (i.e., rarely separated by crossing over) from the fact that the two traits segregate together. It was hoped that from these kinds of observations a human gene map could be created.

Difficulties arise, however, when genes of interest are separated on a chromosome such that recombinant gametes are formed, obscuring linkage in a pedigree. In such cases, the demonstration of linkage is enhanced by an approach that relies on probability, called the **lod score method.** First devised by J. B. S. Haldane and C. A. Smith in 1947, and refined by Newton Morton in 1955, the lod score (*l*og of the *od*ds favoring linkage) assesses the probability that a particular pedigree involving two traits reflects linkage or does not. First, the probability is calculated that family data (pedigrees) concerning two or more traits conform to the transmission of traits without linkage. Then the probability is calculated that the identical family data following these same traits result from linkage with a specified recombination frequency. These probability calculations factor in the statistical significance at the $p = 0.05$ level. The ratio of these probability values is then calculated and converted to the logarithm of this value, which reflects the "odds" for, and against, linkage. Traditionally, a value of 3.0 or higher strongly indicates linkage, whereas a value of <2.0 argues strongly against linkage.

Accuracy using the lod score method is limited by the extent of the family data, but nevertheless represents an important advance in assigning human genes to specific chromosomes and constructing preliminary human chromosome maps. The initial results were discouraging because of the method's inherent limitations and the relatively high haploid number of human chromosomes (23), and by 1960, almost no linkage or mapping information had become available. Today, however, in contrast to its restricted impact when originally developed, this elegant technique has become important in human linkage analysis, owing to the discovery of countless molecular *DNA markers* along every human chromosome. The discovery of these markers, which behave just as genes do in occupying a particular locus along a chromosome, was a result of recombinant DNA techniques (Chapter 17) and genomic analysis (Chapter 18). Any human trait may now be tested for linkage with such markers.

In the 1960s, a new technique, **somatic cell hybridization,** proved to be an immense aid in assigning human genes to their respective chromosomes. This technique, first discovered by Georges Barsky, relies on the fact that two cells in culture can be induced to fuse into a single hybrid cell. Barsky used two mouse-cell lines, but it soon became evident that cells from different organisms could also be fused. When fusion occurs, an initial cell type called a **heterokaryon** is produced. The hybrid cell contains two nuclei in a common cytoplasm. By using the proper techniques, it is possible to fuse human and mouse cells, for example, and isolate the hybrids from the parental cells.

As the heterokaryons are cultured *in vitro,* two interesting changes occur. The nuclei eventually fuse, creating a **synkaryon.** Then, as culturing is continued for many generations, chromosomes from one of the two parental species are gradually lost. In the case of the human–mouse hybrid, human chromosomes are lost randomly until eventually the synkaryon has a full complement of mouse chromosomes and only a few human chromosomes. As we shall see, it is the preferential loss of human chromosomes (rather than mouse chromosomes) that makes possible the assignment of human genes to the chromosomes on which they reside.

The experimental rationale is straightforward. If a specific human gene product is synthesized in a synkaryon containing one to three human chromosomes, then the gene responsible for that product must reside on one of the human chromosomes remaining in the hybrid cell. On the other hand, if the human gene product is not synthesized in a synkaryon, the gene responsible is not present on those human chromosomes that remain in this hybrid cell. Ideally, a panel of 23 hybrid-cell lines, each with just one unique human chromosome, would allow the immediate assignment of any human gene for which the product could be characterized.

In practice, a panel of cell lines, each with several remaining human chromosomes, is most often used. The correlation of the presence or absence of each chromosome with the presence or absence of each gene product is called **synteny testing.**

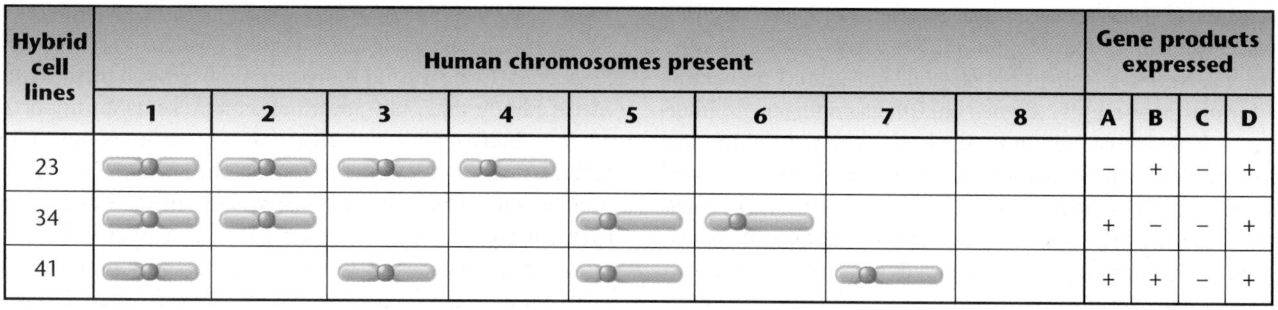
Hybrid cell lines
Human chromosomes present
Gene products expressed
1
2
3
4
5
6
7
8
A
B
C
D
23
– + – +
34
+ – – +
41
+ + – +

FIGURE 7–13 A hypothetical grid of data used in synteny testing to assign genes to their appropriate human chromosomes. Three somatic hybrid-cell lines, designated 23, 34, and 41, have each been scored for the presence or absence of human chromosomes 1–8, as well as for their ability to produce the hypothetical human gene products A, B, C, and D.

Consider, for example, the hypothetical data provided in Figure 7–13, where four gene products (A, B, C, and D) are tested in relationship to eight human chromosomes. Let's analyze the gene that produces product A.

1. Product A is not produced by cell line 23, but chromosomes 1, 2, 3, and 4 are present in cell line 23. Therefore, we rule out the presence of gene *A* on those four chromosomes and conclude that it must be on chromosome 5, 6, 7, or 8.
2. Product A is produced by cell line 34, which contains chromosomes 5 and 6 but not 7 and 8. Therefore, gene *A* is on chromosome 5 or 6, but cannot be on 7 or 8 because they are absent even though product A is produced.
3. Product A is also produced by cell line 41, which contains chromosome 5 but not chromosome 6. Using similar reasoning, we see that gene *A* must be on chromosome 5.

Using a similar approach, we can assign gene *B* to chromosome 3. Perform this analysis for yourself to demonstrate that this is correct.

Gene *C* presents a unique situation. The data indicate that it is not present on chromosomes 1–7. While it might be on chromosome 8, no direct evidence supports this conclusion, and other panels are needed. We leave gene *D* for you to analyze—on what chromosome does it reside?

Using this technique, researchers were able to assign hundreds of human genes to one chromosome or another. To map genes for which the products are not known, however, researchers have to rely on other more modern approaches.

7.6 Chromosome Mapping Is Now Possible Using DNA Markers and Annotated Computer Databases

The development of technology allowing direct analysis of DNA has greatly enhanced mapping in those organisms. We will address this topic using humans as an example. Progress has initially relied on the discovery of **DNA markers** (mentioned earlier) that have been identified during recombinant DNA and genomic studies. These markers are short segments of DNA whose sequence and location are known, making them useful *landmarks* for mapping purposes. The analysis of human genes in relation to these markers has extended our knowledge of the location within the genome of countless genes, which is the ultimate goal of mapping.

The earliest examples are the DNA markers referred to as **restriction fragment length polymorphisms (RFLPs)** (see Chapter 19) and **microsatellites** (see Chapter 11). RFLPs are polymorphic sites generated when specific DNA sequences are recognized and cut by restriction enzymes. Microsatellites are short repetitive sequences that are found throughout the genome, and they vary in the number of repeats at any given site. For example, the two-nucleotide sequence CA is repeated 5 to 50 times per site [$(CA)_n$] and appears throughout the genome approximately every 10,000 bases, on average. Microsatellites may be identified not only by the number of repeats but by the DNA sequences that flank them. More recently, variation in single nucleotides (called **single-nucleotide polymorphisms** or **SNPs**) has been utilized. Found throughout the genome, up to several million of these variations may be screened for an association with a disease or trait of interest, thus providing geneticists with a means to identify and locate related genes.

Cystic fibrosis offers an early example of a gene located by using DNA markers. This condition is a life-shortening autosomal recessive exocrine disorder resulting in excessive, thick mucus that impedes the function of organs such as the lung and pancreas. After scientists established that the gene causing this disorder is located on chromosome 7, they were then able to pinpoint its exact location on the long arm (the q arm) of that chromosome.

Several years ago (June 2007) using SNPs as DNA markers, associations between 24 genomic locations were established with seven common human diseases: Type 1 (insulin dependent) and Type 2 diabetes, Crohn's disease (inflammatory bowel disease), hypertension, coronary artery disease, bipolar (manic-depressive) disorder, and rheumatoid arthritis. In each case, an inherited susceptibility effect has been mapped to a specific location on a specific chromosome within the genome. In some cases, this has either confirmed or led to the identification of a specific gene involved in the cause of the disease. In other cases, new genes will no doubt soon be discovered as a result of the identification of their location. We will return to this topic in much greater detail in Chapters 18 and 19.

The many Human Genome Project databases that have been completed now make it possible to map genes along a human chromosome in base-pair distances rather than recombination frequency. This distinguishes what is referred to as a **physical map** of the genome from the genetic maps described above. Distances can then be determined relative to other genes and to features such as the DNA markers discussed above. The power of this approach is that geneticists will soon be able to construct chromosome maps for individuals that designate specific allele combinations at each gene site.

ESSENTIAL POINT

Human linkage studies, initially relying on pedigree and lod score analysis and subsequently on somatic cell hybridization techniques, are now enhanced by the use of newly discovered molecular DNA markers.

7.7 Linkage and Mapping Studies Can Be Performed in Haploid Organisms

Many of the single-celled eukaryotes are haploid during the vegetative stages of their life cycle. The alga *Chlamydomonas* and the mold *Neurospora* demonstrate this genetic condition. But these organisms do form reproductive cells that fuse during fertilization, producing a diploid zygote. However, this structure soon undergoes meiosis, resulting in haploid vegetative cells that then propagate by mitotic divisions. In genetic studies, small haploid organisms have several important advantages compared with diploid eukaryotes. Haploid organisms can be cultured and manipulated in genetic crosses much more easily. In addition, a haploid organism contains only a single allele of each gene, which is expressed directly in the phenotype. This greatly simplifies genetic analysis. As a result, organisms such as *Chlamydomonas* and *Neurospora* serve as the subjects of research investigations in many areas of genetics, including linkage and mapping studies.

In order to perform genetics experiments with such organisms, crosses are made, and following fertilization, the meiotic structures may be isolated. Because all four meiotic products give rise to spores, the structures bearing these products (asci) are called **tetrads.** *The term tetrad has a different meaning here than earlier when it was used to describe a precise chromatid configuration in meiosis.* Individual tetrads are isolated, and the resultant cells are grown and analyzed separately from those of other tetrads. In the results we are about to describe, the data reflect the proportion of tetrads that show one combination of genotypes, the proportion that show another combination, and so on. Such experimentation is called **tetrad analysis.**

Gene-to-Centromere Mapping

When a single gene in *Neurospora* is analyzed (Figure 7–14, p. 150), the data can be used to calculate the map distance between the gene and the centromere. This process is sometimes referred to as **mapping the centromere.** It is accomplished by experimentally determining the frequency of recombination using tetrad data. Note that once the four meiotic products of the tetrad are formed, a mitotic division occurs, resulting in eight ordered products (ascospores). If no crossover event occurs between the gene under study and the centromere, the pattern of ascospores contained within an ascus (pl., asci) appears as shown in Figure 7–14(a), *aaaa*++++.*

This pattern represents **first-division segregation** because the two alleles are separated during the first meiotic division. However, a crossover event will alter this pattern, as shown in Figure 7–14(b), *aa*++*aa*++, and Figure 7–14(c), ++*aaaa*++. Two other recombinant patterns occur but are not shown: ++*aa*++*aa* and *aa*++++*aa*. These depend on the chromatid orientation during the second meiotic division. These four patterns reflect **second-division segregation** because the two alleles are not separated until the second meiotic division. Usually, the ordered tetrad data are condensed to reflect the genotypes of the identical ascospore pairs, and six combinations are possible.

First-Division Segregation	Second-Division Segregation	
aa++	*a*+*a*+	+*aa*+
++*aa*	+*a*+*a*	*a*++*a*

To calculate the distance between the gene and the centromere, a large number of asci resulting from a controlled cross are counted. We then use these data to calculate the distance (d):

$$d = \frac{1/2\ (\text{second-division segregant asci})}{\text{total asci scored}}$$

The distance (d) reflects the percentage of recombination and is only half the number of second-division segregant asci. This is because crossing over in each occurs in only two of the four chromatids during meiosis.

To illustrate, we use a for albino and + for wild type in *Neurospora.* In crosses between the two genetic types, suppose we observe 65 first-division segregants, and 70 second-division segregants. The distance between a and the centromere is

$$d = \frac{(1/2)\,(70)}{135} = 0.259$$

or about 26 mu.

As the distance increases to 50 mu, all asci should reflect second-division segregation. However, numerous factors prevent this. As in diploid organisms, mapping accuracy based on crossover events is greatest when the gene and centromere are relatively close together.

We can also analyze haploid organisms to distinguish between linkage and independent assortment of two genes—mapping distances between gene loci are calculated once linkage is established. As a result, detailed maps of organisms such as *Neurospora* and *Chlamydomonas* are now available.

* The pattern (++++*aaaa*) can also be formed. However, it is indistinguishable from (*aaaa*++++).

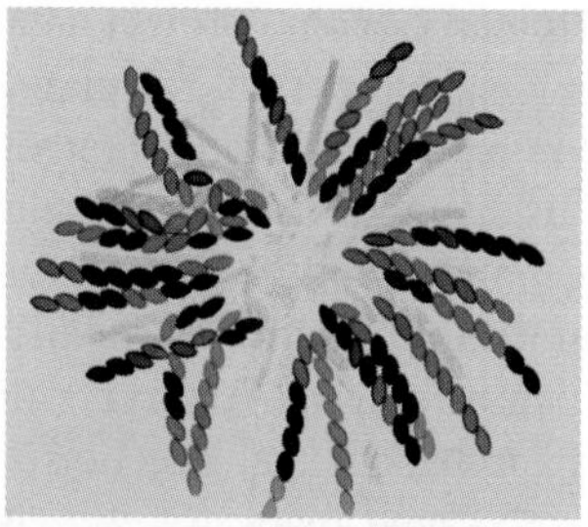

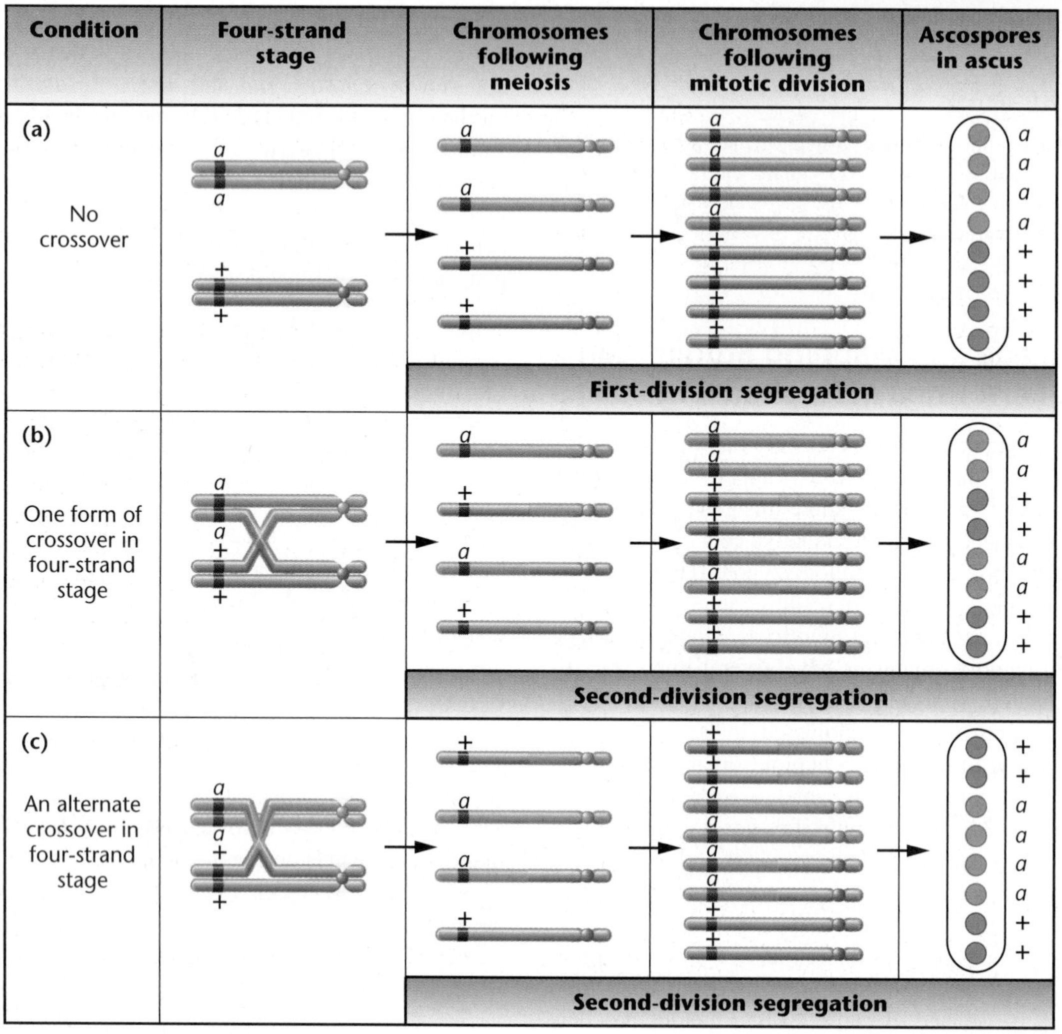

FIGURE 7–14 Three ways in which different ascospore patterns can be generated in *Neurospora.* Analysis of these patterns is the basis of gene-to-centromere mapping. The photograph shows a variety of ascospore arrangements within *Neurospora* asci.

ESSENTIAL POINT

Linkage analysis and chromosome mapping are possible in haploid eukaryotes, relying on the direct analysis of meitotic products such as ascospores in *Neurospora.*

7.8 Other Aspects of Genetic Exchange

Careful analysis of crossing over during gamete formation allows us to construct chromosome maps in both diploid and haploid organisms. However, we should not lose sight of the real biological significance of crossing over, which is to generate genetic variation in gametes and, subsequently, in the offspring derived from the resultant eggs and sperm. Many unanswered questions remain, which we consider next.

Cytological Evidence for Crossing Over

Once genetic mapping was understood, it was of great interest to investigate the relationship between chiasmata observed in meiotic prophase I and crossing over. For example, are chiasmata visible manifestations of crossover events? If so, then crossing over in higher organisms appears to result from an actual physical exchange between homologous chromosomes. That this is the case was demonstrated independently in the 1930s by Harriet Creighton and Barbara McClintock in *Zea mays* and by Curt Stern in *Drosophila.*

Since the experiments are similar, we will consider only the work with maize. Creighton and McClintock studied two linked genes on chromosome 9. At one locus, the alleles *colorless* (*c*) and *colored* (*C*) control endosperm coloration. At the other locus, the alleles *starchy* (*Wx*) and *waxy* (*wx*) control the carbohydrate characteristics of the endosperm.

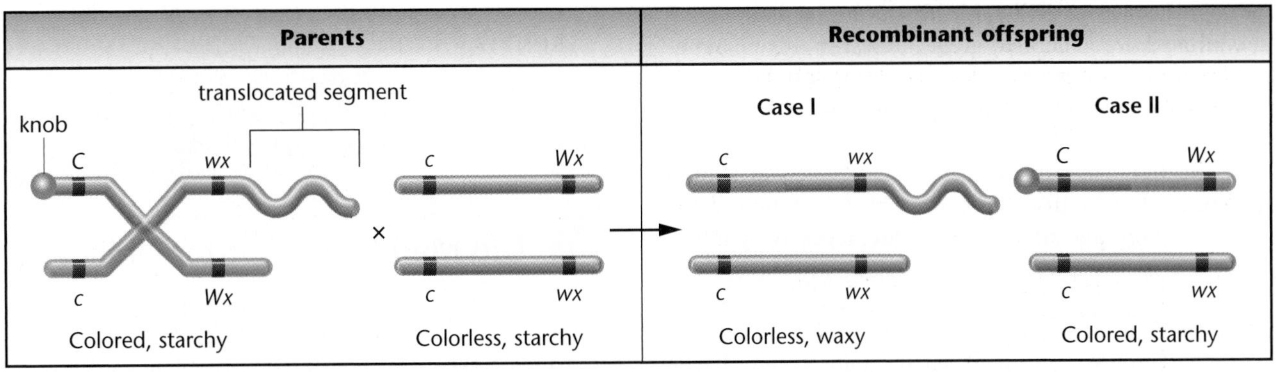

FIGURE 7–15 The phenotypes and chromosome compositions of parents and recombinant offspring in Creighton and McClintock's experiment in maize. The knob and translocated segment are the cytological markers that established that crossing over involves an actual exchange of chromosome arms.

The maize plant studied is heterozygous at both loci. The key to this experiment is that one of the homologs contains two unique cytological markers. The markers consist of a densely stained knob at one end of the chromosome and a translocated piece of another chromosome (8) at the other end. The arrangements of alleles and cytological markers can be detected cytologically and are shown in Figure 7–15.

Creighton and McClintock crossed this plant to one homozygous for the *colored* allele (*c*) and heterozygous for the endosperm alleles. They obtained a variety of different phenotypes in the offspring, but they were most interested in a crossover result involving the chromosome with the unique cytological markers. They examined the chromosomes of this plant with the colorless, waxy phenotype (Case I in Figure 7–15) for the presence of the cytological markers. If physical exchange between homologs accompanies genetic crossing over, the translocated chromosome will still be present, but the knob will not—this is exactly what happened. In a second plant (Case II), the phenotype colored, starchy should result from either nonrecombinant gametes or crossing over. Some of the plants then ought to contain chromosomes with the dense knob but not the translocated chromosome. This condition was also found, and the conclusion that a physical exchange takes place was again supported. Along with Stern's findings with *Drosophila,* this work clearly established that crossing over has a cytological basis.

ESSENTIAL POINT

Cytological investigations of both maize and *Drosophila* reveal that crossing over involves a physical exchange of segments between nonsister chromatids.

Sister Chromatid Exchanges between Mitotic Chromosomes

Considering that crossing over occurs between synapsed homologs in meiosis, we might ask whether a similar physical exchange occurs between homologs during mitosis. While homologous chromosomes do not usually pair up or synapse in somatic cells (*Drosophila* is an exception), each individual chromosome in prophase and metaphase of mitosis consists of two identical sister chromatids, joined at a common centromere. Surprisingly, several experimental approaches have demonstrated that reciprocal exchanges similar to crossing over occur between sister chromatids. These **sister chromatid exchanges (SCEs)** do not produce new allelic combinations, but evidence is accumulating that attaches significance to these events.

Identification and study of SCEs are facilitated by several modern staining techniques. In one technique, cells replicate for two generations in the presence of the thymidine analog **bromodeoxyuridine (BrdU).** Following two rounds of replication, each pair of sister chromatids has one member with one strand of DNA "labeled" with BrdU and the other member with both strands labeled with BrdU. Using a differential stain, chromatids with the analog in both strands stain less brightly than chromatids with BrdU in only one strand. As a result, SCEs are readily detectable if they occur. In Figure 7–16, numerous instances of SCE events are clearly evident. These sister chromatids are sometimes referred to as **harlequin chromosomes** because of their patchlike appearance.

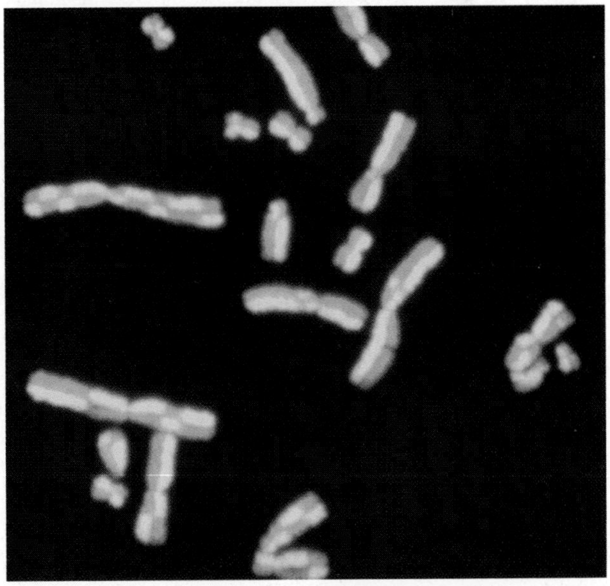

FIGURE 7–16 Light micrograph of sister chromatid exchanges (SCEs) in mitotic chromosomes. Sometimes called harlequin chromosomes because of their patchlike appearance, chromatids with the thymidine analog BrdU in both DNA strands fluoresce *less* brightly than do those with the analog in only one strand. These chromosomes were stained with 33258-Hoechst reagent and acridine orange and then viewed using fluorescence microscopy.

The significance of SCEs is still uncertain, but several observations have generated great interest in this phenomenon. We know, for example, that agents that induce chromosome damage (viruses, X rays, ultraviolet light, and certain chemical mutagens) increase the frequency of SCEs. The frequency of SCEs is also elevated in **Bloom syndrome,** a human disorder caused by a mutation in the *BLM* gene on chromosome 15. This rare, recessively inherited disease is characterized by prenatal and postnatal retardation of growth, a great sensitivity of the facial skin to the sun, immune deficiency, a predisposition to malignant and benign tumors, and abnormal behavior patterns. The chromosomes from cultured leukocytes, bone marrow cells, and fibroblasts derived from homozygotes are very fragile and unstable compared to those of homozygous and heterozygous normal individuals. Increased breaks and rearrangements between nonhomologous chromosomes are observed in addition to excessive amounts of sister chromatid exchanges. Work by James German and colleagues suggests that the *BLM* gene encodes an enzyme called DNA helicase, which is best known for its role in DNA replication (see Chapter 10).

ESSENTIAL POINT

Recombination between sister chromatids in mitosis, referred to as sister chromatid exchanges (SCEs), occur at an elevated frequency in the human disorder, Bloom syndrome.

7.9 Did Mendel Encounter Linkage?

We conclude this chapter by examining a modern-day interpretation of the experiments that form the cornerstone of transmission genetics—Mendel's crosses with garden peas.

Some observers believe that Mendel had extremely good fortune in his classic experiments. He did not encounter apparent linkage relationships between the seven mutant characters in any of his crosses. Had Mendel obtained highly variable data characteristic of linkage and crossing over, these unorthodox ratios might have hindered his successful analysis and interpretation.

The article by Stig Blixt, reprinted in its entirety in the following box, demonstrates the inadequacy of this hypothesis. As we shall see, some of Mendel's genes were indeed linked. We leave it to Stig Blixt to enlighten you as to why Mendel did not detect linkage.

Why Didn't Gregor Mendel Find Linkage?

It is quite often said that Mendel was very fortunate not to run into the complication of linkage during his experiments. He used seven genes, and the pea has only seven chromosomes. Some have said that had he taken just one more, he would have had problems. This, however, is a gross oversimplification. The actual situation, most probably, is shown in Table 7.1. This shows that Mendel worked with three genes in chromosome 4, two genes in chromosome 1, and one gene in each of chromosomes 5 and 7. It seems at first glance that, out of the 21 dihybrid combinations Mendel theoretically could have studied, no fewer than four (that is, *a–i, v–fa, v–le, fa–le,*) ought to have resulted in linkages. However, as found in hundreds of crosses and shown by the genetic map of the pea, *a* and *i* in chromosome 1 are so distantly located on the chromosome that no linkage is normally detected. The same is true for *v* and *le* on the one hand, and *fa* on the other, in chromosome 4. This leaves *v–le*, which ought to have shown linkage.

Mendel, however, seems not to have published this particular combination and thus, presumably, never made the appropriate cross to obtain both genes segregating simultaneously. It is therefore not so astonishing that Mendel did not run into the complication of linkage, although he did not avoid it by choosing one gene from each chromosome.

Stig Blixt
Weibullsholm Plant Breeding Institute,
Landskrona, Sweden,
and Centro Energia Nucleate na
Agricultura, Piracicaba, SP, Brazil

Source: Reprinted by permission from *Nature,* Vol. 256, p. 206. © 1975 Macmillan Magazines Limited.

TABLE 7.1 Relationship between Modern Genetic Terminology and Character Pairs Used by Mendel

Character Pair Used by Mendel	Alleles in Modern Terminology	Located in Chromosome
Seed color, yellow–green	*I–i*	1
Seed coat and flowers, colored–white	*A–a*	1
Mature pods, smooth expanded–wrinkled indented	*V–v*	4
Inflorescences, from leaf axis–umbellate in top of plant	*Fa–fa*	4
Plant height, 0.5–1m	*Le–le*	4
Unripe pods, green–yellow	*Gp–gp*	5
Mature seeds, smooth–wrinkled	*R–r*	7

EXPLORING GENOMICS

Human Chromosome Maps on the Internet

In this chapter we discussed how recombination data can be analyzed to develop chromosome maps based on linkage. Although linkage analysis and chromosome mapping continue to be important approaches in genetics, chromosome maps are increasingly being developed for many species using genomics techniques to sequence entire chromosomes. As a result of the Human Genome Project, maps of human chromosomes are now freely available on the Internet. With the click of a mouse you can have immediate access to an incredible wealth of information. In this exercise we will explore the **National Center for Biotechnology Information (NCBI) Genes and Disease** Web site to learn more about human chromosome maps.

■ Exercise I: NCBI Genes and Disease

The NCBI Web site is an outstanding resource for genome data. Here we explore the Genes and Disease site, which presents human chromosome maps that show the locations of specific disease genes.

1. Access the Genes and Disease site at **http://www.ncbi.nlm.nih.gov/books/bv.fcgi?rid=gnd&ref=toc.**
2. Click on any of the chromosomes at the top of the page to see a map showing selected genes on that chromosome that are associated with genetic diseases. For example, click on chromosome 7; then click on the OB gene label to learn about the role of this gene in obesity. Notice that the top of each chromosome page displays the number of genes on the chromosome and the number of base pairs the chromosome contains.
3. Look again at chromosome 7. At first you might think there are only five disease genes on this chromosome because the initial view shows only selected disease genes. However, if you click the "MapViewer" link for the chromosome (just above the drawing), you will see detailed information about the chromosome, including a complete "Master Map" of the genes it contains, including the symbols used in naming genes:.

   Gene Symbols: Clicking on the gene symbols takes you to the **NCBI Entrez Gene database,** a searchable tool for information on genes in the NCBI database.

   Links: The items in the "Links" column provide access to OMIM (Online Mendelian Inheritance in Man, discussed in the Exploring Genomics feature for Chapter 3) data for a particular gene, as well as to protein information (*pn*) and lists of homologous genes (*hm;* these are other genes that have similar sequences).
4. Click on the links in MapViewer to learn more about a gene of interest.
5. Scan the chromosome maps in MapViewer until you see one of the genes listed as a "hypothetical gene or protein."

   a. What does it mean if a gene or protein is referred to as hypothetical?

   b. What information do you think genome scientists use to assign a gene locus for a gene encoding a hypothetical protein?

Visit the **NCBI Map Viewer** home page **(www.ncbi.nlm.nih.gov/mapview/)** for an excellent database containing chromosome maps for a wide variety of different organisms. Search this database to learn more about chromosome maps for an organism you are interested in.

CASE STUDY Links to autism

As parents of an autistic child, entering a research study seemed to be a way of not only educating themselves about their son's condition, but also of furthering research into this complex, behaviorally defined disorder. Researchers told the couple that there is a strong genetic influence for autism since the concordance rate in identical twins is about 75 percent and only about 5 percent in fraternal twins. In addition, researchers have identified interactions among at least ten candidate genes distributed among chromosomes 3, 7, 15, and 17. Generally unaware of the principles of basic genetics, the couple asked a number of interesting questions. How would you respond to them?

1. How can there be ten candidate genes on only four chromosomes?
2. Since several candidate genes must be on the same chromosome, will they be transmitted as a block of harmful genetic information to future offspring?
3. With such an apparently complex genetic condition, what is the likelihood that our next child will also be autistic?
4. Is prenatal diagnosis possible during future pregnancies?

INSIGHTS AND SOLUTIONS

1. In rabbits, black color (B) is dominant to brown (b), while full color (C) is dominant to *chinchilla* (c^{ch}). The genes controlling these traits are linked. Rabbits that are heterozygous for both traits and express black, full color are crossed to rabbits that express brown, chinchilla with the following results:

31	brown, chinchilla
34	black, full
16	brown, full
19	black, chinchilla

Determine the arrangement of alleles in the heterozygous parents and the map distance between the two genes.

Solution: This is a two-point map problem, where the two most prevalent reciprocal phenotypes are the noncrossovers. The less frequent reciprocal phenotypes arise from a single crossover. The arrangement of alleles is derived from the noncrossover phenotypes because they enter gametes intact.

The single crossovers give rise to 35/100 offspring (35 percent). Therefore, the distance between the two genes is 35 mu.

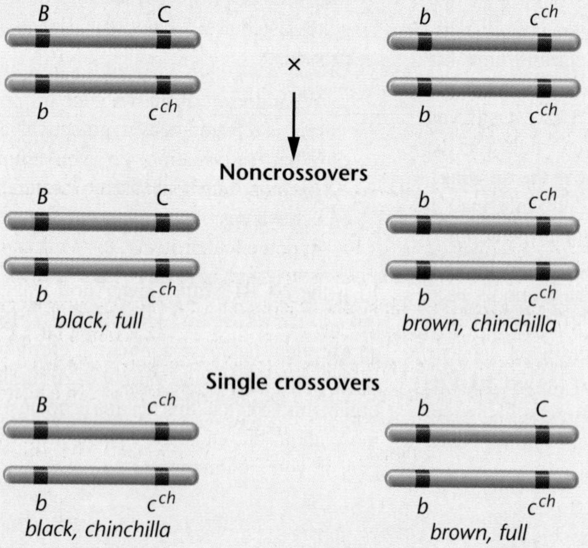

2. In *Drosophila*, *Lyra* (Ly) and *Stubble* (Sb) are dominant mutations located at locus 40 and 58, respectively, on chromosome III. A recessive mutation with bright red eyes is discovered and shown also to be located on chromosome III. A map is obtained by crossing a female who is heterozygous for all three mutations to a male that is homozygous for the *bright-red* mutation (which we will call *br*), and the data in the table are generated. Determine the location of the *br* mutation on chromosome III.

Phenotype			Number
(1) *Ly*	*Sb*	*br*	404
(2) +	+	+	422
(3) *Ly*	+	+	18
(4) +	*Sb*	*br*	16
(5) *Ly*	+	*br*	75
(6) +	*Sb*	+	59
(7) *Ly*	*Sb*	+	4
(8) +	+	*br*	2
		Total =	1000

Solution: First, determine the arrangement of the alleles on the homologs of the heterozygous crossover parent (the female in this case). To do this, locate the most frequent reciprocal phenotypes, which arise from the noncrossover gametes—these are phenotypes (1) and (2). Each phenotype represents the arrangement of alleles on one of the homologs. Therefore, the arrangement is

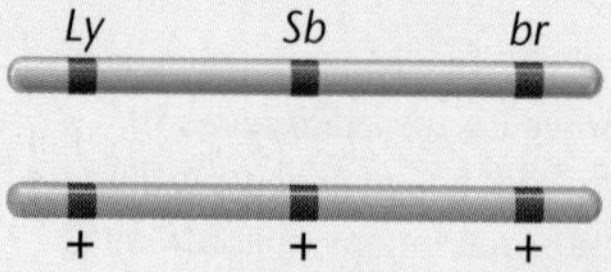

Second, find the correct sequence of the three loci along the chromosome. This is done by determining which sequence yields the observed double-crossover phenotypes, which are the least frequent reciprocal phenotypes (7 and 8). If the sequence is correct as written, then the double crossover depicted here,

will yield $Ly + br$ and $+ Sb +$ as phenotypes. Inspection shows that these categories (5 and 6) are actually single crossovers, not double crossovers. Therefore, the sequence is incorrect as written. Only two other sequences are possible: The *br* gene is either to the left of *Ly* (Case A), or it is between *Ly* and *Sb* (Case B).

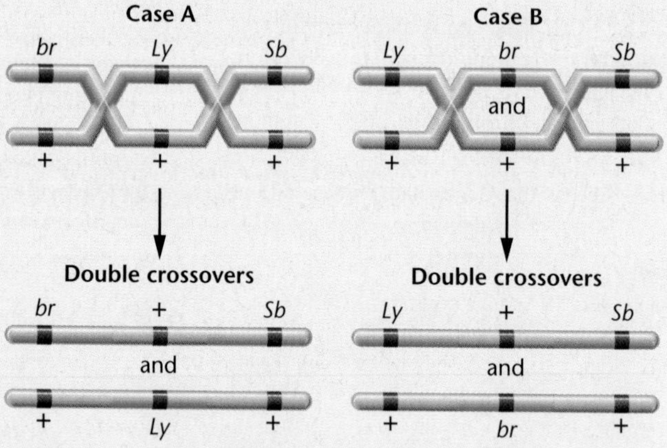

Comparison with the actual data shows that Case B is correct. The double-crossover gametes yield flies that express *Ly* and *Sb* but not *br*, or express *br* but not *Ly* and *Sb*. Therefore, the correct arrangement and sequence are as follows.

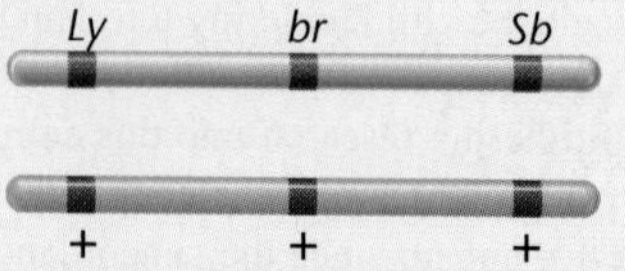

Once this sequence is found, determine the location of *br* relative to *Ly* and *Sb*. A single crossover between *Ly* and *br*, as shown here,

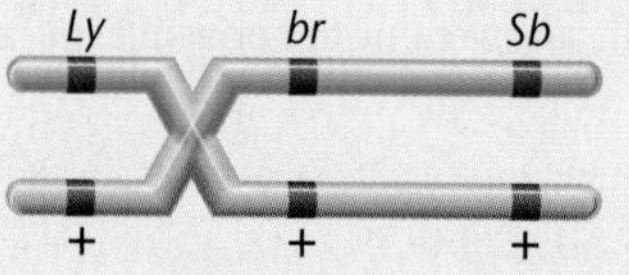

yields flies that are Ly + + and $+br\ Sb$ (phenotypes 3 and 4). Therefore, the distance between the Ly and br loci is equal to

$$(18 + 16 + 4 + 2)/1000 = 40/1000 = 0.04 = 4 \text{ mu}$$

Remember to add the double crossovers because they represent two single crossovers occurring simultaneously. You need to know the frequency of all crossovers between Ly and br, so they must be included.

Similarly, the distance between the br and Sb loci is derived mainly from single crossovers between them.

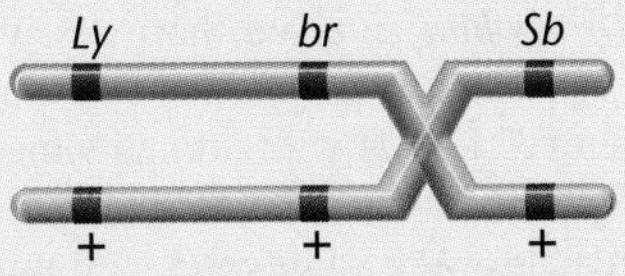

This event yields $Ly\ br+$ and + + Sb phenotypes (phenotypes 5 and 6). Therefore, the distance equals

$$(75 + 59 + 4 + 2)/1000 = 140/1000 = 0.14 = 14 \text{ mu}$$

The final map shows that *br* is located at locus 44, since *Lyra* and *Stubble* are known.

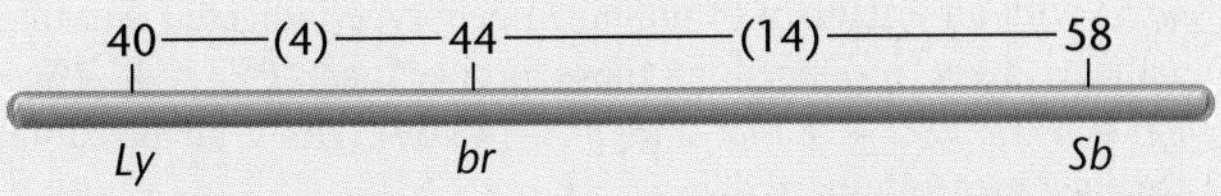

3. In reference to Problem 2, a student predicted that the mutation was actually the known mutation *scarlet* located at locus 44.0. Suggest an experimental cross that would confirm this prediction.

Solution: Since the *scarlet* locus is identical to the experimental assignment, it is reasonable to hypothesize that the *bright-red* eye mutation is an allele at the *scarlet* locus.

To test this hypothesis, you could perform complementation analysis (see Chapter 4) by crossing females expressing the *bright-red* mutation with known *scarlet* males. If the two mutations are alleles, no complementation will occur and all progeny will reveal a *bright-red* mutant eye phenotype. If complementation occurs, all progeny will express normal brick-red (wild-type) eyes, since the bright-red mutation and *scarlet* are at different loci (they are probably very close together). In such a case, all progeny will be heterozygous at both the *bright-red* eye and the *scarlet* loci and will not express either mutation because they are both recessive. This type of complementation analysis is called an **allelism test.**

Problems and Discussion Questions

1. What is the significance of crossing over (which leads to genetic recombination) to the process of evolution?
2. Describe the cytological observation that suggests that crossing over occurs during the first meiotic prophase.
3. Why does more crossing over occur between two distantly linked genes than between two genes that are very close together on the same chromosome?
4. Why is a 50 percent recovery of single-crossover products the upper limit, even when crossing over *always* occurs between two linked genes?
5. Why are double-crossover events expected less frequently than single-crossover events?
6. What is the proposed basis for positive interference?
7. What three essential criteria must be met in order to execute a successful mapping cross?
8. The genes *dumpy* wings (*dp*), *clot* eyes (*cl*), and *apterous* wings (*ap*) are linked on chromosome II of *Drosophila.* In a series of two-point mapping crosses, the genetic distances shown below were determined. What is the sequence of the three genes?

dp–ap	42
dp–cl	3
ap–cl	39

9. Consider two hypothetical recessive autosomal genes *a* and *b,* where a heterozygote is testcrossed to a double-homozygous mutant. Predict the phenotypic ratios under the following conditions:
(a) *a* and *b* are located on separate autosomes.
(b) *a* and *b* are linked on the same autosome but are so far apart that a crossover always occurs between them.
(c) *a* and *b* are linked on the same autosome but are so close together that a crossover almost never occurs.
(d) *a* and *b* are linked on the same autosome about 10 mu apart.

See **Now Solve This** on page 134.

10. Colored aleurone in the kernels of corn is due to the dominant allele *R*. The recessive allele *r*, when homozygous, produces colorless aleurone. The plant color (not kernel color) is controlled by another gene with two alleles, *Y* and *y*. The dominant *Y* allele results in green color, whereas the homozygous presence of the recessive *y* allele causes the plant to appear yellow. In a testcross between a plant of unknown genotype and phenotype and a plant that is homozygous recessive for both traits, the following progeny were obtained:

colored, green	88
colored, yellow	12
colorless, green	8
colorless, yellow	92

Explain how these results were obtained by determining the exact genotype and phenotype of the unknown plant, including the precise association of the two genes on the homologs (i.e., the arrangement).

11. In the cross shown here, involving two linked genes, *ebony* (*e*) and *claret* (*ca*), in *Drosophila,* where crossing over does not occur in males, offspring were produced in a (2+:1 *ca*:1 *e*) phenotypic ratio:

$$\overset{\female}{\frac{e \quad ca^+}{e^+ \quad ca}} \times \overset{\male}{\frac{e \quad ca^+}{e^+ \quad ca}}$$

These genes are 30 mu apart on chromosome III. What did crossing over in the female contribute to these phenotypes?

12. With two pairs of genes involved (*P, p* and *Z, z*), a testcross (to *ppzz*) with an organism of unknown genotype indicated that the gametes were produced in these proportions: $PZ = 42.4\%$; $Pz = 6.9\%$; $pZ = 7.1\%$; and $pz = 43.6\%$. Draw all possible conclusions from these data.

13. In a series of two-point map crosses involving five genes located on chromosome II in *Drosophila,* the following recombinant (single-crossover) frequencies were observed:

pr–adp	29
pr–vg	13
pr–c	21
pr–b	6
adp–b	35
adp–c	8
adp–vg	16
vg–b	19
vg–c	8
c–b	27

(a) If the *adp* gene is present near the end of chromosome II (locus 83), construct a map of these genes.
(b) In another set of experiments, a sixth gene (*d*) was tested against *b* and *pr,* and the results were $d - b = 17\%$ and $d - pr = 23\%$. Predict the results of two-point maps between *d* and *c, d* and *vg,* and *d* and *adp.*

14. Two different female *Drosophila* were isolated, each heterozygous for the autosomally linked genes *black* body (*b*), *dachs* tarsus (*d*), and *curved* wings (*c*). These genes are in the order *d–b–c*, with *b* closer to *d* than to *c*. Shown in the following table is the genotypic arrangement for each female, along with the various gametes formed by both. Identify which categories are noncrossovers (NCO), single crossovers (SCO), and double crossovers (DCO) in each case. Then, indicate the relative frequency with which each will be produced. See **Now Solve This** on page 140.

Female A			Female B	
d b + / *+ + c* ↓		Gamete formation	*d + +* / *+ b c* ↓	
(1) *d b c*	(5) *d + +*		(1) *d b +*	(5) *d b c*
(2) *+ + +*	(6) *+ b c*		(2) *+ + c*	(6) *+ + +*
(3) *+ + c*	(7) *d + c*		(3) *d + c*	(7) *d + +*
(4) *d b +*	(8) *+ b +*		(4) *+ b +*	(8) *+ b c*

15. In *Drosophila,* a cross was made between females expressing the three X-linked recessive traits, *scute* bristles (*sc*), *sable* body (*s*), and *vermilion* eyes (*v*), and wild-type males. All females were wild type in the F_1, while all males expressed all three mutant traits. The cross was carried to the F_2 generation, and 1000 offspring were counted, with the results shown in the following table. No determination of sex was made in the F_2 data. (a) Using proper nomenclature, determine the genotypes of the P_1 and F_1 parents. (b) Determine the sequence of the three genes and the map distance between them. (c) Are there more or fewer double crossovers than expected? (d) Calculate the coefficient of coincidence; does this represent positive or negative interference?

Phenotype			Offspring
sc	*s*	*v*	314
+	+	+	280
+	*s*	*v*	150
sc	+	+	156
sc	+	*v*	46
+	*s*	+	30
sc	*s*	+	10
+	+	*v*	14

16. A cross in *Drosophila* involved the recessive, X-linked genes *yellow* body (*y*), *white* eyes (*w*), and *cut* wings (*ct*). A yellow-bodied, white-eyed female with normal wings was crossed to a male whose eyes and body were normal, but whose wings were cut. The F_1 females were wild type for all three traits, while the F_1 males expressed the yellow-body, white-eye traits. The cross was carried to F_2 progeny, and only male offspring were tallied. On the basis of the data shown here, a genetic map was constructed. (a) Diagram the genotypes of the F_1 parents. (b) Construct a map, assuming that *w* is at locus 1.5 on the X chromosome. (c) Were any double-crossover offspring expected? (d) Could the F_2 female offspring be used to construct the map? Why or why not?

Phenotype			Male Offspring
y	+	*ct*	9
+	*w*	+	6
y	*w*	*ct*	90
+	+	+	95
+	+	*ct*	424
y	*w*	+	376
y	+	+	0
+	*w*	*ct*	0

17. In *Drosophila, Dichaete* (*D*) is a mutation on chromosome III with a dominant effect on wing shape. It is lethal when homozygous. The genes *ebony* body (*e*) and *pink* eye (*p*) are recessive mutations on chromosome III. Flies from a Dichaete stock were crossed to homozygous ebony, pink flies, and the F_1 progeny with a Dichaete phenotype were backcrossed to the ebony, pink homozygotes. Using the results of this backcross shown in the following table, (a) diagram the cross, showing the genotypes of the parents and offspring of both crosses. (b) What is the sequence and interlocus distance between these three genes?

Phenotype	Number
Dichaete	401
ebony, pink	389
Dichaete, ebony	84
pink	96
Dichaete, pink	2
ebony	3
Dichaete, ebony, pink	12
wild type	13

18. *Drosophila* females homozygous for the third chromosomal genes *pink* eye (*p*) and *ebony* body (*e*) were crossed with males homozygous for the second chromosomal gene *dumpy* wings (*dp*). Because these genes are recessive, all offspring were wild type (normal). F_1 females were testcrossed to triply recessive males. If we assume that the two linked genes (*p* and *e*) are 20 mu apart, predict the results of this cross. If the reciprocal cross were made (F_1 males—where no crossing over occurs—with triply recessive females), how would the results vary, if at all?

19. In *Drosophila,* the two mutations *Stubble* bristles (*Sb*) and *curled* wings (*cu*) are linked on chromosome III. *Sb* is a dominant gene that is lethal in a homozygous state, and *cu* is a recessive gene. If a female of the genotype

$$\frac{Sb \quad cu}{+ \quad +}$$

is to be mated to detect recombinants among her offspring, what male genotype would you choose as her mate?

20. In *Drosophila,* a heterozygous female for the X-linked recessive traits *a, b,* and *c* was crossed to a male that phenotypically expressed *a, b,* and *c.* The offspring occurred in the phenotypic ratios in the following table, and no other phenotypes were observed. (a) What is the genotypic arrangement of the alleles of these genes on the X chromosome of the female? (b) Determine the correct sequence, and construct a map of these genes on the X chromosome. (c) What progeny phenotypes are missing, and why? See **Now Solve This** on page 145.

+	*b*	*c*	460
a	+	+	450
a	*b*	*c*	32
+	+	+	38
a	+	*c*	11
+	*b*	+	9

21. Are sister chromatid exchanges effective in producing genetic variability in an individual? in the offspring of individuals?

22. What conclusion can be drawn from the observations that in male *Drosophila,* no crossing over occurs and that during meiosis, synaptonemal complexes (ultrastructural components found between synapsed homologs in meiosis) are not seen in males but are observed in females where crossing over occurs?

23. An organism of the genotype *AaBbCc* was testcrossed to a triply recessive organism (*aabbcc*). The genotypes of the progeny are in the following table.

AaBbCc	20	*AaBbcc*	20
aabbCc	20	*aabbcc*	20
AabbCc	5	*Aabbcc*	5
aaBbCc	5	*aaBbcc*	5

(a) Assuming simple dominance and recessiveness in each gene pair, if these three genes were all assorting independently, how many genotypic and phenotypic classes would result in the offspring, and in what proportion?

(b) Answer part (a) again, assuming the three genes are so tightly linked on a single chromosome that no crossover gametes were recovered in the sample of offspring.

(c) What can you conclude from the *actual* data about the location of the three genes in relation to one another?

24. Based on our discussion of the potential inaccuracy of mapping (see Figure 7–12), would you revise your answer to Problem 23(c)? If so, how?

25. In a plant, fruit color is either red or yellow, and fruit shape is either oval or long. Red and oval are the dominant traits. Two plants, both heterozygous for these traits, were testcrossed, with the results shown in the following table. Determine the location of the genes relative to one another and the genotypes of the two parental plants.

	Progeny	
Phenotype	**Plant A**	**Plant B**
red, long	46	4
yellow, oval	44	6
red, oval	5	43
yellow, long	5	47
Total	100	100

26. In a plant heterozygous for two gene pairs (*Ab*/*aB*), where the two loci are linked and 25 mu apart, two such individuals were crossed together. Assuming that crossing over occurs during the formation of both male and female gametes and that the *A* and *B* alleles are dominant, determine the phenotypic ratio of the offspring.

27. In a cross in *Neurospora* involving two alleles, *B* and *b,* the tetrad patterns in the following table were observed. Calculate the distance between the gene and the centromere.

Tetrad Pattern	Number
BBbb	36
bbBB	44
BbBb	4
bBbB	6
BbbB	3
bBBb	7

28. In Creighton and McClintock's experiment demonstrating that crossing over involves physical exchange between chromosomes (see Section 7.8), explain the importance of the cytological markers (the translocated segment and the chromosome knob) in the experimental rationale.

29. A number of human–mouse somatic cell hybrid clones were examined for the expression of specific human genes and the presence of human chromosomes; the results are summarized in this table. Assign each gene to the chromosome upon which it is located.

	Hybrid-Cell Clone					
	A	**B**	**C**	**D**	**E**	**F**
Genes (expressed or not)						
ENO1 (*enolase-1*)	−	+	−	+	+	−
MDH1 (*malate dehydrogenase-1*)	+	+	−	+	−	+
PEPS (*peptidase S*)	+	−	+	−	−	−
PGM1 (*phosphoglucomutase-1*)	−	+	−	+	+	−
Chromosomes (present or absent)						
1	−	+	−	+	+	−
2	+	+	−	+	−	+
3	+	+	−	−	+	−
4	+	−	+	−	−	−
5	−	+	+	+	+	+

30. A female of genotype

$$\frac{a \quad b \quad c}{+ \quad + \quad +}$$

produces 100 meiotic tetrads. Of these, 68 show no crossover events. Of the remaining 32, 20 show a crossover between *a* and *b*, 10 show a crossover between *b* and *c*, and 2 show a double crossover between *a* and *b* and between *b* and *c*. Of the 400 gametes produced, how many of each of the eight different genotypes will be produced? Assuming the order *a–b–c* and the allele arrangement shown above, what is the map distance between these loci?

31. *Drosophila melanogaster* has one pair of sex chromosomes (XX or XY) and three autosomes (chromosomes II, III, and IV). A genetics student discovered a male fly with very short (*sh*) legs. Using this male, the student was able to establish a pure-breeding stock of this mutant and found that it was recessive. She then incorporated the mutant into a stock containing the recessive gene *black* (*b*, body color, located on chromosome II) and the recessive gene *pink* (*p*, eye color, located on chromosome III). A female from the homozygous black, pink, short stock was then mated to a wild-type male. The F_1 males of this cross were all wild type and were then backcrossed to the homozygous *b, p, sh* females. The F_2 results appeared as shown in the following table, and no other phenotypes were observed. (a) Based on these results, the student was able to assign *sh* to a linkage group (a chromosome). Determine which chromosome, and include step-by-step reasoning. (b) The student repeated the experiment, making the reciprocal cross: F_1 females backcrossed to homozygous *b, p, sh* males. She observed that 85 percent of the offspring fell into the given classes, but that 15 percent of the offspring were equally divided among *b*+*p*, *b*++,+*shp*, and +*sh*+ phenotypic males and females. How can these results be explained, and what information can be derived from these data?

Phenotype	Female	Male
wild	63	59
pink*	58	65
black, short	55	51
black, pink, short	69	60

*Pink indicates that the other two traits are wild type (normal). Similarly, black, short offspring are wild type for eye color.

32. In *Drosophila*, a female fly is heterozygous for three mutations, *Bar* eyes (*B*), *miniature* wings (*m*), and *ebony* body (*e*). (Note that *Bar* is a dominant mutation.) The fly is crossed to a male with normal eyes, miniature wings, and ebony body. The results of the cross are shown in the following list. Interpret the results of this cross. If you conclude that linkage is involved between any of the genes, determine the map distance(s) between them.

miniature	111	Bar	117
wild	29	Bar, miniature	26
Bar, ebony	101	ebony	35
Bar, miniature, ebony	31	miniature, ebony	115

33. Because of the relatively high frequency of meiotic errors that lead to developmental abnormalities in humans, many research efforts have focused on identifying correlations between error frequency and chromosome morphology and behavior. Tease et al. (2002) studied human fetal oocytes of chromosomes 21, 18, and 13 using an immunocytological approach that allowed a direct estimate of the frequency and position of meiotic recombination. Following is a summary of information (modified from Tease et al., 2002) that compares recombination frequency with the frequency of trisomy for chromosomes 21, 18, and 13. (*Note:* You may want to read appropriate portions of Chapter 8 for descriptions of these trisomic conditions.)

Trisomic	Mean Recombination Frequency	Live-born Frequency
Chromosome 21	1.23	1/700
Chromosome 18	2.36	1/3000–1/8000
Chromosome 13	2.50	1/5000–1/19000

(a) What conclusions can be drawn from these data in terms of recombination and nondisjunction frequencies? How might recombination frequencies influence trisomic frequencies?

(b) Other studies indicate that the number of crossovers per oocyte is somewhat constant, and it has been suggested that positive chromosomal interference acts to spread out a limited number of crossovers among as many chromosomes as possible. Considering information in part (a), speculate on the selective advantage positive chromosomal interference might confer.

An electron micrograph showing the sex pilus between two conjugating *E. coli* cells.

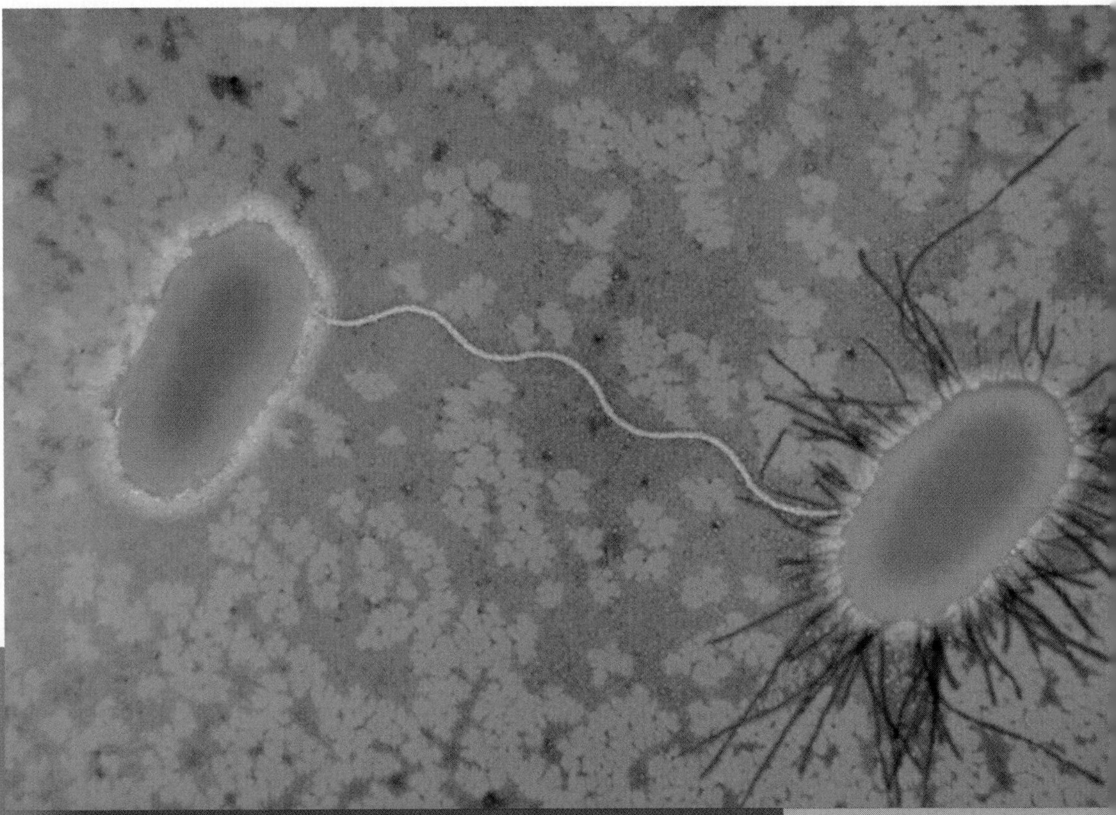

8

Genetic Analysis and Mapping in Bacteria and Bacteriophages

Chapter Concepts

- Bacterial genomes are most often contained in a single circular chromosome.
- Bacteria have developed numerous ways in which they can exchange and recombine genetic information between individual cells, including conjugation, transformation, and transduction.
- The ability to undergo conjugation and to transfer the bacterial chromosome from one cell to another is governed by the presence of genetic information contained in the DNA of a "fertility," or F factor.
- The F factor can exist autonomously in the bacterial cytoplasm as a plasmid, or it can integrate into the bacterial chromosome, where it facilitates the transfer of the host chromosome to the recipient cell, leading to genetic recombination.
- Genetic recombination during conjugation provides a means of mapping bacterial genes.
- Bacteriophages are viruses that have bacteria as their hosts.
- During infection of the bacterial host, bacteriophage DNA is injected into the host cell, where it is replicated and directs the reproduction of the bacteriophage.

In this chapter, we shift from consideration of mapping genetic information in eukaryotes to discussion of the analysis and mapping of genes in **bacteria** (prokaryotes) and **bacteriophages,** viruses that use bacteria as their hosts. The study of bacteria and bacteriophages has been essential to the accumulation of knowledge in many areas of genetic study. For example, much of what we know about molecular genetics, recombinational phenomena, and gene structure was initially derived from experimental work with them. Furthermore, our extensive knowledge of bacteria and their resident plasmids has led to their widespread use in DNA cloning and other recombinant DNA studies.

Bacteria and their viruses are especially useful research organisms in genetics for several reasons. They have extremely short reproductive cycles—literally hundreds of generations, giving rise to billions of genetically identical bacteria or phages, can be produced in short periods of time. Furthermore, they can be studied in pure cultures. That is, a single species or mutant strain of bacteria or one type of virus can be isolated and investigated independently of other similar organisms.

In this chapter, we focus on genetic recombination and chromosome mapping. Complex processes have evolved in bacteria and bacteriophages that facilitate the transfer of genetic information between individual cells within populations. As we shall see, these processes are the basis for the chromosome mapping analysis that forms the cornerstone of molecular genetic investigations of bacteria and the viruses that invade them.

How Do We Know?

In this chapter, we will focus on genetic systems present in bacteria and the viruses that use bacteria as hosts (bacteriophages). In particular, we will discuss mechanisms by which bacteria and their phages undergo genetic recombination, the basis of chromosome mapping. As you study these topics, you should try to answer several fundamental questions:

1. How do we know that bacteria undergo genetic recombination, allowing the transfer of genes from one organism to another?
2. How do we know that conjugation leading to genetic recombination between bacteria involves cell contact, which precedes the transfer of genes from one bacterium to another?
3. How do we know that during transduction bacterial cell-to-cell contact is not essential?
4. How do we know that intergenic exchange occurs in bacteriophages?

BACTERIAL GENETICS WEB TUTORIAL

8.1 Bacteria Mutate Spontaneously and Grow at an Exponential Rate

It has long been known that pure cultures of bacteria give rise to cells that exhibit heritable variation, particularly with respect to growth under unique environmental conditions. Prior to 1943, the source of this variation was hotly debated. The majority of bacteriologists believed that environmental factors induced changes in certain bacteria that led to their adaptation to the new conditions. For example, strains of *Escherichia coli* are known to be sensitive to infection by the bacteriophage T1. Infection by this bacteriophage leads to the virus reproducing at the expense of the bacterial cell, from which new phages are released as the host cell is disrupted, or lysed. If a plate of *E. coli* is uniformly sprayed with T1, almost all cells are lysed. Rare *E. coli* cells, however, survive infection and are not lysed. If these cells are isolated and established in pure culture, all of the descendants are resistant to T1 infection. The **adaptation hypothesis,** put forth to explain this type of observation, implies that the interaction of the phage and bacterium is essential to the acquisition of immunity. In other words, the phage "induces" resistance in the bacteria.

On the other hand, the existence of **spontaneous mutations,** which occur in the presence or the absence of bacteriophage T1, suggested an alternative model to explain the origin of resistance in *E. coli.* In 1943, Salvador Luria and Max Delbruck presented the first convincing evidence that bacteria, like eukaryotic organisms, are capable of spontaneous mutation. This experiment, referred to as the **fluctuation test,** marks the initiation of modern bacterial genetic study. Spontaneous mutation is now considered the primary source of genetic variation in bacteria.

Mutant cells that arise spontaneously in otherwise pure cultures can be isolated and established independently from the parent strain by using established selection techniques. As a result, mutations for almost any desired characteristic can now be isolated. Because bacteria and viruses usually contain only a single chromosome and are therefore haploid, all mutations are expressed directly in the descendants of mutant cells, adding to the ease with which these microorganisms can be studied.

Bacteria are grown in a liquid culture medium or in a petri dish on a semisolid agar surface. If the nutrient components of the growth medium are simple and consist only of an organic carbon source (such as glucose or lactose) and a variety of ions, including Na^+, K^+, Mg^{2+}, Ca^{2+}, and NH_4^+, present as inorganic salts, it is called **minimal medium.** To grow on such a medium, a bacterium must be able to synthesize all essential organic compounds (e.g., amino acids, purines, pyrimidines, sugars, vitamins, and fatty acids). A bacterium that can accomplish this remarkable biosynthetic feat—one that we ourselves cannot duplicate—is a **prototroph.** It is said to be wild type for all growth requirements. On the other hand, if a bacterium loses, through mutation, the ability to synthesize one or more organic components, it is an **auxotroph.** For example, if a bacterium loses the ability to make histidine, then this amino acid must be added as a supplement to the minimal medium for growth to occur. The resulting bacterium is designated as a his^- auxotroph, in contrast to its prototrophic his^+ counterpart.

To study bacterial growth quantitatively, an inoculum of bacteria is placed in liquid culture medium. A graph of the characteristic growth pattern is shown in Figure 8–1. Initially, during the **lag phase,** growth is slow. Then, a period of rapid growth, called the **logarithmic (log) phase,** ensues. During this phase, cells divide continually with a fixed time interval

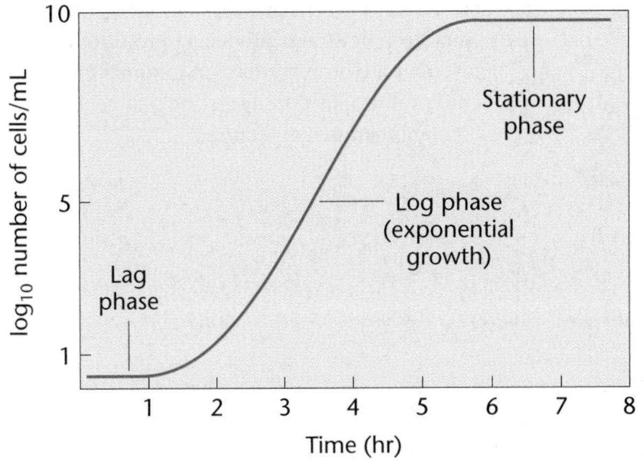

FIGURE 8–1 Typical bacterial population growth curve showing the initial lag phase, the subsequent log phase where exponential growth occurs, and the stationary phase that occurs when nutrients are exhausted.

between cell divisions, resulting in exponential growth. When a cell density of about 10^9cells/mL is reached, nutrients and oxygen become limiting and the cells cease dividing; at this point, the cells enter the **stationary phase.** The doubling time during the log phase can be as short as 20 minutes. Thus, an initial inoculum of a few thousand cells easily achieves maximum cell density during an overnight incubation.

Cells grown in liquid medium can be quantified by transferring them to a semisolid medium in a petri dish. Following incubation and many divisions, each cell gives rise to a visible colony on the surface of the medium. If the number of colonies is too great to count, then a series of successive dilutions (a technique called **serial dilution**) of the original liquid culture is made and plated, until the colony number is reduced to the point where it can be counted (Figure 8–2). This technique allows the number of bacteria present in the original culture to be calculated.

As an example, let's assume that the three dishes in Figure 8–2 represent serial dilutions of 10^{-3}, 10^{-4}, and 10^{-5} (from left to right). We need only select the dish in which the number of colonies can be counted accurately. Because each colony arose from a single bacterium, the number of colonies multiplied by the dilution factor represents the number of bacteria in each milliliter of the initial inoculum used to start the serial dilutions. In Figure 8–2, the rightmost dish has 15 colonies. The dilution factor for a 10^{-5} dilution is 10^5. Therefore, the initial number of bacteria was 15×10^5 per mL.

8.2 Conjugation Is One Means of Genetic Recombination in Bacteria

Development of techniques that allowed the identification and study of bacterial mutations led to detailed investigations of the arrangement of genes on the bacterial chromosome. Such studies began in 1946 when Joshua Lederberg and Edward Tatum showed that bacteria undergo **conjugation,** a parasexual process in which the genetic information from one bacterium is transferred to and recombined with that of another bacterium (see the chapter opening photograph). Like meiotic crossing over in eukaryotes, genetic recombination in bacteria enabled the development of methodology for chromosome mapping. Note that the term **genetic recombination,** as applied to bacteria and bacteriophages, leads to the *replacement* of one or more genes present in one strain with those from a genetically distinct strain. Although this is somewhat different from our use of genetic recombination in eukaryotes, where the term describes crossing over that results in *reciprocal exchange events,* the overall effect is the same: Genetic information is transferred from one chromosome to another, resulting in an altered genotype. Two other phenomena that result in the transfer of genetic information from one bacterium to another, *transformation* and *transduction,* have helped us determine the arrangement of genes on the bacterial chromosome. We will discuss these processes in later sections of this chapter.

Lederberg and Tatum's initial experiments were performed with two multiple auxotrophic strains of *E. coli* K12. Strain A required methionine and biotin in order to grow, whereas strain B required threonine, leucine, and thiamine (Figure 8–3, p. 162). Therefore, neither strain would grow on minimal medium. The two strains were first grown separately in supplemented media, and cells from both were mixed and grown together for several more generations and then plated on minimal medium. Any bacterial cells that grew on minimal medium were prototrophs. It is highly improbable that any of the cells that contained two or three mutant genes underwent spontaneous mutation simultaneously at two or three independent locations, leading to wild-type cells. Therefore, the researchers assumed that any prototrophs recovered arose as a result of some form of genetic exchange and recombination between the two mutant strains.

In this experiment, prototrophs were recovered at a rate of $1/10^7$ (or 10^{-7}) cells plated. The controls for this experiment involved separate plating of cells from strains A and B on

FIGURE 8–2 Results of the serial dilution technique and subsequent culture of bacteria. Each dilution varies by a factor of 10. Each colony is derived from a single bacterial cell.

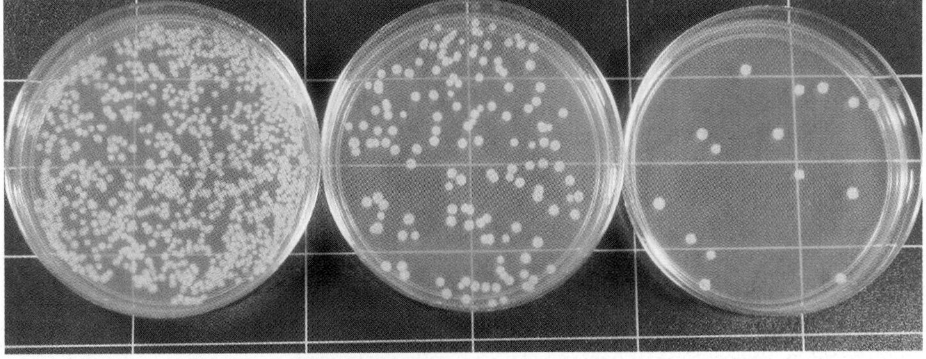

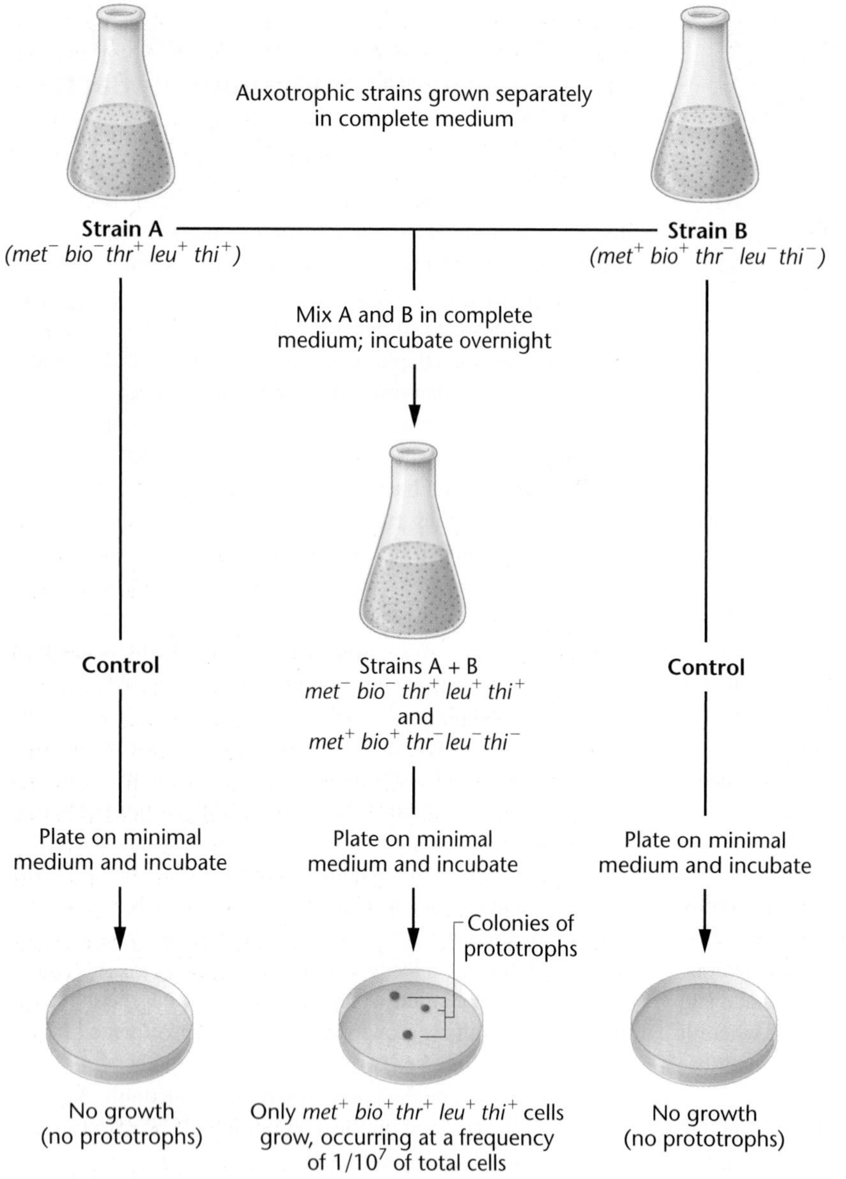

FIGURE 8–3 Genetic recombination of two auxotrophic strains producing prototrophs. Neither auxotroph grows on minimal medium, but prototrophs do, suggesting that genetic recombination has occurred.

minimal medium. No prototrophs were recovered. Based on these observations, Lederberg and Tatum proposed that genetic exchange had occurred.

F^+ and F^- Bacteria

Lederberg and Tatum's findings were soon followed by numerous experiments that elucidated the genetic basis of conjugation. It quickly became evident that different strains of bacteria are involved in a unidirectional transfer of genetic material. When cells serve as donors of parts of their chromosomes, they are designated as **F^+ cells** (F for "fertility"). Recipient bacteria receive the donor chromosome material (now known to be DNA), and recombine it with part of their own chromosome. They are designated as **F^- cells.**

Experimentation subsequently established that cell contact is essential for chromosome transfer to occur. Support for this concept was provided by Bernard Davis, who designed the Davis U-tube for growing F^+ and F^- cells shown in Figure 8–4. At the base of the tube is a sintered glass filter with a pore size that allows passage of the liquid medium but that is too small to allow the passage of bacteria. The F^+ cells are placed on one side of the filter, and F^- cells on the other side. The medium is moved back and forth across the filter so that the cells share a common medium during bacterial incubation. When Davis plated samples from both sides of the tube on minimal medium, no prototrophs were found, and he logically concluded that *physical contact between cells of the two strains* is essential to genetic recombination. We now know that this physical interaction is the initial step in the process of conjugation established by a structure called the **F pilus** (or **sex pilus;** pl., pili). Bacteria often have many pili, which are tubular extensions of the cell. After contact is initiated between mating pairs (see the chapter opening photograph), chromosome transfer is then possible.

Later evidence established that F^+ cells contain a **fertility factor** (**F factor**) that confers the ability to donate part of their chromosome during conjugation. Experiments by Joshua and Esther Lederberg and by William Hayes and Luca Cavalli-Sforza show that certain conditions eliminate the F factor in otherwise fertile cells. However, if these "infertile" cells are then grown with fertile donor cells, the F factor is regained.

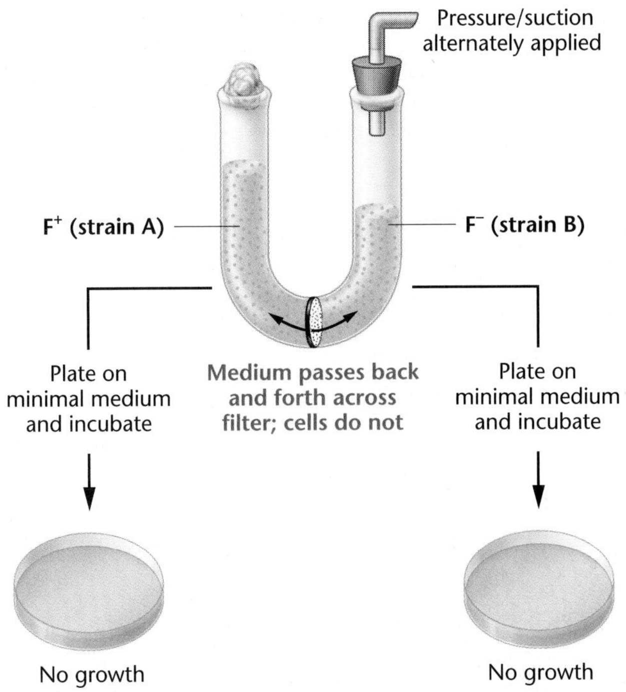

FIGURE 8–4 When strain A and B auxotrophs are grown in a common medium but separated by a filter, as in this Davis U-tube apparatus, no genetic recombination occurs and no prototrophs are produced.

The conclusion that the F factor is a mobile element is further supported by the observation that, following conjugation and genetic recombination, recipient cells always become F^+. Thus, in addition to the *rare* cases of gene transfer from the bacterial chromosome (genetic recombination), the F factor itself is passed to *all* recipient cells. On this basis, the initial cross of Lederberg and Tatum (see Figure 8–3) is as follows:

Strain A		*Strain B*
F^+	×	F^-
Donor		*Recipient*

Characterization of the F factor confirmed these conclusions. Like the bacterial chromosome, though distinct from it, the F factor has been shown to consist of a circular, double-stranded DNA molecule, equivalent to about 2 percent of the bacterial chromosome (about 100,000 nucleotide pairs). There are 19 genes contained within the F factor, whose products are involved in the transfer of genetic information, including those essential to the formation of the sex pilus.

Geneticists believe that transfer of the F factor during conjugation involves separation of the two strands of its double helix and movement of one of the two strands into the recipient cell. Both strands, one moving across the conjugation tube and one remaining in the donor cell, are replicated. The result is that both the donor *and* the recipient cells become F^+. This process is diagrammed in Figure 8–5, p. 164.

To summarize, an *E. coli* cell may or may not contain the F factor. When this factor is present, the cell is able to form a sex pilus and potentially serve as a donor of genetic information. During conjugation, a copy of the F factor is almost always transferred from the F^+ cell to the F^- recipient, converting the recipient to the F^+ state. The question remained as to exactly why such a low proportion of cells involved in these matings (10^{-7}) also results in genetic recombination. The answer awaited further experimentation.

As you soon shall see, the F factor is in reality an autonomous genetic unit called a *plasmid.* However, in our historical coverage of its discovery, we will continue to refer to it as a factor.

ESSENTIAL POINT

Conjugation may be initiated by a bacterium housing a plasmid called the F factor in its cytoplasm, making it a donor (F^+) cell. Following conjugation, the recipient (F^-) cell receives a copy of the F factor and is converted to the F^+ status.

Hfr Bacteria and Chromosome Mapping

Subsequent discoveries not only clarified how genetic recombination occurs but also defined a mechanism by which the *E. coli* chromosome could be mapped. Let's address chromosome mapping first.

In 1950, Cavalli-Sforza treated an F^+ strain of *E. coli* K12 with nitrogen mustard, a chemical known to induce mutations. From these treated cells, he recovered a genetically altered strain of donor bacteria that underwent recombination at a rate of $1/10^4$ (or 10^{-4})—1000 times more frequently than the original F^+ strains. In 1953, Hayes isolated another strain that demonstrated an elevated frequency. Both strains were designated **Hfr,** for **high-frequency recombination.** Because Hfr cells behave as donors, they are a special class of F^+ cells.

Another important difference was noted between Hfr strains and the original F^+ strains. If the donor is from an Hfr strain, recipient cells, though sometimes displaying genetic recombination, never become Hfr; that is, they remain F^-. In comparison, then,

$$F^+ \times F^- \longrightarrow F^+ \quad \text{(low rate of recombination)}$$

$$\text{Hfr} \times F^- \longrightarrow F^- \quad \text{(higher rate of recombination)}$$

Perhaps the most significant characteristic of Hfr strains is the *nature of recombination.* In any given strain, certain genes are more frequently recombined than others, and some not at all. This *nonrandom* pattern was shown to vary between Hfr strains. Although these results were puzzling, Hayes interpreted them to mean that some physiological alteration of the F factor had occurred, resulting in the production of Hfr strains of *E. coli.*

In the mid-1950s, experimentation by Ellie Wollman and François Jacob elucidated the difference between Hfr and F^+ strains and showed how Hfr strains allow genetic mapping of the *E. coli* chromosome. In their experiments, Hfr and F^- strains with suitable marker genes were mixed, and recombination of specific genes was assayed at different times. To accomplish this, a culture containing a mixture of an Hfr and an F^- strain was first incubated, and samples were removed at various intervals and placed in a blender. The shear forces in the blender separated conjugating bacteria so that the transfer of the chromosome was terminated. The cells were then assayed for genetic recombination.

This process, called the **interrupted mating technique,** demonstrated that specific genes of a given Hfr strain were

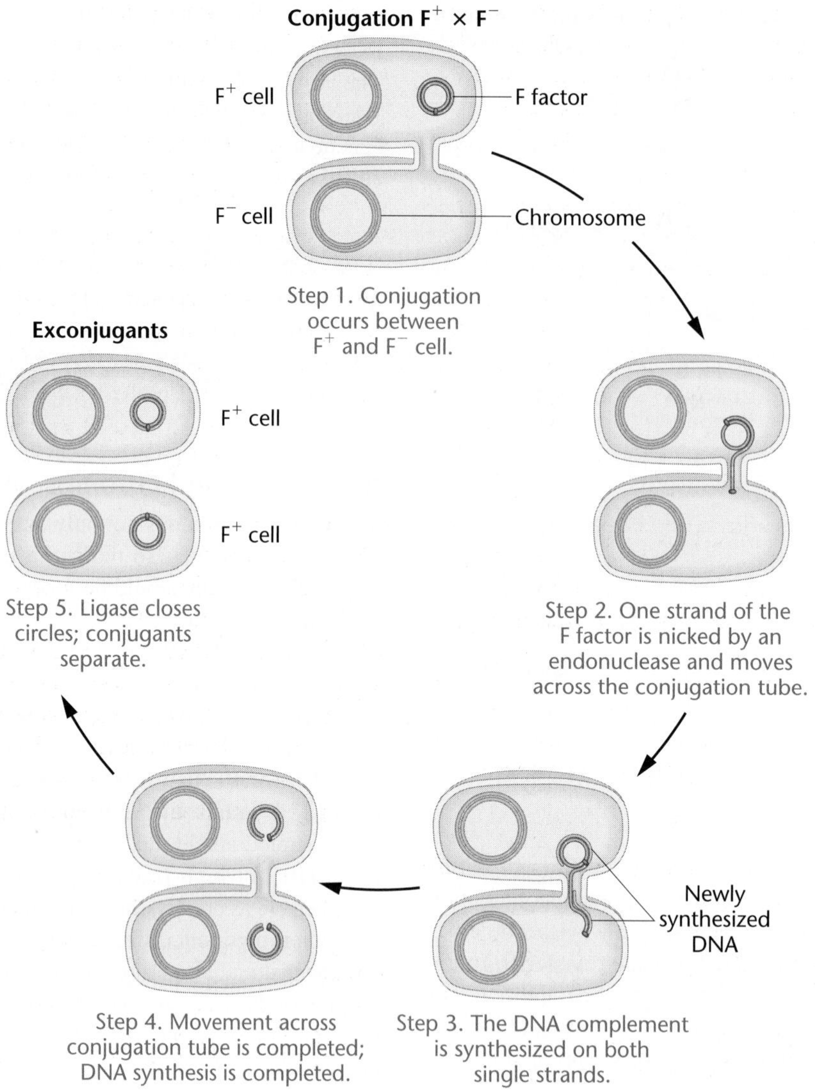

FIGURE 8–5 An $F^+ \times F^-$ mating demonstrating how the recipient F^- cell converts to F^+. During conjugation, the DNA of the F factor replicates with one new copy entering the recipient cell, converting it to F^+. The bars added to the F factors follow their clockwise rotation during replication. Newly replicated DNA is depicted by a lighter shade of blue as the F factor is transferred.

transferred and recombined sooner than others. The graph in Figure 8–6 illustrates this point. During the first 8 minutes after the two strains were mixed, no genetic recombination was detected. At about 10 minutes, recombination of the *azi*^{R} gene was detected, but no transfer of the *ton*^{s}, *lac*$^+$, or *gal*$^+$ genes was noted. By 15 minutes, 50 percent of the recombinants were *azi*^{R}, and 15 percent were *ton*^{s}; but none were *lac*$^+$ or *gal*$^+$. Within 20 minutes, the *lac*$^+$ was found among the recombinants; and within 25 minutes, *gal*$^+$ was also being transferred. Wollman and Jacob had demonstrated *an ordered transfer of genes* that correlated with the length of time conjugation proceeded.

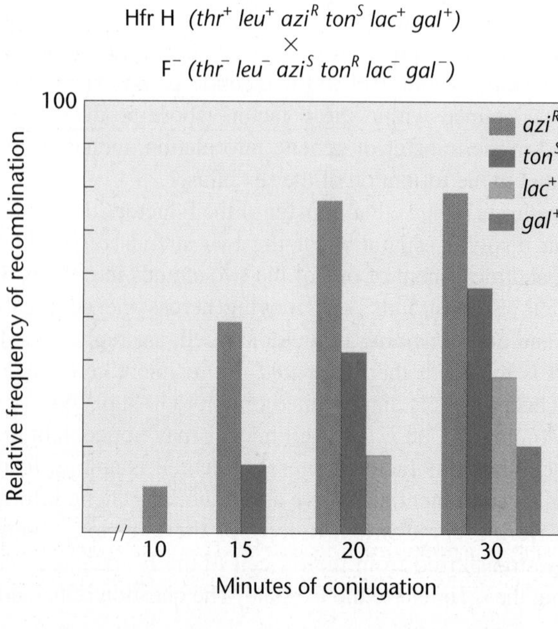

FIGURE 8–6 The progressive transfer during conjugation of various genes from a specific Hfr strain of *E. coli* to an F^- strain. Certain genes (*azi* and *ton*) transfer more quickly than others and recombine more frequently. Others (*lac* and *gal*) take longer to transfer, and recombinants are found at a lower frequency.

It appeared that the chromosome of the Hfr bacterium was transferred linearly and that the gene order and distance between genes, as measured in minutes, could be predicted from such experiments (Figure 8–7). This information served as the basis for the first genetic map of the *E. coli* chromosome. Minutes in bacterial mapping are equivalent to map units in eukaryotes.

Wollman and Jacob then repeated the same type of experiment with other Hfr strains, obtaining similar results with one important difference. Although genes were always transferred linearly with time, as in their original experiment, the order in which genes entered seemed to vary from Hfr strain to Hfr strain [see Figure 8–8(a)]. When they reexamined the entry rate of genes, and thus the genetic maps for each strain, a definite pattern emerged. The major difference between each strain was simply the point of origin (*O*) and the direction in which entry proceeded from that point [Figure 8–8(b)].

To explain these results, Wollman and Jacob postulated that the *E. coli* chromosome is circular (a closed circle, with no free ends). If the point of origin (*O*) varies from strain to strain, a different sequence of genes will be transferred in each case. But what determines *O*? They proposed that in various Hfr strains, the F factor integrates into the chromosome at different points and that its position determines the site of *O*. A case of integration is shown in step 1 of Figure 8–9, p. 166. During conjugation between an Hfr and an F^- cell, the position of the F factor determines the initial point of transfer (steps 2 and 3). Those genes adjacent to *O* are transferred first, and the F factor becomes the last part that can be transferred (step 4). However, conjugation rarely, if ever, lasts long enough to allow the entire chromosome to pass across the conjugation tube (step 5). *This proposal explains why recipient cells, when mated with Hfr cells, remain F^-.*

Figure 8–9 also depicts the way in which the two strands making up a DNA molecule behave during transfer, allowing for the entry of one strand of DNA into the recipient (see step 3). Following replication, the entering DNA now has the potential to recombine with its homologous region of the host chromosome. The DNA strand that remains in the donor also undergoes replication.

Use of the interrupted mating technique with different Hfr strains allowed researchers to map the entire *E. coli* chromosome. Mapped in time units, strain K12 (or *E. coli* K12) was shown to be 100 minutes long. While modern genome analysis of the *E. coli* chromosome has now established the presence of just over 4000 protein-coding sequences, this original mapping procedure established the location of approximately 1000 genes.

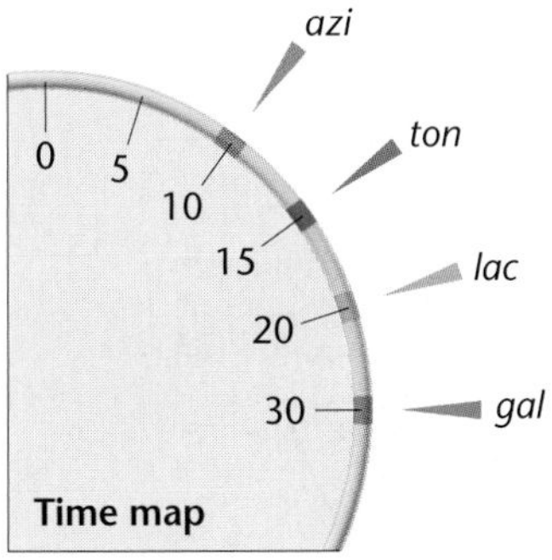

FIGURE 8–7 A time map of the genes studied in the experiment depicted in Figure 8–6.

ESSENTIAL POINT

When the F factor is integrated into the donor cell chromosome (making it Hfr), the donor chromosome moves unidirectionally into the recipient, initiating recombination and providing the basis for time mapping of the bacterial chromosome.

(a)

Hfr strain	(earliest)				Order of transfer											(latest)
H	*thr*	–	*leu*	–	*azi*	–	*ton*	–	*pro*	–	*lac*	–	*gal*	–	*thi*	
1	*leu*	–	*thr*	–	*thi*	–	*gal*	–	*lac*	–	*pro*	–	*ton*	–	*azi*	
2	*pro*	–	*ton*	–	*azi*	–	*leu*	–	*thr*	–	*thi*	–	*gal*	–	*lac*	
7	*ton*	–	*azi*	–	*leu*	–	*thr*	–	*thi*	–	*gal*	–	*lac*	–	*pro*	

(b)

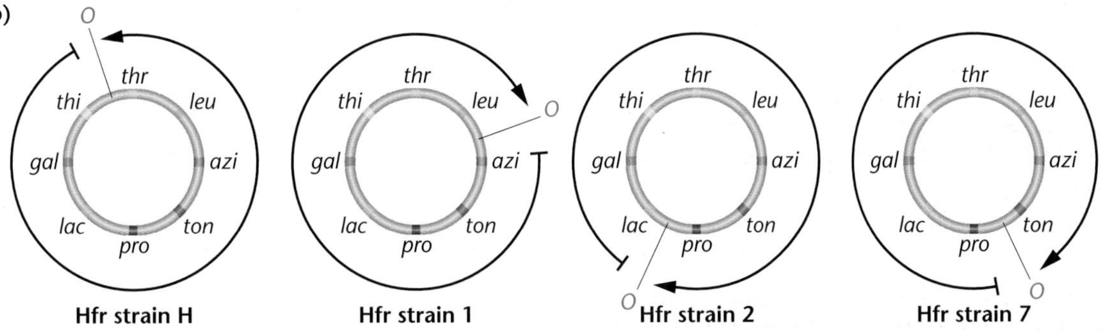

FIGURE 8–8 (a) The order of gene transfer in four Hfr strains, suggesting that the *E. coli* chromosome is circular. (b) The point where transfer originates (*O*) is identified in each strain. The origin is determined by the point of integration into the chromosome of the F factor, and the direction of transfer is determined by the orientation of the F factor as it integrates. The arrowheads indicate the points of initial transfer.

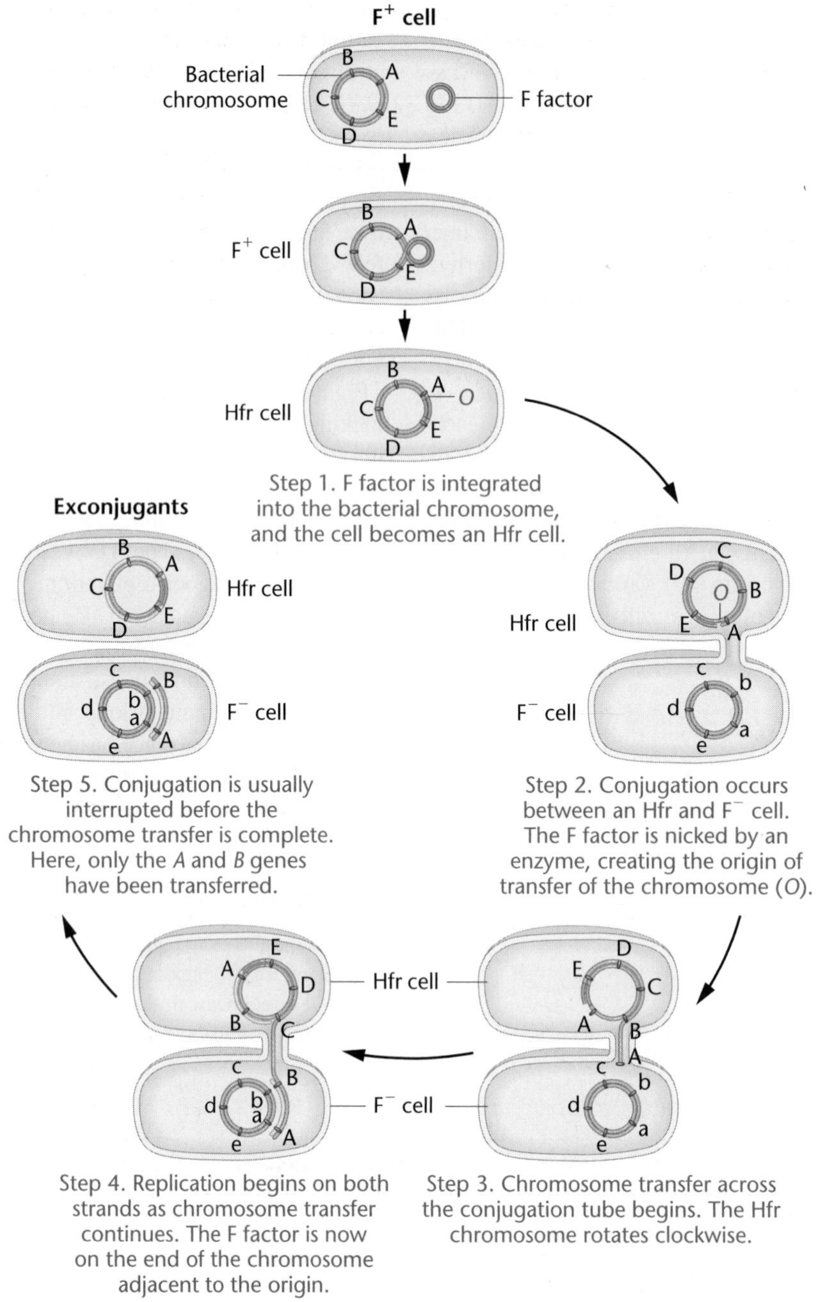

FIGURE 8–9 Conversion of F^+ to an Hfr state occurs by integrating the F factor into the bacterial chromosome. The point of integration determines the origin (O) of transfer. During conjugation, an enzyme nicks the F factor, now integrated into the host chromosome, initiating transfer of the chromosome at that point. Conjugation is usually interrupted prior to complete transfer. Only the *A* and *B* genes are transferred to the F^- cell, which may recombine with the host chromosome. Newly replicated DNA of the chromosome is depicted by a lighter shade of orange.

NOW SOLVE THIS

Problem 25 on page 179 involves an understanding of how the bacterial chromosome is transferred during conjugation, leading to recombination and mapping. You are asked to draw a map of the bacterial chromosome.

Hint: Chromosome transfer is strain-specific and depends on the position within the chromosome where the F factor is integrated.

Recombination in $F^+ \times F^-$ Matings: A Reexamination

The preceding experiment helped geneticists better understand how genetic recombination occurs during $F^+ \times F^-$ matings. Recall that recombination occurs much less frequently than in Hfr $\times$ F^- matings and that random gene transfer is involved. The current belief is that when F^+ and F^- cells are mixed, conjugation occurs readily and each F^- cell involved in conjugation with an F^+ cell receives a copy of the F factor, *but no genetic recombination occurs.* However, at an extremely low frequency in a population of F^+ cells, the F factor integrates spontaneously from the cytoplasm to a random point in the bacterial chromosome, converting the F^+ cell to the Hfr state as we saw in Figure 8–9. Therefore, in $F^+ \times F^-$ matings, the extremely low frequency of genetic recombination (10^{-7}) is attributed to the rare, newly formed Hfr cells, which then undergo conjugation with F^- cells. Because the point of integration of the F factor is random, the gene or genes that are transferred by any newly formed Hfr donor *will also appear to be random within the larger F^+/F^- population.* The recipient bacterium will appear as a recombinant but will remain F^-. If it subsequently undergoes conjugation with an F^+ cell, it will then be converted to F^+.

The F′ State and Merozygotes

In 1959, during experiments with Hfr strains of *E. coli,* Edward Adelberg discovered that the F factor could lose its integrated status, causing the cell to revert to the F^+ state (Figure 8–10, step 1). When this occurs, the F factor frequently carries

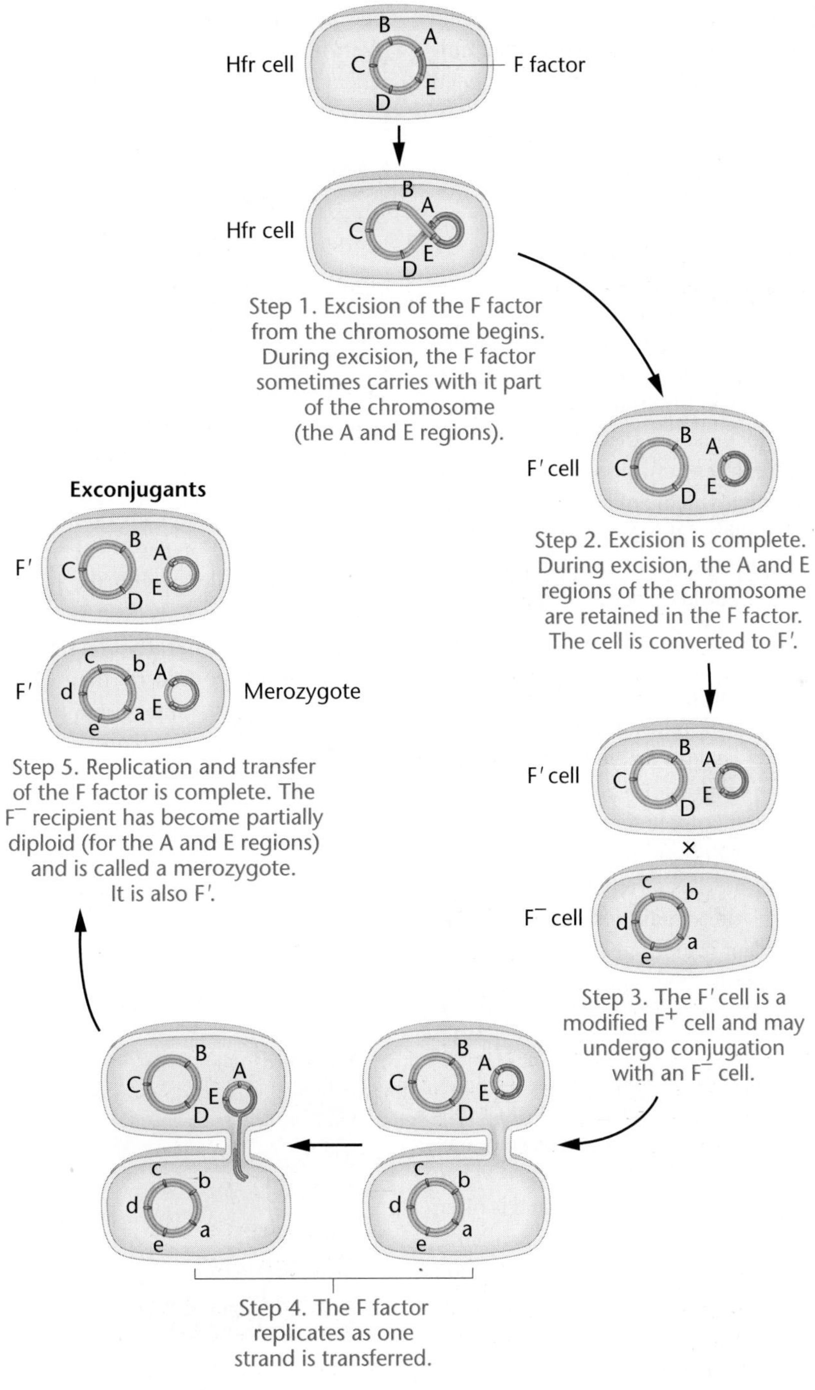

FIGURE 8–10 Conversion of an Hfr bacterium to F′ and its subsequent mating with an F^- cell. The conversion occurs when the F factor loses its integrated status. During excision from the chromosome, it carries with it one or more chromosomal genes (*A* and *E*). Following conjugation with an F^- cell, the recipient cell becomes partially diploid and is called a merozygote; it also behaves as an F^+ donor cell.

several adjacent bacterial genes along with it (step 2). Adelberg labeled this condition **F′** to distinguish it from F^+ and Hfr. F′, like Hfr, is thus another special case of F^+, but this conversion is from Hfr to F′.

The presence of bacterial genes within a cytoplasmic F factor creates an interesting situation. An F′ bacterium behaves like an F^+ cell by initiating conjugation with F^- cells (Figure 8–10, step 3). When this occurs, the F factor, containing chromosomal genes, is transferred to the F^- cell (step 4). As a result, whatever chromosomal genes are part of the F factor are now present as duplicates in the recipient cell (step 5) because the recipient still has a complete chromosome. This creates a partially diploid cell called a **merozygote.** Pure cultures of F′ merozygotes can be established. They have been extremely useful in the study of bacterial genetics, particularly in genetic regulation.

8.3 Rec Proteins Are Essential to Bacterial Recombination

Once researchers established that a unidirectional transfer of DNA occurs between bacteria, they became interested in determining how the actual recombination event occurs in the recipient cell. Just how does the donor DNA replace the homologous region in the recipient chromosome? As with many systems, the biochemical mechanism by which recombination occurs was deciphered through genetic studies. Major insights were gained as a result of the isolation of a group of mutations that impaired the process of recombination and led to the discovery of *rec* (for recombination) genes.

The first relevant observation in this case involved a series of mutant genes labeled *recA, recB, recC,* and *recD.* The first mutant gene, *recA,* diminished genetic recombination in bacteria 1000-fold, nearly eliminating it altogether; each of the other *rec* mutations reduced recombination by about 100 times. Clearly, the normal wild-type products of these genes play some essential role in the process of recombination.

Researchers looked for, and subsequently isolated, several functional gene products present in normal cells but missing in *rec* mutant cells and showed that they played a role in genetic recombination. The first product is called the **RecA protein.*** This protein plays an important role in recombination involving either a single-stranded DNA molecule or the linear end of a double-stranded DNA molecule that has unwound. As it turns out, **single-strand displacement** is a common form of recombination in many bacterial species. When double-stranded DNA enters a recipient cell, one strand is often degraded, leaving the complementary strand as the only source of recombination. This strand must find its homologous region along the host chromosome, and once it does, RecA facilitates recombination.

The second related gene product is a more complex protein called the **RecBCD protein,** an enzyme consisting of polypeptide subunits encoded by three other *rec* genes. This protein is important when double-stranded DNA serves as the source of genetic recombination. RecBCD unwinds the helix, facilitating recombination that involves RecA. These discoveries have extended our knowledge of the process of recombination considerably and underscore the value of isolating mutations, establishing their phenotypes, and determining the biological role of the normal, wild-type genes. The model of recombination based on the *rec* discoveries also applies to eukaryotes: eukaryotic proteins similar to RecA have been isolated and studied. We will return to this topic in Chapter 10 where we will discuss a detailed model of DNA recombination.

8.4 The F Factor Is an Example of a Plasmid

In the preceding sections, we examined the extrachromosomal heredity unit called the F factor. When it exists autonomously in the bacterial cytoplasm, the F factor is composed of a double-stranded closed circle of DNA. These characteristics place the F factor in the more general category of genetic structures called **plasmids** [Figure 8–11(a)]. These structures contain one or more genes and often, quite a few. Their replication depends on the same enzymes that replicate the chromosome of the host cell, and they are distributed to daughter cells along with the host chromosome during cell division.

Plasmids are generally classified according to the genetic information specified by their DNA. The F factor plasmid confers fertility and contains the genes essential for sex pilus formation, on which genetic recombination depends. Other examples of plasmids include the R and Col plasmids.

Most **R plasmids** consist of two components: the **resistance transfer factor (RTF)** and one or more **r-determinants** [Figure 8–11(b)]. The RTF encodes genetic information essential to transferring the plasmid between bacteria, and

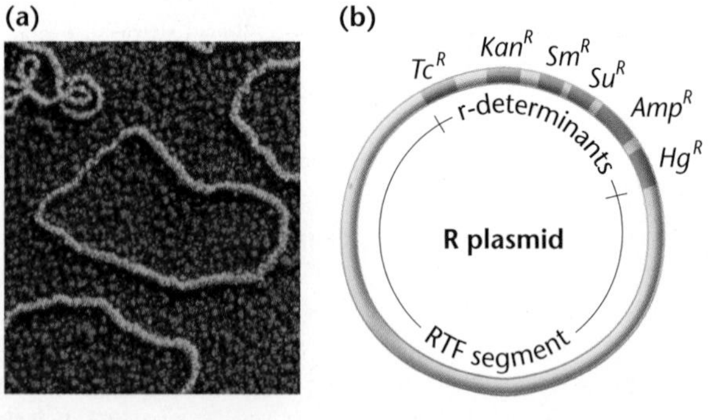

FIGURE 8–11 (a) Electron micrograph of a plasmid isolated from *E. coli.* (b) An R plasmid containing resistance transfer factors (RTFs) and multiple r-determinants (Tc, tetracycline; Kan, kanamycin; Sm, streptomycin; Su, sulfonamide; Amp, ampicillin; and Hg, mercury).

* Note that the names of bacterial genes use lowercase letters and are italicized, while the names of the corresponding gene products begin with capital letters and are not italicized. For example, the *recA* gene encodes the RecA protein.

the r-determinants are genes that confer resistance to antibiotics or mercury. While RTFs are similar in a variety of plasmids from different bacterial species, r-determinants are specific for resistance to one class of antibiotic and vary widely.

Resistance to tetracycline, streptomycin, ampicillin, sulfonamide, kanamycin, or chloramphenicol is most frequently encountered. Sometimes several r-determinants occur in a single plasmid, conferring multiple resistance to several antibiotics [Figure 8–11(b)]. Bacteria bearing these plasmids are of great medical significance not only because of their multiple resistance but because of the ease with which the plasmids can be transferred to other bacteria. Sometimes a bacterial cell contains r-determinant plasmids but no RTF—the cell is resistant but cannot transfer the genetic information for resistance to recipient cells. The most commonly studied plasmids, however, contain the RTF as well as one or more r-determinants.

The **Col plasmid,** ColE1 (derived from *E. coli*), is clearly distinct from the R plasmid. It encodes one or more proteins that are highly toxic to bacterial strains that do not harbor the same plasmid. These proteins, called **colicins,** can kill neighboring bacteria, and bacteria that carry the plasmid are said to be *colicinogenic*. Present in 10 to 20 copies per cell, a gene in the Col plasmid encodes an immunity protein that protects the host cell from the toxin. Unlike an R plasmid, the Col plasmid is not usually transmissible to other cells.

Interest in plasmids has increased dramatically because of their role in recombinant DNA research. As we will see in Chapter 17, specific genes from any source can be inserted into a plasmid, which may then be inserted into a bacterial cell. As the altered cell replicates its DNA and undergoes division, the foreign gene is also replicated, thus cloning the foreign genes.

ESSENTIAL POINT

Plasmids, such as the F factor, are autonomously replicating DNA molecules found in the bacterial cytoplasm, sometimes containing unique genes conferring antibiotic resistance as well as the genes necessary for plasmid transfer during conjugation.

8.5 Transformation Is Another Process Leading to Genetic Recombination in Bacteria

Transformation provides another mechanism for recombining genetic information in some bacteria. Small pieces of extracellular DNA are taken up by a living bacterium, potentially leading to a stable genetic change in the recipient cell. We discuss transformation in this chapter because in those bacterial species where it occurs, the process can be used to map bacterial genes, though in a more limited way than conjugation. Transformation has also played a central role in experiments proving that DNA is the genetic material.

The process of transformation (Figure 8–12, p. 170) consists of numerous steps divided into two categories: (1) entry of DNA into a recipient cell and (2) recombination of the donor DNA with its homologous region in the recipient chromosome. In a population of bacterial cells, only those in the particular physiological state of **competence** take up DNA. Entry is thought to occur at a limited number of receptor sites on the surface of the bacterial cell (Figure 8–12, step 1). Passage into the cell is an active process that requires energy and specific transport molecules. This model is supported by the fact that substances that inhibit energy production or protein synthesis in the recipient cell also inhibit the transformation process.

Soon after entry, one of the two strands of the double helix is digested by nucleases, leaving only a single strand to participate in transformation (Figure 8–12, steps 2 and 3). The surviving strand of DNA then aligns with its complementary region of the bacterial chromosome. In a process involving several enzymes, the segment replaces its counterpart in the chromosome, which is excised and degraded (step 4).

For recombination to be detected, the transforming DNA must be derived from a different strain of bacteria that bears some genetic variation, such as a mutation. Once it is integrated into the chromosome, the recombinant region contains one host strand (present originally) and one mutant strand. Because these strands are from different sources, this helical region is referred to as a **heteroduplex.** Following one round of DNA replication, one chromosome is restored to its original configuration, and the other contains the mutant gene. Following cell division, one untransformed cell (nonmutant) and one transformed cell (mutant) are produced (step 5).

Transformation and Linked Genes

For DNA to be effective in transformation, it must include between 10,000 and 20,000 nucleotide pairs, a length equal to about 1/200 of the *E. coli* chromosome—this size is sufficient to encode several genes. Genes adjacent to or very close to one another on the bacterial chromosome can be carried on a single segment of this size. Consequently, a single event can result in the **cotransformation** of several genes simultaneously. Genes that are close enough to each other to be cotransformed are *linked.* In contrast to *linkage groups* in eukaryotes, which indicate all genes on a single chromosome, linkage here refers to the close proximity of genes.

If two genes are not linked, simultaneous transformation occurs only as a result of two independent events involving two distinct segments of DNA. As in double crossovers in eukaryotes, the probability of two independent events occurring simultaneously is equal to the product of the individual probabilities. Thus, the frequency of two unlinked genes being transformed simultaneously is much lower than if they are linked.

Studies have shown that a variety of bacteria readily undergo transformation (e.g., *Hemophilus influenzae, Bacillus subtilis, Shigella paradysenteriae, Diplococcus pneumoniae,* and *E. coli*). Under certain conditions, the relative distances between linked genes can be determined from transformation data. Although analysis is more complex, such data are interpreted in a manner analogous to chromosome mapping in eukaryotes.

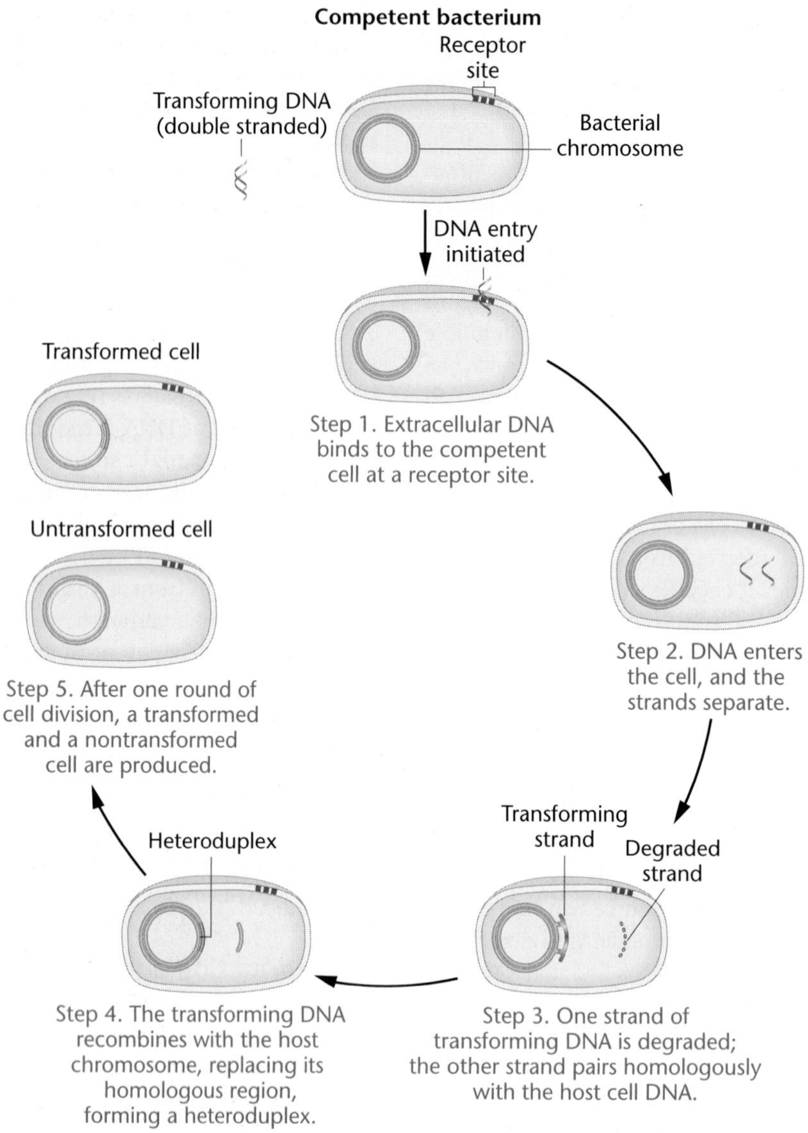
Competent bacterium
Receptor site
Transforming DNA (double stranded)
Bacterial chromosome
DNA entry initiated
Step 1. Extracellular DNA binds to the competent cell at a receptor site.
Step 2. DNA enters the cell, and the strands separate.
Transforming strand
Degraded strand
Step 3. One strand of transforming DNA is degraded; the other strand pairs homologously with the host cell DNA.
Heteroduplex
Step 4. The transforming DNA recombines with the host chromosome, replacing its homologous region, forming a heteroduplex.
Transformed cell
Untransformed cell
Step 5. After one round of cell division, a transformed and a nontransformed cell are produced.

FIGURE 8–12 Proposed steps for transforming a bacterial cell by exogenous DNA. Only one of the two entering DNA strands is involved in the transformation event, which is completed following cell division.

ESSENTIAL POINT

Transformation in bacteria, which does not require cell-to-cell contact, involves exogenous DNA that enters a recipient bacterium and recombines with the host's chromosome. Linkage mapping of closely aligned genes is possible during the analysis of transformation.

PHAGE GENETICS

WEB TUTORIAL

NOW SOLVE THIS

Problem 9 on page 179 involves an understanding of how transformation can be used to determine if bacterial genes are "linked" in close proximity to one another. You are asked to predict the location of two genes, relative to one another.

Hint: Cotransformation occurs according to the laws of probability. Two "unlinked" genes are transformed as a result of two separate events. In such a case, the probability of that occurrence is equal to the product of the individual probabilities.

8.6 Bacteriophages Are Bacterial Viruses

Bacteriophages, or **phages** as they are commonly known, are viruses that have bacteria as their hosts. During their reproduction, phages can be involved in still another mode of bacterial genetic recombination called **transduction.** To understand this process, we must consider the genetics of bacteriophages, which themselves undergo recombination.

A great deal of genetic research has been done using bacteriophages as a model system, making them a worthy subject of discussion. In this section, we will first examine the structure and life cycle of one type of bacteriophage. We then discuss how these phages are studied during their infection of bacteria. Finally, we contrast two possible modes of behavior once the initial phage infection occurs. This information is background for our discussion of *transduction* and *bacteriophage recombination.*

Phage T4: Structure and Life Cycle

Bacteriophage T4 is one of a group of related bacterial viruses referred to as T-even phages. It exhibits the intricate structure shown in Figure 8–13. The phage T4's genetic material (DNA) is contained within an icosahedral (a polyhedron with 20 faces) protein coat, making up the head of the virus. The DNA is sufficient in quantity to encode more than 150 average-sized genes. The head is connected to a tail that contains a collar and a contractile sheath surrounding a central core. Tail fibers, which protrude from the tail, contain binding sites in their tips that specifically recognize unique areas of the outer surface of the cell wall of the bacterial host, *E. coli*.

The life cycle of phage T4 (Figure 8–14) is initiated when the virus binds by adsorption to the bacterial host cell. Then, an ATP-driven contraction of the tail sheath causes the central core to penetrate the cell wall. The DNA in the head is extruded, and it moves across the cell membrane into the bacterial cytoplasm. Within minutes, all bacterial DNA, RNA, and protein synthesis in the host cell is inhibited, and synthesis of viral molecules begins. At the same time, degradation of the host DNA is initiated.

A period of intensive viral gene activity characterizes infection. Initially, phage DNA replication occurs, leading to a pool of viral DNA molecules. Then, the components of the head, tail, and tail fibers are synthesized. The assembly of mature viruses is a complex process that has been well studied by William Wood, Robert Edgar, and others. Three sequential pathways occur: (1) DNA packaging as the viral heads are assembled, (2) tail assembly, and (3) tail fiber assembly. Once DNA is packaged into the head, it combines with the tail components, to which tail fibers are added. Total construction is a combination of self-assembly and enzyme-directed processes.

When approximately 200 new viruses have been constructed, the bacterial cell is ruptured by the action of the enzyme lysozyme (a phage gene product), and the mature phages are released from the host cell. The new phages infect other available bacterial cells, and the process repeats itself over and over again.

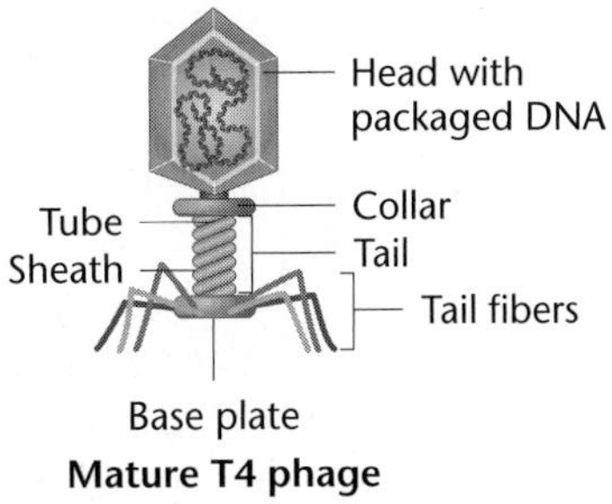

FIGURE 8–13 The structure of bacteriophage T4 includes an icosahedral head filled with DNA, a tail consisting of a collar, tube, sheath, base plate, and tail fibers. During assembly, the tail components are added to the head, and then tail fibers are added.

The Plaque Assay

Bacteriophages and other viruses have played a critical role in our understanding of molecular genetics. During infection of bacteria, enormous quantities of bacteriophages can be obtained for investigation. Often, over 10^{10} viruses are produced per milliliter of culture medium. Many genetic studies rely on

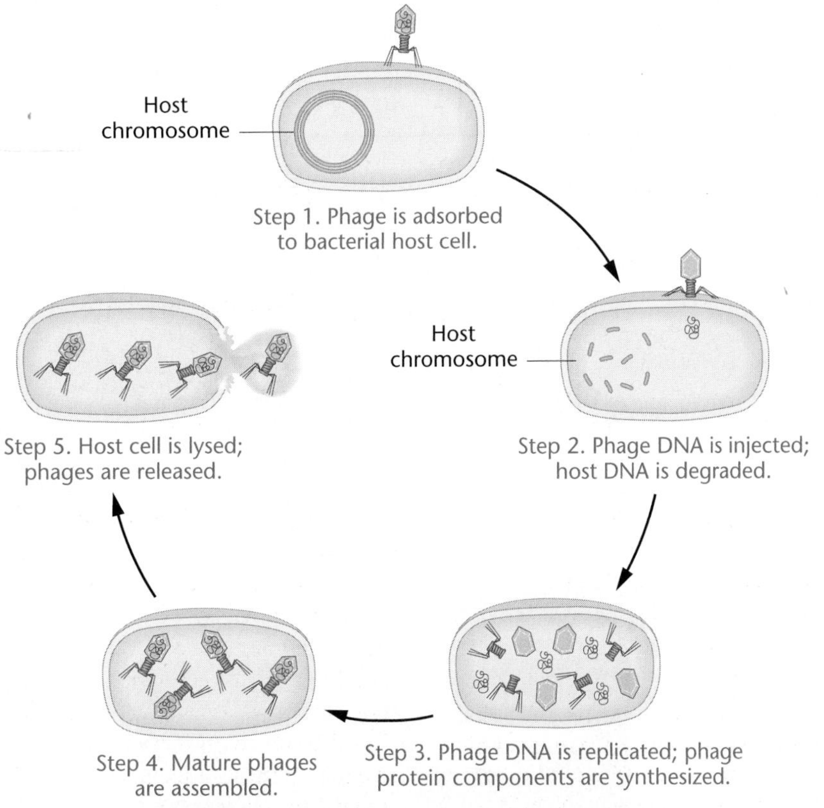

FIGURE 8–14 Life cycle of bacteriophage T4.

the ability to quantify the number of phages produced following infection under specific culture conditions. The **plaque assay** is a routinely used technique, which is invaluable in mutational and recombinational studies of bacteriophages.

This assay is shown in Figure 8–15, where actual plaque morphology is also illustrated. A serial dilution of the original virally infected bacterial culture is performed first. Then, a 0.1-mL sample (an *aliquot*) from a dilution is added to melted nutrient agar (about 3 mL) into which a few drops of a healthy bacterial culture have been added. The solution is then poured evenly over a base of solid nutrient agar in a petri dish and allowed to solidify before incubation. A clear area called a **plaque** occurs wherever a single virus initially infected one bacterium in the culture (the lawn) that has grown up during incubation. The plaque represents clones of the single infecting bacteriophage, created as reproduction cycles are repeated. If the dilution factor is too low, the plaques are plentiful, and they will fuse, lysing the entire lawn—which has occurred in the 10^{-3} dilution of Figure 8–15. On the other hand, if the dilution factor is increased, plaques can be counted and the density of viruses in the initial culture can be estimated as:

$$\text{initial phage density} = (\text{plaque number/mL}) \times (\text{dilution factor})$$

Using the results shown in Figure 8–15, 23 phage plaques are derived from the 0.1-mL aliquot of the 10^{-5} dilution. Therefore, we estimate that there are 230 phages/mL *at this dilution* (since the initial aliquot was 0.1 mL). The initial phage density in the undiluted sample, factoring in the 10^{-5} dilution, is then calculated as

$$\text{initial phage density} = (230/\text{mL}) \times (10^5) = 230 \times 10^5/\text{mL}$$

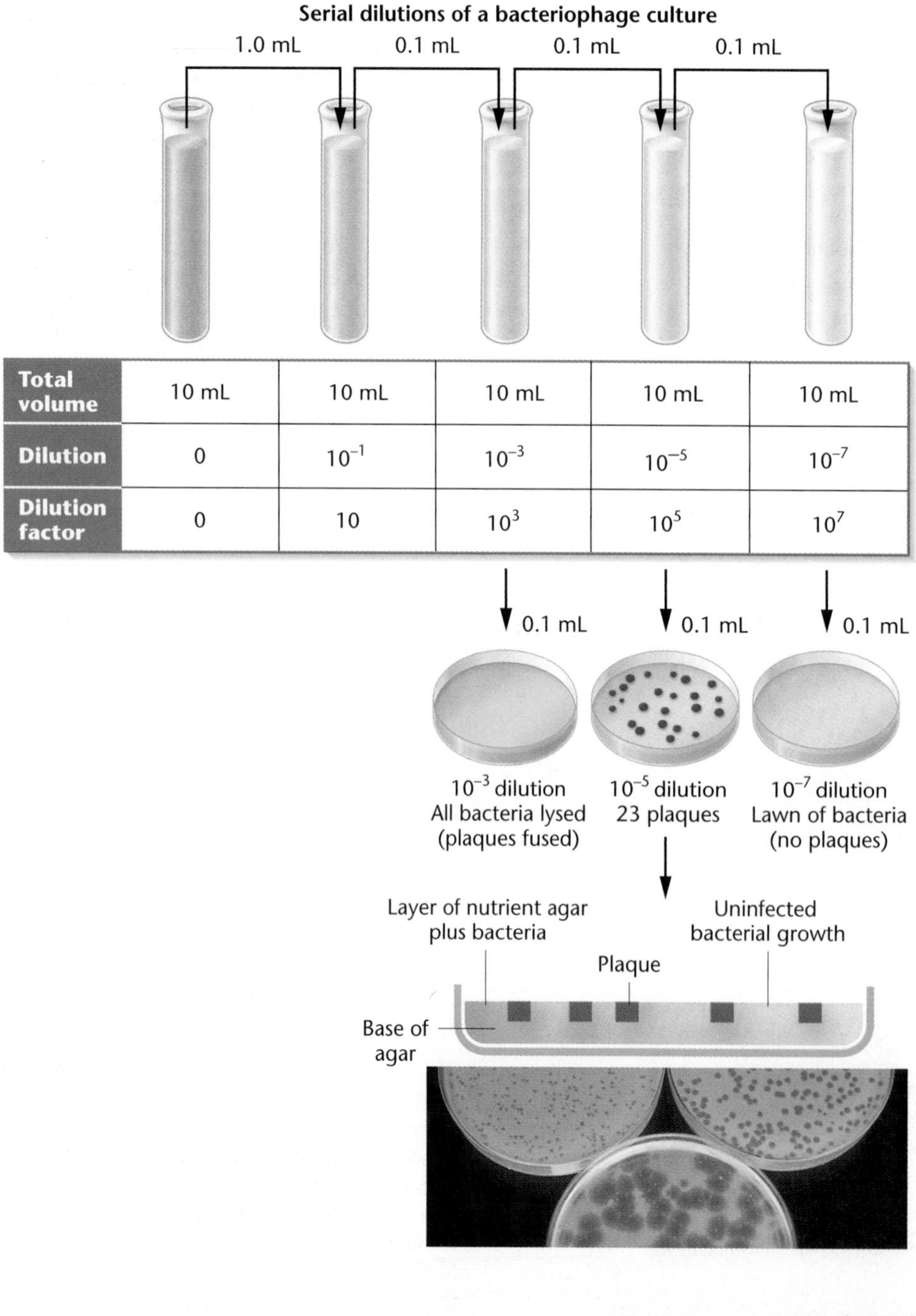

FIGURE 8–15 A plaque assay for bacteriophage analysis. Serial dilutions of a bacterial culture infected with bacteriophages are first made. Then three of the dilutions (10^{-3}, 10^{-5}, and 10^{-7}) are analyzed using the plaque assay technique. Each plaque represents the initial infection of one bacterial cell by one bacteriophage. In the 10^{-3} dilution, so many phages are present that all bacteria are lysed. In the 10^{-5} dilution, 23 plaques are produced. In the 10^{-7} dilution, the dilution factor is so great that no phages are present in the 0.1-mL sample, and thus no plaques form. From the 0.1-mL sample of the 10^{-5} dilution, the original bacteriophage density is calculated to be $23 \times 10 \times 10^5$ phages/mL (23×10^6, or 2.3×10^7). The photograph shows phage T2 plaques on lawns of *E. coli.*

Because this figure is derived from the 10^{-5} dilution, we can also estimate that there will be only 0.23 phage/0.1 mL in the 10^{-7} dilution. Thus, when 0.1 mL from this tube is assayed, it is predicted that no phage particles will be present. This possibility is borne out in Figure 8–15, which depicts an intact lawn of bacteria lacking any plaques. The dilution factor is simply too great.

ESSENTIAL POINT ■ ■ ■

Bacteriophages (viruses that infect bacteria) demonstrate a well-defined life cycle where they reproduce within the host cell and can be studied using the plaque assay.

Lysogeny

The relationship between a virus and a bacterium does not always result in viral reproduction and lysis. As early as the 1920s, it was known that a virus can enter a bacterial cell and coexist with it. The precise molecular basis of this relationship is now well understood. Upon entry, the viral DNA is integrated into the bacterial chromosome instead of replicating in the bacterial cytoplasm, a step that characterizes the developmental stage referred to as **lysogeny.** Subsequently, each time the bacterial chromosome is replicated, the viral DNA is also replicated and passed to daughter bacterial cells following division. No new viruses are produced, and no lysis of the bacterial cell occurs. However, under certain stimuli, such as chemical or ultraviolet light treatment, the viral DNA loses its integrated status and initiates replication, phage reproduction, and lysis of the bacterium.

Several terms are used to describe this relationship. The viral DNA that integrates into the bacterial chromosome is a **prophage.** Viruses that either lyse the cell or behave as a prophage are **temperate phages.** Those that only lyse the cell are referred to as **virulent phages.** A bacterium harboring a prophage is **lysogenic;** that is, it is capable of being lysed as a result of induced viral reproduction. The viral DNA, which can either replicate in the bacterial cytoplasm or become integrated into the bacterial chromosome, is thus classified as an **episome.**

ESSENTIAL POINT ■ ■ ■

Bacteriophages can be lytic, meaning they infect the host cell, reproduce, and then lyse it, or in contrast, they can lysogenize the host cell, where they infect it and integrate their DNA into the host chromosome, but do not reproduce.

8.7 Transduction Is Virus-Mediated Bacterial DNA Transfer

In 1952, Norton Zinder and Joshua Lederberg were investigating possible recombination in the bacterium *Salmonella typhimurium.* Although they recovered prototrophs from mixed cultures of two different auxotrophic strains, investigation revealed that recombination was occurring in a manner different from that attributable to the presence of an F factor, as in *E. coli.* What they had discovered was a process of bacterial recombination mediated by bacteriophages and now called **transduction.**

The Lederberg–Zinder Experiment

Lederberg and Zinder mixed the *Salmonella* auxotrophic strains LA-22 and LA-2 together, and when the mixture was plated on minimal medium, they recovered prototrophic cells. The LA-22 strain was unable to synthesize the amino acids phenylalanine and tryptophan (phe^-, trp^-), and LA-2 could not synthesize the amino acids methionine and histidine (met^-, his^-). Prototrophs (phe^+, trp^+, met^+, his^+) were recovered at a rate of about $1/10^5$ (10^{-5}) cells.

Although these observations at first suggested that the recombination involved was the type observed earlier in conjugative strains of *E. coli,* experiments using the Davis U-tube soon showed otherwise (Figure 8–16). The two auxotrophic strains were separated by a sintered glass filter, thus preventing cell contact but allowing growth to occur in a common medium. Surprisingly, when samples were removed from both sides of the filter and plated independently on minimal medium, prototrophs *were* recovered, but only from the side of the tube containing LA-22 bacteria. Recall that if conjugation were responsible, the conditions in the Davis U-tube would be expected to *prevent* recombination altogether (see Figure 8–4).

Since LA-2 cells appeared to be the source of the new genetic information (phe^+and trp^+), how that information crossed the filter from the LA-2 cells to the LA-22 cells, allowing

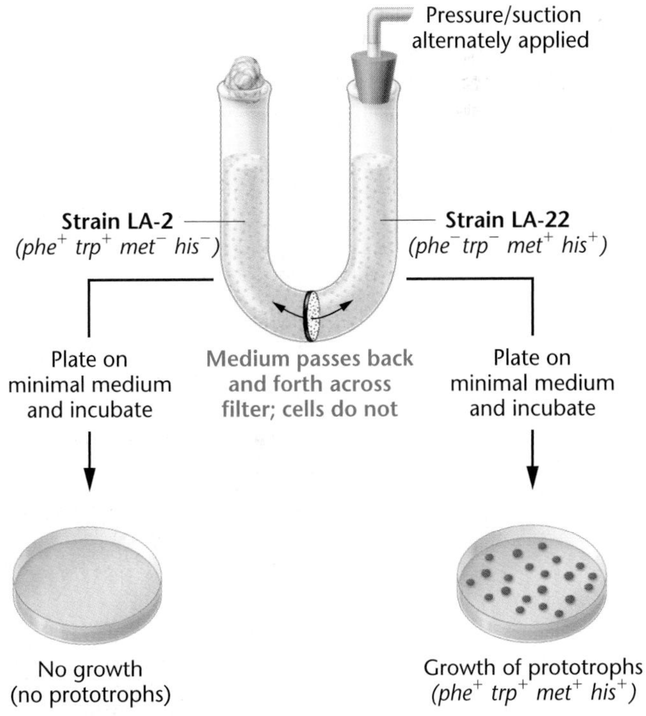

FIGURE 8–16 The Lederberg–Zinder experiment using *Salmonella.* After placing two auxotrophic strains on opposite sides of a Davis U-tube, Lederberg and Zinder recovered prototrophs from the side with the LA-22 strain, but not from the side containing the LA-2 strain.

recombination to occur, was a mystery. The unknown source was designated simply as a **filterable agent (FA).**

Three observations were used to identify the FA:

1. The FA was produced by the LA-2 cells only when they were grown in association with LA-22 cells. If LA-2 cells were grown independently and that culture medium was then added to LA-22 cells, recombination did not occur. Therefore, LA-22 cells play some role in the production of FA by LA-2 cells and do so only when they share a common growth medium.
2. The addition of DNase, which enzymatically digests DNA, did not render the FA ineffective. Therefore, the FA is not naked DNA, ruling out transformation.
3. The FA could not pass across the filter of the Davis U-tube when the pore size was reduced below the size of bacteriophages.

Aided by these observations and aware that temperate phages can lysogenize *Salmonella,* researchers proposed that the genetic recombination event is mediated by bacteriophage P22, present initially as a prophage in the chromosome of the LA-22 *Salmonella* cells. They hypothesized that P22 prophages rarely enter the vegetative or lytic phase, reproduce, and are released by the LA-22 cells. Such P22 phages, being much smaller than a bacterium, then cross the filter of the U-tube and subsequently infect and lyse some of the LA-2 cells. In the process of lysis of LA-2, these P22 phages occasionally package a region of the LA-2 chromosome in their heads. If this region contains the phe^+ and trp^+ genes and the phages subsequently pass back across the filter and infect LA-22 cells, these newly lysogenized cells will behave as prototrophs. This process of transduction, whereby bacterial recombination is mediated by bacteriophage P22, is diagrammed in Figure 8–17.

ESSENTIAL POINT

Transduction is virus-mediated bacterial DNA recombination.

The Nature of Transduction

Further studies have revealed the existence of transducing phages in other species of bacteria. For example, *E. coli* can be transduced by phages P1 and λ, and *B. subtilis* and *Pseudomonas aeruginosa* can be transduced by phages SPO1 and F116, respectively. The details of several different modes of transduction have also been established. Even though the initial discovery of transduction involved a temperate phage and a lysogenized bacterium, the same process can occur during the normal lytic cycle. Sometimes a small piece of bacterial DNA is packaged *along with* the viral chromosome so that the transducing phage contains both viral and bacterial DNA.

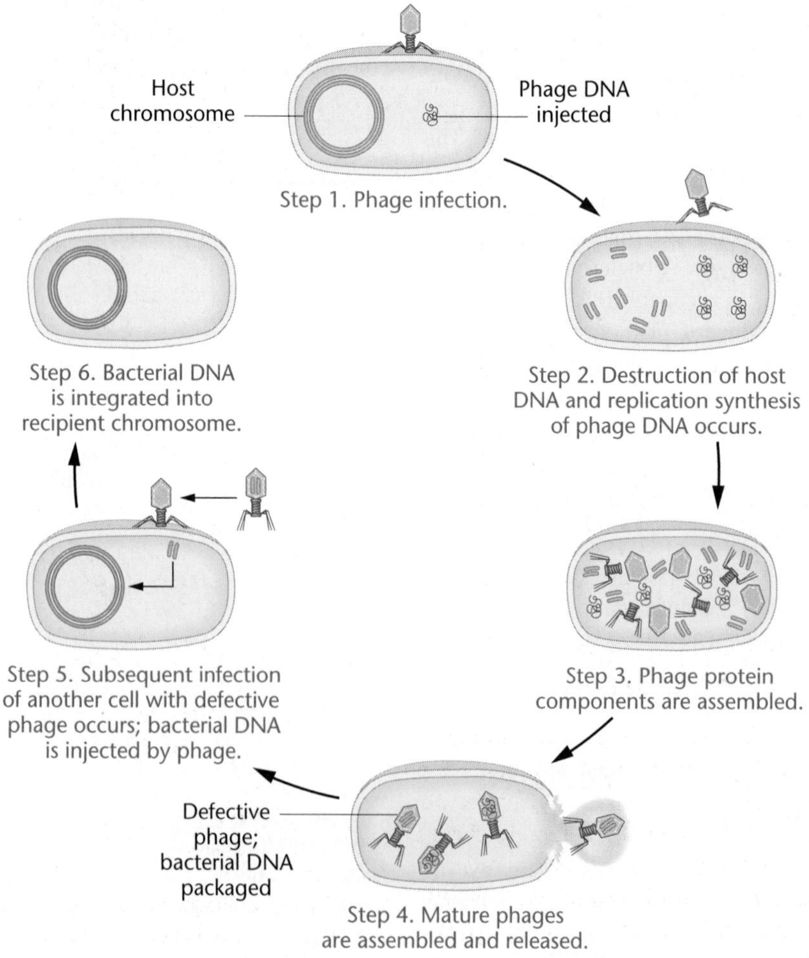
Host chromosome
Phage DNA injected
Step 1. Phage infection.
Step 2. Destruction of host DNA and replication synthesis of phage DNA occurs.
Step 3. Phage protein components are assembled.
Step 4. Mature phages are assembled and released.
Defective phage; bacterial DNA packaged
Step 5. Subsequent infection of another cell with defective phage occurs; bacterial DNA is injected by phage.
Step 6. Bacterial DNA is integrated into recipient chromosome.

FIGURE 8–17 Generalized transduction.

In such cases, only a few bacterial genes are present in the transducing phage. However, when *only* bacterial DNA is packaged, regions as large as 1 percent of the bacterial chromosome can become enclosed in the viral head. In either case, the ability to infect a host cell is unrelated to the type of DNA in the phage head, making transduction possible.

When bacterial rather than viral DNA is injected into the bacterium, it either remains in the cytoplasm or recombines with the homologous region of the bacterial chromosome. If the bacterial DNA remains in the cytoplasm, it does not replicate but is transmitted to one progeny cell following each division. When this happens, only a single cell, partially diploid for the transduced genes, is produced—a phenomenon called *abortive transduction.* If the bacterial DNA recombines with its homologous region of the bacterial chromosome, *complete transduction* occurs, where the transduced genes are replicated as part of the chromosome and passed to all daughter cells.

Both abortive and complete transduction are subclasses of the broader category of *generalized transduction,* which is characterized by the random nature of DNA fragments and genes that are transduced. Each fragment of the bacterial chromosome has a finite but small chance of being packaged in the phage head. Most cases of generalized transduction are of the abortive type; some data suggest that complete transduction occurs 10 to 20 times less frequently.

Transduction and Mapping

Like transformation, generalized transduction was used in linkage and mapping studies of the bacterial chromosome. The fragment of bacterial DNA involved in a transduction event is large enough to include numerous genes. As a result, two genes that closely align (are linked) on the bacterial chromosome can be simultaneously transduced, a process called **cotransduction.** Two genes that are not close enough to one another along the chromosome to be included on a single DNA fragment require two independent events to be transduced into a single cell. Since this occurs with a much lower probability than cotransduction, linkage can be determined.

By concentrating on two or three linked genes, transduction studies can also determine the precise order of these genes. The closer linked genes are to each other, the greater the frequency of cotransduction. Mapping studies involving three closely aligned genes can thus be executed, and the analysis of such an experiment is predicated on the same rationale underlying other mapping techniques.

8.8 Bacteriophages Undergo Intergenic Recombination

Around 1947, several research teams demonstrated that genetic recombination can be detected in bacteriophages. This led to the discovery that gene mapping can be performed in these viruses. Such studies relied on finding numerous phage mutations that could be visualized or assayed. As in bacteria and eukaryotes, these mutations allow genes to be identified and followed in mapping experiments. Thus, before considering recombination and mapping in these bacterial viruses, we briefly introduce several of the mutations that were studied.

Bacteriophage Mutations

Phage mutations often affect the morphology of the plaques formed following lysis of bacterial cells. For example, in 1946, Alfred Hershey observed unusual T2 plaques on plates of *E. coli* strain B. Normal T2 plaques are small and have a clear center surrounded by a diffuse (nearly invisible) halo, but the unusual plaques were larger and possessed a distinctive outer perimeter (Figure 8–18). When the viruses were isolated from these plaques and replated on *E. coli* B cells, the resulting plaque appearance was identical. Thus, the plaque phenotype was an inherited trait resulting from the reproduction of mutant phages. Hershey named the mutant *rapid lysis* (*r*) because the plaques were larger, apparently resulting from a more rapid or more efficient life cycle of the phage. We now know that in wild-type phages, reproduction is inhibited once a particular-sized plaque has been formed. The *r* mutant T2 phages overcome this inhibition, producing larger plaques.

Salvador Luria discovered another bacteriophage mutation, *host range* (*h*). This mutation extends the range of bacterial hosts that the phage can infect. Although wild-type T2 phages can infect *E. coli* B (a unique strain), they normally cannot attach or be adsorbed to the surface of *E. coli* B-2 (a different strain). The *h* mutation, however, provides the basis for

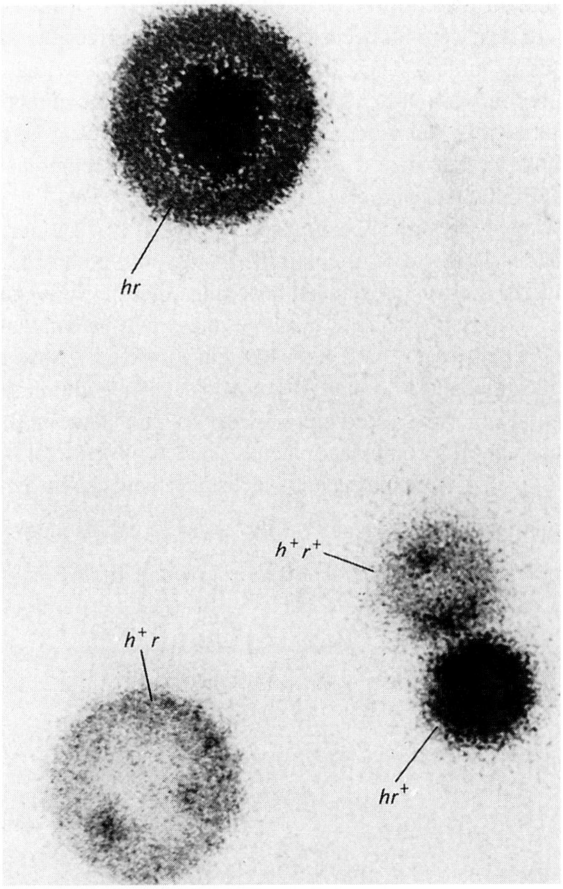

FIGURE 8–18 Plaque morphology phenotypes observed following simultaneous infection of *E. coli* by two strains of phage T2, h^+r and $h\,r^+$. In addition to the parental genotypes, recombinant plaques $h\,r$ and h^+r^+ are shown.

TABLE 8.1 Some Mutant Types of T-even Phages

Name	Description
minute	Small plaques
turbid	Turbid plaques on *E. coli* B
star	Irregular plaques
UV-sensitive	Alters UV sensitivity
acriflavin-resistant	Forms plaques on acriflavin agar
osmotic shock	Withstands rapid dilution into distilled water
lysozyme	Does not produce lysozyme
amber	Grows in *E. coli* K12, but not B
temperature-sensitive	Grows at 25°C, but not at 42°C

adsorption and subsequent infection of *E. coli* B-2. When grown on a mixture of *E. coli* B and B-2, the center of the h plaque appears much darker than the h^+ plaque (Figure 8–18).

Table 8.1 lists other types of mutations that have been isolated and studied in the T-even series of bacteriophages (e.g., T2, T4, T6). These mutations are important to the study of genetic phenomena in bacteriophages.

Mapping in Bacteriophages

Genetic recombination in bacteriophages was discovered during **mixed infection experiments** in which two distinct mutant strains were allowed to *simultaneously* infect the same bacterial culture. These studies were designed so that the number of viral particles sufficiently exceeded the number of bacterial cells to ensure simultaneous infection of most cells by both viral strains. If two loci are involved, recombination is referred to as intergenic.

For example, in one study using the T2/*E. coli* system, the parental viruses were of either the h^+r (wild-type host range, rapid lysis) or $h\,r^+$ (extended host range, normal lysis) genotype. If no recombination occurred, these two parental genotypes would be the only expected phage progeny. However, the recombinants h^+r^+ and $h\,r$ were detected in addition to the parental genotypes (see Figure 8–18). As with eukaryotes, the percentage of recombinant plaques divided by the total number of plaques reflects the relative distance between the genes.

$$\text{recombinational frequency} = (h^+r^+ + h\,r)/\text{total plaques} \times 100$$

Sample data for the h and r loci are shown in Table 8.2.

TABLE 8.2 Results of a Cross Involving the *h* and *r* Genes in Phage T2 ($hr^+ \times h^+r$)

Genotype	Plaques	Designation
$h\ r^+$	42	Parental progeny 76%
$h^+\ r$	34	
$h^+\ r^+$	12	Recombinants 24%
$h\ r$	12	

Source: Data derived from Hershey and Rotman (1949).

Similar recombinational studies have been conducted with numerous mutant genes in a variety of bacteriophages. Data are analyzed in much the same way as in eukaryotic mapping experiments. Two- and three-point mapping crosses are possible, and the percentage of recombinants in the total number of phage progeny is calculated. This value is proportional to the relative distance between two genes along the DNA molecule constituting the chromosome.

Investigations into phage recombination support a model similar to that of eukaryotic crossing over—a breakage and reunion process between the viral chromosomes. A fairly clear picture of the dynamics of viral recombination has emerged. After the early phase of infection, the chromosomes of the phages begin replication. As this stage progresses, a pool of phage chromosomes accumulates in the bacterial cytoplasm. If double infection by phages of two genotypes has occurred, then the pool of chromosomes initially consists of the two parental types. Genetic exchange between these two types will occur before, during, and after replication, producing recombinant chromosomes.

In the case of the h^+r and $h\,r^+$ example discussed here, recombinant h^+r^+ and $h\,r$ chromosomes are produced. Each of these chromosomes can undergo replication, with new replicates exchanging with each other and with parental chromosomes. Furthermore, recombination is not restricted to exchanges between two chromosomes—three or more can be involved simultaneously. As phage development progresses, chromosomes are randomly removed from the pool and packed into the phage head, forming mature phage particles. Thus, a variety of parental and recombinant genotypes are represented in progeny phages.

ESSENTIAL POINT

Various mutant phenotypes, including mutations in plaque morphology and host range, have been studied in bacteriophages and have served as the basis for mapping in these viruses.

NOW SOLVE THIS

Problem 21 on page 179 involves an understanding of intergenic mapping as performed in bacteriophages. You are asked to construct a map from recombination data.

Hint: Mapping is based on recombination occurring during simultaneous infection of a bacterium by two or more strain-specific bacteriophages. Mapping in phages is similar to three-point mapping in eukaryotes. The probability of recombination varies directly with the distance between two genes along the bacteriophage chromosome.

GENETICS, TECHNOLOGY, AND SOCIETY

From Cholera Genes to Edible Vaccines

Using an expanding toolbox of molecular genetic tools, scientists are tackling some of the most serious bacterial diseases affecting our species. Our ability to clone bacterial genes and transfer them into other organisms is leading directly to exciting new treatments. The story of edible vaccines for the treatment of cholera illustrates how genetic engineering is being applied to control a serious human disease.

Cholera is caused by *Vibrio cholerae,* a curved, rod-shaped bacterium found in rivers and oceans. Most genetic strains of *V. cholerae* are harmless; only a few are pathogenic. Infection occurs when a person drinks water or eats food contaminated with pathogenic *V. cholerae.* Once in the digestive system, these bacteria colonize the small intestine and produce proteins called enterotoxins that invade the mucosal cells lining the intestine. This triggers a massive secretion of water and dissolved salts resulting in violent diarrhea, severe dehydration, muscle cramps, lethargy, and often death. The enterotoxin consists of two polypeptides, called the A and B subunits, encoded by two separate genes.

Cholera remains a leading cause of human deaths throughout the Third World, where basic sanitation is lacking and water supplies are often contaminated. For example, in July 1994, 70,000 cases of cholera leading to 12,000 fatalities were reported among the Rwandans crowded into refugee camps in Goma, Zaire. And after an absence of over 100 years, cholera reappeared in Latin America in 1991, spreading from Peru to Mexico and claiming more than 10,000 lives. In 2001, more than 40 cholera outbreaks in 28 countries were reported to the World Health Organization.

A new gene-based technology is emerging to attack cholera. This technology centers on genetically engineered plants that act as vaccines. Scientists introduce a cloned gene—such as a gene encoding a bacterial protein—into the plant genome. The transgenic plant produces the new gene product, and immunity is acquired when an animal eats the plant. The gene product in the plant acts as an antigen, stimulating the production of antibodies to protect against bacterial infection or the effects of their toxins. Since the B subunit of the cholera enterotoxin binds to intestinal cells, research has focused on using this polypeptide as the antigen, with the hope that antibodies against it will prevent toxin binding and render the bacteria harmless.

Leading the efforts to develop an edible vaccine are Charles Arntzen and associates at Cornell University. To test the system, they are using the B subunit of an *E. coli* enterotoxin, which is similar in structure and immunological properties to the cholera protein. Their first step was to obtain the DNA clone of the gene encoding the B subunit and to attach it to a promoter that would induce transcription in all tissues of the plant. Second, the researchers introduced the hybrid gene into potato plants by means of *Agrobacterium*-mediated transformation. The engineered plants expressed their new gene and produced the enterotoxin B subunit. Third, they fed mice a few grams of the genetically engineered tubers. Arntzen's group found that the mice produced specific antibodies against the B subunit and secreted them into the small intestine. When they fed purified enterotoxin to the mice, the mice were protected from its effects and did not develop the symptoms of cholera. In clinical trials conducted using humans in 1998, almost all of the volunteers developed an immune response, and none experienced adverse side effects.

Arntzen's experiments have served as models for other research efforts involving edible vaccines. Currently, scientists are developing edible vaccines against bacterial diseases such as anthrax and tetanus, as well as viral diseases such as rabies, AIDS, and measles.

Your Turn

Take time, individually or in groups, to answer the following questions. Investigate the references and links to help you understand some of the issues that surround the development and uses of edible vaccines.

1. What are the latest research developments on edible vaccines for cholera?

A source of information is the PubMed Web site (**http://www.ncbi.nlm.nih.gov/sites/entrez?db=PubMed**)*, as described in the Exploring Genomics feature in Chapter 2.*

2. Several oral vaccines against cholera are currently available. Given the availability of these vaccines, why do you think that scientists are also developing edible vaccines for cholera? Which vaccine type would you choose for vaccinating populations at risk for cholera?

Read about these cholera vaccines on the World Health Organization Web site at **http://www.who.int/topics/cholera/vaccines/current/en/index.html**.

3. Cholera is spread through ingestion of contaminated water and food. Despite its severity, cholera patients can be effectively treated by oral rehydration. Cholera becomes a major health problem only when proper sanitation and medical treatment are lacking. Given these facts, how much research funding do you think we should spend to develop cholera vaccines, relative to funds spent on improved sanitation, water treatment, and education about treatments?

A discussion of cholera prevention and treatment can be found at **http://en.wikipedia.org/wiki/Cholera**.

4. One of the problems associated with edible vaccines is the public's concern about genetically modified organisms (GMOs). Attitudes vary from outright moral opposition to GMOs to concern about the potential environmental hazards associated with growing transgenic plants. How do you feel about the use of GMOs? What do you think are the most valid arguments for and against them, and why?

Scientists also debate these issues. One such debate is presented in the article Arntzen, C. J., et al., 2003, GM crops: Science, politics and communication, *Nature Rev Genet* 4: 839–843.

CASE STUDY To treat or not to treat

A 4-month-old infant had been running a moderate fever for 36 hours, and a nervous mother made a call to her pediatrician. Examination and testing revealed no outward signs of infection or cause of the fever. The anxious mother asked the pediatrician about antibiotics, but the pediatrician recommended watching the infant carefully for two days before making a decision. He explained that decades of rampant use of antibiotics in medicine and agriculture had caused a worldwide surge in bacteria that are now resistant to such drugs. He also said that the reproductive behavior of bacteria allows them to exchange antibiotic resistance traits with a wide range of other disease-causing bacteria, and that many strains are now resistant to multiple antibiotics. The physician's information raises several interesting questions.

1. Was the physician correct in saying that bacteria can share resistance?
2. Where do bacteria carry antibiotic resistance genes, and how are they exchanged?
3. If the infant was given an antibiotic as a precaution, how might it contribute to the production of resistant bacteria?
4. Aside from hospitals, where else would infants and children come in contact with antibiotic-resistant strains of bacteria? Does the presence of such bacteria in the body always mean an infection?

INSIGHTS AND SOLUTIONS

1. Time mapping is performed in a cross involving the genes *his, leu, mal,* and *xyl.* The recipient cells are auxotrophic for all four genes. After 25 minutes, mating is interrupted, with the results in recipient cells shown below. Diagram the positions of these genes relative to the origin (*O*) of the F factor and to one another.

(a) 90% are xyl^+.

(b) 80% are mal^+.

(c) 20% are his^+.

(d) None are leu^+.

Solution: The *xyl* gene is transferred most frequently, so it is closest to *O* (very close). The *mal* gene is next and reasonably close to *xyl,* followed by the more distant *his* gene. The *leu* gene is far beyond these three, since no recovered recombinants include it. The diagram shows these relative locations along a piece of the circular chromosome.

2. In four Hfr strains of bacteria, all derived from an original F^+ culture grown over several months, a group of hypothetical genes is studied and shown to transfer in the orders shown in the following table. (a) Assuming *b* is the first gene along the chromosome, determine the sequence of all genes shown. (b) One strain creates an apparent dilemma. Which one is it? Explain why the dilemma is only apparent, not real.

Hfr Strain	Order of Transfer					
1	*e*	*r*	*i*	*u*	*m*	*b*
2	*u*	*m*	*b*	*a*	*c*	*t*
3	*c*	*t*	*e*	*r*	*i*	*u*
4	*r*	*e*	*t*	*c*	*a*	*b*

Solution:

(a) The sequence is found by overlapping the genes in each strain.

Strain 2	*u*	*m*	***b***	***a***	***c***	*t*				
Strain 3					*c*	***t***	***e***	***r***	*i*	*u*
Strain 1							*e*	*r*	***i***	***u***	***m***	*b*

Starting with *b* in strain 2, the gene sequence is *bacterium.*

(b) Strain 4 creates a dilemma, which is resolved when we realize that the F factor is integrated in the opposite orientation. Thus, the genes enter in the opposite sequence, starting with gene *r.*

$$\xrightarrow{retcab}$$

3. Three strains of bacteria, each bearing a separate mutation, a^-, b^-, or c^-, are the sources of donor DNA in a transformation experiment. Recipient cells are wild type for those genes but express the mutation d^-.

(a) Based on the following data and assuming that the location of the *d* gene precedes the *a, b,* and *c* genes, propose a linkage map for these four genes.

DNA Donor	Recipient	Transformants	Frequency of Transformants
a^-d^+	a^+d^-	a^+d^+	0.21
b^-d^+	b^+d^-	b^+d^+	0.18
c^-d^+	c^+d^-	c^+d^+	0.63

(b) If the donor DNA is wild type and the recipient cells are either a^-b^-, a^-c^-, or b^-c^-, in which case would wild-type transformants be expected most frequently?

Solution:

(a) These data reflect the relative distances between each of the *a, b,* and *c* genes and the *d* gene. The *a* and *b* genes are about the same distance from the *d* gene and are thus tightly linked to one another. The *c* gene is more distant. Assuming that the *d* gene precedes the others, the map looks like this:

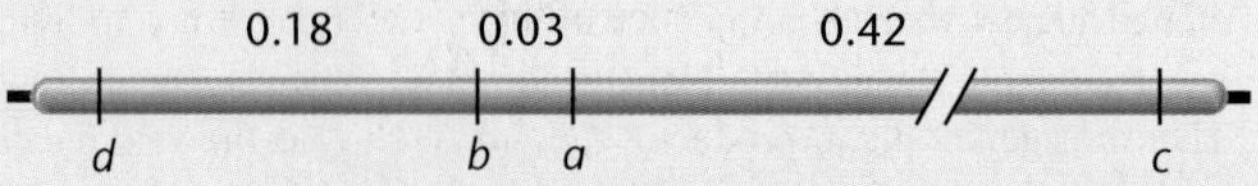

(b) Because the *a* and *b* genes are closely linked, they most likely cotransform in a single event. Thus, recipient cells of a^-b^- are most likely to convert to wild type.

Problems and Discussion Questions

1. Distinguish among the three modes of recombination in bacteria.
2. With respect to F^+ and F^- bacterial matings,
   (a) How was it established that physical contact was necessary?
   (b) How was it established that chromosome transfer was unidirectional?
   (c) What is the genetic basis of a bacterium being F^+?
3. List all of the differences between $F^+ \times F^-$ and Hfr $\times F^-$ bacterial crosses.
4. List all of the differences between F^+, F^-, Hfr, and F′ bacteria.
5. Describe the basis for chromosome mapping in the Hfr $\times F^-$ crosses.
6. Why are the recombinants produced from an Hfr $\times F^-$ cross rarely, if ever, F^+?
7. Describe the origin of F′ bacteria and merozygotes.
8. Describe the mechanism of transformation.
9. In a transformation experiment involving a recipient bacterial strain of genotype a^-b^- the results below were obtained. What can you conclude about the location of the *a* and *b* genes relative to each other? See **Now Solve This** on page 170.

	Transformants (%)		
Transforming DNA	a^+b^-	a^-b^+	a^+b^+
a^+b^+	3.1	1.2	0.04
a^+b^- and a^-b^+	2.4	1.4	0.03

10. In a transformation experiment, donor DNA was obtained from a prototroph bacterial strain ($a^+b^+c^+$), and the recipient was a triple auxotroph ($a^-b^-c^-$). What general conclusions can you draw about the linkage relationships among the three genes from the following transformant classes that were recovered?

$a^+b^-c^-$	180
$a^-b^+c^-$	150
$a^+b^+c^-$	210
$a^-b^-c^+$	179
$a^+b^-c^+$	2
$a^-b^+c^+$	1
$a^+b^+c^+$	3

11. The bacteriophage genome consists primarily of genes encoding proteins that make up the head, collar and tail, and tail fibers. When these genes are transcribed following phage infection, how are these proteins synthesized, since the phage genome lacks genes essential to ribosome structure?
12. Describe the temporal sequence of the bacteriophage life cycle.
13. In the plaque assay, what is the precise origin of a single plaque?
14. In the plaque assay, exactly what makes up a single plaque?
15. A plaque assay is performed beginning with 1.0 mL of a solution containing bacteriophages. This solution is serially diluted three times by taking 0.1 mL and adding it to 9.9 mL of liquid medium. The final dilution is plated and yields 17 plaques. What is the initial density of bacteriophages in the original 1.0 mL?
16. Describe the difference between the lytic cycle and lysogeny when bacteriophage infection occurs.
17. Define prophage.
18. Explain the observations that led Zinder and Lederberg to conclude that the prototrophs recovered in their transduction experiments were not the result of Hfr-mediated conjugation.
19. Describe generalized transduction and distinguish between abortive and complete transduction.
20. Describe how generalized transduction can be used to map the bacterial chromosome. How does cotransduction play a role in mapping?
21. Two theoretical genetic strains of a virus ($a^-b^-c^-$ and $a^+b^+c^+$) are used to simultaneously infect a culture of host bacteria. Of 10,000 plaques scored, the genotypes in the following table were observed. Determine the genetic map of these three genes on the viral chromosome. See **Now Solve This** on page 176.

$a^+b^+c^+$	4100	$a^-b^+c^-$	160
$a^-b^-c^-$	3990	$a^+b^-c^+$	140
$a^+b^-c^-$	740	$a^-b^-c^+$	90
$a^-b^+c^+$	670	$a^+b^+c^-$	110

22. Describe the conditions under which genetic recombination may occur in bacteriophages.
23. If a single bacteriophage infects one *E. coli* cell present in a culture of bacteria and, upon lysis, yields 200 viable viruses, how many phages will exist in a single plaque if three more lytic cycles occur?
24. A phage-infected bacterial culture was subjected to a series of dilutions, and a plaque assay was performed in each case, with the following results. What conclusion can be drawn in the case of each dilution?

Dilution Factor	Assay Results
10^4	All bacteria lysed
10^5	14 plaques
10^6	0 plaques

25. When the interrupted mating technique was used with five different strains of Hfr bacteria, the order of gene entry during recombination shown in the following table was observed. On the basis of these data, draw a map of the bacterial chromosome. Do the data support the concept of circularity? See **Now Solve This** on page 166.

Hfr Strain	Order				
1	*t*	*c*	*h*	*r*	*o*
2	*h*	*r*	*o*	*m*	*b*
3	*m*	*o*	*r*	*h*	*c*
4	*m*	*b*	*a*	*k*	*t*
5	*c*	*t*	*k*	*a*	*b*

26. In *B. subtilis,* linkage analysis of two mutant genes affecting the synthesis of the two amino acids, tryptophan (trp_2^-) and tyrosine (tyr_1^-), was performed using transformation (E. Nester, M. Schafer, and J. Lederberg (1963) *Genetics* 48: 529–551). Examine the data in the following table, and draw all possible conclusions regarding linkage.

	Donor DNA	Recipient Cell	Transformants	Number of Transformants
			$trp_2^+ tyr_1^-$	196
Part A	$trp_2^+ tyr_1^+$	$trp_2^- tyr_1^-$	$trp_2^- tyr_1^+$	328
			$trp_2^+ tyr_1^+$	367
	$trp_2^+ tyr_1^-$		$trp_2^+ tyr_1^-$	190
Part B	and	$trp_2^- tyr_1^-$	$trp_2^- tyr_1^+$	256
	$trp_2^- tyr_1^+$		$trp_2^+ tyr_1^+$	2

27. What is the role of part B in the experiment shown in Problem 26?
28. An Hfr strain is used to map three genes in an interrupted mating experiment. The cross is $Hfr/a^+b^+c^+rif^s \times F^-/a^-b^-c^-rif^r$ (no map order is implied in the listing of the alleles; *rif* = the antibiotic rifampicin). The a^+ gene is required for biosynthesis of nutrient A, the b^+ gene for nutrient B, and the c^+ gene for nutrient C. The minus alleles are auxotrophs for these nutrients. The cross is initiated at time = 0, and, at various intervals, the mating mixture is plated on three types of medium. Each plate contains minimal medium (MM) *plus* rifampicin *plus* the specific supplements indicated in the following table (results for each time point are shown as number of colonies growing on each plate).
   (a) What is the purpose of rifampicin in the experiment?
   (b) Based on these data, determine the approximate location on the chromosome of the *a, b,* and *c* genes relative to one another and to the F factor.
   (c) Can the location of the *rif* gene be determined in this experiment? If not, design an experiment to determine the location of *rif* relative to the F factor and to gene *b.*

	Time of Interruption (min)			
Supplements Added to MM	**5**	**10**	**15**	**20**
Nutrients A and B	0	0	4	21
Nutrients B and C	0	5	23	40
Nutrients A and C	4	25	60	82

29. In a cotransformation experiment using various combinations of genes, two at a time, the following data were produced. Determine which genes are linked and to which others.

Successful Cotransformation	Unsuccessful Cotransformation
a and *d*; *b* and *c*; *b* and *f*	*a* and *b*; *a* and *c*; *a* and *f*
	d and *b*; *d* and *c*; *d* and *f*
	a and *e*; *b* and *e*; *c* and *e*
	d and *e*; *e* and *f*

30. In Problem 29, another gene, *g,* was also studied. It demonstrated positive cotransformation when tested with gene *f.* Predict the results of experiments when it was tested with genes *a, b, c, d,* and *e.*

Computer-generated space-filling models of alternative forms of DNA.

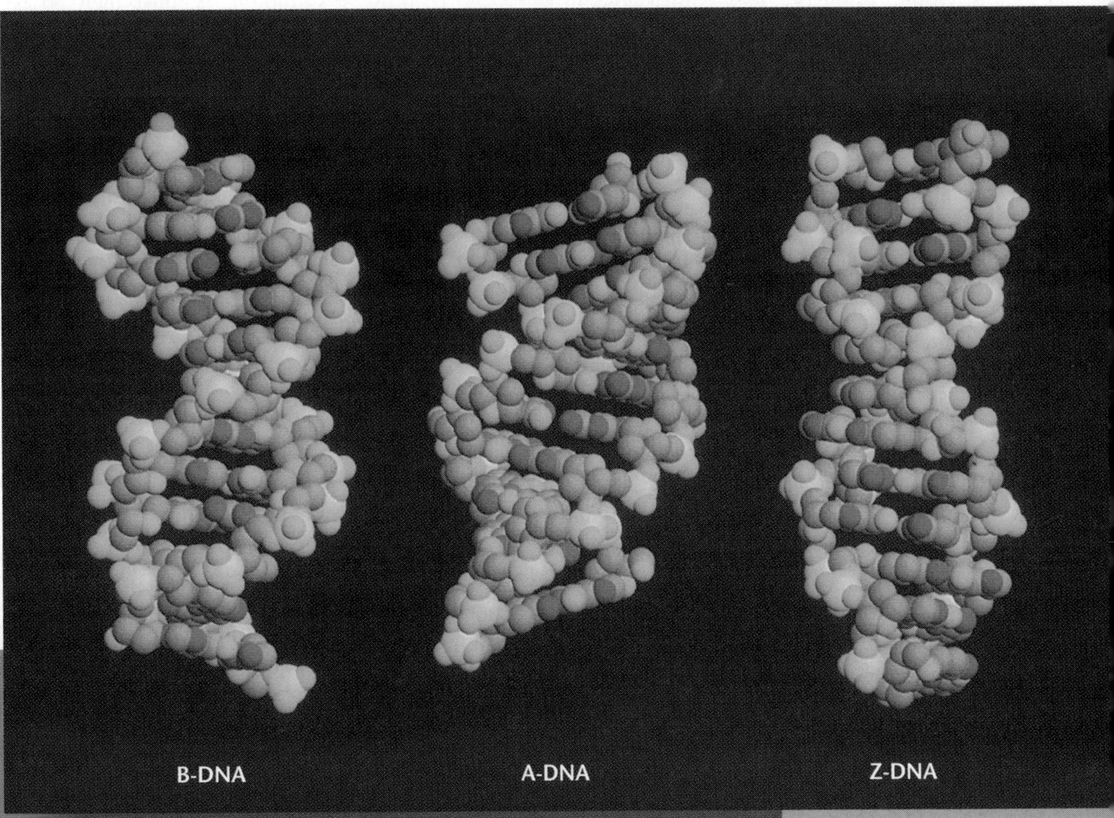

9

DNA Structure and Analysis

Chapter Concepts

- With the exception of some viruses, DNA serves as the genetic material in all living organisms on Earth.
- According to the Watson–Crick model, DNA exists in the form of the right-handed double helix.
- The strands of the double helix are antiparallel and held together by hydrogen bonding between complementary nitrogenous bases.
- The structure of DNA provides the basis for storing and expressing genetic information.
- RNA has many similarities to DNA but exists mostly as a single-stranded molecule.
- In some viruses, RNA serves as the genetic material.

Up to this point in the text, we have described chromosomes as containing genes that control phenotypic traits that are transmitted through gametes to future offspring. Logically, genes must contain some sort of information that, when passed to a new generation, influences the form and characteristics of each individual. We refer to that information as the **genetic material.** Logic also suggests that this same information in some way directs the many complex processes that lead to an organism's adult form.

Until 1944, it was not clear what chemical component of the chromosome makes up genes and constitutes the genetic material. Because chromosomes were known to have both a nucleic acid and a protein component, both were candidates. In 1944, however, there emerged direct experimental evidence that the nucleic acid DNA serves as the informational basis for heredity.

Once the importance of DNA in genetic processes was realized, work intensified with the hope of discerning not only the structural basis of this molecule but also the relationship of its structure to its function. Between 1944 and 1953, many scientists sought information that might answer the most significant and intriguing question in the history of biology: How does DNA serve as the genetic basis for the living process? Researchers believed the answer depended strongly on the chemical structure of the DNA molecule, given the complex but orderly functions ascribed to it.

These efforts were rewarded in 1953 when James Watson and Francis Crick set forth their hypothesis for the double-helical nature of DNA. The assumption that the molecule's functions would be clarified more easily once its general structure was determined proved to be correct. In this chapter, we initially review the evidence that DNA is the genetic material and then discuss the elucidation of its structure.

How Do We Know?

In this chapter, we will focus on DNA, the molecule that stores genetic information in all living things. We shall define its structure and delve into how we analyze this molecule. As you study this topic, you should try to answer several fundamental questions:

1. How were we able to determine that DNA, and not some other molecule, serves as the genetic material in bacteria, bacteriophages, and eukaryotes?
2. How do we know that the structure of DNA is in the form of a right-handed double-helical molecule?
3. How do we know that in DNA G pairs with C and that A pairs with T as complementary strands are formed?
4. How do we know that repetitive DNA sequences exist in eukaryotes?

9.1 The Genetic Material Must Exhibit Four Characteristics

For a molecule to serve as the genetic material, it must possess four major characteristics: **replication, storage of information, expression of information,** and **variation by mutation.** *Replication* of the genetic material is one facet of the cell cycle, a fundamental property of all living organisms. Once the genetic material of cells replicates and is doubled in amount, it must then be partitioned equally into daughter cells. During the formation of gametes, the genetic material is also replicated but is partitioned so that each cell gets only one-half of the original amount of genetic material—the process of meiosis. Although the products of mitosis and meiosis differ, both of these processes are part of the more general phenomenon of cellular reproduction.

Storage of information requires the molecule to act as a repository of genetic information that may or may not be expressed by the cell in which it resides. It is clear that while most cells contain a complete copy of the organism's genome, at any point in time they express only a part of this genetic potential. For example, in bacteria many genes "turn on" in response to specific environmental conditions and "turn off" when conditions change. In vertebrates, skin cells may display active melanin genes but never activate their hemoglobin genes; in contrast, digestive cells activate many genes specific to their function but do not activate their melanin genes.

Expression of the stored genetic information is the basis of the process of **information flow** within the cell (Figure 9–1). The initial event is the **transcription** of DNA, in which three main types of RNA molecules are synthesized: messenger RNA (mRNA), ribosomal RNA (rRNA), and transfer RNA (tRNA). Of these, mRNAs are translated into proteins. Each mRNA is the product of a specific gene and directs the synthesis of a different protein. **Translation** occurs in conjunction with ribosomes and involves tRNA, which adapts the chemical information in mRNA to the amino acids that make up proteins. Collectively, these processes form the **central dogma of molecular genetics:** "DNA makes RNA, which makes proteins."

The genetic material is also the source of *variation* among organisms through the process of mutation. If a mutation—a

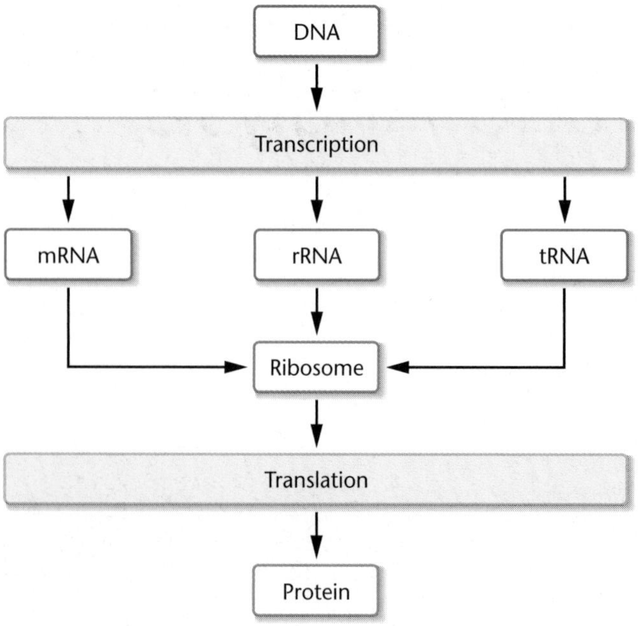

FIGURE 9–1 Simplified view of information flow (the central dogma) involving DNA, RNA, and proteins within cells.

change in the chemical composition of DNA—occurs, the alteration is reflected during transcription and translation, affecting the specific protein. If a mutation is present in gametes, it may be passed to future generations and, with time, become distributed throughout the population. Genetic variation, which also includes alterations of chromosome number and rearrangements within and between chromosomes, provides the raw material for the process of evolution.

9.2 Until 1944, Observations Favored Protein as the Genetic Material

The idea that genetic material is physically transmitted from parent to offspring has been accepted for as long as the concept of inheritance has existed. Beginning in the late nineteenth century, research into the structure of biomolecules progressed considerably, setting the stage for describing the genetic material in chemical terms. Although both proteins and nucleic acid were major candidates for the role of the genetic material, until the 1940s many geneticists favored proteins. This is not surprising because a diversity of proteins was known to be abundant in cells, and much more was known about protein chemistry.

DNA was first studied in 1868 by a Swiss chemist, Friedrich Miescher. He isolated cell nuclei and derived an acid substance containing DNA that he called **nuclein.** As investigations progressed, however, DNA, which was shown to be present in chromosomes, seemed to lack the chemical diversity necessary to store extensive genetic information. This conclusion was based largely on Phoebus A. Levene's observations in 1910 that DNA contained approximately equal amounts of four similar molecules called *nucleotides.* Levene postulated incorrectly that identical groups of these four components were repeated over and over, which was the basis of his **tetranucleotide hypothesis** for DNA structure. Attention was thus directed away from DNA, favoring proteins. However, in the 1940s, Erwin Chargaff showed that Levene's proposal was incorrect when he demonstrated that most organisms do not contain precisely equal proportions of the four nucleotides. We shall see later that the structure of DNA accounts for Chargaff's observations.

ESSENTIAL POINT

Although both proteins and nucleic acids were initially considered as possible candidates for genetic material, proteins were initially favored.

TABLE 9.1 Strains of *Diplococcus pneumoniae* Used by Frederick Griffith in His Original Transformation Experiments

Serotype	Colony Morphology	Capsule	Virulence
II*R*	Rough	Absent	Avirulent
III*S*	Smooth	Present	Virulent

9.3 Evidence Favoring DNA as the Genetic Material Was First Obtained during the Study of Bacteria and Bacteriophages

Oswald Avery, Colin MacLeod, and Maclyn McCarty's 1944 publication on the chemical nature of a "transforming principle" in bacteria was the initial event that led to the acceptance of DNA as the genetic material. Their work, along with subsequent findings of other research teams, constituted the first direct experimental proof that DNA, and not protein, is the biomolecule responsible for heredity. It marked the beginning of the era of molecular genetics, a period of discovery in biology that made biotechnology feasible and has moved us closer to understanding the basis of life. The impact of the initial findings on future research and thinking paralleled that of the publication of Darwin's theory of evolution and the subsequent rediscovery of Mendel's postulates of transmission genetics. Together, these events constitute the three great revolutions in biology.

Transformation Studies

The research that provided the foundation for Avery, MacLeod, and McCarty's work was initiated in 1927 by Frederick Griffith, a medical officer in the British Ministry of Health. He experimented with several different strains of the bacterium *Diplococcus pneumoniae.** Some were **virulent strains,** which cause pneumonia in certain vertebrates (notably humans and mice), while others were **avirulent strains,** which do not cause illness.

The difference in virulence depends on the existence of a polysaccharide capsule; virulent strains have this capsule, whereas avirulent strains do not. The nonencapsulated bacteria are readily engulfed and destroyed by phagocytic cells in the animal's circulatory system. Virulent bacteria, which possess the polysaccharide coat, are not easily engulfed; they multiply and cause pneumonia.

The presence or absence of the capsule causes a visible difference between colonies of virulent and avirulent strains. Encapsulated bacteria form **smooth colonies** (*S*) with a shiny surface when grown on an agar culture plate; nonencapsulated strains produce **rough colonies** (*R*) (Figure 9–2, p. 184). Thus, virulent and avirulent strains are easily distinguished by standard microbiological culture techniques.

Each strain of *Diplococcus* may be one of dozens of different types called **serotypes.** The specificity of the serotype is due to the detailed chemical structure of the polysaccharide constituent of the thick, slimy capsule. Serotypes are identified by immunological techniques and are usually designated by Roman numerals. Griffith used the avirulent type II*R* and the virulent type III*S* in his critical experiments. Table 9.1 summarizes the characteristics of these strains.

Griffith knew from the work of others that only living virulent cells produced pneumonia in mice. If heat-killed virulent

* This organism is now named *Streptococcus pneumoniae.*

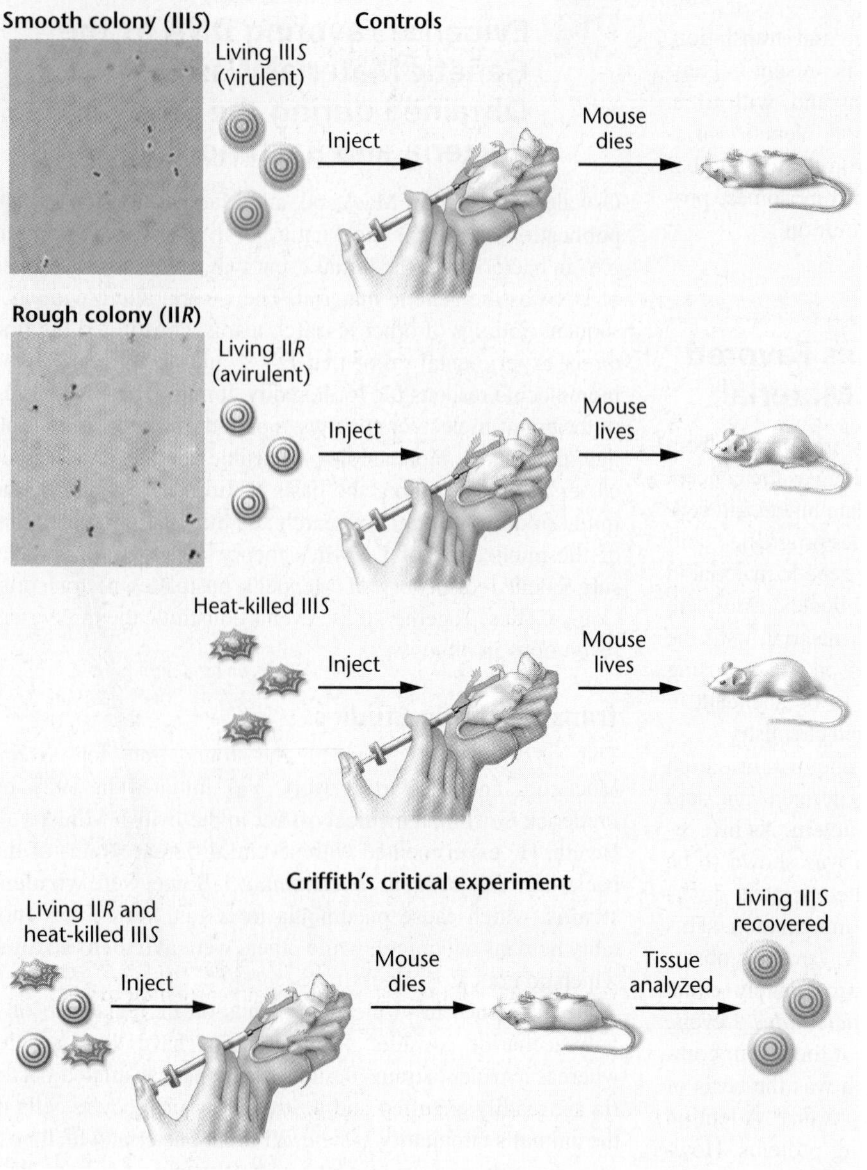

FIGURE 9–2 Griffith's transformation experiment. The photographs show bacterial cells that exhibit capsules (type III*S*) or not (type II*R*).

bacteria were injected into mice, no pneumonia resulted, just as living avirulent bacteria failed to produce the disease. Griffith's critical experiment (Figure 9–2) involved injecting mice with living II*R* (avirulent) cells combined with heat-killed III*S* (virulent) cells. Since neither cell type caused death in mice when injected alone, Griffith expected that the double injection would not kill the mice. But, after five days, all of the mice that had received both types of cells were dead. Paradoxically, analysis of their blood revealed large numbers of living type III*S* bacteria.

As far as could be determined, these III*S* bacteria were identical to the III*S* strain from which the heat-killed cell preparation had been made. Control mice, injected only with living avirulent II*R* bacteria, did not develop pneumonia and remained healthy. This ruled out the possibility that the avirulent II*R* cells simply changed (or mutated) to virulent III*S* cells in the absence of the heat-killed III*S* bacteria. Instead, some type of interaction had taken place between living II*R* and heat-killed III*S* cells.

Griffith concluded that the heat-killed III*S* bacteria somehow converted live avirulent II*R* cells into virulent III*S* cells. Calling the phenomenon **transformation,** he suggested that the **transforming principle** might be some part of the polysaccharide capsule or a compound required for capsule synthesis, although the capsule alone did not cause pneumonia. To use Griffith's term, the transforming principle from the dead III*S* cells served as a "pabulum" for the II*R* cells.

Griffith's work led bacteriologists and other physicians to explore the phenomenon of transformation. By 1931, Henry Dawson and his coworkers showed that transformation could occur *in vitro* (in a test tube containing only bacterial cells). That is, injection into mice was not necessary for transformation to occur. By 1933, Lionel J. Alloway had refined the *in vitro* experiments using extracts from *S* cells added to living *R* cells. The soluble filtrate from the heat-killed III*S* cells was as effective in inducing transformation as were the intact cells. Alloway and others did not view transformation as a

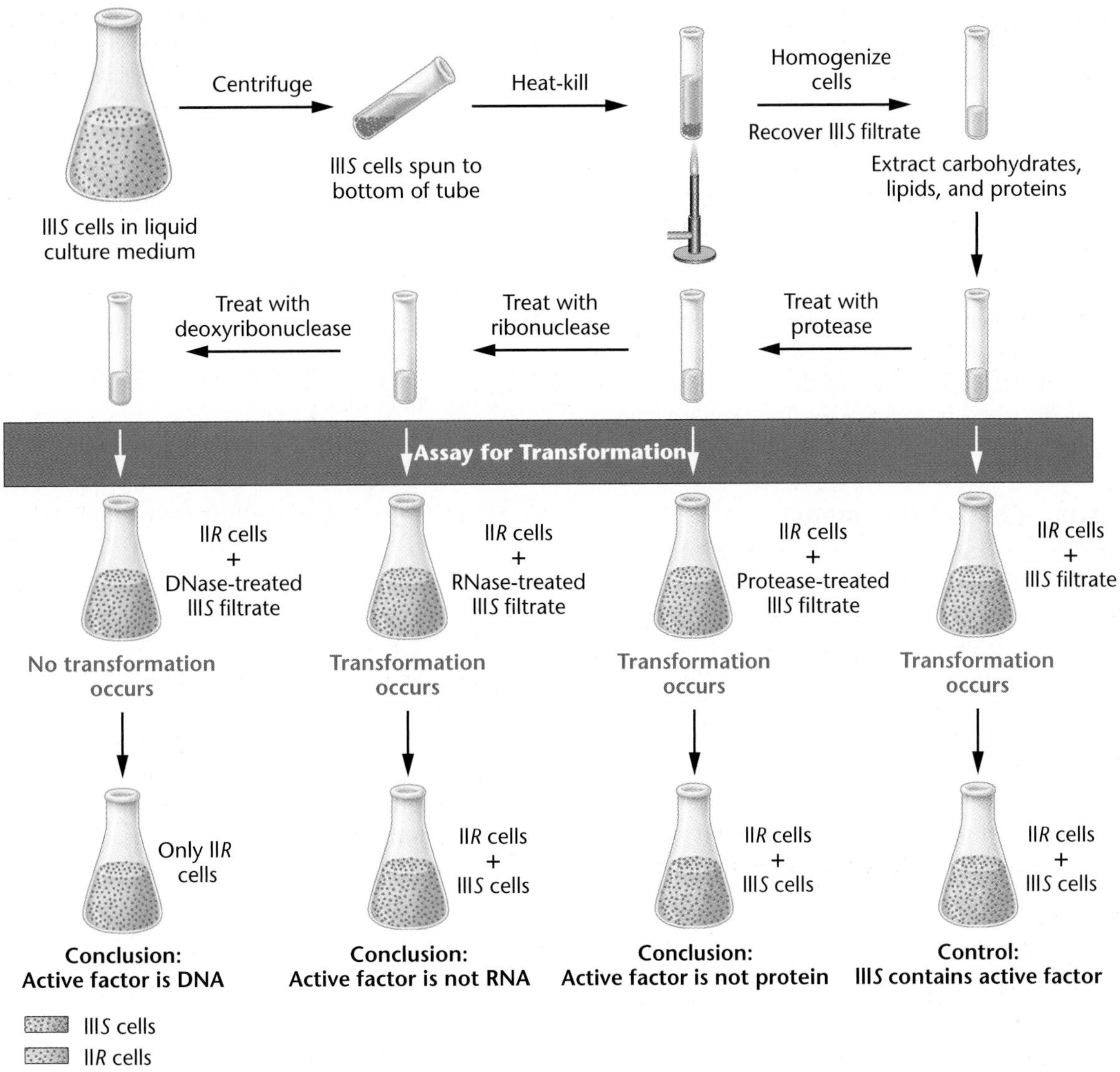

FIGURE 9–3 Summary of Avery, MacLeod, and McCarty's experiment demonstrating that DNA is the transforming principle.

genetic event, but rather as a physiological modification of some sort. Nevertheless, the experimental evidence that a chemical substance was responsible for transformation was quite convincing.

Then, in 1944, after ten years of work, Avery, MacLeod, and McCarty published their results in what is now regarded as a classic paper in the field of molecular genetics. They reported that they had obtained the transforming principle in a highly purified state and that beyond reasonable doubt it was DNA.

The details of their work are illustrated in Figure 9–3. The researchers began their isolation procedure with large quantities (50–75 L) of liquid cultures of type III*S* virulent cells. The cells were centrifuged, collected, and heat-killed. Following various chemical treatments, a soluble filtrate was derived from these cells, which retained the ability to induce transformation of type II*R* avirulent cells. The soluble filtrate was treated with a protein-digesting enzyme, called a protease, and an RNA-digesting enzyme, called **ribonuclease.** Such treatment destroyed the activity of any remaining protein and RNA. Nevertheless, transforming activity still remained. They concluded that neither protein nor RNA was responsible for transformation. The final confirmation came with experiments using crude samples of the DNA-digesting enzyme **deoxyribonuclease,** isolated from dog and rabbit sera. Digestion with this enzyme destroyed transforming activity present in the filtrate; thus, Avery and his coworkers were certain that the active transforming principle in these experiments was DNA.

The great amount of work, the confirmation and reconfirmation of the conclusions, and the logic of the experimental design involved in the research of these three scientists are truly impressive. Their conclusion in the 1944 publication, however, was stated very simply: "The evidence presented supports the belief that a nucleic acid of the desoxyribose* type is the fundamental unit of the transforming principle of *Pneumococcus* type III."

The researchers also immediately recognized the genetic and biochemical implications of their work. They suggested that the transforming principle interacts with the II*R* cell and

* Desoxyribose is now spelled deoxyribose.

gives rise to a coordinated series of enzymatic reactions that culminates in the synthesis of the type III*S* capsular polysaccharide. They emphasized that, once transformation occurs, the capsular polysaccharide is produced in successive generations. Transformation is therefore heritable, and the process affects the genetic material.

Transformation, originally introduced in Chapter 8, has now been shown to occur in *Hemophilus influenzae, Bacillus subtilis, Shigella paradysenteriae,* and *Escherichia coli,* among many other microorganisms. Transformation of numerous genetic traits other than colony morphology has been demonstrated, including those that resist antibiotics. These observations further strengthened the belief that transformation by DNA is primarily a genetic event rather than simply a physiological change.

The Hershey–Chase Experiment

The second major piece of evidence supporting DNA as the genetic material was provided during the study of the bacterium *E. coli* and one of its infecting viruses, **bacteriophage T2.** Often referred to simply as a **phage,** the virus consists of a protein coat surrounding a core of DNA. Electron micrographs reveal that the phage's external structure is composed of a hexagonal head plus a tail. Figure 9–4 shows the life cycle of a T-even bacteriophage such as T2, as it was known in 1952. Recall that the phage adsorbs to the bacterial cell and that some component of the phage enters the bacterial cell. Following infection, the viral information "commandeers" the cellular machinery of the host and undergoes viral reproduction. In a reasonably short time, many new phages are constructed and the bacterial cell is lysed, releasing the progeny viruses.

In 1952, Alfred Hershey and Martha Chase published the results of experiments designed to clarify the events leading to phage reproduction. Several of the experiments clearly established the independent functions of phage protein and nucleic acid in the reproduction process of the bacterial cell. Hershey and Chase knew from this existing data that:

1. T2 phages consist of approximately 50 percent protein and 50 percent DNA.
2. Infection is initiated by adsorption of the phage by its tail fibers to the bacterial cell.
3. The production of new viruses occurs within the bacterial cell.

It appeared that some molecular component of the phage, DNA and/or protein, entered the bacterial cell and directed viral reproduction. Which was it?

Hershey and Chase used radioisotopes to follow the molecular components of phages during infection. Both ^{32}P and ^{35}S, radioactive forms of phosphorus and sulfur, respectively, were used. DNA contains phosphorus but not sulfur, so ^{32}P effectively labels DNA. Because proteins contain sulfur but not phosphorus, ^{35}S labels protein. *This is a key*

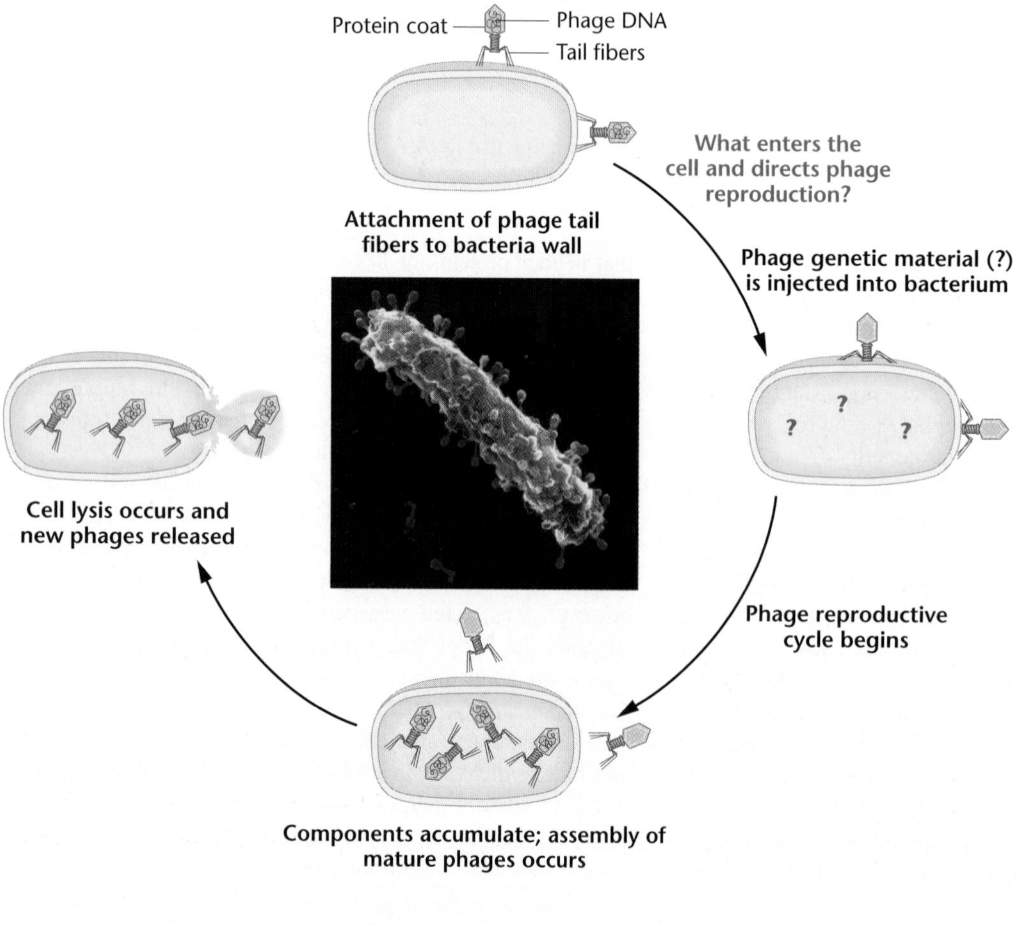

Protein coat
Phage DNA
Tail fibers
What enters the cell and directs phage reproduction?
Attachment of phage tail fibers to bacteria wall
Phage genetic material (?) is injected into bacterium
?
?
?
Phage reproductive cycle begins
Components accumulate; assembly of mature phages occurs
Cell lysis occurs and new phages released

FIGURE 9–4 Life cycle of a T-even bacteriophage. The electron micrograph shows an *E. coli* cell during infection by numerous T2 phages (shown in blue).

point in the experiment. If *E. coli* cells are first grown in the presence of *either* ^{32}P *or* ^{35}S and then infected with T2 viruses, the progeny phage will have *either* a labeled DNA core *or* a labeled protein coat, respectively. These radioactive phages can be isolated and used to infect unlabeled bacteria (Figure 9–5).

When labeled phage and unlabeled bacteria were mixed, an *adsorption complex* was formed as the phages attached their tail fibers to the bacterial wall. These complexes were isolated and subjected to a high shear force by placing them in a blender. This force stripped off the attached phages, which were then analyzed separately (Figure 9–5). By tracing the

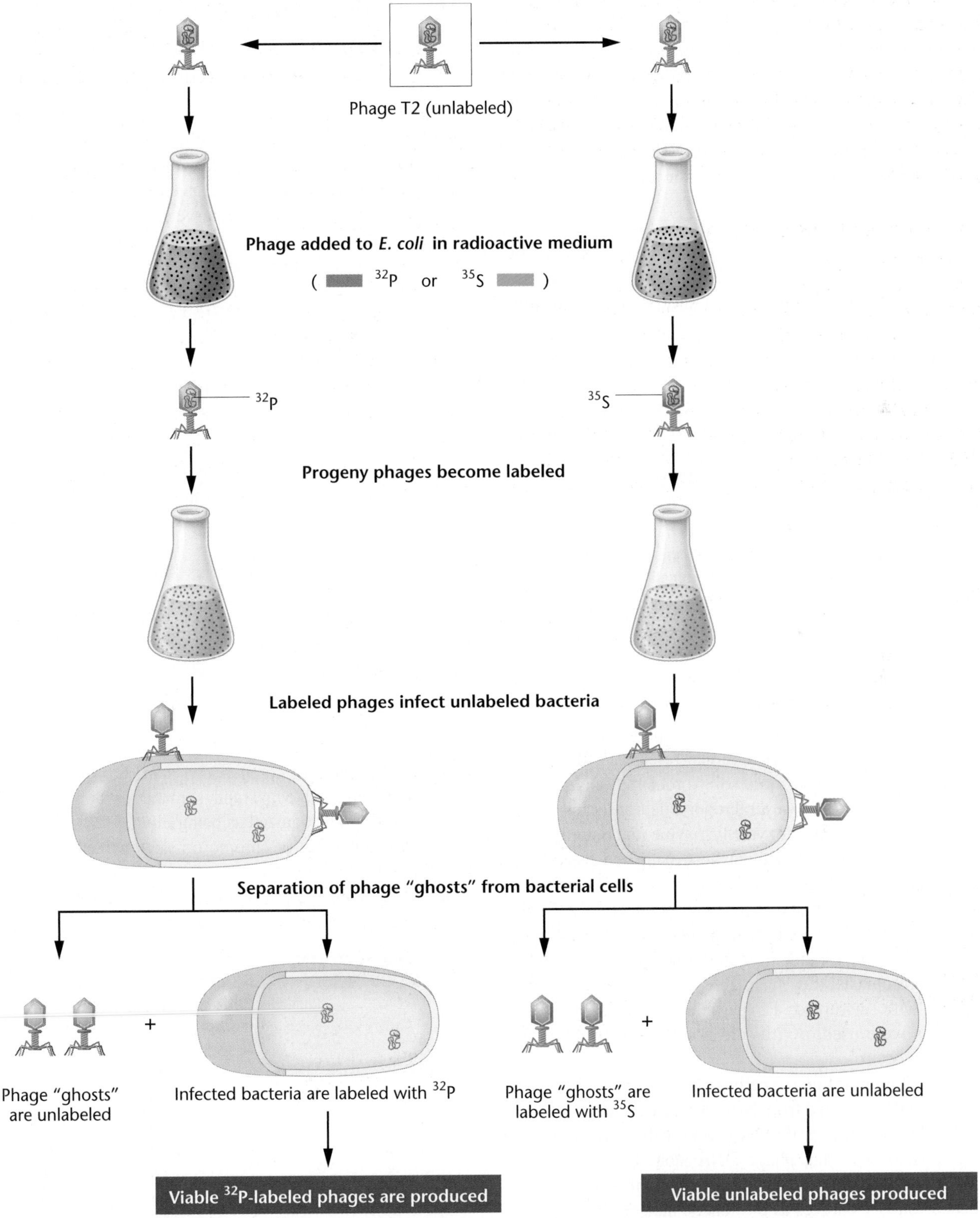
Phage T2 (unlabeled)
Phage added to E. coli in radioactive medium
(32P or 35S)
32P
35S
Progeny phages become labeled
Labeled phages infect unlabeled bacteria
Separation of phage "ghosts" from bacterial cells
+
+
Phage "ghosts" are unlabeled
Infected bacteria are labeled with 32P
Phage "ghosts" are labeled with 35S
Infected bacteria are unlabeled
Viable 32P-labeled phages are produced
Viable unlabeled phages produced

FIGURE 9–5 Summary of the Hershey–Chase experiment demonstrating that DNA, not protein, is responsible for directing the reproduction of phage T2 during the infection of *E. coli*.

radioisotopes, Hershey and Chase were able to demonstrate that most of the ^{32}P-labeled DNA had transferred into the bacterial cell following adsorption; on the other hand, almost all of the ^{35}S-labeled protein remained outside the bacterial cell and was recovered in the phage "ghosts" (empty phage coats) after the blender treatment. After separation, the bacterial cells, which now contained viral DNA, were eventually lysed as new phages were produced. These progeny contained ^{32}P but not ^{35}S.

Hershey and Chase interpreted these results as indicating that the protein of the phage coat remains outside the host cell and is not involved in the production of new phages. On the other hand, and most important, phage DNA enters the host cell and directs phage reproduction. Hershey and Chase had demonstrated that the genetic material in phage T2 is DNA, not protein.

These experiments, along with those of Avery and his colleagues, provided convincing evidence that DNA is the molecule responsible for heredity. This conclusion has since served as the cornerstone of the field of molecular genetics.

NOW SOLVE THIS

Problem 7 on page 201 asks about applying the protocol of the Hershey–Chase experiment to the investigation of transformation. You are asked about the possible success of attempting to do so.

Hint: As you attempt to apply the Hershey–Chase protocol, remember that in transformation, exogenous DNA enters the soon-to-be transformed cell and no cell-to-cell contact is involved in the process.

Transfection Experiments

During the eight years following the publication of the Hershey–Chase experiment, additional research with bacterial viruses provided even more solid proof that DNA is the genetic material. In 1957, several reports demonstrated that if *E. coli* is treated with the enzyme lysozyme, the outer wall of the cell can be removed without destroying the bacterium. Enzymatically treated cells are naked, so to speak, and contain only the cell membrane as the outer boundary of the cell; these structures are called **protoplasts** (or **spheroplasts**). John Spizizen and Dean Fraser independently reported that by using protoplasts, they were able to initiate phage multiplication with disrupted T2 particles. That is, provided protoplasts were used, a virus did not have to be intact for infection to occur.

Similar but more refined experiments were reported in 1960 using only DNA purified from bacteriophages. This process of infection by only the viral nucleic acid, called **transfection,** proves conclusively that phage DNA alone contains all the necessary information for producing mature viruses. Thus, the evidence that DNA serves as the genetic material in all organisms was further strengthened, even though all direct evidence had been obtained from bacterial and viral studies.

ESSENTIAL POINT

By 1952, transformation studies and experiments using bacteria infected with bacteriophages strongly suggested that DNA is the genetic material in bacteria and most viruses.

9.4 Indirect and Direct Evidence Supports the Concept that DNA Is the Genetic Material in Eukaryotes

In 1950, eukaryotic organisms were not amenable to the types of experiments that used bacteria and viruses to demonstrate that DNA is the genetic material. Nevertheless, it was generally assumed that the genetic material would be a universal substance and also serve this role in eukaryotes. Initially, support for this assumption relied on several circumstantial observations that, taken together, indicated that DNA is also the genetic material in eukaryotes. Subsequently, direct evidence established unequivocally the central role of DNA in genetic processes.

Indirect Evidence: Distribution of DNA

The genetic material should be found where it functions—in the nucleus as part of chromosomes. Both DNA and protein fit this criterion. However, protein is also abundant in the cytoplasm, whereas DNA is not. Both mitochondria and chloroplasts are known to perform genetic functions, and DNA is also present in these organelles. Thus, DNA is found only where primary genetic function is known to occur. Protein, however, is found everywhere in the cell. These observations are consistent with the interpretation favoring DNA over protein as the genetic material.

Because it had been established earlier that chromosomes within the nucleus contain the genetic material, a correlation was expected between the ploidy (n, $2n$, etc.) of cells and the quantity of the molecule that functions as the genetic material. Meaningful comparisons can be made between gametes (sperm and eggs) and somatic or body cells. The somatic cells are recognized as being diploid ($2n$) and containing twice the number of chromosomes as gametes, which are haploid (n).

Table 9.2 compares the amount of DNA found in haploid sperm and the diploid nucleated precursors of red blood cells

TABLE 9.2 DNA Content of Haploid Versus Diploid Cells of Various Species (in picograms)

Organism*	n	$2n$
Human	3.25	7.30
Chicken	1.26	2.49
Trout	2.67	5.79
Carp	1.65	3.49
Shad	0.91	1.97

*Sperm (n) and nucleated precursors to red blood cells ($2n$) were used to contrast ploidy levels.

from a variety of organisms. The amount of DNA and the number of sets of chromosomes are closely correlated. No consistent correlation can be observed between gametes and diploid cells for proteins, thus again favoring DNA over proteins as the genetic material of eukaryotes.

Indirect Evidence: Mutagenesis

Ultraviolet (UV) light is one of many agents capable of inducing mutations in the genetic material. Simple organisms such as yeast and other fungi can be irradiated with various wavelengths of UV light, and the effectiveness of each wavelength can then be measured by the number of mutations it induces. When the data are plotted, an **action spectrum** of UV light as a mutagenic agent is obtained. This action spectrum can then be compared with the **absorption spectrum** of any molecule suspected to be genetic material (Figure 9–6). *The molecule serving as the genetic material is expected to absorb at the wavelengths shown to be mutagenic.*

UV light is most mutagenic at the wavelength (λ) of about 260 nanometers (nm), and both DNA and RNA strongly absorb UV light at 260 nm. On the other hand, protein absorbs most strongly around 280 nm, yet no significant mutagenic effects are observed at this wavelength. This indirect evidence also supports the idea that a nucleic acid is the genetic material and tends to exclude protein.

Direct Evidence: Recombinant DNA Studies

Although the circumstantial evidence described earlier does not constitute direct proof that DNA is the genetic material in eukaryotes, these observations spurred researchers to forge ahead, basing their work on this hypothesis. Today, there is no doubt of the validity of this conclusion. DNA *is* the genetic material in eukaryotes. The strongest evidence is provided by molecular analysis utilizing **recombinant DNA technology.** In this procedure, segments of eukaryotic DNA corresponding to specific genes are isolated and spliced into bacterial DNA. This complex can then be inserted into a bacterial cell, and its genetic expression is monitored. If a eukaryotic gene is introduced, the presence of the corresponding eukaryotic protein product demonstrates directly that this DNA is present and functional in the bacterial cell. This has been shown to be the case in countless instances. For example, the products of the human genes specifying insulin and interferon are produced by bacteria after the human genes that encode these proteins are inserted. As the bacterium divides, the eukaryotic DNA replicates along with the host DNA and is distributed to the daughter cells, which also express the human genes and synthesize the corresponding proteins.

The availability of vast amounts of DNA coding for specific genes, derived from recombinant DNA research, has led to other direct evidence that DNA serves as the genetic material. Work done in the laboratory of Beatrice Mintz demonstrated that DNA encoding the human β-globin gene, when microinjected into a fertilized mouse egg, is later found to be present and expressed in adult mouse tissue, and it is transmitted to and expressed in that mouse's progeny. These mice are examples of **transgenic animals,** now commonplace in genetic research. They clearly demonstrate that DNA meets the requirement of expression of genetic information in eukaryotes.

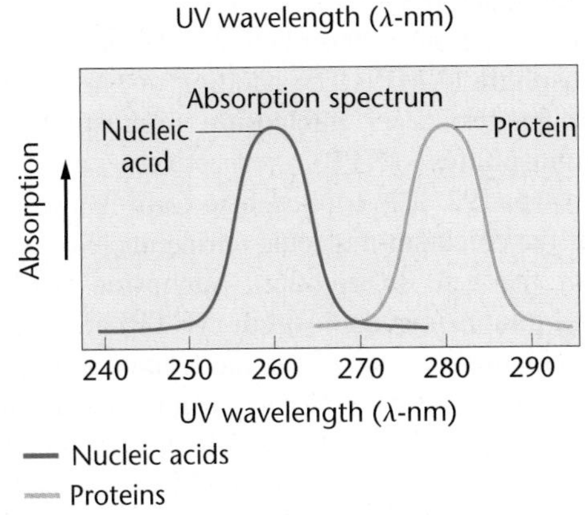

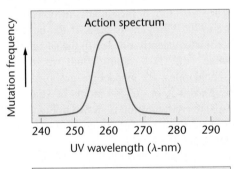

FIGURE 9–6 Comparison of the action spectrum, which determines the most effective mutagenic UV wavelength, and the absorption spectrum, which shows the range of wavelengths where nucleic acids and proteins absorb UV light.

ESSENTIAL POINT

Although initially only indirect observations supported the hypothesis that DNA controls inheritance in eukaryotes, subsequent studies involving recombinant DNA techniques and transgenic mice provided direct experimental evidence that the eukaryotic genetic material is DNA.

9.5 RNA Serves as the Genetic Material in Some Viruses

Some viruses contain an RNA core rather than a DNA core. In these viruses, it appears that RNA serves as the genetic material—an exception to the general rule that DNA performs this function. In 1956, it was demonstrated that when purified RNA from **tobacco mosaic virus (TMV)** was spread on tobacco leaves, the characteristic lesions caused by viral infection subsequently appeared on the leaves. Thus, it was concluded that RNA is the genetic material of this virus.

In 1965 and 1966, Norman Pace and Sol Spiegelman demonstrated further that RNA from the phage Qβ can be isolated and replicated *in vitro.* Replication depends on an enzyme, **RNA replicase,** which is isolated from host *E. coli* cells following normal infection. When the RNA replicated *in vitro* is added to *E. coli* protoplasts, infection and viral multiplication (transfection) occur. Thus, RNA synthesized in a test tube serves as the genetic material in these phages by directing the production of all the components necessary for viral reproduction.

One other group of RNA-containing viruses bears mention. These are the **retroviruses,** which replicate in an unusual way. Their RNA serves as a template for the synthesis of the complementary DNA molecule. The process, **reverse transcription,** occurs under the direction of an RNA-dependent DNA polymerase enzyme called **reverse transcriptase.** This DNA intermediate can be incorporated into the genome of the host cell, and when the host DNA is transcribed, copies of the original retroviral RNA chromosomes are produced. Retroviruses include the human immunodeficiency virus (HIV), which causes AIDS, as well as RNA tumor viruses.

ESSENTIAL POINT

RNA serves as the genetic material in some bacteriophages as well as some plant and animal viruses.

9.6 The Structure of DNA Holds the Key to Understanding Its Function

Having established that DNA is the genetic material in all living organisms (except certain viruses), we turn now to the structure of this nucleic acid. In 1953, James Watson and Francis Crick proposed that the structure of DNA is in the form of a double helix. Their proposal was published in a short paper in the journal *Nature* (reprinted in its entirety on p. 195). In a sense, this publication was the finish of a highly competitive scientific race to obtain what some consider to be the most significant finding in the history of biology. This race, as recounted in Watson's book *The Double Helix,* demonstrates the human interaction, genius, frailty, and intensity involved in the scientific effort that eventually led to the elucidation of DNA structure.

The data available to Watson and Crick, crucial to the development of their proposal, came primarily from two sources: (1) base composition analysis of hydrolyzed samples of DNA and (2) X-ray diffraction studies of DNA. Watson and Crick's analytical success can be attributed to model building that conformed to the existing data. If the correct solution to the structure of DNA is viewed as a puzzle, Watson and Crick, working in the Cavendish Laboratory in Cambridge, England, were the first to fit the pieces together successfully.

DNA STRUCTURE WEB TUTORIAL

Nucleic Acid Chemistry

Before turning to this work, a brief introduction to nucleic acid chemistry is in order. This chemical information was well known to Watson and Crick during their investigation and served as the basis of their model building.

DNA is a nucleic acid, and nucleotides are the building blocks of all nucleic acid molecules. Sometimes called mononucleotides, these structural units have three essential components: a **nitrogenous base,** a **pentose sugar** (a five-carbon sugar), and a **phosphate group.** There are two kinds of nitrogenous bases: the nine-member double-ring **purines** and the six-member single-ring **pyrimidines.**

Two types of purines and three types of pyrimidines are found in nucleic acids. The two purines are **adenine** and **guanine,** abbreviated A and G, respectively. The three pyrimidines are **cytosine, thymine,** and **uracil** (respectively, C, T, and U). The chemical structures of the five bases are shown in Figure 9–7(a). Both DNA and RNA contain A, G, and C, but only DNA contains the base T and only RNA contains the base U. Each nitrogen or carbon atom of the ring structures of purines and pyrimidines is designated by a number. Note that corresponding atoms in the purine and pyrimidine rings are numbered differently.

The pentose sugars found in nucleic acids give them their names. **Ribonucleic acids (RNA)** contain **ribose,** while **deoxyribonucleic acids (DNA)** contain **deoxyribose.** Figure 9–7(b) shows the chemical structures for these two pentose sugars. Each carbon atom is distinguished by a number with a prime sign ($'$). As you can see in Figure 9–7(b), compared with ribose, deoxyribose has a hydrogen atom rather than a hydroxyl group at the C-2$'$ position. The absence of a hydroxyl group at the C-2$'$ position thus distinguishes DNA from RNA. In the absence of the C-2$'$ hydroxyl group, the sugar is more specifically named **2-deoxyribose.**

If a molecule is composed of a purine or pyrimidine base and a ribose or deoxyribose sugar, the chemical unit is called a **nucleoside.** If a phosphate group is added to the nucleoside, the molecule is now called a **nucleotide.** Nucleosides and nucleotides are named according to the specific nitrogenous base (A, G, C, T, or U) that is part of the molecule. The structure of a nucleotide and the nomenclature used in naming DNA nucleotides and nucleosides are shown in Figure 9–8.

The bonding between components of a nucleotide is highly specific. The C-1$'$ atom of the sugar is involved in the chemical linkage to the nitrogenous base. If the base is a purine, the N-9 atom is covalently bonded to the sugar; if the base is a pyrimidine, the N-1 atom bonds to the sugar. In deoxyribonucleotides, the phosphate group may be bonded to the C-2$'$, C-3$'$, or C-5$'$ atom of the sugar. The C-5$'$-phosphate configuration is shown in Figure 9–8. It is by far the most prevalent one in biological systems and the one found in DNA and RNA.

Nucleotides are also described by the term **nucleoside monophosphate (NMP)**. The addition of one or two phosphate groups results in **nucleoside diphosphates (NDP)** and **triphosphates (NTP)**, respectively, as shown in Figure 9–9, p. 192. The triphosphate form is significant because it is the precursor molecule during nucleic acid synthesis within the cell. In addition, **adenosine triphosphate (ATP)** and **guanosine triphosphate (GTP)** are important in cell bioenergetics because of the large amount of energy involved in adding or removing the terminal phosphate group. The hydrolysis of ATP or GTP to ADP or GDP and inorganic phosphate (P_i) is accompanied by the release of a large amount of energy in the cell. When these chemical conversions are coupled to other reactions, the energy produced is used to drive them. As a result, ATP and GTP are involved in many cellular activities.

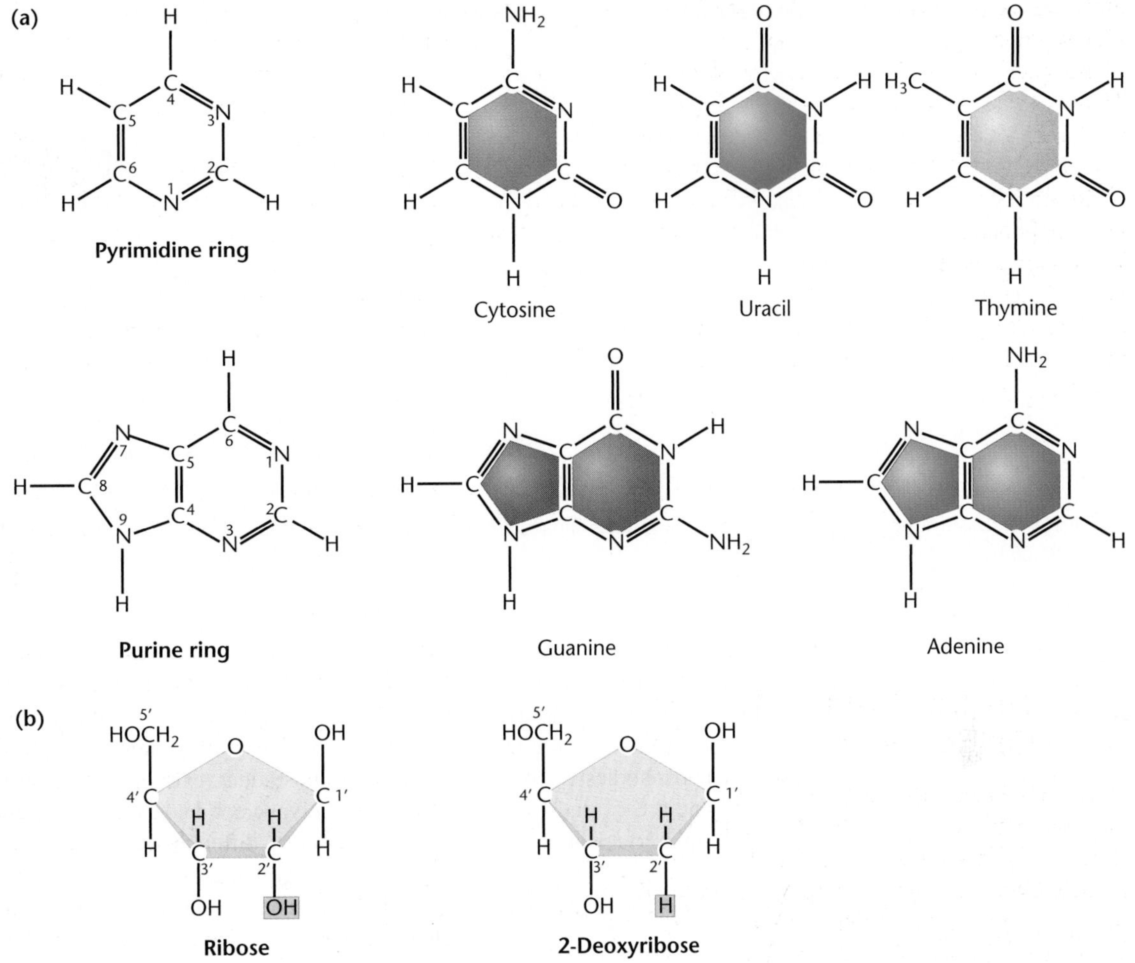

FIGURE 9–7 (a) Chemical structures of the pyrimidines and purines that serve as the nitrogenous bases in RNA and DNA. (b) Chemical ring structures of ribose and 2-deoxyribose, which serve as the pentose sugars in RNA and DNA, respectively.

FIGURE 9–8 Structures and names of the nucleosides and nucleotides of RNA and DNA.

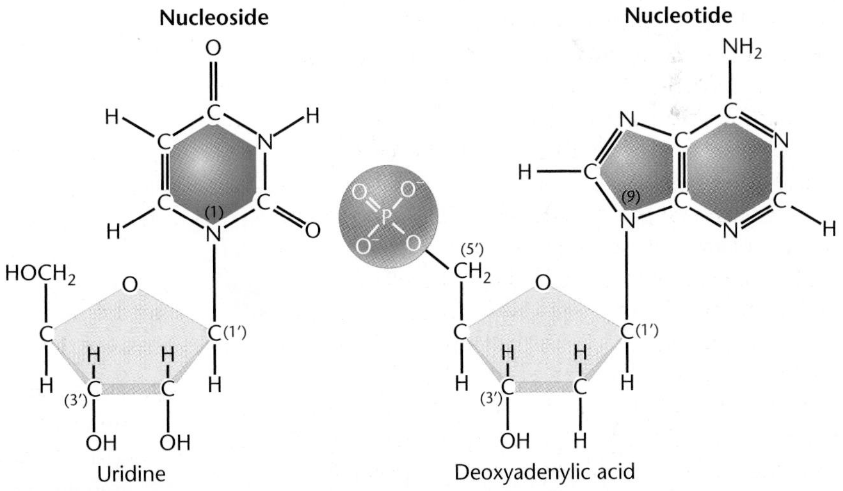

Ribonucleosides	Ribonucleotides
Adenosine Cytidine Guanosine Uridine	Adenylic acid Cytidylic acid Guanylic acid Uridylic acid
Deoxyribonucleosides	**Deoxyribonucleotides**
Deoxyadenosine Deoxycytidine Deoxyguanosine Deoxythymidine	Deoxyadenylic acid Deoxycytidylic acid Deoxyguanylic acid Deoxythymidylic acid

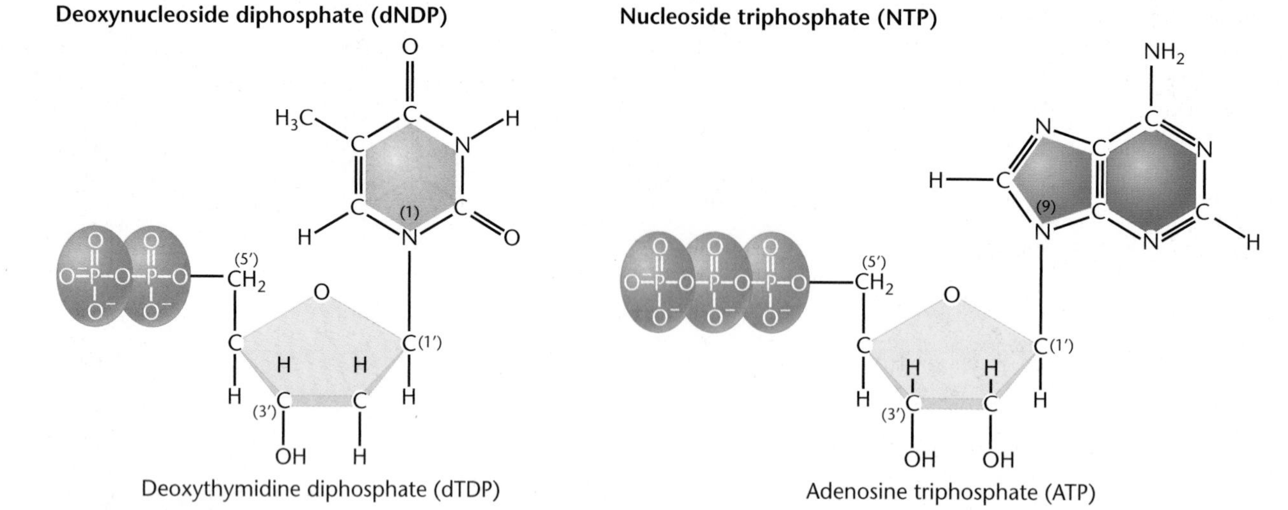

FIGURE 9–9 Basic structures of nucleoside diphosphates and triphosphates. Deoxythymidine diphosphate and adenosine triphosphate are diagrammed here.

The linkage between two mononucleotides involves a phosphate group linked to two sugars. A **phosphodiester bond** is formed as phosphoric acid is joined to two alcohols (the hydroxyl groups on the two sugars) by an ester linkage on both sides. Figure 9–10 shows the resultant phosphodiester bond in DNA. Each structure has a C-3′ end and a C-5′ end. Two joined nucleotides form a dinucleotide; three nucleotides, a trinucleotide; and so forth. Short chains consisting of up to 20 nucleotides or so are called **oligonucleotides;** longer chains are **polynucleotides.**

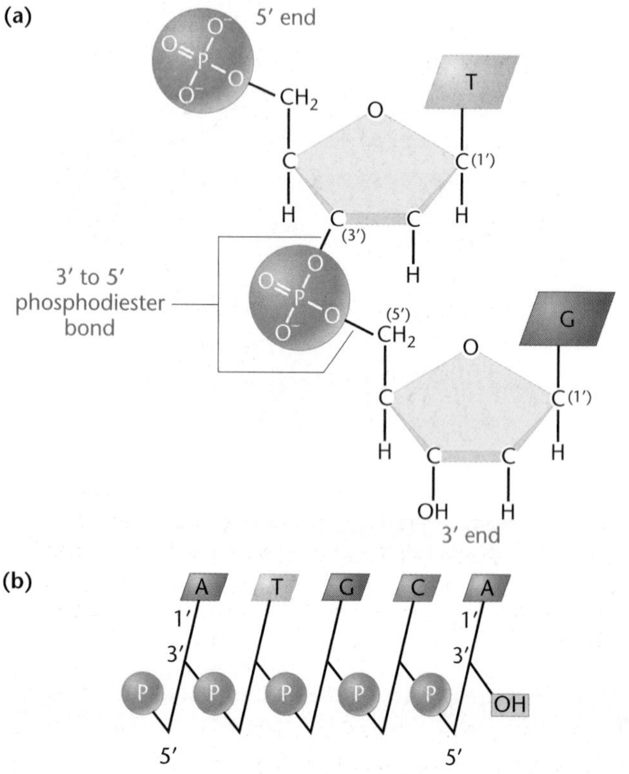

FIGURE 9–10 (a) Linkage of two nucleotides by the formation of a C-3′–C-5′ (3′–5′) phosphodiester bond, producing a dinucleotide. (b) Shorthand notation for a polynucleotide chain.

Long polynucleotide chains account for the large molecular weight of DNA and explain its most important property—storage of vast quantities of genetic information. If each nucleotide position in this long chain can be occupied by any one of four nucleotides, extraordinary variation is possible. For example, a polynucleotide only 1000 nucleotides in length can be arranged 4^{1000} different ways, each one different from all other possible sequences. This potential variation in molecular structure is essential if DNA is to store the vast amounts of chemical information necessary to direct cellular activities.

Base-Composition Studies

Between 1949 and 1953, Erwin Chargaff and his colleagues used chromatographic methods to separate the four bases in DNA samples from various organisms. Quantitative methods were then used to determine the amounts of the four nitrogenous bases from each source. Table 9.3(a) lists some of Chargaff's original data. Parts (b) and (c) show more recently derived information that reinforces Chargaff's findings. As we shall see, Chargaff's data were critical to the successful model of DNA put forward by Watson and Crick. On the basis of these data, the following conclusions may be drawn.

1. The amount of adenine residues is proportional to the amount of thymine residues in DNA (columns 1, 2, and 5). Also, the amount of guanine residues is proportional to the amount of cytosine residues (columns 3, 4, and 6).
2. Based on this proportionality, the sum of the purines (A + G) equals the sum of the pyrimidines (C + T), as shown in column 7.
3. The percentage of (G + C) does not necessarily equal the percentage of (A + T). As you can see, this ratio varies greatly between different organisms, as shown in column 8, and in Part (c).

These conclusions indicate a definite pattern of base composition of DNA molecules. The data provided the initial clue to the DNA puzzle. They also directly refuted Levene's tetra-

TABLE 9.3 DNA Base-Composition Data

(a) Chargaff's Data

	Molar Proportions†			
	1	2	3	4
Source	*A*	*T*	*G*	*C*
Ox thymus	26	25	21	16
Ox spleen	25	24	20	15
Yeast	24	25	14	13
Avian tubercle bacilli	12	11	28	26
Human sperm	29	31	18	18

(b) Base Compositions of DNAs from Various Sources

	Base Composition				Base Ratio		Combined Base Ratios	
	1	2	3	4	5	6	7	8
Source	*A*	*T*	*G*	*C*	*A*/*T*	*G*/*C*	(*A* + *G*)/(*C* + *T*)	(*A* + *T*)/(*C* + *G*)
Human	30.9	29.4	19.9	19.8	1.05	1.00	1.04	1.52
Sea urchin	32.8	32.1	17.7	17.3	1.02	1.02	1.02	1.58
E. coli	24.7	23.6	26.0	25.7	1.04	1.01	1.03	0.93
Sarcina lutea	13.4	12.4	37.1	37.1	1.08	1.00	1.04	0.35
T7 bacteriophage	26.0	26.0	24.0	24.0	1.00	1.00	1.00	1.08

†Moles of nitrogenous constituent per mole of P (often, the recovery was less than 100%).

(c) *G* + *C* Content in Several Organisms

Organism	*G* + *C* (%)
Phage T2	36.0
Drosophila	45.0
Maize	49.1
Euglena	53.5
Neurospora	53.7

nucleotide hypothesis, which stated that all four bases are present in equal amounts.

X-Ray Diffraction Analysis

When fibers of a DNA molecule are subjected to X-ray bombardment, the X rays scatter according to the molecule's atomic structure. The pattern of scatter can be captured as spots on photographic film and analyzed, particularly for the overall shape of and regularities within the molecule. This process, **X-ray diffraction analysis,** was applied successfully to the study of protein structure by Linus Pauling and other chemists. The technique had been attempted on DNA as early as 1938 by William Astbury. By 1947, he had detected a periodicity of 3.4 angstroms (Å)* within the structure of the molecule, which suggested to him that the bases were stacked like coins on top of one another.

Between 1950 and 1953, Rosalind Franklin, working in the laboratory of Maurice Wilkins, obtained improved X-ray data from more purified samples of DNA (Figure 9–11). Her work confirmed the 3.4 Å periodicity seen by Astbury and suggested that the structure of DNA was some sort of helix. However, she did not propose a definitive model. Pauling had analyzed the work of Astbury and others and proposed incorrectly that DNA is a triple helix.

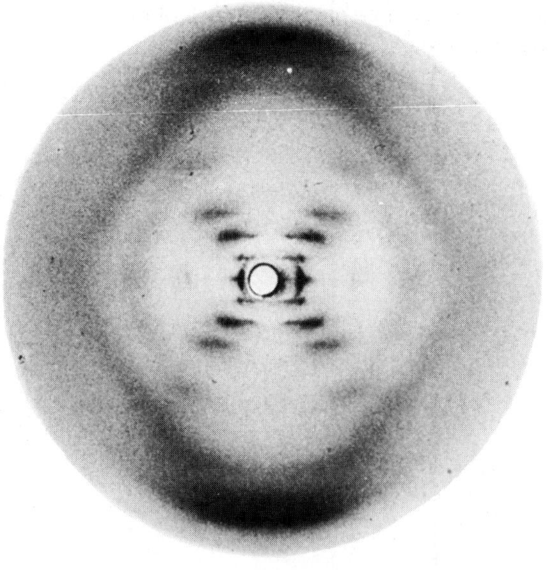

FIGURE 9–11 X-ray diffraction photograph of purified DNA fibers. The strong arcs on the periphery show closely spaced aspects of the molecule, providing an estimate of the periodicity of nitrogenous bases, which are 3.4 Å apart. The inner cross pattern of spots shows the grosser aspect of the molecule, indicating its helical nature.

The Watson–Crick Model

Watson and Crick published their analysis of DNA structure in 1953 (see p. 195). By building models under the constraints of the information just discussed, they proposed the double-helical form of DNA shown in Figure 9–12(a), p. 194. This model has these major features:

1. Two long polynucleotide chains are coiled around a central axis, forming a right-handed double helix.

* Today, measurement in nanometers (nm) is favored (1 nm = 10 Å).

2. The two chains are antiparallel; that is, their C-5′ to C-3′ orientations run in opposite directions.
3. The bases of both chains are flat structures, lying perpendicular to the axis; they are "stacked" on one another, 3.4 Å (0.34 nm) apart, and located on the inside of the structure.
4. The nitrogenous bases of opposite chains are *paired* as the result of hydrogen bonds; in DNA, only A=T and G≡C pairs occur.
5. Each complete turn of the helix is 34 Å (3.4 nm) long; thus, 10 bases exist per turn in each chain.
6. In any segment of the molecule, alternating larger **major grooves** and smaller **minor grooves** are apparent along the axis.
7. The double helix measures 20 Å (2.0 nm) in diameter.

The nature of *base pairing* (point 4 above) is the most genetically significant feature of the model. Before we discuss it in detail, several other important features warrant emphasis. First, the antiparallel nature of the two chains is a key part of the double helix model. While one chain runs in the 5′ to 3′ orientation (what seems right side up to us), the other chain is in the 3′ to 5′ orientation (and thus appears upside down). This is illustrated in Figure 9–12(b) and (c). Given the constraints of the bond angles of the various nucleotide components, the double helix could not be constructed easily if both chains ran parallel to one another.

The key to the model proposed by Watson and Crick is the specificity of base pairing. Chargaff's data suggested that the amounts of A equaled T and that the amounts of G equaled C. Watson and Crick realized that if A pairs with T and C pairs with G, this would account for these proportions and that such pairing could occur as a result of hydrogen bonding between base pairs [Figure 9–12(c)], providing the chemical stability necessary to hold the two chains together. Arranged in this way, both major and minor grooves become apparent along the axis. Further, a purine (A or G) opposite a pyrimidine (T or C) on each "rung of the spiral staircase" of the proposed double helix accounts for the 20 Å (2 nm) diameter suggested by X-ray diffraction studies.

The specific A=T and G≡C base pairing is the basis for **complementarity.** This term describes the chemical affinity provided by hydrogen bonding between the bases. As we shall see, complementarity is very important in DNA replication and gene expression.

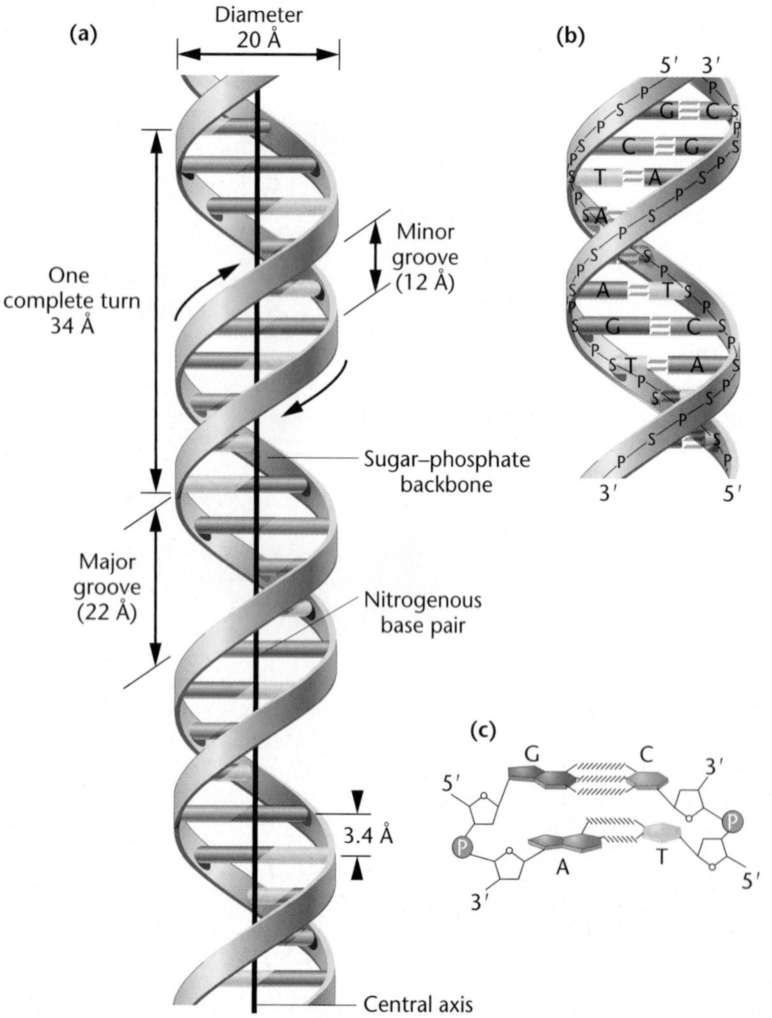

FIGURE 9–12 (a) The DNA double helix as proposed by Watson and Crick. The ribbonlike strands constitute the sugar–phosphate backbones, and the horizontal rungs constitute the nitrogenous base pairs, of which there are 10 per complete turn. The major and minor grooves are shown. The solid vertical bar represents the central axis. (b) A detailed view labeled with the bases, sugars, phosphates, and hydrogen bonds of the helix. (c) A demonstration of the antiparallel nature of the helix and the horizontal stacking of the bases.

Molecular Structure of Nucleic Acids: A Structure for Deoxyribose Nucleic Acid

We wish to suggest a structure for the salt of deoxyribose nucleic acid (DNA). This structure has novel features which are of considerable biological interest. A structure for nucleic acid has already been proposed by Pauling and Corey.[1] They kindly made their manuscript available to us in advance of publication.

Their model consists of three intertwined chains, with the phosphates near the fibre axis, and the bases on the outside. In our opinion, this structure is unsatisfactory for two reasons: (1) We believe that the material which gives the X-ray diagrams is the salt, not the free acid. Without the acidic hydrogen atoms it is not clear what forces would hold the structure together, especially as the negatively charged phosphates near the axis will repel each other. (2) Some of the van der Waals distances appear to be too small.

Another three-chain structure has also been suggested by Fraser (in the press). In his model the phosphates are on the outside and the bases on the inside, linked together by hydrogen bonds. This structure as described is rather ill-defined, and for this reason we shall not comment on it.

We wish to put forward a radically different structure for the salt of deoxyribose nucleic acid. This structure has two helical chains each coiled round the same axis. We have made the usual chemical assumptions, namely, that each chain consists of phosphate diester groups joining β-D-deoxyribofuranose residues with $3', 5'$ linkages. The two chains (but not their bases) are related by a dyad perpendicular to the fibre axis. Both chains follow right-handed helices, but owing to the dyad the sequences of the atoms in the two chains run in opposite directions. Each chain loosely resembles Furberg's[2] model No. 1; that is, the bases are on the inside of the helix and the phosphates on the outside. The configuration of the sugar and the atoms near it is close to Furberg's "standard configuration," the sugar being roughly perpendicular to the attached base. There is a residue on each chain every 3.4 Å in the z-direction. We have assumed an angle of 36° between adjacent residues in the same chain, so that the structure repeats after 10 residues on each chain, that is, after 34 Å. The distance of a phosphate atom from the fibre axis is 10 Å. As the phosphates are on the outside, cations have easy access to them.

The structure is an open one, and its water content is rather high. At lower water content we would expect the bases to tilt so that the structure could become more compact.

The novel feature of the structure is the manner in which the two chains are held together by the purine and pyrimidine bases. The planes of the bases are perpendicular to the fibre axis. They are joined together in pairs, a single base from one chain being hydrogen-bonded to a single base from the other chain, so that the two lie side by side with identical z-coordinates. One of the pair must be a purine and the other a pyrimidine for bonding to occur. The hydrogen bonds are made as follows: purine position 1 to pyrimidine position 1; purine position 6 to pyrimidine position 6.

If it is assumed that the bases only occur in the structure in the most plausible tautomeric forms (that is, with the keto rather than the enol configuration), it is found that only specific pairs of bases can bond together. These pairs are: adenine (purine) with thymine (pyrimidine), and guanine (purine) with cytosine (pyrimidine).

In other words, if an adenine forms one member of a pair, on either chain, then on these assumptions the other member must be thymine; similarly for guanine and cytosine. The sequence of bases on a single chain does not appear to be restricted in any way. However, if only specific pairs of bases can be formed, it follows that if the sequence of bases on one chain is given, then the sequence on the other chain is automatically determined.

It has been found experimentally[3, 4] that the ratio of the amounts of adenine to thymine, and the ratio of guanine to cytosine, are always very close to unity for deoxyribose nucleic acid.

It is probably impossible to build this structure with a ribose sugar in place of deoxyribose, as the extra oxygen atom would make too close a van der Waals contact.

The previously published X-ray data[5, 6] on deoxyribose nucleic acid are insufficient for a rigorous test of our structure. So far as we can tell, it is roughly compatible with the experimental data, but it must be regarded as unproved until it has been checked against more exact results. Some of these are given in the following communications. We were not aware of the details of the results presented there when we devised our structure, which rests mainly though not entirely on published experimental data and stereochemical arguments.

It has not escaped our notice that the specific pairing we have postulated immediately suggests a possible copying mechanism for the genetic material. Full details of the structure, including the conditions assumed in building it, together with a set of co-ordinates for the atoms, will be published elsewhere.

We are much indebted to Dr. Jerry Donohue for constant advice and criticism, especially on interatomic distances. We have also been stimulated by a knowledge of the general nature of the unpublished experimental results and ideas of Dr. M. H. F. Wilkins, Dr. R. E. Franklin and their co-workers at King's College, London. One of us (J. D. W.) has been aided by a fellowship from the National Foundation for Infantile Paralysis.

J. D. Watson
F. H. C. Crick
Medical Research Council Unit for the Study of the Molecular Structure of Biological Systems, Cavendish Laboratory, Cambridge, England

[1]Pauling L., and Corey, R. B., *Nature,* 171, 346 (1953); *Proc. U.S. Nat. Acad. Sci.,* 39, 84 (1953).

[2]Furberg, S., *Acta Chem. Scand.,* 6, 634 (1952).

[3]Chargaff, E., for references see Zamenhof, S., Brawerman, G., and Chargaff, E., *Biochim. et Biophys. Acta,* 9, 402 (1952).

[4]Wyatt, G. R., *J. Gen. Physiol.,* 36, 201 (1952).

[5]Astbury, W. T., *Symp. Soc. Exp. Biol. 1, Nucleic Acid,* 66 (Camb. Univ. Press, 1947).

[6]Wilkins, M. H. F., and Randall, J. T., *Biochim. et Biophys. Acta,* 10, 192 (1953).

It is appropriate to inquire into the nature of a hydrogen bond and to ask whether it is strong enough to stabilize the helix. A **hydrogen bond** is a very weak electrostatic attraction between a covalently bonded hydrogen atom and an atom with an unshared electron pair. The hydrogen atom assumes a partial positive charge, while the unshared electron pair—characteristic of covalently bonded oxygen and nitrogen atoms—assumes a partial negative charge. These opposite charges are responsible for the weak chemical attractions. As oriented in the double helix, adenine forms two hydrogen bonds with thymine, and guanine forms three hydrogen bonds with cytosine. Although two or three individual hydrogen bonds are energetically very weak, 2000 to 3000 bonds in tandem (typical of two long polynucleotide chains) provide great stability to the helix.

Another stabilizing factor is the arrangement of sugars and bases along the axis. In the Watson–Crick model, the *hydrophobic* ("water-fearing") nitrogenous bases are stacked almost horizontally on the interior of the axis and are thus shielded from water. The *hydrophilic* ("water-loving") sugar–phosphate backbone is on the outside of the axis, where both components can interact with water. These molecular arrangements provide significant chemical stabilization to the helix.

A more recent and accurate analysis of the form of DNA that served as the basis for the Watson–Crick model reveals a minor structural difference. A precise measurement of the number of base pairs (bp) per turn has demonstrated a value of 10.4 bp rather than the 10.0 bp predicted by Watson and Crick. In the classical model, each base pair is rotated 36° around the helical axis relative to the adjacent base pair, whereas the new finding requires a rotation of 34.6°. Thus, there are slightly more than 10 bp per turn.

The Watson–Crick model had an immediate effect on the emerging discipline of molecular biology. Even in their initial 1953 article, the authors noted, "It has not escaped our notice that the specific pairing we have postulated immediately suggests a possible copying mechanism for the genetic material." Two months later, Watson and Crick pursued this idea in a second article in *Nature,* suggesting a specific mechanism of replication of DNA—the **semiconservative mode of replication.** The second article alluded to two new concepts: (1) the storage of genetic information in the sequence of the bases, and (2) the mutations or genetic changes that would result from an alteration of the bases. These ideas have received vast amounts of experimental support since 1953 and are now universally accepted.

Watson and Crick's synthesis of ideas was highly significant with regard to subsequent studies of genetics and biology. The nature of the gene and its role in genetic mechanisms could now be viewed and studied in biochemical terms. Recognition of their work, along with that of Wilkins, led to their receipt of the Nobel Prize in Physiology or Medicine in 1962. Unfortunately, Rosalind Franklin had died in 1958 at the age of 37, making her contributions ineligible for consideration since the award is not given posthumously. The Nobel Prize was to be one of many such awards bestowed for work in the field of molecular genetics.

ESSENTIAL POINT

As proposed by Watson and Crick, DNA exists in the form of a right-handed double helix composed of two long antiparallel polynucleotide chains held together by hydrogen bonds formed between complementary, nitrogenous base pairs.

NOW SOLVE THIS

Problem 27 on page 202 asks you to solve the structure of the nucleic acid serving as the genetic material of a hypothetical organism.

Hint: Compare what is known about the Watson–Crick model of DNA when analyzed using the techniques utilized in the problem. Then construct a model of the hypothetical nucleic acid based on the differences that are described.

9.7 Alternative Forms of DNA Exist

Under different conditions of isolation, several conformational forms of DNA have been recognized. At the time Watson and Crick performed their analysis, two forms—**A-DNA** and **B-DNA**—were known. Watson and Crick's analysis was based on X-ray studies of B-DNA performed by Franklin, which is present under aqueous, low-salt conditions and is believed to be the biologically significant conformation.

While DNA studies around 1950 relied on the use of X-ray diffraction, more recent investigations have been performed using **single-crystal X-ray analysis.** The earlier studies achieved limited resolution of about 5 Å, but single crystals diffract X rays at about 1 Å, near atomic resolution. As a result, every atom is "visible" and much greater structural detail is available during analysis.

With this modern technique, A-DNA, which is prevalent under high-salt or dehydration conditions, has now been scrutinized. In comparison to B-DNA (Figure 9–13), A-DNA is slightly more compact, with 11 bp in each complete turn of the helix, which is 23 Å (2.3 nm) in diameter. It is also a right-handed helix, but the orientation of the bases is somewhat different—they are tilted and displaced laterally in relation to the axis of the helix. As a result, the appearance of the major and minor grooves is modified. It seems doubtful that A-DNA occurs *in vivo* (under physiological conditions).

Other forms of DNA (e.g., C-, D-, E-, and most recently, P-DNA) are now known, but it is **Z-DNA** that has drawn the most attention. Discovered by Andrew Wang, Alexander Rich, and their colleagues in 1979 when they examined a small synthetic DNA fragment containing only G≡C base pairs, Z-DNA takes on the rather remarkable configuration of a *left-handed double helix.* Like A- and B-DNA, Z-DNA consists of two antiparallel chains held together by Watson–Crick base pairs. The three forms of DNA are shown in the chapter opening photograph on page 181. Beyond these characteristics, Z-DNA is quite different. The

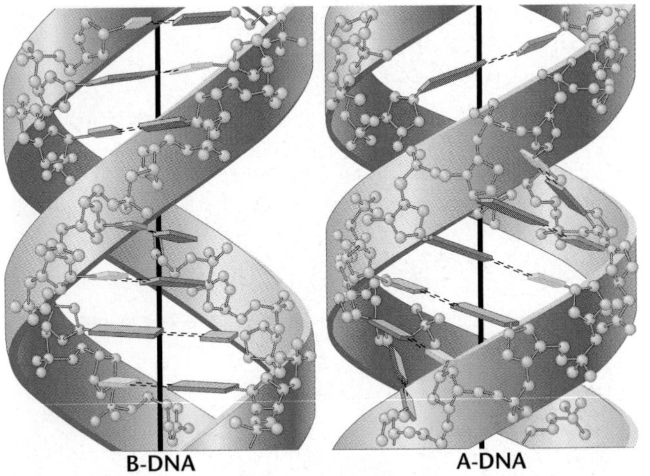

FIGURE 9–13 An artist's rendering showing the orientation of the base pairs of B-DNA and A-DNA. Note that in B-DNA the base pairs are perpendicular to the helix, while they are tilted and pulled away from the helix in A-DNA. In comparison to B-DNA, A-DNA also displays a slightly greater diameter and contains one more base pair per turn.

left-handed helix is 18 Å (1.8 nm) in diameter, contains 12 bp per turn, and assumes a zigzag conformation (hence its name). The major groove present in B-DNA is nearly eliminated in Z-DNA.

Speculation abounds over the possibility that regions of Z-DNA exist in the chromosomes of living organisms. The unique helical arrangement could provide an important recognition point for the interaction with other molecules. However, it is still not clear whether Z-DNA occurs *in vivo.*

Still other forms of DNA have been studied, including P-DNA, named after Linus Pauling. It is produced by artificial "stretching" of DNA, creating a longer, more narrow version with the phosphate groups on the interior.

9.8 The Structure of RNA Is Chemically Similar to DNA, but Single-Stranded

The structure of RNA molecules resembles DNA, with several important exceptions. Although RNA also has nucleotides linked with polynucleotide chains, the sugar ribose replaces deoxyribose, and the nitrogenous base uracil replaces thymine. Another important difference is that most RNA is single-stranded, although there are two important exceptions. First, RNA molecules sometimes fold back on themselves to form double-stranded regions of complementary base pairs. Second, some animal viruses that have RNA as their genetic material contain double-stranded helices.

As established earlier (see Figure 9–1), three major classes of cellular RNA molecules function during the expression of genetic information: **ribosomal RNA (rRNA), messenger RNA (mRNA),** and **transfer RNA (tRNA).** These molecules all originate as complementary copies of deoxyribonucleotide sequences of DNA. Because uracil replaces thymine in RNA, uracil is complementary to adenine during transcription and RNA base pairing.

Different RNAs are distinguished by their sedimentation behavior in a centrifugal field. Sedimentation behavior depends on a molecule's density, mass, and shape, and its measure is called the **Svedberg coefficient** (*S*). Although higher *S* values almost always designate molecules of greater molecular weight, the correlation is not direct; that is, a two-fold increase in molecular weight does not lead to a two-fold increase in *S*.

Ribosomal RNAs are generally the largest of these molecules and usually constitute about 80 percent of all RNA in the cell. Ribosomal RNAs are important structural components of **ribosomes,** which function as nonspecific workbenches where proteins are synthesized during translation. The various forms of rRNA found in prokaryotes and eukaryotes differ distinctly in size.

Messenger RNA molecules carry genetic information from the DNA of the gene to the ribosome. The mRNA molecules vary considerably in size, which reflects the variation in the size of the protein encoded by the mRNA as well as the gene serving as the template for transcription of mRNA.

Transfer RNA, the smallest class of these RNA molecules, carries amino acids to the ribosome during translation. Since more than one tRNA molecule interacts simultaneously with the ribosome, the molecule's smaller size facilitates these interactions.

Other unique RNAs exist that perform various roles, especially in eukaryotes. For example, **small nuclear RNA (snRNA)** participates in processing mRNAs. **Telomerase RNA** is involved in DNA replication at the ends of chromosomes (the telomeres), and **microRNA (miRNA),** and **short interfering RNA (siRNA)** are involved in gene regulation (Chapter 15).

ESSENTIAL POINT

The second category of nucleic acids important in genetic function is RNA, which is similar to DNA with the exceptions that it is usually single stranded, the sugar ribose replaces the deoxyribose, and the pyrimidine uracil replaces thymine.

9.9 Many Analytical Techniques Have Been Useful during the Investigation of DNA and RNA

Since 1953, the role of DNA as the genetic material and the role of RNA in transcription and translation have been clarified through detailed analysis of nucleic acids. Several important methods of analysis are based on the unique nature of the hydrogen bond that is so integral to the structure of nucleic acids. For example, if DNA is subjected to heat, the double helix is denatured and unwinds. During unwinding, the viscosity of DNA decreases and UV absorption increases (called the **hyperchromic shift**). A melting profile, in which OD_{260} is plotted against temperature, is shown for

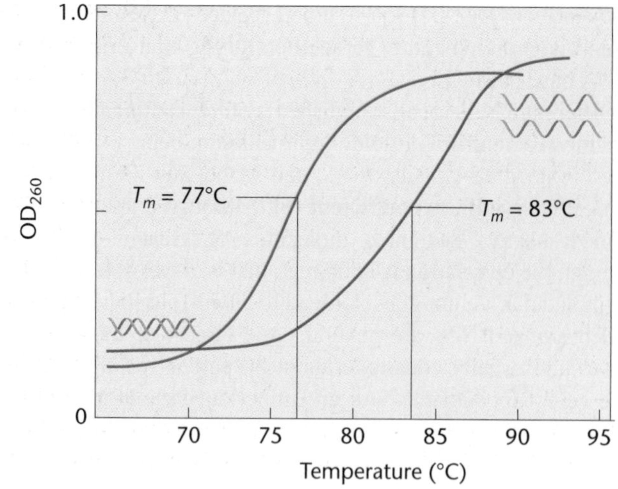

FIGURE 9–14 A graph of increase in UV absorption versus temperature (the hyperchromic effect) for two DNA molecules with different G≡C contents. The molecule with a melting point (T_m) of 83°C has a greater G≡C content than the molecule with T_m = 77°C.

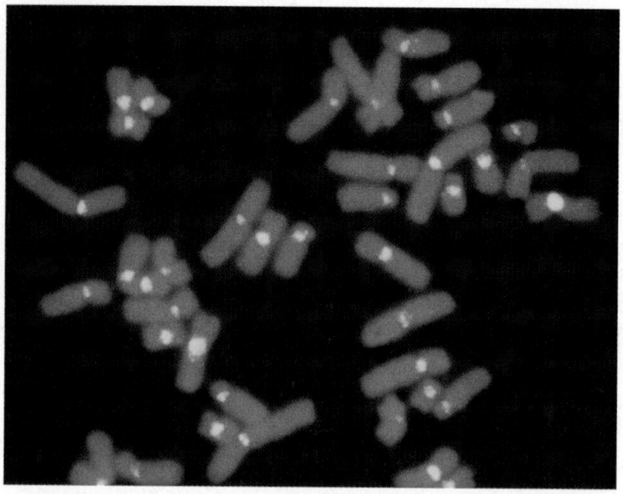

FIGURE 9–15 Fluorescent *in situ* hybridization (FISH) of human metaphase chromosomes. The probe, specific to centromeric DNA, produces a yellow fluorescence signal indicating hybridization. The red fluorescence is produced by propidium iodide counterstaining of chromosomal DNA.

two DNA molecules in Figure 9–14. The midpoint of each curve is called the **melting temperature** (T_m), where 50 percent of the strands have unwound. The molecule with a higher T_m has a higher percentage of G≡C base pairs than A=T base pairs since G≡C pairs share three hydrogen bonds compared to the two bonds between A=T pairs.

The denaturation/renaturation of nucleic acids is the basis for one of the most useful techniques in molecular genetics—**molecular hybridization.** Provided that a reasonable degree of base complementarity exists between any two nucleic acid strands, denaturation can be reversed whereby molecular hybridization is possible. Duplexes can be re-formed between DNA strands, even from different organisms, and between DNA and RNA strands. For example, an RNA molecule will hybridize with the segment of DNA from which it was transcribed. As a result, nucleic acid **probes** are often used to identify complementary sequences.

The technique can even be performed using the DNA present in chromosomal preparations as the "target" for hybrid formation. This process is called ***in situ* molecular hybridization.** Mitotic cells are first fixed to slides and then subjected to hybridization conditions. Single-stranded DNA or RNA is added (a probe), and hybridization is monitored. The nucleic acid that is added may be either radioactive or contain a fluorescent label to allow its detection. In the former case, autoradiography is used.

Figure 9–15 illustrates the use of a fluorescent label. A short fragment of DNA that is complementary to DNA in the chromosomes' centromere regions has been hybridized. Fluorescence occurs only in the centromere regions and thus identifies each one along its chromosome. Because fluorescence is used, the technique is known by the acronym **FISH (fluorescent *in situ* hybridization).** The use of this technique to identify chromosomal locations housing specific genetic information has been a valuable addition to geneticists' repertoire of experimental techniques.

Reassociation Kinetics and Repetitive DNA

In one extension of molecular hybridization procedures, the *rate of reassociation* of complementary single DNA strands is analyzed. This technique, **reassociation kinetics,** was first refined and studied by Roy Britten and David Kohne.

The DNA used in such studies is first fragmented into small pieces by shearing forces introduced during its isolation. The resultant DNA fragments have an average size of several hundred base pairs. The fragments are then dissociated into single strands by heating (denatured), and when the temperature is lowered, reassociation is monitored. During reassociation, pieces of single-stranded DNA randomly collide. If they are complementary, a stable double strand is formed; if not, they separate and are free to encounter other DNA fragments. The process continues until all possible matches are made.

A great deal of information can be obtained from studies that compare the reassociation of DNA of different organisms. As genome size increases and there is more DNA, reassociation time is extended. Reassociation occurs more slowly in larger genomes because with random collisions it takes more time for all correct matches to be made.

When reassociation kinetics in eukaryotic organisms with much larger genome sizes was first studied, a surprising observation was made. Rather than requiring an extended reassociation time, *some* eukaryotic DNA reassociated even more rapidly than those derived from bacteria. The remaining DNA, as expected because of its complexity, took longer to reassociate.

Based on this observation, Britten and Kohne hypothesized that the rapidly reassociating fraction might represent **repetitive DNA sequences.** This interpretation would explain why these segments reassociate so rapidly—multiple copies of the same sequence are much more likely to make matches, thus reassociating more quickly than single copies. On the other hand, the remaining DNA segments consist of **unique DNA sequences,** present only once in the genome.

It is now known that repetitive DNA is prevalent in eukaryotic genomes and is key to our understanding of how genetic information is organized in chromosomes. Careful study has shown that various levels of repetition exist. In some cases, short DNA sequences are repeated over a million times. In other cases, longer sequences are repeated only a few times, or intermediate levels of sequence redundancy are present. We will return to this topic in Chapter 11, where we will discuss the organization of DNA in genes and chromosomes. For now, we will simply point out that the discovery of repetitive DNA was one of the first clues that much of the DNA in eukaryotes is not contained in genes that encode proteins. We will develop and elaborate on this concept as we proceed with our coverage of the molecular basis of heredity.

Electrophoresis

Another technique essential to the analysis of nucleic acids is **electrophoresis.** This technique may be adapted to separate different-sized fragments of DNA and RNA chains and is invaluable in current research investigations in molecular genetics.

Electrophoresis separates the molecules in a mixture by causing them to migrate under the influence of an electric field. A sample is placed on a porous substance, such as a semisolid gel, which is then placed in a solution that conducts electricity. Mixtures of molecules with a similar charge–mass ratio but of different sizes will migrate at different rates through the gel based on their size. For example, two polynucleotide chains of different lengths, such as 10 versus 20 nucleotides, are both negatively charged (based on the phosphate groups of the nucleotides) and will both move to the positively charged pole (the anode), but at different rates. Using a medium such as a **polyacrylamide gel** or an **agarose gel,** which can be prepared with various pore sizes, the *shorter chains migrate at a faster rate through the gel than larger chains* (Figure 9–16). Once electrophoresis is complete, bands representing the variously sized molecules are identified either by autoradiography (if a component of the molecule is radioactive) or by use of a fluorescent dye that binds to nucleic acids. The resolving power is so great that polynucleotides that vary by just one nucleotide in length may be separated.

Electrophoretic separation of nucleic acids is at the heart of a variety of other commonly used research techniques. Of particular note are the various "blotting" techniques (e.g., Southern blots and Northern blots), as well as DNA sequencing methods, which we will discuss in detail later in the text (see Chapter 17).

> ESSENTIAL POINT
>
> Various methods of analysis of nucleic acids, particularly molecular hybridization and electrophoresis, have led to studies essential to our understanding of genetic mechanisms.

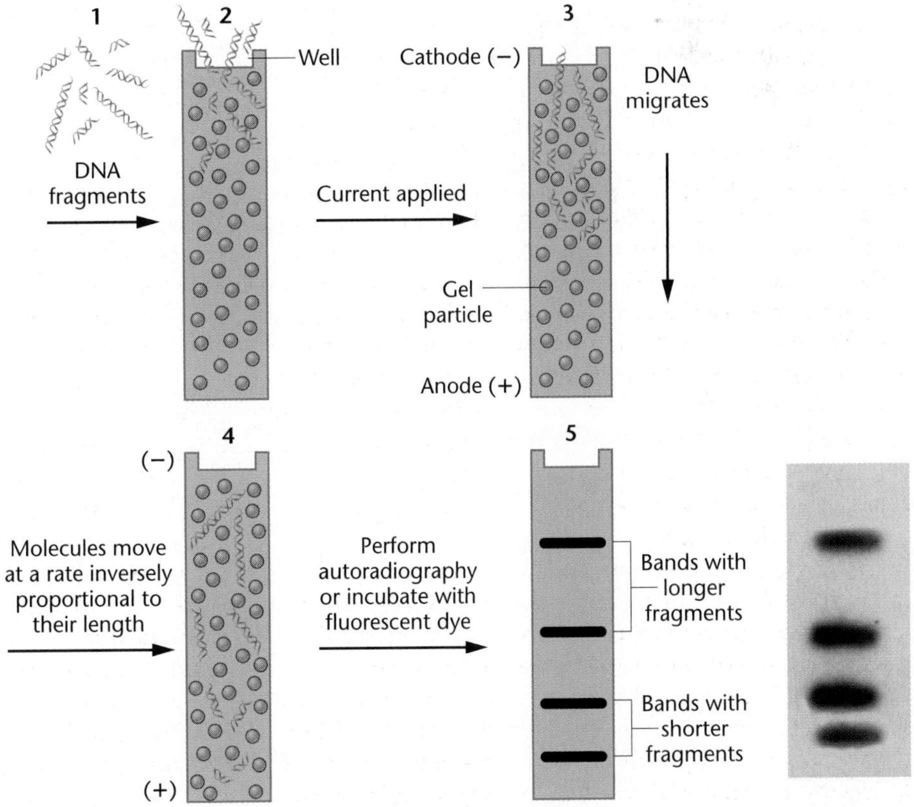

FIGURE 9–16 Electrophoretic separation of a mixture of DNA fragments that vary in length. The photograph at the bottom right shows an agarose gel with DNA bands.

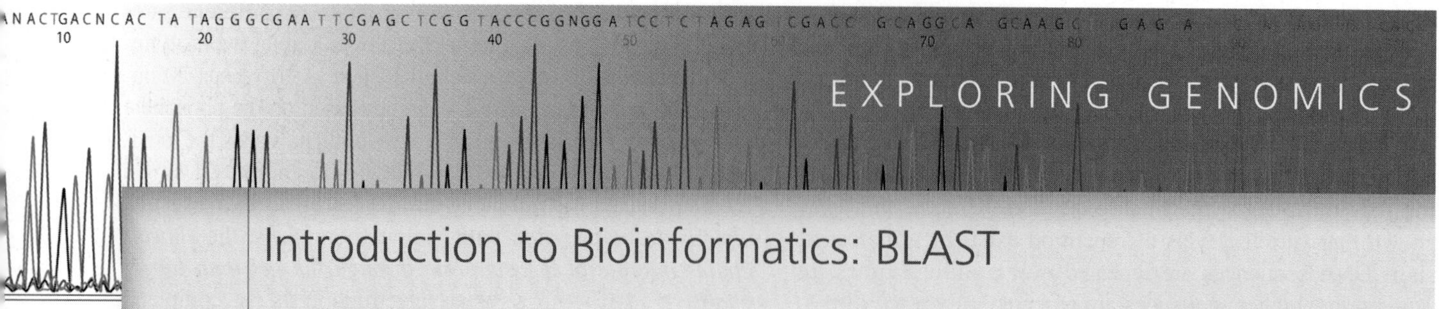

Introduction to Bioinformatics: BLAST

In this chapter, we focused on the structural details of DNA, the genetic material for living organisms. In Chapter 17, you will learn how scientists can clone and sequence DNA—a routine technique in molecular biology and genetics laboratories. The explosion of DNA and protein sequence data that has occurred in the last 15 years has launched the field of *bioinformatics,* an interdisciplinary science that applies mathematics and computing technology to develop hardware and software for storing, sharing, comparing, and analyzing nucleic acid and protein sequence data.

A large number of sequence databases that make use of bioinformatics have been developed. An example is **GenBank (www.ncbi.nlm.nih.gov/Genbank/index.html)**, which is the National Institutes of Health sequence database. This global resource, with access to databases in Europe and Japan, currently contains more than 65 billion base pairs of sequence data!

In the Exploring Genomics exercises for Chapter 7, you were introduced to the National Center for Biotechnology Information (NCBI) Genes and Disease site. Now we will use an NCBI application called **BLAST, Basic Local Alignment Search Tool.** BLAST is an invaluable program for searching through GenBank and other databases to find DNA- and protein-sequence similarities between cloned substances. It has many additional functions that we will explore in other exercises.

■ Exercise I: Introduction to BLAST

1. Access BLAST from the NCBI Web site at **www.ncbi.nlm.nih.gov/BLAST/.**
2. Click on "nucleotide blast." This feature allows you to search DNA databases to look for a similarity between a sequence you enter and other sequences in the database. Do a nucleotide search with the following sequence:
   CCAGAGTCCAGCTGCTGCTCATA
   CTACT GATACTGCTGGG
3. Imagine that this sequence is a short part of a gene you cloned in your laboratory. You want to know if this gene or others with similar sequences have been discovered. Enter this sequence into the "Enter Query Sequence" text box at the top of the page. Near the bottom of the page, under the "Program Selection" category, choose "blastn"; then click on the "BLAST" button at the bottom of the page to run the search. It may take several minutes for results to be available because BLAST is using powerful algorithms to scroll through billions of bases of sequence data! A new page will appear with the results of your search.
4. Near the top of this page you will see a table showing significant matches to the sequence you searched with (called the query sequence). BLAST determines significant matches based on statistical measures that consider the length of the query sequence, the number of matches with sequences in the database, and other factors. Significant *alignments,* regions of significant similarity in the query and subject sequences, typically have E values less than 1.0.
5. The top part of the table lists matches to transcripts (mRNA sequences), and the lower part lists matches to genomic DNA sequences, in order of highest to lowest number of matches. Use the "Links" column to the far right of this table to explore gene and chromosome databases relevant to the matched sequences.
6. Alignments are indicated by horizontal lines. BLAST adjusts for gaps in the sequences, that is, for areas that may not align precisely because of missing bases in otherwise similar sequences. Scroll below the table to see the aligned sequences from this search, and then answer the following questions:
   a. What were the top three matches to your query sequence?
   b. For each alignment, BLAST also indicates the percent *identity* and the number of gaps in the match between the query and subject sequences (shown in the column under "Max ident"). What was the percent identity for the top three matches? What percentage of each aligned sequence showed gaps indicating sequence differences?
   c. Click on the links for the first matched sequence (far-right column). These will take you to a wealth of information, including the size of the sequence; the species it was derived from; a PubMed-linked chronology of research publications pertaining to this sequence; the complete sequence; and if the sequence encodes a polypeptide, the predicted amino acid sequence coded by the gene. Skim through the information presented for this gene. What is the gene's function?
7. A BLAST search can also be done by entering the *accession number* for a sequence, which is a unique identifying number assigned to a sequence before it can be put into a database. For example, search with the accession number NM_007305. What did you find?
8. Run a BLAST search using the sequences or accession numbers listed below. In each case, after entering the accession number or sequence in the "Enter Query Sequence" box, go to the "Choose Search Set" box and click on the "Others" button for database. Then go to the "Program Selection" box and click "megablast" before running your search. These features will allow you to align the query sequence with similar genes from a number of other species. When each search is completed, explore the information BLAST provides so that you can identify and learn about the gene encoded by the sequence.
   a. NM_001006650. What is the top sequence that aligns with the query sequence of this accession number and shows 100 percent sequence identity?
   b. DQ991619. What gene is encoded by this sequence?
   c. NC_007596. What living animal has a sequence similar to this one?

CASE STUDY Zigs and zags of the smallpox virus

Smallpox, a once highly lethal contagious disease, has been eradicated worldwide. However, research continues with stored samples of variola, the smallpox virus, because it is a potential weapon in bioterrorism. Human cells protect themselves from the variola virus (and other viruses) by activating genes that encode protective proteins. It has recently been discovered that in response to variola, human cells create small transitory stretches of Z-DNA at sites that regulate these genes. The smallpox virus can bypass this cellular defense mechanism by specifically targeting the segments of Z-DNA and inhibiting the synthesis of the protective proteins. This discovery raises some interesting questions:

1. What is unique about Z-DNA that might make it a specific target during viral infection?
2. How might the virus target host-cell Z-DNA formation to block the synthesis of antiviral proteins?
3. To study the interaction between viral proteins and Z-DNA, how could Z DNA-forming DNA be synthesized in the lab?
4. How could this research lead to the development of drugs to combat infection by variola and related viruses?

INSIGHTS AND SOLUTIONS

In contrast to preceding chapters, this chapter does not emphasize genetic problem solving. Instead, it recounts some of the initial experimental analyses that launched the era of molecular genetics. Quite fittingly, then, our "Insights and Solutions" section shifts its emphasis to an experimental rationale and analytical thinking, an approach that will continue throughout the remainder of the text, whenever appropriate.

1. Based strictly on the transformation analysis of Avery, MacLeod, and McCarty, what objection might be made to the conclusion that DNA is the genetic material? What other conclusion might be considered?

Solution: Based solely on their results, we could conclude that DNA is essential for transformation. However, DNA might have been a substance that caused capsular formation by converting nonencapsulated cells *directly* to cells with a capsule. That is, DNA may simply have played a catalytic role in capsular synthesis, leading to cells that display smooth, type III colonies.

2. What observations argue against this objection?

Solution: First, transformed cells pass the trait on to their progeny cells, thus supporting the conclusion that DNA is responsible for heredity, not for the direct production of polysaccharide coats. Second, subsequent transformation studies over the next five years showed that other traits, such as antibiotic resistance, could be transformed. Therefore, the transforming factor has a broad general effect, not one specific to polysaccharide synthesis.

3. If RNA were the universal genetic material, how would this have affected the Avery experiment and the Hershey–Chase experiment?

Solution: In the Avery experiment, ribonuclease (RNase), rather than deoxyribonuclease (DNase), would have eliminated transformation. Had this occurred, Avery and his colleagues would have concluded that RNA was the transforming factor. Hershey and Chase would have obtained identical results, since ^{32}P would also label RNA but not protein.

4. Sea urchin DNA, which is double stranded, contains 17.5 percent of its bases in the form of cytosine (C). What percentages of the other three bases are expected to be present in this DNA?

Solution: The amount of C equals G, so guanine is also present at 17.5 percent. The remaining bases, A and T, are present in equal amounts, and together they represent the remaining bases (100–35). Therefore, $A = T = 65/2 = 32.5$ percent.

PROBLEMS AND DISCUSSION QUESTIONS

1. The functions ascribed to the genetic material are replication, expression, storage, and mutation. What does each of these terms mean?
2. Discuss the reasons why proteins were generally favored over DNA as the genetic material before 1940. What was the role of the tetranucleotide hypothesis in this controversy?
3. Contrast the various contributions made to our understanding of transformation by Griffith, Alloway, and Avery.
4. When Avery and his colleagues had obtained what was concluded to be the transforming factor from the III*S* virulent cells, they treated the fraction with proteases, ribonuclease, and deoxyribonuclease, followed by the assay for retention or loss of transforming ability. What were the purpose and results of these experiments? What conclusions were drawn?
5. Why were ^{32}P and ^{35}S chosen in the Hershey–Chase experiment? Discuss the rationale and conclusions of this experiment.
6. Does the design of the Hershey–Chase experiment distinguish between DNA and RNA as the molecule serving as the genetic material? Why or why not?
7. Would an experiment similar to that performed by Hershey and Chase work if the basic design were applied to the phenomenon of transformation? Explain why or why not. See **Now Solve This** on page 188.
8. What observations are consistent with the conclusion that DNA serves as the genetic material in eukaryotes? List and discuss them.
9. What are the exceptions to the general rule that DNA is the genetic material in all organisms? What evidence supports these exceptions?
10. Draw the chemical structure of the three components of a nucleotide, and then link them together. What atoms are removed from the structures when the linkages are formed?
11. How are the carbon and nitrogen atoms of the sugars, purines, and pyrimidines numbered?
12. Adenine may also be named 6-amino purine. How would you name the other four nitrogenous bases, using this alternative system? (O is oxy, and CH_3 is methyl.)
13. Draw the chemical structure of a dinucleotide composed of A and G. Opposite this structure, draw the dinucleotide composed

of T and C in an antiparallel (or upside-down) fashion. Form the possible hydrogen bonds.

14. Describe the various characteristics of the Watson–Crick double helix model for DNA.
15. What evidence did Watson and Crick have at their disposal in 1953? What was their approach in arriving at the structure of DNA?
16. What might Watson and Crick have concluded, had Chargaff's data from a single source indicated the following base composition?

	A	*T*	*G*	*C*
%	29	19	21	31

Why would this conclusion be contradictory to Wilkins and Franklin's data?

17. How do covalent bonds differ from hydrogen bonds? Define base complementarity.
18. List three main differences between DNA and RNA.
19. What are the three major types of RNA molecules? How is each related to the concept of information flow?
20. What component of the nucleotide is responsible for the absorption of ultraviolet light? How is this technique important in the analysis of nucleic acids?
21. What is the physical state of DNA after being denatured by heat?
22. What is the hyperchromic effect? How is it measured? What does T_m imply?
23. Why is T_m related to base composition?
24. What is the chemical basis of molecular hybridization?
25. What did the Watson–Crick model suggest about the replication of DNA?
26. A genetics student was asked to draw the chemical structure of an adenine- and thymine-containing dinucleotide derived from DNA. His answer is shown below. The student made more than six major errors. One of them is circled, numbered 1, and explained. Find five others. Circle them, number them 2–6, and briefly explain each by following the example given.

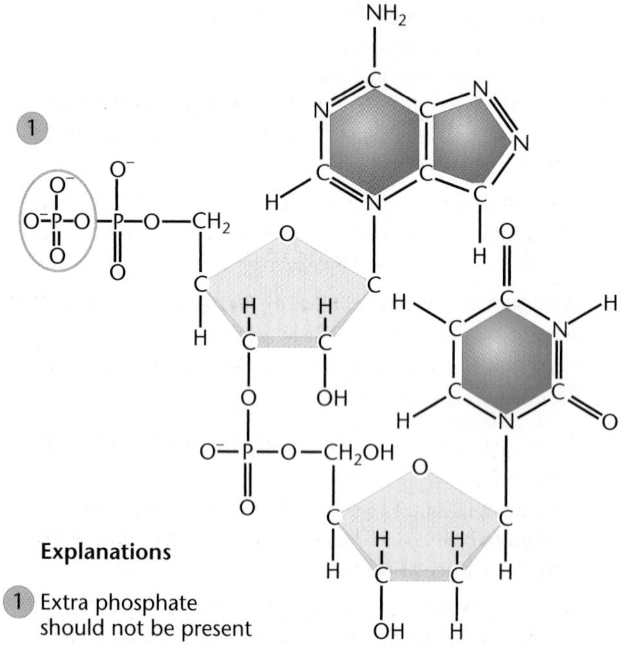

27. A primitive eukaryote was discovered that displayed a unique nucleic acid as its genetic material. Analysis revealed the following observations:
    (a) X-ray diffraction studies display a general pattern similar to DNA, but with somewhat different dimensions and more irregularity.
    (b) A major hyperchromic shift is evident upon heating and monitoring UV absorption at 260 nm.
    (c) Base-composition analysis reveals four bases in the following proportions:

Adenine	= 8%	Hypoxanthine	= 18%
Guanine	= 37%	Xanthine	= 37%

    (d) About 75 percent of the sugars are deoxyribose, whereas 25 percent are ribose. Attempt to solve the structure of this molecule by postulating a model that is consistent with the foregoing observations. See **Now Solve This** on page 196.

28. One of the most common spontaneous lesions that occurs in DNA under physiological conditions is the hydrolysis of the amino group of cytosine, converting it to uracil. What would be the effect on DNA structure if a uracil group replaced cytosine?
29. In some organisms, cytosine is methylated at carbon 5 of the pyrimidine ring after it is incorporated into DNA. If a 5-methyl cytosine is then hydrolyzed, as described in Problem 28, what base will be generated?
30. *Newsdate: March 1, 2015.* A unique creature has been discovered during exploration of outer space. Recently, its genetic material has been isolated and analyzed, and has been found to be similar in some ways to DNA in chemical makeup. It contains in abundance the 4-carbon sugar erythrose and a molar equivalent of phosphate groups. In addition, it contains six nitrogenous bases: adenine (A), guanine (G), thymine (T), cytosine (C), hypoxanthine (H), and xanthine (X). These bases exist in the following relative proportions:

$$A = T = H \quad \text{and} \quad C = G = X$$

X-ray diffraction studies have established a regularity in the molecule and a constant diameter of about 30 Å.

Together, these data have suggested a model for the structure of this molecule. (a) Propose a general model of this molecule, and briefly describe it. (b) What base-pairing properties must exist for H and for X in the model? (c) Given the constant diameter of 30 Å, do you think *either* (i) both H and X are purines or both pyrimidines, *or* (ii) one is a purine and one is a pyrimidine?

31. You are provided with DNA samples from two newly discovered bacterial viruses. Based on the various analytical techniques discussed in this chapter, construct a research protocol that would be useful in characterizing and contrasting the DNA of both viruses. Indicate the type of information you hope to obtain for each technique included in the protocol.
32. During electrophoresis, DNA molecules can easily be separated according to size because all DNA molecules have the same charge–mass ratio and the same shape (long rod). Would you expect RNA molecules to behave in the same manner as DNA during electrophoresis? Why or why not?
33. Considering the information in this chapter on B- and Z-DNA and right- and left-handed helices, carefully analyze structures (a) and (b) below and draw conclusions about their helical nature. Which is right handed and which is left handed?

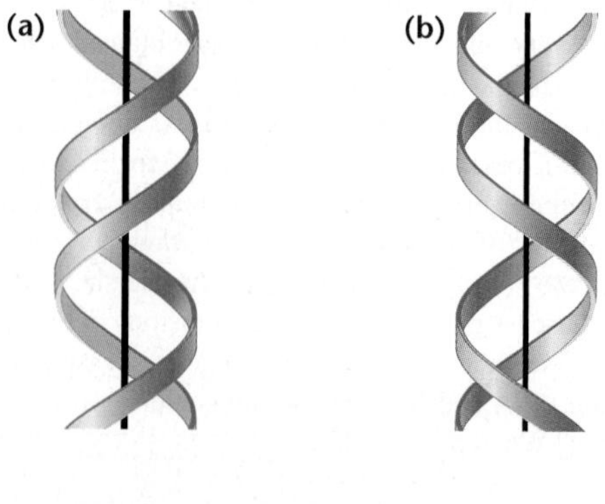

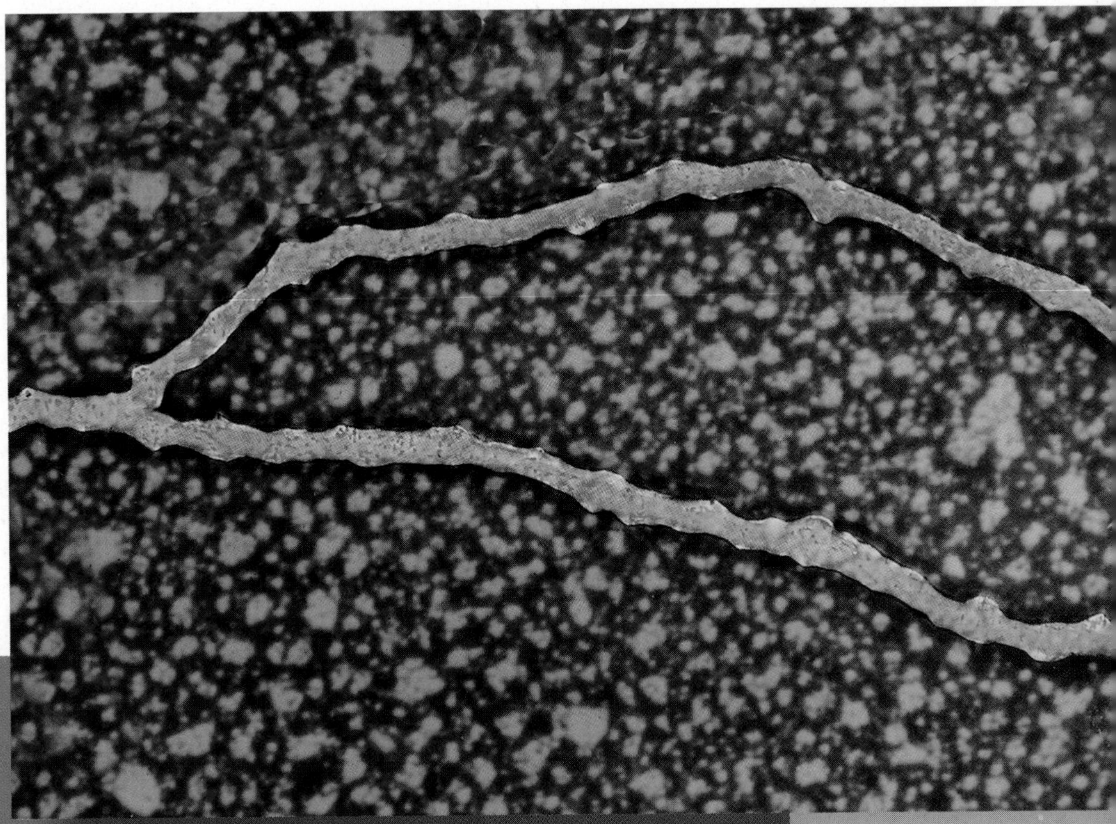

Transmission electron micrograph of human DNA from a HeLa cell, illustrating a replication fork characteristic of active DNA replication.

10

DNA Replication and Recombination

Chapter Concepts

- Genetic continuity between parental and progeny cells is maintained by semiconservative replication of DNA, as predicted by the Watson–Crick model.
- Semiconservative replication uses each strand of the parent double helix as a template, and each newly replicated double helix includes one "old" and one "new" strand of DNA.
- DNA synthesis is a complex but orderly process, occurring under the direction of a myriad of enzymes and other proteins.
- DNA synthesis involves the polymerization of nucleotides into polynucleotide chains.
- DNA synthesis is similar in prokaryotes and eukaryotes, but more complex in eukaryotes.
- In eukaryotes, DNA synthesis at the ends of chromosomes (telomeres) poses a special problem, overcome by a unique RNA-containing enzyme, telomerase.
- Genetic recombination, an important process leading to the exchange of segments between DNA molecules, occurs under the direction of a group of enzymes.

Following Watson and Crick's proposal for the structure of DNA, scientists focused their attention on how this molecule is replicated. Replication is an essential function of the genetic material and must be executed precisely if genetic continuity between cells is to be maintained following cell division. It is an enormous, complex task. Consider for a moment that more than 3×10^9 (3 billion) base pairs exist within the human genome. To duplicate faithfully the DNA of just one of these chromosomes requires a mechanism of extreme precision. Even an error rate of only 10^{-6} (one in a million) will still create 3000 errors (obviously an excessive number) during each replication cycle of the genome. Although it is not error free, and much of evolution would not have occurred if it were, an extremely accurate system of DNA replication has evolved in all organisms.

As Watson and Crick noted in the concluding paragraph of their 1953 paper (reprinted on page 195, their proposed model of the double helix provided the initial insight into how replication occurs. Called *semiconservative replication,* this mode of DNA duplication was soon to receive strong support from numerous studies of viruses, prokaryotes, and eukaryotes. Once the general *mode* of replication was clarified, research to determine the precise details of *DNA synthesis* intensified. What has since been discovered is that numerous enzymes and other proteins are needed to copy a DNA helix. Because of the complexity of the chemical events during synthesis, this subject remains an extremely active area of research.

In this chapter, we will discuss the general mode of replication, as well as the specific details of DNA synthesis. The research leading to such knowledge is another link in our understanding of life processes at the molecular level.

How Do We Know?

In this chapter, we will focus on how DNA is replicated and synthesized. We shall elucidate the general mechanism of replication and describe how DNA is synthesized when it is copied. As you study this topic, you should try to answer several fundamental questions:

1. What is the experimental basis for concluding that DNA replicates semiconservatively in both prokaryotes and eukaryotes?
2. How was it demonstrated that DNA synthesis occurs under the direction of DNA polymerase III and not polymerase I?
3. How do we know that *in vivo* DNA synthesis occurs in the 5′ to 3′ direction?
4. How do we know that DNA synthesis is discontinuous on one of the two template strands?
5. What observations reveal that a "telomere problem" exists during eukaryotic DNA replication, and how did we learn of the solution to this problem?

10.1 DNA Is Reproduced by Semiconservative Replication

Watson and Crick recognized that, because of the arrangement and nature of the nitrogenous bases, each strand of a DNA double helix could serve as a template for the synthesis of its complement (Figure 10–1). They proposed that, if the helix were unwound, each nucleotide along the two parent strands would have an affinity for its complementary nucleotide. As we learned in Chapter 9, the complementarity is due to the potential hydrogen bonds that can be formed. If thymidylic acid (T) were present, it would "attract" adenylic acid (A); if guanidylic acid (G) were present, it would attract cytidylic acid (C); likewise, A would attract T, and C would attract G. If these nucleotides were then covalently linked into polynucleotide chains along both templates, the result would be the production of two identical double strands of DNA. Each replicated DNA molecule would consist of one "old" and one "new" strand, hence the reason for the name **semiconservative replication.**

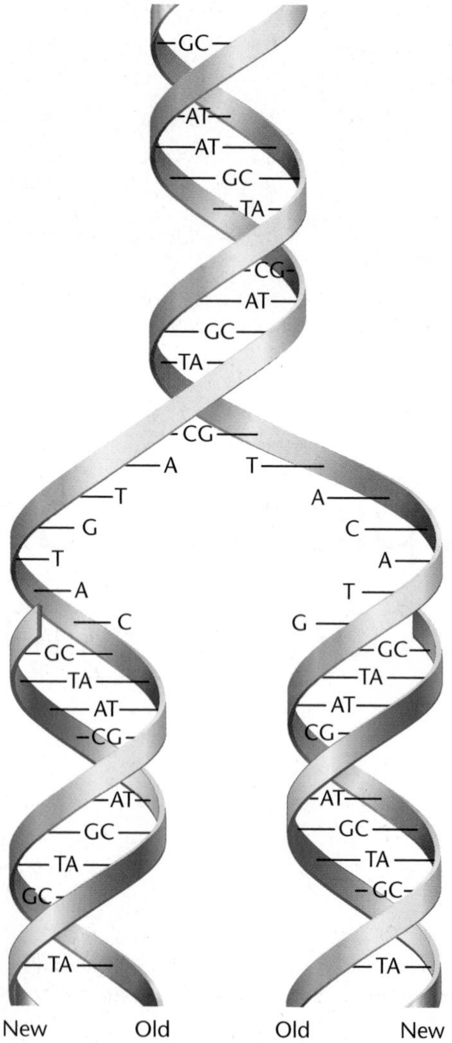

FIGURE 10–1 Generalized model of semiconservative replication of DNA. New synthesis is shown in blue.

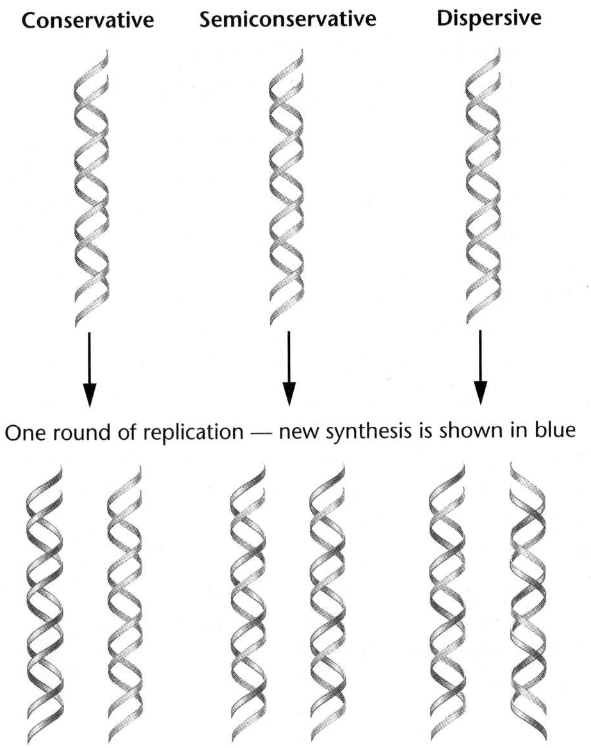

FIGURE 10–2 Results of one round of replication of DNA for each of the three possible modes by which replication could be accomplished.

Two other theoretical modes of replication are possible that also rely on the parental strands as a template (Figure 10–2). In **conservative replication,** complementary polynucleotide chains are synthesized as described earlier. Following synthesis, however, the two newly created strands then come together and the parental strands reassociate. The original helix is thus "conserved."

In the second alternative mode, called **dispersive replication,** the parental strands are dispersed into two new double helices following replication. Hence, each strand consists of both old and new DNA. This mode would involve cleavage of the parental strands during replication. It is the most complex of the three possibilities and is therefore considered to be least likely to occur. It could not, however, be ruled out as an experimental model. Figure 10–2 shows the theoretical results of a single round of replication by each of the three different modes.

The Meselson–Stahl Experiment

In 1958, Matthew Meselson and Franklin Stahl published the results of an experiment providing strong evidence that semiconservative replication is the mode used by bacterial cells to produce new DNA molecules. They grew *E. coli* cells for many generations in a medium that had $^{15}NH_4Cl$ (ammonium chloride) as the only nitrogen source. A "heavy" isotope of nitrogen, ^{15}N contains one more neutron than the naturally occurring ^{14}N isotope; thus, molecules containing ^{15}N are more dense than those containing ^{14}N. Unlike radioactive isotopes, ^{15}N is stable. After many generations in this medium, almost all nitrogen-containing molecules in the *E. coli* cells, including the nitrogenous bases of DNA, contained the heavier isotope.

Critical to the success of this experiment, DNA containing ^{15}N can be distinguished from DNA containing ^{14}N. The experimental procedure involves the use of a technique referred to as **sedimentation equilibrium centrifugation,** or as it is also called, *buoyant density gradient centrifugation.* Samples are forced by centrifugation through a density gradient of a heavy metal salt, such as cesium chloride. Molecules of DNA will reach equilibrium when their density equals the density of the gradient medium. In this case, ^{15}N-DNA will reach this point at a position closer to the bottom of the tube than will ^{14}N-DNA.

In this experiment (Figure 10–3, p. 206), uniformly labeled ^{15}N cells were transferred to a medium containing only $^{14}NH_4Cl$. Thus, all "new" synthesis of DNA during replication contained only the "lighter" isotope of nitrogen. The time of transfer to the new medium was taken as time zero ($t = 0$). The *E. coli* cells were allowed to replicate over several generations, with cell samples removed after each replication cycle. DNA was isolated from each sample and subjected to sedimentation equilibrium centrifugation.

After one generation, the isolated DNA was present in only a single band of intermediate density—the expected result for semiconservative replication in which each replicated molecule was composed of one new ^{14}N-strand and one old ^{15}N-strand (Figure 10–4, p. 206). This result was not consistent with the prediction of conservative replication, in which two distinct bands would occur; thus this mode may be rejected.

After two cell divisions, DNA samples showed two density bands—one intermediate band and one lighter band corresponding to the ^{14}N position in the gradient. Similar results occurred after a third generation, except that the proportion of the lighter band increased. This was again consistent with the interpretation that replication is semiconservative.

You may have realized that a molecule exhibiting intermediate density is also consistent with dispersive replication. However, Meselson and Stahl ruled out this mode of replication on the basis of two observations. First, after the first generation of replication in an ^{14}N-containing medium, they isolated the hybrid molecule and heat denatured it. Recall from Chapter 9 that heating will separate a duplex into single strands. When the densities of the single strands of the hybrid were determined, they exhibited *either* an ^{15}N profile *or* an ^{14}N profile, but *not* an intermediate density. This observation is consistent with the semiconservative mode but inconsistent with the dispersive mode.

Furthermore, if replication were dispersive, *all* generations after $t = 0$ would demonstrate DNA of an intermediate density. In each generation after the first, the ratio of $^{15}N/^{14}N$ would decrease, and the hybrid band would become lighter and lighter, eventually approaching the ^{14}N band. This result was not observed. The Meselson–Stahl experiment provided conclusive support for semiconservative replication in bacteria and tended to rule out both the conservative and dispersive modes.

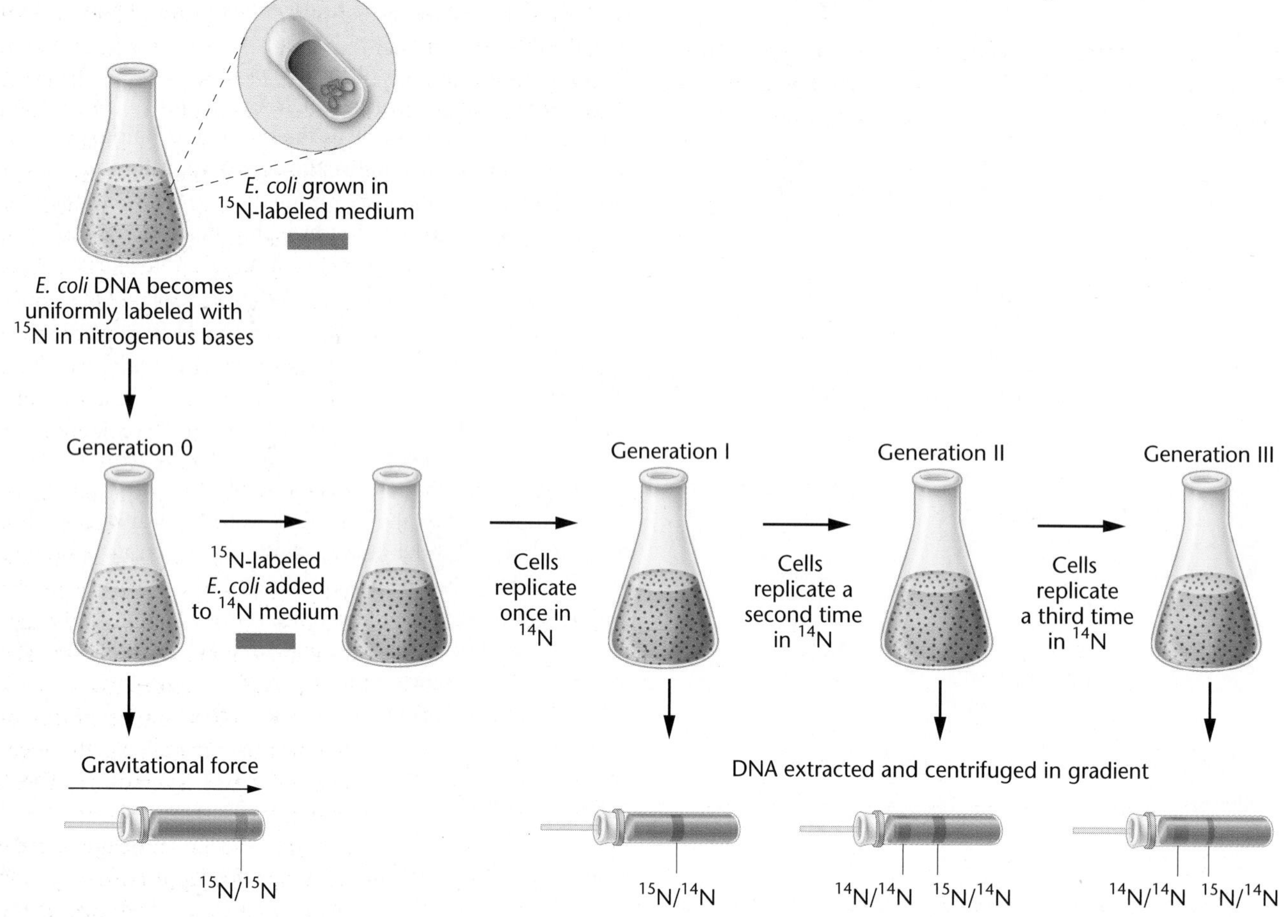

FIGURE 10–3 The Meselson–Stahl experiment.

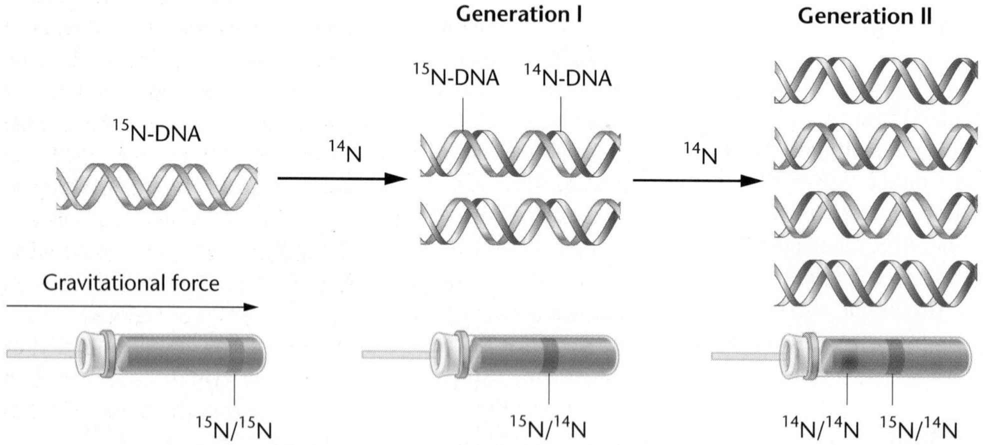

FIGURE 10–4 The expected results of two generations of semiconservative replication in the Meselson–Stahl experiment.

ESSENTIAL POINT

In 1958, Meselson and Stahl resolved the question of which of three potential modes of replication is utilized by *E. coli* during the duplication of DNA in favor of semiconservative replication, showing that newly synthesized DNA consists of one old strand and one new strand.

NOW SOLVE THIS

Problem 2 on page 222 asks you to describe the role of ^{15}N during the Meselson–Stahl experiment.

Hint: Remember that ^{15}N is a stable, heavy isotope of nitrogen that can be distinguished from ^{14}N using centrifugation techniques.

Semiconservative Replication in Eukaryotes

In 1957, the year before the work of Meselson and Stahl was published, J. Herbert Taylor, Philip Woods, and Walter Hughes presented evidence that semiconservative replication also occurs in eukaryotic organisms. They experimented with root tips of the broad bean *Vicia faba,* which are an excellent source of dividing cells. These researchers were able to monitor the process of replication by labeling DNA with ^{3}H-thymidine, a radioactive precursor of DNA, and performing autoradiography.

Autoradiography is a common technique that, when applied cytologically, pinpoints the location of a radioisotope in a cell. In this procedure, a photographic emulsion is placed over a histological preparation containing cellular material (root tips, in this experiment), and the preparation is stored in the dark. The slide is then developed, much as photographic film is processed. Because the radioisotope emits energy, following development the emulsion turns black at the approximate point of emission. The end result is the presence of dark spots or "grains" on the surface of the section, identifying the location of newly synthesized DNA within the cell.

Taylor and his colleagues grew root tips for approximately one generation in the presence of the radioisotope and then placed them in unlabeled medium in which cell division continued. At the conclusion of each generation, they arrested the cultures at metaphase by adding colchicine (a chemical derived from the crocus plant that poisons the spindle fibers) and then examined the chromosomes by autoradiography. They found radioactive thymidine only in association with chromatids that contained newly synthesized DNA. Figure 10–5 illustrates the replication of a single chromosome over two division cycles, including the distribution of grains.

These results are compatible with the semiconservative mode of replication. After the first replication cycle in the presence of the isotope, both sister chromatids show radioactivity, indicating that each chromatid contains one new radioactive DNA strand and one old unlabeled strand. After the second replication cycle, *which*

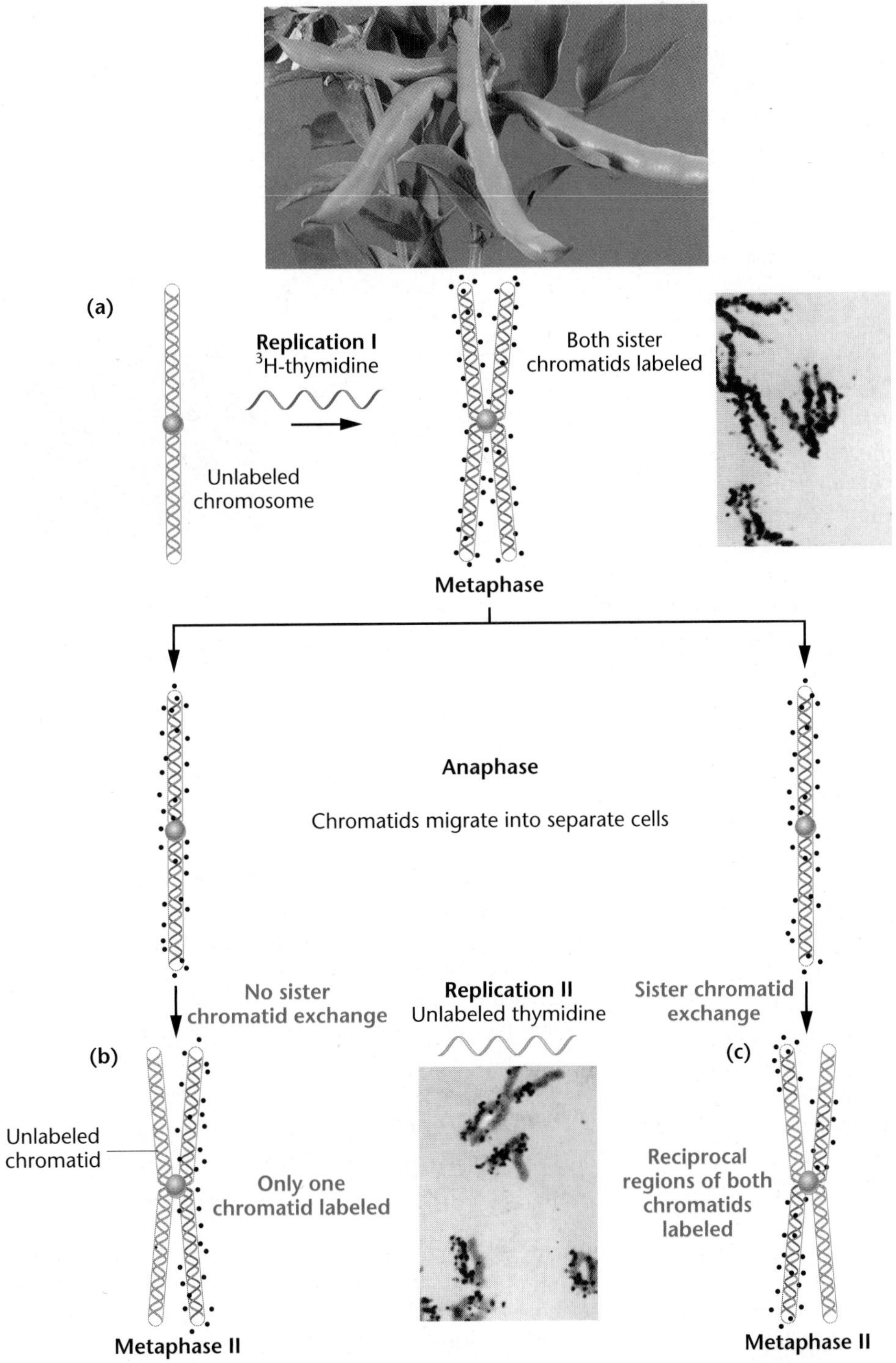

FIGURE 10–5 The Taylor–Woods–Hughes experiment, demonstrating the semiconservative mode of replication of DNA in root tips of *Vicia faba.* A portion of the plant is shown in the top photograph. (a) An unlabeled chromosome proceeds through the cell cycle in the presence of ^{3}H-thymidine. As it enters mitosis, both sister chromatids of the chromosome are labeled, as shown, by autoradiography. After a second round of replication (b), this time in the absence of ^{3}H-thymidine, only one chromatid of each chromosome is expected to be surrounded by grains. Except where a reciprocal exchange has occurred between sister chromatids (c), the expectation was upheld. The micrographs are of the actual autoradiograms obtained in the experiment.

takes place in unlabeled medium, only one of the two sister chromatids of each chromosome should be radioactive because half of the parent strands are unlabeled. With only the minor exceptions of *sister chromatid exchanges* (discussed in Chapter 7), this result was observed.

Together, the Meselson–Stahl experiment and the experiment by Taylor, Woods, and Hughes soon led to the general acceptance of the semiconservative mode of replication. Later studies with other organisms reached the same conclusion and also strongly supported Watson and Crick's proposal for the double helix model of DNA.

ESSENTIAL POINT

Taylor, Woods, and Hughes demonstrated semiconservative replication in eukaryotes using the root tips of the broad bean as the source of dividing cells.

Origins, Forks, and Units of Replication

To enhance our understanding of semiconservative replication, let's briefly consider a number of relevant issues. The first concerns the **origin of replication.** Where along the chromosome is DNA replication initiated? Is there only a single origin, or does DNA synthesis begin at more than one point? Is any given point of origin random, or is it located at a specific region along the chromosome? Second, once replication begins, does it proceed in a single direction or in both directions away from the origin? In other words, is replication **unidirectional** or **bidirectional**?

To address these issues, we need to introduce two terms. First, at each point along the chromosome where replication is occurring, the strands of the helix are unwound, creating what is called a **replication fork.** Such a fork will initially appear at the point of origin of synthesis and then move along the DNA duplex as replication proceeds. If replication is bidirectional, two such forks will be present, migrating in opposite directions away from the origin. The second term refers to the length of DNA that is replicated following one initiation event at a single origin. This is a unit referred to as the **replicon.**

The evidence is clear regarding the origin and direction of replication. John Cairns tracked replication in *E. coli,* using radioactive precursors of DNA synthesis and autoradiography. He was able to demonstrate that in *E. coli* there is only a single region, called *oriC,* where replication is initiated. The presence of only a single origin is characteristic of bacteria, which have only one circular chromosome. Since DNA synthesis in bacteriophages and bacteria originates at a single point, the entire chromosome constitutes one replicon. In *E. coli,* the replicon consists of the entire genome of 4.6 Mb (4.6 million base pairs).

Figure 10–6 illustrates Cairns's interpretation of DNA replication in *E. coli.* This interpretation (and the accompanying micrograph) does not answer the question of unidirectional versus bidirectional synthesis. However, other results, derived from studies of bacteriophage lambda, demonstrated that replication is bidirectional, moving away from *oriC* in both directions. Figure 10–6 therefore interprets Cairns's work with that understanding. Bidirectional replication creates two replication forks that migrate farther and farther apart as replication proceeds. These forks eventually merge, as semiconservative replication of the entire chromosome is completed, at a termination region, called *ter.*

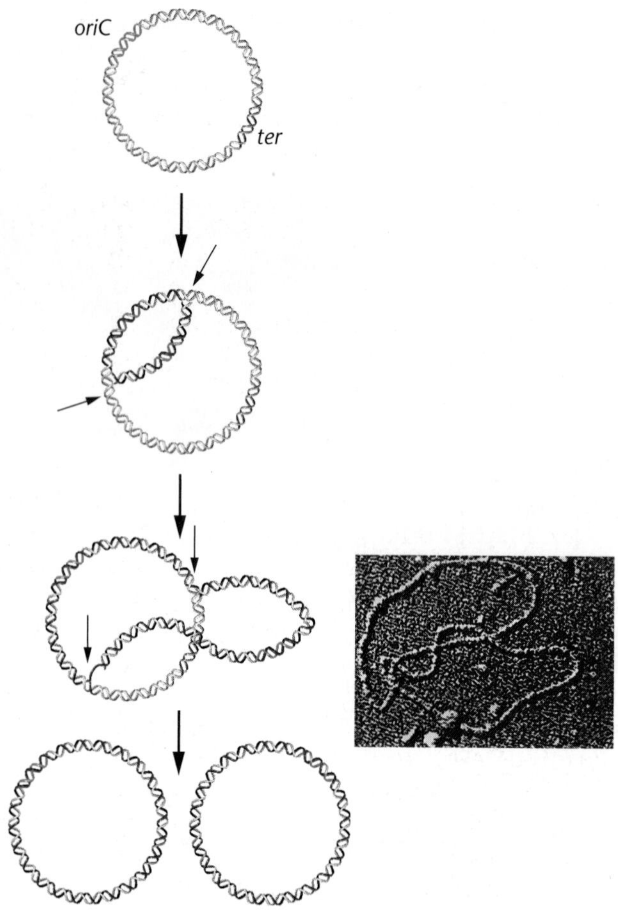

FIGURE 10–6 Bidirectional replication of the *E. coli* chromosome. The thin black arrows identify the advancing replication forks. The micrograph is of a bacterial chromosome in the process of replication, comparable to the figure next to it.

Later in this chapter we will see that in eukaryotes, each chromosome contains multiple points of origin.

10.2 DNA Synthesis in Bacteria Involves Five Polymerases, as Well as Other Enzymes

To say that replication is semiconservative and bidirectional describes the overall *pattern* of DNA duplication and the association of finished strands with one another once synthesis is completed. However, it says little about the more complex issue of how the actual *synthesis* of long complementary polynucleotide chains occurs on a DNA template. Like most questions in molecular biology, this one was first studied using microorganisms. Research on DNA synthesis began about the same time as the Meselson–Stahl work, and the topic is still an active area of investigation. What is most apparent in this research is the tremendous complexity of the biological synthesis of DNA.

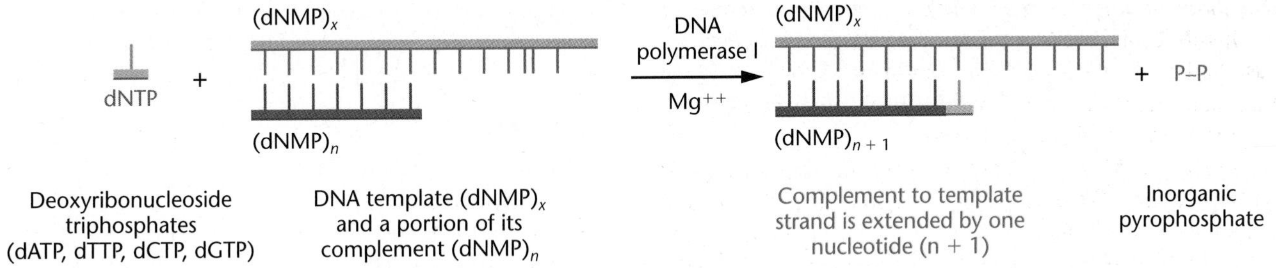

FIGURE 10–7 The chemical reaction catalyzed by DNA polymerase I. During each step, a single nucleotide is added to the growing complement of the DNA template, using a nucleoside triphosphate as the substrate. The release of inorganic pyrophosphate drives the reaction energetically.

DNA Polymerase I

Studies of the enzymology of DNA replication were first reported by Arthur Kornberg and colleagues in 1957. They isolated an enzyme from *E. coli* that was able to direct DNA synthesis in a cell-free (*in vitro*) system. The enzyme is called **DNA polymerase I,** because it was the first of several similar enzymes to be isolated.

Kornberg determined that there were two major requirements for *in vitro* DNA synthesis under the direction of DNA polymerase I: (1) all four deoxyribonucleoside triphosphates (dNTPs) and (2) template DNA. If any one of the four deoxyribonucleoside triphosphates was omitted from the reaction, no measurable synthesis occurred. If derivatives of these precursor molecules other than the nucleoside triphosphate were used (nucleotides or nucleoside diphosphates), synthesis also did not occur. If no template DNA was added, synthesis of DNA occurred but was reduced greatly.

Most of the synthesis directed by Kornberg's enzyme appeared to be exactly the type required for semiconservative replication. The reaction is summarized in Figure 10–7, which depicts the addition of a single nucleotide. The enzyme has since been shown to consist of a single polypeptide containing 928 amino acids.

The way in which each nucleotide is added to the growing chain is a function of the specificity of DNA polymerase I. As shown in Figure 10–8, the precursor dNTP contains the three phosphate groups attached to the 5′-carbon of deoxyribose. As the two terminal phosphates are cleaved during synthesis, the remaining phosphate attached to the 5′-carbon is covalently linked to the 3′-OH group of the deoxyribose to which it is added. Thus, **chain elongation** occurs in the **5′ to 3′ direction** by the addition of one nucleotide at a time to the growing 3′ end. Each step provides a newly exposed 3′-OH group that can participate in the next addition of a nucleotide as DNA synthesis proceeds.

Having isolated DNA polymerase I and demonstrated its catalytic activity, Kornberg next sought to demonstrate the accuracy, or fidelity, with which the enzyme replicated the DNA template. Because technology for ascertaining the nucleotide sequences of the template and newly synthesized strand was not yet available in 1957, he initially had to rely on several indirect methods.

One of Kornberg's approaches was to compare the nitrogenous base compositions of the DNA template with those of the recovered DNA product. Table 10.1 shows Kornberg's base-composition analysis of three DNA templates. Within

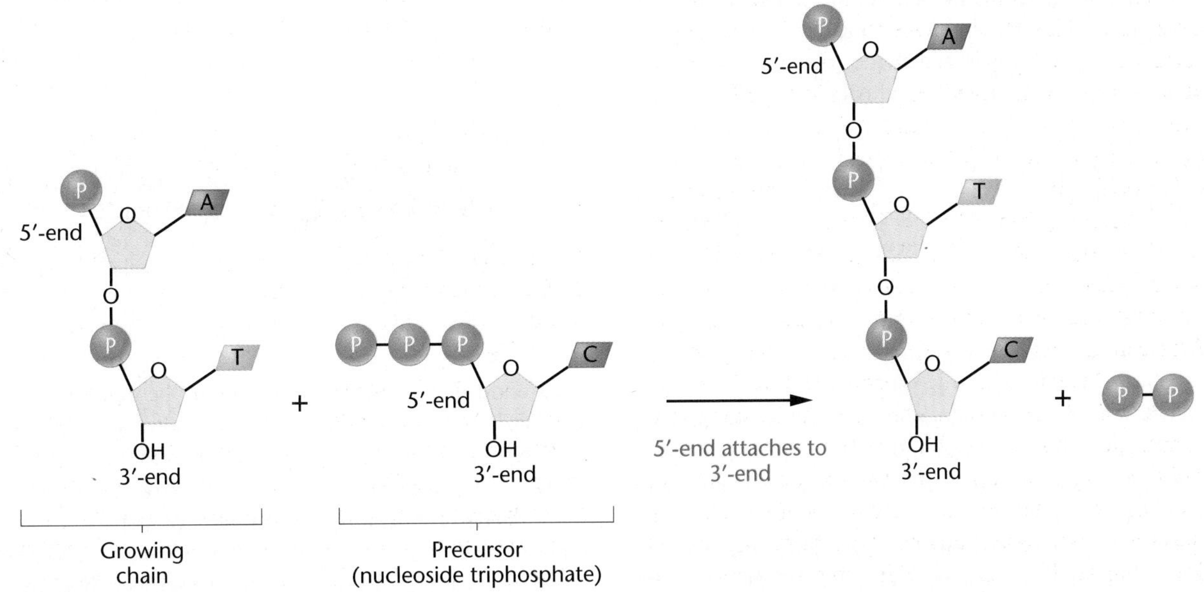

FIGURE 10–8 Demonstration of 5′ to 3′ synthesis of DNA.

TABLE 10.1 Base Composition of the DNA Template and the Product of Replication in Kornberg's Early Work

Organism	Template or Product	%A	%T	%G	%C
T2	Template	32.7	33.0	16.8	17.5
	Product	33.2	32.1	17.2	17.5
E. coli	Template	25.0	24.3	24.5	26.2
	Product	26.1	25.1	24.3	24.5
Calf	Template	28.9	26.7	22.8	21.6
	Product	28.7	27.7	21.8	21.8

Source: Kornberg (1960).

experimental error, the base composition of each product agreed with the template DNAs used. These data, along with other types of comparisons of template and product, suggested that the templates were replicated faithfully.

ESSENTIAL POINT

Arthur Kornberg isolated the enzyme DNA polymerase I from *E. coli* and showed that it is capable of directing *in vitro* DNA synthesis, provided that a template and precursor nucleoside triphosphates are supplied.

DNA Polymerase II, III, IV, and V

While DNA polymerase I clearly directs the synthesis of DNA, a serious reservation about the enzyme's true biological role was raised in 1969. Paula DeLucia and John Cairns discovered a mutant strain of *E. coli* that was deficient in polymerase I activity. The mutation was designated *polA1*. In the absence of the functional enzyme, this mutant strain of *E. coli* still duplicated its DNA and successfully reproduced. However, the cells were deficient in their ability to repair DNA. For example, the mutant strain is highly sensitive to ultraviolet light (UV) and radiation, both of which damage DNA and are mutagenic. Nonmutant bacteria are able to repair a great deal of UV-induced damage.

These observations led to two conclusions:

1. At least one other enzyme that is responsible for replicating DNA *in vivo* is present in *E. coli* cells.
2. DNA polymerase I serves a secondary function *in vivo*, now believed to be critical to the maintenance of fidelity of DNA synthesis.

To date, four other unique DNA polymerases have been isolated from cells lacking polymerase I activity and from normal cells that contain polymerase I. Table 10.2 contrasts several characteristics of DNA polymerase I with **DNA polymerase II** and **III.** Although none of the three can *initiate* DNA synthesis on a template, all three can *elongate* an existing DNA strand, called a **primer.**

All the DNA polymerase enzymes are large proteins exhibiting a molecular weight in excess of 100,000 Daltons (Da). All three possess 3′ to 5′ exonuclease activity, which

TABLE 10.2 Properties of Bacterial DNA Polymerases I, II, and III

Properties	I	II	III
Initiation of chain synthesis	−	−	−
5′–3′ polymerization	+	+	+
3′–5′ exonuclease activity	+	+	+
5′–3′ exonuclease activity	+	−	−
Molecules of polymerase/cell	400	?	15

means that they have the potential to polymerize in one direction and then pause, reverse their direction, and excise nucleotides just added. As we will discuss later in the chapter, this activity provides a capacity to proofread newly synthesized DNA and to remove and replace incorrect nucleotides.

DNA polymerase I also demonstrates 5′ to 3′ exonuclease activity. This activity allows the enzyme to excise nucleotides, starting at the end at which synthesis begins and proceeding in the same direction of synthesis. Two final observations probably explain why Kornberg isolated polymerase I and not polymerase III: polymerase I is present in greater amounts than is polymerase III, and it is also much more stable.

What then are the roles of the polymerases *in vivo*? Polymerase III is the enzyme responsible for the 5′ to 3′ polymerization essential to *in vivo* replication. Its 3′ to 5′ exonuclease activity also provides a proofreading function that is activated when it inserts an incorrect nucleotide. When this occurs, synthesis stalls and the polymerase "reverses course," excising the incorrect nucleotide. Then, it proceeds back in the 5′ to 3′ direction, synthesizing the complement of the template strand. Polymerase I is believed to be responsible for removing the primer, as well as for the synthesis that fills gaps produced after this removal. Its exonuclease activities also allow for its participation in DNA repair. Polymerase II, as well as **polymerase IV** and **V,** are involved in various aspects of repair of DNA that has been damaged by external forces, such as ultraviolet light. Polymerase II is encoded by a gene activated by disruption of DNA synthesis at the replication fork.

ESSENTIAL POINT

The discovery of the *polA1* mutant strain of *E. coli,* capable of DNA replication despite its lack of polymerase I activity, cast doubt on the enzyme's hypothesized *in vivo* replicative function. Polymerase III has been identified as the enzyme responsible for DNA replication *in vivo.*

We end this section by emphasizing the complexity of the DNA polymerase III molecule. In contrast to DNA polymerase I, which is but a single polypeptide, the active form of DNA polymerase III is a **holoenzyme**—a complex enzyme made up of multiple subunits. Polymerase III consists of ten kinds of polypeptide subunits (Table 10.3) and has a molecular weight of 900,000 Da. The largest subunit, α, has a molecular weight of 140,000 Da and, along with subunits ε

TABLE 10.3 Subunits of the DNA Polymerase III Holoenzyme

Subunit	Function	Groupings
α	5′–3′ polymerization	Core enzyme: Elongates polynucleotide chain and proofreads
ε	3′–5′ exonuclease	
θ	Core assembly	
γ	Loads enzyme on template (serves as clamp loader)	γ complex
δ		
δ'		
χ		
ψ		
β	Sliding clamp structure (processivity factor)	
τ	Dimerizes core complex	

and θ, constitutes a **core enzyme** responsible for the polymerization activity. The α subunit is responsible for nucleotide polymerization on the template strands, whereas the ε subunit of the core enzyme possesses the 3′ to 5′ exonuclease activity. A single DNA polymerase III holoenzyme contains, along with other components, two core enzymes combined into a dimer.

A second group of five subunits (γ, δ, δ', χ, and ψ) forms what is called the γ complex, which is involved in "loading" the enzyme onto the template at the replication fork. This enzymatic function requires energy and is dependent on the hydrolysis of ATP. The β subunit serves as a "clamp" and prevents the core enzyme from falling off the template during polymerization. Finally, the τ subunit functions to dimerize two core polymerases facilitating simultaneous synthesis of both strands of the helix at the replication fork. The holoenzyme and several other proteins at the replication fork together form a huge complex (nearly as large as a ribosome) known as the **replisome.** We consider the function of DNA polymerase III in more detail later in this chapter.

10.3 Many Complex Issues Must Be Resolved during DNA Replication

We have thus far established that in bacteria and viruses replication is semiconservative and bidirectional along a single replicon. We also know that synthesis is catalyzed by DNA polymerase III and occurs in the 5′ to 3′ direction. Bidirectional synthesis creates two replication forks that move in opposite directions away from the origin of synthesis. As we can see from the following list, many issues remain to be resolved in order to provide a comprehensive understanding of DNA replication:

1. The helix must undergo localized unwinding, and the resulting "open" configuration must be stabilized so that synthesis may proceed along both strands.
2. As unwinding and subsequent DNA synthesis proceed, increased coiling creates tension further down the helix, which must be reduced.
3. A primer of some sort must be synthesized so that polymerization can commence under the direction of DNA polymerase III. Surprisingly, RNA, not DNA, serves as the primer.
4. Once the RNA primers have been synthesized, DNA polymerase III begins to synthesize the DNA complement of both strands of the parent molecule. Because the two strands are antiparallel to one another, continuous synthesis in the direction that the replication fork moves is possible along only one of the two strands. On the other strand, synthesis must be discontinuous and thus involves a somewhat different process.
5. The RNA primers must be removed prior to completion of replication. The gaps that are temporarily created must be filled with DNA complementary to the template at each location.
6. The newly synthesized DNA strand that fills each temporary gap must be joined to the adjacent strand of DNA.
7. While DNA polymerases accurately insert complementary bases during replication, they are not perfect, and, occasionally, incorrect nucleotides are added to the growing strand. A proofreading mechanism that also corrects errors is an integral process during DNA synthesis.

As we consider these points, examine Figures 10–9, 10–10, 10–11, and 10–12 to see how each issue is resolved. Figure 10–13 summarizes the model of DNA synthesis.

Unwinding the DNA Helix

As discussed earlier, there is a single point of origin along the circular chromosome of most bacteria and viruses at which DNA synthesis is initiated. This region of the *E. coli* chromosome has been particularly well studied. Called ***oriC,*** it consists of 245 nucleotide pairs characterized by repeating sequences of 9 and 13 bases (called **9mers** and **13mers**). As shown in Figure 10–9, p. 212, one particular protein, called **DnaA** (because it is encoded by the gene called *dnaA*), is responsible for the initial step in unwinding the helix. A number of subunits of the DnaA protein bind to each of several 9mers. This step facilitates the subsequent binding of **DnaB** and **DnaC** proteins that further open and destabilize the helix. Proteins such as these, which require the energy supplied by the hydrolysis of ATP in order to break hydrogen bonds and denature the double helix, are called **helicases.** Other proteins, called **single-stranded binding proteins (SSBPs),** stabilize this open conformation.

As unwinding proceeds, a coiling tension is created ahead of the replication fork, often producing **supercoiling.** In circular molecules, supercoiling may take the form of added twists and turns of the DNA, much like the coiling you can create in a rubber band by stretching it out and then twisting one end. Such supercoiling can be relaxed by **DNA gyrase,** a member of a larger group of enzymes referred to as **DNA topoisomerases.** The gyrase makes either single- or double-stranded "cuts" and also catalyzes localized movements that have the

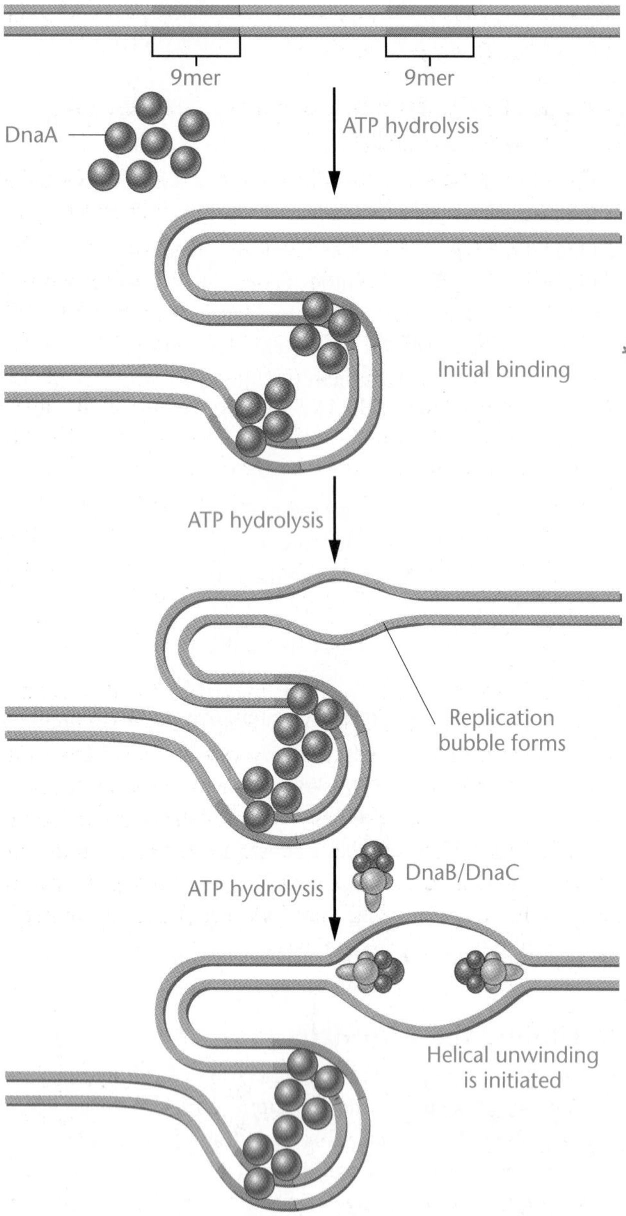

FIGURE 10–9 Helical unwinding of DNA during replication as accomplished by DnaA, DnaB, and DnaC proteins. Initial binding of many monomers of DnaA occurs at DNA sites containing repeating sequences of 9 nucleotides, called 9mers. Not illustrated are 13mers, which are also involved.

effect of "undoing" the twists and knots created during supercoiling. The strands are then resealed. These various reactions are driven by the energy released during ATP hydrolysis.

Together, the DNA, the polymerase complex, and associated enzymes make up an array of molecules that participate in DNA synthesis and are part of what we have previously called the *replisome.*

ESSENTIAL POINT

During the initiation of DNA synthesis, the double helix unwinds, forming a replication fork at which synthesis begins. Proteins stabilize the unwound helix and assist in relaxing the coiling tension created ahead of the replication fork.

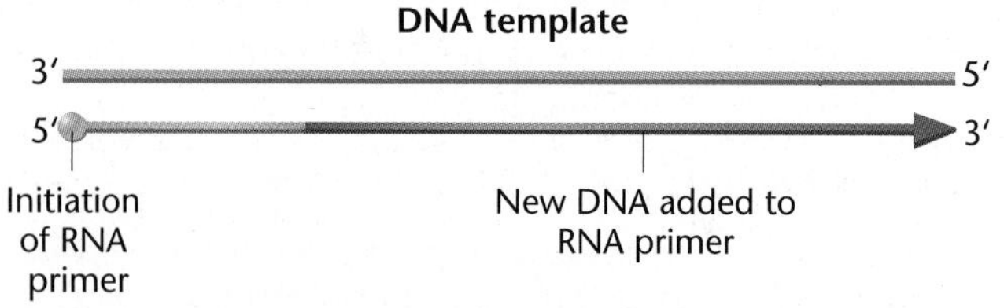

FIGURE 10–10 The initiation of DNA synthesis. A complementary RNA primer is first synthesized, to which DNA is added. All synthesis is in the 5′ to 3′ direction. Eventually, the RNA primer is replaced with DNA under the direction of DNA polymerase I.

Initiation of DNA Synthesis Using an RNA Primer

Once a small portion of the helix is unwound, what else is needed to initiate synthesis? As we have seen, DNA polymerase III requires a primer with a free 3′-hydroxyl group in order to elongate a polynucleotide chain. Since none is available in a circular chromosome, this absence prompted researchers to investigate how the first nucleotide could be added. It is now clear that RNA serves as the primer that initiates DNA synthesis.

A short segment of RNA (about 10 to 12 nucleotides long), complementary to DNA, is first synthesized on the DNA template. Synthesis of the RNA is directed by a form of RNA polymerase called **primase,** which does not require a free 3′ end to initiate synthesis. It is to this short segment of RNA that DNA polymerase III begins to add deoxyribonucleotides, initiating DNA synthesis. A conceptual diagram of initiation on a DNA template is shown in Figure 10–10. Later, the RNA primer is clipped out and replaced with DNA. This is thought to occur under the direction of DNA polymerase I. Recognized in viruses, bacteria, and several eukaryotic organisms, RNA priming is a universal phenomenon during the initiation of DNA synthesis.

ESSENTIAL POINT

DNA synthesis is initiated at specific sites along each template strand by the enzyme primase, resulting in short segments of RNA that provide suitable 3′ ends upon which DNA polymerase III can begin polymerization.

Continuous and Discontinuous DNA Synthesis

We must now revisit the fact that the two strands of a double helix are **antiparallel** to each other—that is, one runs in the 5′−3′ direction, while the other has the opposite 3′−5′ polarity. Because DNA polymerase III synthesizes DNA in only the 5′−3′ direction, synthesis along an advancing replication fork occurs in one direction on one strand and in the opposite direction on the other.

As a result, as the strands unwind and the replication fork progresses down the helix (Figure 10–11), only one strand can serve as a template for **continuous DNA synthesis.** This newly synthesized DNA is called the **leading strand.** As the fork progresses, many points of initiation are necessary on the opposite DNA template, resulting in **discontinuous DNA synthesis** of the **lagging strand.***

Evidence supporting the occurrence of discontinuous DNA synthesis was first provided by Reiji and Tuneko Okazaki.

* Because DNA synthesis is continuous on one strand and discontinuous on the other, the term **semidiscontinuous synthesis** is sometimes used to describe the overall process.

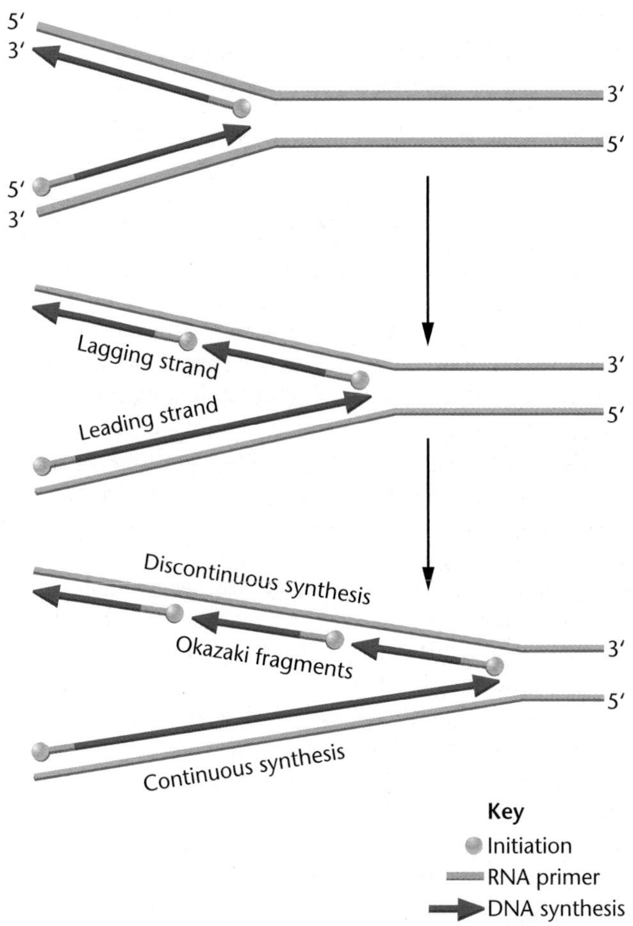

FIGURE 10–11 Opposite polarity of synthesis along the two strands of DNA is necessary because they run antiparallel to one another, and because DNA polymerase III synthesizes in only one direction (5′ to 3′). On the lagging strand, synthesis must be discontinuous, resulting in the production of Okazaki fragments. On the leading strand, synthesis is continuous. RNA primers are used to initiate synthesis on both strands.

They discovered that when bacteriophage DNA is replicated in *E. coli,* some of the newly formed DNA that is hydrogen bonded to the template strand is present as small fragments containing 1000 to 2000 nucleotides. RNA primers are part of each such fragment. These pieces, now called **Okazaki fragments,** are converted into longer and longer DNA strands of higher molecular weight as synthesis proceeds.

Discontinuous synthesis of DNA requires enzymes that both remove the RNA primers and unite the Okazaki fragments into the lagging strand. As we have noted, DNA polymerase I removes the primers and replaces the missing nucleotides. Joining the fragments is the work of **DNA ligase,** which is capable of catalyzing the formation of the phosphodiester bond that seals the nick between the discontinuously synthesized strands. The evidence that DNA ligase performs this function during DNA synthesis is strengthened by the observation of a ligase-deficient mutant strain (*lig*) of *E. coli,* in which a large number of unjoined Okazaki fragments accumulate.

Concurrent Synthesis Occurs on the Leading and Lagging Strands

Given the model just discussed, we might ask how DNA polymerase III synthesizes DNA on both the leading and lagging strands. Can both strands be replicated simultaneously at the same replication fork, or are the events distinct, involving two separate copies of the enzyme? Evidence suggests that both strands can be replicated simultaneously. As Figure 10–12 illustrates, if the lagging strand forms a loop, nucleotide polymerization can occur on both template strands under the direction of a dimer of the enzyme. After the synthesis of 1000 to 2000 nucleotides, the monomer of the enzyme on the lagging strand will encounter a completed Okazaki fragment, at which point it releases the lagging strand. A new loop is then formed with the lagging template strand, and the process is repeated. Looping inverts the orientation of the template but not the direction of actual synthesis on the lagging strand, which is always in the 5′ to 3′ direction.

Another important feature of the holoenzyme that facilitates synthesis at the replication fork is a dimer of the β subunit that forms a clamplike structure around the newly formed DNA duplex. This β-subunit clamp prevents the **core enzyme** (the α, ε, and θ subunits that are responsible for catalysis of nucleotide addition) from falling off the template as polymerization proceeds. Because the entire holoenzyme moves along the parent duplex, advancing the replication fork, the β-subunit dimer is often referred to as a *sliding clamp.*

ESSENTIAL POINT

Concurrent DNA synthesis occurs continuously on the leading strand and discontinuously on the opposite lagging strand, resulting in short Okazaki fragments that are later joined by DNA ligase.

NOW SOLVE THIS

In Problem 28 on page 223, a hypothetical organism is observed in which no Okazaki fragments are produced. You are asked to suggest a DNA model consistent with this information.

Hint: The lack of Okazaki fragments suggests that DNA synthesis is continuous on both strands.

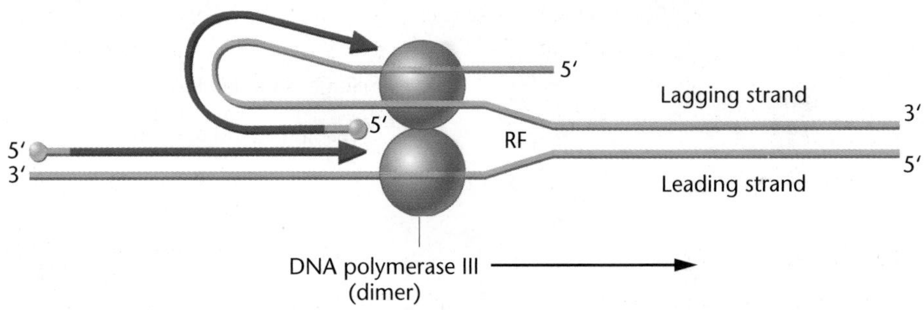

FIGURE 10–12 Illustration of how concurrent DNA synthesis may be achieved on both the leading and lagging strands at a single replication fork (RF). The lagging template strand is "looped" in order to invert the physical direction of synthesis, but not the biochemical direction. The enzyme functions as a dimer, with each core enzyme achieving synthesis on one or the other strand.

Proofreading and Error Correction Occurs during DNA Replication

The immediate purpose of DNA replication is the synthesis of a new strand that is precisely complementary to the template strand at each nucleotide position. Although the action of DNA polymerases is very accurate, synthesis is not perfect and a noncomplementary nucleotide is occasionally inserted erroneously. To compensate for such inaccuracies, the DNA polymerases all possess 3′ to 5′ exonuclease activity. This property imparts the potential for them to detect and excise a mismatched nucleotide (in the 3′−5′ direction). Once the mismatched nucleotide is removed, 5′ to 3′ synthesis can again proceed. This process, called **proofreading,** increases the fidelity of synthesis by a factor of about 100. In the case of the holoenzyme form of DNA polymerase III, the epsilon (ε) subunit is directly involved in the proofreading step. In strains of *E. coli* with a mutation that has rendered the ε subunit nonfunctional, the error rate (the mutation rate) during DNA synthesis is increased substantially.

10.4 A Coherent Model Summarizes DNA Replication

We can now combine the various aspects of DNA replication occurring at a single replication fork into a coherent model, as shown in Figure 10–13. At the advancing fork, a helicase is unwinding the double helix. Once unwound, single-stranded binding proteins associate with the strands, preventing the reformation of the helix. In advance of the replication fork, DNA gyrase functions to diminish the tension created as the helix supercoils. Each half of the dimeric polymerase is a core enzyme bound to one of the template strands by a β-subunit sliding clamp. Continuous synthesis occurs on the leading strand, while the lagging strand must loop around in order for simultaneous (concurrent) synthesis to occur on both strands. Not shown in the figure, but essential to replication on the lagging strand, is the action of DNA polymerase I and DNA ligase, which together replace the RNA primers with DNA and join the Okazaki fragments, respectively.

Because the investigation of DNA synthesis is still an extremely active area of research, this model will no doubt be extended in the future. In the meantime, it provides a summary of DNA synthesis against which genetic phenomena can be interpreted.

10.5 Replication Is Controlled by a Variety of Genes

Much of what we know about DNA replication in viruses and bacteria is based on genetic analysis of the process. For example, we have already discussed studies involving the *polA1* mutation, which revealed that DNA polymerase I is not the major enzyme responsible for replication. Many other mutations interrupt or seriously impair some aspect of replication, such as the ligase-deficient and the proofreading-deficient mutations mentioned previously. Because such mutations are lethal ones, genetic analysis frequently uses **conditional mutations,** which are expressed under one condition but not under a different condition. For example, a **temperature-sensitive mutation** may not be expressed at a particular *permissive* temperature. When mutant cells are grown at a *restrictive* temperature, the mutant phenotype is expressed and can be studied. By examining the effect of the loss of function associated with the mutation, the investigation of such temperature-sensitive mutants can provide insight into the product and the associated function of the normal, nonmutant gene.

As shown in Table 10.4, a variety of genes in *E. coli* specify the subunits of the DNA polymerases and encode products involved in specification of the origin of synthesis, helix unwinding and stabilization, initiation and priming, relaxation of supercoiling, repair, and ligation. The discovery of such a large group of genes attests to the complexity of the process of replication, even in the relatively simple prokaryote. Given the enormous quantity of DNA that must be unerringly replicated in a very brief time, this level of complexity is not unexpected. As we will see in the next section, the process is even more involved and therefore more difficult to investigate in eukaryotes.

NOW SOLVE THIS

Problem 21 on page 223 involves several temperature-sensitive mutations in *E. coli,* and you are asked to interpret the action of the gene that has mutated based on the phenotype that results.

Hint: Each mutation has disrupted one of the many steps essential to DNA synthesis. In each case, the mutant phenotype provides the clue as to which enzyme or function is affected.

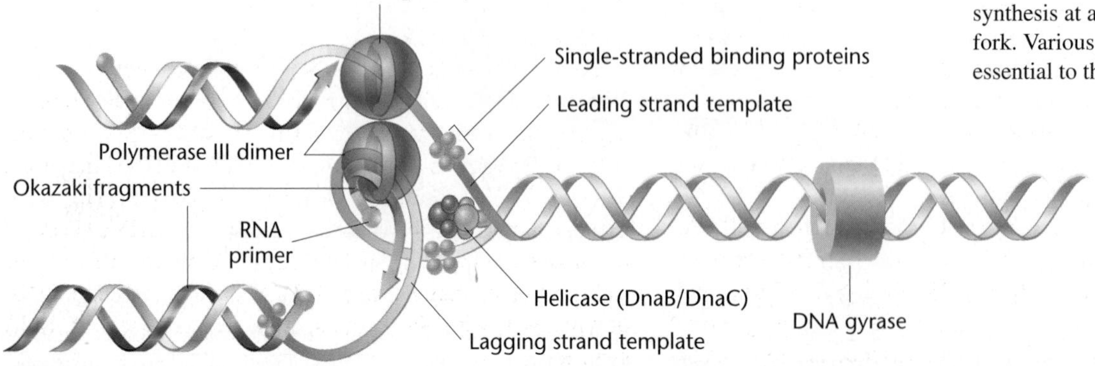

FIGURE 10–13 Summary of DNA synthesis at a single replication fork. Various enzymes and proteins essential to the process are shown.

TABLE 10.4 Some of the Various *E. coli* Genes and Their Products or Role in Replication

Gene	Product or Role
polA	DNA polymerase I
polB	DNA polymerase II
dnaE, N, Q, X, Z	DNA polymerase III subunits
dnaG	Primase
dnaA, I, P	Initiation
dnaB, C	Helicase at *oriC*
gyrA, B	Gyrase subunits
lig	DNA ligase
rep	DNA helicase
ssb	Single-stranded binding proteins
rpoB	RNA polymerase subunit

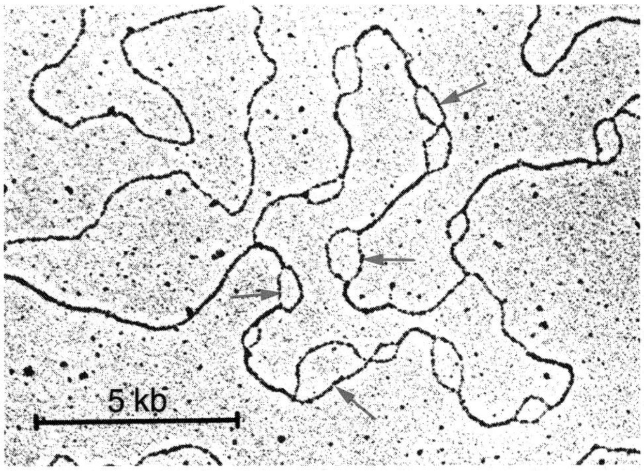

FIGURE 10–14 A demonstration of the multiple origins of replication along a eukaryotic chromosome. Each origin is apparent as a replication bubble along the axis of the chromosome. Arrows identify some of these replication bubbles.

10.6 Eukaryotic DNA Replication Is Similar to Replication in Prokaryotes, but Is More Complex

Eukaryotic DNA replication shares many features with replication in bacteria. In both systems, double-stranded DNA is unwound at replication origins, replication forks are formed, and bidirectional DNA synthesis creates leading and lagging strands from single-stranded DNA templates under the direction of DNA polymerase. Eukaryotic polymerases have the same fundamental requirements for DNA synthesis as do bacterial polymerases: four deoxyribonucleoside triphosphates, a template, and a primer. However, because eukaryotic cells contain much more DNA, this DNA is complexed with nucleosomes, and because eukaryotic chromosomes are linear rather than circular, eukaryotic DNA synthesis is more complicated. In this section, we will describe some of the ways in which eukaryotes deal with this added complexity.

Initiation of Replication at Multiple Replication Origins

The most obvious difference between eukaryotic and prokaryotic DNA replication is that eukaryotic replication must deal with greater amounts of DNA. For example, yeast cells contain three times as much DNA, and *Drosophila* cells contain 40 times as much as *E. coli* cells. In addition, eukaryotic DNA polymerases synthesize DNA at a rate 25 times slower (about 2000 nucleotides per minute) than that in prokaryotes. Under these conditions, replication from a single origin on a typical eukaryotic chromosome would take days to complete! However, replication of entire eukaryotic genomes is usually accomplished in a matter of minutes to hours.

To facilitate the rapid synthesis of large quantities of DNA, eukaryotic chromosomes contain multiple replication origins. Yeast genomes contain between 250 and 400 origins, and mammalian genomes have as many as 25,000. Multiple origins are visible under the electron microscope as "replication bubbles" that form as the DNA helix opens up, each bubble providing two potential replication forks (Figure 10–14). Origins in yeast, called **autonomously replicating sequences (ARSs),** consist of an approximately 120 base-pair unit containing a **consensus sequence** (meaning a sequence that is the same, or nearly the same, in all yeast ARSs) of 11 base pairs. Origins in mammalian cells appear to be unrelated to specific sequence motifs and may be defined more by chromatin structure over a 6–55 kb region.

Eukaryotic replication origins are not only the sites of replication initiation, but also control the timing of DNA replication. These regulatory functions are carried out by a complex of more than 20 proteins, called the **prereplication complex (pre-RC),** which assembles at replication origins. In the early G1 phase of the cell cycle, replication origins are recognized by a six-protein complex known as an **origin recognition complex (ORC),** which tags the origin as a site of initiation. Throughout the G1 phase of the cell cycle, other proteins associate with the ORC to form the pre-RC. The presence of a pre-RC at an origin "licenses" that origin for replication. Once DNA polymerases initiate synthesis at the origin, the pre-RC is disrupted and does not reassemble again until the G1 phase of the next cell cycle. This is an important mechanism because it distinguishes segments of DNA that have completed replication from segments of unreplicated DNA, thus maintaining orderly and efficient replication. It ensures that replication occurs only once along each stretch of DNA during each cell cycle.

The initiation of DNA replication is also regulated at the pre-RC. A number of cell-cycle kinases that phosphorylate replication proteins, along with helicases that unwind DNA, associate with the pre-RC and are essential for initiation. The kinases are activated in S phase, at which time they phosphorylate other proteins that trigger the initiation of DNA replication. The end result is the unwinding of DNA at the replication forks, the stabilization of single-stranded DNA, the association of DNA polymerases with the origins, and the initiation of DNA synthesis.

DNA Synthesis by Multiple Eukaryotic DNA Polymerases

To accommodate the increased number of replicons, eukaryotic cells contain many more DNA polymerase molecules than do bacterial cells. For example, a single *E. coli* cell contains about 15 molecules of DNA polymerase III, but mammalian cells contain tens of thousands of DNA polymerase molecules.

Eukaryotes also utilize a larger array of different DNA polymerase types than do prokaryotes. The human genome contains genes that encode at least 14 different DNA polymerases, only three of which are involved in the majority of nuclear genome DNA replication. The nomenclature and characteristics of these DNA polymerases are summarized in Table 10.5.

Pol α and δ, as well as ε, are the major forms of the enzyme involved in initiation and elongation during nuclear DNA synthesis, so we will concentrate our discussion on these. Two of the four subunits of the Pol α enzyme synthesize RNA primers on both the leading and lagging strands. After the RNA primer reaches a length of about 10 ribonucleotides, another subunit adds 20 to 30 complementary deoxyribonucleotides. Pol α is said to possess low **processivity,** a term that refers to the strength of the association between the enzyme and its substrate, and thus the length of DNA that is synthesized before the enzyme dissociates from the template. Once the primer is in place, an event known as **polymerase switching** occurs, whereby Pol α dissociates from the template and is replaced by Pol δ and ε. These enzymes extend the primers on opposite strands of DNA, possess much greater processivity, and exhibit 3′ to 5′ exonuclease activity, thus having the potential to proofread. Pol ε synthesizes DNA on the leading strand, and Pol δ synthesizes the lagging strand. Both Pol δ and ε participate in other DNA synthesizing events in the cell, including several types of DNA repair and recombination. All three enzymes are essential for viability.

As in prokaryotic DNA replication, the final stages in eukaryotic DNA replication involve replacing the RNA primers with DNA and ligating the Okazaki fragments on the lagging strand. In eukaryotes, the Okazaki fragments are about ten times smaller (100 to 150 nucleotides) than in prokaryotes.

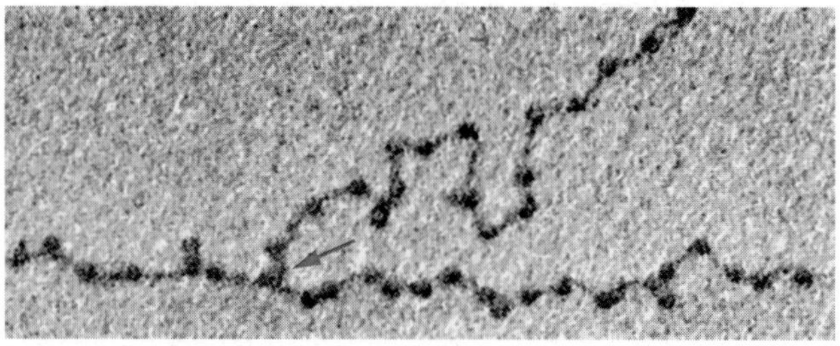

FIGURE 10–15 An electron micrograph of a eukaryotic replicating fork demonstrating the presence of histone-protein-containing nucleosomes on both branches.

Included in the remainder of DNA-replicating enzymes is Pol γ, which is found exclusively in mitochondria, synthesizing the DNA present in that organelle. DNA polymerases are involved in DNA repair and replication through regions of the DNA template that contain damage or distortions (called **translesion synthesis,** or **TLS**). Although these translesion DNA polymerases are less faithful in copying DNA and thus make more errors than Pol α and δ, and ε, they are able to bypass the distortions, leaving behind gaps that may then be repaired.

Replication through Nucleosomal Chromatin

One of the major differences between prokaryotic and eukaryotic DNA is that eukaryotic DNA is complexed with DNA-binding proteins, including the histones and nonhistone proteins. As we will discuss in Chapter 11, these DNA-protein complexes are known as **chromatin** (those specifically complexed with nucleosomes are called *nucleosomal chromatin*). Before polymerases can begin synthesis, nucleosomes and other DNA-binding proteins must be stripped away or otherwise modified to allow the passage of replication proteins. As DNA synthesis proceeds, the histones and nonhistone proteins must rapidly reassociate with the newly formed duplexes, reestablishing the characteristic nucleosome pattern (Figure 10–15).

In order to re-create nucleosomal chromatin on replicated DNA, the synthesis of new histone proteins is tightly coupled to DNA synthesis during the S phase of the cell cycle. Research data suggest that nucleosomes are disrupted just ahead of the replication fork and that the preexisting histone proteins assemble with newly synthesized histone proteins into new nucleosomes. After assembly, the new nucleosomes

TABLE 10.5 Properties of Eukaryotic DNA Polymerases

DNA Polymerase	Subunits	3′ → 5′ Exonuclease	Function
α (alpha)	4	No	RNA/DNA primers, initiation of DNA synthesis
δ (delta)	4	Yes	Lagging strand synthesis, DNA repair, proofreading
ε (epsilon)	4	Yes	Leading strand synthesis, proofreading
γ (gamma)	2	Yes	Mitochondrial DNA replication and repair
β (beta)	1	No	Base-excision DNA repair
η (eta)	1	No	Translesion DNA synthesis
ζ (zeta)	2	No	Translesion DNA synthesis
κ (kappa)	1	No	Translesion DNA synthesis
ι (iota)	1	No	Translesion DNA synthesis
θ (theta)	1	No	DNA repair
λ (lambda)	1	No	DNA repair
μ (mu)	1	No	DNA repair
ν (nu)	1	No	Unknown
Rev 1	1	No	DNA repair

are redistributed at random to the two daughter strands of DNA. The assembly of new nucleosomes is carried out by **chromatin assembly factors (CAFs)** that move along with the replication fork.

ESSENTIAL POINT

DNA replication is more complex than replication in prokaryotes, using multiple replication origins, multiple forms of DNA polymerases, and factors that disrupt and assemble nucleosomal chromatin.

10.7 The Ends of Linear Chromosomes Are Problematic during Replication

A final difference between prokaryotic and eukaryotic DNA synthesis stems from the structural differences in their chromosomes. Unlike the closed, circular DNA of bacteria and most bacteriophages, eukaryotic chromosomes are linear. During replication, a special problem arises at the "ends" of these linear molecules.

Eukaryotic chromosomes end in distinctive sequences called **telomeres** that help preserve the integrity and stability of the chromosome. Telomeres are necessary because the double-stranded "ends" of DNA molecules at the termini of linear chromosomes potentially resemble the **double-stranded breaks (DSBs)** that can occur when a chromosome becomes fragmented internally. In such cases, the double-stranded loose ends can fuse to other such ends; if they don't fuse, they are vulnerable to degradation by nucleases. Either outcome can lead to problems. Telomeres are believed to create inert chromosome ends, protecting intact eukaryotic chromosomes from improper fusion or degradation.

Telomere Structure

We could speculate that there must be something unique about the DNA sequence or the proteins that bind to it that confers this protective property to telomeres. Indeed, this has been shown to be the case. First discovered by Elizabeth Blackburn and Joe Gall in their study of micronuclei—the smaller of two nuclei in the ciliated protozoan *Tetrahymena*—the DNA at the protozoan's chromosome ends consists of the short tandem repeating sequence TTGGGG. This sequence is present many times on one of the two helical strands making up each telomere. This strand is referred to as the G-rich strand, in contrast to its complementary strand, the so-called C-rich strand, which displays the repeated sequence AACCCC. In a similar way, all vertebrates contain the sequence TTAGGG at the ends of G-rich strands, repeated several thousand times in somatic cells. Since each linear chromosome ends with two helical DNA strands running antiparallel to one another, one strand has a 3′-ending and the other has a 5′-ending. It is the 3′-strand that is the G-rich one. This has special significance during telomere replication.

But first, let's describe how this tandemly repeated DNA confers inertness to the chromosome ends. One model is based on the discovery that the 3′-ending G-rich strand extends as an overhang, lacking a complement, and thus forms a single-stranded tail at the terminus of each telomere. In *Tetrahymena,* this tail is only 12 to 16 nucleotides long. However, in vertebrates, it may be several hundred nucleotides long. The final conformation of these tails has been correlated with chromosome inertness. Though not considered complementary in the same way as A-T and G-C base pairs are, G-containing nucleotides are nevertheless capable of base pairing with one another when several are aligned opposite another G-rich sequence. Thus, the G-rich single-stranded tails are capable of looping back on themselves, forming multiple G-G hydrogen bonds to create what are referred to as **G-quartets.** The resulting loops at the chromosome ends (called **t-loops**) are much like those created when you tie your shoelaces into a bow. It is believed that these structures, in combination with specific proteins that bind to them, essentially close off the ends of chromosomes and make them inert.

Replication at the Telomere

Now let's consider the problem that semiconservative replication poses at the end of a double-stranded DNA molecule. Although 5′ to 3′ synthesis on the leading-strand template may proceed to the end, a difficulty arises on the lagging strand once the final RNA primer is removed (Figure 10–16).

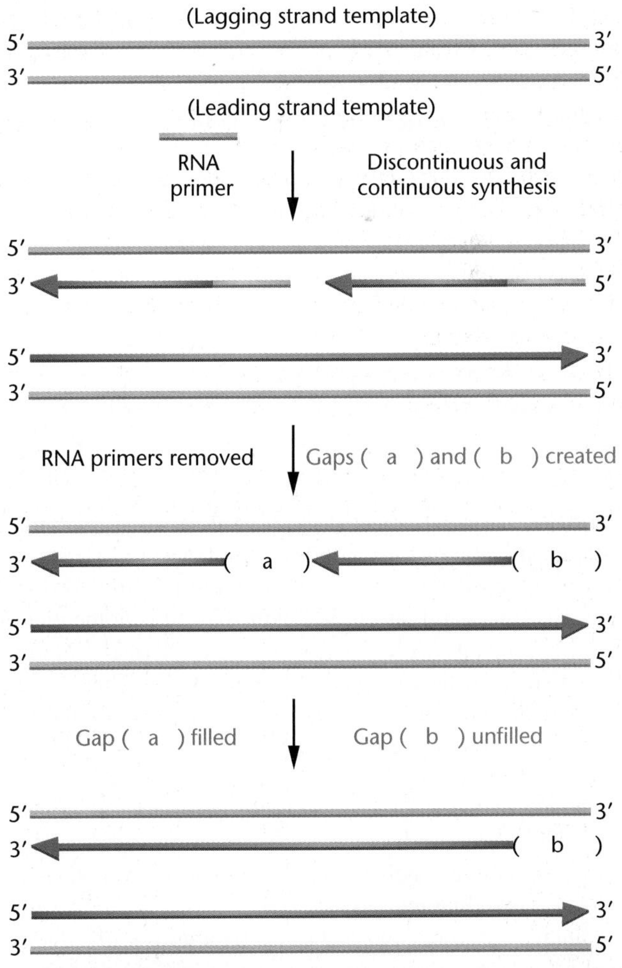

FIGURE 10–16 Diagram illustrating the difficulty encountered during the replication of the ends of linear chromosomes. A gap (- -b- -) is left following synthesis on the lagging strand.

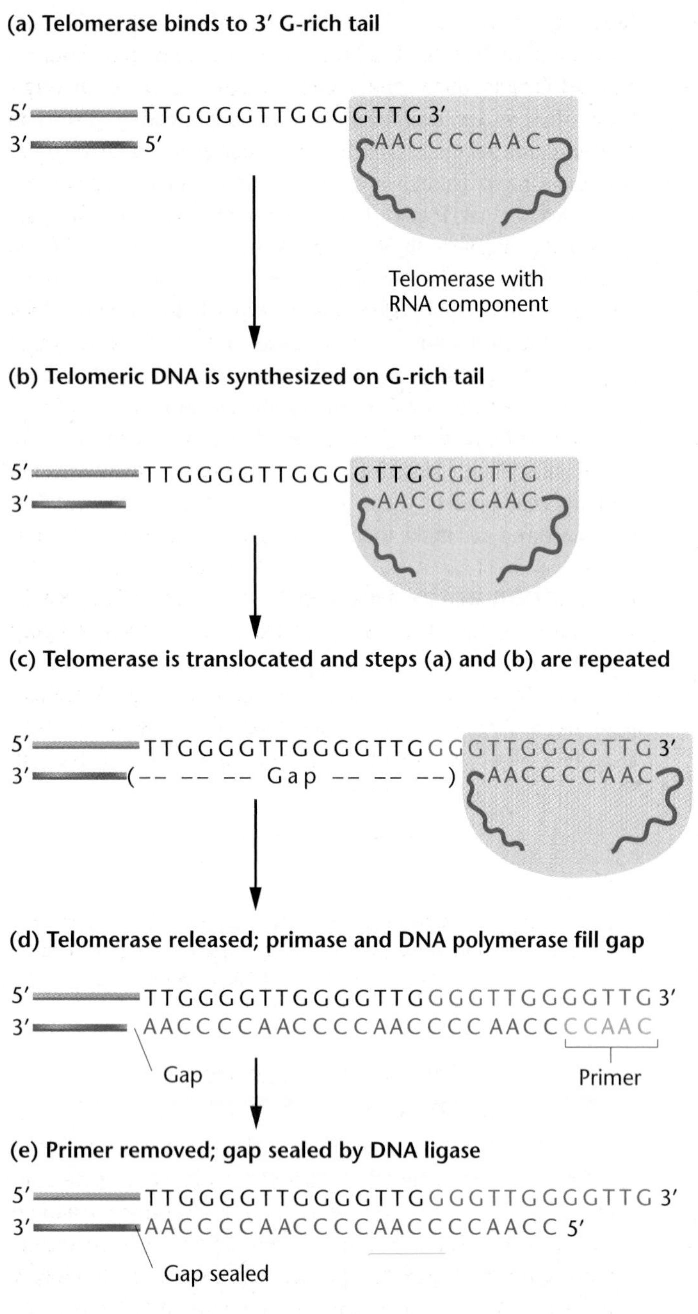

FIGURE 10–17 The predicted solution to the problem posed in Figure 10–16. The enzyme telomerase (with its RNA component shown in green) directs synthesis of repeated TTGGGG sequences, resulting in the formation of an extended 3′-overhang. This facilitates DNA synthesis on the opposite strand, filling in the gap that would otherwise be created on the ends of linear chromosomes during each replication cycle.

Normally, the newly created gap would be filled in starting with the addition of a nucleotide to the adjacent 3′-OH group [the group to the right of gap (a) in Figure 10–16]. However, since the final gap [gap (b) in Figure 10–16] is at the end of the strand being synthesized, there is no Okazaki fragment present to provide the needed 3′-OH group. Thus, in the situation depicted in Figure 10–16, a gap remains on the lagging strand produced in each successive round of synthesis, shortening the double-stranded end of the chromosome by the length of the RNA primer. With each round of replication, the shortening becomes more severe in each daughter cell, eventually extending beyond the telomere to potentially delete gene-coding regions.

A unique eukaryotic enzyme called **telomerase,** first discovered by Elizabeth Blackburn and Carol Greider in studies of *Tetrahymena,* has helped us understand the solution to the problem of telomere shortening. As noted earlier, telomeric DNA in eukaryotes is always found to consist of many short, repeated nucleotide sequences, with the G-rich strand overhanging in the form of a single-stranded tail. In *Tetrahymena* the tail contains several repeats of the sequence 5′-TTGGGG-3′. As we will see, telomerase is capable of adding several more repeats of this six-nucleotide sequence to the 3′-end of the G-rich strand (using 5′−3′ synthesis). Detailed investigation by Blackburn and Greider of how the *Tetrahymena* telomerase enzyme accomplishes this synthesis yielded an extraordinary finding. The enzyme is highly unusual in that it is a *ribonucleoprotein,* containing within its molecular structure a short piece of RNA that is essential to its catalytic activity. The RNA component serves as both a "guide" (to proper attachment of the enzyme to the telomere) and a template for the synthesis of its DNA complement, the latter being a process called **reverse transcription.** In *Tetrahymena,* the RNA contains the sequence AACCCCAAC, within which is found the complement of the repeating telomeric DNA sequence that must be synthesized (TTGGGG).

Figure 10–17 shows one model of how researchers envision the enzyme working. Part of the RNA sequence of the enzyme (shown in green) base pairs with the ending sequence of the single-stranded overhanging DNA, while the remainder of the RNA extends beyond the overhang. Next, reverse transcription of this extending RNA sequence—synthesizing DNA on an RNA template—extends the length of the G-rich lagging strand. It is believed that the enzyme is then translocated toward the (newly formed) end of the strand, and the same events are repeated, continuing the extension process.

Note that the newly formed extension of the overhanging G-rich strand by telomerase has not filled in the "gap" that represents the "telomere problem" during replication. However, it is felt that the DNA extension created by telomerase facilitates DNA synthesis on the opposite C-rich strand. One model suggests that the single-stranded extension loops back on itself, providing the 3′-OH group necessary for the initiation of synthesis to fill the gap. A second model, shown in Figure 10–17 (d) and (e), suggests that conventional DNA synthesis involving primase, DNA polymerase, and DNA ligase occurs, using the overhang as a template, filling in and closing the gap.

ESSENTIAL POINT

Replication at the ends of linear chromosomes in eukaryotes poses a special problem that can be solved by the presence of telomeres and by a unique RNA-containing enzyme called telomerase.

10.8 DNA Recombination, Like DNA Replication, Is Directed by Specific Enzymes

We now turn to a topic previously discussed in Chapter 5—genetic recombination. There, we pointed out that the process of crossing over between homologs depends on the breakage and rejoining of the DNA strands, and results in the exchange of genetic information between DNA molecules. Now that we have discussed the chemistry and replication of DNA, we can consider how recombination occurs at the molecular level. In general, our discussion pertains to genetic exchange between any two homologous double-stranded DNA molecules, whether they be viral or bacterial chromosomes or eukaryotic homologs during meiosis. Genetic exchange at equivalent positions along two chromosomes with substantial DNA sequence homology is referred to as **general,** or **homologous recombination.**

Several models attempt to explain homologous recombination, but they all have certain features in common. First, all are based on proposals first put forth independently by Robin Holliday and Harold L. K. Whitehouse in 1964. Second, they all depend on the complementarity between DNA strands to explain the precision of the exchange. Finally, each model relies on a series of enzymatic processes in order to accomplish genetic recombination.

One such model is shown in Figure 10–18, p. 220. It begins with two paired DNA duplexes, or homologs [Step (a)], in each of which an endonuclease introduces a single-stranded nick at an identical position [Step (b)]. The internal strand endings produced by these cuts are then displaced and subsequently pair with their complements on the opposite duplex [Step (c)]. Next, a ligase seals the loose ends [Step (d)], creating hybrid duplexes called **heteroduplex DNA molecules,** held together by a cross-bridge structure. The position of this cross bridge can then move down the chromosome by a process referred to as **branch migration** [Step (e)], which occurs as a result of a zipperlike action as hydrogen bonds are broken and then reformed between complementary bases of the displaced strands of each duplex. This migration yields an increased length of heteroduplex DNA on both homologs.

If the duplexes bend [Step (f)] and the bottom portion shown in the figure rotates 180° [Step (g)], an intermediate planar structure called a χ (chi) form—or **Holliday structure**—is created. If the two strands on opposite homologs previously uninvolved in the exchange are now nicked by an endonuclease [Step (h)] and ligation occurs as in Step (i), two recombinant duplexes are created. Note that the arrangement of alleles is altered as a result of this recombination.

Whereas the model above involves *single-stranded breaks,* other recombination models have been proposed that involve *double-stranded breaks* in one of the DNA double helices. In these models, endonucleases remove nucleotides at the breakpoint, creating 3′ overhangs on each strand. One of the broken strands invades the intact double helix of the other homolog, and both strands line up with the intact homolog. DNA repair synthesis then fills all gaps, and two Holliday junctions are formed. Endonuclease cleavages and ligations finalize the exchange. The end result is the same as our original model: genetic exchange occurs during crossing over in meiotic tetrads. A similar mechanism is thought to occur when cells repair double-stranded breaks in chromosomes. Such damage can occur from numerous causes, including the energy of ionizing radiation. We discuss this topic again in Chapter 14.

WEB TUTORIAL DNA RECOMBINATION

As with DNA replication, the processes involved in DNA recombination require the activities of numerous enzymes and other proteins. Mutations in genes encoding these proteins may cause defects in recombination, as well as in DNA repair and replication. One of the key proteins involved in *E. coli* recombination is the **RecA protein.** This molecule promotes the exchange of reciprocal single-stranded DNA molecules as occurs in Step (c) of the model. RecA also enhances the hydrogen-bond formation during strand displacement, thus initiating heteroduplex formation. The **RecB, RecC,** and **RecD proteins** can cleave DNA strands and help unwind the duplex. Other proteins are involved in branch migration and resolution of Holliday structures. DNA replication proteins, such as DNA polymerases, DNA ligase, gyrases, and single-stranded binding proteins, are also involved in DNA recombination and repair.

ESSENTIAL POINT

Homologous recombination between DNA molecules relies on precise alignment of homologs and the actions of a series of enzymes that can cut, realign, and reseal DNA strands.

Gene Conversion, a Consequence of DNA Recombination

A modification of the preceding model has helped us to better understand a unique genetic phenomenon known as **gene conversion.** Initially found in yeast by Carl Lindegren and in *Neurospora* by Mary Mitchell, gene conversion is characterized by a *nonreciprocal* genetic exchange between two closely linked genes. For example, if we were to cross two *Neurospora* strains, each bearing a separate mutation ($a+ \times +b$), a *reciprocal* recombination between the genes would yield spore pairs of the ++ and the *ab* genotypes. However, a nonreciprocal exchange yields one pair without the other. Working with pyridoxine mutants, Mitchell observed several asci-containing spore pairs with the ++ genotype, but not the reciprocal product (*ab*). Because the frequency of these events was higher than the predicted mutation rate and consequently could not be accounted for by mutation, they were called *gene conversions.* They were so named because it appeared that one allele had somehow been "converted" to another in which genetic exchange had also occurred. Similar findings come from studies of other fungi as well.

Gene conversion is now considered to be a consequence of the process of DNA recombination. One possible explanation interprets conversion as a mismatch of base pairs during heteroduplex formation. Mismatched regions of hybrid strands

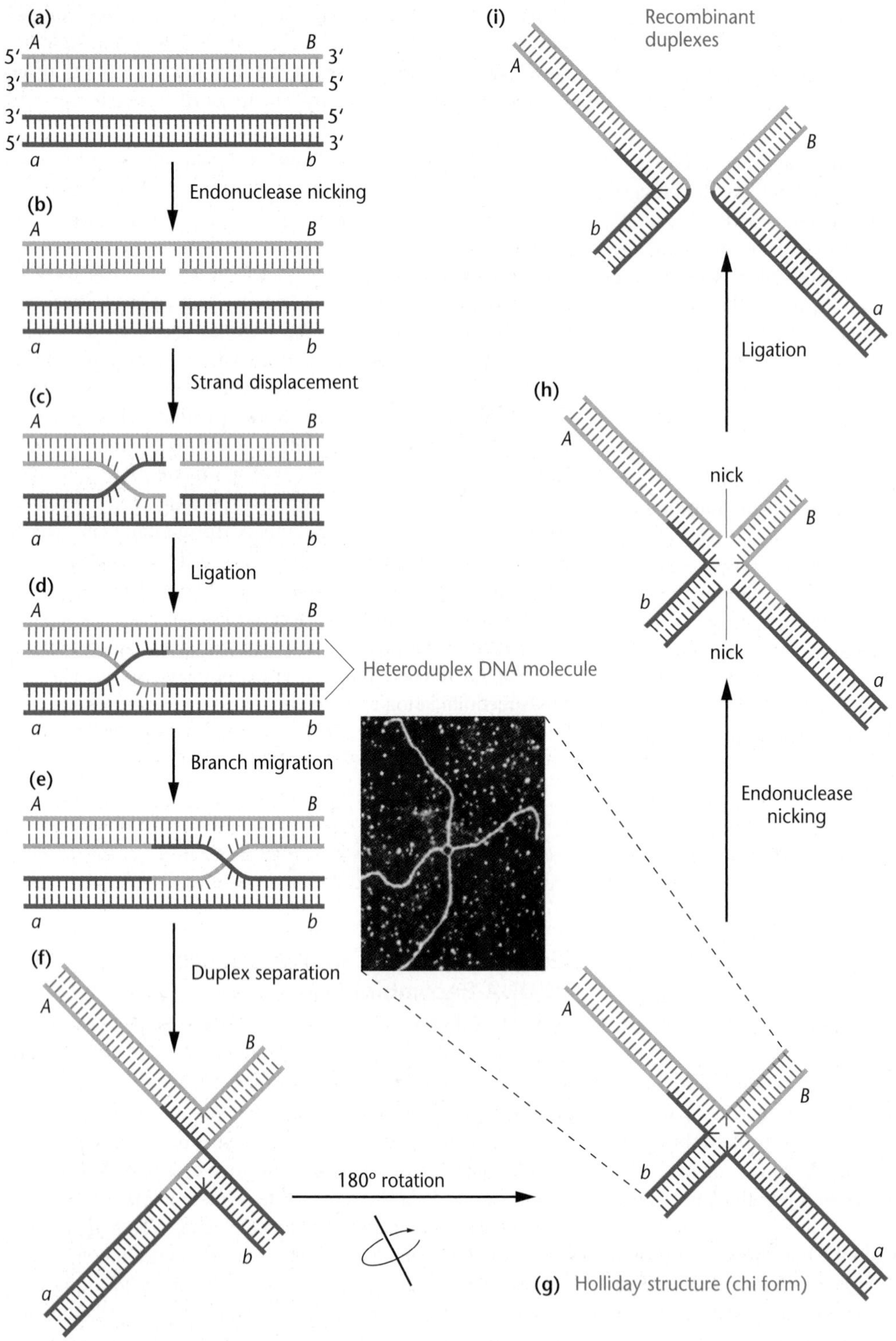

FIGURE 10–18 Model depicting how genetic recombination can occur as a result of the breakage and rejoining of heterologous DNA strands. Each stage is described in the text. The electron micrograph shows DNA in a χ-form structure similar to the diagram in (g); the DNA is an extended Holliday structure, derived from the *Col*E1 plasmid of *E. coli*. *David Dressler, Oxford University, England.*

can be repaired by the excision of one of the strands and the synthesis of the complement by using the remaining strand as a template. Excision may occur in either one of the strands, yielding two possible "corrections." One repairs the mismatched base pair and "converts" it to restore the original sequence. The other also corrects the mismatch but does so by copying the altered strand, creating a base-pair substitution. Conversion may have the effect of creating identical alleles on the two homologs that were different initially.

Gene-conversion events have helped to explain other puzzling genetic phenomena in fungi. For example, when mutant and wild-type alleles of a single gene are crossed, asci should yield equal numbers of mutant and wild-type spores. However, exceptional asci with 3:1 or 1:3 ratios are sometimes observed. These ratios can be understood in terms of gene conversion. The phenomenon also has been detected during mitotic events in fungi, as well as during the study of unique compound chromosomes in *Drosophila.*

GENETICS, TECHNOLOGY, AND SOCIETY

Telomeres: The Key to Immortality?

Humans, like all multicellular organisms, grow old and die. As we age, our immune systems become less efficient, wound healing is impaired, and tissues lose resilience. It has always been a mystery why we go through these age-related declines and why each species has a characteristic finite life span. Why do we grow old? Can we reverse this march to mortality? Some recent discoveries suggest that the answers to these questions may lie at the ends of our chromosomes.

The study of human aging begins with a study of human cells growing in culture dishes. Like the organisms from which the cells are taken, cells in culture have a finite life span. This *replicative senescence* was first noted by Leonard Hayflick in the 1960s. He reported that normal human fibroblasts lose their ability to grow and divide after about 50 cell divisions. These senescent cells remain metabolically active but can no longer proliferate. Eventually, they die. Although we don't know whether cellular senescence directly causes organismal aging, the evidence is suggestive. For example, cells derived from young people undergo more divisions than those from older people; cells from short-lived species stop growing after fewer divisions than those from longer-lived species; and cells from patients with premature aging syndromes undergo fewer divisions than those from normal patients.

Another characteristic of aging cells involves their telomeres. In most mammalian somatic cells, telomeres become shorter with each DNA replication because DNA polymerase cannot synthesize new DNA at the ends of each parent strand. However, as discussed in detail in this chapter, cells that undergo extensive proliferation, like embryonic cells, germ cells, and adult stem cells, maintain their telomere length by using *telomerase*—a remarkable RNA-containing enzyme that adds telomeric DNA sequences onto the ends of linear chromosomes. However, most somatic cells in adult organisms do not proliferate and do not contain active telomerase.

Could we gain perpetual youth and vitality by increasing our telomere lengths? Studies suggest that it may be possible to reverse senescence by artificially increasing the amount of telomerase in our cells. When investigators introduced cloned telomerase genes into normal human cells in culture, telomeres lengthened, and the cells continued to grow past their typical senescence point. These studies suggest that some of the atrophy of tissues that accompanies old age could be reversed by activating telomerase genes. However, before we use telomerase to achieve immortality, we need to consider a potential serious side effect: cancer.

Although normal cells shorten their telomeres and undergo senescence after a specific number of cell divisions, cancer cells do not. More than 80 percent of human tumor cells contain telomerase activity, maintain telomeres, and achieve immortality. Those that do not contain active telomerase use a less well understood mechanism known as ALT (for "alternative lengthening of telomeres").

These observations have motivated scientists to devise new cancer therapies based on the idea that agents that inhibit telomerase might destroy cancer cells by allowing telomeres to shorten, thereby forcing the cells into senescence. Because most normal human cells do not express telomerase, such a therapy might target tumor cells and be less toxic than most current anticancer drugs. Many such anti-telomerase drugs are currently under development, and some are in clinical trials.

Will a deeper understanding of telomeres allow us to both arrest cancers *and* reverse the descent into old age? Time will tell.

Your Turn

Take time, individually or in groups, to answer the following questions. Investigate the references and links to help you understand some of the research on telomeres, aging, and cancer.

1. How might our knowledge about telomeres and telomerase be applied to anti-aging strategies? Are such strategies or therapies being developed?

   Sources of information can be obtained by using the PubMed Web site (**http://www.ncbi.nlm.nih.gov/sites/entrez?db=PubMed**).

2. One anti-telomerase drug, called GRN163, is being developed by Geron Corporation as a treatment for cancer. How does GRN163 work? What is the current status of GRN163 clinical trials? What are some possible side effects for anti-telomerase drugs?

   Read about this drug and its clinical trials on the Geron Web site at **http://www.geron.com**. *Search on PubMed for scientific papers dealing with GRN163's anticancer effects.*

3. People suffering from chronic stress appear to have more health problems and to age prematurely. Is there any evidence that chronic stress, poor health, and telomere length are linked? How might stress affect telomere length or vice versa?

   Some recent papers suggest how these phenomena may be linked. One such paper is Epel, E. S. et al. 2004. Accelerated telomere shortening in response to life stress. *Proc. Natl. Acad. Sci. USA* **101**(49): 17312–17315.

4. In 2006, the Lasker Award for Basic Medical Research was awarded to Drs. Elizabeth Blackburn, Carol Greider, and Jack Szostak. How did the unexpected intersections of people, ideas, and good fortune lead to their discovery of telomerase and its role in aging and cancer? What is the future for this research?

   Listen to a series of fascinating interviews with these scientists, in which they tell their stories about their research and where they see the field going, at **http://www.laskerfoundation.org/2006videoawards/**

CASE STUDY At loose ends

A researcher was asked if his work on human telomere replication was related to any genetic disorders. He replied that one might think that any such mutation would be lethal during early development, but in fact a rare human genetic disorder affecting telomeres is known. This disorder, dyskeratosis congenita (DKC), is associated with mutations in the protein subunits of telomerase, the enzyme responsible for replicating the ends of eukaryotic chromosomes. Initial symptoms appear in tissues derived from rapidly dividing cells, including the skin, nails, and bone marrow, and first affect children between the ages of 5 and 15 years.

This disorder raises several interesting questions.

1. How could such individuals survive?
2. Why are the tissues derived from rapidly dividing cells initially affected?
3. Is this disorder likely to impact the life span?
4. Would you predict that mutations in the RNA component of telomerase might also cause DKC?

INSIGHTS AND SOLUTIONS

1. Predict the theoretical results of conservative and dispersive replication of DNA under the conditions of the Meselson–Stahl experiment. Follow the results through two generations of replication after cells have been shifted to an ^{14}N-containing medium, using the following sedimentation pattern.

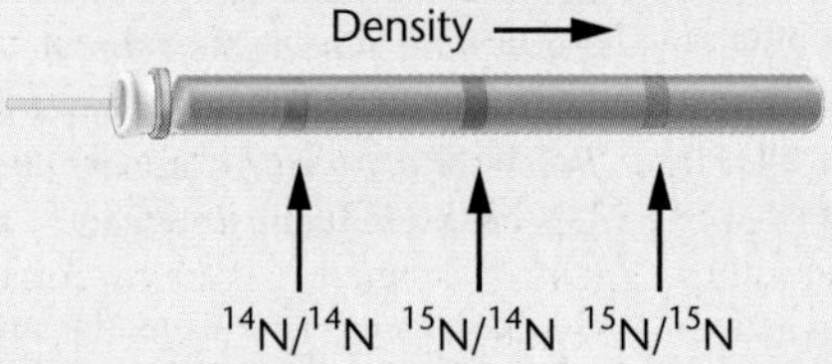

Solution:

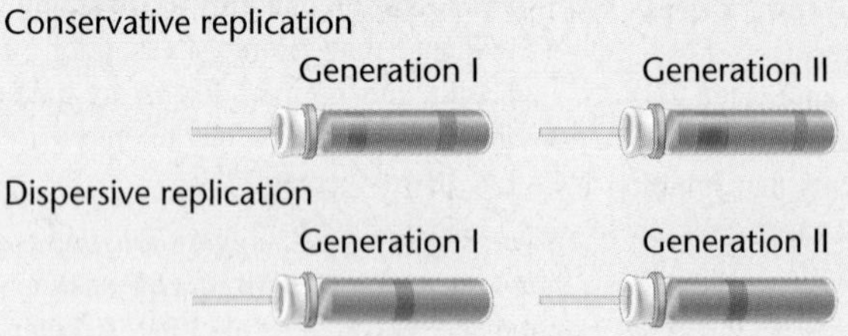

2. Mutations in the *dnaA* gene of *E. coli* are lethal and can only be studied following the isolation of conditional, temperature-sensitive mutations. Such mutant strains grow nicely and replicate their DNA at the permissive temperature of 18°C, but they do not grow or replicate their DNA at the restrictive temperature of 37°C. Two observations were useful in determining the function of the DnaA protein product. First, *in vitro* studies using DNA templates that have unwound do not require the DnaA protein. Second, if intact cells are grown at 18°C and are then shifted to 37°C, DNA synthesis continues at this temperature until one round of replication is completed and then stops. What do these observations suggest about the role of the *dnaA* gene product?

Solution: At 18°C (the permissive temperature), the mutation is not expressed and DNA synthesis begins. Following the shift to the restrictive temperature, the already initiated DNA synthesis continues, but no new synthesis can begin. Because the DnaA protein is not required for synthesis of unwound DNA, these observations suggest that, *in vivo,* the DnaA protein plays an essential role in DNA synthesis by interacting with the intact helix and somehow facilitating the localized denaturation necessary for synthesis to proceed.

PROBLEMS AND DISCUSSION QUESTIONS

1. Compare conservative, semiconservative, and dispersive modes of DNA replication.
2. Describe the role of ^{15}N in the Meselson–Stahl experiment. See **Now Solve This** on page 206.
3. In the Meselson–Stahl experiment, which of the three modes of replication could be ruled out after one round of replication? after two rounds?
4. Predict the results of the experiment by Taylor, Woods, and Hughes if replication were (a) conservative and (b) dispersive.
5. Reconsider Problem 30 in Chapter 9. In the model you proposed, could the molecule be replicated semiconservatively? Why? Would other modes of replication work?
6. What are the requirements for *in vitro* synthesis of DNA under the direction of DNA polymerase I?
7. In Kornberg's initial experiments, it was rumored that he grew *E. coli* in Anheuser-Busch beer vats. (He was working at Washington University in St. Louis.) Why do you think this might have been helpful to the experiment?
8. How did Kornberg assess the fidelity of DNA polymerase I in copying a DNA template?
9. Which characteristics of DNA polymerase I raised doubts that its *in vivo* function is the synthesis of DNA leading to complete replication?
10. Kornberg showed that nucleotides are added to the 3′-end of each growing DNA strand. In what way does an exposed 3′-OH group participate in strand elongation?
11. What was the significance of the *polA1* mutation?
12. Summarize and compare the properties of DNA polymerase I, II, and III.
13. List and describe the function of the ten subunits constituting DNA polymerase III. Distinguish between the holoenzyme and the core enzyme.

14. Distinguish between (a) unidirectional and bidirectional synthesis, and (b) continuous and discontinuous synthesis of DNA.
15. List the proteins that unwind DNA during *in vivo* DNA synthesis. How do they function?
16. Define and indicate the significance of (a) Okazaki fragments, (b) DNA ligase, and (c) primer RNA during DNA replication.
17. Outline the current model for DNA synthesis.
18. Why is DNA synthesis expected to be more complex in eukaryotes than in bacteria? How is DNA synthesis similar in the two types of organisms?
19. If the analysis of DNA from two different microorganisms demonstrated very similar base compositions, are the DNA sequences of the two organisms also nearly identical?
20. Suppose that *E. coli* synthesizes DNA at a rate of 100,000 nucleotides per minute and takes 40 minutes to replicate its chromosome. (a) How many base pairs are present in the entire *E. coli* chromosome? (b) What is the physical length of the chromosome in its helical configuration—that is, what is the circumference of the chromosome if it were opened into a circle?
21. Several temperature-sensitive mutant strains of *E. coli* display the following characteristics. Predict what enzyme or function is being affected by each mutation.
    (a) Newly synthesized DNA contains many mismatched base pairs.
    (b) Okazaki fragments accumulate, and DNA synthesis is never completed.
    (c) No initiation occurs.
    (d) Synthesis is very slow.
    (e) Supercoiled strands remain after replication, which is never completed. See **Now Solve This** on page 214.
22. Define gene conversion, and describe how this phenomenon is related to genetic recombination.
23. Many of the gene products involved in DNA synthesis were initially defined by studying mutant *E. coli* strains that could not synthesize DNA. (a) The *dnaE* gene encodes the α subunit of DNA polymerase III. What effect is expected from a mutation in this gene? How could the mutant strain be maintained? (b) The *dnaQ* gene encodes the ε subunit of DNA polymerase. What effect is expected from a mutation in this gene?
24. In 1994, telomerase activity was discovered in human cancer cell lines. Although telomerase is not active in human somatic tissue, this discovery indicated that humans do contain the genes for telomerase proteins and telomerase RNA. Since inappropriate activation of telomerase can cause cancer, why do you think the genes coding for this enzyme have been maintained in the human genome throughout evolution? Are there any types of human body cells where telomerase activation would be advantageous or even necessary? Explain.
25. The genome of *D. melanogaster* consists of approximately 1.7×10^8 base pairs. DNA synthesis occurs at a rate of 30 base pairs per second. In the early embryo, the entire genome is replicated in five minutes. How many bidirectional origins of synthesis are required to accomplish this feat?
26. Assume a hypothetical organism in which DNA replication is conservative. Design an experiment similar to that of Taylor, Woods, and Hughes that will unequivocally establish this fact. Using the format established in Figure 10–5, draw sister chromatids and illustrate the expected results establishing this mode of replication.
27. DNA polymerases in all organisms add only 5′ nucleotides to the 3′ end of a growing DNA strand, never to the 5′ end. One possible reason for this is the fact that most DNA polymerases have a proofreading function that would not be *energetically* possible if DNA synthesis occurred in the 3′ to 5′ direction. (a) Sketch the reaction that DNA polymerase would have to catalyze if DNA synthesis occurred in the 3′ to 5′ direction. (b) Consider the information in your sketch and speculate as to why proofreading would be problematic.
28. An alien organism was investigated. It displayed characteristics of eukaryotes. When DNA replication was examined, two unique features were apparent: (1) no Okazaki fragments were observed; and (2) there was a telomere problem (i.e., telomeres were shortened) but only on one end of the chromosome. Create a model of DNA that is consistent with both of these observations. See **Now Solve This** on page 213.
29. Assume that the sequence of bases given in this problem is present on one nucleotide chain of a DNA duplex and that the chain has opened up at a replication fork. Synthesis of an RNA primer occurs on this template starting at the base that is underlined. (a) If the RNA primer consists of eight nucleotides, what is its base sequence? (b) In the intact RNA primer, which nucleotide has a free 3′-OH terminus?

    3′.....GGCTACC**T**GGATTCA......5′

30. Given the following diagram, assume that the phase G1 chromosome on the left underwent one round of replication in ^{3}H-thymidine and that the metaphase chromosome on the right had both chromatids labeled. Which of the replicative models (conservative, dispersive, semiconservative) could be eliminated by this observation?

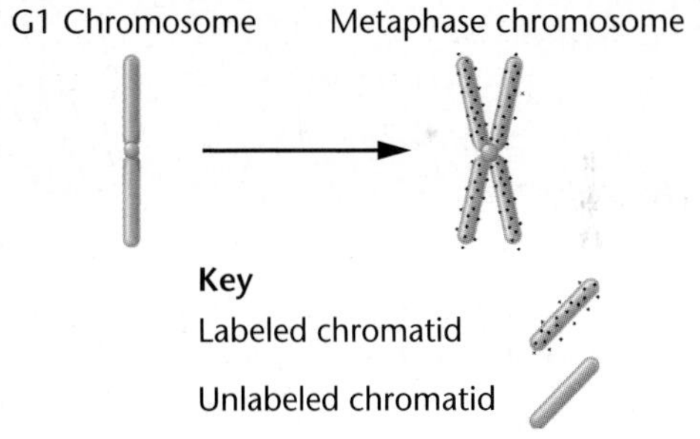

31. DNA is allowed to replicate in moderately radioactive ^{3}H-thymidine for several minutes and is then switched to a highly radioactive medium for several more minutes. Synthesis is stopped, and the DNA is subjected to autoradiography and electron microscopy. Interpret as much as you can regarding DNA replication from the drawing of the electron micrograph presented here.

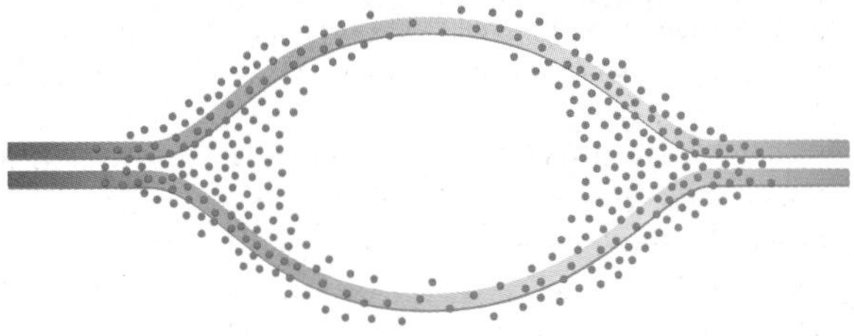

A chromatin fiber viewed using a scanning transmission electron microscope (STEM)

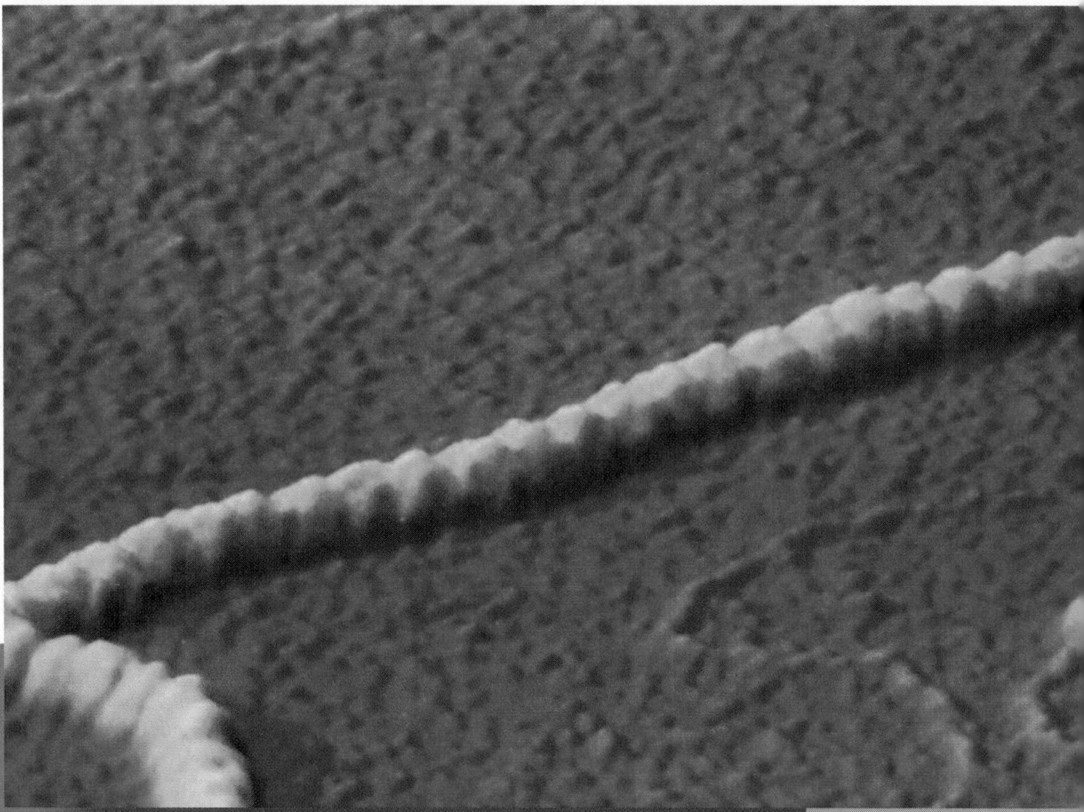

11

Chromosome Structure and DNA Sequence Organization

Chapter Concepts

- Genetic information in viruses, bacteria, mitochondria, and chloroplasts is most often contained in a short, circular DNA molecule, relatively free of associated proteins.
- Eukaryotic cells, in contrast to viruses and bacteria, contain relatively large amounts of DNA organized into nucleosomes and present during most of the cell cycle as chromatin fibers.
- Uncoiled chromatin fibers characteristic of interphase coil up and condense into chromosomes during eukaryotic cell division.
- Eukaryotic genomes are characterized by both unique and repetitive DNA sequences.
- Eukaryotic genomes consist mostly of noncoding DNA sequences.

Once geneticists understood that DNA houses genetic information, it became very important to determine how DNA is organized into genes and how these basic units of genetic function are organized into chromosomes. In short, the major question had to do with how the genetic material was organized as it makes up the genome of organisms. There has been much interest in this question because knowledge of the organization of the genetic material and associated molecules is important to understanding many other areas of genetics. For example, the way in which the genetic information is stored, expressed, and regulated must be related to the molecular organization of the genetic molecule, DNA. How genomic organization varies in different organisms—from viruses to bacteria to eukaryotes—will undoubtedly provide a better understanding of the evolution of organisms on Earth.

In this chapter, we focus on the various ways DNA is organized into chromosomes. These structures have been studied using numerous techniques, instruments, and approaches, including analysis by light microscopy and electron microscopy. More recently, molecular analysis has provided significant insights into chromosome organization. In the first half of the chapter, after surveying what we know about chromosomes in viruses and bacteria, we examine the large specialized eukaryotic structures called polytene and lampbrush chromosomes. Then, in the second half, we discuss how eukaryotic chromosomes are organized at the molecular level—for example, how DNA is complexed with proteins to form chromatin and how the chromatin fibers characteristic of interphase are condensed into chromosome structures visible during mitosis and meiosis. We conclude the chapter by examining certain aspects of DNA sequence organization in eukaryotic genomes.

How Do We Know?

In this chapter, we will focus on chromosome structure and the way DNA is organized within chromosomes. As you study this topic, you should try to answer several fundamental questions:

1. What is the experimental basis for concluding that DNA contained within puffs in polytene chromosomes and loops in lampbrush chromosomes are areas of intense transcription of RNA?
2. How do we know that mitochondria and chloroplasts contain DNA?
3. What experimental evidence supports the idea that eukaryotic chromatin exists in the form of repeating nucleosomes, each consisting of about 200 base pairs and an octamer of histones?
4. How do we know that satellite DNA consists of repetitive sequences and has been derived from regions of the centromere?

11.1 Viral and Bacterial Chromosomes Are Relatively Simple DNA Molecules

The chromosomes of viruses and bacteria are much less complicated than those in eukaryotes. They usually consist of a single nucleic acid molecule, unlike the multiple chromosomes comprising the genome of higher forms. Compared to eukaryotes, the chromosomes contain much less genetic information and the DNA is not as extensively bound to proteins. These characteristics have greatly simplified analysis, and we now have a fairly comprehensive view of the structure of viral and bacterial chromosomes.

The chromosomes of viruses consist of a nucleic acid molecule—either DNA or RNA—that can be either single- or double-stranded. They can exist as circular structures (closed loops), or they can take the form of linear molecules. For example, the single-stranded DNA of the **ϕX174 bacteriophage** and the double-stranded DNA of the **polyoma virus** are closed loops housed within the protein coat of the mature viruses. The **bacteriophage lambda (λ),** on the other hand, possesses a linear double-stranded DNA molecule prior to infection, which closes to form a ring upon its infection of the host cell. Still other viruses, such as the T-even series of bacteriophages, have linear double-stranded chromosomes of DNA that do not form circles inside the bacterial host. Thus, circularity is not an absolute requirement for replication in viruses.

Viral nucleic acid molecules have been seen with the electron microscope. Figure 11–1, p. 226, shows a mature bacteriophage λ and its double-stranded DNA molecule in the circular configuration. One constant feature shared by viruses, bacteria, and eukaryotic cells is the ability to package an exceedingly long DNA molecule into a relatively small volume. In λ, the DNA is 17 μm long and must fit into the phage head, which is less than 0.1 μm on any side. Table 11.1 compares the length of the chromosomes of several viruses to the size of their head structure. In each case, a similar packaging feat must be accomplished. Compare the dimensions given for phage T2 with the micrograph of both the DNA and the viral particle shown in Figure 11–2, p. 227. Seldom does the space available in the head of a virus exceed the chromosome volume by more than a factor of two. In many cases, almost all of the space is filled, indicating nearly perfect packing. Once packed within the head, the genetic material is functionally inert until it is released into a host cell.

Bacterial chromosomes are also relatively simple in form. They always consist of a double-stranded DNA molecule, compacted into a structure sometimes referred to as the **nucleoid.** *Escherichia coli,* the most extensively studied bacterium, has a large circular chromosome measuring approximately 1200 μm (1.2 mm) in length that may occupy up to one-third of the volume of the cell. When the cell is gently lysed and the chromosome is released, it can be visualized under the electron microscope (Figure 11–3, p. 227.)

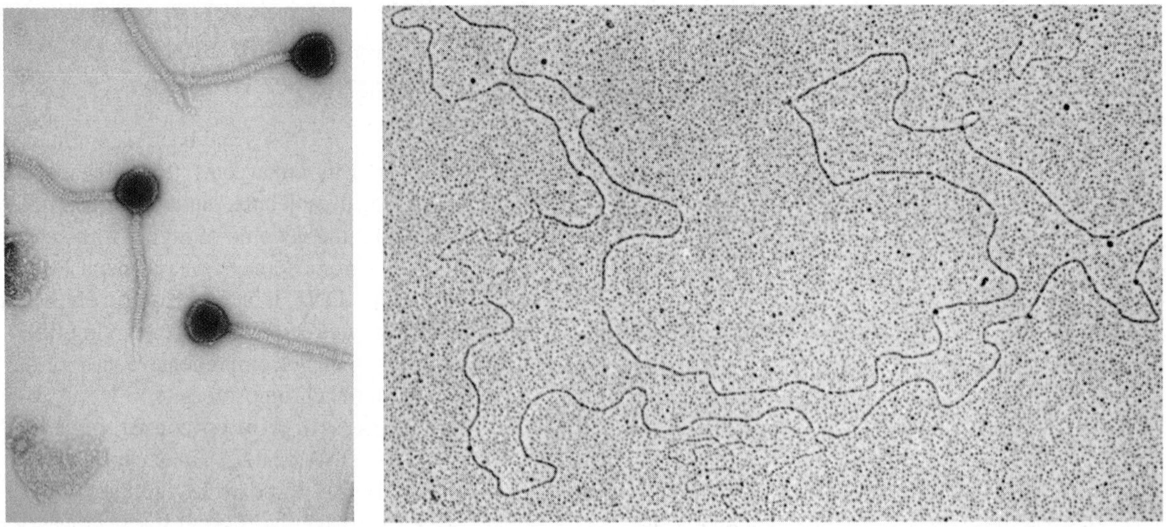

FIGURE 11–1 Electron micrographs of phage λ (a) and the DNA that was isolated from it (b). The chromosome is 17 μm long. Note that the phages are magnified about five times more than the DNA.

TABLE 11.1 The Genetic Material of Representative Viruses and Bacteria

			Nucleic Acid		
	Organism	Type	SS or DS*	Length (μm)	Overall Size of Viral Head or Bacteria (μm)
Viruses	φX174	DNA	SS	2.0	0.025 × 0.025
	Tobacco mosaic virus	RNA	SS	3.3	0.30 × 0.02
	Phage λ	DNA	DS	17.0	0.07 × 0.07
	T2 phage	DNA	DS	52.0	0.07 × 0.10
Bacteria	*Haemophilus influenzae*	DNA	DS	832.0	1.00 × 0.30
	Escherichia coli	DNA	DS	1200.0	2.00 × 0.50

SS = single-stranded, DS = double-stranded.

DNA in bacterial chromosomes is associated with several types of **DNA-binding proteins.** Two, called **HU** and **H1 proteins,** are small but abundant in the cell and contain a high percentage of positively charged amino acids that can bond ionically to the negative charges of the phosphate groups in DNA. Although these proteins are structurally similar to molecules called histones that are associated with eukaryotic DNA, they clearly do not compact DNA to the extent of histones, since the bacterial chromosome is *not* functionally inert and can be readily replicated and transcribed. Nor are these proteins essential since null mutations in the genes encoding them are not lethal.

ESSENTIAL POINT

Bacteriophage and bacterial chromosomes, in contrast to eukaryotes, are largely devoid of associated proteins, are of much smaller size, and most often consist of circular DNA.

NOW SOLVE THIS

Problem 2 on page 238 involves viral chromosomes that are linear in the bacteriophage but circularize after they enter the bacterial host cell. You are asked to consider the advantages of circular DNA molecules versus linear molecules during replication.

Hint: Recall from Chapter 10 that the enzyme telomerase is essential to replication in eukaryotes.

11.2 Mitochondria and Chloroplasts Contain DNA Similar to Bacteria and Viruses

Numerous studies demonstrate that both **mitochondria** and **chloroplasts** contain their own DNA and genetic system for expressing this information. This was first suggested by the discovery of mutations in yeast, other fungi, and plants that

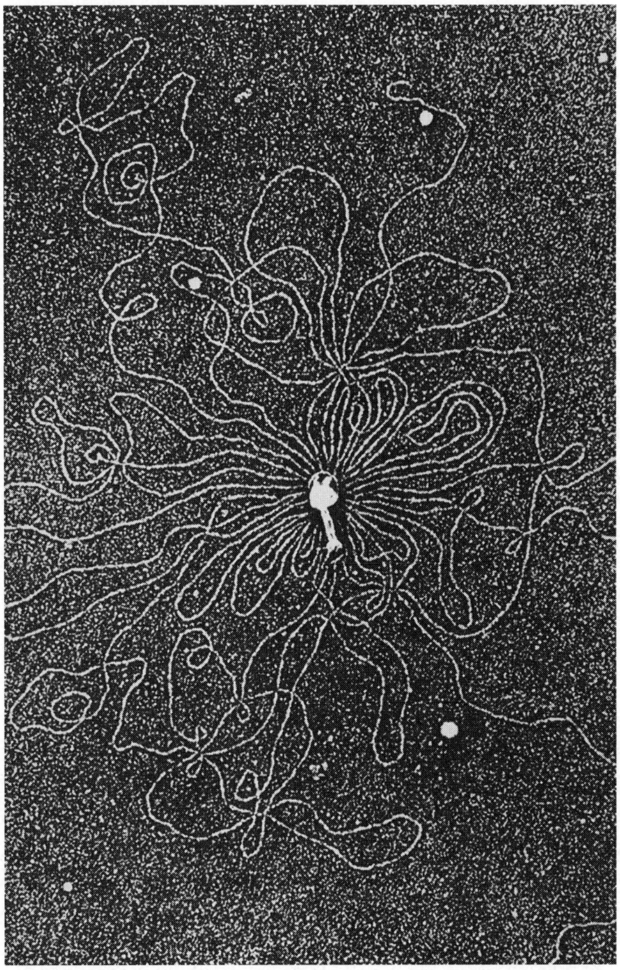

FIGURE 11–2 Electron micrograph of bacteriophage T2, which has had its DNA released by osmotic shock. The chromosome is 52 μm long.

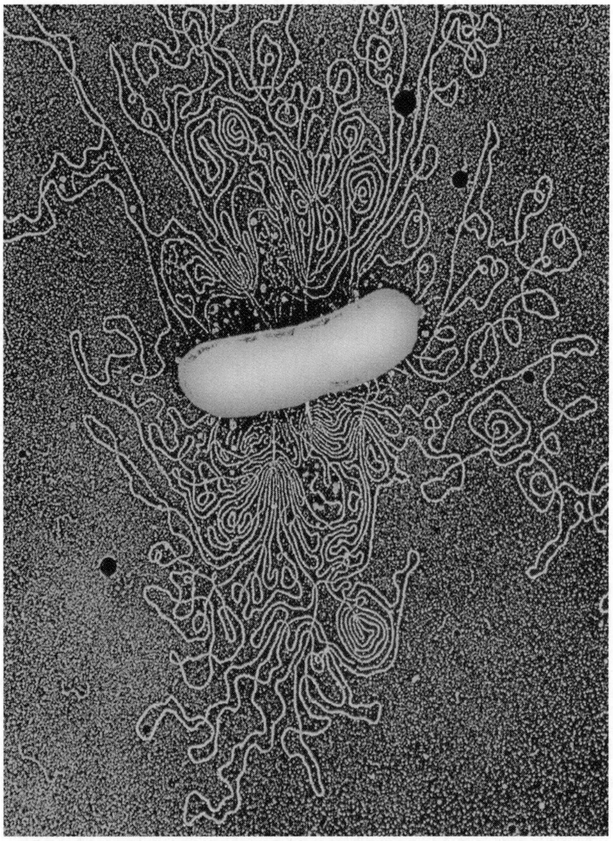

FIGURE 11–3 Electron micrograph of the bacterium *E. coli,* which has had its DNA released by osmotic shock. The chromosome is 1200 μm long.

produce altered phenotypes that can be linked to these organelles (see Chapter 4). Thus, geneticists set out to look for more direct evidence of DNA in these organelles. Electron microscopists not only documented the presence of DNA in both organelles, they also saw DNA in a form quite unlike that seen in the nucleus of the eukaryotic cells that house these organelles. This DNA looked remarkably similar to that seen in viruses and bacteria. This similarity, along with other observations, led to the idea that mitochondria and chloroplasts arose independently more than a billion years ago from free-living, prokaryote-like organisms that possessed the ability to undergo aerobic respiration (mitochondria) or photosynthesis (chloroplasts). This theory, called the **endosymbiotic hypothesis,** was championed by Lynn Margulis and others. They proposed that the prokaryotes were engulfed by larger primitive eukaryotic cells, which lacked these bioenergetic functions. A symbiotic relationship developed whereby the prokaryotic organisms eventually lost their ability to function independently, while the eukaryotic host cells gained the ability to either respire aerobically or undergo photosynthesis. The basic tenets of this theory are widely accepted.

Molecular Organization and Gene Products of Mitochondrial DNA

Extensive information is now available on the molecular aspects and gene products of **mitochondrial DNA (mtDNA).** In most eukaryotes, mtDNA is a double-stranded closed circle (Figure 11–4) that replicates semiconservatively and is free of the chromosomal proteins characteristic of eukaryotic DNA. An exception is found in some ciliated protozoans, in

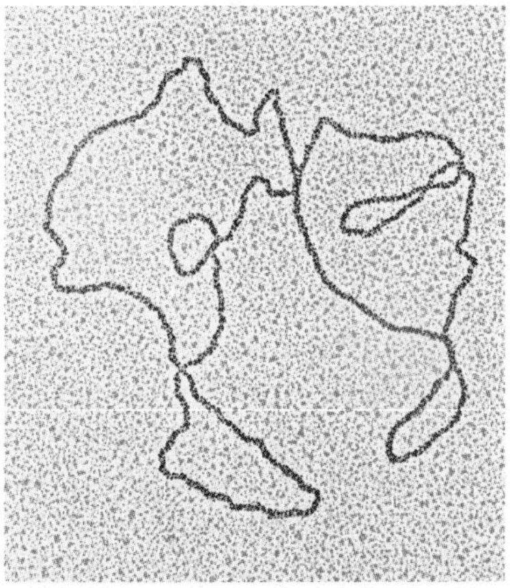

FIGURE 11–4 Electron micrograph of mitochondrial DNA (mtDNA) derived from *Xenopus laevis.*

which the DNA is linear. In size, mtDNA differs greatly among organisms. In a variety of animals, including humans, mtDNA consists of about 16,000 to 18,000 bp (16–18 kb). However, yeast (*Saccharomyces*) contains 75 kb of mtDNA. Plants typically exceed this amount—367 kb is present in mitochondria in the mustard plant, *Arabidopsis*. Vertebrates have 5 to 10 such DNA molecules per organelle, while plants have 20 to 40 copies per organelle.

There are several other noteworthy aspects of mtDNA. With only rare exceptions, *introns,* the noncoding regions of genes, are absent from mitochondrial genes, and there are few or no gene repetitions. Nor is there usually much in the way of intergenic spacer DNA. This is particularly true in species whose mtDNA is fairly small in size, such as humans. In *Saccharomyces,* with a much larger mtDNA molecule, much of the excess DNA is accounted for by introns and intergenic spacer DNA. As will be discussed in Chapter 12, the expression of mitochondrial genes uses several modifications of the otherwise standard genetic code. Also of interest is the fact that replication in mitochondria is dependent on enzymes encoded by nuclear DNA.

In humans, mtDNA encodes 2 ribosomal RNAs (rRNAs), 22 transfer RNAs (tRNAs), as well as 13 polypeptides essential to the oxidative respiratory functions of the organelles. In almost every case, the polypeptides are part of multichain proteins, and the other polypeptides of each protein are encoded in the nucleus, synthesized in the cytoplasm, and then transported into the organelle.

Another interesting observation is that in vertebrate mtDNA, the two strands vary in density, as revealed by centrifugation. This provides researchers with a way to isolate the strands for study, designating one heavy (H) and the other light (L). Although most of the mitochondrial genes are encoded by the H strand, several are encoded by the complementary L strand.

As might be predicted by the endosymbiotic theory, ribosomes found in the organelle are different from those present in the neighboring cytoplasm. Mitochondrial ribosomes of different species vary considerably in their sedimentation coefficients, ranging from 55*S* to 80*S,* while cytoplasmic ribosomes are uniformly 80 (*S* refers to the Svedberg coefficient, described in Chapter 9, and is related to the molecule's size and shape).

The majority of proteins that function in mitochondria are encoded by nuclear genes. In fact, over 1000 nuclear-coded gene products are essential to biological activity in the organelle. They include, for example, DNA and RNA polymerases, initiation and elongation factors essential for translation, ribosomal proteins, aminoacyl tRNA synthetases, and several tRNA species. These imported components are distinct from their cytoplasmic counterparts, even though both sets are coded by nuclear genes. For example, the synthetase enzymes essential for charging mitochondrial tRNA molecules (a process essential to translation) show a distinct affinity for the mitochondrial tRNA species as compared with the cytoplasmic tRNAs. Similar affinity has been shown for the initiation and elongation factors. Furthermore, although bacterial and nuclear RNA polymerases are known to be composed of numerous subunits, the mitochondrial variety consists of only one polypeptide chain. This polymerase is generally sensitive to antibiotics that inhibit bacterial RNA synthesis, but not to eukaryotic inhibitors.

Molecular Organization and Gene Products of Chloroplast DNA

Chloroplasts, like mitochondria, contain an autonomous genetic system distinct from that found in the nucleus and cytoplasm. This system includes DNA as a source of genetic information and a complete protein-synthesizing apparatus. Also similar to mitochondria, the molecular components of the chloroplast translation apparatus are jointly derived from both nuclear and organelle genetic information.

Chloroplast DNA (cpDNA), shown in Figure 11–5, is fairly uniform in size among different organisms, ranging between 100 and 225 kb in length. It shares many similarities to DNA found in prokaryotic cells. It is circular and double stranded, and it is free of the associated proteins characteristic of eukaryotic DNA.

The size of cpDNA is much larger than that of mtDNA. To some extent, this can be accounted for by a larger number of genes. However, most of the difference appears to be due to the presence in cpDNA of many long noncoding nucleotide sequences both between and within genes, the latter being called **introns.** Duplications of many DNA sequences are also present.

In the green alga *Chlamydomonas,* there are about 75 copies of the chloroplast DNA molecule per organelle. Each copy consists of a length of DNA that contains 195,000 base pairs (195 kb). In higher plants, such as the sweet pea, multiple copies of the DNA molecule are also present in each organelle, but the molecule (134 kb) is considerably smaller than that in *Chlamydomonas.*

Numerous gene products encoded by chloroplast DNA function during translation within the organelle. Present on cpDNA of the moss, which is representative of a variety of higher plants, are two sets each of the genes coding for the

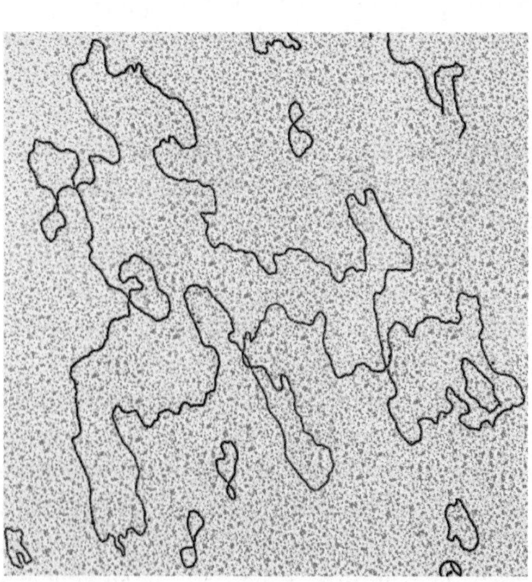

FIGURE 11–5 Electron micrograph of chloroplast DNA obtained from lettuce.

ribosomal RNAs—16*S*, and 23*S* rRNA. In addition, cpDNA encodes numerous transfer RNAs (tRNAs), as well as many ribosomal proteins specific to the chloroplast ribosomes. For example, in the liverwort, whose cpDNA was the first to be sequenced, there are genes encoding 30 tRNAs, RNA polymerase, multiple rRNAs, and numerous ribosomal proteins.

Still other chloroplast genes specific to the photosynthetic function have been identified. For example, in the moss, there are 92 chloroplast genes encoding proteins that are part of the thylakoid membrane, a cellular component integral to the light-dependent reactions of photosynthesis. Mutations in these genes may inactivate photosynthesis. A typical distribution of genes between the nucleus and the chloroplast is illustrated by one of the major photosynthetic enzymes, **ribulose-1-5-bisphosphate carboxylase** (known as **Rubisco**). This enzyme has its small subunit encoded by a nuclear gene, whereas the large subunit is encoded by cpDNA.

ESSENTIAL POINT

Mitochondria and chloroplasts contain DNA that is remarkably similar in form and appearance to some bacterial and bacteriophage DNA, lending support to the endosymbiotic hypothesis.

11.3 Specialized Chromosomes Reveal Variations in the Organization of DNA

We now consider two cases of genetic organization that demonstrate the specialized forms that eukaryotic chromosomes can take. Both types—*polytene chromosomes* and *lampbrush chromosomes*—are so large that their organization was discerned using light microscopy long before we understood how mitotic chromosomes form from interphase chromatin. The study of these chromosomes provided many of our initial insights into the arrangement and function of the genetic information.

Polytene Chromosomes

Giant **polytene chromosomes** are found in various tissues (salivary, midgut, rectal, and malpighian excretory tubules) in the larvae of some flies and in several species of protozoans and plants. Such structures, first observed by E. G. Balbiani in 1881, provided a model system for subsequent investigations of chromosomes. What is particularly intriguing about polytene chromosomes is that they can be seen in the nuclei of interphase cells.

Each polytene chromosome is 200 to 600 μm long, and when they are observed under the light microscope, they reveal a linear series of alternating bands and interbands (Figure 11–6). The banding pattern is distinctive for each chromosome in any given species. Individual bands are sometimes called **chromomeres,** a generalized term describing lateral condensations of material along the axis of a chromosome.

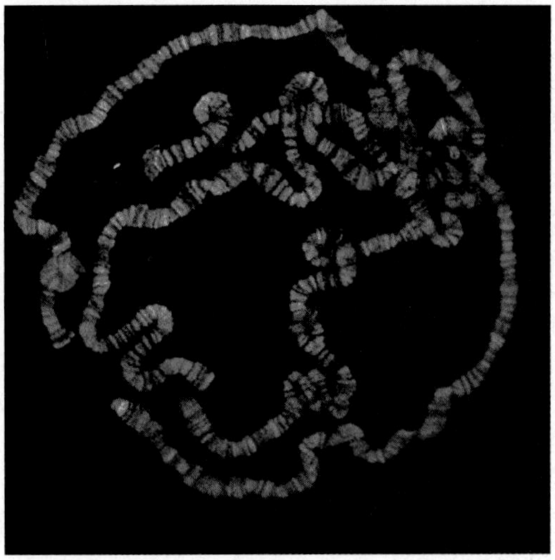

FIGURE 11–6 Polytene chromosomes derived from larval salivary gland cells of *Drosophila.*

Extensive study using electron microscopy and radioactive tracers led to an explanation for the unusual appearance of these chromosomes. First, polytene chromosomes represent paired homologs. This is highly unusual because they are present in somatic cells, where in most organisms, chromosomal material is normally dispersed as chromatin and homologs are not paired. Second, their large size and distinctiveness result from the many DNA strands that compose them. The DNA of these paired homologs undergoes many rounds of replication, *but without strand separation or cytoplasmic division.* As replication proceeds, chromosomes contain 1000 to 5000 DNA strands that remain in precise parallel alignment with one another, giving rise to the distinctive band pattern along the axis of the chromosome.

The presence of bands on polytene chromosomes was initially interpreted as the visible manifestation of individual genes. The discovery that the strands present in bands undergo localized uncoiling during genetic activity further strengthened this view. Each such uncoiling event results in what is called a **puff** because of its appearance (Figure 11–7, p. 230). That puffs are visible manifestations of gene activity (transcription that produces RNA) is evidenced by their high rate of incorporation of radioactively labeled RNA precursors, as assayed by autoradiography. Bands that are not extended into puffs incorporate fewer radioactive precursors or none at all.

The study of bands during development in insects, such as *Drosophila* and the midge fly *Chironomus,* reveals differential gene activity. A characteristic pattern of band formation, which is equated with gene activation, is observed as development proceeds. Despite attempts to resolve the issue, it is not yet clear how many genes are contained in each band; however, we do know that a band can contain up to 10^7 bp of DNA, certainly enough DNA to encode 50 to 100 average-sized genes.

Lampbrush Chromosomes

Another specialized chromosome that has given us insight into chromosomal structure is the **lampbrush chromosome,**

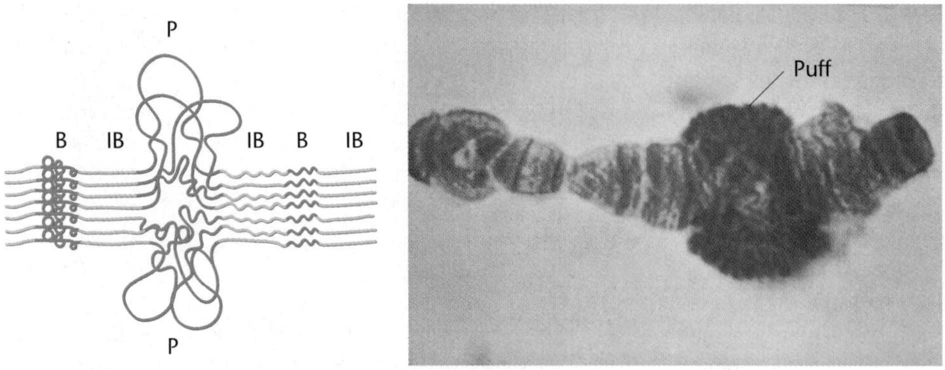

FIGURE 11–7 Photograph of a puff within a polytene chromosome. The diagram depicts the uncoiling of strands within a band (B) region to produce a puff (P) in polytene chromosomes. Interband regions (IB) are also labeled.

so named because it resembles the brushes used to clean kerosene-lamp chimneys in the nineteenth century. Lampbrush chromosomes were first discovered in 1892 in the oocytes of sharks and are now known to be characteristic of most vertebrate oocytes as well as the spermatocytes of some insects. Therefore, they are meiotic chromosomes. Most experimental work has been done with material taken from amphibian oocytes.

These unique chromosomes are easily isolated from oocytes in the diplotene stage of the first prophase of meiosis, where they are active in directing the metabolic activities of the developing cell. The homologs are seen as synapsed pairs held together by chiasmata. However, instead of condensing, as most meiotic chromosomes do, lampbrush chromosomes often extend to lengths of 500 to 800 μm. Later in meiosis, they revert to their normal length of 15 to 20 μm. Based on these observations, lampbrush chromosomes are interpreted as extended, uncoiled versions of the normal meiotic chromosomes.

The two views of lampbrush chromosomes in Figure 11–8 provide significant insights into their morphology. Part (a) shows the meiotic configuration under the light microscope. The linear axis of each structure contains a large number of condensed areas, and as with polytene chromosomes, these are referred to as *chromomeres.* Emanating from each chromomere is a pair of **lateral loops,** which give the chromosome its distinctive appearance. In part (b), the scanning electron micrograph (SEM) reveals adjacent loops present along one of the two axes of the chromosome, providing a clear view of the chromomeres and the chromosomal fibers emanating from them. As with bands in polytene chromosomes, much more DNA is present in each loop than is needed to encode a single gene. Consistent with the belief that each meiotic chromosome is composed of a pair of sister chromatids, each chromosomal loop is thought to be composed of one DNA double helix, whereas the central axis is composed of two DNA helices. Studies using radioactive RNA precursors have revealed that the loops are active in the synthesis of RNA. The lampbrush loops are thus viewed in a manner similar to puffs in polytene chromosomes, representing DNA that has been uncoiled from the central chromomere axis during transcription.

NOW SOLVE THIS

Problem 8 on page 239 involves polytene chromosomes that are cultured in ^{3}H-thymidine and subjected to autoradiography. You are asked to predict the pattern of grains that will result.

Hint: ^{3}H-thymidine will only be incorporated during the synthesis of DNA.

ESSENTIAL POINT

Polytene and lampbrush chromosomes are examples of specialized structures that extended our knowledge of genetic organization and function well in advance of the technology available to the modern-day molecular biologist.

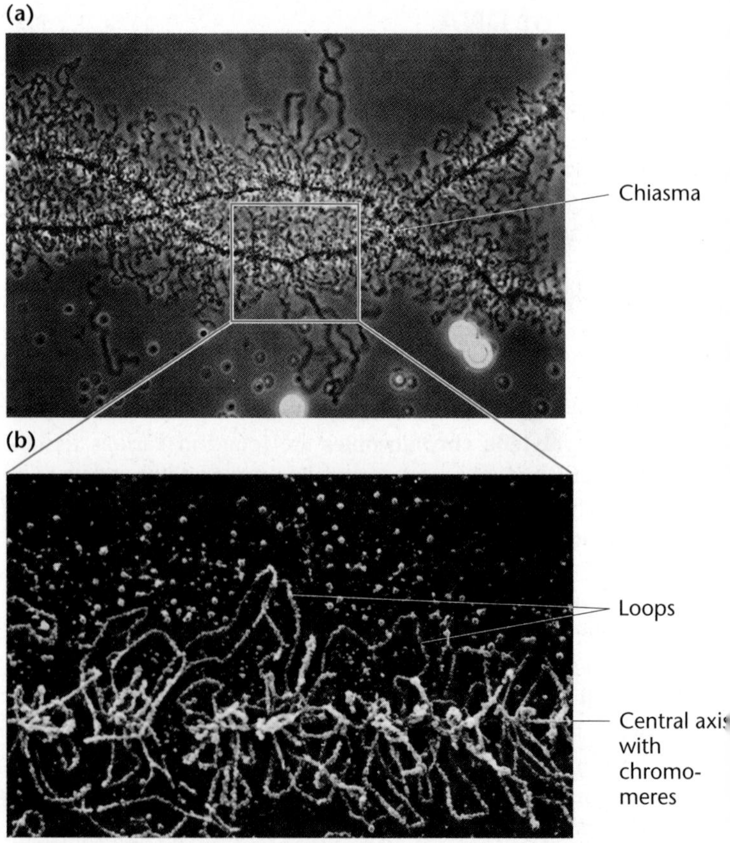

FIGURE 11–8 Lampbrush chromosomes derived from amphibian oocytes. Part (a) is a photomicrograph; part (b) is a scanning electron micrograph.

11.4 DNA Is Organized into Chromatin in Eukaryotes

We now turn our attention to the way DNA is organized in eukaryotic chromosomes. Our focus will be on eukaryotic cells, in which chromosomes are visible only during mitosis. After chromosome separation and cell division, cells enter the interphase stage of the cell cycle, during which time the components of the chromosome uncoil and are present in the form referred to as **chromatin.** While in interphase, the chromatin is dispersed in the nucleus, and the DNA of each chromosome is replicated. As the cell cycle progresses, most cells reenter mitosis, whereupon chromatin coils into visible chromosomes once again. This condensation represents a length contraction of some 10,000 times for each chromatin fiber.

The organization of DNA during the transitions just described is much more intricate and complex than in viruses or bacteria, which never exhibit a process similar to mitosis. This is due to the greater amount of DNA per chromosome, as well as the presence of a large number of proteins associated with eukaryotic DNA. For example, while DNA in the *E. coli* chromosome is 1200 μm long, the DNA in each human chromosome ranges from 19,000 to 73,000 μm in length. In a single human nucleus, all 46 chromosomes contain sufficient DNA to extend to more than 2 meters. This genetic material, along with its associated proteins, is contained within a nucleus that usually measures about 5 to 10 μm in diameter.

Chromatin Structure and Nucleosomes

As we have seen, the genetic material of viruses and bacteria consists of strands of DNA or RNA that are nearly devoid of proteins. In eukaryotic chromatin, a substantial amount of protein is associated with the chromosomal DNA in all phases of the eukaryotic cell cycle. The associated proteins are divided into basic, positively charged **histones** and less positively charged nonhistones. The histones clearly play the most essential structural role of all the proteins associated with DNA. Histones contain large amounts of the positively charged amino acids lysine and arginine, making it possible for them to bond electrostatically to the negatively charged phosphate groups of nucleotides. Recall that a similar interaction has been proposed for several bacterial proteins. The five main types of histones are shown in Table 11.2.

TABLE 11.2 Categories and Properties of Histone Proteins

Histone Type	Lysine-Arginine Content	Molecular Weight (Da)
H1	Lysine-rich	23,000
H2A	Slightly lysine-rich	14,000
H2B	Slightly lysine-rich	13,800
H3	Arginine-rich	15,300
H4	Arginine-rich	11,300

The general model for chromatin structure is based on the assumption that chromatin fibers, composed of DNA and protein, undergo extensive coiling and folding as they are condensed within the cell nucleus. X-ray diffraction studies confirm that histones play an important role in chromatin structure. Chromatin produces regularly spaced diffraction rings, suggesting that repeating structural units occur along the chromatin axis. If the histone molecules are chemically removed from chromatin, the regularity of this diffraction pattern is disrupted.

A basic model for chromatin structure was worked out in the mid-1970s. Several observations were particularly relevant to the development of this model:

1. Digestion of chromatin by certain endonucleases, such as micrococcal nuclease, yields DNA fragments that are approximately 200 bp in length or multiples thereof. This demonstrates that enzymatic digestion is not random, for if it were, we would expect a wide range of fragment sizes. Thus, chromatin consists of some type of repeating unit, each of which is protected from enzymatic cleavage, except where any two units are joined. It is the area between units that is attacked and cleaved by the endonuclease.
2. Electron microscopic observations of chromatin reveal that chromatin fibers are composed of linear arrays of spherical particles (Figure 11–9). Discovered by Ada and Donald Olins, the particles occur regularly along the axis of a chromatin strand and resemble beads on a string. These particles, initially referred to as ν-bodies (ν is the Greek letter nu), are now called **nucleosomes.** This conforms nicely to the earlier observation, which suggests the existence of repeating units.
3. Studies of precise interactions of histone molecules and DNA in the nucleosomes constituting chromatin show that histones H2A, H2B, H3, and H4 occur as two types of tetramers, $(H2A)_2 \cdot (H2B)_2$ and $(H3)_2 \cdot (H4)_2$. Roger Kornberg predicted that each repeating nucleosome unit consists of one of each tetramer (creating an octamer) in association with about 200 bp of DNA. Such a structure is consistent with previous observations and provides the basis for a model that explains the interaction of histones and DNA in chromatin.

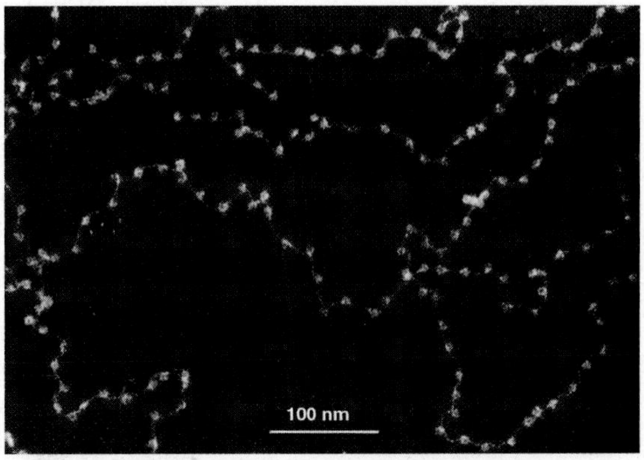

FIGURE 11–9 Dark-field electron micrograph of nucleosomes present in chromatin derived from a chicken erythrocyte nucleus.

4. When nuclease digestion time is extended, some of the 200 bp of DNA are removed from the nucleosome, creating a **nucleosome core particle** consisting of 147 bp. The DNA lost in this prolonged digestion is responsible for linking nucleosomes together. This linker DNA is associated with the fifth histone, H1.

On the basis of this information, as well as on X-ray and neutron-scattering analyses of crystallized core particles by John T. Finch, Aaron Klug, and others, a detailed model of the nucleosome was put forward in 1984, providing a basis for predicting chromatin structure and its condensation into chromosomes. In this model, illustrated in Figure 11–10, a 147-bp length of the 2-nm-diameter DNA molecule coils around an octamer of histones in a left-handed superhelix that completes about 1.7 turns per nucleosome. Each nucleosome, ellipsoidal in shape, measures about 11 nm at its longest point [Figure 11–10(a)]. Significantly, the formation of the nucleosome represents the first level of packing, whereby the DNA helix is reduced to about one-third of its original length by winding around the histones.

In the nucleus, the chromatin fiber seldom, if ever, exists in the extended form described here. Instead, the 11-nm chromatin fiber is further packed into a thicker 30-nm fiber, initially called a **solenoid** [Figure 11–10(b)]. This thicker fiber consists of numerous closely coiled nucleosomes, creating the second level of packing. Solenoids condense the eukaryotic fiber by a factor of five. The exact details of this structure are not completely clear, but 30-nm chromatin fibers are characteristically seen under the electron microscope.

In the transition to the mitotic chromosome, still another level of packing occurs. The 30-nm fiber forms a series of looped domains that condense the structure into the chromatin fiber, which is 300 nm in diameter [Figure 11–10(c)]. The fibers are then coiled into the chromosome arms that constitute a chromatid, which is part of the metaphase chromosome [Figure 11–10(d)]. While this figure shows the chromatid arms to be 700 nm in diameter, this value undoubtedly varies among different organisms. At a value of 700 nm, a pair of sister chromatids comprising a chromosome measures about 1400 nm.

The importance of the organization of DNA into chromatin and chromatin into mitotic chromosomes can be illustrated by considering a human cell that stores its genetic material in a nucleus that is about 5 to 10 μm in diameter. The haploid genome contains 3.2×10^9 base pairs of DNA distributed among 23 chromosomes. The diploid cell contains twice that amount. At 0.34 nm per base pair, this amounts to an enormous length of DNA (as stated earlier, to more than 2 m). One estimate is that the DNA inside a typical human nucleus is complexed with roughly 25×10^6 nucleosomes.

In the overall transition from a fully extended DNA helix to the extremely condensed status of the mitotic chromosome, a packing ratio (the ratio of DNA length to the length of the struc-

CHROMOSOME STRUCTURE

WEB TUTORIAL

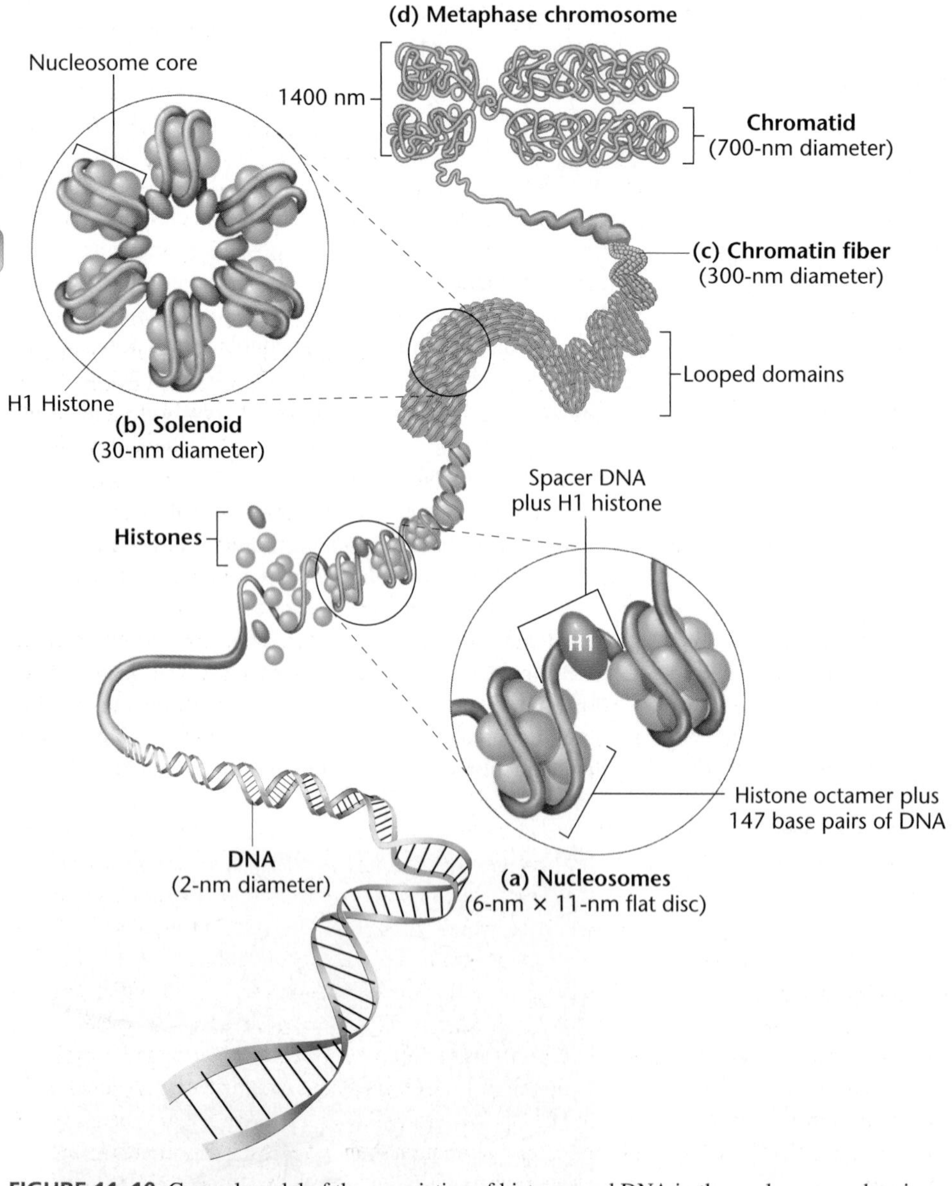

FIGURE 11–10 General model of the association of histones and DNA in the nucleosome, showing how the chromatin fiber can coil into a more condensed structure, ultimately producing a metaphase chromosome.

ture containing it) of about 500:1 must be achieved. In fact, our model accounts for a ratio only one-tenth that. Obviously, the larger fiber can be further bent, coiled, and packed, as even greater condensation occurs during the formation of a mitotic chromosome.

ESSENTIAL POINT

Eukaryotic chromatin is a nucleoprotein organized into repeating units called nucleosomes, which are composed of 200 base pairs of DNA, an octamer of four types of histones, plus one linker histone.

NOW SOLVE THIS

Problem 20 on page 239 involves the extent to which the nucleus is filled by all of the diploid content of human DNA.

Hint: Assuming the nucleus is a perfect sphere, start with the formula $V = (4/3)\pi r^3$.

Chromatin Remodeling

As with many significant findings in genetics, the study of nucleosomes has answered some important questions, but at the same time has also led us to new ones. For example, in the preceding discussion, we established that histone proteins play an important structural role in packaging DNA into the nucleosomes that make up chromatin. While solving the structural problem of how to organize a huge amount of DNA within the eukaryotic nucleus, a new problem was apparent: *the chromatin fiber, when complexed with histones and folded into various levels of compaction, makes the DNA inaccessible to interaction with important nonhistone proteins.* The variety of proteins that function in enzymatic and regulatory roles during the processes of replication and gene expression must interact directly with DNA. To accommodate these protein–DNA interactions, chromatin must be induced to change its structure, a process called **chromatin remodeling.** In the case of replication and gene expression, chromatin must relax its compact structure but be able to reverse the process during periods of inactivity.

Insights into how different states of chromatin structure may be achieved were forthcoming in 1997, when Timothy Richmond and members of his research team were able to significantly improve the level of resolution in X-ray diffraction studies of nucleosome crystals (from 7 Å in the 1984 studies to 2.8 Å in the 1997 studies). At this resolution, most atoms are visible, thus revealing the subtle twists and turns of the superhelix of DNA that encircles the histones. Recall that the double-helical ribbon represents 147 bp of DNA surrounding four pairs of histone proteins. This configuration is repeated over and over in the chromatin fiber and is the principal packaging unit of DNA in the eukaryotic nucleus.

The work of Richmond and colleagues, extended to a resolution of 1.9 Å in 2003, has revealed the details of the location of each histone entity within the nucleosome. Of particular interest to chromatin remodeling is that unstructured **histone tails** are not packed into the folded histone domains within the core of the nucleosome. For example, tails devoid of any secondary structure extending from histones H3 and H2B protrude through the minor groove channels of the DNA helix. The tails of histone H4 appear to make a connection with adjacent nucleosomes. Histone tails also provide potential targets for a variety of chemical modifications that may be linked to genetic functions along the chromatin fiber, including the regulation of gene expression.

Potential modifications of histones, now recognized as important to genetic function, include the chemical processes of **acetylation, methylation,** and **phosphorylation** of amino acids that are part of histones. Such chemical modifications are believed to be related to gene regulation as well as the cycle of chromatin unfolding and condensation that occurs during and after DNA replication. Although a great deal more work must be done to elucidate the specific involvement of chromatin remodeling during genetic processes, it is now clear that the dynamic forms in which chromatin exists are vitally important to the way that all genetic processes directly involving DNA are executed. We will return to a more detailed discussion of the role of chromatin remodeling when we consider the regulation of eukaryotic gene expression in Chapter 15.

Heterochromatin

Although we know that the DNA of the eukaryotic chromosome consists of one continuous double-helical fiber along its entire length, we also know that the whole chromosome is not structurally uniform from end to end. In the early part of the twentieth century, it was observed that some parts of the chromosome remain condensed and stain deeply during interphase, while most parts are uncoiled and do not stain. In 1928, the terms **euchromatin** and **heterochromatin** were coined to describe the parts of chromosomes that are uncoiled and those that remain condensed, respectively.

Subsequent investigation revealed a number of characteristics that distinguish heterochromatin from euchromatin. Heterochromatic areas are genetically inactive because they either lack genes or contain genes that are repressed. Also, heterochromatin replicates later during the S phase of the cell cycle than euchromatin does. The discovery of heterochromatin provided the first clues that parts of eukaryotic chromosomes do not always encode proteins. Instead, some chromosome regions are thought to be involved in maintenance of the chromosome's structural integrity and in other functions, such as chromosome movement during cell division.

The presence of heterochromatin is unique to and characteristic of the genetic material of eukaryotes. In some cases, whole chromosomes are heterochromatic. A case in point is the mammalian Y chromosome, much of which is genetically inert. And, as we discussed in Chapter 5, the inactivated X chromosome in mammalian females is condensed into an inert heterochromatic Barr body. In some species, such as mealy bugs, all chromosomes of one entire haploid set are heterochromatic.

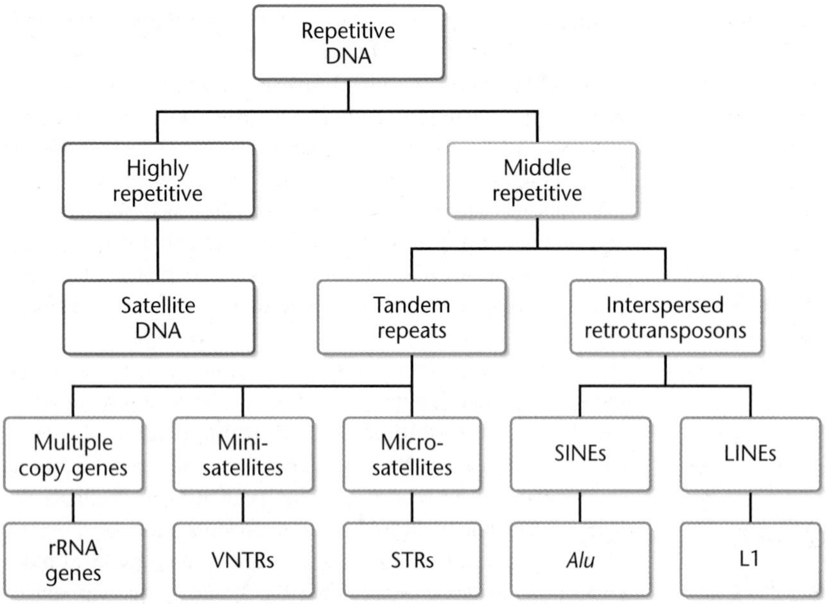

FIGURE 11–11 An overview of the categories of repetitive DNA.

When certain heterochromatic areas from one chromosome are translocated to a new site on the same or another nonhomologous chromosome, genetically active areas sometimes become genetically inert if they lie adjacent to the translocated heterochromatin. This influence on existing euchromatin is one example of what is more generally referred to as a **position effect.** That is, the position of a gene or group of genes relative to all other genetic material may affect their expression.

ESSENTIAL POINT

Heterochromatin, prematurely condensed in interphase and for the most part genetically inert, is illustrated by centromeric and telomeric regions of eukaryotic chromosomes, the Y chromosome, and the Barr body.

11.5 Eukaryotic Genomes Demonstrate Complex Sequence Organization Characterized by Repetitive DNA

Thus far, we have looked at how DNA is organized into chromosomes in bacteriophages, bacteria, and eukaryotes. We now begin an examination of what we know about the organization of DNA sequences within the chromosomes making up an organism's genome, placing our emphasis on eukaryotes. Once we establish the pattern of genome organization, in a later chapter we will focus on how the genes themselves are organized within chromosomes (see Chapter 18).

We have already established that, in addition to single copies of unique DNA sequences that comprise genes, a great deal of the DNA sequences within chromosomes is repetitive in nature and that various levels of repetition occur within the genome of organisms. Many studies have now provided insights into **repetitive DNA,** demonstrating various classes of these sequences and their organization within the genome. Figure 11–11 outlines the various categories of repetitive DNA. Some functional genes are present in more than one copy and are therefore repetitive in nature. However, the majority of repetitive sequences are nongenic, and in fact, most serve no known function. We explore three main categories: (1) heterochromatin found associated with centromeres and making up telomeres, (2) tandem repeats of both short and long DNA sequences, and (3) transposable sequences that are interspersed throughout the genome of eukaryotes.

Repetitive DNA and Satellite DNA

The nucleotide composition of the DNA (e.g., the percentage of G≡C versus A=T pairs) of a particular species is reflected in its density, which can be measured with sedimentation equilibrium centrifugation. When eukaryotic DNA is analyzed in this way, the majority is present as a single main band, or peak, of fairly uniform density. However, one or more additional peaks represent DNA that differs slightly in density. This component, called **satellite DNA,** represents a variable proportion of the total DNA, depending on the species. A profile of main-band and satellite DNA from the mouse is shown in Figure 11–12. By contrast, prokaryotes contain only main-band DNA.

The significance of satellite DNA remained an enigma until the mid-1960s, when Roy Britten and David Kohne developed the technique for measuring the reassociation kinetics of DNA that had previously been dissociated into single strands. They demonstrated that certain portions of DNA reassociated more rapidly than others. They concluded that rapid reassociation was characteristic of multiple DNA fragments composed of identical or nearly identical nucleotide

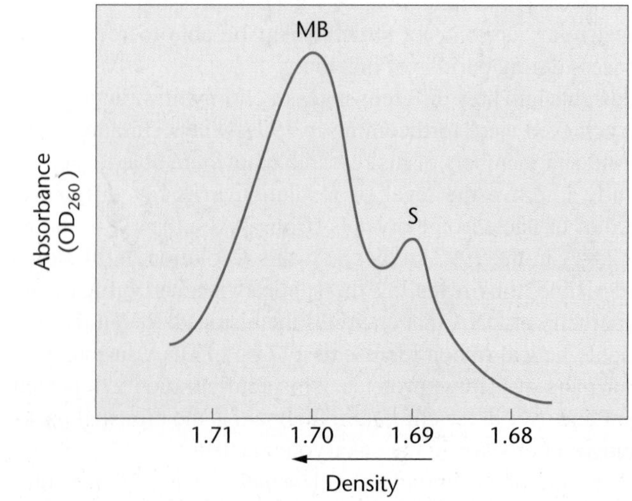

FIGURE 11–12 Separation of main-band (MB) and satellite (S) DNA from the mouse, using ultracentrifugation in a CsCl gradient.

sequences—the basis for the descriptive term repetitive DNA (see Chapter 9).

When satellite DNA is subjected to analysis by reassociation kinetics, it falls into the category of **highly repetitive DNA,** which is known to consist of relatively short sequences repeated a large number of times. Further evidence suggests that these sequences are present as tandem repeats clustered in very specific chromosomal areas known to be heterochromatic—the regions flanking centromeres. This was discovered in 1969 when several researchers, including Mary Lou Pardue and Joe Gall, applied ***in situ* molecular hybridization** to the study of satellite DNA. This technique involves the molecular hybridization between an isolated fraction of radioactively labeled DNA or RNA probes and the DNA contained in the chromosomes of a cytological preparation. Following the hybridization procedure, autoradiography is performed to locate the chromosome areas complementary to the fraction of DNA or RNA.

Pardue and Gall demonstrated that radioactive probes made from mouse satellite DNA hybridize with the DNA of centromeric regions of mouse mitotic chromosomes (Figure 11–13). Several conclusions were drawn: Satellite DNA differs from main-band DNA in its molecular composition, as established by buoyant density studies. It is composed of repetitive sequences. Finally, satellite DNA is found in the heterochromatic centromeric regions of chromosomes.

Centromeric DNA Sequences

The separation of homologs during mitosis and meiosis depends on **centromeres,** described cytologically in the late nineteenth century as the *primary constrictions* along eukaryotic chromosomes (see Chapter 2). In this role, it is believed that the DNA sequence contained within the centromere is critical to this role. Careful analysis has confirmed this prediction. The minimal region of the centromere that supports the function of chromosomal segregation is designated the **CEN region.** Within this heterochromatic region of the chromosome, the DNA binds a platform of proteins, which in multicellular organisms includes the **kinetochore** that binds to the spindle fiber during division.

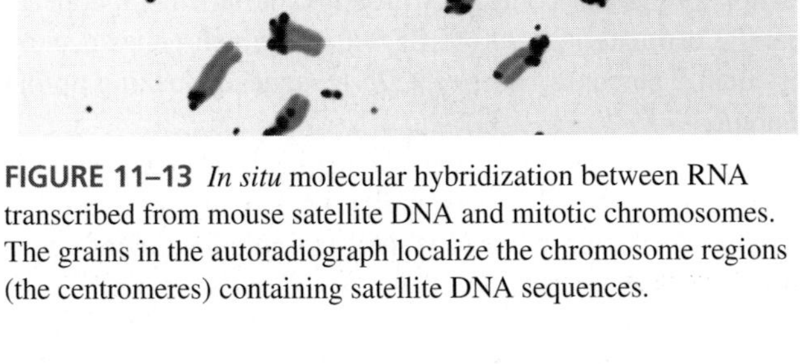

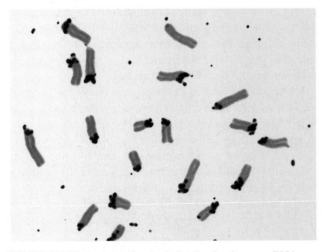

FIGURE 11–13 *In situ* molecular hybridization between RNA transcribed from mouse satellite DNA and mitotic chromosomes. The grains in the autoradiograph localize the chromosome regions (the centromeres) containing satellite DNA sequences.

The CEN regions of the yeast *Saccharomyces cerevisiae* were the first to be studied. Each centromere serves an identical function, so it is not surprising that CENs from different chromosomes were found to be remarkably similar in their organization. The CEN region of yeast chromosomes consists of about 120 bp. Mutational analysis suggests that portions near the 3′-end of this DNA region are most critical to centromere function since mutations in them, but not those nearer the 5′-end, disrupt centromere function. Thus, the DNA of this region appears to be essential to the eventual binding to the spindle fiber (yeast do not have kinetechores).

Centromere sequences of multicellular eukaryotes are much more extensive than in yeast and vary considerably in size. Such sequences, absent from yeast but characteristic of most multicellular organisms, vary considerably in size. For example, in *Drosophila* the CEN region is found within some 200 to 600 kb of DNA, much of which is highly repetitive. Recall from our prior discussion that highly repetitive satellite DNA is localized in the centromere regions of mice. In humans, one of the most recognized satellite DNA sequences is the **alphoid family,** found mainly in the centromere regions. Alphoid sequences, each about 170 bp in length, are present in tandem arrays of up to 1 million base pairs. Embedded within this repetitive DNA are more specific sequences that are critical to centromere function.

Middle Repetitive Sequences: VNTRs and STRs

A brief look at still another prominent category of repetitive DNA sheds additional light on the organization of the eukaryotic genome. In addition to highly repetitive DNA, which constitutes about 5 percent of the human genome (and 10 percent of the mouse genome), a second category, **middle** (or **moderately**) **repetitive DNA,** recognized by C_0t analysis, is fairly well characterized. Because we now know a great deal about the human genome, we will use our own species to illustrate this category of DNA in genome organization.

Although middle repetitive DNA does include some duplicated genes (such as those encoding ribosomal RNA), most prominent in this category are either noncoding tandemly repeated sequences or noncoding interspersed sequences. No function has been ascribed to these components of the genome. An example is DNA described as **variable number tandem repeats (VNTRs).** These repeating DNA sequences may be 15 to 100 bp long and are found within and between genes. Many such clusters are dispersed throughout the genome, and they are often referred to as **minisatellites.**

The number of tandem copies of each specific sequence at each location varies from one individual to the next, creating localized regions of 1,000 to 20,000 bp (1–20 kb) in length. As we will see in Chapter 19, the variation in size (length) of these regions between individual humans was originally the basis for the forensic technique referred to as **DNA fingerprinting.**

Another group of tandemly repeated sequences consists of di-, tri-, tetra-, and pentanucleotides, also referred to as **microsatellites** or **short tandem repeats (STRs).** Like VNTRs, they are dispersed throughout the genome and vary among individuals in the number of repeats present at any site. For example, in humans, the most common microsatellite is the dinucleotide $(CA)_n$, where n equals the number of repeats. Most commonly, n is between 5 and 50. These clusters have served as useful molecular markers for genome analysis.

Repetitive Transposed Sequences: SINEs and LINEs

Still another category of repetitive DNA consists of sequences that are interspersed individually throughout the genome, rather than being tandemly repeated. They can be either short or long, and many have the added distinction of being **transposable sequences,** which are mobile and can potentially move to different locations within the genome. A large portion of the human genome is composed of such sequences.

For example, **short interspersed elements,** called **SINEs,** are less than 500 base pairs long and may be present 500,000 times or more in the human genome. The best characterized human SINE is a set of closely related sequences called the ***Alu* family** (the name is based on the presence of DNA sequences recognized by the restriction endonuclease *Alu*I). Members of this DNA family, also found in other mammals, are 200 to 300 base pairs long and are dispersed rather uniformly throughout the genome, both between and within genes. In humans, this family encompasses more than 5 percent of the entire genome.

Alu sequences are particularly interesting, although their function, if any, is yet undefined. Members of the *Alu* family are sometimes transcribed. The role of this RNA is not certain, but it is thought to be related to its mobility in the genome. *Alu* sequences are thought to have arisen from an RNA element whose DNA complement was dispersed throughout the genome as a result of the activity of reverse transcriptase (an enzyme that synthesizes DNA on an RNA template).

The group of **long interspersed elements (LINEs)** represents yet another category of repetitive transposable DNA sequences. LINEs are usually about 6 kb in length and in the human genome are present 850,000 times. The most prominent example in humans is the **L1 family.** Members of this sequence family are about 6400 base pairs long and are present up to 100,000 times. Their 5′-end is highly variable, and their role within the genome has yet to be defined.

The general mechanism for transposition of L1 elements is now clear. The L1 DNA sequence is first transcribed into an RNA molecule. The RNA then serves as the template for the synthesis of the DNA complement using the enzyme reverse transcriptase. This enzyme is encoded by a portion of the L1 sequence. The new L1 copy then integrates into the DNA of the chromosome at a new site. Because of the similarity of this transposition mechanism to that used by retroviruses, LINEs are referred to as **retrotransposons.**

SINEs and LINEs represent a significant portion of human DNA. SINEs constitute about 13 percent of the human genome, whereas LINEs constitute up to 21 percent. Within both types of elements, repeating sequences of DNA are present in combination with unique sequences.

Middle Repetitive Multiple-Copy Genes

In some cases, middle repetitive DNA includes functional genes present tandemly in multiple copies. For example, many copies exist of the genes encoding ribosomal RNA. *Drosophila* has 120 copies per haploid genome. Single genetic units encode a large precursor molecule that is processed into the 5.8*S*, 18*S*, and 28*S* rRNA components. In humans, multiple copies of this gene are clustered on the p arm of the acrocentric chromosomes 13, 14, 15, 21, and 22. Multiple copies of the genes encoding 5*S* rRNA are transcribed separately from multiple clusters found together on the terminal portion of the p arm of chromosome 1.

ESSENTIAL POINT

Eukaryotic genomes demonstrate complex sequence organization characterized by numerous categories of repetitive DNA, consisting of either tandem repeats clustered in various regions of the genome or single sequences repeatedly interspersed at random throughout the genome.

11.6 The Vast Majority of a Eukaryotic Genome Does Not Encode Functional Genes

Given the preceding information concerning various forms of repetitive DNA in eukaryotes, it is of interest to pose an important question: *What proportion of the eukaryotic genome actually encodes functional genes?*

We have seen that, taken together, the various forms of highly repetitive and moderately repetitive DNA comprise a substantial portion of the human genome. In addition to repetitive DNA, a large amount of the DNA consists of single-copy sequences as defined by C_0t analysis (Chapter 9) that appear to be noncoding. Included are many instances of what we call **pseudogenes.** These are DNA sequences representing evolutionary vestiges of duplicated copies of genes that have undergone significant mutational alteration. As a result, although they show some homology to their parent gene, they are usually not transcribed because of insertions and deletions throughout their structure.

Although the proportion of the genome consisting of repetitive DNA varies among organisms, one feature seems to be shared: *Only a very small part of the genome actually codes for proteins.* For example, the 20,000 to 30,000 genes encoding proteins in sea urchin occupy less than 10 percent of the genome. In *Drosophila,* only 5 to 10 percent of the genome is occupied by genes coding for proteins. In humans, it appears that the estimated 20,000 to 25,000 functional genes occupy less than 2 percent of the total DNA sequence making up the genome.

Study of the various forms of repetitive DNA has significantly enhanced our understanding of genome organization. In the next chapter, we will explore the organization of genes within chromosomes.

EXPLORING GENOMICS

Database of Genomic Variants: Structural Variations in the Human Genome

In this chapter, we focused on structural details of chromosomes and DNA sequence organization in chromosomes. As discussed in Chapter 6, we have learned that large segments of DNA and a number of genes can vary greatly in copy number due to duplications, creating **copy number variations (CNVs).** Many studies are underway to identify and map CNVs and to find possible disease conditions associated with them.

To date, approximately 2000 CNVs have been identified in the human genome, and estimates suggest there may be thousands more within human populations. In this Exploring Genomics exercise we will visit the **Database of Genomic Variants (DGV),** which provides a quickly expanding summary of structural variations in the human genome including CNVs.

■ Exercise I: Database of Genomic Variants

1. Access the DGV at **http://projects.tcag.ca/variation/**. Click the "About the Project" tab to learn more about the purpose of the DGV.
2. Information in the DGV is easily viewed by clicking on a chromosome of interest using the "View Data by Chromosome" feature or by clicking on a chromosome using the "View Data by Genome" feature.
3. Click on a chromosome of interest to you using the "View Data by Chromosome" feature. A table will appear showing several columns of data including:
   - Locus: Shows the locus for the CNV, including the base pairs that span the variation.
   - Landmark: Shows different variations of CNVs for a particular locus.
   - Variation ID: Provides a unique identifying number for each variation.
   - Variation Type: Listed as "copy number." Most variations in this database are CNVs. Variations known to be insertions or deletions based on relatively small changes (a few bases) are labeled "InDel." Inversions labeled "Inv."
   - Cytoband: Indicates the chromosomal banding location for the variation.
   - Position (Mb): Shows the relative location in megabases (Mb) on the chromosome.
   - Known Genes in the Locus: Lists genes that are located in a particular CNV.
4. Let's analyze a particular group of CNVs. Many CNVs are unlikely to affect phenotype because they involve large areas of non–protein-coding or nonregulatory sequences. But gene-containing CNVs have been identified, including variants containing genes associated with Alzheimer's disease, Parkinson's disease, and other conditions.

Defensin (*DEF*) genes are part of a large family of highly duplicated genes. To learn more about *DEF* genes and CNVs, use the Keyword Search box to search for *DEF* genes (click "No" for the exact match button). A results page for the search will appear with a listing of CNVs. Click on one of the links shown. The top part of each report has graphs indicating the position of each CNV on a chromosome. Scroll down to the bottom of this page until you see "Known Genes." Click on the different *DEF* genes listed in the known genes category, which will take you to the National Center for Biotechnology Information (NCBI) Entrez site with a wealth of information about these genes so that you can answer the questions below. Do this for several *DEF*-containing CNVs on different chromosomes to find the information you will need for your answers.

On what chromosome(s) did you find CNVs containing *DEF* genes?

a. What did you learn about the function of *DEF* gene products? What do DEF proteins do?

b. Variations in *DEF* genotypes and *DEF* gene expression in humans have been implicated in a number of different human disease conditions. Give examples of the kinds of disorders affected by variations in *DEF* genotypes.

c. Explore the DGV to search a chromosome of interest to you and learn more about CNVs that have been mapped to that chromosome. Try the "View Data by Genome" feature that will show you maps of each chromosome indicating different variations. For CNVs (shown in blue), clicking on the CNV will take you to its locus on the chromosome.

CASE STUDY Art inspires learning

A genetics student visiting a museum saw a painting by Goya showing a woman with a newborn baby in her lap that had very short arms and legs along with some facial abnormalities. Wondering whether this condition might be a genetic disorder, the student went online, learning that the baby might have Roberts syndrome (RBS), a rare autosomal recessive trait. She read that cells in RBS have mitotic chromosome abnormalities, including centromeres and other heterochromatic regions of homologs that separate prematurely in metaphase instead of anaphase, and as a result, metaphase chromosomes have a rigid, or "railroad track" appearance. RBS has been shown to be caused by mutant alleles of the *ESCO2* gene, which functions during cell division.

The student wrote a list of questions to investigate in an attempt to better understand this condition.

1. What do centromeres and other heterochromatic regions have in common that might cause this appearance?
2. What might be the role of the protein encoded by *ESCO2,* which in mutant form could cause these changes in mitotic chromosomes?
3. How could premature separation of centromeres cause the problems seen in RBS?

INSIGHTS AND SOLUTIONS

A previously undiscovered single-cell organism was found living at a great depth on the ocean floor. Its nucleus contained only a single linear chromosome with 7×10^6 *nucleotide pairs of DNA coalesced with three types of histonelike proteins. Consider the following questions:*

1. A short micrococcal nuclease digestion yielded DNA fractions of 700, 1400, and 2100 bp. Predict what these fractions represent. What conclusions can be drawn?

Solution: The chromatin fiber may consist of a variation of nucleosomes containing 700 bp of DNA. The 1400- and 2100-bp fractions, respectively, represent two and three linked nucleosomes. Enzymatic digestion may have been incomplete, leading to the latter two fractions.

2. The analysis of individual nucleosomes reveals that each unit contained one copy of each protein and that the short linker DNA contained no protein bound to it. If the entire chromosome consists of nucleosomes (discounting any linker DNA), how many are there, and how many total proteins are needed to form them?

Solution: Since the chromosome contains 7×10^6 bp of DNA, the number of nucleosomes, each containing 7×10^2 bp, is equal to

$$7 \times 10^6/7 \times 10^2 = 10^4 \text{ nucleosomes}$$

The chromosome contains 10^4 copies of each of the three proteins, for a total of 3×10^4 proteins.

3. Further analysis revealed the organism's DNA to be a double helix similar to the Watson–Crick model but containing 20 bp per complete turn of the right-handed helix. The physical size of the nucleosome was exactly double the volume occupied by that found in all other known eukaryotes, by virtue of increasing the distance along the fiber axis by a factor of two. Compare the degree of compaction of this organism's nucleosome to that found in other eukaryotes.

Solution: The unique organism compacts a length of DNA consisting of 35 complete turns of the helix (700 bp per nucleosome/20 bp per turn) into each nucleosome. The normal eukaryote compacts a length of DNA consisting of 20 complete turns of the helix (200 bp per nucleosome/10 bp per turn) into a nucleosome one-half the volume of that in the unique organism. The degree of compaction is therefore less in the unique organism.

4. No further coiling or compaction of this unique chromosome occurs in the unique organism. Compare this to a eukaryotic chromosome. Do you think an interphase human chromosome 7×10^6 bp in length would be a shorter or longer chromatin fiber?

Solution: The length of the unique chromosome is compacted into 10^4 nucleosomes, each with an axis length twice that of the eukaryotic fiber. The eukaryotic fiber consists of

$$7 \times 10^6/2 \times 10^2 = 3.5 \times 10^4 \text{ nucleosomes}$$

which is 3.5 times more than the unique organism. However, they are compacted by a factor of five in each solenoid. Therefore, the chromosome of the unique organism is a longer chromatin fiber.

PROBLEMS AND DISCUSSION QUESTIONS

1. Contrast the sizes of the chromosomes of bacteriophage λ and T2 with that of *E. coli.* How does this relate to the relative size and complexity of the phages and bacteria?
2. Bacteriophages and bacteria almost always contain their DNA as circular (closed loops) chromosomes. Phage λ is an exception, maintaining its DNA in a linear chromosome within the viral particle. However, as soon as it is injected into a host cell, it circularizes before replication begins. Taking into account the information in Chapter 10, what advantage exists in replicating circular DNA molecules compared to linear molecules?

See **Now Solve This** on page 226.

3. Contrast the appearance of the DNA associated with mitochondria and chloroplasts.

4. Compare the size of DNA and the encoded gene products in mitochondria and chloroplasts.
5. While protein synthesis occurs within mitochondria and chloroplasts, not all necessary gene products are encoded by the organellar DNA. Explain the origin of the genetic machinery within the organelles.
6. Mitochondria and chloroplasts contain ribosomal RNA molecules that are unlike those found in the adjoining cytoplasm. How does this observation relate to the endosymbiotic hypothesis?
7. Describe how giant polytene chromosomes are formed.
8. Salivary gland cells from *Drosophila* are isolated and placed in the presence of radioactive thymidylic acid. Autoradiography is performed, revealing polytene chromosomes. Predict the distribution of the grains along the chromosomes. See **Now Solve This** on page 230.
9. What genetic process is occurring in a puff of a polytene chromosome? How do we know this experimentally?
10. Describe the structure of LINE sequences. Why are LINEs referred to as retrotransposons?
11. During what genetic process are lampbrush chromosomes present in vertebrates?
12. Why might we predict that the organization of eukaryotic genetic material will be more complex than that of viruses or bacteria?
13. Describe the sequence of research findings that led to the development of the model of chromatin structure.
14. What is the molecular composition and arrangement of the components in the nucleosome?
15. Describe the transitions that occur as nucleosomes are coiled and folded, ultimately forming a chromatid.
16. Provide a comprehensive definition of heterochromatin, and list as many examples as you can.
17. Mammals contain a diploid genome consisting of at least 10^9 bp. If this amount of DNA is present as chromatin fibers, where each group of 200 bp of DNA is combined with 9 histones into a nucleosome and each group of 6 nucleosomes is combined into a solenoid, achieving a final packing ratio of 50, determine (a) the total number of nucleosomes in all fibers, (b) the total number of histone molecules combined with DNA in the diploid genome, and (c) the combined length of all fibers.
18. Assume that a viral DNA molecule is a 50-μm-long circular strand of a uniform 20 Å diameter. If this molecule is contained in a viral head that is a 0.08-μm-diameter sphere, will the DNA molecule fit into the viral head, assuming complete flexibility of the molecule? Justify your answer mathematically.
19. How many base pairs are in a molecule of phage T2 DNA 52 μm long?
20. If a human nucleus is 10 μm in diameter, and it must hold as much as 2 m of DNA, which is complexed into nucleosomes that are 11 nm in diameter during full extension, what percentage of the volume of the nucleus is occupied by the genetic material? See **Now Solve This** on page 233.
21. Sun and others (2002. *Proc. Natl. Acad. Sci. (USA)* 99: 8695–8700) studied *Drosophila* in which two normally active genes, w^+ (wild-type allele of the *white*-eye gene) and *hsp*26 (a heat-shock gene), were introduced (using a plasmid vector) into euchromatic and heterochromatic chromosomal regions. The relative activity of each gene was assessed, and an approximation of the data obtained is shown below. Considering three characteristics of heterochromatin, which is/are supported by the experimental data?

	Activity (relative percentage)	
Gene	**Euchromatin**	**Heterochromatin**
*hsp*26	100%	31%
w^+	100%	8%

22. While much remains to be learned about the role of nucleosomes and chromatin structure and function, recent research indicates that *in vivo* chemical modification of histones is associated with changes in gene activity. For example, Bernstein and others (2000. *Proc. Natl. Acad. Sci. (USA)* 97: 5340–5345) determined that acetylation of H3 and H4 is associated with 21.1 percent and 13.8 percent increase in yeast gene activity, respectively, and that yeast heterochromatin is hypomethylated relative to the genome average. Speculate on the significance of these findings in terms of nucleosome–DNA interactions and gene activity.
23. In an article entitled "*Nucleosome Positioning at the Replication Fork*," Lucchini and others (2002. *EMBO* 20: 7294–7302) state, "both the 'old' randomly segregated nucleosomes as well as the 'new' assembled histone octamers rapidly position themselves (within seconds) on the newly replicated DNA strands." Given this statement, how would one compare the distribution of nucleosomes and DNA in newly replicated chromatin? How could one experimentally test the distribution of nucleosomes on newly replicated chromosomes?
24. The human genome contains approximately 10^6 copies of an *Alu* sequence, one of the best-studied classes of short interspersed elements (SINEs), per haploid genome. Individual *Alu*s share a 282-nucleotide consensus sequence followed by a 3′-adenine-rich tail region (Schmid. 1998. *Nuc. Acids Res.* 26: 4541–4550). Given that there are approximately 3×10^9 bp per human haploid genome, about how many base pairs are spaced between each *Alu* sequence?
25. Below is a diagram of the general structure of the bacteriophage λ chromosome. Speculate on the mechanism by which it forms a closed ring upon infection of the host cell.

5′GGGCGGCGACCT—double-stranded region——3′
3′—double-stranded region——CCCGCCGCTGGA5′

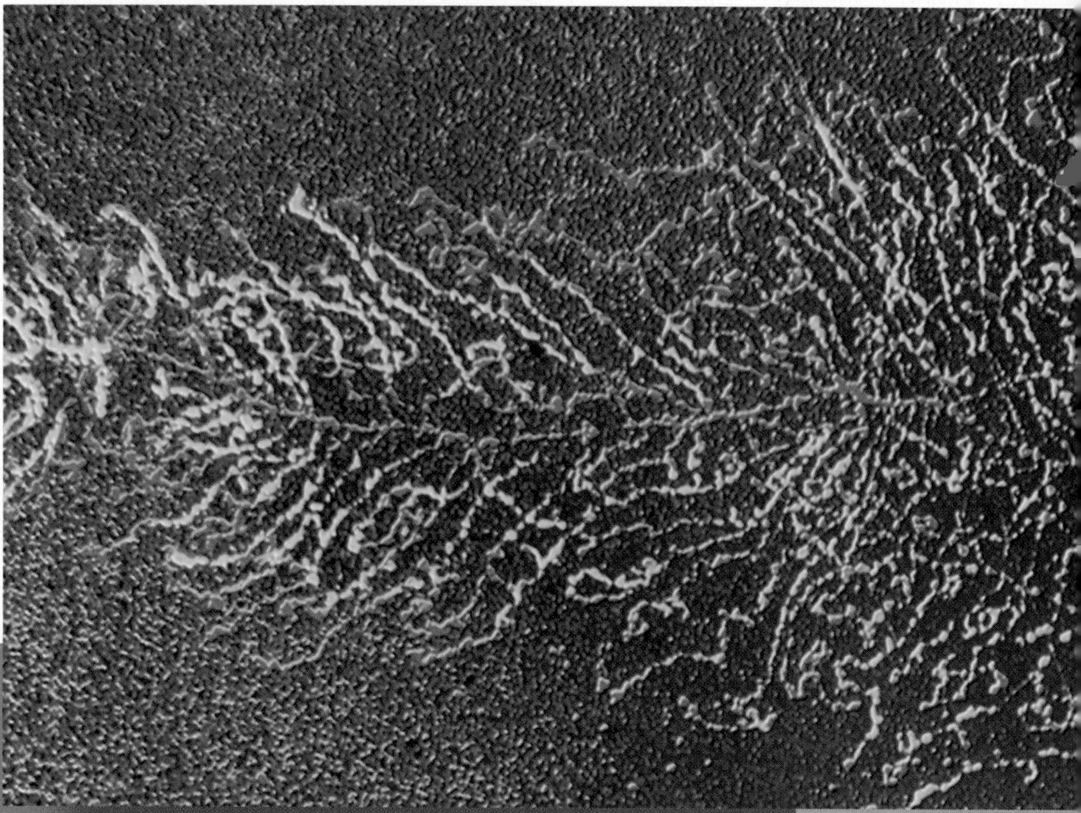

Electron micrograph visualizing the process of transcription.

12

The Genetic Code and Transcription

Chapter Concepts

- Genetic information is stored in DNA using a triplet code that is nearly universal to all living things on Earth.
- The genetic code is initially transferred from DNA to RNA during the process of transcription.
- Once transferred from DNA to RNA, the genetic code exists as triplet codons, using the four ribonucleotides in RNA as the letters composing it.
- By using four different letters taken three at a time, 64 triplet sequences are possible. Most encode one of the 20 amino acids present in proteins, which serve as the end products of most genes.
- Several codons provide signals that initiate or terminate protein synthesis.
- The process of transcription is similar but more complex in eukaryotes compared to prokaryotes and bacteriophages that infect them.

The linear sequence of deoxyribonucleotides making up DNA ultimately dictates the components constituting proteins, the end product of most genes. The central question is how such information stored as a nucleic acid is decoded into a protein. Figure 12–1 gives a simplified overview of how this transfer of information occurs. In the first step in gene expression, information on one of the two strands of DNA (the template strand) is transferred into an RNA complement through transcription. Once synthesized, this RNA acts as a "messenger" molecule bearing the coded information—hence its name, **messenger RNA (mRNA).** The mRNAs then associate with ribosomes, where decoding into proteins takes place.

In this chapter, we focus on the initial phases of gene expression by addressing two major questions. First, how is genetic information encoded? Second, how does the transfer from DNA to RNA occur, thus defining the process of transcription? As you shall see, ingenious analytical research established that the genetic code is written in units of three letters—ribonucleotides present in mRNA that reflect the stored information in genes. Most all triplet code words direct the incorporation of a specific amino acid into a protein as it is synthesized. As we can predict based on our prior discussion of the replication of DNA, transcription is also a complex process dependent on a major polymerase enzyme and a cast of supporting proteins. We will explore what is known about transcription in bacteria and then contrast this prokaryotic model with the differences found in eukaryotes. Together, the information in this and the next chapter provides a comprehensive picture of molecular genetics, which serves as the most basic foundation for understanding living organisms. In Chapter 13, we will address how translation occurs and discuss the structure and function of proteins.

How Do We Know?

In this chapter, we will focus on how the genetic information stored in DNA encodes proteins, the end products of most genes. We shall also elucidate the general mechanism by which genetic information in DNA is transfered to RNA, the process of transcription. As you study this topic, you should try to answer several fundamental questions:

1. Why did geneticists believe, even before experimental evidence was obtained, that the genetic code is a triplet?
2. What experimental evidence provided the *compositions* of triplet codons encoding specific amino acids, and subsequently, how were the *specific sequences* of triplet codes determined?
3. How do we know that expression of the information encoded in DNA involves an RNA intermediate?
4. How were the experimentally derived triplet codon assignments verified in studies using bacteriophage MS2?
5. How do we know that the initial transcript of a eukaryotic gene contains noncoding sequences that must be removed before accurate translation into proteins can occur?

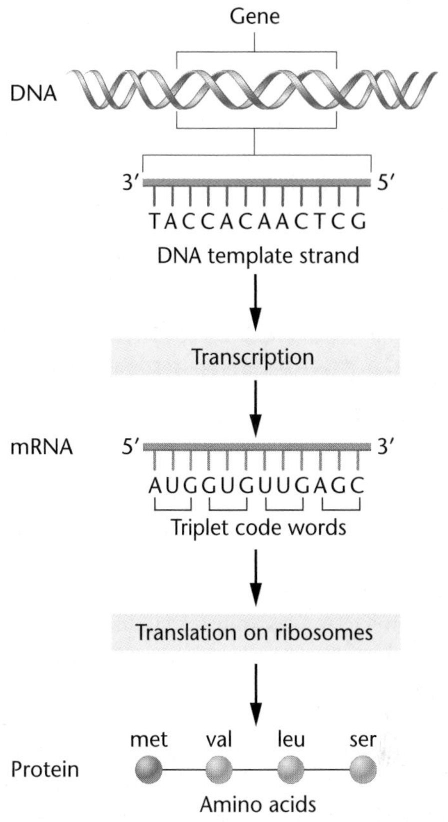

FIGURE 12–1 Flowchart illustrating how genetic information encoded in DNA produces protein.

12.1 The Genetic Code Exhibits a Number of Characteristics

WEB TUTORIAL TRANSCRIPTION

Before we consider the various analytical approaches that led to our current understanding of the genetic code, let's summarize the general features that characterize it.

1. The genetic code is written in linear form, using the ribonucleotide bases that compose mRNA molecules as "letters." The ribonucleotide sequence is derived from the complementary nucleotide bases in DNA.
2. Each "word" within the mRNA contains three ribonucleotide letters. With only three exceptions, each group of three ribonucleotides, called a **codon,** specifies one amino acid; the code is thus a **triplet codon.**
3. The code is **unambiguous**—each triplet specifies only a single amino acid.
4. The code is **degenerate;** that is, a given amino acid can be specified by more than one triplet codon. This is the case for 18 of the 20 amino acids.
5. The code contains one "start" and three "stop" signals, triplets that **initiate** and **terminate** translation.
6. No internal punctuation (such as a comma) is used in the code. Thus, the code is said to be **commaless.** Once translation of mRNA begins, the codons are read one after the other, with no breaks between them.

7. The code is **nonoverlapping.** Once translation commences, any single ribonucleotide at a specific location within the mRNA is part of only one triplet.
8. The code is nearly **universal.** With only minor exceptions, almost all viruses, prokaryotes, archaea, and eukaryotes use a single coding dictionary.

12.2 Early Studies Established the Basic Operational Patterns of the Code

In the late 1950s, before it became clear that mRNA is the intermediate that transfers genetic information from DNA to proteins, researchers thought that DNA itself might directly encode proteins during their synthesis. Because ribosomes had already been identified, the initial thinking was that information in DNA was transferred in the nucleus to the RNA of the ribosome, which served as the template for protein synthesis in the cytoplasm. This concept soon became untenable as accumulating evidence indicated the existence of an unstable intermediate template. The RNA of ribosomes, on the other hand, was extremely stable. As a result, in 1961 François Jacob and Jacques Monod postulated the existence of **messenger RNA (mRNA).** Once mRNA was discovered, it was clear that even though genetic information is stored in DNA, the code that is translated into proteins resides in RNA. The central question then was how only four letters—the four nucleotides—could specify 20 words—the amino acids?

The Triplet Nature of the Code

In the early 1960s, Sidney Brenner argued on theoretical grounds that the code had to be a triplet since three-letter words represent the minimal use of four letters to specify 20 amino acids. A code of four nucleotides, taken two at a time, for example, provides only 16 unique code words (4^2). A triplet code yields 64 words (4^3)—clearly more than the 20 needed—and is much simpler than a four-letter code (4^4), which specifies 256 words.

Experimental evidence supporting the triplet nature of the code was subsequently derived from research by Francis Crick and his colleagues. Using phage T4, they studied **frameshift mutations,** which result from the addition or deletion of one or more nucleotides within a gene and subsequently the mRNA transcribed by it. The gain or loss of letters shifts the *frame of reading* during translation. Crick and his colleagues found that the gain or loss of one or two nucleotides caused a frameshift mutation, but when three nucleotides were involved, the frame of reading was reestablished (Figure 12–2). This would not occur if the code was anything other than a triplet. This work also suggested that most triplet codes are not blank, but rather encode amino acids, supporting the concept of a degenerate code.

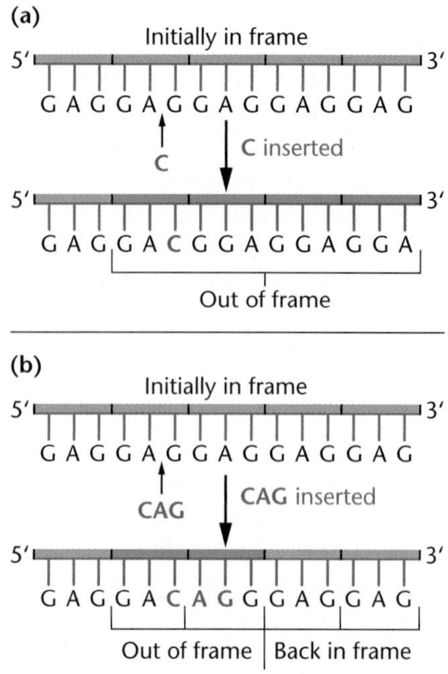

FIGURE 12–2 The effect of frameshift mutations on a DNA sequence with the repeating triplet sequence GAG. (a) The insertion of a single nucleotide shifts all subsequent triplet reading frames. (b) The insertion of three nucleotides changes only two triplets, but the frame of reading is then reestablished to the original sequence.

12.3 Studies by Nirenberg, Matthaei, and Others Deciphered the Code

In 1961, Marshall Nirenberg and J. Heinrich Matthaei deciphered the first specific coding sequences, which served as a cornerstone for the complete analysis of the genetic code. Their success, as well as that of others who made important contributions to breaking the code, was dependent on the use of two experimental tools—an ***in vitro*** **(cell-free) protein-synthesizing system** and an enzyme, **polynucleotide phosphorylase,** which enabled the production of synthetic mRNAs. These mRNAs are templates for polypeptide synthesis in the cell-free system.

Synthesizing Polypeptides in a Cell-Free System

In the cell-free system, amino acids are incorporated into polypeptide chains. This *in vitro* mixture must contain the essential factors for protein synthesis in the cell: ribosomes, tRNAs, amino acids, and other molecules essential to translation (see Chapter 13). In order to follow (or trace) protein synthesis, one or more of the amino acids must be radioactive. Finally, an mRNA must be added, which serves as the template that will be translated.

In 1961, mRNA had yet to be isolated. However, use of the enzyme polynucleotide phosphorylase allowed the artificial synthesis of RNA templates, which could be added to the cell-free system. This enzyme, isolated from bacteria,

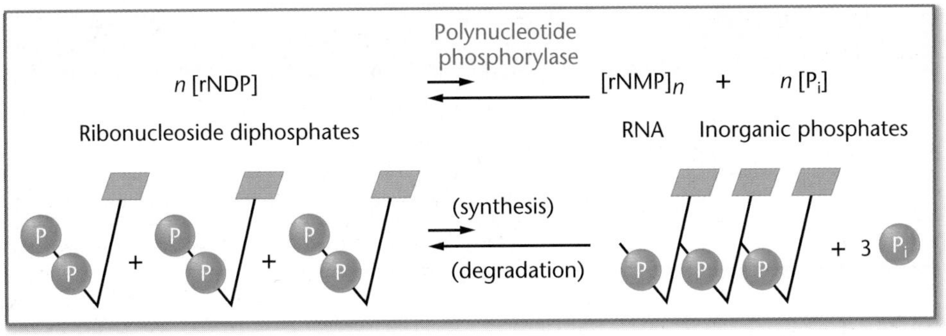

FIGURE 12–3 The reaction catalyzed by the enzyme polynucleotide phosphorylase. Note that the equilibrium of the reaction favors the degradation of RNA but can be "forced" in the direction favoring synthesis.

catalyzes the reaction shown in Figure 12–3. Discovered in 1955 by Marianne Grunberg-Manago and Severo Ochoa, the enzyme functions metabolically in bacterial cells to degrade RNA. However, *in vitro,* with high concentrations of ribonucleoside diphosphates, the reaction can be "forced" in the opposite direction to synthesize RNA, as shown.

In contrast to RNA polymerase, polynucleotide phosphorylase does not require a DNA template. As a result, each addition of a ribonucleotide is random, based on the relative concentration of the four ribonucleoside diphosphates added to the reaction mixtures. The probability of the insertion of a specific ribonucleotide is proportional to the availability of that molecule, relative to other available ribonucleotides. *This point is absolutely critical to understanding the work of Nirenberg and others in the ensuing discussion.*

Together, the cell-free system and the availability of synthetic mRNAs provided a means of deciphering the ribonucleotide composition of various triplets encoding specific amino acids.

The Use of Homopolymers

In their initial experiments, Nirenberg and Matthaei synthesized **RNA homopolymers,** each with only one type of ribonucleotide. Therefore, the mRNA added to the *in vitro* system was either UUUUUU . . . , AAAAAA . . . , CCC CCC . . . , or GGGGGG . . . . They tested each mRNA and were able to determine which, if any, amino acids were incorporated into newly synthesized proteins. To do this, the researchers labeled 1 of the 20 amino acids added to the *in vitro* system and conducted a series of experiments, each with a different radioactively labeled amino acid.

For example, in experiments using ^{14}C-phenylalanine (Table 12.1), Nirenberg and Matthaei concluded that the message poly U (polyuridylic acid) directed the incorporation of only phenylalanine into the homopolymer polyphenylalanine. Assuming the validity of a triplet code, they determined the first specific codon assignment—UUU codes for phenylalanine. Using similar experiments, they quickly found that AAA codes for lysine and CCC codes for proline. Poly G was not an adequate template, probably because the molecule folds back upon itself. Thus, the assignment for GGG had to await other approaches.

TABLE 12.1 Incorporation of ^{14}C-phenylalanine into Protein

Artificial mRNA	Radioactivity (counts/min)
None	44
Poly U	39,800
Poly A	50
Poly C	38

Source: After Nirenberg and Matthaei (1961).

Note that the specific triplet codon assignments were possible only because homopolymers were used. This method yields only the composition of triplets, but since three identical letters can have only one possible sequence (e.g., UUU), the actual codons were identified.

Mixed Copolymers

With these techniques in hand, Nirenberg and Matthaei, and Ochoa and coworkers turned to the use of **RNA heteropolymers.** In this type of experiment, two or more different ribonucleoside diphosphates are added in combination to form the artificial message. The researchers reasoned that if they knew the relative proportion of each type of ribonucleoside diphosphate, they could predict the frequency of any particular triplet codon occurring in the synthetic mRNA. If they then added the mRNA to the cell-free system and ascertained the percentage of any particular amino acid present in the new protein, they could analyze the results and predict the *composition* of triplets specifying particular amino acids.

This approach is shown in Figure 12–4, p. 244. Suppose that A and C are added in a ratio of 1A:5C. The insertion of a ribonucleotide at any position along the RNA molecule during its synthesis is determined by the ratio of A:C. Therefore, there is a $1/6$ chance for an A and a $5/6$ chance for a C to occupy each position. On this basis, we can calculate the frequency of any given triplet appearing in the message.

For AAA, the frequency is $(1/6)^3$, or about 0.4 percent. For AAC, ACA, and CAA, the frequencies are identical—that is, $(1/6)^2(5/6)$, or about 2.3 percent for each triplet. Together, all three 2A:1C triplets account for 6.9 percent of the total three-letter sequences. In the same way, each of three 1A:2C triplets accounts for $(1/6)(5/6)^2$, or 11.6 percent (or a total of 34.8 percent); CCC is represented by $(5/6)^3$, or 57.9 percent of the triplets.

By examining the percentages of any given amino acid incorporated into the protein synthesized under the direction of this message, we can propose probable base compositions for each amino acid (Figure 12–4). Since proline appears 69 percent of the time, we could propose that proline is encoded by CCC (57.9 percent) and one triplet of 2C:1A (11.6 percent). Histidine, at 14 percent, is probably coded by one 2C:1A (11.6 percent) and one 1C:2A (2.3 percent). Threonine, at 12 percent, is likely coded by only one 2C:1A. Asparagine and glutamine each appear to be coded by one of the 1C:2A triplets, and lysine appears to be coded by AAA.

Using as many as all four ribonucleotides to construct the mRNA, the researchers conducted many similar experiments. Although determining the *composition* of the triplet code words for all 20 amino acids represented a significant breakthrough, the *specific sequences* of triplets were still unknown—other approaches were needed.

Possible compositions	Possible triplets	Probability of occurrence of any triplet	Final %
3A	AAA	$(1/6)^3 = 1/216 = 0.4\%$	0.4
1C:2A	AAC ACA CAA	$(5/6)(1/6)^2 = 5/216 = 2.3\%$	$3 \times 2.3 = 6.9$
2C:1A	ACC CAC CCA	$(5/6)^2(1/6) = 25/216 = 11.6\%$	$3 \times 11.6 = 34.8$
3C	CCC	$(5/6)^3 = 125/216 = 57.9\%$	57.9
			100.0

Chemical synthesis of message ↓

CCCCCCCCACCCCCCAACCACCCCCACCCCCACCCAA — RNA

Translation of message ↓

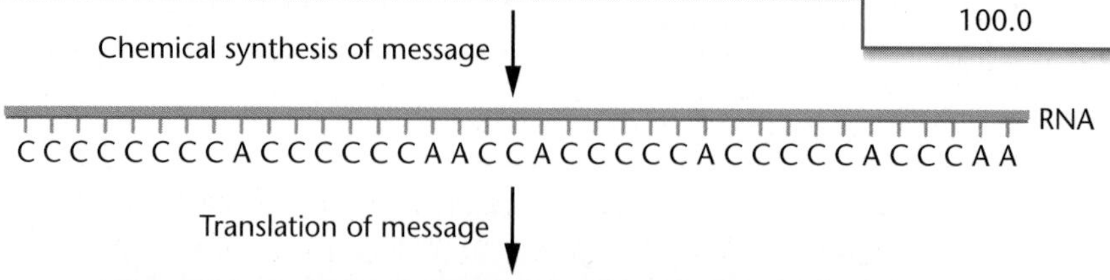

Percentage of amino acids in protein		Probable base-composition assignments
Lysine	<1	AAA
Glutamine	2	1C:2A
Asparagine	2	1C:2A
Threonine	12	2C:1A
Histidine	14	2C:1A, 1C:2A
Proline	69	CCC, 2C:1A

FIGURE 12–4 Results and interpretation of a mixed copolymer experiment where a ratio of 1A:5C is used (1/6A:5/6C).

ESSENTIAL POINT ▪ ▪ ▪

The use of RNA homopolymers and mixed copolymers in a cell-free system allowed the determination of the composition, but not the sequence, of triplet codons designating specific amino acids.

NOW SOLVE THIS

Problem 24 on page 260 asks you to analyze a reciprocal pair of mixed copolymer experiments and to predict codon compositions to the amino acids they encode.

Hint: The data sets generated in the reciprocal experiments are essential to solving the problem because analysis of the initial data set will provide you with more than one possible answer. However, only one answer is consistent with both data sets.

The Triplet Binding Assay

It was not long before more advanced techniques were developed. In 1964, Nirenberg and Philip Leder developed the **triplet binding assay,** which led to specific assignments of triplets. The technique took advantage of the observation that ribosomes, when presented *in vitro* with an RNA sequence as short as three ribonucleotides, will bind to it and form a complex similar to that found *in vivo*. The triplet acts like a

codon in mRNA, attracting the complementary sequence within tRNA (Figure 12–5). The triplet sequence in tRNA that is complementary to a codon of mRNA is an **anticodon.**

Although it was not yet feasible to chemically synthesize long stretches of RNA, triplets of known sequence could be synthesized in the laboratory to serve as templates. All that was needed was a method to determine which tRNA–amino acid was bound to the triplet RNA–ribosome complex. The test system Nirenberg and Leder devised was quite simple. The amino acid to be tested was made radioactive, and a charged tRNA was produced. Because codon compositions were known, researchers could narrow the range of amino acids that should be tested for each specific triplet.

The radioactively charged tRNA, the RNA triplet, and ribosomes were incubated together and then passed through a nitrocellulose filter, which retains the larger ribosomes but not the other smaller components, such as unbound charged tRNA. If radioactivity is not retained on the filter, an incorrect amino acid has been tested. But if radioactivity remains on the filter, it is retained because the charged tRNA has bound to the triplet associated with the ribosome. When this occurs, a specific codon assignment can be made.

Work proceeded in several laboratories, and in many cases clear-cut, unambiguous results were obtained. Table 12.2, for example, shows 26 triplets assigned to 9 amino acids. However, in some cases, the degree of triplet binding was inefficient and assignments were not possible. Eventually, about 50 of the

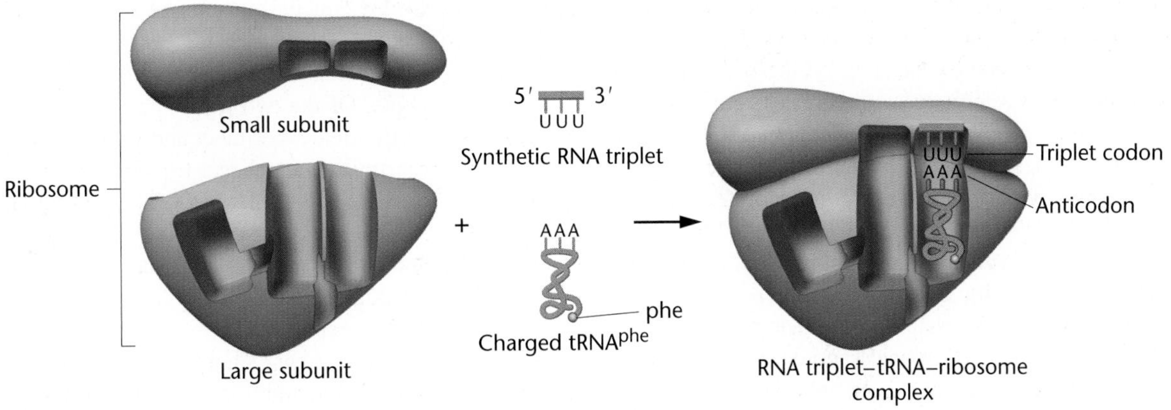

FIGURE 12–5 Illustration of the behavior of the components during the triplet binding assay. The UUU triplet RNA sequence acts as a codon, attracting the complementary AAA anticodon of the charged $tRNA^{phe}$, which together are bound by the subunits of the ribosome.

TABLE 12.2 Amino Acid Assignments to Specific Trinucleotides Derived from the Triplet Binding Assay

Trinucleotides	Amino Acid
AAA AAG	Lysine
AUG	Methionine
AUU AUG AUA	Isoleucine
CCG CCA CCU CCC	Proline
CUC CUA CUG CUU	Leucine
GAA GAG	Glutamic acid
UCA UCG UCU UCC	Serine
UGU UGC	Cysteine
UUA UUG	Leucine
UUU UUC	Phenylalanine

64 triplets were assigned. These specific assignments of triplets to amino acids led to two major conclusions. First, the genetic code is *degenerate*; that is, one amino acid can be specified by more than one triplet. Second, the code is *unambiguous*. That is, a single triplet specifies only one amino acid. As you shall see later in this chapter, these conclusions have been upheld with only minor exceptions. The triplet binding technique was a major innovation in deciphering the genetic code.

Repeating Copolymers

Yet another innovative technique used to decipher the genetic code was developed in the early 1960s by Gobind Khorana, who chemically synthesized long RNA molecules consisting of short sequences repeated many times. First, he created shorter sequences (e.g., di-, tri-, and tetranucleotides), which were then replicated many times and finally joined enzymatically to form the long polynucleotides. As shown in Figure 12–6, a

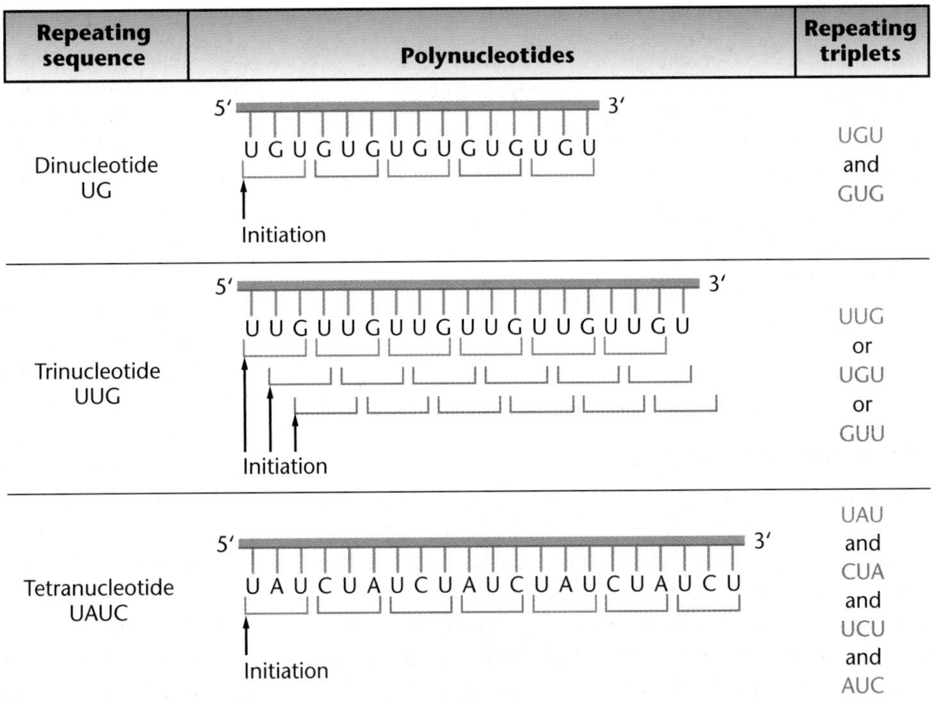

FIGURE 12–6 The conversion of di-, tri-, and tetranucleotides into repeating copolymers. The triplet codons that are produced in each case are shown.

dinucleotide made in this way is converted to an mRNA with two repeating triplets. A trinucleotide is converted to an mRNA with three potential triplets, depending on the point at which initiation occurs, and a tetranucleotide creates four repeating triplets.

When synthetic messages were added to a cell-free system, the predicted number of amino acids incorporated was upheld. Several examples are shown in Table 12.3. When the data were combined with those on composition assignment and triplet binding, specific assignments were possible.

One example of specific assignments made from this data demonstrates the value of Khorana's approach. Consider the following three experiments in concert with one another. The repeating trinucleotide sequence UUCUUCUUC . . . produces three possible triplets: UUC, UCU, and CUU, depending on the initiation point. When placed in a cell-free translation system, the polypeptides containing phenylalanine (phe), serine (ser), and leucine (leu) are produced. On the other hand, the repeating dinucleotide sequence UCUCUCUC . . . produces the triplets UCU and CUC with the incorporation of leucine and serine into the polypeptide. These results indicate that the triplets UCU and CUC specify leucine and serine, but they don't indicate which triplet specifies which amino acid. We can further conclude that *either* the CUU *or* the UUC triplet also encodes leucine or serine, while the other encodes phenylalanine.

To derive more specific information, let's examine the results of using the repeating tetranucleotide sequence UUAC, which produces the triplets UUA, UAC, ACU, and CUU. The CUU triplet is one of the two that interest us. Three amino acids are incorporated: leucine, threonine, and tyrosine. Because CUU must specify only serine or leucine and because, of these two, only leucine appears, we can conclude that CUU specifies leucine.

Once this is established, we can logically determine all other assignments. Of the two triplet pairs remaining (UUC and UCU from the first experiment and UCU and CUC from the second experiment), whichever triplet is common to both must encode serine. This triplet is UCU. By elimination, we find that UUC encodes phenylalanine and CUC encodes leucine. While the logic must be followed carefully, four specific triplets encoding three different amino acids have been assigned from these experiments.

From these interpretations, Khorana reaffirmed triplets that were already deciphered and filled in gaps left from other approaches. For example, the use of two tetranucleotide sequences, GAUA and GUAA, suggested that at least two triplets were *termination codons*. He reached this conclusion because neither of these repeating sequences directed the incorporation of more than a few amino acids into a polypeptide. Because no triplets are common to both messages, he predicted that each repeating sequence would contain at least one triplet that terminates protein synthesis. Table 12.3 lists the possible triplets for the poly-(GAUA) sequence, of which UAG is a termination codon.

TABLE 12.3 Amino Acids Incorporated Using Repeated Synthetic Copolymers of RNA

Repeating Copolymer	Codons Produced	Amino Acids in Polypeptides
UG	UGU	Cysteine
	GUG	Valine
AC	ACA	Threonine
	CAC	Histidine
UUC	UUC	Phenylalanine
	UCU	Serine
	CUU	Leucine
AUC	AUC	Isoleucine
	UCA	Serine
	CAU	Histidine
UAUC	UAU	Tyrosine
	CUA	Leucine
	UCU	Serine
	AUC	Isoleucine
GAUA	GAU	None
	AGA	None
	UAG	None
	AUA	None

ESSENTIAL POINT

Use of the triplet-binding assay and of repeating copolymers allowed the determination of the specific sequences of triplet codons designating specific amino acids.

NOW SOLVE THIS

Problem 4 on page 259 asks you to consider several outcomes of repeating copolymer experiments.

Hint: On a repeating copolymer of RNA, translation can be initiated at different ribonucleotides. Determining the number of triplet codons produced by each possible initiation point is the key to solving this problem.

12.4 The Coding Dictionary Reveals the Function of the 64 Triplets

The various techniques used to decipher the genetic code have yielded a dictionary of 61 triplet codon–amino acid assignments. The remaining three triplets are termination signals and do not specify any amino acid.

Degeneracy and the Wobble Hypothesis

A general pattern of triplet codon assignments becomes apparent when we look at the genetic coding dictionary. Figure 12–7 designates the assignments in a particularly illustrative form first suggested by Francis Crick.

Most evident is that the code is degenerate, as the early researchers predicted. That is, almost all amino acids are specified by two, three, or four different codons. Three amino

Second position

First position (5′-end)	U	C	A	G	Third position (3′-end)
U	UUU phe	UCU ser	UAU tyr	UGU cys	U
U	UUC phe	UCC ser	UAC tyr	UGC cys	C
U	UUA leu	UCA ser	UAA Stop	UGA Stop	A
U	UUG leu	UCG ser	UAG Stop	UGG trp	G
C	CUU leu	CCU pro	CAU his	CGU arg	U
C	CUC leu	CCC pro	CAC his	CGC arg	C
C	CUA leu	CCA pro	CAA gln	CGA arg	A
C	CUG leu	CCG pro	CAG gln	CGG arg	G
A	AUU ile	ACU thr	AAU asn	AGU ser	U
A	AUC ile	ACC thr	AAC asn	AGC ser	C
A	AUA ile	ACA thr	AAA lys	AGA arg	A
A	AUG met	ACG thr	AAG lys	AGG arg	G
G	GUU val	GCU ala	GAU asp	GGU gly	U
G	GUC val	GCC ala	GAC asp	GGC gly	C
G	GUA val	GCA ala	GAA glu	GGA gly	A
G	GUG val	GCG ala	GAG glu	GGG gly	G

Initiation Termination

FIGURE 12–7 The coding dictionary. AUG encodes methionine, which initiates most polypeptide chains. All other amino acids except tryptophan, which is encoded only by UGG, are encoded by two to six triplets. The triplets UAA, UAG, and UGA are termination signals and do not encode any amino acids.

acids (serine, arginine, and leucine) are each encoded by six different codons. Only tryptophan and methionine are encoded by single codons.

Also evident is the *pattern* of degeneracy. Most often, in a set of codons specifying the same amino acid, the first two letters are the same, with only the third differing. Crick discerned a pattern in the degeneracy at the third position, and in 1966, he postulated the **wobble hypothesis.**

Crick's hypothesis first predicted that the initial two ribonucleotides of triplet codes are more critical than the third in attracting the correct tRNA. He postulated that hydrogen bonding at the third position of the codon–anticodon interaction is less constrained and need not adhere as specifically to the established base-pairing rules. The wobble hypothesis thus proposes a more flexible set of base-pairing rules at the third position of the codon (Table 12.4).

This relaxed base-pairing requirement, or "wobble," allows the anticodon of a single form of tRNA to pair with more than one triplet in mRNA. Consistent with the wobble hypothesis and degeneracy, U at the first position (the 5′ end) of the tRNA anticodon may pair with A or G at the third position (the 3′ end) of the mRNA codon, and G may likewise pair with U or C. Inosine (I), one of the modified bases found in tRNA, may pair with C, U, or A. Applying these wobble rules, a minimum of about 30 different tRNA species is necessary to accommodate the 61 triplets specifying an amino acid. If nothing more, wobble can be considered a potential economy measure, provided that the fidelity of translation is not compromised. Current estimates are that 30 to 40 tRNA

TABLE 12.4 Codon–Anticodon Base-Pairing Rules

Base at first position (5′ end) of tRNA	*Base at third position (3′ end) of mRNA*
A	U
C	G
G	C or U
U	A or G
I	A, U, or C

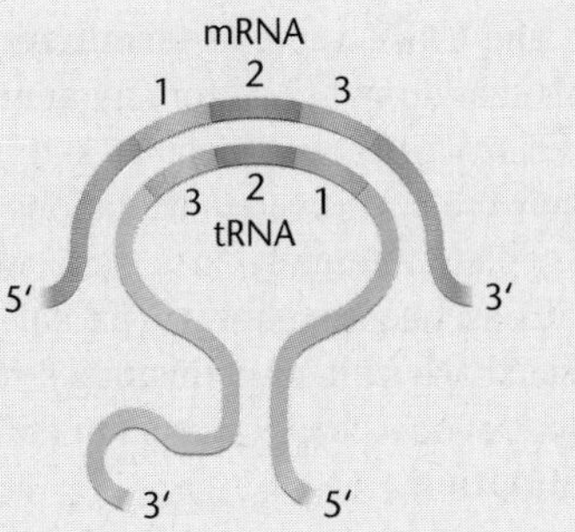

species are present in bacteria and up to 50 tRNA species exist in animal and plant cells.

The Ordered Nature of the Code

Still another observation has become apparent in the pattern of codon sequences and their corresponding amino acids, leading to the description referred to as an **ordered genetic code.** Chemically similar amino acids often share one or two "middle" bases in the different triplets encoding them. For example, either U or C is often present in the second position of triplets that specify hydrophobic amino acids, including valine and alanine, among others. Two codons (AAA and AAG) specify the positively charged amino acid lysine. If only the middle letter of these codons is changed from A to G (AGA and AGG), the positively charged amino acid arginine is specified. Hydrophilic amino acids, such as serine and threonine, are specified by triplet codons, with G or C in the second position.

The chemical properties of amino acids will be discussed in more detail in Chapter 13. The end result of an "ordered" code is that it buffers the potential effect of mutation on protein function. While many mutations of the second base of triplet codons result in a change of one amino acid to another, the change is often to an amino acid with similar chemical properties. In such cases, protein function may not be noticeably altered.

Initiation and Termination

In contrast to the *in vitro* experiments discussed earlier, initiation of protein synthesis *in vivo* is a highly specific process. In bacteria, the initial amino acid inserted into all polypeptide chains is a modified form of methionine—***N*-formylmethionine (fmet).** Only one codon, AUG, codes for methionine, and it is sometimes called the **initiator codon.** However, when AUG appears internally in mRNA, rather than at an initiating position, unformylated methionine is inserted into the polypeptide chain. Rarely, another codon, GUG, specifies methionine during initiation, though it is not clear why this happens, since GUG normally encodes valine.

In bacteria, either the formyl group is removed from the initial methionine upon the completion of protein synthesis or the entire formylmethionine residue is removed. In eukaryotes, methionine is also the initial amino acid during polypeptide synthesis. However, it is not formylated.

As mentioned in the preceding section, three other triplets (UAG, UAA, and UGA) serve as **termination codons,** punctuation signals that do not code for any amino acid. They are not recognized by a tRNA molecule, and translation terminates when they are encountered. Mutations that produce any of the three triplets internally in a gene will also result in termination. Consequently, only a partial polypeptide has been synthesized when it is prematurely released from the ribosome. When such a change occurs in the DNA, it is called a **nonsense mutation.**

ESSENTIAL POINT

The complete coding dictionary reveals that of the 64 possible triplet codons, 61 encode the 20 amino acids found in proteins, while three triplets terminate translation.

12.5 The Genetic Code Has Been Confirmed in Studies of Bacteriophage MS2

The various aspects of the genetic code discussed thus far yield a fairly complete picture. The code is triplet in nature, degenerate, unambiguous, and commaless, but it contains punctuation with respect to start and stop signals. These individual principles have been confirmed by a detailed analysis of the RNA-containing **bacteriophage MS2** by Walter Fiers and his coworkers.

MS2 is a bacteriophage that infects *E. coli.* Its nucleic acid (RNA) contains only about 3500 ribonucleotides, making up only three genes. These genes specify a coat protein, an RNA-directed replicase, and a maturation protein (the A protein). This simple system of a small genome and a few gene products enabled Fiers and his colleagues to sequence the genes and their products. The amino acid sequence of the coat protein was completed in 1970, and the nucleotide sequence of the gene and several nucleotides on each of its ends was reported in 1972.

When the chemical constitution of this gene and its encoded protein are compared, they are found to exhibit **colinearity.** That is, based on the coding dictionary, the linear sequence of nucleotides, and thus the sequence of triplet codons, correspond precisely with the linear sequence of amino acids in the protein. Furthermore, the codon for the first amino acid is AUG, the common initiator codon. The codon for the last amino acid is followed by two consecutive termination codons, UAA and UAG.

By 1976, the other two genes and their protein products were sequenced. The analysis clearly showed that the genetic code in this virus was identical to that established in bacterial systems. Other evidence suggested that the code was also identical in eukaryotes, thus providing confirmation of what seemed to be a universal genetic code.

12.6 The Genetic Code Is Nearly Universal

Between 1960 and 1978, it was generally assumed that the genetic code would be found to be universal, applying equally to viruses, bacteria, archaea, and eukaryotes. Certainly, the nature of mRNA and the translation machinery seemed to be very similar in these organisms. For example, cell-free systems derived from bacteria can translate eukaryotic mRNAs. Poly U stimulates synthesis of polyphenylalanine in cell-free systems when the components are derived from eukaryotes. Many recent studies involving recombinant DNA technology (see Chapter 17) reveal that eukaryotic genes can be inserted into bacterial cells, which are then transcribed and translated. Within eukaryotes, mRNAs from mice and rabbits have been injected into amphibian eggs and efficiently translated. For the many eukaryotic genes that have been sequenced, notably those for hemoglobin molecules, the amino acid sequence of the encoded proteins adheres to the coding dictionary established from bacterial studies.

However, several 1979 reports on the coding properties of DNA derived from mitochondria (**mtDNA**) of yeast and humans undermined the principle of the universality of the genetic language. Since then, mtDNA has been examined in many other organisms.

Cloned mtDNA fragments have been sequenced and compared with the amino acid sequences of various mitochondrial proteins, revealing several exceptions to the coding dictionary (Table 12.5). Most surprising is that the codon UGA, normally specifying termination, specifies the insertion of tryptophan during translation in yeast and human mitochondria. In yeast mitochondria, threonine is inserted instead of leucine when CUA is encountered in mRNA. In human mitochon-

TABLE 12.5 Exceptions to the Universal Code

Triplet	Normal Code Word	Altered Code Word	Source
UGA	Termination	Tryptophan	Human and yeast mitochondria; *Mycoplasma*
CUA	Leucine	Threonine	Yeast mitochondria
AUA	Isoleucine	Methionine	Human mitochondria
AGA	Arginine	Termination	Human mitochondria
AGG	Arginine	Termination	Human mitochondria
UAA	Termination	Glutamine	*Paramecium; Tetrahymena; Stylonychia*
UAG	Termination	Glutamine	*Paramecium*

dria, AUA, which normally specifies isoleucine, directs the internal insertion of methionine.

In 1985, several other exceptions to the standard coding dictionary were discovered in the bacterium *Mycoplasma capricolum* and in the nuclear genes of the protozoan ciliates *Paramecium, Tetrahymena,* and *Stylonychia.* For example, as shown in Table 12.5, one alteration converts the termination codon UGA to tryptophan, yet several others convert the normal termination codons UAA and UAG to glutamine. These changes are significant because both a prokaryote and several eukaryotes are involved, representing distinct species that have evolved separately over a long period of time.

Note the apparent pattern in several of the altered codon assignments. The change in coding capacity involves only a shift in recognition of the third, or wobble, position. For example, AUA specifies isoleucine in the cytoplasm and methionine in the mitochondrion, but in the cytoplasm, methionine is specified by AUG. Similarly, UGA calls for termination in the cytoplasm, but it specifies tryptophan in the mitochondrion; in the cytoplasm, tryptophan is specified by UGG. It has been suggested that such changes in codon recognition may represent an evolutionary trend toward reducing the number of tRNAs needed in mitochondria; only 22 tRNA species are encoded in human mitochondria, for example. However, until more examples are found, the differences must be considered to be exceptions to the previously established general coding rules.

12.7 Different Initiation Points Create Overlapping Genes

Earlier we stated that the genetic code is nonoverlapping—each ribonucleotide in an mRNA is part of only one codon. However, this characteristic of the code does not rule out the possibility that a single mRNA may have multiple initiation points for translation. If so, these points could theoretically create several different reading frames within the same mRNA, thus specifying more than one polypeptide and leading to the concept of **overlapping genes.**

That this might actually occur in some viruses was suspected when phage ϕX174 was carefully investigated. The circular DNA chromosome consists of 5386 nucleotides, which should encode a maximum of 1795 amino acids, sufficient for five or six proteins. However, this small virus in fact synthesizes 11 proteins consisting of more than 2300 amino acids. A comparison of the nucleotide sequence of the DNA and the amino acid sequences of the polypeptides synthesized has clarified the apparent paradox. At least four cases of multiple initiation have been discovered, creating overlapping genes.

For example, in one case, the coding sequences for the initiation of two polypeptides are found at separate positions within the reading frame that specifies the sequence of a third polypeptide. In one case, seven different polypeptides may be created from a DNA sequence that might otherwise have specified only three polypeptides.

A similar situation has been observed in other viruses, including phage G4 and the animal virus SV40. Like ϕX174, phage G4 contains a circular single-stranded DNA molecule. The use of overlapping reading frames optimizes the use of a limited amount of DNA present in these small viruses. However, such an approach to storing information has a distinct disadvantage in that a single mutation may affect more than one protein and thus increase the chances that the change will be deleterious or lethal.

12.8 Transcription Synthesizes RNA on a DNA Template

Even while the genetic code was being studied, it was quite clear that proteins were the end products of many genes. Thus, while some geneticists attempted to elucidate the code, other research efforts focused on the nature of genetic expression. The central question was how DNA, a nucleic acid, could specify a protein composed of amino acids.

The complex multistep process begins with the transfer of genetic information stored in DNA to RNA. The process by which RNA molecules are synthesized on a DNA template is called **transcription.** It results in an mRNA molecule complementary to the gene sequence of one of the double helix's two strands. Each triplet codon in the mRNA is, in turn, complementary to the anticodon region of its corresponding tRNA as the amino acid is correctly inserted into the polypeptide chain during translation. The significance of transcription is enormous, for it is the initial step in the process of **information flow** within the cell. The idea that RNA is involved as an intermediate molecule in the process of information flow between DNA and protein was suggested by the following observations:

1. DNA is, for the most part, associated with chromosomes in the nucleus of the eukaryotic cell. However, protein synthesis occurs in association with ribosomes located outside the nucleus in the cytoplasm. Therefore, DNA does not appear to participate directly in protein synthesis.
2. RNA is synthesized in the nucleus of eukaryotic cells, where DNA is found, and is chemically similar to DNA.
3. Following its synthesis, most RNA migrates to the cytoplasm, where protein synthesis (translation) occurs.
4. The amount of RNA is generally proportional to the amount of protein in a cell.

Collectively, these observations suggested that genetic information, stored in DNA, is transferred to an RNA intermediate, which directs the synthesis of proteins. As with most new ideas in molecular genetics, the initial supporting experimental evidence was based on studies of bacteria and their phages. It was clearly established that during initial infection, RNA synthesis preceded phage protein synthesis and that the RNA is complementary to phage DNA.

The results of these experiments agree with the concept of a messenger RNA (mRNA) being made on a DNA template and then directing the synthesis of specific proteins in association with ribosomes. This concept was formally proposed by François Jacob and Jacques Monod in 1961 as part of a model for gene regulation in bacteria. Since then, mRNA has been isolated and studied thoroughly. There is no longer any question about its role in genetic processes.

12.9 RNA Polymerase Directs RNA Synthesis

To prove that RNA can be synthesized on a DNA template, it was necessary to demonstrate that there is an enzyme capable of directing this synthesis. By 1959, several investigators, including Samuel Weiss, had independently isolated such a molecule from rat liver. Called **RNA polymerase,** it has the same general substrate requirements as does DNA polymerase, the major exception being that the substrate nucleotides contain the ribose rather than the deoxyribose form of the sugar. Unlike DNA polymerase, no primer is required to initiate synthesis; the initial base remains as a nucleoside triphosphate (NTP). The overall reaction summarizing the synthesis of RNA on a DNA template can be expressed as

$$n(\mathrm{NTP}) \xrightarrow[\text{enzyme}]{\text{DNA}} (\mathrm{NMP})_n + n(\mathrm{PP_i})$$

As this equation reveals, nucleoside triphosphates (NTPs) are substrates for the enzyme, which catalyzes the polymerization of nucleoside monophosphates (NMPs), or nucleotides, into a polynucleotide chain $(\mathrm{NMP})_n$. Nucleotides are linked during synthesis by 3′-to-5′ phosphodiester bonds (see Figure 9–10). The energy created by cleaving the triphosphate precursor into the monophosphate form drives the reaction, and inorganic pyrophosphates ($\mathrm{PP_i}$) are produced.

A second equation summarizes the sequential addition of each ribonucleotide as the process of transcription progresses:

$$(\mathrm{NMP})_n + \mathrm{NTP} \xrightarrow[\text{enzyme}]{\text{DNA}} (\mathrm{NMP})_{n+1} + \mathrm{PP_i}$$

As this equation shows, each step of transcription involves the addition of one ribonucleotide (NMP) to the growing polyribonucleotide chain $(\mathrm{NMP})_{n+1}$, using a nucleoside triphosphate (NTP) as the precursor.

RNA polymerase from *E. coli* has been extensively characterized and shown to consist of subunits designated α, β, β', and σ. The active form of the enzyme, the **holoenzyme,** contains the subunits α_2, β, β', σ and has a molecular weight of almost 500,000 Da. Of these subunits, it is the β and β' polypeptides that provide the catalytic basis and active site for transcription. As we shall see, another subunit, called the **σ (sigma) factor,** plays a regulatory function in the initiation of RNA transcription.

Although there is but a single form of the enzyme in *E. coli,* there are several different σ factors, creating variations of the polymerase holoenzyme. On the other hand, eukaryotes display three distinct forms of RNA polymerase, each consisting of a greater number of polypeptide subunits than in bacteria.

ESSENTIAL POINT

Transcription—the initial step in gene expression—is the synthesis, under the direction of RNA polymerase, of a strand of RNA complementary to a DNA template.

NOW SOLVE THIS

Problem 23 on page 260 asks you to consider the outcome of the transfer of complementary information from DNA to RNA and the amino acids encoded by this information.

Hint: In RNA, uracil is complementary to adenine, and while DNA stores genetic information in the cell, the code that is translated is contained in the RNA complementary to the template strand of DNA making up a gene.

Promoters, Template Binding, and the σ Subunit

Transcription results in the synthesis of a single-stranded RNA molecule complementary to a region along only one strand of the DNA double helix. For simplicity, let's call the transcribed DNA strand the **template strand** and its complement the **partner strand.**

The initial step is **template binding** (Figure 12–8). In bacteria, the site of this initial binding is established when the RNA polymerase σ subunit recognizes specific DNA sequences called **promoters.** These regions are located in the region upstream (5′) from the point of initial transcription of a gene. It is believed that the enzyme "explores" a length of DNA until it recognizes the promoter region and binds to about 60 nucleotide pairs of the helix, 40 of which are upstream from the point of initial transcription. Once this occurs, the helix is denatured or unwound locally, making the DNA template accessible to the action of the enzyme. The point at which transcription actually begins is called the **transcription start site.**

The importance of promoter sequences cannot be overemphasized. They govern the efficiency of the initiation of transcription. In bacteria, both strong promoters and weak promoters have been discovered. Because the interaction of promoters with RNA polymerase governs transcription, the nature of the binding between them is at the heart of discussions concerning genetic regulation, the subject of Chapter 15. While we will pursue more detailed information involving promoter–enzyme interactions, we must address two points here.

The first point is the concept of **consensus sequences** of DNA. These sequences are similar (homologous) in different genes of the same organism or in one or more genes of related organisms. Their conservation throughout evolution attests to the critical nature of their role in biological processes. Two such sequences have been found in bacterial promoters. One, TATAAT, is located 10 nucleotides upstream from the site of initial transcription (the −10 region, or **Pribnow box**). The other, TTGACA, is located 35 nucleotides upstream (the −35 region). Mutations in either region diminish transcription, often severely.

Sequences such as these are said to be ***cis*-acting elements.** Use of the term *cis* is drawn from organic chemistry nomenclature, meaning "next to" or on the same side as, in contrast to being "across from," or *trans,* to other functional groups. In molecular genetics, then, *cis*-elements are adjacent parts of the same DNA molecule. This is in contrast to ***trans*-acting factors,** molecules that bind to these DNA elements. As we will soon see, in most eukaryotic genes studied, a consensus sequence comparable to that in the −10 region has been recognized. Because it is rich in adenine and thymine residues, it is called the **TATA box.**

The second point involves the σ subunit in bacteria. The major form is designated as σ^{70}, based on its molecular weight of 70 kilodaltons (kDa). The promoters of most bacterial genes are recognized by this form; however, several alternative forms of RNA polymerase in *E. coli* have unique σ subunits associated with them (e.g., σ^{32}, σ^{54}, σ^{S}, and σ^{E}). Each form recognizes different promoter sequences, which in turn provides specificity to the initiation of transcription.

Initiation, Elongation, and Termination of RNA Synthesis

Once it has recognized and bound to the promoter [Figure 12–8(b)], RNA polymerase catalyzes **initiation,** the insertion of the first 5′-ribonucleoside triphosphate, which is complementary to the first nucleotide at the start site of the DNA template strand. As we noted earlier, no primer is required. Subsequent ribonucleotide complements are inserted and linked by phosphodiester bonds as RNA polymerization proceeds. This process of **chain elongation** [Figure 12–8(c)], continues in a 5′ to 3′ extension, creating a temporary DNA/RNA duplex whose chains run antiparallel to one another.

After a few ribonucleotides have been added to the growing RNA chain, the σ subunit dissociates from the holoenzyme and elongation proceeds under the direction of the core enzyme. In *E. coli,* this process proceeds at the rate of about 50 nucleotides/second at 37°C.

Eventually, the enzyme traverses the entire gene until it encounters a specific nucleotide sequence that acts as a

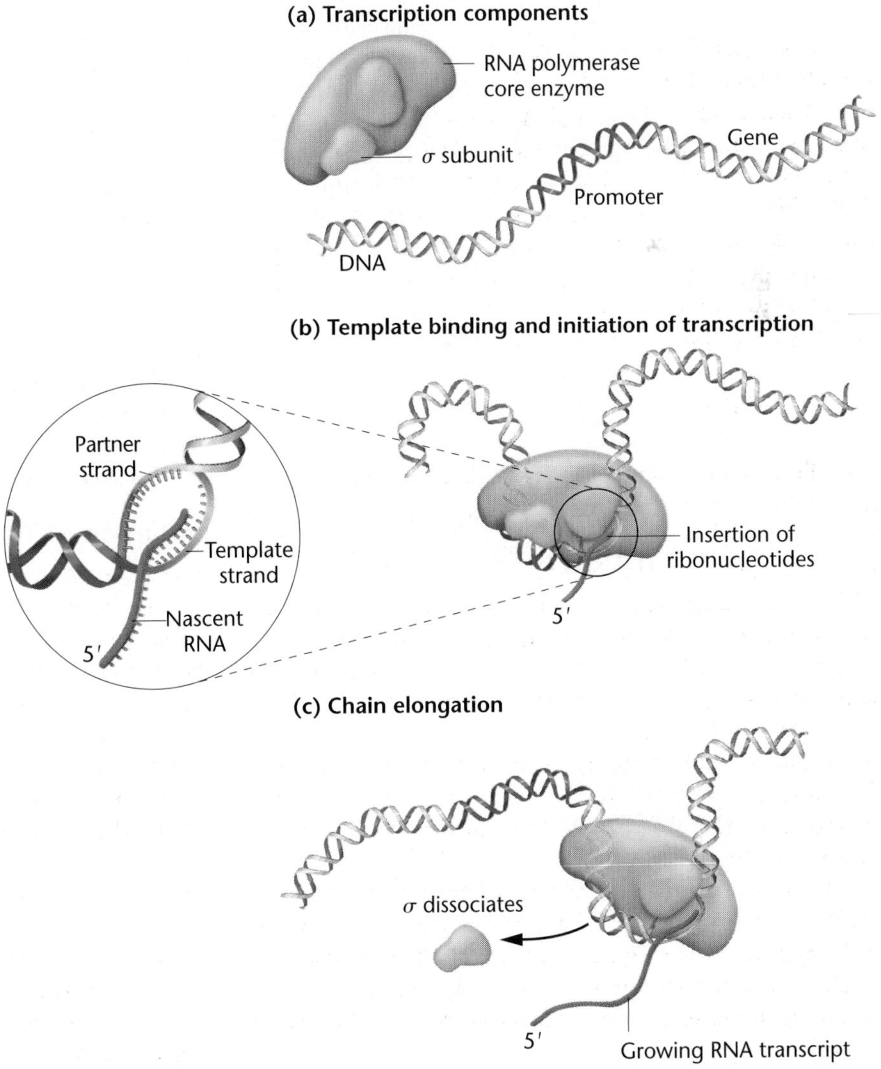

FIGURE 12–8 The early stages of transcription in prokaryotes, showing (a) the components of the process; (b) template binding at the −10 site involving the σ subunit of RNA polymerase and subsequent initiation of RNA synthesis; and (c) chain elongation, after the σ subunit has dissociated from the transcription complex and the enzyme moves along the DNA template.

termination signal. The termination sequences, about 40 base pairs in length, are extremely important in prokaryotes because of the close proximity of the end of one gene and the upstream sequences of the adjacent gene. An interesting aspect of termination in bacteria is that the termination sequence alluded to above is actually transcribed into RNA. The unique sequence of nucleotides in this termination region causes the newly formed transcript to fold back on itself, forming what is called a **hairpin secondary structure,** held together by hydrogen bonds. The hairpin is important to termination. In some cases, the termination of synthesis is dependent on the **termination factor, ρ (rho)**—a large hexameric protein that physically interacts with the growing RNA transcript.

At the point of termination, the transcribed RNA molecule is released from the DNA template and the core polymerase enzyme dissociates. The synthesized RNA molecule is precisely complementary to a DNA sequence representing the template strand of a gene. Wherever an A, T, C, or G residue existed, a corresponding U, A, G, or C residue, respectively, is incorporated into the RNA molecule. These RNA molecules ultimately provide the information leading to the synthesis of all proteins present in the cell.

In bacteria, groups of genes whose products are related are often clustered along the chromosome. In many such cases, they are contiguous, and all but the last gene lack the encoded signals for termination. The result is that during transcription, a large mRNA is produced that encodes more than one protein. Since genes in bacteria are sometimes called **cistrons,** the RNA is called a **polycistronic mRNA.** The products of genes transcribed in this fashion are usually all needed by the cell at the same time, so this is an efficient way to transcribe and subsequently translate the needed genetic information. In eukaryotes, **monocistronic mRNAs** are the rule.

12.10 Transcription in Eukaryotes Differs from Prokaryotic Transcription in Several Ways

Much of our knowledge of transcription has been derived from studies of prokaryotes. The general aspects of the mechanics of these processes are mostly similar in eukaryotes, but there are several notable differences:

1. Transcription in eukaryotes occurs within the nucleus under the direction of three separate forms of RNA polymerase. Unlike prokaryotes, in eukaryotes the RNA transcript is not free to associate with ribosomes prior to the completion of transcription. For the mRNA to be translated, it must move out of the nucleus into the cytoplasm.
2. Initiation of transcription of eukaryotic genes requires that the compact chromatin fiber, characterized by nucleosome coiling (Chapter 11), must be uncoiled and the DNA made accessible to RNA polymerase and other regulatory proteins. This transition is referred to as **chromatin remodeling,** reflecting the dynamics involved in the conformational change that occurs as the DNA helix is opened.
3. Initiation and regulation of transcription involve a more extensive interaction between *cis*-acting upstream DNA sequences and *trans*-acting protein factors involved in stimulating and initiating transcription. In addition to promoters, other control units, called **enhancers,** may be located in the 5′ regulatory region upstream from the initiation point, but they have also been found within the gene or even in the 3′ downstream region beyond the coding sequence.
4. Alteration of the primary RNA transcript to produce mature eukaryotic mRNA involves many complex stages referred to generally as "processing." An initial processing step involves the addition of a 5′ cap and a 3′ tail to most transcripts destined to become mRNAs. Other extensive modifications occur to the internal nucleotide sequence of eukaryotic RNA transcripts that eventually serve as mRNAs. The initial (or primary) transcripts are most often much larger than those that are eventually translated. Sometimes called **pre-mRNAs,** they are part of a group of molecules found only in the nucleus—a group referred to collectively as **heterogeneous nuclear RNA (hnRNA).** Such RNA molecules are of variable but large size and are complexed with proteins, forming **heterogeneous nuclear ribonucleoprotein particles (hnRNPs).** Only about 25 percent of hnRNA molecules are converted to mRNA. In those that are converted, substantial amounts of the ribonucleotide sequence are excised, and the remaining segments are spliced back together prior to nuclear export and translation. This phenomenon has given rise to the concepts of **split genes** and **splicing** in eukaryotes.

In the remainder of this chapter we will look at the basic details of transcription in eukaryotic cells. The process of transcription is highly regulated, determining which DNA sequences are copied into RNA and when and how frequently they are transcribed. We will return to topics directly related to regulation of eukaryotic transcription in Chapter 15.

Initiation of Transcription in Eukaryotes

The recognition of certain highly specific DNA regions by RNA polymerase is the basis of orderly genetic function in all cells. Eukaryotic RNA polymerase exists in three unique forms, each of which transcribes different types of genes, as indicated in Table 12.6. Each enzyme is larger and more complex than the single prokaryotic polymerase. For example, in

TABLE 12.6 RNA Polymerases in Eukaryotes

Form	Product	Location
I	rRNA	Nucleolus
II	mRNA, snRNA	Nucleoplasm
III	5*S* rRNA, tRNA	Nucleoplasm

yeast, the holoenzyme consists of two large subunits and ten smaller subunits.

In regard to the initial template-binding step and promoter regions, most is known about **RNA polymerase II (RNP II),** which is responsible for the production of all mRNAs in eukaryotes. The activity of RNP II is dependent on both *cis*-acting elements in the gene itself and a number of *trans*-acting transciption factors that bind to these DNA elements (we will return to the topic of transcription factors below). At least three *cis*-acting DNA elements regulate the initiation of transcription by RNP II. The first of these sequences, called a **core-promoter element,** determines where RNP II binds to the DNA and where it begins copying the DNA into RNA. The other two types of regulatory DNA sequences, called **promoter** and **enhancer elements,** influence the *efficiency* as the process proceeds from the core promoter element. Recall that in prokaryotes, the DNA sequence recognized by RNA polymerase is also called the promoter. In eukaryotes, however, transcriptional initiation is controlled by this group of *cis*-acting DNA elements, and they are *collectively* referred to as the promoter. Thus, in eukaryotes the term *promoter* refers to both the core-promoter sequence where RNP II binds and to the promoter and enhancer elements in the DNA that influence RNP II activity.

In most, if not all, genes studied, the *cis*-acting core-promoter element is the **Goldberg–Hogness,** or **TATA, box** present in almost all eukaryotic genes. Located about 35 nucleotide pairs upstream (-35) from the start point of transcription, the sequence and function are analogous to that found in the -10 promoter region of prokaryotic genes. The TATA box is thought to be nonspecific and to be responsible only for fixing the site of transcription initiation by facilitating denaturation of the helix.

Another *cis*-acting DNA sequence that is part of the eukaryotic promoter is the **CAAT box.** This specific DNA sequence, which contains the consensus sequence GGCCAATCT, is typically located upstream (in the $5'$ region) of the gene at about 80 nucleotides from the start of transcription (-80). Still other upstream regulatory regions have been found, and most genes contain one or more of them. They influence the efficiency of the promoter, along with the TATA box and CAAT box. The locations of these elements have been identified through studies of deletions of particular regions of promoters that reduce the efficiency of transcription.

DNA regions called enhancers represent another *cis*-acting element. Although their locations can vary, enhancers are often found farther upstream than the regions already mentioned, or even downstream, or within the gene. Thus they can modulate transcription from a distance. Although they may not participate directly in RNP II binding to the core promoter, they are essential to highly efficient initiation of transcription.

Complementing the *cis*-acting regulatory sequences are various *trans*-acting factors that facilitate RNP II binding and, therefore, the initiation of transcription. These are proteins referred to as **transcription factors.** There are two broad categories of transcription factors: the **general transcription factors** that are absolutely required for all RNP II–mediated transcription, and the **specific transcription factors** that influence the efficiency or the rate of RNP II transcription. Transcription factors are discussed in more detail in Chapter 15.

Heterogeneous Nuclear RNA and Its Processing: Caps and Tails

Our discussion continues from the point at which an initial transcript has been produced. The genetic code that is now contained in the ribonucleotide sequence of this newly synthesized chain of RNA originated in the template strand of a DNA molecule, where complementary sequences of deoxyribonucleotides are stored. In bacteria, the relationship between DNA and mRNA appears to be quite direct. The DNA base sequence is transcribed into an mRNA sequence, which is then immediately translated into an amino acid sequence according to the genetic code. In eukaryotes, by contrast, the RNA must undergo significant processing before being transported to the cytoplasm as mRNA to participate in translation.

By 1970, accumulating evidence showed that eukaryotic mRNA is transcribed initially as a precursor molecule much larger than that which is translated. This notion was based on the observation by James Darnell and his coworkers of heterogeneous nuclear RNA (hnRNA) in mammalian nuclei that contained nucleotide sequences common to the smaller mRNA molecules present in the cytoplasm. They proposed that the initial transcript of a gene results in a large RNA molecule that must first be processed in the nucleus before it appears in the cytoplasm as a mature mRNA molecule. The various processing steps are summarized in Figure 12–9, p. 254.

The initial **posttranscriptional modification** of eukaryotic RNA transcripts destined to become mRNAs involves the $5'$ end of these molecules, where a **7-methylguanosine (7-mG) cap** is added (Figure 12–9, step 2). The cap, which is added even before the initial transcript is complete, appears to be important to subsequent processing within the nucleus, perhaps by protecting the $5'$ end of the molecule from nuclease attack. Subsequently, this cap may be involved in the transport of mature mRNAs across the nuclear membrane into the cytoplasm. The cap is fairly complex and distinguished by a unique $5'$ to $5'$ bonding between the cap and the initial ribonucleotide of the RNA. Some eukaryotes also contain a methyl group (CH_3) added to the $2'$-carbon of the ribose sugars of the first two ribonucleotides of the RNA.

Further insights into the processing of RNA transcripts during the maturation of mRNA came from the discovery that both hnRNAs and mRNAs have a stretch of as many as 250 adenylic acid residues at their $3'$ end. Such **poly-A sequences** are added after the 7-mG cap has been added. First, the $3'$ end of the initial transcript is cleaved enzymatically at a point 10 to 35 ribonucleotides from a highly conserved AAUAAA sequence (step 3). Then polyadenylation occurs by the sequential addition of a poly-A sequence (step 4). Poly A has been found at the $3'$ end of almost all mRNAs studied in a variety of eukaryotic organisms. The exceptions seem to be the products of histone genes.

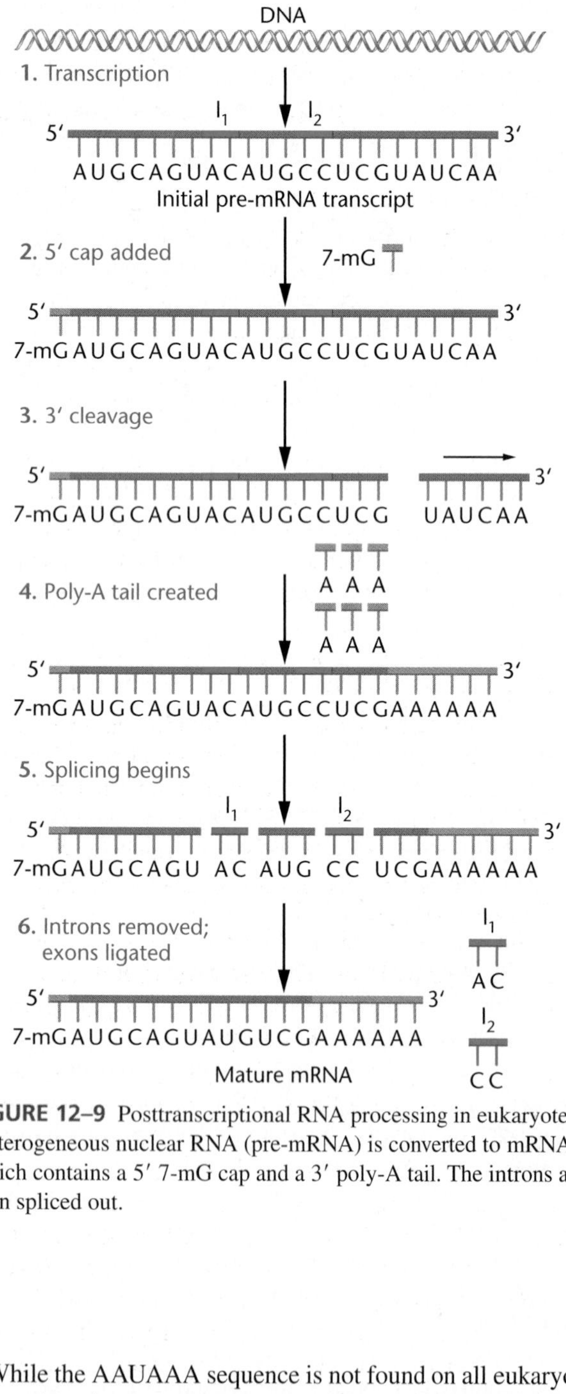

FIGURE 12–9 Posttranscriptional RNA processing in eukaryotes. Heterogeneous nuclear RNA (pre-mRNA) is converted to mRNA, which contains a 5′ 7-mG cap and a 3′ poly-A tail. The introns are then spliced out.

While the AAUAAA sequence is not found on all eukaryotic transcripts, it appears to be essential to those that have it. If the sequence is changed as a result of a mutation, those transcripts that would normally have it cannot add the poly-A tail. In the absence of this tail, these RNA transcripts are rapidly degraded. Therefore, both the 5′ cap and the 3′ poly-A tail are critical if an mRNA transcript is to be further processed and transported to the cytoplasm.

ESSENTIAL POINT

The process of creating the initial transcript during transcription is more complex in eukaryotes than in prokaryotes, including the addition of a 5′ 7-mG cap and a 3′ poly-A tail, to the pre-mRNA.

12.11 The Coding Regions of Eukaryotic Genes Are Interrupted by Intervening Sequences Called Introns

One of the most exciting breakthroughs in the history of molecular genetics occurred in 1977 when Susan Berget, Philip Sharp, and Richard Roberts presented direct evidence that the genes of animal viruses contain *internal* nucleotide sequences that are not expressed in the amino acid sequence of the proteins they encode. These internal DNA sequences are present in initial RNA transcripts, but they are removed before the mature mRNA is translated (Figure 12–9, steps 5 and 6). These nucleotide segments are called **intervening sequences** (I_1 and I_2 in Figure 12–9), and the genes that contain them are **split genes.** DNA sequences that are not represented in the final mRNA product are also called **introns** (*int* for intervening), and those retained and expressed are called **exons** (*ex* for expressed). Splicing involves the removal of the ribonucleotide sequences present in introns as a result of an excision process and the rejoining of exons.

Similar discoveries were soon made in a variety of eukaryotic genes. Two approaches have been most fruitful. The first involves the molecular hybridization of purified, functionally mature mRNAs with DNA containing the genes specifying that mRNA. Hybridization between nucleic acids that are not perfectly complementary results in **heteroduplexes,** in which introns present in the DNA but absent in the mRNA loop out and remain unpaired. Such structures can be visualized with the electron microscope, as shown in Figure 12–10. The chicken ovalbumin shown in the figure is a heteroduplex with seven loops (A–G), representing seven introns whose sequences are present in DNA but not in the final mRNA.

The second approach provides more specific information. It involves a direct comparison of nucleotide sequences of DNA with those of mRNA and the correlation with amino acid sequences. Such an approach allows the precise identification of all intervening sequences.

Thus far, most eukaryotic genes have been shown to contain introns. One of the first to be identified was the **β-globin gene** in mice and rabbits, studied independently by Philip Leder and Richard Flavell. Both the mouse and rabbit genes contain two introns of approximately the same size and same location within the gene. The rabbit gene is diagrammed in Figure 12–11, revealing the intron locations. Similar introns have been found in the β-globin gene in all mammals examined.

The **ovalbumin gene** of chickens has been extensively characterized by Bert O'Malley in the United States and Pierre Chambon in France. As shown in Figure 12–11, the gene contains seven introns. In fact, the majority of the gene's DNA sequence is composed of introns and is thus "silent." The initial RNA transcript is nearly three times the length of the mature mRNA. Compare the ovalbumin gene in Figures 12–10 and 12–11. Can you match the unpaired loops in Figure 12–10 with the sequence of introns specified in Figure 12–11?

The list of genes containing intervening sequences is long. In fact, few eukaryotic genes seem to lack introns. An extreme

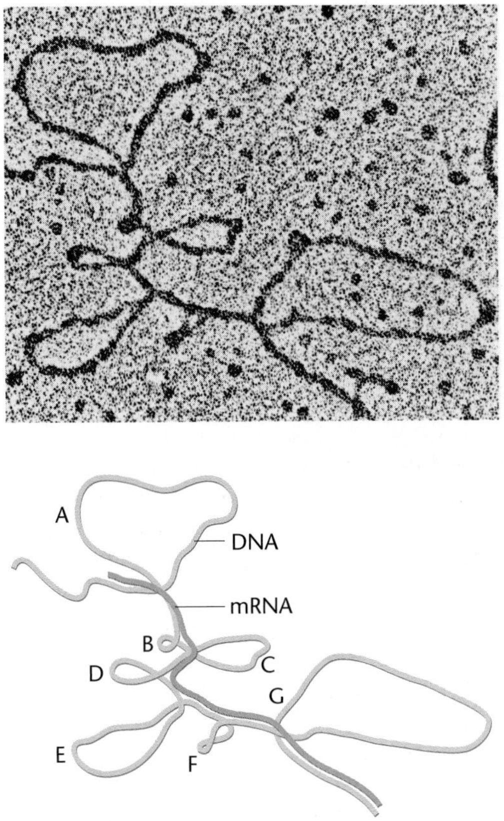

FIGURE 12–10 An electron micrograph and interpretive drawing of the hybrid molecule (heteroduplex) formed between the template DNA strand of the chicken ovalbumin gene and the mature ovalbumin mRNA. Seven DNA introns, labeled A–G, produce unpaired loops.

example of the number of introns in a single gene is found in the gene coding for one subunit of collagen, the major connective tissue protein in vertebrates. The *pro-α-2(1) collagen* gene contains 50 introns. The precision of cutting and splicing that occurs must be extraordinary if errors are not to be introduced into the mature mRNA. Equally noteworthy is a comparison of the size of genes with the size of the final mRNA once the introns are removed. As shown in Table 12.7, about

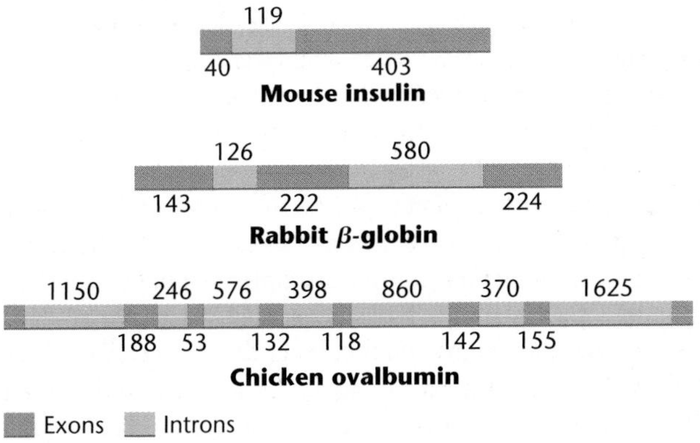

FIGURE 12–11 Intervening sequences in various eukaryotic genes. The numbers indicate the number of nucleotides present in various intron and exon regions.

TABLE 12.7 Comparing Human Gene Size, mRNA Size, and the Number of Introns

Gene	Gene Size (kb)	mRNA Size (kb)	Number of Introns
Insulin	1.7	0.4	2
Collagen [*pro-α-2(1)*]	38.0	5.0	50
Albumin	25.0	2.1	14
Phenylalanine hydroxylase	90.0	2.4	12
Dystrophin	2000.0	17.0	50

15 percent of the collagen gene consists of exons that finally appear in mRNA. For other proteins, an even more extreme picture emerges. Only about 8 percent of the albumin gene remains to be translated, and in the largest human gene known, dystrophin (the missing protein product in Duchenne muscular dystrophy), less than 1 percent of the gene sequence is retained in the mRNA. Two other human genes are also included in Table 12.7.

Although the vast majority of eukaryotic genes examined thus far contain introns, there are several exceptions. Notably, the genes coding for histones and interferon do not appear to contain introns. It is not clear why or how the genes encoding these molecules have been maintained throughout evolution without acquiring the extraneous information characteristic of almost all other genes.

Splicing Mechanisms: Autocatalytic RNAs

The discovery of split genes led to intensive attempts to elucidate the mechanism by which introns of RNA are excised and exons are spliced back together, and a great deal of progress has been made. Interestingly, it appears that somewhat different mechanisms exist for different types of RNA, as well as for RNAs produced in mitochondria and chloroplasts.

Introns can be categorized into several groups based on their splicing mechanisms. Group I, represented by introns that are part of the primary transcript of rRNAs, requires no additional components for intron excision; the intron itself is the source of the enzymatic activity necessary for its own removal. This amazing discovery, which contradicted expectations, was made in 1982 by Thomas Cech and his colleagues during a study of the ciliate protozoan *Tetrahymena.* Reflecting their autocatalytic properties, RNAs that are capable of splicing themselves are sometimes called **ribozymes.**

The **self-excision process** is shown in Figure 12–12, p. 256. Chemically, two nucleophilic reactions (called transesterification reactions) take place. The first is an interaction between guanosine, which acts as a cofactor in the reaction, and the primary transcript [Figure 12–12(a)]. The 3′-OH group of guanosine is transferred to the nucleotide adjacent to the 5′ end of the intron [Figure 12–12(b)]. The second reaction involves the interaction of the newly acquired 3′-OH group on the left-hand exon and the phosphate on the 3′ end of the right intron [Figure 12–12(c)]. The intron is spliced out, and the two exon regions are ligated, leading to the mature RNA [Figure 12–12(d)].

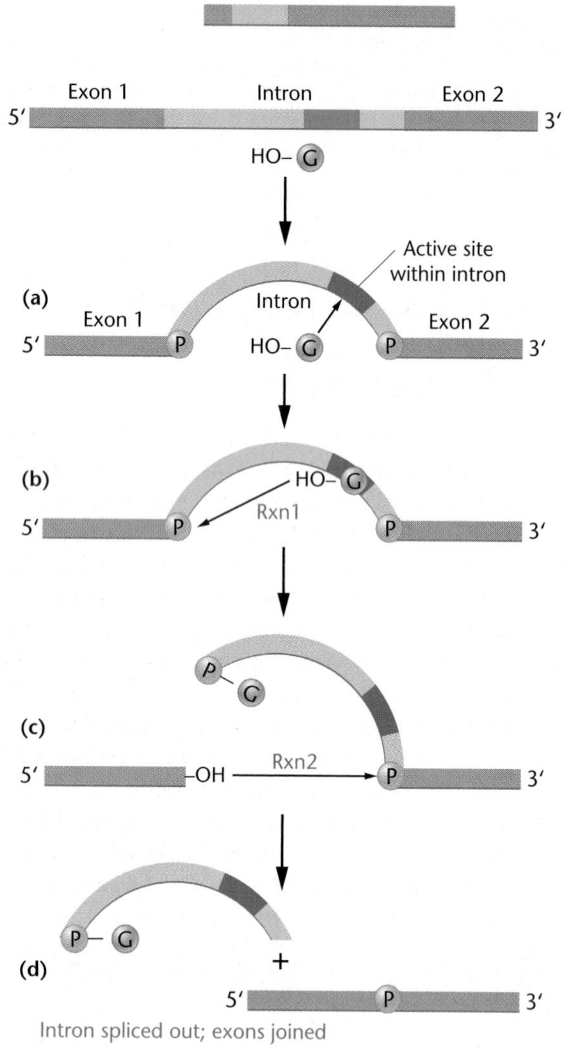

FIGURE 12–12 Splicing mechanism of pre-rRNA involving group I introns that are removed from the initial transcript. The process is one of self-excision involving two transesterification reactions.

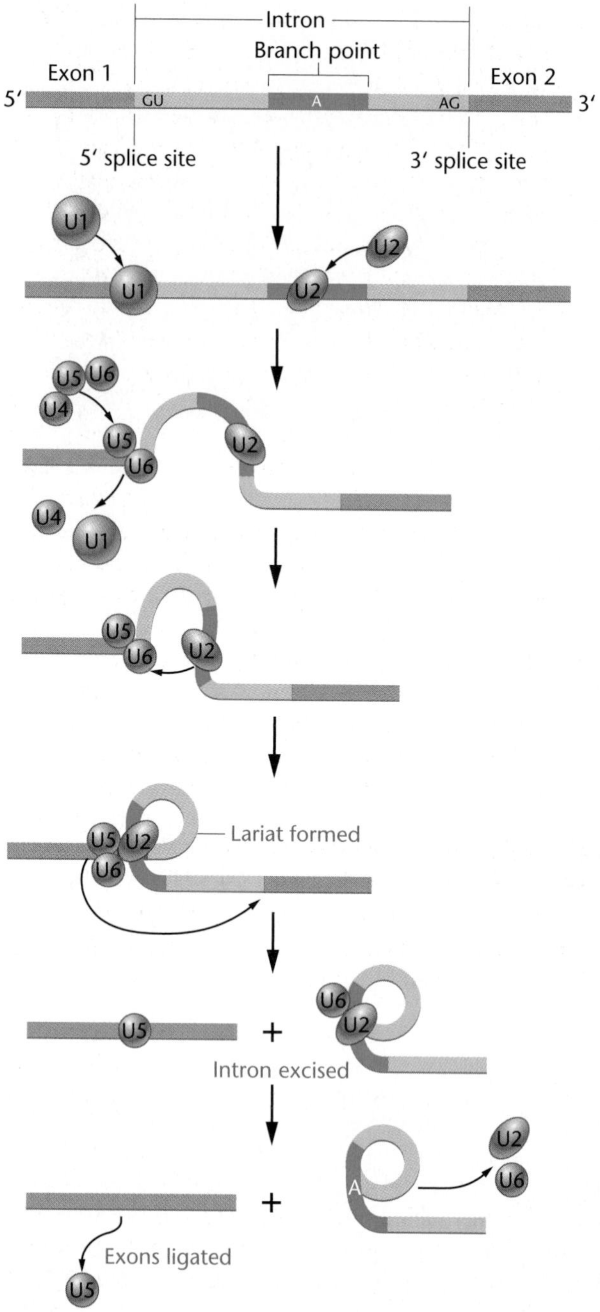

FIGURE 12–13 A model of the splicing mechanism involved with the removal of an intron from a pre-mRNA. Excision is dependent on various snRNAs (U1, U2, . . . , U6) that combine with proteins to form snurps, which are part of the spliceosome. The lariat structure in the intermediate stage is characteristic of this mechanism.

Self-excision of group I introns, as described, is now known to apply to pre-rRNAs from other protozoans. Self-excision also seems to govern the removal of introns present in the primary mRNA and tRNA transcripts produced in the mitochondria and chloroplasts. These are referred to as group II introns. Like group I molecules, splicing involves two autocatalytic reactions leading to the excision of introns. However, guanosine is not involved as a cofactor in group II introns.

Splicing Mechanisms: The Spliceosome

Introns are a major component of nuclear-derived pre-mRNA transcripts. Compared with the other RNAs we have discussed, introns in nuclear-derived mRNA can be much larger—up to 20,000 nucleotides—and they are more plentiful. Their removal appears to require a much more complex mechanism, which has been more difficult to define.

Nevertheless, many clues are now emerging, and the model in Figure 12–13 illustrates the removal of one intron. The nucleotide sequences near the perimeters of this type of intron are often similar. Many begin at the 5′ end with a GU dinucleotide sequence and terminate at the 3′ end with an AG dinucleotide sequence. These, as well as other consensus sequences shared by introns, attract specific molecules that form a molecular complex, a **spliceosome,** essential to splicing. Spliceosomes have been identified in extracts of yeast as well as mammalian cells and are very large, being 40*S* in yeast and 60*S* in mammals. Perhaps the most essential component of spliceosomes is the unique set of **small nuclear RNAs (snRNAs).** These RNAs are usually 100 to 200 nucleotides or less and are often complexed with proteins to form **small nuclear ribonucleoproteins (snRNPs,** or **snurps**). These are found only in the nucleus. Because they are rich in uridine residues, the snRNAs have been arbitrarily designated U1, U2, . . . , U6.

The snRNA of U1 bears a nucleotide sequence that is homologous to the 5′ end of the intron. Base pairing resulting from this homology promotes binding that represents the initial step in formation of the spliceosome. After the other snurps (U2, U4, U5, and U6) are added, splicing commences. As with group I splicing, two transesterification reactions are involved. The first involves the interaction of the 3′-OH group from an adenine (A) residue present within the **branch point** region of the intron. The A residue attacks the 5′ splice site, cutting the RNA chain. In a subsequent step involving several other snurps, an intermediate structure is formed and the second reaction ensues, linking the cut 5′ end of the intron to the A. This results in the formation of the characteristic loop structure called a *lariat,* which contains the excised intron. The exons are then ligated, and the snurps are released.

The processing involved in splicing represents a potential regulatory step during gene expression. For example, several cases are known in which introns present in pre-mRNAs *derived from the same gene* are spliced *in more than one way,* thereby yielding different collections of exons in the mature mRNA. This process of **alternative splicing** yields a group of mRNAs that, upon translation, results in a series of related proteins called **isoforms.** A growing number of examples have been found in organisms ranging from viruses to *Drosophila* to humans. Alternative splicing of pre-mRNAs provides the basis for producing related proteins from a single gene, thus increasing the number of gene products that can be derived from an organism's genome.

ESSENTIAL POINT

The primary transcript in eukaryotes reflects the presence of intervening sequences, or introns, present in DNA, which must be spliced out to create the mature mRNA.

12.12 Transcription Has Been Visualized by Electron Microscopy

We conclude this chapter by referring you back to the chapter opening photograph (p. 240), which is a striking visualization of transcription occurring in the oocyte nucleus of *Xenopus laevis,* the clawed frog. Note the central axis that runs horizontally from left to right and from which threads appear to be emanating vertically. This axis, appearing as a thin thread, is the DNA of most of one gene encoding ribosomal RNA (rDNA). Each of the emanating threads, which grows longer the farther to the right it is found, is an rRNA molecule being transcribed. What is apparent is that multiple copies of RNA polymerase have initiated transcription at a point near the left end and that transcription by each of them has proceeded to the right. Simultaneous transcription by many of these polymerases results in the electron micrograph that has captured an image of the entire process.

It is fascinating to visualize the process and to confirm our expectations based on the biochemical analysis of this process.

GENETICS, TECHNOLOGY, AND SOCIETY

Nucleic Acid-Based Gene Silencing: Attacking the Messenger

Standard chemotherapies for diseases such as cancer and AIDS are often accompanied by toxic side effects. Conventional therapeutic drugs affect both normal and diseased cells, with diseased or infected cells being only slightly more susceptible than the patient's normal cells. Scientists have long wished for a magic bullet that could seek out and destroy viruses or diseased cells, leaving normal cells alive and healthy. Over the last decade, a group of promising candidates has emerged, collectively described as *nucleic acid-based gene-silencing* drugs.

The two chief nucleic acid-based therapies currently being investigated are *antisense oligonucleotides* (*ASOs*) and *RNA interference* (*RNAi*). Both have been developed through an understanding of the molecular biology of gene expression: First, a single-stranded messenger RNA (mRNA) is copied from the template strand of the duplex DNA molecule; and second, the mRNA is complexed with ribosomes, and its coded information is translated into the amino acid sequence of a polypeptide.

Normally, a gene is transcribed into RNA from only one strand of the DNA duplex. The resulting RNA is known as *sense RNA*. However, it is possible for the other DNA strand to be copied into RNA, and this RNA, produced by transcription of the "wrong" strand of DNA, is called *antisense RNA*. When present together, sense and antisense RNA strands can form double-stranded duplex structures, the formation of which may affect the sense RNA in several ways.

In ASO technologies, scientists design single-stranded antisense DNA oligonucleotides (about 20 nucleotides long) of known sequence and then synthesize large amounts of these antisense nucleic acids *in vitro*. It is theoretically possible to treat cells with these synthetic antisense oligonucleotides, so that they enter the cells and bind to precise target mRNAs.

(*Cont. on the next page*)

The binding of antisense DNA to sense mRNA may physically block its translation. Alternatively, the degradation of the RNA may result. In either case, gene expression is blocked.

The antisense approach is exciting because of its potential specificity. Because an ASO has a sequence that specifically binds to a particular sense RNA, it should be possible to inhibit synthesis of the specific protein encoded by the sense RNA. If the protein is necessary for virus reproduction or cancer cell growth (but is not necessary in normal cells), the antisense oligonucleotide should have only therapeutic effects.

One ASO drug, which targets cytomegalovirus infection, is currently on the market. Others, which are designed to counteract cancers, Crohn's disease, HIV-1, and Hepatitis C, are in Phase II clinical trials.

The second gene-silencing approach, called RNAi technology (discussed in more detail in Chapter 15), uses short double-stranded RNA molecules (~20–25 nucleotides long) with sequences complementary to specific mRNAs within cells. These are known as *short interfering RNAs* (siRNAs). These siRNA molecules may be synthesized *in vitro* or may be transcribed within a cell from cloned vectors that are introduced into cells.

Once within a cell's cytoplasm, the siRNA associates with an enzyme complex called an *RNA-induced silencing complex* (*RISC*), which is found within the cell's cytoplasm. One strand of the siRNA is degraded within the RISC, and the other strand binds to a target mRNA that contains the complementary RNA sequence. When RISC and the siRNA are bound to the target mRNA, the RISC may degrade the target mRNA or may interfere with its translation.

RNAi clinical trials are being conducted to study its use in combating the eye disease macular degeneration, with encouraging results. Other areas of high interest for RNAi-based treatments are cancers, diseases of the nervous system, and viral infections such as hepatitis B and HIV-1.

Your Turn

Take time, individually or in groups, to answer the following questions. Investigate the references and links to help you discuss some of the issues that surround the development and uses of antisense therapies.

1. What are some of the challenges in the use of ASOs as therapeutics? Do siRNAs share these challenges?

A balanced discussion of antisense oligonucleotide drugs is presented in: Lebedeva, I. and Stein, C.A. 2001. Antisense oligonucleotides: Promise and reality. *Annu. Rev. Pharmacol.* **41**:403–419. RNAi drugs are critiqued in Dykxhoorn, D. M. and Lieberman, J. 2006. Running interference: Prospects and obstacles to using small interfering RNAs as small molecule drugs. *Annu. Rev. Biomed.* Eng. **8**:377–402.

2. Have any RNAi-based therapeutic drugs reached the market? What clinical trials for RNAi drugs are currently in progress?

Information about clinical trials can be found at **http://www.ClinicalTrials.gov**. *A number of biotechnology companies, including Sirna Therapeutics, Alnylam Pharmaceuticals, and Opko Health, are developing RNAi-based therapeutics. Their Web sites also contain information about the RNAi drug pipelines and clinical trials.*

3. Studies in model organisms show that RNAi is effective in silencing genes involved in a wide range of infections and diseases. What do you think is the most promising use of RNAi as a therapeutic?

To read about animal studies using siRNA gene silencing, see Dykxhoorn, D. M., et al. 2006. The silent treatment: siRNAs as small molecule drugs. *Gene Therapy* **13**:541–552.

CASE STUDY A drug that sometimes works

A 30-year-old woman with β-thalassemia, a recessively inherited genetic disorder caused by absence of the hemoglobin β chain, had been treated with blood transfusions since the age of 7. However, in spite of the transfusions, her health was declining. As an alternative treatment, her physician administered 5-azacytidine to induce transcription of the fetal β hemoglobin chain to replace her missing β chain. This drug activates gene transcription by removing methyl groups from DNA. Addition of methyl groups silences genes. However, the physician expressed concern that approximately 40 percent of all human genes are normally silenced by methylation. Nevertheless, after several weeks of 5-azacytidine treatment, the patient's condition improved dramatically. Although the treatment was successful, use of this drug raises several important questions.

1. Why was her physician concerned that a high percentage of human genes are transcriptionally silenced by methylation?
2. What genes might raise the greatest concern?
3. What criteria would you use when deciding to administer a drug such as 5-azacytidine?

INSIGHTS AND SOLUTIONS

1. Calculate how many triplet codons would be possible had evolution seized on six bases (three complementary base pairs) rather than four bases within the structure of DNA. Would six bases accommodate a two-letter code, assuming 20 amino acids and start and stop codons?

Solution: Six things taken three at a time will produce $(6)^3$ or 216 triplet codes. If the code was a doublet, there would be $(6)^2$ or 36 two-letter codes, more than enough to accommodate 20 amino acids and start and stop signals.

2. In a heteropolymer experiment using 1/2C:1/4A:1/4G, how many different triplets will occur in the synthetic RNA molecule? How frequently will the most frequent triplet occur?

Solution: There will be $(3)^3$ or 27 triplets produced. The most frequent will be CCC, present $(1/2)^3$ or 1/8 of the time.

3. In a regular copolymer experiment, where UUAC is repeated over and over, how many different triplets will occur in the synthetic RNA, and how many amino acids will occur in the polypeptide when this RNA is translated? (Consult Figure 12–7.)

Solution: The synthetic RNA will repeat four triplets—UUA, CUU, ACU, UAC—over and over. Because both UUA and CUU encode leucine, while ACU and UAC encode threonine and tyrosine, respectively, the polypeptides synthesized under the directions of this RNA would contain three amino acids in the repeating sequence leu-leu-thr-tyr.

4. Actinomycin D inhibits DNA-dependent RNA synthesis. This antibiotic is added to a bacterial culture where a specific protein is being monitored. Compared to a control culture, where no antibiotic is added, translation of the protein declines over a period of 20 minutes, until no further protein is made. Explain these results.

Solution: The mRNA, which is the basis for translation of the protein, has a lifetime of about 20 minutes. When actinomycin D is added, transcription is inhibited and no new mRNAs are made. Those already present support the translation of the protein for up to 20 minutes.

Problems and Discussion Questions

1. Early proposals regarding the genetic code considered the possibility that DNA served directly as the template for polypeptide synthesis. In eukaryotes, what difficulties would such a system pose? What observations and theoretical considerations argue against such a proposal?
2. In studies of frameshift mutations, Crick, Barnett, Brenner, and Watts–Tobin found that either three nucleotide insertions or deletions restored the correct reading frame. (a) Assuming the code is a triplet, what effect would the addition or loss of six nucleotides have on the reading frame? (b) If the code were a sextuplet (consisting of six nucleotides), would the reading frame be restored by the addition or loss of three, six, or nine nucleotides?
3. In a mixed copolymer experiment using polynucleotide phosphorylase, 3/4G:1/4C was added to form the synthetic message. Using the resulting amino acid composition of the ensuing protein shown below, (a) indicate the percentage of time each possible triplet will occur in the message, and (b) determine one consistent base-composition assignment for the amino acids present. (c) Considering the wobble hypothesis, predict as many specific triplet assignments as possible.

Glycine	36/64	(56%)
Alanine	12/64	(19%)
Arginine	12/64	(19%)
Proline	4/64	(6%)

4. When repeating copolymers are used to form synthetic mRNAs, dinucleotides produce a single type of polypeptide that contains only two different amino acids. On the other hand, a trinucleotide sequence produces three different polypeptides, each consisting of only a single amino acid. Why? What will be produced when a repeating tetranucleotide is used? See **Now Solve This** on page 246.
5. The mRNA formed from the repeating tetranucleotide UUAC incorporates only three amino acids, but the use of UAUC incorporates four amino acids. Why?
6. In studies using repeating copolymers, AC . . . incorporates threonine and histidine, and CAACAA . . . incorporates glutamine, asparagine, and threonine. What triplet code can definitely be assigned to threonine?
7. In a coding experiment using repeating copolymers (as shown in Table 12.3), the following data were obtained. AGG is known to code for arginine. Taking into account the wobble hypothesis, assign each of the four remaining different triplet codes to its correct amino acid.

Copolymer	Codons Produced	Amino Acids in Polypeptide
AG	AGA, GAG	arg, glu
AAG	AGA, AAG, GAA	lys, arg, glu

8. In the triplet-binding assay technique, radioactivity remains on the filter when the amino acid corresponding to the experimental triplet is labeled. Explain the basis of this technique.
9. When the amino acid sequences of insulin isolated from different organisms were determined, some differences were noted. For example, alanine was substituted for threonine, serine was substituted for glycine, and valine was substituted for isoleucine at corresponding positions in the protein. List the single-base changes that could occur in triplets to produce these amino acid changes.
10. In studies of the amino acid sequence of wild-type and mutant forms of tryptophan synthetase in *E. coli*, the following changes have been observed:

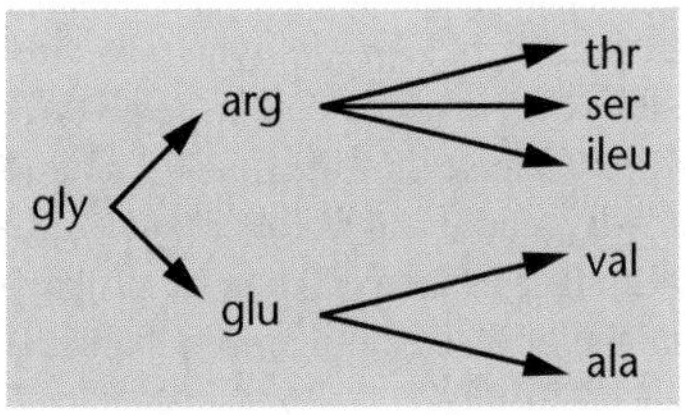

Determine a set of triplet codes in which only a single nucleotide change produces each amino acid change.

11. Why doesn't polynucleotide phosphorylase (Ochoa's enzyme) synthesize RNA *in vivo*?
12. Refer to Table 12.1. Can you hypothesize why a mixture of (Poly U) + (Poly A) would not stimulate incorporation of ^{14}C-phenylalanine into protein?
13. Predict the amino acid sequence produced during translation of the short theoretical mRNA sequences below. (Note that the second sequence was formed from the first by a deletion of only one nucleotide.) What type of mutation gave rise to sequence 2?

Sequence 1: 5′-AUGCCGGAUUAUAGUUGA-3′
Sequence 2: 5′-AUGCCGGAUUAAGUUGA-3′

14. A short RNA molecule was isolated that demonstrated a hyperchromic shift indicating secondary structure. Its sequence was determined to be

5′-AGGCGCCGACUCUACU-3′

(a) Propose a two-dimensional model for this molecule.
(b) What DNA sequence would give rise to this RNA molecule through transcription?
(c) If the molecule were a tRNA fragment containing a CGA anticodon, what would the corresponding codon be?
(d) If the molecule were an internal part of a message, what amino acid sequence would result from it following translation? (Refer to the code chart in Figure 12–7.)

15. A glycine residue exists at position 210 of the tryptophan synthetase enzyme of wild-type *E. coli.* If the codon specifying glycine is GGA, how many single-base substitutions will result in an amino acid substitution at position 210, and what are they? How many will result if the wild-type codon is GGU?

16. Shown here is a theoretical viral mRNA sequence:

5′-AUGCAUACCUAUGAGACCCUUGGA-3′

(a) Assuming that it could arise from overlapping genes, how many different polypeptide sequences can be produced? Using the chart in Figure 12–7, what are the sequences?
(b) A base substitution mutation that altered the sequence in part (a) eliminated the synthesis of all but one polypeptide. The altered sequence is shown here. Use Figure 12–7 to determine why it was altered.

5′-AUGCAUACCUAUGUGACCCUUGGA-3′

17. Most proteins have more leucine than histidine residues but more histidine than tryptophan residues. Correlate the number of codons for these three amino acids with this information.

18. Define the process of transcription. Where does this process fit into the central dogma of molecular genetics?

19. What was the initial evidence for the existence of mRNA?

20. Describe the structure of RNA polymerase in bacteria. What is the core enzyme? What is the role of the σ subunit?

21. In a written paragraph, describe the abbreviated chemical reactions that summarize RNA polymerase-directed transcription.

22. Messenger RNA molecules are very difficult to isolate from prokaryotes because they are quickly degraded. Can you suggest a reason why this occurs? Eukaryotic mRNAs are more stable and exist longer in the cell than do prokaryotic mRNAs. Is this an advantage or disadvantage for a pancreatic cell making large quantities of insulin?

23. Deoxyribonucleotide sequences derived from a template strand of DNA are shown below.
(a) Determine the mRNA sequence that would be derived from transcription of each DNA sequence.
(b) Using Figure 12–7, determine the amino acid sequence that is encoded by each mRNA.
(c) For sequence 1, what is the sequence of the partner DNA strand?

Sequence 1:	5′-CTTTTTTGCCAT-3′
Sequence 2:	5′-ACATCAATAACT-3′
Sequence 3:	5′-TACAAGGGTTCT-3′

See **Now Solve This** on page 250.

24. In a mixed copolymer experiment, messages were created with either 4/5C:1/5A or 4/5A:1/5C. These messages yielded proteins with the amino acid compositions shown in the following table. Using these data, predict the most specific coding composition for each amino acid.

4/5C:1/5A		4/5A:1/5C	
Proline	63.0%	Proline	3.5%
Histidine	13.0%	Histidine	3.0%
Threonine	16.0%	Threonine	16.6%
Glutamine	3.0%	Glutamine	13.0%
Asparagine	3.0%	Asparagine	13.0%
Lysine	0.5%	Lysine	50.0%
	98.5%		99.1%

See **Now Solve This** on page 244.

25. Shown in this problem are the amino acid sequences of the wild type and three mutant forms of a short protein.
(a) Using Figure 12–7, predict the type of mutation that created each altered protein.
(b) Determine the specific ribonucleotide change that led to the synthesis of each mutant protein.
(c) The wild-type RNA consists of nine triplets. What is the role of the ninth triplet?
(d) For the first eight wild-type triplets, which, if any, can you determine specifically from an analysis of the mutant proteins? In each case, explain why or why not.
(e) Another mutation (mutant 4) is isolated. Its amino acid sequence is unchanged, but mutant cells produce abnormally low amounts of the wild-type proteins. As specifically as you can, predict where this mutation exists in the gene.

Wild type:	met-trp-tyr-arg-gly-ser-pro-thr
Mutant 1:	met-trp
Mutant 2:	met-trp-his-arg-gly-ser-pro-thr
Mutant 3:	met-cys-ile-val-val-val-gln-his

26. The concept of consensus sequences of DNA was introduced in this chapter as those sequences that are similar (homologous) in different genes of the same organism or in genes of different organisms. Examples were the Pribnow box and the −35 region in prokaryotes and the TATA-box region in eukaryotes. Work by Novitsky and colleagues (2002. Virology. *J. Virol.* 76: 5435–5451) indicates that among 73 isolates of HIV-Type 1C (a major contributor to the AIDS epidemic), a GGGNNNNNCC consensus sequence exists (where N equals any nitrogenous base) in the promoter-enhancer region of the NF-κB transcription factor, a *cis*-acting motif that is critical in initiating HIV transcription in human macrophages. The authors contend that finding this and other conserved sequences may be of value in designing an AIDS vaccine. What advantages would finding these consensus sequences confer? What disadvantages?

27. Recent observations indicate that alternative splicing is a common mechanism for eukaryotes to expand their repertoire of gene functions. Studies by Xu and colleagues (2002. *Nuc. Acids Res.* 30: 3754–3766) indicate that approximately 50 percent of human genes use alternative splicing, and approximately 15 percent of disease-causing mutations involve aberrant alternative splicing. Different tissues show remarkably different frequencies of alternative splicing, with the brain accounting for approximately 18 percent of such events.
(a) What does alternative splicing mean, and what is an isoform?
(b) What evolutionary strategy does alternative splicing offer, and why might some tissues engage in more alternative splicing than others?

Crystal structure of a *Thermus thermophilus* 70S ribosome containing three bound transfer RNAs.

(Reprinted from the front cover of Science, Vol. 292, May 4, 2001. Crystal structure of a Thermus thermophilus *70S ribosome containing three bound transfer RNAs (top) and exploded views showing its different molecular components (middle and bottom). Image provided by Dr. Albion Baucom (baucom@biology.ucsc.edu). Copyright American Association for the Advancement of Science.)*

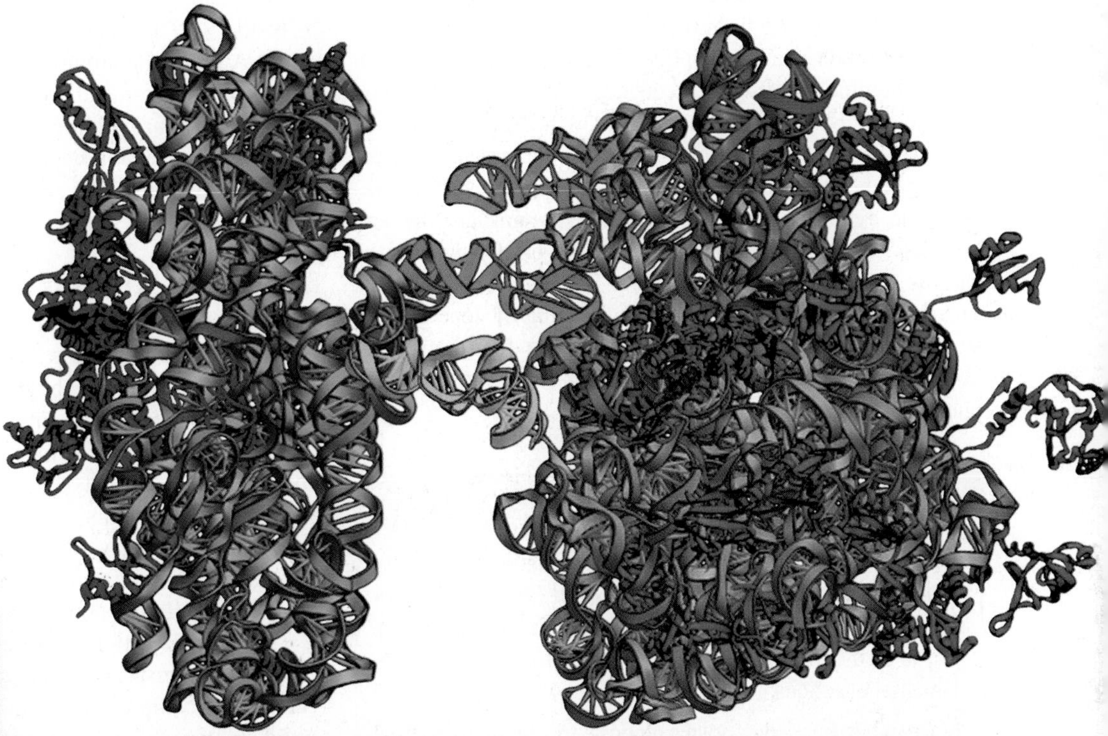

13

Translation and Proteins

Chapter Concepts

- The ribonucleotide sequence of messenger RNA (mRNA) reflects genetic information stored in DNA that makes up genes and corresponds to the amino acid sequences in proteins encoded by those genes.
- The process of translation decodes the information in mRNA, leading to the synthesis of polypeptide chains.
- Translation involves the interactions of mRNA, tRNA, ribosomes, and a variety of translation factors essential to the initiation, elongation, and termination of the polypeptide chain.
- Proteins, the final product of most genes, achieve a three-dimensional conformation that is based on the primary amino acid sequences of the polypeptide chains making up each protein.
- The function of any protein is closely tied to its three-dimensional structure, which can be disrupted by mutation.

We have already learned that a genetic code exists that stores information in the form of triplet nucleotides in DNA and that this information is initially expressed through the process of transcription into a messenger RNA that is complementary to one strand of the DNA helix. However, in most instances, the final product of gene expression is a polypeptide chain consisting of a linear series of amino acids whose sequence has been prescribed by the genetic code. In this chapter, we will examine how the information present in mRNA is processed to create polypeptides, which then fold into protein molecules. We will also review the evidence that confirms that proteins are the end products of gene expression, and we will briefly discuss the various levels of protein structure, diversity, and function. This information extends our understanding of gene expression and provides an important foundation for interpreting how the mutations that arise in DNA can result in the diverse phenotypic effects observed in organisms.

How Do We Know?

In this chapter, we will focus on how genetic information, transferred from DNA to mRNA, is expressed through the production of proteins, the end product of most gene expression. As you study this topic, you should try to answer several fundamental questions:

1. On what basis did Holley propose the two-dimensional cloverleaf model of tRNA?
2. How do we know that the process of translation occurs in association with ribosomes?
3. How do we know that proteins are the end products of genetic expression?
4. How did we learn that most genes ultimately encode a single polypeptide chain?
5. How do we know that the structure of a protein is intimately related to the function of that protein?

13.1 Translation of mRNA Depends on Ribosomes and Transfer RNAs

Translation of mRNA is the biological polymerization of amino acids into polypeptide chains. This process, alluded to in our discussion of the genetic code in Chapter 12, occurs only in association with ribosomes, which serve as nonspecific workbenches. The central question in translation is how triplet ribonucleotides of mRNA direct specific amino acids into their correct position in the polypeptide. This question was answered once **transfer RNA (tRNA)** was discovered. This class of molecules adapts specific triplet codons in mRNA to their correct amino acids. The *adaptor hypothesis* for the role of tRNA was postulated by Francis Crick in 1957.

In association with a ribosome, mRNA presents a triplet codon that calls for a specific amino acid. A specific tRNA molecule contains within its nucleotide sequence three consecutive ribonucleotides complementary to the codon, called the **anticodon,** which can base-pair with the codon. Another region of this tRNA is covalently bonded to its corresponding amino acid.

Inside the ribosome, hydrogen bonding of tRNAs to mRNA holds the amino acids in proximity so that a peptide bond can be formed. This process occurs over and over as mRNA runs through the ribosome and amino acids are polymerized into a polypeptide. Before we discuss the actual process of translation, let's first consider the structures of the ribosome and tRNA.

Ribosomal Structure

Because of its essential role in the expression of genetic information, the **ribosome** has been extensively analyzed. One bacterial cell contains about 10,000 ribosomes, and a eukaryotic cell contains many times more. Electron microscopy reveals that the bacterial ribosome is about 25 μm at its largest diameter and consists of two subunits, one large and one small. Both subunits consist of one or more molecules of rRNA and an array of **ribosomal proteins.** When the two subunits are associated with each other in a single ribosome, the structure is sometimes called a **monosome.**

The specific differences between prokaryotic and eukaryotic ribosomes are summarized in Figure 13–1. The subunit and rRNA components are most easily isolated and characterized on the basis of their sedimentation behavior in sucrose gradients (their rate of migration, or Svedberg coefficient *S,* which reflects their density, mass, and shape). In prokaryotes, the monosome is a 70*S* particle; in eukaryotes, it is approximately 80*S*. Sedimentation coefficients, which reflect the variable rate of migration of different-sized particles and molecules, are not additive. For example, the prokaryotic 70*S* monosome consists of a 50*S* and a 30*S* subunit, and the eukaryotic 80*S* monosome consists of a 60*S* and a 40*S* subunit.

The larger subunit in prokaryotes consists of a 23*S* rRNA molecule, a 5*S* rRNA molecule, and 31 ribosomal proteins. In the eukaryotic equivalent, a 28*S* rRNA molecule is accompanied by a 5.8*S* and 5*S* rRNA molecule and 49 proteins. The smaller prokaryotic subunits consist of a 16*S* rRNA component and 21 proteins. In the eukaryotic equivalent, an 18*S* rRNA component and 33 proteins are found. The approximate molecular weights (MWs) and the numbers of nucleotides of these components are also shown in Figure 13–1.

It is now clear that the RNA molecules perform the all-important catalytic functions associated with translation. The many proteins, whose functions were long a mystery, are thought to promote the binding of the various molecules involved in translation and, in general, to fine-tune the process. This conclusion is based on the observation that some of the catalytic functions in ribosomes still occur in experiments involving "ribosomal protein-depleted" ribosomes.

Molecular hybridization studies have established the degree of redundancy of the genes coding for the rRNA components. The *E. coli* genome contains seven copies of a single sequence that encodes all three components—23*S*, 16*S*, and 5*S*. The initial transcript of each set of these genes produces a

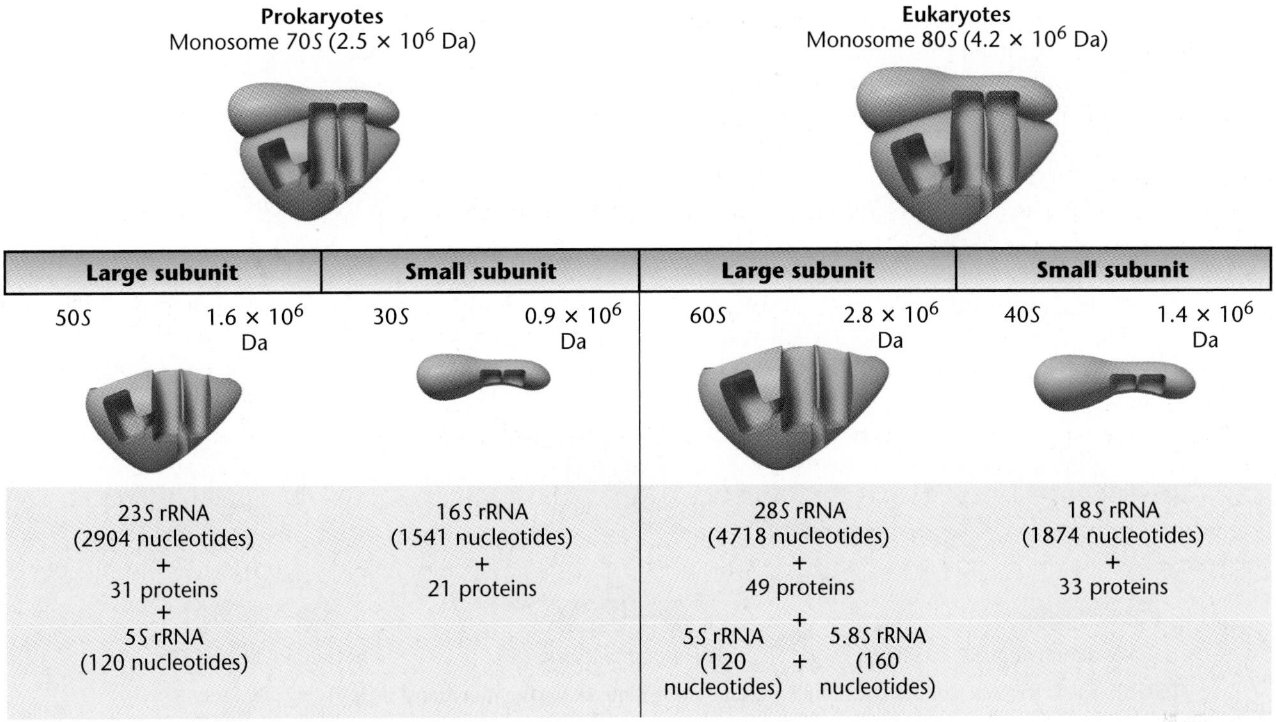

FIGURE 13–1 A comparison of the components of prokaryotic and eukaryotic ribosomes.

30*S* RNA molecule that is enzymatically cleaved into these smaller components. Coupling of the genetic information encoding these three rRNA components ensures that after multiple transcription events, equal quantities of all three will be present as ribosomes are assembled.

In eukaryotes, many more copies of a sequence encoding the 28*S*, 18*S*, and 5.8*S* components are present. In *Drosophila,* approximately 120 copies per haploid genome are each transcribed into a molecule of about 34*S*. This molecule is then processed into the 28*S*, 18*S*, and 5.8*S* rRNA species. These species are homologous to the three rRNA components of *E. coli.* In *Xenopus laevis,* over 500 copies of the 34*S* component are present per haploid genome. In mammalian cells, the initial transcript is even larger at 45*S*.

The rRNA genes, called **rDNA,** are part of the moderately repetitive DNA fraction and are present in clusters at various chromosomal sites. Each cluster in eukaryotes consists of tandem repeats, with each unit separated by a noncoding spacer DNA sequence. In humans, these gene clusters have been localized near the ends of chromosomes 13, 14, 15, 21, and 22. The unique 5*S* rRNA component of eukaryotes is not part of this larger transcript. Instead, genes coding for this ribosomal component are distinct and located separately. In humans, a gene cluster encoding the 5*S* rRNA has been located on chromosome 1.

Despite the detailed knowledge available about the structure and genetic origin of the ribosomal components, a complete understanding of the function of these components has thus far eluded geneticists. This is not surprising; the ribosome is the largest and perhaps the most intricate of all cellular structures. For example, the bacterial monosome has a combined molecular weight of 2.5 million Da!

ESSENTIAL POINT

Translation is the synthesis of polypeptide chains under the direction of mRNA in association with ribosomes.

tRNA Structure

Because of their small size and stability in the cell, transfer RNAs (tRNAs) have been investigated extensively and are the best-characterized RNA molecules. They are composed of only 75 to 90 nucleotides, displaying a nearly identical structure in bacteria and eukaryotes. In both types of organisms, tRNAs are transcribed as larger precursors, which are cleaved into mature 4*S* tRNA molecules. In *E. coli,* for example, $tRNA^{tyr}$ (the superscript identifies the specific tRNA and the cognate amino acid that binds to it) is composed of 77 nucleotides, yet its precursor contains 126 nucleotides.

In 1965, Robert Holley and his colleagues reported the complete sequence of $tRNA^{ala}$ isolated from yeast. They found that a number of nucleotides are unique to tRNA. As shown in Figure 13–2, p. 264, each nucleotide is a modification of one of the four nitrogenous bases normally present in RNA (G, C, A, and U). These include inosinic acid (which contains the purine hypoxanthine), ribothymidylic acid, and pseudouridine, among others. These modified structures, variously referred to as *unusual, rare,* or *odd bases,* are created *after* transcription, illustrating the more general concept of **posttranscriptional modification.** In this case, the unmodified base is inserted during transcription of tRNA, and subsequently, enzymatic reactions catalyze the chemical modifications to the base.

Holley's sequence analysis led him to propose the two-dimensional **cloverleaf model of tRNA.** It was known that

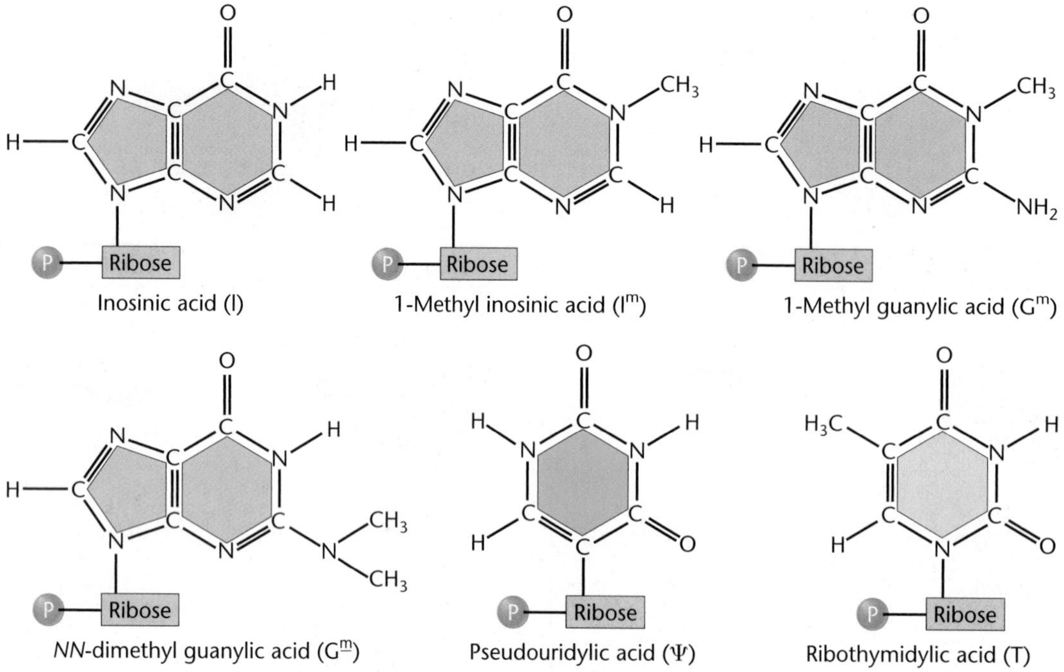

FIGURE 13–2 Ribonucleotides containing unusual nitrogenous bases found in transfer RNA.

tRNA demonstrates a secondary structure due to base pairing. Holley discovered that he could arrange the linear model in such a way that several stretches of base pairing would result. This arrangement created a series of paired stems and unpaired loops resembling the shape of a cloverleaf. Loops consistently contained modified bases that did not generally form base pairs. Holley's model is shown in Figure 13–3.

The triplets GCU, GCC, and GCA specify alanine; therefore, Holley looked for an anticodon sequence complementary to one of these codons in his $tRNA^{ala}$ molecule. He found it in the form of CGI (the 3′ to 5′ direction) in one loop of the cloverleaf. The nitrogenous base I (inosinic acid) can form hydrogen bonds with U, C, or A, the third members of the alanine triplets. Thus, the **anticodon loop** was established.

Studies of other tRNA species reveal many constant features. At the 3′ end, all tRNAs contain the sequence (… pCpCpA-3′). This is the end of the molecule where the amino acid is covalently joined to the terminal adenosine residue. All tRNAs contain the nucleotide (5′-Gp…) at the other end of the molecule. In addition, the lengths of various stems and loops are very similar. Each tRNA that has been examined also contains an anticodon complementary to the known amino acid codon for which it is specific, and all anticodon loops are present in the same position of the cloverleaf.

Because the cloverleaf model was predicted strictly on the basis of nucleotide sequence, there was great interest in the X-ray crystallographic examination of tRNA, which reveals a three-dimensional structure. By 1974, Alexander Rich and his colleagues in the United States, and J. Roberts, B. Clark, Aaron Klug, and their colleagues in England had succeeded in crystallizing tRNA and performing X-ray crystallography at a resolution of 3 Å. At this resolution, the pattern formed by individual nucleotides is discernible.

As a result of these studies, a complete three-dimensional model of tRNA is shown in Figure 13–4. At one end of the molecule is the anticodon loop and stem, and at the other end is the 3′-acceptor region where the amino acid is bound. Geneticists speculate that the shapes of the intervening loops may be recognized by the specific enzymes responsible for adding amino acids to tRNAs—a subject to which we now turn our attention.

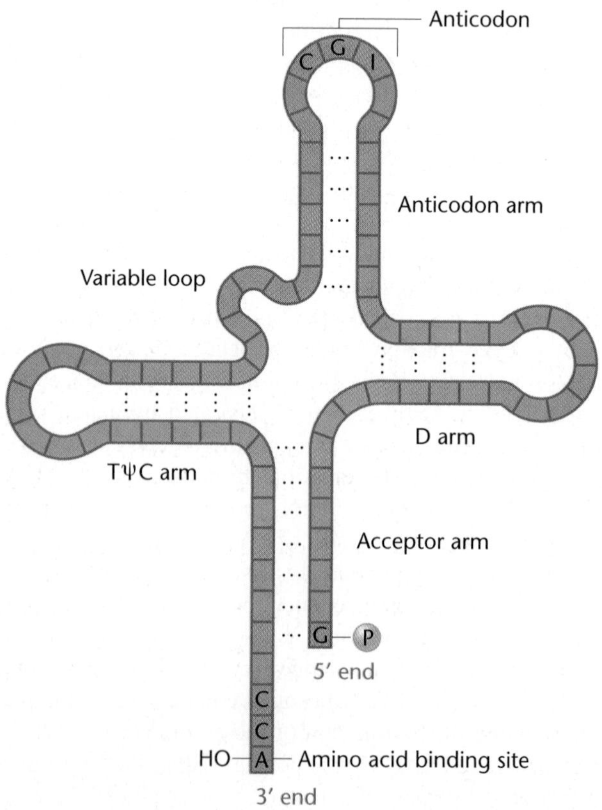

FIGURE 13–3 Holley's two-dimensional cloverleaf model of transfer RNA. Blocks represent nitrogenous bases.

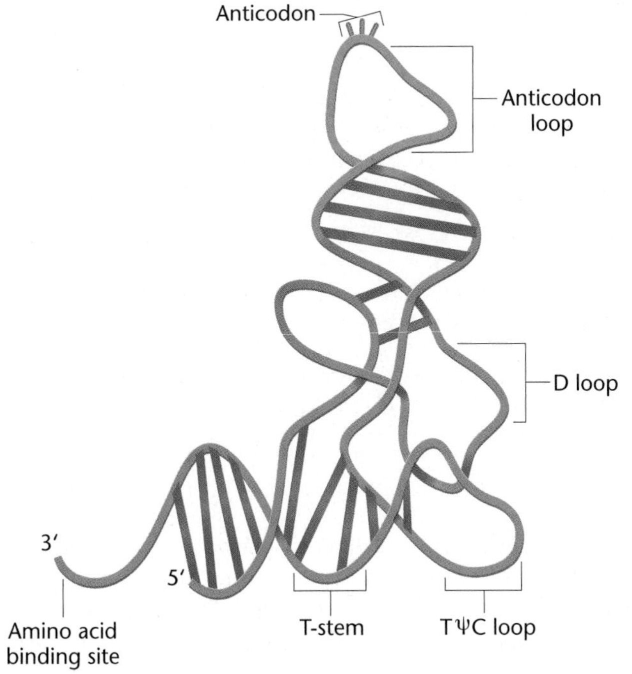

FIGURE 13–4 A three-dimensional model of transfer RNA.

Charging tRNA

Before translation can proceed, the tRNA molecules must be chemically linked to their respective amino acids. This activation process, called **charging,** occurs under the direction of enzymes called **aminoacyl tRNA synthetases.** There are 20 different amino acids, so there must be at least 20 different tRNA molecules and as many different enzymes. In theory, because there are 61 triplets that encode amino acids, there could be 61 specific tRNAs and enzymes. However, because of the ability of the third member of a triplet code to "wobble," it is now thought that there are only 31 different tRNAs. It is also believed that there are only 20 synthetases, one for each amino acid, regardless of the greater number of corresponding tRNAs.

The charging process is outlined in Figure 13–5. In the initial step, the amino acid is converted to an activated form, reacting with ATP to create an **aminoacyladenylic acid.** A covalent linkage is formed between the 5′-phosphate group of ATP and the carboxyl end of the amino acid. This molecule remains associated with the synthetase enzyme, forming a complex that then reacts with a specific tRNA molecule. During this next step, the amino acid is transferred to the appropriate tRNA and bonded covalently to the adenine residue at the 3′ end. The charged tRNA may now participate directly in protein synthesis. Aminoacyl tRNA synthetases are highly specific enzymes because they recognize only one amino acid and the subset of corresponding tRNAs called **isoaccepting tRNAs.** Accurate charging is crucial if fidelity of translation is to be maintained.

ESSENTIAL POINT

Translation depends on tRNA molecules that serve as adaptors between triplet codons in mRNA and the corresponding amino acids.

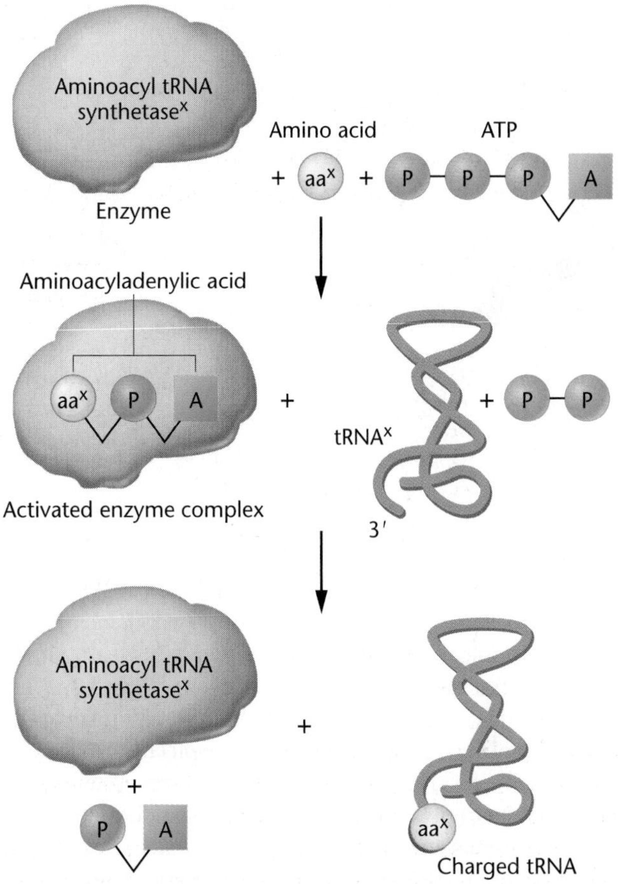

FIGURE 13–5 Steps involved in charging tRNA. The superscript x denotes that only the corresponding specific tRNA and specific aminoacyl tRNA synthetase enzyme are involved in the charging process for each amino acid.

NOW SOLVE THIS

Problem 27 on page 283 asks you to consider an experiment attempting to determine whether the anticodon of tRNA or the amino acid added to the tRNA during charging is more critical to the accuracy of translation.

Hint: In this experiment, when the triplet codon in mRNA calls for cysteine, alanine is inserted during translation, even though it is the "incorrect" amino acid.

13.2 Translation of mRNA Can Be Divided into Three Steps

In a way similar to transcription, the process of translation can be best described by breaking it into discrete phases. We will consider three phases, each with its own illustration, but keep in mind that translation is a dynamic, continuous process. You should correlate the following discussion with the step-by-step characterization in the figures. Many of the protein factors and their roles in translation are summarized in Table 13.1.

TABLE 13.1 Various Protein Factors Involved during Translation in *E. coli*

Process	Factor	Role
Initiation of translation	IF1	Stabilizes 30*S* subunit
	IF2	Binds fmet-tRNA to 30*S*-mRNA complex; binds to GTP
	IF3	Binds 30*S* subunit to mRNA
Elongation of polypeptide	EF-Tu	Binds GTP; mediates aminoacyl-tRNA entry to the A site of ribosome
	EF-Ts	Generates active EF-Tu
	EF-G	Stimulates translocation; GTP-dependent
Termination of translation and release of polypeptide	RF1	Catalyzes release of the polypeptide chain from tRNA and dissociation of the translocation complex; specific for UAA and UAG termination codons
	RF2	Behaves like RF1; specific for UGA and UAA codons
	RF3	Stimulates RF1 and RF2

Initiation

Initiation of translation is depicted in Figure 13–6. Recall that the ribosome serves as a nonspecific workbench for the translation process. Most ribosomes, when they are not involved in translation, are dissociated into their large and small subunits. Initiation of translation in *E. coli* involves the small ribosome subunit, an mRNA molecule, a specific charged tRNA, GTP, Mg^{2+}, and at least three proteinaceous **initiation factors (IFs)** that enhance the binding affinity of the various translational components. In prokaryotes, the initiation codon of mRNA (AUG) calls for the modified amino acid **formylmethionine (fmet).**

The small ribosomal subunit binds to several initiation factors, and this complex then binds to mRNA (Step 1). In bacteria, this binding involves a sequence of up to six ribonucleotides (AGGAGG, not shown), which *precedes* the initial AUG start codon of mRNA. This sequence (containing only purines and called the **Shine-Dalgarno sequence**) base-pairs with a region of the 16*S* rRNA of the small ribosomal subunit, facilitating initiation.

Another initiation protein then enhances the binding of charged fmet-tRNA to the small subunit in response to the AUG triplet (Step 2). This step sets the reading frame so that all subsequent groups of three ribonucleotides are translated accurately. This aggregate represents the **initiation complex,** which then combines with the large ribosomal subunit. At this point, a molecule of GTP is hydrolyzed, providing the required energy for the release of the initiation factors (Step 3).

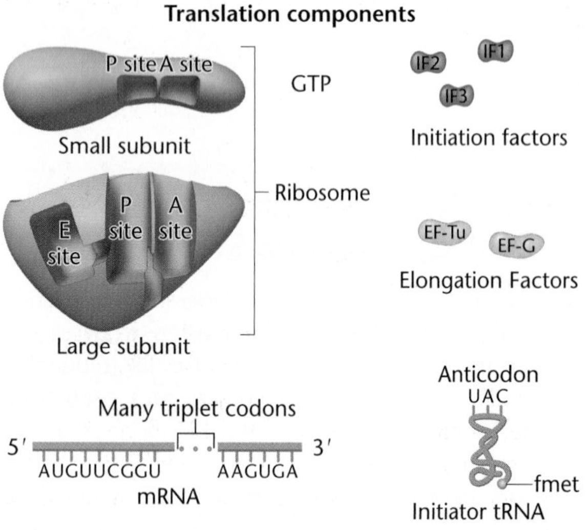

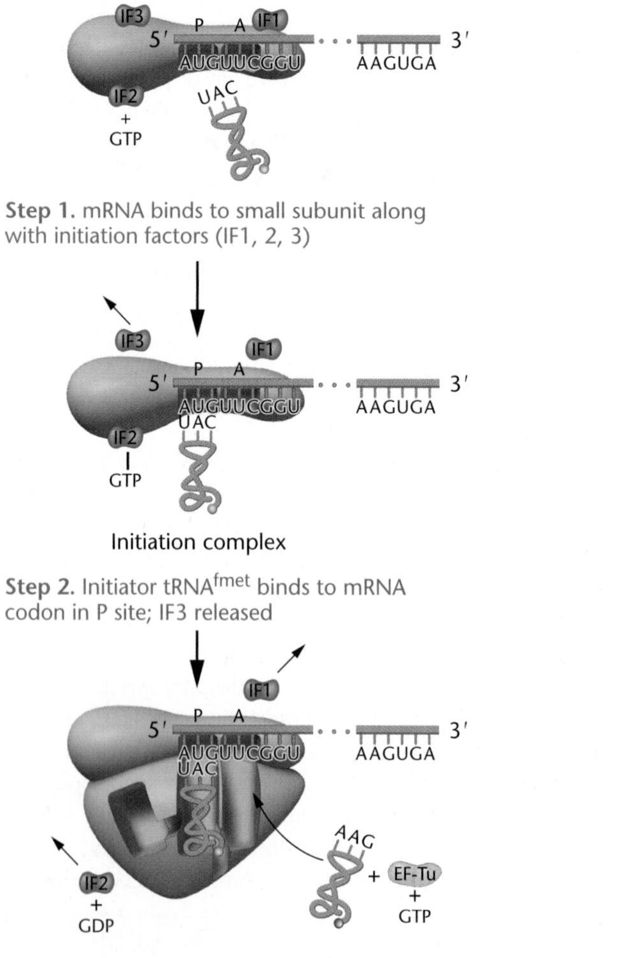

FIGURE 13–6 Initiation of translation. The components are depicted at the left of the figure.

Elongation

The second phase of translation, elongation, is depicted in Figure 13–7. Once both subunits of the ribosome are assembled with the mRNA, binding sites for two charged tRNA molecules are formed. These are the **P (peptidyl) site** and the **A (aminoacyl) site.** The charged initiator tRNA binds to the P site, provided that the AUG triplet of mRNA is in the corresponding position of the small subunit.

The lengthening of the growing polypeptide chain by one amino acid is called **elongation.** The sequence of the second triplet in mRNA dictates which charged tRNA molecule will become positioned at the A site (Step 1). Once tRNA is present, an enzymatic reaction occurs within the large subunit of the ribosome, catalyzing the formation of the peptide bond that links the two amino acids together. At the same time, the covalent bond between the tRNA occupying the P site and its amino acid is hydrolyzed (broken). The dipeptide remains attached to the end of the tRNA still residing in the A site (Step 2). These reactions were initially believed to be catalyzed by an enzyme called **peptidyl transferase,** embedded in but never isolated from the large subunit of the ribosome. However, it is now clear that this catalytic activity is a function of the 23*S* rRNA of the large subunit, perhaps in conjunction with one or more of the ribosomal proteins. In such a case, as we saw with splicing of pre-mRNAs in Chapter 12, we refer to the complex as a **ribozyme,** recognizing the catalytic role that RNA plays in the process.

Before elongation can be repeated, the tRNA attached to the P site, which is now uncharged, must be released from the large subunit. The uncharged tRNA moves transiently through a third site on the ribosome, called the **E (exit) site.** The entire *mRNA–tRNA–*aa_2*–*aa_1 *complex* then shifts in the direction of the P site by a distance of three nucleotides (Step 3). This event requires several protein **elongation factors (EFs)** as well as the energy derived from hydrolysis of GTP. The result is that the third triplet of mRNA is now in a position to accept another specific charged tRNA into the A site (Step 4). One simple way to distinguish the two sites is to remember that, *following the shift,* the P site (P for peptide) contains a tRNA attached to a peptide chain, whereas the A site (A for amino acid) contains a tRNA with an amino acid attached.

The sequence of elongation is repeated over and over (Steps 4 and 5). An additional amino acid is added to the growing polypeptide chain each time the mRNA advances through the ribosome. Once a polypeptide chain of reasonable size is assembled (about 30 amino acids), it begins to emerge from the base of the large subunit, as illustrated in Step 6. A tunnel exists within the large subunit, from which the elongating polypeptide emerges.

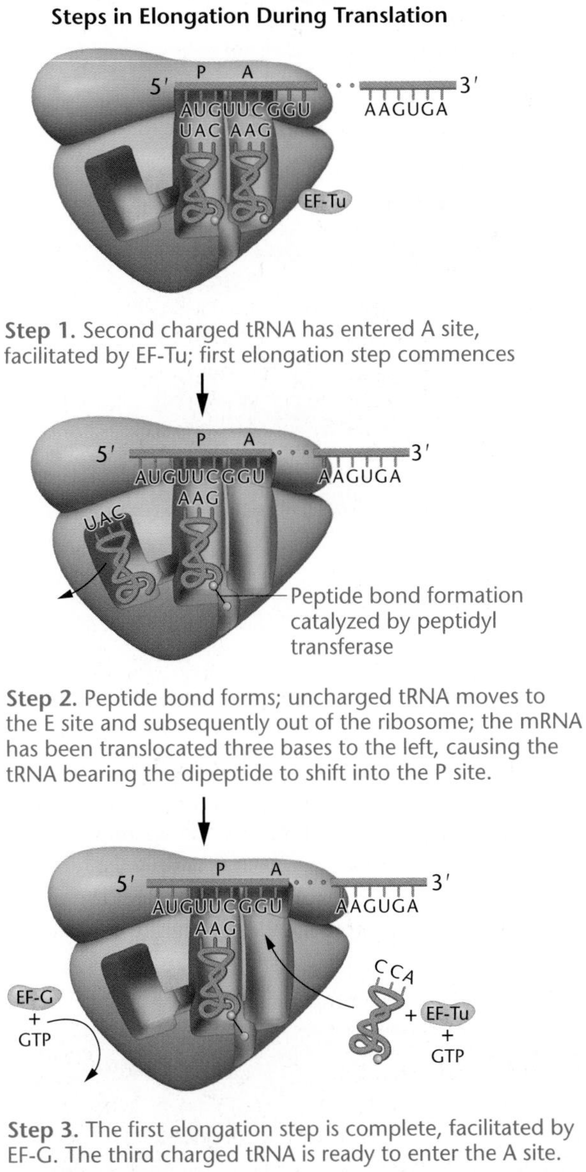

Steps in Elongation During Translation
P A
5′
3′
AUGUUCGGU
AAGUGA
UAC AAG
EF-Tu
Step 1. Second charged tRNA has entered A site, facilitated by EF-Tu; first elongation step commences
AAG
UAC
Peptide bond formation catalyzed by peptidyl transferase
Step 2. Peptide bond forms; uncharged tRNA moves to the E site and subsequently out of the ribosome; the mRNA has been translocated three bases to the left, causing the tRNA bearing the dipeptide to shift into the P site.
EF-G
+
GTP
CCA
+ EF-Tu
+
GTP
Step 3. The first elongation step is complete, facilitated by EF-G. The third charged tRNA is ready to enter the A site.

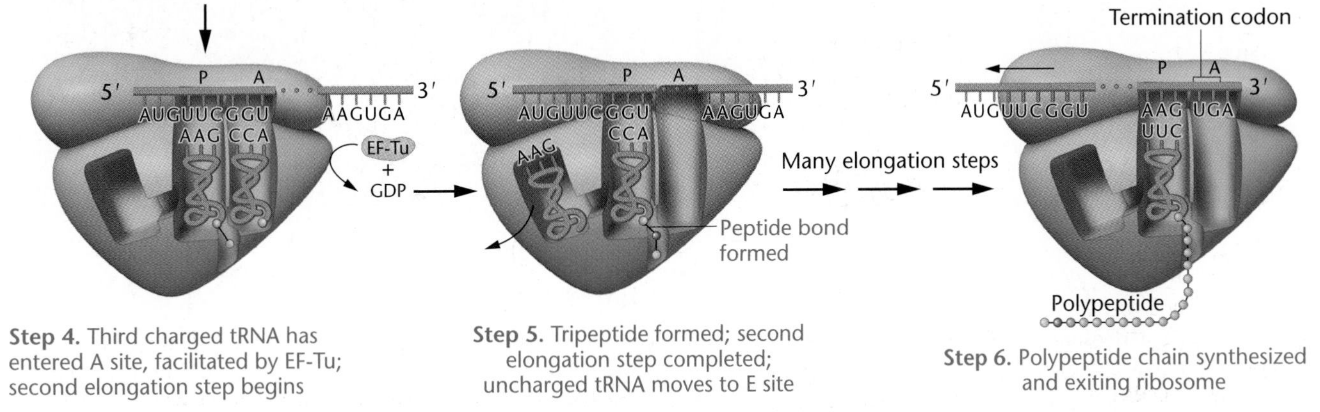

FIGURE 13–7 Elongation of the growing polypeptide chain during translation.

As we have seen, the role of the small subunit during elongation is one of "decoding" the triplets present in mRNA, whereas the role of the large subunit is peptide bond synthesis. The efficiency of the process is remarkably high. The observed error rate is only about 10^{-4}. At this rate, an incorrect amino acid will occur only once in every 20 polypeptides of an average length of 500 amino acids. In *E. coli,* elongation occurs at a rate of about 15 amino acids per second at 37°C.

Termination

Termination, the third phase of translation, is depicted in Figure 13–8. The process is signaled by one or more of three triplet codes in the A site: UAG, UAA, or UGA. These codons do not specify an amino acid, nor do they call for a tRNA in the A site. They are called **stop codons, termination codons,** or **nonsense codons.** Often, several such consecutive codons are part of an mRNA. The finished polypeptide is therefore still attached to the terminal tRNA at the P site, and the A site is empty. The termination codon signals the action of **GTP-dependent release factors,** which cleave the polypeptide chain from the terminal tRNA, releasing it from the translation complex (Step 1). Then, the tRNA is released from the ribosome, which then dissociates into its subunits (Step 2). If a termination codon should appear in the middle of an mRNA molecule as a result of mutation, the same process occurs, and the polypeptide chain is prematurely terminated.

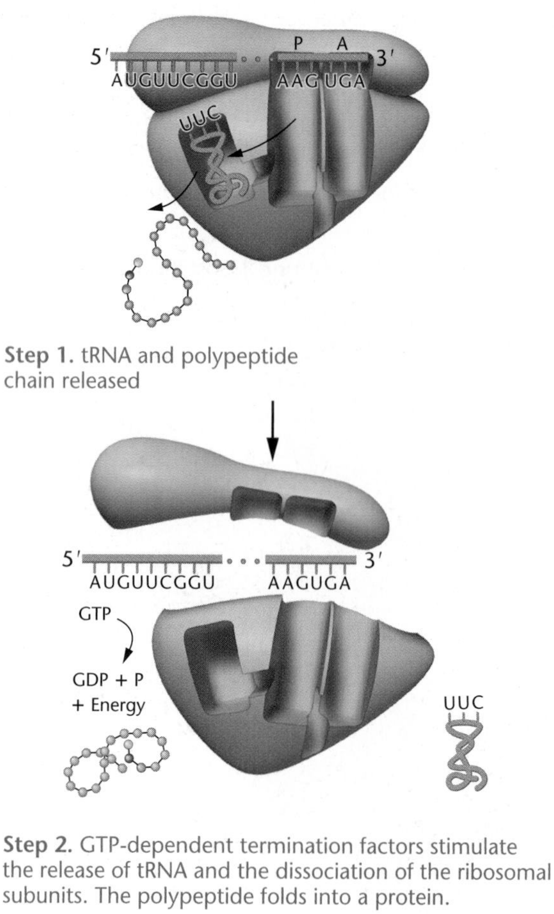

FIGURE 13–8 Termination of the process of translation.

Polyribosomes

As elongation proceeds and the initial portion of mRNA has passed through the ribosome, this mRNA is free to associate with another small subunit to form a second initiation complex. This process can be repeated several times with a single mRNA and results in what are called **polyribosomes,** or just **polysomes.**

Polyribosomes can be isolated and analyzed following a gentle lysis of cells. The photos in Figure 13–9 show these complexes as seen under an electron microscope. In Figure 13–9(a), you can see the thin lines of mRNA between the individual ribosomes. The micrograph in Figure 13–9(b) is even more remarkable, for it shows the polypeptide chains emerging from the ribosomes during translation. The formation of polysome complexes represents an efficient use of the components available for protein synthesis during a particular

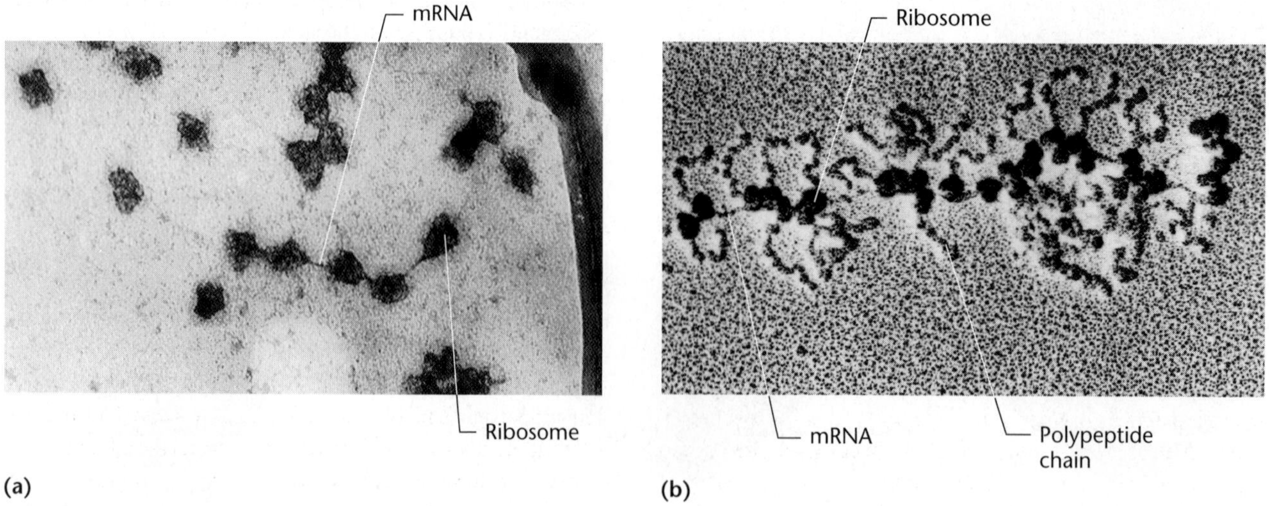

FIGURE 13–9 Polyribosomes as seen under the electron microscope. Those in (a) were derived from rabbit reticulocytes engaged in the translation of hemoglobin mRNA. The polyribosomes in (b) were taken from giant salivary gland cells of the midgefly, *Chironomus thummi.* Note that the nascent polypeptide chains are apparent as they emerge from each ribosome. Their length increases as translation proceeds from left (5′) to right (3′) along the mRNA.

unit of time. Using the analogy of a song recorded on a tape and a tape recorder, in polysome complexes one tape (mRNA) would be played simultaneously by several recorders (the ribosomes), but at any given moment, each song (the polypeptide being synthesized in each ribosome) would be at a different point in the lyrics.

ESSENTIAL POINT

Translation, like transcription, is subdivided into the stages of initiation, elongation, and termination and relies on base-pairing affinities between complementary nucleotides.

13.3 Crystallographic Analysis Has Revealed Many Details about the Functional Prokaryotic Ribosome

Our knowledge of the process of translation and the structure of the ribosome, as described in the previous sections, is based primarily on biochemical and genetic observations, in addition to visualization of ribosomes under the electron microscope. Because of the tremendous size and complexity of the functional ribosome during active translation, obtaining the crystals needed to perform X-ray diffraction studies was extremely difficult. Nevertheless, great strides have been made in the past several years. First, the individual ribosomal subunits were crystallized and examined in several laboratories, most prominently that of V. Ramakrishnan. Then, in 2001, the crystal structure of the intact 70*S* ribosome, complete with associated mRNA and tRNAs, was examined by Harry Noller and his colleagues—in essence, the entire translational complex was seen at the atomic level. Both Ramakrishnan and Noller derived the ribosomes from the bacterium *Thermus thermophilus.*

These investigations have produced many noteworthy observations. One of the models based on Noller's findings is shown as the opening photograph of this chapter (p. 261). For example, the sizes and shapes of the subunits, measured at atomic dimensions, are in agreement with earlier estimates based on high-resolution electron microscopy. Furthermore, the shape of the ribosome changes during different functional states, attesting to the dynamic nature of the process of translation. A great deal has also been learned about the prominence and location of the RNA components of the subunits. For example, about one-third of the 16*S* rRNA is responsible for producing a flat projection within the smaller 30*S* subunit referred to as the "platform," which modulates movement of the mRNA–tRNA complex during translocation.

Information also supports the concept that RNA is the major "player" in the ribosome during translation. The interface between the two subunits, considered to be the location in the ribosome where polymerization of amino acids occurs, is composed almost exclusively of RNA. In contrast, the numerous ribosomal proteins are found mostly on the periphery of the ribosome. These observations confirm what has been predicted on genetic grounds—the catalytic steps that join amino acids during translation occur under the direction of RNA, not proteins.

Another interesting finding involves the actual location of the three sites predicted to house tRNAs during translation. All three sites (A, P, and E) have been identified, and in each case, the RNA of the ribosome makes direct contact with the various loops and domains of the tRNA molecule. This observation points to the importance of the different regions of tRNA and helps us understand why the specific three-dimensional conformation of all tRNA molecules has been preserved throughout evolution.

A final observation takes us back almost 50 years, to when Francis Crick proposed the wobble hypothesis. The Ramakrishnan research group has identified the precise location along the 16*S* rRNA of the 30*S* subunit involved in the decoding step between mRNA and tRNA. At this location, two particular nucleotides of the 16*S* rRNA actually flip out and probe the codon–anticodon region and are also believed to check the accuracy of base pairing during this interaction. According to the wobble hypothesis, the stringency of this step is high for the first two base pairs but less stringent for the third (or wobble) base pair.

These landmark studies provide us with a much better picture of the dynamic changes that must occur within the ribosome during translation. However, numerous questions still remain about ribosome structure and function. In particular, the role of the many ribosomal proteins has yet to be clarified. Nevertheless, the models that are emerging based on the work of Noller, Ramakrishnan, and their many colleagues provide us with a much better understanding of the mechanism of translation.

13.4 Translation Is More Complex in Eukaryotes

The general features of the model we just discussed were initially derived from investigations of the translation process in bacteria. As we have seen, one main difference between translation in prokaryotes and eukaryotes is that in the eukaryotes, translation occurs on larger ribosomes whose rRNA and protein components are more complex than those of prokaryotes (see Figure 13–1).

Several other differences are also important. Eukaryotic mRNAs are much longer-lived than their prokaryotic counterparts. Most exist for hours rather than minutes prior to degradation by nucleases in the cell; thus they remain available much longer to orchestrate protein synthesis. Another significant distinction is that whereas transcription and translation are coupled in prokaryotes, in eukaryotes these two processes are separated both spatially and temporally. In eukaryotic cells transcription occurs in the nucleus and translation in the cytoplasm. This separation provides multiple opportunities for regulation of genetic expression in eukaryotic cells.

Several aspects involving the initiation of translation are also different in eukaryotes. First, as we discussed in Chapter 12, the 5′ end of mRNA is capped with a 7-methylguanosine (7mG) residue at maturation. The presence of the 7mG cap, absent in prokaryotes, is essential to efficient translation, since RNAs that lack the cap are translated poorly. In addition, most eukaryotic mRNAs contain a short recognition sequence that surrounds the initiating AUG codon—5′-ACCAUGG. Named after Marilyn Kozak, who discovered it, this **Kozak sequence** appears to function during initiation in the same way that the Shine–Dalgarno sequence functions in prokaryotic mRNA. Both greatly facilitate the initial binding of mRNA to the small subunit of the ribosome.

Another difference is that the amino acid formylmethionine is not required to initiate eukaryotic translation. However, as in prokaryotes, the AUG triplet, which encodes methionine, is essential to the formation of the translational complex, and a unique transfer RNA ($tRNA_i^{met}$) is used during initiation.

Protein factors similar to those in prokaryotes guide the initiation, elongation, and termination of translation in eukaryotes. Many of these eukaryotic factors are clearly homologous to their counterparts in prokaryotes. However, a greater number of factors are usually required during each step, and some are more complex than in prokaryotes.

Finally, recall that in eukaryotes a large proportion of the ribosomes are found in association with the membranes that make up the endoplasmic reticulum (forming the rough ER). Such membranes are absent from the cytoplasm of prokaryotic cells. This association in eukaryotes facilitates the secretion of newly synthesized proteins from the ribosomes directly into the channels of the endoplasmic reticulum. Recent studies using cryo-electron microscopy have established how this occurs. A **tunnel** in the large subunit of ribosomes begins near the point where the two subunits interface and exits near the back of the large subunit. The location of the tunnel within the large subunit is the basis for the belief that it provides the conduit for the movement of the newly synthesized polypeptide chain out of the ribosome. In studies in yeast, newly synthesized polypeptides enter the ER through a membrane channel formed by a specific protein, Sec61. This channel is perfectly aligned with the exit point of the ribosomal tunnel. In prokaryotes, the polypeptides are released by the ribosome directly into the cytoplasm.

13.5 The Initial Insight that Proteins Are Important in Heredity Was Provided by the Study of Inborn Errors of Metabolism

Let's consider how we know that proteins are the end products of genetic expression. The first insight into the role of proteins in genetic processes was provided by observations made by Sir Archibald Garrod and William Bateson early in the twentieth century. Garrod was born into an English family of medical scientists. His father was a physician with a strong interest in the chemical basis of rheumatoid arthritis, and his eldest brother was a leading zoologist in London. It is not surprising, then, that as a practicing physician, Garrod became interested in several human disorders that seemed to be inherited. Although he also studied albinism and cystinuria, we shall describe his investigation of the disorder **alkaptonuria.** Individuals afflicted with this disorder have an important metabolic pathway blocked (Figure 13–10). As a result, they cannot metabolize the alkapton 2,5-dihydroxyphenylacetic acid, also known as homogentisic acid. Homogentisic acid accumulates in cells and tissues and is excreted in the urine. The molecule's oxidation products are black and easily detectable in the diapers of newborns. The products tend to accumulate in cartilaginous areas, causing the ears and nose to darken. The deposition of homogentisic acid in joints leads to a benign arthritic condition. This rare disease is not serious, but it persists throughout an individual's life.

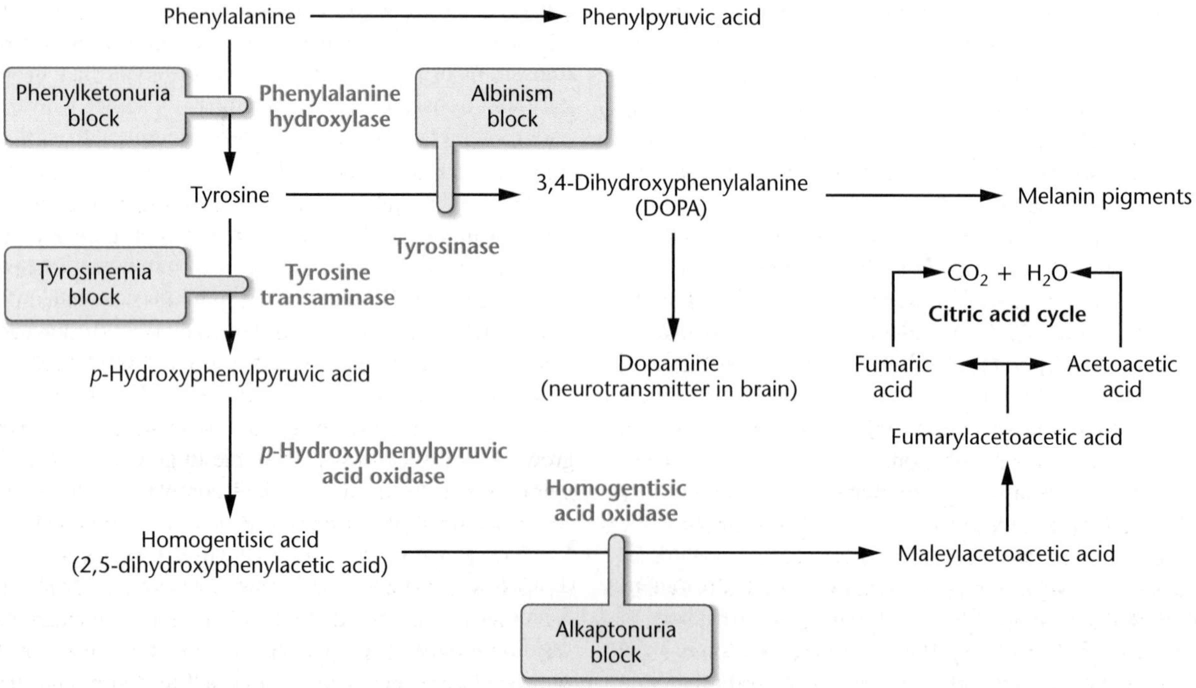

FIGURE 13–10 Metabolic pathway involving phenylalanine and tyrosine. Various metabolic blocks resulting from mutations lead to the disorders phenylketonuria, alkaptonuria, albinism, and tyrosinemia.

Garrod studied alkaptonuria by looking for patterns of inheritance of this benign trait. Eventually he concluded that it was genetic in nature. Of 32 known cases, he ascertained that 19 were confined to seven families, with one family having four affected siblings. In several instances, the parents were unaffected but known to be related as first cousins, and therefore **consanguine,** a term describing relatives having a common recent ancestor. Parents who are so related have a higher probability than unrelated parents of producing offspring that express recessive traits because such parents are both more likely to be heterozygous for some of the same recessive traits (see Chapter 27). Garrod concluded that this inherited condition was the result of an alternative mode of metabolism, thus implying that hereditary information controls chemical reactions in the body. While *genes* and *enzymes* were not familiar terms during Garrod's time, he used the corresponding concepts of *unit factors* and *ferments.* Garrod published his initial observations in 1902.

Only a few geneticists, including Bateson, were familiar with or referred to Garrod's work. Garrod's ideas fit nicely with Bateson's belief that inherited conditions are caused by the lack of some critical substance. In 1909, Bateson published *Mendel's Principles of Heredity,* in which he linked Garrod's ferments with heredity. However, for almost 30 years, most geneticists failed to see the relationship between genes and enzymes. Garrod and Bateson, like Mendel, were ahead of their time.

Phenylketonuria

The inherited human metabolic disorder, **phenylketonuria (PKU),** results when another reaction in the pathway shown in Figure 13–10 is blocked. Described first in 1934, this disorder can result in mental retardation and is transmitted as an autosomal recessive disease. Afflicted individuals are unable to convert the amino acid phenylalanine to the amino acid tyrosine. These molecules differ by only a single hydroxyl group (OH) that is present in tyrosine but absent in phenylalanine. The reaction is catalyzed by the enzyme **phenylalanine hydroxylase,** which is inactive in affected individuals and active at about a 30 percent level in heterozygotes. The enzyme functions in the liver. The normal blood level of phenylalanine is about 1 mg/100 mL; phenylketonurics show levels as high as 50 mg/100 mL.

As phenylalanine accumulates, it can be converted to phenylpyruvic acid and subsequently to other derivatives. These are less efficiently resorbed by the kidney and tend to spill into the urine more quickly than phenylalanine. Both phenylalanine and its derivatives subsequently enter the cerebrospinal fluid, resulting in elevated levels in the brain. The presence of these substances during early development is thought to cause mental retardation.

Phenylketonuria occurs in approximately 1 in 11,000 births, and newborns are routinely screened for PKU throughout the United States. When the condition is detected in the analysis of an infant's blood, a strict dietary regimen is instituted in time to prevent retardation. A low-phenylalanine diet can reduce byproducts such as phenylpyruvic acid, and the development of abnormalities characterizing the disease can be diminished.

Our knowledge of inherited metabolic disorders such as alkaptonuria and phenylketonuria has caused a revolution in medical thinking and practice. Human diseases, once believed to be attributable solely to the action of invading microorganisms, viruses, or parasites, clearly can have a genetic basis. We now know that hundreds of medical conditions are caused by errors in metabolism resulting from mutant genes. These human biochemical disorders include all classes of organic biomolecules.

ESSENTIAL POINT

Inherited metabolic disorders are most often due to the loss of enzyme activity resulting from mutations in genes encoding those proteins.

13.6 Studies of *Neurospora* Led to the One-Gene:One-Enzyme Hypothesis

In two separate investigations beginning in 1933, George Beadle provided the first convincing experimental evidence that genes are directly responsible for the synthesis of enzymes. The first investigation, conducted in collaboration with Boris Ephrussi, involved *Drosophila* eye pigments. Together, they confirmed that mutant genes that alter the eye color of fruit flies could be linked to biochemical errors that, in all likelihood, involved the loss of enzyme function. Encouraged by these findings, Beadle then joined with Edward Tatum to investigate nutritional mutations in the pink bread mold *Neurospora crassa.* This investigation led to the **one-gene:one-enzyme hypothesis.**

Analysis of *Neurospora* Mutants by Beadle and Tatum

In the early 1940s, Beadle and Tatum chose to work with *Neurospora* because much was known about its biochemistry and because mutations could be induced and isolated with relative ease. By inducing mutations, they produced strains that had genetic blocks of reactions essential to the growth of the organism.

Beadle and Tatum knew that this mold could manufacture nearly everything necessary for normal development. For example, using rudimentary carbon and nitrogen sources, this organism can synthesize nine water-soluble vitamins, 20 amino acids, numerous carotenoid pigments, and all essential purines and pyrimidines. Beadle and Tatum irradiated asexual conidia (spores) with X rays to increase the frequency of mutations and allowed them to be grown on "complete" medium containing all the necessary growth factors (e.g., vitamins and amino acids). Under such growth conditions, a mutant strain unable to grow on minimal medium was able to grow by virtue of supplements present in the enriched complete medium. All the cultures were then transferred to minimal medium. If growth occurred on the minimal medium, the organisms were able to synthesize all the necessary growth factors themselves, and the researchers concluded that the culture did not contain a nutritional mutation. If no growth occurred, then they concluded that the culture contained a nutritional mutation, and the only task remaining was to determine its type. These results are shown in Figure 13–11(a), p. 272.

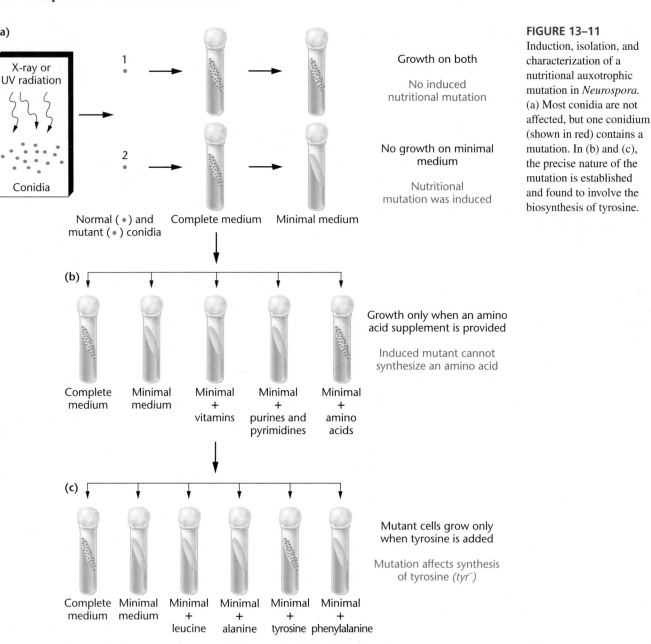

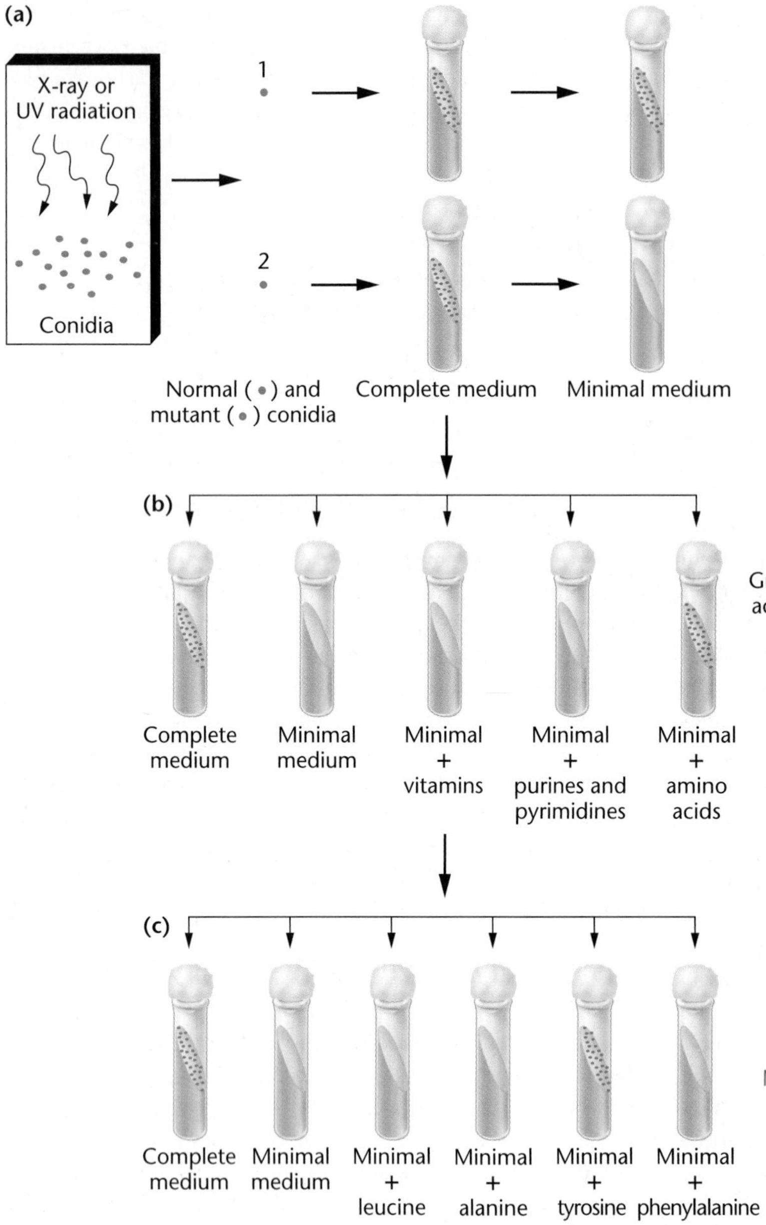

FIGURE 13–11 Induction, isolation, and characterization of a nutritional auxotrophic mutation in *Neurospora.* (a) Most conidia are not affected, but one conidium (shown in red) contains a mutation. In (b) and (c), the precise nature of the mutation is established and found to involve the biosynthesis of tyrosine.

Many thousands of individual spores from this procedure were isolated and grown on complete medium. In subsequent tests on minimal medium, many cultures failed to grow, indicating that a nutritional mutation had been induced. To identify the mutant type, the mutant strains were then tested on a series of different minimal media [Figure 13–11(b)], each containing groups of supplements, and subsequently on media containing single vitamins, purines, pyrimidines, or amino acids [Figure 13–11(c)] until one specific supplement that permitted growth was found. Beadle and Tatum reasoned that the supplement that restored growth would be the molecule that the mutant strain could not synthesize.

The first mutant strain they isolated required vitamin B_6 (pyridoxine) in the medium, and the second required vitamin B_1 (thiamine). Using the same procedure, Beadle and Tatum eventually isolated and studied hundreds of mutants deficient in the ability to synthesize other vitamins, amino acids, or other substances.

The findings derived from testing over 80,000 spores convinced Beadle and Tatum that genetics and biochemistry have much in common. It seemed likely that each nutritional mutation caused the loss of the enzymatic activity that facilitated an essential reaction in wild-type organisms. It also appeared that a mutation could be found for nearly any enzymatically controlled reaction. Beadle and Tatum had thus provided sound experimental evidence for the hypothesis that *one gene specifies one enzyme,* an idea alluded to over 30 years earlier by Garrod and Bateson. With modifications, this concept was to become another major principle of genetics.

ESSENTIAL POINT

Beadle and Tatum's work with nutritional mutations in *Neurospora* led them to propose that one gene encodes one enzyme.

Genes and Enzymes: Analysis of Biochemical Pathways

The one-gene:one-enzyme concept and its attendant methods have been used over the years to work out many details of metabolism in *Neurospora, E. coli,* and a number of other microorganisms. One of the first metabolic pathways to be investigated in detail was that leading to the synthesis of the amino acid arginine in *Neurospora.* By studying seven mutant strains, each requiring arginine for growth (arg^-), Adrian Srb and Norman Horowitz ascertained a partial biochemical pathway that leads to the synthesis of this molecule. Their work demonstrates how genetic analysis can be used to establish biochemical information.

Srb and Horowitz tested each mutant strain's ability to grow if either citrulline or ornithine, two compounds with close chemical similarity to arginine, was used to supplement a minimal medium. If either compound was able to substitute for arginine, they reasoned that it must be involved in the biosynthetic pathway of arginine. They found that both molecules could be substituted in one or more strains.

Of the seven mutant strains, four of them (*arg4–7*) grew if supplied with either citrulline, ornithine, or arginine. Two of them (*arg2* and *arg3*) grew if supplied with citrulline or arginine. One strain (*arg1*) would grow only if arginine was supplied—neither citrulline nor ornithine could substitute for it. From these experimental observations, the following pathway and metabolic blocks for each mutation were deduced:

$$\text{Precursor} \xrightarrow[\text{Enzyme A}]{arg\ 4-7} \text{Ornithine} \xrightarrow[\text{Enzyme B}]{arg\ 2 \text{ and } arg\ 3} \text{Citrulline} \xrightarrow[\text{Enzyme C}]{arg\ 1} \text{Arginine}$$

The logic supporting these conclusions is as follows: If mutants *arg4* through *arg7* can grow regardless of which of the three molecules is supplied as a supplement to minimal medium, the mutations preventing growth must cause a metabolic block that occurs *prior* to the involvement of ornithine, citrulline, or arginine in the pathway. When any one of these three molecules is added, *its presence bypasses the block.* As a result, both citrulline and ornithine appear to be involved in the biosynthesis of arginine. However, the sequence of their participation in the pathway cannot be determined on the basis of these data.

On the other hand, the *arg2* and *arg3* mutations grow if supplied with citrulline, but not if they are supplied with only ornithine. Therefore, ornithine must be synthesized in the pathway *prior to the block.* Its presence will not overcome the block. Citrulline, however, *does overcome the block,* so it must be synthesized beyond the point of blockage. Therefore, the conversion of ornithine to citrulline represents the correct sequence in the pathway.

Finally, we can conclude that *arg1* represents a mutation preventing the conversion of citrulline to arginine. Neither ornithine nor citrulline can overcome the metabolic block because both participate earlier in the pathway.

Taken together, the analysis as described above supports the sequence of biosynthesis shown in Figure 13–12. Since Srb and Horowitz's experiments in 1944, the detailed pathway has been worked out and the enzymes controlling each step have been characterized.

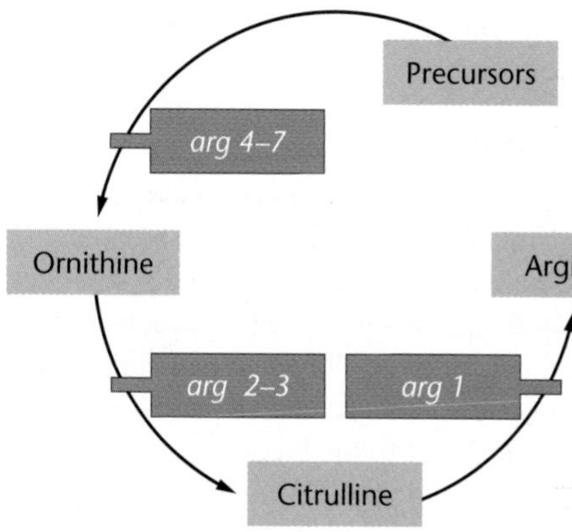

FIGURE 13–12 Abbreviated pathway resulting in the biosynthesis of arginine in *Neurospora.*

NOW SOLVE THIS

Problem 12 on page 282 asks you to analyze data to establish a biochemical pathway in the bacterium *Salmonella typhimurium.*

Hint: Apply the same principles and approach used to decipher biochemical pathways in *Neurospora.*

13.7 Studies of Human Hemoglobin Established that One Gene Encodes One Polypeptide

The concept of the one-gene:one-enzyme hypothesis that developed in the early 1940s was not immediately accepted by all geneticists. This is not surprising because it was not yet clear how mutant enzymes could cause variation in many phenotypic traits. For example, *Drosophila* mutants demonstrate altered eye size, wing shape, wing-vein pattern, and so on. Plants exhibit mutant varieties of seed texture, height, and fruit size. How an inactive mutant enzyme could result in such phenotypes puzzled many geneticists.

Two factors soon modified the one-gene:one-enzyme hypothesis. First, although *nearly all enzymes are proteins, not all proteins are enzymes.* As the study of biochemical genetics progressed, it became clear that all proteins are specified by the information stored in genes, leading to the more accurate phraseology, **one-gene:one-protein hypothesis.** Second, proteins often show a substructure consisting of two or more polypeptide chains. This is the basis of the quaternary protein structure, which we will discuss later in this chapter. Because each distinct polypeptide chain is encoded by a separate gene, a more accurate statement of Beadle and Tatum's basic tenet is **one-gene:one-polypeptide chain hypothesis.** These modifications of the original hypothesis became apparent during the analysis of hemoglobin structure in individuals afflicted with sickle-cell anemia.

Sickle-cell Anemia

The first direct evidence that genes specify proteins other than enzymes came from work on mutant hemoglobin molecules found in humans afflicted with the disorder **sickle-cell anemia.** Affected individuals have erythrocytes that, under low oxygen tension, become elongated and curved because of the polymerization of hemoglobin. The sickle shape of these erythrocytes is in contrast to the biconcave disc shape characteristic in unaffected individuals (Figure 13–13). Those with the disease suffer attacks when red blood cells aggregate in the venous side of capillary systems, where oxygen tension is very low. As a result, a variety of tissues are deprived of oxygen and suffer severe damage. When this occurs, an individual is said to experience a sickle-cell crisis. If left untreated, a crisis can be fatal. The kidneys, muscles, joints, brain, gastrointestinal tract, and lungs can be affected.

In addition to suffering crises, these individuals are anemic because their erythrocytes are destroyed more rapidly than are normal red blood cells. Compensatory physiological mechanisms include increased red blood cell production by bone marrow, along with accentuated heart action. These mechanisms lead to abnormal bone size and shape, as well as dilation of the heart.

In 1949, James Neel and E. A. Beet demonstrated that the disease is inherited as a Mendelian trait. Pedigree analysis revealed three genotypes and phenotypes controlled by a single pair of alleles, Hb^A and Hb^S. Unaffected and affected individuals result from the homozygous genotypes $Hb^A\ Hb^A$ and $Hb^S\ Hb^S$, respectively. The red blood cells of heterozygotes, who exhibit the **sickle-cell trait** but not the disease, undergo much less sickling because over half of their hemoglobin is normal. Although they are largely unaffected, heterozygotes are "carriers" of the defective gene, which is transmitted on average to 50 percent of their offspring.

Also in 1949, Linus Pauling and his coworkers provided the first insight into the molecular basis of the disease. They showed that hemoglobins isolated from diseased and normal individuals differ in their rates of electrophoretic migration. In this technique, charged molecules migrate in an electric field. If the net charge of two molecules is different, their rates of migration will be different. On this basis, Pauling and his colleagues concluded that a chemical difference exists between normal **(HbA)** and sickle-cell **(HbS)** hemoglobin.

Figure 13–14(a) shows the migration pattern of hemoglobin derived from individuals of all three possible genotypes when it was subjected to **starch gel electrophoresis.** The gel provides the supporting medium for the molecules during migration. In this experiment, samples were placed at a point of origin between a cathode (−) and an anode (+) and an electric field was applied. The migration pattern revealed that all of the molecules moved toward the anode, indicating a net negative charge. However, HbA migrated farther than HbS, suggesting that its net negative charge was greater. The electrophoretic pattern of hemoglobin derived from carriers revealed the presence of both HbA and HbS and confirmed their heterozygous genotype.

Pauling's findings suggested two possibilities. It was known that hemoglobin consists of four nonproteinaceous, iron-containing *heme groups* and a *globin portion* that contains four polypeptide chains. The alteration in net charge in HbS had to be due, theoretically, to a chemical change in one of these components.

Work carried out between 1954 and 1957 by Vernon Ingram resolved this question. He demonstrated that the chemical change occurs in the primary structure of the globin portion of the hemoglobin molecule. Using the **protein fingerprinting technique** shown in Figure 13–14(b), Ingram showed that HbS differs in amino acid composition compared to HbA. Human adult hemoglobin contains two identical α chains of 141 amino acids and two identical β chains of 146 amino acids in its quaternary structure.

The fingerprinting technique involves enzymatic digestion of the protein into peptide fragments. The mixture is then placed on absorbent paper and exposed to an electric field, where migration occurs according to net charge. The paper is then turned at a right angle to its first exposure and placed in a solvent, where chromatographic action causes the migration of the peptides in the second direction. The end result is a two-dimensional separation of the peptide fragments into a

(a)

(b)

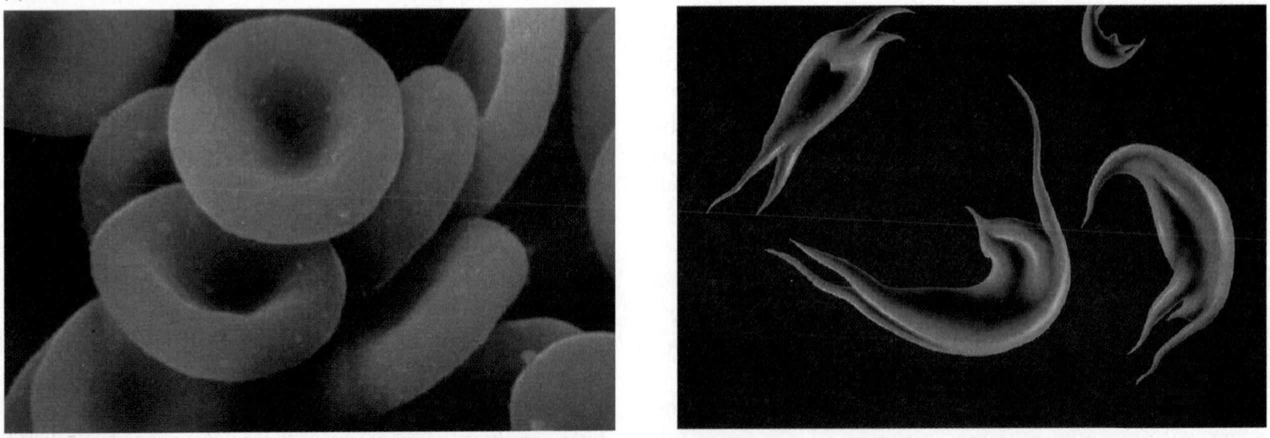

FIGURE 13–13 A comparison of erythrocytes from (a) healthy individuals and (b) those with sickle-cell anemia.

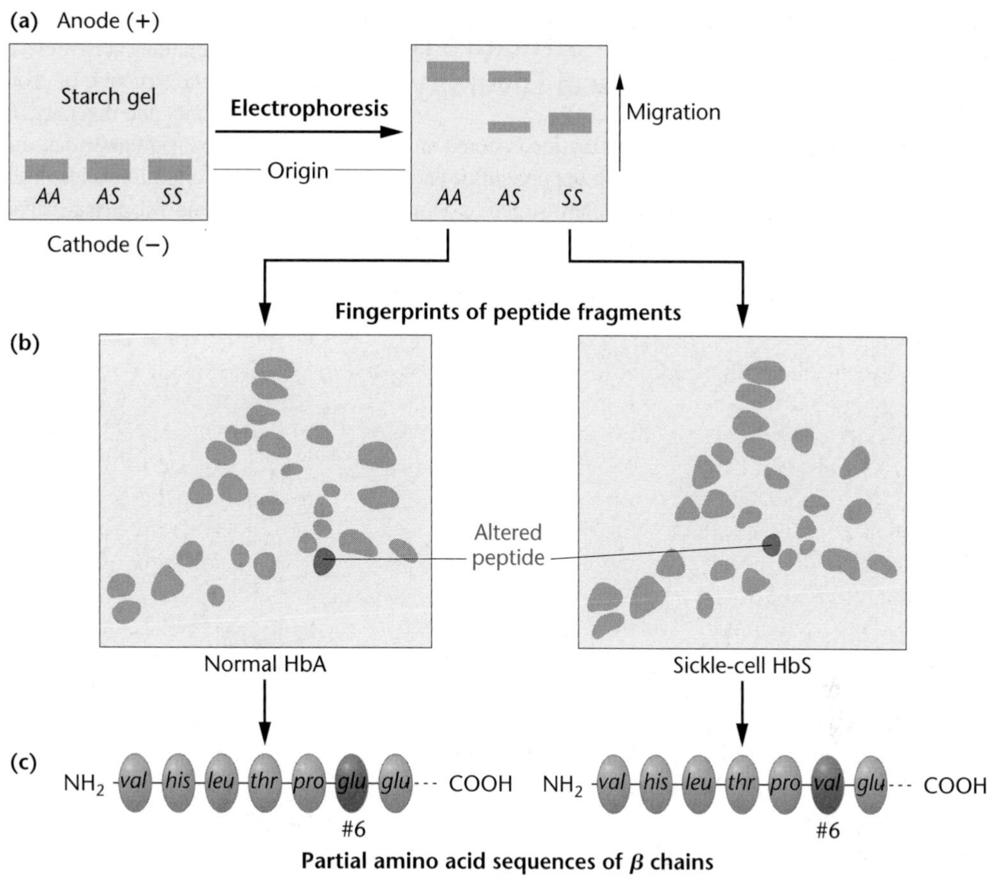

FIGURE 13–14 Investigation of hemoglobin derived from $Hb^A\ Hb^A$, $Hb^A Hb^S$, and $Hb^S\ Hb^S$ individuals by using electrophoresis, protein fingerprinting, and amino acid analysis. Hemoglobin from individuals with sickle-cell anemia ($Hb^S\ Hb^S$) (a) migrates differently in an electrophoretic field, (b) shows an altered peptide in fingerprint analysis, and (c) shows an altered amino acid, valine, at the sixth position in the β chain. During electrophoresis, heterozygotes ($Hb^A\ Hb^S$) reveal both forms of hemoglobin.

distinctive pattern of spots or a "fingerprint." Ingram's work revealed that HbS and HbA differed by only a single peptide fragment. Further analysis revealed just a single amino acid change: valine was substituted for glutamic acid at the sixth position of the β chain, thus accounting for the peptide difference [Figure 13–14(c)].

The significance of this discovery has been multifaceted. It clearly establishes that a single gene provides the genetic information for a single polypeptide chain. Studies of HbS also demonstrate that a mutation can affect the phenotype by directing a single amino acid substitution. Also, by providing the explanation for sickle-cell anemia, the concept of **inherited molecular disease** was firmly established. Finally, this work has led to a thorough study of human hemoglobins, which has provided valuable genetic insights.

In the United States, sickle-cell anemia is found almost exclusively in the African-American population. It affects about 1 in every 625 African-American infants. Currently, about 50,000 to 75,000 individuals are afflicted. In 1 of about every 145 African-American married couples, both partners are heterozygous carriers. In these cases, each of their children has a 25 percent chance of having the disease.

ESSENTIAL POINT

Pauling and Ingram's investigations of hemoglobin from patients with sickle-cell anemia led to the modification of the one-gene:one-enzyme hypothesis to indicate that one gene encodes one polypeptide chain.

Human Hemoglobins

Having introduced human hemoglobins in an historical context, it may be useful to extend our discussion about these molecules in our species. Molecular analysis reveals that a variety of hemoglobin molecules are produced in humans at different stages of the life cycle. All are tetramers consisting of numerous combinations of seven distinct polypeptide chains, each encoded by a separate gene. The expression of these various genes is developmentally regulated.

Almost all adult hemoglobin consists of **HbA,** which contains two α and two β chains. Recall that the mutation in sickle-cell anemia involves the β chain. HbA represents about 98 percent of all hemoglobin found in an adult's erythrocytes after the age of six months. The remaining 2 percent consists of **HbA$_2$,** a minor adult component. This molecule contains two α chains and two **delta (δ) chains.** The δ chain is very similar to the β chain, consisting of 146 amino acids.

During embryonic and fetal development, a completely different set of polypeptides are found in hemoglobin. The earliest set detected is in embryos and is called **Gower 1.** It contains two **zeta (ζ) chains,** which are similar to α chains, and two **epsilon (ε) chains,** which are similar to β chains. By eight weeks gestation, the embryonic form is gradually replaced by another hemoglobin molecule with still different polypeptides. This molecule is called **HbF,** or **fetal hemoglobin,** and consists of two alpha chains and two **gamma (γ) chains.** There are two different types of γ chains, designated $^G\gamma$ and $^A\gamma$. Both are similar to β chains and differ from each other by only a single amino acid. These persist until birth, and gradually, HbF is replaced with HbA and HbA$_2$.

13.8 Variation in Protein Structure Is the Basis of Biological Diversity

Having established that the genetic information is stored in DNA and influences cellular activities through the proteins it encodes, we turn now to a brief discussion of protein structure. How can these molecules play such a critical role in determining the complexity of cellular activities? As we shall see, the fundamental aspects of the structure of proteins provide the basis for incredible complexity and diversity. At the outset, we should differentiate between **polypeptides** and **proteins.** Both are molecules composed of amino acids. They differ, however, in their state of assembly and functional capacity. Polypeptides are the precur-

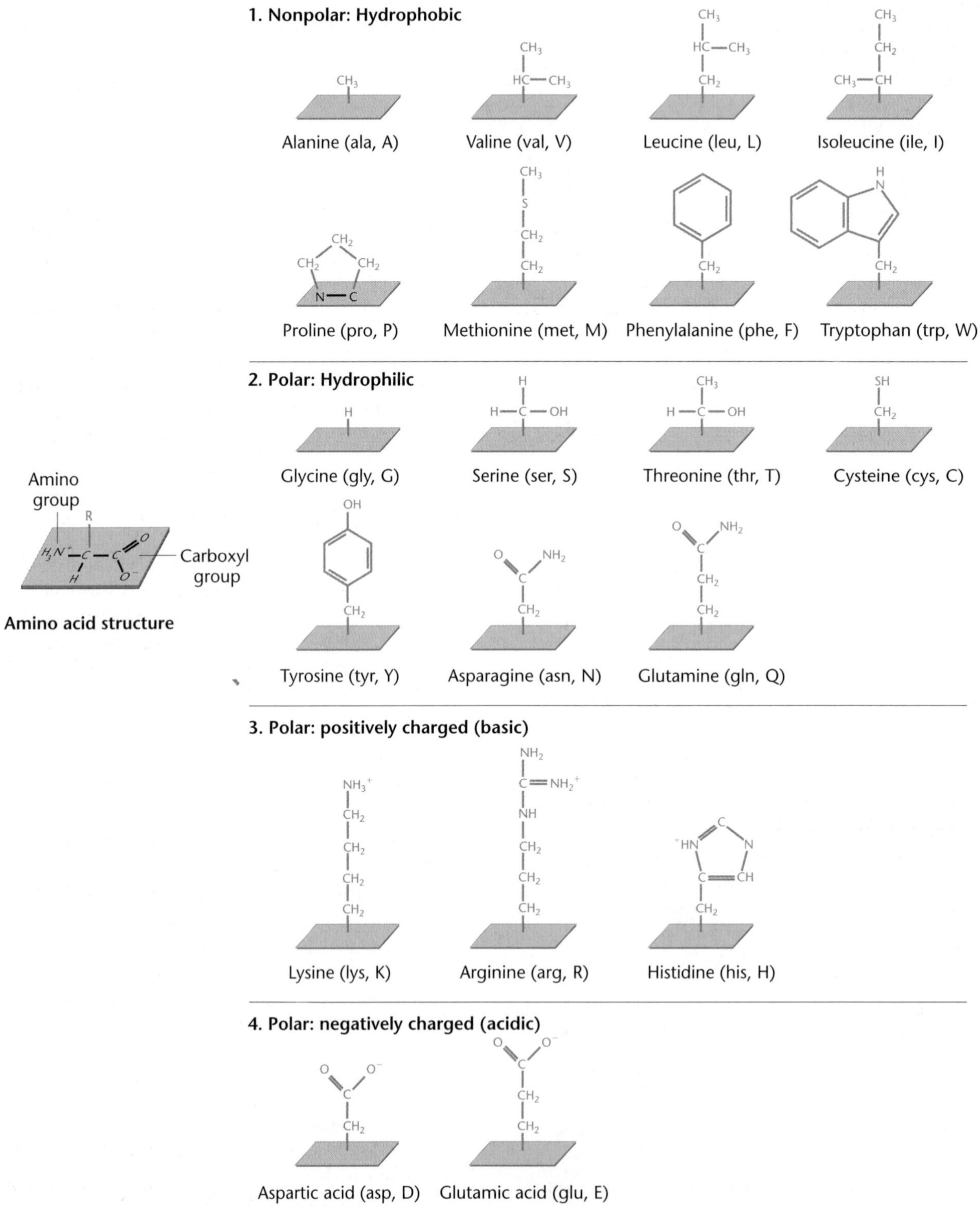

FIGURE 13–15 Chemical structures and designations of the 20 amino acids found in living organisms, divided into four major categories. Each amino acid has two abbreviations; that is, alanine is designated either ala or A (a universal nomenclature).

sors of proteins. As it is assembled on the ribosome during translation, the molecule is called a *polypeptide.* When released from the ribosome following translation, a polypeptide folds up and assumes a higher order of structure. When this occurs, a three-dimensional conformation emerges. In many cases, several polypeptides interact to produce this conformation. When the final conformation is achieved, the molecule is now fully functional and is appropriately called a *protein.* Its three-dimensional conformation is essential to the function of the molecule.

The polypeptide chains of proteins, like nucleic acids, are linear nonbranched polymers. There are 20 commonly occurring amino acids that serve as the subunits (the building blocks) of proteins. Each amino acid has a **carboxyl group,** an **amino group,** and an **R (radical) group** (a side chain) bound covalently to a **central carbon (C) atom.** The R group gives each amino acid its chemical identity. Figure 13–15 shows the 20 R groups, which exhibit a variety of configurations and can be divided into four main classes: (1) *nonpolar* (hydrophobic), (2) *polar* (hydrophilic), (3) *positively charged,* and (4) *negatively charged.* Because polypeptides are often long polymers and because each position may be occupied by any 1 of the 20 amino acids with their unique chemical properties, enormous variation in chemical conformation and activity is possible. For example, if an average polypeptide is composed of 200 amino acids (molecular weight of about 20,000 Da), 20^{200} different molecules, each with a unique sequence, can be created using the 20 different building blocks.

Around 1900, German chemist Emil Fischer determined the manner in which the amino acids are bonded together. He showed that the amino group of one amino acid reacts with the carboxyl group of another amino acid during a dehydration reaction, releasing a molecule of H_2O. The resulting covalent bond is a **peptide bond** (Figure 13–16). Two amino acids linked together constitute a **dipeptide,** three a **tripeptide,** and so on. Once 10 or more amino acids are linked by peptide bonds, the chain is referred to as a polypeptide. Generally, no matter how long a polypeptide is, it will contain a free amino group at one end (the N-terminus) and a free carboxyl group at the other end (the C-terminus).

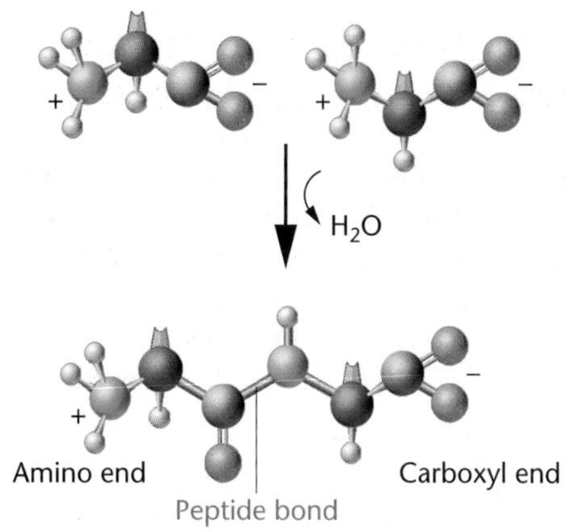

FIGURE 13–16 Peptide bond formation between two amino acids, resulting from a dehydration reaction.

Four levels of protein structure are recognized: primary, secondary, tertiary, and quaternary. The sequence of amino acids in the linear backbone of the polypeptide constitutes its **primary structure.** It is specified by the sequence of deoxyribonucleotides in DNA via an mRNA intermediate. The primary structure of a polypeptide helps determine the specific characteristics of the higher orders of organization as a protein is formed.

Secondary structures are certain regular or repeating configurations in space assumed by amino acids lying close to one another in the polypeptide chain. In 1951, Linus Pauling and Robert Corey predicted, on theoretical grounds, an **α helix** as one type of secondary structure. The α-helix model [Figure 13–17(a)] has since been confirmed by X-ray crystallographic studies. It is rodlike and has the greatest possible theoretical stability. The helix is composed of a spiral chain of amino acids stabilized by hydrogen bonds.

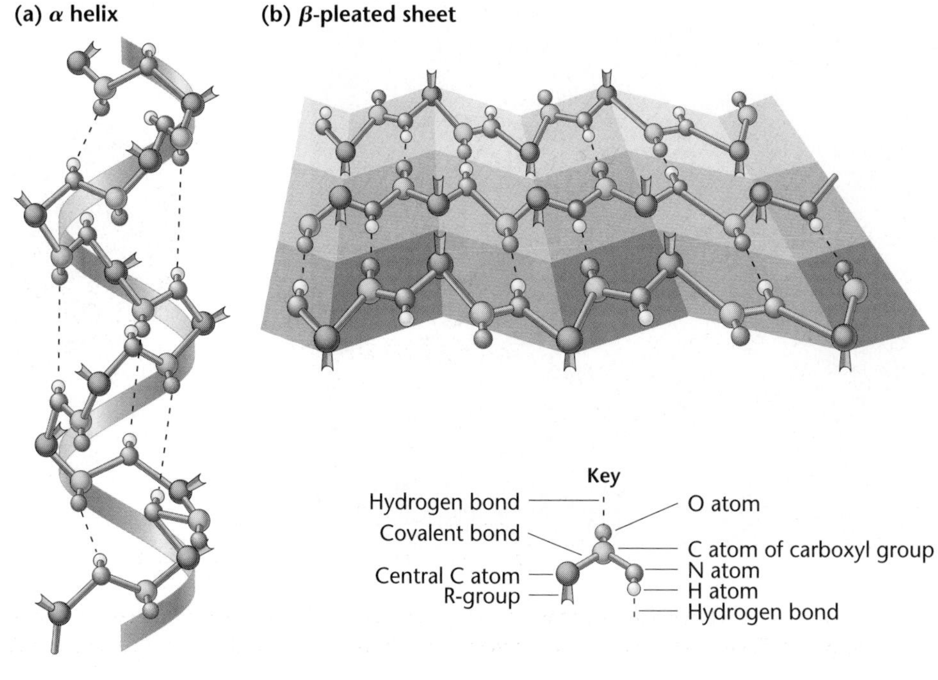

FIGURE 13–17 (a) The right-handed α helix, which represents one form of secondary structure of a polypeptide chain. (b) The β-pleated sheet, an alternative form of secondary structure of polypeptide chains. To maintain clarity, not all atoms are shown.

The side chains (the R groups) of amino acids extend outward from the helix, and each amino acid residue occupies a vertical distance of 1.5 Å in the helix. There are 3.6 residues per turn. While left-handed helices are theoretically possible, all proteins seen with an α helix are right-handed.

Also in 1951, Pauling and Corey proposed a second structure, the **β-pleated sheet.** In this model, a single-polypeptide chain folds back on itself, or several chains run in either parallel or antiparallel fashion next to one another. Each such structure is stabilized by hydrogen bonds formed between atoms on adjacent chains [Figure 13–17(b)]. A zigzagging plane is formed in space with adjacent amino acids 3.5 Å apart.

As a general rule, most proteins demonstrate a mixture of α-helix and β-pleated-sheet structures. Globular proteins, most of which are round in shape and water soluble, usually contain a core of β-pleated-sheet structure as well as many areas with α-helical structures. The more structurally rigid proteins, many of which are water insoluble, rely on more extensive β-pleated-sheet regions for their rigidity. For example, **fibroin,** the protein made by the silk moth, depends extensively on this form of secondary structure.

The secondary structure describes the arrangement of amino acids within certain areas of a polypeptide chain, but the **tertiary structure** defines the three-dimensional conformation of the entire chain in space. Each protein twists and turns and loops around itself in a very particular fashion, characteristic of the specific protein. A model of the three-dimensional tertiary structure of the respiratory pigment myoglobin is shown in Figure 13–18. Three aspects of this level of structure are most important in stabilizing the molecule and in determining its conformation.

1. Covalent disulfide bonds form between closely aligned cysteine residues to make the unique amino acid cystine.
2. Usually, the polar hydrophilic R groups are located on the surface, where they can interact with water.
3. The nonpolar hydrophobic R groups are usually located on the inside of the molecule, where they interact with one another, avoiding interaction with water.

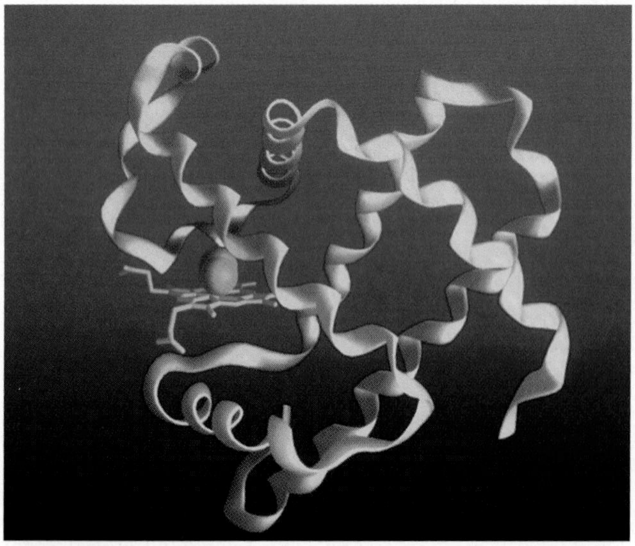

FIGURE 13–18 The tertiary level of protein structure in a respiratory pigment, myoglobin. The bound oxygen atom is shown in red.

The three-dimensional conformation achieved by any protein is a product of the *primary structure* of the polypeptide. The three stabilizing factors depend on the location of each amino acid relative to all others in the chain. As the polypeptide is folded, the most thermodynamically stable conformation is created. This level of organization is essential because the specific function of any protein is directly related to its tertiary structure.

The concept of **quaternary structure** of proteins applies only to those composed of more than one polypeptide chain and indicates the position of the various chains in relation to one another. This type of protein is *oligomeric,* and each chain is a *protomer* or, less formally, a *subunit.* Protomers have conformations that facilitate their fitting together in a specific complementary fashion. Hemoglobin, an oligomeric protein consisting of four polypeptide chains, has been studied in great detail—its quaternary protein structure is shown in Figure 13–19. Most enzymes, including DNA and RNA polymerase, demonstrate quaternary structure.

ESSENTIAL POINT

Proteins, the end products of genes, demonstrate four levels of structural organization that together describe their three-dimensional conformation, which is the basis of each molecule's function.

Protein Folding and Misfolding

It was long thought that **protein folding** was a spontaneous process whereby a linear molecule exiting the ribosome achieved a three-dimensional, thermodynamically stable conformation based solely on the combined chemical properties inherent in the amino acid sequence. This indeed is the case for many proteins. However, numerous studies have shown that for other proteins, correct folding is dependent

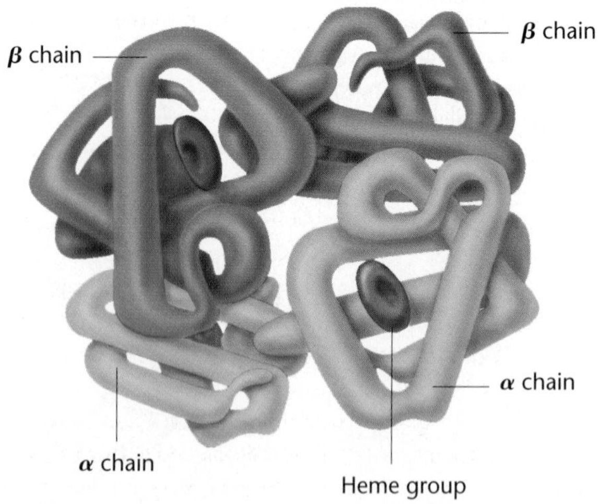

FIGURE 13–19 The quaternary level of protein structure as seen in hemoglobin. Four chains (two α and two β) interact with four heme groups to form the functional molecule.

on members of a family of molecules called **chaperones.** Chaperones are themselves proteins (sometimes called *molecular chaperones* or *chaperonins*) that function by mediating the folding process by excluding the formation of alternative, incorrect patterns. While they may bind to the protein in question, like enzymes, they do not become part of the final product. Initially discovered in *Drosophila,* in which they are called **heat-shock proteins,** chaperones are ubiquitous, having now been discovered in all organisms. They are even present in mitochondria and chloroplasts.

In eukaryotic cells, chaperones are particularly important when translation occurs on membrane-bound ribosomes, where the newly translated polypeptide is extruded into the lumen of the endoplasmic reticulum. Even in their presence, misfolding may still occur, and one more system of "quality control' exists. As misfolded proteins are transported out of the endoplasmic reticulum to the cytoplasm, they are "tagged" by another class of small proteins called **ubiquitins.** The protein–ubiquitin complex moves to a cellular structure called the **proteasome,** within which the ubiquitin is released and the misfolded proteins are degraded by proteases.

Protein folding is a critically important process, not only because misfolded proteins may be nonfunctional, but also because improperly folded proteins can accumulate and be detrimental to cells and the organisms that contain them. For example, a group of transmissible brain disorders in mammals—**scrapie** in sheep, **bovine spongioform encephalopathy (mad cow disease)** in cattle, and **Creutzfeldt–Jakob disease** in humans—are caused by the presence in the brain of **prions,** which are aggregates of a misfolded protein. The misfolded protein (called PrPsc) is an altered version of a normal cellular protein (called PrPc) synthesized in neurons and found in the brains of all adult animals. The difference between PrPc and PrPsc lies in their secondary protein structures. Normal, noninfectious PrPc folds into an α helix, whereas infectious PrPsc folds into a β-pleated sheet. When an abnormal PrPsc molecule contacts a PrPc molecule, the normal protein refolds into the abnormal conformation. The process continues as a chain reaction, with potentially devastating results—the formation of prion particles that eventually destroy the brain. Hence, this group of disorders can be considered diseases of secondary protein structure.

Currently, many laboratories are studying protein folding and misfolding, particularly as related to genetics. Numerous inherited human disorders are caused by misfolded proteins that form abnormal aggregates. **Sickle-cell anemia,** discussed earlier in this chapter, is a case in point, where the β chains of hemoglobin are altered as the result of a single amino acid change, causing the molecules to aggregate within erythrocytes, with devastating results. An autosomal dominant inherited form of **Creutzfeldt–Jakob disease** is known in which the mutation alters the PrP amino acid sequence, leading to prion formation. And various progressive neurodegenerative diseases such as **Huntington disease, Alzheimer disease,** and **Parkinson disease** are linked to the formation of abnormal protein aggregate in the brain. Huntington disease is inherited as an autosomal dominant trait, whereas less clearly defined genetic components are associated with Alzheimer and Parkinson diseases.

NOW SOLVE THIS

Problem 30 on page 283 asks you to consider the potential impact of several amino acid substitutions that result due to mutations in one of the genes encoding one of the chains making up human hemoglobin.

Hint: When considering the three amino acids (glutamic acid, lysine, and valine), consider the net charge of each R group.

13.9 Proteins Function in Many Diverse Roles

The essence of life on Earth rests at the level of diverse cellular function. One can argue that DNA and RNA simply serve as vehicles to store and express genetic information. However, proteins are at the heart of cellular function. And it is the capability of cells to assume diverse structures and functions that distinguishes most eukaryotes from less evolutionarily advanced organisms such as bacteria. Therefore, an introductory understanding of protein function is critical to a complete view of genetic processes.

Proteins are the most abundant macromolecules found in cells. As the end products of genes, they play many diverse roles. For example, the respiratory pigments **hemoglobin** and **myoglobin** transport oxygen, which is essential for cellular metabolism. **Collagen** and **keratin** are structural proteins associated with the skin, connective tissue, and hair of organisms. **Actin** and **myosin** are contractile proteins, found in abundance in muscle tissue. Still other examples are the **immunoglobulins,** which function in the immune system of vertebrates; **transport proteins,** involved in movement of molecules across membranes; some of the **hormones** and their **receptors,** which regulate various types of chemical activity; and **histones,** which bind to DNA in eukaryotic organisms.

The largest group of proteins with a related function are the **enzymes.** Since we have referred to these molecules throughout this chapter, it may be useful to extend our discussion and include a more detailed description of their biological role.

Enzymes specialize in catalyzing chemical reactions within living cells. They increase the rate at which a chemical reaction reaches equilibrium but do not alter the end-point of the chemical equilibrium. Their remarkable, highly specific catalytic properties largely determine the metabolic capacity of any cell type. The specific functions of many enzymes involved in the genetic and cellular processes of cells are described throughout this text.

Biological catalysis is a process whereby the **energy of activation** (E_a) for a given reaction is lowered. The energy of activation is the increased kinetic energy state that molecules usually must reach before they react with one another. This state can be attained as a result of elevated temperatures, but enzymes allow biological reactions to occur at lower physiological temperatures. In this way, enzymes make life as we know it possible.

The catalytic properties and specificity of an enzyme are determined by the chemical configuration of the molecule's **active site.** This site is associated with a crevice, a cleft, or a pit on the surface of the enzyme, which binds the reactants, or substrates, facilitating their interaction. Enzymatically catalyzed reactions control metabolic activities in the cell. Each reaction is either catabolic or anabolic. **Catabolism** is the degradation of large molecules into smaller, simpler ones with the release of chemical energy. **Anabolism** is the synthetic phase of metabolism and yields the various components that make up nucleic acids, proteins, lipids, and carbohydrates.

ESSENTIAL POINT

Of the myriad functions performed by proteins, the most influential role belongs to enzymes, which serve as highly specific biological catalysts that play a central role in the production of all classes of molecules in living systems.

Protein Domains Impart Function

We conclude this chapter by briefly discussing the important finding that regions made up of specific amino acid sequences are associated with specific functions in protein molecules. Such sequences, usually between 50 and 300 amino acids, constitute **protein domains** and represent modular portions of the protein that fold into stable, unique conformations independently of the rest of the molecule. Different domains impart different functional capabilities. Some proteins contain only a single domain, while others contain two or more.

The significance of domains resides in the tertiary structures of proteins. Each domain can contain a mixture of secondary structures, including α helices and β-pleated sheets. The unique conformation of a given domain imparts a specific function to the protein. For example, a domain may serve as the catalytic site of an enzyme, or it may impart an ability to bind to a specific ligand. Thus, discussions of proteins may mention *catalytic domains, DNA-binding domains,* and so on. In short, a protein must be seen as being composed of a series of structural and functional modules. Obviously, the presence of multiple domains in a single protein increases the versatility of each molecule and adds to its functional complexity.

EXPLORING GENOMICS

Translation Tools and Swiss-Prot for Studying Protein Sequences

Many of the databases and bioinformatics programs we have used for Exploring Genomics exercises have focused on manipulating and analyzing DNA and RNA sequences. However, scientists working on various aspects of translation and protein structure and function also have a wide range of bioinformatics tools and databases at their disposal via the Internet. Many of these sites were developed as a consequence of the newly emerging discipline of *proteomics,* the study of all the proteins expressed in a cell or tissue. We will discuss proteomics in more detail in Chapter 18.

In this Exploring Genomics exercise, we will use a program from **ExPASy (Expert Protein Analysis System)** to translate a segment of a gene into a possible polypeptide. We will then explore databases for learning more about this polypeptide.

■ Exercise I: Translating a Nucleotide Sequence and Analyzing a Polypeptide

ExPASy (Expert Protein Analysis System), hosted by the Swiss Institute of Bioinformatics, provides a wealth of resources for studies in proteomics. In this exercise we will use a program from ExPASy called **Translate Tool** to translate a nucleotide sequence to a polypeptide sequence. Although many other programs are available on the Web for this purpose, ExPASy is one of the more student-friendly tools. Translate Tool allows you to make a predicted polypeptide sequence from a cloned gene and then look for open reading frames and variations in possible polypeptides.

1. Below is a partial sequence for a human gene based on a complementary DNA (cDNA) sequence. In Chapter 17 you will learn that cDNA sequences are DNA copies complementary to mRNA molecules expressed in a cell. Before you translate this sequence in ExPASy, run a nucleotide–nucleotide BLAST search from the NCBI Web site (**http://www. ncbi.nlm.nih.gov/BLAST/**) to identify the gene corresponding to this sequence. Refer to the Exploring Genomics exercise in Chapter 9 if you need help with BLAST searches.

ACATTTGCTTCTGACACAATTGTGT
TCACTAGCAACCTCAAACAGACAC
CATGGTGCATCTGACTCCTGAGGAG
AAGTCTGCCGTTACTGCCCTGTGGG
GCAAGGTGAACGTGGATGAAGTTG
GTGGTGAGGCCCTGGGCAG

2. Access the ExPASy Translate Tool program at **http://us.expasy.org/tools/dna.html**. Copy and paste the cDNA sequence into Translate Tool and click "Translate Sequence" to generate possible polypeptide sequences encoded by this cDNA.
3. Review the translation results and then answer the following questions:
   a. Did Translate Tool provide one or multiple possible polypeptide sequences?
   b. If the translation results showed multiple polypeptide sequences, what does this mean? Explain.
   c. Refer to Figure 13–14 in the chapter. Based on this figure, which reading frame generated by Translate Tool appears to be correct?
4. ExPASy also provides access to a wealth of information about this polypeptide by connecting to a large number of different databases such as **UniProt KB/Swiss-Prot,** a protein sequence database maintained by the Swiss Institute for Bioinformatics (SIB) and the European Bioinformatics

Institute (EBI), and a database called the Protein Data Bank. UniProt KB/Swiss-Prot is widely used by scientists around the world. Visit UniProt KB/Swiss-Prot (**http://ca.expasy.org/sprot/hpi/**) for a wealth of information on the human genome and proteomics.

5. To learn more about the features of this polypeptide, it is best to work with a complete sequence. To obtain a complete sequence, click on the reading frame you believe is correct.
6. To retrieve the amino acid sequence for the entire polypeptide, click "methionine (m)" as the first amino acid in the polypeptide and then use the "BLAST" link near the bottom of the next page to run a BLAST search. From the BLAST results page, locate UniProtKB/Swiss-Prot entry P68871; this is an accession number for the protein sequence, similar to the accession numbers assigned to DNA and RNA sequences, and it is the correct match for this sequence. Click on this link to reveal a comprehensive report about the protein. Be sure the identity of this protein agrees with what you discovered in Step 1.
7. Scroll past the references in the Swiss-Prot report; then explore the following features:
   a. 3D Structure Databases—presents 3D modeling representations showing polypeptide folding arrangements.
   b. Under 2D Gel Databases, use the Swiss-2D PAGE link to view 2D gels of this polypeptide from different tissue samples. Refer to Figure 18–23 for a representation of 2D gels. When viewing a 2D gel image with this feature, click on links under "Map Locations" to identify specific spots on a gel that correspond to this polypeptide.
   c. The Family and Domain databases links will take you to a wealth of information about this polypeptide and related polypeptides and proteins.
   d. Explore the "Other" category and visit the DrugBank link that provides information on drugs that bind to and affect this polypeptide.
   e. At the very bottom of the page, under "Sequence analysis tools," use the "ProtParam" feature to learn more about predicted secondary structures formed by this polypeptide.

CASE STUDY Lost in translation

A recessively inherited brain disorder called vanishing white matter (VWM) was first described in the 1990s using magnetic resonance imaging (MRI). Affected individuals show neurological deterioration early in childhood, dying soon after diagnosis, or they may express a slow progressive form of the disease. VWM is caused by mutations in any of the five genes that encode the protein subunits of the translation initiation factor 2B. This factor helps position the ribosome on the mRNA during the initiation of translation. It is known that other cells which rapidly synthesize large amounts of protein, such as insulin-secreting cells, are also affected. Many questions about this disorder remain to be answered.

1. Given the two forms of the disease, discuss the possible nature of VWM mutations. Why do you think that this is a recessive rather than a dominant mutation?
2. Why might some cells in the body be more susceptible to mutations in genes encoding factor 2B than other cells?
3. How could we study the cellular and molecular aspects of VWM?

INSIGHTS AND SOLUTIONS

1. The growth responses in the following chart were obtained using four mutant strains of *Neurospora* and the related compounds A, B, C, and D. None of the mutants grow on minimal medium. Draw all possible conclusions from this data.

	Growth Supplement			
Mutation	**A**	**B**	**C**	**D**
1	–	–	–	–
2	+	+	–	+
3	+	+	–	–
4	–	+	–	–

Solution: Nothing can be concluded about mutation *1* except that it lacks some essential growth factor, perhaps even unrelated to the biochemical pathway represented by mutations *2, 3,* and *4.* Nor can anything be concluded about compound C. If it is involved in the pathway, it is a product that was synthesized prior to compounds A, B, and D.

We now analyze these three compounds and the control of their synthesis by the enzymes encoded by mutations *2, 3,* and *4.* Because product B allows growth in all three cases, it may be considered the "end product"—it bypasses the block in all three instances. Using similar reasoning, product A precedes B in the pathway because it bypasses the block in two of the three steps, and product D precedes B yielding a partial solution:

$$C(?) \longrightarrow D \longrightarrow A \longrightarrow B$$

Now let's determine which mutations control which steps. Since mutation *2* can be alleviated by products D, B, and A, it must control a step prior to all three products, perhaps the direct conversion to D (although we cannot be certain). Mutation *3* is alleviated by B and A, so its effect must precede them in the pathway. Thus, we assign it as controlling the conversion of D to A. Likewise, we can assign mutation *4* to the conversion of A to B, leading to a more complete solution:

$$C(?) \xrightarrow{2(?)} D \xrightarrow{3} A \xrightarrow{4} B$$

PROBLEMS AND DISCUSSION QUESTIONS

1. List and describe the role of all molecular constituents present in a functional polyribosome.
2. Contrast the roles of tRNA and mRNA during translation, and list all enzymes that participate in the transcription and translation process.
3. Francis Crick proposed the adaptor hypothesis for the function of tRNA. Why did he choose that description?
4. During translation, what molecule bears the anticodon? the codon?
5. The α chain of eukaryotic hemoglobin is composed of 141 amino acids. What is the minimum number of nucleotides in an mRNA coding for this polypeptide chain?
6. Summarize the steps involved in charging tRNAs with their appropriate amino acids.
7. Each transfer RNA requires at least four specific recognition sites that must be inherent in its tertiary protein structure in order for it to carry out its role. What are these sites?
8. Discuss the potential difficulties involved in designing a diet to alleviate the symptoms of phenylketonuria.
9. Phenylketonurics cannot convert phenylalanine to tyrosine. Why don't these individuals exhibit a deficiency of tyrosine?
10. Phenylketonurics are often more lightly pigmented than are normal individuals. Can you suggest a reason why this is so?
11. The synthesis of flower pigments is known to be dependent on enzymatically controlled biosynthetic pathways. Postulate the role of mutant genes and their products in producing the observed phenotypes in the crosses shown.

(a)	P_1:	white strain A × white strain B
	F_1:	all purple
	F_2:	9/16 purple: 7/16 white
(b)	P_1:	white × pink
	F_1:	all purple
	F_2:	9/16 purple: 3/16 pink: 4/16 white

12. A series of mutations in the bacterium *Salmonella typhimurium* results in the requirement of either tryptophan or some related molecule in order for growth to occur. From the data shown here, suggest a biosynthetic pathway for tryptophan.
See **Now Solve This** on page 273.

	Growth Supplement				
Mutation	Minimal Medium	Anthranilic Acid	Indole Glycerol Phosphate	Indole	Tryptophan
trp-8	–	+	+	+	+
trp-2	–	–	+	+	+
trp-3	–	–	–	+	+
trp-1	–	–	–	–	+

13. The study of biochemical mutants in organisms such as *Neurospora* has demonstrated that some pathways are branched. The following data illustrate the branched nature of the pathway resulting in the synthesis of thiamine. Why don't the data support a linear pathway? Can you postulate a pathway for the synthesis of thiamine in *Neurospora*?

	Growth Supplement			
Mutation	Minimal Medium	Pyrimidine	Thiazole	Thiamine
thi-1	–	–	+	+
thi-2	–	+	+	+
thi-3	–	–	–	+

14. Explain why the one-gene:one-enzyme hypothesis is no longer considered to be totally accurate.
15. Why is an alteration of electrophoretic mobility interpreted as a change in the primary structure of the protein under study?
16. Hemoglobin is a tetramer consisting of two α and two β chains. How does this information relate to each of the four levels of protein structure?
17. Using sickle-cell anemia as a basis, describe what is meant by a genetic or inherited molecular disease. What are the similarities and dissimilarities between this type of a disorder and a disease caused by an invading microorganism?
18. Contrast the contributions Pauling and Ingram made to our understanding of the genetic basis for sickle-cell anemia.
19. Hemoglobins from two individuals are compared by starch gel electrophoresis and with protein fingerprinting. Electrophoresis reveals no difference in migration, but fingerprinting shows an amino acid difference. How is this possible?
20. Assuming that each nucleotide is 0.34 nm long in mRNA, how many triplet codes can simultaneously occupy space in a ribosome that is 20 nm in diameter?
21. Review the concept of colinearity in Section 12.5 (p. 248) and consider the following question: Certain mutations called *amber* in bacteria and viruses result in premature termination of polypeptide chains during translation. Many *amber* mutations have been detected at different points along the gene that codes for a head protein in phage T4. How might this system be further investigated to demonstrate and support the concept of colinearity?
22. In your opinion, which of the four levels of protein organization is the most critical to a protein's function? Defend your choice.
23. List and describe the function of as many nonenzymatic proteins as you can that are unique to eukaryotes.
24. How does an enzyme function? Why are enzymes essential for living organisms?
25. Shown in the following table are several amino acid substitutions in the α and β chains of human hemoglobin. Use the genetic code table in Figure 12–7 to determine how many of them can occur as a result of a single nucleotide change.

Hb Type	Normal Amino Acid	Substituted Amino Acid
HbJ Toronto	ala	asp (α-5)
HbJ Oxford	gly	asp (α-15)
Hb Mexico	gln	glu (α-54)
Hb Bethesda	tyr	his (β-145)
Hb Sydney	val	ala (β-67)
HbM Saskatoon	his	tyr (β-63)

26. Early detection and adherence to a strict dietary regime has relieved much of the mental retardation that once occurred in people afflicted with phenylketonuria (PKU). Now, affected individuals often lead normal lives and have families. For various reasons, some individuals adhere less rigorously to their diet as they get older. Predict the effect of such dietary neglect on the newborns of mothers with PKU.
27. In 1962, F. Chapeville and others (1962. *Proc. Natl. Acad. Sci. (USA)* 48:1086–1093) reported an experiment in which they isolated radioactive ^{14}C-cysteinyl-tRNA^{cys} (charged tRNA^{cys} + cysteine). They then removed the sulfur group from the cysteine, creating alanyl-tRNA^{cys} (charged tRNA^{cys} + alanine). When alanyl-tRNA^{cys} was added to a synthetic mRNA calling for cysteine, but not alanine, a polypeptide chain was synthesized containing alanine. What can you conclude from this experiment? See **Now Solve This** on page 265.
28. Three independently assorting genes are known to control the biochemical pathway here that provides the basis for flower color in a hypothetical plant:

$$\text{colorless} \xrightarrow{A-} \text{yellow} \xrightarrow{B-} \text{green} \xrightarrow{C-} \text{speckled}$$

Homozygous recessive mutations, which interrupt each step, are known. Determine the phenotypic results in the F_1 and F_2 generations resulting from the P_1 crosses involving true-breeding plants given here.

(a) speckled	(*AABBCC*)	×	yellow	(*AAbbCC*)
(b) yellow	(*AAbbCC*)	×	green	(*AABBcc*)
(c) colorless	(*aaBBCC*)	×	green	(*AABBcc*)

29. How would the results in cross (a) of Problem 28 vary if genes *A* and *B* were linked with no crossing over between them? How would the results of cross (a) vary if genes *A* and *B* were linked and 20 map units apart?
30. HbS results from the amino acid change of glutamic acid to valine at the number 6 position in the β chain of human hemoglobin. HbC is the result of a change at the same position in the β chain, but lysine replaces glutamic acid. Return to the genetic code table in Figure 12–7, and determine whether single-nucleotide changes can account for these mutations. Then turn to Figure 13–15, and examine the R groups in the amino acids glutamic acid, valine, and lysine. Describe the chemical differences between the three amino acids and predict how the changes might alter the structure of the molecule and lead to altered hemoglobin function. See **Now Solve This** on page 279.
31. HbS results in anemia and resistance to malaria, whereas in those with HbA, the parasite *Plasmodium falciparum* invades red blood cells and causes the disease. Predict whether those with HbC are likely to be anemic and whether they would be resistant to malaria.
32. Deep in a previously unexplored South American rain forest, a species of plant was discovered with true-breeding varieties whose flowers were either pink, rose, orange, or purple. A very astute plant geneticist made a single cross, carried to the F_2 generation, as shown here. Based solely on these data, he was able to propose both a mode of inheritance for flower pigmentation and a biochemical pathway for the synthesis of these pigments. Carefully study the data. Create your own hypothesis to explain the mode of inheritance, and then propose a biochemical pathway consistent with your hypothesis. What other crosses would enable you to test the hypothesis?

P_1:	**purple × pink**
F_1:	all purple
F_2:	27/64 purple
	16/64 pink
	12/64 rose
	9/64 orange

33. The emergence of antibiotic-resistant strains of *Enterococci* and transfer of resistant genes to other bacterial pathogens have highlighted the need for new generations of antibiotics to combat serious infections. To grasp the range of potential sites for the action of existing antibiotics, sketch the components of the translation machinery (e.g., see Step 3 of Figure 13–6), and using a series of numbered pointers, indicate the specific location for the action of the antibiotics shown in the following table.

Antibiotic	**Action**
1. Streptomycin	Binds to 30*S* ribosomal subunit
2. Chloramphenicol	Inhibits peptidyl transferase of 70*S* ribosome
3. Tetracycline	Inhibits binding of charged tRNA to ribosome
4. Erythromycin	Binds to free 50*S* particle and prevents formation of 70*S* ribosome
5. Kasugamycin	Inhibits binding of tRNA^{fmet}
6. Thiostrepton	Prevents translocation by inhibiting EF-G

34. Development of antibiotic resistance by pathogenic bacteria represents a major health concern. One potential new antibiotic is evernimicin, which was isolated from *Micromonospora carbonaceae.* Evernimicin is an oligosaccharide antibiotic with activity against a broad range of gram-positive pathogenic bacteria. To determine the mode of action of this drug, Adrian and others (2000. *Antimicrob. Ag. and Chemo.* 44:3101–3106) analyzed 23*S* ribosomal DNA mutants that showed reduced sensitivity to evernimicin. They and others discovered two classes of mutants that conferred resistance: 23*S* rRNA nucleotides 2475–2483 and ribosomal protein L16. This suggests that these two ribosomal components are structurally and functionally linked. It turns out that the tRNA anticodon stem-loop appears to bind to the A site of the ribosome at rRNA bases 2465–2485. This finding conforms to the proposed function of L16, which appears to be involved in attracting the aminoacyl stem of the tRNA to the ribosome at its A site. Using your sketch of the translation machinery from Problem 33 along with this information, designate where the proposed antibacterial action of evernimicin is likely to occur.

Pigment mutations within an ear of corn, caused by transposition of the Ds element.

14

Gene Mutation, Transposition, and DNA Repair

Chapter Concepts

- Mutations comprise any change in the base-pair sequence of DNA.
- Mutation is a source of genetic variation and provides the raw material for natural selection. It is also the source of genetic damage that contributes to cell death, genetic diseases, and cancer.
- Mutations have a wide range of effects on organisms depending on the type of base-pair alteration, the location of the mutation within the chromosome, and the function of the affected gene product.
- Mutations can occur spontaneously as a result of natural biological and chemical processes, or they can be induced by external factors, such as chemicals or radiation.
- Organisms rely on a number of DNA repair mechanisms to counteract mutations. These mechanisms range from proofreading and correction of replication errors to base excision and homologous recombination repair.
- Mutations in genes whose products control DNA repair lead to genome hypermutability and human DNA repair diseases.
- Geneticists induce gene mutations as the first step in classical genetic analysis.
- Transposable elements create mutations by moving into and out of chromosomes, inducing mutations within coding regions and in gene-regulatory regions, and causing chromosome breaks.

The ability of DNA molecules to store, replicate, transmit, and decode information is the basis of genetic function. But equally important is the capacity of DNA to make mistakes. Without the variation that arises from changes in DNA sequences, there would be no phenotypic variability, no adaptation to environmental changes, and no evolution. Gene mutations are the source of most new alleles and are the origin of genetic variation within populations. On the downside, they are also the source of genetic changes that can lead to cell death, genetic diseases, and cancer.

Mutations also provide the basis for genetic analysis. The phenotypic variability resulting from mutations allows geneticists to identify and study the genes responsible for the modified trait. In genetic investigations, mutations act as identifying "markers" for genes so that they can be followed during their transmission from parents to offspring. Without phenotypic variability, classical genetic analysis would be impossible. For example, if all pea plants displayed a uniform phenotype, Mendel would have had no foundation for his research.

In Chapter 6, we examined mutations in large regions of chromosomes—chromosomal mutations. In contrast, the mutations we will now explore are those occurring primarily in the base-pair sequence of DNA within individual genes—**gene mutations.** We will also describe how the cell defends itself from such mutations using various mechanisms of DNA repair. The chapter also describes how geneticists use mutations to identify genes and analyze gene functions in humans and other organisms.

How Do We Know?

In this chapter, we will focus on how mutations arise and how the cell repairs mutations and damaged DNA. We will find many opportunities to consider the methods and reasoning by which much of this information was acquired. From the explanations given in the chapter, you should answer the following fundamental questions:

1. How do we know that X rays are mutagenic and that the dose of X rays is directly related to the rate of mutation?
2. How do we know whether a substance or chemical compound is mutagenic?
3. How do we know that mutant phenotypes are sometimes due to transposable elements?
4. How do we know whether two or more mutations that cause the same phenotype occur in the same gene, or in different genes?
5. How do we know that one step in the repair of damaged DNA involves DNA synthesis?

14.1 Gene Mutations Are Classified in Various Ways

A gene mutation can be defined as an alteration in DNA sequence. Any base-pair change in any part of a DNA molecule can be considered a mutation. A mutation may comprise a single base-pair substitution, a deletion or insertion of one or more base pairs, or a major alteration in the structure of a chromosome.

Mutations may occur within regions of a gene that code for protein or within noncoding regions of a gene such as introns and regulatory sequences. Mutations may or may not bring about a detectable change in phenotype. The extent to which a mutation changes the characteristics of an organism depends on where the mutation occurs and the degree to which the mutation alters the function of the gene product.

Mutations can occur in somatic cells or within germ cells. Those that occur in germ cells are heritable and are the basis for the transmission of genetic diversity and evolution, as well as genetic diseases. Those that occur in somatic cells are not transmitted to the next generation but may lead to altered cellular function or tumors.

Because of the wide range of types and effects of mutations, geneticists classify mutations according to several different schemes. These organizational schemes are not mutually exclusive. In this section, we outline some of the ways in which gene mutations are classified.

Spontaneous and Induced Mutations

Mutations can be classified as either spontaneous or induced, although these two categories overlap to some degree. **Spontaneous mutations** are changes in the nucleotide sequence of genes that appear to have no known cause. No specific agents are associated with their occurrence, and they are generally assumed to be accidental. Many of these mutations arise as a result of normal biological or chemical processes in the organism that alter the structure of nitrogenous bases. Often, spontaneous mutations occur during the enzymatic process of DNA replication, as we discuss later in this chapter.

In contrast to spontaneous mutations, mutations that result from the influence of extraneous factors are considered to be **induced mutations.** Induced mutations may be the result of either natural or artificial agents. For example, radiation from cosmic and mineral sources and ultraviolet radiation from the sun are energy sources to which most organisms are exposed and, as such, may be factors that cause induced mutations. The earliest demonstration of the artificial induction of mutations occurred in 1927, when Hermann J. Muller reported that X rays could cause mutations in *Drosophila.* In 1928, Lewis J. Stadler reported that X rays had the same effect on barley. In addition to various forms of radiation, numerous natural and synthetic chemical agents are also mutagenic.

Several generalizations can be made regarding spontaneous mutation rates in organisms. First, the rate of spontaneous mutation is exceedingly low for all organisms. Second, the rate varies considerably between different organisms. Third, even within the same species, the spontaneous mutation rate varies from gene to gene.

Viral and bacterial genes undergo spontaneous mutation at an average of about 1 in 100 million (10^{-8}) replications. *Neurospora* exhibits a similar rate, but maize, *Drosophila,* and humans demonstrate rates several orders of magnitude higher. The genes studied in these groups average between 1/1,000,000 and 1/100,000 (10^{-6} and 10^{-5}) mutations per gamete formed. Mouse genes are another order of magnitude higher in their spontaneous mutation rate, 1/100,000 to 1/10,000 (10^{-5} to 10^{-4}). It is not clear why such a large variation occurs in mutation rates. The variation between organisms may reflect the relative efficiencies of their DNA proofreading and repair systems. We will discuss these systems later in the chapter.

Classification Based on Location of Mutation

MUTATIONS AT THE DNA LEVEL — WEB TUTORIAL

Mutations may be classified according to the cell type or chromosomal locations in which they occur. **Somatic mutations** are those occurring in any cell in the body except germ cells. **Germ-line mutations** are those occurring in gametes. **Autosomal mutations** are mutations within genes located on the autosomes, whereas **X-linked mutations** are those within genes located on the X chromosome.

Mutations arising in somatic cells are not transmitted to future generations. When a recessive autosomal mutation occurs in a somatic cell of a diploid organism, it is unlikely to result in a detectable phenotype. The expression of most such mutations is likely to be masked by expression of the wild-type allele within that cell. Somatic mutations will have a greater impact if they are dominant or, in males, if they are X-linked, since such mutations are most likely to be immediately expressed. Similarly, the impact of dominant or X-linked somatic mutations will be more noticeable if they occur early in development, when a small number of undifferentiated cells replicate to give rise to several differentiated tissues or organs. Dominant mutations that occur in cells of adult tissues are often masked by the activity of thousands upon thousands of nonmutant cells in the same tissue that perform the nonmutant function.

Mutations in gametes are of greater significance because they are transmitted to offspring as part of the germ line. They have the potential of being expressed in all cells of an offspring. Inherited dominant autosomal mutations will be expressed phenotypically in the first generation. X-linked recessive mutations arising in the gametes of a **homogametic** female may be expressed in hemizygous male offspring. This will occur provided that the male offspring receives the affected X chromosome. Because of heterozygosity, the occurrence of an autosomal recessive mutation in the gametes of either males or females (even one resulting in a lethal allele) may go unnoticed for many generations, until the resultant allele has become widespread in the population. Usually, the new allele will become evident only when a chance mating brings two copies of it together into the homozygous condition.

Classification Based on Type of Molecular Change

Geneticists often classify gene mutations in terms of the nucleotide changes that constitute the mutation. A change of one base pair to another in a DNA molecule is known as a **point mutation,** or **base substitution** (Figure 14–1). A change of one nucleotide of a triplet within a protein-coding portion of a gene may result in the creation of a new triplet that codes for a different amino acid in the protein product. If this occurs, the mutation is known as a **missense mutation.** A second possible outcome is that the triplet will be changed into a stop codon, resulting in the termination of translation of the protein. This is known as a **nonsense mutation.** If the point mutation alters a codon but does not result in a change in the amino acid at that position in the protein (due to degeneracy of the genetic code), it can be considered a **silent mutation.**

You will often see two other terms used to describe base substitutions. If a pyrimidine replaces a pyrimidine or a purine replaces a purine, a **transition** has occurred. If a purine replaces a pyrimidine, or vice versa, a **transversion** has occurred.

Another type of change is the insertion or deletion of one or more nucleotides at any point within the gene. As illustrated in Figure 14–1, the loss or addition of a single nucleotide causes all of the subsequent three-letter codons to be changed. These are called **frameshift mutations** because the frame of triplet reading during translation is altered. A frameshift mutation will occur when any number of bases are added or deleted, except multiples of three, which would reestablish the initial frame of reading. It is possible that one of the many altered triplets will be UAA, UAG, or UGA, the translation termination codons. When one of these triplets is encountered during translation, polypeptide synthesis is terminated at that point. Obviously, the results of frameshift mutations can be very severe, especially if they occur early in the coding sequence.

Classification Based on Phenotypic Effects

Depending on their type and location, mutations can have a wide range of phenotypic effects, from none to severe.

As discussed in Chapter 4, a **loss-of-function mutation** is one that reduces or eliminates the function of the gene prod-

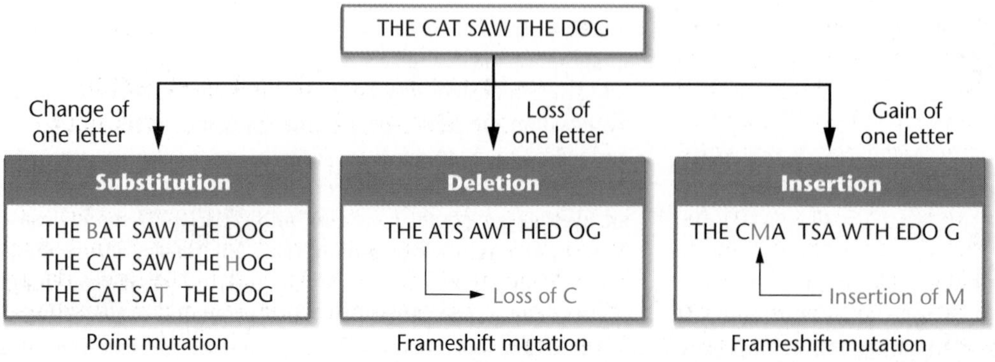

FIGURE 14–1 Analogy showing the effects of substitution, deletion, and insertion of one letter in a sentence composed of three-letter words to demonstrate point and frameshift mutations.

uct. Any type of mutation, from a point mutation to deletion of the entire gene, may lead to a loss of function. Mutations that result in complete loss of function are known as **null mutations.** It is possible for a loss-of-function mutation to be either dominant or recessive. A dominant loss-of-function mutation may result from the presence of a defective protein product that binds to, or inhibits the action of, the normal gene product, which is also present in the same organism. A **gain-of-function** mutation results in a gene product with enhanced or new functions. This may be due to a change in the amino acid sequence of the protein that confers a new activity, or it may result from a mutation in a regulatory region of the gene, leading to expression of the gene at higher levels, or the synthesis of the gene product at abnormal times, or places. Most gain-of-function mutations are dominant.

The most easily observed mutations are those affecting a morphological trait. These mutations are known as **visible mutations** and are recognized by their ability to alter a normal or wild-type visible phenotype. For example, all of Mendel's pea characteristics and many genetic variations encountered in *Drosophila* fit this designation, since they cause obvious changes to the morphology of the organism.

Some mutations exhibit nutritional or biochemical effects. In bacteria and fungi, a typical **nutritional mutation** results in a loss of ability to synthesize an amino acid or vitamin. In humans, sickle-cell anemia and hemophilia are examples of diseases resulting from **biochemical mutations.** Although such mutations do not always affect morphological characters, they can have an effect on the well-being and survival of the affected individual.

Still another category consists of mutations that affect the behavior patterns of an organism. The primary effect of **behavioral mutations** is often difficult to analyze. For example, the mating behavior of a fruit fly may be impaired if it cannot beat its wings. However, the defect may be in the flight muscles, the nerves leading to them, or the brain, where the nerve impulses that initiate wing movements originate.

Another group of mutations may affect the regulation of gene expression. For example, as we will see with the *lac* operon discussed in Chapter 15, a regulatory gene can produce a product that controls the transcription of other genes. A mutation in a regulatory gene or a gene control region can disrupt normal regulatory processes and inappropriately activate or inactivate expression of a gene. Our knowledge of genetic regulation has been dependent on the study of such **regulatory mutations.**

It is also possible that a mutation may interrupt a process that is essential to the survival of the organism. In this case, it is referred to as a **lethal mutation.** For example, a mutant bacterium that has lost the ability to synthesize an essential amino acid will cease to grow and eventually will die when placed in a medium lacking that amino acid. Various inherited human biochemical disorders are also examples of lethal mutations. For example, Tay-Sachs disease and Huntington disease are caused by mutations that result in lethality, but at different points in the life cycle of humans.

Another interesting class of mutations are those whose expression depends on the environment in which the organism finds itself. Such mutations are called **conditional mutations** because the mutation is present in the genome of an organism but can be detected only under certain conditions. Among the best examples of conditional mutations are **temperature-sensitive mutations.** At a "permissive" temperature, the mutant gene product functions normally, but it loses its function at a different, "restrictive" temperature. Therefore, when the organism is shifted from the permissive to the restrictive temperature, the impact of the mutation becomes apparent.

Because eukaryotic genomes consist mainly of noncoding regions, the vast majority of mutations are likely to occur in the large portions of the genome that do not contain genes. These are considered to be **neutral mutations,** if they do not affect gene products or gene expression. Although it is interesting to consider the rates at which neutral mutations occur, it is of even greater interest to consider the rates at which deleterious mutations occur, especially in our own species. Recent molecular techniques, including DNA sequencing, have allowed geneticists to calculate the rates of deleterious mutations. The rate of deleterious mutations in humans is surprisingly high—at least 1.6 deleterious genetic changes per individual per generation.

ESSENTIAL POINT

Mutations can be spontaneous or induced, somatic or germline, autosomal or X-linked. They can have many different effects on gene function, depending on the type of nucleotide changes that comprise the mutation. Phenotypic effects can range from neutral or silent, to loss of function or gain of function, to lethality.

NOW SOLVE THIS

Problem 17 on page 306 provides data concerning a bacterial mutation that interrupts the biosynthesis of the amino acid leucine. You are asked to calculate the spontaneous mutation rate.

Hint: You must first determine the number of bacteria per milliliter in the original cultures grown under the two conditions by converting the "dilution" data back to the "undiluted" data. For example, if six colonies were observed in 1 mL after a 1000-fold dilution, then the original cell concentration was 6×10^3 bacteria/mL.

14.2 Spontaneous Mutations Arise from Replication Errors and Base Modifications

In this section, we will outline some of the processes that lead to spontaneous mutations. It is useful to keep in mind, however, that many of the base modifications that occur during spontaneous mutagenesis also occur, at a higher rate, during induced mutagenesis.

DNA Replication Errors

As we learned in Chapter 10, the process of DNA replication is imperfect. Occasionally, DNA polymerases insert incorrect nucleotides during replication of a strand of DNA. Although DNA polymerases can correct most of these replication errors using their inherent 3′ to 5′ exonuclease proofreading capacity, misincorporated nucleotides may persist after replication. If these errors are not detected and corrected by DNA repair mechanisms, they may lead to mutations. Replication errors due to mispairing predominantly lead to point mutations. The fact that bases can take several forms, known as **tautomers,** increases the chance of mispairing during DNA replication, as we explain next.

Replication Slippage

In addition to mispairing and point mutations, DNA replication can lead to the introduction of small insertions or deletions. These mutations can occur when one strand of the DNA template loops out and becomes displaced during replication, or when DNA polymerase slips or stutters during replication. If a loop occurs in the template strand during replication, DNA polymerase may miss the looped-out nucleotides, and a small deletion in the new strand will be introduced. If DNA polymerase repeatedly introduces nucleotides that are not present in the template strand, an insertion of one or more nucleotides will occur, creating an unpaired loop on the newly synthesized strand. Insertions and deletions may lead to frameshift mutations, or amino acid insertions or deletions in the gene product.

Replication slippage can occur anywhere in the DNA but seems distinctly more common in regions containing repeated sequences. Repeat sequences are hot spots for DNA mutation and in some cases contribute to hereditary diseases, such as Fragile X syndrome and Huntington disease. The hypermutability of repeat sequences in noncoding regions of the genome is the basis for current methods of forensic DNA analysis.

Tautomeric Shifts

Purines and pyrimidines can exist in tautomeric forms—that is, in alternate chemical forms that differ by only a single proton shift in the molecule. The biologically important tautomers are the keto–enol forms of thymine and guanine and the amino–imino forms of cytosine and adenine. These shifts change the bonding structure of the molecule, allowing hydrogen bonding with noncomplementary bases. Hence, **tautomeric shifts** may lead to permanent base-pair changes and mutations. Figure 14–2 compares normal base-pairing relationships with rare unorthodox pairings. Anomalous $T{\equiv}G$ and $C{=}A$ pairs, among others, may be formed.

A mutation occurs during DNA replication when a transiently formed tautomer in the template strand pairs with a noncomplementary base. In the next round of replication, the "mismatched" members of the base pair are separated, and each becomes the template for its normal complementary base. The end result is a point mutation (Figure 14–3).

Depurination and Deamination

Some of the most common causes of spontaneous mutations are two forms of DNA base damage: depurination and deamination. **Depurination** is the loss of one of the nitrogenous bases in an intact double-helical DNA molecule. Most frequently, the base is a purine—either guanine or adenine. These bases may be lost if the glycosidic bond linking the 1′-C of the deoxyribose and the number 9 position of the purine ring is broken, leaving an **apurinic site** on one strand of the DNA. Geneticists estimate that thousands of such spontaneous lesions are formed daily in the DNA of mammalian cells in culture. If apurinic sites are not repaired, there will be no base at that position to act as a template during DNA replication. As a result, DNA polymerase may introduce a nucleotide at random at that site.

In **deamination,** an amino group in cytosine or adenine is converted to a keto group (Figure 14–4). In these cases, cytosine is converted to uracil, and adenine is changed to hypoxanthine. The major effect of these changes is an alteration in the base-pairing specificities of these two bases during DNA replication. For example, cytosine normally pairs with guanine. Following its conversion to uracil, which pairs with adenine, the original

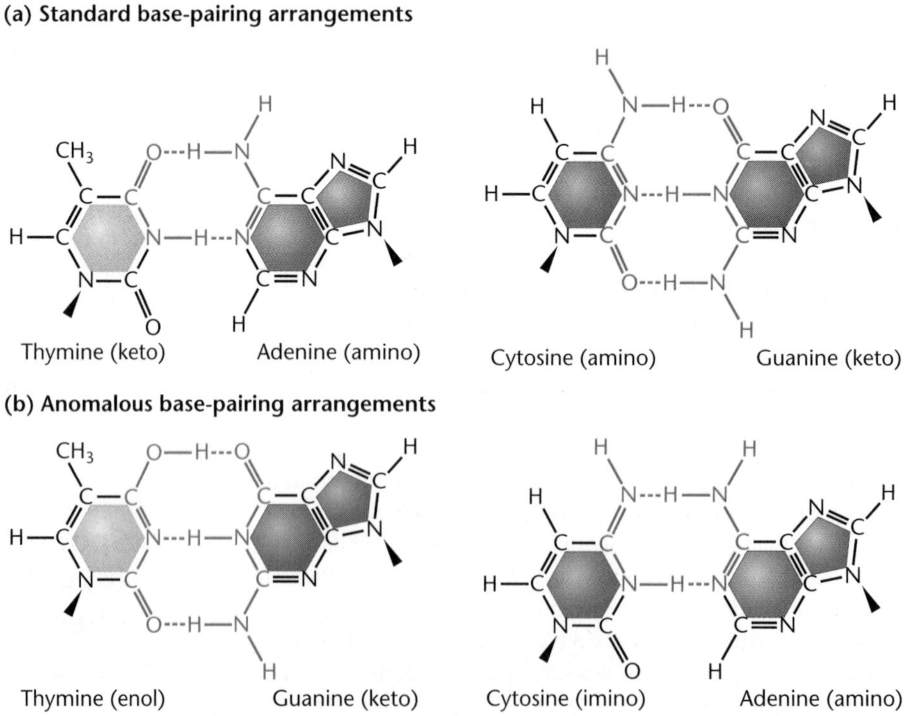

FIGURE 14–2 Standard base-pairing relationships (a) compared with examples of the anomalous base-pairing that occurs as a result of tautomeric shifts (b). The long triangle indicates the point at which the base bonds to the pentose sugar.

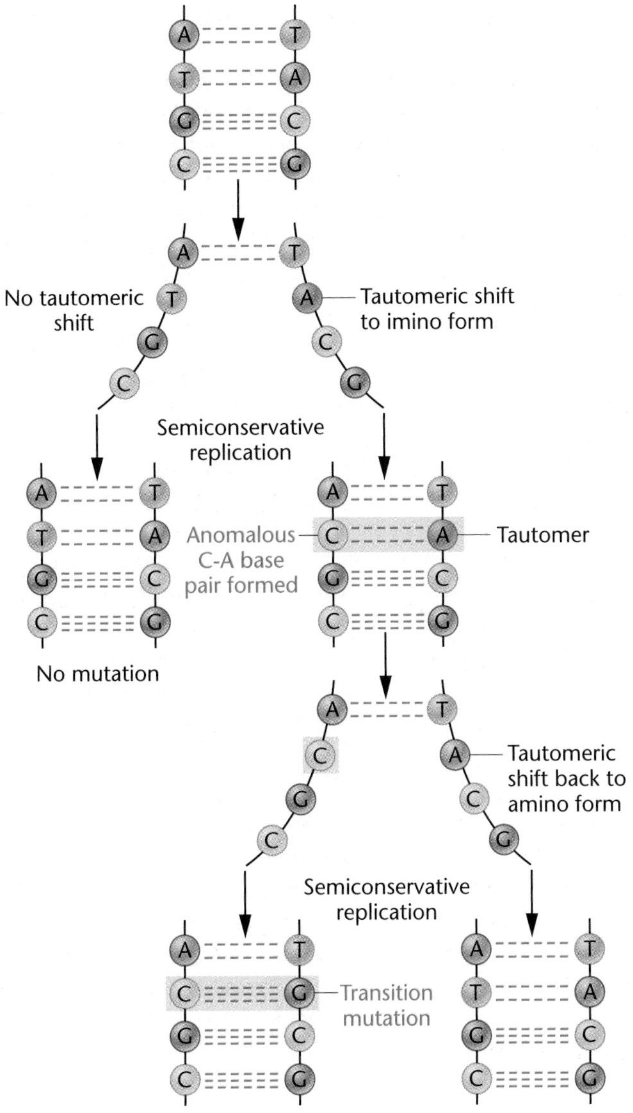

FIGURE 14–3 Formation of an A═T to G≡C transition mutation as a result of a tautomeric shift in adenine.

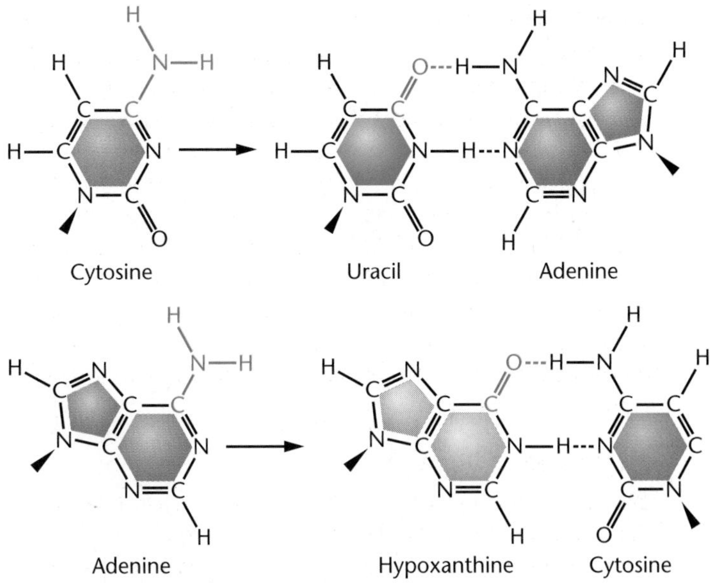

FIGURE 14–4 Deamination of cytosine and adenine, leading to new base pairing and mutation. Cytosine is converted to uracil, which base-pairs with adenine. Adenine is converted to hypoxanthine, which base-pairs with cytosine.

G≡C pair is converted to an A═U pair and then, in the next replication, is converted to an A═T pair. When adenine is deaminated, the original A═T pair is converted to a G≡C pair because hypoxanthine pairs naturally with cytosine. Deamination may occur spontaneously or as a result of treatment with chemical mutagens such as nitrous acid (HNO_2).

Oxidative Damage

DNA may also suffer damage from the by-products of normal cellular processes. These by-products include reactive oxygen species (electrophilic oxidants) that are generated during normal aerobic respiration. For example, superoxides ($O_2^{\cdot -}$), hydroxyl radicals (•OH), and hydrogen peroxide (H_2O_2) are created during cellular metabolism and are constant threats to the integrity of DNA. Such **reactive oxidants,** also generated by exposure to high-energy radiation, can produce more than 100 different types of chemical modifications in DNA, including modifications to bases that lead to mispairing during replication.

Transposons

Transposable genetic elements, or **transposons,** are DNA elements that can move within, or between, genomes. These elements are present in the genomes of all organisms, from bacteria to humans, and often comprise large portions of these genomes. Transposons can act as naturally occurring mutagens. If in moving to a new location they insert themselves into the coding region of a gene, they can alter the reading frame or introduce stop codons. If they insert into the regulatory region of a gene, they can disrupt proper expression of the gene. Transposons can also create chromosomal damage, including double-stranded breaks, inversions, and translocations. Transposable genetic elements are described in detail later in this chapter (Section 14.8).

ESSENTIAL POINT

Spontaneous mutations occur in many ways, ranging from errors during DNA replication to changes in DNA base pairing caused by tautomeric shifts, depurination and deamination, and reactive oxidant damage.

14.3 Induced Mutations Arise from DNA Damage Caused by Chemicals and Radiation

All cells on Earth are exposed to a plethora of agents called **mutagens,** which have the potential to damage DNA and cause mutations. Some of these agents, such as some fungal toxins, cosmic rays, and ultraviolet light, are natural components of our environment. Others, including some industrial

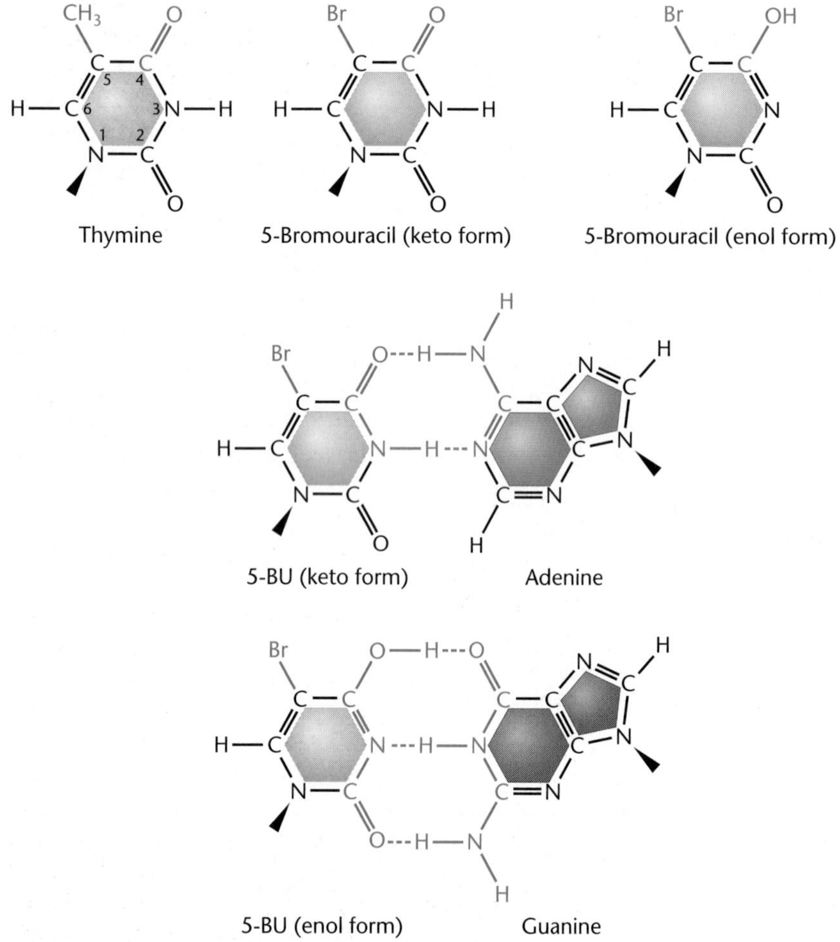

FIGURE 14–5 Similarity of 5-bromouracil (5-BU) structure to thymine structure. In the common keto form, 5-BU base-pairs normally with adenine, behaving as a thymine analog. In the rare enol form, it pairs anomalously with guanine.

pollutants, medical X rays, and chemicals within tobacco smoke, can be considered as unnatural or human-made additions to our modern world. On the positive side, geneticists harness some mutagens for use in analyzing genes and gene functions, as discussed later in this chapter (Section 14.7). The mechanisms by which some of these natural and unnatural agents lead to mutations are outlined in this section.

Base Analogs

One category of mutagenic chemicals are **base analogs,** compounds that can substitute for purines or pyrimidines during nucleic acid biosynthesis. For example, **5-bromouracil (5-BU),** a derivative of uracil, behaves as a thymine analog but is halogenated at the number 5 position of the pyrimidine ring. If 5-BU is chemically linked to deoxyribose, the nucleoside analog **bromodeoxyuridine (BrdU)** is formed. Figure 14–5 compares the structure of this analog with that of thymine. The presence of the bromine atom in place of the methyl group increases the probability that a tautomeric shift will occur. If 5-BU is incorporated into DNA in place of thymine and a tautomeric shift to the enol form occurs, 5-BU base-pairs with guanine. After one round of replication, an A═T to G≡C transition results. Furthermore, the presence of 5-BU within DNA increases the sensitivity of the molecule to ultraviolet (UV) light, which itself is mutagenic.

There are other base analogs that are mutagenic. For example, **2-amino purine (2-AP)** can act as an analog of adenine. In addition to its base-pairing affinity with thymine, 2-AP can also base-pair with cytosine, leading to possible transitions from A═T to G≡C following replication.

Alkylating Agents and Acridine Dyes

The sulfur-containing mustard gases, discovered during World War I, were some of the first chemical mutagens identified in chemical warfare studies. Mustard gases are **alkylating agents**—that is, they donate an alkyl group, such as CH_3 or CH_3CH_2, to amino or keto groups in nucleotides. Ethylmethane sulfonate (EMS), for example, alkylates the keto groups in the number 6 position of guanine and in the number 4 position of thymine. As with base analogs, base-pairing affinities are altered, and transition mutations result. For example, 6-ethylguanine acts as an analog of adenine and pairs with thymine (Figure 14–6).

Chemical mutagens called **acridine dyes** cause frameshift mutations. As illustrated in Figure 14–1, frameshift mutations result from the addition or removal of one or more base pairs in the polynucleotide sequence of the gene. Acridine dyes such as proflavin and acridine orange are about the same dimensions as nitrogenous base pairs and **intercalate,** or wedge, between the purines and pyrimidines of intact DNA. Interca-

FIGURE 14–6 Conversion of guanine to 6-ethylguanine by the alkylating agent ethylmethane sulfonate (EMS). The 6-ethylguanine base-pairs with thymine.

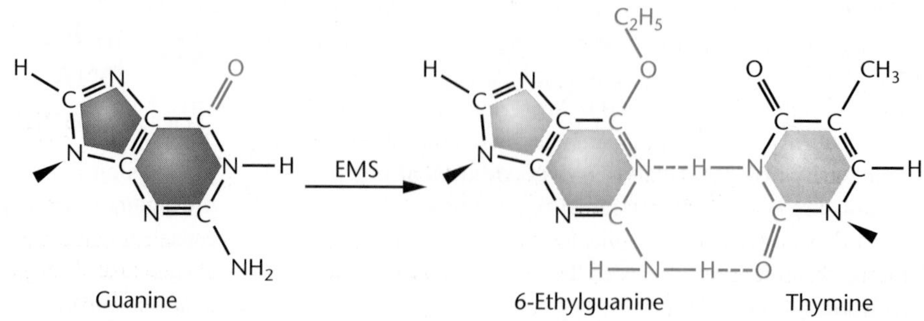

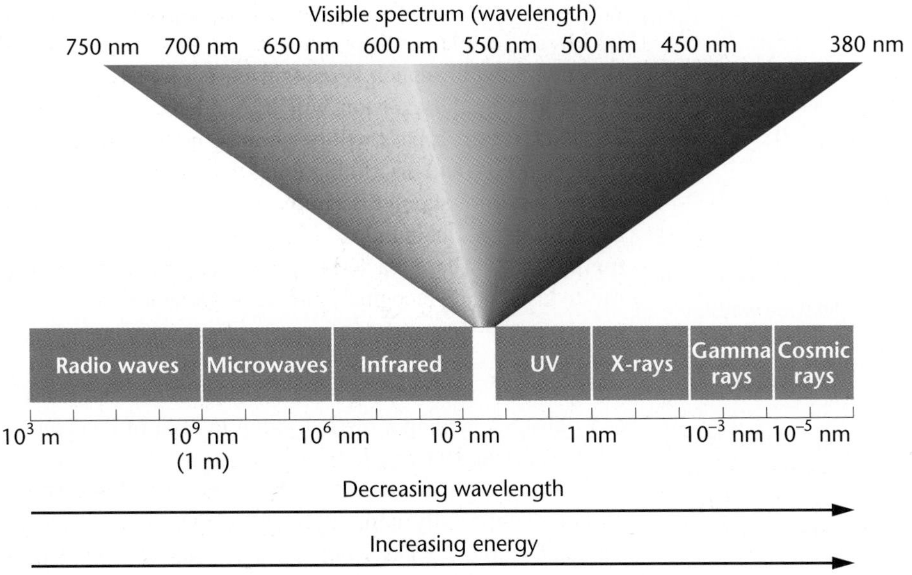

FIGURE 14–7 The regions of the electromagnetic spectrum and their associated wavelengths.

lation introduces contortions in the DNA helix that may cause deletions and insertions during DNA replication or repair, leading to frameshift mutations.

Ultraviolet Light

All electromagnetic radiation consists of energetic waves that we define by their different wavelengths (Figure 14–7). The full range of wavelengths is referred to as the **electromagnetic spectrum,** and the energy of any radiation in the spectrum varies inversely with its wavelength. Waves in the range of visible light and longer are benign when they interact with most organic molecules. However, waves of shorter length than visible light, being inherently more energetic, have the potential to disrupt organic molecules. As we know, purines and pyrimidines absorb **ultraviolet (UV) radiation** most intensely at a wavelength of about 260 nm. One major effect of UV radiation on DNA is the creation of **pyrimidine dimers**—chemical species consisting of two identical pyrimidines—particularly ones consisting of two thymine residues (Figure 14–8). The dimers distort the DNA conformation and inhibit normal replication. As a result, errors can be introduced in the base sequence of DNA during replication. When UV-induced dimerization is extensive, it is responsible (at least in part) for the killing effects of UV radiation on cells.

Ionizing Radiation

As noted above, the energy of radiation varies inversely with wavelength. Therefore, **X rays, gamma rays,** and **cosmic rays** are more energetic than UV radiation (Figure 14–7). As a result, they penetrate deeply into tissues, causing ionization of the molecules encountered along the way.

As X rays penetrate cells, electrons are ejected from the atoms of molecules encountered by the radiation. Thus, stable molecules and atoms are transformed into **free radicals**—chemical species containing one or more unpaired electrons. Free radicals can directly or indirectly affect the genetic material, altering purines and pyrimidines in DNA and resulting in point mutations. **Ionizing radiation,** radiation energetic enough to produce ions and other high-energy particles, is also capable of breaking phosphodiester bonds, disrupting the integrity of chromosomes, and producing a variety of chromosomal aberrations, such as deletions, translocations, and chromosomal fragmentation.

Figure 14–9 shows a graph of the percentage of induced X-linked recessive lethal mutations versus the dose of X rays administered. There is a linear relationship between X-ray dose and the induction of mutation; for each doubling of the dose, twice as many mutations are induced. Because the line intersects near the zero axis, this graph suggests that even very small doses of radiation are mutagenic.

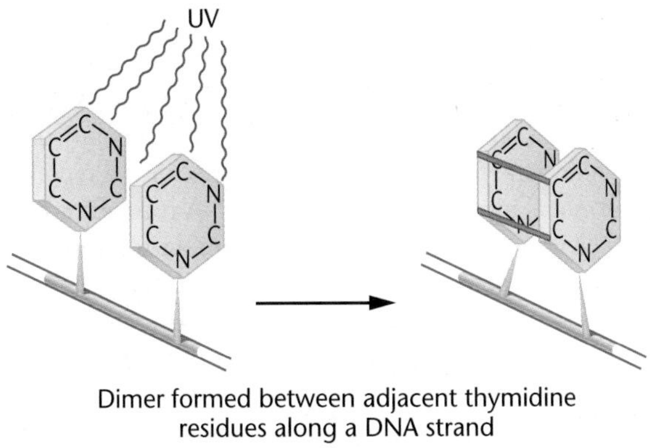

FIGURE 14–8 Induction of a thymine dimer by UV radiation, leading to distortion of the DNA. The covalent crosslinks occur between the atoms of the pyrimidine ring.

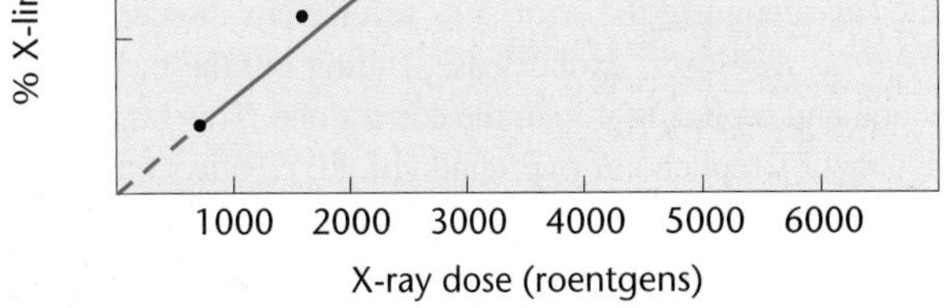

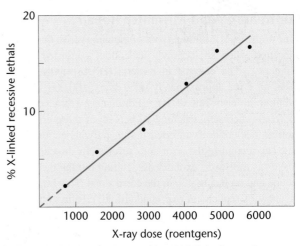

FIGURE 14–9 Plot of the percentage of X-linked recessive mutations induced in *Drosophila* by increasing doses of X rays. If extrapolated, the graph intersects the zero axis as shown by the dashed line.

ESSENTIAL POINT

Mutations can be induced by many types of chemicals and radiation. These agents can damage both DNA bases and the sugar-phosphate backbone of DNA molecules.

NOW SOLVE THIS

Problem 19 on page 306 describes the cancer drug melphalan and its effects on DNA. You are asked to describe ways in which the chemotherapy drug melphalan kills cancer cells and how these cells might repair the DNA damage caused by this drug.

Hint: To answer this question, consider the effect of the alkylation of guanine on base pairing during DNA replication. In Section 14.4, you will learn about the ways in which cells repair the types of mutations introduced by alkylating agents.

14.4 Organisms Use DNA Repair Systems to Counteract Mutations

Living systems have evolved a variety of elaborate repair systems that counteract both spontaneous and induced DNA damage. These **DNA repair** systems are absolutely essential to the maintenance of the genetic integrity of organisms and, as such, to the survival of organisms on Earth. The balance between mutation and repair results in the observed mutation rates of individual genes and organisms. Of foremost interest in humans is the ability of these systems to counteract genetic damage that would otherwise result in genetic diseases and cancer. The link between defective DNA repair and cancer susceptibility is described in Chapter 16.

We now embark on a review of some systems of DNA repair. Since the field is expanding rapidly, our goal here is merely to survey the major approaches that organisms use to counteract genetic damage.

Proofreading and Mismatch Repair

Some of the most common types of mutations arise during DNA replication when an incorrect nucleotide is inserted by DNA polymerase. The enzyme in bacteria (**DNA polymerase III**) makes an error approximately once every 100,000 insertions, leading to an error rate of 10^{-5}. Fortunately, DNA polymerase proofreads each step, catching 99 percent of those errors. If an incorrect nucleotide is inserted during polymerization, the enzyme can recognize the error and "reverse" its direction. It then behaves as a 3′ to 5′ exonuclease, cutting out the incorrect nucleotide and replacing it with the correct one. This improves the efficiency of replication one hundredfold, creating only 1 mismatch in every 10^7 insertions, for a final error rate of 10^{-7}.

WEB TUTORIAL DNA REPAIR: PROOFREADING, MISMATCH, AND SOS

To cope with those errors that remain after proofreading, another mechanism, called **mismatch repair,** may be activated. During mismatch repair, the mismatches are detected, the incorrect nucleotide is removed, and the correct nucleotide is inserted in its place. But how does the repair system recognize which nucleotide is correct (on the template strand) and which nucleotide is incorrect (on the newly synthesized strand)? If the mismatch is recognized but no such discrimination occurs, the excision will be random, and the strand bearing the correct base will be clipped out 50 percent of the time. Hence, strand discrimination is a critical step.

The process of strand discrimination has been elucidated in some bacteria, including *E. coli,* and is based on **DNA methylation.** These bacteria contain an enzyme, **adenine methylase,** which recognizes the DNA sequence

5′—GATC—3′
3′—CTAG—5′

as a substrate, adding a methyl group to each of the adenine residues during DNA replication.

Following replication, the newly synthesized DNA strand remains temporarily unmethylated, as the adenine methylase lags behind the DNA polymerase. Prior to methylation, the repair enzyme recognizes the mismatch and binds to the unmethylated (newly synthesized) DNA strand. An **endonuclease** enzyme creates a nick in the backbone of the unmethylated DNA strand, either 5′ or 3′ to the mismatch. An **exonuclease** unwinds and degrades the nicked DNA strand, until the region of the mismatch is reached. Finally, DNA polymerase fills in the gap created by the exonuclease, using the correct DNA strand as a template. DNA ligase then seals the gap.

A series of *E. coli* gene products, Mut H, L, and S, are involved in mismatch repair. Mutations in any of these genes result in bacterial strains deficient in mismatch repair. While the preceding mechanism is based on studies of *E. coli,* similar mechanisms involving homologous proteins exist in yeast and in mammals.

Postreplication Repair and the SOS Repair System

Another DNA repair system, called **postreplication repair,** responds *after* damaged DNA has escaped repair and has failed to be completely replicated. As illustrated in Figure 14–10, when DNA bearing a lesion of some sort (such as a pyrimidine dimer) is being replicated, DNA polymerase may stall at the lesion and then skip over it, leaving an unreplicated gap on the newly synthesized strand. To correct the gap, the RecA protein directs a recombinational exchange with the corresponding region on the undamaged parental strand of the same polarity (the "donor" strand). When the undamaged segment of the donor strand DNA replaces the gapped segment, a gap is created on the donor strand. The gap can be filled by repair synthesis as replication proceeds. Because a recombinational event is involved in this type of DNA repair, it is considered to be a form of **homologous recombination repair.**

Still another repair pathway, the *E. coli* **SOS repair system,** also responds to damaged DNA, but in a different way. In the presence of a large number of unrepaired DNA mismatches and gaps, bacteria can induce the expression of about 20 genes (including *lexA, recA,* and *uvr*) whose products allow DNA replication to occur even in the presence of these lesions. This type of repair is a last resort to minimize DNA damage, hence its name. During SOS repair, DNA synthesis

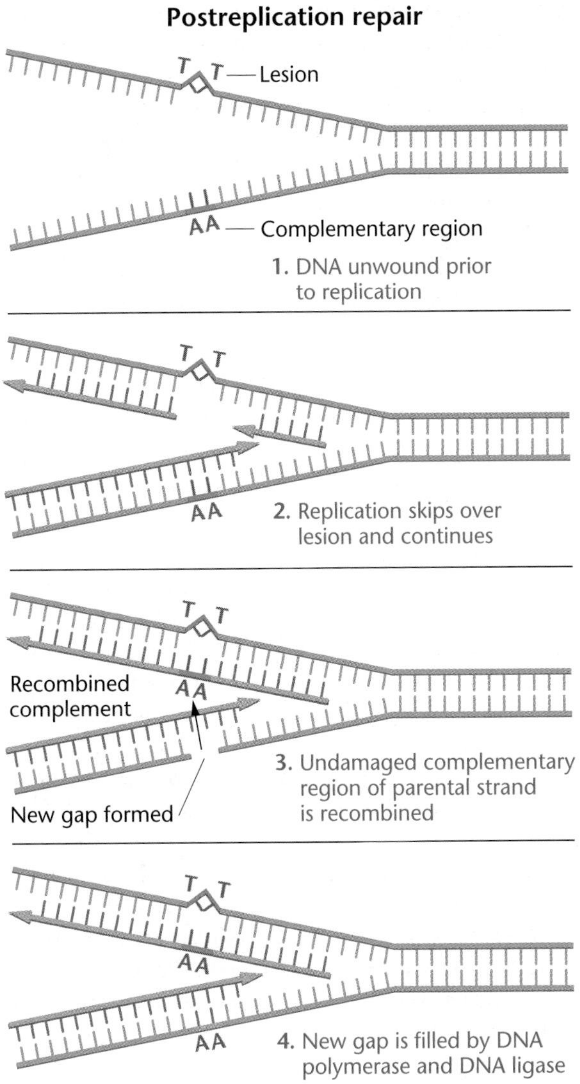

FIGURE 14–10 Postreplication repair occurs if DNA replication has skipped over a lesion such as a thymine dimer. Through the process of recombination, the correct complementary sequence is recruited from the parental strand and inserted into the gap opposite the lesion. The new gap is filled by DNA polymerase and DNA ligase.

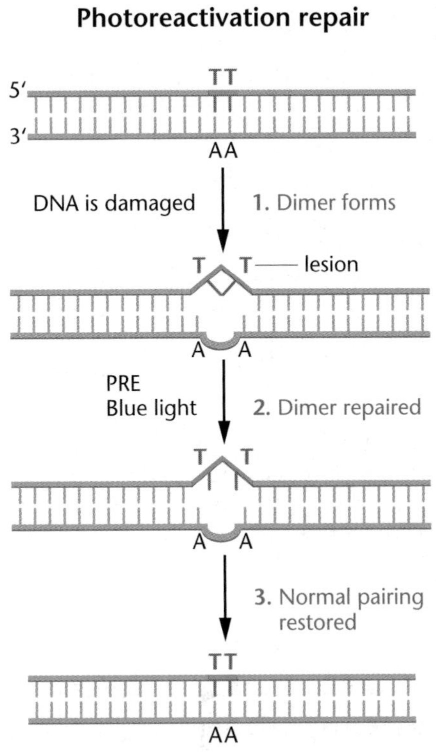

FIGURE 14–11 Damaged DNA repaired by photoreactivation repair. The bond creating the thymine dimer is cleaved by the photoreactivation enzyme (PRE), which must be activated by blue light in the visible spectrum.

becomes error-prone, inserting random and possibly incorrect nucleotides in places that would normally stall DNA replication. As a result, SOS repair itself becomes mutagenic—although it may allow the cell to survive DNA damage that would otherwise kill it.

Photoreactivation Repair: Reversal of UV Damage

As illustrated in Figure 14–8, UV light is mutagenic as a result of the creation of pyrimidine dimers. UV-induced damage to *E. coli* DNA can be partially reversed if, following irradiation, the cells are exposed briefly to light in the blue range of the visible spectrum. The process is dependent on the activity of a protein called **photoreactivation enzyme (PRE).** The enzyme's mode of action is to cleave the bonds between thymine dimers, thus directly reversing the effect of UV radiation on DNA (Figure 14–11). Although the enzyme will associate with a dimer in the dark, it must absorb a photon of light to cleave the dimer. In spite of its ability to reduce the number of UV-induced mutations, **photoreactivation repair** is not absolutely essential in *E. coli*; we know this because a mutation creating a null allele in the gene coding for PRE is not lethal. Nonetheless, the enzyme is detectable in many organisms, including bacteria, fungi, plants, and some vertebrates—although not in humans. Humans and other organisms that lack photoreactivation repair must rely on other repair mechanisms to reverse the effects of UV radiation.

Base and Nucleotide Excision Repair

A number of light-independent DNA repair systems exist in all prokaryotes and eukaryotes. The basic mechanisms involved in these types of repair—collectively referred to as **excision repair** or cut-and-paste mechanisms—consist of the following three steps.

WEB TUTORIAL
DNA REPAIR: PHOTOREACTIVATION AND EXCISION

1. The distortion or error present on one of the two strands of the DNA helix is recognized and enzymatically clipped out by an endonuclease. Excisions in the phosphodiester backbone usually include a number of nucleotides adjacent to the error as well, leaving a gap on one strand of the helix.
2. A DNA polymerase fills in the gap by inserting nucleotides complementary to those on the intact strand, which it uses as a replicative template. The enzyme adds these nucleotides to the free 3′-OH end of the clipped DNA. In *E. coli,* this step is usually performed by DNA polymerase I.
3. DNA ligase seals the final "nick" that remains at the 3′-OH end of the last nucleotide inserted, closing the gap.

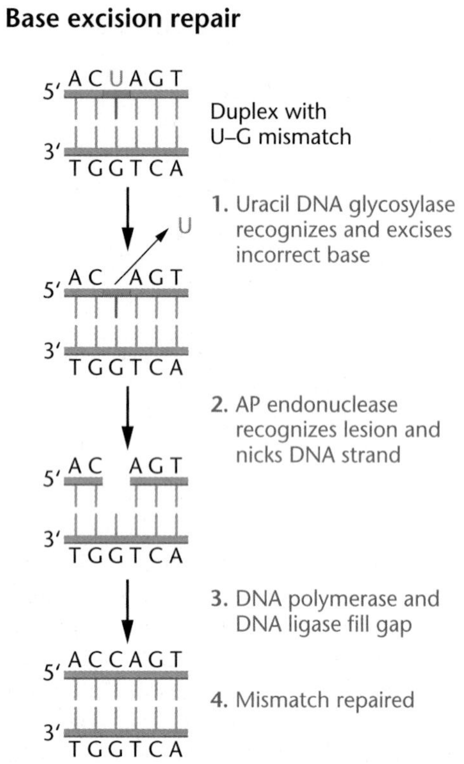

FIGURE 14–12 Base excision repair (BER) accomplished by uracil DNA glycosylase, AP endonuclease, DNA polymerase, and DNA ligase. Uracil is recognized as a noncomplementary base, excised, and replaced with the complementary base (C).

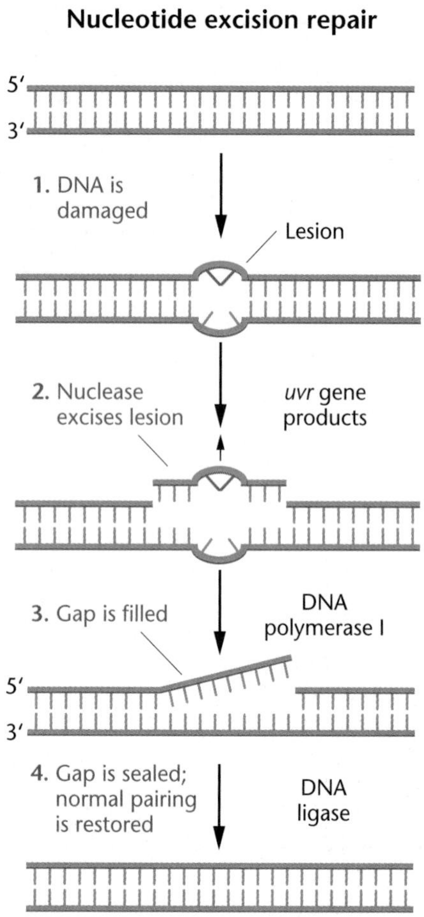

FIGURE 14–13 Nucleotide excision repair (NER) of a UV-induced thymine dimer. During repair, 13 nucleotides are excised in prokaryotes, and 28 nucleotides are excised in eukaryotes.

There are two types of excision repair: base excision repair and nucleotide excision repair. **Base excision repair (BER)** corrects DNA that contains a damaged DNA base. The first step in the BER pathway in *E. coli* involves the recognition of the altered base by an enzyme called **DNA glycosylase.** There are a number of DNA glycosylases, each of which recognizes a specific base (Figure 14–12). For example, the enzyme uracil DNA glycosylase recognizes the presence of uracil in DNA. DNA glycosylases first cut the glycosidic bond between the base and the sugar, creating an **apyrimidinic** or **apurinic site.** The sugar with the missing base is then recognized by an enzyme called **AP endonuclease.** The AP endonuclease makes a cut in the phosphodiester backbone at the apyrimidinic or apurinic site. Endonucleases then remove the deoxyribose sugar and the gap is filled by DNA polymerase and DNA ligase.

Although much has been learned about the mechanisms of BER in *E. coli,* BER systems have also been detected in eukaryotes from yeast to humans. Experimental evidence shows that both mouse and human cells that are defective in BER activity are hypersensitive to the killing effects of gamma rays and oxidizing agents.

Nucleotide excision repair (NER) pathways repair "bulky" lesions in DNA that alter or distort the double helix, such as the UV-induced pyrimidine dimers discussed previously. The NER pathway (Figure 14–13) was first discovered in *E. coli* by Paul Howard-Flanders and coworkers, who isolated several independent mutants that are sensitive to UV radiation. One group of genes was designated *uvr* (ultraviolet repair) and included the *uvrA, uvrB,* and *uvrC* mutations. In the NER pathway, the *uvr* gene products are involved in recognizing and clipping out lesions in the DNA. Usually, a specific number of nucleotides is clipped out around both sides of the lesion. In *E. coli,* usually a total of 13 nucleotides are removed, including the lesion. The repair is then completed by DNA polymerase I and DNA ligase, in a manner similar to that occurring in BER. The undamaged strand opposite the lesion is used as a template for the replication, resulting in repair.

Nucleotide Excision Repair and Xeroderma Pigmentosum in Humans

The mechanism of NER in eukaryotes is much more complicated than that in prokaryotes and involves many more proteins, encoded by about 30 genes. Much of what is known about the system in humans has come from detailed studies of individuals with **xeroderma pigmentosum (XP),** a rare recessive genetic disorder that predisposes individuals to severe skin abnormalities and skin cancers, particularly when they are exposed to UV radiation in sunlight. The condition is severe and may be lethal, although early detection and protection from sunlight can arrest it (Figure 14–14).

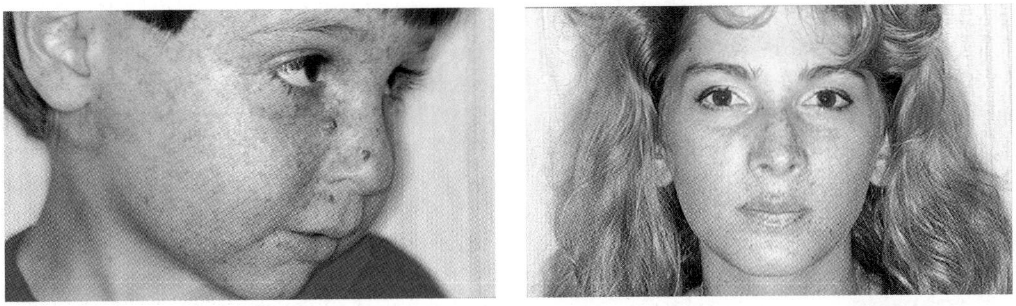

FIGURE 14–14 Two individuals with xeroderma pigmentosum. The 4-year-old boy on the left shows marked skin lesions induced by sunlight. Mottled redness (erythema) and irregular pigment changes in response to cellular injury are apparent. Two nodular cancers are present on his nose. The 18-year-old girl on the right has been carefully protected from sunlight since her diagnosis of xeroderma pigmentosum in infancy. Several cancers have been removed, and she has worked as a successful model.

The repair of UV-induced lesions in XP has been investigated *in vitro,* using human fibroblast cell cultures derived from normal individuals and those with XP. (Fibroblasts are undifferentiated connective tissue cells.) The results of these studies suggest that the XP phenotype is caused by defects in NER pathways and by mutations in more than one gene.

In 1968, James Cleaver showed that cells from XP patients were deficient in DNA synthesis other than that occurring during chromosome replication—a phenomenon known as **unscheduled DNA synthesis.** Unscheduled DNA synthesis is elicited in normal cells by UV radiation. Because this type of synthesis is thought to represent the activity of DNA polymerization during NER, the lack of unscheduled DNA synthesis in XP patients suggested that XP may be a deficiency in NER.

The involvement of multiple genes in NER and XP was further investigated by studies using **somatic cell hybridization.** Fibroblast cells from any two unrelated XP patients, when grown together in tissue culture, can fuse together, forming heterokaryons. A **heterokaryon** is a single cell with two nuclei from different organisms but a common cytoplasm. NER in the heterokaryon can be measured by the level of unscheduled DNA synthesis. If the mutation in each of the two XP cells occurs in the same gene, the heterokaryon, like the cells that fused to form it, will still be unable to undergo NER. This is because there is no normal copy of the relevant gene present in the heterokaryon. However, if NER does occur in the heterokaryon, the mutations in the two XP cells must have been present in two different genes. Hence, the two mutants are said to demonstrate **complementation,** a concept also discussed in Chapter 4. Complementation occurs because the heterokaryon has at least one normal copy of each gene in the fused cell. By fusing XP cells from a large number of XP patients, researchers were able to determine how many genes contribute to the XP phenotype.

Based on these and other studies, XP patients have been divided into seven complementation groups, indicating that at least seven different genes are involved in excision repair in humans. A gene representing each of these complementation groups, *XPA* to *XPG* (*X*eroderma *P*igmentosum gene *A* to *G*), have now been identified, and a homologous gene for each has been identified in yeast. Approximately 20 percent of XP patients do not fall into any of the seven complementation groups. These patients manifest similar symptoms, but their fibroblasts do not demonstrate defective NER. There is some evidence that these XP cells are less efficient in normal DNA replication.

As a result of the study of defective genes in XP, a great deal is now known about how NER counteracts DNA damage in normal cells. The first step in humans is recognition of the damaged DNA by proteins encoded by the *XPC, XPE,* and *XPA* genes. These proteins then recruit the remainder of the repair proteins to the site of DNA damage. The *XPB* and *XPD* genes encode helicases, and the *XPF* and *XPG* genes encode nucleases. The excision repair complex containing these and other factors is responsible for the excision of a 28-nucleotide-long fragment from the DNA strand that contains the lesion.

Double-Strand Break Repair in Eukaryotes

Thus far, we have discussed repair pathways that deal with damage or errors within one strand of DNA. We conclude our discussion of DNA repair by considering what happens when both strands of the DNA helix are cleaved—as a result of exposure to ionizing radiation, for example. These types of damage are extremely dangerous to cells, leading to chromosome rearrangements, cancer, or cell death. In this section, we will discuss double-strand breaks in eukaryotic cells.

Specialized forms of DNA repair, the **DNA double-strand break repair (DSB repair)** pathways, are activated and are responsible for reattaching two broken DNA strands. Recently, interest in DSB repair has grown because defects in these pathways are associated with X-ray hypersensitivity and immune deficiency. Such defects may also underlie familial disposition to breast and ovarian cancer. Several human disease syndromes, such as Fanconi's anemia and ataxia telangiectasia, result from defects in DSB repair.

One pathway involved in double-strand break repair is **homologous recombination repair.** The first step in this process involves the activity of an enzyme that recognizes the double-strand break, then digests back the 5′ ends of the

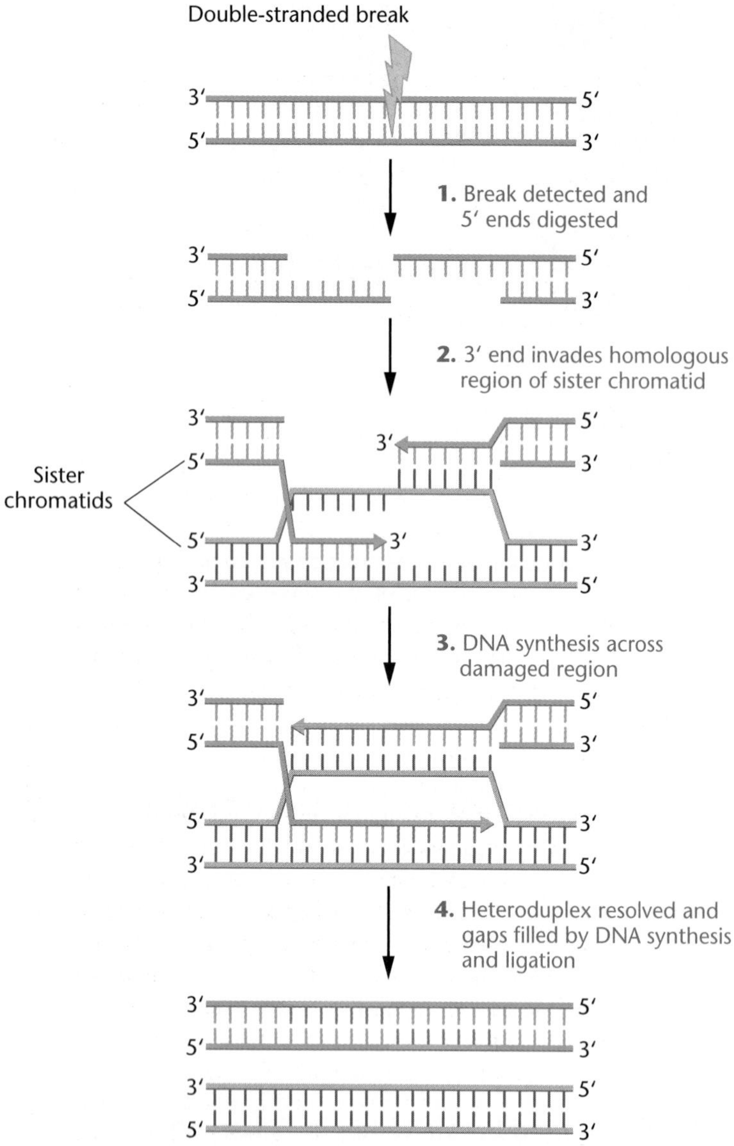
Double-stranded break
3′
5′
5′
3′
1. Break detected and 5′ ends digested
2. 3′ end invades homologous region of sister chromatid
Sister chromatids
3. DNA synthesis across damaged region
4. Heteroduplex resolved and gaps filled by DNA synthesis and ligation

FIGURE 14–15 Steps in homologous recombination repair of double-stranded breaks.

broken DNA helix, leaving overhanging 3′ ends (Figure 14–15). One 3′ overhanging end searches for a region of sequence complementarity on the sister chromatid and then invades the homologous DNA duplex, aligning the complementary sequences. Once aligned, DNA synthesis proceeds from the 3′ overhanging ends, using the undamaged DNA strands as templates. The interaction of two sister chromatids is necessary because, when both strands of one helix are broken, there is no undamaged parental DNA strand available to use as a source of the complementary template DNA sequence during repair. After DNA repair synthesis, the resulting heteroduplex molecule is resolved and the two chromatids separate, as previously discussed in Chapter 10.

DSB repair usually occurs during the late S or early G2 phase of the cell cycle, after DNA replication, a time when sister chromatids are available to be used as repair templates. Because an undamaged template is used during repair synthesis, homologous recombination repair is an accurate process.

A second pathway, called **nonhomologous end joining,** also repairs double-strand breaks. However, as the name implies, the mechanism does not recruit a homologous region of DNA during repair. This system is activated in G1, prior to DNA replication. End joining involves a complex of three proteins, including DNA-dependent protein kinase. These proteins bind to the free ends of the broken DNA, trim the ends, and ligate them back together. Because some nucleotide sequences are lost in the process of end joining, it is an error-prone repair system. In addition, if more than one chromosome suffers a double-strand break, the wrong ends could be joined together, leading to abnormal chromosome structures, such as those discussed in Chapter 6.

ESSENTIAL POINT

Organisms counteract mutations by using a range of DNA repair systems. Errors in DNA synthesis can be repaired by proofreading, mismatch repair, and postreplication repair. DNA damage can be repaired by photoreactivation repair, SOS repair, base excision repair, nucleotide excision repair, and double-strand break repair.

NOW SOLVE THIS

Problem 23 on page 306 examines the results of heterokaryon analysis of seven cell lines derived from XP patients. You are asked to determine the number of complementation groups represented by these data.

Hint: Complementation (+) results only when the two mutations being examined are in different complementation groups (genes).

14.5 The Ames Test Is Used to Assess the Mutagenicity of Compounds

There is great concern about the possible mutagenic properties of any chemical that enters the human body, whether through the skin, the digestive system, or the respiratory tract. Examples of synthetic chemicals that concern us are those found in air and water pollution, food preservatives, artificial sweeteners, herbicides, pesticides, and pharmaceutical products. Mutagenicity can be tested in various organisms, including fungi, plants, and cultured mammalian cells; however, one of the most common tests, which we describe here, uses bacteria.

The **Ames test** uses a number of different strains of the bacterium *Salmonella typhimurium* that have been selected for their ability to reveal the presence of specific types of mutations. For example, some strains are used to detect base-pair substitutions, and other strains detect various frameshift mutations. Each strain contains a mutation in one of the genes of the histidine operon. The mutant strains are unable to synthesize histidine (*his*$^-$ strains) and therefore require histidine for

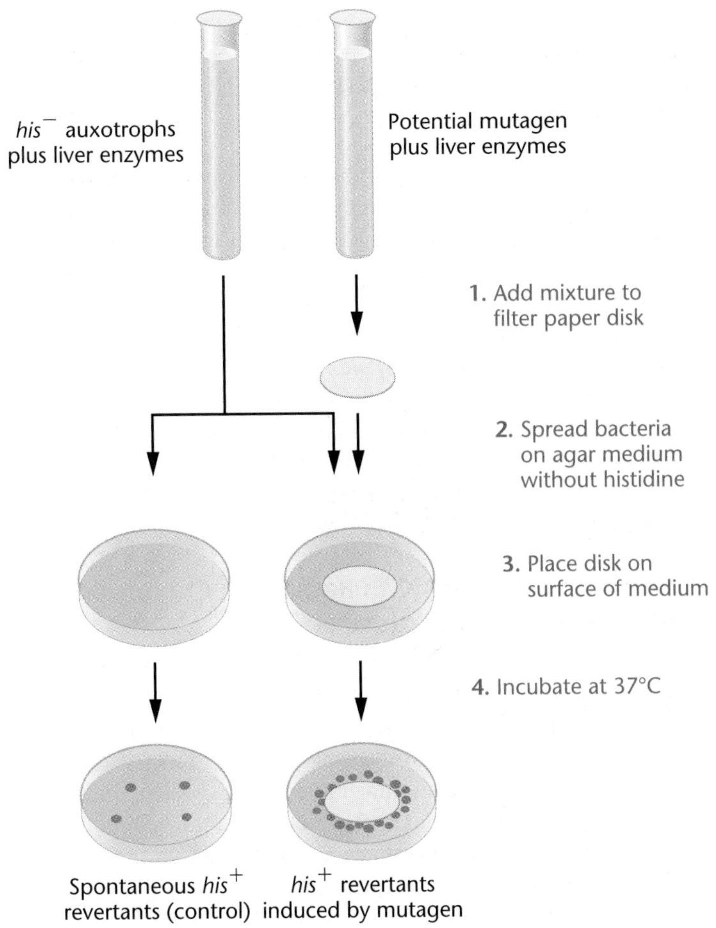

his^- auxotrophs plus liver enzymes
Potential mutagen plus liver enzymes
1. Add mixture to filter paper disk
2. Spread bacteria on agar medium without histidine
3. Place disk on surface of medium
4. Incubate at 37°C
Spontaneous his^+ revertants (control)
his^+ revertants induced by mutagen

FIGURE 14–16 The Ames test, which screens compounds for potential mutagenicity.

growth (Figure 14–16). The assay measures the frequency of reverse mutations that occur within the mutant gene, yielding wild-type bacteria (his^+ revertants). These *Salmonella* strains also have an increased sensitivity to mutagens due to the presence of mutations in genes involved in DNA damage repair and synthesis of the lipopolysaccharide barrier that coats bacteria and protects them from external substances.

Many substances entering the human body are relatively innocuous until activated metabolically, usually in the liver, to a more chemically reactive product. Thus, the Ames test includes a step in which the test compound is incubated *in vitro* in the presence of a mammalian liver extract. Alternatively, test compounds may be injected into a mouse where they are modified by liver enzymes and then recovered for use in the Ames test.

In the initial use of Ames testing in the 1970s, a large number of known **carcinogens,** or cancer-causing agents, were examined, and more than 80 percent of these were shown to be strong mutagens. This is not surprising, as the transformation of cells to the malignant state occurs as a result of mutations. Although a positive response in the Ames test does not prove that a compound is carcinogenic, the Ames test is useful as a preliminary screening device. The Ames test is used extensively during the development of industrial and pharmaceutical chemical compounds.

ESSENTIAL POINT

The Ames test allows scientists to estimate the mutagenicity and cancer-causing potential of chemical agents by following the rate of mutation in specific strains of bacteria.

14.6 DNA Sequencing Has Enhanced Our Understanding of Mutations in Humans

Historically, scientists have learned about mutations by analyzing the amino acid sequences of proteins isolated from individuals with certain diseases or traits. These amino acid sequences are a reflection of changes in the triplet codons following substitution, insertion, or deletion of one or more nucleotides in the DNA sequences of protein-coding genes. As our ability to analyze DNA more directly increases, we are able to examine the actual nucleotide sequence of genes and surrounding DNA regions, and to gain greater insights into the nature of mutations. In this section, we describe several human genetic traits and what we have learned about the mutations responsible for them.

WEB TUTORIAL CHEMICAL MUTAGENESIS: THE AMES TEST

Mutations Creating the ABO Blood Groups

The **ABO blood group system** is based on a series of antigenic determinants found on erythrocytes and other cells, particularly epithelial cells. As we discussed in Chapter 4, three alleles of the *I* gene encode the **glycosyltransferase** enzyme. The H substance is modified to either the A or B antigen as a result of the activity of this enzyme, depending on whether the enzyme was encoded by the I^A or I^B allele, respectively. Failure to modify the H substance results from the presence of the null, I^O, allele.

When the DNA sequences of the I^A and I^B alleles are compared, four consistent single-nucleotide substitutions are found. The resulting changes in the amino acid sequence of the glycosyltransferase gene product lead to altered gene product function and, consequently, to different modifications of the H substance.

The I^O allele situation is unique and interesting. Individuals homozygous for this allele have type O blood, lack glycosyltransferase activity, and fail to modify the H substance. Analysis of the DNA of the I^O allele shows one consistent change that is unique compared with the sequences of the other alleles—the deletion of a single nucleotide early in the coding sequence, causing a frameshift mutation. A complete messenger RNA is transcribed, but at translation, the reading frame shifts at the point of the deletion and continues out of frame for about 100 nucleotides before a stop codon is encountered. At this point, the glycosyltransferase polypeptide chain terminates prematurely, resulting in a nonfunctional product.

These findings provide a direct molecular explanation of the ABO allele system and the basis for the biosynthesis of the corresponding antigens. The molecular basis for the antigenic phenotypes is clearly a matter of base substitutions and frameshift mutations within the gene encoding the glycosyltransferase enzyme.

Mutations Associated with Muscular Dystrophies

Muscular dystrophies are genetic diseases characterized by progressive muscle weakness and degeneration. There are many types of muscular dystrophy, and they differ in their severities, onsets, and genetic causes. Two related forms of muscular dystrophy—**Duchenne muscular dystrophy (DMD)** and **Becker muscular dystrophy (BMD)**—are recessive, X-linked conditions. DMD is the more severe of the two diseases, with a rapid progression of muscle degeneration and involvement of the heart and lungs. Males with DMD usually lose the ability to walk by the age of 12 and may die in their early 20s. The incidence of 1 in 3500 live male births makes DMD one of the most common life-shortening hereditary diseases. In contrast, BMD does not involve the heart or lungs and progresses slowly, from adolescence to the age of 50 or more.

The gene responsible for DMD and BMD—the *dystrophin* gene—is unusually large, consisting of about 2.5 million base pairs. In normal (unaffected) individuals, transcription and subsequent processing of the *dystrophin* initial transcript results in a messenger RNA containing only about 14,000 bases (14 kb). It is translated into the protein **dystrophin,** which consists of 3685 amino acids.

In recent years, geneticists have determined the molecular basis of mutations leading to BMD and DMD. Studies have shown that more than 70 percent of mutations in the *dystrophin* gene that lead to DMD and BMD are deletions and insertions. The remainder are a collection of small insertions, and point mutations. With few exceptions, DMD mutations change the reading frame of the *dystrophin* gene, whereas BMD mutations usually do not. The majority of DMD frameshift mutations lead to premature termination of translation. This, in turn, leads to degradation of the improperly translated *dystrophin* transcript as well as the truncated protein. In contrast, the majority of BMD gene mutations alter the internal sequence of the *dystrophin* transcript and protein, but do not alter the translation reading frame. As a result, a modified but somewhat functional dystrophin protein is produced, preventing the severe consequences of DMD.

ESSENTIAL POINT

The human ABO blood groups arise from point mutations or frameshift mutations within the gene encoding the enzyme that modifies the H substance. Muscular dystrophies are caused by insertions, deletions, and point mutations.

NOW SOLVE THIS

Problem 20 on page 306 describes muscular dystrophies as some of the most common inherited diseases. You are asked to speculate on why the *dystrophin* gene appears to suffer a large number of mutations.

Hint: In answering this question, you may want to think about the size of the gene and the organization of its introns and exons.

14.7 Geneticists Use Mutations to Identify Genes and Study Gene Function

In order to dissect the genes and processes that regulate biological functions, geneticists study the effects of mutations in **model organisms.** A good model organism must be easy to grow, have a short generation time, produce abundant progeny, and be readily mutagenized and crossed. The most extensively used model organisms are bacteria (*E. coli*), budding yeasts (*Saccharomyces cerevisiae*), fruit flies (*Drosophila melanogaster*), nematodes (*Caenorhabditis elegans*), mustard plants (*Arabidopsis thaliana*), and mice (*Mus musculus*).

In this section, we will describe some of the methods that geneticists use to generate and detect gene mutations, as the first step toward a full genetic analysis.

Inducing Mutations with Radiation, Chemicals, and Transposon Insertion

Geneticists sometimes begin their analyses of genes by examining spontaneous, naturally occurring mutations. However, as mutations are generally rare in nature, researchers need to induce mutations in model organisms in order to increase the chances of detecting a relevant mutant. The goal of mutagenesis is to create one mutation at random in the genome of each individual in the experimental population, so that one gene product is disrupted in each individual, leaving the rest of the genome wild type.

During genetic analyses, researchers use a wide range of different mutagens, depending on the type of mutation desired. For example, ionizing radiation can be used to create chromosome breaks, deletions, and translocations. Though useful for some types of studies, such mutations often have severe effects on the phenotype, making further genetic analysis difficult. In contrast, chemicals such as ethyl methane sulfonate (EMS) and nitrosoguanidine cause single base-pair changes and small deletions and insertions. With these mutagens, a range of mild to severe mutant phenotypes can be generated. It is more likely that single base-pair mutations will result in the creation of **conditional mutations,** such as temperature-sensitive mutations, which are particularly useful for the study of essential gene functions. Geneticists also use transposons to create mutations. If a transposon, such as a *Drosophila P* element, inserts into a gene's coding or regulatory regions, it can disrupt the gene's function.

Because humans are obviously not suitable experimental organisms (for both practical and ethical reasons), techniques developed for the detection of mutations in organisms such as bacteria and fruit flies are not available for human genetics research. To determine the mutational basis for any human characteristic or disorder, geneticists may analyze a **pedigree** that traces the family history as far back as possible. If a trait is shown to be inherited, the pedigree may also be useful in determining whether the mutant allele is behaving as a dominant or a recessive mutation and whether it is X-linked or autosomal. Pedigree analysis is described in more detail in Chapter 3.

More recently, genomics and reverse genetic techniques have expanded the methods available for studying human mutations. Once a mutant gene has been identified and cloned, sequence analysis of mutant and normal genes from affected and unaffected individuals may reveal the molecular basis of the disease or mutant phenotype. In addition, knowledge of the mutant gene sequence may open the way to developing specific genetic tests and gene-based therapeutics. These new genomics-based methods are described in Chapters 18 and 19.

Screening and Selecting for Mutations

The next step in a genetic analysis involves detecting those individuals, within the mutagenized population, who display mutations in the gene or genes affecting the phenotype of interest. The most frequently used method to detect mutants is a **genetic screen.** A genetic screen may involve the visual or biochemical examination of large numbers of mutagenized organisms. For example, Mendel formed the scientific basis for transmission genetics by screening thousands of individual pea plants for visible phenotypic features. Similarly, scientists used modern biochemical assays of the proteins in maize endosperm, in order to detect the *opaque-2* mutant strain. This strain contains high levels of lysine, which improves the nutritional value of the plant.

It is easy to see how dominant mutations can be detected during a genetic screen, in either haploid or diploid organisms. As long as the dominant mutation is not lethal at an early stage of development, the phenotype will be visible immediately in any organism that bears a dominant mutation in the relevant gene. The mutant can then be crossed, and heterozygous or homozygous stocks can be maintained.

Most mutations, however, result in a loss of gene function, and most of these loss-of-function mutations are recessive. Organisms that have a haploid phase in their life cycle, such as yeast, have a significant advantage for recessive mutant detection, as the mutant phenotype will be immediately evident in the mutated haploid individual. The mutation can then be crossed and analyzed in diploid phases of the life cycle.

Diploid organisms that are heterozygous for a recessive allele will not show the mutant phenotype. To deal with this situation, geneticists have devised some intricate strategies to detect and recover recessive mutations in diploid organisms. An example of one of these methods is the **attached-X** method for detecting recessive mutations on the X chromosome of *Drosophila.* This method was developed by Hermann Muller in the 1920s, in order to demonstrate that X rays cause mutations in *Drosophila.*

Attached-X females have two X chromosomes attached to a single centromere and one Y chromosome, in addition to the normal diploid complement of autosomes. When attached-X females are mated to males with normal sex chromosomes (XY), four types of progeny result: triplo-X females that die, viable attached-X females, YY males that also die, and viable XY males. Figure 14–17 shows how P_1 males that have been treated with a mutagenic agent produce F_1 male offspring that express any recessive mutation induced on the X chromosome. In this screening procedure, the mutant phenotype is expressed in the first generation.

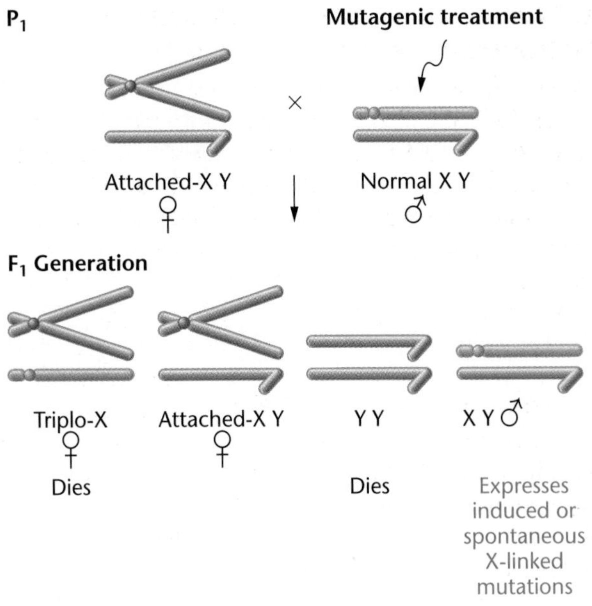

FIGURE 14–17 The attached-X method for detection of induced morphological mutations in *Drosophila.*

Geneticists have devised similar techniques to detect recessive mutations on autosomes of *Drosophila* and other diploid organisms. These methods involve crossing mutagenized individuals with special strains bearing a number of visible and lethal gene markers or chromosomes containing specific deletions. After three or more generations, the recessive lethal mutation may be revealed, but the mutant strain can be maintained in a heterozygous state. Although these techniques are cumbersome to perform, they are fairly efficient. Many of the developmental mutations in *Drosophila* that are discussed in Chapter 20 were identified using these techniques.

ESSENTIAL POINT

A range of genetic and biochemical techniques are used to induce and analyze mutations in model organisms. In contrast, the study of human mutations may begin with pedigree analysis or modern genomics methods.

14.8 Transposable Elements Move within the Genome and May Create Mutations

Transposable elements, also known as **transposons** or "jumping genes," can move or transpose within and between chromosomes, inserting themselves into various locations within the genome.

Transposons are present in the genomes of all organisms from bacteria to humans. Not only are they ubiquitous, but they also comprise large portions of some eukaryotic genomes. For example, almost 50 percent of the human genome is derived from transposable elements. Some organisms with unusually large genomes, such as salamanders and barley, contain hundreds of thousands of copies of various types of transposable elements. Although the function of these elements is unknown, data from human genome sequencing suggest that some genes may have evolved from transposons and that transposons may help to modify and reshape the genome. Transposable elements are also valuable tools in genetic research. Geneticists harness transposons as mutagens, as cloning tags, and as vehicles for introducing foreign DNA into model organisms.

In this section, we discuss transposable elements as naturally occurring mutagens. The movement of transposons from one place in the genome to another has the capacity to disrupt genes and cause mutations, as well as to create chromosomal damage such as double-strand breaks.

Insertion Sequences

There are two types of transposable elements in bacteria: insertion sequences and bacterial transposons. **Insertion sequences (IS elements)** can move from one location to another and, if they insert into a gene or gene-regulatory region, may cause mutations.

IS elements were first identified during analyses of mutations in the *gal* operon of *E. coli.* Researchers discovered that certain mutations in this operon were due to the presence of several hundred base pairs of extra DNA inserted into the beginning of the operon. Surprisingly, the segment of mutagenic DNA could spontaneously excise from this location, restoring wild-type function to the *gal* operon. Subsequent research revealed that several other DNA elements could behave in a similar fashion, inserting into bacterial chromosomes and affecting gene function.

IS elements are relatively short, not exceeding 2000 bp (2 kb). The first insertion sequence to be characterized in *E. coli,* IS1, is about 800 bp long. Other IS elements such as IS2, 3, 4, and 5 are about 1250 to 1400 bp in length. IS elements are present in multiple copies in bacterial genomes. For example, the *E. coli* chromosome contains five to eight copies of IS1, five copies each of IS2 and IS3, as well as copies of IS elements on plasmids such as F factors.

All IS elements contain two features that are essential for their movement. First, they contain a gene that encodes an enzyme called **transposase.** This enzyme is responsible for making staggered cuts in chromosomal DNA, into which the IS element can insert. Second, the ends of IS elements contain **inverted terminal repeats (ITRs).** ITRs are short segments of DNA that have the same nucleotide sequence as each other but are oriented in the opposite direction (Figure 14–18).

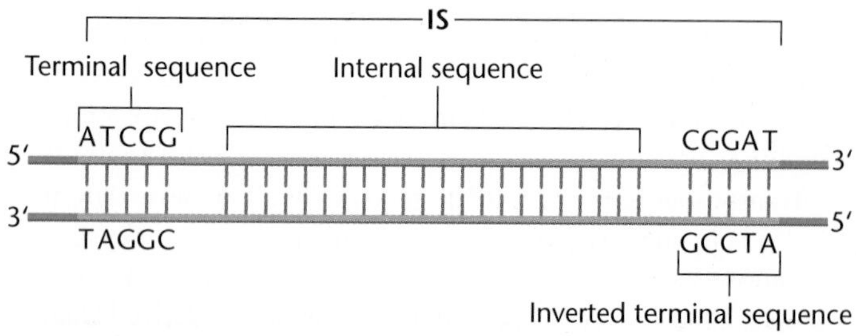

FIGURE 14–18 An insertion sequence (IS), shown in purple. The terminal sequences are perfect inverted repeats of one another.

Although Figure 14–18 shows the ITRs to consist of only a few nucleotides, IS ITRs usually contain about 20 to 40 nucleotide pairs. ITRs are essential for transposition and act as recognition sites for the binding of the transposase enzyme.

Bacterial Transposons

Bacterial transposons (**Tn elements**) are larger than IS elements and contain protein-coding genes that are unrelated to their transposition. Some Tn elements, such as Tn10, are comprised of a drug-resistance gene flanked by two IS elements present in opposite orientations. The IS elements encode the transposase enzyme that is necessary for transposition of the Tn element. Other types of Tn elements, such as Tn3, have shorter inverted repeat sequences at their ends and encode their transposase enzyme from a transposase gene located in the middle of the Tn element. Like IS elements, Tn elements are mobile in both bacterial chromosomes and in plasmids, and can cause mutations if they insert into genes or gene-regulatory regions.

Tn elements are currently of interest because they can introduce multiple drug resistance onto bacterial plasmids. These plasmids, called **R factors,** may contain many Tn elements conferring simultaneous resistance to heavy metals, antibiotics, and other drugs. These elements can move from plasmids onto bacterial chromosomes and can spread multiple drug resistance between different strains of bacteria.

The *Ac–Ds* System in Maize

About 20 years before the discovery of transposons in bacteria, Barbara McClintock discovered mobile genetic elements in corn plants (maize). She did this by analyzing the genetic behavior of two mutations, ***dissociation (Ds)*** and ***activator (Ac),*** expressed in either the endosperm or aleurone layers. She then correlated her genetic observations with cytological examinations of the maize chromosomes. Initially, McClintock determined that *Ds* was located on chromosome 9. If *Ac* was also present in the genome, *Ds* induced breakage at a point on the chromosome adjacent to its own location. If chromosome breakage occurred in somatic cells during their development, progeny cells often lost part of the broken chromosome, causing a variety of phenotypic effects.

Subsequent analysis suggested to McClintock that both *Ds* and *Ac* elements sometimes moved to new chromosomal locations. While *Ds* moved only if *Ac* was also present, *Ac* was capable of autonomous movement. Where *Ds* came to reside determined its genetic effects—that is, it might cause chromosome breakage, or it might inhibit expression of a certain gene. In cells in which *Ds* caused a gene mutation, *Ds* might move again, restoring the gene mutation to wild type.

Figure 14–19 illustrates the types of movements and effects brought about by *Ds* and *Ac* elements. In McClintock's original observation, pigment synthesis was restored in cells in which the *Ds* element jumped out of chromosome 9. McClintock concluded that the *Ds* and *Ac* genes were **mobile controlling elements.** We now commonly refer to them as transposable elements, a term coined by another great maize geneticist, Alexander Brink.

Several *Ac* and *Ds* elements have now been analyzed, and the relationship between the two elements has been clarified

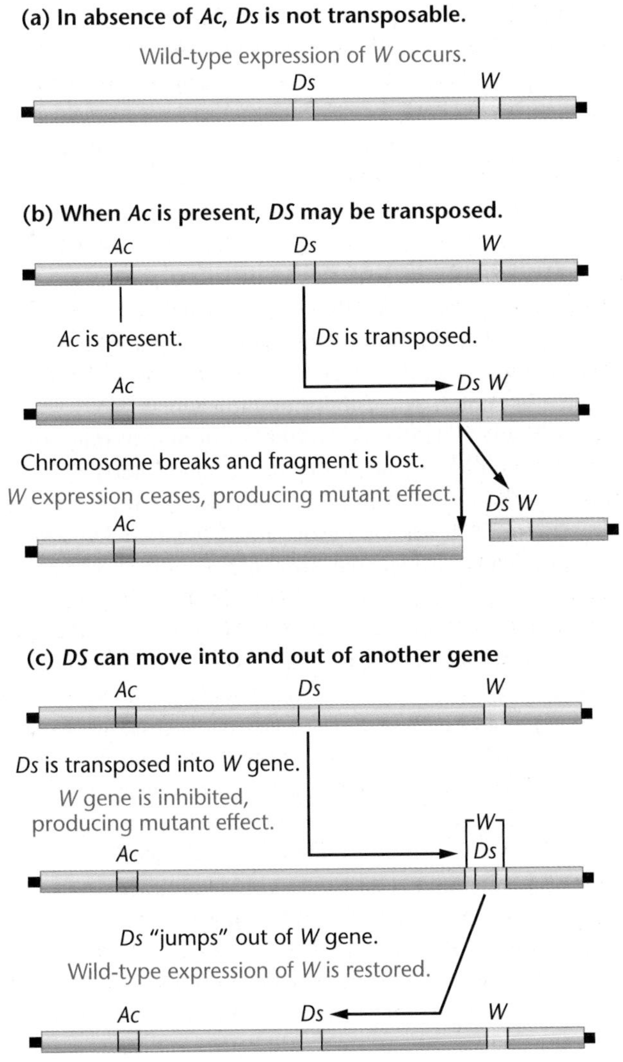

FIGURE 14–19 Effects of *Ac* and *Ds* elements on gene expression. (a) If *Ds* is present in the absence of *Ac,* there is normal expression of a distantly located hypothetical gene *W.* (b) In the presence of *Ac, Ds* may transpose to a region adjacent to *W. Ds* can induce chromosome breakage, which may lead to loss of a chromosome fragment bearing the *W* gene. (c) In the presence of *Ac, Ds* may transpose into the *W* gene, disrupting *W*-gene expression. If *Ds* subsequently transposes out of the *W* gene, *W*-gene expression may return to normal.

(Figure 14–20). The *Ac* element is 4563 nucleotides long, and its structure is strikingly similar to that of bacterial transposons. The *Ac* sequence contains two 11-base-pair imperfect ITRs, two open reading frames (ORFs), and three noncoding regions. One of the two ORFs encodes the *Ac* transposase enzyme. The first *Ds* element studied (*Ds-a*) is nearly identical to *Ac* except for a 194-bp deletion within the transposase gene. The deletion of part of the transposase gene in the *Ds-a* element explains its dependence on the *Ac* element for transposition. Several other *Ds* elements have also been sequenced, and each contains an even larger deletion within the transposase gene. In each case, however, the ITRs are retained.

Although the significance of Barbara McClintock's mobile controlling elements was not fully appreciated following her initial observations, molecular analysis has since verified her conclusions. She was awarded the Nobel Prize in Physiology or Medicine in 1983.

Copia Elements in *Drosophila*

In 1975, David Hogness and his colleagues David Finnegan, Gerald Rubin, and Michael Young identified a class of DNA elements in *Drosophila melanogaster* that they designated as ***copia.*** These elements are transcribed into "copious" amounts of RNA (hence their name). *Copia* elements are present in 10 to 100 copies in the genomes of *Drosophila* cells. Mapping studies show that they are transposable to different chromosomal locations and are dispersed throughout the genome.

Each *copia* element consists of approximately 5000 to 8000 bp of DNA, including a long **direct terminal repeat (DTR)** sequence of 267 bp at each end. Within each DTR is an inverted terminal repeat (ITR) of 17 bp (Figure 14–21, p. 302). The short ITR sequences are characteristic of *copia* elements. The DTR sequences are found in other transposons in other organisms, but they are not universal.

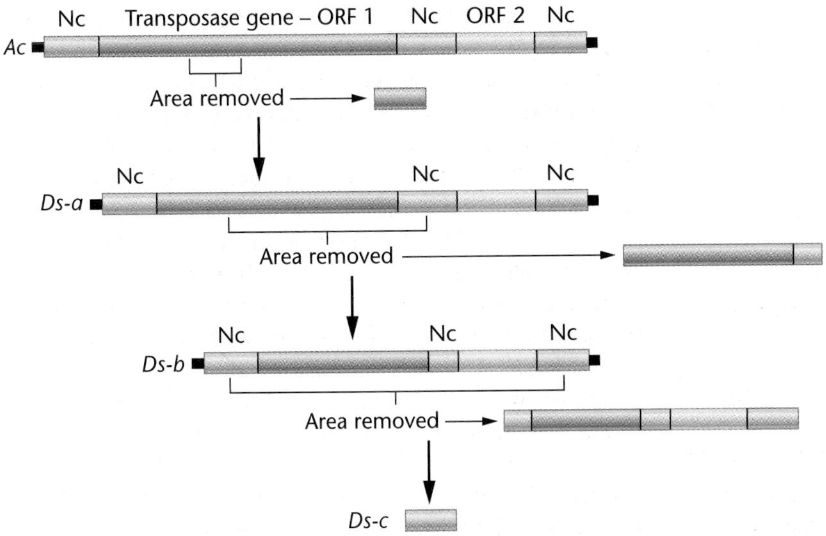

FIGURE 14–20 A comparison of the structures of an *Ac* element and three *Ds* elements. The transposase gene is in the open reading frame designated ORF 1. No function has yet been assigned to ORF 2. (Noncoding regions are designated Nc.) As this illustration shows, *Ds-a* appears to be an *Ac* element that has a small deletion in the gene encoding the transposase enzyme.

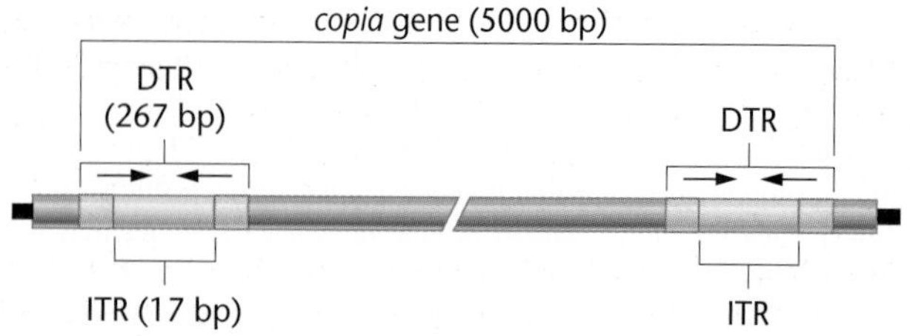

FIGURE 14–21 Structural organization of a *copia* transposable element in *Drosophila melanogaster,* showing the terminal repeats.

Insertion of *copia* is dependent on the presence of the ITR sequences and seems to occur preferentially at specific target sites in the genome. The *copia*-like elements demonstrate regulatory effects at the point of their insertion in the chromosome. Certain mutations, including those affecting eye color and segment formation, are due to *copia* insertions within genes. For example, the eye-color mutation *white-apricot,* is caused by an allele of the *white* gene which contains a *copia* element within the gene. Transposition of the *copia* element out of the *white-apricot* allele can restore the allele to wild type.

Copia elements are only one of approximately 30 families of transposable elements in *Drosophila,* each of which is present in 20 to 50 copies in the genome. Together, these families constitute about 5 percent of the *Drosophila* genome and over half of the middle repetitive DNA of this organism. One study suggests that 50 percent of all visible mutations in *Drosophila* are the result of the insertion of transposons into otherwise wild-type genes.

P Element Transposons in *Drosophila*

Perhaps the most significant *Drosophila* transposable elements are the ***P* elements.** These were discovered while studying the phenomenon of **hybrid dysgenesis,** a condition characterized by sterility, elevated mutation rates, and chromosome rearrangements in the offspring of crosses between certain strains of fruit flies. Hybrid dysgenesis is caused by high rates of *P* element transposition in the germ line, in which transposons insert themselves into or near genes, thereby causing mutations. *P* elements range from 0.5 to 2.9 kb long, with 31-bp ITRs. Full-length *P* elements encode at least two proteins, one of which is the transposase enzyme that is required for transposition, and another is a repressor protein that inhibits transposition. The transposase gene is expressed only in the germ line, accounting for the tissue specificity of *P* element transposition. Strains of flies that contain full-length *P* elements inserted into their genomes are resistant to further transpositions due to the presence of the repressor protein encoded by the *P* elements.

Mutations can arise from several kinds of insertional events. If a *P* element inserts into the coding region of a gene, it can terminate transcription of the gene and destroy normal gene expression. If it inserts into the promoter region of a gene, it can affect the level of expression of the gene. Insertions into introns can affect splicing or cause the premature termination of transcription.

Geneticists have harnessed *P* elements as tools for genetic analysis. One of the most useful applications of *P* elements is as vectors to introduce transgenes into *Drosophila*—a technique known as **germ-line transformation.** *P* elements are also used to generate mutations and to clone mutant genes. In addition, researchers are perfecting methods to target *P* element insertions to precise single-chromosomal sites, which should increase the precision of germ-line transformation in the analysis of gene activity.

Transposable Elements in Humans

The human genome, like that of other eukaryotes, is riddled with DNA derived from transposons. Recent genomic sequencing data reveal that approximately half of the human genome is comprised of transposable element DNA. As we saw in Chapter 11, the major families of human transposons are the long interspersed elements and short interspersed elements (**LINES** and **SINES,** respectively). Together, they comprise over 30 percent of the human genome. Other families of transposable elements account for a further 11 percent. As coding sequences comprise only about 1 percent of the human genome, there is about 40 to 50 times more transposable element DNA in the human genome than DNA in functional genes.

Although most human transposons appear to be inactive, the potential mobility and mutagenic effects of transposable elements have far-reaching implications for human genetics, as can be seen in a recent example of a transposon "caught in the act." The case involves a male child with hemophilia. One cause of hemophilia is a defect in blood-clotting factor VIII, the product of an X-linked gene. Haig Kazazian and his colleagues found LINES inserted at two points within the gene. Researchers were interested in determining if one of the mother's X chromosomes also contained this specific LINE. If so, the unaffected mother would be heterozygous and pass the LINE-containing chromosome to her son. The surprising finding was that the LINE sequence was *not* present on either of her X chromosomes but *was* detected on chromosome 22 of both parents. This suggests that this mobile element may have transposed from one chromosome to another in the gamete-forming cells of the mother, prior to being transmitted to the son.

LINE insertions into the human *dystrophin* gene have resulted in at least two separate cases of Duchenne muscular dystrophy. In one case, a transposon inserted into exon 48 and in another case, a transposon inserted into exon 44, both leading to frameshift mutations and premature termination of translation of the dystrophin protein. There are also reports that LINES have inserted into the *APC* and *c-myc* genes, leading to mutations that may have contributed to the development of some colon and breast cancers. In the latter cases, the transposition had occurred within one or a few somatic cells.

SINE insertions are also responsible for a number of cases of human disease. In one case, an ***Alu* element** integrated into the *BRCA2* gene, inactivating this tumor suppressor gene and leading to a familial case of breast cancer. Other genes that have been mutated by *Alu* integrations are the *factor IX* gene (leading to hemophilia B), the *ChE* gene (leading to acholinesterasemia), and the *NF1* gene (leading to neurofibromatosis).

Transposons, Mutations, and Evolution

Transposons can have a wide range of effects on genes. The insertion of a transposon into the coding region of a gene may disrupt the gene's normal translation reading frame or may induce premature termination of translation of the mRNA transcribed from the gene. Many transposons contain their own transcription promoters and enhancers, as well as splice sites and polyadenylation signals. The presence of these transposon regulatory sequences can have effects on nearby genes. The insertion of a transposon containing polyadenylation or transcription termination signals into a gene's intron may bring about termination of the gene's transcription within the transposon. In addition, it can cause aberrant splicing of an RNA transcribed from the gene. Insertions of a transposon into a gene's transcription regulatory region may disrupt the gene's normal regulation or may cause the gene to be expressed differently as a result of the presence of the transposon's own transcription promoter or enhancer sequences. The presence of two or more identical transposons in a genome creates the potential for recombination between the transposons, leading to duplications, deletions, inversions, or chromosome translocations. Any of these rearrangements may bring about phenotypic changes or disease.

New germ-line transpositions are estimated to occur once in every 50 to 100 human births. Most of these do not cause disease or a change in phenotype; however, it is thought that about 0.2 percent of detectable human mutations may be due to transposon insertions. Other organisms appear to suffer more damage due to transposition. About 10 percent of new mouse mutations and 50 percent of *Drosophila* mutations are caused by insertions of transposons in or near genes.

Because of their ability to alter genes and chromosomes, transposons may contribute to the variability that underlies evolution. For example, the Tn elements of bacteria carry antibiotic resistance genes between organisms, conferring a survival advantage to the bacteria under certain conditions. Another example of a transposon's contribution to evolution is provided by *Drosophila* telomeres. LINE-like elements are present at the ends of *Drosophila* chromosomes, and these elements act as telomeres, maintaining the length of *Drosophila* chromosomes over successive cell divisions. Other examples of evolved transposons are the *RAG1* and *RAG2* genes in humans. These genes encode **recombinase** enzymes that are essential to the development of the immune system. These two genes appear to have evolved from transposons.

Transposons may also affect the evolution of genomes by altering gene-expression patterns in ways that are subsequently retained by the host. For example, the human *amylase* gene contains an enhancer that causes the gene to be expressed in the parotid gland. This enhancer evolved from transposon sequences that were inserted into the gene-regulatory region early in primate evolution. Other examples of gene expression patterns that were affected by the presence of transposon sequences are T-cell-specific expression of the *CD8* gene and placenta-specific expression of the *leptin* and *CYP19* genes.

ESSENTIAL POINT

Transposable elements can move within a genome, creating mutations and altering gene expression. Besides creating mutations, transposons may contribute to evolution. Geneticists use transposons as a research tool to create mutations, clone genes, and introduce genes into model organisms.

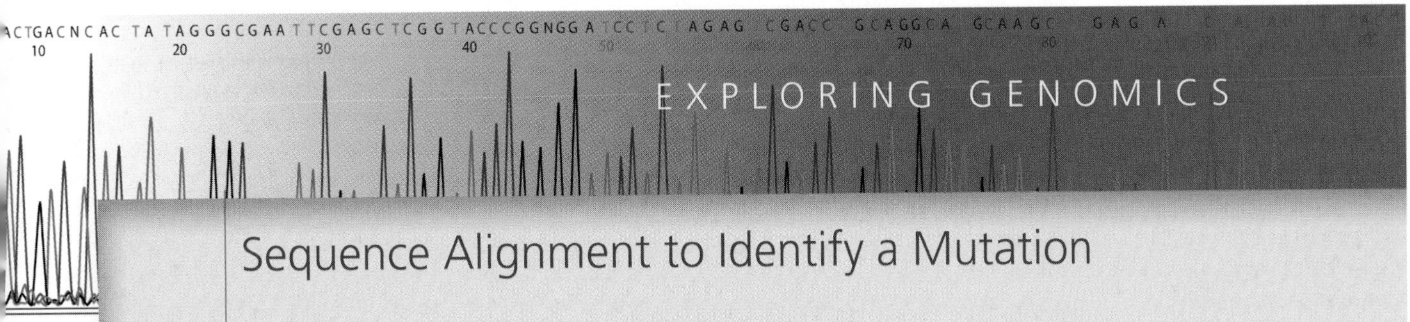

EXPLORING GENOMICS

Sequence Alignment to Identify a Mutation

In this chapter, we examined the causes of different types of mutations and how mutations affect phenotype by altering the structure and function of proteins. The emergence of genomics, bioinformatics, and proteomics as key disciplines in modern genetics has provided geneticists with an unprecedented set of tools for identifying and analyzing mutations in gene and protein sequences.

In this exercise we will use the **ExPASy (Expert Protein Analysis System)** site, which is hosted by the Swiss Institute for Bioinformatics and provides a wealth of resources for studying proteins. Here we will use an ExPASy program called SIM (for "similarity" in sequence) to compare two polypeptide sequences so as to pinpoint a mutation. Once the mutation has been identified, you will learn more about the gene encoding these polypeptides and about a human disease condition associated with this gene.

■ Exercise I: Identifying a Missense Mutation Affecting a Protein in Humans

1. Begin this exercise by accessing the ExPASy site at **http://www.expasy.ch/tools/sim-prot.html**. The SIM feature is an algorithm-based software program that allows us to compare multiple polypeptide sequences by looking for amino acid similarity in the sequences.

(*Cont. on the next page*)

2. Below are amino acid sequences for polypeptides expressed in two different people.

Person A

MGAPACALALCVAVAIVAGASSESLGTEQ
RVVGRAAEVPGPEPGQQEQLVFGSGDAV
ELSCPPPGGGPMGPTVWVKDGTGLVPSE
RVLVGPQRLQVLNASHEDSGAYSCRQRLT
QRVLCHFSVRVTDAPSSGDDEDGEDEA
EDTGVDTGAPYWTRPERMDKKLLAVPA
ANTVRFRCPAAGNPTPSISWLKNGREFR
GEHRIGGIKLRHQQWSLVMESVVPSDRG
NYTCVVENKFGSIRQTYTLDVLERS
PHRPILQAGLPANQTAVLGSDVEFHC
KVYSDAQPHIQWLKHVEVNGSKVG
PDGTPYVTVLKTAGANTTDKELEVLSLH
NVTFEDAGEYTCLAGNSIGFSHHSAWLVV
LPAEEELVEADEAGSVYAGILSYGVGFFL
FILVVAAVTLCRLRSPPKKGLGSPTVHK
ISRFPLKRQVSLESNASMSSNTPLVRIARL
SSGEGPTLANVSELELPADPKWELSRARL
TLGKPLGEGCFGQVVMAEAIGIDKDRAA
KPVTVAVKMLKDDATDKDLSDLVSEMEM
MKMIGKHKNIINLLGACTQGGPLYVLVEY
AAKGNLREFLRARRPPGLDYSFDTCKPPE
EQLTFKDLVSCAYQVARGMEYLASQKCI
HRDLAARNVLVTEDNVMKIADFGLARD
VHNLDYYKKTTNGRLPVKWMAPEALFD
RVYTHQSDVWSFGVLLWEIFTLGGSPYPG
IPVEELFKLLKEGHRMDKPANCTHDLYMI
MRECWHAAPSQRPTFKQLVEDLDRVLT
VTSTDEYLDLSAPFEQYSPGGQDTPSSSS
GDDSVFAHDLLPPAPPSSGGSRT

Person B

MGAPACALALCVAVAIVAGASSESLGTEQ
RVVGRAAEVPGPEPGQQEQLVFGSGDAV
ELSCPPPGGGPMGPTVWVKDGTGLVPSE
RVLVGPQRLQVLNASHEDSGAYSCRQRLT
QRVLCHFSVRVTDAPSSGDDEDGEDEAE
DTGVDTGAPYWTRPERMDKKLLAVPAA
NTVRFRCPAAGNPTPSISWLKNGREFRGE
HRIGGIKLRHQQWSLVMESVVPSDRGNY
TCVVENKFGSIRQTYTLDVLERSPHRPILQ
AGLPANQTAVLGSDVEFHCKVYSDAQPHI
QWLKHVEVNGSKVGPDGTPYVTVLKTA
GANTTDKELEVLSLHNVTFEDAGEYTCL
AGNSIGFSHHSAWLVVLPAEEELVEADEA
GSVYAGILSYRVGFFLFILVVAAVTLCRLR
SPPKKGLGSPTVHKISRFPLKRQVSLESNA
SMSSNTPLVRIARLSSGEGPTLANVSELEL
PADPKWELSRARLTLGKPLGEGCFGQVV
MAEAIGIDKDRAAKPVTVAVKMLKDDAT
DKDLSDLVSEMEMMKMIGKHKNIINLL
GACTQGGPLYVLVEYAAKGNLREFLRAR
RPPGLDYSFDTCKPPEEQLTFKDLVSCAY
QVARGMEYLASQKCIHRDLAARNVLVT
EDNVMKIADFGLARDVHNLDYYKKTTN
GRLPVKWMAPEALFDRVYTHQSDVWSF
GVLLWEIFTLGGSPYPGIPVEELFKLLKEG
HRMDKPANCTHDLYMIMRECWHAAPSQ
RPTFKQLVEDLDRVLTVTSTDEYLDLSAP
FEQYSPGGQDTPSSSSGDDSVFAHDLLP
PAPPSSGGSRT

3. Copy and paste each sequence into the "SEQUENCE" text boxes in SIM. (*Hint:* Access these sequences from the Companion Web site so that you can copy and paste the sequence into SIM.) Use the Person A sequence for sequence 1 and the Person B sequence for sequence 2. Click the "User-entered sequence" button for each. Name the sequences Person A and Person B as appropriate. Submit the sequences for comparison and then answer the following questions:
   a. How many amino acids are in each polypeptide sequence that was analyzed?
   b. Look carefully at the alignment results. Can you find any differences in amino acid sequence when comparing these two polypeptides? What did you find?

■ Exercise II: Identifying the Genetic Basis for a Human Genetic Disease Condition

1. Go to the ExPASy home page and find the BLAST link. Run a protein BLAST (blastp) search to identify which polypeptide you have been studying. Explore the BLAST reports for the top three protein sequences that aligned with your query sequence by clicking on the link for each sequence. Pay particular attention to the "Comment" section of each report to help you answer the questions in the following part.
2. Now that you know what gene you are working with, go to PubMed (**http://www.ncbi.nlm.nih.gov/entrez/query.fcgi?db=PubMed**) and search for a review article from the authors Vajo, Z., Francomano, C. A., and Wilkin, D. J.
3. Answer the following questions:
   a. What gene codes for the polypeptides you have been studying?
   b. What is the function of this protein?
   c. Based on what you learned from the alignment results you analyzed in Exercise I, the BLAST reports, and your PubMed search, what human disease is caused by the mutation you identified in Exercise I? Explain your answer and briefly describe phenotypes associated with the disease.

CASE STUDY Genetic dwarfism

Seven months pregnant, an expectant mother was undergoing a routine ultrasound. While prior tests had been normal, this one showed that the limbs of the fetus were unusually short. The doctor suspected that the baby might have a genetic form of dwarfism called achondroplasia. He told her that the disorder was due to an autosomal dominant mutation and occurred with a frequency of about 1 in 25,000 births. The expectant mother had studied genetics in college and immediately raised several questions. How would you answer them?

1. How could her baby have a dominantly inherited disorder if there was no history of this condition on either side of the family?
2. Is the mutation more likely to have come from the mother or the father?
3. If this child has achondroplasia, would the chances increase that their next child would also have this disorder?
4. Could this disorder have been caused by X rays or ultrasounds she had earlier in pregnancy?

INSIGHTS AND SOLUTIONS

1. The base analog 2-amino purine (2-AP) substitutes for adenine during DNA replication, but it may base-pair with cytosine. The base analog 5-bromouracil (5-BU) substitutes for thymidine, but it may base-pair with guanine. Follow the double-stranded trinucleotide sequence shown here through three rounds of replication, assuming that, in the first round, both analogs are present and become incorporated wherever possible. Before the second and third round of replication, any unincorporated base analogs are removed. What final sequences occur?

Solution:

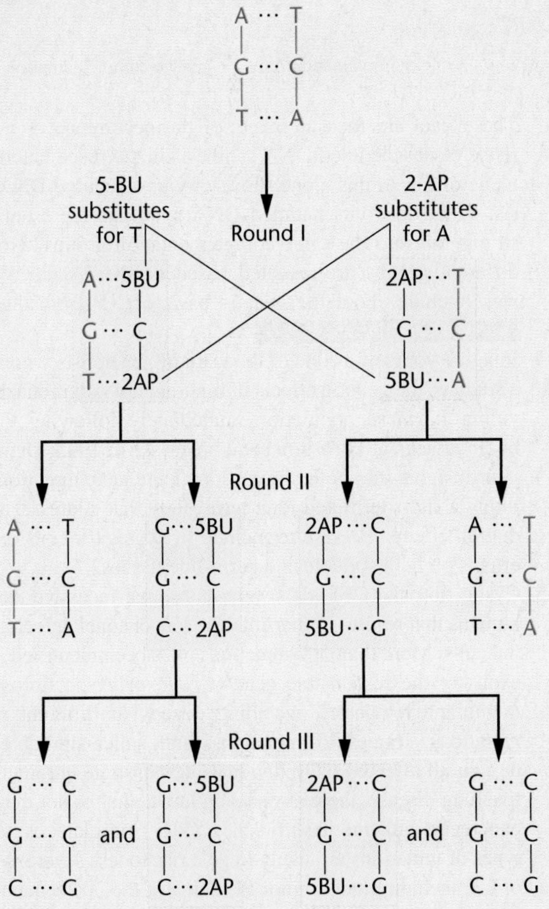

2. A rare dominant mutation expressed at birth was studied in humans. Records showed that six cases were discovered in 40,000 live births. Family histories revealed that in two cases, the mutation was already present in one of the parents. Calculate the spontaneous mutation rate for this mutation. What are some underlying assumptions that may affect our conclusions?

Solution: Only four cases represent a new mutation. Because each live birth represents two gametes, the sample size is from 80,000 meiotic events. The rate is equal to

$$\frac{4}{80{,}000} = \frac{1}{20{,}000} = 5 \times 10^{-5}$$

We have assumed that the mutant gene is fully penetrant and is expressed in each individual bearing it. If it is not fully penetrant, our calculation may be an underestimate because one or more mutations may have gone undetected. We have also assumed that the screening was 100 percent accurate. One or more mutant individuals may have been "missed," again leading to an underestimate. Finally, we assumed that the viability of the mutant and nonmutant individuals is equivalent and that they survive equally *in utero*. Therefore, our assumption is that the number of mutant individuals at birth is equal to the number at conception. If this were not true, our calculation would again be an underestimate.

3. Consider the following estimates:

(a) There are 5.5×10^9 humans living on this planet.

(b) Each individual has about 30,000 (0.3×10^5) genes.

(c) The average mutation rate at each locus is 10^{-5}.

How many spontaneous mutations are currently present in the human population? Assuming that these mutations are equally distributed among all genes, how many new mutations have arisen in each gene in the human population?

Solution: First, since each individual is diploid, there are two copies of each gene per person, each arising from a separate gamete. Therefore, the total number of spontaneous mutations is

$$(2 \times 0.3 \times 10^5 \text{ genes/individual})$$
$$\times (5.5 \times 10^9) \text{ individuals} \times (10^{-5} \text{ mutations/gene})$$
$$= (0.6 \times 10^5) \times (5.5 \times 10^9) \times (10^{-5}) \text{ mutations}$$
$$= 3.3 \times 10^9 \text{ mutations in the population}$$

$$3.3 \times 10^9 \text{ mutations}/0.3 \times 10^5 \text{ genes}$$
$$= 11 \times 10^4 \text{ mutations per gene in the population}$$

PROBLEMS AND DISCUSSION QUESTIONS

1. Discuss the importance of mutations in genetic studies.
2. Why would a mutation in a somatic cell of a multicellular organism escape detection?
3. Most mutations are thought to be deleterious. Why, then, is it reasonable to state that mutations are essential to the evolutionary process?
4. Why is a random mutation more likely to be deleterious than beneficial?
5. Most mutations in a diploid organism are recessive. Why?
6. What is meant by a conditional mutation?
7. Describe a tautomeric shift and how it may lead to a mutation.
8. Contrast and compare the mutagenic effects of deaminating agents, alkylating agents, and base analogs.
9. Acridine dyes induce frameshift mutations. Why are frameshift mutations likely to be more detrimental than point mutations, in which a single pyrimidine or purine has been substituted?

10. Why are X rays more potent mutagens than UV radiation?
11. DNA damage brought on by a variety of natural and artificial agents elicits a wide variety of cellular responses involving numerous signaling pathways. In addition to the activation of DNA repair mechanisms, there can be activation of pathways leading to apoptosis (programmed cell death) and cell-cycle arrest. Why would apoptosis and cell-cycle arrest often be part of a cellular response to DNA damage?
12. Contrast the various types of DNA repair mechanisms known to counteract the effects of UV radiation. What is the role of visible light in repairing UV-induced mutations?
13. Mammography is an accurate screening technique for the early detection of breast cancer in humans. Because this technique uses X rays diagnostically, it has been highly controversial. Can you explain why? What reasons justify the use of X rays for such a medical screening technique?
14. Explain the molecular basis of Duchenne muscular dystrophy and Becker muscular dystrophy, and how different types of mutations affect the severity of the disorders.
15. Describe how the Ames test screens for potential environmental mutagens. Why is it thought that a compound that tests positively in the Ames test may also be carcinogenic?
16. What genetic defects result in the disorder xeroderma pigmentosum (XP) in humans? How do these defects create the phenotypes associated with the disorder?
17. In a bacterial culture in which all cells are unable to synthesize leucine (leu^-), a potent mutagen is added, and the cells are allowed to undergo one round of replication. At that point, samples are taken, a series of dilutions is made, and the cells are plated on either minimal medium or minimal medium containing leucine. The first culture condition (minimal medium) allows the growth of only leu^+ cells, while the second culture condition (minimum medium with leucine added) allows the growth of all cells. The results of the experiment are as follows:

Culture Condition	Dilution	Colonies
minimal medium	10^{-1}	18
minimal + leucine	10^{-7}	6

What is the rate of mutation at the locus associated with leucine biosynthesis? See Now Solve This on page 287.
18. Speculate on how improved living conditions and medical care in the developed nations might affect human mutation rates, both neutral and deleterious.
19. The cancer drug melphalan is an alkylating agent of the mustard gas family. It acts in two ways: by causing alkylation of guanine bases and by crosslinking DNA strands together. Describe two ways in which melphalan might kill cancer cells. What are two ways in which cancer cells could repair the DNA-damaging effects of melphalan? See Now Solve This on page 292.
20. Muscular dystrophies are some of the most common inherited diseases in humans, resulting from a large number of different mutations in the *dystrophin* gene. Speculate on why this gene appears to suffer so many mutations. See Now Solve This on page 298.
21. As described in Chapter 6, some mutations that lead to diseases such as Huntington disease are caused by the insertion of trinucleotide repeats. Describe how the process of DNA replication can lead to expansions of trinucleotide repeat regions.
22. The origin of the mutation that led to X-linked hemophilia in Queen Victoria's family is controversial. Her father did not have X-linked hemophilia, and there is no evidence that any members of her mother's family had the condition. What are some possible explanations of how the mutation arose? What types of mutations could lead to the disease?
23. Presented here are hypothetical findings from studies of heterokaryons formed from seven human xeroderma pigmentosum cell strains:

	XP1	XP2	XP3	XP4	XP5	XP6	XP7
XP1	–						
XP2	–	–					
XP3	–	–	–				
XP4	+	+	+	–			
XP5	+	+	+	+	–		
XP6	+	+	+	+	–	–	
XP7	+	+	+	+	–	–	–

Note: " + " = complementation; " – " = no complementation

These data are measurements of the occurrence or nonoccurrence of unscheduled DNA synthesis in the fused heterokaryon. None of the strains alone shows any unscheduled DNA synthesis. What does unscheduled DNA synthesis represent? Which strains fall into the same complementation groups? How many different groups are revealed based on these data? What can we conclude about the genetic basis of XP from these data? See Now Solve This on page 296.
24. Imagine yourself as one of the team of geneticists who launches a study of the genetic effects of high-energy radiation on the surviving Japanese population immediately following the atom-bomb attacks at Hiroshima and Nagasaki in 1945. Demonstrate your insights into both chromosomal and gene mutation by outlining a short-term and long-term study that addresses these radiation effects. Be sure to include strategies for considering the effects on both somatic and germ-line tissues.
25. Cystic fibrosis (CF) is a severe autosomal recessive disorder in humans that results from a chloride ion–channel defect in epithelial cells. More than 500 mutations have been identified in the 24 exons of the responsible gene (*CFTR*, or cystic fibrosis transmembrane regulator), including dozens of different missense mutations, frameshift mutations, and splice-site defects. Although all affected CF individuals demonstrate chronic obstructive lung disease, there is variation in whether or not they exhibit pancreatic enzyme insufficiency (PI). Speculate as to which types of mutations are likely to give rise to less severe symptoms of CF, including only minor PI. Some of the 300 sequence alterations that have been detected within the exon regions of the *CFTR* gene do not give rise to cystic fibrosis. Taking into account your knowledge of the genetic code, gene expression, protein function, and mutation, describe why this might be so, as if you were explaining it to a freshman biology major.
26. Electrophilic oxidants are known to create the modified base named 7,8-dihydro-8-oxoguanine (oxoG) in DNA. Whereas guanine base-pairs with cytosine, oxoG base-pairs with either cytosine or adenine.
    (a) What are the sources of reactive oxidants within cells that cause this type of base alteration?
    (b) Drawing on your knowledge of nucleotide chemistry, draw the structure of oxoG, and, below it, draw guanine. Opposite guanine, draw cytosine, including the hydrogen bonds that allow these two molecules to base-pair. Does the structure of oxoG, in contrast to guanine, provide any hint as to why it base-pairs with adenine?

(c) Assume that an unrepaired oxoG lesion is present in the helix of DNA opposite cytosine. Predict the type of mutation that will occur following several rounds of replication.

(d) Which DNA repair mechanisms might work to counteract an oxoG lesion? Which of these is likely to be most effective?

27. Among Betazoids in the world of *Star Trek*®, the ability to read minds is under the control of a gene called *mindreader* (abbreviated *mr*). Most Betazoids can read minds, but rare recessive mutations in the *mr* gene result in two alternative phenotypes: *delayed-receivers* and *insensitives.* Delayed-receivers have some mind-reading ability but perform the task much more slowly than normal Betazoids. Insensitives cannot read minds at all. Betazoid genes do not have introns, so the gene only contains coding DNA. It is 3332 nucleotides in length, and Betazoids use a four-letter genetic code.

The following table shows some data from five unrelated *mr* mutations.

Description of Mutation		Phenotype
mr-1	Nonsense mutation in codon 829	delayed-receiver
mr-2	Missense mutation in codon 52	delayed-receiver
mr-3	Deletion of nucleotides 83–150	delayed-receiver
mr-4	Missense mutation in codon 192	insensitive
mr-5	Deletion of nucleotides 83–93	insensitive

For each mutation, provide a plausible explanation for why it gives rise to its associated phenotype and not to the other phenotype. For example, hypothesize why the *mr-1* nonsense mutation in codon 829 gives rise to the milder delayed-receiver phenotype rather than the more severe insensitive phenotype. Then repeat this type of analysis for the other mutations. (More than one explanation is possible, so be creative within *plausible* bounds!)

28. Skin cancer carries a lifetime risk nearly equal to that of all other cancers combined. Following is a graph (modified from Kraemer, 1997. *Proc. Natl. Acad. Sci. (USA)* 94: 11–14) depicting the age of onset of skin cancers in patients with or without XP, where cumulative percentage of skin cancer is plotted against age. The non-XP curve is based on 29,757 cancers surveyed by the National Cancer Institute, and the curve representing those with XP is based on 63 skin cancers from the Xeroderma Pigmentosum Registry.

(a) Provide an overview of the information contained in the graph.

(b) Explain why individuals with XP show such an early age of onset.

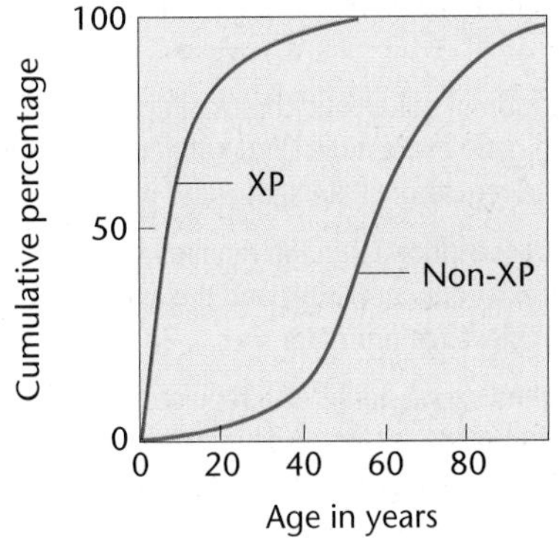

29. The initial discovery of IS elements in bacteria revealed the presence of an element upstream (5′) of three genes controlling galactose metabolism. All three genes were affected simultaneously, although there was only one IS insertion. Offer an explanation as to why this might occur.

30. It has been noted that most transposons in humans and other organisms are located in noncoding regions of the genome—regions such as introns, pseudogenes, and stretches of particular types of repetitive DNA. There are several ways to interpret this observation. Describe two possible interpretations. Which interpretation do you favor? Why?

31. Mutations in the *IL2RG* gene cause approximately 30 percent of severe combined immunodeficiency disorder (SCID) cases. These mutations result in alterations to a protein component of cytokine receptors that are essential for proper development of the immune system. The *IL2RG* gene is composed of eight exons and contains upstream and downstream sequences that are necessary for proper transcription and translation. Below are some of the mutations observed. For each, explain its likely influence on the *IL2RG* gene product (assume its length to be 375 amino acids).

(a) Nonsense mutation in coding regions
(b) Insertion in Exon 1, causing frameshift
(c) Insertion in Exon 7, causing frameshift
(d) Missense mutation
(e) Deletion in Exon 2, causing frameshift
(f) Deletion in Exon 2, in frame
(g) Large deletion covering Exons 2 and 3

32. A variety of neural and muscular disorders are associated with expansions of trinucleotide repeat sequences. The following table lists several disorders, the repeat motifs, their locations, the normal number of repeats, and the number of repeats in the full mutations.

Disorder	Repeat	Location	Normal Number	Full Mutation
Fragile X	CCG	5′ untranslated region	6–230	>230
Huntington	CAG	Exon	6–35	36–120
Myotonic Dystrophy	CTG	3′untranslated region	5–37	37–1500
Oculopharyngeal Muscular Dystrophy	GCG	Exon	6	8–13
Friedreich Ataxia	AAG	Intron	20	200–900

(a) Most disorders attributable to trinucleotide repeats result from expansion of the repeats. Two mechanisms are often proposed to explain repeat expansion: (1) unequal synapsis and crossing over and (2) errors in DNA replication where single-stranded, base-paired loops are formed that conflict with linear replication. Present a simple sketch of each mechanism.

(b) Notice that some of the repeats occur in areas of the gene that are not translated. How can a mutation occur if the alteration is not reflected in an altered amino acid sequence?

(c) In the two cases where the repeat expansions occur in exons, the extent of expansion is considerably less than when the expansion occurs outside exons. Present an explanation for this observation.

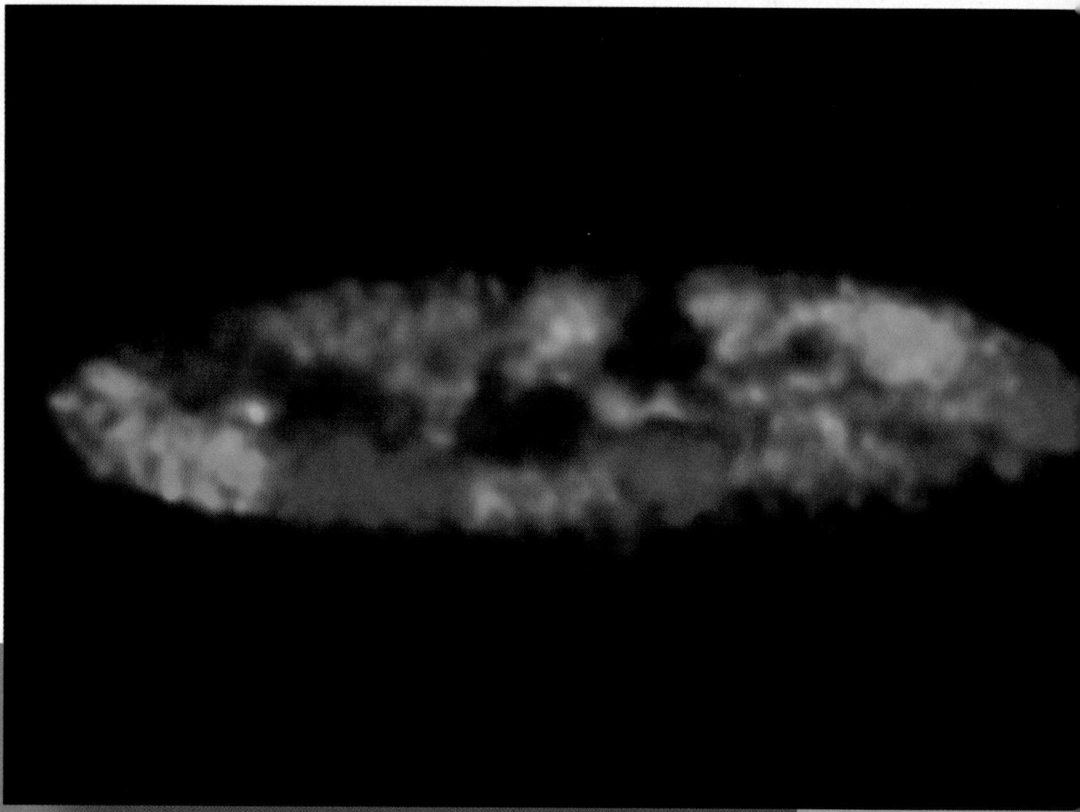

Chromosome territories in an interphase chicken cell nucleus. Each chromosome is stained with a different-colored probe.

15 Regulation of Gene Expression

Chapter Concepts

- Expression of genetic information is regulated by intricate regulatory mechanisms that exert control over transcription, mRNA stability, translation, and posttranslational modifications.
- In prokaryotes, genes that encode proteins with related functions tend to be organized in clusters and are often under coordinated control. Such clusters, including their associated regulatory sequences, are called operons.
- Transcription within operons is either inducible or repressible and is often regulated by the metabolic substrate or end product of the pathway.
- Eukaryotic gene regulation is more complex than prokaryotic gene regulation.
- The organization of eukaryotic chromatin in the nucleus plays a role in regulating gene expression. Chromatin must be remodeled to provide access to regulatory DNA sequences within it.
- Eukaryotic transcription initiation requires the assembly of transcription regulators at enhancer sites and the assembly of basal transcription complexes at promoter sites.
- Eukaryotic gene expression is also regulated at multiple post-transcriptional steps, including alternative splicing of pre-mRNA, control of mRNA stability, translation, and posttranslational processing.

In previous chapters, we have described how DNA is organized into genes, how genes store genetic information, and how this information is expressed through the processes of transcription and translation. We now consider one of the most fundamental questions in molecular genetics: *How is gene expression regulated?*

It is clear that not all genes are expressed at all times in all situations. For example, some proteins in the bacterium *E. coli* are present in as few as 5 to 10 molecules per cell, whereas others, such as ribosomal proteins and the many proteins involved in the glycolytic pathway, are present in as many as 100,000 copies per cell. Although many prokaryotic gene products are present continuously at low levels, the level of these products can increase dramatically when required. In multicellular eukaryotes, differential gene expression is also essential and is at the heart of embryonic development and maintenance of the adult state. Cells of the pancreas, for example, do not synthesize retinal pigment, and retinal cells do not make insulin.

The activation and repression of gene expression is a delicate balancing act for an organism; expression of a gene at the wrong time, in the wrong cell type, or in abnormal amounts can lead to a deleterious phenotype, cancer, or cell death—even when the gene itself is normal.

In this chapter, we will explore the ways in which prokaryotic and eukaryotic organisms regulate gene expression. We will describe some of the fundamental components of gene regulation, including the *cis*-acting DNA elements and *trans*-acting factors that regulate transcription initiation. We will then explain how these components interact with each other and with other factors such as activators, repressors, and chromatin proteins. We will also consider the roles that posttranscriptional mechanisms play in the regulation of eukaryotic gene expression.

How Do We Know?

In this chapter, we will focus on how prokaryotic and eukaryotic organisms regulate the expression of genetic information stored in DNA. We will find many opportunities to consider the methods and reasoning by which much of this information was acquired. From the explanations given in the chapter, you should answer the following fundamental questions:

1. How do we know that bacteria regulate the expression of certain genes in response to the environment?
2. How do we know that bacterial gene clusters are often coordinately regulated by a regulatory region that must be located adjacent to the cluster?
3. What led researchers to conclude that a *trans*-acting repressor molecule regulates the *lac* operon?
4. How do we know that promoters and enhancers regulate transcription of eukaryotic genes?
5. How do we know that DNA methylation plays a role in the regulation of eukaryotic gene expression?

15.1 Prokaryotes Regulate Gene Expression in Response to Both External and Internal Conditions

Not only do bacteria respond metabolically to changes in their environment, but they also regulate gene expression in order to synthesize products required for a variety of normal cellular activities, including DNA replication, recombination, repair, and cell division. In the following sections, we will focus on prokaryotic gene regulation at the level of transcription, which is the predominant level of regulation in prokaryotes. Keep in mind, however, that posttranscriptional regulation also occurs in bacteria. We will defer discussion of posttranscriptional gene-regulatory mechanisms to subsequent sections dealing with eukaryotic gene expression.

The idea that microorganisms regulate the synthesis of gene products is not a new one. As early as 1900, it was shown that when lactose (a galactose and glucose-containing disaccharide) is present in the growth medium of yeast, the organisms synthesize enzymes required for lactose metabolism. When lactose is absent, the enzymes are not manufactured. Soon thereafter, investigators were able to generalize that bacteria also adapt to their environment, producing certain enzymes only when specific chemical substrates are present. Such enzymes are referred to as **inducible,** reflecting the role of the substrate, which serves as the **inducer** in enzyme production. In contrast, enzymes that are produced continuously, regardless of the chemical makeup of the environment, are called **constitutive.**

More recent investigation has revealed a contrasting system whereby the presence of a specific molecule inhibits gene expression. This is usually true for molecules that are end products of anabolic biosynthetic pathways. For example, the amino acid tryptophan can be synthesized by bacterial cells. If a sufficient supply of tryptophan is present in the environment or culture medium, it is energetically inefficient for the organism to synthesize the enzymes necessary for tryptophan production. A mechanism has evolved whereby tryptophan plays a role in repressing transcription of genes that encode the appropriate biosynthetic enzymes. In contrast to the inducible system controlling lactose metabolism, the system governing tryptophan expression is said to be **repressible.**

Regulation, whether it is inducible or repressible, may be under either **negative** or **positive control.** Under negative control, gene expression occurs *unless it is shut off by some form of a regulator molecule.* In contrast, under positive control, transcription occurs *only if a regulator molecule directly stimulates RNA production.* In theory, either type of control or a combination of the two can govern inducible or repressible systems.

15.2 Lactose Metabolism in *E. coli* Is Regulated by an Inducible System

Beginning in 1946, the studies of Jacques Monod (with later contributions by Joshua Lederberg, François Jacob, and André Lwoff) revealed genetic and biochemical insights into the mechanisms of lactose metabolism in bacteria. These studies explained how gene expression is repressed when lactose is absent, but induced when it is available. In the presence of lactose, concentrations of the enzymes responsible for lactose metabolism increase rapidly from a few molecules to thousands per cell. The enzymes responsible for lactose metabolism are thus *inducible,* and lactose serves as the *inducer.*

In prokaryotes, genes that code for enzymes with related functions (in this case, the genes involved with lactose metabolism) tend to be organized in clusters on the bacterial chromosome. In addition, transcription of these genes is often under the coordinated control of a single transcription regulatory region. The location of this regulatory region is almost always upstream of the gene cluster it controls, and we refer to the regulatory region as a ***cis*-acting site.** *Cis*-acting regulatory regions bind molecules that control transcription of the gene cluster. Such molecules are called ***trans*-acting molecules.** Actions at the *cis*-acting regulatory site determine whether the genes are transcribed into RNA and thus whether the corresponding enzymes or other protein products are synthesized from the mRNA. Binding of a *trans*-acting molecule at a *cis*-acting site can regulate the gene cluster either negatively (by turning off transcription) or positively (by turning on transcription of genes in the cluster). In this section, we discuss how transcription of such bacterial gene clusters is coordinately regulated.

ESSENTIAL POINT

Research on the *lac* operon in *E. coli* pioneered our understanding of gene regulation in bacteria.

Structural Genes

As illustrated in Figure 15–1, three genes and an adjacent regulatory region constitute the **lactose,** or ***lac,* operon.** Together, the entire gene cluster functions in an integrated fashion to provide a rapid response to the presence or absence of lactose.

Genes coding for the primary structure of the enzymes are called **structural genes.** There are three structural genes in the *lac* operon. The *lacZ* gene encodes **β-galactosidase,** an enzyme whose role is to convert the disaccharide lactose to the monosaccharides glucose and galactose (Figure 15–2). This conversion is essential if lactose is to serve as an energy source in glycolysis. The second gene, *lacY,* encodes the amino acid sequence of **permease,** an enzyme that facilitates the entry of lactose into the bacterial cell. The third gene, *lacA,* codes for the enzyme **transacetylase.** Although its physiological role is still not completely clear, it may be involved in the removal of toxic by-products of lactose digestion from the cell.

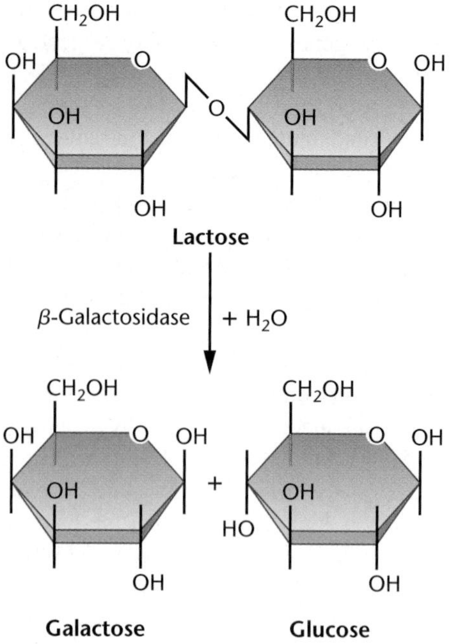

FIGURE 15–2 The catabolic conversion of the disaccharide lactose into its monosaccharide units, galactose and glucose.

To study the genes encoding these three enzymes, researchers isolated numerous mutants, each of which eliminated the function of one of the enzymes. These mutants were first isolated and studied by Joshua Lederberg. Mutant cells that fail to produce active β-galactosidase ($lacZ^-$) or permease ($lacY^-$) are unable to use lactose as an energy source and are collectively known as lac^- mutants. Mapping studies by Lederberg established that all three genes are closely linked or contiguous to one another on the bacterial chromosome, in the order Z–Y–A. (See Figure 15–1.) All three genes are transcribed as a single unit, resulting in a polycistronic mRNA (Figure 15–3). This results in the coordinate regulation of all three genes, since a single message RNA is simultaneously translated into all three gene products.

The Discovery of Regulatory Mutations

How does lactose stimulate transcription of the *lac* operon and induce the synthesis of the related enzymes? A partial answer comes from studies using **gratuitous inducers,** chemical analogs of lactose such as the sulfur analog

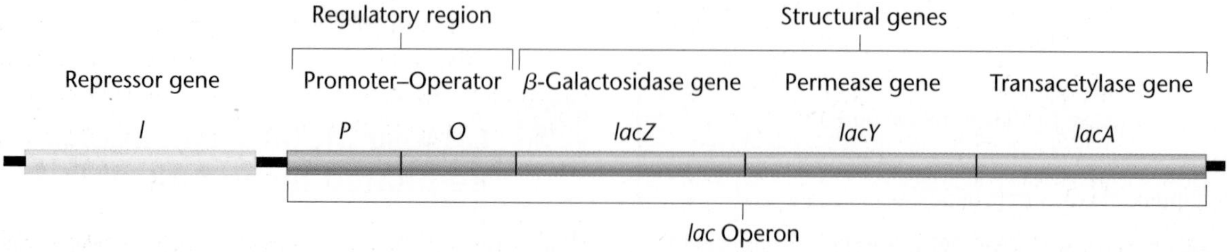

FIGURE 15–1 A simplified overview of the genes and regulatory units involved in the control of lactose metabolism. (This region of DNA is not drawn to scale.) A more detailed model is described later in this chapter.

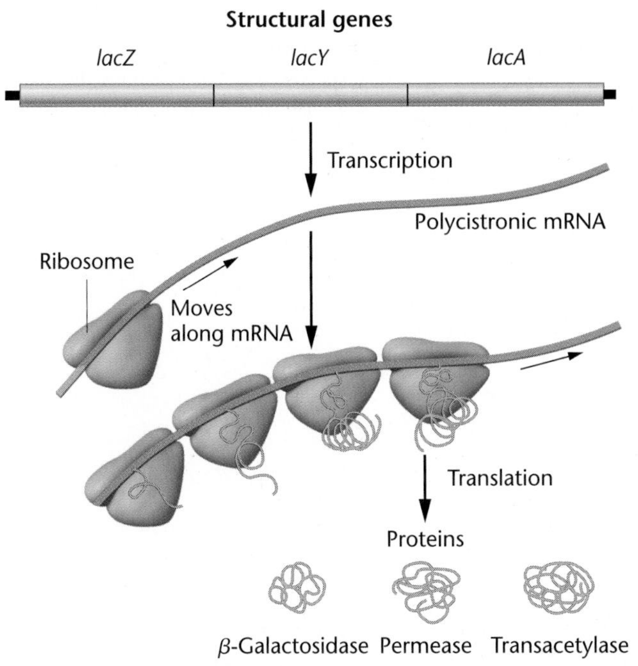

FIGURE 15–3 The structural genes of the *lac* operon are transcribed into a single polycistronic mRNA, which is translated simultaneously by several ribosomes into the three enzymes encoded by the operon.

isopropylthiogalactoside (IPTG), shown in Figure 15–4. Gratuitous inducers behave like natural inducers, but they do not serve as substrates for the enzymes that are subsequently synthesized.

What, then, is the role of lactose in gene regulation? The answer to this question required the study of a class of mutants called **constitutive mutants.** In cells bearing constitutive mutations, enzymes are produced regardless of the presence or absence of lactose. Studies of the constitutive mutation *lacI*$^-$ mapped the mutation to a site on the bacterial chromosome close to, but distinct from, the *lacZ, lacY,* and *lacA* genes. This mutation defined the *lacI* gene, which is appropriately called a **repressor gene.** Another set of constitutive mutations that produce identical effects to those of *lacI*$^-$ occur in a region immediately adjacent to the structural genes. This class of mutations, designated *lacO*^{C}, occur in the **operator region** of the operon. In both types of constitutive mutants, the enzymes are continually produced, inducibility is eliminated, and gene regulation is lost.

The Operon Model: Negative Control

Around 1960, Jacob and Monod proposed a scheme involving negative control called the **operon model,** whereby a group of genes is regulated and expressed together as a unit. As we saw in Figure 15–1, the *lac* operon they proposed consists of the *Z, Y,* and *A* structural genes, as well as the adjacent sequences of DNA referred to as the *operator region.* They argued that the *lacI* gene regulates the transcription of the structural genes by producing a **repressor molecule,** and that the repressor is **allosteric,** meaning that it reversibly interacts with another molecule, causing both a conformational change in the repressor's three-dimensional shape and a change in its chemical activity. Figure 15–5, p. 312, illustrates the components of the *lac* operon as well as the action of the *lac* repressor in the presence and absence of lactose.

Jacob and Monod suggested that the repressor normally binds to the DNA sequence of the operator region. When it does so, it inhibits the action of RNA polymerase, effectively repressing the transcription of the structural genes [Figure 15–5(b)]. However, when lactose is present, this sugar binds to the repressor molecule and causes an allosteric conformational change. This change renders the repressor incapable of interacting with operator DNA [Figure 15–5(c)]. In the absence of the repressor–operator interaction, RNA polymerase transcribes the structural genes, and the enzymes necessary for lactose metabolism are produced. Because transcription occurs only when the repressor *fails* to bind to the operator region, regulation is said to be under *negative control.*

The operon model invokes a series of molecular interactions between proteins, inducers, and DNA to explain the efficient regulation of structural gene expression. In the absence of lactose, the enzymes encoded by the genes are not needed, and expression of genes encoding these enzymes is repressed. When lactose is present, it indirectly induces the transcription of the structural genes by interacting with the repressor.* If all lactose is metabolized, none is available to bind to the repressor, which is again free to bind to operator DNA and repress transcription.

Both the I^- and O^C constitutive mutations interfere with these molecular interactions, allowing continuous transcription of the structural genes. In the case of the I^- mutant, seen in Figure 15–6(a), p. 313, the repressor protein is altered or absent and cannot bind to the operator region, so the structural genes are always transcribed. In the case of the O^C mutant [Figure 15–6(b)], the nucleotide sequence of the operator DNA is altered and will not bind with a normal repressor molecule. The result is the same: the structural genes are always transcribed.

Genetic Proof of the Operon Model

The operon model leads to three major predictions that can be tested to determine its validity. The major predictions to be tested are that (1) the *I* gene produces a diffusible product; (2) the *O* region is involved in regulation but does not produce a product; and (3) the *O* region must be adjacent to the structural genes in order to regulate transcription.

The construction of partially diploid bacteria allows us to assess these assumptions, particularly those that predict

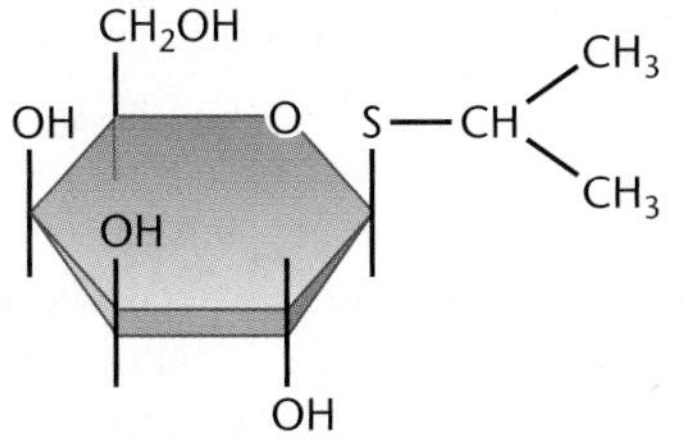

FIGURE 15–4 The gratuitous inducer isopropylthiogalactoside (IPTG).

* Technically, allolactose, an isomer of lactose, is the inducer. When lactose enters the bacterial cell, some of it is converted to allolactose by the β-galactosidase enzyme.

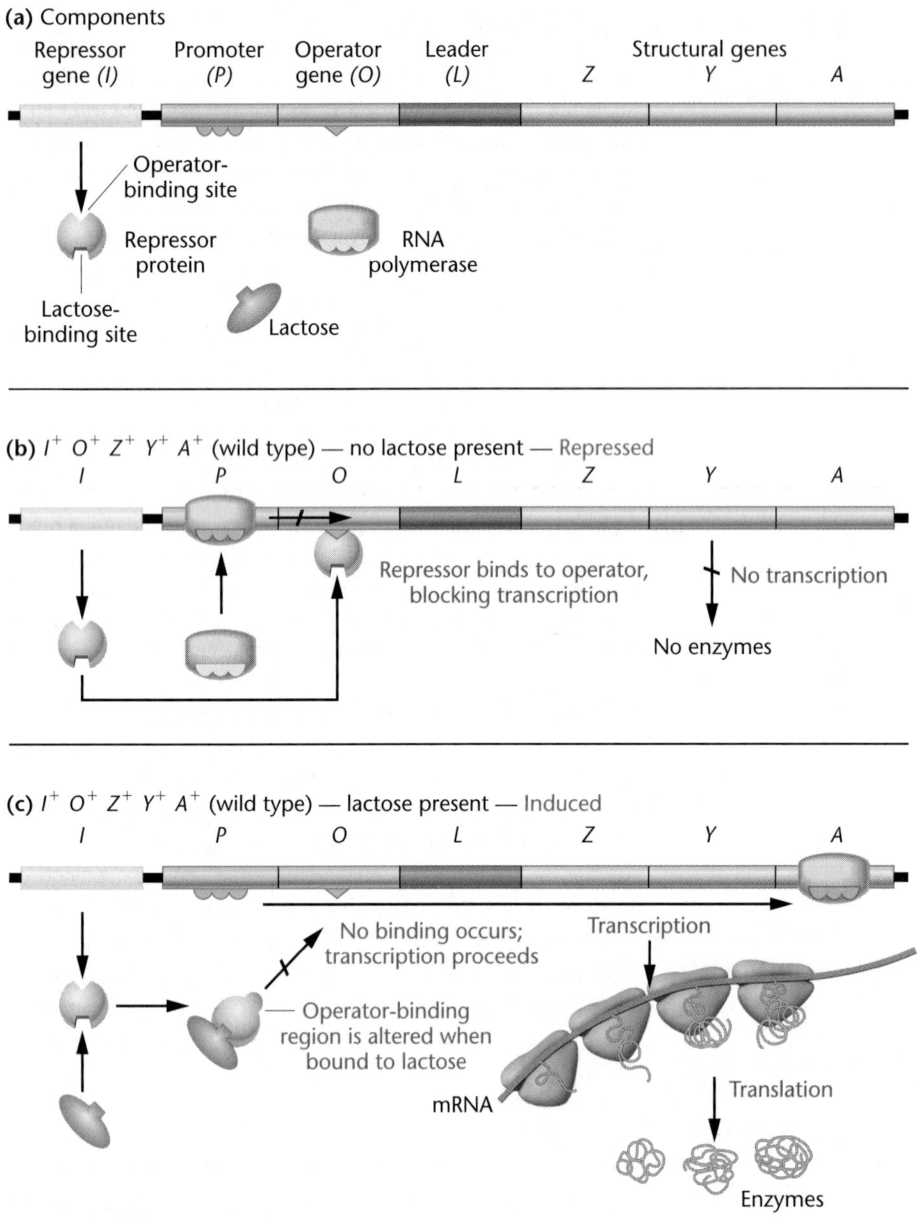

FIGURE 15–5 The components of the wild-type *lac* operon (a) and the response of the *lac* operon to the absence (b) and presence (c) of lactose.

trans-acting regulatory molecules. For example, as introduced in Chapter 8, the F plasmid may contain chromosomal genes, in which case it is designated F′. When an F^- cell acquires such a plasmid, it contains its own chromosome plus one or more additional genes present in the plasmid. This creates a host cell, called a **merozygote,** that is diploid for those genes. The use of such a plasmid makes it possible, for example, to introduce an I^+ gene into a host cell whose genotype is I^- or to introduce an O^+ region into a host cell of genotype O^C. The Jacob–Monod operon model predicts how regulation should be affected in such cells. Adding an I^+ gene to an I^- cell should restore inducibility because the normal wild-type repressor, which is a *trans*-acting factor, would be produced by the inserted I^+ gene. Adding an O^+ region to an O^C cell should have no effect on constitutive enzyme production, since regulation depends on the presence of an O^+ region immediately adjacent to the structural genes—that is, O^+ is a *cis*-acting regulator.

Results of these experiments are shown in Table 15.1, where *Z* represents the structural genes. The inserted genes are listed after the designation F′. In both cases described here, the Jacob–Monod model is upheld (part B of Table 15.1). Part C shows the reverse experiments, where either an I^- gene or an O^C region is added to cells of normal inducible genotypes. As the model predicts, inducibility is maintained in these partial diploids.

Another prediction of the operon model is that certain mutations in the *I* gene should have the opposite effect of I^-. That is, instead of being constitutive because the repressor can't bind the operator, mutant repressor molecules should be produced that cannot interact with the inducer, lactose. As a result, the repressor would always bind to the operator sequence, and the structural genes would be permanently repressed (Figure 15–7, p. 314). If this were the case, the presence of an additional I^+ gene would have little or no effect on repression.

FIGURE 15–6 The response of the *lac* operon in the absence of lactose when a cell bears either the I^- (a) or the O^C (b) mutation.

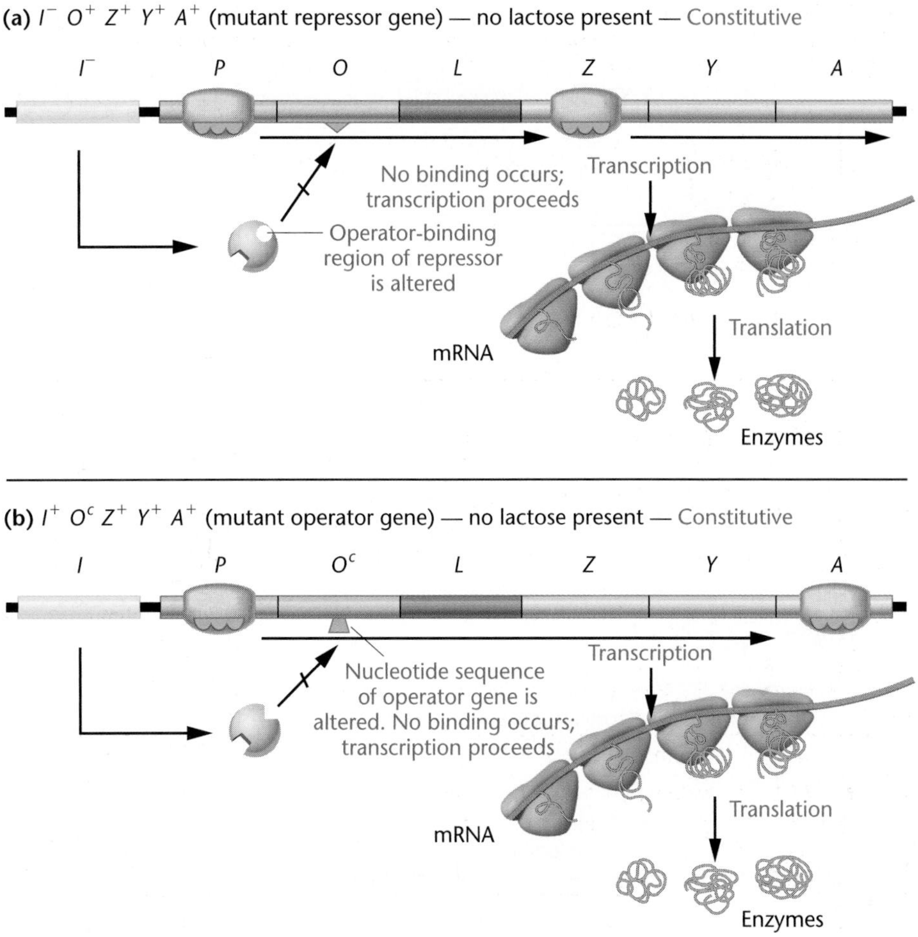

TABLE 15.1 A Comparison of Gene Activity (+ or −) in the Presence or Absence of Lactose for Various *E. coli* Genotypes

	Genotype	Presence of β-Galactosidase Activity: Lactose Present	Lactose Absent
	$I^+O^+Z^+$	+	−
A.	$I^+O^+Z^-$	−	−
	$I^-O^+Z^+$	+	+
	$I^+O^CZ^+$	+	+
B.	$I^-O^+Z^+/F'I^+$	+	−
	$I^+O^CZ^+/F'O^+$	+	+
C.	$I^+O^+Z^+/F'I^-$	+	−
	$I^+O^+Z^+/F'O^C$	+	−
D.	$I^SO^+Z^+$	−	−
	$I^SO^+Z^+/F'I^+$	−	−

Note: In parts B to D, most genotypes are partially diploid, containing an F factor plus attached genes (F′).

In fact, such a mutation, I^S, was discovered wherein the operon is "superrepressed," as shown in part D of Table 15.1. An additional I^+ gene does not effectively relieve repression of gene activity. These observations are consistent with the idea that the repressor contains separate DNA-binding domains and inducer binding domains. The binding of lactose to the inducer binding domain causes an allosteric change in the DNA-binding domain.

ESSENTIAL POINT

Genes involved in the metabolism of lactose are coordinately regulated by a negative control system that responds to the presence or absence of lactose.

Isolation of the Repressor

Although Jacob and Monod's operon theory succeeded in explaining many aspects of genetic regulation in prokaryotes, the nature of the repressor molecule was not known when their landmark paper was published in 1961. While they had assumed that the allosteric repressor was a protein, RNA was also a candidate because activity of the molecule required the ability to bind to DNA. Despite many attempts to isolate and characterize the hypothetical repressor molecule, no direct chemical evidence was immediately forthcoming. A single *E. coli* cell contains no more than ten or so molecules of the *lac* repressor; therefore, direct chemical identification of ten molecules in a population of millions of proteins and RNAs in a single cell presented a tremendous challenge. Nevertheless, in 1966, Walter Gilbert and Benno Müller-Hill reported the

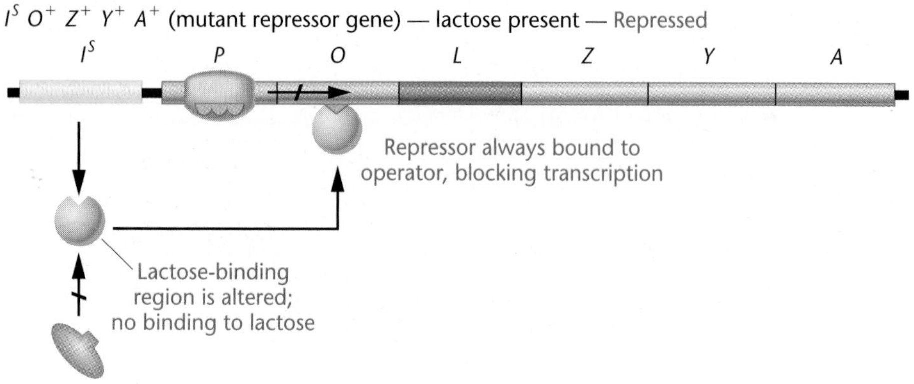

FIGURE 15–7 The response of the *lac* operon in the presence of lactose in a cell bearing the I^S mutation.

isolation of the *lac* repressor. Once the repressor was purified, it was shown to have various characteristics of a protein. The isolation of the repressor thus confirmed the operon model, which had been put forward strictly on genetic grounds.

NOW SOLVE THIS

Problem 6 on page 332 lists a number of *lac* operon genotypes and growth conditions of lactose presence or absence. You are asked to predict the outcome of gene expression under the various conditions listed.

Hint: Determine initially whether the repressor is active or inactive based on whether the gene encoding it is wild type or mutant. Then consider the impact of the presence or absence of lactose.

15.3 The Catabolite-Activating Protein (CAP) Exerts Positive Control over the *lac* Operon

As we discussed at the beginning of this section, the role of β-galactosidase is to cleave lactose into its components, glucose and galactose. Then, for galactose to be used by the cell, it also is converted to glucose. What if the cell found itself in an environment that contained an ample amount of lactose *and* glucose? Given that glucose is the preferred carbon source for *E. coli,* it would not be energetically efficient for a cell to induce transcription of the *lac* operon, make β-galactosidase, and metabolize lactose, since what it really needs—glucose—is already present. As we shall see next, a molecule called the **catabolite-activating protein (CAP)** helps activate expression of the *lac* operon, but is able to inhibit expression when glucose is present. This inhibition is called **catabolite repression.**

To understand CAP and its role in regulation, let's backtrack for a moment. When the *lac* repressor is bound to the inducer, the *lac* operon is activated and RNA polymerase transcribes the structural genes. As stated in Chapter 12, transcription is initiated as a result of the binding that occurs between RNA polymerase and the nucleotide sequence of the promoter region, found upstream (5′) from the initial coding sequences. Within the *lac* operon, the promoter is found between the *I* gene and the operator region (*O*). (See Figure 15–1.) Careful examination has revealed that RNA polymerase binding is never very efficient unless CAP is also present to facilitate the process.

The mechanism is summarized in Figure 15–8. In the absence of glucose and under inducible conditions, CAP exerts positive control by binding to the CAP site, facilitating RNA polymerase binding at the promoter and thus transcription. Therefore, for maximal transcription of the structural genes, the repressor must be bound by lactose (so as not to repress operon expression), *and* CAP must be bound to the CAP-binding site.

This leads us to the central question about CAP. What role does glucose play in inhibiting CAP binding when it is present? The answer involves still another molecule, **cyclic adenosine monophosphate (cAMP),** upon which CAP binding is dependent. In order to bind to the *lac* operon promoter, CAP must be bound to cAMP. The level of cAMP is itself dependent on an enzyme, **adenyl cyclase,** which catalyzes the conversion of ATP to cAMP.

The role of glucose in catabolite repression is to inhibit the activity of adenyl cyclase, causing a decline in the level of cAMP in the cell. Under this condition, CAP cannot form the CAP–cAMP complex essential to the positive control of transcription of the *lac* operon.

The structures of CAP and cAMP–CAP have been examined by using X-ray crystallography. CAP is a dimer that binds adjacent regions of a specific nucleotide sequence of the DNA making up the *lac* promoter. The cAMP–CAP complex, when bound to DNA, bends it, causing it to assume a new conformation.

Binding studies in solution further clarify the mechanism of gene activation. Alone, neither cAMP–CAP nor RNA polymerase has a strong affinity to bind to *lac* promoter DNA, nor does either molecule have a strong affinity to bind to the other. However, when both are together in the presence of the *lac* promoter DNA, a tightly bound complex is formed, an example of what is called **cooperative binding.** The control conferred by the cAMP-CAP provides another illustration of how the regulation of one small group of genes can be fine-tuned by several simultaneous influences.

In contrast to the negative regulation conferred by the *lac* repressor, the action of cAMP–CAP constitutes positive

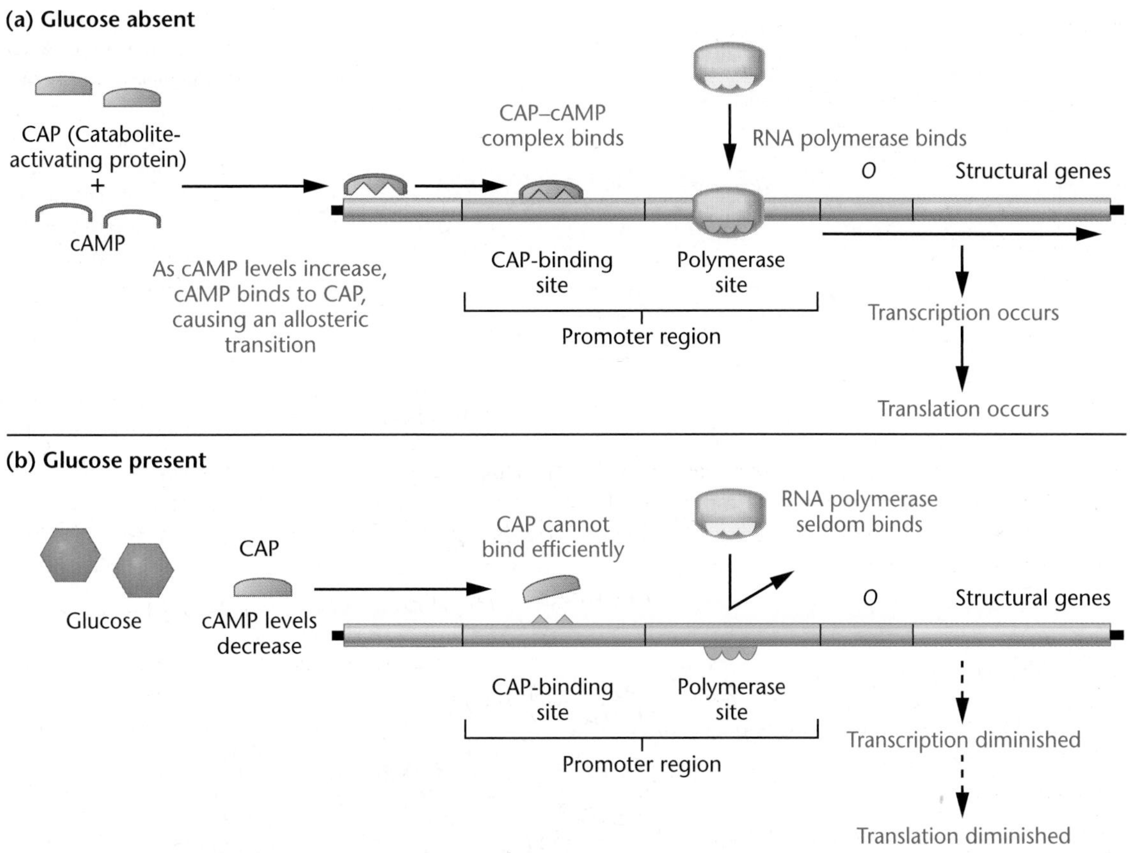

FIGURE 15–8 Catabolite repression. (a) In the absence of glucose, cAMP levels increase, resulting in the formation of a CAP–cAMP complex, which binds to the CAP site of the promoter, stimulating transcription. (b) In the presence of glucose, cAMP levels decrease, CAP–cAMP complexes are not formed, and transcription is not stimulated.

regulation. Thus, a combination of positive and negative regulatory mechanisms determine transcription levels of the *lac* operon. Catabolite repression involving CAP has also been observed for other inducible operons, including those controlling the metabolism of galactose and arabinose.

NOW SOLVE THIS

Problem 10 on page 332 describes four growth conditions for *E. coli* cells. You are asked to assess the level of gene activity in the *lac* operon for cells growing in the presence or absence of lactose and/or glucose.

Hint: You must keep in mind that regulation involving lactose is a negative control system, while regulation involving glucose is a positive control system.

ESSENTIAL POINT

The catabolite-activating protein (CAP) exerts positive control over *lac* gene expression by interacting with RNA polymerase at the *lac* promoter and by responding to the levels of cyclic AMP in the bacterial cell.

15.4 The Tryptophan (*trp*) Operon in *E. coli* Is a Repressible Gene System

Although inducible gene regulation had been known for some time, it was not until 1953 that Monod and colleagues discovered a repressible system. Studies on the biosynthesis of the essential amino acid tryptophan revealed that, if tryptophan is present in sufficient quantity in the growth medium, the enzymes necessary for its synthesis (such as **tryptophan synthase**) are not produced. It is energetically advantageous for bacteria to repress expression of genes involved in tryptophan synthesis when ample tryptophan is present in the growth medium.

Further investigation showed that a series of enzymes encoded by five contiguous genes on the *E. coli* chromosome is involved in tryptophan synthesis. These genes are part of an operon, and in the presence of tryptophan, all are coordinately repressed, and none of the enzymes is produced. Because of the great similarity between this repression and the induction of enzymes for lactose metabolism, Jacob and Monod proposed a model of gene regulation resembling that of the *lac* system (Figure 15–9, p. 316).

The model suggests the presence of a *normally inactive repressor* that alone cannot interact with the operator region of

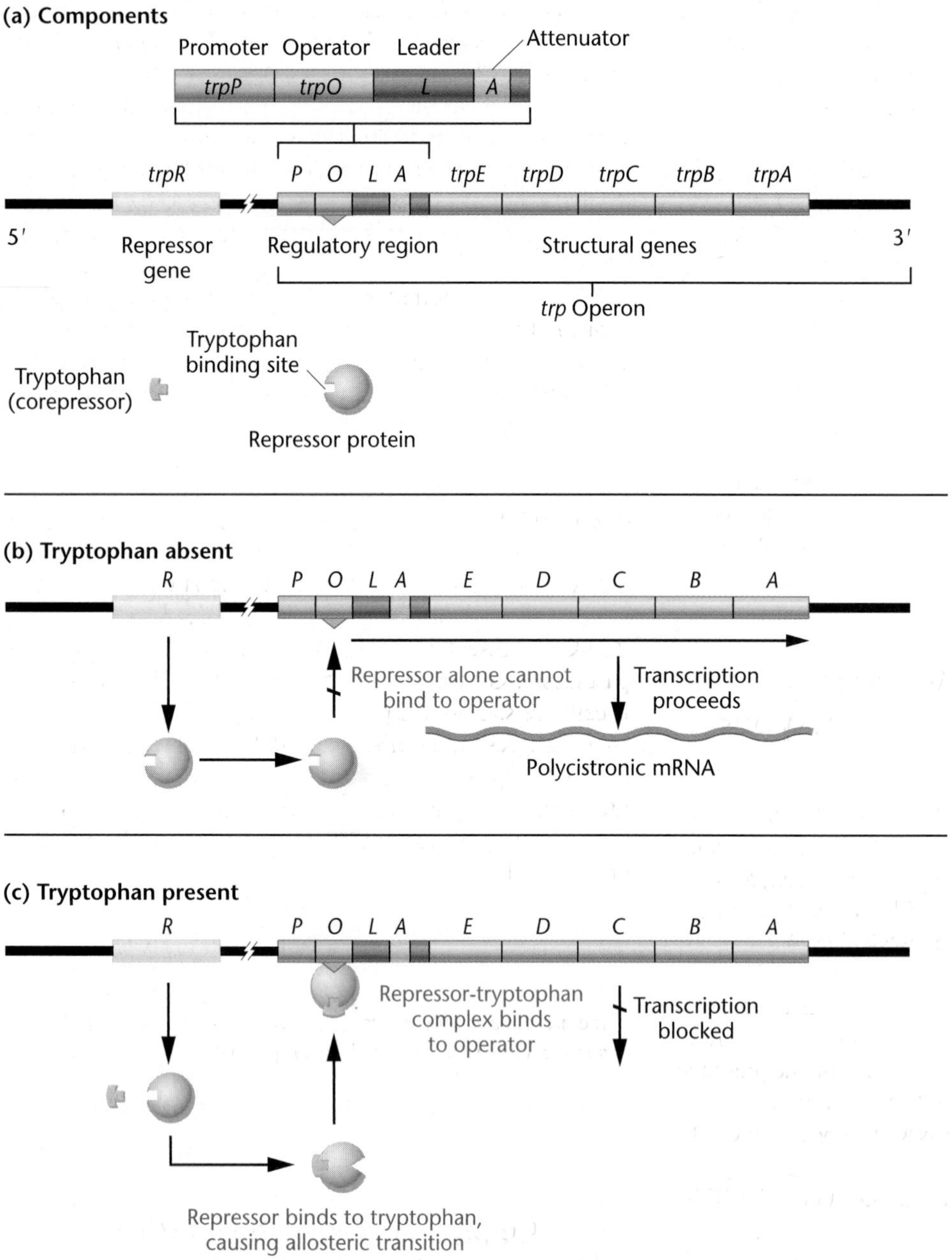

FIGURE 15–9 (a) The components involved in the regulation of the tryptophan operon. (b) Regulatory conditions are depicted that involve either activation or (c) repression of the structural genes. In the absence of tryptophan, an inactive repressor is made that cannot bind to the operator (O), thus allowing transcription to proceed. In the presence of tryptophan, it binds to the repressor, causing an allosteric transition to occur. This complex binds to the operator region, leading to repression of the operon.

the operon. However, the repressor is an allosteric molecule that can bind to tryptophan. When tryptophan is present, the resultant complex of repressor and tryptophan attains a new conformation that binds to the operator, repressing transcription. Thus, when tryptophan, the end product of this anabolic pathway, is present, the system is repressed and enzymes are not made. Since the regulatory complex inhibits transcription of the operon, this repressible system is under negative control. And as tryptophan participates in repression, it is referred to as a **corepressor** in this regulatory scheme.

Evidence for the *trp* Operon

Support for the concept of a repressible operon is based primarily on the isolation of two distinct categories of constitutive mutations. The first class, $trpR^-$, maps at a considerable distance from the structural genes. This locus represents the gene coding for the repressor. Presumably, the mutation either inhibits the interaction of the repressor with tryptophan or inhibits repressor formation entirely. Whichever the case, no repression is present in cells with the $trpR^-$ mutation. As expected, if the $trpR^+$ gene encodes a functional repressor molecule, the presence of a copy of this gene will restore repressibility.

The second constitutive mutant is analogous to the O^C mutant of the lactose operon because it maps immediately adjacent to the structural genes. Furthermore, the addition of a wild-type operator gene into mutant cells (as an external element) does not restore repression. This is predictable if the mutant operator no longer interacts with the repressor–tryptophan complex.

The entire *trp* operon has now been well defined, as shown in Figure 15–9. Five contiguous structural genes (*trp E, D, C,*

B, and *A*) are transcribed as a polycistronic message directing translation of the enzymes that catalyze the biosynthesis of tryptophan. As in the *lac* operon, a promoter region (*trpP*) represents the binding site for RNA polymerase, and an operator region (*trpO*) is the binding site for the repressor. In the absence of binding, transcription is initiated within the overlapping *trpP–trpO* region and proceeds along a **leader sequence** 162 nucleotides prior to the first structural gene (*trpE*). Within that leader sequence, still another regulatory site exists, called an *attenuator*—the subject of the next section of this chapter. As we shall see, this regulatory unit is an integral part of the control mechanism of the operon.

ESSENTIAL POINT

Unlike the inducible *lac* operon, the *trp* operon is repressible. In the presence of tryptophan, the repressor binds to the regulatory region of the *trp* operon and represses transcription initiation.

15.5 Attenuation Is a Regulatory Mechanism in Some Prokaryotic Operons

Many bacterial operons utilize another level of control over transcription—a mechanism known as attenuation. Attenuation regulates the *E. coli* operons that encode enzymes involved in the biosynthesis of amino acids such as tryptophan, threonine, histidine, leucine, and phenylalanine. As with the *trp* operon, attenuation occurs in a leader sequence that contains an attenuator region (see Figure 15–9). The attenuators in these operons contain multiple codons specifying the amino acid being regulated. As described in this section, the presence of these codons in the attenuator regulates the progress of cotranscriptional translation of the mRNA. The presence or absence of the amino acid end product affects ribosome progress, resulting in elongation or premature termination of mRNA transcription.

Charles Yanofsky, his coworker Kevin Bertrand, and their colleagues defined the mechanisms of bacterial attenuation. They observed that, even when tryptophan is present and the *trp* operon is repressed, initiation of transcription still occurs, but it is terminated at a point about 140 nucleotides along the transcript. This finding suggested that there must be a mechanism by which tryptophan inhibits transcription of the entire operon. This process was called **attenuation,** indicative of its effect in diminishing genetic expression of the operon. When tryptophan is absent, or present in very low concentrations, transcription is initiated and *not* subsequently terminated, instead continuing beyond the leader sequence into the structural genes. As a result, a polycistronic mRNA is produced, and the enzymes essential to the biosynthesis of tryptophan are translated.

Identification of the site involved in attenuation was made possible by the isolation of various deletion mutations in the region 115 to 140 nucleotides into the leader sequence. Such mutations abolish attenuation. This region is referred to as the **attenuator.**

Yanofsky and colleagues presented a model to explain how attenuation occurs and is regulated. The initial DNA sequence that is transcribed gives rise to an mRNA sequence that has the potential to fold into two mutually exclusive stem-loop structures referred to as "hairpins." In the presence of excess tryptophan, the mRNA hairpin that is formed behaves as a **terminator** structure, and transcription is almost always terminated prematurely, just beyond the attenuator. On the other hand, if tryptophan is scarce, an alternative mRNA hairpin referred to as the **antiterminator hairpin** is formed. Transcription proceeds past the antiterminator hairpin region, and the entire mRNA is subsequently produced.

The question is how the absence (or a low concentration) of tryptophan allows attenuation to be bypassed. A key point in Yanofsky's model is that the leader transcript must be translated in order for the antiterminator hairpin to form. Yanofsky discovered that the leader transcript includes two triplets (UGG) that encode tryptophan preceded upstream by the initial AUG sequence that prompts the initiation of translation by ribosomes. When adequate tryptophan is present, charged $tRNA^{trp}$ is present in the cell. As a result, translation proceeds past these UGG triplets, and the *terminator hairpin* is formed. If cells are starved of tryptophan, charged $tRNA^{trp}$ is unavailable. As a result, the ribosome "stalls" during translation of the UGG triplets. This event allows the formation of the antiterminator hairpin within the leader transcript. As a result, transcription proceeds, leading to expression of the entire set of structural genes.

ESSENTIAL POINT

The process of attenuation, which regulates operons based on the presence of an end product, involves alterations to mRNA secondary structure, leading to premature termination of transcription.

15.6 Eukaryotic Gene Regulation Differs from That in Prokaryotes

In eukaryotes, gene expression is finely tuned in specific cell types and in response to complex changes in the cellular environment. For example, some white blood cells express genes encoding certain immunoglobulins, allowing these cells to synthesize antibodies that defend the organism from infection and foreign agents. However, skin, kidney, and liver cells do not express immunoglobulin genes. Pancreatic islet cells synthesize and secrete insulin in response to the presence of blood sugars; however, they do not manufacture immunoglobulins. In addition, they do not synthesize insulin when it is not required. Eukaryotic cells, as part of multicellular organisms, do not grow solely in response to the availability of nutrients. Instead, they regulate their growth and division to occur at appropriate places in the body and at appropriate times during development. The loss of gene regulation that controls normal cell growth and division may lead to developmental defects or cancer.

To achieve this degree of fine tuning, eukaryotes employ a wide range of mechanisms for altering the expression of genes. In contrast to prokaryotic gene regulation, which occurs primarily at the level of transcription initiation, regulation of gene expression in eukaryotes can occur at many different levels. These include the initiation of transcription, mRNA modifications and stability, and synthesis, modification, and stability of the protein product (Figure 15–10).

Several features of eukaryotic cells make it possible for them to use more types of gene regulation than are possible in prokaryotic cells:

- Eukaryotic cells contain a much greater amount of DNA than do prokaryotic cells, and this DNA is complexed with histones and other proteins to form highly compact chromatin structures within an enclosed nucleus. Eukaryotic cells modify this structural organization in order to influence gene expression.
- The mRNAs of most eukaryotic genes must be spliced, capped, and polyadenylated prior to transport from the nucleus. Each of these processes can be regulated in order to influence the numbers and types of mRNAs available for translation.
- Genetic information in eukaryotes is carried on many chromosomes (rather than just one), and these chromosomes are enclosed within a double-membrane-bound nucleus. After transcription, transport of RNAs into the cytoplasm can be regulated in order to modulate the availability of mRNAs for translation.
- Eukaryotic mRNAs can have a wide range of half-lives ($t_{1/2}$). In contrast, the majority of prokaryotic mRNAs decay very rapidly. Rapid turnover of mRNAs allows prokaryotic cells to rapidly respond to environmental changes. In eukaryotes, the complement of mRNAs in each cell type can be more subtly manipulated by altering mRNA decay rates over a larger range.
- In eukaryotes, translation rates can be modulated, as well as the way proteins are processed, modified, and degraded.

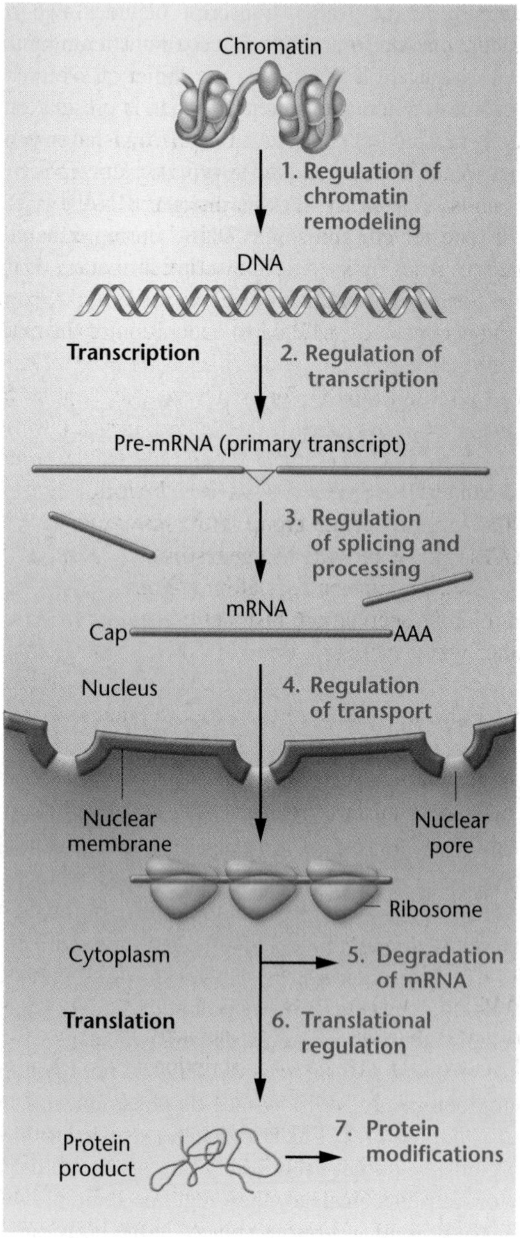

FIGURE 15–10 Regulation can occur at any stage in the expression of genetic material in eukaryotes. All these forms of regulation affect the degree to which a gene is expressed.

REGULATION OF GENE EXPRESSION: EUKARYOTES

WEB TUTORIAL

In the following sections, we examine some of the major ways in which eukaryotic gene expression is regulated. As most eukaryotic genes are regulated, at least in part, at the transcriptional level, we will emphasize transcriptional control, although we will discuss other levels of control as well. In addition, we will limit our discussion to regulation of genes transcribed by RNA polymerase II. As we previously described in Chapter 12, eukaryotic genes use three RNA polymerases for transcription. In contrast, all prokaryotic genes are transcribed by a single RNA polymerase. In eukaryotes, RNA polymerase II transcribes all mRNAs and some small nuclear RNAs, whereas RNA polymerases I and III transcribe ribosomal RNAs, some small nuclear RNAs, and transfer RNAs. The promoter for each type of polymerase has a different nucleotide sequence and binds different transcription factors.

15.7 Eukaryotic Gene Expression Is Influenced by Chromosome Organization and Chromatin Modifications

Two structural features of eukaryotic genes distinguish them from the genes of prokaryotes. First, eukaryotic genes are situated on chromosomes that occupy a distinct location within the cell—the nucleus. This sequestering of genetic information in a discrete compartment allows the proteins that directly regulate transcription to be kept apart from those involved with translation and other aspects of cellular metabolism. Second, as described in Chapter 11, eukaryotic DNA is combined with histones and nonhistone proteins to form chromatin. Chromatin's basic structure is characterized by repeating units called nucleosomes that are wound into 30-nm fibers, which in turn form other, even more compact structures. The compactness of these chromatin structures is inhibitory to many processes, including transcription, replication, and DNA repair. In this section, we outline some of the ways in which eukaryotic cells use these structural features of eukaryotic genes to regulate their expression.

Chromosome Territories and Transcription Factories

During interphase of the cell cycle, chromosomes are unwound and cannot be seen as intact structures by light microscopy. The development of chromosome-painting techniques has revealed that the interphase nucleus is not a bag of tangled chromosome arms, but has a highly organized structure. In the interphase nucleus, each chromosome occupies a discrete domain called a **chromosome territory** and stays separate from other chromosomes. Channels between chromosomes contain little or no DNA and are called **interchromosomal domains.**

Chromosome organization appears to be continuously rearranging so that transcriptionally active genes are cycled to the edge of chromosome territories at the border of the interchromosomal domain channels. Although evidence suggests that transcription of many, if not most, genes occurs when they are in direct contact with the interchromosomal compartment, other evidence suggests that transcription can also take place within chromosomal territories. The physiological role of gene relocation within territories is not known. However, it is hypothesized that this organization may bring actively expressed genes into closer association with transcription factors, or with other actively expressed genes, thereby facilitating their coordinated expression.

Another feature within the nucleus—the **transcription factory**—may also contribute to regulating gene expression. Transcription factories are specific nuclear sites at which most RNA polymerase II transcription occurs. These sites also contain the majority of active RNA polymerase and other transcription factors. By concentrating transcription proteins and actively transcribed genes in specific locations in the nucleus, the cell may enhance the expression of these genes.

Chromatin Remodeling

The ability of the cell to alter the association of DNA with other chromatin components is essential to allow regulatory proteins to access DNA. Hence, chromatin modification, referred to as **chromatin remodeling,** is an important step in gene regulation. Chromatin remodeling appears to be a prerequisite for transcription of some eukaryotic genes, although it can occur simultaneously with transcription of other genes.

Chromatin can be remodeled in two general ways. The first involves changes to nucleosomes, and the second involves modifications to DNA. In this section, we will discuss changes to the nucleosomal component of chromatin. In the next subsection, we present DNA modifications, specifically DNA methylation.

Nucleosomal chromatin can be remodeled in three ways: by altering nucleosome composition, by adding or removing covalent modifications to or from histones, and by repositioning the nucleosome on a gene region. These remodeling functions are performed by specific chromatin remodeling complexes, all of which require ATP hydrolysis. Remodeling complexes are recruited to specific genes that are tagged for transcription by the presence of certain transcription activators or repressors on their promoter regions, or by the presence of modified histones or methylated DNA in the genes' regulatory regions.

Changes in nucleosome composition can affect gene transcription. For example, most nucleosomes contain the normal histone H2A. However, the promoter regions of transcriptionally active and potentially active genes are often flanked by nucleosomes containing variant histones, such as H2A.Z. These variant nucleosomes help keep promoter regions free of repressive nucleosomes, thereby facilitating gene transcription.

A second mechanism of chromatin alteration involves histone modification. One such modification is acetylation, a chemical alteration of the histone component of nucleosomes that is catalyzed by **histone acetyltransferase enzymes (HATs).** When an acetate group is added to specific basic amino acids on the histone tails, the attraction between the basic histone protein and acidic DNA is lessened. HATs are recruited to genes by specific transcription factors. The loosening of histones from DNA facilitates further chromatin remodeling catalyzed by ATP-dependent chromatin remodeling complexes, as described below. These modifications make promoter regions available for binding to transcription factors that initiate the chain of events leading to gene transcription, as well as to RNA polymerase. Of course, what can be opened can also be closed. In that case, **histone deacetylases (HDACs)** remove acetate groups from histone tails. HDACs, like HATs, can be recruited to genes by the presence of certain repressor proteins on regulatory regions.

In addition to acetylation, histones can be modified in several other ways, including phosphorylation and methylation. These structural alterations occur at specific amino acid residues in histones. It has been proposed that such specific, reversible patterns of covalent histone modifications bring about gene activation or gene silencing.

The third mechanism for chromatin remodeling involves the repositioning of nucleosomes on DNA. Chromatin remodeling complexes that reposition nucleosomes make different regions of the chromosome accessible to transcription proteins, including transcription activators, transcription repressors, and RNA polymerase II. One of the best-studied remodeling complexes is the **SWI/SNF** complex. Proteins in the SWI/SNF complex were originally identified as transcriptional activators, since their actions lead to increases in gene transcription. Remodelers such as SWI/SNF can act in several different ways (Figure 15–11, p. 320). They may loosen the attachment between histones and DNA, resulting in the nucleosome sliding along the DNA and exposing regulatory regions. Alternatively, they may loosen the DNA strand from the nucleosome core, or they may cause reorganization of the internal nucleosome components. In all cases, the DNA is left transiently exposed to association with transcription factors and RNA polymerase.

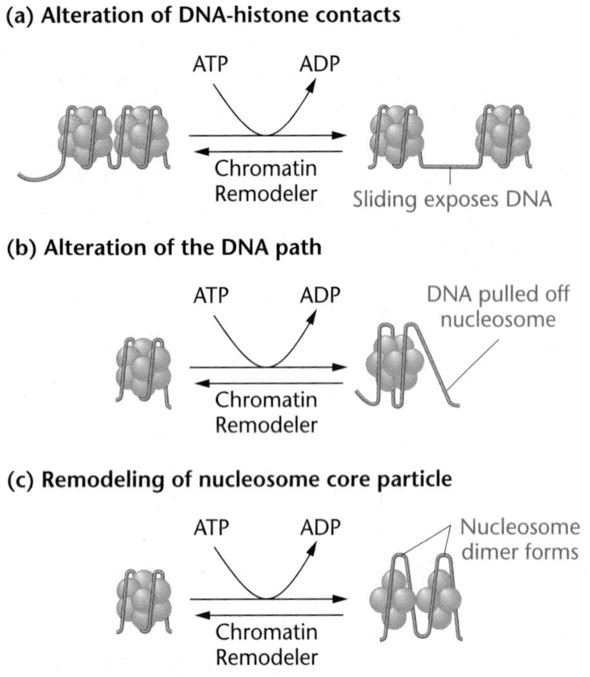

FIGURE 15–11 Three ways by which chromatin remodelers, such as the SWI/SNF complex, alter the association of nucleosomes with DNA. (a) The DNA–histone contacts may be loosened, allowing the nucleosomes to slide along the DNA, exposing DNA regulatory regions. (b) The path of the DNA around a nucleosome core particle may be altered. (c) Components of the core nucleosome particle may be rearranged, resulting in a modified nucleosome structure.

DNA Methylation

Another type of change in chromatin that plays a role in gene regulation is the addition or removal of methyl groups to or from bases in DNA. The DNA of most eukaryotic organisms can be modified after DNA replication by the enzyme-mediated addition of methyl groups to bases and sugars. **DNA methylation** most often involves cytosine. In the genome of any given eukaryotic species, approximately 5 percent of the cytosine residues are methylated. However, the extent of methylation can be tissue specific and can vary from less than 2 percent to more than 7 percent of cytosine residues.

Evidence of a role for methylation in eukaryotic gene expression is based on a number of observations. First, an inverse relationship exists between the degree of methylation and the degree of expression. That is, low amounts of methylation are associated with high levels of gene expression, and high levels of methylation are associated with low levels of gene expression. Large transcriptionally inert regions of the genome are often heavily methylated. In mammalian females, the inactivated X chromosome, which is almost totally transcriptionally inactive, has a higher level of methylation than does the active X chromosome. Within the inactive X chromosome, regions that escape inactivation have much lower levels of methylation than those in adjacent inactive regions.

Second, methylation patterns are tissue specific and, once established, are heritable for all cells of that tissue. It appears that proper patterns of DNA methylation are essential for normal mammalian development. Undifferentiated embryonic cells that are not able to methylate DNA die when they are required to differentiate into specialized cell types. Transgenic mice that are unable to methylate DNA die soon after birth. A loss of normal DNA methylation appears to occur in cancer cells that have lost the ability to regulate their growth and division.

Perhaps the most direct evidence for the role of methylation in gene expression comes from studies using base analogs. The nucleotide **5-azacytidine** can be incorporated into DNA in place of cytidine during DNA replication. This analog cannot be methylated, causing the undermethylation of the sites where it is incorporated. The incorporation of 5-azacytidine into DNA changes the pattern of gene expression and stimulates expression of alleles on inactivated X chromosomes. In addition, the presence of 5-azacytidine in DNA can induce the expression of genes that would normally be silent in certain differentiated cells.

How might methylation affect gene regulation? Data from *in vitro* studies suggest that methylation can repress transcription by inhibiting the binding of transcription factors to DNA. Methylated DNA may also recruit repressive chromatin remodeling complexes to gene-regulatory regions. Another observation is that certain proteins bind to 5-methyl cytosine without regard to the DNA sequence. It is possible that these proteins recruit transcription repressor proteins or repressive chromatin remodeling complexes.

NOW SOLVE THIS

Problem 28 on page 333 presents data from experiments designed to measure the activity of a reporter gene when various regions upstream of the gene are methylated. You are asked to interpret the effect of methylation of DNA on the expression of a gene.

Hint: Remember that the location of various regulatory sequences outside the gene will affect the results.

ESSENTIAL POINT

Eukaryotic gene regulation at the level of chromatin may involve gene-specific chromatin remodeling, histone modifications, or DNA modifications. These modifications may either allow or inhibit access of promoters and enhancers to transcription factors, resulting in increased or decreased levels of transcription initiation.

15.8 Eukaryotic Transcription Is Regulated at Specific *Cis*-Acting Sites

Eukaryotic transcription regulation requires the interactions of many regulatory factors with specific DNA sequences located in and around genes, as well as with sequences located at great distances. In this section, we will discuss some of these *cis*-acting DNA sequences, including promoters, enhancers, and silencers (Figure 15–12).

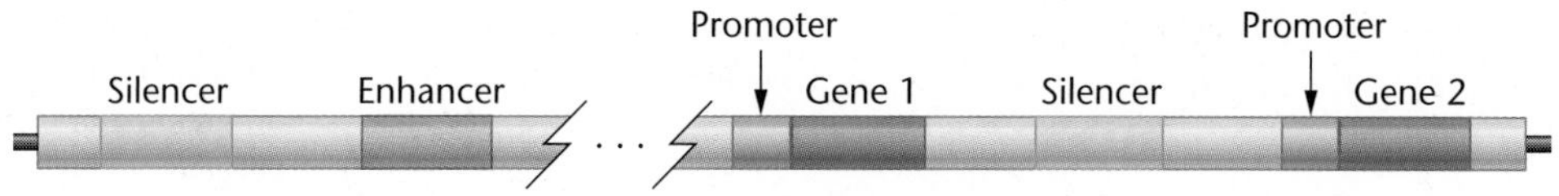

FIGURE 15–12 Transcription of eukaryotic genes is controlled by regulatory elements directly adjacent to the gene (promoters) and by others located at a distance (enhancers and silencers).

Promoters

Promoters are nucleotide sequences that serve as recognition sites for transcription proteins. They are necessary in order for transcription to be initiated at a basal level. Promoters are located immediately adjacent to the genes they regulate. These regions are usually several hundred nucleotides in length and specify the site at which transcription begins and the direction of transcription along the DNA.

The promoters of most eukaryotic genes that are transcribed by RNA polymerase II contain one or more elements, including TATA, CAAT, and GC boxes, as well as the transcription start site (Figure 15–13). Some genes utilize elements downstream of the start site of transcription (to about +30) as part of their promoters.

Most, but not all, eukaryotic gene promoters contain TATA boxes. Located about 25 to 30 bases upstream from the transcription start site (a location designated as −25 to −30), the **TATA box** consists of a 7- to 8-bp consensus sequence composed of AT base pairs. The TATA box is often flanked on either side by GC-rich regions. Genetic analysis shows that mutations within TATA sequences reduce transcription (Figure 15–14, p. 322), and that mutations may also alter the initiation point of transcription.

Many promoter regions also contain **CAAT boxes.** These elements have the consensus sequence CAAT or CCAAT. The CAAT box frequently appears 70 to 80 bp upstream from the start site. Mutational analysis suggests that CAAT boxes (when present) are critical to the promoter's ability to initiate transcription. Mutations on either side of this element have no effect on transcription, whereas mutations within the CAAT sequence dramatically lower the rate of transcription (Figure 15–14). The **GC box,** another element often found in promoter regions, has the consensus sequence GGGCGG and is located, in one or more copies, at about position −110. The CAAT and GC elements function somewhat like enhancers, which we will cover in the next section.

In summary, each eukaryotic gene contains a number of promoter elements that specify the basal level of transcription from that gene. Different genes may have different subsets of these promoter elements, and they may vary in location and organization from gene to gene.

Enhancers and Silencers

Transcription of eukaryotic genes is regulated not only by promoters but also by DNA sequences called **enhancers.** Enhancers can be located on either side of a gene, at some distance from the gene, or even within the gene. They are called ***cis* regulators** because they function when adjacent to the structural genes they regulate, as opposed to ***trans* regulators** (such as DNA-binding proteins), which can regulate a gene on any chromosome. While promoter sequences are essential for basal-level transcription, enhancers are necessary for achieving the maximum level of transcription. In addition, enhancers are responsible for time- and tissue-specific gene expression. Thus, there is some degree of analogy between enhancers and operator regions in prokaryotes. However, enhancers are more complex in both structure and function.

Scientists have studied promoters and enhancers by analyzing the effects that specific mutations have on the transcription of cloned genes introduced into cultured cells. In addition, they have moved these elements into new positions relative to the coding genes to which the elements are attached. These studies have revealed several features that distinguish promoters from enhancers:

1. The position of an enhancer need not be fixed; it will function whether it is upstream, downstream, or within the gene it regulates.

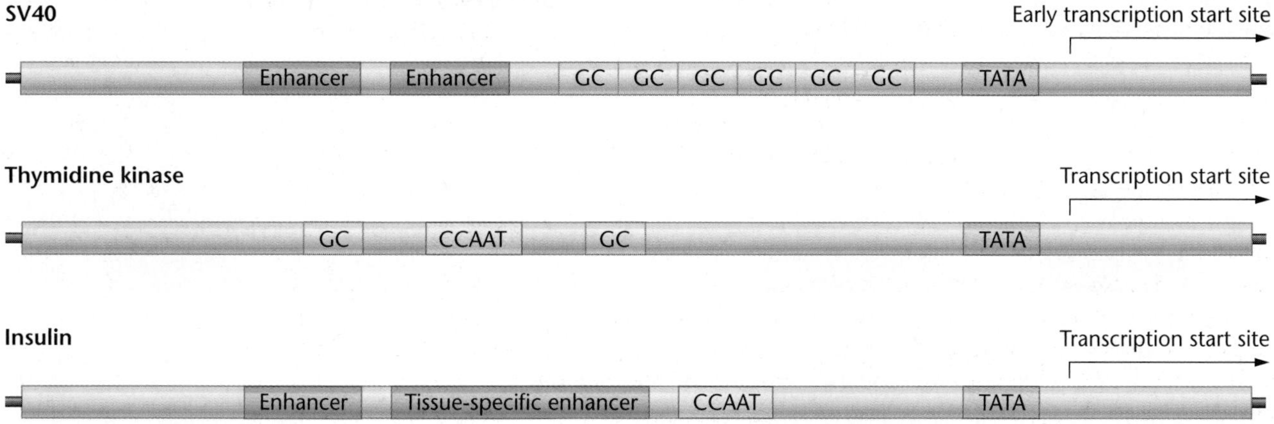

FIGURE 15–13 Organization of transcription regulatory regions in promoters of several genes expressed in eukaryotic cells, illustrating the variable nature, number, and arrangement of controlling elements.

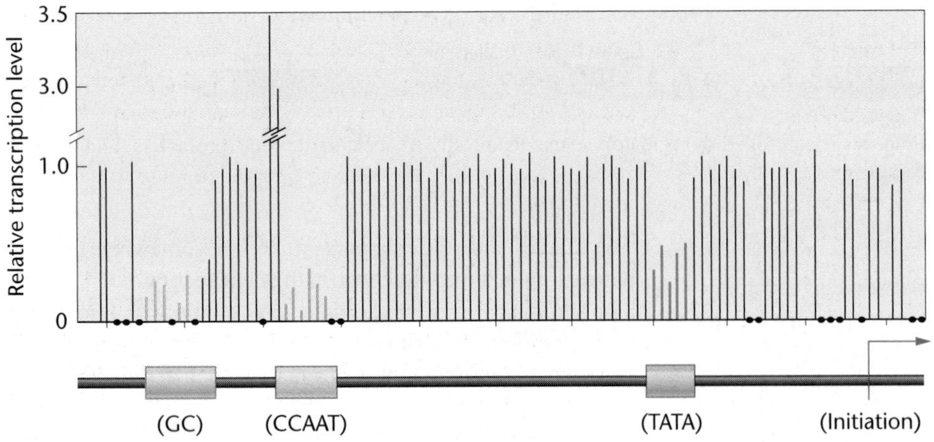

FIGURE 15–14 Summary of the effects on transcription levels of different point mutations in the promoter region of the β-globin gene. Each line represents the level of transcription produced in a separate experiment by a single-nucleotide mutation (relative to wild-type) at a particular location. Dots represent nucleotides for which no mutation was obtained. Note that mutations within specific elements of the promoter have the greatest effects on the level of transcription.

2. The orientation of an enhancer can be inverted without significant effect on its action.
3. If an enhancer is experimentally moved adjacent to a gene elsewhere in the genome, or if an unrelated gene is placed near an enhancer, the transcription of the newly adjacent gene is enhanced.

An example of an enhancer located *within* the gene it regulates is the immunoglobulin heavy-chain gene enhancer, which is located in an intron between two exons. This enhancer is active only in cells expressing the immunoglobulin genes, indicating that tissue-specific gene expression can be modulated through the enhancer. Internal enhancers have been discovered in other eukaryotic genes, including the immunoglobulin light-chain gene. An example of a downstream enhancer is the β-globin gene enhancer. In chickens, an enhancer located between the β-globin gene and the ε-globin gene works in one direction to control transcription of the ε-globin gene during embryonic development and in the opposite direction to regulate expression of the β-globin gene during adult life.

Enhancers are modular and often contain several different short DNA sequences. For example, the enhancer of the SV40 virus (which is transcribed inside a eukaryotic cell) has a complex structure consisting of two adjacent sequences of approximately 100-bp each, located some 200 bp upstream from a transcriptional start point. One of these sequences is shown in Figure 15–15. Each of the two 100-bp regions contains multiple sequence motifs that contribute to achieving the maximum rate of transcription. If one or the other of these regions is deleted, there is no effect on transcription; but if both are deleted, *in vivo* transcription is greatly reduced.

Another type of *cis*-acting transcription regulatory element, the **silencer,** acts upon eukaryotic genes to repress the level of transcription initiation. Silencers, like enhancers, are short DNA sequence elements that affect the rate of transcription initiated from an associated promoter. They often act in tissue- or temporal-specific ways to control gene expression. An example of a silencer element is found in the human thyrotropin-β gene. The thyrotropin-β gene encodes one subunit of the thyrotropin hormone, and this gene is expressed only in thyrotrophs (thyrotropin-producing cells) of the pituitary gland. Transcription of this gene is restricted to thyrotrophs due to the actions of a silencer element located -140 bp upstream from the transcription start site. The silencer binds a cellular factor known as Oct-1, which represses transcription in all cell types except thyrotrophs. In thyrotrophs, the action of the silencer is overcome by an enhancer element located over 1.2 kb upstream of the promoter.

ESSENTIAL POINT

Eukaryotic transcription is regulated at gene-specific promoter, enhancer, and silencer elements. These *cis*-acting regulatory sites may act constitutively or may be active in tissue- or temporal-specific ways.

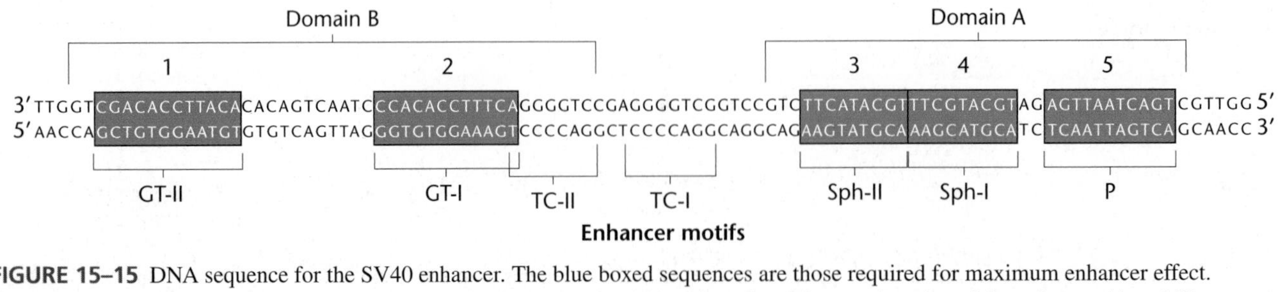

FIGURE 15–15 DNA sequence for the SV40 enhancer. The blue boxed sequences are those required for maximum enhancer effect. The brackets below the sequence name the various sequence motifs within this region. The two domains of the enhancer (A and B) are indicated.

15.9 Eukaryotic Transcription Is Regulated by Transcription Factors that Bind to *Cis*-Acting Sites

It is generally accepted that *cis*-acting regulatory sites—including promoters, enhancers, and silencers—influence transcription initiation by acting as binding sites for transcription regulatory proteins. These transcription regulatory proteins, known as **transcription factors,** can have diverse and complicated effects on transcription. Some transcription factors are expressed in tissue-specific ways, thereby regulating their target genes for tissue-specific levels of expression. In addition, some transcription factors are expressed in cells at certain times during development or in response to external physiological signals. In some cases, a transcription factor that binds to a *cis*-acting site and regulates a certain gene may be present in a cell and may even bind to its appropriate *cis*-acting site but will only become active when modified structurally (for example, by phosphorylation or by binding to a coactivator such as a hormone). These modifications to transcription factors can also be regulated in tissue- or temporal-specific ways. In addition, different transcription factors may compete for binding to a given DNA sequence or to one of two overlapping sequences. In these cases, transcription factor concentrations and the strength with which each factor binds to the DNA will dictate which factor binds. The same site may also bind different factors in different tissues. Finally, multiple transcription factors that bind to several different enhancers and promoter elements within a gene-regulatory region can interact with each other to fine-tune the levels and timing of transcription initiation.

The Human Metallothionein IIA Gene: Multiple *Cis*-Acting Elements and Transcription Factors

The **human metallothionein IIA gene** (*hMTIIA*) provides an example of how one gene can be transcriptionally regulated through the interplay of multiple promoter and enhancer elements and the transcription factors that bind to them. The product of the *hMTIIA* gene is a protein that binds to heavy metals such as zinc and cadmium, thereby protecting cells from the toxic effects of high levels of these metals. The gene is expressed at low levels in all cells but is transcriptionally induced to express at high levels when cells are exposed to heavy metals and steroid hormones such as glucocorticoids.

The *cis*-acting regulatory elements controlling transcription of the *hMTIIA* gene include promoter, enhancer, and silencer elements (Figure 15–16). Each *cis*-acting element is a short DNA sequence that has specificity for binding to one or more transcription factors.

The *hMTIIA* gene contains the promoter elements TATA box and start site, which specify the start of transcription. The promoter element, GC, binds the SP1 factor, which is present in most eukaryotic cells and stimulates transcription at low levels in most cells. Basal levels of expression are also regulated by the BLE (basal element) and ARE (AP factor response element) regions. These *cis*-elements bind the activator proteins 1, 2, and 4 (AP1, AP2, and AP4), which are present in various levels in different cell types and can be activated in response to extracellular growth signals. The BLE element contains overlapping binding sites for the AP1 and AP4 factors, providing some degree of selectivity in how these factors stimulate transcription of *hMTIIA* when bound to the BLE in different cell types. High levels of transcription induction are conferred by the MRE (metal response element) and GRE (glucocorticoid response element). The metal-inducible transcription factor (MTF1) binds to the MRE element in response to the presence of heavy metals. The glucocorticoid receptor protein binds to the GRE, but only when the receptor protein is also bound to the glucocorticoid steroid hormone. The glucocorticoid receptor is normally located in the cytoplasm of the cell; however, when glucocorticoid hormone enters the cytoplasm, it binds to the receptor and causes a conformational change that allows the receptor to enter the nucleus, bind to the GRE, and stimulate *hMTIIA* gene transcription. In addition to induction, transcription of the *hMTIIA* gene can be repressed by the actions of the repressor protein PZ120, which binds over the transcription start region.

The presence of multiple regulatory elements and transcription factors that bind to them allows the *hMTIIA* gene to be transcriptionally induced or repressed in response to subtle changes in both extracellular and intracellular conditions.

Functional Domains of Eukaryotic Transcription Factors

We have described transcription factors as proteins that bind to DNA and activate or repress transcription initiation. These

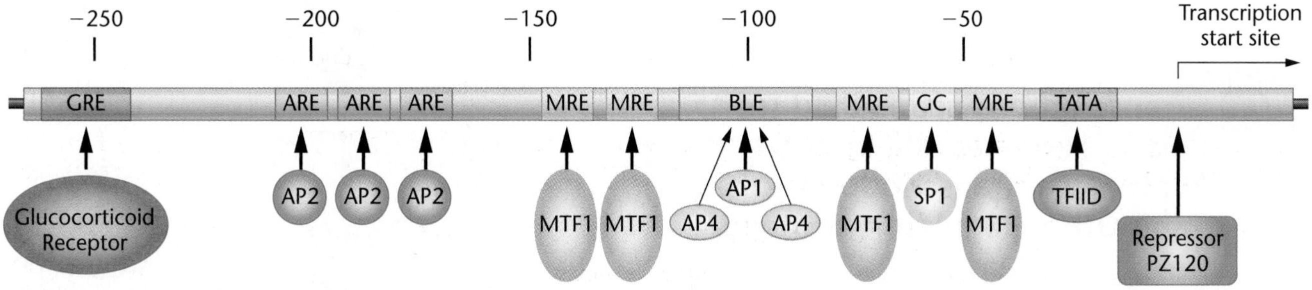

FIGURE 15–16 The human metallothionein IIA gene promoter and enhancer regions, containing multiple *cis*-acting regulatory sites. The transcription factors controlling both basal and induced levels of MTIIA transcription, and their binding sites, are indicated below the gene and are described in the text.

actions are achieved through the presence of two functional domains (clusters of amino acids that carry out a specific function). One domain, the **DNA-binding domain,** binds to specific DNA sequences present in the *cis*-acting regulatory site; the other, the ***trans*-activating** or ***trans*-repression domain,** activates or represses transcription. The *trans*-activating and *trans*-repression domains bring about their effects by interacting with other transcription factors or RNA polymerase, as we will discuss in the next section.

The DNA-binding domains of eukaryotic transcription factors have various characteristic three-dimensional structural motifs. Examples include the helix–turn–helix (HTH), zinc-finger, and basic leucine zipper (bZIP) motifs.

The **helix–turn–helix (HTH)** motif, present in both prokaryotic and eukaryotic transcription factors, is characterized by a certain geometric conformation rather than a distinctive amino acid sequence. The presence of two adjacent α-helices separated by a "turn" of several amino acids (hence the name of the motif) enables the protein to bind to DNA. The **zinc-finger** motif is found in a wide range of transcription factors that regulate gene expression related to cell growth, development, and differentiation. A typical zinc-finger protein contains clusters of two cysteines and two histidines at repeating intervals. These clusters bind zinc atoms, fold into loops, and interact with specific DNA sequences. The **basic leucine zipper (bZIP)** motif contains a region called a leucine zipper that allows protein–protein dimerization. When two bZIP-containing molecules dimerize, the leucine residues "zip" together. The resulting dimer contains two basic α-helical regions adjacent to the zipper that bind to phosphate residues and specific bases in DNA.

Transcription factors also contain *trans*-activating or *trans*-repressing domains that are distinct from their DNA-binding domains. These domains can occupy from 30 to 100 amino acids. As we will see in the next section, these domains interact with the RNA polymerase II transcription apparatus. In addition, many transcription factors contain domains that bind to chromatin remodeling proteins or to **coactivators,** small molecules such as hormones or metabolites that regulate the transcription factor's activity.

ESSENTIAL POINT

Transcription factors influence transcription rates by binding to *cis*-acting regulatory sites within or adjacent to a gene promoter. They contain DNA-binding domains as well as *trans*-activating or *trans*-repression domains.

15.10 Transcription Factors Bind to *Cis*-Acting Sites and Interact with Basal Transcription Factors and Other Regulatory Proteins

We have now discussed the first steps in eukaryotic transcription regulation: first, chromatin must be remodeled and modified in such a way that transcription proteins can bind to their specific *cis*-acting sites; second, transcription factors bind to *cis*-acting sites and bring about positive and negative effects on the transcription initiation rate—often in response to extracellular signals or in tissue- or time-specific ways. The next question for us to consider is, how do these *cis*-acting regulatory elements and their DNA-binding factors act to influence transcription initiation? To answer this question, we must first discuss how eukaryotic RNA polymerase II and its basal transcription factors assemble at promoters.

Formation of the Transcription Initiation Complex

A number of proteins called **basal** or **general transcription factors** are needed to initiate either basal-level or enhanced levels of transcription. These proteins assemble at the promoter in a specific order, forming a transcriptional **pre-initiation complex (PIC)** that in turn provides a platform for RNA polymerase to recognize and bind to the promoter (Figure 15–17). To initiate formation of the PIC, a complex called **TFIID** binds to the TATA box through one of its subunits, called **TBP** (*T*ATA *B*inding *P*rotein). The TFIID complex is composed of TBP plus approximately 13 proteins called *TAF*s (*T*ATA *A*ssociated *F*actors). TFIID then binds additional

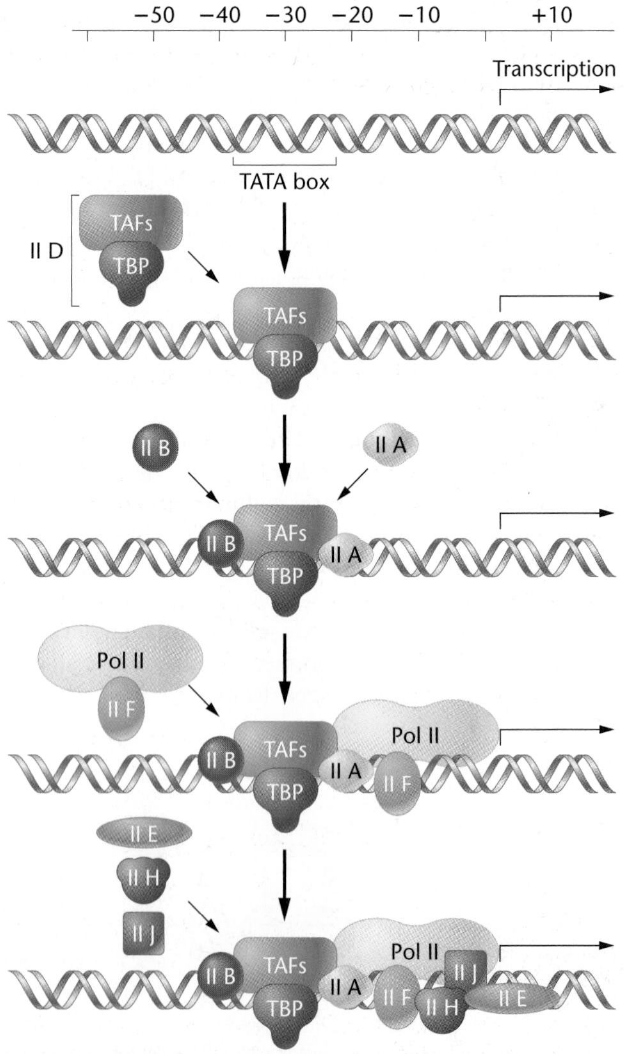

FIGURE 15–17 The assembly of general transcription factors required for the initiation of transcription by RNA polymerase II.

general factors, such as **TFIIB, TFIIA, TFIIF, TFIIH,** and **TFIIJ.** RNA polymerase II is recruited to the complex along with TFIIF. Once all of the general transcription factors have assembled at a promoter, and RNA polymerase II has made stable contacts with the start site of transcription, the DNA double helix is unwound at the start site and RNA polymerase begins transcription.

Interactions of the General Transcription Factors with Transcription Activators

Transcription activators alter the rate of transcription initiation in several ways. One way is for certain activators to bind to closed chromatin near a gene's promoter. These activators may recruit chromatin remodeling complexes that open regions of the promoter for further interactions with the transcription machinery. A second way involves interactions of the activators with general transcription factors. The first step is the interaction of transcription activators, which are bound to their DNA regulatory elements, with other proteins such as coactivators in a complex known as an **enhanceosome.** The second step is the interaction of the enhanceosome components with one or more general transcription factors in the PIC (Figure 15–18). In order for proteins bound to the promoter and enhancer regions to interact, the DNA loops out or bends between the promoter and enhancer. By enhancing the rate of PIC assembly, or by influencing its stability, transcription activators stimulate the rate of transcription initiation.

In addition to chromatin remodeling and PIC formation, transcription activators may also increase the rate of DNA unwinding within the gene and accelerate the release of RNA polymerase from the promoter into the transcribed region of the gene. Conversely, gene transcription can be repressed by the actions of repressor proteins bound at silencer DNA elements. Repressors inhibit the formation of a PIC, block the association of a gene's regulatory elements with activator transcription factors, or stimulate the actions of chromatin remodeling proteins that create repressive chromatin structures.

In summary, the picture of transcription regulation in eukaryotes is complex, but several important generalizations

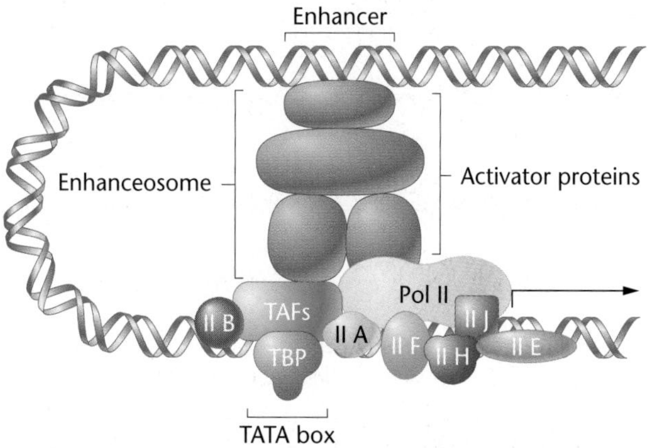

FIGURE 15–18 Formation of DNA loops allows factors that bind to an enhancer (or silencer) at a distance from a promoter to interact with general transcription factors in the pre-initiation complex and to regulate the level of transcription.

can be drawn. First, alterations in chromatin structure allow (or repress) the binding of transcription factors and RNA polymerase. Second, the interactions of multiple activator or repressor proteins with general transcription factors affect the rate of formation of the PIC and the release of RNA polymerase from the promoter. In addition, different transcription factors may compete for binding at a given DNA sequence or at one of two overlapping sequences. The same DNA sequence may bind different transcription factors in different tissues or at different times or in response to different extracellular signals. Transcription factor concentrations or the efficiency with which each factor binds to the DNA may dictate which factor binds and the effect on transcription levels. Finally, the rate of transcription initiation can be affected simultaneously by multiple enhancers, silencers, and a multiplicity of transcription factors that may bind to these elements in different circumstances.

ESSENTIAL POINT

Transcription factors are thought to act by enhancing or repressing the association of general transcription factors at the promoter. They may also assist in chromatin remodeling.

15.11 Posttranscriptional Gene Regulation Occurs at All the Steps from RNA Processing to Protein Modification

Although transcriptional control is a major type of gene regulation in eukaryotes, **posttranscriptional regulation** also contributes to control of gene expression. Modification of eukaryotic nuclear RNA transcripts prior to translation includes the removal of noncoding introns, the precise splicing together of the remaining exons, and the addition of a cap at the mRNA's 5′ end and a poly-A tail at its 3′ end. The messenger RNA is then exported to the cytoplasm, where it is translated and degraded. Each of the mRNA processing steps can be regulated to control the quantity of functional mRNA available for synthesis of a protein product. In addition, the rate of translation, as well as the stability and activity of protein products, can be regulated. We will examine several mechanisms of posttranscriptional gene regulation that are especially important in eukaryotes—alternative splicing, mRNA stability, translation, and protein stability. In a subsequent section, we will discuss a newly discovered method of posttranscriptional gene regulation—RNA silencing.

Alternative Splicing of mRNA

Alternative splicing can generate different forms of mRNA from identical pre-mRNA molecules, so that expression of one gene can give rise to a number of proteins with similar or different functions. Changes in splicing patterns can have many different effects on the translated protein. They can alter the protein's enzymatic activity, receptor-binding capacity, or

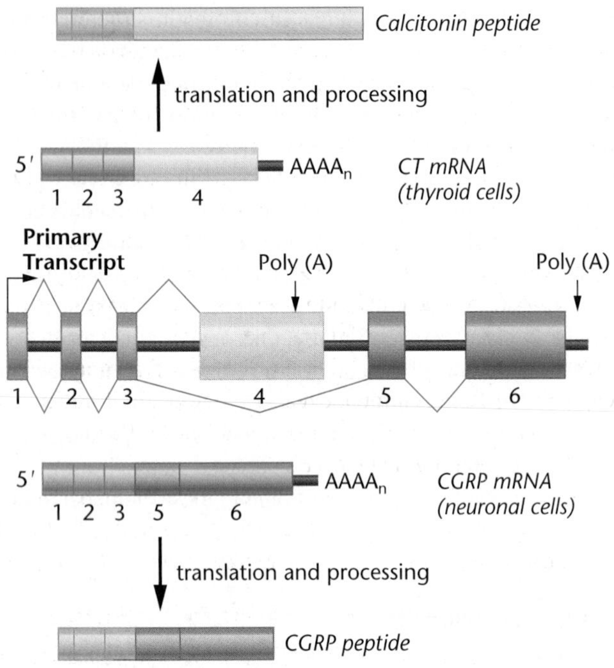

FIGURE 15–19 Alternative splicing of the *CT/CGRP* gene transcript. The primary transcript, which is shown in the middle of the diagram, contains six exons. The primary transcript can be spliced into two different mRNAs, both containing the first three exons but differing in their final exons. The *CT* mRNA contains exon 4, with polyadenylation occurring at the end of the fourth exon. The *CGRP* mRNA contains exons 5 and 6, and polyadenylation occurs at the end of exon 6. The *CT* mRNA is produced in thyroid cells. After translation, the resulting protein is processed into the calcitonin peptide. In contrast, the *CGRP* mRNA is produced in neuronal cells, and after translation, its protein product is processed into the CGRP peptide.

protein localization in the cell. Changes in splicing patterns are therefore important regulatory events that help control many aspects of multicellular development and function.

Figure 15–19 presents an example of alternative splicing of the pre-mRNA transcribed from the **calcitonin/calcitonin gene-related peptide gene (*CT/CGRP* gene).** In thyroid cells, the *CT/CGRP* primary transcript is spliced in such a way that the mature mRNA contains the first four exons only. In these cells, the exon 4 polyadenylation signal is used to process the mRNA and add the poly-A tail. This mRNA is translated into the calcitonin peptide, a 32-amino acid peptide hormone that is involved in regulating calcium. In the brain and peripheral nervous system, the *CT/CGRP* primary transcript is spliced to include exons 5 and 6, but not exon 4. In these cells, the exon 6 polyadenylation site is recognized. The *CGRP* mRNA encodes a 37-amino acid peptide with hormonal activities in a wide range of tissues. Through alternative splicing, two peptide hormones with different structures, locations, and functions are synthesized from the same gene. Even more complex alternative splicing patterns occur in some genes, such as the example in Figure 15–20.

Alternative splicing increases the number of proteins that can be made from each gene. As a result, the number of proteins that an organism can make—its **proteome**—is not the same as the number of genes in the genome, and protein diversity can exceed gene number by an order of magnitude. Alternative splicing is found in all metazoans but is especially common in vertebrates, including humans. It has been estimated that 30 to 60 percent of the genes in the human genome can undergo alternative splicing. Thus, humans can produce several hundred thousand different proteins (or perhaps more) from the approximately 25,000 genes in the haploid genome.

Mutations that affect regulation of splicing contribute to several genetic disorders. One of these disorders, **myotonic dystrophy,** provides an example of how defects in alternative RNA splicing can lead to a wide range of symptoms. Myotonic dystrophy (DM) is the most common form of adult muscular dystrophy, affecting 1 in 8000 individuals. It is an autosomal dominant disorder that occurs in two forms—DM1 and DM2. Both of these diseases show a wide range of symptoms, including muscle wasting, **myotonia** (difficulty relaxing muscles), insulin resistance, cataracts, testicular atrophy, behavior and cognitive defects, cardiac muscle problems, and hair follicle tumors.

DM1 is caused by the expansion of the trinucleotide repeat CTG in the 3′-untranslated region of the ***DMPK* gene.** In unaffected individuals, the *DMPK* gene contains between 5 and 35 copies of the CTG repeat sequence, whereas in DM1 patients, the gene contains between 150 and 2000 copies. The severity of the symptoms are directly related to the number of copies of the repeat sequence. DM2 is caused by an expansion of the repeat sequence CCTG within the first intron of the ***ZNF9* gene.** Affected individuals may have up to 11,000 copies of the repeat sequence in the *ZNF9* intron. In DM2, the severity of symptoms is not related to the number of repeats.

Recently, scientists have discovered that DM1 and DM2 are caused not by changes in the protein products of the *DMPK* or *ZNF9* genes, but by the toxic effects of their repeat-containing RNAs. These RNAs accumulate and form inclusions within the nucleus. In the case of *ZNF9,* only the CCUG sequence repeat itself accumulates in the nucleus, as the remainder of the

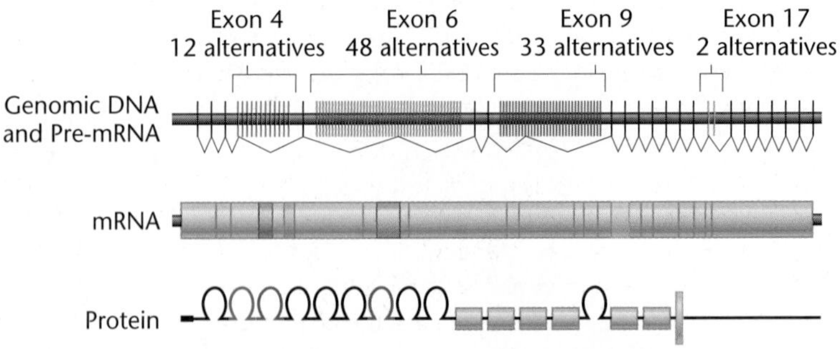

FIGURE 15–20 Alternative splicing of the *Dscam* gene mRNA. (Top) Organization of the *Dscam* gene in *Drosophila melanogaster* and the transcribed pre-mRNA. The *Dscam* gene encodes a protein that guides axon growth during development. Each mRNA will contain one of the 12 possible exons for exon 4 (red), one of the 48 possible exons for exon 6 (blue), one of the 33 possible exons for exon 9 (green), and one of the 2 possible exons for exon 17 (yellow). Counting all possible combinations of these exons, the *Dscam* gene could encode 38,016 different versions of the DSCAM protein.

intron is degraded after splicing of the gene. It appears that the accumulated RNAs bind to, and sequester, proteins that would normally be involved in regulating the alternative splicing patterns of a large number of other RNAs. These RNAs include those whose products are required for the proper functioning of muscle and neural tissue. So far, scientists have discovered over 20 genes that are inappropriately spliced in DM1 muscle, heart, and brain. Often, the fetal splicing patterns occur in DM1 and DM2 patients, and the normal transitions to adult splicing patterns are lacking. Such defects in the regulation of RNA splicing are known as **spliceopathies.**

Other human diseases are caused by expansions of trinucleotide repeats. Two of these, Fragile X syndrome and Huntington disease, are discussed in Chapter 4 and Chapter 21, respectively.

NOW SOLVE THIS

Problem 27 on page 333 describes cases of Becker muscular dystrophy caused by deletions of both exons 45 and 46 of the *dystrophin* gene. You are asked to explain how a deletion of two exons could lead to this disease.

Hint: To answer this question, consider how the deletion or addition of certain numbers of base pairs in DNA can affect reading frames in the resulting mRNA. You might also consider the various outcomes for both the mRNA and the encoded protein that could result from frameshift mutations.

Control of mRNA Stability

The **steady-state level** of an mRNA is its amount in the cell as determined by a combination of the rate at which the gene is transcribed and the rate at which the mRNA is degraded. In turn, the steady-state level determines the amount of mRNA that is available for translation. All mRNA molecules are degraded at some point after their synthesis, but the lifetime of an mRNA, defined in terms of its **half-life,** or $t_{1/2}$, can vary widely between different mRNAs and can be regulated in response to the needs of the cell. For example, the abundance of some proteins involved in regulating gene transcription, cell growth, and differentiation is determined more by controlling the rate of degradation of the mRNA for those proteins than by regulating the rate of gene transcription. Some mRNAs are degraded within minutes after their synthesis, whereas others can remain stable for hours, months, or even years (in the case of mRNAs stored in oocytes).

Regulation of mRNA stability is often linked with the process of translation. Several observations demonstrate this link between translation and mRNA stability. First, most mRNA molecules become stable in cells that are treated with translation inhibitors. Second, the presence of premature stop codons in the body of an mRNA, as well as premature translation termination, causes rapid degradation of mRNAs. Third, many of the ribonucleases and mRNA-binding proteins that affect mRNA stability associate with ribosomes.

One of the best studied examples of mRNA stability linked to translation is the synthesis of α- and β-tubulins, the subunit components of eukaryotic microtubules. Treatment of a cell with the drug colchicine leads to a rapid disassembly of its microtubules and an increase in the concentration of free α- and β-subunits. Under these conditions, the synthesis of α- and β-tubulins drops dramatically. At low concentrations of free subunits, the synthesis of tubulins is stimulated, whereas at higher concentrations, synthesis is inhibited.

Research by Don Cleveland and his colleagues has uncovered how the stability of tubulin mRNA is regulated during translation. They propose that the first four amino acids of the tubulin protein form a recognition site to which protein factors—the α and β-tubulins—bind during the process of tubulin translation. The interaction of free tubulin subunits with the nascent protein chain activates an RNase that may be either a ribosomal component or a nonspecific cytoplasmic RNase. The RNase degrades the tubulin mRNA in the act of translation, shutting down tubulin subunit biosynthesis.

Translation-coupled mRNA turnover has been proposed as a regulatory mechanism for other genes, including histones, some transcription factors, lymphokines, and cytokines.

Translational and Post-translational Controls

The ultimate end-point of gene expression is the action or presence of a gene's protein product. In some cases, the translation of an mRNA can be regulated by the extent of the cell's requirement for the gene product. In some cases, specific regulatory proteins can bind to mRNA molecules, blocking their translation. The binding of these regulatory proteins can be controlled by their abundance in the cell or by their molecular modifications.

Protein levels and activities can also be controlled by regulation of their stabilities and by protein modifications. An example of regulated stability and modification is offered by the **p53 protein.** The p53 protein is essential to protect normal cells from the effects of DNA damage and other stresses. It is a transcription factor that increases the transcription of a number of genes whose products are involved in cell-cycle arrest, DNA repair, and programmed cell death. Under normal conditions, the levels of p53 protein are extremely low in cells, and the p53 that is present is inactive. When cells suffer DNA damage or metabolic stresses, the levels of p53 protein increase dramatically, and p53 becomes an active transcription factor.

The changes in the abundance and activity of p53 are due to a combination of increased protein stability and modifications to the protein. In unstressed cells, p53 is bound by another protein called **Mdm2.** The Mdm2 protein binds to the transcriptional activation domain of the p53 molecule, blocking its ability to induce transcription. In addition, Mdm2 adds ubiquitin residues onto the p53 protein. **Ubiquitin** is a small protein that tags other proteins for degradation by proteolytic enzymes. The presence of ubiquitin on p53 results in p53 degradation. When cells are stressed, Mdm2 and p53 become modified by phosphorylation and acetylation, resulting in the release of Mdm2 from p53. As a consequence, p53 proteins become stable, the levels of p53 increase, and the protein is able to act as a transcription factor. An added level of control is that p53 is a transcription factor that induces the transcription of the *Mdm2* gene. Hence, the presence of active p53 triggers a negative feedback loop that creates more Mdm2 protein, which rapidly returns p53 to its rare and inactive state.

ESSENTIAL POINT

Posttranscriptional gene regulation can involve alternative splicing of nascent RNA, changes in mRNA stability, translational control, and posttranslational modifications. These mechanisms may alter the type, quantity, or activity of a gene's protein product.

15.12 RNA-induced Gene Silencing Controls Gene Expression in Several Ways

In the last several years, the discovery that small RNA molecules control gene expression has given rise to a new field of research. First discovered in plants, short RNA molecules, ~21 nucleotides long, are now known to regulate gene expression in the cytoplasm of both plants and animals by repressing translation and triggering the degradation of mRNAs. This form of sequence-specific posttranscriptional regulation is known as **RNA interference (RNAi).** More recently, short RNAs have been shown to act in the nucleus to alter chromatin structure and bring about repression of transcription. Together, these phenomena are known as **RNA-induced gene silencing.** We will outline some basic features of RNA-induced gene silencing, while keeping in mind that this field is rapidly expanding and advancing. In addition, we will discuss some of the ways in which RNAi is being used in biotechnology and medicine.

RNAi was first discovered during laboratory research, in studies of plant and animal gene expression. In one research project, Andrew Fire and Craig Mello injected roundworm (*Caenorhabditis elegans*) cells with either single-stranded or double-stranded RNA molecules—both containing sequences complementary to the mRNA of the *unc-22* gene. Although they expected that the single-stranded antisense RNA molecules would suppress *unc-22* gene expression by binding to the endogenous sense mRNA, they were surprised to discover that the injection of double-stranded *unc-22* RNA was 10- to 100-fold more powerful in repressing expression of the *unc-22* mRNA. They studied the phenomenon further and published their results in the journal *Nature* in 1998. They reported that the presence of double-stranded RNA acts to degrade the mRNA if the mRNA is complementary in sequence to one strand of the double-stranded RNA. Only a few molecules of double-stranded RNA are needed to bring about the degradation of large amounts of mRNA. Fire and Mello's research opened up an entirely new and surprising branch of molecular biology, with far-reaching implications for practical applications. For their insights into RNAi, they were awarded the Nobel Prize in Physiology or Medicine for 2006.

The Molecular Mechanisms of RNA-induced Gene Silencing

Two types of short RNA molecules are involved in RNA-induced gene silencing: The **small interfering RNAs (siRNAs)** and the **microRNAs (miRNAs).** Although they arise from different sources, their mechanisms of action are similar. Both types of RNA are short, double-stranded molecules, between 21 and 24 ribonucleotides long.

The siRNAs are derived from longer RNA molecules that are linear, double-stranded, and located in the cell cytoplasm. In nature, these siRNA precursors arise within cells as a result of virus infection or the expression of transposons—both of which synthesize double-stranded RNA molecules as part of their life cycles. RNAi may be a method by which cells recognize these double-stranded RNAs and inactivate them, protecting the organism from external or internal assaults. Another source of siRNA molecules is in the research lab. Scientists are now able to introduce double-stranded RNAs into cells for research or therapeutic purposes. In the cytoplasm, double-stranded RNA molecules are recognized by an enzyme complex known as **Dicer** and are cleaved by Dicer into siRNAs.

The miRNAs are derived from single-stranded RNAs that are transcribed within the nucleus from the cell's own genome and that contain a double-stranded stem-loop structure. Nuclease enzymes within the nucleus recognize these stem-loop structures and cleave them from the longer single-stranded RNA. The stem-loop RNA fragments are exported from the nucleus into the cytoplasm where they are further processed by the Dicer complex into short, linear, double-stranded miRNAs. Over the last few years, scientists have discovered that significant amounts of eukaryotic genomes are transcribed by RNA polymerase II into RNA products that contain no open reading frames and are not translated into protein products. These RNAs are transcribed either from sequences within the introns of other protein-coding genes or from their own promoters. So far, more than 400 of these noncoding RNA genes have been discovered in the human genome. *Arabidopsis* has more than 130, and *C. elegans* has more than 100. This is probably an underestimate, and scientists speculate that eukaryotic genomes may contain thousands of genes that are transcribed into short noncoding RNAs, which may regulate the expression of more than half of all protein-coding genes.

How do these noncoding RNAs work to negatively regulate gene expression? The several different pathways involved in RNA-induced gene silencing are outlined in Figure 15–21.

The RNAi pathway takes several steps. First, siRNA or miRNA molecules associate with an enzyme complex called the **RNA-induced silencing complex (RISC).** Second, within the RISC, the short double-stranded RNA is denatured and the sense strand is degraded. Third, the RNA/RISC complex becomes a functional and highly specific agent of RNAi, seeking out mRNA molecules that are complementary to the antisense RNA contained in the RISC. At this point, RNAi can take one of two different pathways. If the antisense RNA in the RISC is perfectly complementary to the mRNA, the RISC will cleave the mRNA. The cleaved mRNA is then degraded by ribonucleases. If the antisense RNA within the RISC is not exactly complementary to the mRNA, the RISC complex stays bound to the mRNA, interfering with the ability of ribosomes to translate the mRNA. Hence, RNAi can silence gene expression by affecting either mRNA stability or translation.

In addition to repressing mRNA translation and triggering mRNA degradation, siRNAs and miRNAs can also repress the transcription of specific genes and larger regions of the genome. They do this by associating with a different complex—the **RNA-induced initiation of transcription silencing complex (RITS).** The antisense RNA strand within the RITS targets the

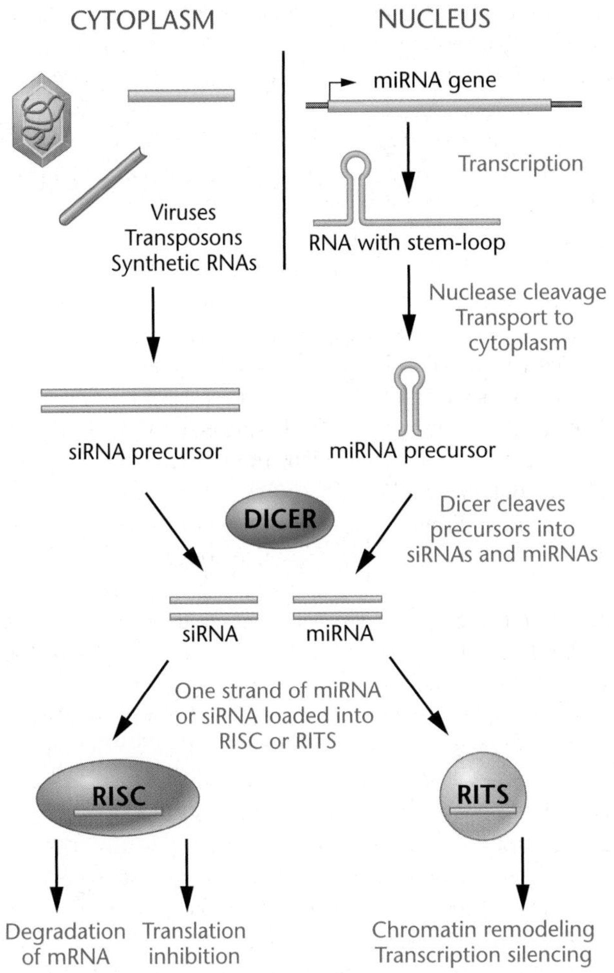

FIGURE 15–21 Mechanisms of gene regulation by RNA-induced gene silencing. The siRNA or miRNA precursors are processed into short double-stranded RNA molecules by the Dicer complex in the cytoplasm. They are then recognized by either the RISC complex or the RITS complex, and one strand is degraded. In the RNAi pathway, the RISC complex, guided by the antisense single-stranded RNA, recognizes target mRNA substrates, marking them for degradation or translation inhibition. In the transcription silencing pathway, the RITS complex acts in the nucleus, by recognizing genomic DNA that is complementary to the single strands of the miRNAs or siRNAs. The RITS complex recruits chromatin remodeling proteins that modify chromatin and repress transcription.

RITS complex to specific gene promoters or larger regions of chromatin. RITS then recruits chromatin remodeling enzymes to these regions. These enzymes methylate histones and DNA, resulting in heterochromatin formation and subsequent transcriptional silencing.

RNAi pathways are also able to repress transcription in indirect ways. Transcription factor mRNAs are frequent targets for RNAi-mediated silencing. When the levels of specific transcription factors are reduced in a cell, transcription of genes whose expression depends on these factors is also repressed.

Recent studies are demonstrating that RNA-induced gene-silencing mechanisms operate during normal development and control the expression of batteries of genes involved in tissue-specific cellular differentiation. In addition, scientists have discovered that abnormal activities of miRNAs contribute to the occurence of cancers, diabetes, and heart disease.

ESSENTIAL POINT

RNA-induced gene silencing affects the translatability or stability of mRNA, as well as transcription. It acts through the hybridization of small antisense RNAs to specific regions of an mRNA or a DNA region.

RNA-Induced Gene Silencing in Biotechnology and Medicine

Recently, geneticists have applied RNAi as a powerful research tool. RNAi technology allows investigators to specifically create single-gene defects without having to induce inherited gene mutations. RNAi-mediated gene silencing is relatively specific and inexpensive, and it allows scientists to rapidly analyze gene function. Several dozen scientific supply companies now manufacture synthetic siRNA molecules of specific ribonucleotide sequence for use in research. These molecules can be introduced into cultured cells to knock out specific gene products.

In addition to its use in laboratory research, RNAi is being developed as a potential pharmaceutical agent. In theory, any disease caused by overexpression of a specific gene, or even normal expression of an abnormal gene product, could be attacked by therapeutic RNAi. Viral infections are an obvious target, and scientists have had promising results using RNAi in tissue cultures to reduce the severity of infection by several types of viruses such as HIV, influenza, and polio. In animal models, siRNA molecules have successfully treated virus infections, eye diseases, cancers, and inflammatory bowel disease. In these studies, the synthetic siRNA molecules were applied to the surfaces of mucous membranes, or into eyes, brain tissues, or the lower intestine. Research into RNAi pharmaceuticals has made rapid progress, and the first human clinical trials have completed their Phase I (toxicity) stages. Phase II (efficacy) clinical trials have begun to test the use of siRNAs directed against the product of the vascular endothelial growth factor gene, as a drug to treat age-related macular degeneration. Clinical trials have also been started to test RNAi in the treatment of respiratory syncytial virus infections. Other clinical trials are planned for siRNA treatment of influenza and hepatitis C virus infections.

Scientists are particularly enthusiastic about the potential uses of RNAi in the diagnosis and treatment of cancers. Recent studies show that the expression profiles of miRNA genes are characteristic of each tumor type. This observation may lead to more precise methods of diagnosing tumors, predicting their course, and planning treatments. In addition, certain cancers appear to have defects in miRNA gene expression. Treatment of these tumors with synthetic siRNAs may be able to correct these defects and reverse the cancer phenotype. Finally, some cancers are characterized by overexpression or abnormal expression of one or several key proto-oncogenes. If RNAi methods could target these specific gene products, they might help treat cancers that have become resistant to other methods such as radiation or chemotherapy.

New as it is, the science of RNAi holds powerful promise for molecular medicine. Analysts expect that the first RNAi drugs may be available within the next decade.

EXPLORING GENOMICS

Tissue-Specific Gene Expression

In this chapter, we discussed how gene expression can be regulated in many complex ways. Recall that one aspect of gene expression regulation we considered is the way promoter, enhancer, and silencer sequences can govern transcriptional initiation of genes to allow for tissue-specific gene expression. All cells and tissues of an organism possess the same genome, and many genes are expressed in all cell and tissue types. However, muscle cells, blood cells, and all other tissue types express genes that are largely tissue-specific (i.e., they have limited or no expression in other tissue types). In this exercise, we use BLAST to learn more about tissue-specific gene expression patterns.

■ Exercise: Tissue-Specific Gene Expression

In this exercise, we return to the National Center for Biotechnology Information site (NCBI) and use the search tool **BLAST, Basic Local Alignment Search Tool,** which you were introduced to in the Exploring Genomics exercise for Chapter 9.

1. Access BLAST from the NCBI Web site at **http://www.ncbi.nlm.nih.gov/BLAST/**.
2. The following are GenBank accession numbers for four different genes that show tissue-specific expression patterns. You will perform your searches on these genes.

   NM_021588.1
   NM_00739.1
   AY260853.1
   NM_004917
3. For each gene, carry out a nucleotide BLAST search using the accession numbers for your sequence query. Refer to the Exploring Genomics feature in Chapter 9 to refresh your memory on BLAST searches. Because the accession numbers are for nucleotide sequences, be sure to use the "nucleotide blast" (blastn) program when running your searches. Once you enter "blastn" under the "Choose Search Set" category, you should set the database to "Others (nr etc.)," so that you are not searching an organism-specific database.
4. For the top alignments for each gene, the "Links" column (far right) contains colored boxes labeled U (for UniGene expression data), E (Gene Expression Profiles), and G (Gene Information). Each of these boxes will link you to information about the gene. The UniGene link will show you a UniGene report. For some genes, upon entering UniGene you may need to click a link above the gene name before retrieving a UniGene report. Be sure to explore the "Expression Profile" link under the "Gene Expression" category in each UniGene report. Expression profiles will show a table of gene expression patterns in different tissues.

Also explore the "GEO profiles" link under the "Gene Expression" category of the UniGene reports, when available. These links will take you to a number of gene expression studies related to each gene of interest. Explore these resources for each gene, and then answer the following questions:

a. What is the identity of each sequence, based on sequence alignment? How do you know this?
b. What species was each gene cloned from?
c. Which tissue(s) are known to express each gene?
d. Does this gene show regulated expression during different times of development?
e. Which gene shows the most restricted pattern of expression by being expressed in the fewest tissues?

CASE STUDY A mysterious muscular dystrophy

A man in his early 30s suddenly developed weakness in his hands and neck, followed a few weeks later by burning muscle pain—all symptoms of late-onset muscular dystrophy. His internist ordered genetic tests to determine whether he had one of the inherited muscular dystrophies, focusing on Becker muscular dystrophy, myotonic dystrophy Type I, and myotonic dystrophy Type II. These tests were designed to detect mutations in the related *dystrophin, DMPK,* and *ZNF9* genes. The testing ruled out Becker muscular dystrophy. While awaiting the results of the *DMPK,* and *ZNF9* gene tests, the internist explained that the possible mutations were due to expanded tri- and tetranucleotide repeats, but not in the protein-coding portion of the genes. She went on to say that the resulting disorders were due not to changes in the encoded proteins, which appear to be normal, but instead to altered RNA splicing patterns, whereby the RNA splicing remnants containing the nucleotide repeats disrupt normal splicing of the transcripts of other genes. This discussion raises several interesting questions about the diagnosis and genetic basis of the disorders.

1. Why did the physician immediately rule out Duchenne muscular dystrophy, a muscular dystrophy which is also caused by mutations in the *dystrophin* gene?
2. What is alternative splicing, where does it occur, and how could disrupting it affect the expression of the affected gene(s)?
3. What role might the expanded tri- and tetranucleotide repeats play in the altered splicing?
4. How does this contrast with the cause of Becker muscular dystrophy?

Insights and Solutions

1. A theoretical operon (*theo*) in *E. coli* contains several structural genes encoding enzymes that are involved sequentially in the biosynthesis of an amino acid. Unlike the *lac* operon, in which the repressor gene is separate from the operon, the gene encoding the regulator molecule is contained within the *theo* operon. When the end product (the amino acid) is present, it combines with the regulator molecule, and this complex binds to the operator, repressing the operon. In the absence of the amino acid, the regulatory molecule fails to bind to the operator, and transcription proceeds.

Characterize this operon, then consider the following mutations, as well as the situation in which the wild-type gene is present along with the mutant gene in partially diploid cells (F′):

(a) Mutation in the operator region.

(b) Mutation in the promoter region.

(c) Mutation in the regulator gene.

In each case, will the operon be active or inactive in transcription, assuming that the mutation affects the regulation of the *theo* operon? Compare each response with the equivalent situation of the *lac* operon.

Solution: The *theo* operon is repressible and under negative control. When there is no amino acid present in the medium (or the environment), the product of the regulatory gene cannot bind to the operator region, and transcription proceeds under the direction of RNA polymerase. The enzymes necessary for the synthesis of the amino acid are produced, as is the regulator molecule. If the amino acid *is* present, or is present after sufficient synthesis occurs, the amino acid binds to the regulator, forming a complex that interacts with the operator region, causing repression of transcription of the genes within the operon.

The *theo* operon is similar to the tryptophan system, except that the regulator gene is within the operon rather than separate from it. Therefore, in the *theo* operon, the regulator gene is itself regulated by the presence or absence of the amino acid.

(a) As in the *lac* operon, a mutation in the *theo* operator region inhibits binding with the repressor complex, and transcription occurs constitutively. The presence of an F′ plasmid bearing the wild-type allele would have no effect, since it is not adjacent to the structural genes.

(b) A mutation in the *theo* promoter region would no doubt inhibit binding to RNA polymerase and therefore inhibit transcription. This would also happen in the *lac* operon. A wild-type allele present in an F′ plasmid would have no effect.

(c) A mutation in the *theo* regulator gene, as in the *lac* system, may inhibit either its binding to the repressor or its binding to the operator gene. In both cases, transcription will be constitutive because the *theo* system is repressible. Both cases result in the failure of the regulator to bind to the operator, allowing transcription to proceed. In the *lac* system, failure to bind the corepressor lactose would permanently repress the system. The addition of a wild-type allele would restore repressibility, provided that this gene was transcribed constitutively.

2. Regulatory sites for eukaryotic genes are usually located within a few hundred nucleotides of the transcription start site, but they can be located up to several kilobases away. DNA sequence-specific binding assays have been used to detect and isolate protein factors present at low concentrations in nuclear extracts. In these experiments, short DNA molecules containing DNA-binding sequences are attached to material that is packed into a glass column, and nuclear extracts are passed over the column. The idea is that if proteins that specifically bind to the DNA sequence are present in the nuclear extract, they will bind to the DNA, and they can be recovered from the column after all other nonbinding material has been washed away. Once a DNA-binding protein has been isolated and identified, the problem is to devise a general method for screening cloned libraries for the genes encoding the DNA-binding factors. Determining the amino acid sequence of the protein and constructing synthetic oligonucleotide probes are time consuming and useful for only one factor at a time. Knowing the strong affinity for binding between the protein and its DNA-recognition sequence, how would you screen for genes encoding binding factors?

Solution: Several general strategies have been developed, and one of the most promising was devised by Steve McKnight's laboratory at the Fred Hutchinson Cancer Center. The cDNA isolated from cells expressing the binding factor is cloned into the lambda vector, gt11. Plaques of this library, containing proteins derived from expression of cDNA inserts, are adsorbed onto nitrocellulose filters and probed with double-stranded radioactive DNA corresponding to the binding site. If a fusion protein corresponding to the binding factor is present, it will bind to the DNA probe. After the unbound probe is washed off, the filter is subjected to autoradiography and the plaques corresponding to the DNA-binding proteins can be identified. An added advantage of this strategy is filter recycling by washing the bound DNA from the filters prior to their reuse. Such an ingenious procedure is similar to the colony-hybridization and plaque-hybridization procedures described for library screening in Chapter 17, and it provides a general method for isolating genes encoding DNA-binding factors.

Problems and Discussion Questions

1. Describe which enzymes are required for lactose and tryptophan metabolism in bacteria when lactose and tryptophan, respectively, are (a) present and (b) absent.
2. Contrast positive versus negative regulation of gene expression.
3. Contrast the role of the repressor in an inducible system and in a repressible system.
4. Even though the *lac Z, Y,* and *A* structural genes are transcribed as a single polycistronic mRNA, each gene contains the appropriate initiation and termination signals essential for translation. Predict what will happen when a cell growing in the presence of lactose contains a deletion of one nucleotide (a) early in the *Z* gene and (b) early in the *A* gene.

5. For the *lac* genotypes shown in the accompanying table, predict whether the structural gene (*Z*) is constitutive, permanently repressed, or inducible in the presence of lactose.

Genotype	Constitutive	Repressed	Inducible
$I^+O^+Z^+$			X
$I^-O^+Z^+$			
$I^+O^CZ^+$			
$I^-O^+Z^+/F'I^+$			
$I^+O^CZ^+/F'O^+$			
$I^SO^+Z^+$			
$I^SO^+Z^+/F'I^+$			

6. For the genotypes and conditions (lactose present or absent) shown in the accompanying table, predict whether functional enzymes, nonfunctional enzymes, or no enzymes are made.

Genotype	Condition	Functional Enzyme Made	Nonfunctional Enzyme Made	No Enzyme Made
$I^+O^+Z^+$	No lactose			X
$I^+O^CZ^+$	Lactose			
$I^-O^+Z^-$	No lactose			
$I^-O^+Z^-$	Lactose			
$I^-O^+Z^+/F'I^+$	No lactose			
$I^+O^CZ^+/F'O^+$	Lactose			
$I^+O^+Z^-/F'I^+O^+Z^+$	Lactose			
$I^-O^+Z^-/F'I^+O^+Z^+$	No lactose			
$I^SO^+Z^+/F'O^+$	No lactose			
$I^+O^CZ^+/F'O^+Z^+$	Lactose			

See **Now Solve This** on page 314.

7. The locations of numerous *lacI*$^-$ and *lacI*^{S} mutations have been determined within the DNA sequence of the *lacI* gene. Among these, *lacI*$^-$ mutations were found to occur in the 5′-upstream region of the gene, while *lacI*^{S} mutations were found to occur farther downstream in the gene. Are the locations of the two types of mutations within the gene consistent with what is known about the function of the repressor that is the product of the *lacI* gene?
8. Explain why catabolite repression is used in regulating the *lac* operon and describe how it fine-tunes β-galactosidase synthesis.
9. Describe experiments that would confirm whether or not two transcription regulatory molecules act through the mechanism of cooperative binding.
10. Predict the level of genetic activity of the *lac* operon as well as the status of the *lac* repressor and the CAP protein under the cellular conditions listed in the accompanying table.

	Lactose	Glucose
(a)	−	−
(b)	+	−
(c)	−	+
(d)	+	+

See **Now Solve This** on page 315.

11. Predict the effect on the inducibility of the *lac* operon of a mutation that disrupts the function of (a) the *crp* gene, which encodes the CAP protein, and (b) the CAP-binding site within the promoter.
12. Describe the role of attenuation in the regulation of tryptophan biosynthesis.
13. Attenuation of the *trp* operon was viewed as a relatively inefficient way to achieve genetic regulation when it was first discovered in the 1970s. Since then, however, attenuation has been found to be a relatively common regulatory strategy. Assuming that attenuation is a relatively inefficient way to achieve genetic regulation, what might explain its widespread use?
14. In a theoretical operon, genes *A, B, C,* and *D* represent the repressor gene, the promoter sequence, the operator gene, and the structural gene, *but not necessarily in that order.* This operon is concerned with the metabolism of a theoretic molecule (tm). From the data provided in the accompanying table, first decide whether the operon is inducible or repressible. Then assign *A, B, C,* and *D* to the four parts of the operon. Explain your rationale. (AE = active enzyme; IE = inactive enzyme; NE = no enzyme)

Genotype	tm Present	tm Absent
$A^+B^+C^+D^+$	AE	NE
$A^-B^+C^+D^+$	AE	AE
$A^+B^-C^+D^+$	NE	NE
$A^+B^+C^-D^+$	IE	NE
$A^+B^+C^+D^-$	AE	AE
$A^-B^+C^+D^+/F'A^+B^+C^+D^+$	AE	AE
$A^+B^-C^+D^+/F'A^+B^+C^+D^+$	AE	NE
$A^+B^+C^-D^+/F'A^+B^+C^+D^+$	AE + IE	NE
$A^-B^+C^+D^-/F'A^+B^+C^+D^+$	AE	NE

15. A bacterial operon is responsible for the production of the biosynthetic enzymes needed to make the theoretical amino acid tisophane (tis). The operon is regulated by a separate gene, *R*, deletion of which causes the loss of enzyme synthesis. In the wild-type condition, when tis is present, no enzymes are made; in the absence of tis, the enzymes are made. Mutations in the operator gene (O^-) result in repression regardless of the presence of tis.

    Is the operon under positive or negative control? Propose a model for (a) repression of the genes in the presence of tis in wild-type cells and (b) the mutations.
16. A marine bacterium is isolated and is shown to contain an inducible operon whose genetic products metabolize oil when it is encountered in the environment. Investigation demonstrates that the operon is under positive control and that there is a *reg* gene whose product interacts with an operator region (*o*) to regulate the structural genes designated *sg*.

    In an attempt to understand how the operon functions, a constitutive mutant strain and several partial diploid strains were isolated and tested with the results shown here:

Host Chromosome	F′ Factor	Phenotype
wild type	none	inducible
wild type	*reg* gene from mutant strain	inducible
wild type	operon from mutant strain	constitutive
mutant strain	*reg* gene from wild type	constitutive

    Draw all possible conclusions about the mutation as well as the nature of regulation of the operon. Is the constitutive mutation in the *trans*-acting *reg* element or in the *cis*-acting *o* operator element?
17. Why is gene regulation more complex in a multicellular eukaryote than in a prokaryote? Why is the study of this phenomenon in eukaryotes more difficult?

18. List and define the levels of gene regulation discussed in this chapter.
19. Distinguish between the *cis*-acting regulatory elements referred to as promoters and enhancers.
20. Is the binding of a transcription factor to its DNA recognition sequence necessary and sufficient for an initiation of transcription at a regulated gene? What else plays a role in this process?
21. Compare the control of gene regulation in eukaryotes and prokaryotes at the level of initiation of transcription. How do the regulatory mechanisms work? What are the similarities and differences in these two types of organisms in terms of the specific components of the regulatory mechanisms? Address how the differences or similarities relate to the biological context of the control of gene expression.
22. Many promoter regions contain CAAT boxes with consensus sequences CAAT or CCAAT approximately 70 to 80 bases upstream from the transcription start site. How might one determine the influence of CAAT boxes on the transcription rate of a given gene?
23. Present an overview of RNA silencing achieved through RNA interference (RNAi) and microRNAs (miRNAs). How do the silencing processes begin, and what major components participate?
24. Although it is customary to consider transcriptional regulation in eukaryotes as resulting from the positive or negative influence of different factors binding to DNA, a more complex picture is emerging. For instance, researchers have described the action of a transcriptional repressor (Net) that is regulated by nuclear export (Ducret et al., 1999. *Mol. and Cell. Biol.* 19: 7076–7087). Under neutral conditions, Net inhibits transcription of target genes; however, when phosphorylated, Net stimulates transcription of target genes. When stress conditions exist in a cell (for example, from ultraviolet light or heat shock), Net is excluded from the nucleus, and target genes are transcribed. Devise a model (using diagrams) that provides a consistent explanation of these three conditions.
25. DNA supercoiling, which occurs when coiling tension is generated ahead of the replication fork, is relieved by DNA gyrase. Supercoiling may also be involved in transcription regulation. Researchers discovered that transcriptional enhancers operating over a long distance (2500 base pairs) are dependent on DNA supercoiling, while enhancers operating over shorter distances (110 base pairs) are not so dependent (Liu et al., 2001. *Proc. Natl. Acad. Sci. [USA]* 98: 14,883–14,888). Using a diagram, suggest a way in which supercoiling may positively influence enhancer activity over long distances.
26. In some organisms, including mammals, there is an inverse relationship between the presence of 5-methylcytosine (m^5C) in CpG sequences and gene activity. In addition, m^5C may be involved in the recruitment of proteins that convert chromatin regions from transcriptionally active to inactive states. Overall, genomic DNA is relatively poor in CpG sequences due to the conversion of m^5C to thymine; however, unmethylated CpG islands are often associated with active genes. Researchers have determined that patterns of DNA methylation in spermatozoa vary with age in rats and suggest that such age-related alterations in DNA methylation may be one mechanism underlying age-related abnormalities in mammals (Oakes et al., 2003. *Proc. Natl. Acad. Sci.* 100: 1775–1780). In light of the above information, provide an explanation that relates paternal age-related alterations in DNA methylation to birth abnormalities.
27. Scientists estimate that approximately 15 percent of disease-causing mutations involve errors in alternative splicing (Philips & Cooper, 2000. *Cell. Mol. Life Sci.* 57: 235–249). However, there is an interesting case where an exon deletion appears to enhance dystrophin production in muscle cells of Duchenne muscular dystrophy (DMD) patients. It turns out that a deletion of exon 45 is the most frequent DMD-causing mutation. But some individuals with Becker muscular dystrophy (BMD), a milder form of muscular dystrophy, have deletions in exons 45 *and* 46 (van Deutekom & van Ommen, 2003. *Nat. Rev. Genetics* 4: 774–783). Recalling that deletions often cause frameshift mutations, provide a possible explanation for why BMD patients, with exon 45 *and* 46 deletions, produce more dystrophin than DMD patients do. See **Now Solve This** on page 327.
28. DNA methylation is commonly associated with a reduction of transcription. The following data come from a study of the impact of the location and extent of DNA methylation on gene activity in human cells. A bacterial gene, luciferase, was cloned next to eukaryotic promoter fragments that were methylated to various degrees, *in vitro*. The chimeric plasmids were then introduced into tissue culture cells, and the luciferase activity was assayed. These data compare the degree of expression of luciferase with differences in the location of DNA methylation (Irvine et al., 2002. *Mol. and Cell. Biol.* 22: 6689–6696). What general conclusions can be drawn from these data? See **Now Solve This** on page 320.

DNA Segment	Patch Size of Methylation (kb)	Number of Methylated CpGs	Relative Luciferase Expression
Outside transcription unit (0–7.6 kb away)	0.0	0	490X
	2.0	100	290X
	3.1	102	250X
	12.1	593	2X
Inside transcription unit	0.0	0	490X
	1.9	108	80X
	2.4	134	5X
	12.1	593	2X

29. The interphase nucleus appears to be a highly structured organelle with chromosome territories, interchromosomal compartments, and transcription factories. In cultured human cells, researchers have identified approximately 8000 transcription factories per cell, each containing an average of eight tightly associated RNA polymerase II molecules actively transcribing RNA. If each RNA polymerase II molecule is transcribing a different gene, how might such a transcription factory appear? Provide a simple diagram that shows eight different genes being transcribed in a transcription factory and include the promoters, structural genes, and nascent transcripts in your presentation.
30. It has been estimated that approximately half of human genes produce alternatively spliced mRNA isoforms. In some cases, incorrectly spliced RNAs lead to human pathologies. Scientists have examined human cancer cells for splice-specific changes and found that many of the changes disrupt tumor-suppressor gene function (Xu and Lee, 2003. *Nucl. Acids Res.* 31: 5635–5643). In general, what would be the effects of splicing changes on these RNAs and the function of tumor-suppressor gene function? How might loss of splicing specificity be associated with cancer?

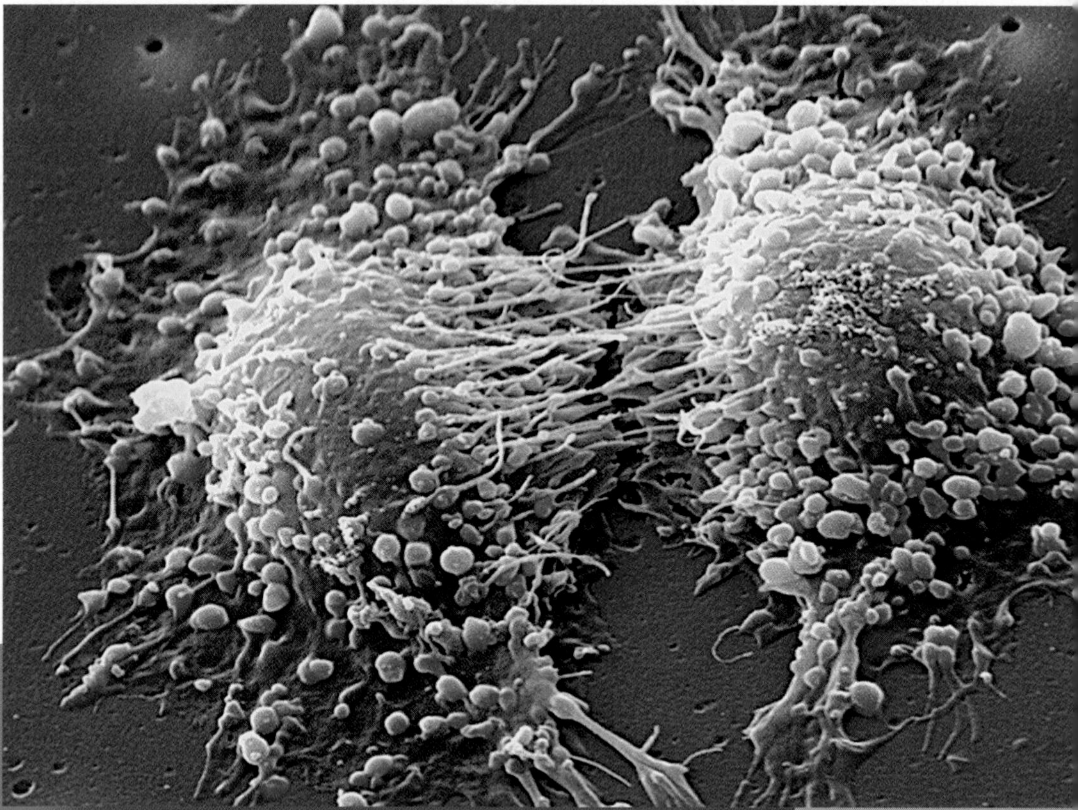

Colored scanning electron micrograph of two prostate cancer cells in the final stages of cell division (cytokinesis). The cells are still joined by strands of cytoplasm.

16

Cancer and Regulation of the Cell Cycle

Chapter Concepts

- Cancer is a group of genetic diseases affecting fundamental aspects of cellular function, including DNA repair, the cell cycle, apoptosis, differentiation, cell migration, and cell–cell contact.
- Most cancer-causing mutations occur in somatic cells; only about 1 percent of cancers have a hereditary component.
- Mutations in cancer-related genes lead to abnormal proliferation and loss of control over how cells spread and invade surrounding tissues.
- The development of cancer is a multistep process requiring mutations in genes controlling many aspects of cell proliferation and metastasis.
- Cancer cells show high levels of genomic instability, leading to the accumulation of multiple mutations in cancer-related genes.
- Epigenetic effects such as DNA methylation and histone modifications may play significant roles in the development of cancers.
- Mutations in proto-oncogenes and tumor-suppressor genes contribute to the development of cancers.
- Oncogenic viruses introduce oncogenes into infected cells and stimulate cell proliferation.
- Environmental agents contribute to cancer by damaging DNA.

Cancer is the second leading cause of death in Western countries, surpassed only by heart disease. It strikes people of all ages, and one out of three people will experience a cancer diagnosis sometime in his or her lifetime. Each year, more than 1 million cases of cancer are diagnosed in the United States, and more than 500,000 people die from the disease.

Over the last 30 years, scientists have discovered that cancer is a genetic disease, characterized by an interplay of mutant forms of oncogenes and tumor-suppressor genes leading to the uncontrolled growth and spread of cancer cells. Although some of these mutations may be inherited, as we will see, most mutations that lead to cancer occur in somatic cells that then divide and lead to tumors. Completion of the Human Genome Project is opening the door to a wealth of new information about the mutations that trigger a cell to become cancerous. This new understanding of cancer genetics is also leading to new gene-specific treatments, some of which are now entering clinical trials. Some scientists predict that gene therapies will replace chemotherapies within the next 25 years.

The goal of this chapter is to highlight our current understanding of the nature and causes of cancer. As we will see, cancer is a genetic disease that arises from mutations in genes controlling many basic aspects of cellular function. We will examine the relationship between genes and cancer, and consider how mutations, chromosomal changes, and environmental agents play roles in the development of cancer.

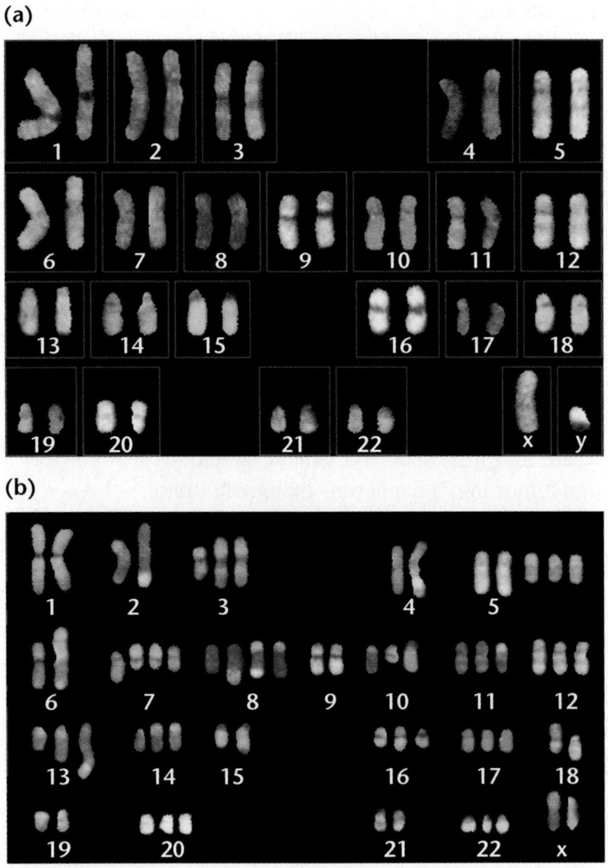

FIGURE 16–1 (a) Spectral karyotype of a normal cell. (b) Karyotype of a cancer cell showing translocations, deletions, and aneuploidy—characteristic features of cancer cells.

How Do We Know?

In this chapter, we will focus on cancer as a genetic disease, with an emphasis on the relationship between cancer, the cell cycle, and DNA damage, as well as on the multiple steps that lead to cancer. We will find many opportunities to consider the methods and reasoning by which much of this information was acquired. From the explanations given in the chapter, you should answer the following fundamental questions:

1. How do we know that malignant tumors arise from a single cell that contains mutations?
2. How do we know that cancer development requires more than one mutation?
3. How do we know that cancer cells contain defects in DNA repair?
4. How do we know that most cancers are not hereditary?

16.1 Cancer Is a Genetic Disease at the Level of Somatic Cells

Perhaps the most significant development in understanding the causes of cancer is the realization that cancer is a genetic disease. Genomic alterations that are associated with cancer range from single-nucleotide substitutions to large-scale chromosome rearrangements, amplifications, and deletions (Figure 16–1). However, unlike other genetic diseases, cancer is caused by mutations that occur predominantly in somatic cells. Only about 1 percent of cancers are associated with germ-line mutations that increase a person's susceptibility to certain types of cancer. Another important difference between cancer and other genetic diseases is that cancers rarely arise from a single mutation, but from the accumulation of many mutations—as many as six to twelve. The mutations that lead to cancer affect multiple cellular functions, including repair of DNA damage, cell division, apoptosis, cellular differentiation, migratory behavior, and cell–cell contact.

What Is Cancer?

Clinically, cancer is defined as a large number of complex diseases, up to a hundred, that behave differently depending on the cell types from which they originate. Cancers vary in their ages of onset, growth rates, invasiveness, prognoses, and responsiveness to treatments. However, at the molecular level, all cancers exhibit common characteristics that unite them as a family.

All cancer cells share two fundamental properties: (1) abnormal cell growth and division **(cell proliferation),** and (2) abnormalities in the normal restraints that keep cells from spreading and invading other parts of the body **(metastasis).**

In normal cells, these functions are tightly controlled by genes that are expressed appropriately in time and place. In cancer cells, these genes are either mutated or are expressed inappropriately.

It is this combination of uncontrolled cell proliferation and metastatic spread that makes cancer cells dangerous. When a cell simply loses genetic control over cell growth, it may grow into a multicellular mass, a **benign tumor.** Such a tumor can often be removed by surgery and may cause no serious harm. However, if cells in the tumor also acquire the ability to break loose, enter the bloodstream, invade other tissues, and form secondary tumors **(metastases),** they become malignant. **Malignant tumors** are difficult to treat and may become life threatening. As we will see later in the chapter, there are multiple steps and genetic mutations that convert a benign tumor into a dangerous malignant tumor.

ESSENTIAL POINT ■ ■ ■

Cancer cells show two fundamental properties: abnormal cell proliferation and a propensity to spread and invade other parts of the body (metastasis).

The Clonal Origin of Cancer Cells

Although malignant tumors may contain billions of cells, and may invade and grow in numerous parts of the body, all cancer cells in the primary and secondary tumors are clonal, meaning that they originated from a common ancestral cell that accumulated numerous specific mutations. This is an important concept in understanding the molecular causes of cancer and has implications for its diagnosis.

Numerous data support the concept of cancer clonality. For example, reciprocal chromosomal translocations are characteristic of many cancers, including leukemias and lymphomas (two cancers involving white blood cells). Cancer cells from patients with **Burkitt's lymphoma** show reciprocal translocations between chromosome 8 (with translocation breakpoints at or near the *c-myc* gene) and chromosomes 2, 14, or 22 (with translocation breakpoints at or near one of the immunoglobulin genes). Each Burkitt's lymphoma patient exhibits unique breakpoints in his or her *c-myc* and immunoglobulin gene DNA sequences; however, all lymphoma cells within that patient contain identical translocation breakpoints. This demonstrates that all cancer cells in each case of Burkitt's lymphoma arise from a single cell, and this cell passes on its genetic aberrations to its progeny.

Another demonstration that cancer cells are clonal is their pattern of X-chromosome inactivation. As explained in Chapter 5, female humans are mosaic, with some cells containing an inactivated paternal X chromosome and other cells containing an inactivated maternal X chromosome. X-chromosome inactivation occurs early in development and takes place at random. All cancer cells within a tumor, both primary and metastatic, within one female individual, contain the same inactivated X chromosome. This supports the concept that all the cancer cells in that patient arose from a common ancestral cell.

ESSENTIAL POINT ■ ■ ■

Cancers are clonal, meaning that all cells within a tumor originate from a single cell that contained a number of mutations.

Cancer As a Multistep Process, Requiring Multiple Mutations

Although we know that cancer is a genetic disease initiated by mutations that lead to uncontrolled cell proliferation and metastasis, a single mutation is not sufficient to transform a normal cell into a tumor-forming (tumorigenic), malignant cell. If it were sufficient, then cancer would be far more prevalent than it is. In humans, mutations occur spontaneously at a rate of about 10^{-6} mutations per gene, per cell division, mainly due to the intrinsic error rates of DNA replication. Because there are approximately 10^{16} cell divisions in a human body during a lifetime, a person might suffer up to 10^{10} mutations per gene somewhere in the body, during his or her lifetime. However, only about one person in three will suffer from cancer.

The phenomenon of age-related cancer is another indication that cancer develops from the accumulation of several mutagenic events in a single cell. The incidence of most cancers rises exponentially with age. If a single mutation were sufficient to convert a normal cell to a malignant one, then cancer incidence would appear to be independent of age. The age-related incidence of cancer suggests that many independent mutations, occurring randomly, and with a low probability, are necessary before a cell is transformed into a malignant cancer cell. Another indication that cancer is a multistep process is the delay that occurs between exposure to **carcinogens** (cancer-causing agents) and the appearance of the cancer. For example, an incubation period of five to eight years separated exposure of people to the radiation of the atomic explosions at Hiroshima and Nagasaki and the onset of leukemias.

The multistep nature of cancer development is supported by the observation that cancers often develop in progressive steps, beginning with mildly aberrant cells and progressing to cells that are increasingly tumorigenic and malignant. This progressive nature of cancer is illustrated by the development of colon cancer, as discussed in Section 16.6.

Each step in **tumorigenesis** (the development of a malignant tumor) appears to be the result of one or more genetic alterations that release the cells progressively from the controls that normally operate on proliferation and malignancy. As we will discover in subsequent sections of this chapter, the genes that undergo mutations leading to cancer (called oncogenes and tumor-suppressor genes) are those that control DNA damage repair, the cell cycle, cell–cell contact, and programmed cell death.

We will now investigate each of these fundamental processes, the genes that control them, and how mutations in these genes may lead to cancer.

ESSENTIAL POINT ■ ■ ■

The development of cancer is a multistep process, requiring mutations in several cancer-related genes.

16.2 Cancer Cells Contain Genetic Defects Affecting Genomic Stability, DNA Repair, and Chromatin Modifications

Cancer cells show higher than normal rates of mutation, chromosomal abnormalities, and genomic instability. Many researchers believe that the fundamental defect in cancer cells is a derangement of the cells' normal ability to repair DNA damage. This loss of genomic integrity leads to a general increase in the mutation rate in every gene, including specific genes that control aspects of cell proliferation, programmed cell death, and cell–cell contact. In turn, the accumulation of mutations in genes controlling these processes leads to cancer. The high level of genomic instability seen in cancer cells is known as the **mutator phenotype.** In addition, recent research has revealed that cancer cells contain aberrations in the types and locations of chromatin modifications, particularly DNA methylation patterns.

Genomic Instability and Defective DNA Repair

Genomic instability in cancer cells is characterized by the presence of gross defects such as translocations, aneuploidy, chromosome loss, DNA amplification, and chromosome deletions (Figures 16–1 and 16–2). Cancer cells that are grown in cultures in the lab also show a great deal of genomic instability—duplicating, losing, and translocating chromosomes or parts of chromosomes. Often cancer cells show specific chromosomal defects that are used to diagnose the type and stage of the cancer. For example, leukemic white blood cells from patients with **chronic myelogenous leukemia (CML)** bear a specific translocation, in which the *C-ABL* gene on chromosome 9 is translocated into the *BCR* gene on chromosome 22. This translocation creates a structure known as the **Philadelphia chromosome** (Figure 16–3). The *BCR-ABL* fusion gene codes for a chimeric BCR-ABL protein. The normal ABL protein is a **protein kinase** that acts within signal transduction pathways, transferring growth factor signals from the external environment to the nucleus. The BCR-ABL protein is an abnormal signal transduction molecule in CML cells, constantly stimulating these cells to proliferate even in the absence of external growth signals.

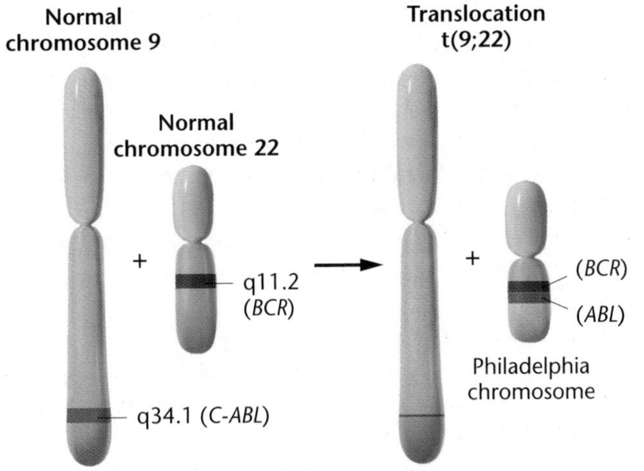

FIGURE 16–3 A reciprocal translocation involving the long arms of chromosomes 9 and 22 results in the formation of a characteristic chromosome, the Philadelphia chromosome, which is associated with chronic myelogenous leukemia (CML). The t(9;22) translocation results in the fusion of the *C-ABL* proto-oncogene on chromosome 9 with the *BCR* gene on chromosome 22. The fusion protein is a powerful hybrid molecule that allows cells to escape control of the cell cycle, contributing to the development of CML.

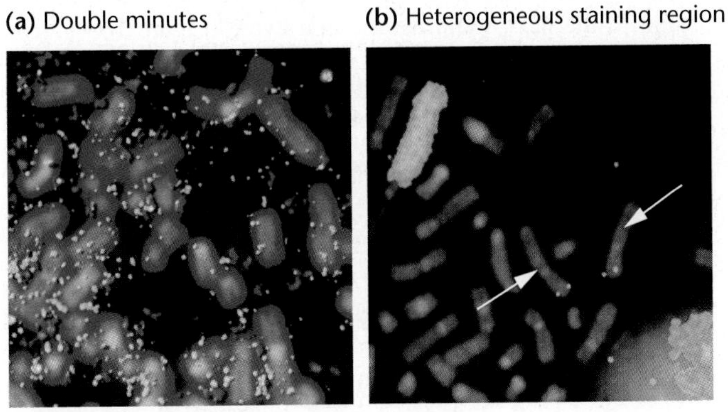

FIGURE 16–2 DNA amplifications in neuroblastoma cells. (a) Two cancer genes (*MYCN* in red and *MDM2* in green) are amplified as small DNA fragments that remain separate from chromosomal DNA within the nucleus. These units of amplified DNA are known as double minute chromosomes. Normal chromosomes are stained blue. (b) Multiple copies of the *MYCN* gene are amplified within one large region called a heterogeneous staining region (green). Single copies of the *MYCN* gene are visible as green dots at the ends of the normal parental chromosomes (white arrows). Normal chromosomes are stained red.

In keeping with the concept of the cancer mutator phenotype, a number of inherited cancers are caused by defects in genes that control DNA repair. For example, xeroderma pigmentosum (XP) is a rare hereditary disorder that is characterized by extreme sensitivity to ultraviolet light and other carcinogens. Patients with XP often develop skin cancer. Cells from patients with XP are defective in nucleotide excision repair, with mutations appearing in any one of seven genes whose products are necessary to carry out DNA repair. XP cells are impaired in their ability to repair DNA lesions such as thymine dimers induced by UV light. The relationship between XP and genes controlling nucleotide excision repair is also described in Chapter 14.

Another hereditary cancer, **hereditary nonpolyposis colorectal cancer (HNPCC),** is also caused by mutations in genes controlling DNA repair. HNPCC is an autosomal dominant syndrome, affecting about one in every 200 to 1000 people. Patients affected by HNPCC have an increased risk of developing colon, ovary, uterine, and kidney cancers. Cells from patients with HNPCC show higher than normal mutation rates and genomic instability. At least eight genes are associated with HNPCC, and four of these genes control aspects of DNA mismatch repair. Inactivation of any of these four genes—*MSH2, MSH6, MLH1,* and *MLH3*—causes a rapid accumulation of genomewide mutations and the subsequent development of colorectal and other cancers.

The observation that hereditary defects in genes controlling nucleotide excision repair and DNA mismatch repair lead to high rates of cancer lends support to the idea that the mutator phenotype is a significant contributor to the development of cancer.

Chromatin Modifications and Cancer Epigenetics

The field of cancer epigenetics is providing new perspectives on the genetics of cancer. **Epigenetics** is the study of factors that affect gene expression in a heritable way but that do not alter the nucleotide sequence of DNA. DNA methylation and histone modifications such as acetylation and phosphorylation are examples of epigenetic modifications. The genomic patterns and locations of these modifications can be inherited and affect gene expression. For example, DNA methylation is thought to be responsible for the gene silencing associated with parental imprinting, heterochromatin gene repression, and X-chromosome inactivation.

Cancer cells contain major alterations in DNA methylation. Overall, there is much less DNA methylation in cancer cells than in normal cells. At the same time, the promoters of some genes are hypermethylated in cancer cells. These changes are thought to result in the release of transcription repression over the bulk of genes that would be silent in normal cells—including cancer-causing proto-oncogenes (introduced in Section 16.4), while at the same time repressing transcription of genes that would regulate normal cellular functions such as DNA repair and cell-cycle control. Methylation profiles in cancer cells are now being used to help diagnose tumors and predict their course.

Histone modifications are also disrupted in cancer cells. Genes that encode histone acetylases, deacetylases, methyltransferases, and demethylases are often mutated or aberrantly expressed in cancer cells. The large numbers of epigenetic abnormalities in tumors have prompted some scientists to speculate that there may be more epigenetic defects in cancer cells than there are gene mutations. In addition, because epigenetic modifications are reversible, cancers may lend themselves to epigenetic-based therapies. Although the field of cancer epigenetics is still in its infancy, it has already provided major insights into tumorigenesis as well as new clinical applications.

ESSENTIAL POINT

Cancer cells show high rates of mutation, chromosomal abnormalities, genomic instability, and abnormal patterns of chromatin modifications.

NOW SOLVE THIS

Problem 17 on page 349 describes how leukemic cells from CML patients contain the BCR-ABL hybrid fusion protein. You are asked to explain how this characteristic could be used as a target for cancer therapy.

Hint: Consider that the BCR-ABL fusion protein is found only in CML white blood cells. As you answer this problem, you may wish to learn more about the drug Gleevec (see http://www.nci.nih.gov/newscenter/qandagleevec).

16.3 Cancer Cells Contain Genetic Defects Affecting Cell-Cycle Regulation

One of the fundamental aberrations in all cancer cells is a loss of control over cell proliferation. Cell proliferation is the process of cell growth and division that is essential for all development and tissue repair in multicellular organisms. Although some cells, such as epidermal cells of the skin or blood-forming cells in the bone marrow, continue to grow and divide throughout the organism's lifetime, most cells in adult multicellular organisms remain in a nondividing, quiescent, and differentiated state. **Differentiated cells** are those that are specialized for a specific function, such as photoreceptor cells of the retina or muscle cells of the heart. The most extreme examples of nonproliferating cells are nerve cells, which divide little, if at all, even to replace damaged tissue. In contrast, many differentiated cells, such as those in the liver and kidney, are able to grow and divide when stimulated by extracellular signals and growth factors. In this way, multicellular organisms are able to replace dead and damaged tissue. However, the growth and differentiation of cells must be strictly regulated; otherwise, the integrity of organs and tissues would be compromised by the presence of inappropriate types and quantities of cells. Normal regulation over cell proliferation involves a large number of gene products that control steps in the cell cycle, programmed cell death, and the response of cells to external growth signals. In cancer cells, many of the genes that control these functions are mutated or aberrantly expressed, leading to uncontrolled cell proliferation.

In this section, we will review steps in the cell cycle, some of the genes that control the cell cycle, and how these genes, when mutated, lead to cancer.

The Cell Cycle and Signal Transduction

The cellular events that occur in sequence from one cell division to the next comprise the **cell cycle** (Figure 16–4). The

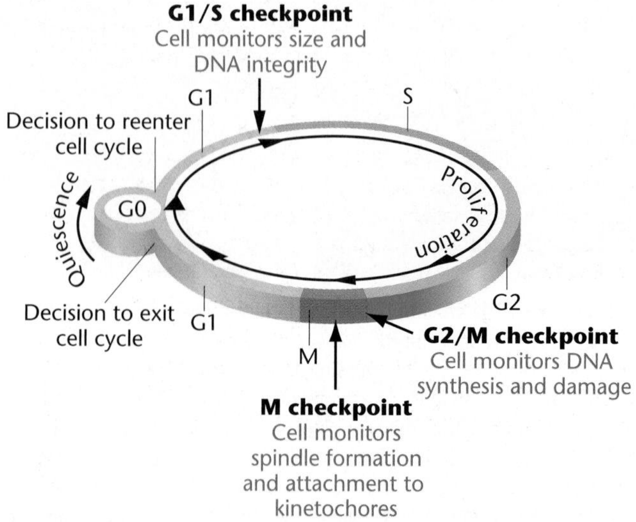

FIGURE 16–4 Checkpoints and proliferation decision points monitor the progress of the cell through the cell cycle.

interphase stage of the cell cycle is the interval between mitotic divisions. During this time, the cell grows and replicates its DNA. During **G1,** the cell prepares for DNA synthesis by accumulating the enzymes and molecules required for DNA replication. G1 is followed by **S phase,** during which the cell's chromosomal DNA is replicated. During **G2,** the cell continues to grow and prepare for division. During **M phase,** the duplicated chromosomes condense, sister chromosomes separate to opposite poles, and the cell divides in two.

In early to mid-G1, the cell makes a decision either to enter the next cell cycle or to withdraw from the cell cycle into quiescence. Continuously dividing cells do not exit the cell cycle but proceed through G1, S, G2, and M phases; however, if the cell receives signals to stop growing, it enters the **G0** phase of the cell cycle. During G0, the cell remains metabolically active but does not grow or divide. Most differentiated cells in multicellular organisms can remain in this G0 phase indefinitely. Some, such as neurons, never reenter the cell cycle. In contrast, cancer cells are unable to enter G0, and instead, they continuously cycle. Their *rate* of proliferation is not necessarily any greater than that of normal proliferating cells; however, they are not able to become quiescent at the appropriate time or place.

Cells in G0 can often be stimulated to reenter the cell cycle by external growth signals. These signals are delivered to the cell by molecules such as growth factors and hormones that bind to cell-surface receptors, which then relay the signal from the plasma membrane to the cytoplasm. The process of transmitting growth signals from the external environment to the cell nucleus is known as **signal transduction.** Ultimately, signal transduction initiates a program of gene expression that propels the cell out of G0 back into the cell cycle. Cancer cells often have defects in signal transduction pathways. Sometimes, abnormal signal transduction molecules send continuous growth signals to the nucleus even in the absence of external growth signals. In addition, malignant cells may not respond to external signals from surrounding cells—signals that would normally inhibit cell proliferation within a mature tissue.

Cell-Cycle Control and Checkpoints

In normal cells, progress through the cell cycle is tightly regulated, and each step must be completed before the next step can begin. There are at least three distinct points in the cell cycle at which the cell monitors external signals and internal equilibrium before proceeding to the next stage. These are the **G1/S,** the **G2/M,** and **M checkpoints** (Figure 16–4). At the G1/S checkpoint, the cell monitors its size and determines whether its DNA has been damaged. If the cell has not achieved an adequate size, or if the DNA has been damaged, further progress through the cell cycle is halted until these conditions are corrected. If cell size and DNA integrity are normal, the G1/S checkpoint is traversed, and the cell proceeds to S phase. The second important checkpoint is the G2/M checkpoint, where physiological conditions in the cell are monitored prior to mitosis. If DNA replication or repair of any DNA damage has not been completed, the cell cycle arrests until these processes are complete. The third major checkpoint occurs during mitosis and is called the M checkpoint. At this checkpoint, both the successful formation of the spindle-fiber system and the attachment of spindle fibers to the kinetochores associated with the centromeres are monitored. If spindle fibers are not properly formed or attachment is inadequate, mitosis is arrested.

In addition to regulating the cell cycle at checkpoints, the cell controls progress through the cell cycle by means of two classes of proteins: **cyclins** and **cyclin-dependent kinases (CDKs).** The cell synthesizes and destroys cyclin proteins in a precise pattern during the cell cycle (Figure 16–5). When a cyclin is present, it binds to a specific CDK, triggering activity of the CDK/cyclin complex. The CDK/cyclin complex then selectively phosphorylates and activates other proteins that in turn bring about the changes necessary to advance the cell through the cell cycle. For example, in G1 phase, CDK4/cyclin D complexes activate proteins that stimulate transcription of genes whose products (such as DNA polymerase δ and DNA ligase) are required for DNA replication during S phase. Another CDK/cyclin complex, CDK1/cyclin B, phosphorylates a number of proteins that bring about the events of early mitosis, such as nuclear membrane breakdown, chromosome condensation, and cytoskeletal reorganization. Mitosis can only be completed, however, when cyclin B is degraded and the protein phosphorylations characteristic of M phase are reversed. Although a large number of different protein kinases exist in cells, only a few are involved in cell-cycle regulation.

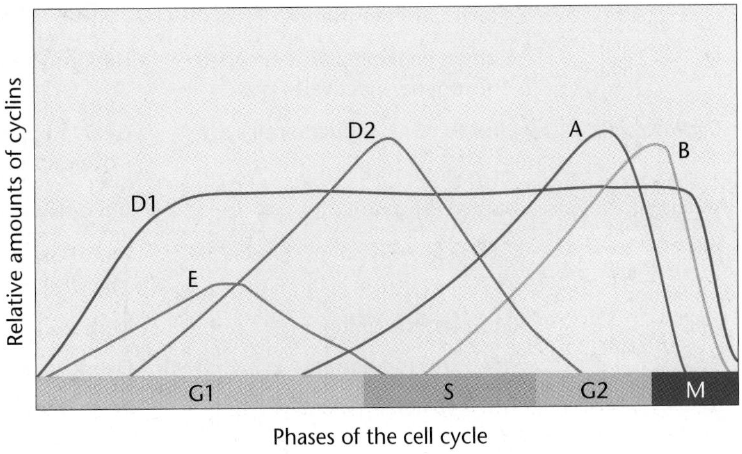

FIGURE 16–5 Relative expression times and amounts of cyclins during the cell cycle. Cyclin D1 accumulates early in G1 and is expressed at a constant level through most of the cycle. Cyclin E accumulates in G1, reaches a peak, and declines by mid-S phase. Cyclin D2 begins accumulating in the last half of G1, reaches a peak just after the beginning of S, and then declines by early G2. Cyclin A appears in late G1, accumulates through S phase, peaks at the G2/M transition, and is rapidly degraded. Cyclin B peaks at the G2/M transition and declines rapidly in M phase.

The cell cycle is regulated by an interplay of genes whose products either promote or suppress cell division. Mutation or misexpression of any of the genes controlling the cell cycle can contribute to the development of cancer. For example, if genes that control the G1/S or G2/M checkpoints are mutated, the cell may continue to grow and divide without repairing DNA damage. As these cells continue to divide, they accumulate mutations in genes whose products control cell proliferation or metastasis. Similarly, if genes that control progress through the cell cycle, such as those that encode the cyclins, are expressed at the wrong time or at incorrect levels, the cell may grow and divide continuously and may be unable to exit the cell cycle into G0. The result in both cases is that the cell loses control over proliferation and is on its way to becoming cancerous.

Control of Apoptosis

As already described, if DNA replication, repair, or chromosome assembly is aberrant, normal cells halt their progress through the cell cycle until the condition is corrected. This reduces the number of mutations and chromosomal abnormalities that accumulate in normal proliferating cells. However, if DNA or chromosomal damage is so severe that repair is impossible, the cell may initiate a second line of defense—a process called **apoptosis,** or **programmed cell death.** Apoptosis is a genetically controlled process whereby the cell commits suicide. Besides its role in preventing cancer, apoptosis is also initiated during normal multicellular development in order to eliminate certain cells that do not contribute to the final adult organism. The steps in apoptosis are the same for damaged cells and for cells being eliminated during development: nuclear DNA becomes fragmented, internal cellular structures are disrupted, and the cell dissolves into small spherical structures known as apoptotic bodies. In the final step, the apoptotic bodies are engulfed by the immune system's phagocytic cells. A series of proteases called **caspases** are responsible for initiating apoptosis and for digesting intracellular components.

By removing damaged cells, programmed cell death reduces the number of mutations that are passed to the next generation, including those in cancer-causing genes. The same genes that control cell-cycle checkpoints can trigger apoptosis. These genes are mutated in many cancers. As a result of the mutation or inactivation of these checkpoint genes, the cell is unable to repair its DNA or undergo apoptosis. This inability leads to the accumulation of even more mutations in genes that control growth, division, and metastasis.

ESSENTIAL POINT

Cancer cells have defects in cell-cycle progression, checkpoint controls, and programmed cell death.

16.4 Proto-oncogenes and Tumor-suppressor Genes Are Altered in Cancer Cells

Two general categories of cancer-causing genes are mutated or misexpressed in cancer cells—the proto-oncogenes and the tumor-suppressor genes (Table 16.1). **Proto-oncogenes** encode transcription factors that stimulate expression of other genes, signal transduction molecules that stimulate cell division, and cell-cycle regulators that move the cell through the cell cycle. Their products are important for normal cell functions, especially cell growth and division. When normal cells become quiescent and cease division, they repress the expression of most proto-oncogenes or the activity of their products. In cancer cells, one or more proto-oncogenes are altered in such a way that their

TABLE 16.1 Some Proto-oncogenes and Tumor-suppressor Genes

Proto-oncogene	Normal Function	Alteration in Cancer	Associated Cancers
c-myc	Transcription factor, regulates cell cycle, differentiation, apoptosis	Translocation, amplification, point mutations	Lymphomas, leukemias, lung cancer, many types
c-kit	Tyrosine kinase, signal transduction	Mutation	Sarcomas
RARα	Hormone-dependent transcription factor, differentiation	Chromosomal translocations with PML gene, fusion product	Acute promyelocytic leukemia
E6	Human papillomavirus encoded oncogene, inactivates p53	HPV infection	Cervical cancer
Cyclins	Bind to CDKs, regulate cell cycle	Gene amplification, overexpression	Lung, esophagus, many types
Tumor Suppressor	**Normal Function**	**Alteration in Cancer**	**Associated Cancers**
RB1	Cell-cycle checkpoints, binds E2F	Mutation, deletion, inactivation by viral oncogene products	Retinoblastoma, osteosarcoma, many types
APC	Cell–cell interaction	Mutation	Colorectal cancers, brain, thyroid
Bcl2	Apoptosis regulation	Overexpression blocks apoptosis	Lymphomas, leukemias
BRCA1, BRCA2	DNA repair	Point mutations	Breast, ovarian, prostate cancers

activities cannot be controlled in a normal fashion. This is sometimes due to a mutation in the proto-oncogene resulting in a protein product that acts abnormally. In other cases, proto-oncogenes are overexpressed or expressed at an incorrect time. If the proto-oncogene is continually in an "on" state, its product may constantly stimulate the cell to divide. When a proto-oncogene is mutated or aberrantly expressed and contributes to the development of cancer, it is known as an **oncogene**—a cancer-causing gene. Oncogenes are proto-oncogenes that have experienced a gain-of-function alteration. As a result, only one allele of a proto-oncogene needs to be mutated or misexpressed in order to trigger uncontrolled growth. Hence, oncogenes confer a dominant cancer phenotype.

Tumor-suppressor genes are genes whose products normally regulate cell-cycle checkpoints or initiate the process of apoptosis. In normal cells, proteins encoded by tumor-suppressor genes halt progress through the cell cycle in response to DNA damage or growth-suppression signals from the extracellular environment. When tumor-suppressor genes are mutated or inactivated, cells are unable to respond normally to cell-cycle checkpoints, or are unable to undergo programmed cell death if DNA damage is extensive. This leads to a further increase in mutations and to the inability of the cell to leave the cell cycle when it should become quiescent. When both alleles of a tumor-suppressor gene are inactivated, and other changes in the cell keep it growing and dividing, cells may become tumorigenic.

The following are examples of proto-oncogenes and tumor-suppressor genes that contribute to cancer when mutated. More than 300 oncogenes and tumor-suppressor genes are now known, and more will likely be discovered as cancer research continues.

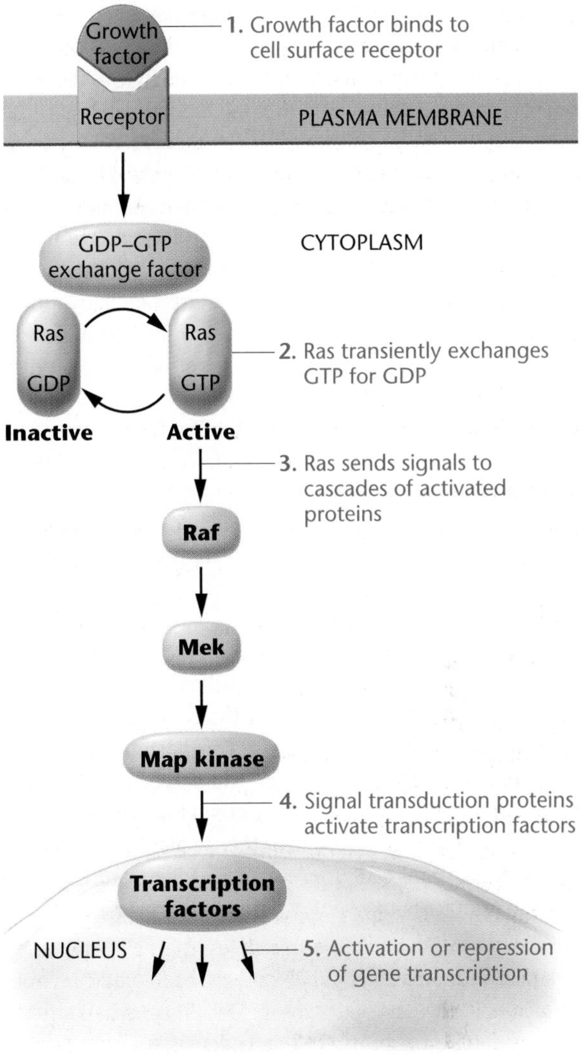

FIGURE 16–6 A signal transduction pathway mediated by Ras.

The *ras* Proto-oncogenes

Some of the most frequently mutated genes in human tumors are those in the ***ras* gene family.** These genes are mutated in more than 30 percent of human tumors. The *ras* gene family encodes signal transduction molecules that are associated with the cell membrane and regulate cell growth and division. Ras proteins normally transmit signals from the cell membrane to the nucleus, stimulating the cell to divide in response to external growth factors (Figure 16–6). Ras proteins alternate between an inactive (switched off) and an active (switched on) state by binding either guanosine diphosphate (GDP) or guanosine triphosphate (GTP). When a cell encounters a growth factor (such as platelet-derived growth factor or epidermal growth factor), growth factor receptors on the cell membrane bind to the growth factor, resulting in autophosphorylation of the cytoplasmic portion of the growth factor receptor. This causes recruitment of proteins known as nucleotide exchange factors to the plasma membrane. These nucleotide exchange factors cause Ras to release GDP and bind GTP, thereby activating Ras. The active, GTP-bound form of Ras then sends its signals through cascades of protein phosphorylations in the cytoplasm. The endpoint of these cascades is activation of nuclear transcription factors that stimulate expression of genes whose products drive the cell from quiescence into the cell cycle. Once Ras has sent its signals to the nucleus, it hydrolyzes GTP to GDP and becomes inactive. Mutations that convert the proto-oncogene *ras* to an oncogene prevent the Ras protein from hydrolyzing GTP to GDP and hence freeze the Ras protein into its "on" conformation, constantly stimulating the cell to divide.

ESSENTIAL POINT

Proto-oncogenes are normal genes that promote cell growth and division. When proto-oncogenes are mutated or misexpressed in cancer cells, they are known as oncogenes.

The *p53* Tumor-suppressor Gene

The most frequently mutated gene in human cancers—mutated in more than 50 percent of all cancers—is the ***p53* gene.** This gene encodes a nuclear protein that acts as a transcription factor, repressing or stimulating transcription of more than 50 different genes.

Normally, the p53 protein is continuously synthesized but is rapidly degraded and therefore is present in cells at low levels. In addition, the p53 protein is normally bound to another protein called **Mdm2,** which has several effects on p53. The presence of Mdm2 on the p53 protein tags p53 for degradation and sequesters the transcriptional activation domain of p53. It also prevents the phosphorylations and acetylations that convert the p53 protein from an inactive to an active form. Several types of events bring about rapid increases in the nuclear levels of activated p53 protein. These include chemical damage to DNA, double-stranded breaks in DNA induced by ionizing radiation, and the presence of DNA-repair intermediates generated by exposure of cells to ultraviolet light. In response to these signals, Mdm2 dissociates from p53, making p53 more stable and unmasking its transcription activation domain. Increases in the levels of activated p53 protein also result from increases in protein phosphorylation, acetylation, and other posttranslational modifications.

The p53 protein initiates two different responses to DNA damage: either arrest of the cell cycle followed by DNA repair, or apoptosis and cell death if DNA cannot be repaired. Both of these responses are accomplished by p53 acting as a transcription factor that stimulates or represses the expression of genes involved in each response.

In normal cells, p53 can arrest the cell cycle at the G1/S and G2/M checkpoints, as well as retarding the progression of the cell through S phase. It accomplishes this by inhibiting cyclin/CDK complexes and regulating the transcription of other genes involved in these phases of the cell cycle.

Activated p53 can also instruct a damaged cell to commit suicide by apoptosis. It does so by activating the transcription of genes whose products control this process. In cancer cells that lack functional p53, these gene products are not synthesized and apoptosis may not occur.

Hence, cells lacking functional p53 are unable to arrest at cell-cycle checkpoints or to enter apoptosis in response to DNA damage. As a result, they move unchecked through the cell cycle, regardless of the condition of the cell's DNA. Cells lacking p53 have high mutation rates and accumulate the types of mutations that lead to cancer. Because of the importance of the *p53* gene to genomic integrity, it is often referred to as the "guardian of the genome."

The *RB1* Tumor-suppressor Gene

The loss or mutation of the ***RB1* (retinoblastoma 1)** tumor-suppressor gene contributes to the development of many cancers, including those of the breast, bone, lung, and bladder. The *RB1* gene was originally identified as a result of studies on **retinoblastoma,** an inherited disorder in which tumors develop in the eyes of young children. Retinoblastoma occurs with a frequency of about 1 in 15,000 individuals. In the familial form of the disease, individuals inherit one mutated allele of the *RB1* gene and have an 85 percent chance of developing retinoblastomas as well as an increased chance of developing other cancers. All somatic cells of patients with hereditary retinoblastoma contain one mutated allele of the *RB1* gene. However, it is only when the second normal allele of the *RB1* gene is lost or mutated in certain retinal cells that retinoblastoma develops. In individuals who do not have this hereditary condition, retinoblastoma is extremely rare, as it requires at least two separate somatic mutations in a retinal cell in order to inactivate both copies of the *RB1* gene (Figure 16–7).

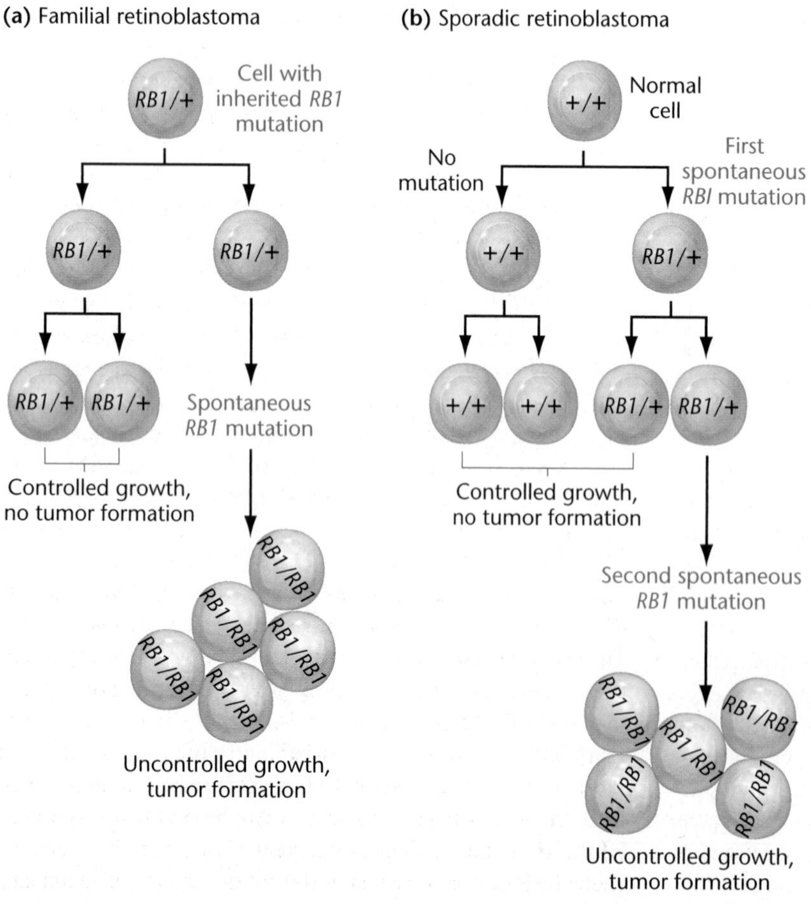

FIGURE 16–7 (a) In familial retinoblastoma, one mutation (designated as *RB1*) is inherited and present in all cells. A second mutation at the retinoblastoma locus in any retinal cell contributes to uncontrolled cell growth and tumor formation. (b) In sporadic retinoblastoma, independent mutations in both alleles of the retinoblastoma gene within a single cell are acquired sequentially, also leading to tumor formation.

The **retinoblastoma protein (pRB)** is a tumor-suppressor protein that controls the G1/S cell-cycle checkpoint. The pRB protein is found in the nuclei of all cell types at all stages of the cell cycle. However, its activity varies throughout the cell cycle, depending on its phosphorylation state. When cells are in the G0 phase of the cell cycle, the pRB protein is nonphosphorylated and binds to transcription factors such as E2F, inactivating them (Figure 16–8). When the cell is stimulated by growth factors, it enters G1 and approaches S phase. Throughout the G1 phase, the pRB protein becomes phosphorylated by the CDK4/cyclin D1 complex. Phosphorylated pRB releases its bound regulatory proteins. When E2F and other regulators are released by pRB, they are free to induce the expression of over 30 genes whose products are required for the transition from G1 into S phase. After cells traverse S, G2, and M phases, pRB reverts to a nonphosphorylated state, binds to regulatory proteins such as E2F, and keeps them sequestered until required for the next cell cycle. In normal quiescent cells, the presence of the pRB protein prevents passage into S phase. In many cancer cells, including retinoblastoma cells, both copies of the *RB1* gene are defective, inactive, or absent, and progression through the cell cycle is not regulated.

ESSENTIAL POINT

Tumor-suppressor genes normally regulate cell-cycle checkpoints and apoptosis. When tumor-suppressor genes are mutated or inactivated, cells cannot correct DNA damage. This leads to accumulations of mutations that may cause cancer.

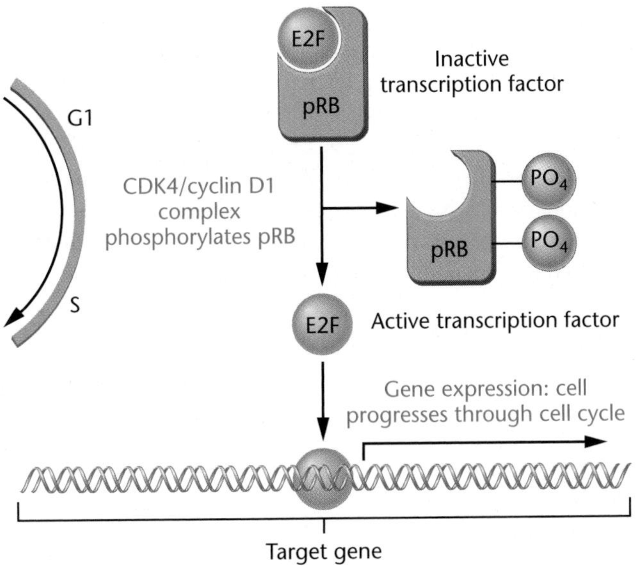

FIGURE 16–8 In the nucleus during G0 and early G1, pRB interacts with and inactivates transcription factor E2F. As the cell moves from G1 to S phase, a CDK4/cyclinD1 complex forms and adds phosphate groups to pRB. As pRB becomes phosphorylated, E2F is released and becomes transcriptionally active, allowing the cell to pass through S phase. Phosphorylation of pRB is transitory; as CDK/cyclin complexes are degraded and the cell moves through the cell cycle to early G1, pRB phosphorylation declines, allowing pRB to reassociate with E2F.

NOW SOLVE THIS

Problem 23 on page 349 describes how people with Li-Fraumeni syndrome have a mutated allele of the *p53* gene and susceptibility to a wide variety of cancers. You are asked to explain how a mutation in one gene can give rise to a number of different cancers.

Hint: Consider which cellular functions are regulated by p53 protein and how the absence of p53 could affect each of these functions. Also, read about loss of heterozygosity in Section 16.6.

16.5 Cancer Cells Metastasize, Invading Other Tissues

As discussed at the beginning of this chapter, uncontrolled growth alone is insufficient to create a malignant and life-threatening cancer. Cancer cells must also acquire the features of metastasis, which include the ability to disengage from the original tumor site, to enter the blood or lymphatic system, to invade surrounding tissues, and to develop into secondary tumors. In order to leave the site of the primary tumor and invade other tissues, tumor cells must dissociate from other cells and digest components of the **extracellular matrix** and **basal lamina,** which normally surround and separate the body's tissues. The extracellular matrix and basal lamina are composed of proteins and carbohydrates. They form the scaffold for tissue growth and inhibit the migration of cells.

The ability to invade the extracellular matrix is also a property of some normal cell types. For example, implantation of the embryo in the uterine wall during pregnancy requires cell migration across the extracellular matrix. In addition, white blood cells reach sites of infection by penetrating capillary walls. The mechanisms of invasion are probably similar in these normal cells and in cancer cells. The difference is that, in normal cells, the invasive ability is tightly regulated, whereas in tumor cells, this regulation has been lost.

Although less is known about the genes that control metastasis than about those controlling the cell cycle, it is likely that metastasis is controlled by a large number of genes, including those that encode cell-adhesion molecules, cytoskeleton regulators, and proteolytic enzymes. For example, epithelial tumors have a lower than normal level of the **E-cadherin glycoprotein,** which is responsible for cell–cell adhesion in normal tissues. Also, proteolytic enzymes such as **metalloproteinases** are present at higher than normal levels in highly malignant tumors and are not susceptible to the normal controls conferred by regulatory molecules such as **tissue inhibitors of metalloproteinases (TIMPs).** It has been shown that the level of aggressiveness of a tumor correlates positively with the levels of proteolytic enzymes expressed by the tumor. Hence, inappropriately expressed cell adhesion and proteinase enzymes may assist malignant tumor cells to metastasize by loosening the normal constraints on cell location and creating holes through which the tumor cells can pass into and out of the circulatory system.

Like the tumor-suppressor genes of primary cancers, **metastasis-suppressor genes** control the growth of secondary tumors. Less than a dozen of these metastasis-suppressor genes have been identified so far, but all appear to affect the growth of metastatic tumors and not the primary tumor. The expression of these genes is often reduced by epigenetic mechanisms rather than by mutation. This observation provides hope that researchers can develop antimetastasis therapies that target the epigenetic silencing of metastasis-suppressor genes.

ESSENTIAL POINT

The ability of cancer cells to metastasize requires defects in gene products that control a number of functions such as cell adhesion, proteolysis, and tissue invasion.

16.6 Predisposition to Some Cancers Can Be Inherited

Although the vast majority of human cancers are sporadic, a small fraction (1 to 2 percent) have a hereditary or familial component. At present, about 50 forms of hereditary cancer are known (Table 16.2).

Most inherited cancer-susceptibility alleles, though transmitted in a Mendelian dominant fashion, are not sufficient in themselves to trigger development of a cancer. At least one other somatic mutation in the other copy of the gene must occur in order to drive a cell toward tumorigenesis. In addition, mutations in still other genes are usually necessary to fully express the cancer phenotype. As mentioned earlier, inherited mutations in the *RB1* gene predispose individuals to developing various cancers. Although the normal somatic cells of these patients are heterozygous for the *RB1* mutation, cells within their tumors contain mutations in both copies of the gene. The phenomenon whereby the second, wild-type, allele is mutated in a tumor is known as **loss of heterozygosity.** Although loss of heterozygosity is an essential first step in expression of these inherited cancers, further mutations in other proto-oncogenes and tumor-suppressor genes are necessary for the tumor cells to become fully malignant.

TABLE 16.2 Inherited Predispositions to Cancer

Tumor Predisposition Syndromes	Chromosome
Early-onset familial breast cancer	17q
Familial adenomatous polyposis	5q
Familial melanoma	9p
Gorlin syndrome	9q
Hereditary nonpolyposis colon cancer	2p
Li-Fraumeni syndrome	17p
Multiple endocrine neoplasia, type 1	11q
Multiple endocrine neoplasia, type 2	22q
Neurofibromatosis, type 1	17q
Neurofibromatosis, type 2	22q
Retinoblastoma	13q
Von Hippel–Lindau syndrome	3p
Wilms tumor	11p

The development of hereditary colon cancer illustrates how inherited mutations in one allele of a gene contribute only one step in the multistep pathway leading to malignancy.

About 1 percent of colon cancer cases result from a genetic predisposition to cancer known as **familial adenomatous polyposis (FAP).** In FAP, individuals inherit one mutant copy of the ***APC*** **(adenomatous polyposis) gene** located on the long arm of chromosome 5. Mutations include deletions, frameshift, and point mutations. The normal function of the *APC* gene product is to act as a tumor suppressor controlling cell–cell contact and growth inhibition by interacting with the β-catenin protein. The presence of a heterozygous *APC* mutation causes the epithelial cells of the colon to partially escape cell-cycle control, and the cells divide to form small clusters of cells called **polyps** or adenomas. People who are heterozygous for this condition develop hundreds to thousands of colon and rectal polyps early in life. Although it is not necessary for the second allele of the *APC* gene to be mutated in polyps at this stage, in the majority of cases, the second *APC* allele becomes mutant in a later stage of cancer development. The relative order of mutations in the development of FAP is shown in Figure 16–9.

The second mutation in polyp cells that contain an *APC* gene mutation occurs in the *ras* proto-oncogene. The combined *APC* and *ras* gene mutations bring about the development of intermediate adenomas. Cells within these adenomas have defects in normal cell differentiation. In addition, these cells will grow in culture and are not growth-inhibited by contact

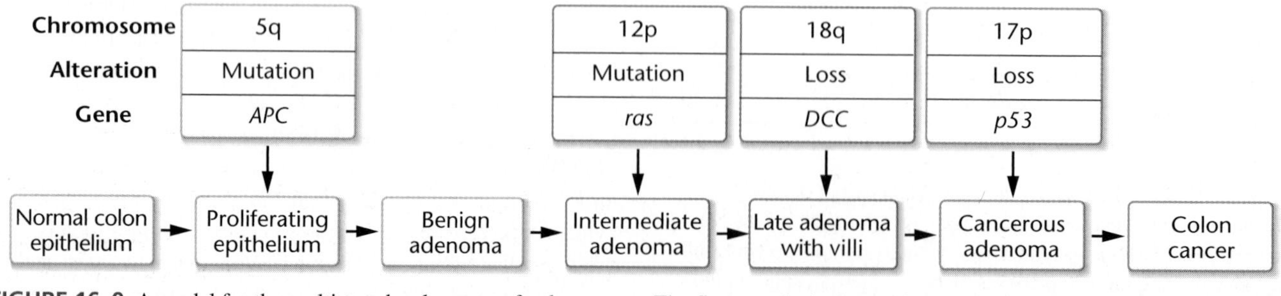

FIGURE 16–9 A model for the multistep development of colon cancer. The first step is the loss or inactivation of one allele of the *APC* gene on chromosome 5. In FAP cases, one mutant *APC* allele is inherited. Subsequent mutations involving genes on chromosomes 12, 17, and 18 in cells of benign adenomas can lead to a malignant transformation that results in colon cancer. Although the mutations on chromosomes 12, 17, and 18 usually occur at a later stage than those involving chromosome 5, the sum of changes is more important than the order in which they occur.

with other cells—a process known as **transformation.** The third step toward malignancy requires loss of function of both alleles of the *DCC* (*d*eleted in *c*olon *c*ancer) gene. The *DCC* gene product is thought to be involved with cell adhesion and differentiation. Mutations in both *DCC* alleles result in the formation of late-stage adenomas with a number of finger-like outgrowths (villi). When late adenomas progress to cancerous adenomas, they usually suffer loss of functional *p53* genes. The final steps toward malignancy involve mutations in an unknown number of genes associated with metastasis.

ESSENTIAL POINT

Inherited mutations in cancer-susceptibility genes are not sufficient to trigger cancer. Other somatic mutations in proto-oncogenes or tumor-suppressor genes are necessary for the development of hereditary cancers.

NOW SOLVE THIS

Problem 7 on page 349 presents the observation that some tobacco smokers and some people with inherited mutations in cancer-related genes never develop cancer. You are asked to explain this phenomenon.

Hint: Consider the steps involved in the development of cancer and the number of abnormal functions in cancer cells. Also, consider how genetics may affect DNA repair functions.

16.7 Viruses Contribute to Cancer in Both Humans and Animals

Viruses that cause cancer in animals have played a significant role in the search for knowledge about the genetics of human cancer. Most cancer-causing animal viruses are RNA viruses known as **retroviruses.** In humans, most of the known cancer viruses are DNA viruses.

To understand how retroviruses cause cancer in animals, it is necessary to know how these viruses replicate in cells. When a retrovirus infects a cell, its RNA genome is copied into DNA by the **reverse transcriptase** enzyme, which is brought into the cell with the infecting virus. The DNA copy then enters the nucleus of the infected cell, where it integrates at random into the host cell's genome. The integrated DNA copy of the retroviral RNA is called a **provirus.** The proviral DNA contains powerful enhancer and promoter elements in its U5 and U3 sequences at the ends of the provirus (Figure 16–10). The proviral promoter uses the host cell's transcription proteins, directing transcription of the viral genes (*gag, pol,* and *env*). The products of these genes are the proteins and RNA genomes that make up the new retroviral particles. Because the provirus is integrated into the host genome, it is replicated along with the host's DNA during the cell's normal cell cycle.

A retrovirus may not kill a cell, but it may continue to use the cell as a factory to replicate more viruses that will then infect surrounding cells.

A retrovirus may cause cancer in three different ways. First, the proviral DNA may integrate by chance near one of the cell's normal proto-oncogenes. The strong promoters and enhancers in the provirus then stimulate high levels or inappropriate timing of transcription of the proto-oncogene, leading to stimulation of host-cell proliferation. Second, a retrovirus may pick up a copy of a host proto-oncogene and integrate it into its genome (Figure 16–10). The new viral oncogene may be mutated during the process of transfer into the virus, or it may be expressed at abnormal levels because it is now under the control of viral promoters. Retroviruses that carry these oncogenes can infect and transform normal cells into tumor cells, and are known as **acute transforming retroviruses.** Through the study of many acute transforming viruses of animals, scientists have identified dozens of proto-oncogenes. Third, a retrovirus may contain a viral gene whose product can either stimulate the cell cycle or act as a gene expression regulator for both cellular and viral genes. As a result, expression of such a viral gene may lead to inappropriate cell growth or to abnormal expression of cancer-related cellular genes.

So far, no acute transforming retroviruses have been identified in humans; however, several human retroviruses, such as **human immunodeficiency virus (HIV)** and the **human T-cell leukemia virus (HTLV-1)** are associated with human cancers. These retroviruses are thought to stimulate cancer development through the third mechanism, described in the previous paragraph.

DNA viruses also contribute to the development of human cancers in a variety of ways. Because viruses are comprised solely of a nucleic acid genome surrounded by a protein coat, they must utilize the host cell's biosynthetic machinery in order to reproduce themselves. To access the host's DNA-synthesizing enzymes, viruses require the host cell to be in an actively growing state. Thus, many DNA viruses, as well as retroviruses, contain genes encoding products that stimulate the cell cycle. These products often interact with tumor-suppressor

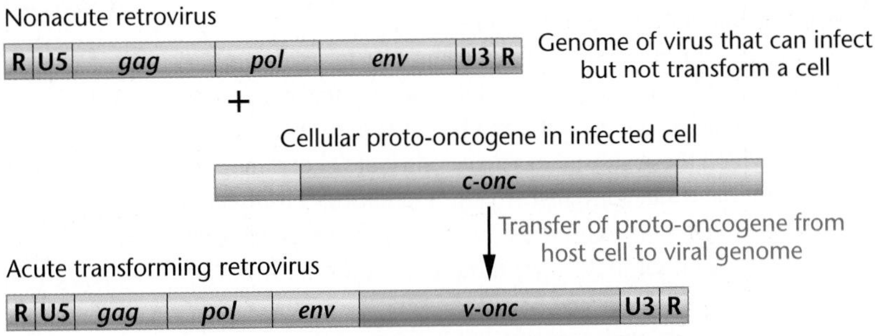

FIGURE 16–10 The genome of a typical retrovirus is shown at the top of the diagram. The genome contains repeats at the termini (R), the U5 and U3 regions that contain promoter and enhancer elements, and the three major genes that encode viral structural proteins (*gag* and *env*) and the viral reverse transcriptase (*pol*). RNA transcripts of the entire viral genome comprise the new viral genomes. If the retrovirus acquires all or part of a host-cell proto-oncogene (*c-onc*), this gene (now known as a *v-onc*) is expressed along with the viral genes, leading to overexpression or inappropriate expression of the *v-onc* gene. The *v-onc* gene may also acquire mutations that enhance its transforming ability.

proteins, inactivating them. If the host cell survives the infection, it may lose control of the cell cycle and begin its journey to carcinogenesis.

It is thought that, worldwide, about 15 percent of human cancers are associated with viruses, making virus infection the second greatest risk factor for cancer, next to tobacco smoking. The most significant contributors to virus-induced cancers are the **papillomaviruses (HPV 16** and **18), human T-cell leukemia virus (HTLV-1), hepatitis B virus, human herpesvirus 8,** and **Epstein-Barr virus.** Like other risk factors for cancer, including hereditary predisposition to certain cancers, virus infection alone is not sufficient to trigger human cancers. Other factors, including DNA damage or the accumulation of mutations in one or more of a cell's oncogenes and tumor-suppressor genes, are required to move a cell down the multistep pathway to cancer.

ESSENTIAL POINT

Tumor viruses contribute to cancers by introducing oncogenes, interfering with tumor-suppressor proteins, or altering expression of a cell's proto-oncogenes.

16.8 Environmental Agents Contribute to Human Cancers

Any substance or event that damages DNA has the potential to be carcinogenic. Unrepaired or inaccurately repaired DNA introduces mutations, which, if they occur in proto-oncogenes or tumor-suppressor genes, can lead to abnormal regulation of the cell cycle or disruption of controls over cell contact and invasion.

Our environment, both natural and human-made, contains abundant carcinogens. These include chemicals, radiation, some viruses, and chronic infections. Perhaps the most significant carcinogen in our environment is tobacco smoke, which contains a number of cancer-causing chemicals. Epidemiologists estimate that about 30 percent of human cancer deaths are associated with cigarette smoking.

Diet is often implicated in the development of cancer. Consumption of red meat and animal fat is associated with some cancers, such as colon, prostate, and breast cancer. The mechanisms by which these substances may contribute to carcinogenesis are not clear but may involve stimulation of cell division through hormones or creation of carcinogenic chemicals during cooking. Alcohol may cause inflammation of the liver and contribute to liver cancer.

Although most people perceive the human-made, industrial environment to be a highly significant contributor to cancer, it may account for only a small percentage of total cancers, and only in special situations. Some of the most mutagenic agents, and hence potentially the most carcinogenic, are natural substances and natural processes. For example, **aflatoxin,** a component of a mold that grows on peanuts and corn, is one of the most carcinogenic chemicals known. Most chemical carcinogens, such as **nitrosamines,** are components of synthetic substances and are found in some preserved meats; however, many are naturally occurring. For example, natural pesticides and antibiotics found in plants may be carcinogenic, and the human body itself creates alkylating agents in the acidic environment of the gut. Nevertheless, these observations do not diminish the serious cancer risks to specific populations who are exposed to human-made carcinogens such as synthetic pesticides or asbestos.

DNA lesions brought about by natural radiation (X rays, ultraviolet light), natural dietary substances, and substances in the external environment contribute the majority of environmentally caused mutations that lead to cancer. In addition, normal metabolism creates oxidative end products that can damage DNA, proteins, and lipids. It is estimated that the human body suffers about 10,000 damaging DNA lesions per day due to the actions of oxygen free radicals. DNA repair enzymes deal successfully with most of this damage; however, some damage may lead to permanent mutations. The process of DNA replication itself is mutagenic. Hence, substances such as growth factors or hormones that stimulate cell division are ultimately mutagenic and perhaps carcinogenic. Chronic inflammation due to infection also stimulates tissue repair and cell division, resulting in DNA lesions accumulating during replication. These mutations may persist, particularly if cell-cycle checkpoints are compromised due to mutations or inactivation of tumor-suppressor genes such as *p53* or *RB1*.

Both ultraviolet (UV) light and ionizing radiation (such as X rays and gamma rays) induce DNA damage. UV in sunlight is well accepted as an inducer of skin cancers. Ionizing radiation has clearly shown itself to be a carcinogen in studies of populations exposed to neutron and gamma radiation from atomic blasts such as those in Hiroshima and Nagasaki. Another significant environmental component, radon gas, may be responsible for up to 50 percent of the ionizing radiation exposure of the U.S. population and could contribute to lung cancers in some populations.

ESSENTIAL POINT

Environmental agents such as chemicals, radiation, viruses, and chronic infections contribute to the development of cancer. The most significant environmental factors that affect human cancers are tobacco smoke, diet, and natural radiation.

NOW SOLVE THIS

Problem 16 on page 349 deals with the concept that cancer can occur spontaneously and in response to environmental factors. You are asked to provide an estimate of what percentage of money spent on cancer research should be devoted to research and education aimed at preventing cancer and what percentage should be devoted to research in pursuit of cancer cures.

Hint: In answering this question, think about the relative rates of environmentally induced and spontaneous cancers. (An interesting source of information on this topic is Ames, B. N. et al. 1995. The causes and prevention of cancer. *Proc. Natl. Acad. Sci. USA 92: 5258–5265.*)

GENETICS, TECHNOLOGY, AND SOCIETY

Breast Cancer: The Double-Edged Sword of Genetic Testing

These are exhilarating times for genetics and biotechnology. The prospect of using genetics to prevent and cure a wide range of diseases is exciting. However, in our enthusiasm, we often forget that these new technologies still have significant limitations and profound ethical complexities. The story of genetic testing for breast cancer illustrates how we must temper our high expectations with respect for uncertainty.

Breast cancer is the most common cancer among women. A woman's lifetime risk of developing breast cancer is about 12 percent. Each year, more than 190,000 new cases are diagnosed in the United States. Breast cancer is not limited to women; about 1400 men are also diagnosed with the disease each year.

Approximately 5 to 10 percent of breast cancers are familial, a category defined by the early onset of the disease and the appearance of several cases of breast or ovarian cancer among near blood relatives. In 1994, two genes were identified that show linkage to familial breast cancers: *BRCA1* and *BRCA2*. In normal cells, these two genes appear to encode tumor-suppressor proteins that are involved in repairing damaged DNA. Women with germ-line mutations in *BRCA1* or *BRCA2* have a 36 to 85 percent lifetime risk of developing breast cancer and a 16 to 60 percent risk of developing ovarian cancer. Men with germ-line mutations in *BRCA2* have a 6 percent lifetime breast cancer risk—a hundredfold increase over the general male population.

BRCA1 and *BRCA2* genetic tests are available, and these detect over 2000 different mutations that are known to occur within the coding regions of these genes. Many patients at risk for familial breast cancer opt to undergo genetic testing. These patients feel that test results could motivate them to take steps to prevent breast or ovarian cancers, guide them in childbearing decisions, and provide information concerning the risk to close relatives. But all these potential benefits are fraught with uncertainties.

A woman whose *BRCA* test results are negative may feel relieved and assume that she is not subject to familial breast cancer. However, her risk of developing breast cancer is still 12 percent (the population risk), and she should continue to monitor herself for the disease. Also, a negative *BRCA* genetic test does not eliminate the possibility that she carries an inherited mutation in another gene that increases breast cancer risk or that *BRCA1* or *BRCA2* gene mutations exist in regions of the genes that are inaccessible to current genetic tests.

A woman whose test results are positive faces difficult choices. Her treatment options consist of close monitoring, prophylactic mastectomy or oophorectomy (removal of breasts and ovaries, respectively), and taking prophylactic drugs such as tamoxifen. Prophylactic surgery reduces her risk but does not eliminate it, as cancers can still occur in tissues that remain after surgery. Drugs such as tamoxifen are helpful but have serious side effects. Genetic tests also affect the patient's entire family. People often experience fear, anxiety, and guilt on learning that they are carriers of a genetic disease. Confidentiality is also a major concern. Patients fear that their genetic test results may be leaked to insurance companies or employers, jeopardizing their prospects for jobs or affordable health and life insurance.

The unanswered scientific and ethical questions about *BRCA1* and *BRCA2* genetic testing are many and important. As we develop genetic tests for more and more diseases over the next few decades, our struggle with these kinds of questions will continue.

Your Turn

Take time, individually or in groups, to answer the following questions. Investigate the references and links, to help you understand some of the issues that surround genetic testing for breast cancer.

1. New genomics research is rapidly identifying genes linked to human diseases, including breast cancer. How many genes are now thought to be involved in familial breast cancer? Are genetic tests available to detect mutations in these genes?

Search for recent scientific data on breast cancer susceptibility genes by using the PubMed Web site **(http://www.ncbi.nlm.nih.gov/sites/entrez?db=PubMed)**, *as described in the Exploring Genomics box in Chapter 2.*

2. In 2008, the U.S. Congress passed a bill designed to eliminate discrimination by employers or health insurance companies, based on a person's genetic test results. Do you think that this act will be effective? What problems exist with the act, and how might they be solved?

Read news articles about the Genetic Information Nondiscrimination Act by searching the New York Times Web site at: **http://www.nytimes.com.**

3. What is the rate of "false negatives" in *BRCA1/2* gene tests—that is, the number of high-risk patients who do not show *BRCA1/2* gene mutations in genetic tests but still have genetic defects that lead to breast cancer?

This topic is discussed in Stokstad, E. 2006. Genetic screen misses mutations in women at high risk of breast cancer. *Science* **311**: 1847.

4. If your family was at risk for familial breast cancer, would you opt to take the *BRCA1/2* genetic tests? What actions would you take if you received a positive test result? How would you feel about such a result?

Two helpful sources are: (a) Surbone, A. 2001. Ethical implications of genetic testing for breast cancer susceptibility. *Crit. Rev. in Onc./Hem.* **40**: 149–157. *(b) Genetic Testing for BRCA1 and BRCA2: It's your choice. National Cancer Institute Fact Sheet.* **http://www.cancer.gov/cancertopics/factsheet/risk/brca.**

CASE STUDY I thought it was safe

A middle-aged woman taking the breast cancer drug Tamoxifen for ten years became concerned when she saw a news report with disturbing information. In some women, the drug made their cancer more aggressive and more likely to spread. Other women with breast cancer, the report stated, do not respond to Tamoxifen at all, and 30 to 40 percent of women who take the drug eventually become resistant to chemotherapy. The woman contacted her oncologist to ask some questions:

1. How can some people react one way to a cancer treatment and others react a different way?
2. Why do most cancers eventually become resistant to a specific chemotherapeutic drug?
3. Why does it seem that some drugs are thought to be safe one day and declared unsafe the next day?

INSIGHTS AND SOLUTIONS

1. In disorders such as retinoblastoma, a mutation in one allele of the *RB1* gene can be inherited from the germ line, causing an autosomal dominant predisposition to the development of eye tumors. To develop tumors, a somatic mutation in the second copy of the *RB1* gene is necessary, indicating that the mutation itself acts as a recessive trait. Given that the first mutation can be inherited, in what ways can a second mutational event occur?

Solution: In considering how this second mutation arises, we must look at several types of mutational events, including changes in nucleotide sequence and events that involve whole chromosomes or chromosome parts. Retinoblastoma results when both copies of the *RB1* locus are lost or inactivated. With this in mind, you must first list the phenomena that can result in a mutational loss or the inactivation of a gene.

One way the second *RB1* mutation can occur is by a nucleotide alteration that converts the remaining normal *RB1* allele to a mutant form. This alteration can occur through a nucleotide substitution or through a frameshift mutation caused by the insertion or deletion of nucleotides during replication. A second mechanism involves the loss of the chromosome carrying the normal allele. This event would take place during mitosis, resulting in chromosome 13 monosomy and leaving the mutant copy of the gene as the only *RB1* allele. This mechanism does not necessarily involve loss of the entire chromosome; deletion of the long arm (*RB1* is on 13q) or an interstitial deletion involving the *RB1* locus and some surrounding material would have the same result. Alternatively, a chromosome aberration involving loss of the normal copy of the *RB1* gene might be followed by duplication of the chromosome carrying the mutant allele. Two copies of chromosome 13 would be restored to the cell, but the normal *RB1* allele would not be present. Finally, a recombination event followed by chromosome segregation could produce a homozygous combination of mutant *RB1* alleles.

2. Proto-oncogenes can be converted to oncogenes in a number of different ways. In some cases, the proto-oncogene itself becomes amplified up to hundreds of times in a cancer cell. An example is the *cyclin D1* gene, which is amplified in some cancers. In other cases, the proto-oncogene may be mutated in a limited number of specific ways leading to alterations in the gene product's structure. The *ras* gene is an example of a proto-oncogene that becomes oncogenic after suffering point mutations in specific regions of the gene. Explain why these two proto-oncogenes (*cyclin D1* and *ras*) undergo such different alterations in order to convert them into oncogenes.

Solution: The first step in solving this question is to understand the normal functions of these proto-oncogenes and to think about how either amplification or mutation would affect each of these functions.

The cyclin D1 protein regulates progression of the cell cycle from G1 into S phase, by binding to CDK4 and activating this kinase. The cyclin D1/CDK4 complex phosphorylates a number of proteins including pRB, which in turn activates other proteins in a cascade that results in transcription of genes whose products are necessary for DNA replication in S phase. The simplest way to increase the activity of cyclin D1 would be to increase the number of cyclin D1 molecules available for binding to the cell's endogenous CDK4 molecules. This can be accomplished by several mechanisms, including amplification of the *cyclin D1* gene. In contrast, a point mutation in the *cyclin D1* gene would most likely interfere with the ability of the cyclin D1 protein to bind to CDK4; hence, mutations within the gene would probably repress cell-cycle progression rather than stimulate it.

The *ras* gene product is a signal transduction protein that operates as an on/off switch in response to external stimulation by growth factors. It does so by binding either GTP (the "on" state) or GDP (the "off" state). Oncogenic mutations in the *ras* gene occur in specific regions that alter the ability of the Ras protein to exchange GDP for GTP. Oncogenic Ras proteins are locked in the "on" conformation, bound to GTP. In this way, they constantly stimulate the cell to divide. An amplification of the *ras* gene would simply provide more molecules of normal Ras protein, which would still be capable of on/off regulation. Hence, simple amplification of *ras* would less likely be oncogenic.

Problems and Discussion Questions

1. As a genetic counselor, you are asked to assess the risk for a couple with a family history of retinoblastoma who are thinking about having children. Both the husband and wife are phenotypically normal, but the husband has a sister with familial retinoblastoma in both eyes. What is the probability that this couple will have a child with retinoblastoma? Are there any tests that you could recommend to help in this assessment?
2. What events occur in each phase of the cell cycle? Which phase is most variable in length?
3. Where are the major regulatory points in the cell cycle?
4. List the functions of kinases and cyclins, and describe how they interact to cause cells to move through the cell cycle.
5. (a) How does pRB function to keep cells at the G1 checkpoint? (b) How do cells get past the G1 checkpoint to move into S phase?
6. What is the difference between saying that cancer is inherited and saying that the predisposition to cancer is inherited?
7. Although tobacco smoking is responsible for a large number of human cancers, not all smokers develop cancer. Similarly, some people who inherit mutations in the tumor-suppressor genes *p53* or *RB1* never develop cancer. Describe some reasons for these observations. See **Now Solve This** on page 345.
8. What is apoptosis, and under what circumstances do cells undergo this process?
9. Define tumor-suppressor genes. Why is a mutation in a single copy of a tumor-suppressor gene expected to behave as a recessive gene?
10. A genetic variant of the retinoblastoma protein, called PSM-RB (phosphorylation site mutated RB), is not able to be phosphorylated by the action of CDK4/cyclinD1 complex. Explain why PSM-RB is said to have a constitutive growth-suppressing action on the cell cycle.
11. Part of the Ras protein is associated with the plasma membrane, and part extends into the cytoplasm. How does the Ras protein transmit a signal from outside the cell into the cytoplasm? What happens in cases where the *ras* gene is mutated?
12. If a cell suffers damage to its DNA while in S phase, how can this damage be repaired before the cell enters mitosis?
13. Distinguish between oncogenes and proto-oncogenes. In what ways can proto-oncogenes be converted to oncogenes?
14. Of the two classes of genes associated with cancer, tumor-suppressor genes and oncogenes, mutations in which group can be considered gain-of-function mutations? In which group are the loss-of-function mutations? Explain.
15. How do translocations such as the Philadelphia chromosome contribute to cancer?
16. Given that cancers can be environmentally induced and that some environmental factors are the result of lifestyle choices such as smoking, sun exposure, and diet, what percentage of the money spent on cancer research do you think should be devoted to research and education on preventing cancer rather than on finding cancer cures? See **Now Solve This** on page 346.
17. In CML, leukemic blood cells can be distinguished from other cells of the body by the presence of a functional BCR-ABL hybrid protein. Explain how this characteristic provides an opportunity to develop a therapeutic approach to a treatment for CML. See **Now Solve This** on page 338.
18. How do normal cells protect themselves from accumulating mutations in genes that could lead to cancer? How do cancer cells differ from normal cells in these processes?
19. Describe the difference between an acute transforming virus and a virus that does not cause tumors.
20. Explain how environmental agents such as chemicals and radiation cause cancer.
21. Radiotherapy (treatment with ionizing radiation) is one of the most effective current cancer treatments. It works by damaging DNA and other cellular components. In which ways could radiotherapy control or cure cancer, and why does radiotherapy often have significant side effects?
22. Genetic tests that detect mutations in the *BRCA1* and *BRCA2* oncogenes are widely available. These tests reveal a number of mutations in these genes—mutations that have been linked to familial breast cancer. Assume that a young woman in a suspected breast cancer family takes the *BRCA1* and *BRCA2* genetic tests and receives negative results. That is, she does not test positive for the mutant alleles of *BRCA1* or *BRCA2*. Can she consider herself free of risk for breast cancer?
23. People with a genetic condition known as Li–Fraumeni syndrome inherit one mutant copy of the *p53* gene. These people have a high risk of developing a number of different cancers, such as breast cancer, leukemias, bone cancers, adrenocortical tumors, and brain tumors. Explain how mutations in one cancer-related gene can give rise to such a diverse range of tumors. See **Now Solve This** on page 343.
24. Explain the differences between a benign and malignant tumor.
25. As part of a cancer research project, you have discovered a gene that is mutated in many metastatic tumors. After determining the DNA sequence of this gene, you compare the sequence with those of other genes in the human genome sequence database. Your gene appears to code for an amino acid sequence that resembles sequences found in some serine proteases. Conjecture how your new gene might contribute to the development of highly invasive cancers.
26. A study by Bose and colleagues (1998. *Blood* 92: 3362–3367) and a previous study by Biernaux and others (1996. *Bone Marrow Transplant* 17: (Suppl. 3) S45–S47) showed that *BCR-ABL* fusion gene transcripts can be detected in 25 to 30 percent of healthy adults who do not develop chronic myelogenous leukemia (CML). Explain how these individuals can carry a fusion gene that is transcriptionally active and yet do not develop CML.
27. Those who inherit a mutant allele of the *RB1* gene are at risk for developing a bone cancer called osteosarcoma. You suspect that in these cases, osteosarcoma requires a mutation in the second *RB1* allele, and you have cultured some osteosarcoma cells and obtained a cDNA clone of a normal human *RB1* gene. A colleague sends you a research paper revealing that a strain of cancer-prone mice develop malignant tumors when injected with osteosarcoma cells, and you obtain these mice. Using these three resources, what experiments would you perform to determine (a) whether osteosarcoma cells carry two *RB1* mutations, (b) whether osteosarcoma cells produce any pRB protein, and (c) if the addition of a normal *RB1* gene will change the cancer-causing potential of osteosarcoma cells?
28. Explain the apparent paradox that both hypermethylation and hypomethylation of DNA are often associated with various cancers.

29. The table in this problem summarizes some of the data that have been collected on *BRCA1* mutations in families with a high incidence of both early-onset breast cancer and ovarian cancer.
    (a) Note the coding effect of the mutation found in kindred group 2082. This results from a single base-pair substitution. Draw the normal double-stranded DNA sequence for this codon (with the 5′ and 3′ ends labeled), and show the sequence of events that generated this mutation, assuming that it resulted from an uncorrected mismatch event during DNA replication.
    (b) Examine the types of mutations that are listed in the table and determine if the *BRCA1* gene is likely to be a tumor-suppressor gene or an oncogene.
    (c) Although the mutations listed in the table are clearly deleterious and cause breast cancer in women at very young ages, each of the kindred groups had at least one woman who carried the mutation but lived until age 80 without developing cancer. Name at least two different mechanisms (or variables) that could underlie variation in the expression of a mutant phenotype and propose an explanation for the incomplete penetrance of this mutation. How do these mechanisms or variables relate to this explanation?

Predisposing Mutations in *BRCA1*

Kindred	Codon	Nucleotide Change	Coding Effect	Frequency in Control Chromosomes
1901	24	−11 bp	Frameshift or splice	0/180
2082	1313	C → T	Gln → Stop	0/170
1910	1756	Extra C	Frameshift	0/162
2099	1775	T → G	Met → Arg	0/120
2035	NA*	?	Loss of transcript	NA*

Source: 1994. *Science* 266: 66–71.

*NA indicates not applicable, as the regulatory mutation is inferred, and the position has not been identified.

30. The following table shows neutral polymorphisms found in control families (those with no increased frequency of breast and ovarian cancer).
    Examine the data in the table and answer the following questions:
    (a) What is meant by a neutral polymorphism?
    (b) What is the significance of this table in the context of examining a family or population for *BRCA1* mutations that predispose an individual to cancer?
    (c) Is the PM2 polymorphism likely to result in a neutral missense mutation or a silent mutation?
    (d) Answer part (c) for the PM3 polymorphism.

Neutral Polymorphisms in *BRCA1*

Name	Codon Location	Base in Codon†	Frequency in Control Chromosomes* A	C	G	T
PM1	317	2	152	0	10	0
PM6	878	2	0	55	0	100
PM7	1190	2	109	0	53	0
PM2	1443	3	0	115	0	58
PM3	1619	1	116	0	52	0

*The number of chromosomes with a particular base at the indicated polymorphic site (A, C, G, or T) is shown.

†Position 1, 2, or 3 of the codon.

31. Explain why many oncogenic viruses contain genes whose products interact with tumor-suppressor proteins.
32. Prostate cancer is a major cause of cancer-related deaths among men. Epigenetic changes that regulate gene expression are involved in both the initiation and progression of such cancers. Following is a table that lists the number of genes known to be hypermethylated in prostate cancer cells (modified from Long-Cheng, L. et al., 2005. *J. Natl. Cancer Inst.* 97: 103–115). For each category of genes, speculate on the mechanism(s) by which cancer initiation or progression might be influenced by hypermethylation.

DNA Hypermethylation of:	Number of Known Genes
Hormonal response genes	5
Cell-cycle control genes	2
Tumor cell invasion genes	8
DNA damage repair genes	2
Signal transduction genes	4

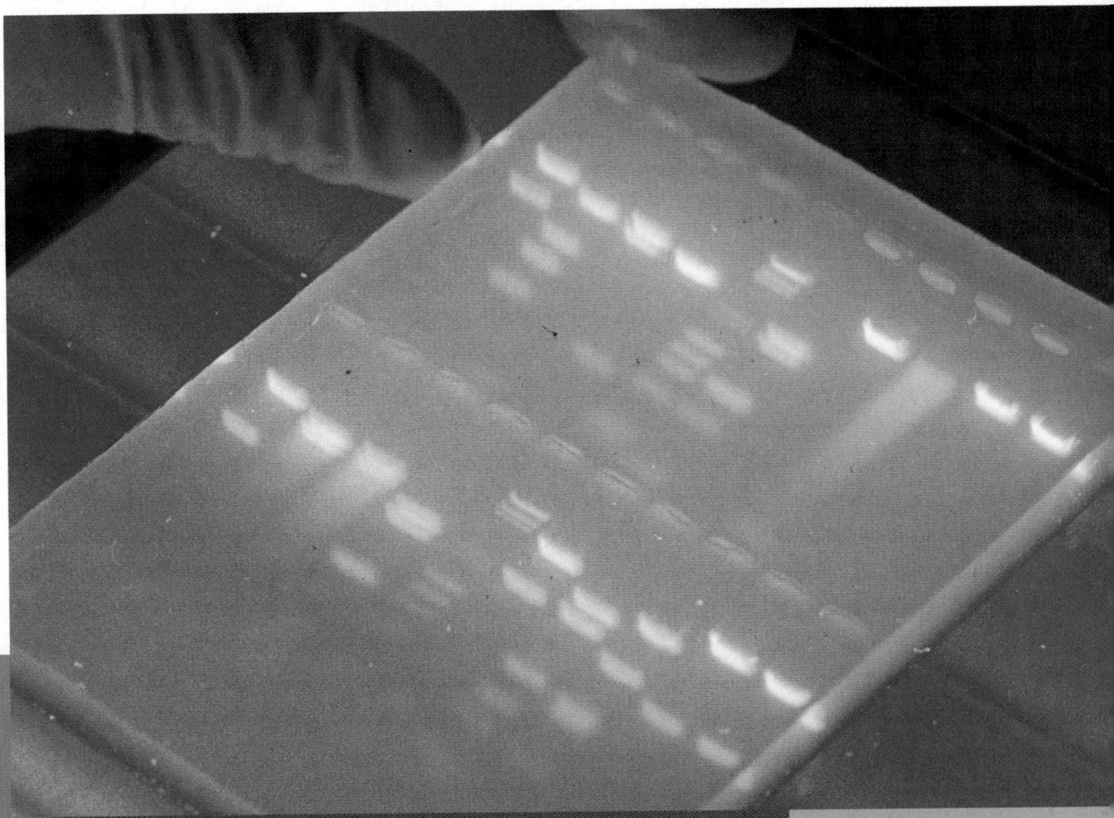

An agarose gel containing separated DNA fragments stained with a dye (ethidium bromide) and visualized under ultraviolet light.

17 Recombinant DNA Technology and Gene Cloning

Chapter Concepts

- Recombinant DNA technology creates artificial combinations of DNA molecules, usually from two different sources, most often from different species.
- Recombinant DNA technology depends in part on the ability to cleave and rejoin DNA segments at specific base sequences.
- The most useful application of recombinant DNA technology is to clone a DNA segment of interest.
- For cloning, specific DNA segments are inserted into vectors (such as plasmids), which are transferred into host cells (such as bacterial cells), where the recombinant molecules replicate as the host cells divide.
- Recombinant molecules can be cloned in prokaryotic or eukaryotic host cells.
- DNA segments can be quickly and efficiently amplified millions of times using the polymerase chain reaction (PCR).
- Cloned DNA segments are analyzed in several ways, the most specific being DNA sequencing.
- Recombinant DNA technology has revolutionized our ability to investigate the genomes of diverse species.

RECOMBINANT DNA TECHNOLOGY: RESTRICTION ENZYMES

WEB TUTORIAL

The term **recombinant DNA** has two meanings in genetics. The more specific of the two refers to a DNA molecule formed in the laboratory by joining together DNA sequences from different biological sources. Such **recombinant DNA molecules** are artificial laboratory creations and are not found in nature. The term *recombinant DNA* is also used more loosely to refer to the technology that is utilized to create and study these hybrid molecules. The power of recombinant DNA technology is astonishing, enabling geneticists to identify and isolate a single gene or DNA segment of interest from the thousands or tens of thousands present in a genome. Subsequently, through cloning, millions of identical copies of this specific DNA molecule can be produced. These identical copies, or **clones,** can then be manipulated for numerous purposes, including research or for the commercial production of its encoded protein. In this chapter, we review the basic methods of recombinant DNA technology used to isolate, replicate, and analyze genes. In Chapters 18 and 19, we will discuss some applications of this technology to genomics, research, medicine, the legal system, agriculture, and industry.

How Do We Know?

In this chapter, we will focus on the essential techniques of recombinant DNA technology, including the construction and cloning of recombinant DNA molecules, and the ways in which clones of interest can be identified and analyzed. As you study this topic, you should try to answer several fundamental questions:

1. How did we discover that restriction enzymes recognize and cut specific DNA sequences?
2. How can we know if DNA fragments have been successfully incorporated into vectors such as plasmids?
3. How did we discover how to make many copies of a DNA molecule without using a vector or host cell?
4. How can a specific DNA sequence be recovered from a large collection of cloned DNA sequences?

17.1 An Overview of Recombinant DNA Technology

Although natural genetic processes such as crossing over produce recombined DNA molecules, the term *recombinant DNA* is generally reserved for molecules produced by artificially joining DNA obtained from different biological sources. The methods used to create these molecules are largely derived from nucleic acid biochemistry, coupled with genetic techniques developed for the study of bacteria and viruses. The technology is used to isolate and to clone DNA sequences (i.e., make many copies of these sequences), including genes. The process involves the following steps:

1. DNA to be cloned is purified from cells or tissues.
2. Proteins called **restriction enzymes** are used to generate specific DNA fragments. These enzymes recognize and cut DNA molecules at specific nucleotide sequences.
3. The fragments produced by restriction enzymes are joined to other DNA molecules that serve as vectors, or carrier molecules. A vector joined to a DNA fragment is a recombinant DNA molecule.
4. The recombinant DNA molecule is transferred to a host cell. Within the host cell, the recombinant molecule replicates, producing dozens of identical copies, or clones, of the recombinant molecule.
5. As host cells replicate, the recombinant DNA molecules within them are passed on to all their progeny, creating a population of host cells, each of which carries copies of the cloned DNA sequence.
6. The cloned DNA can be recovered from host cells, purified, and analyzed.
7. The cloned DNA can also be transcribed, its mRNA translated, and the encoded gene product isolated and used for research or commercial purposes.

ESSENTIAL POINT

The development of recombinant DNA technology was made possible by the discovery of proteins called restriction enzymes, which cut DNA at specific recognition sites, producing fragments that can be joined with other DNA fragments to form recombinant DNA molecules.

17.2 Constructing Recombinant DNA Molecules Requires Several Steps

Recombinant DNA technology allows the preparation of large quantities of specific DNA sequences, including genes. We will begin our discussion of this technology by considering the steps used to make specific cuts in DNA and link the fragments produced to carrier DNA molecules.

Restriction Enzymes

Restriction enzymes are produced by bacteria as a defense mechanism against viral infection by degrading the DNA of invading viruses. More than 3500 restriction enzymes have been identified, and about 150 of these are commonly used by researchers. A restriction enzyme binds to DNA and recognizes a specific nucleotide sequence called a **recognition sequence.** The enzyme then cuts both strands of the DNA within that sequence. The usefulness of restriction enzymes in cloning derives from their ability to accurately and reproducibly cut genomic DNA into fragments called **restriction fragments.**

The size of restriction fragments is determined by the number of times a given restriction enzyme cuts the DNA. Enzymes with a four-base recognition sequence—such as the enzyme *Alu*I, which recognizes the sequence AGCT—will

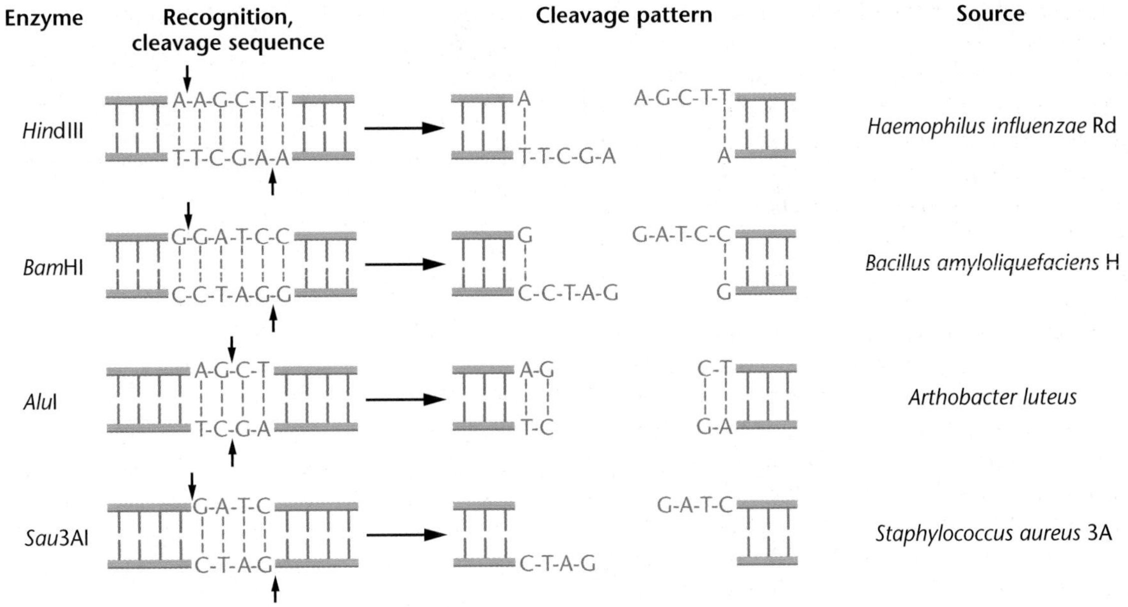

FIGURE 17–1 Some common restriction enzymes, with their recognition sequences, cleavage patterns, and sources. The arrows indicate the cutting sites for each enzyme.

cut, on average, every 256 base pairs ($4^n = 4^4 = 256$) if all four nucleotides are present in equal proportions, producing many small fragments. Enzymes such as *Not*I have an eight-base recognition sequence (GCGGCCGC) and cut the DNA on average every 65,500 base pairs ($4^n = 4^8$), producing fewer, larger fragments. The actual fragment sizes produced by cutting DNA with a given restriction enzyme vary because the number and location of recognition sequences are not always distributed randomly in DNA. Enzymes such as *Eco*RI and *Bam*HI make offset or staggered cuts in the DNA strands, producing fragments with short, single-stranded tails. Other enzymes such as *Alu*I cut both strands at the same nucleotide pair, producing blunt ends (Figure 17–1).

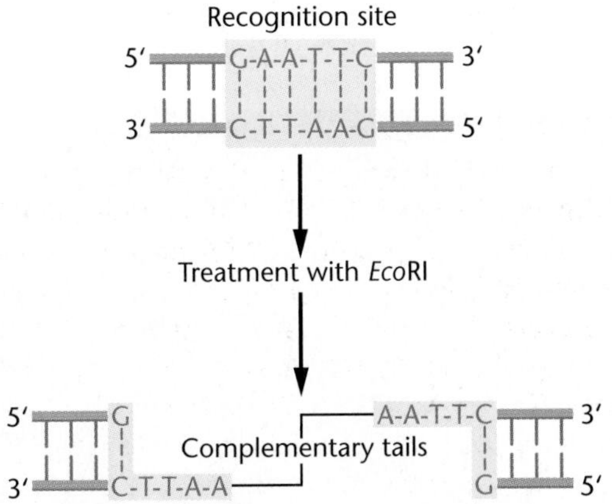

FIGURE 17–2 The restriction enzyme *Eco*RI recognizes and binds to the nucleotide sequence GAATTC. Cleavage of the DNA at this site produces complementary single-stranded tails (often called sticky ends). These single-stranded tails can anneal with single-stranded tails from other DNA fragments that, after ligation, form recombinant DNA molecules.

One of the first restriction enzymes to be identified was isolated from *Escherichia coli* strain R and is designated *Eco*RI (pronounced echo-r-one or eeko-r-one). DNA fragments produced by *Eco*RI digestion (Figure 17–2) have overhanging single-stranded tails ("sticky ends") that can base-pair with complementary single-stranded tails on other DNA fragments. Complementary single-stranded ends of DNA fragments from different sources can thus **anneal,** or stick together, by complementary base pairing at their single-stranded ends. Adding the enzyme **DNA ligase** to the solution closes the gaps in the sugar-phosphate backbone groups, covalently linking the fragments to form recombinant DNA molecules (Figure 17–3, p. 354).

Vectors

WEB TUTORIAL RECOMBINANT DNA TECHNOLOGY: VECTORS

The fragments of DNA produced by restriction enzyme digestion are often copied (cloned) inside bacteria or some other host cell, but they cannot directly enter the bacterial cells for cloning without first being joined to a carrier DNA molecule, or vector. **Vectors** transfer and help replicate inserted DNA fragments. Many different vectors are available for cloning; they differ in which host cells they are able to enter, in the size of inserts they can carry, and in other properties, such as the number of copies that can be produced, the number of recognition sequences available for inserting DNA fragments, and the number and type of selectable marker genes they contain.

To serve as a vector, a DNA molecule must:

- replicate independently along with any DNA fragment it carries, once inside the host cell
- contain several restriction-enzyme cleavage sites that allow insertion of DNA fragments to be cloned
- carry a selectable marker gene to identify host cells that contain recombinant vectors (usually an antibiotic resistance gene or the gene for an enzyme absent from the host cell)

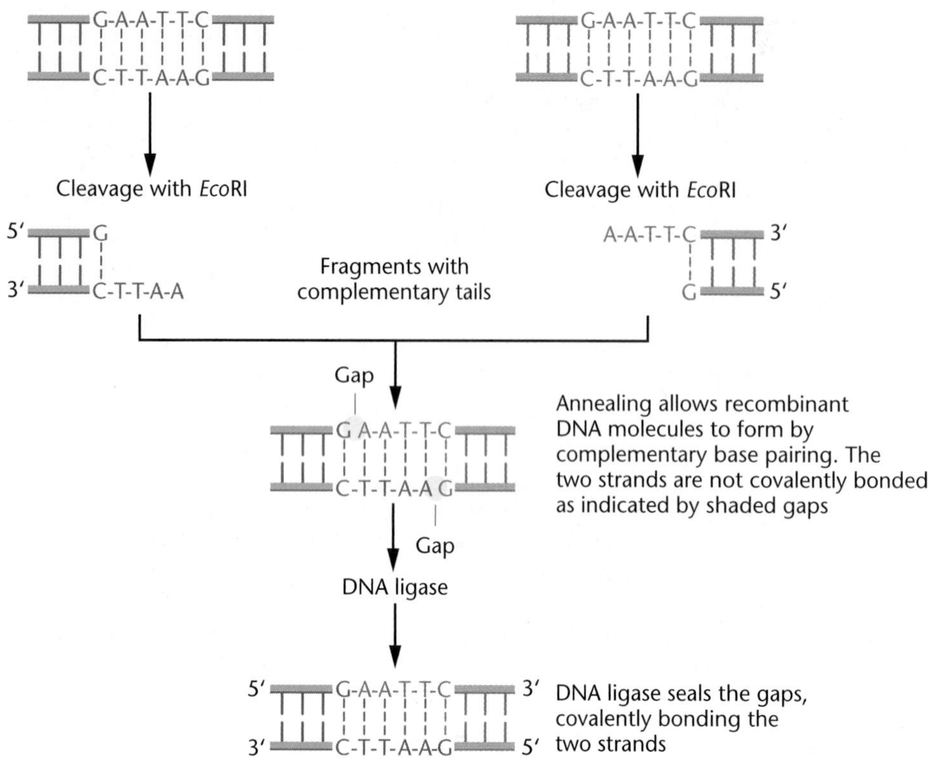

FIGURE 17–3 DNA from different sources is cleaved with *Eco*RI and mixed to allow annealing. The enzyme DNA ligase then chemically bonds these annealed fragments into an intact recombinant DNA molecule.

In addition, the vector and its inserted DNA fragment should be easy to recover from the host cell.

To prepare a vector for use in cloning, a preparation of purified vector DNA is added to a tube containing buffer and a restriction enzyme, and incubated at the optimum temperature for the enzyme, usually 37°C. Over the course of one or more hours, the enzyme binds to and cuts the DNA at all its specific recognition sites. For insertion of a DNA fragment, the cut vector is mixed with a collection of DNA fragments produced by cutting with the same enzyme and treated with the enzyme DNA ligase. Vectors carrying an inserted fragment are called **recombinant vectors,** and each is an example of a recombinant DNA molecule, produced by joining DNA from two different sources.

Genetically modified plasmids were the first vectors developed and are still widely used for cloning. These vectors were derived from naturally occurring plasmids (Figure 17–4), the extrachromosomal, double-stranded self-replicating DNA molecules found in certain bacterial strains. The genetics of plasmids and their host bacterial cells were introduced in Chapter 8. Here, we will focus on the use of plasmids as vectors for cloning DNA.

Plasmids have been extensively modified by genetic engineering to serve as cloning vectors. Many are now available with a range of useful features. For example, although only a single plasmid generally enters a bacterial host cell, once inside, some plasmids can increase their number so that several hundred copies of the original plasmid are present. When used as vectors, these plasmids greatly enhance the number of DNA clones that can be produced. For added convenience, these vectors have also been genetically engineered to contain a number of convenient restriction-enzyme recognition sequences, as well as marker genes that reveal their presence in host cells.

A plasmid with several useful features as a vector is **pUC18** (Figure 17–5).

1. pUC18 is small (2686 base pairs), so it can carry relatively large DNA inserts.
2. It has an origin of replication, necessary for initiating DNA synthesis, and it can produce up to 500 copies of inserted DNA fragments per cell.
3. A large number of restriction-enzyme recognition sequences have been engineered into pUC18, conveniently clustered in one region called a **polylinker site.**

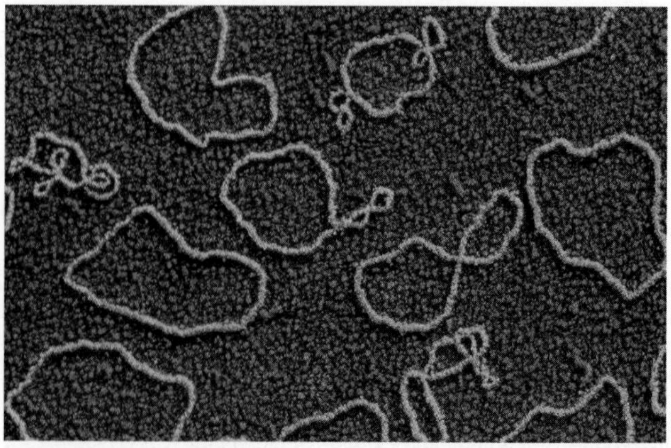

FIGURE 17–4 A color-enhanced electron micrograph of circular plasmid molecules isolated from *E. coli.* Genetically engineered plasmids are used as vectors for cloning DNA.

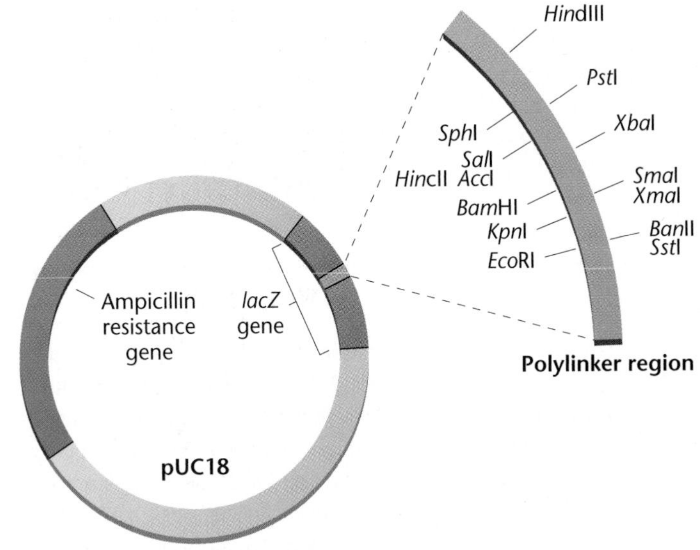

FIGURE 17–5 A diagram of the plasmid pUC18 showing the polylinker region located within a *lacZ* gene. DNA inserted into the polylinker region disrupts the *lacZ* gene, resulting in white colonies that allow direct identification of bacterial colonies carrying cloned DNA inserts.

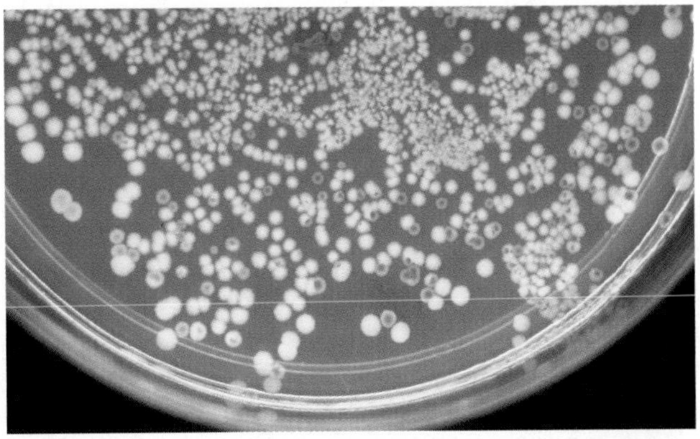

FIGURE 17–6 A Petri dish showing the growth of bacterial cells after uptake of recombinant plasmids. The medium on the plate contains a compound called Xgal. DNA inserted into the pUC18 vector disrupts the gene responsible for the formation of blue colonies. As a result, it is easy to distinguish colonies carrying cloned DNA inserts. Cells in blue colonies contain vectors without cloned DNA inserts, whereas cells in white colonies contain vectors carrying DNA inserts.

4. Recombinant pUC18 plasmids are easily identified. For example, pUC18 carries a fragment of the bacterial *lacZ* gene as a selectable marker, and the polylinker site is inserted into this fragment. Expression of *lacZ* causes bacterial host cells carrying pUC18 to produce blue colonies when grown on medium containing a compound known as Xgal. If a DNA fragment is inserted anywhere in the polylinker site, the *lacZ* gene is disrupted and becomes inactive. Thus, a bacterial cell carrying pUC18 with an inserted DNA fragment forms white colonies on Xgal medium, whereas colonies carrying pUC18 plasmids without inserted DNA fragments form blue colonies (Figure 17–6).

The steps for generating a recombinant DNA molecule using pUC18 are illustrated in Figure 17–7.

Although many types of plasmid cloning vectors are available, other more specialized vectors have been developed from viral and bacterial genomes and from yeast plasmids.

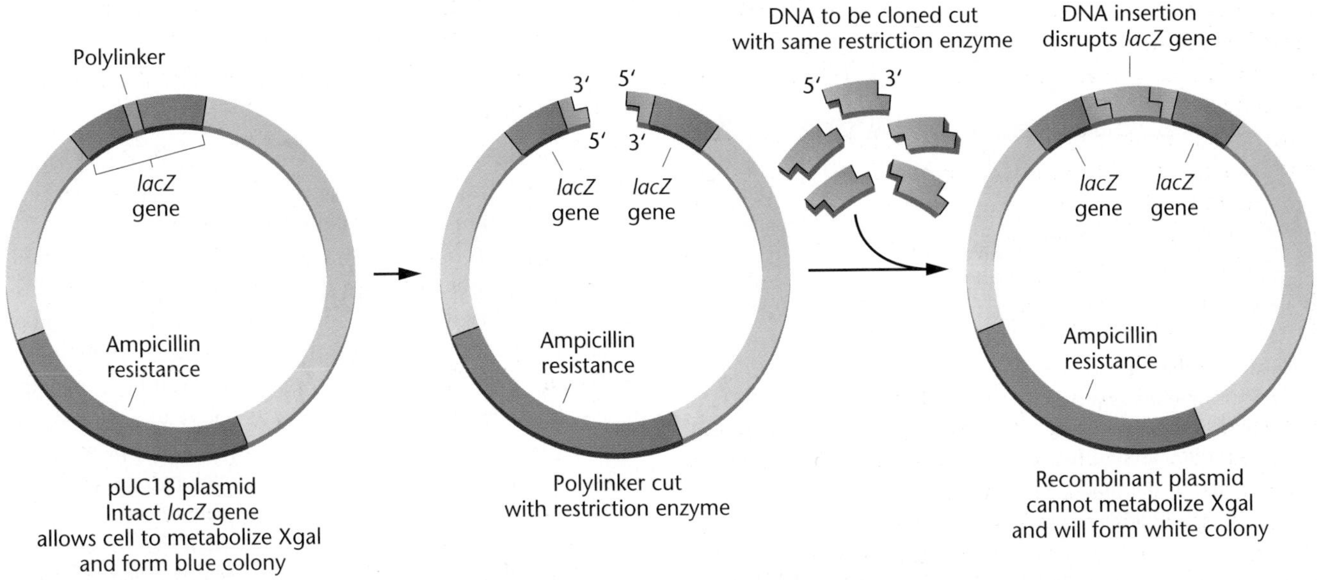

FIGURE 17–7 The plasmid vector pUC18 carries unique restriction cleavage sites in a polylinker region within the *lacZ* gene. Cleavage of the plasmid and DNA to be cloned with a restriction enzyme, followed by insertion of a DNA fragment into the polylinker region, disrupts the *lacZ* gene, so that cells with plasmids carrying inserts are unable to metabolize Xgal. These plasmids therefore form white colonies on nutrient plates containing Xgal.

Some specialized vectors are **expression vectors,** engineered with control sequences that allow expression of inserted genes. Other vectors, such as **bacterial artificial chromosomes** (**BACs,** pronounced "backs") or **yeast artificial chromosomes** (**YACs,** pronounced as "yaks"), are designed to clone very long segments of DNA.

Like natural chromosomes, a YAC (Figure 17–8) has telomeres at each end, an origin of replication (which initiates DNA synthesis), and a centromere. These components are joined to selectable marker genes (*TRP1, SUP4,* and *URA3*) and a cluster of restriction sites for DNA inserts. Yeast chromosomes range from 230 kb to over 1900 kb, making it possible to clone DNA inserts from 100 to 1000 kb in YACs. The ability to clone large pieces of DNA into these vectors makes them an important tool in genome projects, including the Human Genome Project, which we will discuss in Chapter 18.

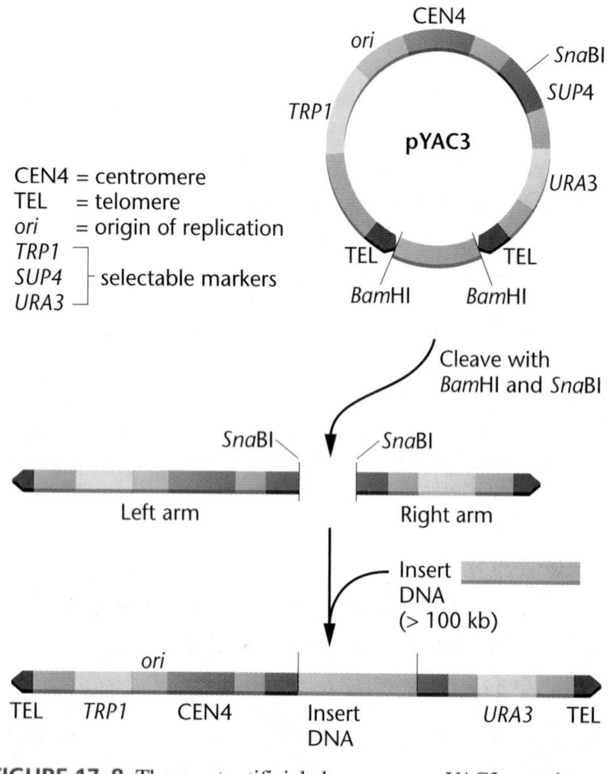

FIGURE 17–8 The yeast artificial chromosome pYAC3 contains telomere sequences (TEL), a centromere (CEN4) derived from yeast chromosome 4, and an origin of replication (*ori*). These elements give the cloning vector the properties of a chromosome. *TRP1* and *URA3* are yeast genes that are selectable markers for the left and right arms of the chromosome. Within the *SUP4* gene is a restriction enzyme recognition sequence for the enzyme *Sna*B1. Two *Bam*H1 recognition sequences flank a spacer segment. Cleavage with *Sna*B1 and *Bam*H1 breaks the artificial chromosome into two arms. The DNA to be cloned is treated with *Sna*B1, producing a collection of fragments. The arms and fragments are ligated together, and the artificial chromosome is inserted into yeast host cells. Because naturally occurring yeast chromosomes range in size from 230 kb to over 1900 kb, it is possible to insert large DNA fragments in YACs. Insert sizes can be in the million base-pair range.

ESSENTIAL POINT

Vectors replicate autonomously in host cells and facilitate the manipulation of the newly created recombinant DNA molecules. Vector components come from many sources, including bacterial plasmids and phages.

17.3 Cloning DNA in Host Cells

As discussed earlier, scientists use recombinant DNA technology to construct *and* replicate recombinant DNA molecules to clone specific DNA sequences. Replication takes place after transfer of recombinant molecules into host cells. This cloning method, known as cell-based cloning, was the first method developed and was carried out using first prokaryotic host cells and later eukaryotic host cells. It is still widely used to make cloned DNA.

One of the most commonly used prokaryotic hosts is a laboratory strain of the bacterium *E. coli* known as JM109. This and other genetically-modified strains of *E. coli* are genetically well characterized and can host a wide range of vectors.

As an example, the following steps can be used to create recombinant DNA molecules and transfer them to an *E. coli* host cell (Figure 17–9) where they are cloned:

1. The DNA to be cloned is isolated and treated with a restriction enzyme to create fragments ending in specific single-stranded tails.
2. The fragments are then linked to plasmid molecules that have been cut with the same restriction enzyme, creating a collection of recombinant vectors.
3. The recombinant vectors are transferred into *E. coli* host cells. Inside the host cell, a vector replicates to form many copies, or clones.
4. The bacteria are plated on nutrient medium, where they form colonies.
5. The colonies are screened to identify those that have taken up recombinant plasmids.

Because the cells in each colony are derived from a single ancestral cell, all the cells in the colony as well as the plasmids they contain are genetically identical clones.

Although *E. coli* is widely used as a prokaryotic host cell, for several reasons, the yeast *Saccharomyces cerevisiae* is extensively used as a host cell for the cloning and expression of eukaryotic genes. These reasons include the following: (1) Although yeast is a eukaryotic organism, it can be grown and manipulated in much the same way as bacterial cells. (2) The genetics of yeast have been intensively studied, providing a large catalog of mutants and a highly developed genetic map. (3) The entire yeast genome has been sequenced, and most genes in the organism have been identified. (4) To study the function of some eukaryotic proteins, it is necessary to use a host cell that can modify the protein after it has been synthesized, to convert it to a functional form (bacteria cannot carry

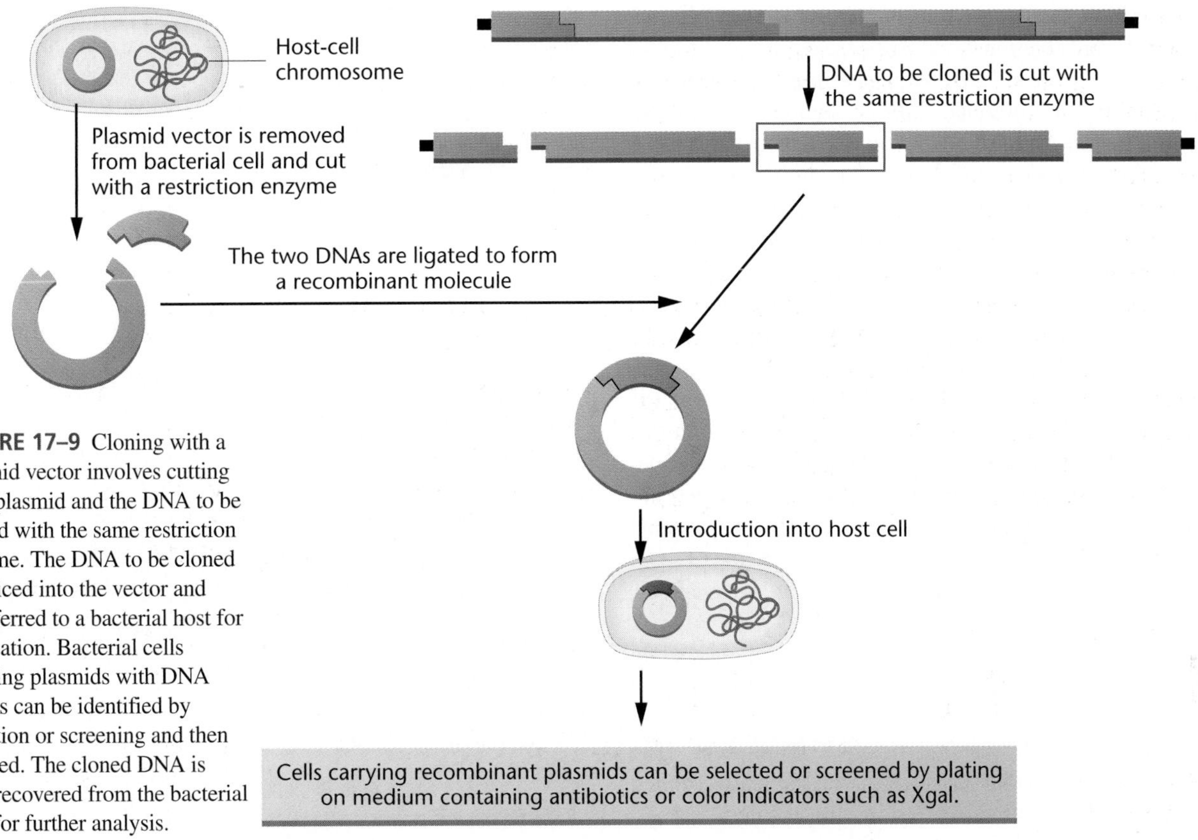

Host-cell chromosome
DNA to be cloned is cut with the same restriction enzyme
Plasmid vector is removed from bacterial cell and cut with a restriction enzyme
The two DNAs are ligated to form a recombinant molecule
Introduction into host cell
Cells carrying recombinant plasmids can be selected or screened by plating on medium containing antibiotics or color indicators such as Xgal.

FIGURE 17–9 Cloning with a plasmid vector involves cutting both plasmid and the DNA to be cloned with the same restriction enzyme. The DNA to be cloned is spliced into the vector and transferred to a bacterial host for replication. Bacterial cells carrying plasmids with DNA inserts can be identified by selection or screening and then isolated. The cloned DNA is then recovered from the bacterial host for further analysis.

out some of these modifications). (5) Yeast has been used for centuries in the baking and brewing industries and is considered to be a safe organism for producing proteins for vaccines and therapeutic agents. Table 17.1 lists some of the medically useful products of cloning in yeast.

NOW SOLVE THIS

Question 12 on page 372 describes the preparation of a genomic library using *Drosophila* DNA. You are asked to determine several specific aspects of the procedure.

Hint: Inserting foreign DNA into vectors carrying antibiotic resistance genes can be used to distinguish vectors that carry inserted DNA from those that do not carry inserts.

TABLE 17.1 Recombinant Proteins Synthesized in Yeast Cells

Hepatitis B virus surface protein
Malaria parasite protein
Epidermal growth factor
Platelet-derived growth factor
α_1-antitrypsin
Clotting factor XIIIA

ESSENTIAL POINT

Recombinant DNA molecules are transferred into any of several types of host cells, where cloned copies are produced during host-cell replication.

17.4 The Polymerase Chain Reaction Makes DNA Copies without Host Cells

WEB TUTORIAL: POLYMERASE CHAIN REACTION

Recombinant DNA techniques were developed in the early 1970s and revolutionized research in genetics and molecular biology. There methods also gave birth to the booming biotechnology industry. However, cloning DNA using vectors and host cells can be labor intensive and time-consuming. In 1986, another technique, called the **polymerase chain reaction (PCR),** was developed. This advance further accelerated the use of recombinant DNA methodology in biological research. The significance of this method was underscored by the awarding of the 1993 Nobel Prize in Chemistry to Kary Mullis, who developed the technique.

PCR is a rapid method of copying DNA that extends the power of recombinant DNA research and eliminates the need to use host cells for cloning. Although cell-based cloning is still widely used, PCR is the method of choice for many applications in molecular biology, human genetics, evolution, development, conservation, and forensics.

PCR copies a specific DNA sequence through a series of *in vitro* reactions and produces billions of copies of this sequence in a matter of hours. PCR can amplify specific DNA sequences that are present in very small quantities even when they are mixed in with many other DNA molecules. As a prerequisite for PCR, three things are needed: (1) single-stranded DNA templates to be copied, (2) primers for each strand that can be used to synthesize a complementary strand on each template, and (3) the enzyme DNA polymerase. The primers are short oligonucleotides about 15–20 base pairs in length. Each primer is complementary to the 3′ end of the denatured single-stranded templates. When added to a sample of DNA that has been heated and denatured to form single-stranded templates, the primers bind to complementary nucleotides flanking the sequence to be amplified. A heat-stable DNA polymerase is then added. Beginning at the 3′ OH group of the primer, the enzyme synthesizes a new DNA strand complementary to the template DNA (Figure 17–10). Repetition of this process using both strands of the newly copied DNA as templates produces large numbers of copies of the DNA very quickly.

In practice, the PCR reaction involves three steps. The amount of amplified DNA produced is theoretically limited only by the number of times these steps are repeated.

1. The DNA to be amplified is *denatured* to form single strands. The DNA can come from many sources, including genomic DNA, mummified remains, fossils, or forensic samples such as dried blood or semen, single hairs, or dried samples from medical records. Heating to 90–95°C denatures the hydrogen bonds between the DNA strands, converting them into single-stranded templates (usually in about 1 minute).
2. The temperature of the reaction is lowered to an *annealing temperature* between 50°C and 70°C, which allows the primers to bind to the single-stranded DNA. The primers serve as starting points for synthesizing new DNA strands complementary to the target DNA.
3. A heat-stable form of DNA polymerase (such as *Taq* polymerase) is added to the reaction mixture, and DNA synthesis is carried out at temperatures between 70°C and 75°C (the optimal temperature for this form of polymerase). The *Taq* polymerase *extends* the primers by adding nucleotides in the 5′ to 3′ direction, making a double-stranded copy of the target DNA.

Each set of three steps—*denaturation* of the double-stranded DNA, *primer annealing,* and *extension* by polymerase—is a cycle. PCR is a chain reaction because the number of new DNA strands is doubled in each cycle, and the new strands, along with the old strands, serve as templates in the next cycle. Each cycle takes 2 to 5 minutes and can be repeated immediately, so that in less than 3 hours, 25 to 30 cycles result in a many million-fold increase in the amount of DNA (Figure 17–10). This process is automated by machines called *thermalcyclers* that can be programmed to carry out a

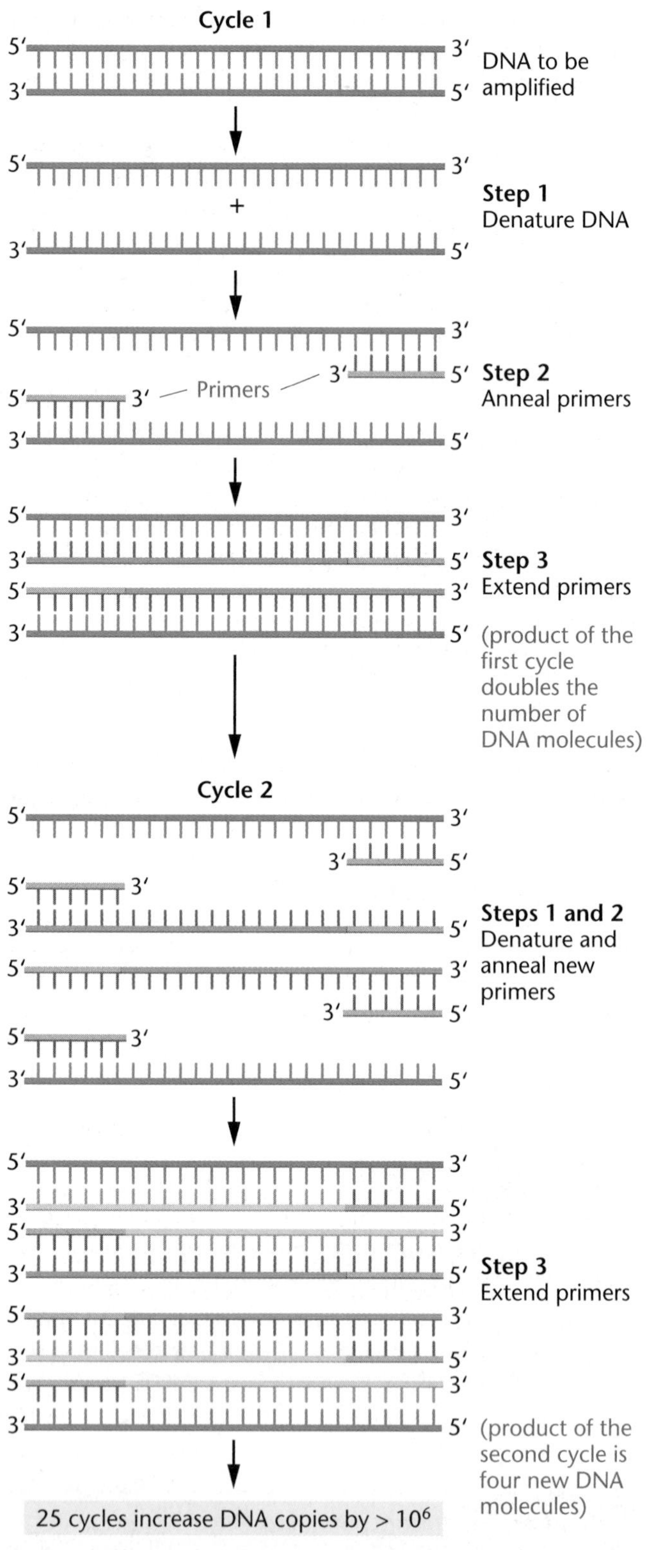

FIGURE 17–10 In the polymerase chain reaction (PCR), the target DNA is denatured into single strands; each strand is then annealed to a short, complementary primer. DNA polymerase extends the primers in the 5′ to 3′ direction, using the single-stranded DNA as a template. The result is two newly synthesized double-stranded DNA molecules, with the primers incorporated into them. Repeated cycles of PCR can quickly amplify the original DNA sequence more than a million-fold.

predetermined number of cycles, yielding large amounts of a specific DNA sequence that can be used for many purposes, including cloning into plasmid vectors, DNA sequencing, clinical diagnosis, and genetic screening.

PCR-based DNA cloning has several advantages over host-cell cloning. PCR is rapid and can be carried out in a few hours, rather than the days required for cell-based cloning. In addition, the design of PCR primers can be done rapidly using the appropriate software, and the commercial synthesis of the oligonucleotides is also fast and economical. If desired, the products of PCR can be cloned into plasmid vectors for further use.

PCR is also very sensitive and amplifies specific DNA sequences from vanishingly small DNA samples, including the DNA in a single cell. This feature of PCR is invaluable in several kinds of applications, including genetic testing, forensics, and molecular paleontology. With carefully designed primers, DNA samples that have been partially degraded, contaminated with other materials, or embedded in a matrix (such as amber) can be recovered and amplified, when conventional cloning would be difficult or impossible.

Limitations of PCR

Although PCR is a valuable technique, it does have its limitations: some information about the nucleotide sequence of the target DNA must be known, and even minor contamination of the sample with DNA from other sources can cause problems. For example, cells shed from the researcher's skin can contaminate samples gathered from a crime scene or taken from fossils, making it difficult to obtain accurate results. PCR reactions must always be performed in parallel with carefully designed and appropriate controls.

Other Applications of PCR

Today, PCR is one of the most widely used techniques in genetics and molecular biology. PCR and its variations have many other applications as well. They quickly identify restriction-enzyme recognition sequence variants as well as variations in tandemly repeated DNA sequences that can be used as genetic markers in gene-mapping studies and forensic identification. Gene-specific primers provide a way of screening for mutations in genetic disorders, allowing the location and nature of mutation to be determined quickly. Primers can be designed to distinguish between target sequences that differ by only a single nucleotide. (This makes it possible to synthesize allele-specific probes for genetic testing.) Random primers indiscriminately amplify DNA and are particularly advantageous when studying samples from single cells, fossils, or a crime scene, where a single hair or even a saliva-moistened postage stamp is the source of the DNA. Using PCR, researchers can also explore uncharacterized DNA regions adjacent to known regions and even sequence DNA. PCR has been used to enforce the worldwide ban on the sale of certain whale products and to settle arguments about the pedigree background of purebred dogs. In short, PCR is one of the most versatile techniques in modern genetics.

ESSENTIAL POINT

PCR allows DNA to be amplified without host cells and is a rapid, sensitive method with wide-ranging applications.

17.5 Recombinant Libraries Are Collections of Cloned Sequences

Only relatively small DNA segments—representing only a single gene or even a portion of a gene—are produced by cloning in conventional plasmid vectors. As a result, a large collection of clones is needed for exploring even a small fraction of an organism's genome. A set of DNA clones derived from a single source (an individual or a single cell) is called a cloned *library* or DNA library. These libraries can represent an entire genome, a single chromosome, or a set of genes that are expressed in a single cell type.

Genomic Libraries

Ideally, a **genomic library** contains at least one copy of every sequence in an organism's genome. Genomic libraries are constructed using host-cell cloning methods, since PCR-cloned DNA fragments are relatively small. In making a genomic library, DNA is extracted from cells or tissues, cut with a restriction enzyme for a short time to produce large, partially cut fragments that are inserted into vectors. Since some vectors (such as plasmids) carry only a few thousand base pairs of inserted DNA, selecting the vector so that the library contains the whole genome in the smallest number of clones is an important consideration.

How big does a genomic library have to be to have a 95 or 99 percent chance of containing all the sequences in a genome? The number of clones required to contain a genome depends on several factors, including the average size of the cloned inserts, the size of the genome to be cloned, and the level of probability desired. The number of clones in a library can be calculated as

$$N = \frac{\ln(1 - P)}{\ln(1 - f)}$$

where N is the number of required clones, P is the probability of recovering a given sequence, and f is the fraction of the genome in each clone.

Suppose we wish to prepare a human genome library large enough to have a 99 percent chance of containing all the sequences in the genome. Because the human genome is so large, the choice of vector is a primary consideration in making this library. If we construct the library using a plasmid vector with an average insert size of 5 kb, more than 2.4 million clones would be required for a 99 percent probability of recovering any given sequence from the genome. Because of its size, this library would be difficult to use efficiently. However, if the library was constructed in a YAC vector with an average insert size of 1 Mb, then the library would only need to contain about 14,000 YACs, making it relatively easy to use. Vectors with large cloning capacities such as YACs are essential tools for the Human Genome Project and many other genome projects.

NOW SOLVE THIS

Problem 13 on page 372 asks about the number of clones needed to contain a *Drosophila* genomic library using a plasmid vector.

Hint: Remember, there are three parameters in this calculation: the size of the genome, the average size of the cloned inserts, and the probability of having a gene included in the library.

Subgenomic Libraries

A **chromosome-specific library** prepared from a single chromosome can be used to isolate and clone specific genes and to study chromosome organization. A technique known as flow cytometry is used to make cloned libraries from single human chromosomes. Metaphase chromosomes are stained with fluorescent dyes and a laser is used to fractionate the chromosomes by light scattering and dye binding. DNA is extracted from the isolated chromosomes, cut with restriction enzymes and inserted into a vector and cloned. Libraries are available for each of the human chromosomes.

To study specific events in development, cell death, cancer, and other biological processes, a library of the subset of the genome that is expressed in a given cell type at a given time can be a valuable tool. Genomic libraries and chromosome libraries contain all the genes in a genome or on a chromosome, but these collections cannot be directly used to find genes that are actively expressed in a cell.

A **cDNA library** contains DNA copied from messenger RNA (mRNA) molecules present in a cell population at a given time, and therefore represents the genes being expressed in the cells at the time the library was made. (Recall from Chapter 1 that genes are transcribed to produce single-stranded messenger RNA molecules.) It is called a cDNA library because the DNA it contains—known as **complementary DNA,** or **cDNA**—is complementary to the nucleotide sequence of the mRNA.

Clones in a cDNA library are not the same as those in a genomic library. Eukaryotic mRNA transcripts are processed to remove some sequences (see Chapter 12), and a poly-A tail is added to the end of the mature mRNA molecule. Moreover, an mRNA molecule does not include the sequences adjacent to the gene that regulate its activity.

A cDNA library is prepared by first isolating mRNAs from a population of cells and using the mRNAs as templates for the synthesis of double-stranded cDNA molecules. The cDNA molecules are then inserted into vectors and cloned to produce a library that is a snapshot of genes that were transcriptionally active at a given time.

The first step in making a cDNA library is to mix mRNAs having poly-A tails with oligo-dT primers, which anneal to the poly-A, forming a partially double-stranded product (Figure 17–11). The enzyme **reverse transcriptase** extends the primer and synthesizes a complementary DNA copy of the mRNA sequence. The product of this reaction is an mRNA–DNA double-stranded hybrid molecule. Action of the enzyme **RNase H** introduces nicks in the RNA strand by partially digesting the RNA. The remaining RNA fragments serve as primers for the enzyme DNA polymerase I. (This situation is similar to synthesis of the lagging strand of DNA in prokaryotes.) DNA polymerase I is then used to synthesize a second DNA strand and removes the RNA primers, producing double-stranded cDNA.

The cDNA can be inserted into a vector by attaching linker sequences to the blunt ends of the cDNA. Linker sequences are short double-stranded oligonucleotides containing a restriction-enzyme recognition sequence (e.g., *Eco*RI). After attachment

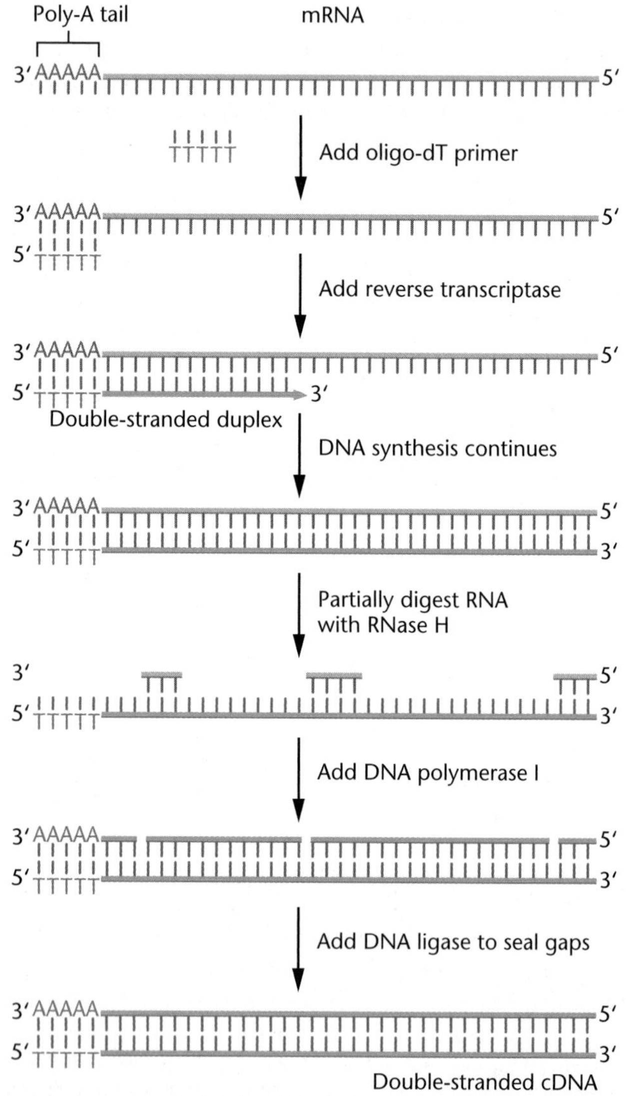

FIGURE 17–11 Producing cDNA from mRNA. Because many eukaryotic mRNAs have a polyadenylated (poly-A) tail of variable length at one end, a short oligo-dT molecule annealed to this tail serves as a primer for the enzyme reverse transcriptase. Reverse transcriptase uses the mRNA as a template to synthesize a complementary DNA strand (cDNA) and forms an mRNA/cDNA double-stranded duplex. The mRNA is digested with the enzyme RNase H, producing gaps in the RNA strand. The 3′ ends of the remaining RNA serve as primers for DNA polymerase, which synthesizes a second DNA strand. The result is a double-stranded cDNA molecule that can be cloned into a suitable vector or used directly as a probe for library screening.

to the cDNAs, the linkers are cut with *Eco*RI, producing fragments with single-stranded tails. These fragments are then ligated to vectors treated with the same enzyme. Transfer of vectors carrying cDNA molecules to host cells and cloning is the final step in making a cDNA library. Many different cDNA libraries are available from cells and tissues in specific stages of development, or different organs such as brain, muscle, and kidney. These libraries provide an instant catalog of all the genes active in a cell at a specific time.

A cDNA library can also be prepared using a variation of PCR called **reverse transcriptase PCR (RT-PCR).** In this procedure, reverse transcriptase is used to generate single-stranded cDNA copies of mRNA molecules as described earlier. This reaction is followed by PCR to copy the single-stranded DNA into double-stranded molecules and then amplify these into many copies. *Taq* polymerase and random DNA primers (instead of primers specific for a given gene) are added to the single-stranded cDNA. After primer binding, the *Taq* polymerase extends the primers, making double-stranded cDNA. Additional cycles of PCR make many copies of the cDNA. The amplified cDNA can be inserted into plasmid vectors, which are replicated to produce a cDNA library. RT-PCR is more sensitive than conventional cDNA preparation and is a powerful tool for identifying mRNAs that may be present in only one or two copies per cell.

17.6 Specific Clones Can Be Recovered from a Library

A genomic library often consists of several hundred thousand different clones. To find a specific gene, we need to identify and isolate only the clone or clones containing that gene. We must also determine whether a given clone contains all or only part of the gene we are studying. Several methods allow us to sort through a library—called *screening* the library—to recover clones of interest. The choice of method often depends on the circumstances and available information about the gene being sought.

Probes Identify Specific Clones

Probes are used to screen a library to recover clones of a specific gene. A **probe** is any DNA or RNA sequence that has been labeled in some way and is complementary to some part of a cloned sequence present in the library.

When used in a hybridization reaction, the probe binds to any complementary DNA sequences present in one or more clones. Probes can be labeled with radioactive isotopes, or with compounds that undergo chemical or color reactions to indicate the location of a specific clone in a library.

Probes are derived from a variety of sources—even related genes isolated from other species can be used if enough of the DNA sequence is conserved. For example, extrachromosomal copies of the ribosomal RNA genes of the African clawed frog *Xenopus laevis* can be isolated, inserted into vectors, and cloned. Because ribosomal gene sequences are highly conserved (that is, are similar in different organisms), clones carrying human ribosomal genes can be recovered from a genomic library using cloned fragments of *Xenopus* ribosomal DNA as probes.

If the gene to be selected from a genomic library is expressed in certain cell types, a cDNA probe can be used. This technique is particularly helpful when purified or enriched mRNA for a gene product can be obtained. For example, β-globin mRNA is present in high concentrations in certain stages of red blood cell development. The mRNA purified from these cells can be copied by reverse transcriptase into a cDNA molecule for use as a probe. In fact, a cDNA probe was originally used to recover the structural gene for human β-globin from a cloned genomic library.

Screening a Library

To screen a library constructed using a plasmid vector (this can be a genomic library or a cDNA library), clones from the library are grown on nutrient agar plates, where they form hundreds or thousands of colonies (Figure 17–12, p. 362). A replica of the colonies on each plate is made by gently pressing a nylon or nitrocellulose filter onto the plate's surface; this transfers the pattern of bacterial colonies from the plate to the filter. The filter is processed to lyse the bacterial cells, denature the double-stranded DNA released from the cells into single strands, and bind these strands to the filter.

The DNA on the filter is screened by incubation with a labeled nucleic acid probe. First, the probe is heated and quickly cooled to form single-stranded molecules, and then added to a solution containing the filter. If the nucleotide sequence of any of the DNA on the filter is complementary to that of the probe, a double-stranded DNA–DNA hybrid molecule will form (one strand from the probe and the other from the cloned DNA on the filter). For example, if a β-globin cDNA probe is used, it will bind to cloned DNA sequences encoding the β-globin gene. After incubation of the probe and the filter, unbound probe molecules are washed away, and the filter is assayed to detect the hybrid molecules that remain. If a radioactive probe has been used, the filter is overlaid with a piece of X-ray film. Radioactive decay in the probe molecules bound to DNA on the filter will expose the film, producing dark spots on the film. These spots represent colonies on the plate containing the cloned gene of interest (Figure 17–12). The positions of spots on the film are used as a guide to identify and recover the corresponding colonies on the plate. The cloned DNA they contain can be used in further experiments. With some nonradioactive probes, a chemical reaction emits photons of light (chemiluminescence) to expose the photographic film and reveal the location of colonies carrying the gene of interest.

NOW SOLVE THIS

Question 18 on page 372 involves selecting a cloned gene from a cDNA library. You are asked to identify a specific clone in this library using a probe from yeast.

Hint: cDNA clones do not have all the sequences of a genomic library but do have the coding sequences, and they can be selected with the proper probe.

CLONING LIBRARIES

WEB TUTORIAL

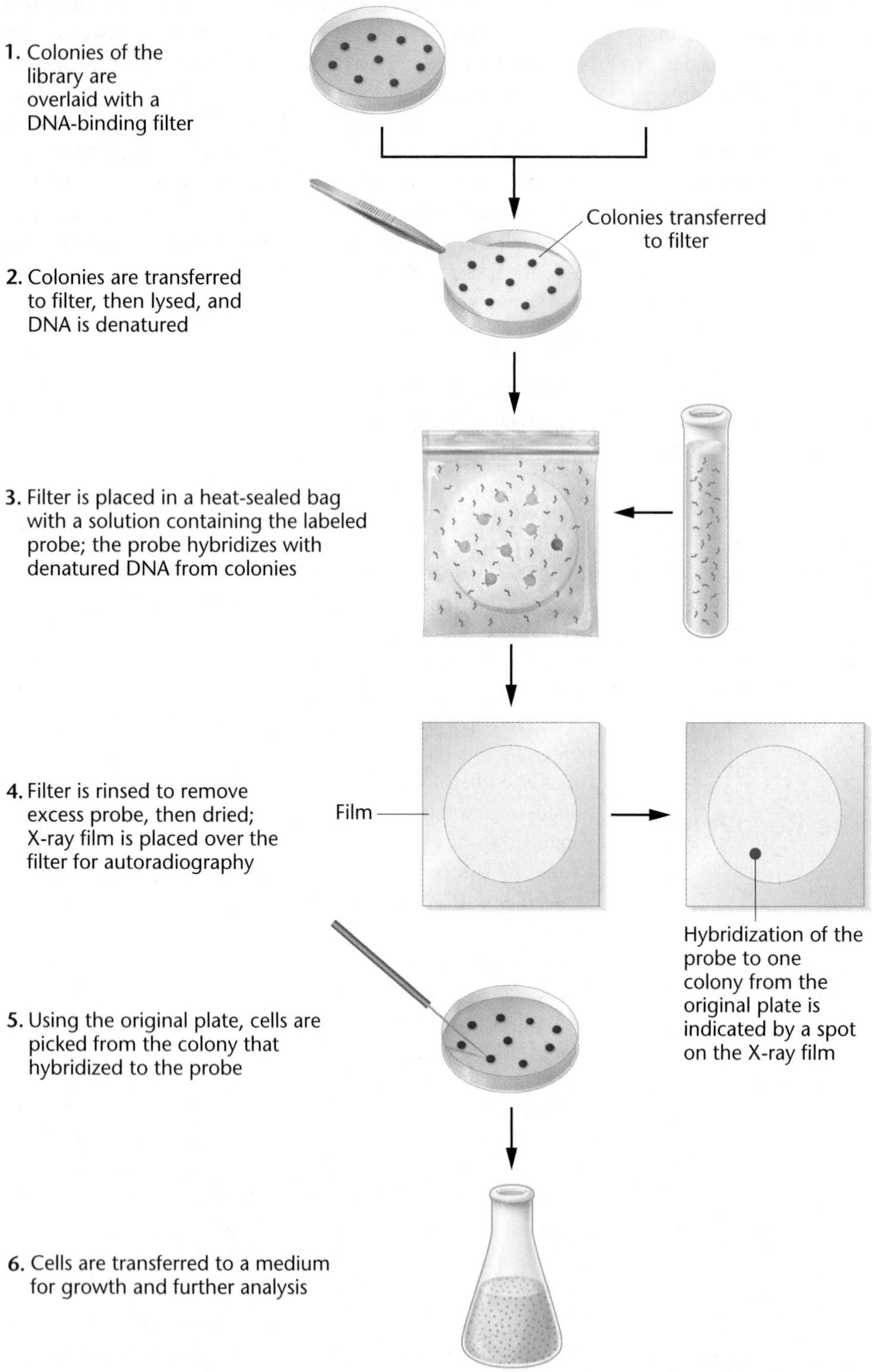

1. Colonies of the library are overlaid with a DNA-binding filter
2. Colonies are transferred to filter, then lysed, and DNA is denatured
Colonies transferred to filter
3. Filter is placed in a heat-sealed bag with a solution containing the labeled probe; the probe hybridizes with denatured DNA from colonies
4. Filter is rinsed to remove excess probe, then dried; X-ray film is placed over the filter for autoradiography
Film
Hybridization of the probe to one colony from the original plate is indicated by a spot on the X-ray film
5. Using the original plate, cells are picked from the colony that hybridized to the probe
6. Cells are transferred to a medium for growth and further analysis

FIGURE 17–12 Screening a library constructed using a plasmid vector to recover a specific gene. The library, present in bacteria on Petri plates, is overlaid with a DNA-binding filter, and colonies are transferred to the filter. Colonies on the filter are lysed, and the DNA is denatured to single strands. The filter is placed in a hybridization bag along with buffer and a labeled single-stranded DNA probe. During incubation, the probe forms a double-stranded hybrid with any complementary sequences on the filter. The filter is removed from the bag and washed to remove excess probe. Hybrids are detected by placing a piece of X-ray film over the filter and exposing it for a short time. The film is developed, and hybridization events are visualized as spots on the film. Colonies containing the insert that hybridized to the probe are identified from the orientation of the spots. Cells are picked from this colony for growth and further analysis.

17.7 Cloned Sequences Can Be Analyzed in Several Ways

The recovery and identification of genes and other DNA sequences by cloning or by PCR is a powerful tool for analyzing genomic structure and function. In fact, much of the Human Genome Project data was acquired through such techniques. In the following sections, we consider some ways these methods are used to provide information about the organization and function of cloned sequences.

Restriction Mapping

One of the first steps in characterizing a DNA clone is the construction of a **restriction map.** A restriction map establishes the number, order, and distance between restriction-enzyme cleavage sites along a cloned segment of DNA. This map provides information about the length of the cloned insert and the location of restriction-enzyme cleavage sites within the cloned DNA. The map units are expressed in base pairs (bp) or, for longer lengths, kilobase (kb) pairs. Restriction maps for different cloned DNAs are usually different enough to serve as identity tags for those clones. The data the maps provide can be used to reclone fragments of a gene or compare its internal organization with that of other cloned sequences.

Fragments generated by cutting DNA with restriction enzymes can be separated by gel electrophoresis, a method that separates fragments by size, with the smallest pieces moving farthest through the gel (see Chapter 9). The fragments form a series of bands that can be visualized by staining the DNA with ethidium bromide and illuminating it with ultraviolet light (see chapter opening photo).

Figure 17–13 shows the construction of a restriction map from a cloned DNA segment. For the sake of this example, let's say that the cloned DNA segment is 7.0 kb in length. Three samples of the cloned DNA are digested with restriction enzymes—one with *Hin*dIII, one with *Sal*I, and one with both *Hin*dIII and *Sal*I. The fragments are separated by gel electrophoresis and stained with ethidium bromide, producing a series of bands on the gel. The resulting bands are visualized with ultraviolet light and photographed or scanned for analysis. The molecular weights of the fragments are measured

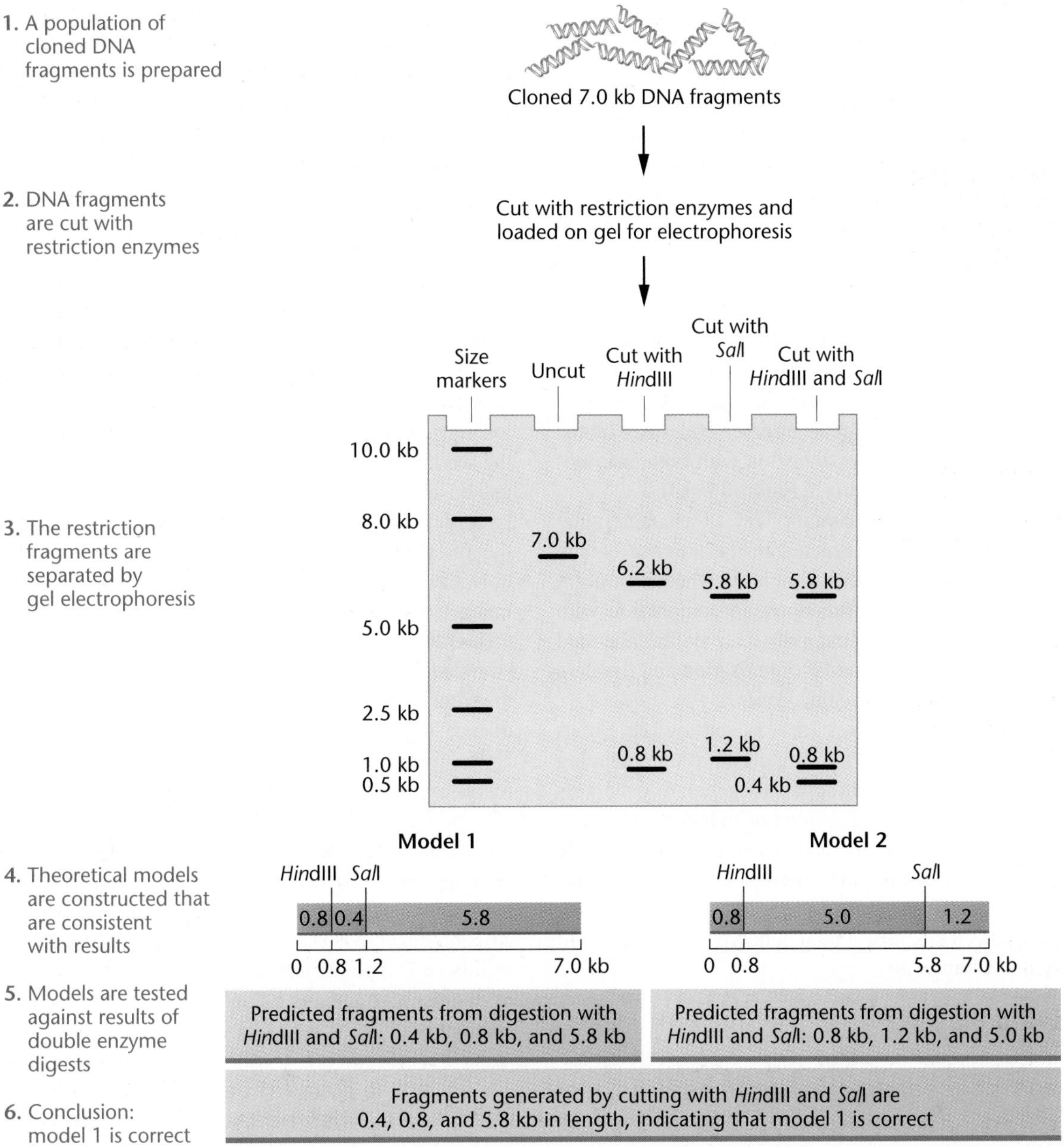

FIGURE 17–13 Constructing a restriction map. Samples of the 7.0-kb DNA fragments are digested with restriction enzymes: one sample is digested with *Hin*dIII, one with *Sal*I, and one with both *Hin*dIII and *Sal*I. The resulting fragments are separated by gel electrophoresis. The separated fragments are measured by comparing them with molecular-weight standards in an adjacent lane. Cutting the DNA with *Hin*dIII generates two fragments: 0.8 kb and 6.2 kb. Cutting with *Sal*I produces two fragments: 1.2 kb and 5.8 kb. Models are constructed showing the possible locations of restriction sites and are used to predict the fragment sizes generated by cutting with both *Hin*dIII and *Sal*I together. Model 1 predicts that 0.4-, 0.8-, and 5.8-kb fragments will result from cutting with both enzymes. Model 2 predicts that 0.8-, 1.2-, and 5.0-kb fragments will result. Comparing the predicted fragments with those observed on the gel indicates that model 1 is the correct restriction map.

WEB TUTORIAL
RESTRICTION MAPPING

by comparing their location on the gel to a set of molecular-weight standards run in an adjacent lane. The restriction map is constructed by analyzing the number and length of the fragments.

When the DNA is cut with *Hin*dIII, two fragments (0.8 and 6.2 kb) are produced, confirming that the cloned insert is 7.0 kb in length and showing that it contains only one cutting site for this enzyme (located 0.8 kb from one end). When the DNA is cut with *Sal*I, two fragments (1.2 and 5.8 kb) result, indicating that the insert also has only one cutting site for this enzyme, located 1.2 kb from one end of the cloned DNA segment.

These results show that each enzyme has one restriction site in the cloned DNA, but the relative positions of the two restriction-enzyme cleavage sites are unknown. From these data, two different maps are possible. One map, model 1 in Figure 17–13, shows the *Hin*dIII recognition sequence located 0.8 kb from one end and the *Sal*I recognition sequence 1.2 kb from the same end. The alternative map, model 2, locates the *Hin*dIII recognition sequence 0.8 kb from one end and the *Sal*I sequence 1.2 kb from the other end.

The correct model is determined by analyzing the results from the sample digested with both *Hin*dIII and *Sal*I. Model 1 predicts that digestion with both enzymes will generate three fragments: 0.4, 0.8, and 5.8 kb in length; model 2 predicts that there will also be three fragments, but with different lengths: 0.8, 5.0, and 1.2 kb. The number and sizes of the fragments seen on the gel after digestion with both enzymes indicate that model 1 is correct (see Figure 17–13).

Restriction maps are an important way of characterizing cloned DNA and can be constructed in the absence of any other information about the DNA, including whether or not it encodes a gene or has other functions. In conjunction with other techniques, restriction mapping can define the end points of a gene, dissect its internal organization and flanking regions, and locate mutations within genes.

Restriction digestion of clones (that is, cutting the clones with restriction enzymes) plays an important role in mapping genes to specific human chromosomes by generating markers that can be mapped to defined regions of individual chromosomes. If one of these markers maps close to a mutant allele that causes a genetic disorder, the marker can be used in genetic testing to identify carriers of recessively inherited disorders or to prenatally diagnose a fetal genotype. This topic will be discussed in Chapter 19.

Nucleic Acid Blotting

Many of the techniques described in this chapter rely on hybridization between complementary nucleic acid (DNA or RNA) molecules, a topic introduced in Chapter 9. One of the most widely used methods for detecting such hybrids is called Southern blotting (after Edwin Southern, who devised it). The **Southern blot** method can be used to identify which clones in a library contain a given DNA sequence (such as ribosomal DNA or a β-globin gene) and to characterize the size of the fragments, thus permitting a restriction map to be made. Southern blots can also be used to identify fragments carrying specific genes in genomic DNA digested with a restriction enzyme.

Southern blotting has three components: separation of DNA fragments by gel electrophoresis, transfer of the DNA to a nylon or nitrocellulose membrane, and hybridization of the DNA fragments on the membrane using labeled probes. Gel electrophoresis can be used, as shown above, to characterize the number of fragments produced by restriction digestion and to estimate their molecular weights. However, restriction-enzyme digestion of large genomes—such as the human genome, with more than 3 billion nucleotides—will produce so many different fragments that they will run together on a gel to produce a continuous smear. The identification of specific fragments in these cases is accomplished in the third step: hybridization, which characterizes the DNA sequences present in the fragments. The DNA to be analyzed by Southern blot hybridization can come from several sources, including clones selected from a library or genomic DNA. Our discussion will use examples from both sources to show how Southern blots are used to characterize the number, size, organization, and base sequence of DNA fragments.

To make a Southern blot, DNA is cut into fragments with one or more restriction enzymes, and the fragments are separated by gel electrophoresis (Figure 17–14). In preparation for hybridization, the DNA in the gel is denatured with alkaline treatment to form single-stranded fragments. The gel is then overlaid with a DNA-binding membrane, usually nitrocellulose or a nylon derivative. Transfer of the DNA fragments to the membrane is accomplished by placing the membrane and gel on a wick (often a sponge) sitting in a buffer solution. Layers of paper towels or blotting paper are placed on top of the filter and held in place with a weight. Capillary action draws buffer up through the gel, transferring the DNA fragments from the gel to the membrane.

The filter is placed in a heat-sealed bag with a labeled, single-stranded DNA probe for hybridization. DNA fragments on the filter that are complementary to the probe's nucleotide sequence bind to the probe to form double-stranded hybrids. Excess probe is then washed away, and the hybridized fragments are visualized on a piece of film (Figure 17–14).

To produce the results shown in Figure 17–15, p. 366, researchers cut samples of genomic DNA with several restriction enzymes, and the resulting fragments are separated in adjacent lanes on the gel. The pattern of fragments obtained for each restriction enzyme is shown in Figure 17–15(a). A Southern blot of this gel is shown in Figure 17–15(b). The probe hybridized to fragments containing complementary sequences, identifying fragments of interest, while other fragments, visible on the stained gel [Figure 17–15(a)], did not hybridize to the probe.

In addition to characterizing cloned DNAs, Southern blots can be used to create restriction maps of segments within and near a gene and to identify DNA fragments carrying all or parts of a single gene in a mixture of fragments. By comparing the pattern of bands in normal cells to those from patients with genetic disorders or cancer, Southern blots also detect rearrangements, deletions, and duplications in genes associated with human genetic disorders and cancers.

To determine whether a gene is actively being expressed in a given cell or tissue type, a related blotting technique probes for the presence of mRNA complementary to a cloned gene.

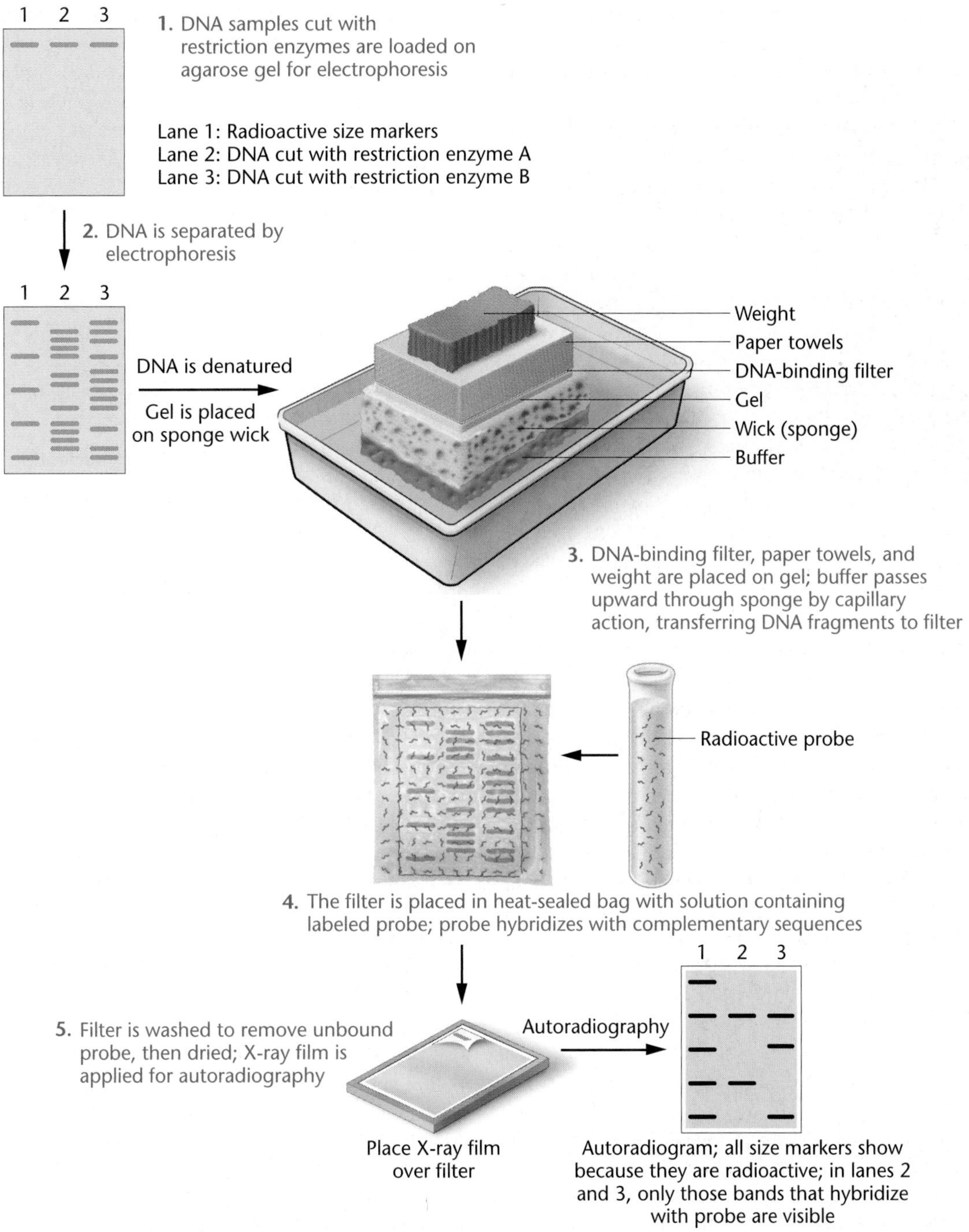
1 2 3
1. DNA samples cut with restriction enzymes are loaded on agarose gel for electrophoresis
Lane 1: Radioactive size markers
Lane 2: DNA cut with restriction enzyme A
Lane 3: DNA cut with restriction enzyme B
2. DNA is separated by electrophoresis
1 2 3
DNA is denatured
Gel is placed on sponge wick
Weight
Paper towels
DNA-binding filter
Gel
Wick (sponge)
Buffer
3. DNA-binding filter, paper towels, and weight are placed on gel; buffer passes upward through sponge by capillary action, transferring DNA fragments to filter
Radioactive probe
4. The filter is placed in heat-sealed bag with solution containing labeled probe; probe hybridizes with complementary sequences
1 2 3
5. Filter is washed to remove unbound probe, then dried; X-ray film is applied for autoradiography
Autoradiography
Place X-ray film over filter
Autoradiogram; all size markers show because they are radioactive; in lanes 2 and 3, only those bands that hybridize with probe are visible

FIGURE 17–14 In the Southern blotting technique, samples of the DNA to be probed are cut with restriction enzymes and the fragments are separated by gel electrophoresis. The pattern of fragments is visualized and photographed under ultraviolet illumination. The DNA in the gel is denatured, then the gel is placed on a sponge wick that is in contact with a buffer solution and covered with a DNA-binding filter. Layers of paper towels or blotting paper are placed on top of the filter and held in place with a weight. Capillary action draws the buffer through the gel, transferring the pattern of DNA fragments from the gel to the filter. The DNA fragments on the filter are then hybridized with a labeled DNA probe. The filter is washed to remove excess probe and overlaid with a piece of X-ray film for autoradiography. The hybridized fragments show up as bands on the X-ray film.

To do this, mRNA is extracted from a specific cell or tissue type and separated by gel electrophoresis. The resulting pattern of RNA bands is transferred to a membrane, as in Southern blotting. The membrane is then exposed to a labeled single-stranded DNA probe derived from a cloned copy of the gene. If mRNA complementary to the DNA probe is present, the complementary sequences will hybridize and be detected as a band on the film. Because the original procedure (DNA bound to a filter) is known as a Southern blot, this variant procedure (RNA bound to a filter) is called a **Northern blot.** (Following this somewhat perverse logic, another procedure involving the visualization of specific proteins bound to a filter is known as a **Western blot.**)

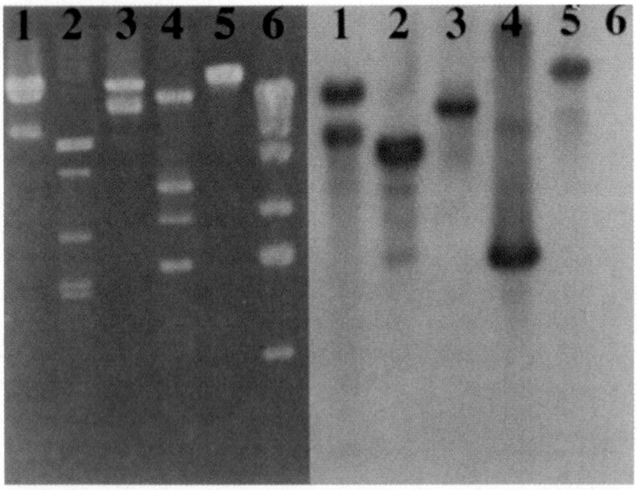

FIGURE 17–15 (a) Agarose gel stained with ethidium bromide to show DNA fragments. (b) Exposed X-ray film of a Southern blot prepared from the gel in part (a). Only those bands containing DNA sequences complementary to the probe show hybridization.

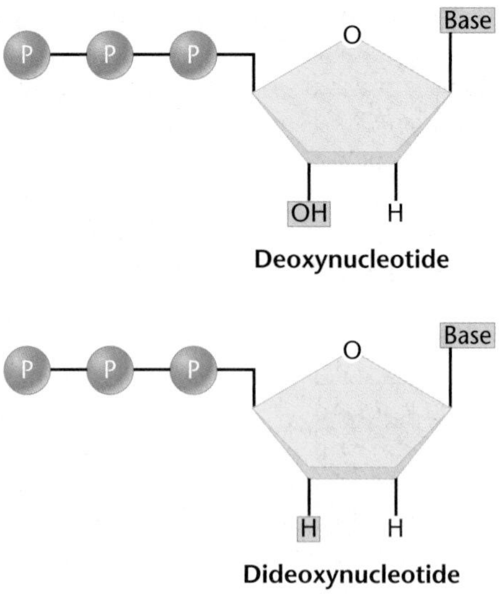

FIGURE 17–16 Deoxynucleotides (top) have an OH group at the 3′ position in the deoxyribose molecule. Dideoxynucleotides (bottom) lack an OH group and have only hydrogen (H) at this position. Dideoxynucleotides can be incorporated into a growing DNA strand, but the lack of a 3′-OH group prevents formation of a phosphodiester bond with another nucleotide, terminating further elongation of the template strand.

Northern blots provide information about the expression of specific genes and are used to study patterns of gene expression in embryonic tissues, cancer, and genetic disorders. Northern blots also detect alternatively spliced mRNAs (multiple types of transcripts derived from a single gene) and are used to derive other information about transcribed mRNAs. If RNAs of known sizes are run on the same gel, Northern blots can be used to measure the size of a gene's mRNA transcripts. Measuring the density of bands gives an estimate of the relative transcriptional activity of the gene. Thus, Northern blots characterize and quantify the transcriptional activity of genes in different cells, tissues, and organisms.

ESSENTIAL POINT

Once cloned, DNA sequences are analyzed through a variety of methods, including restriction mapping, and Southern blotting, which are used to identify genes and flanking regulatory regions within the cloned sequences.

17.8 DNA Sequencing Is the Ultimate Way to Characterize a Clone

In a sense, a cloned DNA molecule or any DNA, from a clone to a genome, is completely characterized only when its nucleotide sequence is known. The ability to sequence DNA has greatly enhanced our understanding of genome organization and increased our knowledge of gene structure, function, and mechanisms of regulation.

The most commonly used method of DNA sequencing was developed by Fred Sanger and his colleagues. In this procedure, DNA to be sequenced is converted to single strands that are used as templates for synthesizing a series of complementary strands. Each of these complementary strands randomly terminates at a different, specific nucleotide. The resulting series of DNA fragments are separated by electrophoresis and analyzed to reveal the sequence of the DNA.

In the first step of the Sanger method, DNA is heated to denature it and convert it into single strands. The single-stranded DNA is mixed with primers designed to anneal to the 3′ end of each single strand. Samples of the primer-bound single-stranded DNA are distributed into four tubes. In the next step, DNA polymerase and the four dNTPs (dATP, dCTP, dGTP, and dTTP) are added to each tube. Each of these deoxynucleotides has a 3′ OH group that covalently links the next nucleotide to the growing complementary strand during DNA synthesis. In addition, each tube receives a small amount of one modified deoxynucleotide (Figure 17–16) called a dideoxynucleotide (i.e., ddATP, ddCTP, ddGTP, or ddTTP). One of the deoxynucleotides is labeled with radioactivity for later analysis of the sequence. DNA polymerase is added to each tube, and the primers are elongated in the 5′ to 3′ direction, forming a complementary strand on the template.

As DNA synthesis takes place, the polymerase occasionally inserts a dideoxynucleotide instead of a deoxynucleotide into a growing DNA strand. Because the dideoxynucleotide does not have a 3′-OH, once it is incorporated into the growing complementary strand, it cannot form a covalent 3′ bond with another nucleotide, and synthesis of that strand terminates. For example, as shown in

Figure 17–17, in the tube with added ddATP, the polymerase inserts ddATP instead of dATP at the first T on the template strand, causing termination of the growing strand. On other strands in the tube, dATP will be inserted at this position, and termination will occur at other T nucleotides along the template strand. As the reaction proceeds, the tube with ddATP will eventually accumulate partially replicated DNA molecules that terminate at all A positions on the newly replicated strand. In the other tubes, reactions terminate at all the G, C, and T, respectively. The DNA fragments from each reaction tube (one for each dideoxynucleotide) are separated in adjacent lanes by gel electrophoresis. The result is a series of bands forming a ladderlike pattern that is visualized by developing film exposed to the gel (Figure 17–18). The nucleotide sequence revealed by the bands is read directly from bottom to top, corresponding to the sequence of the DNA strand complementary to the template.

DNA sequencing is now largely automated and uses machines that can sequence several hundred thousand nucleotides per day. In the automated procedure, each of the four dideoxynucleotide analogs is labeled with a different colored fluorescent dye (Figure 17–19, p. 368) so that chains terminating in adenosine are labeled with one color, those ending in cytosine with another color, and so forth. All four labeled dideoxynucleotides are added to a single tube, and after primer extension by DNA polymerase, the reaction products are loaded into one lane on a gel. The gel is scanned with a laser, causing each band to fluoresce a different color. A detector in the sequencing machine reads the color of each band and determines whether it represents an A, T, C, or G. The data are represented as a series of colored peaks, each corresponding to one nucleotide in the sequence (Figure 17–20, p. 368).

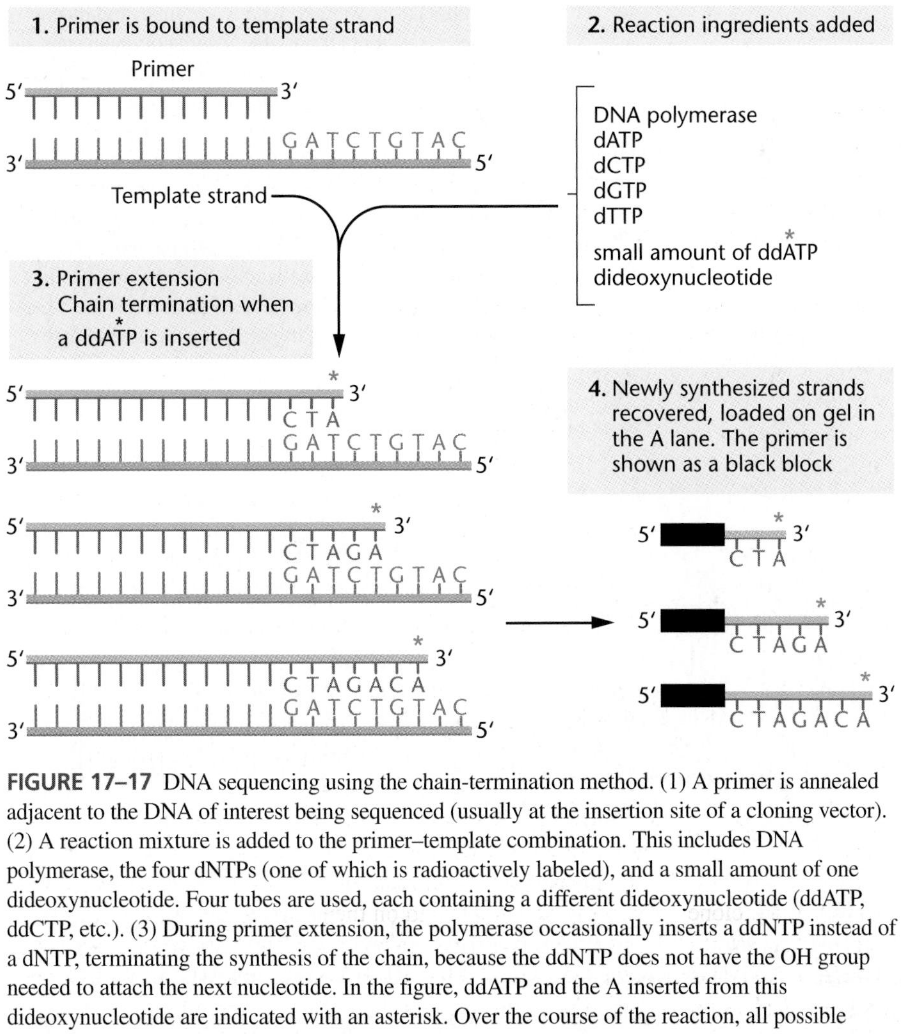

FIGURE 17–17 DNA sequencing using the chain-termination method. (1) A primer is annealed adjacent to the DNA of interest being sequenced (usually at the insertion site of a cloning vector). (2) A reaction mixture is added to the primer–template combination. This includes DNA polymerase, the four dNTPs (one of which is radioactively labeled), and a small amount of one dideoxynucleotide. Four tubes are used, each containing a different dideoxynucleotide (ddATP, ddCTP, etc.). (3) During primer extension, the polymerase occasionally inserts a ddNTP instead of a dNTP, terminating the synthesis of the chain, because the ddNTP does not have the OH group needed to attach the next nucleotide. In the figure, ddATP and the A inserted from this dideoxynucleotide are indicated with an asterisk. Over the course of the reaction, all possible termination sites will have a ddNTP inserted. (4) The newly synthesized strands are removed from the template, and the mixture is placed on a gel. DNA fragments from the reaction tube containing ddATP and terminating with A are loaded in the A lane, those ending in C are loaded in the C lane, and so forth.

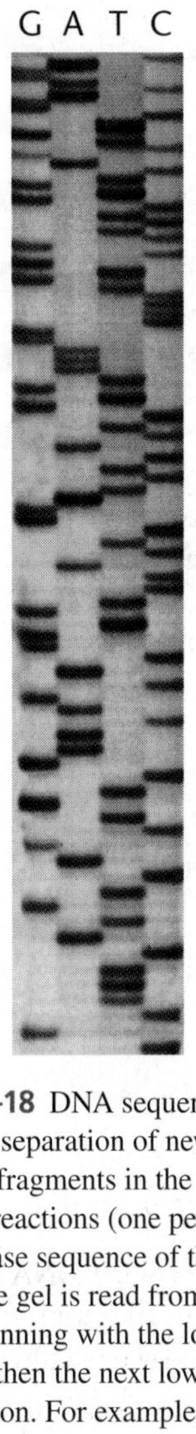

FIGURE 17–18 DNA sequencing gel showing the separation of newly synthesized fragments in the four sequencing reactions (one per lane). To obtain the base sequence of the DNA fragment, the gel is read from the bottom, beginning with the lowest band in any lane, then the next lowest, then the next, and so on. For example, the sequence of the DNA on this gel begins with 5′-CG at the very bottom of the gel, proceeds upward as 5′-CGCTTTCATGTCA, and so forth.

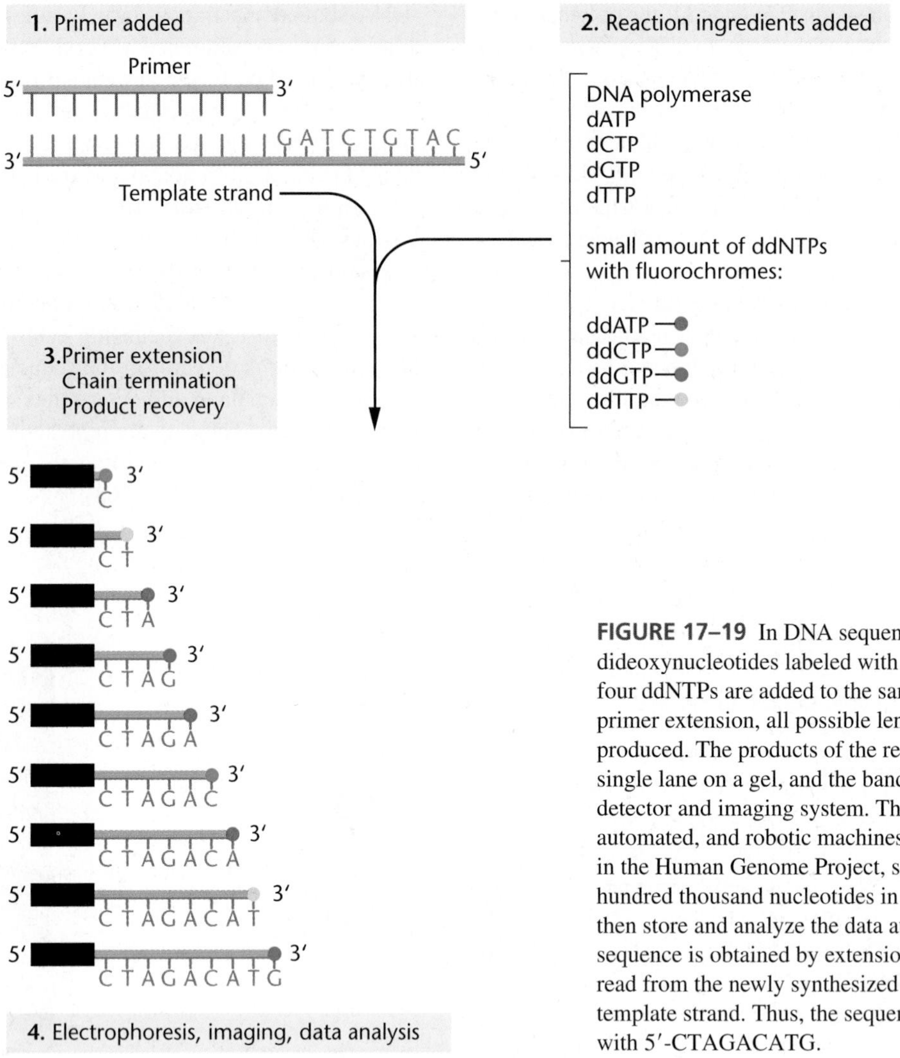

FIGURE 17–19 In DNA sequencing using dideoxynucleotides labeled with fluorescent dyes, all four ddNTPs are added to the same tube, and during primer extension, all possible lengths of chains are produced. The products of the reaction are added to a single lane on a gel, and the bands are read by a detector and imaging system. This process is now automated, and robotic machines, such as those used in the Human Genome Project, sequence several hundred thousand nucleotides in a 24-hour period and then store and analyze the data automatically. The sequence is obtained by extension of the primer and is read from the newly synthesized strand, not the template strand. Thus, the sequence obtained begins with 5′-CTAGACATG.

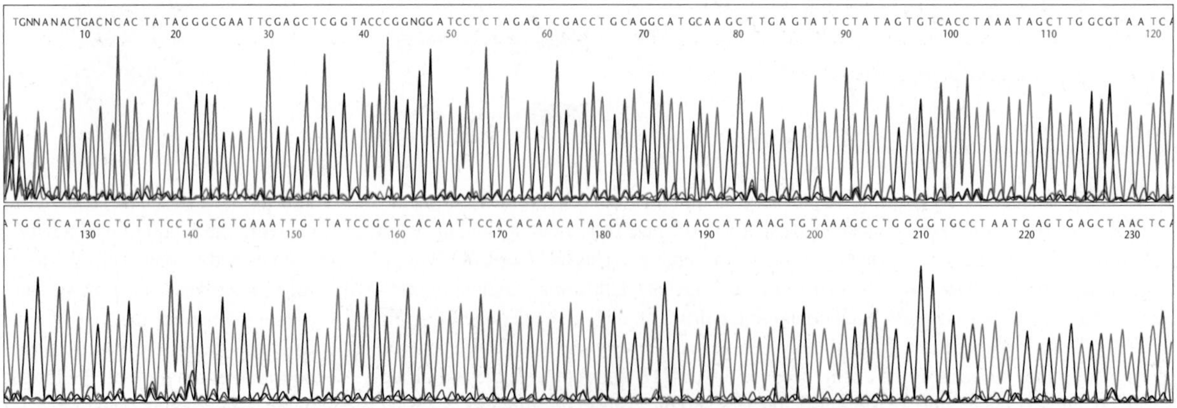

FIGURE 17–20 Automated DNA sequencing using fluorescent dyes, one for each base. Each peak represents the correct nucleotide in the sequence. The sequence extending from the primer (which is not shown here) starts at the upper left of the diagram and extends rightward. The bases labeled as N are ambiguous and cannot be identified with certainty. These ambiguous base readings are more likely to occur near the primer because the quality of sequence determination deteriorates the closer the sequence is to the primer. The separated bases are read in order along the axis from left to right. Thus, this sequence begins as 5′-TGNNANACTGACNCAC. Numbers below the bases indicate length of the sequence in base pairs.

Recombinant DNA Technology and Genomics

DNA sequencing is one of the technologies that make genome projects possible. Through a combination of recombinant DNA techniques and nucleotide sequencing, the genomes of more than 500 species have been sequenced, and almost 2000 additional projects are underway. The Human Genome Project sponsored by the U.S. Department of Energy and the National Institutes of Health and a private project sponsored by the biotechnology company Celera used a combination of genomic and chromosome-specific libraries to finish sequencing the coding portion of the human genome in 2003. Newer methods of DNA sequencing now under development will increase the speed, capacity, and accuracy of sequencing at a lower cost than present technology. Once these new methods of sequencing are available, it will be feasible to sequence the genomes of many more species.

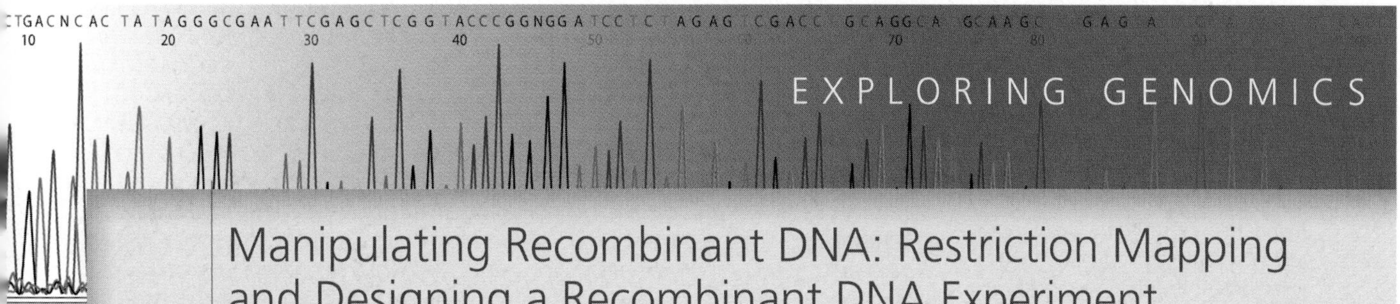

EXPLORING GENOMICS

Manipulating Recombinant DNA: Restriction Mapping and Designing a Recombinant DNA Experiment

As you learned in this chapter, restriction enzymes are sophisticated "scissors" that molecular biologists use to cut DNA, and they are routinely used in genetics and molecular biology laboratories for recombinant DNA experiments. Yet another advantage of the genomics revolution has been the development of a wide variety of online tools to assist scientists working with restriction enzymes and manipulating recombinant DNA for different applications, such as restriction mapping and designing primers for PCR experiments. Here we explore **Webcutter,** a site that makes recombinant DNA experiments much easier.

■ Exercise I: Creating a Restriction Map in Webcutter

Suppose you had cloned and sequenced a gene and you wanted to design a probe approximately 600 bp long that could be used to analyze the expression of this gene in different human tissues by northern blot analysis. Not too long ago, you had primarily two ways to approach this task. You could digest the cloned DNA with whatever restriction enzymes were in your freezer, and then run agarose gels and develop restriction maps in the hope of identifying cutting sites that would give you the size fragment you wanted. Or you could scan the sequence with your eyes, looking for restriction sites of interest—a very time-consuming and eye-straining effort! Internet sites such as Webcutter take the guesswork out of developing restriction maps and make it relatively easy to design experiments for manipulating recombinant DNA. In this exercise, you will use Webcutter to create a restriction map of human DNA with the enzymes *Eco*RI, *Bam*HI, and *Pst*I.

1. Access Webcutter at **http://rna.lundberg.gu.se/cutter2**. Copy the sequence of cloned human DNA shown below and paste it into the text box in Webcutter. (*Hint*: Access this sequence from the Companion Web site so that you can copy and paste the sequence into Webcutter.)

Human DNA Sequence

CCCCAGGAGACCTGGTTGTGGAATTCTG
TGTGTGAGTGGTTGACCTTCCTCCATCC
CCTGGTCCTTCCCTTCCCTTCCCGAGGC
ACAGAGAGACAGGGCAGGATCCACGTG
CCCATTGTGGAGGCAGAGAAAAGAGAA
AGTGTTTTATATACGGTACTTATTTAATAT
CCCTTTTTAATTAGAAATTAAAACAGTTA
ATTTAATTAAAGAGTAGGGTTTTTTTTCA
GTATTCTTGGTTAATATTTAATTTCAACTA
TTTATGAGATGTATCTTTTGCTCTCTCTT
GCTCTCTTATTTGTACCGGTTTTTGTATAT
AAAATTCATGTTTCCAATCTCTCTCTCCC
TGATCGGTGACAGTCACTAGCTTATCTTG
AACAGATATTTAATTTTGCTAACACTCAG
CTCTGCCCTCCCCGATCCCCTGGCTCCC
CAGCACACATTCCTTTGAAATAAGGTTT
CAATATACATCTACATACTATATATATATTT
GGCAACTTGTATTTGTGTGTATATATATAT
ATATATGTTTATGTATATATGTGATTCTGA
TAAAATAGACATTGCTATTCTGTTTTTTA
TATGTAAAAACAAAACAAGAAAAAAATA
GAGAATTTACATACTAAATCTCTCTCCTT
TTTTAATTTTAATATTTGTTATCATTTATTT
ATTGGTGCTACTGTTTATCCGTAATAATT
GTGGGGAAAAGATATTAACATCACGTCT
TTGTCTCTAGTGCAGTTTTTCGAGATATT
CCGTAGTACATATTTATTTTTAAACAACG
ACAAAGAAATACAGATATATCTTAAAAA
AAAAAAAGCATTTTGTATTAAA GAATTT
AATTCTGATCTGCAGCTCAAAAAAAAA
AAAA

2. Scroll down to "Please indicate which enzymes to include in the analysis." Click the button indicating "Use only the following enzymes." Select the restriction enzymes *Eco*RI, *Bam*HI, and *Pst*I from the list provided, then click "Analyze sequence." (*Note*: Use the command, control, or shift key to select multiple restriction enzymes.)
3. After examining the results provided by Webcutter, create a table showing the number of cutting sites for each enzyme and the fragment sizes that would be generated by digesting with each enzyme. Draw a restriction map indicating cutting sites for each enzyme with distances between each site and the total size of this piece of human DNA.

■ Exercise II: Designing a Recombinant DNA Experiment

Now that you have created a restriction map of your piece of human DNA, you

(*Cont. on the next page*)

need to ligate the DNA into a plasmid DNA vector that you can use to make your probe (molecular biologists often refer to this procedure as subcloning). To do this you will need to determine which restriction enzymes would best be suited for cutting both the plasmid and the human DNA.

1. Below is a plasmid DNA sequence. Copy this sequence into the text box in Webcutter and identify cutting sites for the same enzymes you used in Exercise I. Then answer the following questions:
   a. What is the total size of the plasmid DNA analyzed in Webcutter?
   b. Which enzyme(s) could be used in a recombinant DNA experiment to ligate the plasmid to the *largest* DNA fragment from the human gene? Briefly explain your answer.
   c. What size recombinant DNA molecule will be created by ligating these fragments?
   d. Draw a simple diagram showing the cloned DNA inserted into the plasmid and indicate the restriction-enzyme cutting site(s) used to create this recombinant plasmid.

Plasmid DNA Sequence

TATAAATATAGAATAATGAATCATATAAA
ACATATCATTATTCATTTATTTACATTTAA
AATTATTGTTTCAGTATCTTTAATTTATTA
TGTATATATAAAAATAACTTACAATTTTAT
TAATAAACAATATATGTTTATTAATTCATG
TTTTGTAATTTATGGGATAGCGATTTTTT
TTACTGTCTGTATTTTTCTTTTTTAATTAT
GTTTTAATTGTATTTTATTTTTATTATTGTT
CTTTTTATAGTATTATTTTAAAACAAAAT
GTATTTTCTAAGAACTTATAATAATAATA
AATATAAATTTTAATAAAAATTATATTTAT
CTTTTACAATATGAACATAAAGTACAACA
TTAATATATAGCTTTTAATATTTTTATTCCT
AATCATGTAAATCTTAAATTTTTCTTTTT
AAACATATGTTAAATATTTATTTCTCATTA
TATATAAGAACATATTTATTAAATCTAGAA
TTCTATAGTGAGTCGTATTACAATTCACT
GGCCGTCGTTTTACAACGTCGTGACTGG
GAAAACCCTGGCGTTACCCAACTTAATC
GCCTTGCAGCACATCCCCCTTTCGCCAG
CTGGCGTAATAGCGAAGAGGCCCGCACC
GATCGCCCTTCCCAACAGTTGCGCAGCC
TGAATGGCGAATGGCGCCTGATGCGGTA
TTTTCTCCTTACGCATCTGTGCGGTATTT
CACACCGCATATGGTGCACTCTCAGTAC
AATCTGCTCTGATGCCGCATAGTTAAGC
CAGCCCCGACACCCGCCAACACCCGCT
GACGCGCCCTGACGGGCTTGTCTGCTCC
CGGCATCCGCTTACAGACAAGCTGTGAC
CGTCTCCGGGAGCTGCATGTGTCAGAGG
TTTTCACCGTCATCACCGAAACGCGCGA
GACGAAAGGGCCTCGTGATACGCCTATT
TTTATAGGTTAATGTCATGATAATAATGG
TTTCTTAGACGTCAGGTGGCACTTTTCG
GGGAAATGTGCGCGGAACCCCTATTTGT
TTATTTTTCTAAATACATTCAAATATGTAT
CCGCTCATGAGACAATAACCCTGATAAA
TGCTTCAATAATATTGAAAAAGGAAGAG
TATGAGTATTCAACATTTCCGTGTCGCCC
TTATTCCCTTTTTTGCGGCATTTTGCCTT
CCTGTTTTTGCTCACCCAGAAACGCTG-
GTGAAAGTAAAAGATGCTGAAGATCAG
TTGGGTGCACGAGTGGGTTACATCGAAC
TGGATCTCAACAGCGGTAAGATCCTTGA
GAGTTTTCGCCCCGAAGAACGTTTTCCA
ATGATGAGCACTTTTAAAGTTCTGCTATG
TGGCGCGGTATTATCCCGTATTGACGCC
GGGCAAGAGCAACTCGGTCGCCGCATA
CACTATTCTCAGAATGACTTGGTTGAGT
ACTCACCAGTCACAGAAAAGCATCTTAC
GGATGGCATGACAGTAAGAGAATTATGC
AGTGCTGCCATAACCATGAGTGATAACA
CTGCGGCCAACTTACTTCTGACAACGAT
CGGAGGACCGAAGGAGCTAACCGCTTT
TTTGCACAACATGGGGGATCATGTAACT
CGCCTTGATCGTTGGGAACCGGAGCTG
AATGAAGCCATACCAAACGACGAGCGT
GACACCACGATGCCTGTAGCAATGCCAA
CAACGTTGCGCAAACTATTAACTGGCGA
ACTACTTACTCTAGCTTCCCGGCAACAAT
TAATAGACTGGATGGAGGCGGATAAAGT
TGCAGGACCACTTCTGCGCTCGGCCCTT
CCGGCTGGCTGGTTTATTGCTGATAAAT
CTGGAGCCGGTGAGCGTGGGTCTCGCG
GTATCATTGCAGCACTGGGGCCAGATGG
TAAGCCCTCCCGTATCGTAGTTATCTACA
CGACGGGGAGTCAGGCAACTATGGATG
AACGAAATAGACAGATCGCTGAGATAG
GTGCCTCACTGATTAAGCATTGGTAACT
GTCAGACCAAGTTTACTCATATATACTTT
AGATTGATTTAAAACTTCATTTTTAATTT
AAAAGGATCTAGGTGAAGATCCTTTTTG
ATAATCTCATGACCAAAATCCCTTAACGT
GAGTTTTCGTTCCACTGAGCGTCAGACC
CCGTAGAAAAGATCAAAGGATCTTCTTG
AGATCCTTTTTTTCTGCGCGTAATCTGCT
GCTTGCAAACAAAAAAACCACCGCTAC
CAGCGGTGGTTTGTTTGCCGGATCAAGA
GCTAC

CASE STUDY Should we worry about recombinant DNA technology?

Early in the 1970s, when recombinant DNA research was first developed, scientists realized that there may be unforeseen dangers, and after a self-imposed moratorium on all such research, they developed and implemented a detailed set of safety protocols for the construction, storage, and use of genetically modified organisms. These guidelines then formed the basis of regulations adopted by the federal government. Over time, safer methods were developed, and these stringent guidelines were gradually relaxed or in many cases, eliminated altogether. Now, however, the specter of bioterrorism has re-focused attention on the potential misuses of recombinant DNA technology. For example, individuals or small groups might use the information in genome databases coupled with recombinant DNA technology to construct or reconstruct agents of disease, such as the smallpox virus or the deadly influenza virus.

1. Do you think that the question of recombinant DNA research regulation by university and corporations should be revisited to monitor possible bioterrorist activity?
2. Should freely available access to genetic databases, including genomes, and gene or protein sequences be continued, or should it be restricted to individuals who have been screened and approved for such access?
3. Forty years after its development, the use of recombinant DNA technology is widespread and is found even in many middle school and high school biology courses. Are there some aspects of gene-splicing that might be dangerous in the hands of an amateur?

INSIGHTS AND SOLUTIONS

1. The recognition sequence for the restriction enzyme *Sau*3AI is GATC. In the recognition sequence for the enzyme *Bam*HI—GGATCC—the four internal bases are identical to the *Sau*3AI sequence. The single-stranded ends produced by the two enzymes are identical. Suppose you have a cloning vector that contains a *Bam*HI recognition sequence and you also have foreign DNA that was cut with *Sau*3AI. (a) Can this DNA be ligated into the *Bam*HI site of the vector, and if so, why? (b) Can the DNA segment cloned into this sequence be cut from the vector with *Sau*3AI? With *Bam*HI? What potential problems do you see with the use of *Bam*HI?

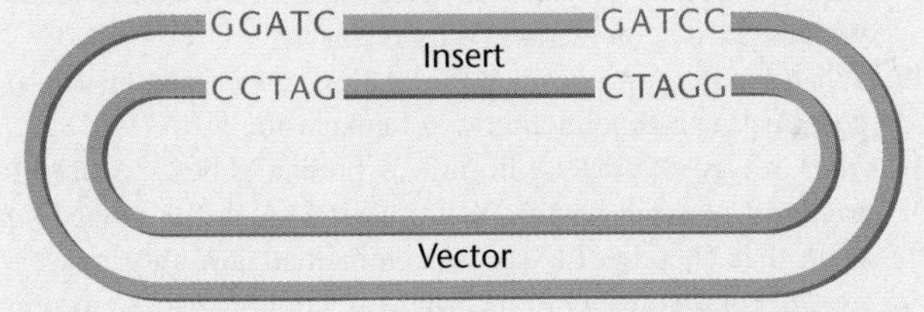

Solution: (a) DNA cut with *Sau*3AI can be ligated into the vector's *Bam*HI cutting site because the single-stranded ends generated by the two enzymes are identical. (b) The DNA can be cut from the vector with *Sau*3AI because the recognition sequence for this enzyme (GATC) is maintained on each side of the insert. Recovering the cloned insert with *Bam*HI is more problematic. In the ligated vector, the conserved sequences are GGATC (left) and GATCC (right). The correct base for recognition by *Bam*HI will *follow* the conserved sequence (to produce GGATCC on the left) only about 25 percent of the time, and the correct base will *precede* the conserved sequence (and produce GGATCC on the right) about 25 percent of the time as well. Thus, *Bam*HI will be able to cut the insert from the vector $(0.25 \times 0.25 = 0.0625)$, or only about 6 percent, of the time.

2. In setting up a PCR reaction, which set of primers (out of the three sets listed in this problem) would you choose to amplify the sequence shown as a series of asterisks?

5′-TTAAGATCCGTTACGTATGC * * ** * *
AACCCGTTCCTACGAACCTT-3′

5′-TTAAGATCCGTTACGTATGC * * ** * *
AACCCGTTCCTACGAACCTT-3′

Primers:

Set 1: 5′-TTAAGATCCGTT-3′ 5′-CGTTCCTACGAA-3′

Set 2: 5′-GATCCGTTACGT-3′ 5′-TTCGTAGGAACG-3′

Set 3: 5′-CGTATGCATTGC-3′ 5′-TTCCAAGCATCC-3′

Solution: A PCR reaction requires two primers, one for each strand. The primers are complementary to sequences near the end of each single-stranded DNA. In the first step of PCR, the DNA to be copied is heated to separate the molecule into single strands. Then the primers are annealed to the single strands, and elongated in the 5′ to 3′ direction by the addition of *Taq* polymerase. The only set of primers that will anneal properly is Set 2. One primer has a complementary sequence four base pairs in from the end of the lower strand of DNA, and the other primer has a complementary sequence four base pairs in from the end of the upper strand.

PROBLEMS AND DISCUSSION QUESTIONS

1. What roles do restriction enzymes, vectors, and host cells play in recombinant DNA studies?
2. Why is oligo-dT an effective primer for reverse transcriptase?
3. The human insulin gene contains a number of sequences that are removed in the processing of the mRNA transcript. In spite of the fact that bacterial cells cannot excise these sequences from mRNA transcripts, explain how a gene like this can be cloned into a bacterial cell and produce insulin.
4. Restriction enzymes recognize palindromic sequences (palindromic sequences read the same in the 5′ and the 3′ direction) in intact DNA molecules and cleave the double-stranded helix at these sequences. Inasmuch as the bases are on the inside of a DNA double helix, how is this recognition accomplished?
5. Although the potential benefits of cloning in higher plants are obvious, the development of this field has lagged behind cloning in bacteria, yeast, and mammalian cells. Can you think of a reason for this?
6. Using DNA sequencing on a cloned DNA segment, you recover the nucleotide sequence shown in this problem. Does this segment contain a palindromic recognition sequence for a restriction enzyme? If so, what is the double-stranded sequence of the palindrome, and what enzyme would cut at this sequence? (Consult Figure 17–1 for a list of restriction-enzyme recognition sequences.)

   CAGTATCCTAGGCAT

7. Restriction-enzyme sequences are palindromic; that is, they read the same in the 5′ to 3′ direction on each strand of DNA. What is the advantage of having restriction recognition sites organized in this way?
8. List the advantages and disadvantages of using plasmids and YACs as cloning vectors.

9. Listed here are several restriction enzymes with four- and six-base recognition sequences. For each of these enzymes, if we assume random distribution and equal amounts of each nucleotide in the DNA to be cut, what is the average distance between each of these recognition sequences?

*Taq*I	TCGA	*Hin*dIII	AAGCTT
*Alu*I	AGCT	*Bal*I	TGGCCA
*Hae*III	GGCC	*Bam*HI	GGATCC

10. Some restriction enzymes have recognition sites that are specific and unambiguous, whereas other recognition sites are ambiguous, meaning that any purine, pyrimidine, or nucleotide can occupy certain positions in the recognition sequence. In the following examples, *Not*I has an unambiguous recognition sequence, *Hin*fI has an ambiguous one (N = any nucleotide), and *Xho*II also has an ambiguous recognition sequence (Pu = any purine; Py = any pyrimidine). Assuming random distribution and equal amounts of each nucleotide in the DNA to be cut, what is the average distance between each of these restriction recognition sequences?

*Not*I	GCGGCCGC
*Hin*fI	GANTC
*Xho*II	PuGATCPy

11. What are the advantages of using a restriction enzyme whose recognition site is relatively rare? When would you use such enzymes?
12. An ampicillin-resistant, tetracycline-resistant plasmid, pBR322, is cleaved with *Pst*I, which cleaves within the ampicillin resistance gene. The cut plasmid is ligated with *Pst*I-digested *Drosophila* DNA to prepare a genomic library, and the mixture is used to transform *E. coli* K12.
    (a) Which antibiotic should be added to the medium to select cells that have incorporated a plasmid?
    (b) What growth pattern should be selected to obtain plasmids containing *Drosophila* inserts?
    (c) How can you explain the presence of colonies that are resistant to both antibiotics? See Now Solve This on page 357.

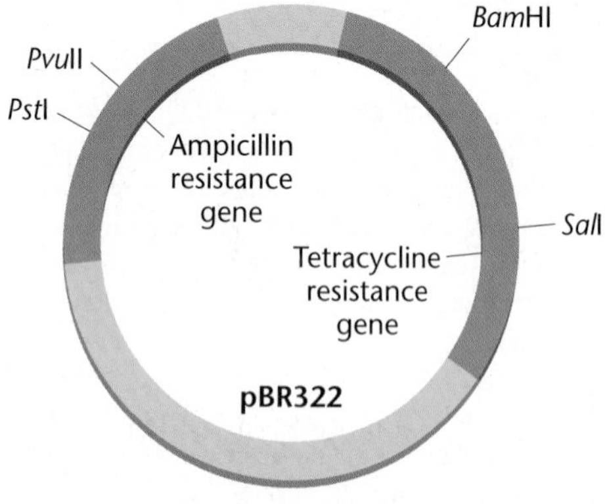

13. Plasmids isolated from the clones in Problem 12 are found to have an average insert length of 5 kb. Given that the *Drosophila* genome is 1.5×10^5 kb long, how many clones would be necessary to give a 99 percent probability that this library contains all genomic sequences? See Now Solve This on page 360.
14. In a control experiment, a plasmid containing a *Hin*dIII recognition sequence within a kanamycin resistance gene is cut with *Hin*dIII, religated, and used to transform *E. coli* K12 cells. Kanamycin-resistant colonies are selected, and plasmid DNA from these colonies is subjected to electrophoresis. Most of the colonies contain plasmids that produce single bands that migrate at the same rate as the original intact plasmid. A few colonies, however, produce two bands, one of original size and one that migrates much higher in the gel. Diagram the origin of this slow band as a product of ligation.
15. You have just created the world's first genomic library from the African okapi, a relative of the giraffe. No genes from this genome have been previously isolated or described. You wish to isolate the gene encoding the oxygen-transporting protein β-globin from the okapi library. This gene has been isolated from humans, and its nucleotide sequence and amino acid sequence are available in databases. Using the information available about the human β-globin gene, what two strategies can you use to isolate this gene from the okapi library?
16. In the production of cDNA, the single-stranded DNA produced by reverse transcriptase can be made double stranded by treatment with DNA polymerase I. However, no primer is required with the DNA polymerase. Why is this?
17. What should you consider in deciding which vector to use in constructing a genomic library of eukaryotic DNA?
18. You are given a cDNA library of human genes prepared in a bacterial plasmid vector. You are also given the cloned yeast gene that encodes EF-1a, a protein that is highly conserved among eukaryotes. Outline how you would use these resources to identify the human cDNA clone encoding EF-1a. See Now Solve This on page 361.
19. Once you have isolated the human cDNA clone for EF-1a in Problem 18, you sequence the clone and find that it is 1384 nucleotide pairs long. Using this cDNA clone as a probe, you isolate the DNA encoding EF-1a from a human genomic library. The genomic clone is sequenced and found to be 5282 nucleotides long. What accounts for the difference in length observed between the cDNA clone and the genomic clone?
20. You have recovered a cloned DNA segment from a vector and determine that the insert is 1300 bp in length. To characterize this cloned segment, you isolate the insert and decide to construct a restriction map. Using enzyme I and enzyme II, followed by gel electrophoresis, you determine the number and size of the fragments produced by enzymes I and II alone and in combination, as recorded in the following table. Construct a restriction map from these data, showing the positions of the restriction-enzyme cutting sites relative to one another and the distance between them in units of base pairs.

Enzymes	Restriction Fragment Sizes
I	350, 950
II	200, 1100
I and II	150, 200, 950

21. To create a cDNA library, cDNA can be inserted into vectors and cloned. In the analysis of cDNA clones, it is often difficult to find clones that are full length—that is, many clones are shorter than the mature mRNA molecules from which they are derived. Why is this so?
22. Although the capture and trading of great apes has been banned in 112 countries since 1973, it is estimated that about 1000 chimpanzees are removed annually from Africa and smuggled into Europe, the United States, and Japan. This illegal trade is often disguised by simulating births in captivity. Until recently, genetic identity tests to uncover these illegal activities were not used because of the lack of highly polymorphic markers (markers that vary from one individual to the next) and the difficulties of obtain-

ing chimpanzee blood samples. Recently, a study was reported in which DNA samples were extracted from freshly plucked chimpanzee hair roots and used as templates for PCR. The primers used in these studies flank highly polymorphic sites in human DNA that result from variable numbers of tandem nucleotide repeats. Several offspring and their putative parents were tested to determine whether the offspring were "legitimate" or the product of illegal trading. The data are shown in the following Southern blot.

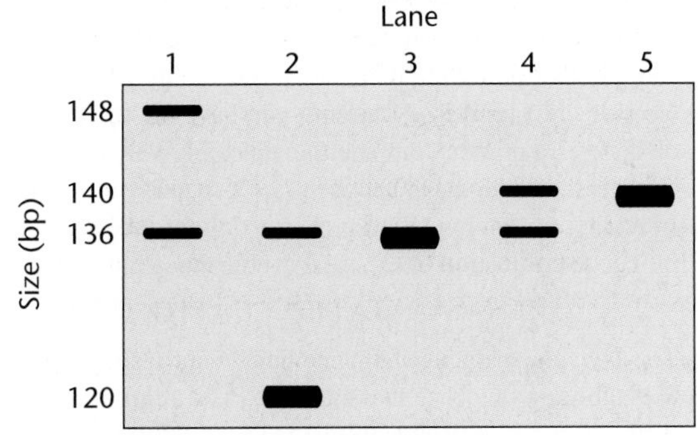

Lane 1: father chimpanzee
Lane 2: mother chimpanzee
Lanes 3–5: putative offspring A, B, C

Examine the data carefully, and choose the best conclusion.

(a) None of the offspring is legitimate.
(b) Offspring B and C are not the products of these parents and were probably purchased on the illegal market. The data are consistent with offspring A being legitimate.
(c) Offspring A and B are products of the parents shown, but C is not and was therefore probably purchased on the illegal market.
(d) There are not enough data to draw any conclusions. Additional polymorphic sites should be examined.
(e) No conclusion can be drawn because "human" primers were used.

23. The following partial restriction map shows a recombinant plasmid, pBIO220, formed by cloning a piece of *Drosophila* DNA (striped segment), including the gene *rosy,* into the vector pBR322, which also contains the penicillin resistance gene, *pen.* The vector part of the plasmid contains only the two E recognition sequences shown and no A or B sequences. The gel (in the next column) shows several restriction digests of pBIO220.

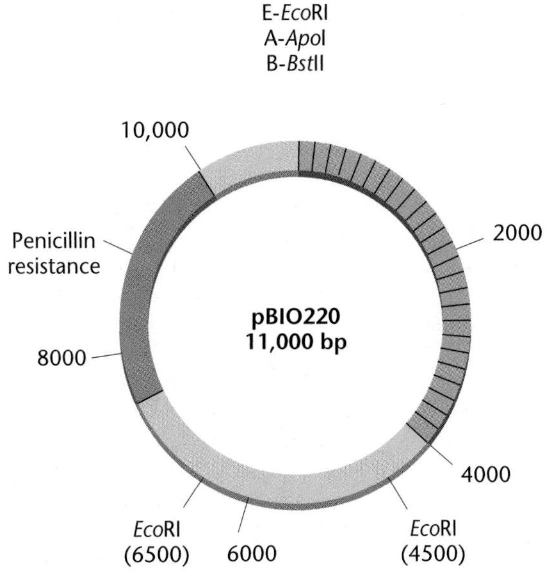

(a) Use the stained gel pattern to deduce where restriction-enzyme recognition sequences are located in the cloned fragment.
(b) A PCR-amplified copy of the entire 2000-bp *rosy* gene was used to probe a Southern blot of the same gel. Use the Southern blot results to deduce the location of *rosy* in the cloned fragment. Redraw the map showing the location of the *rosy* gene.

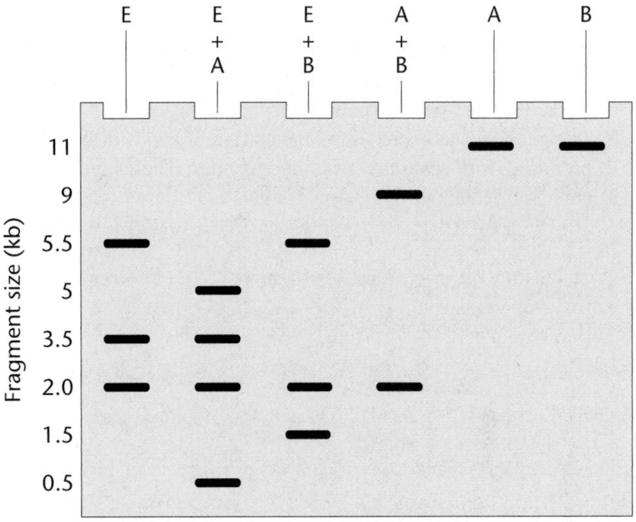

Stained gel

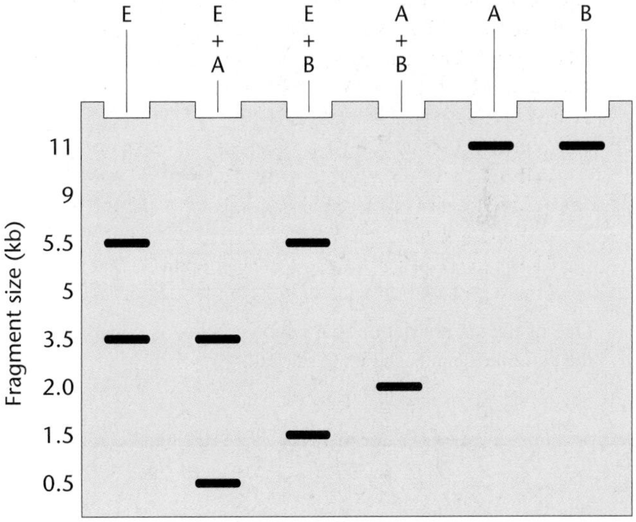

Southern blot

24. List the steps involved in screening a genomic library. What must be known before starting such a procedure? What are the potential problems with such a procedure, and how can they be overcome or minimized?

25. To estimate the number of cleavage sites in a particular piece of DNA with a known size, you can apply the formula $N/4^n$, where N is the number of base pairs in the target DNA and n is the number of bases in the recognition sequence of the restriction enzyme. If the recognition sequence for *Bam*HI is GGATCC and the λ phage DNA contains approximately 48,500 bp, how many cleavage sites would you expect?

26. In a typical PCR reaction, what phenomena are occurring at temperature ranges (a) 90–95°C, (b) 50–70°C, and (c) 70–75°C?

27. We usually think of enzymes as being most active at around 37°C, yet in PCR the DNA polymerase is subjected to multiple exposures of relatively high temperatures and seems to function appropriately at 70–75°C. What is special about the DNA polymerizing enzymes typically used in PCR?
28. How are dideoxynucleotides (ddNTPs) used in the chain-termination method of DNA sequencing?
29. Assume you have conducted a standard DNA sequencing reaction using the chain-termination method. You performed all the steps correctly and electrophoresed the resulting DNA fragments correctly, but when you looked at the sequencing gel, many of the bands were duplicated (in terms of length) in other lanes. What might have happened?
30. The gel presented here shows the pattern of bands of fragments produced with several restriction enzymes. The enzymes used are identified above and below the gel.

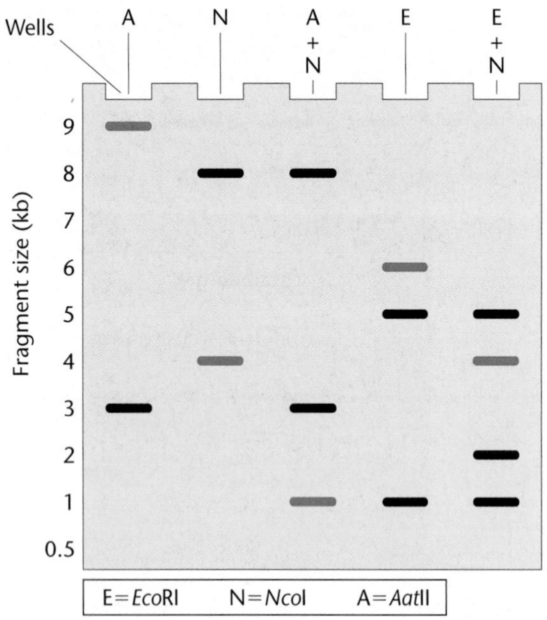

One of the six restriction maps shown below is consistent with the pattern of bands shown in the gel.

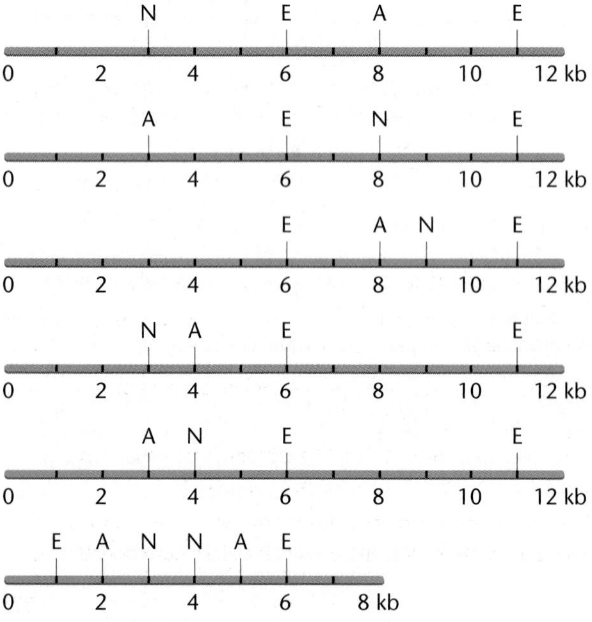

(a) From your analysis of the pattern of bands on the gel, select the correct map and explain your reasoning.

(b) In a Southern blot prepared from this gel, the highlighted bands (pink) hybridized with the gene *pep*. Where is the *pep* gene located?

31. A widely used method for calculating the annealing temperature for a primer used in PCR is 5 degrees below the T_m(°C), which is computed by the equation $81.5 + 0.41\,(\%\,\text{GC}) - (675/N)$, where %GC is the percentage of GC nucleotides in the oligonucleotide and *N* is the length of the oligonucleotide. Notice from the formula that both the GC content *and* the length of the oligonucleotide are variables. Assuming you have the following oligonucleotide as a primer, compute the annealing temperature for PCR. What is the relationship between T_m(°C) and %GC? Why? (*Note:* In reality, this computation provides only a starting point for empirical determination of the most useful annealing temperature.)

5′-TTGAAAATATTTCCCATTGCC-3′

32. The U.S. Department of Justice has recently sponsored research to establish a database that catalogs PCR amplification products from short tandem repeats of the Y (Y-STRs) chromosome in humans. The database contains polymorphisms of five U.S. ethnic groups (African Americans, European Americans, Hispanics, Native Americans, and Asian Americans) as well as worldwide population.

(a) Given that STRs are repeats of varying lengths, for example $(\text{TCTG})_{9-17}$ or $(\text{TAT})_{6-14}$, explain how PCR could reveal differences (polymorphisms) among individuals. How could the Department of Justice make use of those differences?

(b) Y-STRs from the nonrecombining region of the Y chromosome (NRY) have special relevance for forensic purposes. Why?

(c) What would be the value of knowing the ethnic population differences for Y-STR polymorphisms?

(d) For forensic applications, the probability of a "match" for a crime scene DNA sample and a suspect's DNA often culminates in a guilty or innocent verdict. How is a "match" determined, and what are the uses and limitations of such probabilities?

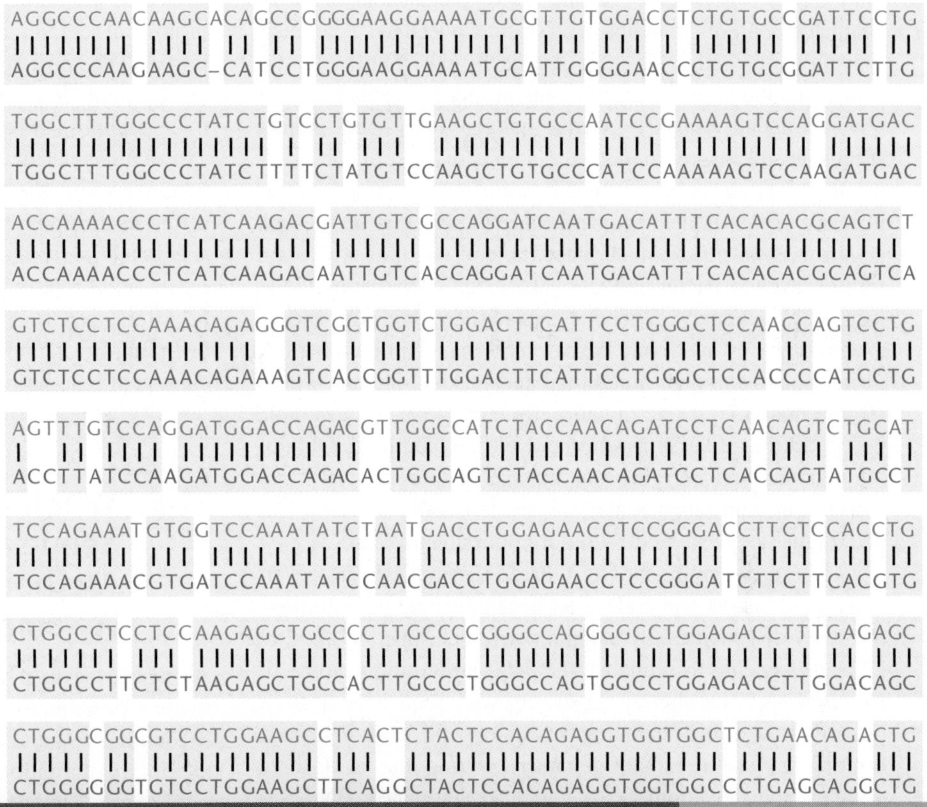

Alignment comparing DNA sequence for the leptin gene from dogs (top) and humans (bottom). Vertical lines and shaded boxes indicate identical bases. LEP encodes a hormone that functions to suppress appetite. This type of analysis is a common application of bioinformatics and a good demonstration of comparative genomics.

18

Genomics, Bioinformatics, and Proteomics

Chapter Concepts

- Genomics applies recombinant DNA, DNA sequencing methods and bioinformatics to sequence, assemble, and analyze genomes.
- Disciplines in genomics encompass several areas of study, including structural and functional genomics, comparative genomics, and metagenomics, and have led to an "omics" revolution in modern biology.
- Bioinformatics is a discipline that merges information technology with biology and mathematics to store, share, compare, and analyze nucleic acid and protein sequence data.
- The Human Genome Project has greatly advanced our understanding of the organization, size, and function of the human genome.
- Genomic analysis of model prokaryotes and eukaryotes has revealed similarities and differences in genome size and organization.
- Metagenomics is the study of genomes from environmental samples and is valuable for identifying microbial genomes.
- Transcriptome analysis provides insight into patterns of gene expression and gene-regulatory activity of a genome.
- Proteomics focuses on the protein content of cells and on the structures, functions, and interactions of proteins.
- Systems biology approaches attempt to uncover complex interactions among genes, proteins, and other cellular components.

The term **genome,** meaning the complete set of DNA in a single cell of an organism, was coined in 1920, at a time when geneticists began to turn from the study of individual genes to a focus on the larger picture. To begin to characterize all of the genes in an organism's DNA, geneticists typically followed a two-part approach: (1) identify spontaneous mutations or collect mutants produced by chemical or physical agents, and (2) generate linkage maps using mutant strains as discussed in Chapter 7. These effective but extremely time-consuming strategies were used to identify genes in many of the model organisms discussed in this book, such as *Drosophila,* maize, mice, bacteria, and yeast, as well as in viruses, such as bacteriophages. These approaches formed the technical backbone of genetic analysis and are still widely used today; however, they have several major limitations. For instance, conventional mutational analysis and linkage would require that at least one mutation for each gene was available before all the genes in a genome could be identified. Obtaining mutants and carrying out linkage studies is very time-consuming, and when mutations are lethal or have no clear phenotype, they can be difficult or impossible to map.

In 1977, as recombinant DNA–based techniques were developed, Fred Sanger and colleagues began the field of **genomics,** the study of genomes, by using a newly developed method of DNA sequencing to sequence the 5400-nucleotide genome of the virus ϕX174. Other viral genomes were sequenced in short order, but even this technology was slow and labor-intensive, limiting its use to small genomes. During the next three decades, the development of computer-automated DNA sequencing methods made it possible to consider sequencing the larger and more complex genomes of eukaryotes, including the 3.1 billion nucleotides that comprise the human genome.

How Do We Know?

In this chapter, we will focus on the analysis of genomes, transcriptomes, and proteomes and consider important applications and findings from these endeavors. We will find many opportunities to consider the methods and reasoning by which much of this information was acquired. From the explanations given in the chapter, you should answer the following fundamental questions:

1. How do we know the sequence of entire chromosomes based on assembling the order of thousands of DNA fragments from a genome?
2. How do we know if a genomic DNA sequence contains a protein-coding gene?
3. How do we know that humans share similarities in gene sequence and function with model organisms?
4. How do we know that gene families result from gene-duplication events?
5. How have microarrays demonstrated that, although all cells of an organism have the same genome, some genes are expressed in almost all cells whereas other genes show cell- and tissue-specific expression?

The development of recombinant DNA technologies coupled with the advent of computer-automated DNA sequencing methods is responsible for accelerating the field of genomics. Genomic technologies have developed so quickly that modern biological research is currently experiencing a genomics revolution. The most recent new subdisciplines of genomics to emerge include *structural* and *functional genomics, comparative genomics,* and *metagenomics.*

In this chapter, we will examine basic technologies used in genomics and then discuss examples of genome data and different disciplines of genomics. We will also discuss *transcriptome analysis,* the study of genes expressed in a cell or tissue (the "transcriptome"), and conclude the chapter with a brief look at *proteomics,* the study of proteins present in a cell or tissue. In Chapter 19, we will discuss many modern applications of recombinant DNA and genomics.

18.1 Whole-Genome Shotgun Sequencing Is a Widely Used Method for Sequencing and Assembling Entire Genomes

As discussed in Chapter 17, recombinant DNA technology made it possible to generate DNA libraries that could be used to identify, clone, and sequence specific genes of interest. But a primary limitation of library screening and even of most polymerase chain reaction (PCR) approaches is that they typically can identify only relatively small numbers of genes at a time. Genomics allows the sequencing of entire genomes. **Structural genomics** focuses on sequencing genomes and analyzing nucleotide sequences to identify genes and other important sequences such as gene-regulatory regions.

Currently, the most widely used strategy for sequencing and assembling an entire genome involves variations of a method called **whole-genome shotgun sequencing.** In simple terms, this technique is analogous to you and a friend taking your respective copies of this genetics textbook and randomly ripping the pages into strips about 5 to 7 inches long. Each chapter represents a chromosome, and all of the letters in the entire book are the "genome." Then you and your friend would go through the painstaking task of comparing the pieces of paper to find places that match, overlapping sentences—areas where there are similar sentences on different pieces of paper. Eventually, in theory, many of the strips containing matching sentences would overlap in ways that you could use to reconstruct the pages and assemble the order of the entire text.

Figure 18–1 shows a basic overview of whole-genome shotgun sequencing. First, an entire chromosome is cut into short, overlapping fragments, either by mechanically shearing the DNA in various ways or, more commonly, by using restriction enzymes to cleave the DNA at different locations. Recall from Chapter 17 that **restriction enzymes** are DNA-digesting enzymes that cut the phosphodiester backbone of DNA at specific sequences.

Different restriction enzymes can be used so that chromosomes are cut at different sites; or sometimes, **partial digests**

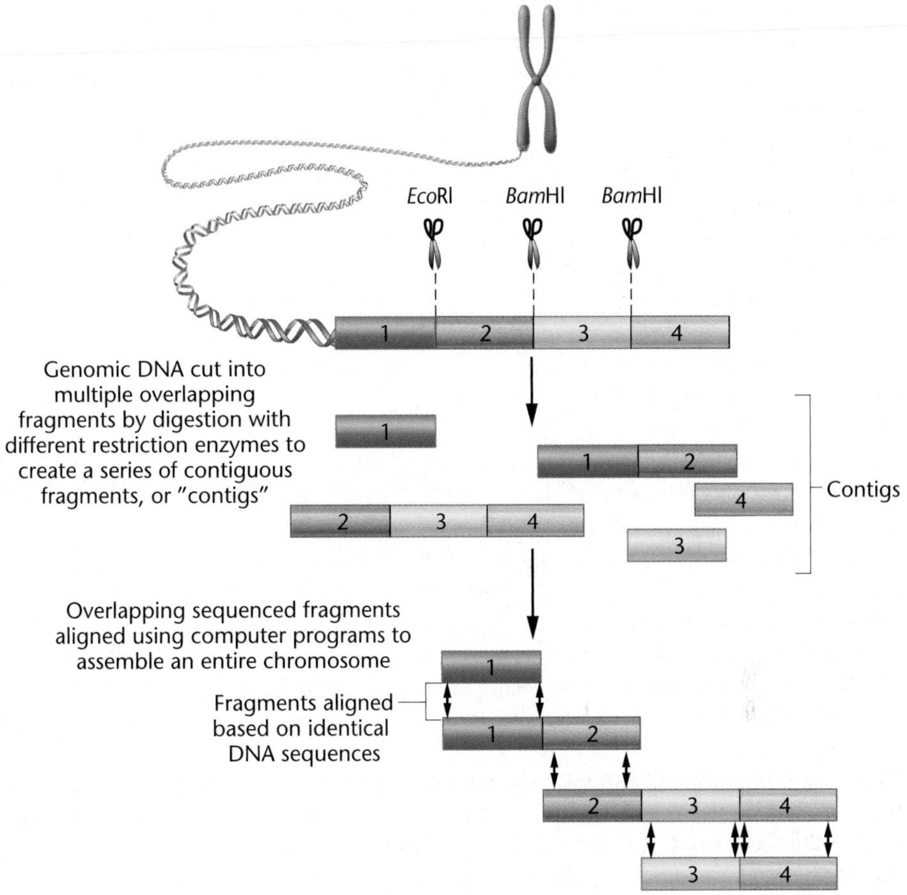
EcoRI
BamHI
BamHI
1
2
3
4
Genomic DNA cut into multiple overlapping fragments by digestion with different restriction enzymes to create a series of contiguous fragments, or "contigs"
Contigs
Overlapping sequenced fragments aligned using computer programs to assemble an entire chromosome
Fragments aligned based on identical DNA sequences

FIGURE 18–1 An overview of whole-genome shotgun sequencing and assembly.

of DNA using the same restriction enzyme are used. With partial digests, DNA is incubated with restriction enzymes for only a short period of time, so that not every site in a particular sequence is cut to completion by an individual enzyme. Either way, restriction digests of whole chromosomes generate thousands to millions of overlapping DNA fragments. For example, a 6-bp cutter such as *Eco*RI creates about 700,000 fragments when used to digest the human genome! Because these overlapping fragments are adjoining segments that collectively form one continuous DNA molecule within a chromosome, they are called **contiguous fragments,** or "**contigs.**"

Cutting a genome into fragments is not particularly difficult; however, a primary hurdle that initially prevented whole-genome sequencing was the question of how to sequence millions or billions of base pairs in a timely and cost-effective way. The Sanger sequencing method discussed in Chapter 17 was the predominant sequencing technique for a long time; however, its major limitation was that even the best sequencing gels would typically yield only several hundred base pairs in each run. Obviously, it would be very time-consuming to manually sequence an entire genome by the Sanger method. The major technological breakthrough that made genomics possible was the development of high-throughput computer-automated sequencers, also discussed in Chapter 17.

High-Throughput Sequencing

Many **computer-automated DNA sequencing instruments** utilize dideoxynucleotides (ddNTPs) labeled with fluorescent dyes (refer to Figure 17–19). A single reaction tube is used, and the original manual approach that employed polyacrylamide gels to separate reaction mixtures has been replaced by a single lane of an ultra-thin-diameter polyacrylamide tube gel called a **capillary gel.** As DNA fragments move through the gel, they are scanned with a laser beam. The laser stimulates the fluorescent dye on each DNA fragment, causing each ddNTP to emit different wavelengths of light. The emitted light is collected by a detector, which amplifies and then feeds this information to a computer to process and convert into the DNA sequence.

Certain types of computer-automated sequencers, designed for so-called **high-throughput sequencing,** can process millions of base pairs in a day. These sequencers contain multiple capillary gels that are several feet long. Some run as many as 96 capillary gels at a time, each producing around 900 bases of sequence. Because these sequencers are computer automated, they can work around the clock, generating over 2 million bases of sequence in a day. In the past 10 years, high-throughput sequencing has increased the productivity of DNA-sequencing technology over 500-fold. The total number of bases that could be sequenced in a single reaction was doubling about every 24 months. At the same time, this increase in efficiency brought about a dramatic decrease in cost, from about $1.00 to less than $0.001 per base pair.

In the next section, we will discuss the importance of bioinformatics to genomics. One of the earliest bioinformatics applications to be developed for genomic purposes was the

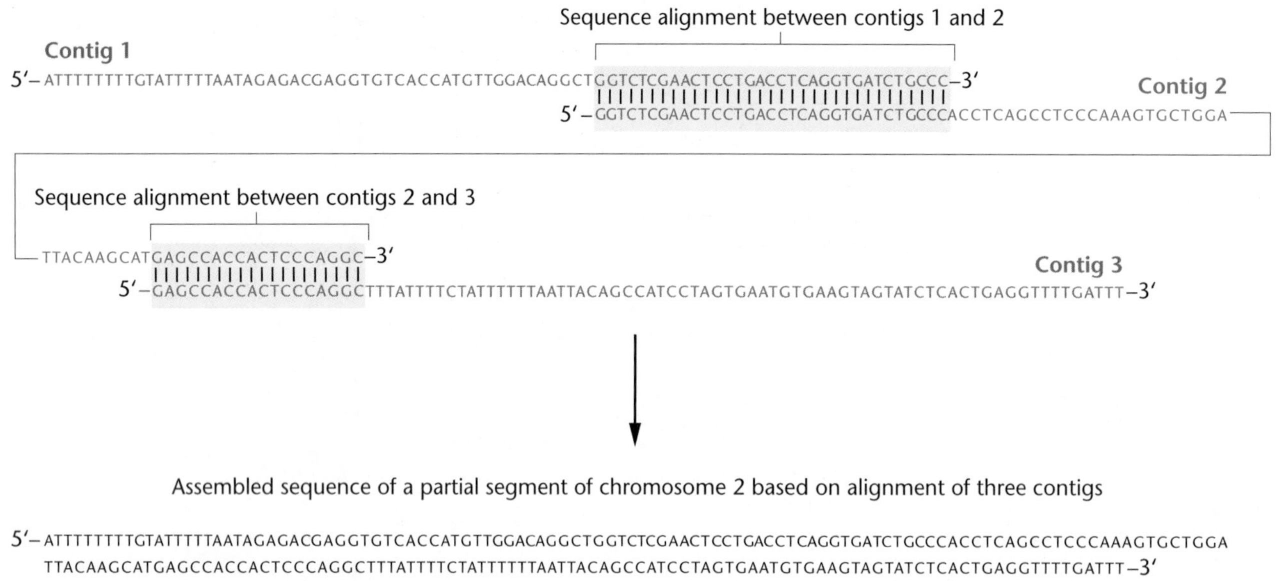

FIGURE 18–2 DNA sequence alignment of contigs on human chromosome 2. Single-stranded DNA for three different contigs from human chromosome 2 is shown in blue, red, or green. The actual sequence from chromosome 2 is shown, but in reality, contig alignment involves fragments that are several thousand bases in length. Alignment of the three contigs allows a portion of chromosome 2 to be assembled. Alignment of all contigs for a particular chromosome would result in assembly of a completely sequenced chromosome.

use of algorithm-based software programs for creating a DNA sequence **alignment,** in which similar sequences of bases, such as contigs, are lined up for comparison. Alignment identifies overlapping sequences, allowing scientists to reconstruct their order in a chromosome. Figure 18–2 shows an example of contig alignment and assembly for a portion of human chromosome 2. For simplicity, this figure shows relatively short sequences for each contig, which in actuality would be much longer. The figure is also simplified in that, in actual alignments, assembled sequences do not always overlap only at their ends.

The whole-genome shotgun sequencing method was developed by J. Craig Venter and colleagues at The Institute for Genome Research (TIGR). In 1995, TIGR scientists used this approach to sequence the 1.83-million-bp genome of the bacterium *Haemophilus influenzae.* This was the first completed genome sequence from a free-living organism, and it demonstrated "proof of concept" that shotgun sequencing could be used to sequence an entire genome. Even after the genome for *H. influenzae* was sequenced, many scientists were skeptical that a shotgun approach would work on the larger genomes of eukaryotes. Variations of shotgun approaches are now the predominant methods for sequencing genomes.

The Clone-by-Clone Approach

Prior to the widespread use of whole-genome sequencing approaches, genomes were being assembled using a **clone-by-clone** approach, also called **map-based cloning** (Figure 18–3). Initial progress on the Human Genome Project was based on this methodology, in which individual DNA fragments from restriction digests of chromosomes are aligned to create the restriction maps of a chromosome. These restriction fragments are then ligated into vectors such as bacterial artificial chromosomes (BACs) or yeast artificial chromosomes (YACs) to create libraries of contigs. Recall from Chapter 17 that BACs and YACs are good cloning vectors for replicating large fragments of DNA. Often, libraries from individual chromosomes were prepared.

Prior to the development of high-throughput approaches capable of sequencing several thousand bases, DNA fragments in BACs and YACs would often be further digested into smaller, more easily manipulated pieces that were then subcloned into cosmids or plasmids so that they could be sequenced in their entirety (Figure 18–3). After each fragment was sequenced and then analyzed for alignment overlaps, a chromosome could be assembled. The bioinformatics approaches we will discuss in the next section would then be used to identify possible protein-coding genes and assign them a location on the chromosome. For example, Figure 18–3 shows the use of map-based cloning to sequence part of chromosome 11, including part of the human β-globin gene.

Compared to whole-genome sequencing, the clone-by-clone approach is cumbersome and time-consuming because of the time required to clone DNA fragments into different vectors, transform bacteria or yeast, select individual clones from the library for sequencing, and then carry out sequence analysis and assembly on relatively short sequences. Essentially, the clone-by-clone approach is the organized sequencing of contigs from a restriction map instead of random sequencing and assembly. As whole-genome sequencing approaches have become the most common method for assembling genomes, map-based cloning approaches are now used primarily to resolve the problems often encountered during whole-genome sequencing. For example, highly repetitive sequences in a chromosome can be difficult to align correctly in order to identify overlaps because with such sequences one cannot be certain whether portions that are nearly identical are overlapping fragments or belong to different parts of a highly repetitive

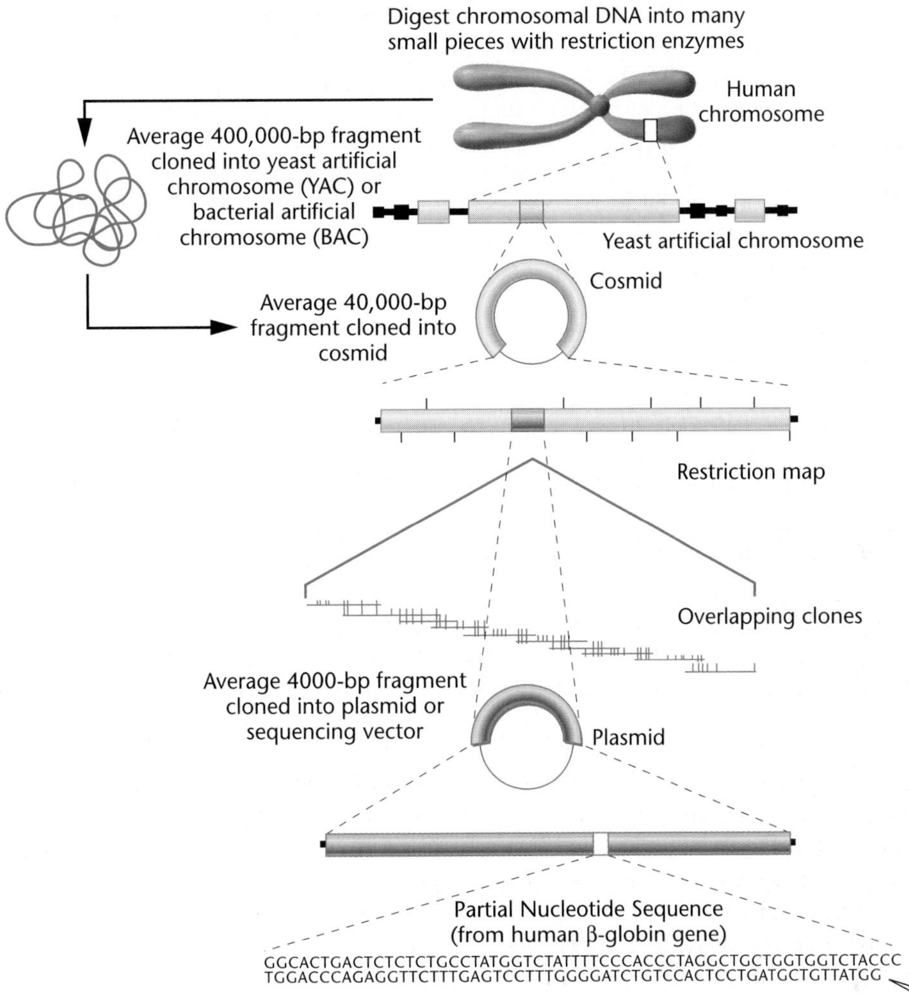

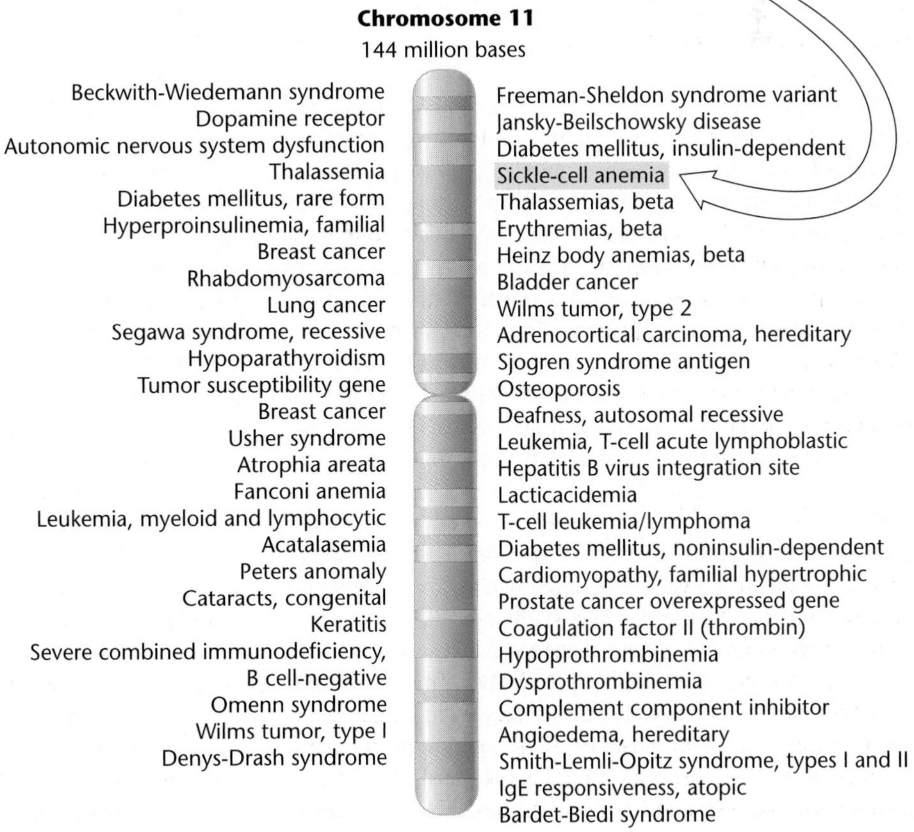

FIGURE 18–3 A clone-by-clone, or map-based, approach to genome sequencing involves cloning overlapping DNA fragments (contigs) into vectors. Different vectors, such as BACs, YACs, cosmids, and plasmids, are used depending on the size of each DNA fragment being analyzed. Overlapping clones are then sequenced and aligned to assemble an entire chromosome.

chromosome. Frequently, there are also gaps between aligned contigs. In the textbook-ripping analogy, after you compare all your pieces of paper, you may be left with some very small ones that contain too few words to be matched with certainty to any others and with some pieces for which you just could not find matches. In these instances, a clone-by-clone approach may enable you to assemble the necessary contigs and complete the chromosome. Thus, whole-genome sequencing and map-based sequencing are commonly combined in the task of sequencing a genome.

It is common for a draft sequence of a genome to be announced several years before a final sequence is released. Draft sequences often contain gaps in areas that, for any number of reasons, may have been difficult to analyze. The decision to designate a sequence as "final" is dictated by the amount of error genome scientists are willing to accept as a cutoff. Once completed, a genome is analyzed to identify gene sequences, regulatory elements, and other features that reveal important information. In the next section we discuss the central role of bioinformatics in this process.

ESSENTIAL POINT

High-throughput computer-automated DNA sequencing methods coupled with bioinformatics enable scientists to assemble sequence maps of entire genomes.

18.2 DNA Sequence Analysis Relies on Bioinformatics Applications and Genome Databases

Genomics necessitated the rapid development of **bioinformatics,** the use of computer hardware and software and mathematics applications to organize, share, and analyze data related to gene structure, gene sequence and expression, and protein structure and function. However, even before whole-genome sequencing projects had been initiated, a large amount of sequence information from a range of different organisms was accumulating as a result of gene cloning by recombinant DNA techniques. Scientists around the world needed databases that could be used to store, share, and obtain the maximum amount of information from protein and DNA sequences. Thus, bioinformatics software was already being widely used to compare and analyze DNA sequences and to create private and public databases. Once genomics emerged as a new approach for analyzing DNA, however, bioinformatics became even more important than before. Today, it is a dynamic area of biological research, providing new career opportunities for anyone interested in merging an understanding of biological data with information technology, mathematics, and statistical analysis.

Among the most important applications of bioinformatics are to compare DNA sequences, as in contig alignment, discussed in the previous section; to identify genes in a genomic DNA sequence; to find gene-regulatory regions, such as promoters and enhancers; to identify structural sequences, such as telomeric sequences, in chromosomes; to predict the amino acid sequence of a putative polypeptide encoded by a cloned gene sequence; to analyze protein structure and predict protein functions on the basis of identified domains and motifs; and to deduce evolutionary relationships between genes and organisms on the basis of sequence information.

High-throughput DNA sequencing techniques were developed almost simultaneously with expansion of the Internet. As genome data accumulated, many DNA-sequence databases became freely available online. Databases are essential for archiving and sharing data with other researchers and with the public. One of the most important genomic databases, called **GenBank,** is maintained by the National Center for Biotechnology Information (NCBI) in Washington, D.C., and is the largest publicly available database of DNA sequences. GenBank shares and acquires data from databases in Japan and Europe; it contains more than 100 billion bases of sequence data from over 100,000 species; and it doubles in size roughly every 14 months! The Human Genome Nomenclature Committee, supported by the NIH, establishes rules for assigning names and symbols to newly cloned human genes. As sequences are identified and genes are named, each sequence deposited into GenBank is provided with an **accession number** that scientists can use to access and retrieve that sequence for analysis.

The NCBI is an invaluable source of public access databases and bioinformatics tools for analyzing genome data. You have already been introduced to NCBI and GenBank through several Exploring Genomics exercises. In the Exploring Genomics exercise for this chapter, you will use NCBI and GenBank to compare and align contigs in order to assemble a chromosome segment.

Annotation to Identify Gene Sequences

One of the fundamental challenges of genomics is that, although genome projects generate tremendous amounts of DNA sequence information, these data are of little use until they have been analyzed and interpreted. Thus, after a genome has been sequenced and compiled, scientists are faced with the task of identifying gene-regulatory sequences and other sequences of interest in the genome so that gene maps can be developed. This process, called **annotation,** relies heavily on bioinformatics, and a wealth of different software tools are available to carry it out.

One initial approach to annotating a sequence is to compare the newly sequenced genomic DNA to the known sequences already stored in various databases. The NCBI provides access to **BLAST (Basic Local Alignment Search Tool),** a very popular software application for searching through banks of DNA and protein sequence data. Using BLAST, we can compare a segment of genomic DNA to sequences throughout major databases such as GenBank to identify portions that align with or are the same as existing sequences.

Figure 18–4 shows a representative example of a sequence alignment based on a BLAST search. Here a 280-bp chromosome 12 contig from the rat was used to search a mouse database to determine whether there was a sequence in the rat contig that matched a known gene in mice. Notice

ref | NT_039455.6 | Mm8_39495_36
Mus musculus chromosome 8 genomic contig, strain C57BL/6J
Features in this part of subject sequence: insulin receptor
Score = 418 bits (226), Expect = 2e-114
Identities = 262/280 (93%), Gaps = 0/280 (0%)

```
Query  1       CAGGCCATCCCGAAAGCGAAGATCCCTTGAAGAGGTGGGCAATGTGACAGCCACTACACC  60
               ||||||||||||||||||||||||||||||||||||||| |||||||||||||| |||| 
Sbjct  174891  CAGGCCATCCCGAAAGCGAAGATCCCTTGAAGAGGTGGGGAATGTGACAGCCACCACACT  174832

Query  61      CACACTTCCAGATTTTCCCAACATCTCCTCCACCATCGCGCCCACAAGCCACGAAGAGCA  120
               |||||||||||||| ||||||| ||||||| ||||| | ||||||||| || || |||||
Sbjct  174831  CACACTTCCAGATTTCCCCAACGTCTCCTCTACCATTGTGCCCACAAGTCAGGAGGAGCA  174772

Query  121     CAGACCATTTGAGAAAGTAGTAAACAAGGAGTCACTTGTCATCTCTGGCCTGAGACACTT  180
               ||| |||||||||||||| || ||||||||||||||||||||||||||||||||||||||
Sbjct  174771  CAGGCCATTTGAGAAAGTGGTGAACAAGGAGTCACTTGTCATCTCTGGCCTGAGACACTT  174712

Query  181     CACTGGGTACCGCATTGAGCTGCAGGCATGCAATCAGGACTCCCCAGAAGAGAGGTGCAG  240
               |||||||||||||||||||||||||||||||||||| || |||||||| |||||||||||
Sbjct  174711  CACTGGGTACCGCATTGAGCTGCAGGCATGCAATCAAGATTCCCCAGATGAGAGGTGCAG  174652

Query  241     CGTGGCTGCCTACGTCAGTGCCCGGACCATGCCTGAAGGT  280
                |||||||||||||||||||||||||||||||||||||||
Sbjct  174651  TGTGGCTGCCTACGTCAGTGCCCGGACCATGCCTGAAGGT  174612
```

FIGURE 18–4 BLAST results showing a 280-base sequence of a chromosome 12 contig from rats (*Rattus norvegicus*, the "query") aligned with a portion of chromosome 8 from mice (*Mus musculus*, the "subject") that contains a partial sequence for the insulin receptor gene. Vertical lines indicate exact matches. The rat contig sequence was used as a query sequence to search a mouse database in GenBank. Notice that the two sequences show 93 percent identity, strong evidence that this rat contig sequence contains a gene for the insulin receptor.

that the rat contig (the query sequence in the BLAST search) aligned with base pairs 174,612 to 174,891 of mouse chromosome 8. BLAST searches compare the query sequence to all of the sequences in the database being analyzed and calculate a **similarity score**—also called the **identity value**—determined by the sum of identical matches between aligned sequences divided by the total number of bases aligned. Similar sequences to the query sequence are rank ordered from highest probable matches to sequences with less similarity. Gaps, indicating missing bases in the two sequences, are usually ignored in calculating similarity scores. The aligned rat and mouse sequences were 93 percent similar and showed no gaps in the alignment. Notice that the BLAST report also provides an "Expect" value, or **E value,** based on the number of matches that would be expected by chance in the aligned sequences. Significant alignments, indicating those DNA sequences that are most likely to be similar to the query sequence, typically have E values less than 1.0.

Because this mouse sequence on chromosome 8 is known to contain an insulin receptor gene (encoding a protein that binds the hormone insulin), it is highly likely that the rat contig sequence also contains an insulin receptor gene. But often predicted amino acid sequence data for protein-coding sequences are compared to confirm gene identity. We will return to the topic of similarity in Sections 18.3 and 18.7, where we consider how similarity between gene sequences can be used to infer function and to identify evolutionarily related genes through comparative genomics.

NOW SOLVE THIS

In problem 18 on page 405, involving annotation analysis of the number of genes on each human chromosome that are expressed in embryos, you are asked to consider this information and suggest explanations for why some aneuploids occur and others do not.

Hint: Consider the relationship between chromosome number and gene number shown in the graph.

Hallmark Characteristics of a Gene Sequence Can Be Recognized During Annotation

A major limitation of this approach to annotation is that it works only if similar gene sequences are already in a database. Fortunately, it is not the only way to identify genes. Whether the genome under study is from a eukaryote or a prokaryote, several hallmark characteristics of genes can be searched for using bioinformatics software (Figure 18–5, p. 382). We discussed many of these characteristics of a "typical" gene in Chapters 12 and 15. For instance, control regions at the beginning of genes are marked by identifiable sequences such as promoters and enhancers. Recall from Chapter 15 that TATA box, GC box, and CAAT box sequences are often present in the promoter regions of eukaryotic genes. Recall also that splice sites between **exons** and **introns** contain a predictable sequence (most introns begin with GT and end with AG). Splice-site sequences are important for determining intron and exon boundaries. There are also well-defined sequences at the end of

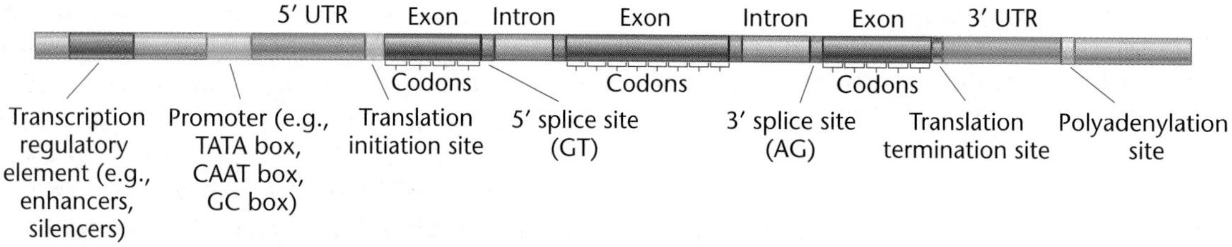

FIGURE 18–5 Characteristics of a gene that can be used during annotation to identify a gene in an unknown sequence of genomic DNA. Most eukaryotic genes are organized into coding segments (exons) and noncoding segments (introns). When annotating a genome sequence to determine whether it contains a gene, it is necessary to distinguish between introns and exons, gene-regulatory sequences, such as promoters and enhancers, untranslated regions (UTRs), and gene termination sequences.

the gene, where a polyadenylation sequence signals the addition of a poly(A) tail to the 3′ end of an mRNA transcript (Figure 18–5). Annotation can sometimes be a little bit easier for prokaryotic genes than for eukaryotic genes because there are no introns in prokaryotic genes.

In addition, protein-coding genes contain one or more **open reading frames (ORFs),** sequences of triplet nucleotides that are translated into the amino acid sequence of a protein. ORFs typically begin with an initiation sequence, usually ATG, which transcribes into the AUG start codon of an mRNA molecule, and end with a termination sequence, TAA, TAG, or TGA, which correspond to the stop codons of UAA, UAG, and UGA in mRNA. Because genetic information is encoded in groups of three nucleotides (triplets), typically ORF searches are done to analyze a sequence starting from the first, second, and third nucleotide to consider all possible triplet reading arrangements. Typically, the sequence adjacent to a promoter is examined for a start (initiation) triplet as a key hallmark of a protein-coding sequence; however, ORFs can be used to identify a gene even when a promoter sequence is not apparent. Software programs can then analyze the ORFs three nucleotides at a time. The discovery of an ORF starting with an ATG followed at some distance by a termination sequence is usually a good indication that the coding region of a gene has been identified.

The way genes are organized in eukaryotic genomes makes direct searching for ORFs more difficult in them than in prokaryotic genomes. First, many eukaryotic genes have introns. As a result, many, if not most, eukaryotic genes are not organized as continuous ORFs; instead, the gene sequences consist of ORFs (exons) interspersed with introns. Second, genes in humans and other eukaryotes are often widely spaced, increasing the chances of finding false ORFs in the regions between gene clusters.

Software designed for ORF analysis of eukaryotic genomes is highly efficient. In addition to the features already mentioned, such software can be used to "translate" ORFs into possible polypeptide sequences as a way to predict the polypeptide encoded by a gene. Shown in Figure 18–6 is a partial sequence for the first exon of the human tubulin alpha 3c gene (*TUBA3C*). Prediction programs scan potential ORFs in both the 5′ to 3′ and 3′ to 5′ direction to predict possible polypeptides from the three possible reading frames in each direction. Figure 18–6 shows the results for the six possible reading frames in the sequence of interest. Amino acids are shown using the single-letter code for each residue. Notice the very different results obtained for each of the six frames. For instance, the 5′ to 3′ ORF 1 contains several stop codons interspersed among amino acids but no methionine residues that are evidence of a start codon. Other ORFs would contain too many methionines to produce a functional polypeptide. For this exon of *TUBA3C,* the 5′ to 3′ ORF 2 is correct.

ESSENTIAL POINT

Annotation is used to identify protein-coding DNA sequences and noncoding sequences such as regulatory elements, while bioinformatics programs are used to identify open reading frames that predict possible polypeptides coded for by a particular sequence.

(a) *Homo sapiens TUBA3C* (bp 1-300)

1 ggttgaggtcaagtagtagcgttgggctgcggcagcggaggagctcaacatgcgtgagtg
61 tatctctatccacgtggggcaggcaggagtccagatcggcaatgcctgctgggaactgta
121 ctgcctggaacatggaattcagcccgatggtcagatgccaagtgataaaaccattggtgg
181 tggggacgactccttcaacacgttcttcagtgagactggagctggcaagcacgtgcccag
241 agcagtgtttgtggacctggagcccactgtggtcgatgaagtgcgcacaggaacctatag (300)

(b) Predicted polypeptides

5'3' Frame 1
G **Stop** G Q V V A L G C G S G G A Q H A **Stop** V Y L Y P R G A G R S P D R Q C L L G T V L P G T W N S A R W S D A K **Stop Stop** N H W W W G R L L Q H V L Q **Stop** D W S W Q A R A Q S S V C G P G A H C G R **Stop** S A H R N L **Stop**

5'3' Frame 2
V E V K **Stop Stop** R W A A A A E E L N **Met** R E C I S I H V G Q A G V Q I G N A C W E L Y C L E H G I Q P D G Q **Met** P S D K T I G G G D D S F N T F F S E T G A G K H V P R A V F V D L E P T V V D E V R T G T Y

5'3' Frame 3
L R S S S S V G L R Q R R S S T C V S V S L S T W G R Q E S R S A **Met** P A G N C T A W N **Met** E F S P **Met** V R C Q V I K P L V V G T T P S T R S S V R L E L A S T C P E Q C L W T W S P L W S **Met** K C A Q E P I

3'5' Frame 1
L **Stop** V P V R T S S T T V G S R S T N T A L G T C L P A P V S L K N V L K E S S P P P **Met** V L S L G I **Stop** P S G **Stop** I P C S R Q Y S S Q Q A L P I W T P A C P T W I E I H S R **Met** L S S S A A A A Q R Y Y L T S T

3'5' Frame 2
Y R F L C A L H R P Q W A P G P Q T L L W A R A C Q L Q S H **Stop** R T C **Stop** R S R P H H Q W F Y H L A S D H R A E F H V P G S T V P S R H C R S G L L P A P R G **Stop** R Y T H A C **Stop** A P P L P Q P N A T T **Stop** P Q

3'5' Frame 3
I G S C A H F I D H S G L Q V H K H C S G H V L A S S S L T E E R V E G V V P T T N G F I T W H L T I G L N S **Met** F Q A V Q F P A G I A D L D S C L P H V D R D T L T H V E L L R C R S P T L L L D L N

FIGURE 18–6 Predicted polypeptide sequences translated from potential ORFs in the human *TUBA3C* gene. (a) Nucleotides 1–300 of the first exon in the human *TUBA3C* gene. (b) A translation program predicts six possible polypeptide sequences from this exon. Which predicted sequence is correct?

NOW SOLVE THIS

In Problem 19 part (d) on page 406, involving a comparison of the annotated genome of *Drosophila,* you are asked to compare differences between two different releases of the genome.

Hint: Consider why continual annotation of a genome can reveal differences following each round (release) of annotation.

18.3 Functional Genomics Attempts to Identify Potential Functions of Genes and Other Elements in a Genome

Reading a genome sequence is a surefire cure for insomnia. What is exciting is not the sequence of the nucleotides but the information that the sequence contains. After a genome has been annotated and ORFs have been identified, the next analytical task is to assign putative functions to all possible genes in the sequence. As the term suggests, **functional genomics** is the study of gene functions, based on the resulting RNAs or possible proteins they encode, and the functions of other components of the genome, such as gene-regulatory elements. Functional genomics also considers how genes are expressed and the regulation of gene expression.

Predicting Gene and Protein Functions by Sequence Analysis

Some newly identified genomic sequences may already have had functions assigned to their genes by classic methods such as mutagenesis and linkage mapping, but many other genes that have been sequenced have not yet been correlated with a function. One approach to assigning functions to genes is to use sequence similarity searches, as described in the previous section. Programs such as BLAST are used to search through databases to find alignments between the newly sequenced genome and genes that have already been identified, either in the same or in different species. You were introduced to this approach for predicting gene function in Figure 18–4, when we demonstrated how sequence similarity to the mouse gene was used to identify a gene in a rat contig as the insulin receptor gene. Inferring gene function from similarity searches is based on a relatively simple idea. If a genome sequence shows statistically significant similarity to the sequence of a gene whose function is known, then it is likely that the genome sequence encodes a protein with a similar or related function. Figure 18–7 indicates functional categories that have been assigned to genes in the *Arabidopsis* genome on the basis of similarity searches.

Another major benefit of similarity searches is that they are often able to identify **homologous genes,** genes that are evolutionarily related. After the human genome was sequenced, many ORFs in it were identified as protein-coding genes based on their alignment with related genes of known function in other species. As an example, Figure 18–8, p. 384 compares portions of the human leptin gene (*LEP*) with its homolog in mice (*ob/Lep*). These two genes are over 85 percent identical in sequence. The leptin gene was first discovered in mice. The match between the *LEP*-containing DNA sequence in humans and the mouse homolog sequence confirms the identity and leptin-coding function of this gene in human genomic DNA.

(a)

(b)

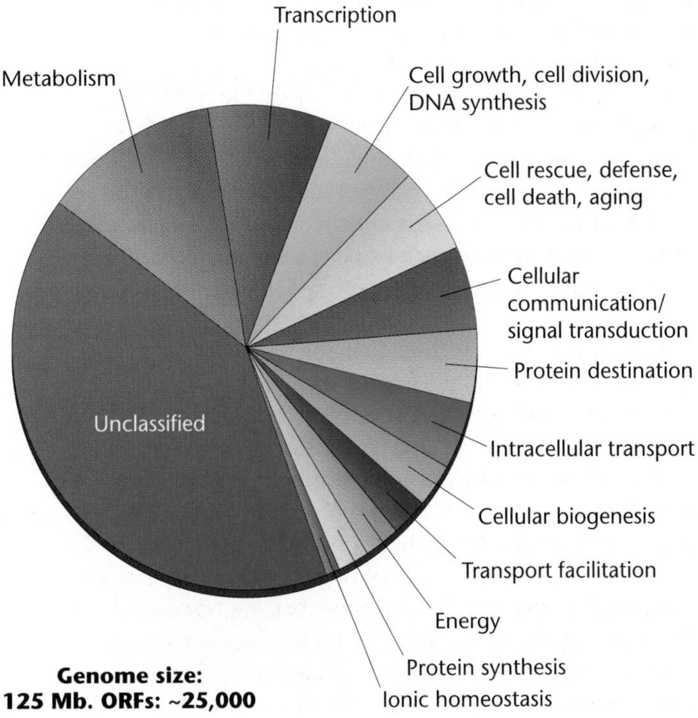

FIGURE 18–7 *Arabidopsis thaliana.* (a) A small plant used as a genetic model for studying genome organization and development of flowering plants. (b) Assignment of *Arabidopsis* genes to functional categories based on homology searches.

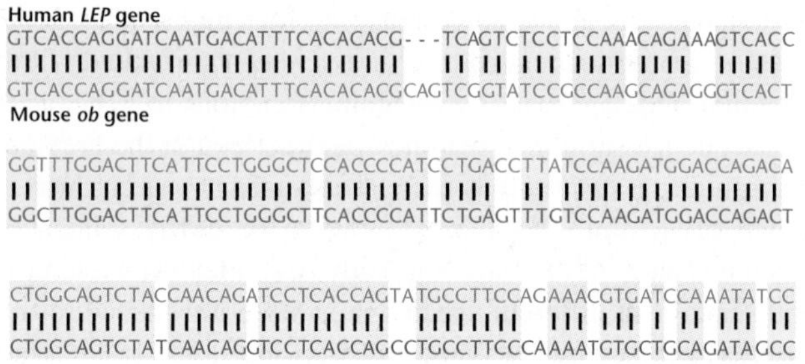

FIGURE 18–8 Comparison of the human *LEP* and mouse *ob/Lep* genes. Partial sequences for these homologs are shown with the human *LEP* gene on top and the mouse *ob/Lep* gene sequence below it. Notice from the number of identical nucleotides, indicated by vertical lines, that the nucleotide sequence for these two genes is very similar. Gaps are indicated by horizontal dashes.

As an interesting aside, the leptin gene (also called *ob,* for obesity, in mice) is highly expressed in fat cells (adipocytes). This gene produces the protein hormone leptin, which targets cells in the brain to suppress appetite. Knockout mice lacking a functional *ob* gene grow dramatically overweight. A similar phenotype has been observed in small numbers of humans with particular mutations in *LEP.* Although it is important to note that weight control is not regulated by a single gene, the discovery of leptin has provided significant insight into lipid metabolism and weight disorders in humans. Further studies on leptin will be important for understanding more about the genetics of weight disorders.

If homologous genes in different species are thought to have descended from a gene in a common ancestor, the genes are known as **orthologs.** In Section 18.7 we will consider the globin gene family. Mouse and human α-globin genes are orthologs evolved from a common ancestor. Homologous genes in the same species are called **paralogs.** The α- and β-globin subunits in humans are paralogs resulting from a gene-duplication event. Paralogs often have similar or identical functions.

Predicting Function from Structural Analysis of Protein Domains and Motifs

When a gene sequence is used to predict a polypeptide sequence, the polypeptide can be analyzed for specific structural domains and motifs. Identification of **protein domains,** such as ion channels, membrane-spanning regions, DNA-binding regions, secretion and export signals, and other structural aspects of a polypeptide that are encoded by a DNA sequence, can in turn be used to predict protein function. Recall from Chapter 15, for example, that the structures of many DNA-binding proteins have characteristic patterns, or **motifs,** such as the helix-turn-helix, leucine zipper, or zinc finger motifs. These motifs can often easily be identified using bioinformatics software, and their presence infers possible functions of a protein.

> ESSENTIAL POINT ■ ■ ■
>
> Functional genomics predicts gene function based on sequence analysis.

18.4 The Human Genome Project Reveals Many Important Aspects of Genome Organization in Humans

Now that you have a general idea of the strategies used for analyzing a genome, let's look at the largest genomics project completed to date. The **Human Genome Project (HGP)** was a coordinated international effort to determine the sequence of the human genome and to identify all the genes it contains. It has produced a plethora of information, much of which is still being analyzed and interpreted. What is clear so far, from all the different kinds of genomes sequenced, is that humans and all other species share a common set of genes essential for cellular function and reproduction, confirming that all living organisms arose from a common ancestor. These evolutionary relationships facilitate the analysis of the human genome and underlie the use of model organisms to study inherited human diseases.

Origins of the Project

The publicly funded Human Genome Project began in 1990 under the direction of James Watson, the co-discoverer of the double-helix structure of DNA. Eventually the public project was led by Dr. Francis Collins, who had previously led a research team involved in identifying the *CFTR* gene as the cause of cystic fibrosis. In the United States, the Collins-led HGP was coordinated by the Department of Energy and the National Center of Human Genome Research, a division of the National Institutes of Health. It established a 15-year plan with a proposed budget of \$3 billion to identify all human genes, originally thought to number between 80,000 and 100,000, to sequence and map them all, and to sequence the approximately 3 billion base pairs thought to comprise the 24 chromosomes (22 autosomes, plus X and Y) in humans. Other primary goals of the HGP included the following:

- To establish functional categories for all human genes
- To analyze genetic variations between humans, including the identification of single-nucleotide polymorphisms (SNPs)
- To map and sequence the genomes of several model organisms used in experimental genetics, including *E. coli, S. cerevisiae, C. elegans, D. melanogaster,* and *M. musculus* (mouse)
- To develop new sequencing technologies, such as high-throughput computer-automated sequencers, in order to facilitate genome analysis
- To disseminate genome information, among both scientists and the general public

Lastly, to deal with the impact that genetic information would have on society, the HGP set up the **ELSI program** (standing for Ethical, Legal, and Social Implications) to consider ethical, legal, and social issues arising from the HGP and to ensure that personal genetic information would be safeguarded and not used in discriminatory ways.

As the HGP grew into an international effort, scientists in 18 countries were involved in the project. Much of the work was carried out by the International Human Genome Sequence Consortium, involving nearly 3000 scientists working at 20 centers in six countries (China, France, Germany, Great Britain, Japan, and the United States).

In 1999, a privately funded human genome project led by J. Craig Venter at Celera Genomics (aptly named from a word meaning "swiftness") was announced. Celera's goal was to use whole-genome shotgun sequencing and computer-automated high-throughput DNA sequencers to sequence the human genome more rapidly than HGP. The public project had proposed using a clone-by-clone approach to sequence the genome. Recall that Venter and colleagues had proven the potential of shotgun sequencing in 1995 when they completed the genome for *H. influenzae.* Celera's announcement set off an intense competition between the two teams, with both aspiring to be first with the human genome sequence. This contest eventually led to the HGP finishing ahead of schedule and under budget after scientists from the public project began to use high-throughput sequencers and whole-genome sequencing strategies as well.

Major Features of the Human Genome

In June 2000, the leaders of the public and private genome projects met at the White House with President Clinton and jointly announced the completion of a draft sequence of the human genome. In February 2001, they each published an analysis covering about 96 percent of the euchromatic region of the genome. The public project sequenced euchromatic portions of the genome 12 times and set a quality control standard of a 0.01 percent error rate for their sequence. Although this error rate may seem very low, it still allows about 600,000 errors in the human genome sequence. Celera sequenced certain areas of the genome more than 35 times when checking sequences for accuracy.

The remaining work of completing the sequence by filling in gaps clustered around centromeres, telomeres, and repetitive sequences, correcting misaligned segments, and re-sequencing portions of the genome to ensure accuracy was completed in 2003. Major features of the human genome are summarized in Table 18.1. As you can see in this table, many unexpected observations have provided us with major new insights. In many ways, the HGP has revealed just how little we know about our genome.

Two of the biggest surprises discovered by the HGP were that less than 2 percent of the genome codes for proteins and that there are only around 20,000 protein-coding genes. Recall that the number of genes had originally been estimated to be about 100,000, based in part on a prediction that human cells produce about 100,000 proteins. At least half of the genes show sequence similarity to genes shared by many other organisms, and as you will learn in Section 18.7, a majority of human genes are similar in sequence to genes from closely related species such as chimpanzees. The exact number of human genes is still not certain. One reason is that it is unclear whether or not many of the presumed genes produce functional proteins. Genome scientists continue to annotate the genome.

The number of genes is much lower than the number of predicted proteins in part because many genes code for multiple proteins through **alternative splicing.** Recall from Chapter 12 that alternative splicing patterns can generate multiple

TABLE 18.1 Major Features of the Human Genome

- The human genome contains 3.1 billion nucleotides, but protein-coding sequences make up only about 2 percent of the genome.
- The genome sequence is ~99.9 percent similar in individuals of all nationalities. Single-nucleotide polymorphisms (SNPs) and copy number variations (CNVs) comprise the majority of sequence differences from person to person.
- At least 50 percent of the genome is derived from transposable elements, such as LINE and *Alu* sequences, and other repetitive DNA sequences.
- The human genome contains approximately 20,000 protein-coding genes, far fewer than the predicted number of 80,000–100,000 genes.
- Many human genes produce more than one protein through alternative splicing, thus enabling human cells to produce a much larger number of proteins (perhaps as many as 200,000) from only ~20,000 genes.
- More than 50 percent of human genes show a high degree of sequence similarity to genes in other organisms; however, the function is unknown for more than 40 percent of the genes identified.
- Genes are not uniformly distributed on the 24 human chromosomes. Gene-rich clusters are separated by gene-poor "deserts" that account for 20 percent of the genome. These deserts correlate with G bands seen in stained chromosomes. Chromosome 19 has the highest gene density, and chromosome 13 and the Y chromosome have the lowest gene densities.
- Chromosome 1 contains the largest number of genes, and the Y chromosome contains the smallest number.
- Human genes are larger and contain more and larger introns than genes in the genomes of invertebrates, such as *Drosophila.* The largest known human gene encodes dystrophin, a muscle protein. This gene, associated in mutant form with muscular dystrophy, is 2.5 Mb in length (Chapter 14), larger than many bacterial chromosomes. Most of this gene is composed of introns.
- The number of introns in human genes ranges from 0 (in histone genes) to 234 (in the titan gene, which encodes a muscle protein).

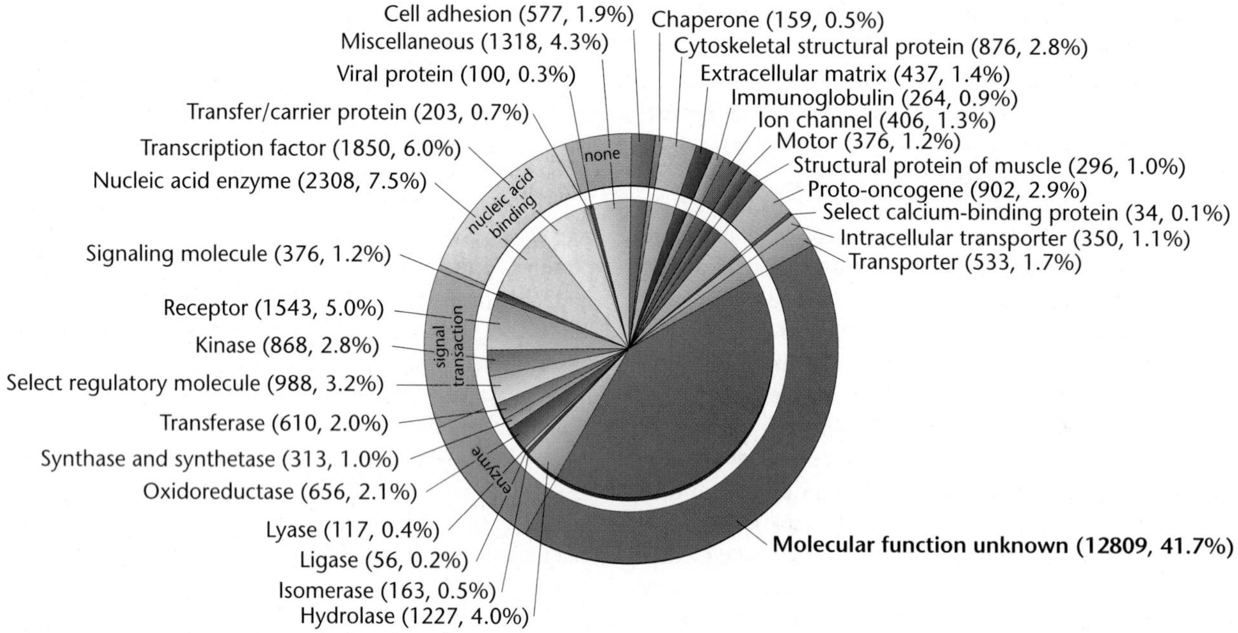

FIGURE 18–9 A preliminary list of the functional categories to which genes in the human genome have been assigned on the basis of similarity to proteins of known function. Among the most common genes are those involved in nucleic acid metabolism (7.5 percent of all genes identified), transcription factors (6.0 percent), receptors (5 percent), hydrolases (4 percent), protein kinases (2.8 percent), and cytoskeletal structural proteins (2.8 percent). A total of 12,809 predicted proteins (41 percent) have unknown functions, indicative of the work that is still needed to fully decipher our genome.

mRNA molecules, and thus multiple proteins, from a single gene, through different combinations of intron–exon splicing arrangements. It is estimated that over 50 percent of human genes undergo alternative splicing to produce multiple transcripts and multiple proteins.

Functional categories have been assigned for many human genes, primarily on the basis of (1) functions determined previously (for example, from recombinant DNA cloning of human genes and known mutations involved in human diseases), (2) comparison to known genes and predicted protein sequences from other species, and (3) predictions based on annotation and analysis of protein functional domains and motifs (Figure 18–9). Although functional categories and assignments continue to be revised, the functions of over 40 percent of human genes remain unknown. Determining human gene functions, deciphering complexities of gene expression regulation and gene interaction, and uncovering the relationships between human genes and phenotypes are among the many challenges for genome scientists.

The HGP has also shown us that in all humans, regardless of racial and ethnic origins, the genomic sequence is approximately 99.9 percent the same. Most genetic differences between individuals result from single-nucleotide polymorphisms (SNPs) and copy number variations (CNVs). Perhaps the most valuable contribution of the HGP will be the identification of disease genes and the development of new treatment strategies as a result. Thus, extensive maps have been developed for genes implicated in human disease conditions. In Chapter 19 we will discuss implications of the HGP for the identification of genes involved in human genetic diseases, and for disease diagnosis, detection, and gene therapy applications.

ESSENTIAL POINT

The Human Genome Project revealed many surprises about human genetics, including gene number, the high degree of DNA sequence similarity between individuals and between humans and other species, and showed that many genes encode multiple proteins.

18.5 The "Omics" Revolution Has Created a New Era of Biological Research Methods

The Human Genome Project and the development of genomics techniques have been largely responsible for launching a new era of biological research—the era of "omics." It seems that every year, more areas of biological research are being described as having an omics connection. Some examples of "omics" are

- proteomics—the analysis of all the proteins in a cell or tissue
- metabolomics—the analysis of proteins and enzymatic pathways involved in cell metabolism
- glycomics—the analysis of the carbohydrates of a cell or tissue
- toxicogenomics—the analysis of the effects of toxic chemicals on genes, including mutations created by toxins and changes in gene expression caused by toxins

- metagenomics—the analysis of genomes of organisms collected from the environment
- pharmacogenomics—the development of customized medicine based on a person's genetic profile for a particular condition
- transcriptomics—the analysis of all expressed genes in a cell or tissue

We will consider several of these genomics disciplines in other parts of this chapter.

As further evidence of the impact of genomics, a new field of nutritional science called nutritional genomics, or **nutrigenomics,** has emerged. Nutrigenomics focuses on understanding the interactions between diet and genes. We have all had routine medical tests for blood pressure, blood sugar levels, and heart rate. Based on these tests, your physician may recommend that you change your diet and exercise more to lose weight, or that you reduce your intake of sodium to help lower your blood pressure. Now several companies claim to provide nutrigenomics tests that analyze your genomes for genes thought to be associated with different medical conditions or aspects of nutrient metabolism. The companies then provide a customized nutrition report, recommending diet changes for improving your health and preventing illness, based on your genes! It remains to be seen whether this approach as currently practiced is of valid scientific or nutritional value.

In yet another example of how genomics has taken over areas of DNA analysis, a number of labs around the world are involved in analyzing "ancient" DNA. These so called **stone-age genomics** studies are generating fascinating data from miniscule amounts of ancient DNA obtained from bone and other tissues that are tens of thousands of years old. Analysis of DNA from a 2400-year-old Egyptian mummy, mammoths, Pleistocene-age cave bears, and Neanderthals are some of the most prominent examples of stone-age genomics. In 2005, researchers from McMaster University in Canada and Pennsylvania State University published about 13 million bp from a 27,000-year-old woolly mammoth. This study revealed a ~98.5 percent sequence identity between mammoths and African elephants. Such work also demonstrates how stable DNA can be under the right conditions, particularly when frozen.

A team of scientists led by Svante Pääbo at the Max Planck Institute for Evolutionary Anthropology in Germany is working to produce a rough draft of the Neanderthal (*Homo neanderthalensis*) genome. In 1997, Pääbo's lab sequenced portions of Neanderthal mitochondrial DNA from a fossil. In late 2006, Pääbo's group along with a number of scientists in the United States reported the first sequence of ~65,000 bp of nuclear DNA isolated from bone of a 38,000-year-old Neanderthal sample from Croatia. Because Neanderthals are close relatives of humans, sequencing the Neanderthal genome is expected to provide an unprecedented opportunity to use comparative genomics (see Section 18.7) to advance our understanding of evolutionary relationships between modern humans and our predecessors.

ESSENTIAL POINT

Genomics has led to a number of other related "omics" disciplines that are rapidly changing how modern biologists study DNA, RNA, and proteins and many aspects of cell function.

18.6 Prokaryotic and Eukaryotic Genomes Display Common Structural and Functional Features and Important Differences

The genomes of over 400 prokaryotic and eukaryotic organisms—including many model organisms and a number of viruses—have been sequenced. Among these organisms are yeast (*Saccharomyces cerevisiae*), bacteria such as *E. coli,* the nematode roundworm (*Caenorhabditis elegans*), the thale cress plant (*Arabidopsis thaliana*), mice (*Mus musculus*), zebrafish (*Danio rerio*), and of course *Drosophila.* In the past few years, genomes for chimpanzees, dogs, chickens, sea urchins, honey bees, pufferfish, rice, and wheat have all been sequenced.

These studies have demonstrated significant differences in genome organization between prokaryotes and eukaryotes but also many similarities between genomes of nearly all species. In this section we provide a basic overview of genome organization in prokaryotes and eukaryotes and discuss interesting aspects of genomes in selected organisms.

Unexpected Features of Prokaryotic Genomes

Since most prokaryotes have small genomes amenable to shotgun cloning and sequencing, many genome projects have focused on prokaryotes, and more than 900 additional projects to sequence prokaryotic genomes are now under way. Many of the prokaryotic genomes already sequenced are from organisms that cause human diseases, such as cholera, tuberculosis, and leprosy. On the basis of genome project results, we can make a number of generalizations about the size of bacterial genomes. Traditionally, the bacterial genome has been thought of as relatively small (less than 5 Mb) and contained within a single circular DNA molecule. *E. coli,* used as the prototypical bacterial model organism in genetics, has a genome with these characteristics. However, the flood of genomic information now available has challenged the validity of this viewpoint for bacteria in general (Table 18.2). Although most prokaryotic genomes are small, their sizes vary across a surprisingly wide range. Gene number in bacterial genomes also demonstrates a wide range, from less than 500 to more than 5000 genes, a tenfold difference. In addition, although many bacteria have a single, circular chromosome, there is substantial variation in chromosome organization and number among bacterial species.

TABLE 18.2 Genome Size and Gene Number in Selected Prokaryotes

	Genome Size (Mb)	Number of Genes
Archaea		
Archaeoglobus fulgidis	2.17	2437
Methanococcus jannaschii	1.66	1783
Thermoplasma acidophilium	1.56	1509
Eubacteria		
Escherichia coli	4.64	4289
Bacillus subtilis	4.21	4779
Haemophilus influenzae	1.83	1738
Aquifex aeolicus	1.55	1749
Rickettsia prowazekii	1.11	834
Mycoplasma pneumoniae	0.82	680
Mycoplasma genitalium	0.58	483

Although most prokaryotic genomes sequenced to date are single, circular DNA molecules, an increasing number of genomes composed of linear DNA molecules are being identified, including the genome of *Borrelia burgdorferi,* the organism that causes Lyme disease. Perhaps more important, genome data are blurring the distinction between plasmids and bacterial chromosomes, thus redefining our view of the bacterial genome as a single DNA molecule (Table 18.3). **Plasmids** are small, extrachromosomal, circular DNA molecules that self-replicate. Because plasmids often contain nonessential genes and can be transferred from one cell to another, and because the same plasmid is often present in bacteria of different species, plasmid genes are usually not considered to be part of a bacterial genome. However, *B. burgdorferi* carries approximately 17 plasmids, containing a total of at least 430 genes. Some of these genes are essential to the bacterium, including genes for purine biosynthesis and membrane proteins, which in most species are carried on the bacterial chromosome.

Other bacteria that have genomes with two or more chromosomes include *Agrobacterium tumefaciens, Deinococcus radiodurans,* and *Rhodobacter sphaeroides* (Table 18.3). The finding that some bacterial species have multiple chromosomes raises questions about how replication and segregation of their chromosomes are coordinated during cell division, and about what undiscovered mechanisms of gene regulation may exist in bacteria. Answers to these and other questions raised by genomic discoveries will redefine some ideas about prokaryotic genomes, the nature of plasmids, and the differences between plasmids and chromosomes, and may provide clues about the evolution of multichromosome eukaryotic genomes.

The genome of *E. coli* strain K12 was sequenced using the shotgun method in 1997 and was the second prokaryotic genome to be sequenced. *E. coli* has a genome of 4.1 Mb, and annotation shows that it contains 4289 ORFs organized as a single circular chromosome. ORFs occupy almost 88 percent of the genome, regulatory sequences about 11 percent, and repetitive sequences only about 0.7 percent. Even though the genetics of *E. coli* had been intensively studied for almost 50 years using classic genetic methods, almost 38 percent of the annotated ORFs were for genes with no known functions.

To reach the goal of assigning functions to all the genes in *E. coli,* and to understand how these genes and their products interact in the biology of the organism, researchers are creating (1) knockout mutations for every gene, (2) a clone for every ORF, (3) a gene expression database for patterns of expression under a variety of physiological conditions, and (4) a three-dimensional model of every protein encoded in the genome. The current classifications of *E. coli* genes are shown in Figure 18–10. Genes with no known homology and no known functions now comprise 19 percent of the 4289 ORFs in the genome.

We can make two generalizations about the organization of protein-coding genes in bacteria. First, gene density is very high, averaging about one gene per kilobase of DNA. For example, *E. coli* has a 4.6-Mb genome containing 4289 protein-coding genes. *Mycoplasma genitalium* has a small genome (0.58 Mb), with 483 genes, and its gene density is also close to one gene per 1000 base pairs. This close packing of genes in prokaryotic genomes means that a very high proportion of the DNA (approximately 85 to 90 percent) serves as coding DNA. Typically, only a small amount of a bacterial genome is noncoding DNA, often in the form of regulatory sequences or

TABLE 18.3 Bacterial Genome Organization

Bacterial Species	Chromosome Number and Organization	Plasmid Number
Agrobacterium tumefaciens	one linear + one circular	two circular
Bacillus subtilis	one circular	
Bacillus thuringiensis	one circular	six
Borrelia burgdorferi	one linear	>17 circular + linear
Buchnera sp.	one circular	two circular
Deinococcus radiodurans	two circular	two circular
Escherichia coli	one circular	
Rhodobacter sphaeroides	two circular	five circular
Sinorhizobium meliloti	one circular	two megaplasmids
Vibrio cholerae	two circular	

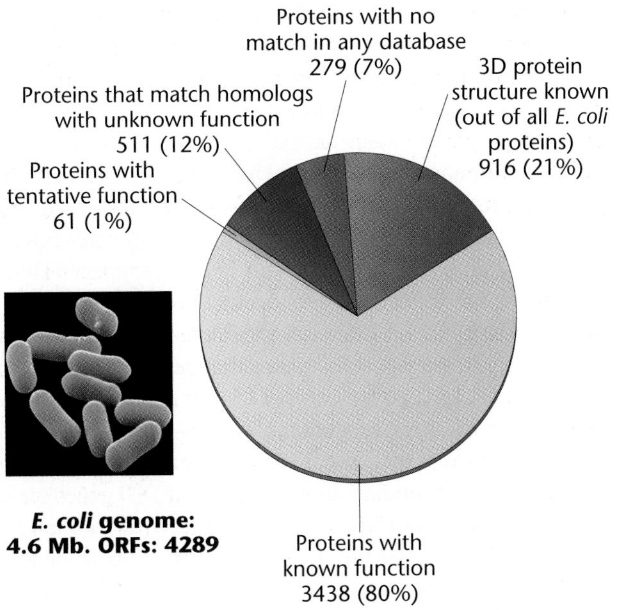

FIGURE 18–10 Status of gene functional assignments for the *E. coli* genome. The total exceeds 100 percent because of overlap between proteins with known functions and those whose 3D structure is known.

of transposable elements that can move from one place to another in the genome.

A second generalization we can make is that bacterial genomes contain operons (recall from Chapter 15 that operons contain multiple genes functioning as a transcriptional unit whose protein products are part of a common biochemical pathway). In *E. coli,* 27 percent of all genes are contained in operons (almost 600 operons). In other bacterial genomes, the organization of genes into transcriptional units is challenging our ideas about the nature of operons.

Organizational Patterns of Eukaryotic Genomes

We now turn our attention to eukaryotic genomes. The human genome has been the most extensively studied eukaryotic genome to date, and we discussed many of its structural features in the previous section when we considered major findings of the Human Genome Project. Here we provide an overview of organizational patterns in eukaryotic genomes and discuss interesting aspects of the genomes for yeast and *Arabidopsis,* two important eukaryotic model organisms. Our discussion on eukaryotic genomes continues in Section 18.7, where we review genomic data for the dog, chimpanzee, and Rhesus monkey.

Although the basic features of eukaryotic genomes are similar in different species, genome size is highly variable (Table 18.4). Genome sizes range from about 10 Mb in fungi to over 100,000 Mb in some flowering plants (a ten-thousandfold range); the number of chromosomes per genome ranges from two into the hundreds (about a hundredfold range), but the number of genes varies much less dramatically than either genome size or chromosome number.

Eukaryotic genomes have several features not found in prokaryotes:

- **Gene density.** In prokaryotes, gene density is close to 1 gene per kilobase. In eukaryotic genomes, there is a wide range of gene density. In yeast, there is about 1 gene/2 kb, in *Drosophila,* about 1 gene/13 kb, and in humans, gene density varies greatly from chromosome to chromosome. Human chromosome 22 has about 1 gene/64 kb, while on chromosome 13 there is 1 gene/155 kb of DNA.

TABLE 18.4 Comparison of Selected Genomes

Organism (Scientific Name)	Approximate Size of Genome (Date Completed)	Number of Genes	Approximate Percentage of Genes Shared with Humans
Bacterium (*Escherichia coli*)	4.1 Mb (1997)	4,403	not determined
Chicken (*Gallus gallus*)	1 Gb (2004)	~20,000–23,000	60%
Dog (*Canis familiaris*)	2.5 Gb (2003)	~18,400	75%
Chimpanzee (*Pan troglodytes*)	~3 Gb (2005)	~20,000–24,000	98%
Fruit fly (*Drosophila melanogaster*)	165 Mb (2000)	~13,600	50%
Human (*Homo sapiens*)	~2.9 Gb (2004)	~20,000	100%
Mouse (*Mus musculus*)	~2.5 Gb (2002)	~30,000	80%
Rat (*Rattus norvegicus*)	~2.75 Gb (2004)	~22,000	80%
Rhesus macaque (*Macaca mulatta*)	2.87 Gb (2007)	~20,000	93%
Rice (*Oryza sativa*)	389 Mb (2005)	~41,000	not determined
Roundworm (*Caenorhabditis elegans*)	97 Mb (1998)	19,099	40%
Sea urchin (*Strongylocentrotus purpuratus*)	814 Mb (2006)	~23,500	60%
Thale cress (plant) (*Arabidopsis thaliana*)	140 Mb (2000)	~27,500	not determined
Yeast (*Saccharomyces cerevisiae*)	12 Mb (1996)	~ 5,700	30%

Adapted from Palladino, M. A. *Understanding the Human Genome Project,* 2nd ed. Benjamin Cummings, 2006.

NOTE: Billion bp (gigabase, Gb).

- **Introns.** Most eukaryotic genes contain introns. There is wide variation among genomes in the number of introns they contain and also wide variation from gene to gene. The entire yeast genome has only 239 introns, whereas just a single gene in the human genome can contain more than 100 introns. Regarding intron size, generally the size in eukaryotes is correlated with genome size. Smaller genomes have smaller average introns, and larger genomes have larger average intron sizes.
- **Repetitive sequences.** The presence of introns and the existence of repetitive sequences are two major reasons for the wide range of genome sizes in eukaryotes. In some plants, such as maize, repetitive sequences are the dominant feature of the genome. The maize genome has about 2500 Mb of DNA, and more than two-thirds of that genome is composed of repetitive DNA. In the human, as discussed previously, about half of the genome is repetitive DNA.

The Yeast Genome

The yeast (*Saccharomyces cerevisiae*) genome was the first eukaryotic genome sequenced. As outlined in Chapter 13, yeast can be grown and manipulated as easily as some prokaryotes. That feature, along with yeast's highly developed genetic map and a large collection of yeast mutants, made it an excellent candidate for genome sequencing.

The yeast genome was sequenced chromosome by chromosome using a map-based approach. The genome contains 12.1 Mb of DNA, distributed over 16 chromosomes. In the years of analysis following publication of the yeast genome, the estimated number of ORFs has changed several times. Originally thought to have about 6200 genes, yeast is currently perceived to have ~5700 genes; the number will probably change again as more evidence is gathered.

Plant Genomes

To geneticists, the flowering plant *Arabidopsis thaliana* is the fruit fly of the plant world. *Arabidopsis* is small (several hundred can be grown in an area the size of this page), has a short generation time, and has a relatively small genome. Because the plant kingdom evolved independently from the animal kingdom, the analysis of *Arabidopsis* can provide insight into the ways in which evolutionary and coevolutionary adaptations have shaped genomes in both kingdoms. The *Arabidopsis* genome was sequenced using map-based sequencing. Its 140-Mb genome is distributed in five chromosomes containing an estimated total of 27,500 genes, with a gene density of 1 gene per 5 kb. At least half the genes in the *Arabidopsis* genome are identical to or closely related to genes found in bacteria and humans, but it also contains genes encoding dozens of protein families that may be unique to plants.

Both genome duplications and gene duplications have played a large role in the evolution of the *Arabidopsis* genome. Many of the 25,000 genes are duplicated, and there are fewer than 15,000 different genes. For example, chromosomes 2 and 4 of *Arabidopsis* contain many tandem gene duplications (239 duplications on chromosome 2, involving 539 genes) as well as larger duplications involving four blocks of DNA sequences, spanning 2.5 Mb. There is some interchromosomal gene duplication, too: genes on chromosome 4 are also present on chromosome 5.

Many crop plants, such as maize, rice, and barley, have much larger genomes than *Arabidopsis,* but some have about the same number of genes. In contrast to *Arabidopsis,* genes in these large-genome plants are clustered in stretches of DNA separated by long stretches of intergenic spacer DNA. Collectively, these gene clusters occupy only about 12 to 24 percent of the genome. In maize, the intergenic DNA is composed mainly of transposons (genes that can shift location in the genome.

Sequencing of the rice genome was completed in 2005. Analysis shows that this cereal crop's genome contains about 389 Mb of DNA, encoding approximately 41,000 genes on 12 chromosomes. About 15 percent of all rice genes are found in duplicated segments of the genome. Close to 90 percent of the genes in *Arabidopsis* are found in rice, but only 70 percent of genes in rice are found in *Arabidopsis,* indicating that in addition to genes found in other flowering plants, cereal crops may also contain unique gene sets. Identification of these genes and their functions in rice and other cereal plants will be critical in helping improve crop yields to feed the Earth's growing population.

Recently, the first genome for a tree, the black cottonwood, a type of poplar, was sequenced. The poplar's 45,555 genes is the highest number found in a genome to date. Scientists anticipate using cottonwood genome data to help the forestry industry make better products, such as biofuels or genetically engineered poplars capable of removing high levels of carbon dioxide from the atmosphere.

ESSENTIAL POINT

Genomic analysis of model prokaryotes and eukaryotes has revealed similarities and important fundamental differences in genome size, gene number, and genome organization.

18.7 Comparative Genomics Analyzes and Compares Genomes from Different Organisms

Analysis of the growing number of genome sequences confirms that all living organisms are related and descended from a common ancestor. Similar gene sets are used in all organisms for basic cellular functions, such as DNA replication, transcription, and translation. These genetic relationships are the rationale for the use of model organisms to study inherited human disorders, the effects of the environment on genes, and interactions of genes in complex diseases, such as cardiovascular disease, diabetes, neurodegenerative conditions, and behavioral disorders.

Comparative genomics compares the genomes of different organisms in order to answer questions about genetics and other aspects of biology. It is a field with many research and practical applications, including gene discovery and the development of model organisms to study human diseases. It also incorporates the study of gene and genome evolution and the relationship between organisms and their environment. Comparative genomics uses a wide range of techniques and resources, such as the construction and use of nucleotide and protein databases containing nucleic acid and amino acid sequences, fluorescent *in situ* hybridization (FISH), and the creation of gene knockout animals. Comparative genomics can reveal genetic differences and similarities between organisms to provide insight into how those differences contribute to differences in phenotype, life cycle, or other attributes, and to ascertain the evolutionary history of those genetic differences.

In this section, we discuss aspects of comparative genomics in model systems and consider how comparative genomics is contributing to identification of evolutionarily conserved gene families. As mentioned earlier, the Human Genome Project sequenced genomes from a number of model nonhuman organisms too, including *E. coli, Arabidopsis thaliana, Saccharomyces cerevisiae, Drosophila melanogaster,* the nematode roundworm *Caenorhabditis elegans,* and the mouse *Mus musculus.* Complete genome sequences of such organisms have been invaluable for comparative genomics studies of gene function in these organisms and in humans. As shown in Table 18.4, the number of genes humans share with other species is very high, ranging from about 30 percent of the genes in yeast to ~80 percent in mice and ~98 percent in chimpanzees. The human genome even contains around 100 genes that are also present in many bacteria. Comparative genomics has also shown us that many mutated genes involved in human disease are also present in model organisms. For instance, approximately 60 percent of genes mutated in nearly 300 human diseases are also found in *Drosophila.* These include genes involved in prostate, colon, and pancreatic cancers; cardiovascular disease; cystic fibrosis; and several other conditions.

The Dog as a Model Organism

Recently, the genome for "man's best friend" was completed, and it revealed that we share about 75 percent of our genes with dogs. A rough draft of the dog (*Canis familiaris*) genome was published in 2003, and a more thorough analysis was completed in 2005, providing one of the most useful models with which to study our own genome. Dogs have a genome that is similar in size to the human genome: about 2.5 billion base pairs with an estimated 18,400 genes (Figure 18–11).

The dog offers several advantages for studying heritable human diseases. Dogs share many genetic disorders with humans, including over 400 single-gene disorders, sex-chromosome aneuploidies, multifactorial diseases (such as epilepsy), behavioral conditions (such as obsessive-compulsive disorder), and genetic predispositions to cancer, blindness, heart disease, and deafness. About 90 conserved blocks of the dog genome can be mapped to human chromosomes by comparative FISH studies. These blocks contain DNA sequences with a high degree of similarity between the dog and the human genomes, reflecting the evolutionary relationship between dogs and our species.

(a)

Dog Genome

Size: 2.47 Gb

Chromosomes: 39

ORFs: 19,300

Repetitive DNA: 31%

(b)

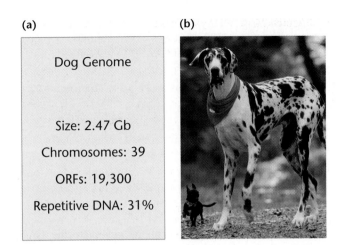

FIGURE 18–11 (a) Basic characteristics of the dog genome. (b) A Chihuahua alongside a Great Dane that is over 50 times its mass. The genetic basis for varied phenotypes in dogs, such as body size, is being revealed by analyzing the dog genome.

Genomic analysis indicates that the molecular causes of at least 60 percent of inherited diseases in dogs, such as point mutations and deletions, are similar or identical to those found in humans. In addition, at least 50 percent of the genetic diseases in dogs are breed-specific, so that the mutant allele segregates in relatively homogeneous genetic backgrounds. Dog breeds resemble isolated human populations in having a small number of founders and a long period of relative genetic isolation. These properties make individual dog breeds useful as models of human genetic disorders.

In addition, differences in biology and behavior among dog breeds are well documented. Domestic dogs show greater variation in body size than all other living terrestrial vertebrates. The size difference between a Chihuahua and a Great Dane is an excellent example (Figure 18–11). Mapping DNA sequence differences (polymorphisms) among breeds may help identify genes that contribute to both physiological and behavioral differences. For instance, in 2007, genome-wide analysis of different large and small dog breeds revealed a locus on chromosome 15 where a single-nucleotide polymorphism in the insulin-like growth factor 1 gene (*Igf1*) is common in all small breeds of dogs but virtually absent from large dogs. It is well known that *Igf1* plays important roles in growth-hormone-regulated increases in muscle mass and bone growth during adolescence in humans. This study provides very strong evidence that mutation of *Igf1* is a primary determinant of body size in small dogs.

Dog breeders are now using genetic tests to screen dogs for inherited disease conditions, for coat color in Labrador retrievers and poodles, and for fur length in Mastiffs. Undoubtedly, we can expect many more genetic tests for dogs in the near future, including DNA analysis for size, type of tail, speed, sense of smell, and other traits deemed important by breeders and owners.

The Chimpanzee Genome

Although the chimpanzee (*Pan troglodytes*) genome was not part of the HGP, its nucleotide sequence was completed in 2004. Overall, the chimp and human sequences differ by less

TABLE 18.5 Comparisons Between Human Chromosome 21 and Chimpanzee Chromosome 22

	Human 21	Chimpanzee 22
Size (bp)	33,127,944	32,799,845
G + C Content	40.94	41.01
CpG Islands	950	885
SINEs (*Alu* elements)	15,137	15,048
Genes	284	272
Pseudogenes	98	89

than 2 percent, and 98 percent of the genes are the same. Comparisons between these genomes offer some interesting insights into what makes some primates humans and others chimpanzees.

The speciation events that separated humans and chimpanzees occurred less than 6.3 million years ago (mya). Recent evidence from genomic analysis indicates that these species initially diverged but then exchanged genes again before separating completely. Their separate evolution after this point is exhibited in such differences as are seen between the sequence of chimpanzee chromosome 22 and its human ortholog, chromosome 21 (Table 18.5; chimps have 48 chromosomes and humans have 46, so the numbering is different). These chromosomes have accumulated nucleotide substitutions that total 1.44 percent of the sequence. The most surprising difference is the discovery of 68,000 nucleotide insertions or deletions, collectively called **indels,** in the chimp and human chromosomes, a frequency of 1 indel every 470 bases. Many of these are *Alu* insertions in human chromosome 21. Although the overall difference in the nucleotide sequence is small, there are significant differences in the encoded genes. Only 17 percent of the genes analyzed encode identical proteins in both chromosomes; the other 83 percent encode genes with one or more amino acid differences.

Differences in the time and place of gene expression also play a major role in differentiating the two primates. Using DNA microarrays (discussed in Section 18.9), researchers compared expression patterns of 202 genes in human and chimp cells from brain and liver. They found more species-specific differences in expression of brain genes than liver genes. To further examine these differences, Svante Pääbo and colleagues compared expression of 10,000 genes in human and chimpanzee brains and found that 10 percent of genes examined differ in expression in one or more regions of the brain. More importantly, these differences are associated with genes in regions of the human genome that have been duplicated subsequent to the divergence of chimps and humans. This finding indicates that genome evolution, speciation, and gene expression are interconnected. Further work on these segmental duplications and the genes they contain may identify genes that help make us human.

The Rhesus Monkey Genome

The Rhesus macaque monkey (*Macaca mulatta*), another primate, has served as one of the most important model organisms in biomedical research. Macaques have played central roles in our understanding of cardiovascular disease, aging, diabetes, cancer, depression, osteoporosis, and many other aspects of human health. They have been essential for research on AIDS vaccines and for the development of polio vaccines.

The macaque's genome is the first monkey genome to have been sequenced. A main reason geneticists are so excited about the completion of this sequencing project is that macaques provide a more distant evolutionary window that is ideally suited for comparing and analyzing human and chimpanzee genomes. As we discussed in the preceding section, humans and chimpanzees shared a common ancestor approximately 6 mya. But macaques split from the ape lineage that led to chimpanzees and humans about 25 mya. The macaque and human genomes have thus diverged farther from one another, as evidenced by the ~93 percent sequence identity between humans and macaques compared to the ~98 percent sequence identity shared by humans and chimpanzees.

The macaque genome was published in 2007, and it was no surprise to learn that it consists of 2.87 billion bp (similar to the size of the human genome) contained in 22 chromosomes (20 autosomes, an X, and a Y) with ~20,000 protein-coding genes. Although comparative analyses of this genome are ongoing, a number of interesting features have been revealed so far. As in humans, about 50 percent of the genome consists of repeat elements (transposons, LINEs, SINEs). Gene duplications and gene families are abundant, including cancer gene families found in humans.

Interesting surprises have also been observed. For instance, the human genetic disorder phenylketonuria (PKU), is an autosomal recessive inherited condition in which individuals cannot metabolize the amino acid phenylalanine due to mutation of the phenylalanine hydroxylase (*PAH*) gene. The histidine substitution encoded by a mutation in the *PAH* gene of humans with PKU appears as the wild-type amino acid in the protein from healthy macaques. Further analysis of the macaque genome and comparison to the human and chimpanzee genomes will be invaluable for geneticists studying genetic variations that played a role in primate evolution.

Evolution and Function of Multigene Families

Comparative genomics has also proven to be valuable for identifying members of **multigene families,** groups of genes that share similar but not identical DNA sequences through duplication and descent from a single ancestral gene. Their gene products frequently have similar functions, and the genes are often, but not always, found at a single chromosomal locus. A group of related multigene families is called a **superfamily.** Sequence data from genome projects is providing evidence that multigene families are present in many, if not all, genomes. To demonstrate how analysis of multigene families provides insight into eukaryotic genome evolution and function, we will now examine the globin gene superfamily, whose members encode very similar but not identical polypeptide chains with closely related functions. Other well-characterized gene superfamilies include the histone, tubulin, actin, and immunoglobulin (antibody) gene superfamilies.

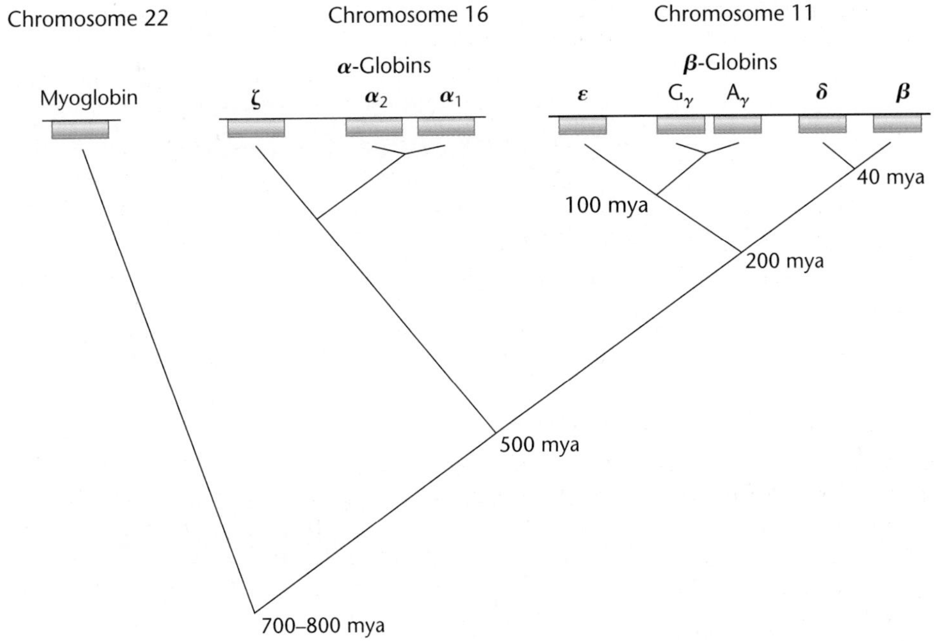

FIGURE 18–12 The evolutionary history of the globin gene superfamily. A duplication event in an ancestral gene gave rise to two lineages about 700 to 800 million years ago (mya). One line led to the myoglobin gene, which is located on chromosome 22 in humans; the other underwent a second duplication event about 500 mya, giving rise to the ancestors of the α-globin and β-globin gene subfamilies. Duplications beginning about 200 mya formed the β-globin gene subfamilies. In humans, the α-globin genes are located on chromosome 16, and the β-globin genes are on chromosome 11.

Recall that paralogs, which we defined in Section 18.3, are homologous genes present in the same single organism, believed to have evolved by gene duplication. One of the best-studied examples of gene family evolution is the **globin gene superfamily** (Figure 18–12). The globin genes that encode the polypeptides in hemoglobin molecules are a paralogous multigene superfamily that arose by duplication and dispersal to occupy different chromosomal sites. In this family, an ancestral gene encoding an oxygen transport protein was duplicated about 800 mya, producing two sister genes, one of which evolved into the modern-day myoglobin gene. **Myoglobin** is an oxygen-carrying protein found in muscle. The other gene underwent further duplication and divergence about 500 mya and formed prototypes of the α-globin and β-globin genes. These genes encode proteins found in **hemoglobin,** the oxygen-carrying molecule in red blood cells. Additional duplications within these genes occurred within the last 200 million years. Events subsequent to each duplication dispersed these gene subfamilies to different chromosomes, and in the human genome, each now resides on a separate chromosome. Similar patterns of evolution are observed in other gene families, including the trypsin–chymotrypsin family of proteases, the homeotic selector genes of animals, and the rhodopsin family of visual pigments.

Adult hemoglobin is a tetramer containing two α- and two β-polypeptides (refer to Figure 13–19 for the structure of hemoglobin). Each polypeptide incorporates a heme group that reversibly binds oxygen. The α-globin gene cluster on chromosome 16 and the β-globin gene cluster on chromosome 11 share nucleotide- and amino acid–sequence similarity (Figure 18–13), but the highest degree of sequence similarity is found within subfamilies.

The ***α*-globin** gene subfamily [Figure 18–14(a), p. 394] contains three genes: the ζ (zeta) gene, expressed only in early embryogenesis, and two copies of the α gene, expressed during the fetal (α_1) and adult stages (α_2). In addition, the cluster contains two pseudogenes (similar to ζ and α_1), which in this family are designated by the prefix ψ (psi) followed by the symbol of the gene they most resemble. Thus, the designation $\psi\alpha_1$ indicates a pseudogene of the fetal α_1 gene.

The organization of the α-globin subfamily members and the locations of their introns and exons demonstrate several characteristic features [Figure 18–14(a)]. First, as is common in eukaryotes, the DNA encoding the three functional α genes

α-globin V – L S P A D K T N V K A A W G K V G A H A G E Y G A E A L E R M F L S F P T T K T Y F P H F – D L S H
β-globin V H L T P E E K S A V T A L W G K V – – N V D E V G G E A L G R L L V V Y P W T Q R F F E S F G D L S T

α-globin – – – G S A Q V K G H G K K V A D A L T N A V A H V D D M P N A L S A L S D L H A H K L R V D P V N
β-globin A V M G N P K V K A H G K K V L G A F S D G L A H L D N L K G T F A T L S E L H C D K L H V D P E N

α-globin L L S H C L L V T L A A H L P A E F T P A V H A S L D K F L A S V S T V L T S K Y R 141 amino acids
β-globin L L G N V L V C V L A H H F G K E F T P P V Q A A Y Q K V V A G V A N A L A H K Y H 146 amino acids

FIGURE 18–13 The amino acid sequences of the α- and β-globin proteins, depicted using the single-letter abbreviations for the amino acids (see Figure 13–15). Shaded areas indicate identical amino acids. The two proteins are slightly different in length. α-globin contains 141 amino acids, while β-globin is 146 amino acids long. Gaps in the two sequences, representing areas that do not align, are indicated by horizontal dashes (–).

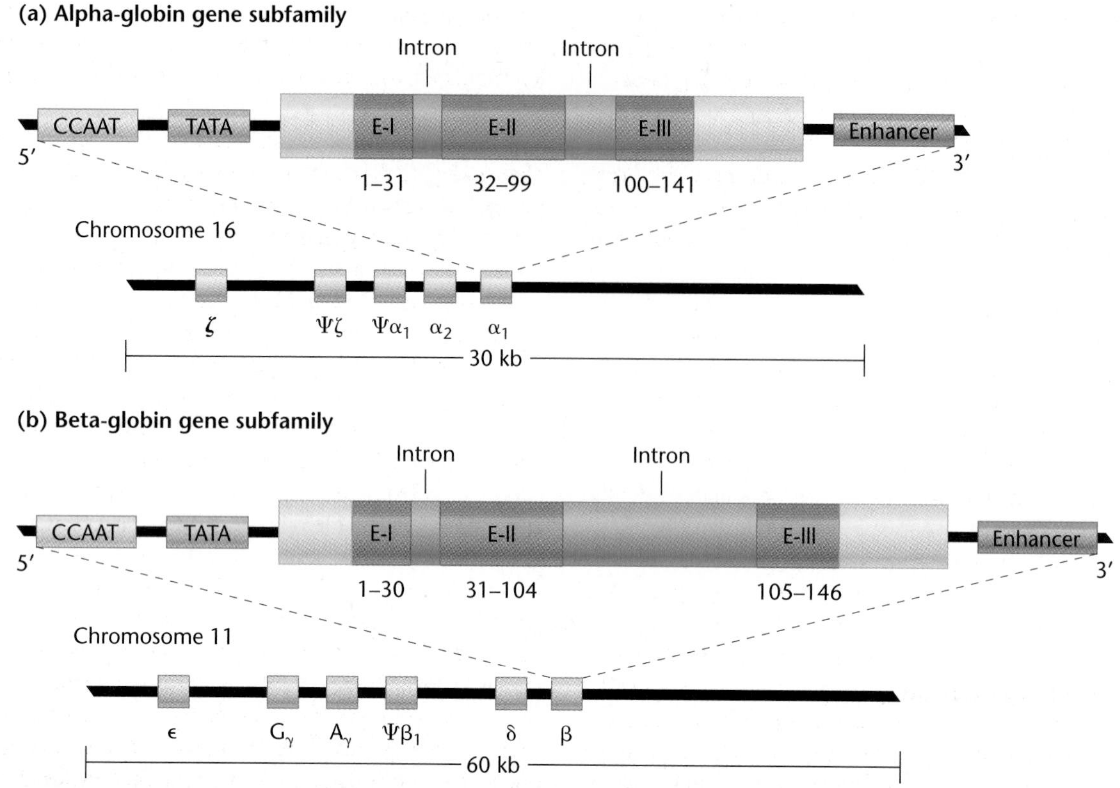

FIGURE 18–14 Organization of (a) the α-globin gene subfamily on chromosome 16 and (b) the β-globin gene subfamily on chromosome 11. Also shown in each case is the internal organization of the α_1 gene and the β gene, respectively. Both of these genes contain three exons (E-I, E-II, E-III) and two introns. The numbers below the exons indicate the location of amino acids in the gene product encoded by each exon.

occupies only a small portion of the 30-kb region containing the subfamily. Most of the DNA in this region is intergenic spacer DNA. Second, each functional gene in this subfamily contains two introns at precisely the same positions. Third, the nucleotide sequences within corresponding exons are nearly identical in the ζ and α genes. Each of these genes encodes a polypeptide chain of 141 amino acids. However, their intron sequences are highly divergent, even though the introns are about the same size. Note that much of the nucleotide sequence of each gene is contained in these noncoding introns.

The human **β-globin** gene cluster contains five genes spaced over 60 kb of DNA [Figure 18–14(b)]. In this and the α-globin gene subfamily, the order of genes on the chromosome parallels their order of expression during development. Three of the five genes are expressed before birth. The ε (epsilon) gene is expressed only during embryogenesis, while the two nearly identical γ genes (G_γ and A_γ) are expressed only during fetal development. The polypeptide products of the two γ genes differ only by a single amino acid. The two remaining genes, δ and β, are expressed after birth and throughout life. A single pseudogene, $\psi\beta_1$, is present in this subfamily. All five functional genes in this cluster encode proteins with 146 amino acids and have two similar-sized introns at exactly the same positions. The second intron in the β-globin subfamily is significantly larger than its counterpart in the functional α-globin subfamily. These features reflect the evolutionary history of each subfamily and the events such as gene duplication, nucleotide substitution, and chromosome translocations that produced the present-day globin superfamily.

ESSENTIAL POINT

Studies in comparative genomics are revealing fascinating similarities and differences in genomes from different organisms, including the identification and analysis of gene families.

18.8 Metagenomics Applies Genomics Techniques to Environmental Samples

Metagenomics, also called **environmental genomics,** is the use of whole-genome shotgun approaches to sequence genomes from entire communities of microbes in environmental samples of water, air, and soil. Oceans, glaciers, deserts, and virtually every other environment on Earth are being sampled for metagenomics projects. Human genome pioneer J. Craig Venter left Celera in 2003 to form the J. Craig Venter Institute, and his group has played a central role in developing metagenomics as an emerging area of genomics research.

One of the institute's major initiatives has been a global expedition to sample marine and terrestrial microorganisms from around the world and to sequence their genomes. Through one expedition of the Institute, called the *Sorcerer II* Global Ocean Sampling (GOS) Expedition, Venter and his researchers traveled the globe by yacht, in a sailing voyage described as a modern-day version of Charles Darwin's famous voyage on the *H.M.S. Beagle.* The Discovery Channel has chronicled this journey.

One of the key benefits of metagenomics is its potential for teaching us more about millions of species of bacteria, of which only a few thousand have been well characterized. Many new viruses, particularly bacteriophages, are also identified through metagenomics studies of water samples. Metagenomics is providing important new information about genetic diversity in microbes that is key to understanding complex interactions between microbial communities and their environment, as well as allowing phylogenetic classification of newly identified microbes. Metagenomics also has great potential for identifying genes with novel functions, some of which have potentially valuable applications in medicine and biotechnology.

The general method used in metagenomics to sequence genomes for all microbes in a given environment involves isolating DNA directly from an environmental sample without requiring cultures of the microbes or viruses. Such an approach is necessary because, often, it is difficult to replicate the complex array of growth conditions needed by the microbes if they are to survive in culture. For the *Sorcerer II* GOS project, samples of water from different layers in the water column were passed through high-density filters of various sizes to capture the microbes. DNA was then isolated from the microbes and subjected to shotgun sequencing and genome assembly. High-throughput sequencers on board the yacht operated nearly around the clock. One of the earliest projects by this group sequenced bacterial genomes from the Sargasso Sea off Bermuda. This project yielded over 1.2 million novel DNA sequences from 1800 microbial species, including 148 previously unknown bacterial species, and identified hundreds of photoreceptor genes. Many aquatic microorganisms rely on photoreceptors for capturing light energy to power photosynthesis. Scientists are interested in learning more about photoreceptors to help develop ways in which photosynthesis may be used to produce hydrogen as a fuel source. Medical researchers are also very interested in photoreceptors because in humans and many other species, photoreceptors in the retina of the eye are key proteins that detect light energy and transduce electrical signals that the brain eventually interprets to create visual images.

By early 2007, the GOS database contained approximately 6 billion bp of DNA from more than 400 uncharacterized microbial species! These sequences included 7.7 million previously uncharacterized sequences, encoding more than 6 million different potential proteins. This is almost twice the total number of previously characterized proteins in all other known databases worldwide. Figure 18–15(a) shows the kingdom assignments for predicted protein sequences in publicly available databases worldwide, such as the NCBI-nonredundant protein database (NCBInr), which accesses GenBank, Ensembl, and other well-known databases. Eukaryotic sequences comprise the majority (63 percent) of predicted proteins in these databases. Reviewing the kingdom assignments of approximately 6 million predicted proteins in the GOS data set shows that, in contrast, the largest majority (90.8 percent) of sequences in this database are from the bacterial kingdom [Figure 18–15(b)].

The GOS Expedition also examined protein families corresponding to the predicted proteins encoded by the genome sequences in the GOS database: 17,067 families were medium (between 20 and 200 proteins) and large-sized (>200 proteins) clusters. These data demonstrate the value of the GOS Expedition and of metagenomics for identifying novel microbial genes and potential proteins.

ESSENTIAL POINT

Metagenomics, or environmental genomics, sequences genomes of microorganisms in environmental samples, often identifying new sequences that encode proteins with novel functions.

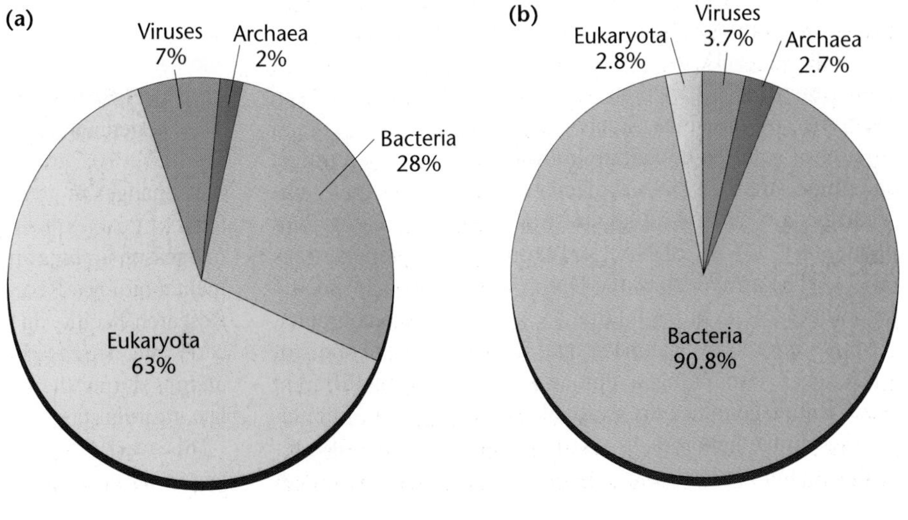

FIGURE 18–15 (a) Kingdom identifications for predicted proteins in NCBInr, NCBI Prokaryotic Genomes, The Institute for Genomics Research Gene Indices, and Ensembl databases. Notice that the publicly available databases of sequenced genomes and the predicted proteins they encode are dominated by eukaryotic sequences. (b) Kingdom identifications for novel predicted proteins in the Global Ocean Sampling (GOS) database. Bacterial sequences dominate this database, demonstrating the value of metagenomics for revealing new information about microbial genomes and microbial communities.

18.9 Transcriptome Analysis Reveals Profiles of Expressed Genes in Cells and Tissues

Sequencing a genome is a major endeavor, and even once any genome has been sequenced and annotated, a formidable challenge still remains: that of understanding genome function by analyzing the genes it contains and the ways the genes expressed by the genome are regulated. **Transcriptome analysis,** also called **transcriptomics** or **global analysis of gene expression,** studies the expression of genes by a genome both qualitatively—by identifying which genes are expressed and which genes are not expressed—and quantitatively—by measuring varying levels of expression for different genes.

As we know, all cells of an organism possess the same genome, but in any cell or tissue type, certain genes will be highly expressed, others expressed at low levels, and some not expressed at all. Transcriptome analysis provides gene expression profiles that for the same genome may vary from cell to cell or from tissue type to tissue type. Identifying genes expressed by a genome is essential for understanding how the genome functions. Transcriptome analysis provides insights into (1) normal patterns of gene expression that are important for understanding how a cell or tissue type differentiates during development, (2) how gene expression dictates and controls the physiology of differentiated cells, and (3) mechanisms of disease development that result from or cause gene expression changes in cells. In Chapter 19, we will consider why gene expression analysis is gradually becoming an important diagnostic tool in certain areas of medicine. For example, examining gene expression profiles in a cancerous tumor can help diagnose tumor type, determine the likelihood of tumor metastasis (spreading), and develop the most effective treatment strategy.

A number of different techniques can be used for transcriptome analysis. PCR-based methods are useful because of their ability to detect genes that are expressed at low levels. **DNA microarray analysis** is widely used because it enables researchers to analyze all of a sample's expressed genes simultaneously.

Most microarrays, also known as **gene chips,** consist of a glass slide onto which single-stranded DNA molecules are attached, or "spotted," using a computer-controlled high-speed robotic arm called an arrayer. Arrayers are fitted with a number of tiny pins. Each pin is immersed in a small amount of solution containing millions of copies of a different single-stranded DNA molecule. For example, many microarrays are made with single-stranded sequences of complementary DNA (cDNA) or expressed sequenced tags (ESTs). The arrayer fixes the DNA onto the slide at specific locations (points, or spots) that are recorded by a computer. A single microarray can have over 20,000 different spots of DNA, each containing a unique sequence for a different gene. Entire genomes are available on microarrays, including the human genome. As you will learn in Chapter 19, researchers are also using microarrays to compare patterns of gene expression in tissues in response to different conditions, to compare gene expression patterns in normal and diseased tissues, and to identify pathogens.

To prepare a microarray for use in transcriptome analysis, scientists typically begin by extracting mRNA from cells or tissues (Figure 18–16). Then the mRNA is usually reverse transcribed to synthesize cDNA tagged with fluorescently labeled nucleotides. The mRNA or cDNA can be labeled in a number of ways, but most methods involve the use of fluorescent dyes. Typically, microarray studies often involve comparing gene expression in different cell or tissue samples. cDNA prepared from one tissue is usually labeled with one color dye, red for example, and cDNA from another tissue is labeled with a different colored dye, such as green. Labeled cDNAs are then denatured and incubated overnight with the microarray so that they will hybridize to spots on the microarray that contain complementary DNA sequences. Next, the microarray is washed, and then it is scanned by a laser that causes the cDNA hybridized to the microarray to fluoresce. The patterns of fluorescent spots reveal which genes are expressed in the tissue of interest, and the intensity of spot fluorescence indicates the relative level of expression. The brighter the spot, the more the particular mRNA is expressed in that tissue. Computerized microarray data analysis programs are essential for organizing gene expression profile data from microarrays to analyze and groups genes according to whether they show increased (upregulated) or decreased (downregulated) expression under the experimental conditions examined.

Microarrays are dramatically changing the way gene expression patterns are analyzed. The biggest advantage of microarrays is that they enable thousands of genes to be studied simultaneously. As a result, however, they can generate an overwhelming amount of gene expression data. In addition, even when properly controlled, microarrays often yield variable results. For example, one experiment under certain conditions may not always yield similar patterns of gene expression as another identical experiment. Some of these differences can be due to real differences in gene expression, but others can be due to variability in chip preparation, cDNA synthesis, probe hybridization, or washing conditions, all of which must be carefully controlled to limit such variability. Commercially available microarrays can reduce the variability that can result when individual researchers make their own arrays.

Recently, the sea urchin genome was sequenced. Sea urchins have a relatively simple body plan [Figure 18–17(a), p. 398]. They contain approximately 1500 cells with only a dozen cell types. Yet sea urchin development progresses through complex changes in gene expression that resemble vertebrate patterns of gene expression changes during development. This is one reason urchins are a valuable model organism for developmental biology. Recently, scientists at NASA's Ames Genome Research Facility in Moffett, California, used microarrays to carry out a transcriptome study on gene expression in sea urchins during the first two days of development (to the mid-late gastrula stage, about 48 hours postfertilization).

This work revealed that approximately 52 percent of all genes in the sea urchin are active during this period of

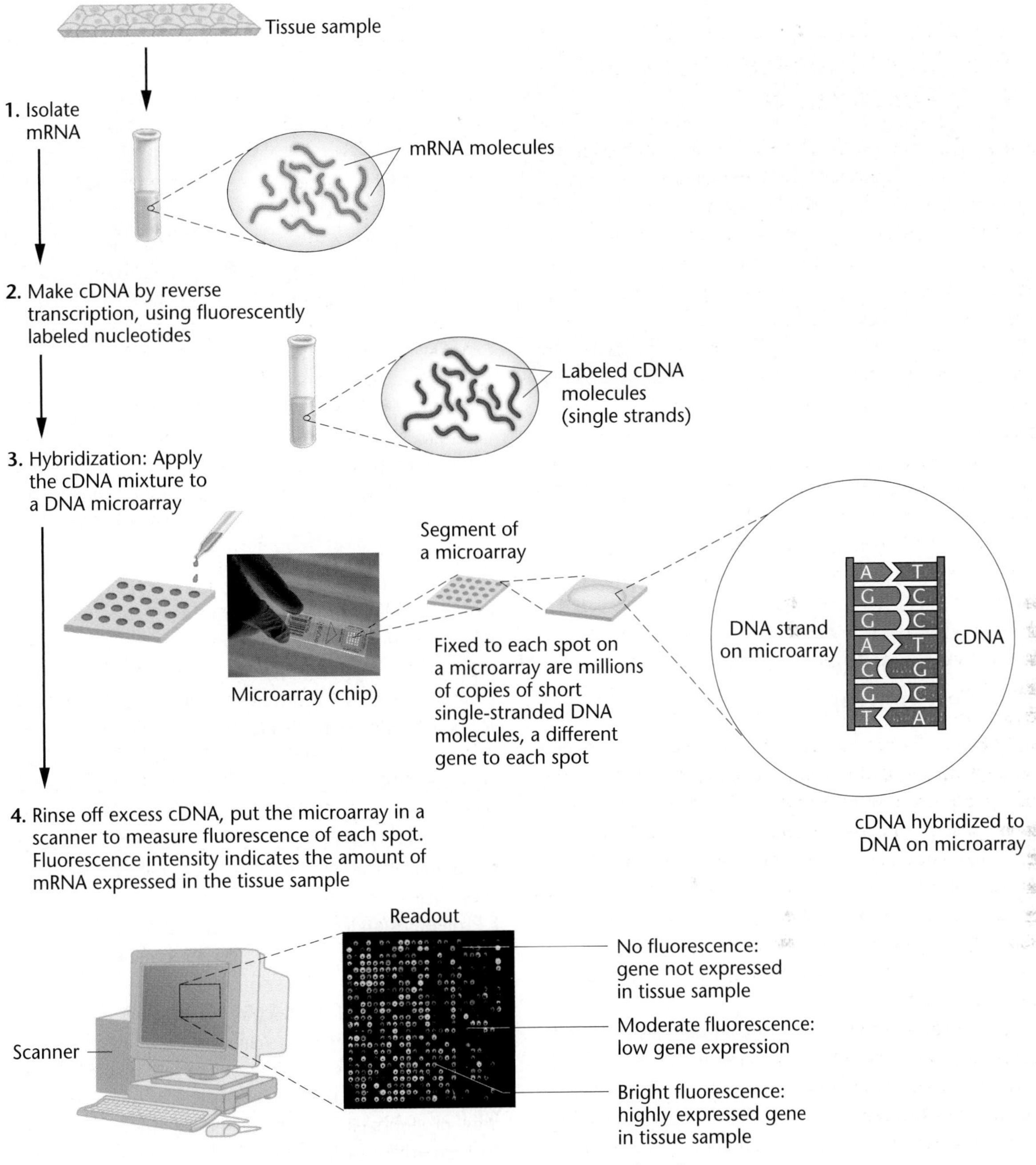

FIGURE 18–16 Microarray analysis for analyzing gene expression patterns in a tissue.

development: 11,500 of the sea urchin's 23,500 known genes were expressed in the embryo. The functional categories of genes expressed in the embryo were diverse, including genes for about 70 percent of the nearly 300 transcription factors in the sea urchin genome, along with genes involved in cell signaling, immunity, fertilization, and metabolism [Figure 18–17(b), p. 398]. Incredibly, 51,000 RNAs of unknown function were also expressed. Further studies are underway to explain the differences between gene number and transcripts expressed, although it is already known that many sea urchin genes are extensively processed through alternative splicing. Further analysis of the sea urchin genome will undoubtedly reveal interesting aspects of gene function during sea urchin development and advance our understanding of the genetics of embryonic development in both invertebrates and vertebrates.

Now that we have considered genomes and transcriptomes, we turn our attention to the ultimate end products of most genes, the proteins encoded by a genome.

ESSENTIAL POINT

DNA chips or microarrays are valuable for transcriptome analysis in studying expression patterns for thousands of genes simultaneously.

(a)

(b)

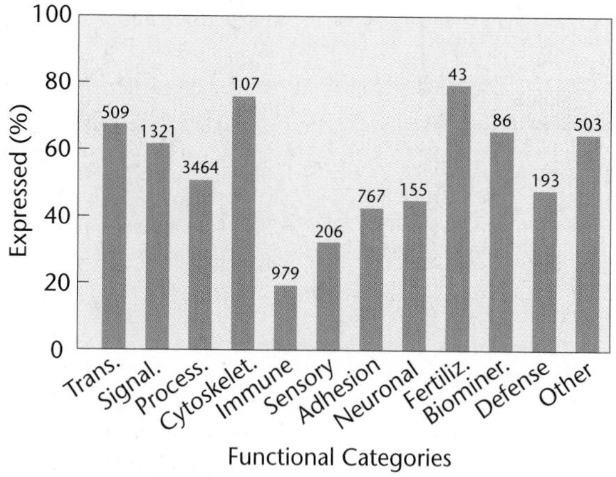

FIGURE 18–17 (a) The sea urchin *Strongylocentrotus purpuratus*. (b) Transcriptome analysis of genes expressed in the sea urchin embryo. The *y*-axis of the histogram displays the percentage of annotated genes in different functional categories expressed in the embryo. The number at the top of each bar represents the total number of annotated genes in the corresponding functional category. Trans., transcription factors; Signal, signaling genes; Process, basic cellular processes, such as metabolism; Cytoskelet., cytoskeletal genes; Fertiliz., fertilization; and Biominer., biomineralization.

18.10 Proteomics Identifies and Analyzes the Protein Composition of Cells

As more genomes have been sequenced and studied, biologists in many different disciplines have focused increasingly on understanding the complex structures and functions of the proteins the genomes encode. This is not surprising given that in most of the genomes sequenced to date, many newly discovered genes and their putative proteins have no known function. Keep in mind, in the ensuing discussion, that although every cell in the body contains an equivalent set of genes, not all cells express the same genes and proteins. **Proteome** is a term that represents the complete set of proteins encoded by a genome, but it is also often used to mean the entire complement of proteins in a cell. This definition would then include proteins that a cell acquired from another cell type.

Proteomics—the identification and characterization of all proteins encoded by the genome of a cell, tissue, or organism—can be used to reconcile differences between the number of genes in a genome and the number of different proteins produced. But equally important, proteomics also provides information about a protein's structure and function; post-translational modifications; protein–protein, protein-nucleic acid, and protein–metabolite interactions; cellular localization of proteins; and relationships (shared domains, evolutionary history) to other proteins. Proteomics is also of clinical interest because it allows comparison of proteins in normal and diseased tissues, which can lead to identifying proteins as biomarkers for disease conditions.

Reconciling the Number of Genes and the Number of Proteins Expressed by a Cell or Tissue

As we have discussed in this chapter and elsewhere, genomics has revealed that the link between gene number and gene product is often quite complex. Genes can have multiple transcription start sites that produce several different transcripts. Alternative splicing of pre-mRNA molecules can generate dozens of different proteins from a single gene. Remember the current estimate that over 50 percent of human genes produce more than one protein by alternative splicing. As a result, proteomes are substantially larger than genomes. For instance, the ~20,000 genes in the human genome encode ~100,000 proteins, but some estimates suggest that the human proteome may be as large as 150,000–200,000 proteins.

Proteomes undergo dynamic changes that are coordinated in part by regulation of gene expression patterns—the transcriptome. However, a number of other factors affect the proteome profile of a cell, further complicating the analysis of protein function. For instance, many proteins are modified by co-translational or posttranslational events, such as cleavage of signal sequences that target a protein for an organelle pathway, propeptides, or initiator methionine residues; by linkage to carbohydrates and lipids; or by the addition of chemical groups through methylation, acetylation, and phosphorylation. Over a hundred different mechanisms of posttranslational modification are known. In addition, many proteins work via elaborate protein–protein interactions or as part of a large molecular complex.

Well before a draft sequence of the human genome was available in 2001, scientists were already discussing the possibility of a "Human Proteome Project." One reason such a project never came to pass is that there is no single human proteome: different tissues produce different sets of proteins. But the idea of such a project has led to a **Protein Structure Initiative** by the National Institute of General Medical Sciences (NIGMS), a division of the National Institutes of Health. Initiated in 2000, PSI is a ten-year project designed to analyze the structures of more than 4000 protein families. Proteins with interesting potential therapeutic properties are a top priority for the PSI. There also are a number of other ongoing projects dedicated to identifying proteome profiles that correlate with diseases such as cancer and diabetes.

Proteomics Technologies: Two-Dimensional Gel Electrophoresis for Separating Proteins

In Chapter 17, we explained that in the early days of DNA cloning and recombinant DNA technology—before genomics—scientists often cloned, sequenced, and studied one or a few genes at a time over a span of several years. Until relatively recently, the same approach typically applied to studying proteins. But now, armed with proteomics technologies, scientists have the ability to study thousands of proteins simultaneously, generating enormous amounts of data quickly and dramatically changing ways of analyzing the protein content of a cell.

The early history of proteomics dates back to 1975 and the development of **two-dimensional gel electrophoresis (2DGE)** as a technique for separating hundreds to thousands of proteins with high resolution. In this technique, proteins isolated from cells or tissues of interest are loaded onto a polyacrylamide tube gel and first separated by **isoelectric focusing,** which causes proteins to migrate according to their electrical charge in a pH gradient. During isoelectric focusing, proteins migrate until they reach the location in the gel where their net charge is zero compared to the pH of the gel (Figure 18–18). Then in a second migration, perpendicular to the first, the proteins are separated by their molecular mass using **sodium dodecyl sulfate polyacrylamide gel electrophoresis (SDS-PAGE).** In this step, the tube gel is rotated 90° and placed on top of an SDS polyacrylamide gel and an electrical current is applied to the gel to separate the proteins by mass.

Proteins in the 2D gel are visualized by staining with Coomassie blue, silver stain, or other dyes that reveal the separated proteins as a series of spots in the gel (Figure 18–19, p. 400). It is not uncommon for a 2D gel loaded with a complex mixture of proteins to show several thousand spots, as in Figure 18–19, which displays the complex mixture of proteins in human platelets (thrombocytes). Particularly abundant protein spots in this figure are labeled with the names of identified proteins. With thousands of different spots on the gel, how are the identities of the proteins ascertained?

In some cases, 2D gel patterns from experimental samples can be compared to gels run with reference standards containing

FIGURE 18–18 Two-dimensional gel electrophoresis (2DGE) is a useful method for separating proteins in a protein extract from cells or tissues that contain a complex mixture of proteins with different biochemical properties.

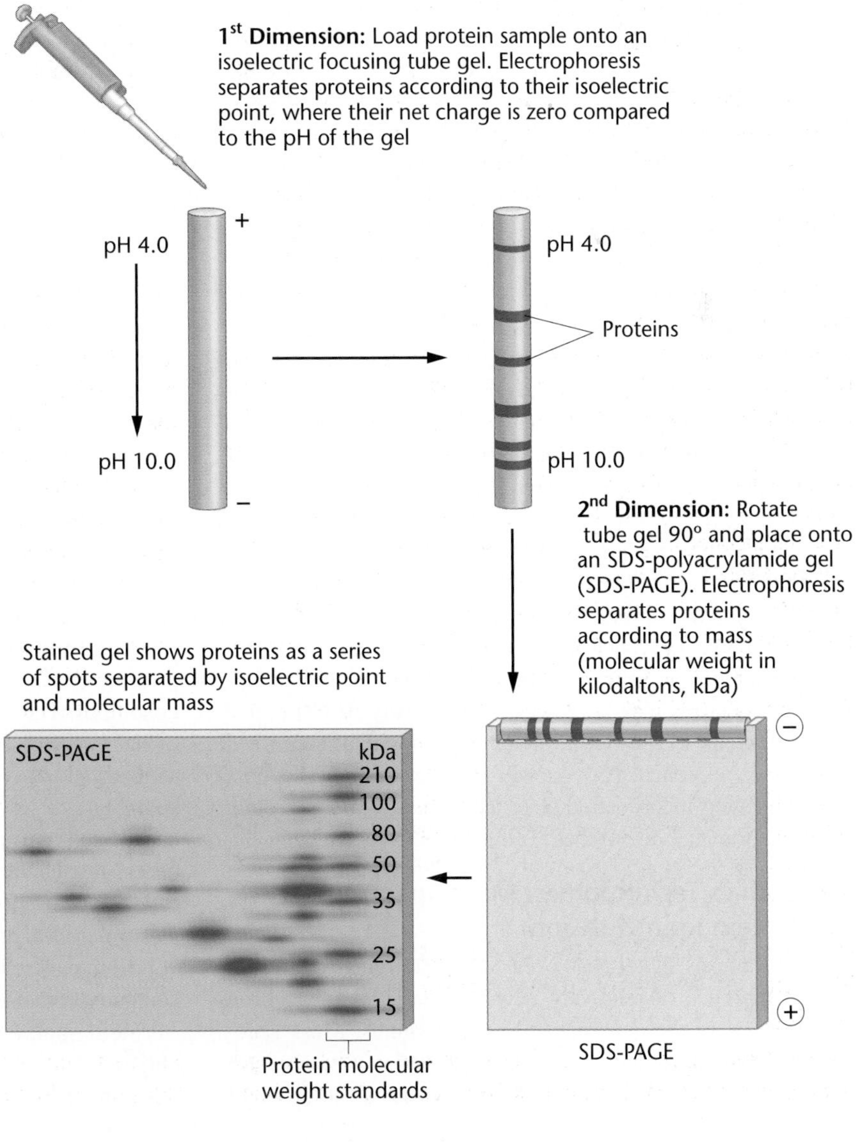
1st Dimension: Load protein sample onto an isoelectric focusing tube gel. Electrophoresis separates proteins according to their isoelectric point, where their net charge is zero compared to the pH of the gel
pH 4.0
pH 10.0
+
–
pH 4.0
Proteins
pH 10.0
2nd Dimension: Rotate tube gel 90° and place onto an SDS-polyacrylamide gel (SDS-PAGE). Electrophoresis separates proteins according to mass (molecular weight in kilodaltons, kDa)
Stained gel shows proteins as a series of spots separated by isoelectric point and molecular mass
SDS-PAGE
kDa
210
100
80
50
35
25
15
Protein molecular weight standards
SDS-PAGE

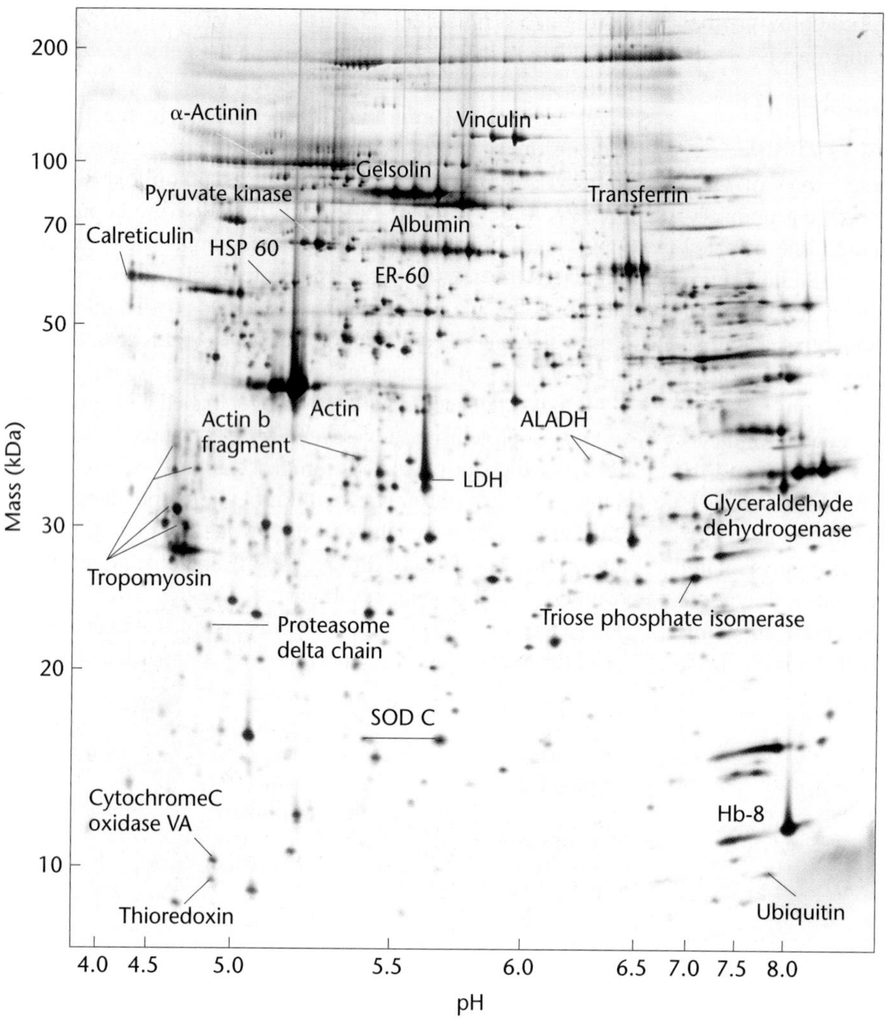

FIGURE 18–19 Two-dimensional gel separations of human platelet proteins. Each spot represents a different polypeptide separated by molecular weight (*y*-axis) and isoelectric point, pH (*x*-axis). Known protein spots are labeled by name based on identification by comparison to a reference gel or by determination of protein sequence using mass spectrometry techniques. Notice that many spots on the gel are unlabeled, indicating proteins of unknown identity.

known proteins with well-characterized migration patterns. Many reference gels for different biological samples such as human plasma, are available, and computer software programs can be used to align and compare the spots from different gels. In the early days of 2DGE, proteins were often identified by cutting spots out of a gel and sequencing the amino acids the spots contained. Only relatively small sequences of amino acids can typically be generated this way; rarely can an entire polypeptide be sequenced using this technique. BLAST and similar programs can be used to search protein databases containing amino acid sequences of known proteins. However, because of alternative splicing or posttranslational modifications, peptide sequences may not always match easily with the final product, and the identity of the protein may have to be confirmed by another approach. As you will learn in the next section, proteomics has incorporated other techniques to aid in protein identification, and one of these is mass spectrometry.

Proteomics Technologies: Mass Spectrometry for Protein Identification

As important as 2DGE has been for protein analysis, **mass spectrometry** is a relatively new technique that has been instrumental in the development of proteomics. Mass spectrometry techniques analyze ionized samples in gaseous form and measure the **mass-to-charge (m/z) ratio** of the different ions in a sample. Proteins analyzed by mass spectra generate m/z spectra that can be correlated with an m/z database containing known protein sequences to discover the protein's identity. Some of the most valuable proteomics applications of this technology are to identify an unknown protein or proteins in a complex mix of proteins, to sequence peptides, to identify posttranslational modifications of proteins, and to characterize multiprotein complexes.

One commonly used mass spectrometry approach is **matrix-assisted laser desorption ionization (MALDI).** MALDI is ideally suited for identifying proteins and is widely used for proteomic analysis of tissue samples treated under different conditions. The proteins are first extracted from cells or tissues of interest and separated by 2DGE, after which MALDI (described below) is used to identify the proteins in the different spots. Figure 18–20 shows an example in which two different sets of cells grown in culture are analyzed for protein differences. Just about any source providing a sufficient number of cells can be used: blood, whole tissues, and organs; tumor samples; microbes; and many other substances. Many proteins involved in cancer have been identified by the use of MALDI to compare protein profiles in normal tissue and tumor samples.

Protein spots are cut out of the 2D gel, and proteins are purified out of each gel spot. Computer-automated high-throughput instruments are available that can pick all of the

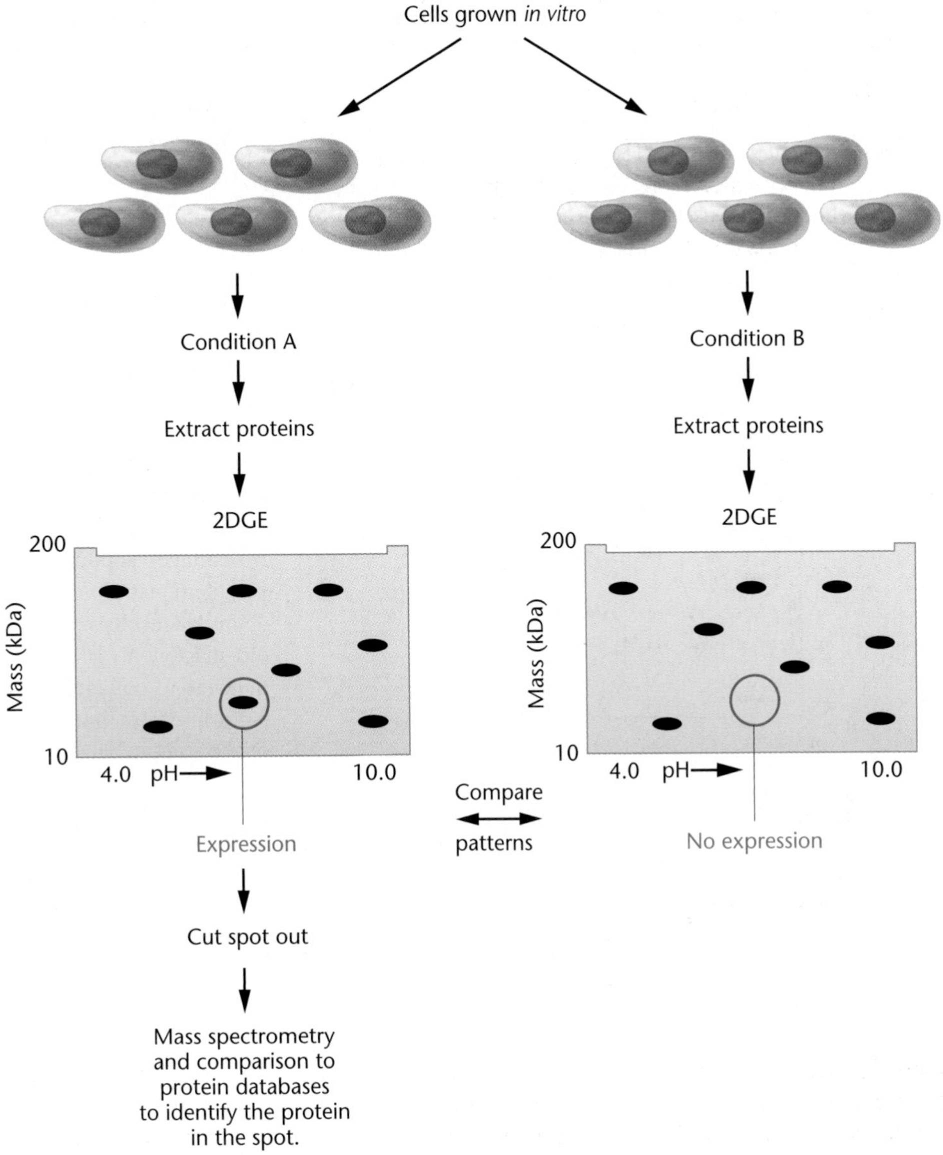

FIGURE 18–20 In a typical proteomic analysis, cells are exposed to different conditions (such as different growth conditions, drugs, or hormones). Then proteins are extracted from these cells and separated by 2DGE, and the resulting patterns of spots are compared for evidence of differential protein expression. Spots of interest are cut out from the gel, digested into peptide fragments, and analyzed by mass spectrometry to identify the protein they contain.

spots out of a 2D gel. Isolated proteins are then enzymatically digested with a protease (a protein-digesting enzyme) such as trypsin to create a series of peptides. This proteolysis produces a complex mixture of peptides determined by the cleavage sites for the protease in the original protein. Each type of protein produces a characteristic set of peptide fragments, and these are identified by MALDI as illustrated in Figure 18–21, p. 402. Because different proteins produce different sets of peptide fragments, MALDI produces a peptide "fingerprint" that is characteristic of the protein being analyzed.

MALDI is often coupled with a protein biochemistry technique for mass analysis called **time of flight (TOF).** TOF moves ionized peptide fragments through an electrical field in a vacuum. Each ion has a speed that varies with its mass. The speed with which each ion crosses the vacuum chamber can be measured, and differences in the ions' kinetic energy can be used to develop a mass-dependent velocity profile—a MALDI-TOF spectrum—that shows each ion's "time of flight." MALDI-TOF spectra can then be compared to databases of spectra for known proteins, as described above for MALDI spectra.

Protein microarrays are also becoming valuable tools for proteomics research. These are designed around the same basic concept as microarrays (gene chips) and are often constructed with antibodies that specifically recognize and bind to different proteins. These microarrays are used, among other applications, for examining protein–protein interactions, for detecting protein markers for disease diagnosis, and for studying in biosensors designed to detect pathogenic microbes and potentially infectious bioweapons.

NOW SOLVE THIS

In Problem 21 on page 406, a database is searched for matches to a query sequence, and the matches prove to correspond to a certain type of protein domain. You are asked to explain why the function of the query sequence is not related to the function of the known proteins, even though they have the same domain.

Hint: Remember that although protein domains may have related functions, proteins can contain several different domains interacting to determine protein function.

Identification of Collagen in *Tyrannosaurus rex* and *Mammut americanum* Fossils

Recently, a team of scientists reported results of mass spectrometry analysis of bone tissue from a *Tyrannosaurus rex* skeleton excavated from the Hell Creek Formation in eastern Montana and estimated to be 68 million years old. As mentioned earlier, DNA has been recovered from fossils, but the general assumption has been that proteins degrade in fossilized materials and cannot be recovered. This study demonstrated that fossilization does not fully destroy all proteins in well-preserved fossils under certain conditions. This research also demonstrates the power and sensitivity of mass spectrometry as a proteomics tool.

In this study, medullary tissue was removed from the inside of the left and right femoral bones. Medullary tissue is porous, spongy bone that contains bone marrow cells, blood vessels, and nerves. *T. rex* proteins extracted from the tissue showed cross-reactivity with antibodies to chicken collagen and were digested by the collagen-specific protease collagenase. These results suggested that the *T. rex* protein samples contained collagen, a major matrix component of bone, ligaments, tendons, and skin. To definitively identify the presence of collagen, tryptic peptides from the *T. rex* samples were analyzed by liquid chromatography and mass spectrometry (LC/MS). The m/z spectra for one of the *T. rex* peptides was identified from a database of m/z spectra as corresponding to collagen. Furthermore, the amino acid sequence of *T. rex* collagen peptide aligned with an isoform of chicken collagen, demonstrating sequence similarity.

Similar results were obtained for 160,000- to 600,000-year-old mastodon (*Mammut americanum*) peptides that showed matches to collagen from extant species, including collagen isoforms from humans, chimps, dogs, cows, chickens, elephants, and mice.

ESSENTIAL POINT

Methods such as two-dimensional gel electrophoresis and mass spectrometry are valuable for analyzing proteomes—the protein content of a cell.

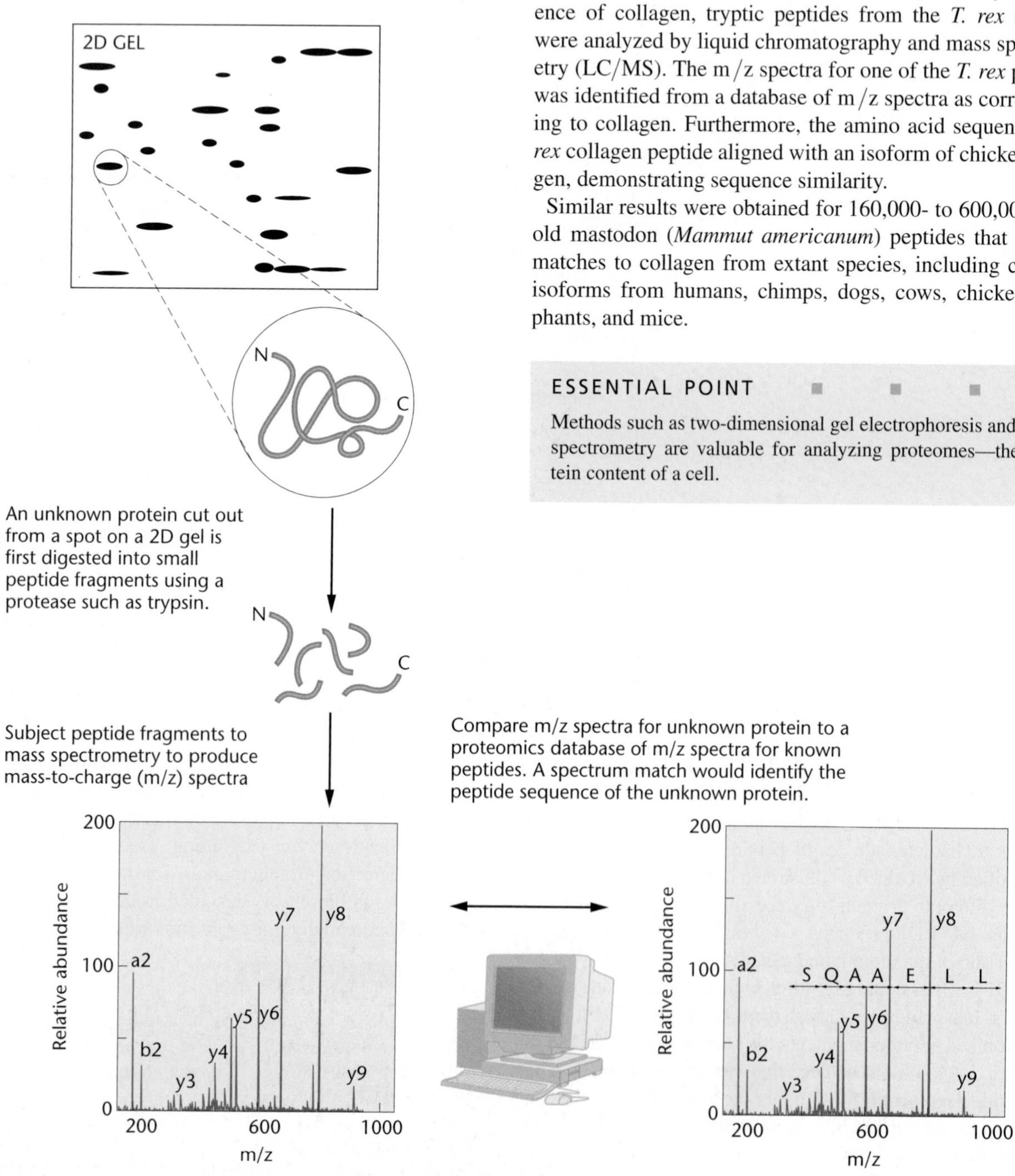

FIGURE 18–21 Mass spectrometry for identifying an unknown protein isolated from a 2D gel. The mass-to-charge spectrum (m/z) (determined, for example, by MALDI) for trypsin-digested peptides from the unknown protein can be compared to a proteomics database for a spectrum match to identify the unknown protein. The peptide in this example was revealed to have the amino acid sequence serine (S)-glutamine (Q)-alanine (A)-alanine (A)-glutamic acid (E)-leucine (L)-leucine (L), shown in single-letter amino acid code.

EXPLORING GENOMICS

Contigs and Shotgun Sequencing

In this chapter, we discussed how whole-genome shotgun sequencing methods can be used to assemble chromosome maps. Recall that in the technique of shotgun cloning, chromosomal DNA is first digested with different restriction enzymes to create a series of overlapping DNA fragments called contiguous sequences, or "contigs." The contigs are then subjected to DNA sequencing, after which bioinformatics-based programs are used to arrange the contigs in their correct order on the basis of short overlapping sequences of nucleotides.

In this Exploring Genomics exercise, you will carry out a simulation of contig alignment to help you understand the underlying logic of this approach to creating sequence maps of a chromosome. For this purpose, you will return to the **National Center for Biotechnology Information BLAST** site that was used in other Exploring Genomics exercises and apply a DNA alignment program called bl2seq.

■ Exercise I: Arranging Contigs to Create a Chromosome Map

1. Access BLAST from the NCBI Web site at **http://www.ncbi.nlm.nih.gov/BLAST/**. Locate and select the "Align two sequences using BLAST (bl2seq)" category at the bottom of the BLAST home page. The bl2seq feature allows you to compare two DNA sequences at a time to check for sequence similarity alignments.
2. Below are eight contig sequences taken from an actual human chromosome sequence deposited in GenBank. For this exercise we have used short fragments; however, in reality contigs are usually several thousand base pairs long. To complete this exercise, copy and paste two sequences into the Align feature of BLAST and then run an alignment (by clicking on "Align"). (*Hint:* Access these sequences from the Companion Web site so that you can copy and paste the sequences easily.) Repeat these steps with other combinations of two sequences to determine which sequences overlap, and then use your findings to create a sequence map that places overlapping contigs in their proper order. A few tips to consider:

- Develop a strategy to be sure that you analyze alignments for all pairs of contigs.
- Only consider alignment overlaps that show 100 percent sequence similarity.

Sequence A

CTAATTTTTTTTGTATTTTTAATAGAGAC
GAGGTGTCACCATGTTGGACAGGCTGGT
CTCGAACTCCTGACCTCAGGTGATCTGC
CCACCTCAGCCTCCCAAAGTGCTGGGAT
TACAAGCATGAGCCACCACTCCCAGGCT
TTATTTTCTATTTTTTAATTACAGCCATCC
TAGTGAATGTGAAGTAGTATCTCACTGA
GGTTTTGATTTGCATTTTTCTATGACAAT
GAACAATGTTTCATGTGCTTGTTGGCT
GTTTGTATATCCTTTTTGGAGAAATACCA
ATTCATGTCCTTTGCCCATTTTTAAAGTG
GATTGCATGTCTTTTTGTTGTTTAGTTGT
AAAGATGTGGGTTTTTCTTTTGAGACGG
AGTCTCGCTGTCGCCTAGGCTGGAGTGC
AGAGGCATGATCTCGGCTGACTGCAATC
CCCACCTCCTGGCATCAAGAAGTTCTCC
TGCCTCAGCCTTCCAAGTAGCTGGGTTT
ACAGATGC

Sequence B

CTTTATCTCAGGACAATGAACCCGCAAG
GAGAGGAAGAGCCAGTAATTCTATAGAG
ACTCGGAGGCGCAGGGGGCACGCTTAG
TTAGAGTGGTGGTGGTATTTTCAGTGTTT
TCTGGTTTTATGATAAACACAAGCATCA
ATGTCTCAAGACTTTCATCTTTATCTTTT
TTTTTTTTTTTTTTTTTTTCTTGAGACAG
GGTTTCCCTCTGTCACCCAGGCTGGAGT
GCATTGGTGGTGTGATCTTGGCTTTCTG
TAACCTCGGGCTTCTGGGCTCAAGCCGT
TCTACTACCTCAGCCTCCCAAATAGCTAG
AACTACAAGCGTGTGCTGCCACACCTG
GCTAATTTGTTGTATTTTATTTATTCATTT
ATTTTTGTGAAGACAAGGTCTTGCCATG
TTGCCCAGGCTGGTCTCAGACTCCTGGG
CTCAAGCAATCCACCCACCTTAGTCTCC
CAAAGTGCTGGGATTACAGGCGTGAGC
CACCACACCCA

Sequence C

GGAATTTCACTCTTGTTGCCCAAGTTGG
AGTGCAATGGCGCGATCTCAGCTCACTG
CAACCTCCGCCTCCCAGGTTCAAACGAT
TCTCCTGCTTCACTCTCCCCAGTAGCTG
GGATTACAGGCTGCACCACCACACCTGG
CTAATTTTTTTTGTATTTTTAATAGAGAC
GAGGTGTCACCATGTTGGACAGGCTGGT
CTCGAACTCCTGACCTCAGGTGATCTGC
CCACCTCAGCCTCCCAAAGTGCTGGGAT
TACAAGCATGAGCCACCACTCCCAGGC

Sequence D

GCTTCATCTTTCTCTTCACCGTAAAACA
GGAAAGTGTGTGGTGACCAGTATTTTAA
GGGAAAGGCACTTACAGAGAATTAAGC
ATTTGACAAAATTTATTTACAGATATTTG
TCTGTGGACCACTTCCGCACCAGCTGTG
CATGAGAGGGCTCATTGCTCTGAATTTG
CCTCCTTGTCTGCACCCAGGAGACCGTT
TCCCAGATCACGCAAACGCTGCCTTCTC
CCCACACCAGGGCCCTCAGCATGGGAA
TGACCTTCCAGCGCTGCACGTTTCCAAT
CCATGCTCTGTTTTTCAGTTCTGGCTCAC
AGAGGACTGCTGGTTGCAAGCAAACTT
GTATCTGGGTCTTCA

Sequence E

CCTTAAGTGATCTACCTGTCTCTGCCTCT
CAAAGTGCTGGGATTGCAGGCATAAGC
CGCCATGCCCGGCCCAAAGTTTCTTTAT
ATGTGCTGGATACTAGGCCCGTAACAGA
TATACAATTTGTAAATATTTTCTCTCATTT
TGAAGATTTTCTTTTCACTTTCTTGATAA
TGTCCTTTGTGTATTTTTTGATAATGTCC
TTTGATACACAAAAGTTTTTAAGTTTGAT
GAAGTTCAATTTACCTATTATTTTCTTTT
GTTGTTCAT

Sequence F

TGTTTGTTTGTTTGTTTTGTTTCATTTTG
TTTTTGAGACAGAGTCTTGTTCTGTCGC
CCAGTGTAGAGTGCAGTGGCATAATCTC

(*Cont. on the next page*)

GGCTCACCGTAACCTCCGCCTCCCGGG
TTCAAGCAACTCTGCCTGCCTCAGCCTC
CCAAGGAGCTGGGATTATAGACGCCCAC
CACCATGCCTGGTTAATTTTTGTAGTTTT
TTTTAGTAGAGATGGGGTTTTGCCATTTT
GGCCAGGCTGGTCTTGAACTCCTGACCT
CAGGTGATCTGCCCACCCTGGCCTCTCA
AAGTGCTGGGATTACAGGTGTGAGCTGC
CACACTCGGCCACAACAAATTTTTGCAC
CAGTTGCTCACA

Sequence G

TCCTTTGATACACAAAAGTTTTTAAGTTT
GATGAAGTTCAATTTACCTATTATTTTCT
TTTGTTGTTCATTCATTTTGTGTCCTATG
TAGGAATCTATTGCCAAATTCAAGGTGA
TAAAGATTTACCCCTATGTTTCCTTCTAA
GAGTTTTATTGTTTTAGCCCTGATATTTA
GCTAAACTTAATTGATTTATTAAGTTTAA
TTTTCCTATGTGGTATGAAGTCATTTATC
TTCTTTAGTTCAGGATCCAAGTGAAAGG
GGCATCTTCTATCTGGGACATGCCATTCT
CATGACAGAGGAAAAAGACAAAAAACT
GACACATACAATGACTTTAAAACTTCAC
TCA

Sequence H

GGGTTTTTCTTTTGAGACGGAGTCTCGC
TGTCGCCTAGGCTGGAGTGCAGAGGCAT
GATCTCGGCTGACTGCAATCCCCACCTC
CTGGCATCAAGAAGTTCTCCTGCCTCAG
CCTTCCAAGTAGCTGGGTTTACAGATGC
CCACCACCATGCCTGGCTGGTTTTTGTA
TTTTTAGTAGACACGGGGTTTTACCATGT
TGGCCGGGCTGGTCTGGAACTCCTAAC
CTTAAGTGATCTACCTGTCTCTGCCTCTC
AAAGTGCTGGGATTGCAGGCATAAGCC
GCCATGCCCGGCCCAAAGTTTCTTTATA
TGTGCTGGATACTAGGCCCGTAACAGAT
ATACAATTTGTAAA

3. On the basis of your alignment results, answer the following questions, referring to the sequences by their letter codes (A through H):

a. What is the correct order of overlapping contigs?
b. What is the length, measured in number of nucleotides, of each sequence overlap between contigs?
c. What is the total size of the chromosome segment that you assembled?
d. Did you find any contigs that do not overlap with any of the others? Explain.

4. Run a nucleotide–nucleotide BLAST search (BLASTn) on any of the overlapping contigs to determine which chromosome these contigs were taken from, and report your answer.

CASE STUDY Bioprospecting in Darwin's wake

The Global Ocean Sampling (GOS) expedition followed the route of Charles Darwin's voyage to chart the genetic diversity of microbes in the marine environment. An approach called metagenomics was used to catalog DNA sequences and their encoded proteins from thousands of previously undescribed organisms present in samples collected in diverse oceanic regions. Although many samples remain to be analyzed, the project has already identified thousands of new species, generated data on more than 1700 previously unknown families of proteins, and assembled information about more than 6 million specific proteins. Some, like those forming light-driven proton pumps and those that function in nitrogen fixation, may have immediate applications. For now, however, the emphasis is on better understanding the distribution and diversity of microbes in the oceans. This massive project has raised several questions about bioprospecting in the waters of coastal nations.

1. Who owns the organisms collected along the coast of countries? Who will own any processes or products developed from these genetic resources?
2. How can the findings from this metagenomics survey of the ocean be applied?
3. Is it surprising that so many previously undiscovered organisms are represented in the samples collected on this expedition?

INSIGHTS AND SOLUTIONS

1. One of the main problems in annotation is deciding how long a putative ORF must be before it is accepted as a gene. Shown here are three different ORF scans of the same *E. coli* genome region—the region containing the *lacY* gene. Regions shaded in brown indicate ORFs. The top scan was set to accept ORFs of 50 nucleotides as genes. The middle and bottom scans accepted ORFs of 100 and 300 nucleotides as genes, respectively. How many putative genes are detected in each scan? The longest ORF covers 1254 bp; the next longest, 234 bp; and the shortest, 54 bp. How can we decide the actual number of genes in this region? In this type of ORF scan, is it more likely that the number of genes in the genome will be overestimated or underestimated? Why?

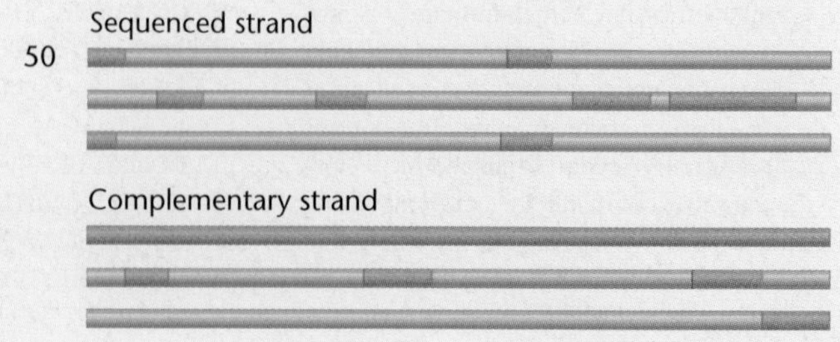

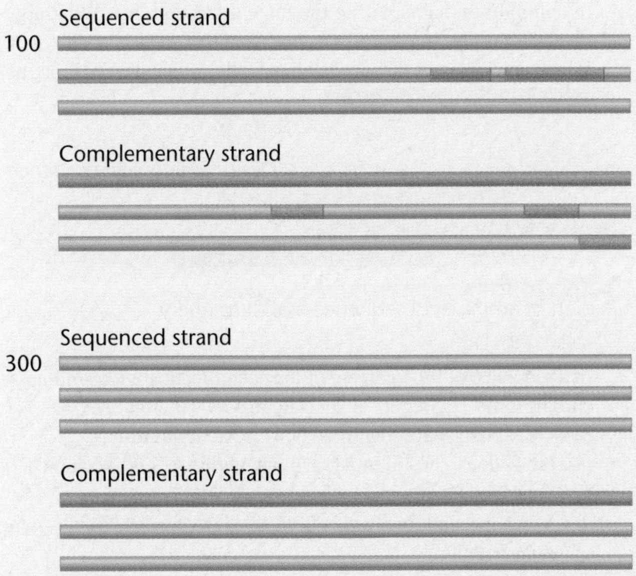

Solution: Generally, one can examine conserved sequences in other organisms to indicate that an ORF is likely a coding region. One can also match a sequence to previously described sequences that are known to code for proteins. The problem is not easily solved—that is, deciding which ORF is actually a gene. The shorter the ORFs scan, the more likely the overestimate of genes because ORFs longer than 200 are less likely to occur by chance. For these scans notice that the 50-bp scans produce the highest number of possible genes, whereas the 300-bp scan produces the lowest number (1) of possible genes.

2. Recent sequencing of the heterochromatic regions (repeat-rich sequences concentrated in centromeric and telomeric areas) of the *Drosophila* genome indicates that within 20.7 Mb, there are 297 protein-coding genes (Bergman et al. 2002. *genomebiology3 (12)@genomebiology.com/2002/3/12/RESEARCH/0086*). Given that the euchromatic regions of the genome contain 13,379 protein-coding genes in 116.8 Mb, what general conclusion is apparent?

Solution: Gene density in euchromatic regions of the *Drosophila* genome is about one gene per 8730 base pairs, whereas gene density in heterochromatic regions is one gene per 70,000 bases (20.7 Mb/297). Clearly, a given region of heterochromatin is much less likely to contain a gene than the same-sized region in euchromatin.

Problems and Discussion Questions

1. What is functional genomics? How does it differ from comparative genomics?
2. Compare and contrast whole-genome shotgun sequencing to a map-based cloning approach.
3. What, if any, features do bacterial genomes share with eukaryotic genomes?
4. Plasmids can be transferred between species of bacteria, and most carry nonessential genes. For these and other reasons, plasmid genes have not been included as part of the genomes of bacterial species. The bacterium *B. burgdorferi* contains 17 plasmids carrying 430 genes, some of which are essential for life. Should plasmids carrying essential genes be considered part of an organism's genome? What about other plasmids that do not carry such genes? In other words, how do we define an organism's genome in these cases?
5. What is bioinformatics, and why is this discipline essential for studying genomes? Provide two examples of bioinformatics applications.
6. List and describe three major goals of the Human Genome Project.
7. How do high-throughput techniques facilitate research in genomics and proteomics? Explain.
8. BLAST searches and related applications are essential for analyzing gene and protein sequences. Define BLAST, describe basic features of this bioinformatics tool, and provide an example of information provided by a BLAST search.
9. What are pseudogenes, and how are they produced?
10. Describe the human genome in terms of genome size, the percentage of the genome that codes for proteins, how much is composed of repetitive sequences, and how many genes it contains. Describe two other features of the human genome.
11. Based on a comparison of general features of eukaryotic and prokaryotic genomes, why might we predict that the organization of eukaryotic genetic material is more complex than that of viruses or bacteria?
12. Compare the organization of bacterial genes to that of eukaryotic genes. What are the major differences?
13. The Human Genome Project has demonstrated that in humans of all races and nationalities approximately 99.9 percent of the sequence is the same, yet different individuals can be identified by DNA fingerprinting techniques. What is one primary variation in the human genome that can be used to distinguish different individuals? Briefly explain your answer.
14. Annotation involves identifying genes and gene-regulatory sequences in a genome. List and describe characteristics of a genome that are hallmarks for identifying genes in an unknown sequence. What characteristics would you look for in a prokaryotic genome? A eukaryotic genome?
15. It can be said that modern biology is experiencing an "omics" revolution. What does this mean? Explain your answer.
16. Metagenomics studies generate very large amounts of sequence data. Provide examples of genetic insight that can be learned from metagenomics.
17. What are gene microarrays? How are microarrays used?
18. In a recent draft annotation and overview of the human genome sequence, F. A. Wright et al. (*Genome Biol.* 2001: 2(7):Research0025) presented a graph similar to the one shown here. The graph details the approximate number of genes from each chromosome that are expressed only in embryos. Review earlier information in the text on human chromosomal aneuploids and correlate that information with the graph. Does this graph

provide insight as to why some aneuploids occur and others do not? See **Now Solve This** on page 381.

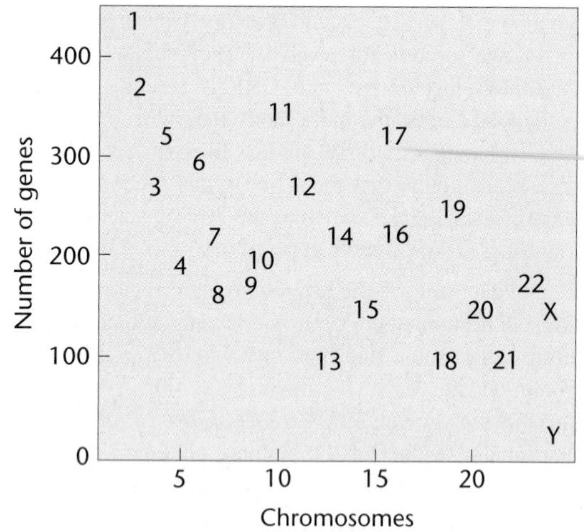

19. The process of annotating a sequenced genome is continual. In March 2000, the first annotated sequence of the *Drosophila* genome was released, which predicted 13,601 protein-coding genes within the euchromatic region of the genome. Shown in the following table are selected data from the next two annotated versions that were released (modified from Misra et al., 2002. *genomebiology3(12)@genomebiology.com/2002/3/12/RESEARCH/0083*). See **Now Solve This** on page 383.
   (a) Assuming a uniform distribution in Release 3, approximately how many base pairs of DNA lie between protein-coding genes in *Drosophila*?
   (b) On average, approximately how many exons are reported per gene in Release 3?
   (c) Approximately how many introns are there per gene?
   (d) What appears to be the most significant difference between Release 2 and Release 3?

Criteria	Release 2	Release 3
Total length of euchromatin	116.2 Mb	116.8 Mb
Total protein-coding genes	13,474	13,379
Protein-coding exons	50,667	54,934
Introns	48,381	48,257
Genes with alternative transcripts	689	2729

20. The β-globin gene family consists of 60 kb of DNA, yet only 5 percent of the DNA encodes gene products. Account for as much of the remaining 95 percent of the DNA as you can.

21. Annotation of a proteome attempts to relate each protein to a function in time and space. Traditionally, protein annotation depended on an amino acid sequence comparison between a query protein and a protein with known function. If the two proteins shared a considerable portion of their sequence, the query would be assumed to share the function of the annotated protein. Following is a representation of this "look-the-same" method of protein annotation involving a query sequence and three different human proteins (modified from Rigoutsos et al. 2002. *Nucl. Acids Res.* 30: 3901–3916). Note that the query sequence aligns to common domains within the three other proteins. What argument might you present to suggest that the function of the query is not related to the function of the other three proteins? See **Now Solve This** on page 401.

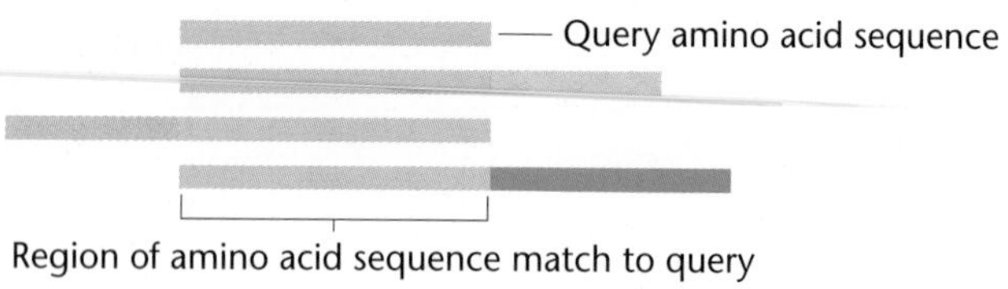

22. In a sequence of 99.4 percent of the euchromatic regions of human chromosome 1, Gregory et al. (Gregory, S. G. et al., *Nature,* 441: 315–321, 2006) have identified 3141 gene structures.
   (a) How does one identify a gene within a raw sequence of bases in DNA?
   (b) What procedures are often used to verify likely gene assignments?
   (c) Given that chromosome 1 contains approximately 8 percent of the human genome, and assuming that there are approximately 20,000 to 25,000 genes, would you consider chromosome 1 to be "gene rich"?

23. M. Stoll and colleagues have compared candidate loci in humans and rats in search of loci in the human genome that are likely to contribute to the constellation of factors leading to hypertension. Through this research, they identified 26 chromosomal regions that they consider likely to contain hypertension genes. How can comparative genomics aid in the identification of genes responsible for such a complex human disease? The researchers state that comparisons of rat and human candidate loci to those in the mouse may help validate their studies. Why might this be so?

24. Comparisons between human and chimpanzee genomes indicate that a gene that may function as a wild type or normal gene in one primate may function as a disease-causing gene in another (The Chimpanzee Sequence and Analysis Consortium, *Nature,* 437: 69–87, 2005). For instance, the *PPARG* locus (regulator of adipocyte differentiation) is associated with type 2 diabetes in humans but functions as a wild-type gene in chimps. What factors might cause this apparent contradiction? Would you consider such apparent contradictions to be rare or common? What impact might such findings have on the use of comparative genomics to identify and design therapies for disease-causing genes in humans?

25. Because of its accessibility and biological significance, the proteome of human plasma has been intensively studied and used to provide biomarkers for such conditions as myocardial infarction (troponin) and congestive heart failure (B-type natriuretic peptide). Polanski and Anderson (Polanski, M., and Anderson, N. L., *Biomarker Insights,* 2: 1–48, 2006) have compiled a list of 1261 proteins, some occurring in plasma, that appear to be differentially expressed in human cancers. Of these 1261 proteins, only 9 have been recognized by the FDA as tumor-associated proteins. First, what advantage should there be in using plasma as a diagnostic screen for cancer? Second, what criteria should be used to validate that a cancerous state can be assessed through the plasma proteome?

Transgenic pigs generated by incorporating a viral vector carrying the jellyfish gene encoding green fluorescent protein into the pig genome.

19

Chapter Concepts

Applications and Ethics of Genetic Engineering and Biotechnology

- Recombinant DNA technology, genetic engineering, and biotechnology have revolutionized medicine and agriculture.
- Genetically modified plants and animals can serve as bioreactors to produce valuable protein products.
- Genetic modifications of plants have resulted in herbicide and pest-resistant crops, and crops with improved nutritional value; similarly, transgenic animals are being created to produce therapeutic proteins and to protect animals from disease.
- Applications of recombinant DNA technology and genomics have become essential for determining genotypes and diagnosing human genetic disorders.
- Pharmacogenomics and rational drug design have led to customized medicines based on a person's genotype.
- Gene therapy by transfer of cloned copies of functional alleles into target tissues is used to treat genetic disorders.
- DNA fingerprinting can identify specific individuals and is widely used for forensics and paternity testing.
- Almost all applications of genetic engineering and biotechnology present ethical dilemmas that involve important moral, social, and legal issues.

Since the dawn of recombinant DNA technology in the 1970s, scientists have harnessed **genetic engineering** not only for biological research, but also for direct applications in medicine, agriculture, and biotechnology. Genetic engineering refers to the alteration of an organism's genome and typically involves the use of recombinant DNA technologies to add genes to a genome, but it can also involve gene removal. The ability to manipulate DNA *in vitro* and to introduce genes into living cells has allowed scientists to generate new varieties of plants, animals, and other organisms with specific gene traits, as well as to manufacture cheaper and more effective therapeutic products. Industry analysts estimate that genetic engineering will lead to U.S. commercial products worth over $45 billion by 2009. Many of these commercial products will be developed by the biotechnology industry.

Biotechnology is the use of living organisms to create a product or a process that helps improve the quality of life for humans or other organisms. Biotechnology as a modern industry began in earnest when recombinant DNA technology developed. But biotechnology is actually a science dating back to ancient civilization when microbes were used to make many products, including wine, beer, vinegar, breads, and cheeses. Modern biotechnology relies heavily on recombinant DNA technology, genetic engineering, and genomics applications. New developments that seemingly occur every day are making biotechnology one of the most rapidly developing industries worldwide, encompassing nearly 5000 companies in 54 countries. Over 350 biotechnology products are currently on the market.

Genetic engineering and biotechnology have the potential to solve major problems globally and to significantly alter how humans deal with the natural world; hence, they are raising unprecedented ethical, social, and economic questions. This chapter presents a selection of applications that illustrate the power of genetic engineering and biotechnology and the complexity of the dilemmas they engender. We begin by explaining how genetic engineering has modified agriculturally important plants and animals. Next, we examine the impact of genetic technologies on the diagnosis and treatment of human diseases. We briefly describe how genetic engineering has affected the production of pharmaceutical products and how DNA technologies are used for forensic applications. Finally, we explore some of the social, ethical, and legal implications of genetic engineering and biotechnology.

How Do We Know?

In this chapter, we will focus on a number of interesting applications of genetic engineering and biotechnology. At the same time, we will find many opportunities to consider the methods and reasoning by which much of this information was acquired. From the explanations given in the chapter, you should answer the following fundamental questions:

1. How do we confirm that we have introduced a useful gene into a transgenic organism and that it is functional?
2. How can we determine whether a human fetus has sickle-cell anemia?
3. How can we identify what genes are expressed in specific tissues?
4. How do we know whether a forensic DNA profile comes from only one individual?

19.1 Genetically Engineered Organisms Synthesize a Wide Range of Biological and Pharmaceutical Products

The most successful and widespread application of recombinant DNA technology has been the production of recombinant proteins as **biopharmaceutical** products—pharmaceutical products produced by biotechnology—particularly, therapeutic proteins to treat diseases. Prior to the recombinant DNA era, therapeutic proteins such as insulin, clotting factors, or growth hormones were purified from blood or organs such as pancreas or pituitary glands. Clearly, these sources were in limited supply, and the purification processes were expensive. In addition, products derived from natural sources could be contaminated by disease agents such as viruses. Now that human genes encoding important therapeutic proteins can be cloned and expressed in nonhuman host-cell types, we have more abundant, safer, and less expensive sources of biopharmaceuticals.

In this section, we outline examples of therapeutic products that are produced by expression of cloned genes in transgenic host cells and organisms. It should not surprise you that cancers, arthritis, diabetes, heart disease, and infectious diseases such as AIDS are among the major diseases that biotechnology companies are targeting for treatment by recombinant therapeutic products. Table 19.1 provides a short list of important recombinant products currently synthesized in transgenic bacteria, plants, yeast, and animals.

Insulin Production in Bacteria

Several therapeutic proteins have been produced by introducing human genes into bacteria. In most cases, the gene is cloned into a plasmid, and the recombinant vector is introduced into the bacterial host. Large quantities of the transformed bacteria are grown, and the recombinant human protein is recovered and purified from bacterial extracts.

The first human gene product manufactured by recombinant DNA technology was insulin, called Humulin, which was licensed for therapeutic use in 1982 by the **U.S. Food and Drug Administration (FDA),** the government agency responsible for regulating the safety of food and drug products and medical devices. In 1977, scientists at Genentech, the San Francisco biotechnology company cofounded in 1976 by Herbert Boyer (a pioneer in the use of plasmids for recombinant DNA technology) and Robert Swanson isolated and cloned the gene for insulin and expressed it in bacterial cells. Genentech, short for "genetic engineering technology," is also generally regarded as the world's first biotechnology company.

TABLE 19.1 Some Genetically Engineered Pharmaceutical Products Now Available or under Development

Gene Product	Condition Treated	Host Type
Erythropoitin	Anemia	*E. coli;* cultured mammalian cells
Interferons	Multiple sclerosis, cancer	*E. coli;* cultured mammalian cells
Tissue plasminogen activator tPA	Heart attack, stroke	Cultured mammalian cells
Human growth hormone	Dwarfism	Cultured mammalian cells
Monoclonal antibodies against vascular endothelial growth factor (VEGF)	Cancers	Cultured mammalian cells
Human clotting factor VIII	Hemophilia A	Transgenic sheep, pigs
C1 inhibitor	Hereditary angioedema	Transgenic rabbits
Recombinant human antithrombin	Hereditary antithrombin deficiency	Transgenic goats
Hepatitis B surface protein vaccine	Hepatitis B infections	Cultured yeast cells, bananas
Immunoglobulin IgG1 to HSV-2	Herpes virus infections	Transgenic soybeans
Recombinant monoclonal antibodies	Passive immunization against rabies (also used in diagnosing rabies), cancer, rheumatoid arthritis	Transgenic tobacco, soybeans, cultured mammalian cells
Norwalk virus capsid protein	Norwalk virus infections	Potato (edible vaccine)
E. coli heat-labile enterotoxin	*E. coli* infections	Potato (edible vaccine)

Previously, insulin was chemically extracted from the pancreas of cows and pigs obtained from slaughterhouses. **Insulin** is a protein hormone that regulates glucose metabolism. Individuals who cannot produce insulin have diabetes, a disease that, in its more severe form (type I), affects more than 2 million individuals in the United States. Although synthetic human insulin can now be produced by another process, a look at the original genetic engineering method is instructive, for it shows both the promise and the difficulty of applying recombinant DNA technology.

Clusters of cells embedded in the pancreas synthesize insulin, which consists of two polypeptide chains (the *A* and *B* chains) joined by disulfide bonds. The *A* subunit contains 21 amino acids, and the *B* subunit contains 30. In the original bioengineering process, synthetic genes that encode the *A* and *B* subunits were constructed and inserted into separate vectors, adjacent to the *lacZ* gene encoding the bacterial form of the enzyme β-galactosidase. When transferred to a bacterial host, the *lacZ* gene and the adjacent insulin oligonucleotide were transcribed and translated as a unit. The product is a **fusion protein**—that is, a hybrid protein consisting of the amino acid sequence for β-galactosidase attached to the amino acid sequence for one of the insulin subunits (Figure 19–1, p. 410). The fusion proteins were purified from bacterial extracts and treated with cyanogen bromide, a chemical that cleaves the fusion protein from the β-galactosidase. When the fusion products were mixed, the two insulin subunits spontaneously united, forming an intact, active insulin molecule. The purified injectable insulin was then packaged for use by diabetics.

Shortly after insulin became available, growth hormone—used to treat children who suffer from a form of dwarfism—was cloned. Soon, recombinant DNA technology made that product readily available too, as well as a wide variety of other medically important proteins that were once difficult to obtain in adequate amounts. Since recombinant insulin ushered in the biotechnology era, well over 200 recombinant protein products have entered the market worldwide.

Transgenic Animal Hosts and Pharmaceutical Products

Although bacteria have been widely used to produce therapeutic proteins, there are some disadvantages in using prokaryotic hosts to synthesize eukaryotic proteins. One problem is that bacterial cells often cannot process and modify eukaryotic proteins. As a result, they frequently cannot add the carbohydrates and phosphate groups to proteins that are needed for full biological activity of the protein. In addition, eukaryotic proteins produced in prokaryotic cells often do not fold into the proper three-dimensional configuration and are therefore inactive. To overcome these difficulties and increase yields, many biopharmaceuticals are now produced in eukaryotic hosts. As seen in Table 19.1, eukaryotic hosts may include cultured plant or animal cells or transgenic farm animals. A herd of goats or cows serve as very effective **bioreactors** or **biofactories**—living factories—that will continuously make milk containing the desired therapeutic protein, which can then be isolated in a noninvasive way.

Regardless of the host, therapeutic proteins may then be purified from the host cells—or when transgenic farm animals are used, isolated from animal products such as milk. An example of a biopharmaceutical product synthesized in transgenic animals is the human protein **α1-antitrypsin.** A deficiency of the enzyme α1-antitrypsin is associated with the heritable form of emphysema, a progressive and fatal respiratory disorder common among people of European ancestry. To produce α1-antitrypsin for therapeutic use, the human gene was cloned into a vector adjacent to a sheep promoter sequence that specifically activates transcription in milk-producing cells of the sheep mammary

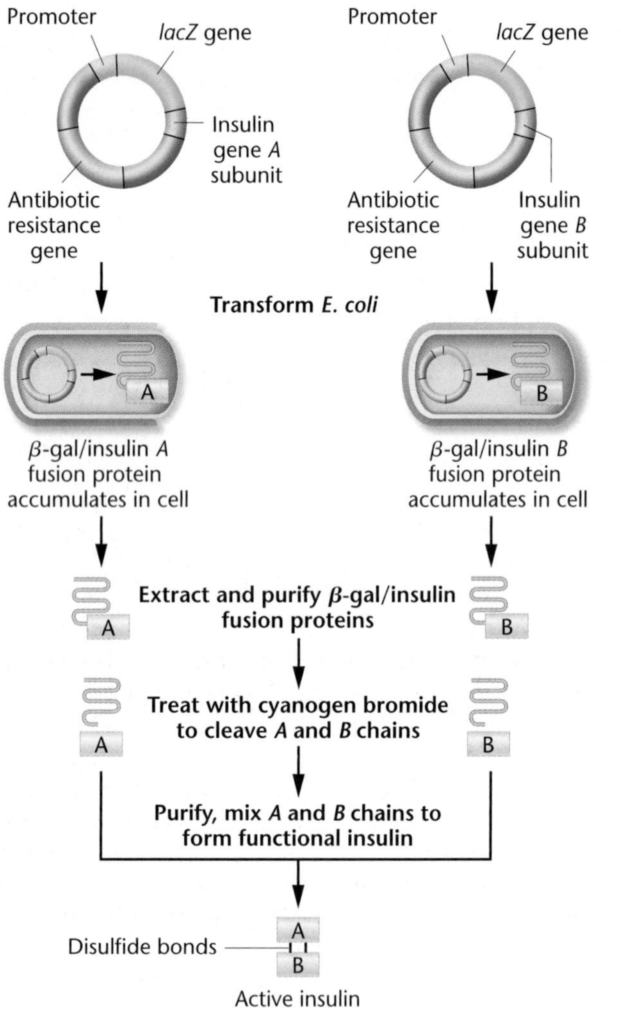
Promoter
lacZ gene
Insulin gene *A* subunit
Antibiotic resistance gene
Promoter
lacZ gene
Antibiotic resistance gene
Insulin gene *B* subunit
Transform *E. coli*
A
B
β-gal/insulin *A* fusion protein accumulates in cell
β-gal/insulin *B* fusion protein accumulates in cell
Extract and purify β-gal/insulin fusion proteins
A
B
Treat with cyanogen bromide to cleave *A* and *B* chains
A
B
Purify, mix *A* and *B* chains to form functional insulin
Disulfide bonds
A
B
Active insulin

FIGURE 19–1 A recombinant form of human insulin was the first therapeutic protein produced by recombinant DNA technology to be approved for use in humans. To synthesize recombinant insulin, synthetic oligonucleotides encoding the insulin *A* and *B* chains were inserted (in separate vectors) adjacent to a cloned *E. coli lacZ* gene. The recombinant plasmids were transformed into *E. coli,* where the β-gal/insulin fusion protein was synthesized and acumulated in the cells. Fusion proteins were then extracted from the host cells and the insulin subunits were purified and mixed to produce a functional insulin molecule.

glands. Genes placed next to this promoter are expressed only in mammary tissue. This fusion gene was microinjected into sheep zygotes fertilized *in vitro.* The fertilized zygotes were transferred to surrogate mothers. Resulting transgenic sheep developed normally and produced milk containing high concentrations of functional human α1-antitrypsin (up to 35 grams per liter of milk) that can be easily extracted and purified. A small herd of lactating transgenic sheep can provide an abundant supply of this protein.

In 2006, recombinant human **antithrombin,** an anticlotting protein, became the world's first therapeutic protein extracted from the milk of farm animals to be approved for use in humans. Scientists at GTC Biotherapeutics of Framingham, Massachusetts, introduced the antithrombin gene into goats. By placing the gene adjacent to a promoter for beta casein, a common protein in milk, GTC scientists were able to target antithrombin expression in the mammary gland, and the protein is highly expressed in the milk. In one year, a single goat will produce the equivalent amount of antithrombin that in the past would have been isolated from ~90,000 blood collections.

In a similar example involving a nonbiopharmaceutical application, transgenic "silk-milk" goats have been generated that express spider-silk proteins in their milk. These goats are a rich source of silk proteins used for applications such as manufacturing bulletproof vests.

Recombinant DNA Approaches for Vaccine Production and Transgenic Plants with Edible Vaccines

One of the most promising applications of recombinant DNA technology for therapeutic purposes may be the production of vaccines. Vaccines stimulate the immune system to produce antibodies against disease-causing agents and thereby confer immunity against specific diseases. Traditionally, two types of vaccines have been used: **inactivated vaccines,** which are prepared from killed samples of the infectious virus or bacteria; and **attenuated vaccines,** which are weakened live bacteria or viruses that can no longer reproduce but can cause a mild form of the disease. Inactivated vaccines include the vaccines for rabies and influenza; vaccines for tuberculosis and chickenpox are examples of attenuated vaccines.

Genetic engineering is being used to produce a relatively new type of vaccine called a **subunit vaccine,** which consists of one or more surface proteins from the virus or bacterium but not the entire virus or bacterium. This surface protein acts as an antigen that stimulates the immune system to make antibodies that act against the pathogen from which it was derived. One of the first subunit vaccines was made against the **hepatitis B virus,** which causes liver damage and cancer. The gene encoding the hepatitis B surface protein was cloned into a yeast expression vector, and it was expressed in yeast host cells. The protein was then extracted, purified, and packaged for use as a vaccine.

In 2005, the FDA approved **Gardasil,** a subunit vaccine produced by the pharmaceutical company Merck and the first cancer vaccine to receive FDA approval. Gardasil targets four strains of **human papillomavirus (HPV)** that cause ~70 percent of cervical cancers. Approximately 70 percent of sexually active women will be infected by an HPV strain during their lifetime. Gardasil is designed to provide immune protection against HPV prior to infection but is not effective against existing infections. You may have heard of Gardasil through media coverage of the legislation pending in several states that would require all adolescent female school children to receive a Gardasil vaccination regardless of whether or not they are sexually active.

Developing countries face serious difficulties in manufacturing, transporting, and storing vaccines. Most vaccines need refrigeration and must be injected under sterile conditions. In many rural areas, refrigeration and sterilization facilities are not available, and in certain cultures people are fearful of

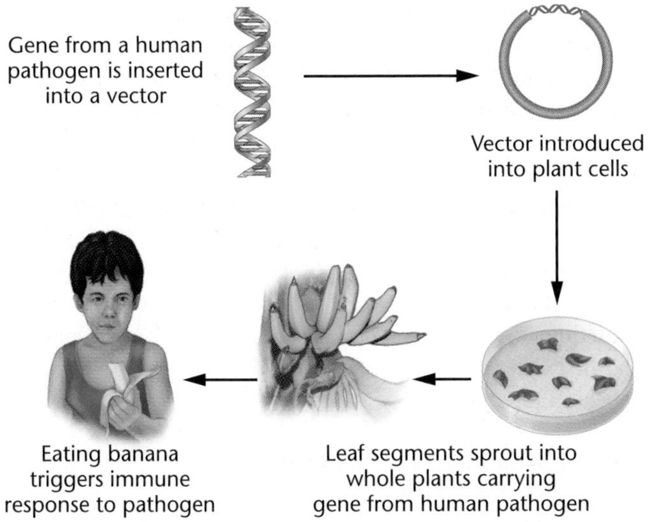
Gene from a human pathogen is inserted into a vector
Vector introduced into plant cells
Eating banana triggers immune response to pathogen
Leaf segments sprout into whole plants carrying gene from human pathogen

FIGURE 19–2 To make an edible vaccine, a gene from a pathogen (a disease-causing agent, such as a virus or bacterium) is placed into a vector, which is then introduced into plant cells. In this example, introduction of the vector into banana plant leaf segments transfers the vector and the pathogen's gene into the nuclei of banana leaf cells. The leaf segments are grown into mature banana trees that express the pathogenic gene. Eating the raw banana produced by these plants triggers an immune response to the protein encoded by the pathogen's gene, conferring immunity to infection by this pathogen.

being injected with needles. To overcome these problems, scientists are attempting to develop vaccines that can be synthesized in edible food plants (Figure 19–2). Successful vaccines would need to be inexpensive to produce, would not require refrigeration, and would not have to be given under sterile conditions by trained personnel.

Plants offer several advantages for expressing recombinant proteins. For instance, once a transgenic plant is made, it can easily be grown and replicated in a greenhouse or field, and it will provide a constant source of recombinant protein. In addition, the cost of expressing a recombinant protein in a transgenic plant is typically much lower than making the same protein in bacteria, yeast, or mammalian cells.

No recombinant proteins expressed in transgenic plants have yet been approved for use by the FDA. In one model system, the gene encoding an antigenic subunit of the hepatitis B virus has been introduced into the tobacco plant and expressed in its leaves. For use as a source of vaccine, however, the gene would have to be inserted into food plants such as grains, vegetables, or fruits. Some edible vaccines are now in clinical trials. For example, a vaccine against the bacterium that causes cholera has been produced in genetically engineered potatoes and used to successfully vaccinate human volunteers. But a number of technical questions about vaccine delivery in plants need to be answered. How can vaccine dose be carefully controlled when fruits and vegetables grow to different sizes and express different amounts of the vaccine? Will vaccine proteins pass through the digestive tract unaltered so that they maintain their ability to provide immune protection? Nevertheless, these are exciting prospects.

ESSENTIAL POINT

Recombinant DNA technology can be used to produce valuable biopharmaceutical protein products such as therapeutic proteins for treating disease.

NOW SOLVE THIS

Problem 5 on page 431 asks you to reflect on questions concerning the development of an edible vaccine.

Hint: Stimulation of antibody formation by the smallest possible portion of a protein is important to ensure vaccine specificity.

19.2 Genetic Engineering of Plants Has Revolutionized Agriculture

For millennia, farmers have manipulated the genetics of plants and animals to enhance food production. Until the advent of genetic engineering, these manipulations were largely restricted to **selective breeding**—the selection and breeding of naturally occurring or mutagen-induced variants. In the last 50 to 100 years, genetic improvement of crop plants through selective breeding and genetic crosses has resulted in dramatic increases in productivity and nutritional enhancement. For example, maize yields have increased fourfold over the last 60 years, and more than half of this increase is due to genetic improvement by selective breeding (Figure 19–3). Modern maize has substantially

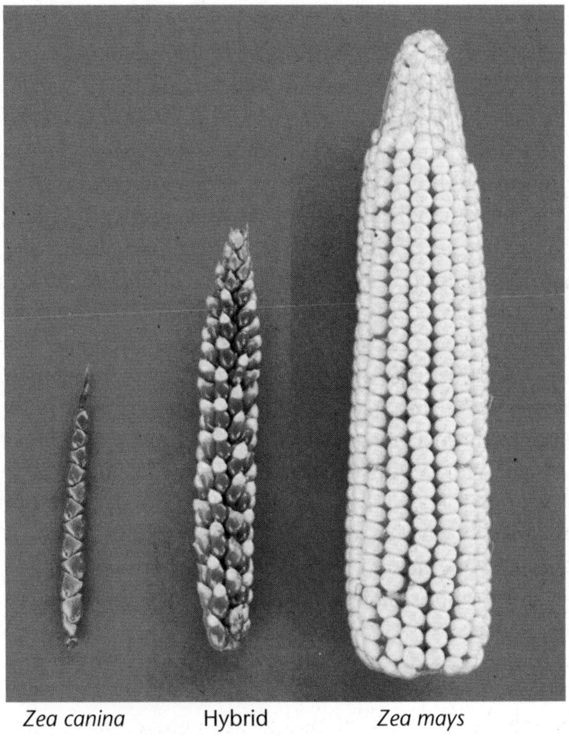
Zea canina
Hybrid
Zea mays

FIGURE 19–3 Selective breeding is one of the oldest methods of genetic alteration of plants. Shown here is teosinte (*Zea canina*, left), a selectively bred hybrid (center), and modern corn (*Zea mays*).

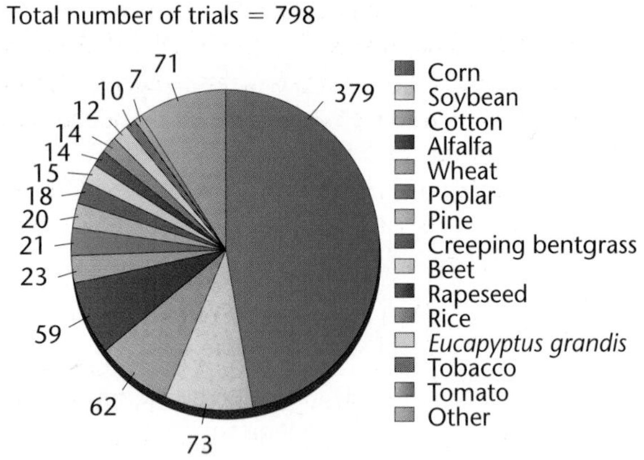

FIGURE 19–4 A recent analysis of nearly 800 transgenic crop trials worldwide shows that GM varieties of corn, soybean, cotton, alfalfa, and wheat are among the most commonly manipulated crops (*Nature Biotechnology,* 23(3), p. 281, March 2005).

larger ears and kernels than predecessor crops, including hybrids from which it was bred.

Recombinant DNA technology provides powerful new tools for altering the genetics of agriculturally important organisms. Scientists can now identify, isolate, and clone genes that confer desired traits, then specifically and efficiently incorporate these into organisms. As a result, it is possible to introduce insect resistance, herbicide resistance, or nutritional characteristics into farm plants and animals, a primary purpose of **agricultural biotechnology.** In this section we consider genetic manipulations to produce transgenic crop plants of agricultural value. In Section 19.3, we discuss examples of genetic manipulations of agriculturally important animals.

Worldwide, over 300 million acres were planted with genetically engineered crops in 2008, particularly herbicide- and pest-resistant soybeans, corn, cotton, and canola; over 50 different transgenic crop varieties have been generated, including alfalfa, corn, rice, potatoes, tomatoes, tobacco, wheat, and cranberries (Figure 19–4). Genetically modified (GM) crops are planted by over 8 million farmers in 17 countries. Both industrialized and developing countries are taking advantage of transgenic crops. Since 1996, there has been a 4000 percent increase in GM crop acreage worldwide. In 2005, the ten-year anniversary of commercial biotech crops, the one-billionth biotech acre was planted. American farmers planted 111 million acres of GM corn, soybeans, and cotton in 2004, a 17 percent increase from the year before. In the United States 86 percent of the soybeans, 78 percent of the cotton, and 46% of corn are genetically engineered to resist pests or herbicides.

Several of the main reasons for generating transgenic crops include:

- Improving the growth characteristics and yield of agriculturally valuable crops
- Increasing the nutritional value of crops
- Providing crop resistance against insect and viral pests, drought, and herbicides

The first commercially available GM food was called the Flavr Savr tomato. The Flavr Savr was designed to increase the shelf life of tomatoes by allowing them to stay ripe for several weeks without softening. Scientists used antisense RNA technology to inhibit an enzyme called polygalacturonase, which digests pectin in the cell wall of tomatoes. Pectin digestion occurs naturally as tomatoes ripen, and this process is a main reason tomatoes soften as they age. This GM approach was generally effective, but the attempts to remedy shipping problems that continued to cause bruising increased the cost of these tomatoes. This and public skepticism about the safety of the first GM food are two reasons Flavr Savr was eventually taken off the market.

Transgenic Crops for Herbicide and Pest Resistance

Damage from weed infestation destroys about 10 percent of crops worldwide. In an attempt to combat this problem, farmers often apply herbicides to the soil to kill weeds prior to seeding a field crop. As the most efficient herbicides also kill crop plants, herbicide uses are limited. The creation of herbicide-resistant crops has opened the way to efficient weed control and increased yields of some major agricultural crops. At present, over 75 percent of soybeans and cotton in the United States are resistant to the herbicide **glyphosate.** You may be familiar with glyphosate because it is the active herbicide in Roundup, which is commonly used to keep sidewalks and patios weed-free (see Figure 19–5).

Glyphosate is effective at very low concentrations, is not toxic to humans, and is rapidly degraded by soil microorganisms. It kills plants by inhibiting the action of a chloroplast enzyme called **EPSP synthase.** This enzyme is important in amino acid biosynthesis in both bacteria and plants. Without the ability to synthesize vital amino acids, plants wither and die.

Agrobacterium tumefaciens is a soil microbe that can infect wounded plants and create crown gall tumors. *A. tumefaciens* contains **Ti plasmids,** so named because they contain tumor-inducing genes. Modified versions of Ti plasmids that lack tumor-inducing genes and contain other features, such as antibiotic resistance, have been widely used as vectors for introducing genes into plants. To produce a glyphosate-resistant

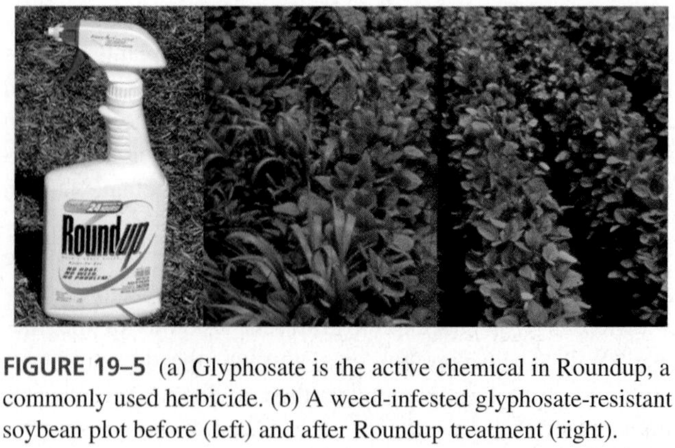

FIGURE 19–5 (a) Glyphosate is the active chemical in Roundup, a commonly used herbicide. (b) A weed-infested glyphosate-resistant soybean plot before (left) and after Roundup treatment (right).

crop plant, researchers began by isolating and cloning an EPSP synthase gene from a glyphosate-resistant strain of *E. coli.* Next, they cloned the *EPSP* gene into a Ti plasmid between promoter sequences derived from a plant virus and transcription termination sequences derived from a plant gene (Figure 19–6). This recombinant vector was introduced into *A. tumefaciens.*

The Ti plasmid-carrying bacteria were then used to infect plant cells derived from plant leaves. The clumps of cells (calluses) that formed after infection with *A. tumefaciens* were grown into transgenic plants and sprayed with glyphosate at concentrations four times higher than that needed to kill wild-type plants. Transgenic plants that expressed EPSP synthase grew and developed, while control plants withered and died. Figure 19–5(b) demonstrates the effectiveness of glyphosate as a herbicide and the resistance of glyphosate-resistant soybeans.

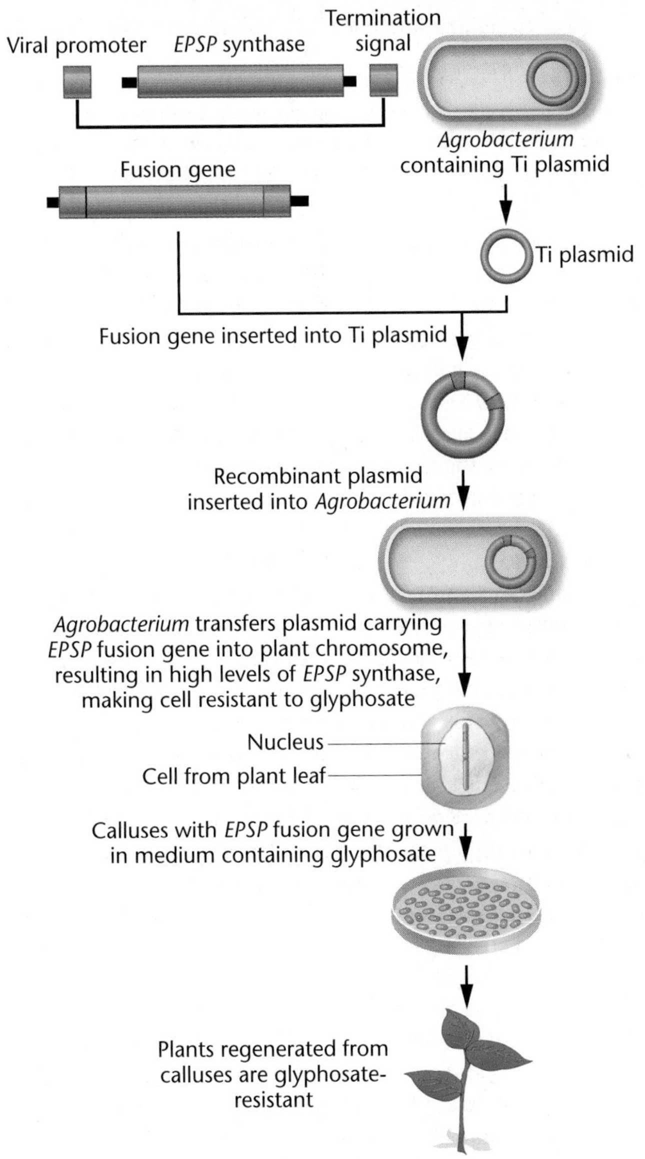
Viral promoter
EPSP synthase
Termination signal
Agrobacterium containing Ti plasmid
Fusion gene
Ti plasmid
Fusion gene inserted into Ti plasmid
Recombinant plasmid inserted into *Agrobacterium*
Agrobacterium transfers plasmid carrying *EPSP* fusion gene into plant chromosome, resulting in high levels of *EPSP* synthase, making cell resistant to glyphosate
Nucleus
Cell from plant leaf
Calluses with *EPSP* fusion gene grown in medium containing glyphosate
Plants regenerated from calluses are glyphosate-resistant

FIGURE 19–6 To create glyphosate-resistant transgenic plants, the EPSP synthase gene from bacteria is fused to a promoter such as the promoter from the cauliflower mosaic virus. This fusion gene is then ligated into a Ti-plasmid vector, and the recombinant vector is transformed into an *Agrobacterium* host. *Agrobacterium* infection of cultured plant cells transfers the EPSP synthase fusion gene into a plant-cell chromosome. Cells that acquire the gene are able to synthesize large quantities of EPSP synthase, making them resistant to the herbicide glyphosate.

Similar techniques have been used to make plants resistant to other herbicides, to pathogens such as viruses, and also to insects. Some of the most well-described and controversial GM crops are **Bt crops,** designed to be resistant to insects. The bacterium *Bacillus thuringiensis* (Bt) produces a protein that when ingested by insects and larvae will crystallize in the gut, killing pests such as corn-borer larvae that are responsible for millions of dollars of crop damage worldwide. Initially, applications of Bt involved spraying these bacteria on crops. But recombinant DNA technology has enabled scientists to produce Bt transgenic crops with built-in insecticide protection. The *cry* genes that encode the Bt crystalline protein have been effectively introduced into a number of different crops, including corn, cotton, tomatoes, and tobacco.

Bt crops have been hailed as a great success story of agricultural biotechnology, but they have also been one of the most controversial. Some studies had suggested a correlation between decreases in Monarch butterfly populations and ingestion of pollen from Bt corn (Monarchs do not feed on the corn itself). More recently, several long-term studies demonstrated that exposure to Bt crops has no apparent effects on the Monarch; however, the possibility of danger to nontarget insect species must be considered whenever pest-resistant plants are used in the wild. Based on the success of Bt crops, many other transgenic crops are under development, including plants with increased tolerance to viral pests, drought, and salty soils.

NOW SOLVE THIS

Problem 4 on page 431 asks you to predict whether the gene conferring glyphosate resistance could escape from a transgenic crop into weed plants.

Hint: Consider the methods by which plants breed and the types of selective pressures on both wild plants and domestic crops.

Nutritional Enhancement of Crop Plants

Because crop plants are deficient in some of the nutrients required in the human diet, biotechnology is being used to produce crops with enhanced nutritional value. One example is the production of **"golden rice,"** with enhanced levels of β-carotene, a precursor to vitamin A (Figure 19–7, p. 414). Vitamin A deficiency is prevalent in many areas of Asia and Africa, and more than 500,000 children a year become permanently blind as a result of this deficiency. Rice is a major staple food in these regions but does not contain vitamin A.

To create golden rice, scientists transferred into the rice genome three genes encoding enzymes required for the biosynthetic pathway leading to β-carotenoid synthesis. Two genes came from the daffodil and one from a bacterium. New

FIGURE 19–7 Golden rice, a strain genetically modified to produce β-carotene, a precursor to vitamin A.

varieties with higher levels of β-carotene production have been developed. One such strain, Golden Rice 2, incorporates the phytoene synthase gene from maize and produces about 20 times more β-carotene than the original golden rice. Acceptance of genetically engineered varieties of rice and efficient distribution of golden rice remain challenges to its wider use. Nonetheless, encouraged by the effectiveness of Golden Rice 2, researchers are working on developing rice with enhanced iron and protein content.

Many other varieties of nutritionally enhanced food crops have been, and are, under development. These include plants with augmented levels of key fatty acids, antioxidants, and other vitamins and minerals. These efforts are directed at addressing nutrient deficiencies affecting more than 40 percent of the world's population. Other expected developments include decaffeinated teas and coffees, as well as crops enhanced for traits affecting taste, growth rates, yields, color, storage, ripening, and similar characteristics.

ESSENTIAL POINT

Genetically modified (GM) plants, designed to improve crop yield and nutritional value and to increase resistance to herbicides, pests, and severe weather, are becoming prevalent worldwide.

19.3 Transgenic Animals with Genetically Enhanced Characteristics Have the Potential to Serve Important Roles in Agriculture and Biotechnology

Although genetically engineered plants are major players in modern agriculture, transgenic animals are less widespread. Nonetheless, some high-profile examples of engineered farm animals have aroused public interest and controversy. Oversize mice containing a human growth hormone transgene were some of the first transgenic animals created. Attempts to create farm animals containing transgenic growth hormone genes have not been particularly successful, probably because growth is a complex, multigene trait. One notable exception is the transgenic Atlantic salmon, bearing copies of a Chinook salmon growth hormone gene adjacent to a constitutive promoter. These salmon mature quickly, grow 400 to 600 percent faster than nontransgenic salmon, weigh approximately ten times more than age-matched nontransgenic salmon, and appear to have no adverse health effects from the added gene (Figure 19–8).

FIGURE 19–8 Transgenic Atlantic salmon (bottom) overexpressing a growth hormone gene display rapidly accelerated rates of growth compared to wild strains and nontransgenic domestic strains (top) and weigh an average of nearly ten times more than nontransgenic strains.

As discussed in Section 19.1, currently, the major uses for transgenic farm animals are as bioreactors to produce useful pharmaceutical products, but other interesting transgenic applications are under development. Several of these applications are designed to increase milk production or enhance the nutritional value of milk. Significant research efforts are also being made to protect farm animals against common pathogens that cause disease and animal loss (including potential bioweapons that could be used in a terrorist attack on food animals) and put the food supply at risk. For instance, controlling mastitis in cattle by creating transgenic cows has shown promise (Figure 19–9). **Mastitis** is an infection of the mammary glands. It is the most costly disease affecting the dairy industry, leading to over $2 billion in losses in the United States. Mastitis can block milk ducts, reducing milk output, and can also contaminate the milk with pathogenic microbes. Infection by the bacterium *Staphylococcus aureus* is the most common cause of mastitis, and most cattle with mastitis typically do not respond well to conventional treatments with antibiotics. As a result, mastitis is a significant cause of herd reduction.

In an attempt to create cattle resistant to mastitis, transgenic cows were generated that possessed the lysostaphin gene from *Staphylococcus simulana.* Lysostaphin is an enzyme that specifically cleaves components of the *S. aureus* cell wall. Transgenic cows expressing this protein in milk produce a natural antibiotic that wards off *S. aureus* infections. These transgenic cows do not completely solve the mastitis problem because lysostaphin is not effective against other microbes that

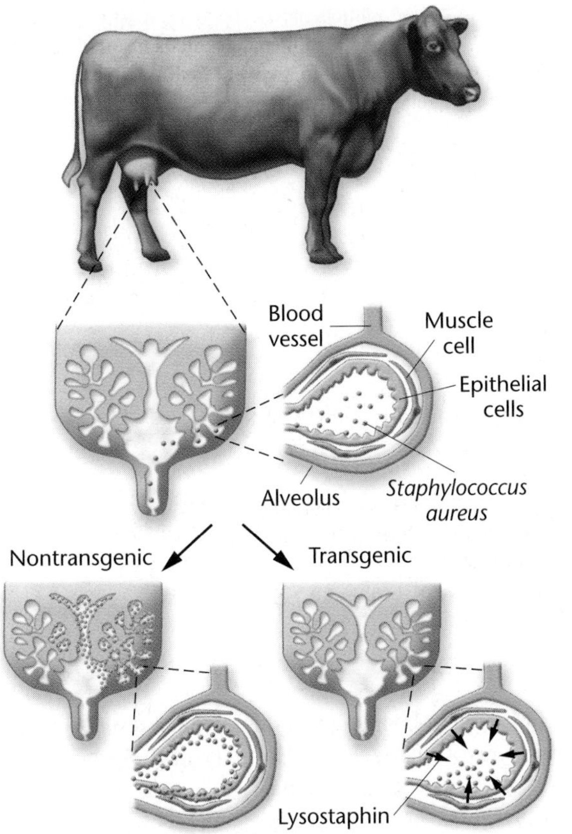

FIGURE 19–9 Transgenic cows for battling mastitis. The mammary glands of nontransgenic cows are highly susceptible to infection by the skin microbe *Staphylococcus aureus.* Transgenic cows express the lysostaphin transgene in milk, where it can kill *S. aureus* before they can multiply in sufficient numbers to cause inflammation and damage mammary tissue.

can cause mastitis; moreover, there is also the potential that *S. aureus* may develop resistance to lysostaphin. Nonetheless, scientists are cautiously optimistic that transgenic approaches have a strong future for providing farm animals with a level of protection against major pathogens. Another successful transgenic farm animal is **EnviroPig,** a pig that expresses the gene encoding the enzyme phytase. These pigs are able to break down dietary phosphorus, thereby reducing amounts of phosphorus in the urine and feces, which is a major environmental pollutant created by pig farming.

Scientists at Yorktown Industries of Austin, Texas, created the **GloFish,** a transgenic strain of zebrafish (*Danio rerio*) containing a red fluorescent protein gene from sea anemones. Marketed as the first GM pet in the United States, GloFish fluoresce bright pink when illuminated by ultraviolet light. GM critics describe these fish as an abuse of genetic technology. However, GloFish may not be as frivolous a use of genetic engineering as some believe. Recently, a variation of this transgenic model, incorporating a heavy-metal-inducible promoter adjacent to the red fluorescent protein gene, has shown promise in a bioassay for heavy metal contamination of water. When these transgenic zebrafish are in water contaminated by mercury and other heavy metals, the promoter becomes activated, inducing transcription of the red fluorescent protein gene. In this way, zebrafish fluorescence can be used as a bioassay to measure heavy metal contamination and uptake by living organisms.

ESSENTIAL POINT

Transgenic animals with improved growth characteristics or desirable phenotypes are being genetically engineered for a number of different applications.

19.4 Genetic Engineering and Genomics Are Transforming Medical Diagnosis

Geneticists are now applying knowledge about the human genome and the genetic basis of many diseases to a wide range of medical applications. Gene-based technologies have already had a major impact on the diagnosis of disease and are revolutionizing medical treatments together with the development of specific and effective pharmaceuticals.

Genetic Tests Based on Restriction Enzyme Analysis

Using DNA-based tests, scientists can directly examine a patient's DNA for mutations associated with disease. Gene testing was one of the first successful applications of recombinant DNA technology, and currently more than 900 gene tests are in use. These tests usually detect DNA mutations associated with single-gene disorders that are inherited in a Mendelian fashion. Examples of such genetic tests are those that detect sickle-cell anemia, cystic fibrosis, Huntington disease, hemophilias, and muscular dystrophies. Other genetic tests have been developed for complex disorders such as breast and colon cancers. Gene tests are used to perform prenatal diagnosis of genetic diseases, to identify carriers, to predict the future development of disease in adults, to confirm the diagnosis of a disease detected by other methods, and to identify genetic diseases in embryos created by *in vitro* fertilization.

For genetic testing of adults, DNA from white blood cells is commonly used. Alternatively, genetic tests are commonly carried out on cheek cells collected by swabbing the inside of the mouth. Some genetic testing can be carried out on gametes. For prenatal diagnosis, fetal cells are obtained by **amniocentesis** or **chorionic villus sampling.** Figure 19–10, p. 416, shows the procedure for amniocentesis, in which a small volume of the amniotic fluid surrounding the fetus is removed. Amniotic fluid contains fetal cells that can be used for karyotyping, genetic testing, and other procedures. For chorionic villus sampling, cells from the fetal portion of the placenta (the chorionic villi) are sampled through a vacuum tube, and analyses can be carried out on this tissue.

A classic method of genetic testing is **restriction fragment length polymorphism (RFLP) analysis.** To illustrate this method, we examine the prenatal diagnosis of **sickle-cell anemia.** As we have discussed before, this disease is an

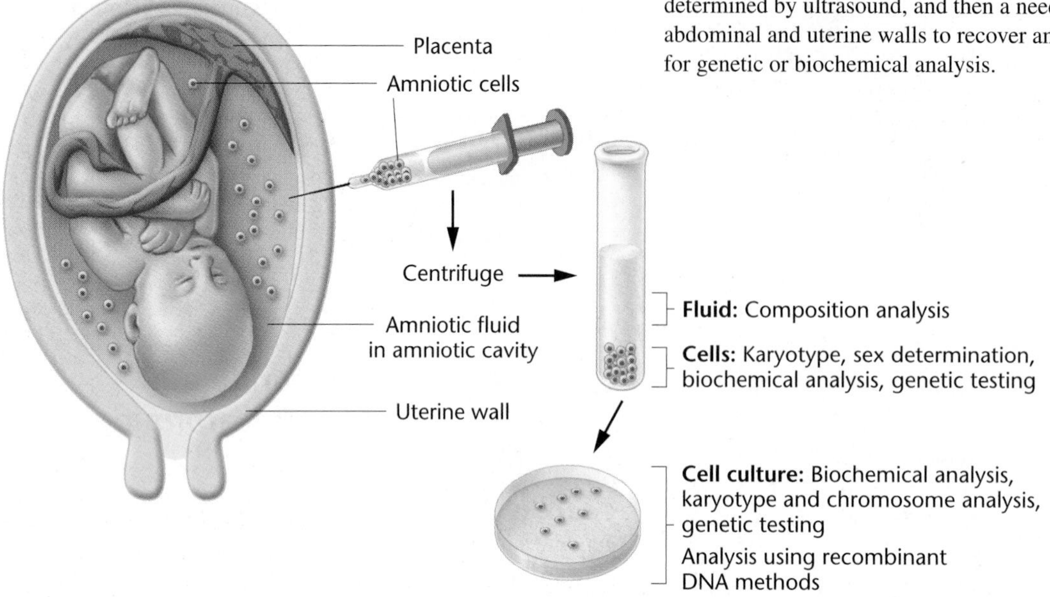

FIGURE 19–10 For amniocentesis, the position of the fetus is first determined by ultrasound, and then a needle is inserted through the abdominal and uterine walls to recover amniotic fluid and fetal cells for genetic or biochemical analysis.

autosomal recessive condition caused by a single amino acid substitution in the β-globin protein, as a consequence of a single-nucleotide substitution in the β-globin gene. This mutation eliminates a cutting site in the gene for the restriction enzymes *Mst*II and *Cvn*I. These differences in restriction cutting sites are used to prenatally diagnose sickle-cell anemia and to establish the parental genotypes and the genotypes of other family members who may be heterozygous carriers of this condition.

DNA is extracted from tissue samples and digested with *Mst*II. This enzyme cuts three times within a region of the normal β-globin gene, producing two small DNA fragments. In the mutant sickle-cell allele, the middle *Mst*II site is destroyed by the mutation, and one large restriction fragment is produced by *Mst*II digestion (Figure 19–11). The restriction-enzyme-digested DNA fragments are separated by gel electrophoresis, transferred to a nylon membrane, and visualized by Southern blot hybridization, using a probe from this region. Alternatively, a gene can be amplified by PCR, subjected to RFLP analysis and analyzed by agarose gel electrophoresis without Southern blotting. This approach is much faster because it eliminates the time required for Southern blot hybridization.

RESTRICTION FRAGMENT LENGTH POLYMORPHISMS — WEB TUTORIAL

Figure 19–11 shows the results of RFLP analysis for sickle-cell anemia in one family. Both parents (I-1 and I-2) are heterozygous carriers of the mutation. *Mst*II digestion of the parents' DNA produces a large band (because of the mutant allele) and two smaller bands (from the normal allele) in each case. The parents' first child (II-1) is homozygous normal because she has only the two smaller bands. The second child (II-2) has sickle-cell anemia; he has only one large band and is homozygous for the mutant allele. The fetus (II-3) has a large band and two small bands and is therefore heterozygous for sickle-cell anemia. He or she will be unaffected but will be a carrier.

Only about 5 to 10 percent of all point mutations can be detected by restriction enzyme analysis because most mutations affecting protein-coding genes do not affect restriction enzyme cutting sites in the genome. However, if the gene of interest has been sequenced and the disease-associated mutations are known, geneticists can employ oligonucleotides to detect these mutations, as described next.

NOW SOLVE THIS

Problem 22 on page 432 asks you to analyze RFLP data to determine if a fetus is affected by β-thalassemia.

Hint: Consider what is known about the genotypes and the RFLP results for each parent.

Genetic Tests Using Allele-Specific Oligonucleotides

Another method of genetic testing involves the use of **allele-specific oligonucleotides (ASOs).** Scientists use these short, synthetic, single-stranded fragments of DNA to identify alleles that differ by as little as a single nucleotide. In contrast to RFLP analysis, which is limited to cases for which a mutation changes a restriction site, ASOs detect single-nucleotide changes (**single-nucleotide polymorphisms or SNPs**), including those that do not affect restriction enzyme cutting sites. As a result, this method offers increased resolution and wider application. Under proper conditions, an ASO will hybridize only with its complementary DNA sequence and not with other sequences, even those that vary by as little as a single nucleotide.

Genetic testing using ASOs and PCR analysis are now available to screen for many disorders, including sickle-cell anemia. In the case of sickle-cell screening, DNA is extracted, and a region of the β-globin gene is amplified by PCR. A small amount of the amplified DNA is spotted onto strips of a DNA-binding membrane, and each strip is

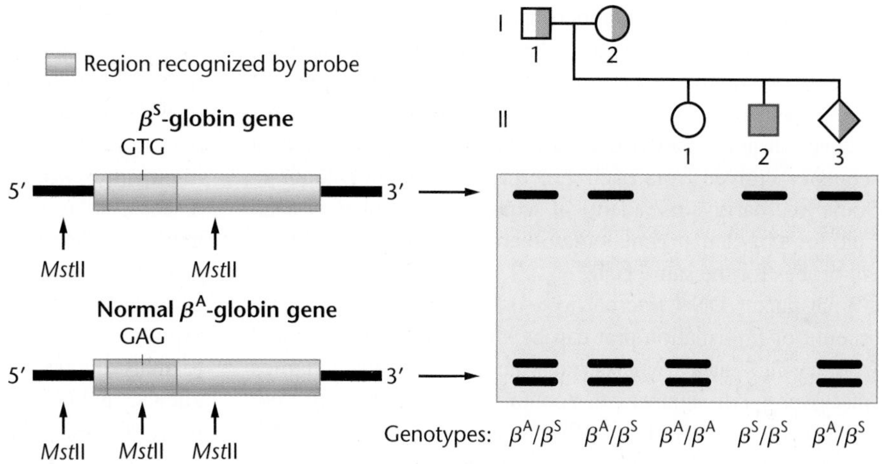

FIGURE 19–11 RFLP diagnosis of sickle-cell anemia. In the mutant β-globin allele (β^S), a point mutation (GAG → GTG) has destroyed a cutting site for the restriction enzyme *Mst*II, resulting in a single large fragment on a Southern blot. In the pedigree, the family has one unaffected homozygous normal daughter (II-1), an affected son (II-2), and an unaffected carrier fetus (II-3). The genotype of each family member can be read directly from the blot and is shown below each lane.

hybridized to an ASO synthesized to resemble the relevant sequence from either a normal or mutant β-globin gene (Figure 19–12). The ASO is tagged with a molecule that is either radioactive or fluorescent, in order to allow visualization of hybridization of the ASO on the membrane. This rapid, inexpensive, and highly accurate technique is used to diagnose a wide range of genetic disorders caused by point mutations. Although highly effective, SNPs can affect probe binding leading to false positive or false negative results that may not reflect a genetic disorder. Sometimes DNA sequencing is carried out on amplified gene segments to confirm identification of a mutation. Because the Human Genome Project has revealed point mutations involved in many human diseases, the use of ASOs for genetic testing is increasing.

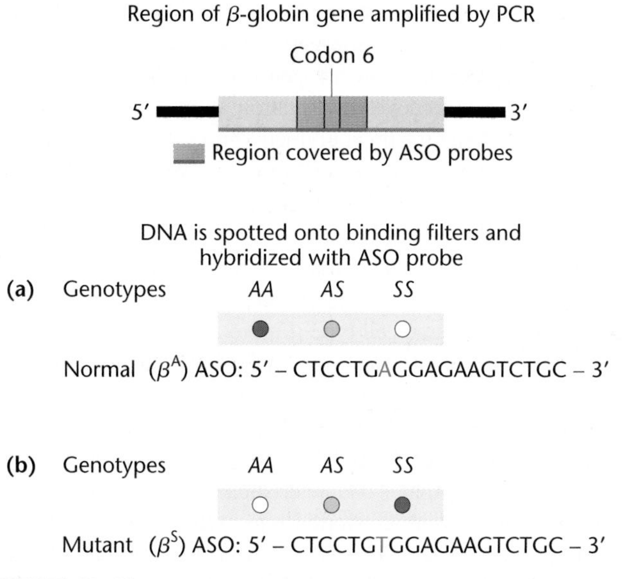

FIGURE 19–12 Allele-specific oligonucleotide (ASO) testing for the β-globin gene and sickle-cell anemia. (a) Results observed when DNA for the three possible genotypes are hybridized to an ASO from the normal β-globin allele: *AA*-homozygous individuals have normal hemoglobin that has two copies of the normal β-globin gene and will show heavy hybridization; *AS*-heterozygous individuals carry one normal β-globin allele and one mutant allele and will show weaker hybridization; *SS*-homozygous sickle-cell individuals carry no normal copy of the β-globin gene and will show no hybridization to the ASO probe for the normal β-globin allele. (b) Results observed when DNA for the three genotypes are hybridized to the probe for the sickle-cell β-globin allele: no hybridization by the *AA* genotype, weak hybridization by the heterozygote (*AS*), and strong hybridization by the homozygous sickle-cell genotype (*SS*).

Because ASO testing makes use of PCR, small amounts of DNA can be analyzed. As a result, ASO testing is ideal for **preimplantation genetic diagnosis (PGD),** the genetic analysis of single cells from embryos created by *in vitro* fertilization. When sperm and eggs are mixed to create zygotes, the early-stage embryos are grown in culture. A single cell can be removed from an early-stage embryo using a vacuum pipette to gently aspirate one cell away from the embryo. The embryo most often continues to divide normally. DNA from the removed cell is then typically analyzed by fluorescent *in situ* hybridization (FISH, for chromosome analysis) or by ASO testing. The genotypes for each cell can then be used to decide which embryos will be implanted into the uterus. Any alleles that can be detected by ASO testing can be used for PGD. Sickle-cell anemia, cystic fibrosis, and dwarfism are often tested for by PGD, but many other alleles can be analyzed.

Genetic Testing Using DNA Microarrays and Genome Scans

Both RFLP and ASO analyses are efficient methods of screening for gene mutations; however, they can only detect the presence of one or a few specific mutations whose identity and locations in the gene are known. There is also a need for genetic tests that detect complex mutation patterns or previously unknown mutations in genes associated with genetic diseases and cancers. Recall from Chapter 18 that one emerging technique is based on the use of **DNA microarrays.**

DNA microarrays (also called DNA or gene chips) are small, solid supports, usually glass or polished quartz-based, on which known fragments of DNA are deposited in a precise pattern. Each spot on a DNA microarray is called a **field.** DNA fragments that are deposited in a microarray field—called probes—are single-stranded and may be oligonucleotides synthesized *in vitro* or longer fragments of DNA created from cloning or PCR amplification. There are typically over a million identical molecules of DNA in each field (see Figure 18–16). The numbers and types of DNA sequences on a microarray are dictated by the type of analysis that is required. For example, each field on a microarray might contain a DNA sequence from each member of a gene family, or sequence variants from one or several genes of interest, or a sequence derived from each gene in an organism's genome. Some microarrays use identical sequences as probes in each field for a particular gene, other microarrays use many different probes for the same gene.

What makes DNA microarrays so amazing is the immense amount of information that can be simultaneously generated from a single array. Microarrays the size of postage stamps (just over 1 cm square) can contain up to 500,000 different fields, each representing a different sequence. In Chapter 18, you learned about the use of microarrays for transcriptome analysis. Scientists are now using DNA microarrays in a range of applications, including the detection of mutations in genomic DNA and the detection of gene expression patterns in diseased tissues.

Most human genes are available on a human genome microarray (Figure 19–13). Geneticists often use a type of DNA microarray known as a **genotyping microarray** to detect mutations in specific genes. Probes on a genotyping microarray consist of short oligonucleotides, about 20 nucleotides long. These probes are designed to methodically scan through the gene of interest, one nucleotide at a time, checking for the presence of a mutation at each position in the gene. Each position in the gene is tested by a set of five oligonucleotides (and hence five fields, arranged in a column) that are identical in sequence except for one nucleotide that differs in each of them, being either A, C, G, T, or a deletion. DNA microarrays have been designed to scan for mutations in many disease-related genes, including the *p53* gene, which is mutated in a majority of human cancers, and the *BRCA1* gene, which, when mutated, predisposes women to breast cancer.

In addition to testing for mutations in single genes, DNA microarrays can contain probes that detect SNPs. SNPs occur randomly about every 100 to 300 nucleotides throughout the human genome. Scientists have discovered that certain SNPs at a specific locus are shared by certain segments of the population. In addition, certain SNPs cosegregate with genes associated with some disease conditions. By correlating the presence or absence of a particular SNP with a genetic disease, scientists are able to use the SNP as a genetic testing marker. SNP microarrays allow scientists to simultaneously screen thousands of genes that might be involved in single-gene diseases as well as those involved in disorders exhibiting multifactorial inheritance. This technique makes it possible to analyze a person's DNA for dozens or hundreds of disease alleles, including those that might predispose the person to heart attacks, asthma, diabetes, Alzheimer disease, and other genetically defined disease subtypes. Perhaps in decades to come, all newborns will undergo SNP testing to determine their lifetime risks for suffering from genetic disorders.

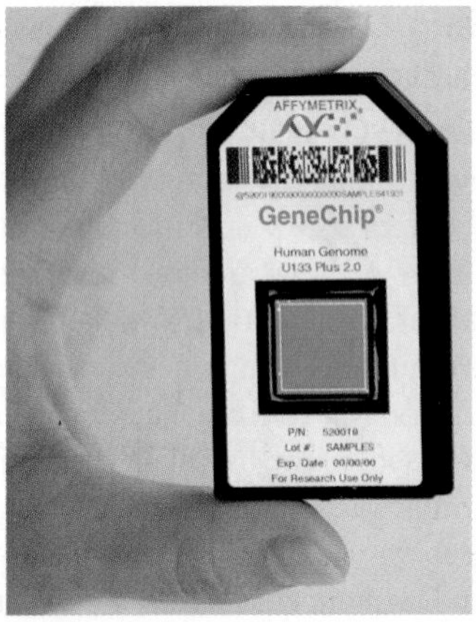

FIGURE 19–13 A commercially available DNA microarray, called a GeneChip, marketed by Affymetrix, Inc. This microarray can be used to analyze expression for approximately 50,000 RNA transcripts. It contains 22 different probes for each transcript and allows scientists to simultaneously assess the expression levels of most of the genes in the human genome.

Genetic Analysis Using Gene Expression Microarrays

Gene expression microarrays are effective for analyzing gene expression patterns in genetic diseases because the progression of a tissue from a healthy to a diseased state is almost always accompanied by changes in expression of hundreds to thousands of genes. These arrays provide a powerful tool for diagnosing genetic disorders and gene expression changes. Expression microarrays may contain probes for only a few specific genes thought to be expressed differently in two cell types or probes representing each gene in the genome. Although microarray techniques provide novel information about gene expression, keep in mind that they do not directly provide us with information about protein levels in a cell or tissue. We often infer what protein levels may be based on mRNA expression patterns, but this may not always be accurate.

In a typical expression microarray analysis, mRNA is isolated from two sources—for example, normal cells and cancer cells arising from the same cell type [Figure 19–14(a)]. The mRNA samples contain transcripts from each gene that is expressed in that cell type. Some genes are expressed more efficiently than others; therefore, each type of mRNA is present at a different level. The level of each mRNA can be used to develop a gene expression profile that is characteristic of the cell type. Isolated mRNA molecules are converted into cDNA molecules, using reverse transcriptase. The cDNAs from the normal cells are tagged with fluorescent dye-labeled nucleotides (for example, green), and the cDNAs from the cancer cells are tagged with a different fluorescent dye-labeled nucleotide (for example, red). The labeled cDNAs are mixed together and applied to a DNA microarray. The cDNA molecules bind to complementary single-stranded probes on the microarray but not to other probes. After washing off the nonbinding cDNAs, scientists scan the microarray with a laser, and a computer captures the fluorescent image pattern for analysis. The pattern of hybridization appears as a series of colored dots, with each dot corresponding to one field of the microarray [Figure 19–14(b)].

The colors on the microarray fields [Figure 19–14(c)] provide a sensitive measure of the relative levels of each cDNA in the mixture. In the example shown here, if an mRNA is

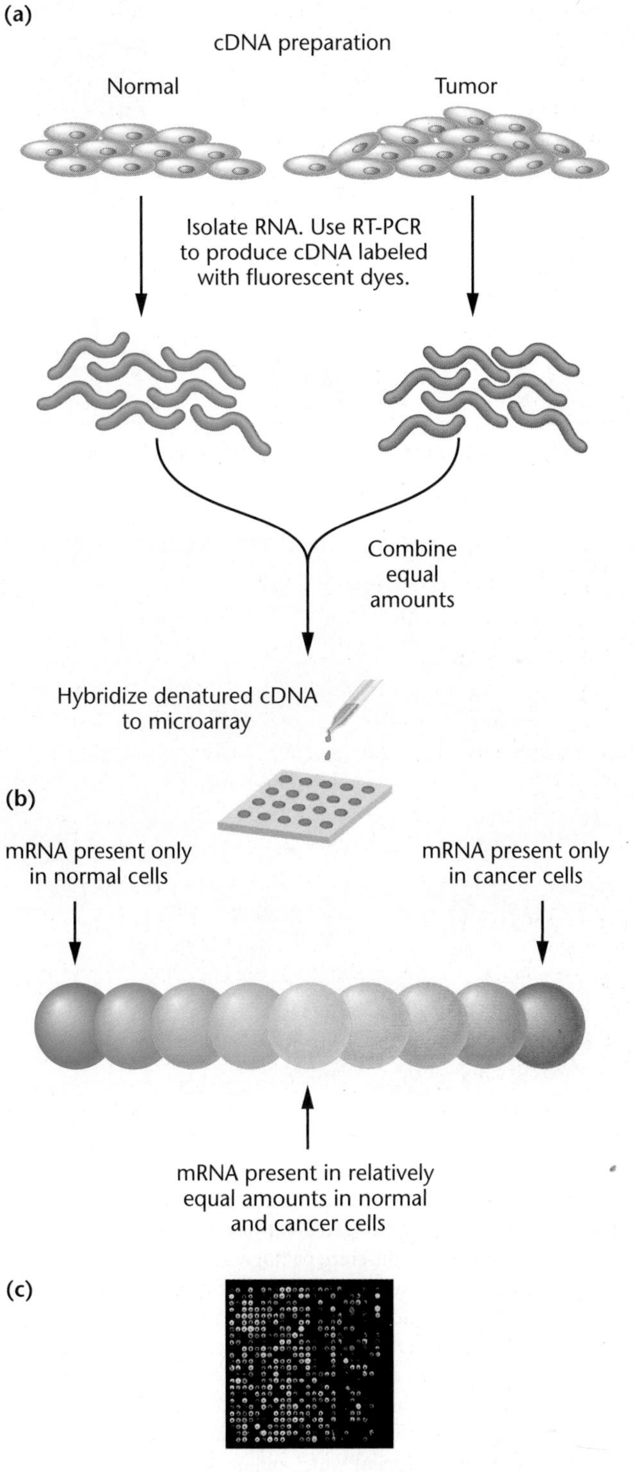
(a)
cDNA preparation
Normal
Tumor
Isolate RNA. Use RT-PCR to produce cDNA labeled with fluorescent dyes.
Combine equal amounts
Hybridize denatured cDNA to microarray
(b)
mRNA present only in normal cells
mRNA present only in cancer cells
mRNA present in relatively equal amounts in normal and cancer cells
(c)

FIGURE 19–14 (a) Microarray procedure for analyzing gene expression in normal and cancer cells. (b) The method shown here is based on a two-channel microarray in which cDNA samples from the two different tissues are competing for binding to the same probe sets. Colors of dots on an expression microarray represent levels of gene expression. In this example, green dots represent genes expressed only in one cell type (e.g., normal cells), and red dots represent genes expressed only in another cell type (e.g., cancer cells). Intermediate colors represent different levels of expression of the same gene in the two cell types. (c) A small portion of a DNA microarray, showing different levels of hybridization to each field.

present only in normal cells, the probe representing the gene encoding that mRNA will appear as a green dot because only "green" cDNAs have hybridized to it. Similarly, if an mRNA is present only in the cancer cells, the probe for that gene will appear as a red dot. If both samples contain the same cDNA, in the same relative amounts, both cDNAs will hybridize to the same field, which will appear yellow [Figure 19–14(b)]. Intermediate colors indicate that the cDNAs are present at different levels in the two samples.

Expression microarray profiling has revealed that certain cancers have distinct patterns of gene expression and that these patterns correlate with factors such as the cancer's stage, clinical course, or response to treatment. In one such experiment, scientists examined gene expression in both normal white blood cells and in cells from a white blood cell cancer known as **diffuse large B-cell lymphoma (DLBCL).** About 40 percent of patients with DLBCL respond well to chemotherapy and have long survival times. The other 60 percent respond poorly to therapy and have short survival. The investigators assayed the expression profiles of 18,000 genes and discovered that there were two types of DLBCL, with almost inverse patterns of gene expression (Figure 19–15, p. 420). One type of DLBCL, called *GC B-like,* had an expression pattern dramatically different from that of a second type, called *activated B-like.* Patients with the activated B-like pattern of gene expression had much lower survival rates than patients with the GC B-like pattern. As a result, it has been possible to adjust therapies for each group of cancer patients and to identify new specific treatments based on gene expression profiles. Similar gene expression profiles have been generated for many other cancers, including breast, prostate, ovarian, and colon cancer. Gene expression microarrays are providing tremendous insight into both substantial and subtle variations in genetic diseases.

Several companies are now promoting "nutrigenomics" services, which claim to use microarrays to identify allele polymorphisms and gene expression patterns for genes involved in nutrient metabolism. For example, polymorphisms in the apolipoprotein A (*APOA1*) gene, involved in lipid metabolism, and the gene for *MTHFR* (methylenetetrahydrofolate reductase), involved in metabolism of folic acid, have been implicated in cardiovascular disease. Nutrigenomics companies claim that microarray analysis of a patient's DNA sample for genes such as these and others enables them to judge whether a patient's allele variations or gene expression profiles warrant dietary changes to potentially improve health and reduce the risk of diet-related diseases.

Application of Microarrays for Gene Expression and Genotype Analysis of Pathogens

Microarrays are also providing infectious disease researchers with powerful new tools for studying pathogens. Genotyping microarrays are being used to identify strains of emergent viruses, such as the virus that causes the highly contagious condition called Severe Acute Respiratory Syndrome (SARS) as well as the H5N1 avian influenza virus, the cause of bird flu, which has killed about two dozen people in Asia, leading to the slaughter of over 80 million chickens and causing concern about possible pandemic outbreaks.

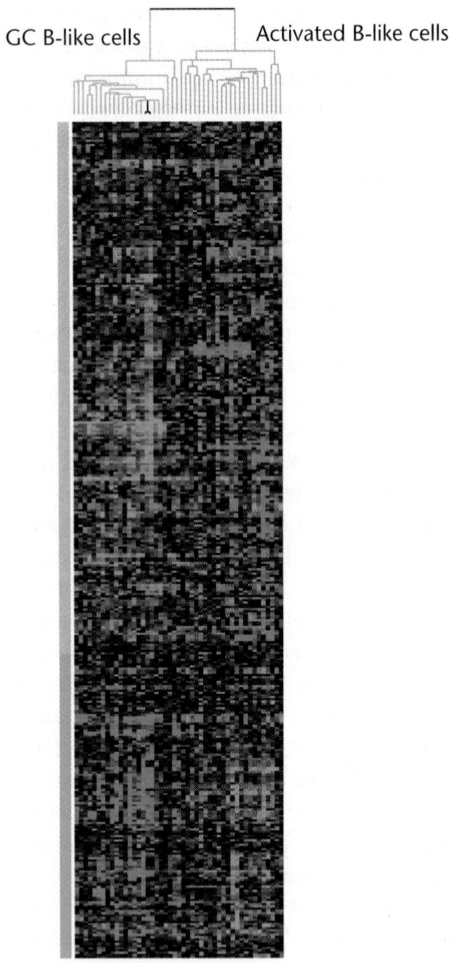

FIGURE 19–15 Gene expression analysis generated from expression DNA microarrays that analyzed 18,000 genes expressed in normal and cancerous lymphocytes. Each row represents a summary of the gene expression from one particular gene; each column represents data from one cancer patient's sample. The colors represent ratios of relative gene expression compared to normal control cells. Red represents expression greater than the mean level in controls, green represents expression lower than in the controls, and the intensity of the color represents the magnitude of difference from the mean. In this summary analysis, the cancer patients' samples are grouped by how closely their gene expression profiles resemble each other. The cluster of cancer patients' samples marked with orange at the top of the figure are GC B-like DLBCL cells. The blue cluster contains samples from cancer patients within the activated B-like DLBCL group. Patients with activated B-like profiles have a much higher rate of death (16 in 21) than those with GC B-like profiles (6 in 19). Data such as these demonstrate the value of microarray analysis for diagnosing disease conditions.

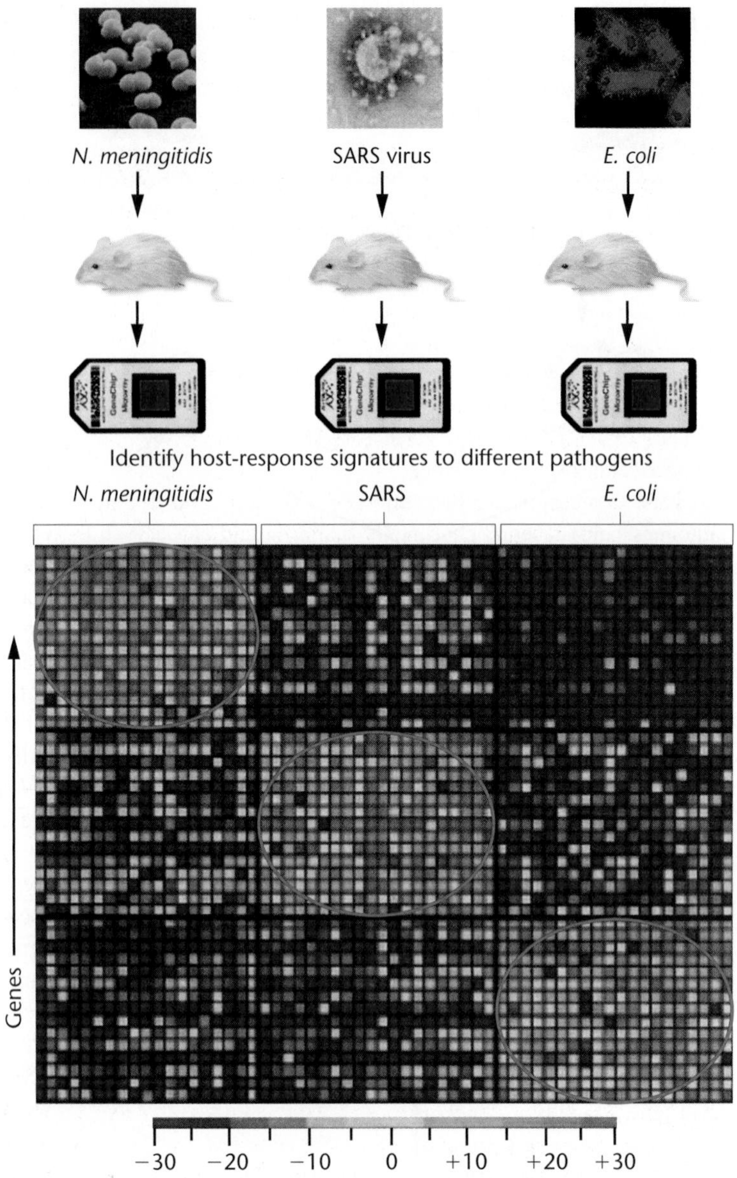

FIGURE 19–16 Gene expression microarrays can reveal host-response signatures for pathogen identification. In this example, mice were infected with different pathogens: *Neisseria meningitidis,* the virus that causes Severe Acute Respiratory Syndrome (SARS), and *E. coli.* Mouse tissues were then used as the source of mRNA for gene expression microarray analysis. Increased expression compared to uninfected control mice is shown in shades of gold. Decreased expression compared to uninfected controls is indicated in shades of blue. Notice that each pathogen elicits a somewhat different response in terms of which major clusters (circled) of host genes are activated by pathogen infection.

Whole-genome transcriptome analysis of pathogens is being used to inform researchers about genes important for pathogen infection and replication. In this approach, bacterial, yeast, protists, or viral pathogens are used to infect host cells *in vitro,* and then expression microarrays are used to analyze pathogen gene expression profiles. Patterns of gene activity during pathogen infection of host cells and replication are useful for identifying pathogens and understanding mechanisms of infection. Gene expression profiling is also valuable for identifying important pathogen genes and the proteins they encode that may be useful targets for subunit vaccine development or for drug treatment strategies to prevent or control disease.

Similarly, researchers are evaluating host responses to pathogens (Figure 19–16). This type of detection has been accelerated in part by the need to develop pathogen-detection strategies for military and civilian use both for detecting outbreaks of naturally emerging pathogens such as SARS and avian influenza and for potential detection of outbreaks such as anthrax (caused by the bacterium *Bacillus anthracis*) that could be the result of a bioterrorism event. Host-response gene expression profiles are developed by exposing a host to a pathogen and then using expression microarrays to analyze host gene expression patterns.

Figure 19–16 shows the different gene expression profiles for mice following exposure to *Neisseria meningitidis,* the SARS virus, or *E. coli.* In this example, although several genes are upregulated or downregulated by each pathogen, notice how each pathogen strongly induces different prominent clusters of genes that reveal a host gene expression response to the pathogen and provide a signature of infection. Comparing host gene expression profiles following exposure to different pathogens helps researchers quickly diagnose and classify infectious diseases. In the future, scientists expect to develop databases of both pathogen and host response expression profile data that can be used to identify pathogens efficiently.

ESSENTIAL POINT

A variety of different techniques, including restriction fragment length polymorphism analysis, allele-specific oligonucleotide tests, and DNA microarrays, can be used to identify genotypes associated with both normal and diseased phenotypes.

19.5 Genetic Engineering and Genomics Promise New, More Targeted Medical Therapies

Recombinant DNA technologies are changing medical diagnosis and allowing scientists to manufacture abundant and effective therapeutic proteins. In the near future, we will see even more transformative medical treatments based on genomics and advanced DNA-based technologies. In this section, we will examine two exciting new methodologies that have the potential to cure genetic diseases and yield specific drug treatments.

Pharmacogenomics and Rational Drug Design

Every year, more than 2 million Americans experience serious side effects of medications, and more than 100,000 die from adverse drug reactions. In addition, most drugs are effective in only about 60 percent of the population. Why are some drugs effective in certain individuals but ineffective or even deadly for others? Until recently, the selection of effective medications for each individual has been a trial-and-error process based largely on using drugs that work in most individuals. The new field of **pharmacogenomics** promises to lead to more specific, effective, and customized drugs that are designed to complement each person's individual genetic makeup.

Many genes affect a person's reaction to drugs. These genes encode products such as cell-surface receptors that bind a drug and allow it to enter a cell, as well as enzymes that metabolize drugs. For example, liver enzymes encoded by the cytochrome *P450* gene family affect the metabolism of many drugs. DNA sequence variations in these genes result in enzymes with different abilities to metabolize and utilize different drugs. Thus, gene variants that encode inactive forms of the cytochrome P450 enzymes are associated with a patient's inability to break down drugs in the body, leading to drug overdoses. A genetic test that recognizes some of these variants is currently being used to screen patients who are recruited into clinical trials for new drugs.

Another example is the reaction of certain people to the thiopurine drugs used to treat childhood leukemias (Figure 19–17, p. 422). Some individuals have sequence variations in the gene encoding the enzyme thiopurine methyltransferase (TPMT), which breaks down thiopurines. Anticancer drugs such as 6-mercaptopurine (6-MP) that are commonly used to treat leukemia have a thiopurine structure. In individuals with mutations in the *TPMT* gene, thiopurine cancer drugs can build up to toxic levels. As a result, although some patients, such as those who are homozygous for the wild-type *TPMT* gene, respond well to 6-MP treatment, others who are heterozygotes or homozygous for mutations in *TPMT* can have severe or even fatal reactions to 6-MP. At first a genetic test was developed to detect *TPMT* gene variants, but now a simple blood test is enough to enable clinicians to tailor the drug dosage to the individual. As a result, toxic effects of 6-MP have decreased, and survival rates for childhood leukemia patients treated with 6-MP have increased—a great example of pharmacogenomics in action.

Several methods are being developed for expanding the uses of pharmacogenomics. One promising method involves the detection of SNPs. Researchers are working to identify shared SNP sequences in the DNA of people who also share a heritable reaction to a drug. If the SNP segregates with a part of the genome containing the gene responsible for the drug reaction, it may be possible to devise gene tests based on the SNP. In the future, DNA microarrays may be used to screen a patient's genome for multiple drug reactions.

Knowledge from genetics and molecular biology is also contributing to the development of new drugs targeted at specific disease-associated molecules. **Rational drug design** involves the synthesis of specific chemical substances that affect specific gene products. **Gleevec** (trade name imatinib) used to treat chronic myelogenous leukemia (CML) is a good example of a rational drug design product. Geneticists had discovered that CML cells contain the Philadelphia chromosome, which results from a reciprocal translocation between chromosomes 9 and 22. Gene cloning revealed that the t(9;22) translocation creates a fusion of the *C-ABL* proto-oncogene with the *BCR* gene. This *BCR-ABL* fusion gene encodes a powerful fusion protein that causes cells to escape cell-cycle control. The fusion protein, which acts as a tyrosine kinase, is not present in noncancer cells from CML patients.

To develop Gleevec, chemists screened chemical libraries to find a molecule that bound to the BCR-ABL enzyme and specifically inhibited BCR-ABL activity. Clinical trials revealed that Gleevec was effective against CML, with minimal side effects and a higher remission rate than that seen with conventional therapies. Gleevec is now used to treat CML and several

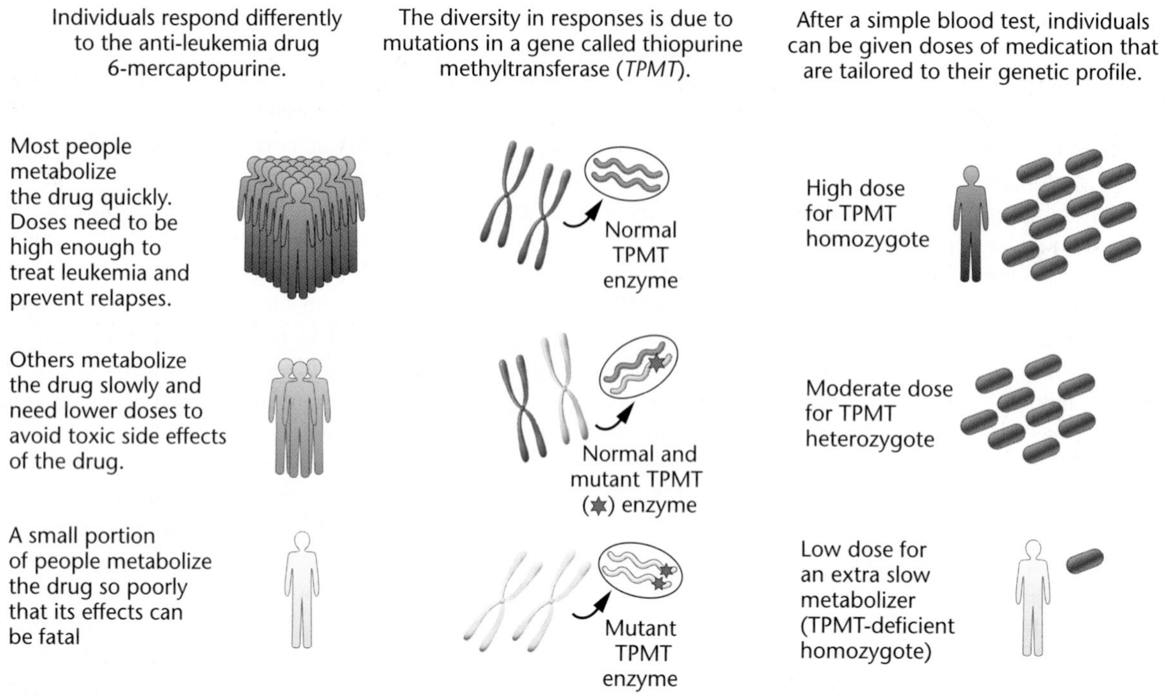

Individuals respond differently to the anti-leukemia drug 6-mercaptopurine.
The diversity in responses is due to mutations in a gene called thiopurine methyltransferase (*TPMT*).
After a simple blood test, individuals can be given doses of medication that are tailored to their genetic profile.
Most people metabolize the drug quickly. Doses need to be high enough to treat leukemia and prevent relapses.
Normal TPMT enzyme
High dose for TPMT homozygote
Others metabolize the drug slowly and need lower doses to avoid toxic side effects of the drug.
Normal and mutant TPMT (✱) enzyme
Moderate dose for TPMT heterozygote
A small portion of people metabolize the drug so poorly that its effects can be fatal
Mutant TPMT enzyme
Low dose for an extra slow metabolizer (TPMT-deficient homozygote)

FIGURE 19–17 Pharmacogenomics approaches to drug development are saving lives. Different individuals with the same disease, in this case childhood leukemia, often respond differently to a drug treatment because of subtle differences in gene expression. The dose of an anticancer drug such as 6-MP that works for one person may be toxic for another person. A simple gene or enzyme test to identify genetic variations can enable physicians to prescribe a drug treatment and dosage based on a person's genetic profile.

other cancers. With scientists discovering more genes and gene products associated with diseases, rational drug design promises to become a powerful technology within the next decade.

ESSENTIAL POINT

Pharmacogenomics provides the basis for customized medical intervention based on an individual's genotype.

Gene Therapy

Although drug treatments are often effective in controlling symptoms of genetic disorders, the ideal outcome of medical treatment is to cure these diseases. In an effort to cure genetic diseases, scientists are actively investigating **gene therapy**—a therapeutic technique that aims to transfer normal genes into a patient's cells. In theory, the normal genes will be transcribed and translated into functional gene products, which, in turn, will bring about a normal phenotype.

In many gene therapy trials, scientists often used genetically modified retroviruses as vectors to deliver therapeutic genes. An example is a vector based on a mouse virus called **Moloney murine leukemia virus (MLV).** The vectors are created by removing a cluster of three genes from the virus and inserting a cloned human gene. After being packaged in a viral protein coat, the recombinant vector is used to infect cells. Once inside a cell, the virus cannot replicate itself because of the missing viral genes. In the cell, the recombinant virus with the inserted human gene moves to the nucleus, and the viral genome integrates into a site on a human chromosome, and becomes part of the genome. If the inserted gene is expressed, it produces a normal gene product that may be able to correct the mutation carried by the affected individual. In initial attempts at gene therapy, several heritable disorders, including severe combined immunodeficiency (SCID), familial hypercholesterolemia, and cystic fibrosis were treated.

Human gene therapy began in 1990 with the treatment of a young girl named Ashanti DeSilva (Figure 19–18), who has a heritable disorder called **severe combined immunodeficiency (SCID).** Individuals with SCID have no functional immune system and usually die from what would normally be minor infections. Ashanti has an autosomal form of SCID caused by a mutation in the gene encoding the enzyme **adenosine deaminase (ADA).** Her gene therapy began when clinicians isolated some of her white blood cells, called T cells (Figure 19–18). These cells, which are key components of the immune system, were mixed with a retroviral vector carrying an inserted copy of the normal *ADA* gene. The virus infected many of the T cells, and a normal copy of the *ADA* gene was inserted into the genome of some T cells which were grown in the laboratory and analyzed to make sure that the transferred *ADA* gene was expressed. Then a billion or so genetically altered T cells were injected into Ashanti's bloodstream. Some of these T cells migrated to her bone marrow and began dividing and producing ADA. She now has ADA protein expression in 25 to 30 percent of her T cells, which is enough to allow her to lead a normal life.

In later trials, attempts were made to transfer the *ADA* gene into the bone marrow cells that form T cells, but these attempts were mostly unsuccessful. To date, gene therapy has successfully restored the health of about 20 children affected by SCID. Although gene therapy was originally developed as a treatment for single-gene inherited diseases, the technique was quickly adapted for trials to treat different forms of can-

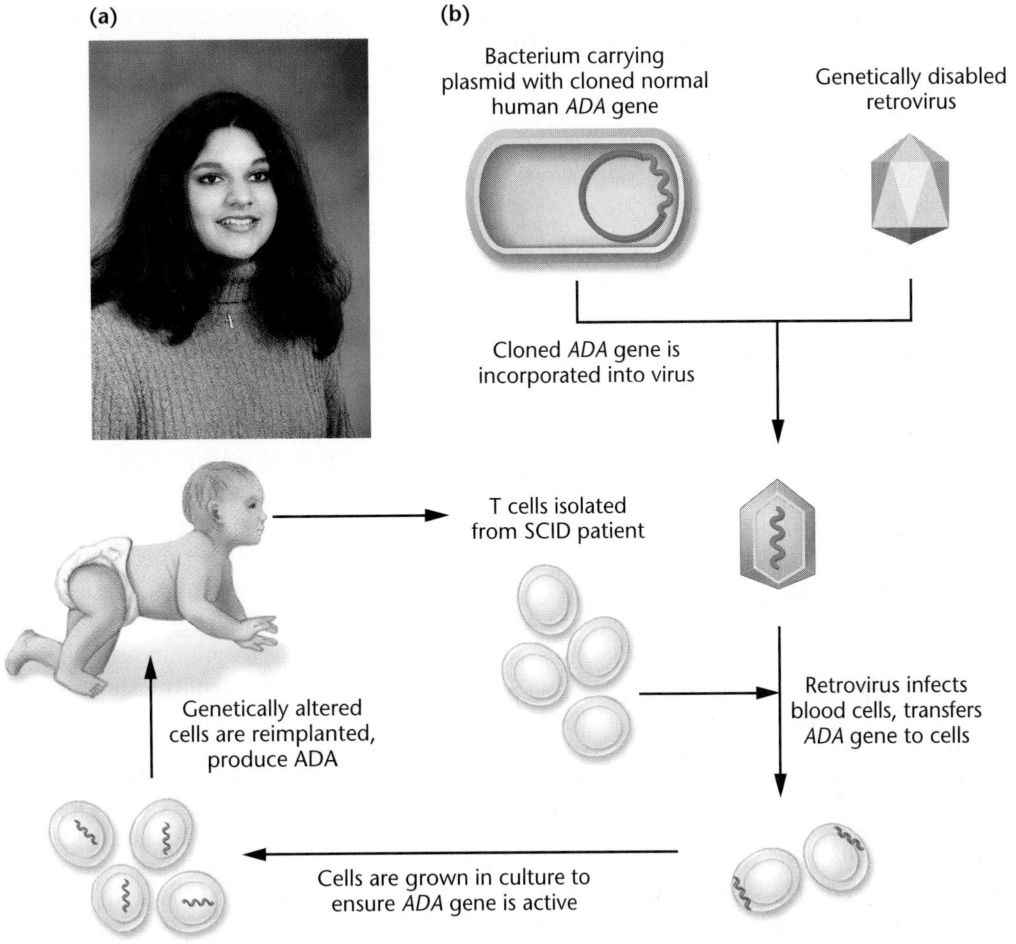
(a)
(b)
Bacterium carrying plasmid with cloned normal human *ADA* gene
Genetically disabled retrovirus
Cloned *ADA* gene is incorporated into virus
T cells isolated from SCID patient
Retrovirus infects blood cells, transfers *ADA* gene to cells
Cells are grown in culture to ensure *ADA* gene is active
Genetically altered cells are reimplanted, produce ADA

FIGURE 19–18 (a) Ashanti DeSilva, the first person to be treated by gene therapy. (b) To treat SCID using gene therapy, a cloned human *ADA* gene is transferred into a viral vector, which is then used to infect white blood cells removed from the patient. The transferred *ADA* gene is incorporated into a chromosome and becomes active. After growth to enhance their numbers, the cells are inserted back into the patient, where they produce ADA, allowing the development of an immune response.

cer, neurodegenerative diseases, cardiovascular disease, and infectious diseases, such as HIV. In addition to retroviral vectors, other viruses, chemically assisted transfer of genes into cells, and fusion of cells with artificial vesicles carrying DNA are being used to transfer genes into human cells.

From 1990 to 1999, more than 4000 people underwent gene therapy for a variety of genetic disorders. These trials often failed and thus led to a loss of confidence in gene therapy. Hopes for gene therapy plummeted even further in September 1999 when teenager Jesse Gelsinger died while undergoing gene therapy. His death was triggered by a massive inflammatory response to the vector, a modified **adenovirus,** one of the viruses that cause colds and respiratory infections. The outlook for gene therapy improved briefly in 2000, when a group of French researchers reported the first large-scale success in gene therapy. Nine children with a fatal X-linked form of SCID developed functional immune systems after being treated with a retroviral vector carrying a normal gene. However, three of these children later developed leukemia, and one died as a result of the treatment. In two of the children, their cancer cells contained the retroviral vector, inserted near or into a gene called *LMO2*. This insertion activated the *LMO2* gene, causing uncontrolled white blood cell proliferation and development of leukemia.

Most problems associated with gene therapy have been traced to the vectors used to transfer therapeutic genes into cells. These vectors, including MLV and adenovirus, have several serious drawbacks. First, integration of retroviral genomes (including the human therapeutic gene) into the host cell's genome occurs only if the host cells are replicating their DNA. Second, most viral vectors are capable of causing an immune response in the patient, as happened in Jesse Gelsinger's case. Third, insertion of viral genomes into host chromosomes can activate or mutate an essential gene, as in the case of the three French patients. (Unfortunately, it is not yet possible to reliably target insertion of therapeutic genes into specific locations in the genome.) Fourth, retroviruses cannot carry DNA sequences much larger than 8 kb. Many human genes exceed this size. Finally, there is a possibility that a fully infectious virus could be created if the vector were to recombine with another viral genome already present in the host cell. To overcome these problems, new viral vectors and strategies for transferring genes into cells are being developed in an attempt to improve the action and safety of vectors. Researchers also hope that new gene delivery systems will allow them to regulate both insertion sites and the levels of gene product produced from the therapeutic genes.

In addition to the vector delivery issues we have already addressed, a number of other barriers must be overcome if gene therapy is to become a viable approach for reliably treating many genetic disorders. Issues include:

- What is the proper route for gene delivery in different disorders? For example, what is the best way to treat brain or muscle tissues?

- What percentage of cells in an organ or a tissue need to express a therapeutic gene to alleviate the effects of a genetic disorder?
- What amount of a therapeutic gene product must be produced to provide lasting improvement of the condition, and how can sufficient production be ensured? Currently, many gene therapy approaches provide only short-lived delivery of the therapeutic gene and its protein.
- Will it be possible to use gene therapy to treat diseases that involve multiple genes?
- Can expression of therapeutic genes be controlled in a patient?

Scientists are also working on gene replacement approaches that involve removing a defective gene from the genome. Encouraging breakthroughs have taken place in this area using model organisms such as mice; however, this technology has not advanced sufficiently for use in humans. Attempts have been made to use antisense oligonucleotides to inhibit translation of mRNAs from defective genes, but this approach to gene therapy has generally not yet proven to be reliable. However, the recent emergence of RNA interference as a powerful gene-silencing tool has reinvigorated gene therapy approaches by gene silencing.

As you learned in Chapter 15, **RNA interference (RNAi)** or **RNA-induced gene silencing** is a form of gene expression regulation (see Figure 15–21). In animals short, double-stranded RNA molecules are delivered into cells where the enzyme dicer chops them into 21-nt long pieces called **small interfering RNAs (siRNAs).** siRNAs then join with an enyzme complex called the **RNA inducing silencing complex (RISC),** which shuttles the siRNAs to their target mRNA, where they bind by complementary base pairing. The RISC complex can block siRNA-bound mRNAs from being translated into protein or can lead to degradation of siRNA-bound mRNAs so that they cannot be translated into protein.

A main challenge to RNAi-based therapeutics so far has been *in vivo* delivery of double-stranded RNA or siRNA. RNAs degrade quickly in the body. It is also hard to get them to penetrate cells and to target the right tissue. Two common delivery approaches are to inject the siRNAs directly or to deliver them via a plasmid vector that is taken in by cells and transcribed to make double-stranded RNA that can be cleaved by Dicer into siRNAs.

Several RNAi clinical trials to treat blindness are underway in the United States. One RNAi strategy to treat a form of blindness called macular degeneration targets a gene called *VEGF.* The VEGF protein promotes blood vessel growth. Overexpression of this gene, causing excessive production of blood vessels in the retina, leads to impaired vision and eventually blindness. Many expect that this disease will soon become the first condition to be treated by RNAi therapy. Other diseases that are candidates for treatment by RNAi include several different cancers, diabetes, multiple sclerosis, and arthritis.

ESSENTIAL POINT

Gene therapy involves the delivery of therapeutic genes or the inhibition of expression of defective genes to treat genetic disorders.

NOW SOLVE THIS

Problem 14 on page 431 asks you to explain why diseases such as muscular dystrophy would be difficult to cure with gene therapy.

Hint: Consider the types of tissues that are affected in muscular dystrophy patients.

19.6 DNA Profiles Identify Individuals

As discussed previously, the presence or absence of restriction sites at specific locations within the human genome can be used as genetic markers. Another type of genetic marker, discovered in the mid-1980s, is based on variations in the length of repetitive DNA sequence clusters. The number of repeats within these clusters varies between individuals and between population groups. These polymorphisms serve as the basis for **DNA profiling**—also called **DNA fingerprinting** or DNA typing. DNA profiling is used in a wide variety of applications, such as paternity testing, forensics, identification of human or animal remains, archaeology, and conservation biology.

DNA Profiling Based on DNA Minisatellites (VNTRs)

The first DNA profiling method was developed in the United Kingdom by Sir Alec Jeffreys and was used in 1986 to convict the murderer of two English schoolgirls. This method involves the analysis of large repeated regions of DNA, called **minisatellites,** or **variable numbers of tandem repeats (VNTRs).**

VNTRs are repeating clusters of about 10 to 100 nucleotides. For example, the VNTR

5′-GACTGCCTGCTAAGAT**GACTGCCTGCTAAGAT**
GACT GCCTGCTAAGAT-3′

is composed of three tandem repeats of the 16-nucleotide sequence GACTGCCTGCTAAGAT. Clusters of such sequences are widely dispersed in the human genome. The number of repeats at each locus ranges from 2 to more than 100, each repeat length representing one allele of that locus. VNTRs are particularly useful for DNA profiling because there are about 30 different possible alleles (repeat lengths) at any VNTR within a population. This creates a large number of possible genotypes. For example, if one examined four different VNTRs within a population, and each VNTR had 20 possible alleles, there would be about 2 billion possible genotypes in this four-locus profile.

To create a VNTR profile, DNA is extracted from the tissue sample and is digested with a restriction enzyme that cleaves on either side of the VNTR repeat region. The digested DNA is separated by gel electrophoresis and subjected to Southern blot analysis using a probe that recognizes DNA sequences within the VNTR. After exposing the membrane to X-ray film, the pattern of bands is determined, with larger VNTR repeat alleles remaining near the top of the gel and smaller VNTRs migrating closer to the bottom (Figure 19–19). The pattern of bands is

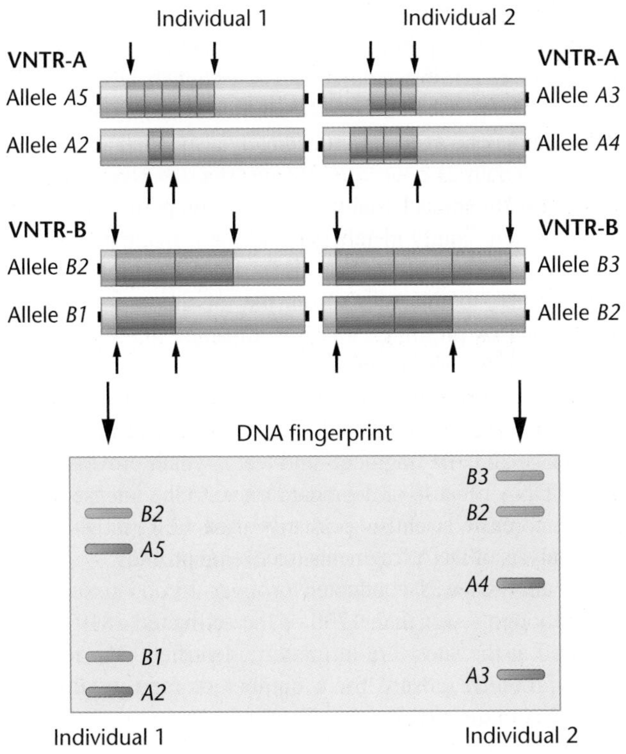

FIGURE 19–19 VNTR alleles at two loci (*A* and *B*) are shown for two different individuals. (Arrows mark restriction enzyme cutting sites flanking the VNTRs.) Restriction enzyme digestion produces a series of fragments that can be detected as bands on a Southern blot (bottom). The number of repeats at each locus is variable, so the overall pattern of bands is distinct for each individual. The DNA fingerprint (profile) shows that these individuals share one allele (*B2*).

always the same for a given individual, no matter what tissue is used as the source of the DNA. If enough VNTRs are analyzed, each person's DNA profile will be unique (except, of course, for identical twins) because of the huge number of possible VNTRs and alleles. In practice, about five or six VNTR loci are analyzed to create a DNA profile.

A significant limitation of VNTR profiling is that it requires a relatively large sample of DNA (10,000 cells or about 50 μg of DNA)—more than is usually found at a typical crime scene—and the DNA must be relatively intact (nondegraded). As a result, VNTR profiling is used most frequently when large tissue samples are available—such as in paternity testing. Blood samples drawn from the child, mother, and alleged father provide abundant fresh, intact cells for DNA extraction and analysis.

DNA Profiling Based on DNA Microsatellites

To overcome the problems of DNA sample size and other limitations of VNTR profiling, scientists have developed alternative techniques based on the use of PCR to amplify shorter polymorphic DNA markers. One of these methods examines sequences called **microsatellites,** or **short tandem repeats (STRs).** Increasingly, STR analysis is replacing VNTR profiling. STRs are similar to VNTRs, but the repeated motif is shorter—between two and nine base pairs. In addition, only about 7 to 40 repeats are found in each STR locus. Although hundreds of STR loci are found in the human genome, only a subset is used for DNA profiling. The FBI and other law enforcement agencies have selected 13 STR loci to be used as a core set for forensic analysis. DNA profiles based on these loci are stored in a national DNA database called the **Combined DNA Index System (CODIS).** Profiles generated in criminal cases, as well as for missing persons and from samples collected from crime scenes, are stored in the CODIS.

To create an STR profile, DNA is extracted from the sample and is amplified using sets of primers that specifically hybridize to regions of DNA flanking each of the STR loci to be analyzed. Each primer set within the mix is labeled with a different fluorescent dye, making the PCR products amplified from each STR locus fluoresce a different color. The amplified mixture is electrophoresed through a thin capillary tube gel (similar to the kind used for computer-automated DNA sequencing discussed in Chapter 17). Larger DNA fragments will be retarded in the capillary gel, and smaller ones will pass through more quickly. This is known as **capillary gel electrophoresis.** At the bottom of the tube, a laser detects when each fluorescent DNA fragment exits the capillary tube and computer software assigns a size to the fragment. All fragments are represented as peaks on a graph [Figure 19–20(a), p. 426]. This entire process is automated—often within self-contained kits that are used in research labs or in mobile crime units.

A major advantage of STR profiles is that they can be generated from trace samples (e.g., single hairs or saliva left on a cigarette butt) and even from samples that are old or degraded (e.g., a skull found in a field or an ancient Egyptian mummy). In addition, STR typing is less expensive, less labor-intensive, and much faster than VNTR analysis, so it is rapidly replacing VNTR typing in many profiling laboratories.

The results of STR analysis are interpreted using statistics, probability, and population genetics [Figure 19–20(b), p. 426]. The population frequency of each STR allele in the standard set has been measured in many groups of people throughout the world. Using this information, scientists can calculate the probability of any combination of these 13 alleles. For example, if an allele of locus 1 is carried by 1 in 333 individuals and an allele of locus 2 is carried by 1 in 83 individuals, the probability that someone will carry both alleles is equal to the product of their individual frequencies, or nearly 1 in 28,000 ($1/333 \times 1/83$). This overall probability may not be especially convincing, but if an allele of a third locus—carried by 1 in 100 individuals—and an allele of a fourth locus—carried by 1 in 25 individuals—are included in the calculations, the combined frequency becomes almost 1 in 70 million. That is, only about four individuals in the U.S. population will carry this combination of alleles by chance. When a profile containing all 13 CODIS STR loci is generated, the chance of anyone having the same profile is 1 in 10 billion. Since the planet's population is only about 6 billion, it is easy to see why STR analysis is so powerful in its ability to discriminate between samples from different individuals.

STR analysis was one of the primary approaches used to identify victims at two mass-casualty disaster sites, the World Trade Center in New York City and in countries affected by the South Asian tsunami of 2004.

(a)

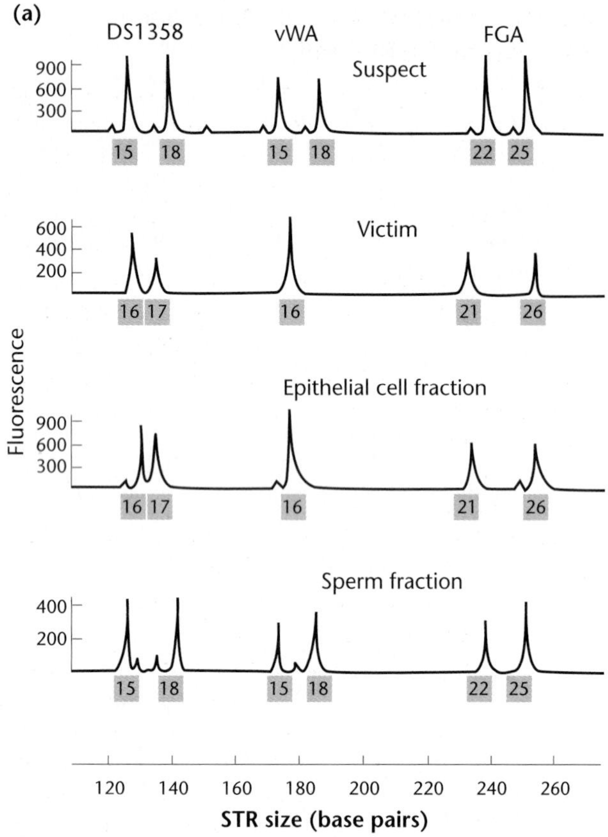

(b)

		Probability of identity		
Locus	*Number of alleles*	*African American*	*US Caucasian*	*US West Coast Hispanic*
DS1358	8	0.12	0.078	0.103
vWA	11	0.058	0.065	0.094
FGA	14	0.035	0.036	0.030
Combined	—	0.00021	0.00018	0.00029

FIGURE 19–20 (a) Capillary electrophoresis readout showing STR profiles of four samples from a rape case. Three STR loci were examined (DS1358, vWA, and FGA) from samples taken from the suspect and victim and from epithelial cells and sperm cells isolated from a vaginal swab taken from the victim. The x-axis shows the size of each STR in bases. The y-axis represents arbitrary values of fluorescence intensity. Notice that the STR profile of the sperm sample taken from the victim matches the STR profile of the suspect. (b) Population frequencies for STR alleles can be calculated to determine the probability of STR allele combinations in different ethnic groups.

Terrorism and Natural Disasters Force Development of New Technologies

World Trade Center Shortly after the twin towers of the World Trade Center in New York City were destroyed by the terror attacks of September 11, 2001, forensic scientists came together and rapidly accelerated efforts to use DNA-based techniques to identify the remains of victims. The tremendous amount of debris at the site, dangerous working conditions, and heat and microbial decomposition of remains, coupled with the hundreds of thousands of tissue samples (primarily bone fragments) at the site from the nearly 3000 individuals lost, made it evident that new strategies would be needed to quickly prepare and organize DNA profiles and compare them to DNA profiles from relatives. How would scientists establish DNA identity for those who perished in over 1.5 million tons of rubble?

State and federal agencies immediately responded to help with this task. Within 24 hours after the disaster, the New York City Police Department had established collection points throughout the city, where family members could file missing person's reports and provide cheek cell swabs for DNA isolation. Personal items (combs, toothbrushes) from the missing were also collected for DNA profiling. Several companies were involved in developing new software programs to help match samples submitted from family members to DNA profiles obtained from victims. Because tissue samples from victims primarily consisted of small bone fragments and teeth, which provided fragmented DNA because of degradation due to the intense heat at the site, forensic scientists primarily used STR, mtDNA, and SNP analysis of DNA fragments to develop profiles.

DNA analysis was conducted on over 15,000 tissue samples, although fewer than 1700 of the estimated 2819 people who died at the site were ultimately identified. The forensic results provided closure for a number of families that lost loved ones in the attack.

South Asian Tsunami The South Asian tsunami of December 2004 was a tragedy that claimed over 225,000 lives and devastated areas of Indonesia, Sri Lanka, and Thailand. The Michigan-based company Gene Codes modified its software system called **Mass Fatality Identification System (M-FISys)** to help with the Thailand Tsunami Victim Identification effort. Because M-FISys was essentially built in response to the 9/11 tragedy, Gene Codes did not have to write entirely new software and was able to customize M-FISys as necessary. In addition to analyzing mitochondrial DNA, M-FISys incorporated male-specific variations in the Y chromosome called Y-STRs to aid in identifying individuals. Within three months approximately 800 victims had been identified.

Recently, Gene Codes established the DNA Shoah Project ("Shoah" is the Hebrew name for the Holocaust), an effort to use M-FISys to establish a genetic database of Nazi-era Holocaust survivors with an overall goal of reuniting an estimated 10,000 postwar orphans around the world.

Forensic Applications of DNA Profiling

Over the last 20 years, DNA profiling has become a powerful tool to help convict the guilty, identify the remains of victims, and exonerate the innocent. Its power emerges from the extreme sensitivity of the technology and its ability to discriminate accurately between crime scene samples. It has rapidly replaced older, more imperfect techniques and is helping to transform the police and justice systems. Since 1992, over 200 convicted persons in the United States have been exonerated of their alleged crimes, based on DNA profile analysis of old crime samples. Many of these convicted individuals had served decades in prison for crimes they did not commit—some were incarcerated on death row, and at least one was wrongly executed prior to having access to DNA profiling evidence. As national and state DNA databases grow, they are becoming important tools in solving so-called **cold cases,** criminal cases in which no arrest has been made and no suspects have been identified. By matching DNA from crime scenes with DNA profiles in the database, more than 1600 cold cases were solved between 2000 and 2003.

Although forensic DNA profiling is a powerful technology, it is important to understand what it can and cannot tell us about guilt and innocence. For example, in a criminal case, when comparing an evidence sample and a suspect's DNA profile, forensic scientists calculate the probability that one individual shares the same DNA profile with another. If a suspect's DNA profile does not match that of the evidence sample, the suspect can be excluded *as the source of that sample.* Whether or not this finding indicates innocence of the suspect would depend on all other evidence in the case. For instance, a suspect in a rape case may not have contributed the semen sample evidence but may have been involved in the crime by restraining the victim.

When all 13 core STR loci are used to create DNA profiles, there is less than one chance in 10 billion that two matching DNA profiles could have come from two different individuals. As these probabilities are based on the assumption of a large unrelated population, there is a greater chance that individuals share the same profile by chance among related individuals—siblings, parents, or other relatives. Certainly, profiles originating from identical twins and therefore identical genomes will be the same. Despite the power to generate individual-specific profiles, the genotypes are only as reliable as the laboratory generating the profiles. Care must be taken and strict protocols must be implemented that separate crime scene samples from the suspect's samples to minimize sample mixup, contamination, or even unintended or deliberate tampering. The very sensitivity of DNA profiling—sufficient to generate a profile from a single hair or from saliva on the back of a postage stamp—makes it highly prone to contamination if sampling and analysis conditions are not stringently controlled.

ESSENTIAL POINT ■ ■ ■

DNA-based profiling methods can be used to identify individuals during paternity testing, forensics analysis, and remains identification.

19.7 Genetic Engineering, Genomics, and Biotechnology Create Ethical, Social, and Legal Questions

Geneticists use recombinant DNA technology to identify genes, diagnose and treat genetic disorders, produce commercial and pharmaceutical products, and solve crimes. However, the applications that arise from genetic engineering raise important ethical, social, and legal issues that must be identified, debated, and resolved. Here we present a brief overview of some current ethical debates concerning the uses of gene technologies.

Concerns about Genetically Modified Organisms and GM Foods

Most GM food products contain an introduced gene encoding a protein that confers a desired trait (for example, herbicide or insect resistance). Much of the concern over GM plants centers on issues of consumer safety and environmental consequences. Are GM plants safe to eat? In general, if the expressed proteins are not found to be toxic or allergenic and do not have other negative physiological effects, they are not considered to be a significant hazard to health. In Europe and Asia, labeling of food containing GM ingredients is mandatory. But in the United States such labeling is not required at the present time, and foods with less than 5 percent of their content from genetically modified organisms (GMOs) can be labeled as GMO-free.

Environmental concerns generally have to do with any risks posed by releasing genetically modified organisms into the environment, which include possible gene transfer by cross breeding with wild plants, toxicity, and invasiveness of the modified plant, resulting in loss of natural species (loss of biodiversity). Although laboratory and field studies suggest that cross-pollination and gene transfer can occur between some genetically engineered plants and wild relatives, there is little evidence that this has occurred in nature. If, for example, glyphosate resistance was transferred from cultivated plants into wild relatives, the herbicide-resistant weeds could make herbicide treatment ineffective. Biotechnology companies have engineered transgenic plants into sterile forms that are unable to transfer their genes into other plants. Built-in sterility was also designed to ensure that farmers could not produce their own seed from genetically modified crops, guaranteeing that biotechnology companies would have exclusive distribution of each year's crop. This, in itself, is an ethical issue, particularly in underdeveloped countries with limited resources to purchase genetically modified seeds.

Genetic Testing and Ethical Dilemmas

We have considered examples of genetic testing and although these technologies are valuable for medical diagnosis, they also present ethical problems that are not easy to resolve. For example, what information should people have before deciding to have a genetic test? How can we protect the information revealed by such tests? How can we define and prevent genetic discrimination? As identification of genetic traits becomes more routine in clinical settings, physicians will need to ensure genetic privacy for their patients. There are significant concerns about how genetic information could be used in negative ways by employers, insurance companies, governmental agencies, or the general public. Genetic privacy and prevention of genetic discrimination will be increasingly important in the coming years. In 2008, the **Genetic Information Nondiscrimination Act (GINA)** was signed into law. GINA prohibits the improper use of genetic information in health insurance and employment.

Many of the potential benefits and consequences of genetic testing are not always clear or certain. For example, we can test for genetic diseases for which there are no effective treatments. But *should* we test people for these disorders? With present technology, a negative result does not necessarily rule out future development of a disease; nor does a positive result always mean that an individual will get the disease. How can we effectively communicate the results of testing and the actual risks to those being tested?

Earlier in this chapter we discussed preimplantation genetic diagnosis (PGD). As we learn more about genes involved in human traits, will other, nondisease-related genes be screened for by PGD? Will couples opt to select embryos with certain genes encoding desirable traits for height, weight, intellect, and other physical or mental characteristics? What do you think of using genetic testing to purposely select for an embryo with a genetic disorder? Recently, there have been several well-publicized cases of couples seeking to use prenatal diagnosis or PGD to select for embryos with dwarfism and deafness.

The Ethical Concerns Surrounding Gene Therapy

Gene therapy raises several ethical concerns, and many forms of therapy are sources of intense debate. At present, all gene therapy trials are restricted to using somatic cells as targets for gene transfer. This form of gene therapy is called **somatic gene therapy;** only one individual is affected, and the therapy is done with the permission and informed consent of the patient or family.

Two other forms of gene therapy have not been approved, primarily because of the unresolved ethical issues surrounding them. The first is called **germ-line therapy,** whereby germ cells (the cells that give rise to the gametes—i.e., sperm and eggs) or mature gametes are used as targets for gene transfer. In this approach, the transferred gene is incorporated into all the future cells of the body, including the germ cells. This means that individuals in future generations will also be affected, without their consent. Is this kind of procedure ethical? Do we have the right to make this decision for future generations?

The second unapproved form of gene therapy—which raises an even greater ethical dilemma—is termed **enhancement gene therapy,** whereby people may be "enhanced" for some desired trait. This use of gene therapy is extremely controversial and is strongly opposed by many people. Should genetic technology be used to enhance human potential? For example, should it be permissible to use gene therapy to increase height, enhance athletic ability, or extend intellectual potential? Presently, the consensus is that enhancement therapy, like germ-line therapy, is an unacceptable use of gene therapy. However, there is an ongoing debate, and many issues are still unresolved. For example, the U.S. FDA permits growth hormone produced by recombinant DNA technology to be used as a growth enhancer, in addition to its medical use for the treatment of growth-associated genetic disorders. Critics charge that the use of a gene product for enhancement will lead to the use of transferred genes for the same purpose. The outcome of these debates may affect not only the fate of individuals but the direction of our society as well.

The Ethical, Legal, and Social Implications (ELSI) Program

When the Human Genome Project was first discussed, scientists and the general public raised concerns about how genome information would be used and how the interests of both individuals and society can be protected. To address these concerns, the **Ethical, Legal, and Social Implications (ELSI) Program** was established as an adjunct to the Human Genome Project. The ELSI Program considers a range of issues, including the impact of genetic information on individuals, the privacy and confidentiality of genetic information, and implications for medical practice, genetic counseling, and reproductive decision making. It is critical that these and other ethical concerns are thoroughly studied and an international consensus developed on appropriate policies and laws.

DNA and Gene Patents

Intellectual property rights are also being debated as an aspect of the ethical implications of genetic engineering, genomics, and biotechnology. Patents on intellectual property (isolated genes, new gene constructs, recombinant cell types, GMOs) can be potentially lucrative for the patent-holders but may also pose ethical and scientific problems. For example, consider the possibilities for a human gene that has been cloned and then patented by the scientists who did the cloning. The person or company holding the patent could require that anyone attempting to do research with the patented gene pay a licensing fee for its use. Should a diagnostic test or therapy result from the research, more fees and royalties may be demanded, and as a result the costs of a genetic test may be too high for many patients to afford. But companies argue that limiting or preventing patents for genes or genetic tools could reduce the incentive for pursuing research that produces such genes and tools. Should scientists and companies be allowed to patent DNA sequences from naturally living organisms? Should there be a lower or an upper limit to the size of those sequences? For example, should patents be awarded for small pieces of genes, such as expressed sequence tags (ESTs), just because some individual or company wants to claim a stake at having cloned a piece of DNA first, even if no one knows whether the DNA sequence has a use? Can or should investigators be allowed to patent the entire genome of any organism they have sequenced?

Since 1980, the U.S. Patent and Trademark Office has granted patents for more than 20,000 genes or gene sequences, including an estimated 20 percent of human genes. Some scientists are concerned that to award a patent for simply cloning a piece of DNA is awarding a patent for too little work. Given that computers do most of the routine work of genome sequencing, who should get the patent? What about individuals who figure out *what* to do with the gene? What if a gene sequence has a role in a disease for which a genetic therapy may be developed? Many scientists believe that it is more appropriate to patent novel technology and applications that make use of gene sequences than to patent the gene sequences themselves.

ESSENTIAL POINT

Applications of genetic engineering and biotechnology involve a wide range of ethical, social, and legal dilemmas with important scientific and societal implications.

GENETICS, TECHNOLOGY, AND SOCIETY

Personal Genome Projects and the Race for the $1000 Genome

It took $3 billion and 15 years to sequence the entire human genome, as part of the Human Genome Project. This ambitious international project yielded the nucleotide sequence of the 3 billion base pairs of DNA that comprise the human genome. The Human Genome Project sequence was not derived from one person's DNA, but from a composite of DNA samples from numerous anonymous donors.

When the Human Genome Project was completed in 2005, the goal of routinely sequencing the genomes of individual humans seemed remote. However, within a few years, the development of high-throughput sequencing technologies, capable of generating long sequence reads at high speeds with great accuracy, has reduced the cost of DNA sequencing dramatically. In 2006, the X Prize Foundation announced the Archon X Prize for Genomics, an award of $10 million to the first private group that develops technology capable of sequencing 100 human genomes with a high degree of accuracy in ten days for under $10,000 per genome. Programs funded by the National Institutes of Health are challenging scientists to develop sequencing technologies to sequence a human genome for $1000 by the year 2014.

The "$1000 genome" came closer to reality with the sequencing of several individual human genomes. In 2007, a Connecticut company called 454 Life Sciences sequenced the entire genome of James D. Watson—the co-discoverer of DNA structure and the first director of the U.S. Human Genome Project. This "Project Jim" was completed in two months at a cost of about $1 million. Once compiled, the company presented Watson with his entire DNA sequence, on two DVDs.

Watson made all of his sequence available to researchers except for that of his apolipoprotein E gene (*ApoE*). *ApoE* variants can indicate a predisposition for Alzheimer's disease. Watson urged others to have their genomes made available to the public so that researchers could correlate physical and mental traits with genes.

Shortly after the announcement of Watson's personal genome sequence, the J. Craig Venter Institute announced the sequencing of another individual genome—that of Craig Venter himself. Venter, whose accomplishments have been discussed in several chapters, had his genome completed and deposited into GenBank in May 2007. He agreed to make his entire genome available, exposing any genetic imperfections for the whole world to see and study, including potentially revealing information about genetic disorders. Shortly after the two personal genome sequences were published, geneticists were comparing the Venter and Watson genomes!

One year after the Watson and Venter sequences were published, two more personal genome sequences were announced. These were sequences of a Yoruba man from Ibadan, Nigeria, and a Han Chinese man. Both projects cost less than U.S. $500,000 each.

As the $1000 genome rapidly approaches, scientists are making plans for even more ambitious projects. In 2008, an international research consortium initiated the "1000 Genomes Project," which aims to sequence the genomes of 1000 volunteers from various backgrounds, including African, Asian, and European volunteers. The goal is to produce an extensive catalog of human DNA sequence variation. Costs are anticipated to be approximately $50 million. George Church of Harvard University and his colleagues have started the Personal Genome Project—a project that aims to sequence the genomes of 100,000 individuals. Volunteers for the Personal Genome Project must provide their DNA samples on the understanding that their genome data will be made publicly available. Church's genome has already been made available online. The goal of the project is to correlate genome sequences with phenotypic characteristics, from height and hair color to disease predisposition.

The race continues to lower the cost and time to sequence an individual genome. Ready or not, we are entering the age of personal genomics.

Your Turn

1. Would you have your genome sequenced if the price was affordable? Why, or why not?

You can find a discussion of the pros and cons of knowing your own genome sequence in a series of articles under the heading of "My Genome. So What? (*Nature* 456: 1, 2008).

2. Would you make your genome sequence publicly available? How might such information be misused?

The issue of genetic privacy is particularly important when considering making genomic sequences available to all. Read about the potential uses and misuses of genome sequences in Taylor, P. 2008. When consent gets in the way. *Nature* 456: 32–33.

3. What information do you think you could obtain by examining your own genome sequence?

The relationship between DNA sequence and predictable phenotype is still a murky one. Read about this in Maher, B. 2008. Personal genomes: The case of the missing heritability. *Nature* 456: 18–21.

4. Private companies are now offering personal DNA sequencing along with interpretation. What services do they offer? Do you think these services should be regulated, and if so, in what way?

Investigate one such company, 23andMe, at **http://www.23andMe.com**. *You can read about ethics and whole-genome sequencing in* McGuire, A. L. et al. 2008. Research ethics and the challenge of whole-genome sequencing. *Nature Rev Genet* 9: 152–155.

CASE STUDY A first for gene therapy

In 1990, a young girl who was born without a functional immune system became the first person to undergo gene therapy. In an attempt to treat her autosomal recessive disorder, known as severe combined immunodeficiency (SCID), cloned copies of the gene encoding the missing enzyme (adenosine deaminase, or ADA) were inserted into some of her white blood cells, which were injected back into her bloodstream. Expression of the normal ADA allele led to the development of a functional immune system, allowing the girl to lead a normal life. An understanding of this spectacular success depends on knowing several details of this process:

1. Is it important that the cloned gene becomes part of a chromosome when inserted into a cell?
2. Does the cloned gene replace the defective copy of the ADA gene?
3. Why were white blood cells chosen as targets for the transferred genes?
4. Would you expect that production of 50% of the normal levels of ADA would be enough to restore immune function?

INSIGHTS AND SOLUTIONS

1. Probes for DNA fingerprinting can be derived from a single locus or multiple loci. Two multiple-loci probes have been widely employed in both criminal and civil cases and are derived from minisatellite loci on chromosome 1 (1cen-q24) and chromosome 7 (7q31.3). These probes, used because they produce a highly individual fingerprint, have determined paternity in thousands of cases. The results of one such DNA fingerprinting of a mother (M), putative father (F), and child (C) are shown in the accompanying figure. The child has 8 maternal bands, 14 paternal bands, 6 bands that are common to both the mother and the alleged father, and 1 band that is not present in either the mother or the alleged father. What are the possible explanations for the presence of the last band? Based on your analysis of the band pattern, which explanation is most likely?

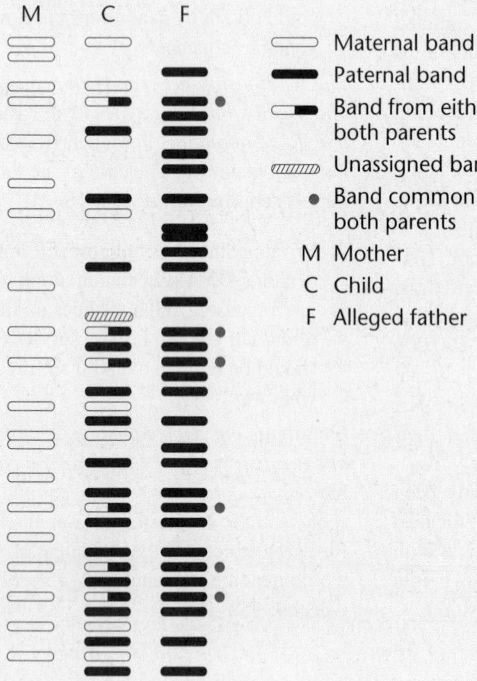

Solution: In this case, one band in the child cannot be assigned to either parent. Two possible explanations are that the child is mutant for one band or that the man tested is not the father. To estimate the probability of paternity, the mean number of resolved bands (n) is determined, and the mean probability (x) that a band in individual A matches that in a second, unrelated individual B is calculated. In this case, because the child and the father share 14 bands in common, the probability that the man tested is not the father is very low (probably 10^{-7} or lower). As a result, the most likely explanation is that the child is mutant for a single band. In fact, in 1419 cases of genuine paternity resolved by the minisatellite probes on chromosomes 1 and 7, single unmatching bands in the children were recorded in 399 cases, accounting for 28 percent of all cases.

2. Infection by HIV-1 (human immunodeficiency virus) weakens the immune system and results in the symptoms of AIDS (acquired immunodeficiency syndrome). Specifically, HIV infects and kills cells of the immune system that carry a cell-surface receptor known as CD4. An HIV surface protein known as gp120 binds to the CD4 receptor and allows the virus to enter the cell. The gene encoding the CD4 protein has been cloned. How might this clone be used along with recombinant DNA techniques to combat HIV infection?

Solution: Researchers hope that clones of the *CD4* gene can be used in the design of systems for the targeted delivery of drugs and toxins to combat the infection. For example, because infection depends on an interaction between the viral gp120 protein and the CD4 protein, the cloned *CD4* gene has been modified to produce a soluble form of the protein (sCD4) that, because of its solubility, would circulate freely in the body. The idea is that HIV might be prevented from infecting cells if the gp120 protein of the virus first encounters and binds to extra molecules of the soluble form of the CD4 protein. Once bound to the extra molecules, the virus would be unable to bind to CD4 proteins on the surface of immune system cells. Studies in cell culture systems indicate that the presence of sCD4 effectively prevents HIV infection of tissue culture cells. However, studies in HIV-positive humans have been somewhat disappointing, mainly because the strains of HIV used in the laboratory are different from those found in infected individuals.

PROBLEMS AND DISCUSSION QUESTIONS

1. What are some of the reasons GM crops are controversial? Describe some of the primary concerns that have been raised about GM foods.
2. Should the United States require mandatory labeling of all foods that contain GMOs? Explain your answer.
3. Provide examples of major questions that need to be answered if gene therapy is to become a safe and reliable treatment for genetic diseases.
4. Outline the steps involved in transferring glyphosate resistance to a crop plant. Do you envision that this trait is likely to escape from the crop plant and make weeds glyphosate-resistant? Why or why not? See **Now Solve This** on page 413.
5. In order to vaccinate people against diseases by having them eat antigens (such as the cholera toxin), the antigen must reach the cells of the small intestine. What are some potential problems of this method? See **Now Solve This** on page 411.
6. What are the advantages of using STRs instead of VNTRs for DNA profiling?
7. Suppose you develop a screening method for cystic fibrosis that allows you to identify the predominant mutation $\Delta 508$ and the next six most prevalent mutations. What must you consider before using this method to screen a population for this disorder?
8. One of the main safety issues associated with genetically modified crops is the potential for allergenicity caused by introducing an allergen or by changing the level of expression of a host allergen. Based on the observation that common allergenic proteins often contain identical stretches of a few (six or seven) amino acids, researchers developed a method for screening transgenic crops to evaluate potential allergenic properties (Kleter & Peijnenburg, 2002. *BMC Struct. Biol.* 2: 8). How do you think they accomplished this? Read the Kleter and Peijnenberg paper to answer this question.
9. Why are most recombinant human proteins produced in animal or plant hosts instead of bacterial host cells?
10. There are more than 1000 cloned farm animals in the United States. In the near future, milk from cloned cows and their offspring (born naturally) may be available in supermarkets. These cloned animals have not been transgenically modified, and they are no different than identical twins. Should milk from such animals and their natural-born offspring be labeled as coming from cloned cows or their descendants? Why?
11. One of the major causes of sickness, death, and economic loss in the cattle industry is *Mannheimia haemolytica,* which causes bovine pasteurellosis, or shipping fever. Noninvasive delivery of a vaccine using transgenic plants expressing immunogens would reduce labor costs and trauma to livestock. An early step toward developing an edible vaccine is to determine whether an injected version of an antigen (usually a derivative of the pathogen) is capable of stimulating the development of antibodies in a test organism. The following table assesses the ability of a transgenic portion of a toxin (Lkt) of *M. haemolytica* to stimulate development of specific antibodies in rabbits.
    (a) What general conclusion can you draw from the data?
    (b) With regards to development of a usable edible vaccine, what work remains to be done?

Immunogen Injected	Antibody Production in Serum
Lkt50*—saline extract	+
Lkt50—column extract	+
Mock injection	−
Pre-injection	−

*Lkt50 is a smaller derivative of Lkt that lacks all hydrophobic regions. + indicates at least 50 percent neutralization of toxicity of Lkt; − indicates no neutralization activity.

Source: Modified from Lee et al. 2001. *Infect. and Immunity* 69: 5786–5793.

12. Recombinant adenoviruses have been used in a number of preclinical studies to determine the efficacy of gene therapy for rheumatoid arthritis and osteoarthritis. In the viruses, genes can be delivered by injection to the tissues that need them. Christopher Evans and colleagues (2001. *Arthritis Res.* 3: 142–146) estimated that approximately 20 percent of all human gene therapy trials have used adenoviruses for gene delivery. The death of a patient in 1999 after infusion of adenoviral vectors has caused concern. As you consider the use of viral vectors as therapy-delivery vehicles for human pathologies, what factors seem of paramount concern?
13. Define somatic gene therapy, germ-line therapy, and enhancement gene therapy. Which of these is currently in use?
14. Gene therapy for human genetic disorders involves transferring a copy of the normal human gene into a vector and using the vector to transfer the cloned human gene into target tissues. Presumably, the gene enters the target tissue and becomes active, and the gene product relieves the symptoms. See **Now Solve This** on page 424.
    (a) Why are disorders such as muscular dystrophy difficult to treat by gene therapy?
    (b) What are the potential problems of using retroviruses as vectors?
    (c) Should gene therapy involve germ-line tissue instead of somatic tissue? What are some of the potential ethical problems associated with the former approach?
15. Sequencing the human genome and the development of microarray technology promise to improve our understanding of normal and abnormal cell behavior. How are microarrays dramatically changing our understanding of complex diseases such as cancer?
16. A couple with European ancestry seeks genetic counseling before having children because of a history of cystic fibrosis (CF) in the husband's family. ASO testing for CF reveals that the husband is heterozygous for the $\Delta 508$ mutation and that the wife is heterozygous for the *R117* mutation. You are the couple's genetic counselor. When consulting with you, they express their conviction that they are not at risk for having an affected child because they each carry different mutations and cannot have a child who is homozygous for either mutation. What would you say to them?

17. The DNA sequence surrounding the site of the sickle-cell mutation in the β-globin gene, for normal and mutant genes, is as follows.

5′-GACTCCTGAGGAGAAGT-3′

3′-CTGAGGACTCCTCTTCA-5′

Normal DNA

5′-GACTCCTGTGGAGAAGT-3′

3′-CTGAGGACACCTCTTCA-5′

Sickle-cell DNA

Each type of DNA is denatured into single strands and applied to a DNA-binding membrane. The membrane containing the two spots is hybridized to an ASO of the sequence

5′-GACTCCTGAGGAGAAGT-3′

Which spot, if either, will hybridize to this probe?

18. The human insulin gene contains introns. Since bacterial cells will not excise introns from mRNA, how can a gene like this be cloned into a bacterial cell that will produce insulin?
19. In mice transfected with the rabbit β-globin gene, the rabbit gene is active in several tissues, including the spleen, brain, and kidney. In addition, some transfected mice suffer from thalassemia (a form of anemia) caused by an imbalance in the coordinate production of α and β-globins. Which problems associated with gene therapy are illustrated by these findings?
20. When genetic testing technologies become widespread, medical records will contain the results of such testing. Who should have access to this information? Should employers, potential employers, or insurance companies be allowed to have this information? Would you favor or oppose having the government establish and maintain a central database containing the results of individuals' genome scans?
21. What limits the use of differences in restriction enzyme sites as a way of detecting point mutations in human genes?
22. You are asked to assist with a prenatal genetic test for a couple, each of whom is found to be a carrier for a deletion in the β-globin gene that produces β-thalassemia when homozygous. The couple already has one child who is unaffected and is not a carrier. The woman is pregnant, and the couple wants to know the status of the fetus. You receive DNA samples obtained from the fetus by amniocentesis and from the rest of the family by extraction from white blood cells. Using a probe for the deletion, you obtain the following blot. Is the fetus affected? What is its genotype for the β-globin gene? See **Now Solve This** on page 416.

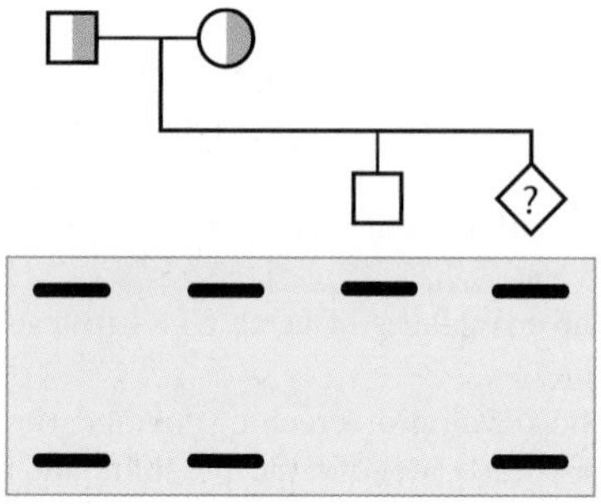

23. Host transgenic mammals are often made by injecting transgenic DNA into a pronucleus, a process that is laborious, technically demanding, and typified by yields of less than 1 percent. Lately, the engineered lentivirus, a retrovirus, has been used to generate transgenic pigs, rats, mice, and cattle with greater than 10 percent efficiency (Whitelaw, 2004). Lentivectors have reverse transcriptase activity, can infect both dividing and nondividing cells, and are replication-defective. A lentivector carrying the reporter gene, which codes a green fluorescent protein, has produced a remarkable set of pigs (refer to the chapter opening figure). Using a labeled diagram, present a strategy for producing a transgenic mammal carrying a pharmaceutically important gene using a lentivector.

This unusual four-winged Drosophila *has developed an extra set of wings as a result of a mutation in a homeotic selector gene.*

20 Developmental Genetics

Chapter Concepts

- Gene action in development is based on differential transcription of selected genes.
- Animals use a small number of signaling systems and regulatory networks to construct adult body forms from the zygote. These shared properties make it possible to use animal models to study human development.
- Differentiation is controlled by cascades of gene action that follow the specification and determination of developmental fate.
- Plants independently evolved developmental mechanisms that parallel those of animals.
- In many organisms, cell–cell signaling programs the developmental fate of adjacent and distant cells.

Over the last two decades genetic analysis and molecular biology have shown that in spite of wide diversity in the size and shape of adult organisms, all multicellular organisms share many genes, genetic pathways, and molecular signaling mechanisms in the processes leading from the zygote to the adult. At the cellular level, development is marked by three important events: **specification**, when the first cues conferring a spatially discrete identity are established, **determination**, the time when a specific developmental fate for a cell becomes fixed, and **differentiation**, the process by which a cell achieves its final form and function. In addition, genomics has shown that higher organisms share many evolutionary and developmental relationships. Genetic analysis has identified genes that regulate developmental processes, and we are beginning to understand how the action and interaction of these genes control basic developmental processes in eukaryotes.

In this chapter, the primary emphasis will be on how genetics has been used to study development. This area, called developmental genetics, has contributed tremendously to our understanding of developmental processes because genetic information is required for the molecular and cellular functions mediating developmental events and contributes to the continually changing phenotype of the newly formed organism.

How Do We Know?

In this chapter, we will focus on both the large-scale and the inter- and intracellular events that take place during embryogenesis and the development of adult structures. As you study this topic, you should try to answer several fundamental questions:

1. How do we know how many genes control development in an organism like *Drosophila*?
2. How do we know that molecular gradients in the egg control development?
3. How did we discover that selector genes specify which adult structures will be formed by body segments?
4. How did we learn about the levels of gene regulation involved in vulval development in *C. elegans*?

20.1 Evolutionary Conservation of Developmental Mechanisms Can Be Studied Using Model Organisms

Genetic analysis of development in a wide range of organisms has demonstrated that all animals use a common set of developmental mechanisms and signaling systems. For example, most of the differences in shape between zebras and zebrafish are controlled by different patterns of expression in one gene set, called homeotic (abbreviated as *Hox*) genes, not by expression of many different gene sets. Genome-sequencing projects have confirmed that genes from a wide range of organisms are conserved. This homology in genes and regulatory mechanisms means that many aspects of normal human embryonic development and associated genetic disorders can be studied in model organisms such as *Drosophila,* which, unlike humans, can be genetically manipulated (see Chapter 1 for a discussion of model organisms in genetics).

Although many developmental mechanisms are similar among all animals, evolution has also generated new and different approaches to transform a zygote into an adult. These evolutionary changes can be attributed to several genetic mechanisms, including mutation, gene duplication and divergence, the assignment of new functions to old genes, and the recruitment of genes to new developmental pathways. The emphasis in this chapter, however, will be on the similarities among species.

Analysis of Developmental Mechanisms

In the space of this chapter, we cannot survey all aspects of development, nor can we explore the genetic analysis of all developmental mechanisms triggered by the fusion of sperm and egg. Instead, we will focus on several general processes in development:

- how the adult body plan of animals is laid down in the embryo
- the program of gene expression that turns undifferentiated cells into differentiated cells
- the role of cell–cell communication in development.

We will use three model systems—*Drosophila melanogaster, Arabidopsis thaliana,* and *Caenorhabditis elegans*—to illustrate these developmental processes and related topics. We will examine the role of differential gene expression in the progressive restriction of developmental options leading to the formation of the adult body plan in the model organisms *Drosophila* and *Arabidopsis.* We will then expand the discussion to include the selection of pathways that result in differentiated cells in plants and animals, and consider the role of cell–cell communication in development in *C. elegans.*

20.2 Genetic Analysis of Embryonic Development in *Drosophila* Reveals How the Body Axis of Animals Is Specified

How a given cell turns specific genes on or off at precisely timed stages of development is a central question in developmental biology. At present, there is no simple answer to this question, but the study of model organisms gives us a starting point. In particular, the genetic and molecular analysis of embryonic development in *Drosophila* highlights the key role of molecular components in the oocyte cytoplasm in controlling gene expression.

Overview of *Drosophila* Development

The life cycle of *Drosophila* is about 10 days long with several distinct phases: the embryo, three larval stages, the pupal

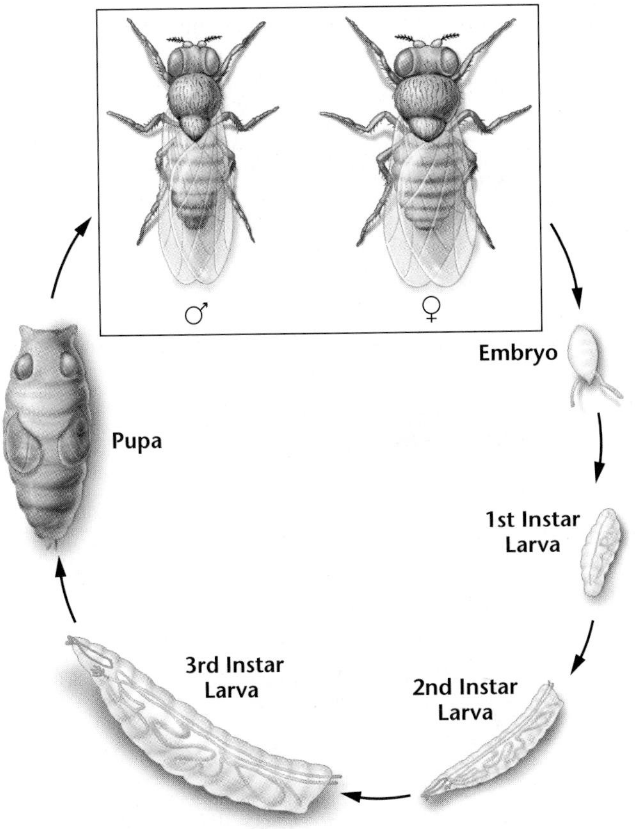

FIGURE 20–1 *Drosophila* life cycle.

stage, and the adult stage (Figure 20–1). Internally, the egg cytoplasm is organized into a series of maternally constructed molecular gradients that play a key role in determining the developmental fates of nuclei located in specific regions of the embryo.

Immediately after fertilization, the zygote nucleus undergoes a series of nuclear divisions without cytokinesis [Figure 20–2 (a) and (b), p. 436], forming a syncytial blastoderm (a syncytium is any cell with more than one nucleus). At about the tenth division, nuclei migrate to the periphery of the egg into cytoplasm containing localized gradients of maternally derived mRNA transcripts and proteins [Figure 20–2 (c)]. After several more divisions, the nuclei become enclosed in plasma membranes [Figure 20–2 (d)].

Germ-cell formation occurs at the posterior pole of the embryo [Figure 20–2 (c) and (d)]. Nuclei transplanted from other regions of the embryo into the posterior cytoplasm will form germ cells, confirming that the posterior pole cytoplasm contains maternal components that direct nuclei to form germ cells.

Transcriptional programs activated in the non–germ-cell nuclei form the embryo's anterior–posterior (front to back) and dorsal–ventral (upper to lower) axes of symmetry, leading to the formation of a segmented embryo [Figure 20–2 (e)]. Under control of the *Hox* gene set (discussed in a later section), these segments give rise to the differentiated structures of the adult fly [Figure 20–2 (f)].

ESSENTIAL POINT

In *Drosophila,* both genetic and molecular studies have confirmed that the egg contains molecular information which initiates a transcriptional cascade that specifies the body plan of the larva, pupa, and adult.

Genetic Analysis of Embryogenesis

Two different gene sets control embryonic development in *Drosophila:* maternal-effect genes and zygotic genes (Figure 20–3, p. 436). Products of maternal-effect genes (mRNA and/or proteins) are placed in the developing egg by the "mother" fly. Many of these products are distributed in a gradient or concentrated in specific regions of the egg cytoplasm. Female flies homozygous for recessive deleterious mutations in maternal-effect genes are sterile: none of their embryos receive wild-type gene products from their mother, so all the embryos develop abnormally. Maternal-effect genes encode transcription factors, receptors, and proteins that regulate gene expression. During embryonic development, these gene products activate or repress expression of the zygotic genome in a temporal and spatial sequence.

Zygotic genes are those transcribed from the nuclei formed after fertilization and expressed in the developing embryo. Flies with homozygous deleterious mutations in zygotic genes exhibit embryonic lethality. In a cross between two flies heterozygous for a recessive zygotic mutation, one-fourth of the embryos (the recessive homozygotes) therefore fail to develop normally and die. In *Drosophila,* many zygotic genes are transcribed in specific regions of the embryo in response to the distribution of maternal-effect proteins.

Much of our knowledge of the genes that regulate *Drosophila* development is based on the work of Christiane Nüsslein-Volhard, Eric Wieschaus, and Ed Lewis, who were awarded the 1995 Nobel Prize for Physiology or Medicine. Ed Lewis initially identified and studied one of these regulatory genes in the 1970s. In the late 1970s, Nüsslein-Volhard and Wieschaus devised a strategy to identify all the genes that control development in *Drosophila.* Their scheme required examining thousands of offspring of randomly mutagenized flies, looking for recessive embryonic lethal mutations with defects in external structures. The parents were thus identified as heterozygous carriers of these mutations, which the researchers grouped into three classes: *gap, pair-rule,* and *segment polarity* genes. In 1980, on the basis of their observations, Nüsslein-Volhard and Wieschaus proposed a model in which embryonic development is initiated by gradients of maternal-effect gene products. The positional information laid down by these molecular gradients is interpreted by two sets of zygotic genes: (1) **segmentation genes** (gap, pair-rule, and segment polarity genes) and (2) **homeotic selector (*Hox*) genes.** The segmentation genes divide the embryo into a series of stripes or segments and define the number, size, and polarity of each segment. The homeotic genes specify the fate of each segment and the adult structures formed from the segments (Figure 20–3).

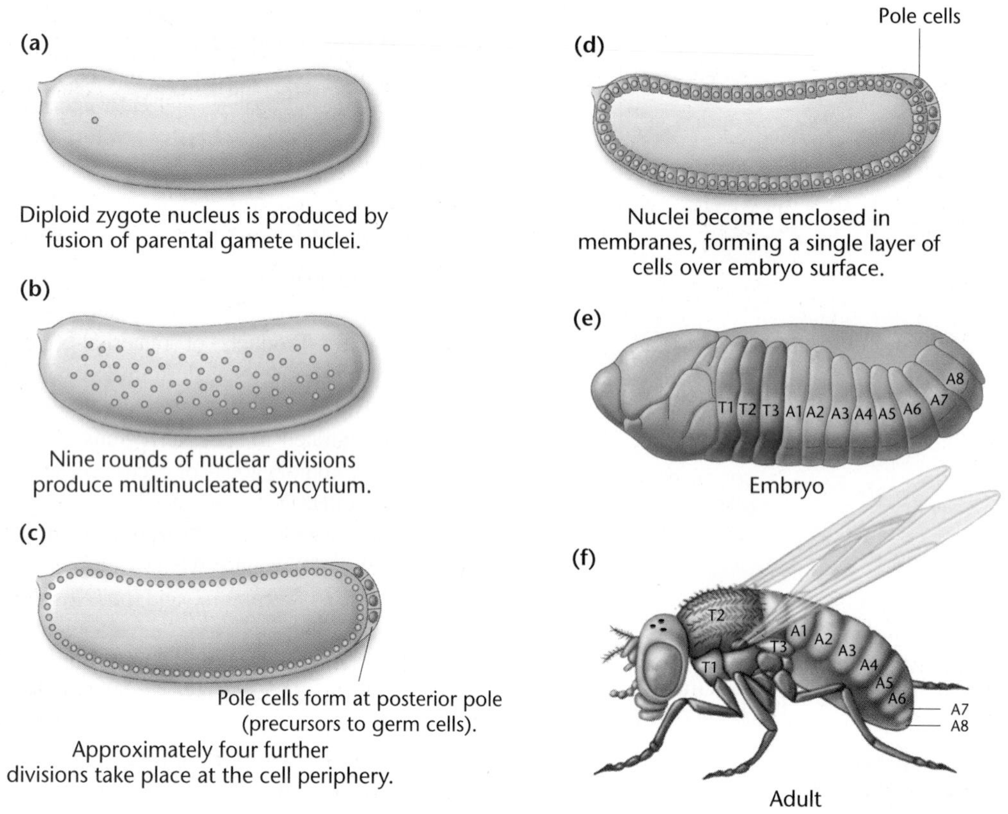

FIGURE 20–2 Early stages of embryonic development in *Drosophila.* (a) Fertilized egg with zygotic nucleus (2*n*), shortly after fertilization. (b) Nuclear divisions occur about every 10 minutes. Nine rounds of division produce a multinucleate cell, the syncytial blastoderm. (c) At the tenth division, the nuclei migrate to the periphery or cortex of the egg, and four additional rounds of nuclear division occur. A small cluster of cells, the pole cells, form at the posterior pole about 2.5 hours after fertilization. These cells will form the germ cells of the adult. (d) About 3 hours after fertilization, the nuclei become enclosed in membranes, forming a single layer of cells over the embryo surface, creating the cellular blastoderm. (e) The embryo at about 10 hours after fertilization. At this stage, the segmentation pattern of the body is clearly established. Behind the segments that will form the head, T1–T3 are thoracic segments, and A1–A8 are abdominal segments. (f) The adult fly showing the structures formed from each segment of the embryo.

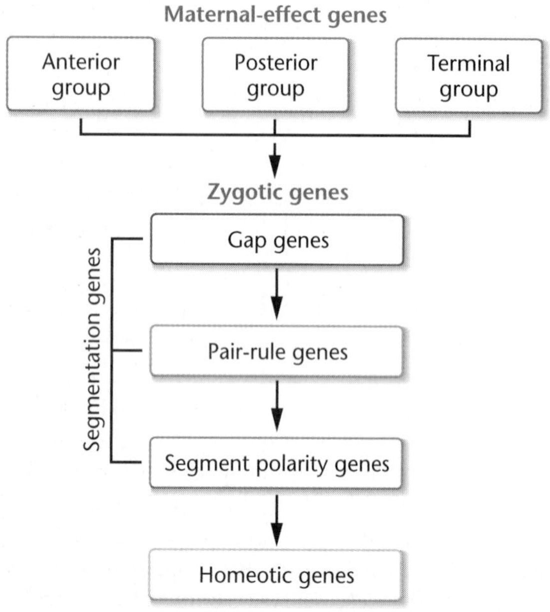

FIGURE 20–3 The hierarchy of genes involved in establishing the segmented body plan in *Drosophila.* Gene products from the maternal genes regulate the expression of the first three groups of zygotic genes (gap, pair-rule, and segment polarity, collectively called the segmentation genes), which in turn control expression of the homeotic genes.

The model is shown in Figure 20–4. Most maternal-effect gene products placed in the egg during oogenesis are activated immediately after fertilization and help establish the anterior–posterior axis of the embryo [Figure 20–4 (a)]. Many maternal gene products encode transcription factors that activate transcription of the gap genes, whose expression divides the embryo into a series of regions corresponding to the head, thorax, and abdomen of the adult [Figure 20–4 (b)]. Gap proteins are transcription factors that activate pair-rule genes, whose products divide the embryo into smaller regions about two segments wide [Figure 20–4 (c)]. The pair-rule genes in turn activate the segment polarity genes, which divide each segment into anterior and posterior regions [Figure 20–4 (d)]. The collective action of the maternal genes that form the anterior–posterior axis and the segmentation genes define the field of action for the homeotic (*Hox*) genes [Figure 20–4 (e)].

ESSENTIAL POINT

Maternal-effect genes lay down the anterior–posterior axis of the embryo and activate genes that specify the location and number of segments, which in turn have their identity determined by homeotic selector genes.

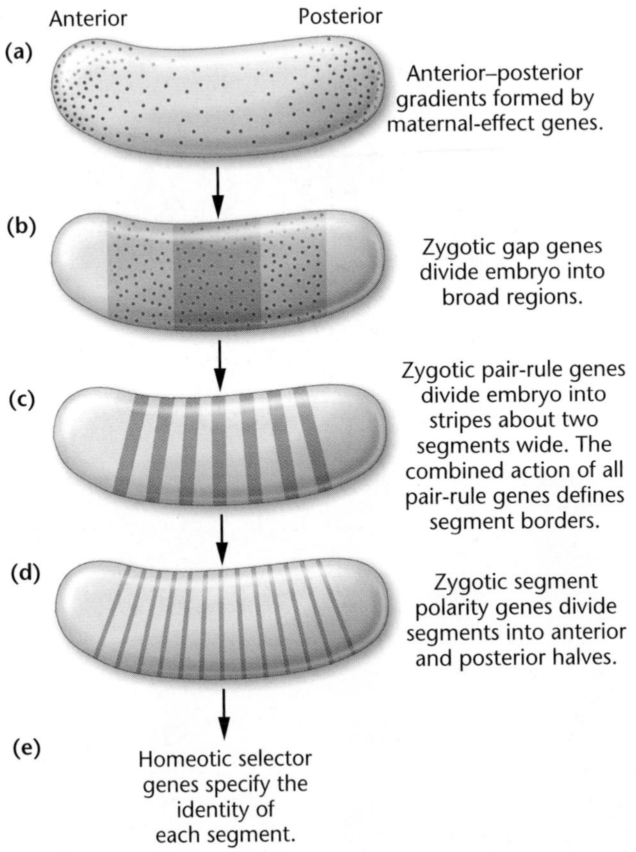
Anterior
Posterior
(a)
Anterior–posterior gradients formed by maternal-effect genes.
(b)
Zygotic gap genes divide embryo into broad regions.
(c)
Zygotic pair-rule genes divide embryo into stripes about two segments wide. The combined action of all pair-rule genes defines segment borders.
(d)
Zygotic segment polarity genes divide segments into anterior and posterior halves.
(e)
Homeotic selector genes specify the identity of each segment.

FIGURE 20–4 (a) Progressive restriction of cell fate during development in *Drosophila*. Gradients of maternal proteins are established along the anterior–posterior axis of the embryo. (b), (c), and (d) Three groups of segmentation genes progressively define the body segments. (e) Individual segments are given identity by the homeotic genes.

NOW SOLVE THIS

Problem 5 on page 448 involves screening for mutants that affect external structures of the embryo. You are asked to reconcile your results with those published in the literature.

Hint: In reconciling the results of your study, remember the differences between genes and alleles.

20.3 Zygotic Genes Program Segment Formation in *Drosophila*

To summarize, zygotic genes are activated or repressed according to a positional gradient of maternal-effect gene products. The expression of three subsets of segmentation genes divides the embryo into a series of segments along its anterior–posterior axis. These segmentation genes are normally transcribed in the developing embryo, and mutations of these genes have embryo-lethal phenotypes.

Over 20 segmentation genes (Table 20.1), have been identified and they are classified on the basis of their mutant phenotypes: (1) mutations in gap genes delete a group of adjacent segments, (2) mutations in pair-rule genes affect every other segment and eliminate a specific part of each affected segment, and (3) mutations in segment polarity genes cause defects in homologous portions of each segment.

TABLE 20.1 Segmentation Genes in *Drosophila*

Gap Genes	Pair-Rule Genes	Segment Polarity Genes
Krüppel	*hairy*	*engrailed*
knirps	*even-skipped*	*wingless*
hunchback	*runt*	*cubitis interruptus*^{D}
giant	*fushi-tarazu*	*hedgehog*
tailless	*paired*	*fused*
buckebein	*odd-paired*	*armadillo*
caudal	*odd-skipped*	*patched*
	sloppy-paired	*gooseberry*
		paired
		naked
		disheveled

In addition to these three sets of genes that determine the anterior–posterior axis of the developing embryo, another set of genes determines the dorsal–ventral axis of the embryo. Our discussion will be limited to the gene sets involved in the anterior–posterior axis. Let us now examine each member of this group in greater detail.

Gap Genes

Transcription of **gap genes** is activated or inactivated by gene products previously expressed along the anterior–posterior axis and by other genes of the maternal gradient system. When mutated, these genes produce large gaps in the embryo's segmentation pattern. *Hunchback* mutants lose head and thorax structures, *Krüppel* mutants lose thoracic and abdominal structures, and *knirps* mutants lose most abdominal structures. Transcription of wild-type gap genes (which encode transcription factors) divides the embryo into a series of broad regions that become the head, thorax, and abdomen. Within these regions, different combinations of gene activity eventually specify both the type of segment that forms and the proper order of segments in the body of the larva, pupa, and adult. Expression domains of the gap genes in different parts of the embryo correlate roughly with the location of their mutant phenotypes: *hunchback* at the anterior, *Krüppel* in the middle (Figure 20–5), and *knirps* at the

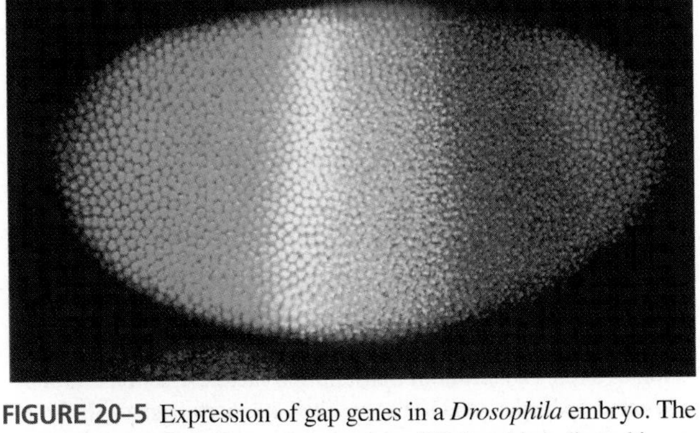

FIGURE 20–5 Expression of gap genes in a *Drosophila* embryo. The hunchback protein is shown in orange, and Krüppel is indicated in green. The yellow stripe is created when cells contain both hunchback and Krüppel proteins. Each dot in the embryo is a nucleus.

posterior. As mentioned earlier, gap genes encode transcription factors that control the expression of pair-rule genes.

Pair-Rule Genes

Pair-rule genes are expressed in a series of seven bands or stripes extending around the embryo. Mutations in pair-rule genes eliminate segment-size sections at every other segment. The pair-rule genes are expressed in narrow bands or stripes of nuclei that extend around the circumference of the embryo. The expression of this gene set first establishes the boundaries of segments and then establishes the developmental fate of the cells within each segment by controlling expression of the segment polarity genes. At least eight pair-rule genes act to divide the embryo into a series of stripes. However, the boundaries of these stripes overlap, so that in each area of overlap, cells express a different combination of pair-rule genes (Figure 20–6). The transcription of the pair-rule genes is mediated by the action of gap gene products and maternal gene products, but the resolution of this segmentation pattern into highly delineated stripes results from the interaction among the gene products of the pair-rule genes themselves (Figure 20–7).

Segment Polarity Genes

Expression of **segment polarity genes** is controlled by transcription factors encoded by pair-rule genes. Within each segment created by pair-rule genes, segment polarity genes become active in a single band of cells that extends around the embryo's circumference (Figure 20–8). This divides the embryo into 14 segments. The products of the segment polarity genes control the cellular identity within each of them and establish the anterior–posterior pattern (the polarity) within each segment.

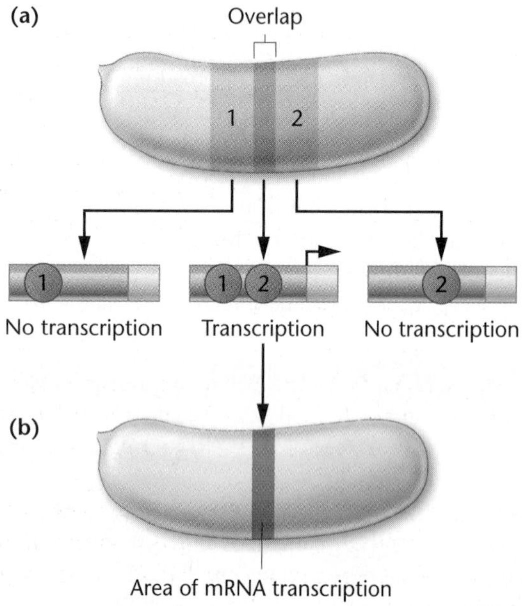

FIGURE 20–6 New patterns of gene expression can be generated by overlapping regions containing two different gene products. (a) Transcription factors 1 and 2 are present in an overlapping region of expression. If both transcription factors must bind to the promoter of a target gene to trigger expression, the gene will be active only in cells containing both factors (most likely in the zone of overlap). (b) The expression of the target gene in the restricted region of the embryo.

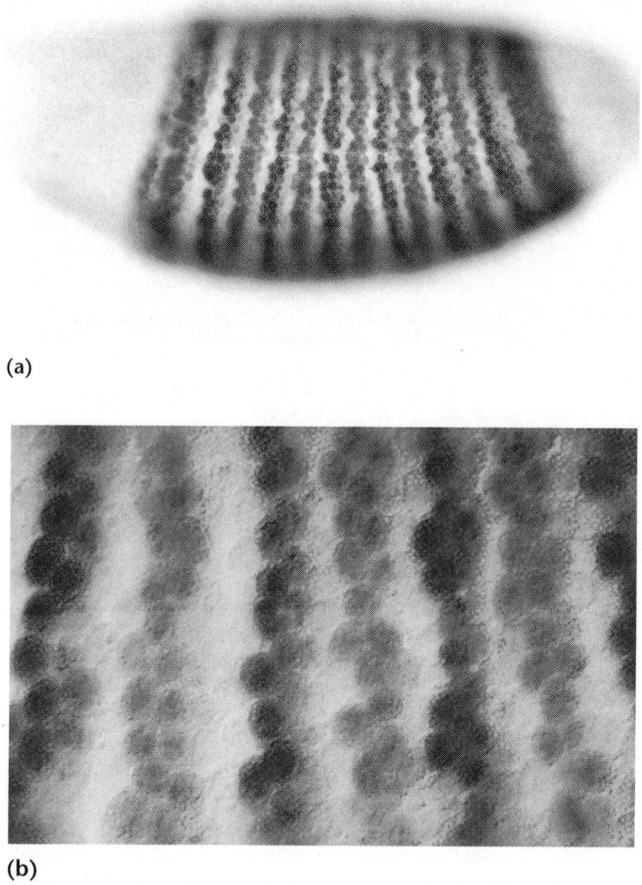

FIGURE 20–7 Stripe pattern of pair-rule gene expression in *Drosophila* embryo. This embryo is stained to show patterns of expression of the genes *even-skipped* and *fushi-tarazu;* (a) low-power view and (b) high-power view of the same embryo.

Segmentation Genes in Mice and Humans

We have seen that segment formation in *Drosophila* depends on the action of three subsets of segmentation genes. Are these genes found in humans and other mammals, and do they

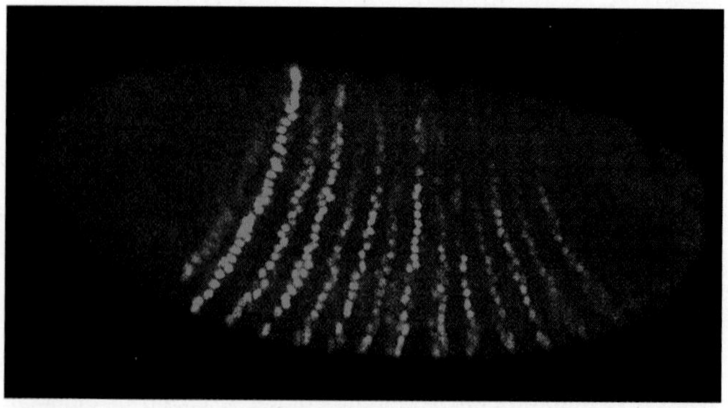

FIGURE 20–8 The 14 stripes of expression of the segment polarity gene *engrailed* in a *Drosophila* embryo.

control aspects of embryonic development in these organisms? To answer this question, let's examine *runt,* one of the pair-rule genes in *Drosophila.* Later in development, it controls aspects of sex determination and formation of the nervous system. The gene encodes a protein that regulates transcription of its target genes. Runt contains a 128-amino-acid DNA-binding region (called the runt domain) that is highly conserved in mouse and human proteins. In fact, *in vitro* experiments show that the *Drosophila* and mouse runt proteins are functionally interchangeable. In mice *runt* is expressed early in development and controls formation of blood cells, bone, and the genital system. In each case, whether in *Drosophila* or mouse, expression of *runt* specifies the fate of uncommitted cells in the embryo by regulating transcription of its target genes.

Mutation in *CBFA,* a human homolog of *runt,* causes cleidocranial dysplasia (CCD), an autosomal dominantly inherited trait. Those affected with CCD have a hole in the top of their skull because their fontanel does not close. Their collar bones (clavicles) do not develop, enabling them to fold their shoulders across their chest (Figure 20–9). Mice with one mutant copy of the *runt* homolog have a phenotype similar to that seen in humans; mice with two mutant copies of the gene have no bones at all. Their skeletons contain only cartilage, much like sharks (Figure 20–10), emphasizing the role of *runt* as an important gene controlling the initiation of bone formation.

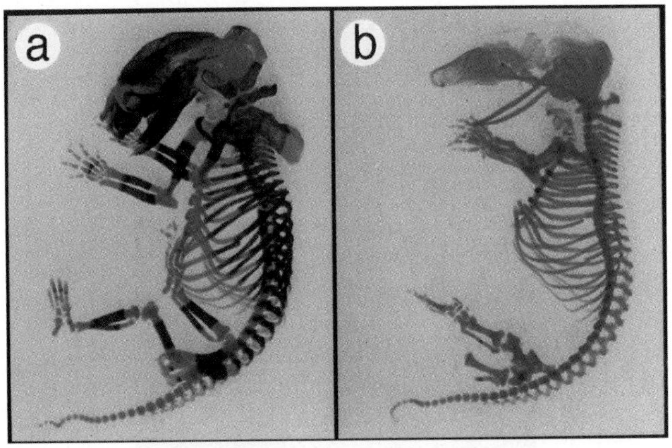

FIGURE 20–10 Bone formation in normal mice and mutants for the *runt* gene *Cbfa1.* (a) Normal mouse embryos at day 17.5 show cartilage (blue) and bone (brown). (b) The skeleton of a 17.5-day homozygous mutant embryo. Only cartilage has formed in the skeleton. There is complete absence of bone formation in the mutant mouse. Expression of a normal copy of the *Cbfa1* gene is essential for specifying the developmental fate of bone-forming osteoblasts.

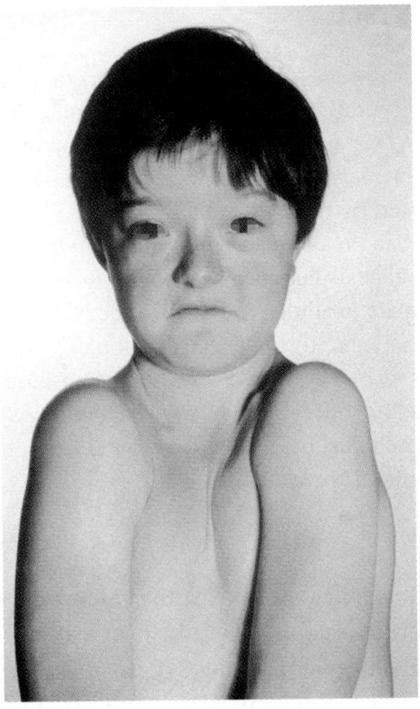

FIGURE 20–9 A boy affected with cleidocranial dysplasia (CCD). This disorder, inherited as an autosomal dominant trait, is caused by mutation in a human *runt* gene, *CBFA.* Affected heterozygotes have a number of skeletal defects, including a hole in the top of the skull where the infant fontanel fails to close, and collar bones that do not develop or form only small stumps. Because the collar bones do not form, CCD individuals can fold their shoulders across their chests. *Reprinted by permission from Macmillan Publishers Ltd.: Fig 1 on p244 from:* British Dental Journal *195: 243–248 2003. Greenwood, M. and Meechan, J. G. "General medicine and surgery for dental practitioners." Copyright © Macmillan Magazines Limited.*

20.4 Homeotic Selector Genes Specify Parts of the Adult Body

As segment boundaries are established by expression of segmentation genes, the homeotic (from the Greek word for "same") genes are activated. Expression of homeotic selector genes determines which adult structures will be formed by each body segment. In *Drosophila,* this includes the antennae, mouth parts, legs, wings, thorax, and abdomen. Mutants of these genes are called **homeotic mutants** because the structure formed by one segment is transformed so that it is the same as that of another segment. For example, the wild-type allele of *Antennapedia (Antp)* specifies formation of a leg on the second segment of the thorax. Dominant gain-of-function *Antp* mutations cause this gene to be expressed in the head as well, and mutant flies have a leg on their head in place of an antenna (Figure 20–11, p. 440).

Hox Genes in *Drosophila*

The *Drosophila* genome contains two clusters of homeotic selector genes (called *Hox* genes) on chromosome 3 that encode transcription factors (Table 20.2). The *Antennapedia (ANT-C)* cluster contains five genes that specify structures in the head and first two segments of the thorax [Figure 20–12 (a), p. 440]. The second cluster, the *bithorax (BX-C)* complex, contains three genes that specify structures in the posterior portion of the second thoracic segment, the entire third thoracic segment, and the abdominal segments [Figure 20–12 (b)].

Hox genes (listed in Table 20.2) have two properties in common. First, each contains a 180-bp domain known as a **homeobox.** (*Hox* is a contraction of homeobox.) The homeobox encodes a DNA-binding sequence of 60 amino acids known as a **homeodomain.** Second, expression of the genes is colinear. Genes at the 3′ end of a cluster are expressed at the anterior end of the embryo, those in the middle

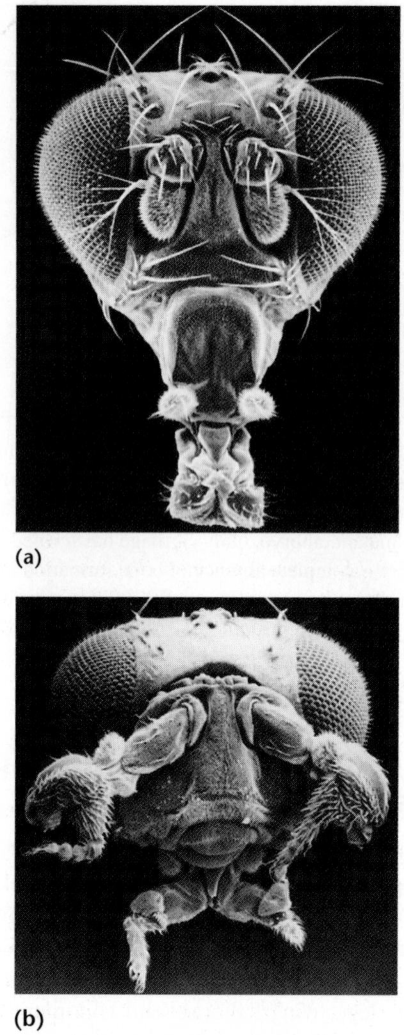

FIGURE 20–11 *Antennapedia (Antp)* mutation in *Drosophila.* (a) Head from wild-type *Drosophila,* showing the antenna and other head parts. (b) Head from an *Antp* mutant, showing the replacement of normal antenna structures with legs. This is caused by activation of the *Antp* gene in the head region.

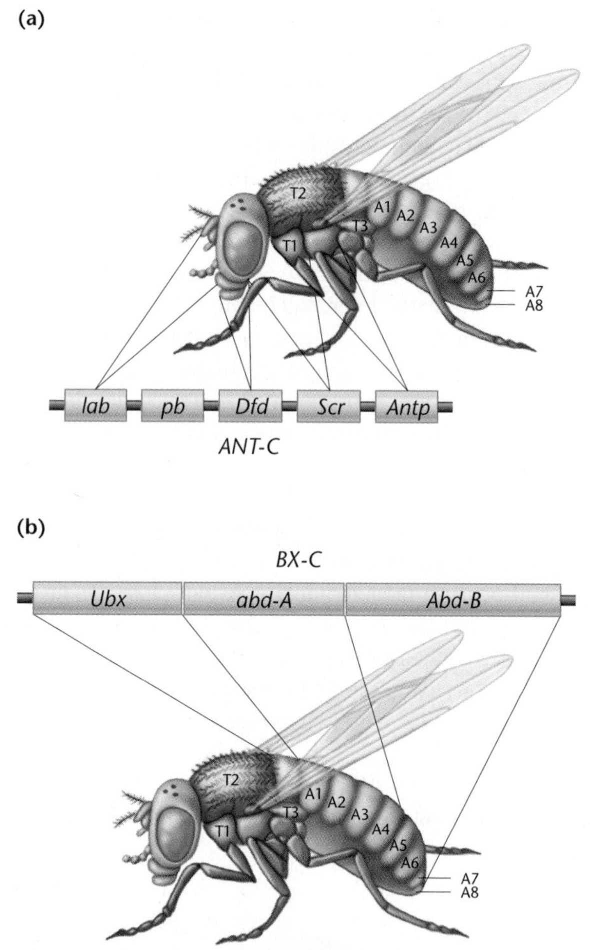

FIGURE 20–12 Genes of the *Antennapedia* complex and the adult structures they specify. (a) In the *ANT-C* complex, the *labial* (*lab*) and *Deformed* (*Dfd*) genes control the formation of head segments. The *Sex comb reduced* (*Scr*) and *Antennapedia* (*Antp*) genes specify the identity of the first two thoracic segments, T1 and T2. The remaining gene in the complex, *proboscipedia* (*pb*), may not act during embryogenesis but may be required to maintain the differentiated state in adults. In mutants, the labial palps are transformed into legs. (b) In the *BX-C* complex, *Ultrabithorax* (*Ubx*) controls formation of structures in the posterior compartment of T2 and structures in T3. The two other genes, *abdominal A* (*abdA*) and *Abdominal B* (*AbdB*), specify the segmental identities of the eight abdominal segments (A1–A8).

TABLE 20.2 *Hox* Genes of *Drosophila*

***Antennapedia* Complex**	***Bithorax* Complex**
labial	*Ultrabithorax*
Antennapedia	*abdominal A*
Sex combs reduced	*Abdominal B*
Deformed	
proboscipedia	

are expressed in the middle of the embryo, and genes at the 5′ end of a cluster are expressed at the embryo's posterior region (Figure 20–13). Although first identified in *Drosophila, Hox* genes are found in the genomes of most eukaryotes with segmented body plans, including zebrafish, *Xenopus* (African clawed frog), chickens, mice, and humans (Figure 20–14).

To summarize, genes that control development in *Drosophila* act in a temporally and spatially ordered cascade, beginning with the genes that establish the anterior–posterior (and dorsal–ventral) axis of the egg and early embryo. Gradients of maternal mRNAs and proteins along the anterior–posterior axis activate gap genes, which subdivide the embryo into broad bands. Gap genes in turn activate pair-rule genes, which divide the embryo into segments. The final group of segmentation genes, the segment polarity genes, divides each segment into anterior and posterior regions arranged linearly along the anterior–posterior axis. The segments are then given identity by the *Hox* genes. Therefore, this progressive restriction of developmental potential of the *Drosophila* embryo's cells (all of which occurs during the first third of embryogenesis) involves a cascade of gene action, with regulatory proteins acting on transcription, translation, and signal transduction.

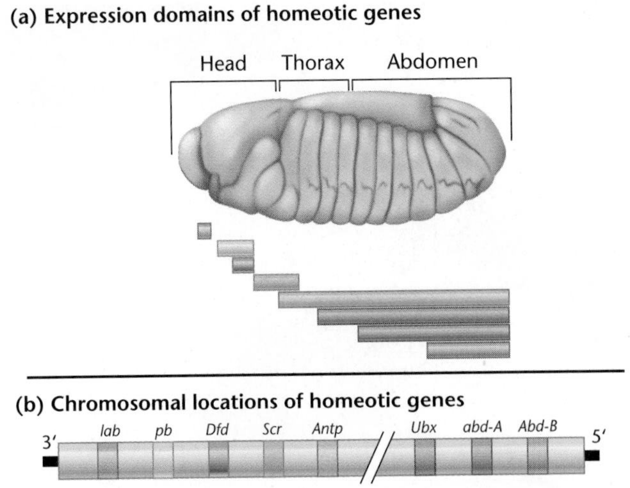

FIGURE 20–13 The colinear relationship between the spatial pattern of expression and chromosomal locations of homeotic genes in *Drosophila.* (a) *Drosophila* embryo and the domains of homeotic gene expression in the embryonic epidermis and central nervous system. (b) Chromosomal location of homeotic selector genes. Note that the order of genes on the chromosome correlates with the sequential anterior borders of their expression domains.

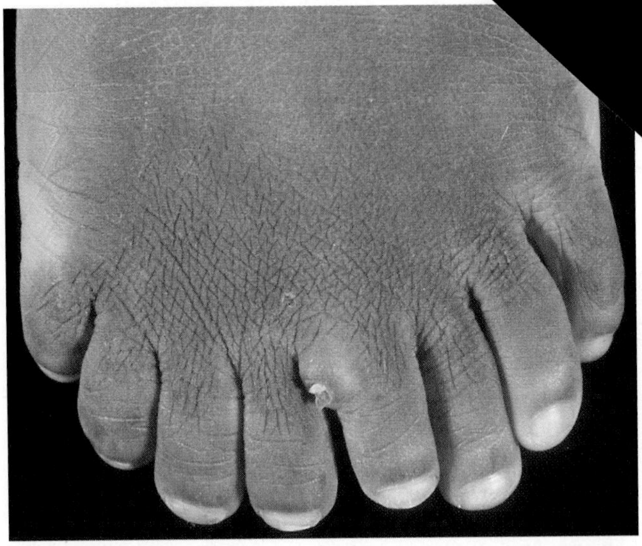

FIGURE 20–15 Mutations in posterior *Hox* genes (*HOXD13* in this case) in humans result in malformations of the limbs, shown here as extra toes. This condition is known as synpolydactyly. Mutations in *HOXD13* are also associated with abnormalities of the bones in the hands and feet.

Hox Genes and Human Genetic Disorders

Although first described in *Drosophila, Hox* genes are found in the genomes of all animals where they play a fundamental role in shaping the body and its appendages. The conservation of sequence, the order of genes in the *Hox* clusters, and their pattern of expression in vertebrates suggests that, as in *Drosophila,* these genes control development along the anterior–posterior and the formation of appendages. This role for *HOXD* genes in humans was confirmed by the discovery that several inherited limb malformations are caused by mutations in *HOXD* genes. For example, mutations in *HOXD13* cause synpolydactyly (SPD), a malformation characterized by extra fingers and toes, and abnormalities in bones of the hands and feet (Figure 20–15).

NOW SOLVE THIS

Problem 15 on page 449 involves analysis of expression patterns for two genes in *Drosophila* embryos with different genetic backgrounds. You are asked to decide which gene regulates the other.

Hint: It is this genetic background that provides the clues for the timing of expression and regulatory patterns of the two genes in question.

FIGURE 20–14 Conservation of organization and patterns of expression in *Hox* genes. (Top) The structures formed in adult *Drosophila* are shown, with the colors corresponding to members of the *Hox* cluster that control their formation. (Bottom) The arrangement of the *Hox* genes in an early human embryo. As in *Drosophila,* genes at the (3′ end) of the cluster form anterior structures, and genes at the (5′ end) of the cluster form posterior structures.

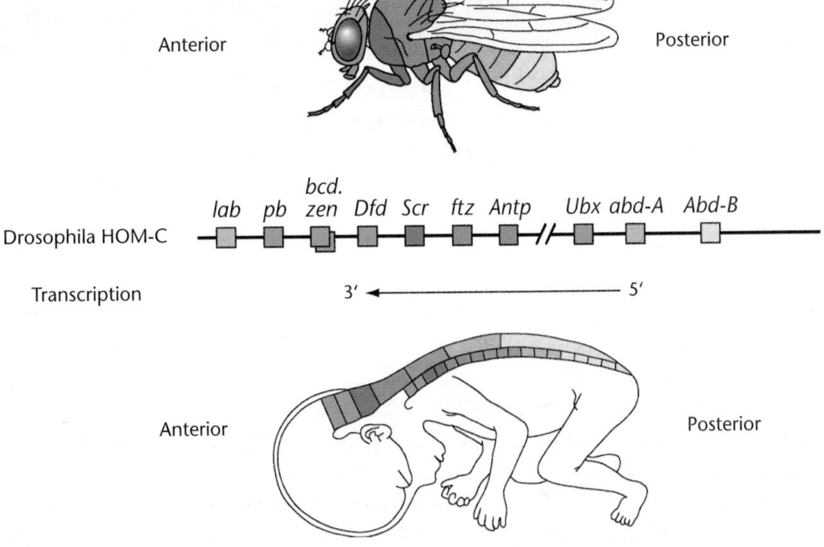

e Evolved ental Systems That ose of Animals

erged from a common unicellular an- years ago, after the origin of eukary- re the rise of multicellular organisms. Genome seq nd genetic analysis of mutants in plants and animals indicate that basic mechanisms of developmental pattern formation have evolved independently in animals and plants. We have already examined the genetic systems that control development and pattern formation in animals, using *Drosophila* as a model organism.

Flower development in *Arabidopsis thaliana* (Figure 20–16), a small plant in the mustard family, has been used to study pattern formation in plants. A cluster of undifferentiated cells, called the *floral meristem,* gives rise to flowers (Figure 20–17). Each flower consists of four organs—sepals, petals, stamens, and carpels—that develop from concentric rings of cells within the meristem [Figure 20–18 (a)]. Each organ develops from a different concentric ring, or whorl of cells.

FIGURE 20–16 The flowering plant *Arabidopsis thaliana,* used as a model organism in plant genetics.

FIGURE 20–17 (a) Parts of the *Arabidopsis* flower. The floral organs are arranged concentrically. The sepals form the outermost ring, followed by petals and stamens, with carpels on the inside. (b) View of the flower from above.

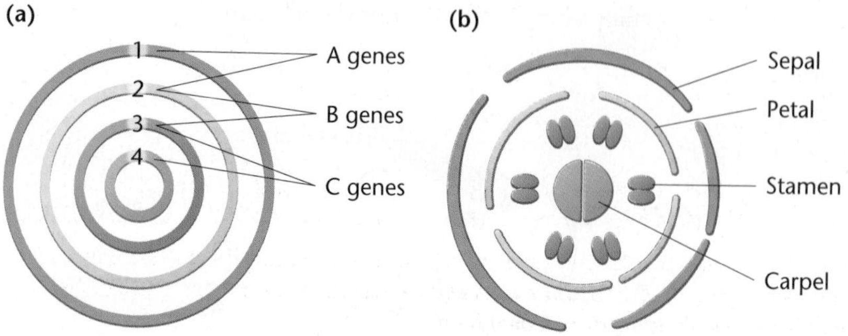

FIGURE 20–18 Cell arrangement in the floral meristem. (a) The four concentric rings, or whorls, labeled 1–4, give rise to (b) arrangement of the sepals, petals, stamens, and carpels, respectively, in the mature flower.

TABLE 20.3	Homeotic Selector Genes in *Arabidopsis**
Class A	*APETALA1 (AP1)*
	APETALA2 (AP2)
Class B	*APETALA3 (AP3)*
	PISTILLATA (P1)
Class C	*AGAMOUS (AG)*

*By convention, wild-type genes in *Arabidopsis* use capital letters.

Homeotic Genes in *Arabidopsis*

Three classes of floral homeotic genes control the development of these organs. Class A genes specify sepals, class A and class B genes specify petals, and class B and class C genes control stamen formation. Class C genes alone specify carpels [Figure 20–18 (b)]. The genes in each class are listed in Table 20.3. Class A genes are active in whorls 1 and 2 (sepals and petals), class B genes are expressed in whorls 2 and 3 (petals and stamens), and class C genes are expressed in whorls 3 and 4 (stamens and carpels). The organ formed depends on the expression pattern of the three gene classes. Expression of class A genes in whorl 1 causes sepals to form. Expression of class A *and* class B genes in whorl 2 leads to petal formation. Expression of class B and class C genes in whorl 3 leads to stamen formation. In whorl 4, expression of class C genes causes carpel formation.

As in *Drosophila,* mutations in homeotic genes cause organs to form in abnormal locations. For example, in *AP2* mutants (mutation of a class A gene), the order of organs is carpel, stamen, stamen, and carpel instead of the normal order, sepal, petal, stamen, and carpel [Figure 20–19 (a) and (b)]. In B loss-of-function mutants, petals become sepals, and stamens are transformed into carpels [Figure 20–19 (c)] and the order of organs becomes sepal, sepal, carpel, carpel. Plants carrying a mutation for the class 3 gene AGAMOUS will have petals in whorl 3 (instead of stamens) and sepals in whorl 4 (instead of carpels), and the order of organs will be sepal, petal, petal, and sepal [Figure 20–19 (d)].

Evolutionary Divergence in Homeotic Genes

Drosophila and *Arabidopsis* use different sets of master regulatory genes to establish the body axis and specify the identity of structures along the axis. In *Drosophila,* this task is accomplished in part by the *Hox* genes, which encode a set of transcription factors sharing a homeobox domain. In *Arabidopsis,* the floral homeotic genes belong to a different family of transcription factors, called the **MADS-box proteins,** characterized by a common sequence of 58 amino acids with no similarity in amino acid sequence or protein structure with the *Hox* genes. Both gene sets encode transcription factors, both sets are master regulators of development expressed in a pattern of overlapping domains, and both specify identity of structures.

Reflecting their evolutionary origin from a common ancestor, the genomes of both *Drosophila* and *Arabidopsis* contain members of the homeobox and MADS-box genes, but these genes have been adapted for different uses in the plant and animal kingdoms, indicating that developmental mechanisms evolved independently in each group.

In both plants and animals, the action of transcription factors depends on changes in chromatin structure that make genes available for expression. Mechanisms of transcription initiation are conserved in plants and animals, as is reflected in the homology of genes in *Drosophila* and *Arabidopsis* that maintain patterns of expression initiated by regulatory gene sets. Action of the floral homeotic genes is controlled by a gene called *CURLY LEAF.* This gene shares significant homology with members of a *Drosophila* gene family called *Polycomb.* This family of regulatory genes controls expression of homeobox genes during development. Both *CURLY LEAF* and *Polycomb* encode proteins that alter chromatin conformation and shut off gene expression. Thus, although different genes are used to control development, both plants and animals use an evolutionarily conserved mechanism to regulate expression of these gene sets.

ESSENTIAL POINT

Flower formation in *Arabidopsis* is controlled by homeotic genes, but these gene sets are from a different gene family than the homeotic selector genes of *Drosophila* and other animals.

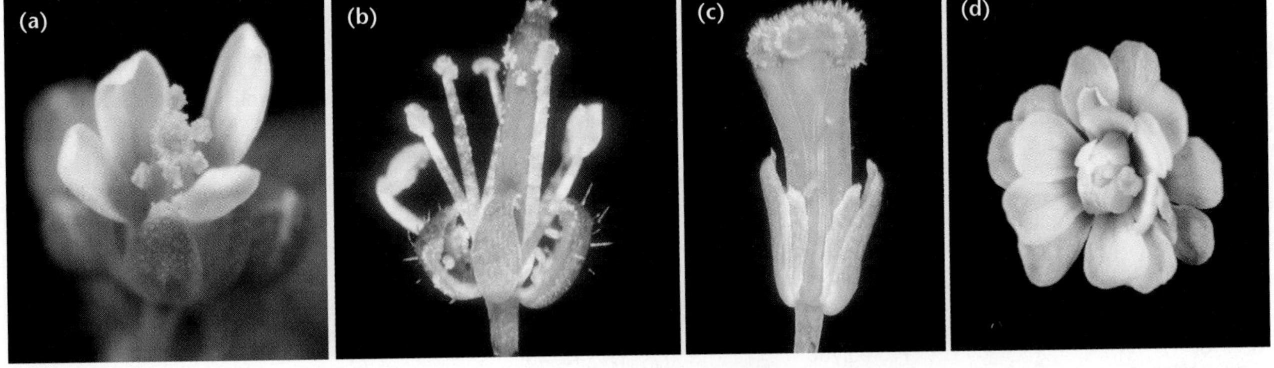

FIGURE 20–19 (a) Wild-type flowers of *Arabidopsis* have (from outside to inside) sepals, petals, stamens, and carpels. (b) Homeotic *APETALA2* mutant flower has carpels, stamens, stamens, and carpels. (c) *PISTILLATA* mutants have sepals, sepals, carpels, and carpels. (d) *AGAMOUS* mutants have petals and sepals at places where stamens and carpels should form.

20.6 Cell–Cell Interactions in Development Are Modeled in *C. elegans*

During development in multicellular organisms, cell–cell interactions influence the transcriptional programs and developmental fate of surrounding cells. Cell–cell interaction is an important process in the embryonic development of most eukaryotic organisms, including *Drosophila,* as well as vertebrates such as clawed frogs *(Xenopus),* mice, and humans.

Signaling Pathways in Development

In early development, animals use a number of signaling pathways to regulate development; after organogenesis begins, other signal pathways are added to those already in use. These newly activated pathways act both independently and in coordinated networks to elicit specific transcriptional responses. The signal networks establish anterior–posterior polarity and body axes, coordinate pattern formation, and direct the differentiation of tissues and organs. The signaling pathways used in early development and some of the developmental processes they control are listed in Table 20.4. After an introduction to the components and interactions of one of these systems—the **Notch signaling pathway**—we will briefly examine its role in the development of the vulva in the nematode, *Caenorhabditis elegans.*

TABLE 20.4 Signaling Pathways Used in Early Embryonic Development

Wnt Pathway
Dorsalization of body
Female reproductive development
Dorsal–ventral differences
TGF-β Pathway
Mesoderm induction
Left–right asymmetry
Bone development
Hedgehog Pathway
Notochord induction
Somitogenesis
Gut/visceral mesoderm
Receptor Tyrosine Kinase Pathway
Mesoderm maintenance
Notch Signaling Pathway
Blood cell development
Neurogenesis
Retina development

*Source: *Taken from* Gerhart, J. 1999. 1998 Warkany lecture: Signaling pathways in development. *Teratology* 60: 226–239.

The Notch Signaling Pathway

The genes in the Notch pathway are named after the *Drosophila* mutants that were used to identify components of this signal transduction system. Notch works through direct cell–cell contact to control the developmental fate of the interacting cells. The *Notch* gene (and the equivalent gene in other organisms) encodes a signal receptor embedded in the plasma membrane (Figure 20–20). The signal is another transmembrane protein encoded by the *Delta* gene (and its equivalents). Because both the signal and receptor are membrane proteins, the Notch signal system works only between adjacent cells. When the Delta protein binds to the Notch receptor protein, the cytoplasmic tail of the Notch protein is cut off and binds to a cytoplasmic protein encoded by the *Su(H)* (suppressor of *Hairless*) gene. This protein complex moves into the nucleus and binds to transcriptional cofactors, activating transcription of a gene set that controls a specific developmental pathway (Figure 20–20).

One of the main roles of the Notch signal system is specifying the fate of equivalent cells in a population. In its simplest form, this interaction involves two neighboring cells that are developmentally equivalent. We will explore the role of the Notch signaling system in development of the vulva in *C. elegans,* after a brief introduction to nematode embryogenesis.

Overview of *C. elegans* Development

The nematode *C. elegans* is widely used to study the genetic control of development. This organism has several advantages for such studies: (1) the genetics of the organism are well

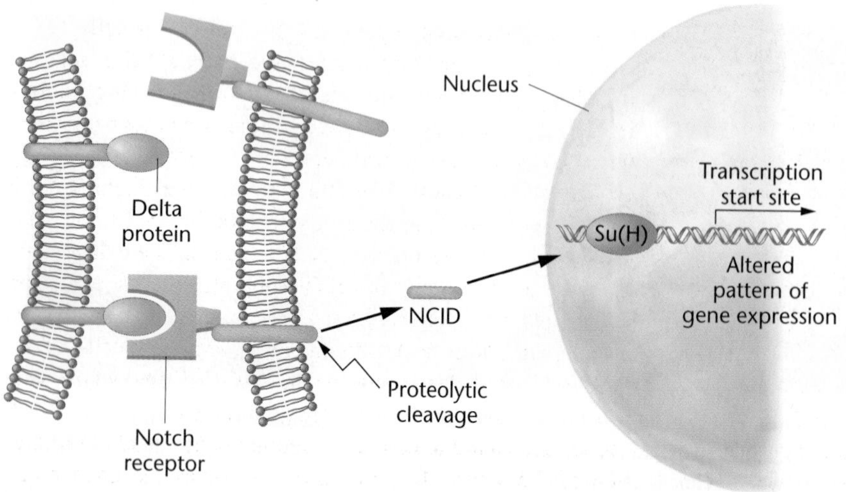

FIGURE 20–20 Components of the Notch signaling pathway in *Drosophila.* The cell carrying the Delta transmembrane protein is the sending cell; the cell carrying the transmembrane Notch protein receives the signal. Binding of Delta to Notch triggers a proteolytic-mediated activation of transcription. The fragment cleaved from the cytoplasmic side of the Notch protein, called the Notch intracellular domain (NCID), combines with the Su(H) protein and moves to the nucleus where it activates a program of gene transcription.

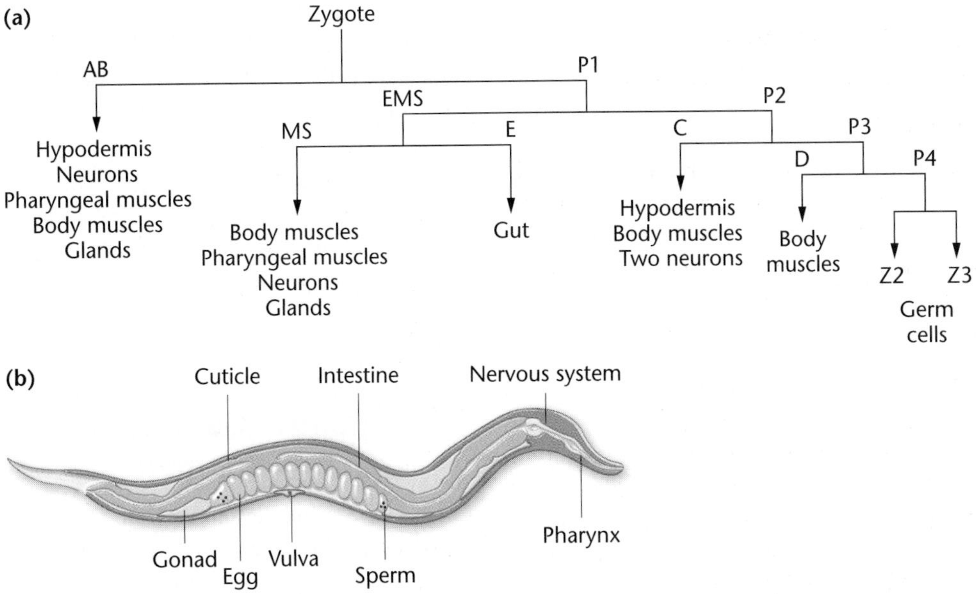

FIGURE 20–21 (a) A truncated cell lineage chart for *C. elegans,* showing early divisions and the tissues and organs formed from these lineages. Each vertical line represents a cell division, and horizontal lines connect the two cells produced. For example, the first division of the zygote creates two new cells, AB and P1. During embryogenesis, cell divisions will produce the 959 somatic cells of the adult hermaphrodite worm. (b) An adult *C. elegans* hermaphrodite. This nematode, about 1 mm in length, consists of 959 cells and is widely used as a model organism to study the genetic control of development.

known, (2) its genome has been sequenced, and (3) adults contain a small number of cells that follow a highly deterministic developmental program. Adult nematodes are about 1 mm long and develop from a fertilized egg in about two days (Figure 20–21). The life cycle includes an embryonic stage (about 16 hours), four larval stages (L1 through L4), and the adult stage. Adults are of two sexes: XX self-fertilizing hermaphrodites that can make both eggs and sperm, and XO males. Self-fertilization of mutagen-treated hermaphrodites is used to develop homozygous stocks of mutant strains, and hundreds of such mutants have been generated, catalogued, and mapped.

Adult hermaphrodites have 959 somatic cells (and about 2000 germ cells). The lineage of each cell, from fertilized egg to adult has been mapped (Figure 20–21) and is invariant from individual to individual. Knowing the lineage of each cell, we can easily follow events caused by mutations that alter cell fate or by killing specific cells with laser microbeams or ultraviolet irradiation. In *C. elegans* hermaphrodites, the developmental fate of cells in the reproductive system is determined by cell–cell interaction, illustrating how gene expression and cell–cell interaction work together to specify developmental outcomes.

NOW SOLVE THIS

Problem 26 on page 449 involves two genes that control sex determination in *C. elegans.* You are asked to analyze a model of sex determination.

Hint: In solving this problem, remember to consider the action of gene products and the effect of loss-of-function mutations on expression of other genes or the action of other proteins.

Genetic Analysis of Vulval Formation

Adult *C. elegans* hermaphrodites lay eggs through the vulva, an opening near the middle of the body (Figure 20–21). The vulva is formed in stages during larval development and involves several rounds of cell–cell interactions.

In *C. elegans,* two neighboring cells, Z1.ppp and Z4.aaa, interact with each other so that one becomes the gonadal anchor cell (from which the vulva forms) and the other becomes a precursor to the uterus (Figure 20–22, p. 446). The determination of which cell becomes which occurs during the second larval stage (L2) and is controlled by the Notch receptor gene, *lin-12.* In recessive *lin-12(0)* mutants (a loss-of-function mutant), both cells become anchor cells. The dominant mutation *lin-12(d)* (a gain-of-function mutation) causes both to become uterine precursors. Thus, it appears that the expression of the *lin-12* gene causes the selection of the uterine pathway, since in the absence of the LIN-12 (Notch) receptor, both cells become anchor cells.

However, the situation is more complex than it first appears. Initially, the two neighboring cells are developmentally equivalent. Each synthesizes low levels of the Notch signal protein (encoded by the *lag-2* gene) *and* the Notch receptor protein. By chance, one cell ends up secreting more of the signal (LAG-2 or Delta protein). This causes its neighboring cell to increase production of the receptor (LIN-12 protein). The cell producing more of the receptor protein becomes the uterine precursor, and the other cell, producing more signal protein, becomes the anchor cell. The critical factor in this first round of cell–cell interaction is the balance between the LAG-2 (Delta) signal gene product and the LIN-12 (Notch) gene product.

A second round of cell–cell communication leads to formation of the vulva. This interaction involves the anchor cell

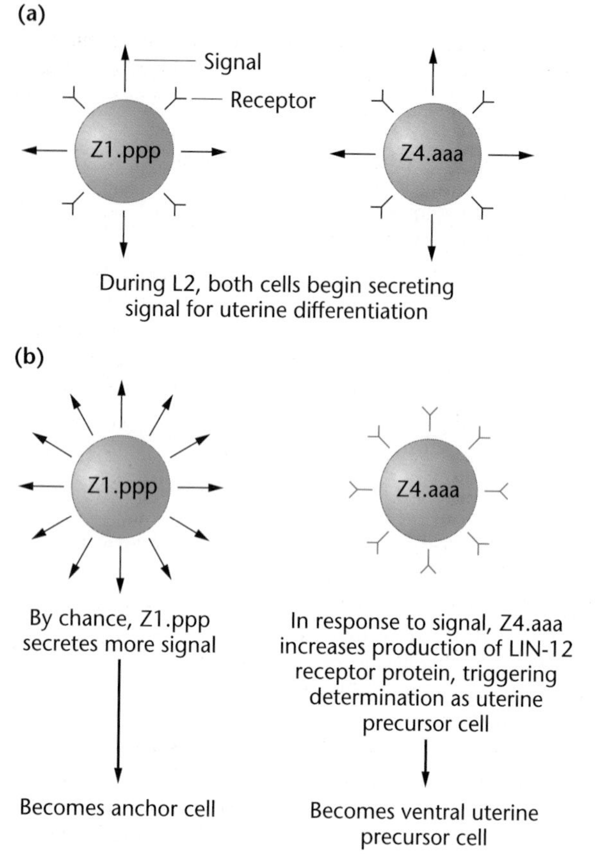
(a)
Signal
Receptor
Z1.ppp
Z4.aaa
During L2, both cells begin secreting signal for uterine differentiation
(b)
Z1.ppp
Z4.aaa
By chance, Z1.ppp secretes more signal
In response to signal, Z4.aaa increases production of LIN-12 receptor protein, triggering determination as uterine precursor cell
Becomes anchor cell
Becomes ventral uterine precursor cell

FIGURE 20–22 Cell–cell interaction in anchor cell determination. (a) During L2, two neighboring cells begin the secretion of chemical signals for the induction of uterine differentiation. (b) By chance, cell Z1.ppp produces more of these signals, causing cell Z4.aaa to increase production of the receptor for signals. The action of increased signals causes Z4.aaa to become the ventral uterine precursor cell and allows Z1.ppp to become the anchor cell.

(located in the gonad) and six of its neighboring cells (called precursor cells) located in the skin. The precursor cells, named P3.p to P8.p, are called Pn.p cells. The fate of each Pn.p cell is specified by its position relative to the anchor cell.

During development, the LIN-3 signal protein is synthesized by the anchor cell, and this signal is received and processed by three adjacent Pn.p precursor cells (Pn.p 5-7). The cell closest to the anchor cell (usually Pn.p 6) becomes the primary vulval precursor cell, and the adjacent cells (Pn.p 5 and 7) become secondary precursor cells. A signal protein from the primary vulval cell activates the *lin-12* gene in the secondary cells, preventing them from becoming primary precursor cells. The other precursor cells (Pn.p 3,4, and 8) receive no signal from the anchor cell and become skin cells.

ESSENTIAL POINT

In *C. elegans,* the well-studied pathway of cell lineage during embryonic development allows developmental biologists to study the cell–cell signaling required for organogenesis.

20.7 Transcriptional Networks Control Gene Expression in Development

As we have seen in examples from *Drosophila* and *C. elegans,* development in higher eukaryotes depends on a closely coordinated pattern of gene expression across both space and time. Understanding development at the level of gene expression will require dissection of the network of transcriptional controls that generates these patterns of expression. Among the key elements coordinating spatiotemporal patterns of gene expression are ***cis*-regulatory elements (CREs),** also called ***cis*-regulatory modules (CRMs).** These DNA sequences can be located upstream from genes, within introns, or downstream from genes. CRMs typically contain binding sites for several transcription factors and usually have multiple binding sites for each of these factors. Binding of transcription factors to sites in CRMs determines whether the gene in question will be activated or repressed. If activated, the basal transcription complex is recruited to the core promoter sequence (see Chapter 15 to review eukaryotic transcriptional regulation). Some of the genes regulated by CRMs encode transcription factors, thus forming an interconnected network of regulatory genes and their targets.

Transcriptional networks have been described for a wide range of organisms, yet all have underlying similarities in organization and function, indicating that the basic regulatory system is conserved by natural selection and is built on the use of CRMs.

ESSENTIAL POINT

Using experimental data and algorithms, researchers are elucidating the transcriptional networks that control the temporal and spatial patterns of gene expression in developing embryos.

GENETICS, TECHNOLOGY, AND SOCIETY

Stem Cell Wars

Stem cell research may be the most controversial research area since the beginning of recombinant DNA technology in the 1970s. Although stem cell research is the focus of presidential proclamations, media campaigns, and ethical debates, few people understand it sufficiently to evaluate its pros and cons.

Stem cells are primitive cells that replicate indefinitely and have the capacity to differentiate into cells with specialized functions, such as the cells of heart, brain, liver, and muscle tissue. All the cells that make up the approximately 200 distinct types of tissues in our bodies are descended from stem cells. Some types of stem cells are defined as *totipotent,* meaning they have the ability to differentiate into any mature cell type in the body. Other types of stem cells are *pluripotent,* meaning that they are able to differentiate into any of a smaller number of mature cell types. In contrast, mature, fully differentiated cells do not replicate or undergo transformations into different cell types.

In the last few years, several research teams have isolated and cultured human pluripotent stem cells. These cells remain undifferentiated and grow indefinitely in culture dishes. When treated with growth factors or hormones, these pluripotent stem cells differentiate into cells that have the characteristics of neural, bone, kidney, liver, heart, or pancreatic cells.

The fact that pluripotent stem cells grow prolifically in culture and differentiate into more specialized cells has created great excitement. Some foresee a day when stem cells may be a cornucopia from which to harvest unlimited numbers of specialized cells to replace cells in damaged and diseased tissues. Hence, stem cells could be used to treat Parkinson disease, type 1 diabetes, chronic heart disease, Alzheimer disease, and spinal cord injuries. Some predict that stem cells will be genetically modified to eliminate transplant rejection or to deliver specific gene products, thereby correcting genetic defects or treating cancers. The excitement about stem cell therapies has been fueled by reports of dramatically successful experiments in animals. For example, mice with spinal cord injuries regained their mobility and bowel and bladder control after they were injected with human stem cells. Both proponents and critics of stem cell research agree that stem cell therapies could be revolutionary. Why, then, should stem cell research be so contentious?

The answer to that question lies in the source of the pluripotent stem cells. Until recently, all pluripotent stem cell lines were derived from five-day-old embryonic blastocysts. Blastocysts at this stage consist of 50–150 cells, most of which will develop into placental and supporting tissues for the early embryo. The inner cell mass of the blastocyst consists of about 30 to 40 pluripotent stem cells that can develop into all the embryo's tissues. *In vitro* fertilization clinics grow fertilized eggs to the five-day blastocyst stage prior to uterine transfer. Embryonic stem cell (ESC) lines are created by taking the inner cell mass out of five-day blastocysts and growing the cells in culture dishes.

The fact that early embryos are destroyed in the process of establishing human ESC lines disturbs people who believe that preimplantation embryos are persons with rights; however, it does not disturb people who believe that these embryos are too primitive to have the status of a human being. Both sides in the debate invoke fundamental questions of what constitutes a human being.

Recently, scientists have developed several types of pluripotent stem cells without using embryos. One of the most promising types—known as *induced pluripotent stem (iPS) cells*—uses adult somatic cells as the source of pluripotent stem cell lines. To prepare iPS cells, scientists isolate somatic cells (such as cells from skin) and infect them with engineered retroviruses that integrate into the cells' DNA. These retroviruses contain several cloned human genes that encode products responsible for converting the somatic cells into immortal, pluripotent stem cells.

The development of iPS cell lines has generated renewed enthusiasm for stem cell research, as these cells bypass the ethical problems associated with the use of human embryos. In addition, they may become sources of patient-specific pluripotent stem cell lines that can be used for transplantation, without immune system rejection.

At the present time, it is unknown whether stem cells of any type will be as miraculous as predicted; however, if stem cell research progresses at its current rapid pace, we won't have long to wait.

Your Turn

1. What, in your opinion, are the scientific and ethical problems that still surround stem cell research? Are these problems solved by the new methods of creating pluripotent stem cells?

You can find descriptions of some new methods of generating pluripotent stem cell lines, and the ethical issues that accompany these methods, in Kastenberg, Z. J. and Odorico, J. S. 2008. Alternative sources of pluripotency: science, ethics, and stem cells. *Transplantation Rev.* **22***:* 215–222.

2. What are the current stem cell research laws in your region, and how do these compare with national regulations?

A starting point for information about stem cell research regulations can be found on the Stem Cell Information Web site of the National Institutes of Health (**http://stemcells.nih.gov**).

3. Do you oppose or support stem cell research? Why, or why not?

An interesting on-line poll, along with arguments for and against stem cell research, is offered by the Public Broadcasting Corporation, at **http://www.pbs.org/wgbh/nova/sciencenow/0305/03-poll-flash.html**.

4. What, in your opinion, is the most significant development in stem cell research in the last year?

Some ideas to start your search are: the PubMed Web site (**http://www.ncbi.nlm.nih.gov/sites/entrez?db=PubMed**) *and the New York Times on-line Stem Cell page* (**http://topics.nytimes.com/top/news/health/diseasesconditionsandhealthtopics/stemcells/**).

CASE STUDY One foot or another

In humans the *HOXD* homeotic gene cluster plays a critical role in limb development. In one large family, 16 of 36 members expressed one of two dominantly inherited malformations of the feet known as rocker bottom foot (CVT) or claw foot (CMT). One individual had one foot with CVT and the other with CMT. Genomic analysis identified a single missense mutation in the *HOXD10* gene, resulting in a single amino acid substitution in the homeodomain of the encoded transcription factor. This region is crucial for making contact and binding to the target genes controlled by this protein. All family members with the foot malformations were heterozygotes; all unaffected members were homozygous for the normal allele.

1. Given that affected heterozygotes carry one normal allele of the *HOXD10* gene, how might a dominant mutation in a gene encoding a transcription factor lead to a developmental malformation?
2. How can two clinically different disorders result from the same mutation?
3. What might we learn about the control of developmental processes from an understanding of how this mutation works?

Insights and Solutions

1. In the slime mold *Dictyostelium,* experimental evidence suggests that cyclic AMP (cAMP) plays a central role in the developmental program leading to spore formation. The genes encoding the cAMP cell-surface receptor have been cloned, and the amino acid sequence of the protein components is known. To form reproductive structures, free-living individual cells aggregate together and then differentiate into one of two cell types, prespore cells or prestalk cells. Aggregating cells secrete waves or oscillations of cAMP to foster the aggregation of cells and then continuously secrete cAMP to activate genes in the aggregated cells at later stages of development. It has been proposed that cAMP controls cell–cell interaction and gene expression. It is important to test this hypothesis by using several experimental techniques. What different approaches can you devise to test this hypothesis, and what specific experimental systems would you employ to test them?

Solution: Two of the most powerful forms of analysis in biology involve the use of biochemical analogs (or inhibitors) to block gene transcription or the action of gene products in a predictable way, and the use of mutations to alter genes and their products. These two approaches can be used to study the role of cAMP in the developmental program of *Dictyostelium.* First, compounds chemically related to cAMP, such as GTP and GDP, can be used to test whether they have any effect on the processes controlled by cAMP. In fact, both GTP and GDP lower the affinity of cell-surface receptors for cAMP, effectively blocking the action of cAMP.

Mutational analysis can be used to dissect components of the cAMP receptor system. One approach is to use transformation with wild-type genes to restore mutant function. Similarly, because the genes for the receptor proteins have been cloned, it is possible to construct mutants with known alterations in the component proteins and transform them into cells to assess their effects.

2. In the sea urchin, early development may occur even in the presence of actinomycin D, which inhibits RNA synthesis. However, if actinomycin D is present early in development but is removed a few hours later, all development stops. In fact, if actinomycin D is present only between the sixth and eleventh hours of development, events that normally occur at the fifteenth hour are arrested. What conclusions can be drawn concerning the role of gene transcription between hours 6 and 15?

Solution: Maternal mRNAs are present in the fertilized sea urchin egg. Thus, a considerable amount of development can take place without transcription of the embryo's genome. Because development past 15 hours is inhibited by prior treatment with actinomycin D, it appears that transcripts from the embryo's genome are required to initiate or maintain these events. This transcription must take place between the sixth and fifteenth hours of development.

3. If it were possible to introduce one of the homeotic genes from *Drosophila* into an *Arabidopsis* embryo homozygous for a homeotic flowering gene, would you expect any of the *Drosophila* genes to negate (rescue) the *Arabidopsis* mutant phenotype? Why or why not?

Solution: The *Drosophila* homeotic genes belong to the *Hox* gene family, whereas *Arabidopsis* homeotic genes belong to the MADS-box protein family. Both gene families are present in *Drosophila* and *Arabidopsis,* but they have evolved different functions in the animal and the plant kingdoms. As a result, it is unlikely that a transferred *Drosophila Hox* gene would rescue the phenotype of a MADS-box mutant, but only an actual experiment would confirm this.

Problems and Discussion Questions

1. Carefully distinguish between the terms *differentiation* and *determination.* Which phenomenon occurs initially during development?
2. Nuclei from almost any source may be injected into *Xenopus* oocytes. Studies have shown that these nuclei remain active in transcription and translation. How can such an experimental system be useful in developmental genetic studies?
3. Distinguish between the syncytial blastoderm stage and the cellular blastoderm stage in *Drosophila* embryogenesis.
4. (a) What are maternal-effect genes? (b) When are gene products from these genes made, and where are they located? (c) What aspects of development do maternal-effect genes control? (d) What is the phenotype of maternal-effect mutations?
5. Suppose you initiate a screen for maternal-effect mutations in *Drosophila* affecting external structures of the embryo and your screen identifies more than 100 mutations that affect external structures. From their screening, researchers concluded that there are about 40 maternal-effect genes. How do you reconcile these different results? See **Now Solve This** on page 437.
6. (a) What are zygotic genes, and when are their gene products made? (b) What is the phenotype associated with zygotic gene mutations? (c) Does the maternal genotype contain zygotic genes?
7. List the main classes of zygotic genes. What is the function of each class of these genes?
8. Experiments have shown that any nuclei placed in the polar cytoplasm at the posterior pole of the *Drosophila* egg will differentiate into germ cells. If polar cytoplasm is transplanted into the anterior end of the egg just after fertilization, what will happen to nuclei that migrate into this cytoplasm at the anterior pole?
9. How can you determine whether a particular gene is being transcribed in different cell types?
10. You observe that a particular gene is being transcribed during development. How can you tell whether the expression of this gene is under transcriptional or translational control?
11. What are *Hox* genes? What properties do they have in common? Are all homeotic genes *Hox* genes?
12. The homeotic mutation *Antennapedia* causes mutant *Drosophila* to have legs in place of antennae and is a dominant gain-of-function mutation. What are the properties of such mutations? How does the *Antennapedia* gene change antennae into legs?
13. The *Drosophila* homeotic mutation *spineless aristapedia* (ss^a) results in the formation of a miniature tarsal structure (normally part of the leg) on the end of the antenna. What insight is provided by (ss^a) concerning the role of genes during determination?

14. Embryogenesis and oncogenesis (generation of cancer) share a number of features including cell proliferation, apoptosis, cell migration and invasion, formation of new blood vessels, and differential gene activity. Embryonic cells are relatively undifferentiated, and cancer cells appear to be undifferentiated or dedifferentiated. Homeotic gene expression directs early development, and mutant expression leads to loss of the differentiated state or an alternative cell identity. M. T. Lewis (2000. *Breast Can. Res.* 2: 158–169) suggested that breast cancer may be caused by the altered expression of homeotic genes. When he examined 11 such genes in cancers, 8 were underexpressed while 3 were overexpressed compared with controls. Given what you know about homeotic genes, could they be involved in oncogenesis?
15. In *Drosophila,* both *fushi tarazu* (*ftz*) and *engrailed (eng)* genes encode homeobox transcription factors and are capable of eliciting the expression of other genes. Both genes work at about the same time during development and in the same region to specify cell fate in body segments. To discover if *ftz* regulates the expression of *engrailed;* if *engrailed* regulates *ftz;* or if both are regulated by another gene, you perform a mutant analysis. In ftz^- embryos (*ftz*/*ftz*) engrailed protein is absent; in $engrailed^-$ embryos (*eng/eng*) *ftz* expression is normal. What does this tell you about the regulation of these two genes—does the *engrailed* gene regulate *ftz,* or does the *ftz* gene regulate *engrailed*? See **Now Solve This** on page 441.
16. Early development depends on the temporal and spatial interplay between maternally supplied material and mRNA and the onset of zygotic gene expression. Maternally encoded mRNAs must be produced, positioned, and degraded (Surdej and Jacobs-Lorena, 1998. *Mol. Cell Biol.* 18: 2892–2900). For example, transcription of the *bicoid* gene that determines anterior–posterior polarity in *Drosophila* is maternal. The mRNA is synthesized in the ovary by nurse cells and then transported to the oocyte, where it localizes to the anterior ends of oocytes. After egg deposition, *bicoid* mRNA is translated and unstable bicoid protein forms a decreasing concentration gradient from the anterior end of the embryo. At the start of gastrulation, *bicoid* mRNA has been degraded. Consider two models to explain the degradation of *bicoid* mRNA: (1) degradation may result from signals within the mRNA (intrinsic model), or (2) degradation may result from the mRNA's position within the egg (extrinsic model). Experimentally, how could one distinguish between these two models?
17. Formation of germ cells in *Drosophila* and many other embryos is dependent on their position in the embryo and their exposure to localized cytoplasmic determinants. Nuclei exposed to cytoplasm in the posterior end of *Drosophila* eggs (the pole plasm) form cells that develop into germ cells under the direction of maternally derived components. R. Amikura et al. (2001. *Proc. Nat. Acad. Sci. (USA)* 98: 9133–9138) consistently found mitochondria-type ribosomes outside mitochondria in the germ plasma of *Drosophila* embryos and postulated that they are intimately related to germ-cell specification. If you were studying this phenomenon, what would you want to know about the activity of these ribosomes?
18. One of the most interesting aspects of early development is the remodeling of the cell cycle from rapid cell divisions, apparently lacking G1 and G2 phases, to slower cell cycles with measurable G1 and G2 phases and checkpoints. During this remodeling, maternal mRNAs that specify cyclins are deadenylated, and zygotic genes are activated to produce cyclins. Audic et al. (2001. *Mol. and Cell. Biol.* 21: 1662–1671) suggest that deadenylation requires transcription of zygotic genes. Present a diagram that captures the significant features of these findings.
19. A number of genes that control expression of *Hox* genes in *Drosophila* have been identified. One of these homozygous mutants is *extra sex combs,* where some of the head and all of the thorax and abdominal segments develop as the last abdominal segment. In other words, all affected segments develop as posterior segments. What does this phenotype tell you about which set of *Hox* genes is controlled by the *extra sex combs* gene?
20. The *apterous* gene in *Drosophila* encodes a protein required for wing patterning and growth. It is also known to function in nerve development, fertility, and viability. When human and mouse genes whose protein products closely resemble *apterous* were used to generate transgenic *Drosophila* (Rincon-Limas et al. 1999. *Proc. Nat. Acad. Sci. [USA]* 96: 2165–2170), the *apterous* mutant phenotype was *rescued.* In addition, the whole-body expression patterns in the transgenic *Drosophila* were similar to normal *apterous.* (a) What is meant by the term *rescued* in this context? (b) What do these results indicate about the molecular nature of development?
21. In *Arabidopsis,* flower development is controlled by sets of homeotic genes. How many classes of these genes are there, and what structures are formed by their individual and combined expression?
22. The floral homeotic genes of *Arabidopsis* belong to the MADS-box gene family, while in *Drosophila,* homeotic genes belong to the homeobox gene family. In both *Arabidopsis* and *Drosophila,* members of the *Polycomb* gene family control expression of these divergent homeotic genes. How do *Polycomb* genes control expression of two very different sets of homeotic genes?
23. The identification and characterization of genes that control sex determination has been another focus of investigators working with *C. elegans.* As with *Drosophila,* sex in this organism is determined by the ratio of X chromosomes to sets of autosomes. A diploid wild-type male has one X chromosome, and a diploid wild-type hermaphrodite has two X chromosomes. Many different mutations have been identified that affect sex determination. Loss-of-function mutations in a gene called *her-1* cause an XO nematode to develop into a hermaphrodite and have no effect on XX development. (That is, XX nematodes are normal hermaphrodites.) In contrast, loss-of-function mutations in a gene called *tra-1* cause an XX nematode to develop into a male. Deduce the roles of these genes in wild-type sex determination from this information.
24. Based on the information in Problem 25 and the analysis of the phenotypes of single- and double-mutant strains, a model for sex determination in *C. elegans* has been generated. This model proposes that the *her-1* gene controls sex determination by establishing the level of activity of the *tra-1* gene, which in turn, controls the expression of genes involved in generating the various sexually dimorphic tissues. Given this information, (a) does the *her-1* gene product have a negative or a positive effect on the activity of the *tra-1* gene? (b) What would be the phenotype of a *tra-1, her-1* double mutant? See **Now Solve This** on page 445.
25. Dominguez et al. (2004) suggest that by studying genes that determine growth and tissue specification in the eye of *Drosophila,* much can be learned about human eye development.
    (a) What evidence suggests that genetic eye determinants in *Drosophila* are also found in humans? Include a discussion of orthologous genes in your answer.
    (b) What evidence indicates that the *eyeless* gene is part of a developmental network?
    (c) Are genetic networks likely to specify developmental processes in general? Explain fully and provide an example.

Genetically engineered Caenorhabditis elegans *roundworms that have turned blue in response to an environmental stress, such as toxins or heat*

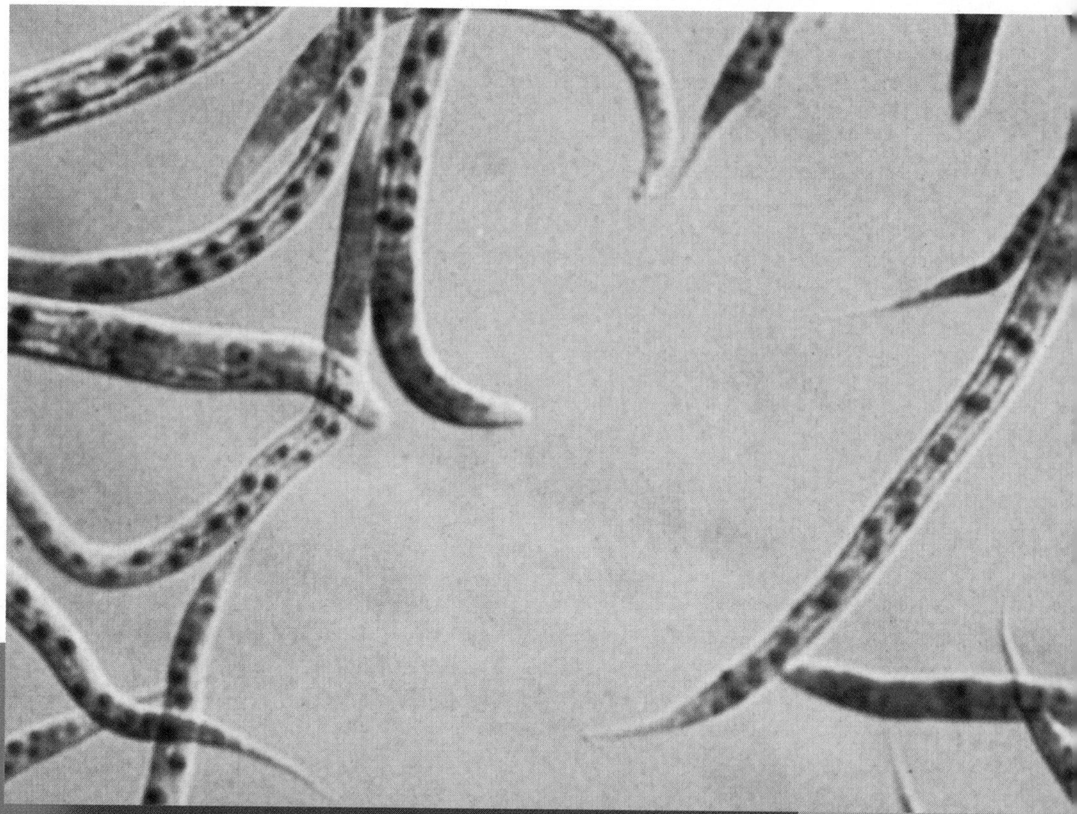

21

Genetics and Behavior

Chapter Concepts

- Behavior is a complex response to stimuli and is mediated both by the genotype and by the environment.
- Bidirectional selection experiments (also called the behavior-first approach) are used to identify heritable behavioral traits. These traits are further studied by genetic crosses to identify the number and chromosomal location of genes contributing to the trait.
- Mutagenesis experiments (also called the gene-first approach) establish strains carrying single-gene mutations that contribute to a behavioral response. Molecular functional analyses of the normal and mutant alleles provide insight into the underlying mechanism controlling the relationship between the gene and a behavioral response.
- The bidirectional and mutagenesis approaches to behavior genetics have been successfully used to dissect behavioral responses in *Drosophila,* making it a useful model organism for the study of nervous system function and human behavioral disorders.
- Many human behavioral disorders are complex responses to environmental stimuli. Mediated by the genotype and polygenically controlled, they may be analyzed by an array of genomic techniques.

Behavior is generally defined as a reaction to stimuli, whether internal or external. In broad terms, every action, reaction, and response represents a type of behavior. Animals run, remain still, or counterattack in the presence of a predator; birds build complex and distinctive nests in response to a combination of internal and external signals; fruit flies execute intricate courtship rituals; plants bend toward light; and humans reflexively avoid painful stimuli as well as "behaving" in a variety of ways as guided by their intellect, emotions, and culture.

Even though clear-cut cases of genetic influence on behavior were known in the early 1900s, the study of behavior was of less interest to the early geneticists than to psychologists, who were primarily concerned with learned or conditioned behavior. Although some behavioral traits were recognized as innate or instinctive, behavior that could be modified by prior experience received the most attention. Such traits or patterns of behavior were thought to reflect events in the environmental setting to the exclusion of the organism's genotype. This philosophy served as the basis of the behaviorist school of psychology.

Such thinking provided a somewhat distorted view of the nature of behavioral patterns. After all, any behavior must rely on the expression of the individual's genotype for its execution; and the expression of that genotype takes place within a hierarchy of environmental settings (that is, gene expression depends on interactions within the cell, the tissue, the organ, the organism, and finally the population and surrounding environment). Without genes and their environments, there could be no behavior. Nevertheless, the nature–nurture controversy flourished well into the 1950s. By that time it was clear that while certain behavioral patterns, particularly in less complex animals, seemed to be innate, others were the result of environmental modifications limited by genetic influences. The latter description is particularly applicable to organisms with more complex nervous systems. It is the nervous system that senses the environment, processes the information, and initiates the response that we perceive as behavior. For that reason, much of the study of behavior in genetics and molecular biology focuses on the development, structure, and function of the nervous system.

Since about 1950, research into the genetic components of behavior has intensified, and appreciation for the importance of genetics in understanding behavior has increased. Behavioral genetics originally employed two approaches to study the genetic basis of behavior. Historically, the first was a top-down, or *behavior-first,* approach, in which, first, a behavioral response to a stimulus was identified in an organism, and then, high-scoring responders were interbred and low-scoring responders were interbred to produce strains that bred true for either a high level or a low level of response. After these strains were established, crosses were made to identify and analyze the genetic components of the behavioral response. The second mode of research was a bottom-up, or *gene-first,* approach in which organisms were subjected to mutagenesis and then screened to identify single-gene mutations associated with abnormal behaviors. Analysis of the molecular mechanism of gene action in these mutant strains often provided a direct explanation of the behavior. As we will discuss in subsequent sections of this chapter, each approach has its advantages and shortcomings. In spite of their differences, both approaches share the same goals: to establish the inherited nature of a specific behavior, to identify and enumerate the genes or gene systems involved in the behavior being studied, to map these genes or gene systems to specific chromosomes, and to elucidate the molecular mechanisms by which these genes influence a behavioral response.

The prevailing view today is that all behavior patterns are influenced by the genotype and by the environment. The genotype provides the physical underpinnings or mental ability (or both) required to execute the behavior and further determines the limitations of environmental influences. Behavioral genetics has grown into an interdisciplinary specialty within genetics as more and more behaviors have been found to be under genetic control.

How Do We Know?

In this chapter, we will focus on the methods of behavioral genetics and the insights gained into the fundamental mechanisms that underlie animal behavior. As you study this topic, you should try to answer several fundamental questions:

1. What observations made it possible for geneticists to conclude that behavioral traits are under genetic control?
2. How did the combined use of behavioral selection and recombinant DNA technology allow the identification of genes that control specific behaviors?
3. What have we learned about human behavioral disorders through the use of transgenic model organisms? What are the advantages and limitations of each method used to study the genetics of behavior?

21.1 Behavioral Differences between Genetic Strains Can Be Identified

One of the classic methods of behavioral genetics is the screening of existing strains of highly inbred organisms to find behavior differences that are strain-specific. This behavior-first approach takes advantage of the fact that genetically distinct strains of experimental organisms are already available and being used in genetic research, and may harbor previously undescribed strain-specific behaviors in response to environmental stimuli. If such genetic differences exist, they can serve as the starting point for analysis of the genes that influence the behavior.

Emotional Behavior Differences in Inbred Mouse Strains

Open-field tests are used to study exploratory and emotional behavior in mice. Open-field testing uses an unfamiliar standard environment much different from the mouse's normal environment to study behaviors, including emotionality or

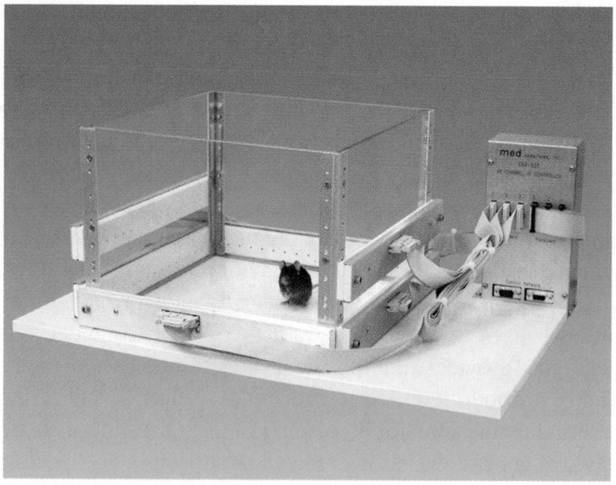

FIGURE 21–1 A setup to measure open-field behavior in mice.

approach/avoidance responses. When mice are placed in a new environment for open-field testing (Figure 21–1), they will normally explore the unfamiliar surroundings, but they are a bit cautious or "nervous" at the same time. The cautious response is evidenced by their elevated rate of defecation and urination. To study these two responses (the exploration and the anxiety) in the laboratory, researchers place the mice in an enclosed, brightly illuminated box; exploration is tracked by recording movements into different areas of the box, and emotion is measured by counting the number of defecations.

Inbred strains of mice differ significantly in their response to the open-field setting. John C. DeFries and his associates focused their work on two contrasting strains, BALB/cJ and C57BL/6J. The BALB strain is homozygous for a coat-color allele, *c*, and is albino, whereas the C57 strain has normal pigmentation (*CC*). BALB mice have low exploratory activity and are highly emotional, whereas C57 mice are active in exploration and relatively unemotional.

To test for genotypic differences, the two strains were crossed, and then the offspring were interbred for several generations. Each generation beyond the F_1 contained albino and nonalbino mice, and these were tested for open-field behavior. In all cases, pigmented mice behaved as strain C57, whereas albino mice behaved as BALB. The general conclusion is that the *c* allele behaves pleiotropically, affecting both coat color and behavior.

Heritability analysis has been used to assess the input of the *c* gene to these behavioral patterns. The results show that this locus accounts for 12 percent of the variance in open-field activity and 26 percent of the variance in defecation-related emotion. These values indicate that the behaviors are polygenically controlled.

To assess the relationship between albinism and behavior, albino and nonalbino mice were tested for open-field behavior under white light and red light. Red light provides little visual stimulation to mice. The behavioral differences between mice with the two types of coat pigmentation disappeared under red light, indicating that the open-field responses of albino mice are visually mediated. This is not surprising; albino mice are photophobic because the lack of pigmentation in albinos extends to the iris as well as to the coat.

Many behaviors are genetically complex and polygenically controlled. Genetic crossing between inbred strains can reveal some aspects of genetic involvement in behavior but has not been a successful strategy for identifying specific genes related to polygenic behavioral phenotypes. Environmental variables make important contributions to many such phenotypes, as illustrated by the disappearance of strain-specific behavior in red as opposed to white light. As a result, phenotypes can be inconsistent and more difficult to define. In addition, pleiotropic effects are common in behavior phenotypes and add to the problems of identifying genes that control complex traits.

Identifying Genes That Control Strain-Specific Behaviors

To narrow the search for genes controlling emotional behavior in mice, quantitative trait loci (QTL) mapping (see Chapter 22 for a discussion of QTLs) has been successfully used. In one experiment, heterozygotes from crosses between C57BL/6J (a nonemotional, low-anxiety strain) and BALB/cJ (an emotional, high-anxiety strain) were intercrossed, and the progeny were selected over multiple generations for extremes of high (H1a) or low (L1a) anxiety. These inbred strains were subjected to several different behavioral tests of anxiety, including the open field, the elevated plus maze (Figure 21–2), and a test called the light–dark box. In this last-named test, mice are placed in an environment in which there is a dark, enclosed box, from which they can emerge into a brightly lit area. Fearful animals prefer to remain in the box and do not explore the lighted areas of the environment. The phenotypically most extreme progeny recovered from these anxiety tests were analyzed for 84 microsatellite markers and four different measures of emotionality. A QTL on chromosome 1 was identified and localized to a small region of the chromosome covering a small region spanning 66 Mb. Over 1600 mice in replicated

FIGURE 21–2 A mouse explores a new environment from the open arm of an elevated plus maze, a standard device for measuring fear in mice. This experiment was one of a series designed to identify genes involved in fear and anxiety in mice.

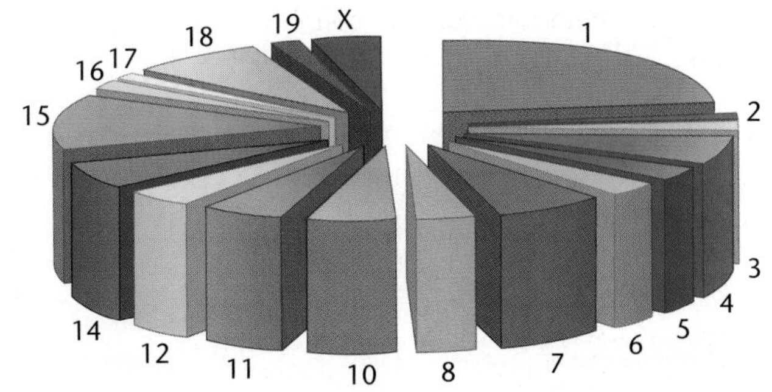

FIGURE 21–3 Loci for genes controlling fear and other forms of emotional behavior are distributed on 17 of 19 autosomes in the mouse genome. Significant numbers of these genes are located on chromosomes 1, 15, and 18.

populations consistently gave the same results. Similar experiments using microsatellite markers have shown that QTLs controlling other aspects of behavior are distributed on all mouse chromosomes (Figure 21–3).

To advance beyond the mapping of a QTL region, which may contain dozens of genes, to identifying the specific loci involved in emotional behavior requires the combined use of genetics and genomics. Genes in the chromosome 1 QTL region are being tested in genetic complementation assays to detect interactions among these loci. Once these genes are identified, their mechanisms of action and relevance to human behavior remain to be assessed.

ESSENTIAL POINT

Both the genotype and environment have been found to have an impact in determining an organism's behavioral response.

NOW SOLVE THIS

Problem 3 on page 463 asks how QTL analysis can be used to identify genes and the molecular mechanisms involved in alcohol preference in mice.

Hint: Review what QTLs are and how they are mapped using chromosome-specific markers such as microsatellites.

21.2 The Behavior-First Approach Can Establish Genetic Strains with Behavioral Differences

The behavior-first approach to studying behavioral genetics selects organisms exhibiting a specific behavior from among a genetically heterogeneous population. If genetic strains can be established that uniformly express this behavior, and if the trait can be transferred by genetic crosses to another strain that initially does not exhibit the behavior under study, genetic involvement in the behavior is confirmed. Often, two groups are selected: one showing high levels and one showing low levels of the behavior under investigation. This bidirectional selection creates two lines with progressively greater differences in behavior. This method maximizes differences in genes that control the trait of interest, and can minimize differences in all other genes. The study of geotaxis in *Drosophila* will be used to illustrate this approach.

Artificial Selection for Geotaxis in *Drosophila*

A **taxis** is a movement response toward or away from an external stimulus. The response may be positive (moving toward the stimulus) or negative (away from it). Many kinds of stimuli elicit this form of behavior, including chemicals (chemotaxis), gravity (geotaxis), and light (phototaxis).

To investigate geotaxis in *Drosophila,* Jerry Hirsch and his colleagues designed a mass-screening device that tests about 200 flies per trial, as shown in Figure 21–4. In this test, flies are lured into a vertical maze that repeatedly requires them to choose between moving up or down. Flies that turn up at each intersection will exit at the top of the maze; those that always turn down will exit at the bottom; and those making both "up" and "down" decisions will exit somewhere in between. Flies can be selected both for positive (going up in the maze) and for negative geotaxis (going down in the maze), establishing the existence of a genetic influence on this behavioral response.

As shown in Figure 21–5, p. 454, mean scores of flies in this maze vary from about +4 to −6, indicating that selection is stronger for negative geotaxis than for positive geotaxis. The two lines have now undergone selection for almost 40 years, encompassing over 1000 generations and the testing of more than 80,000 flies. Throughout the experiment, clear-cut, but fluctuating, differences have been observed. These results indicate that this complex behavior in *Drosophila* is genetically controlled and is a polygenic trait.

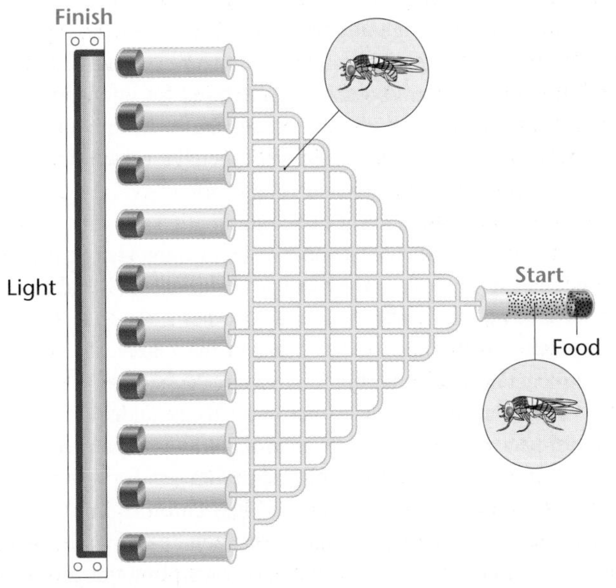

FIGURE 21–4 Schematic drawing of a maze used to study geotaxis in *Drosophila.*

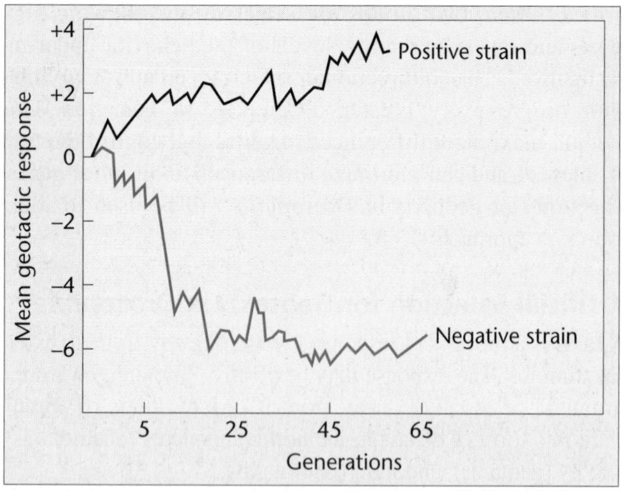

FIGURE 21–5 Selection for positive and negative geotaxis in *Drosophila* over many generations. The scores represent the sum of upward (+) or downward (−) turns in the maze.

Using the lines selected for positive and negative geotaxis, Hirsch and his colleagues analyzed the relative contribution of loci on different chromosomes to geotaxis. Loci on chromosomes 2 and 3 and on the X were identified in an ingenious way. Crosses produced flies that were either heterozygous or homozygous for a given chromosome. Figure 21–6 shows how this is accomplished. From a selected line, a male is crossed to a "tester" female, whose chromosomes are each marked with a dominant mutation (these chromosomes are called balancer chromosomes). Each marked chromosome carries an inversion to suppress the recovery of any crossover products. An F_1 female heterozygous for each chromosome is backcrossed to a male from the original line. The resulting female offspring contain all combinations of the chromosomes. The dominant mutations make it possible to recognize which chromosomes from the selected lines are present in homozygous or heterozygous configurations.

Flies with genotypes 0 and X (Figure 21–6) are homozygous and heterozygous, respectively, for the X chromosome. Similarly, flies with genotypes 0 and 2 are homozygous and heterozygous, respectively, for chromosome 2, and flies with genotypes 0 and 3 are homozygous and heterozygous, respectively, for chromosome 3. By testing these different flies in the maze, it is possible to assess the behavioral influence of loci on any given chromosome.

For negatively geotaxic flies (the ones that go up in the maze), genes on the second chromosome make the largest contribution to the phenotype, followed by loci on chromosome 3 and on the X chromosome (a gradient of chromosomes 2 > 3 > X). For the positive line (flies that go down in the maze), the reverse arrangement (a gradient of chromosomes X > 3 > 2) is seen. The overall results indicate that geotaxis is under polygenic control and that the loci controlling this trait are distributed on all three major chromosomes of *Drosophila*.

Further genetic testing in which each chromosome from a selected line has been isolated in homozygous form has been used to estimate the number of genes controlling the geotactic response in *Drosophila*. Once the chromosome has been placed in a genetic background that has not been selected for

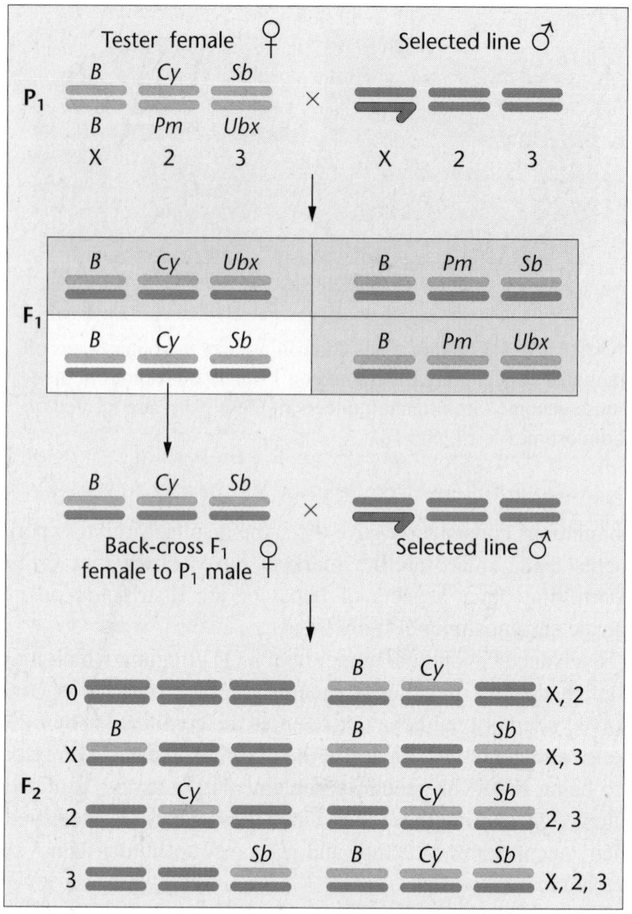

FIGURE 21–6 In this mating scheme in *Drosophila,* the effect of genes located on specific chromosomes that contribute to geotaxis can be assessed. The progeny produced by backcrossing the female contain all combinations of chromosomes. Examining the presence or absence of dominant marker phenotypes in these flies makes it possible to determine which chromosomes from the selected strain are present. Subsequent testing for geotaxis is then performed (*B* = *Bar eyes*; *Cy* = *Curly wing*; *Sb* = *Stubble bristles*; *Ubx* = *ultrabithorax*). The designations alongside each genotype in the F_2 generation (e.g., X, 2, and 3) indicate which chromosomes are heterozygous.

geotactic behavior, it is possible to estimate the number of genes that control the phenotype. This work indicates that a small number of genes, perhaps two to four loci, are responsible for geotaxis in *Drosophila.*

As ingenious as this work is, it is also arduous, illustrating one of the limitations of behavioral selection. Although selection for geotaxis showed that this behavior is genetically controlled and that loci on all major chromosomes are involved, it cannot identify specific genes involved in this behavior. This is partly because the behavior is polygenically controlled and influenced by several to many genes, each of which may make only a small contribution to the phenotype.

The development of new technology, including microarrays (described in Chapters 18 and 19), offers a new approach to identifying genes that influence complex polygenic behavioral traits. Microarrays are one example of the high-throughput experiments that are revolutionizing genetics and molecular biology. Instead of measuring changes in expression one gene at a time, microarrays make it possible to measure changes in dozens, hundreds, or even thousands of genes in a single experiment. Genes used in microarray analysis can be selected on the basis of their expression patterns in specific tissues, in specific cells, or in specific structures.

Researchers employed two steps to identify the genes involved in geotaxis. First, they used cDNA microarrays containing about one-third of the genes in the *Drosophila* genome to analyze mRNA levels from the strains with the highest (*Hi5*) and lowest (*Lo*) geotaxis behavior. A small number of genes showed reproducible differences in expression and were selected for further analysis. To narrow the list of candidate genes further, the researchers selected mutants of these genes that led to neurological defects. Ten mutant lines, including several control lines, were selected for the second step in this gene-identification strategy: analysis of geotaxis behavior.

The geotaxis behavior and mRNA levels of three genes in the mutant flies corresponded to levels of mRNA expression and behavior in *Hi5* and *Lo* flies. These genes each made small, incremental contributions to geotaxis, as would be expected for a polygenically controlled trait. The mechanism by which these three genes, *cryptochrome* (*cry*), *Pendulin* (*Pen*), and *Pigment-dispersing-factor* (*Pdf*), are involved in geotaxis is unknown. However, all three genes are known to be involved in the development and function of the brain and nervous system, and their role in geotaxis is currently being investigated. The use of this two-step strategy is significant because it illustrates the successful integration of classical genetics with genomics to identify genes involved in a polygenic trait.

ESSENTIAL POINT

Using the behavior-first approach to study geotaxis in *Drosophila,* researchers have been able to identify and map genes associated with this behavior, and to calculate their numbers and relative contributions.

NOW SOLVE THIS

Problem 2 on page 463 asks how to determine whether a behavioral trait is genetic or nongenetic.

Hint: Determine whether you will be using the behavior-first or the gene-first method.

21.3 The Gene-First Approach Uses Analysis of Mutant Alleles to Study Molecular Mechanisms That Underlie Behavior

Although we discussed geotaxis in *Drosophila* as an example of the behavior-first approach, much has also been learned using this organism for gene-first studies of behavior. This is not at all surprising in view of our extensive knowledge of *Drosophila's* genetics and the ease with which it can be manipulated experimentally.

Although *Drosophila* and humans are separated by millions of years from a common ancestor, molecular aspects of function in cells of the nervous system are the same in these two organisms. As a result, work with mutant strains of *Drosophila* has provided insight into the fundamental mechanisms that govern nerve impulse transmission.

As the name implies, the gene-first approach to studying behavior first isolates mutant strains with a behavioral abnormality and then investigates the structural and/or functional alterations that underlie this altered behavior.

The gene-first approach has several advantages over the behavior-first approach. Using a bottom-up approach allows the researcher to quantify the impact of single genes on a behavioral response. By rigorous application of this method, all genes contributing to the behavior can eventually be identified, mapped, cloned, and sequenced. The limiting factor is that it requires the use of an organism for which researchers have developed an extensive and sophisticated set of genetic tools, especially marker chromosomes and collections of strains with deleted and duplicated chromosome regions, as well as mutants with cytological markers for mosaic analysis. The fruit fly, *Drosophila,* is the organism that best fits this requirement, and fortunately, it is an excellent model system for studying not only behavior but its underlying basis in the structure and function of the nervous system.

In the following sections, we will explore the use of mutant analysis in dissecting mechanisms of nerve signals and their direct application to human disorders, and then we will discuss two forms of complex behavior, courtship and learning, to illustrate how *Drosophila* serves as a model organism for studying the molecular aspects of behavior.

Genes Involved in Transmission of Nerve Impulses

Figure 21–7, p. 456, depicts a simplified nerve cell (neuron), in which nerve impulses are generated at one end (on a dendrite) and move along the cell to the end of the axon, which transmits the impulse to adjacent neurons. During this process, the impulse is propagated by the movement of sodium and potassium ions across the neuron's plasma membrane. The movement of these electrically charged ions can be monitored by measuring changes in the electrical potential of the membrane. To screen for genes that control the generation and transmission of nerve impulses, Barry Ganetzky and his colleagues screened behavioral mutants of *Drosophila* to identify those with electrophysiological abnormalities in the generation and propagation of nerve impulses. Two general classes of such mutants have been isolated: those with defects in the movement of sodium and those with defects in potassium transport (Table 21.1).

One mutant identified in the screening procedure is a temperature-sensitive allele of the *paralytic* gene. Flies homozygous for this mutant allele become paralyzed when exposed to temperatures at or above 29°C, but they recover rapidly when the temperature is lowered to 25°C. Mosaic

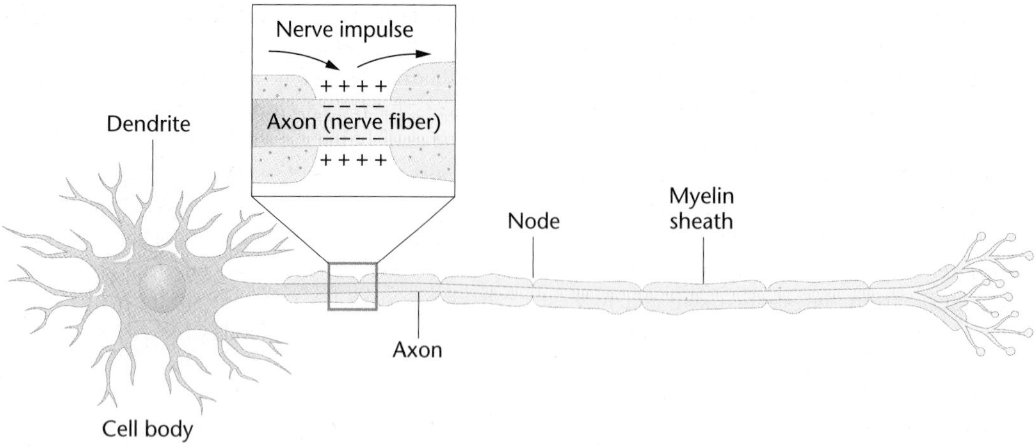

FIGURE 21–7 A nerve cell (neuron) has extensions called dendrites that carry impulses toward the cell body and also has one or more axons that carry impulses away from the cell body. The axon is encased in a sheath of myelin, except at nodes spaced at intervals along the axon. Nerve impulses are accelerated by jumping from node to node. Electrodes placed on either side of the neuron's plasma membrane can record the electric potential across the membrane. When the neuron is at rest, there is more sodium outside the cell and more potassium inside the cell. When a nerve impulse is generated, sodium moves into the cell, and potassium moves out, altering the electric potential. As the impulse moves away, ions are pumped across the membrane to restore the original electric potential.

TABLE 21.1 Behavioral Mutants of *Drosophila* in Which Nerve Impulse Transmission Is Affected

Mutation	Map Location	Ion Channel Affected	Phenotype
nap^{ts}	2-56.2	Sodium	Adults and larvae paralyzed at 37.5°C; reversible at 25°C
$para^{ts}$	1-53.9	Sodium	Adults paralyzed at 29°C, larvae at 37°C; reversible at 25°C
Tip-E	3-13.5	Sodium	Adults and larvae paralyzed at 39–40°C; reversible at 25°C
sei^{ts}	2-10.6	Sodium	Adults paralyzed at 38°C, larvae unaffected; adults recover at 25°C
Sh	1-57.7	Potassium	Aberrant leg shaking in adults exposed to ether
eag	1-50.0	Potassium	Aberrant leg shaking in adults exposed to ether
Hk	1-30.0	Potassium	Ether-induced leg shaking
sio	3-85.0	Potassium	At 22°C, adults are weak fliers; at 38°C, adults are weak, uncoordinated

studies revealed that both the brain and thoracic ganglia are the focus of this abnormal behavior. Electrophysiological studies showed that mutant flies have defective sodium transport associated with the conduction of nerve impulses. Subsequently, Ganetzky and his colleagues mapped, isolated, and cloned the *paralytic* gene. This locus encodes a protein called the sodium channel, which controls the movement of sodium across the membrane of nerve cells in many different organisms, from flies to humans.

A second mutant, called *Shaker,* originally isolated over 40 years ago as a behavioral mutant, encodes a potassium channel protein that has also been cloned and characterized. Because the mechanism of nerve impulse conduction has been highly conserved during animal evolution, the cloned *Drosophila* genes were used as probes to isolate the equivalent human ion channel genes. The cloned human genes are being used to provide new insights into the molecular basis of neuronal activity. One of the human ion channel genes was first identified in *Drosophila* and is defective in a heritable form of cardiac arrhythmia. The identification and cloning of the human gene now make it possible to screen for family members at risk for this potentially fatal condition.

Genetic Control of Courtship

As early as 1915, Alfred Sturtevant observed that, in addition to its effects on body pigmentation, the X-linked recessive gene *yellow* affects mating preference of *Drosophila* females. He found that both wild-type and *yellow* females, when given the choice of wild-type or *yellow* males, prefer to mate with wild types. Both wild-type and *yellow* males prefer to mate with *yellow* females. These conclusions were based on measurements of mating success in all combinations of *yellow* and wild-type (gray-bodied) males and females.

In 1956, Margaret Bastock extended these observations by investigating which, if any, component of courtship behavior was affected by the *yellow* mutant gene. Courtship (the "singing" shown in Figure 21–8) in *Drosophila* is a complex ritual. The male first shows orientation. He follows the female, often circling her, and then orients himself at a right angle to her body and taps her on the abdomen. Once he has her attention, the male begins wing display. He raises the wing closest to the female and vibrates this wing rapidly for several seconds. He then moves behind her and makes contact with her genitalia. If she signals acceptance by remaining in place, he mounts her, and copulation occurs (Figure 21–8). Bastock compared the courtship ritual in wild-type and *yellow* males. The *yellow* males prolong orientation but spend much less time in the vibrating and genital contact phases. Her observations indicate that the *yellow* mutation alters the pattern of male courtship, making these males less successful in mating.

From these beginnings, *Drosophila* geneticists identified, cloned, and sequenced many genes that affect courtship (Table 21.2). Each of the steps in the courtship ritual involves different parts of the nervous system. One gene, *fruitless* (*fru*), affects all aspects of male courtship but has no effect on female behavior. The gene has a complex organization and encodes a transcription factor. The *fru* gene can be transcribed from any of four promoter sites (P1–P4), but only transcripts from the P1 promoter may be spliced in a male-specific manner, and proteins translated from these mRNAs control male courtship. (Transcripts from that promoter may

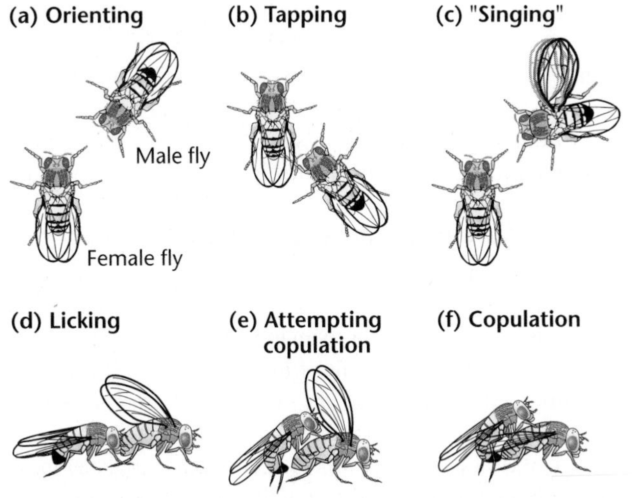

FIGURE 21–8 The stages of male behavior during courtship and mating in *Drosophila.*

TABLE 21.2 Some *Drosophila* Mutations That Affect Courtship and Learning

Behavior Genes	Molecular Function	Behavior Phenotype, Function
Courtship		
fruitless (fru)	Transcription factor	All aspects of male courtship
doublesex (dsx)	Transcription factor	Song defect
dissatisfaction (dsf)	Steroid hormone receptor	Poor sex discrimination, reduced female receptiveness
courtless (crl)	Ubiquitin-conjugating enzyme	Failure to court
slowpoke (slo)	Calcium-activated potassium channel	Song defect
cacophony (cac); nightblind A (nbA)	Voltage-sensitive calcium channel	Song defect, optomotor behavior, photophobia
dissonance (diss); no on-or-off transient (nonA)	RNA binding	Song defect, optomotor behavior
Learning and Memory		
dunce (dnc)	cAMP-specific phosphodiesterase	Locomotor rhythms, ethanol tolerance
rutabaga (rut)	Adenylate cyclase	Courtship learning, ethanol tolerance, grooming
amnesiac (amn); cheapdate (chpd)	Neuropeptide	Ethanol tolerance
latheo (lat)	DNA-replication factor	Larval feeding
Shaker (Sh)	Voltage-sensitive potassium channel	Courtship suppression, gustation defect, ether sensitivity
G protein s α 60 A (*G-s* α60A)	Heterotrimeric G protein	Visual behavior, cocaine sensitivity
DCO; cAMP-dependent protein kinase I (Pka-C1)	Protein Ser/Thr kinase	Locomotor rhythms, ethanol tolerance
cAMP-response-element-binding protein B at 17A	Transcription factor	Locomotor rhythms *(CrebB17A); dCREB*
Calcium/calmodulin-dependent protein kinase II (CaMKII)	Protein Ser/Thr kinase	Courtship suppression
Neurofibromatosis 1 (Nf1)	Ras GTPase activator	Embryonic, nervous system defects

Source: Sokolowski, M. 2001. *Drosophila*: Genetics meets behaviour. *Nature Reviews Genetics 2*: 879–890, Table 1, p. 882.

also be spliced in a female-specific manner.) Male-specific *fru* transcripts are expressed during pupal development in about 20 cells in the central nervous system, including both the brain and the ventral nerve cord. Presumably, *fru* expression results in male-specific neural connections that process sensory information during courtship.

Drosophila Can Learn and Remember

Researchers gain a great advantage when they can study the genetics of a complex behavior, such as learning, in a model organism that is as convenient for them to manipulate as *Drosophila.* However, this is possible only if the model organism can perform the complex behavior the researchers are interested in studying. Thus, before using *Drosophila* to study learning, researchers needed to know whether *Drosophila* could learn.

Seymour Benzer's lab was one of the first to isolate genes that control learning and memory in *Drosophila.* To do so, Benzer's group used an olfactory-based shock-avoidance learning system in which flies are presented with a pair of odors, one of which is associated with an electric shock. Flies quickly learn to avoid the odor associated with the shock. For several reasons, this response is thought to be learned. First, performance is associated with the pairing of a stimulus or response with reinforcement. Second, the response is reversible: flies can be trained to select an odor they previously avoided. Third, flies exhibit short-term memory for the training they have received.

The demonstration that *Drosophila* can learn opened the way to selecting mutants that are defective in learning and memory (Table 21.2). To accomplish this part of the study, males from an inbred wild-type strain were mutagenized and mated to females from the same strain. Their progeny were recovered and mated to produce stocks that carried a mutagenized X chromosome. Mutations that affected learning were selected by testing for responses in the olfactory shock apparatus. A number of learning-deficient mutants, including *dunce, turnip, rutabaga,* and *cabbage,* were isolated in this way. In addition, a memory-deficient mutant, *amnesia,* which learns normally, but forgets four times faster than normal, was recovered. Each of these mutations represents a single-gene defect that affects a specific form of behavior. Because of the method used to recover them, all the mutants found so far are X-linked genes. Presumably, similar genes that control behavior are also located on the autosomes.

Since, in many cases, mutation results in the alteration or abolition of a single protein, the nervous system mutants described here can provide a link to the biochemical and molecular basis of behavior. One group of *Drosophila* learning mutations, for example, involves defects in a cellular signal transduction system that proved to play a role in learning.

In cells of the nervous system, and in many other cell types, cyclic adenosine monophosphate (cAMP) is produced from adenosine 5′ triphosphate (ATP) in the cytoplasm in response to signals received at the cell's surface (see Figure 21–9). Cyclic AMP (cAMP) activates protein kinases, which, in turn, phosphorylate proteins and initiate a cascade of metabolic effects that control gene expression.

Behavioral mutants of *Drosophila* were among the first to show the link between cAMP and learning (Figure 21–9). The

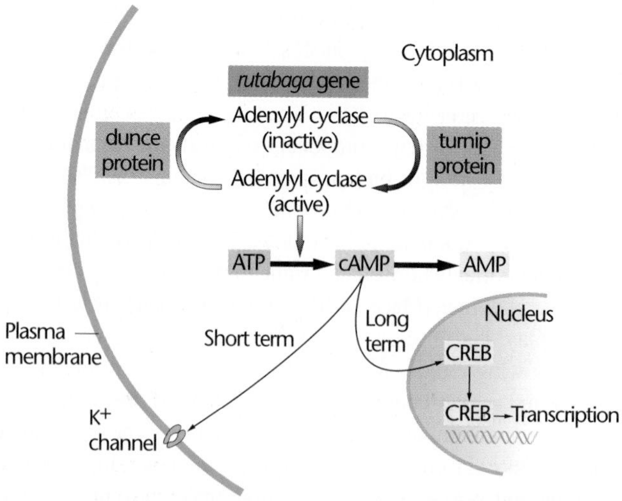

FIGURE 21–9 The pathway for cyclic AMP (cAMP) synthesis and degradation. ATP is converted into cAMP by the enzyme adenylyl cyclase; cAMP affects short-term memory via the potassium channel in the plasma membrane and affects long-term memory by regulating the transcription factor CREB. In *Drosophila,* the *rutabaga* gene encodes adenylyl cyclase. When produced, the enzyme is inactive; it is converted to its active form by a protein encoded by the *turnip* gene. The enzyme is inactivated by a protein encoded by the *dunce* gene. Mutations in any of these genes affect learning and memory in *Drosophila.*

rutabaga (*rut*) locus encodes adenylyl cyclase, the enzyme that synthesizes cAMP from ATP. In the mutant allele of *rutabaga,* a missense mutation destroys the catalytic activity of adenylyl cyclase and is associated with learning deficiency in homozygous flies. The *dunce* (*dnc*) locus encodes the structural gene for the enzyme cAMP phosphodiesterase, which degrades adenylyl cyclase. The *turnip* (*tur*) mutation occurs in a gene encoding a G protein, a class of molecules that bind guanosine 5′-triphosphate (GTP) and, in turn, activate adenylyl cyclase.

Cells in the *Drosophila* nervous system, called mushroom body neurons, integrate and process sensory input from olfactory stimuli and from electric shocks. When both stimuli arrive at a mushroom body simultaneously, adenylyl cyclase is stimulated and activates a G-protein-linked receptor, which leads to elevated cAMP levels. The increase in cAMP leads to changes in the excitability of the mushroom body neuron. Elevated levels of cAMP upregulate expression of a protein kinase, which enhances short-term and long-term memory. Short-term memory is mediated by changes in potassium channels. Long-term memory is associated with changes in mushroom body neuron gene expression controlled by a transcriptional activator, CREB (cAMP response element-binding protein). CREB binds to elements upstream of cAMP-inducible genes and changes the pattern of gene expression, encoding memory.

The biochemical pathways that control learning and memory in *Drosophila* are similar to those in other organisms, including mice. The mammalian homologs of the *Drosophila* genes are active in parts of the mouse brain involved in learning and memory. New genetic screening methods in both mice and flies are being used to identify new genes and other biochemical pathways involved in learning.

ESSENTIAL POINT

Using *Drosophila* as a model organism, researchers have isolated mutant strains with deviations from normal behavior and have used these strains to establish the role of the corresponding wild-type allele in the normal response.

21.4 Human Behavior Has Genetic Components

The genetic control of behavior has proven more difficult to characterize in humans than in other organisms. Not only are humans unavailable as experimental subjects in genetic investigations, but the types of responses considered to be the most interesting forms of human behavior, including aspects of *intelligence, language, personality,* and *emotion,* are difficult to study. Two problems arise in studying such behaviors. First, all behaviors are difficult to define objectively and to measure quantitatively. Second, they are affected by environmental factors. In each case, the environment is extremely important in shaping, limiting, or facilitating the final phenotype for each trait.

Historically, the study of human behavior genetics has been hampered by other factors as well. Many early studies of human behavior were performed by psychologists without adequate input from geneticists. Moreover, traits involving intelligence, personality, and emotion have great social and political significance. Consequently, research findings concerning these traits are likely to be distorted by sensationalism when reported to the public. In fact, because the study of these traits often comes close to infringing upon individual liberties, such as the right to privacy, the studies themselves, much less their conclusions, very frequently stir up controversy.

In lamenting the gulf between psychology and genetics in the study of human behavior, C. C. Darlington wrote in 1963, "Human behavior has thus become a happy hunting ground for literary amateurs. And the reason is that psychology and genetics, whose business it is to explain behavior, have failed to face the task together." Since 1963, some progress has been made in bridging this gap, but the genetics of human behavior remains controversial. Like other areas of behavioral genetics, the study of human behavior depends on the analysis of single genes, as well as on the investigation of complex, polygenic traits with environmental components.

Single Genes and Behavior: Huntington Disease

Neurodegenerative disorders affect millions of people in the United States and across the world. These diseases are associated with the progressive accumulation of misfolded proteins in intra- and extracellular spaces, leading to the death of brain cells, with behavioral consequences. Some of these disorders, such as Alzheimer disease (AD), amyotrophic lateral sclerosis (ALS), and **Huntington disease (HD)** are caused by mutations in single genes. HD is inherited as an autosomal dominant disorder that affects about 1 in 10,000 people, causing cell death in selected brain cells. Symptoms usually appear in the fifth decade of life as a gradual loss of motor function and coordination. As structural degeneration of the brain progresses, personality changes occur. The symptoms are due to brain-cell death taking place in specific brain regions, including the cerebral cortex and the striatum. Most victims die within 10 to 15 years after the onset of the disease.

HD usually appears in a person after they have children, who must then live with the knowledge that they face a 50 percent probability of developing the disorder (affected individuals are usually heterozygotes). The *HD* gene, on chromosome 4, was one of the first genes to be located by using restriction fragment length polymorphisms (RFLPs). The gene encodes a large protein (350 kDa) called huntingtin (Htt) that is essential for the survival of certain neurons in adult brains. Mutant HD alleles have an expanded number of cytosine-adenine-guanine (CAG) trinucleotide repeats in exon 1 (see Chapter 14). Normal individuals have 7 to 34 repeats, encoding multiple copies of glutamine near the amino terminus of the huntingtin protein. In mutant alleles, expansion of CAG repeats leads to additional glutamine residues in the encoded protein. In those with more than 40 CAG repeats, the additional glutamine residues cause Htt to become toxic and produce behavioral abnormalities as brain cells die.

Transgenic animal models of neurodegenerative diseases caused by single-gene mutations, including AD, ALS, and HD have been developed. These model systems are being used to study the molecular events of synthesis, processing, and aggregation of mutant proteins, and the links between protein accumulation, cell death, and behavioral changes.

A Transgenic Mouse Model of Huntington Disease

Several animal models of HD have been developed to study the normal function of huntingtin, to duplicate the disease process and study its effects at the molecular level, and to develop drugs that slow or stop the degeneration and death of brain cells.

To study the relationship between CAG repeat length and disease progression, Danilo Tagle and his colleagues constructed transgenic mice carrying a full-length mutated copy of a human *HD* gene with 16, 48, or 89 copies of the CAG repeat. The vector used for transferring the gene into the mice had the *HD* gene inserted at a site adjacent to a promoter and an enhancer to achieve high levels of expression (see Chapter 19 for a discussion of transgenic organisms).

Mice carrying the transgene were monitored from birth to death to determine the age of onset and progression of abnormal behavioral phenotypes. Animals carrying 48- or 89-repeat *HD* genes showed behavioral abnormalities as early as 8 weeks, and by 20 weeks, these mice showed both behavioral and motor-coordination abnormalities compared with the control animals.

At various ages, brain sections of wild-type mice and of the transgenic animals carrying mutant alleles with 16, 48, and 89 copies of the CAG repeat were examined for changes in

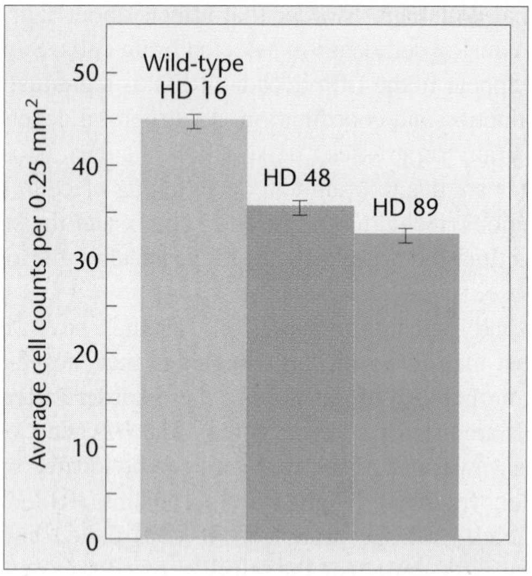

FIGURE 21–10 Relative levels of neuronal loss in HD transgenic mice. Cell counts show a significant reduction of certain neurons (small-medium neurons) in the corpus striatum of the brains of HD48 mutants (middle column) and HD89 mutants (right column) compared to wild-type (left column) mice. Cell loss in this brain tissue is also found in humans with HD, making these transgenic mice valuable models to study the course of this disease.

brain structure. Degenerating neurons and cell loss were evident in mice carrying 48 and 89 repeats, but no changes were seen in brains of wild-type mice or of those carrying a 16-repeat transgene (Figure 21–10).

The behavioral changes and the degeneration in specific brain regions in the transgenic mice parallel the progression of HD in humans. These mice are now used to study early changes in brain structure that occur before the onset of locomotor changes, and in the development of experimental treatments to slow or reverse cell loss. Having a mouse model for the disease allows researchers to administer treatment at specific times in disease progression and to evaluate the outcome of treatments in the presymptomatic stages of HD.

Mechanisms of Huntington Disease

Although the *HD* gene was mapped, isolated, and characterized almost 15 years ago, the mechanism by which mutant Htt causes HD is still unknown. Htt is expressed in many cell types but causes cell death only in striatal cells in the brain. Several mechanisms have been proposed, including abnormal mitochondrial metabolism, increased activity of proteases associated with cell death (apoptosis), misfolding of Htt, and the formation of nuclear inclusions that contain Htt fragments. Recently, attention has focused on the role of mutant Htt in the disruption of transcriptional regulation.

In affected individuals, onset of HD is associated with decreases in mRNA levels for genes encoding certain neurotransmitter receptors. In a transgenic HD mouse model, researchers used cDNA arrays to screen nearly 6000 mRNAs from striatal cells. They found lower levels of mRNA from a small set of genes involved in signaling pathways that are critical to striatal cell function. Similar results were found in another transgenic HD mouse model and are consistent with findings from HD patients, implicating aberrant gene expression as an underlying cause of HD.

Transcriptional regulation involves the action of transcription factors that alter chromatin structure. For example, acetylation of histones increases transcriptional activity, and deacetylation lowers or turns off transcription. The mutant form of Htt interacts with and binds to several transcription factors and may alter patterns of gene expression by making these proteins unavailable. Some proteins that interact with Htt have histone-modifying activity and are found in Htt protein aggregates in brains of transgenic HD mice and patients with HD. In addition, mutant versions of amino terminal Htt fragments have been shown to bind to and dissociate proteins from transcriptional complexes on gene promoters.

Using transgenic HD mice, several groups are studying the use of drugs that inhibit histone deacetylation. Sodium butyrate and phenylbutyrate both increased survival and reduced cell death in the transgenic mice. Drugs that inhibit histone deacetylation are now being used in human clinical trials on HD patients, and other drugs are being screened in mouse models for therapeutic effects.

Multifactorial Behavioral Traits: Schizophrenia

Although RFLP linkage studies have been successful in mapping and cloning a small number of single genes with behavioral phenotypes (such as *HD*), other behavioral disorders are polygenic and have strong environmental components, making their genetics more difficult to dissect. Schizophrenia is a complex brain disorder affecting about 1 percent of the population and is the example we will focus on in this section.

Schizophrenia is a collection of mental disorders characterized by avoidance of social contact and by bizarre and sometimes delusional behavior. Those affected by the disease are unable to lead normal lives and are periodically disabled by the condition. It is clearly a familial disorder, with relatives of schizophrenics having a much higher incidence of the condition than the general population (Figure 21–11). Furthermore, the closer the relationship to an affected individual, the greater is a person's probability of developing the disorder.

Concordance for schizophrenia in monozygotic and dizygotic twins has been the subject of many studies. In almost every investigation, concordance has been higher in monozygotic twins (~50 percent) than in dizygotic twins (~17 percent). Although these results suggest that a genetic component exists, they do not reveal the precise genetic basis of schizophrenia. Other studies implicate the importance of shared environmental risk factors. The current view is that schizophrenia is genetically mediated but not genetically determined. The genetic contributions to schizophrenia have been difficult to identify. Simple monohybrid inheritance, dihybrid inheritance, and multiple-gene control have been proposed for schizophrenia. It does not seem likely that only one or two loci are involved, nor is it likely that the disorder is strictly quantitative, as in polygenic inheritance. Several attempts to identify genes involved in schizophrenia have failed, and new approaches are being used. We will describe one of these new approaches: the use of

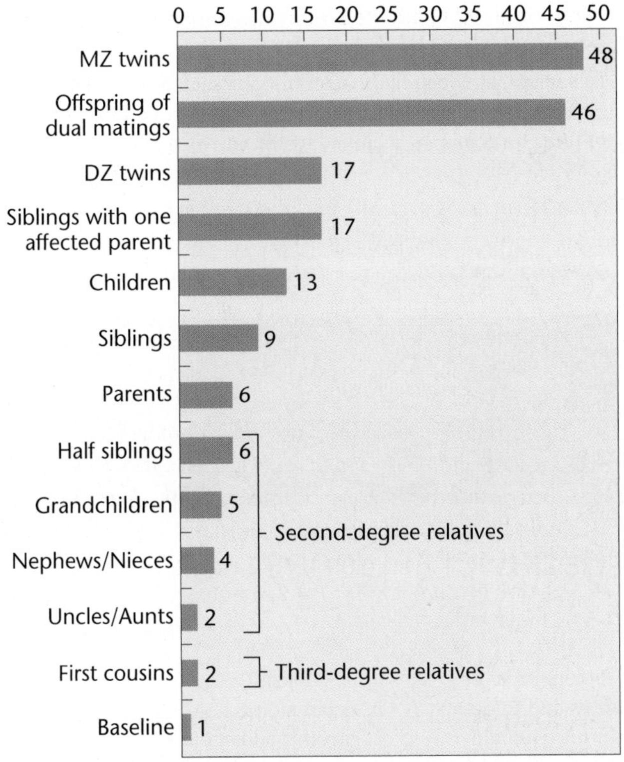

FIGURE 21–11 Relative lifetime risk of schizophrenia for nonrelated individuals and for relatives of a schizophrenic proband.

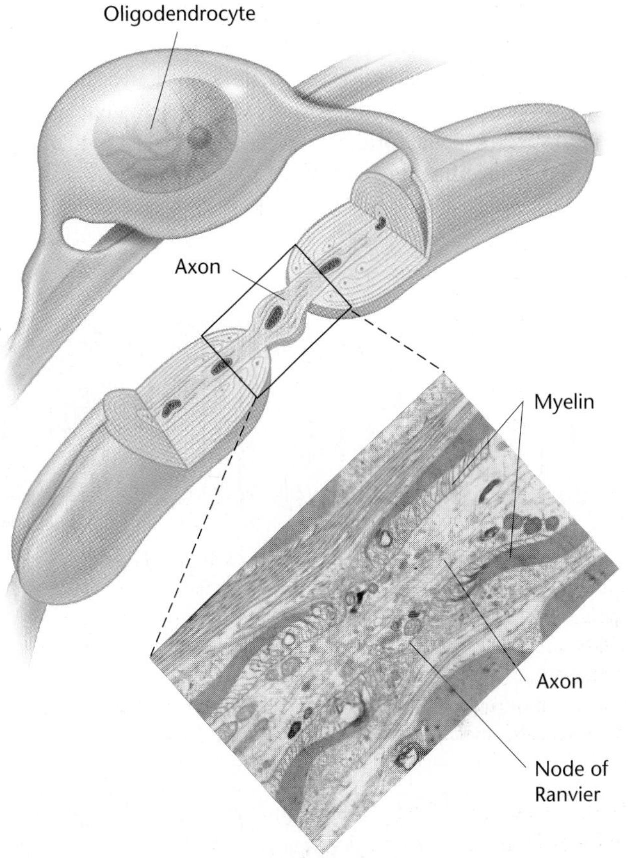

FIGURE 21–13 Oligodendrocytes and the myelin sheath. Oligodendrocytes (left) are cells in the nervous system that synthesize and deposit the myelin sheaths that surround neurons. A cross section of a neuron (right) shows the layers of myelin. Genomic analysis indicates that abnormal myelin synthesis is associated with schizophrenia.

DNA microarrays to carry out genomewide scans to identify genes whose expression is altered in schizophrenia.

In one important study, researchers isolated RNA from the prefrontal cortex (a region of the brain known to be affected in this disease) of normal and schizophrenic individuals obtained at autopsy. The isolated RNA was used to prepare cDNA for hybridization to the microarrays. The microarrays carried over 6000 probes for human genes. Partial results from the study are shown in Figure 21–12. The genes tested have been organized into clusters with similar biological functions. The analysis focused on genes that were previously found to have altered levels of expression in schizophrenia. The cluster of genes involved in myelination had lower levels of expression in schizophrenia, while most of the other genes tested had increased levels of expression. This suggests that schizophrenia is associated with functional disruption in oligodendrocytes, the cells that form the myelin sheath that wraps around nerve cells and ensures rapid conduction of nerve impulses (Figure 21–13). Myelination in the prefrontal cortex occurs during late adolescence and early adulthood, which is usually the time that the behavioral symptoms of schizophrenia first appear.

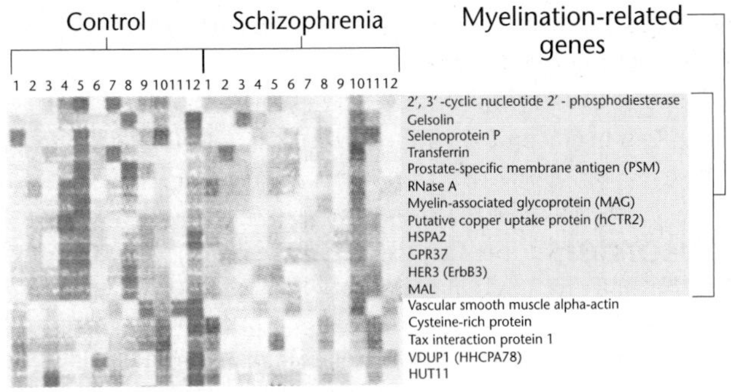

FIGURE 21–12 Relative levels of expression for genes differentially expressed in individuals with schizophrenia. Each column shows the gene expression level (blue is low, red is high) in one individual, and each row shows the levels of expression for a given gene in a number of different individuals. There is an overall significant reduction in expression of myelin-related genes in schizophrenic individuals compared to normal control individuals, leading to the conclusion that there may be a disruption in oligodendrocytes (the cells that make myelin) in schizophrenia.

Various research groups, using over two dozen genomewide microarray scans of gene expression in schizophrenia, have identified several sets of candidate genes whose expression is altered in affected individuals. Although the results of these studies have not been consistent, six myelin-related genes whose expression is specific to oligodendrocytes have been identified. In schizophrenia, expression of all six of these genes is reduced. The emerging evidence indicates that myelination abnormalities are an important part of the disease process in schizophrenia, but that environmental components also play a major role.

Strains of knockout mice, each missing a myelin-related gene, are being developed to connect changes in gene expression to specific behavioral phenotypes and to design therapies targeted at specific gene clusters.

ESSENTIAL POINT

In humans, although multifactorial traits such as schizophrenia have genetic components, studies have shown that expression of these traits may be modified by the environment.

EXPLORING GENOMICS

HomoloGene: Searching for Behavioral Genes

This chapter discussed some of the complexities of understanding behavior—in humans and other species—as a product of both genetics and environment. In this exercise, we will explore the NCBI database **HomoloGene** to search for genes implicated in behavioral conditions. HomoloGene is a database of homologs from several eukaryotic species, and it contains genes that have been implicated in behavioral conditions both in animal models and in humans. Some of these are mutant genes, which, on the basis of single-nucleotide polymorphisms or linkage mapping, are thought to contribute to a behavioral phenotype. Keep in mind that correlation of a gene with a particular behavioral condition does not necessarily mean that the gene is the cause of the condition. Nor does it mean that the gene is the only one involved in the behavioral condition or that the condition is entirely due to genetics. Nonetheless, exploring HomoloGene is an interesting way to search for putative behavioral genes and learn more about them.

■ Exercise I: HomoloGene

1. Access **HomoloGene** from the ENTREZ site at **http://www.ncbi.nlm.nih.gov/gquery/ gquery.fcgi**.
2. Search HomoloGene using the following list of terms related to behavioral conditions in humans. What human genes are listed in the top two categories of gene homologs that result from the HomoloGene search for each condition? Name these genes and identify their chromosomal loci. Notice the list of other species thought to share homologs of each gene.
   a. alcoholism
   b. depression
   c. nicotine addiction
   d. seasonal affective disorder
   e. intelligence
3. For each of the top two categories of homologs, click on the category title to open a report on those genes. Then use the links for individual genes to learn more about the functions of those genes. In addition to other good resources, the "Additional Links" category at the bottom of each gene report page will provide you with access to the OnLine Mendelian Inheritance in Man (OMIM) site that we have used for other exercises.
4. Search HomoloGene for any behavioral conditions you are interested in to see if the database contains genes that may be implicated in those conditions.

CASE STUDY Primate models for human disorders

Huntington disease (HD) is a dominantly inherited, late-onset neurodegenerative disorder that affects about 1 in 10,000 people. Onset is around 50 years of age, there is no treatment, and the disorder inevitably causes dementia, loss of motor control, and is fatal. To study the cellular and molecular mechanisms associated with HD, animal models, including mice and *Drosophila,* have been created by transferring mutant human HD genes into these organisms. However, some of the behavioral changes seen in HD mice and fruit flies are difficult to correlate with those seen in affected humans. To more closely replicate the cognitive and motor changes in HD, researchers have created a primate model using rhesus macaques, which are one of our closest relatives and share many similarities with humans, including life span, cellular metabolism, as well as endocrine and reproductive functions. However, their use raises several important questions.

1. What are the ethical concerns related to research with primate models of human disease?
2. Does it seem likely that results from using rhesus macaques will translate directly to an understanding of the disease process in humans and lead to treatment?
3. If the primate HD model is successful, should it be used for other human behavioral disorders such as Alzheimer or Parkinson disease?

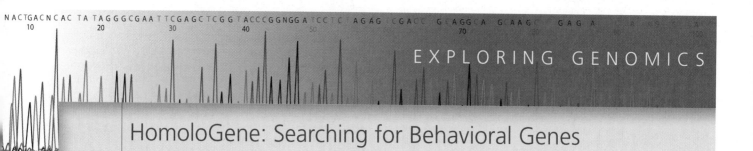

INSIGHTS AND SOLUTIONS

1. Bipolar disorder is a mental illness associated with pervasive and wide mood swings. It is estimated that 1 in 100 individuals has or will suffer from this disorder at least once in his or her lifetime. Genetic studies indicate that bipolar disorder is familial and that mutations in several genes are involved.

In an attempt to identify genes associated with bipolar disorder (this condition was formerly called manic depression), researchers began by using conventional genetic analysis. In 1987, two separate studies using pedigree analysis, RFLP analysis, and a variety of other genetic markers reported a linkage between manic depression and, in one report, markers on the X chromosome and, in the other report, markers on the short arm of chromosome 11. At the time, these reports were hailed as landmark discoveries. It was thought that genes controlling this behavioral disorder could be identified, mapped, and sequenced in much the same way as genes for cystic fibrosis, neurofibromatosis, and other disorders. It was hoped that this discovery would open the way to the development of therapeutic strategies based on knowledge of the nature of the gene product and its mechanism of action. Soon after, however, further work on the same populations (reported in 1989 and 1990) demonstrated that the original results were invalid after individuals in the study who did not carry the markers subsequently developed bipolar disorder. As a result, it became clear that no linkage to markers on the X chromosome or to markers on chromosome 11 could be established. These findings do not exclude the role of major genes on the X chromosome and chromosome 11 in bipolar disorder, but they do exclude linkage to the markers used in the original reports. This setback not only caused embarrassment and confusion, but also forced a reexamination of the validity of the newly developed recombinant DNA-based methods used in mapping other human genes and of the analysis of data from linkage studies involving complex behavioral traits. Several factors have been proposed to explain the flawed conclusions reported in the original studies. What do you suppose some of these factors were?

Solution: Although some of the criticisms were directed at the choice of markers, most of the factors at work in this situation appear to be related to the phenotype of manic depression. At least three confounding factors have been identified. One factor relates to the diagnosis of bipolar disorder itself. The phenotype is complex and not as easily quantified as height or weight. In addition, mood swings are a universal part of everyday life, and it is not always easy to distinguish transient mood alterations and the role of environmental factors from disorders having a biological or genetic basis (or both). A second confounding factor is that, in the populations studied, there may in fact be more than one major X-linked gene and more than one major autosomal gene that trigger bipolar disorder. A third factor is age of onset. There is a positive correlation between age and the appearance of the symptoms of bipolar disorder. Therefore, at the time of a pedigree study, younger individuals who will be affected later in life may not show any signs of bipolar disorder. Taken together, these factors point up the difficulty in researching the genetic basis of complex behavioral traits. Individually or in combinations, the factors described might skew the results enough so that the guidelines for proof that are adequate for other traits are not stringent enough for behavioral traits with complex phenotypes and complex underlying causes.

PROBLEMS AND DISCUSSION QUESTIONS

1. Contrast the two principal approaches used in studying behavioral genetics. What are the advantages of each method with respect to the type of information gained?
2. Assume that you discovered a fruit fly that walked with a limp and was continually off balance as it moved. Describe how you would determine whether this behavior was due to an injury (induced in the environment) or to an inherited trait. See **Now Solve This** on page 455.
3. Certain inbred strains of mice (i.e., C57BL and C3H/2) appear to exhibit a preference for alcohol in free-choice consumption experiments, and a variety of techniques have been applied to assess underlying genetic predispositions. Tarantino et al. (1998) applied QTL analysis to alcohol preference in mice using microsatellite markers. They found three significant QTLs on chromosomes 1, 4, and 9 and three suggestive QTLs on chromosomes 2, 3, and 10. How might QTL analysis lead to an understanding of alcohol preference in mice? See **Now Solve This** on page 453.
4. In humans, the chemical phenylthiocarbamide (PTC) is either tasted or not. When the offspring of various combinations of taster and nontaster parents are examined, the following data are obtained:

Parents	Offspring
Both tasters	All tasters
Both tasters	1/2 tasters 1/2 nontasters
Both tasters	3/4 tasters 1/4 nontasters
One taster One nontaster	All tasters
One taster One nontaster	1/2 tasters 1/2 nontasters
Both nontasters	All nontasters

Based on these data, how is PTC tasting behavior inherited?

5. Jerry Hirsch and colleagues have used the behavior-first approach to study the genes responsible for geotaxis in *Drosophila*. Their elegant genetic analysis has shown that this behavior is under genetic control, and they estimate the number of genes responsible for geotactic behavior. Although the approach has been successful, what are the limitations of the behavior-first approach

as used here? What steps did Hirsch and his colleagues take to overcome those limitations? Has this approach been successful?

6. Various approaches have been applied to study the genetics of problem and pathological gambling (PG), and within-family vulnerability has been well documented. However, family studies, while showing clusters within blood relatives, cannot separate genetic from environmental influences. Eisen (2001) applied "twin studies" using 3359 twin pairs from the Vietnam-era Twin Registry and found that a substantial portion of the variance associated with PG can be attributed to inherited factors. How might twin studies be used to distinguish environmental from genetic factors in complex behavioral traits such as PG?
7. *Caenorhabditis elegans* has become a valuable model organism for the study of development and genetics for a variety of reasons. The developmental fate of each cell (1031 in males and 959 in hermaphrodites) has been mapped. *C. elegans* has only 302 neurons whose pattern of connectivity is known, and it displays a variety of interesting behaviors including chemotaxis, thermotaxis, and mating behavior, to name a few. In addition, isogenic lines have been established. What advantage would the use of *C. elegans* have over other model organisms in the study of animal behavior? What are likely disadvantages?
8. Using the behavioral phenotypes of a series of 18 mutants falling into five phenotypic classes, Thomas (1990) described the genetic program that controls the cyclic defecation motor program in *Caenorhabditis elegans.* The first step in each cycle is contraction of the posterior body muscles, followed by contraction of the anterior body muscles. Finally, anal muscles open and expel the intestinal contents. Below is a list of selected mutants (simplified) that cause defects in defecation.

Gene	Phenotype
exp	failure to open the anus and expel intestinal contents
unc	failure to contract anterior body muscles
pbo	failure to contract posterior body muscles
aex	failure to open the anus and expel intestinal contents and failure to contract anterior body muscles
cha	alteration of cycle periodicity, all other components being completely functional

Present a simple schematic that illustrates genetic involvement in *C. elegans* defecation. Account for the action of the *aex* gene and speculate on the involvement and placement of the *cha* gene.

9. Discuss why the study of human behavior genetics has lagged behind that of other organisms.
10. J. P. Scott and J. L. Fuller studied 50 traits in five pure breeds of dogs. Almost all the traits varied significantly in the five breeds, but very few bred true in crosses. What can you conclude with respect to the genetic control of these behavioral traits?
11. Although not discussed in this chapter, *C. elegans* is a model system whose life cycle makes it an excellent choice for the genetic dissection of many biological processes. *C. elegans* has two natural sexes: hermaphrodite and male. The hermaphrodite is essentially a female that can generate sperm as well as oocytes, so reproduction can occur by hermaphrodite self-fertilization or hermaphrodite–male mating. In the context of studying mutations in the nervous system, what is the advantage of hermaphrodite self-fertilization with respect to the identification of recessive mutations and the propagation of mutant strains?
12. Hypothetical data concerning the genetic effect of *Drosophila* chromosomes on geotaxis are shown here:

	Chromosome		
	X	*2*	*3*
Positive geotaxis	+0.2	+0.1	+3.0
Unselected	+0.1	−0.2	+1.0
Negative geotaxis	−0.1	−2.6	+0.1

What conclusions can you draw?

13. In July 2006, a population of flies, *Drosophila melanogaster,* rode the space shuttle *Discovery* to the International Space Station (ISS) where a number of graviperception experiments and observations are to be conducted over a nine-generation period. Eventually, frozen specimens will be collected by astronauts and returned to Earth. Researchers expect to correlate behavioral and physiological responses to microgravity with changes in gene activity by analyzing RNA and protein profiles. The title of the project is "*Drosophila* Behavior and Gene Expression in Microgravity." If you were in a position to conduct three experiments on the behavioral aspects of these flies, what would they be? How would you go about assaying changes in gene expression in response to microgravity? Given that humans share over half of the genome and proteins of *Drosophila,* how would you justify the expense of such a project in terms of improving human health?
14. Of a variety of investigational approaches that have been applied to the study of schizophrenia, two separate distinct approaches, twin studies and microarray analysis, have provided strong support for a genetic component.
    (a) Provide a summary of the contribution that each has played in understanding schizophrenia.
    (b) Speculate on the value that microarray analysis using samples from monozygotic and dizygotic twins might have for enhancing research on schizophrenia.
15. An interesting and controversial finding by Wedekind and colleagues (1995) suggested that a sense of smell, influenced by various HLA (Human Leukocyte Antigen) haplotypes, plays a role in mate selection in humans. Women preferred T-shirts from men with HLA haplotypes unlike their own. The HLA haplotypes are components of the Major Histocompatibility Complex (MHC) and are known to have significant immunological functions. In addition to humans, mice and fish also condition mate preference on MHC constitution. The three-spined stickleback (*Gasterosteus aculeatus*) has recently been studied to determine whether females use an MHC odor-based system to select males as mates. Examine the following data (modified from Milinski et al. 2005) and provide a summary conclusion. Why might organisms evolve a mate selection scheme for assessing and optimizing MCH diversity?

	Male with optimal MHC	Male with nonoptimal MHC
Gravid female time spent (seconds)	350	250
Number of spawning females	11	3

Gravid females chose different amounts of time (seconds) for exposure to males with optimal MHC alleles versus those with nonoptimal MHC alleles. The number of females spawning when exposed to different males with different MHC complements is also presented.

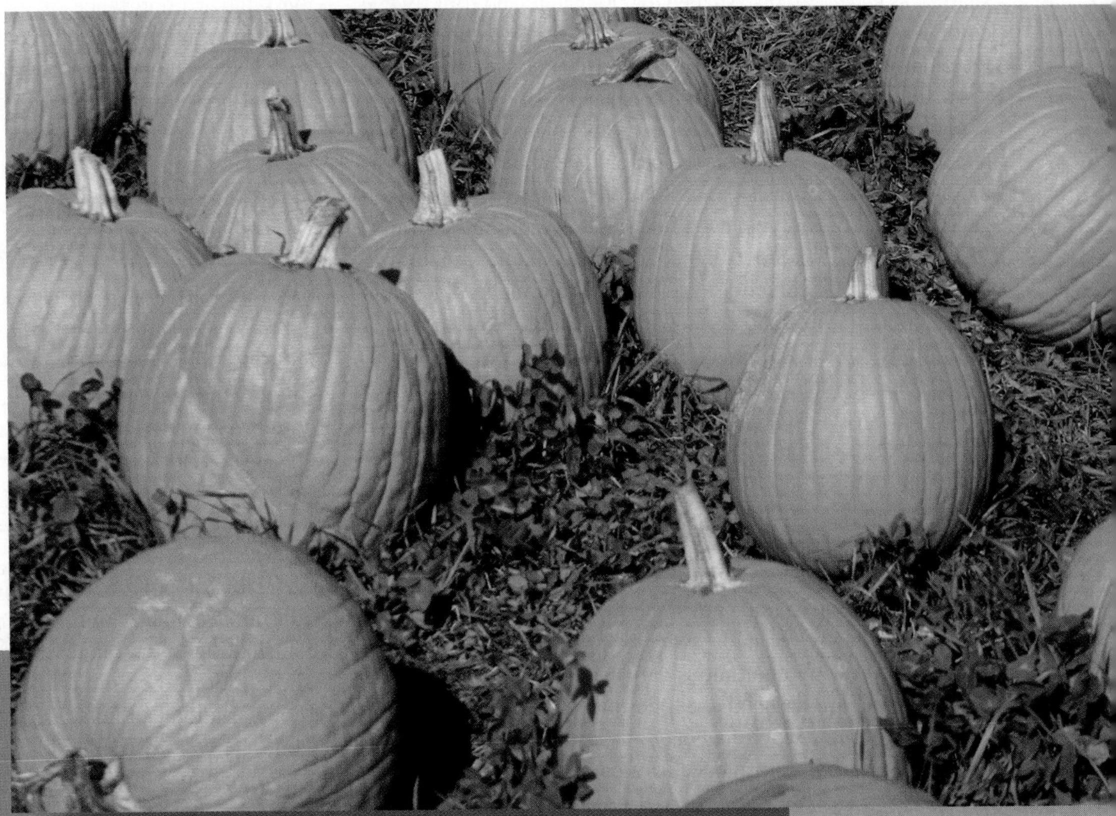

A field of pumpkins, where size is under the influence of additive alleles.

22

Quantitative Genetics and Multifactorial Traits

Chapter Concepts

- Quantitative inheritance results in a range of measurable phenotypes for a polygenic trait.
- With some exceptions, polygenic traits tend to demonstrate continuous variation.
- Quantitative traits can be explained in Mendelian terms whereby certain alleles have an additive effect on the traits under study.
- The study of polygenic traits relies on statistical analysis.
- Heritability values estimate the genetic contribution to phenotypic variability under specific environmental conditions.
- Twin studies allow an estimation of heritability in humans.
- Quantitative trait loci (QTLs) can be mapped and identified.

Up to this point, most of our examples of phenotypic variation have been assigned to distinct and separate categories: pea plants were tall or dwarf; squash fruit shape was spherical, disc shaped, or elongated; and fruit fly eye color was red or white (see Chapter 4). Traits such as these, which have a small number of discrete phenotypes, are said to show **discontinuous variation.** Typically, in these traits, a genotype will produce a single identifiable phenotype, although phenomena such as variable penetrance and expressivity, pleiotropy, and epistasis can obscure the relationship between genotype and phenotype, even for simple discontinuous traits.

Now we will look at traits that are more complex, including many that have medical or agricultural importance. The traits we examine in this chapter show much more variation, with a continuous range of phenotypes that cannot be as easily classified into distinct categories. Examples of traits showing **continuous variation** include human height or weight, milk or meat production in cattle, crop yield, and seed protein content. Continuous variation across a range of phenotypes is measured and described in quantitative terms, so this genetic phenomenon is known as **quantitative inheritance.** Because the varying phenotypes result from the input of genes at multiple loci, quantitative traits are sometimes said to be **polygenic** (literally "of many genes").

For traits showing continuous variation, the genotype generated at fertilization establishes the quantitative range within which a particular individual can fall. However, the final phenotype is often also influenced by environmental factors to which that individual is exposed. Human height, for example, is in part genetically determined, but is also affected by environmental factors such as nutrition. Those phenotypes that result from both gene action and environmental influences are sometimes termed **complex,** or **multifactorial, traits.**

How Do We Know?

In this chapter, we focus on a mode of inheritance referred to as quantitative genetics, as well as many of the statistical parameters utilized to study quantitative traits. Along the way, we will find opportunities to consider the methods and reasoning by which geneticists acquired much of their understanding of quantitative genetics. From the explanations given in the chapter, what answers would you propose to the following fundamental questions:

1. How do we know that threshold traits are actually polygenic even though they may have as few as two discrete phenotypic classes?
2. How can we ascertain the number of polygenes involved in the inheritance of a quantitative trait?
3. What findings led geneticists to postulate the multiple-factor hypothesis that invoked the idea of additive alleles to explain inheritance patterns?
4. How do we assess environmental factors to determine whether they impact the phenotype of a quantitatively inherited trait?

In this chapter, we will examine examples of quantitative inheritance and some of the statistical techniques used to study complex traits. We will also consider how geneticists assess the relative importance of genetic versus environmental factors contributing to continuous phenotypic variation, and we will discuss approaches to identifying and mapping genes that influence quantitative traits.

22.1 Not All Polygenic Traits Show Continuous Variation

When a continuous trait is examined in a population, individual measurements often lie along a continuum of phenotypes with no clear separation into categories. It is possible for a single-gene trait to show such a continuum of phenotypes within a population, if, for example, there are different degrees of gene penetrance among individuals. Single-gene traits can also be multifactorial if the interaction of alleles with the environment produces a range of different phenotypes. More often, however, a continuous quantitative trait is the result of polygenic inheritance. Furthermore, polygenic traits are frequently multifactorial, with environmental factors contributing to the range of phenotypes observed.

In addition to continuous quantitative traits, where phenotypic variation can lie at any point within a range of measurement, there are two other classes of polygenic traits. **Meristic** traits are those in which the phenotypes are described by whole numbers. Examples of meristic traits include the number of seeds in a pod or the number of eggs laid by a chicken in a year. These are quantitative traits, but they do not have an infinite range of phenotypes: for example, a pod may contain 2, 4, or 6 seeds, but not 5.75. **Threshold traits** are polygenic (and frequently, environmental factors affect the phenotypes, making them also multifactorial), but they are distinguished from continuous and meristic traits by having a small number of discrete phenotypic classes. Threshold traits are currently of heightened interest to human geneticists because an increasing number of diseases are now thought to show this pattern of polygenic inheritance. One example is **Type II diabetes,** also known as adult-onset diabetes because it typically affects individuals who are middle aged or older. A population can be divided into just two phenotypic classes for this trait—individuals who have Type II diabetes and those who do not—so at first glance, it may appear to more closely resemble a simple monogenic trait. However, no single adult-onset diabetes gene has been identified. Instead, the combination of alleles present at multiple contributing loci gives an individual a greater or lesser likelihood of developing the disease. These varying levels of liability form a continuous range: at one extreme of the distribution are those at very low risk for Type II diabetes, while at the other end are those whose genotypes make it highly likely they will develop the disease (Figure 22–1). As with many threshold traits, environmental factors also play a role in determining the final phenotype, with diet and lifestyle having significant impact on whether an individual with moderate to high genetic liability will actually develop Type II diabetes.

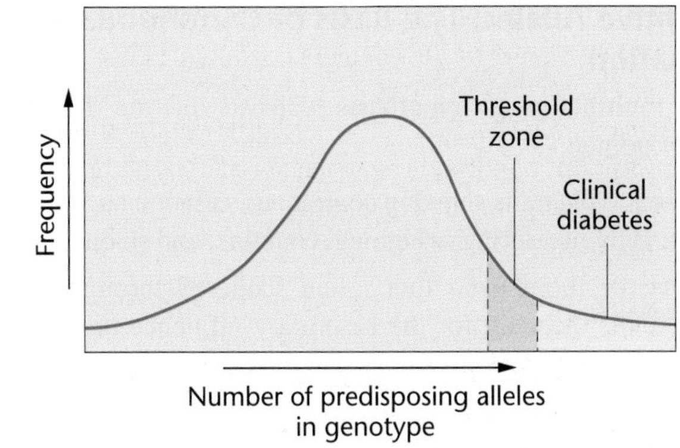

FIGURE 22–1 A graphic depiction of predisposing alleles characteristic of a threshold trait within a population, illustrated by Type II diabetes.

22.2 Quantitative Traits Can Be Explained in Mendelian Terms

The question of whether continuous phenotypic variation could be explained in Mendelian terms caused considerable controversy in the early 1900s. Some scientists argued that, although Mendel's unit factors, or genes, explained patterns of discontinuous segregation with discrete phenotypic classes, they could not also account for the range of phenotypes seen in quantitative patterns of inheritance. However, geneticists William Bateson and Gudny Yule, adhering to a Mendelian explanation, proposed the **multiple-factor** or **multiple-gene hypothesis** in which many genes, each individually behaving in a Mendelian fashion, contribute to the phenotype in a *cumulative* or *quantitative* way.

The Multiple-Gene Hypothesis for Quantitative Inheritance

The **multiple-gene hypothesis** was initially based on a key set of experimental results published by Hermann Nilsson-Ehle in 1909. Nilsson-Ehle used grain color in wheat to test the concept that the cumulative effects of alleles at multiple loci produce the range of phenotypes seen in quantitative traits. In one set of experiments, wheat with red grain was crossed to wheat with white grain (Figure 22–2). The F_1 generation demonstrated an intermediate pink color, which at first sight suggested incomplete dominance of two alleles at a single locus. However, in the F_2 generation, Nilsson-Ehle did not observe the 3:1 segregation typical of a monohybrid cross. Instead, approximately 15/16 of the plants showed some degree of red grain color, while 1/16 of the plants showed white grain color. Careful examination of the F_2 revealed that grain with color could be classified into four different shades of red. Because the F_2 ratio occurred in sixteenths, it appears that two genes, each with two alleles, control the phenotype and that they segregate independently from one another in a Mendelian fashion.

If each gene has one potential **additive allele** that contributes to the red grain color and one potential **nonadditive allele** that fails to produce any red pigment, we can see how the multiple-factor hypothesis could account for the various grain color phenotypes. In the P_1 both parents are homozygous; the red parent contains only additive alleles (*AABB* in Figure 22–2), while the white parent contains only nonadditive alleles (*aabb*). The F_1 plants are heterozygous (*AaBb*), contain two additive (*A* and *B*) and two nonadditive (*a* and *b*) alleles, and express the intermediate pink phenotype. Each of the F_2 plants has 4, 3, 2, 1, or 0 additive alleles. F_2 plants with no additive alleles are white (*aabb*) like one of the P_1 parents, while F_2 plants with 4 additive alleles are red (*AABB*) like the other P_1 parent. Plants with 3, 2, or 1 additive alleles constitute the other three categories of red color observed in the F_2

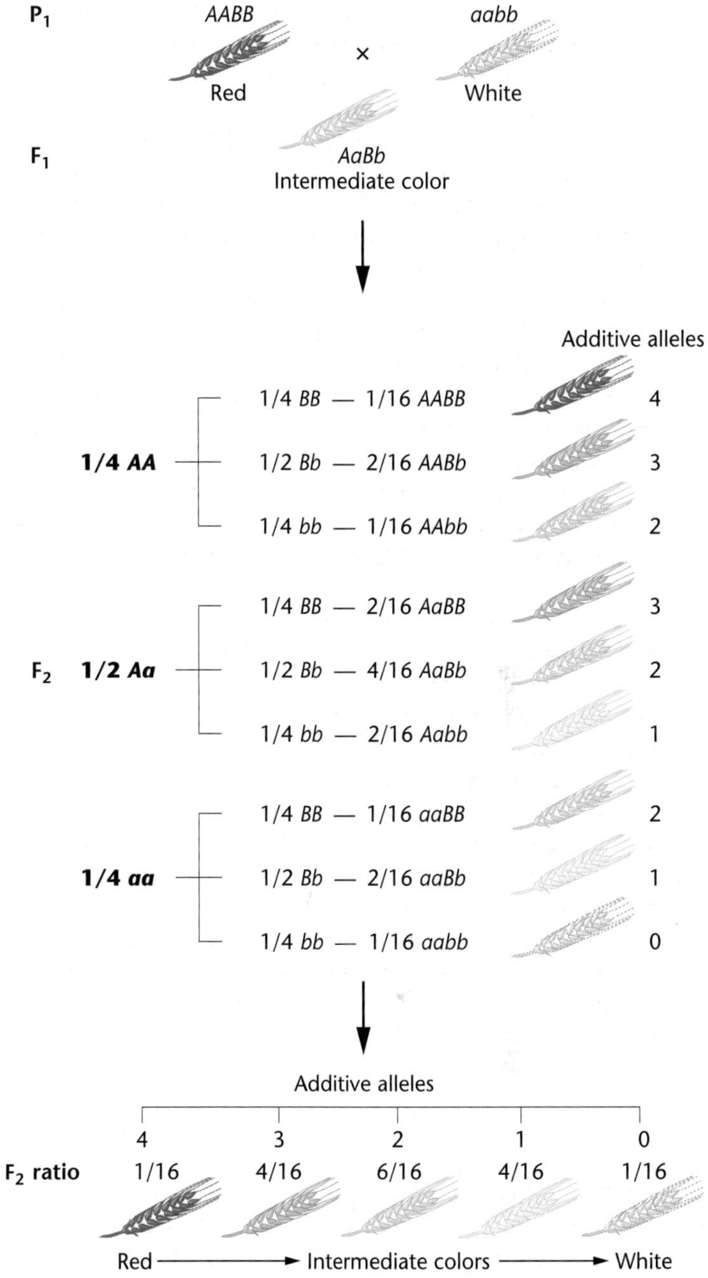

FIGURE 22–2 How the multiple-factor hypothesis accounts for the 1:4:6:4:1 phenotypic ratio of grain color when all alleles designated by an uppercase letter are additive and contribute an equal amount of pigment to the phenotype.

generation. The greater the number of additive alleles in the genotype, the more intense the red color expressed in the phenotype, as each additive allele present contributes equally to the cumulative amount of pigment produced in the grain.

Nilsson-Ehle's results showed how continuous variation could still be explained in a Mendelian fashion, with additive alleles at multiple loci influencing the phenotype in a quantitative manner, but each individual allele segregating according to Mendelian rules. As we saw in Nilsson-Ehle's initial cross, if two loci, each with two alleles, were involved, then five F_2 phenotypic categories in a 1:4:6:4:1 ratio would be expected. However, there is no reason why three, four, or more loci cannot function in a similar fashion in controlling various quantitative phenotypes. As more quantitative loci become involved, greater and greater numbers of classes appear in the F_2 generation in more complex ratios. The number of phenotypes and the expected F_2 ratios for crosses involving up to four gene pairs are illustrated in Figure 22–3.

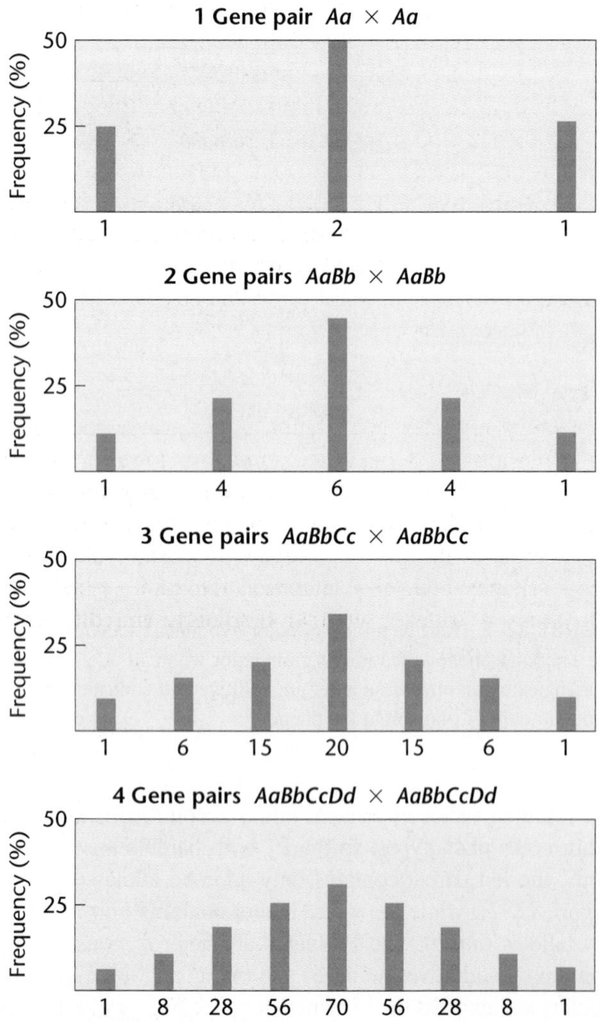

FIGURE 22–3 The genetic ratios (on the *X*-axis) resulting from crossing two heterozygotes when polygenic inheritance is in operation with one to four gene pairs. The histogram bars indicate the distinct F_2 phenotypic classes, ranging from one extreme (left end) to the other extreme (right end).

Additive Alleles: The Basis of Continuous Variation

The multiple-gene hypothesis consists of the following major points:

1. Phenotypic traits showing continuous variation can be quantified by measuring, weighing, counting, and so on.
2. Two or more gene loci, often scattered throughout the genome, account for the hereditary influence on the phenotype in an *additive way.* Because many genes may be involved, inheritance of this type is called *polygenic.*
3. Each gene locus may be occupied by either an *additive* allele, which contributes a constant amount to the phenotype, or a *nonadditive* allele, which does not contribute quantitatively to the phenotype.
4. The contribution to the phenotype of each additive allele, though often small, is approximately equal.
5. Together, the additive alleles contributing to a single quantitative character produce substantial phenotypic variation.

Calculating the Number of Polygenes

Various formulas have been developed for estimating the number of **polygenes,** the genes contributing to a quantitative trait. For example, if the ratio of F_2 individuals resembling *either* of the two extreme P_1 phenotypes can be determined, the number of polygenes involved (n) may be calculated as follows:

$$1/4^n = \text{ratio of } F_2 \text{ individuals expressing either extreme phenotype}$$

In the example of the red and white wheat grain color summarized in Figure 22–2, 1/16 of the progeny are either red *or* white like the P_1 phenotypes. This ratio can be substituted on the right side of the equation to solve for n:

$$\frac{1}{4^n} = \frac{1}{16}$$

$$\frac{1}{4^2} = \frac{1}{16}$$

$$n = 2$$

Table 22.1 lists the ratio and the number of F_2 phenotypic classes produced in crosses involving up to five gene pairs.

For low numbers of polygenes (n), it is sometimes easier to use the equation

$$(2n + 1) = \text{the number of distinct phenotypic categories observed}$$

For example, when there are two polygenes involved ($n = 2$), then $(2n + 1) = 5$ and each phenotype is the result of 4, 3, 2, 1, or 0 additive alleles. If $n = 3$, $2n + 1 = 7$ and each phenotype is the result of 6, 5, 4, 3, 2, 1, or 0 additive alleles. Thus, working backwards with this rule and knowing the number of phenotypes, we can calculate the number of polygenes controlling them.

It should be noted, however, that both of these simple methods for estimating the number of polygenes involved in a

TABLE 22.1 Determination of the Number of Polygenes (*n*) Involved in a Quantitative Trait

n	Individuals Expressing Either Extreme Phenotype	Distinct Phenotypic Classes
1	1/4	3
2	1/16	5
3	1/64	7
4	1/256	9
5	1/1024	11

quantitative trait assume that all the relevant alleles contribute equally and additively, and also that phenotypic expression in the F_2 is not affected significantly by environmental factors. As we will see later, for many quantitative traits, these assumptions may not be true.

ESSENTIAL POINT

Quantitative inheritance results in a range of phenotypes due to the action of additive alleles from two or more genes, as influenced by environmental factors.

NOW SOLVE THIS

Problem 3 on page 480 gives F_1 and F_2 ranges for a quantitative trait and asks you to calculate the number of polygenes involved.

Hint: Remember that unless you know the total number of distinct phenotypes involved, then the ratio (not the number) of parental phenotypes reappearing in the F_2 is the key.

22.3 The Study of Polygenic Traits Relies on Statistical Analysis

Before considering the approaches that geneticists use to dissect how much of the phenotypic variation observed in a population is due to genotypic differences among individuals and how much is due to environmental factors, we need to consider the basic statistical tools they use for the task. Because it is not usually feasible to measure expression of a polygenic trait in every individual in a population, a random subset of individuals is usually selected for measurement to provide a *sample.* It is important to remember that the accuracy of the final results of the measurements depends on whether the sample is truly random and representative of the population from which it was drawn. Suppose, for example, that a student wants to determine the average height of the 100 students in his genetics class, and for his sample he measures the two students sitting next to him, both of whom happen to be centers on the college basketball team. It is unlikely that this sample will provide a good estimate of the average height of the class, for two reasons: first, it is too small; second, it is not a representative subset of the class (unless all 100 students are centers on the basketball team).

If the sample measured for expression of a quantitative trait is sufficiently large and also representative of the population from which it is drawn, we often find that the data form a **normal distribution;** that is, they produce a characteristic bell-shaped curve when plotted as a frequency histogram (Figure 22–4). Several statistical concepts are useful in the analysis of traits that exhibit a normal distribution, including the mean, variance, standard deviation, standard error of the mean, and covariance.

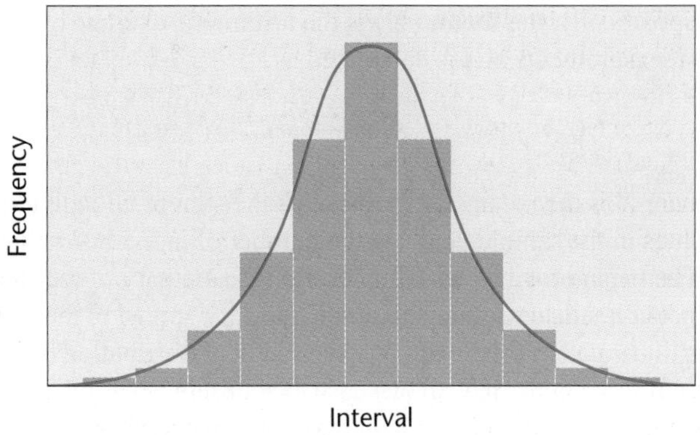

FIGURE 22–4 Normal frequency distribution, characterized by a bell-shaped curve.

The Mean

The mean provides information about where the central point lies along a range of measurements for a quantitative trait. Figure 22–5 shows the distribution curves for two different sets of phenotypic measurements. Each of these sets of measurements clusters around a central value (as it happens, they both cluster around the same value). This clustering is called a **central tendency,** and the central point is the mean.

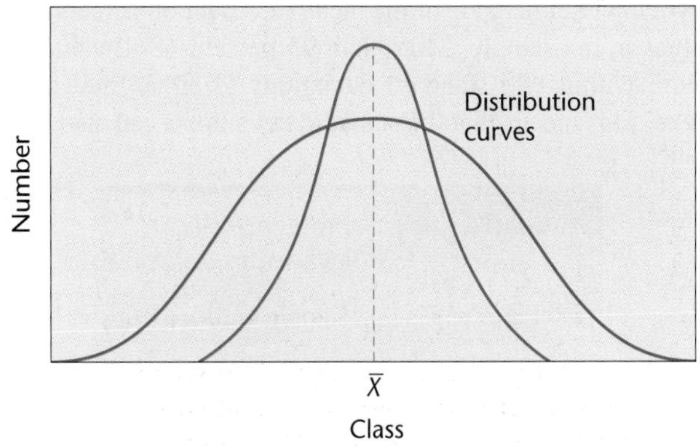

FIGURE 22–5 Two normal frequency distributions with the same mean but different amounts of variation.

Specifically, the **mean** ($\overline{X}$) is the arithmetic average of a set of measurements and is calculated as

$$\overline{X} = \frac{\Sigma X_i}{n}$$

where $\overline{X}$ is the mean, ΣX_i represents the sum of all individual values in the sample, and n is the number of individual values.

The mean provides a useful descriptive summary of the sample, but it tells us nothing about the range or spread of the data. As illustrated in Figure 22–5, a symmetrical distribution of values in the sample may, in one case, be clustered near the mean. Or a set of measurements may have the same mean but be distributed more widely around it. A second statistic, the variance, provides information about the spread of data around the mean.

Variance

The **variance** (s^2) for a sample is the average squared distance of all measurements from the mean. It is calculated as

$$s^2 = \frac{\Sigma(X_i - \overline{X})^2}{n - 1}$$

where the sum (Σ) of the squared differences between each measured value (X_i) and the mean ($\overline{X}$) is divided by one less than the total sample size $n - 1$.

As Figure 22–5 shows, it is possible for two sets of sample measurements for a quantitative trait to have the same mean but a different distribution of values around it. This range will be reflected in different variances. Estimation of variance can be useful in determining the degree of genetic control of traits when the immediate environment also influences the phenotype.

Standard Deviation

Because the variance is a squared value, its unit of measurement is also squared (m^2, g^2, etc.). To express variation around the mean in the original units of measurement, we can use the square root of the variance, a term called the **standard deviation** (*s*):

$$s = \sqrt{s^2}$$

Table 22.2 shows the percentage of individual values within a normal distribution that fall within different multiples of the standard deviation. The values that fall within one standard deviation to either side of the mean represent 68 percent of all values in the sample. More than 95 percent of all values are found within two standard deviations to either side of the mean. This means that the standard deviation *s* can also be interpreted in the form of a probability. For example, a sample measurement picked at random has a 68 percent probability of falling within the range of one standard deviation.

TABLE 22.2 Sample Inclusion for Various *s* Values

Multiples of *s*	Sample Included (%)
$\overline{X} \pm 1s$	68.3
$\overline{X} \pm 1.96s$	95.0
$\overline{X} \pm 2s$	95.5
$\overline{X} \pm 3s$	99.7

Standard Error of the Mean

If multiple samples are taken from a population and measured for the same quantitative trait, we might find that their means vary. Theoretically, larger, truly random samples will represent the population more accurately, and their means will be closer to each other. To measure the accuracy of the sample mean we use the **standard error of the mean ($S_{\overline{X}}$),** calculated as

$$S_{\overline{X}} = \frac{s}{\sqrt{n}}$$

where s is the standard deviation and $\sqrt{n}$ is the square root of the sample size. Because the standard error of the mean is computed by dividing s by $\sqrt{n}$, it is always a smaller value than the standard deviation.

Covariance

Often geneticists working with quantitative traits find they have to consider two phenotypic characters simultaneously. For example, a poultry breeder might investigate the correlation between body weight and egg production in hens: Do heavier birds tend to lay more eggs? The **covariance** statistic measures how much variation is common to both quantitative traits. It is calculated by taking the deviations from the mean for each trait (just as we did for estimating variance) for each individual in the sample. This gives a pair of values for each individual. The two values are multiplied together, and the sum of all these individual products is then divided by one fewer than the number in the sample. Thus, the covariance cov_{XY} of two sets of trait measurements, X and Y, is calculated as

$$\text{cov}_{XY} = \frac{\Sigma\left[(X_i - \overline{X})(Y_i - \overline{Y})\right]}{n - 1}$$

The covariance can then be standardized as yet another statistic, the **correlation coefficient (*r*).** The calculation of r is

$$r = \text{cov}_{XY}/S_X S_Y$$

where S_X is the standard deviation of the first set of quantitative measurements X, and S_Y is the standard deviation of the second set of quantitative measurements Y. Values for the correlation coefficient r can range from -1 to $+1$. Positive r values mean that an increase in measurement for one trait tends to be associated with an increase in measurement for the other, while negative r values mean that increases in one trait are associated with decreases in the other. Therefore, if heavier hens do tend to lay more eggs, a positive r value can be expected. A negative r value, on the other hand, suggests that greater egg production is more likely from less heavy birds. One important point to note about correlation coefficients is that even significant r values—close to $+1$ or -1—do not prove that a cause-and-effect relationship exists between two traits. Correlation simply tells us the extent to which variation in one quantitative trait is associated with variation in another, not what causes that variation.

ESSENTIAL POINT

Numerous statistical methods are essential during the analysis of quantitative traits, including the mean, variance, standard deviation, standard error, covariance, and the correlation coefficient.

NOW SOLVE THIS

Problem 15 on page 481 gives data for two quantitative traits in a flock of sheep. In Part (d), you are asked to determine whether the traits are correlated.

Hint: Once the calculation of the correlation coefficient (*r*) is completed, you must analyze and interpret that value.

Analysis of a Quantitative Character

To apply these statistical concepts, let's consider a genetic experiment that crossed two different homozygous varieties of tomato. One of the tomato varieties produces fruit averaging 18 oz in weight, whereas fruit from the other averages 6 oz. The F_1 obtained by crossing these two varieties has fruit weights ranging from 10 to 14 oz. The F_2 population contains individuals that produce fruit ranging from 6 to 18 oz. The results characterizing both generations are shown in Table 22.3.

The mean value for the fruit weight in the F_1 generation can be calculated as

$$\overline{X} = \frac{\Sigma X_i}{n} = \frac{626}{52} = 12.04$$

The mean value for fruit weight in the F_2 generation is calculated as

$$\overline{X} = \frac{\Sigma X_i}{n} = \frac{872}{72} = 12.11$$

Although these mean values are similar, the frequency distributions in Table 22.3 show more variation in the F_2 generation. The range of variation can be quantified as the sample variance s^2, calculated, as we saw above, as the sum of the squared differences between each value and the mean, divided by one less than the total number of observations.

$$s^2 = \frac{\Sigma(X_i - \overline{X})^2}{n - 1}$$

When the above calculation is made, the variance is found to be 1.29 for the F_1 generation and 4.27 for the F_2 generation. When converted to the standard deviation ($s = \sqrt{s^2}$), the values become 1.13 and 2.06, respectively. Therefore, the distribution of tomato weight in the F_1 generation can be described as 12.04 ± 1.13, and in the F_2 generation it can be described as 12.11 ± 2.06.

Assuming both tomato varieties are homozygous at the loci of interest and that the alleles controlling fruit weight act additively, we can estimate the number of polygenes involved in this trait. Since 1/72 of the F_2 offspring have a phenotype that overlaps one of the parental strains (72 total F_2 offspring; one weighs 6 oz, one weighs 18 oz; see Table 22.3), use of the formula $1/4^n = 1/72$ indicates that n is between 3 and 4, providing evidence of the number of genes that control fruit weight in these tomato strains.

22.4 Heritability Values Estimate the Genetic Contribution to Phenotypic Variability

The question most often asked by geneticists working with multifactorial traits is how much of the observed phenotypic variation in a population is due to genotypic differences among individuals and how much is due to environment. The term **heritability** is used to describe what proportion of total phenotypic variation in a population is due to genetic factors. For a multifactorial trait in a given population, a high heritability estimate indicates that much of the variation can be attributed to genetic factors, with the environment having less impact on expression of the trait. With a low heritability estimate, environmental factors are likely to have a greater impact on phenotypic variation within the population.

The concept of heritability is frequently misunderstood and misused. It should be emphasized that heritability indicates neither how much of a trait is genetically determined nor the extent to which an individual's phenotype is due to genotype. In recent years, such misinterpretations of heritability for human quantitative traits have led to controversy, notably in relation to measurements such as "intelligence quotients," or IQs. Variation in heritability estimates for IQ among different "racial" groups tested led to incorrect suggestions that unalterable genetic factors control differences in intelligence levels among humans of different ancestries. Such suggestions misrepresented the meaning of heritability and ignored the contribution of genotype-by-environment interaction variance (see p. 472) to phenotypic variation in a population. Moreover, heritability is not fixed for a trait. For example, a heritability estimate for egg production in a flock of chickens kept in individual cages might be high, indicating that differences in egg output among individual birds are largely due to genetic differences, as they all have very similar environments.

TABLE 22.3 Distribution of F_1 and F_2 Progeny Derived from a Theoretical Cross Involving Tomatoes

	Weight (oz)												
	6	7	8	9	10	11	12	13	14	15	16	17	18
Number of Individuals F_1					4	14	16	12	6				
Number of Individuals F_2	1	1	2	0	9	13	17	14	7	4	3	0	1

For a different flock kept outdoors, heritability for egg production might be much lower because variation among different birds may also reflect differences in their individual environments. Such differences could include how much food each bird manages to find and whether it competes successfully for a good roosting spot at night. Thus a heritability estimate tells us the proportion of phenotypic variation that can be attributed to genetic variation *within a certain population in a particular environment.*

If we measure heritability for the same trait among different populations in a range of environments, we frequently find that the calculated heritability values have large standard errors. This is an important point to remember when considering heritability estimates for traits in human populations. A mean heritability estimate of 0.65 for human height does not mean that your height is 65 percent due to your genes, but rather that in the populations sampled, on average, *65 percent of the overall variation in height could be explained by genotypic differences among individuals.*

With this subtle, but important, distinction in mind, we will now consider how geneticists divide the phenotypic variation observed in a population into genetic and environmental components. As we saw in the previous section, variation can be quantified as a sample variance: taking measurements of the trait in question from a representative sample of the population and determining the extent of the spread of those measurements around the sample mean. This gives us an estimate of the total **phenotypic variance** in the population (**V_P**). Heritability estimates are obtained by using different experimental and statistical techniques to partition V_P into **genotypic variance (V_G)** and **environmental variance (V_E)** components.

An important factor contributing to overall levels of phenotypic variation is the extent to which individual genotypes affect the phenotype differently depending on the environment. For example, wheat variety A may yield an average of 20 bushels an acre on poor soil, while variety B yields an average of 17 bushels. On good soil, variety A yields 22 bushels, while variety B averages 25 bushels an acre. Because there are differences in yield between the two genotypically distinct varieties, variation in wheat yield has a genetic component. Both varieties yield more on good soil, so yield is also affected by environment. However, we also see that the two varieties do not respond to better soil conditions equally: The genotype of wheat variety B achieves a greater increase in yield on good soil than does variety A. Thus, we have differences in the interaction of genotype with environment contributing to variation for yield in populations of wheat plants. This third component of phenotypic variation is **genotype-by-environment interaction variance ($V_{G\times E}$)** (Figure 22–6).

We can now summarize all the components of total phenotypic variance V_P using the following equation:

$$V_P = V_G + V_E + V_{G\times E}$$

In other words, total phenotypic variance can be subdivided into genotypic variance, environmental variance, and genotype-by-environment interaction variance. When obtaining heritability estimates for a multifactorial trait, researchers often assume that the genotype-by-environment interaction variance is small enough that it can be ignored or combined with the environmental variance. However, it is worth remembering that this kind of approximation is another reason heritability values are *estimates* for a given population in a particular context, not a *fixed attribute* for a trait.

Animal and plant breeders use a range of experimental techniques to estimate heritabilities by partitioning measurements of phenotypic variance into genotypic and environmental components. One approach uses inbred strains containing genetically homogeneous individuals with highly homozygous genotypes. Experiments are then designed to test the effects of a range of environmental conditions on phenotypic variability. Variation *between* different inbred strains reared in a constant environment is due predominantly to genetic factors. Variation *among* members of the same inbred strain reared under different conditions is more likely to be due to environmental factors. Other approaches involve analyzing variance for a quantitative trait among offspring from different crosses, or comparing expression of a trait among offspring and parents reared in the same environment.

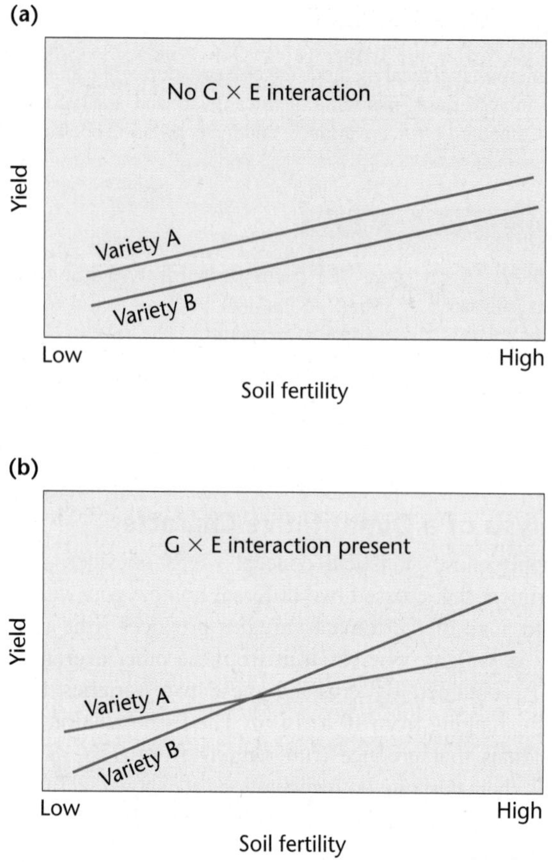

FIGURE 22–6 Differences in yield between two wheat varieties at different soil fertility levels. (a) No genotype-by-environment, or G × E, interaction: The varieties show genetic differences in yield but respond equally to increasing soil fertility. (b) G × E interaction present: Variety A outyields B at low soil fertility, but B yields more than A at high fertility levels.

Broad-Sense Heritability

Broad-sense heritability (represented by the term H^2) measures the contribution of the genotypic variance to the total phenotypic variance. It is estimated as a proportion:

$$H^2 = \frac{V_G}{V_P}$$

Heritability values for a trait in a population range from 0.0 to 1.0. A value approaching 1.0 indicates that the environmental conditions have little impact on phenotypic variance, which is therefore largely due to genotypic differences among individuals in the population. Low values close to 0.0 indicate that environmental factors, not genotypic differences, are largely responsible for the observed phenotypic variation within the population studied. Few quantitative traits have very high or very low heritability estimates, suggesting that both genetics and environment play a part in the expression of most phenotypes for the trait.

The genotypic variance component V_G used in broad-sense heritability estimates includes all types of genetic variation in the population. It does not distinguish between quantitative trait loci with alleles acting additively as opposed to those with epistatic or dominance effects. Broad-sense heritability estimates also assume that the genotype-by-environment variance component is negligible. Although broad-sense heritability estimates for a trait are of general genetic interest, these limitations mean that this kind of heritability is not very useful in breeding programs. Animal or plant breeders wishing to develop improved strains of livestock or higher-yielding crop varieties need more precise heritability estimates for the traits they wish to manipulate in a population. Therefore, another type of estimate, narrow-sense heritability, has been devised that is of more practical use.

Narrow-Sense Heritability

Narrow-sense heritability (h^2) is the proportion of phenotypic variance due to additive genotypic variance alone. Genotypic variance can be divided into subcomponents representing the different modes of action of alleles at quantitative trait loci. As not all the genes involved in a quantitative trait affect the phenotype in the same way, this partitioning distinguishes among three different kinds of gene action contributing to genotypic variance. **Additive variance, V_A,** is the genotypic variance due to the additive action of alleles at quantitative trait loci. **Dominance variance, V_D,** is the deviation from the additive components that results when phenotypic expression in heterozygotes is not precisely intermediate between the two homozygotes. **Interactive variance,** V_I, is the deviation from the additive components that occurs when two or more loci behave epistatically. The amount of interactive variance is often negligible, and so this component is often excluded from calculations of total genotypic variance.

The partitioning of the total genotypic variance V_G is summarized in the equation

$$V_G = V_A + V_D + V_I$$

and a narrow-sense heritability estimate based only on that portion of the genotypic variance due to additive gene action becomes

$$h^2 = \frac{V_A}{V_P}$$

Omitting V_I and separating V_P into genotypic and environmental variance components, we obtain

$$h^2 = \frac{V_A}{V_E + V_A + V_D}$$

Heritability estimates are used in animal and plant breeding to indicate the potential response of a population to artificial selection for a quantitative trait. Narrow-sense heritability, h^2, provides a more accurate prediction of selection response than broad-sense heritability, H^2, and therefore h^2 is more widely used by breeders.

Artificial Selection

Artificial selection is the process of choosing specific individuals with preferred phenotypes from an initially heterogeneous population for future breeding purposes. Theoretically, if artificial selection based on the same trait preferences is repeated over multiple generations, a population can be developed containing a high frequency of individuals with the desired characteristics. If selection is for a simple trait controlled by just one or two genes subject to little environmental influence, generating the desired population of plants or animals is relatively fast and easy. However, many traits of economic importance in crops and livestock, such as grain yield in plants, weight gain or milk yield in cattle, and speed or stamina in horses, are polygenic and frequently multifactorial. Artificial selection for such traits is slower and more complex. Narrow-sense heritability estimates are valuable to the plant or animal breeder because, as we have just seen, they estimate the proportion of total phenotypic variance for the trait that is due to additive genetic variance. Quantitative trait alleles with additive action are those most easily manipulated by the breeder. Alleles at quantitative trait loci that generate dominance effects or interact epistatically (and therefore contribute to V_D or V_I) are less responsive to artificial selection. Thus narrow-sense heritability, h^2, can be used to predict the impact of selection. The higher the estimated value for h^2 in a population, the greater the change in phenotypic range for the trait that the breeder will see in the next generation after artificial selection.

Partitioning the genetic variance components to calculate h^2 and predict response to selection is a complex task requiring careful experimental design and analysis. The simplest approach is to select individuals with superior phenotypes for the desired quantitative trait from a heterogeneous population and to breed offspring from those individuals. The mean score for the trait of those offspring ($M2$) can then be compared to that of: (1) the original population's mean score (M) and (2) the selected individuals used as parents ($M1$). The relationship between these means and h^2 is

$$h^2 = \frac{M2 - M}{M1 - M}$$

This equation can be further simplified by defining $M2 - M$ as the **selection response** (***R***)—the degree of response to mating the selected parents—and $M1 - M$ as the **selection differential** (***S***)—the difference between the mean for the whole population and the mean for the selected population—so h^2 reflects the ratio of the response observed to the total response possible. Thus,

$$h^2 = \frac{R}{S}$$

A narrow-sense heritability value obtained in this way by selective breeding and measuring the response in the offspring is referred to as an estimate of **realized heritability.**

As an example of a realized heritability estimate, suppose that we measure the diameter of corn kernels in a population where the mean diameter M is 20 mm. From this population, we select a group with the smallest diameters, for which the mean $M1$ equals 10 mm. The selected plants are interbred, and the mean diameter $M2$ of the progeny kernels is 13 mm. We can calculate the realized heritability h^2 to estimate the potential for artificial selection on kernel size:

$$h^2 = \frac{M2 - M}{M1 - M}$$

$$h^2 = \frac{13 - 20}{10 - 20}$$

$$= \frac{-7}{-10}$$

$$= 0.70$$

This value for narrow-sense heritability indicates that the selection potential for kernel size is relatively high.

The longest running artificial selection experiment known is still being conducted at the State Agricultural Laboratory in Illinois. Since 1896, corn has been selected for both high and low oil content. After 76 generations, selection continues to result in increased oil content (Figure 22–7). With each cycle of successful selection, more of the corn plants accumulate a higher percentage of additive alleles involved in oil production. Consequently, the narrow-sense heritability h^2 of increased oil content in succeeding generations has declined (see parenthetical values at generations 9, 25, 52, and 76 in Figure 22–7) as artificial selection comes closer and closer to optimizing the genetic potential for oil production. Theoretically, the process will continue until all individuals in the population possess a uniform genotype that includes all the additive alleles responsible for high oil content. At that point, h^2 will be reduced to zero, and response to artificial selection will cease. The decrease in response to selection for low oil content shows that heritability for low oil content is approaching this point.

Table 22.4 lists narrow-sense heritability estimates expressed as percentage values for a variety of quantitative traits in different organisms. As you can see, these h^2 values vary, but heritability tends to be low for quantitative traits that are essential to an organism's survival. Remember, this does not indicate the absence of a genetic contribution to the observed phenotypes for such traits. Instead, the low h^2 values show that natural selection has already largely optimized the genetic component of these traits during evolution. Egg production, litter size, and conception rate are examples of how such physiological limitations on selection have already been reached. Traits that are less critical to survival, such as body weight, tail length, and wing length, have higher heritabilities because more genotypic variation for such traits is still present in the population. Remember also that any single heritability estimate can only provide information about one population in a specific environment. Therefore, narrow-sense heritability is a more valuable predictor of response to selection when estimates

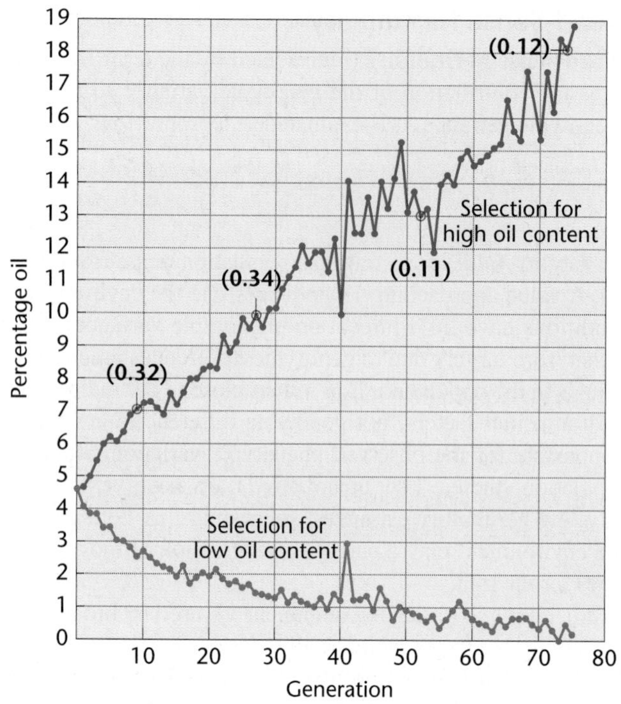

FIGURE 22–7 Response of corn selected for high and low oil content over 76 generations. The numbers in parentheses at generations 9, 25, 52, and 76 for the "high oil" line indicate the calculation of heritability at these points in the continuing experiment.

TABLE 22.4 Estimates of Heritability for Traits in Different Organisms

Trait	Heritability (h^2)
Mice	
Tail length	60%
Body weight	37
Litter size	15
Chickens	
Body weight	50
Egg production	20
Egg hatchability	15
Cattle	
Birth weight	45
Milk yield	44
Conception rate	3

are calculated for many populations and environments and show the presence of a clear trend.

ESSENTIAL POINT

Heritability is an estimate of the relative contribution of genetic versus environmental factors to the range of phenotypic variation seen in a quantitative trait in a particular population and environment.

NOW SOLVE THIS

Problem 12 on page 481 gives data for two quantitative traits in a herd of hogs and asks you to choose the trait that the farmer should select that will respond favorably to selection.

Hint: Consider which variance component's proportion is the best indicator of potential response to selection.

22.5 Twin Studies Allow an Estimation of Heritability in Humans

For obvious reasons, traditional heritability studies are not possible in humans. However, human twins are useful subjects for examining how much phenotypic variance for a multifactorial trait is due to the genotype as opposed to the environment. **Monozygotic (MZ),** or **identical, twins,** derived from the division and splitting of a single egg following fertilization, are genotypically identical. **Dizygotic (DZ),** or **fraternal, twins,** on the other hand, originate from two separate fertilization events and are only as genetically similar as any other two siblings, with about 50 percent of their alleles in common. For a given trait, therefore, phenotypic differences between pairs of identical twins will be equivalent to the environmental variance (V_E) (because the genotypic variance is zero). Phenotypic differences between dizygotic twins, however, represent both environmental variance (V_E) and approximately half the genotypic variance (V_G). Comparison of phenotypic variances for the same trait in monozygotic and dizygotic sets of twins provides an estimate of broad-sense heritability for the trait.

Another approach used in twin studies has been to compare identical twin pairs reared together with pairs that were separated and raised in different settings. For any particular trait, average similarities or differences can be investigated. Twins are said to be **concordant** for a given trait if both express it or neither expresses it. If one expresses the trait and the other does not, the pair is said to be **discordant.** Comparison of concordance values of MZ versus DZ twins reared together illustrates the potential value for heritability assessment. (See the Now Solve This feature at the end of this section.)

Before any conclusions can be drawn, such data must be examined very carefully. If the concordance value approaches 90 to 100 percent in monozygotic twins, we might be inclined to interpret that value as indicating a large genetic contribution to the expression of the trait. In some cases—for example, blood types and eye color—we know that this is indeed true. In the case of contracting measles, however, a high concordance value merely indicates that the trait is almost always induced by a factor in the environment—in this case, a virus.

It is more meaningful to compare the *difference* between the concordance values of monozygotic and dizygotic twins. If these values are significantly higher for monozygotic twins than for dizygotic twins, we suspect a strong genetic component in the determination of the trait. We reach this conclusion because monozygotic twins, with identical genotypes, would be expected to show a greater concordance than genetically related, but not genetically identical, dizygotic twins. In the case of measles, where concordance is high in both types of twins, the environment is assumed to contribute significantly. Such an analysis is useful because phenotypic characteristics that remain similar in different environments are likely to have a strong genetic component.

Interesting as they are, human twin studies contain some unavoidable sources of error. For example, identical twins are often treated more similarly by parents and teachers than are nonidentical twins, especially where the nonidentical siblings are of different sex. This may inflate the environmental variance for nonidentical twins. Another possible error source is genotype-by-environment interaction, which can increase the total phenotypic variance for nonidentical twins compared to identical twins raised in the same environment. Heritability estimates for human traits based on twin studies should therefore be considered approximations and examined very carefully before any conclusions are drawn.

In addition, recent findings in genomics are throwing new light on differences between members of MZ twin pairs and differences between MZ and DZ twins, relationships that affect basic assumptions about concordance and heritability. MZ twins have identical genotypes, but possess epigenetic differences in the form of DNA methylation and histone modification, which may explain the low levels of discordance seen in members of a MZ twin pair. MZ twins are epigenetically identical at birth, but older MZ twins show significant differences in epigenetic modifications and patterns of gene expression. These progressive genomic modifications may represent inherent changes in the genome or exposure to different environmental factors. These findings also indicate that concordance studies in DZ twins must take into account genetic as well as *epigenetic differences* that contribute to discordance in these twin pairs.

ESSENTIAL POINT

Twin studies are used to estimate heritabilities for polygenic traits in humans.

NOW SOLVE THIS

Problem 24 on page 482 lists the percentages of monozygotic (MZ) and dizygotic (DZ) twins expressing the same phenotype for different traits and asks you to evaluate the relative importance of genetic and environmental factors.

Hint: Consider how MZ twins differ genetically from DZ twins. If each pair of twins was raised in the same environment, consider how that would affect expression of genetically determined versus environmentally determined phenotypes.

22.6 Quantitative Trait Loci Can Be Mapped

The pedigree analysis we examined in earlier chapters (see Chapters 3 and 4) is of little use in identifying the multiple genes involved in quantitative traits. Environmental effects, interaction among segregating alleles, and the number of genes contributing to a polygenic phenotype make it difficult to isolate the effect of any one individual gene. However, because many quantitative traits are of economic or medical importance, it is desirable to identify the genes involved and determine their location in the genome. Multiple genes contributing to a quantitative trait are known as **quantitative trait loci (QTLs).** A single quantitative trait locus is designated a QTL. Identifying and studying these loci help geneticists estimate how many genes are involved in a given quantitative trait and whether they all contribute equally to the trait or whether some genes influence the phenotype more strongly than others. Mapping reveals whether the QTLs associated with a trait are grouped on a single chromosome or scattered throughout the genome.

To find and map QTLs, researchers look for associations between DNA marker sequences and phenotypes that fall within a certain range of the quantitative phenotype. One way to do this is to cross homozygous inbred lines that have different phenotypic extremes for the trait of interest to create an F_1 generation whose members will be heterozygous at most of the loci contributing to the trait. Additional crosses, either among F_1 individuals or between the F_1 and the inbred parent lines, result in F_2 generations with a high degree of segregation for different QTL genotypes and associated phenotypes. This segregating F_2 is known as the **QTL mapping population.**

Researchers measure expression of the trait among individuals in the mapping population and identify different genotypes using DNA markers such as restriction fragment length polymorphisms (RFLPs) and microsatellites (see Chapter 19). Computer-based statistical analysis is then used to search for linkage between the markers and an aspect of phenotypic variation for the trait. If a DNA marker *is not* linked to a QTL, then the phenotypic mean score for the trait will not vary among individuals with different genotypes at that marker locus. However, if a DNA marker *is* linked to a QTL, then different genotypes at that marker locus will also differ in their phenotypic expression of the trait. When this occurs, the marker locus and the QTL are said to *cosegregate.* Consistent cosegregation establishes the presence of a QTL at or near the DNA marker along the chromosome; in other words, the marker and QTL are linked. When numerous QTLs for a given trait have been located, a genetic map is created, locating chromosomal regions carrying these traits. Cloning, sequencing, and annotation are used to identify QTLs in these regions.

DNA markers are now available for the genomes of many agriculturally important organisms, making possible systematic mapping of QTLs. For example, hundreds of RFLP markers have been located in the tomato, and its genome is being sequenced. QTL analysis begins with standard genetic crosses between plants with extreme, opposite phenotypes and following the crosses through several generations. Many loci have been identified that are responsible for quantitative traits such as fruit size, shape, soluble solid content, and acidity. These loci are distributed on all 12 chromosomes representing the haploid genome of this plant. Several chromosomes contain loci for more than one of these traits.

Mapping and characterizing QTLs in the tomato have been the focus of a highly successful research effort conducted by Steven Tanksley and his colleagues at Cornell University. Their research shows that approximately 30 QTLs contribute to tomato fruit size and shape. However, these QTLs do not have equal effects: 10 genes scattered among seven chromosomes account for most of the observed phenotypic variation for these traits.

While the cultivated tomato can weigh up to 1000 grams, fruit from the related wild species thought to be the progenitor of the modern tomato weighs only a few grams. QTL mapping has demonstrated that a relatively small number of genes is responsible for this thousand-fold variation in fruit size. Tanksley's group has identified and studied three of these genes: *fw2.2* (on chromosome 2), *fasciated* (*fas*) on chromosome 11, and *locule-number*, also on chromosome 2. The *fw2.2* locus was the first QTL locus to be identified in plants, and alleles at this locus are responsible for about 30 percent of the variation in fruit size. This gene has been isolated, cloned, and transferred between plants, with interesting results. One allele of *fw2.2* is present in all wild small-fruited varieties of tomatoes investigated. Another allele is present in all domesticated large-fruited varieties. When a cloned *fw2.2* allele from small-fruited varieties is transferred to a plant that normally produces large tomatoes, the transformed plant produces fruits that are greatly reduced in weight (Figure 22–8). In the varieties studied by Tanksley's group, the reduction averaged 17 grams, a significant phenotypic change caused by the action of a single QTL.

Further analysis of *fw2.2* revealed that this gene encodes a protein that negatively regulates the cell cycle during fruit development. Differences in the time of gene expression and

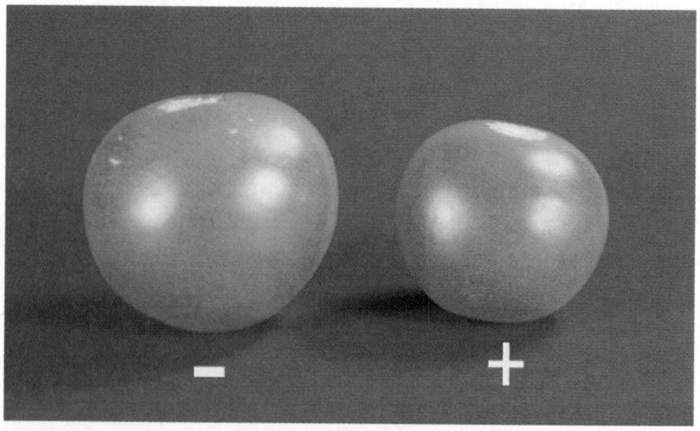

FIGURE 22–8 Phenotypic effect of the *fw2.2* transgene in the tomato. When the allele causing small fruit is transferred to a plant that normally produces large fruit, the fruit is reduced in size (+). A control fruit (−) that has not been transformed is shown for comparison.

differences in the amount of transcript produced result in small or large fruit. Higher levels of expression exert a negative control over cell division, resulting in smaller tomatoes.

Yet, *fw2.2* and other cell-cycle regulatory genes cannot account for all the variation in fruit size. Analysis of the *fas* gene indicates that another step in the development of extreme differences in fruit size involves an increase in the number of compartments in the mature fruit. The small, ancestral stocks produce fruit with two to four seed compartments, but the large-fruited present-day strains have eight or more compartments. Thus, the QTLs that affect fruit size in tomatoes work by controlling at least two developmental processes: the cell cycle and the determination of the number of ovarian compartments.

Marker-based mapping can identify chromosomal regions where QTLs are found, but identifying individual quantitative genes usually requires additional techniques. One approach is to develop DNA markers that are tightly linked to a gene of interest and then use positional cloning (see Chapter 24). Clearly, finding and characterizing all the genes involved in a quantitative trait are part of a long and complex task. At present, the genes controlling tomato size are among a small number of quantitative genes that have been isolated, cloned, and studied in detail. However, mapping QTLs and defining the function of genes present in these regions in agriculturally important plants will help develop strategies for improving crop yields. These studies also pave the way for similar research on animal genes.

ESSENTIAL POINT

Quantitative trait loci, or QTLs, may be identified and mapped using DNA markers.

GENETICS, TECHNOLOGY, AND SOCIETY

The Green Revolution Revisited: Genetic Research with Rice

Of the 6.7 billion people now living on Earth, over 800 million do not have enough to eat. That number is expected to grow by an additional 1 million people each year for the next several decades. How will we be able to feed the estimated 8 billion people on Earth by 2025?

The past gives us some reasons to be optimistic. In the 1950s and 1960s, in the face of looming population increases, plant scientists around the world set about to increase the production of crop plants, including the three most important grains—rice, wheat, and maize. These efforts became known as the *Green Revolution.* The approach was three-pronged: (1) to increase the use of fertilizers, pesticides, and irrigation; (2) to bring more land under cultivation; and (3) to develop improved varieties of crop plants by intensive plant breeding.

The results were dramatic. Developing nations more than doubled their production of rice, wheat, and maize between 1961 and 1985. Nations that were facing widespread famine in the 1960s were able to feed themselves and became major exporters of grain.

The Green Revolution saved millions of people from starvation and improved the quality of life for millions more; however, its effects may be diminishing. The rate of increase in grain yields has slowed since the 1980s, due to slower growth in irrigation development and fertilizer use. If food production is to keep pace with the projected increase in the world's population, we will have to depend more and more on the genetic improvement of crop plants to provide higher yields. Is this possible, or are we approaching the theoretical limits of yield in important crop plants? Recent work with rice suggests that the answer to the second question is a resounding no!

About half of the Earth's population depends on rice for basic nourishment. The Green Revolution for rice began in 1960, aided by the establishment of the International Rice Research Institute (IRRI). One of its major developments was the breeding of a rice variety with improved disease resistance and higher yield. The IRRI research team crossed a Chinese rice variety (*Dee-geo-woo-gen*) and an Indonesian variety (*Peta*) to create a new cultivar known as IR8. IR8 produced a greater number of rice kernels per plant, when grown in the presence of fertilizers and irrigation. Under these cultural practices, IR8 plants were so top-heavy with grain that they tended to fall over—a trait called "lodging." To reduce lodging, IRRI breeders crossed IR8 with a dwarf native variety to create semi-dwarf lines. Owing in part to the adoption of the semi-dwarf IR8 lines, the world production of rice doubled in 25 years.

Predictions suggest that a 40 percent increase in the annual rice harvest may be necessary to keep pace with anticipated population growth during the next 30 years. As land and water resources become scarcer and as the prices of fertilizers and pesticides grow, more emphasis will be placed on creating new rice varieties that have even higher yields and greater disease resistance.

Geneticists are now using several strategies to develop more productive rice. Conventional hybridization and selection techniques, such as those used to create the IR8 strain, continue to produce varieties that contribute to a productivity increase of 1 to 10 percent a year. Several QTLs from wild rice appear to contribute to increased yields, and scientists are attempting to introduce these traits into current dwarf varieties of domestic rice. Genomics and genetic engineering are

(*Cont. on the next page*)

also contributing to the new Green Revolution for rice. In 2002, the rice genome was the first cereal crop genome to be sequenced. Research is now concentrated on assigning a function to all the genes in the rice genome. Once identified and characterized, these genes may be transferred into crop plants, speeding the creation of rice varieties with desirable traits, such as disease resistance, tolerance to drought and salinity, and improved nutritional content.

Your Turn

Take time, individually or in groups, to answer the following questions. Investigate the references and links to help you understand some of the technologies and issues surrounding the new Green Revolution.

1. How do you think that genomics and genetic engineering will contribute to the development of more productive rice varieties?

One source of information on this topic is: Khush, G. S. 2005. What it will take to feed 5.0 billion rice consumers in 2030. *Plant Mol Biol.* 59: 1–6.

2. A serious complication in our efforts to feed the Earth's population is global climate change. Given your assessment of current climate change predictions, what traits should we introduce into rice plants, and what methods can we use to do this?

To learn more about the effects of global climate change on agriculture, read Battisti, D. S. and Naylor, R. L. 2009. Historical warnings of future food insecurity with unprecedented seasonal heat. *Science* 323: 240–244. *Also, search for "climate change" publications on the Web site of the International Rice Research Institute* (**http://www.irri.org**).

3. Despite its benefits, the Green Revolution has been the subject of controversy. What are the main criticisms of the Green Revolution, and how can we mitigate some of the Green Revolution's negative aspects?

A summary of criticisms of the Green Revolution can be found on the Wikipedia Web site (**http://en.wikipedia.org/wiki/Green_revolution**).

CASE STUDY A flip of the genetic coin

On July 11, 2008, twin sons were born to Stephan Gerth from Germany and Addo Gerth from Ghana. Stephan is very fair-skinned with blue eyes and straight hair; Addo is dark-skinned, with brown eyes and curly hair. The first born of the twins, Ryan, is fair-skinned, with blue eyes and straight hair; his brother, Leo, has light brown skin, brown eyes and curly hair. Although the twins' hair texture and eye color were the same as those of one or the other parent, the twins had different skin colors, intermediate to that of their parents. Experts explained that the blending effect of skin color in the twins resulted from quantitative inheritance involving at least three different gene pairs, whereas hair texture and eye color are not quantitatively inherited. Using this as an example of quantitative genetics, we can ask the following questions:

1. What approach is used in estimating how many gene pairs are involved in a quantitative trait? Why would this be extremely difficult in the case of skin color in humans?
2. Would either parent need to have mixed-race ancestry for the twins to be so different?
3. Would twins showing some parental traits (hair texture, eye color) but a blending of other traits (skin color in this case) seem to be a commonplace event, or are we looking at a "one in a million" event?

INSIGHTS AND SOLUTIONS

1. In a certain plant, height varies from 6 to 36 cm. When 6-cm and 36-cm plants were crossed, all F_1 plants were 21 cm. In the F_2 generation, a continuous range of heights was observed. Most were around 21 cm, and 3 of 200 were as short as the 6-cm P_1 parent.

(a) What mode of inheritance does this illustrate, and how many gene pairs are involved?

(b) How much does each additive allele contribute to height?

(c) List all genotypes that give rise to plants that are 31 cm.

Solution:

(a) Polygenic inheritance is illustrated when a trait is continuous and when alleles contribute additively to the phenotype. The 3/200 ratio of F_2 plants is the key to determining the number of gene pairs. This reduces to a ratio of 1/66.7, very close to 1/64. Using the formula $1/4^n = 1/64$ where 1/64 is equal to the proportion of F_2 phenotypes as extreme as either P_1 parent), $n = 3$. Therefore, three gene pairs are involved.

(b) The variation between the two extreme phenotypes is

$$36 - 6 = 30 \text{ cm}$$

Because there are six potential additive alleles (*AABBCC*), each contributes

$$30/6 = 5 \text{ cm}$$

to the base height of 6 cm, which results when no additive alleles (*aabbcc*) are part of the genotype.

(c) All genotypes that include five additive alleles will be 31 cm (5 alleles $\times$ 5 cm/allele + 6 cm base height = 31 cm). Therefore, *AABBCc, AABbCC,* and *AaBBCC* are the genotypes that will result in plants that are 31 cm.

2. In a cross separate from the above-mentioned F_1 crosses, a plant of unknown phenotype and genotype was testcrossed, with the following results:

1/4 11 cm

2/4 16 cm

1/4 21 cm

An astute genetics student realized that the unknown plant could be only one phenotype but could be any of three genotypes. What were they?

Solution: When testcrossed (with *aabbcc*), the unknown plant must be able to contribute either one, two, or three additive alleles in its gametes in order to yield the three phenotypes in the offspring. Since no 6-cm offspring are observed, the unknown plant never contributes all nonadditive alleles (*abc*). Only plants that are homozygous at one locus and heterozygous at the other two loci will meet these criteria. Therefore, the unknown parent can be any of three genotypes, all of which have a phenotype of 26 cm:

AABbCc

AaBbCC

AaBBCc

For example, in the first genotype (*AABbCc*),

AABbCc $\times$ *aabbcc*

yields

1/4 *AaBbCc* 21 cm

1/4 *AaBbcc* 16 cm

1/4 *AabbCc* 16 cm

1/4 *Aabbcc* 11 cm

which is the ratio of phenotypes observed.

3. The results shown in the following table were recorded for ear length in corn:

	Length of Ear in cm																
	5	6	7	8	9	10	11	12	13	14	15	16	17	18	19	20	21
Parent A	4	21	24	8													
Parent B									3	11	12	15	26	15	10	7	2
F_1					1	12	12	14	17	9	4						

For each of the parental strains and the F_1 generation, calculate the mean values for ear length.

Solution: The mean values can be calculated as follows:

$$\overline{X} = \frac{\Sigma X_i}{n}$$

$$P_A : \overline{X} = \frac{\Sigma X_i}{n} = \frac{378}{57} = 6.63$$

$$P_B : \overline{X} = \frac{\Sigma X_i}{n} = \frac{1697}{101} = 16.80$$

$$F_1 : \overline{X} = \frac{\Sigma X_i}{n} = \frac{836}{69} = 12.11$$

4. For the corn plant described in Exercise 3, compare the mean of the F_1 with that of each parental strain. What does this tell you about the type of gene action involved?

Solution: The F_1 mean (12.11) is almost midway between the parental means of 6.63 and 16.80. This indicates that the genes in question may be additive in effect.

5. The mean and variance of corolla length in two highly inbred strains of *Nicotiana* and their progeny are shown in the following table. One parent (P_1) has a short corolla, and the other parent (P_2) has a long corolla. Calculate the broad-sense heritability (H^2) of corolla length in this plant.

Strain	Mean (mm)	Variance (mm)
P_1 short	40.47	3.12
P_2 long	93.75	3.87
F_1 ($P_1 \times P_2$)	63.90	4.74
F_2 ($F_1 \times F_1$)	68.72	47.70

Solution: The formula for estimating heritability is $H^2 = V_G/V_P$, where V_G and V_P are the genetic and phenotypic components of variation, respectively. The main issue in this problem is obtaining some estimate of two components of phenotypic variation: genetic and environmental factors. V_P is the combination of genetic and environmental variance. Because the two parental strains are true breeding, they are assumed to be homozygous, and the variances of 3.12 and 3.87 are considered to be the result of environmental influences. The average of these two values is 3.50. The F_1 is also genetically homogeneous and gives us an additional estimate of the impact of environmental factors. By averaging this value along with that of the parents,

$$\frac{4.74 + 3.50}{2} = 4.12$$

we obtain a relatively good idea of environmental impact on the phenotype. The phenotypic variance in the F_2 is the sum of the genetic (V_G) and environmental (V_E) components. We have estimated the environmental input as 4.12, so 47.70 minus 4.12 gives us an estimated V_G of 43.58. Heritability then becomes 43.58/47.70, or 0.91. This value, when interpreted as a percentage, indicates that about 91 percent of the variation in corolla length is due to genetic influences.

PROBLEMS AND DISCUSSION QUESTIONS

1. What is the difference between continuous and discontinuous variation? Which of the two is most likely to be the result of polygenic inheritance?
2. Define the following: (a) polygene, (b) additive alleles, (c) correlation, (d) monozygotic and dizygotic twins, (e) heritability, and (f) QTL.
3. A homozygous plant with 20-cm diameter flowers is crossed with a homozygous plant of the same species that has 40-cm diameter flowers. The F_1 plants all have flowers 30 cm in diameter. In the F_2 generation of 512 plants, 2 plants have flowers 20 cm in diameter, 2 plants have flowers 40 cm in diameter, and the remaining 508 plants have flowers of a range of sizes in between. See Now Solve This on page 469.
   (a) Assuming that all alleles involved act additively, how many genes control flower size in this plant?
   (b) What frequency distribution of flower diameter would you expect to see in the progeny of a backcross between an F_1 plant and the large-flowered parent?
   (c) Calculate the mean, variance, and standard deviation for flower diameter in the backcross progeny from (b).
4. A dark-red strain and a white strain of wheat are crossed and produce an intermediate, medium-red F_1. When the F_1 plants are interbred, an F_2 generation is produced in a ratio of 1 dark-red : 4 medium-dark-red : 6 medium-red : 4 light-red : 1 white. Further crosses reveal that the dark-red and white F_2 plants are true breeding.
   (a) Based on the ratios in the F_2 population, how many genes are involved in the production of color?
   (b) How many additive alleles are needed to produce each possible phenotype?
   (c) Assign symbols to these alleles and list possible genotypes that give rise to the medium-red and the light-red phenotypes.
   (d) Predict the outcome of the F_1 and F_2 generations in a cross between a true-breeding medium-red plant and a white plant.
5. Height in humans depends on the additive action of genes. Assume that this trait is controlled by the four loci R, S, T, and U and that environmental effects are negligible. Instead of additive versus nonadditive alleles, assume that additive and partially additive alleles exist. Additive alleles contribute two units, and partially additive alleles contribute one unit to height.
   (a) Can two individuals of moderate height produce offspring that are much taller or shorter than either parent? If so, how?
   (b) If an individual with the minimum height specified by these genes marries an individual of intermediate or moderate height, will any of their children be taller than the tall parent? Why or why not?
6. An inbred strain of plants has a mean height of 24 cm. A second strain of the same species from a different geographical region also has a mean height of 24 cm. When plants from the two strains are crossed together, the F_1 plants are the same height as the parent plants. However, the F_2 generation shows a wide range of heights; the majority are like the P_1 and F_1 plants, but approximately 4 of 1000 are only 12 cm high, and about 4 of 1000 are 36 cm high.
   (a) What mode of inheritance is occurring here?
   (b) How many gene pairs are involved?
   (c) How much does each gene contribute to plant height?
   (d) Indicate one possible set of genotypes for the original P_1 parents and the F_1 plants that could account for these results.
   (e) Indicate three possible genotypes that could account for F_2 plants that are 18 cm high and three that account for F_2 plants that are 33 cm high.
7. Erma and Harvey were a compatible barnyard pair but a curious sight. Harvey's tail was only 6 cm long, while Erma's was 30 cm. Their F_1 piglet offspring all grew tails that were 18 cm. When inbred, an F_2 generation resulted in many piglets (Erma and Harvey's grandpigs), whose tails ranged in 4-cm intervals from 6 to 30 cm (6, 10, 14, 18, 22, 26, and 30). Most had 18-cm tails, while 1/64 had 6-cm tails and 1/64 had 30-cm tails.
   (a) Explain how these tail lengths were inherited by describing the mode of inheritance, indicating how many gene pairs were at work, and designating the genotypes of Harvey, Erma, and their 18-cm-tail offspring.
   (b) If one of the 18-cm F_1 pigs is mated with one of the 6-cm F_2 pigs, what phenotypic ratio would be predicted if many offspring resulted? Diagram the cross.
8. In the following table, average differences of height, weight, and fingerprint ridge count between monozygotic twins (reared together and apart), dizygotic twins, and nontwin siblings are compared:

Trait	MZ Reared Together	MZ Reared Apart	DZ Reared Together	Sibs Reared Together
Height (cm)	1.7	1.8	4.4	4.5
Weight (kg)	1.9	4.5	4.5	4.7
Ridge count	0.7	0.6	2.4	2.7

   Based on the data in this table, which of these quantitative traits has the highest heritability values?
9. What kind of heritability estimates (broad sense or narrow sense) are obtained from human twin studies?
10. List as many human traits as you can that are likely to be under the control of a polygenic mode of inheritance.
11. Corn plants from a test plot are measured, and the distribution of heights at 10-cm intervals is recorded in the following table:

Height (cm)	Plants (no.)
100	20
110	60
120	90
130	130
140	180
150	120
160	70
170	50
180	40

   Calculate (a) the mean height, (b) the variance, (c) the standard deviation, and (d) the standard error of the mean. Plot a rough graph of plant height against frequency. Do the values represent a normal distribution? Based on your calculations, how would you assess the variation within this population?

12. The following variances were calculated for two traits in a herd of hogs.

Trait	V_P	V_G	V_A
Back fat	30.6	12.2	8.44
Body length	52.4	26.4	11.7

(a) Calculate broad-sense (H^2) and narrow-sense (h^2) heritabilities for each trait in this herd.
(b) Which of the two traits will respond best to selection by a breeder? Why? See **Now Solve This** on page 475.

13. The mean and variance of plant height of two highly inbred strains (P_1 and P_2) and their progeny (F_1 and F_2) are shown here:

Strain	Mean (cm)	Variance
P_1	34.2	4.2
P_2	55.3	3.8
F_1	44.2	5.6
F_2	46.3	10.3

Calculate the broad-sense heritability (H^2) of plant height in this species.

14. A hypothetical study investigated the vitamin A content and the cholesterol content of eggs from a large population of chickens. The variances (*V*) were calculated as shown here:

	Trait	
Variance	Vitamin A	Cholesterol
V_P	123.5	862.0
V_E	96.2	484.6
V_A	12.0	192.1
V_D	15.3	185.3

(a) Calculate the narrow-sense heritability (h^2) for both traits.
(b) Which trait, if either, is likely to respond to selection?

15. The following table shows measurements for fiber lengths and fleece weight in a small flock of eight sheep.

Sheep	Fiber Length (cm)	Fleece Weight (kg)
1	9.7	7.9
2	5.6	4.5
3	10.7	8.3
4	6.8	5.4
5	11.0	9.1
6	4.5	4.9
7	7.4	6.0
8	5.9	5.1

(a) What are the mean, variance, and standard deviation for each trait in this flock?
(b) What is the covariance of the two traits?
(c) What is the correlation coefficient for fiber length and fleece weight?
(d) Do you think greater fleece weight is caused by an increase in fiber length? Why or why not? See **Now Solve This** on page 471.

16. In a herd of dairy cows, the narrow-sense heritability for milk protein content is 0.76, and for milk butterfat it is 0.82. The correlation coefficient between milk protein content and butterfat is 0.91. If the farmer selects for cows producing more butterfat in their milk, what will be the most likely effect on milk protein content in the next generation?

17. In an assessment of learning in *Drosophila,* flies were trained to avoid certain olfactory cues. In one population, a mean of 8.5 trials was required. A subgroup of this parental population that was trained most quickly (mean = 6.0) was interbred, and their progeny were examined. These flies demonstrated a mean training value of 7.5. Calculate realized heritability for olfactory learning in *Drosophila.*

18. Suppose you want to develop a population of *Drosophila* that would rapidly learn to avoid certain substances the flies could detect by smell. Based on the heritability estimate you obtained in Problem 17, do you think it would be worth doing this by artificial selection? Why or why not?

19. In a population of tomato plants, mean fruit weight is 60 g and (h^2) is 0.3. Predict the mean weight of the progeny if tomato plants whose fruit averaged 80 g were selected from the original population and interbred.

20. In a population of 100 inbred, genotypically identical rice plants, variance for grain yield is 4.67. What is the heritability for yield? Would you advise a rice breeder to improve yield in this strain of rice plants by selection?

21. A mutant strain of *Drosophila* was isolated and shown to be resistant to an experimental insecticide, whereas normal (wild-type) flies were sensitive to the chemical. Following a cross between resistant flies and sensitive flies, isolated populations were derived that had various combinations of chromosomes from the two strains. Each was tested for resistance, as shown in the following histogram. Analyze the data and draw any appropriate conclusion about which chromosome(s) contain a gene responsible for inheritance of resistance to the insecticide.

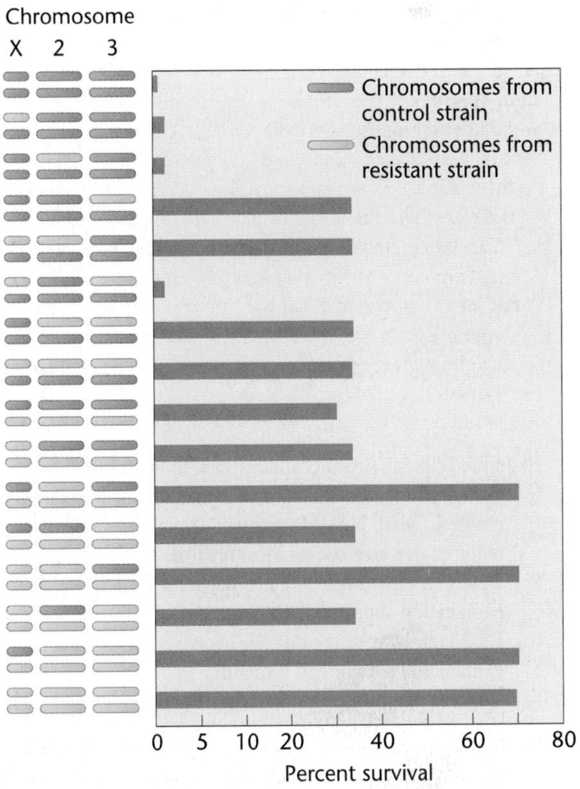

22. In 1988, Horst Wilkens investigated blind cavefish, comparing them with members of a sibling species with normal vision that are found in a lake [Wilkens, H. (1988). *Ecol. Biol.* 23: 271–367]. (We will call them cavefish and lakefish, respectively.) Wilkens found that cavefish eyes are about seven times smaller than lakefish eyes. F_1 hybrids have eyes of intermediate size. These data, as well as the $F_1 \times F_1$ cross and those from backcrosses (F_1 × cavefish and F_1 × lakefish), are depicted as follows:

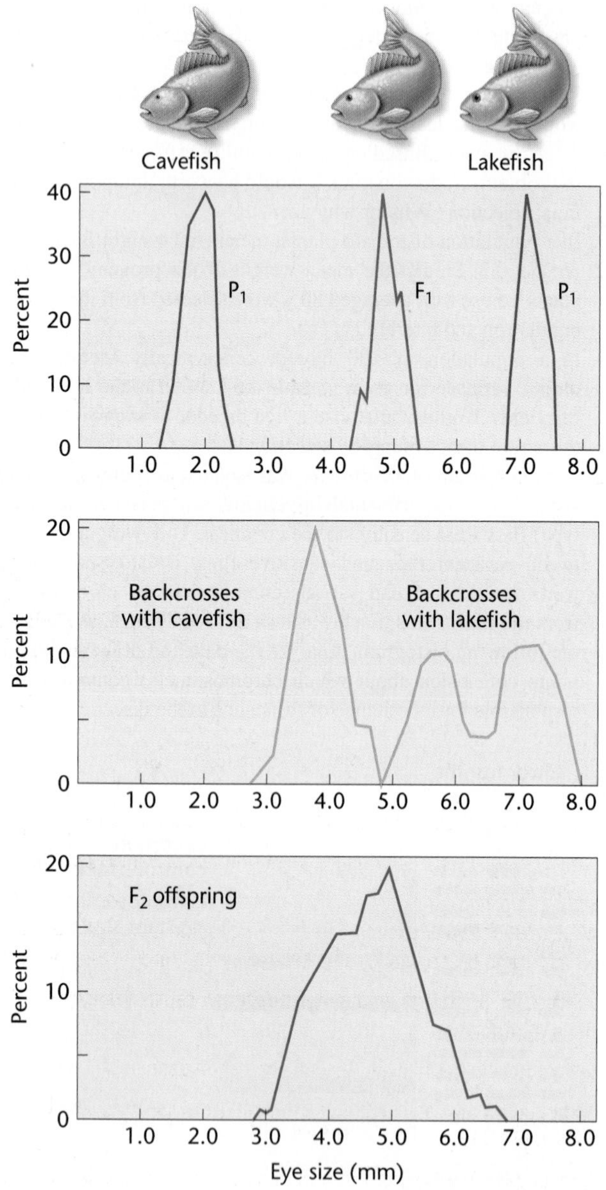

Examine Wilkens's results and respond to the following questions:

(a) Based strictly on the F_1 and F_2 results of Wilkens's initial crosses, what possible explanation concerning the inheritance of eye size seems most feasible?

(b) Based on the results of the F_1 backcross with cavefish, is your explanation supported? Explain.

(c) Based on the results of the F_1 backcross with lakefish, is your explanation supported? Explain.

(d) Wilkens examined about 1000 F_2 progeny and estimated that 6–7 genes are involved in determining eye size. Is the sample size adequate to justify this conclusion? Propose an experimental protocol to test the hypothesis.

(e) A comparison of the embryonic eye in cavefish and lakefish revealed that both reach approximately 4 mm in diameter. However, lakefish eyes continue to grow, while cavefish eye size is greatly reduced. Speculate on the role of the genes involved in this problem.

23. A 3-inch plant was crossed with a 15-inch plant, and all F_1 plants were 9 inches. In the F_2, plants exhibited a "normal distribution," with heights of 3, 4, 5, 6, 7, 8, 9, 10, 11, 12, 13, 14, and 15 inches.

(a) What ratio will constitute the "normal distribution" in the F_2?

(b) What will be the outcome if the F_1 plants are testcrossed with plants that are homozygous for all nonadditive alleles?

24. The following table gives the percentage of twin pairs studied in which both twins expressed the same phenotype for a trait (concordance). Percentages listed are for concordance for each trait in monozygotic (MZ) and dizygotic (DZ) twins. Assuming that both twins in each pair were raised together in the same environment, what do you conclude about the relative importance of genetic versus environmental factors for each trait? See **Now Solve This** on page 475.

Trait	MZ%	DZ%
Blood types	100	66
Eye color	99	28
Mental retardation	97	37
Measles	95	87
Hair color	89	22
Handedness	79	77
Idiopathic epilepsy	72	15
Schizophrenia	69	10
Diabetes	65	18
Identical allergy	59	5
Cleft lip	42	5
Club foot	32	3
Mammary cancer	6	3

25. In a cross between a strain of large guinea pigs and a strain of small guinea pigs, the F_1 are phenotypically uniform, with an average size about intermediate between that of the two parental strains. Among 1014 individuals, 3 are about the same size as the small parental strain and 5 are about the same size as the large parental strain. How many gene pairs are involved in the inheritance of size in these strains of guinea pigs?

26. Type A1B brachydactyly (short middle phalanges) is a genetically determined trait that maps to the short arm of chromosome 5 in humans. If you classify individuals as either having or not having brachydactyly, the trait appears to follow a single-locus, incompletely dominant pattern of inheritance. However, if one examines the fingers and toes of affected individuals, one sees a range of expression from extremely short to only slightly short. What might cause such variation in the expression of brachydactyly?

27. Students in a genetics laboratory began an experiment in an attempt to increase heat tolerance in two strains of *Drosophila melanogaster.* One strain was trapped from the wild six weeks before the experiment was to begin; the other was obtained from a *Drosophila* repository at a university laboratory. In which strain would you expect to see the most rapid and extensive response to heat-tolerance selection, and why?

These lady-bird beetles, from the Chiricahua Mountains in Arizona, show considerable phenotypic variation.

23

Population and Evolutionary Genetics

Chapter Concepts

- Most populations and species harbor considerable genetic variation.
- This variation is reflected in the alleles distributed among populations of a species.
- The relationship between allele frequencies and genotype frequencies in an ideal population is described by the Hardy–Weinberg law.
- Selection, migration, and genetic drift can cause changes in allele frequency.
- Mutation creates new alleles in a population gene pool.
- Nonrandom mating changes population genotype frequency but not allele frequency.
- A reduction in gene flow between populations, accompanied by selection or genetic drift, can lead to reproductive isolation and speciation.
- Genetic differences between populations or species are used to reconstruct evolutionary history.

In the mid-nineteenth century, Alfred Russel Wallace and Charles Darwin identified natural selection as the mechanism of evolution. In his book, *On the Origin of Species,* published in 1859, Darwin provided evidence that populations and species are not fixed, but change, or evolve, over time as a result of natural selection. However, Wallace and Darwin could not explain either the origin of the variations that provide the raw material for evolution or the mechanisms by which such variations are passed from parents to offspring. Gregor Mendel published his work on the inheritance of traits in 1866, but it received little notice at the time. The rediscovery of Mendel's work in 1900 began a 30-year effort to reconcile Mendel's concept of genes and alleles with the theory of evolution by natural selection. As twentieth-century biologists applied the principles of Mendelian genetics to populations, both the source of variation (mutation) and the mechanism of inheritance (segregation of alleles) were explained. We now view evolution as a consequence of changes in genetic material through mutation and changes in allele frequencies in populations over time. This union of population genetics with the theory of natural selection generated a new view of the evolutionary process, called *neo-Darwinism.*

In addition to natural selection, other forces including mutation, migration, and drift, individually and collectively, alter allele frequencies and bring about evolutionary divergence that eventually may result in **speciation,** the formation of new species. Speciation is facilitated by environmental diversity. If a population is spread over a geographic range encompassing a number of ecologically distinct subenvironments with different selection pressures, the populations occupying these areas may gradually adapt and become genetically distinct from one another. Genetically differentiated populations may remain in existence, become extinct, reunite with each other, or continue to diverge until they become reproductively isolated. Populations that are reproductively isolated are regarded as separate species. Genetic changes within populations can modify a species over time, transform it into another species, or cause it to split into two or more species.

Population geneticists investigate patterns of genetic variation within and among groups of interbreeding individuals. As changes in genetic structure form the basis for evolution of a population, population genetics has become an important subdiscipline of evolutionary biology. In this chapter, we examine the population genetics processes of **microevolution**—defined as evolutionary change within populations of a species—and then consider how molecular aspects of these processes can be extended to **macroevolution**—defined as evolutionary events leading to the emergence of new species and other taxonomic groups.

How Do We Know?

Population geneticists study changes in the nature and amount of genetic variation in populations, the distribution of different genotypes, and how forces such as selection and drift act on genetic variation to bring about evolutionary change in populations and the formation of new species. As you read this chapter think about these questions:

1. How do we know how much genetic variation is in a population?
2. How do geneticists detect the presence of genetic variation as different alleles in a population?
3. How do we know whether the genetic structure of a population is static or dynamic?
4. How do we know when populations have diverged to the point that they form two different species?
5. How do we know the age of the last common ancestor shared by two species?

23.1 Genetic Variation Is Present in Most Populations and Species

A **population** is a group of individuals belonging to the same species that live in a defined geographic area and actually or potentially interbreed. In thinking about the human population, we can define it as everyone who lives in the United States, or in Sri Lanka, or we can specify a population as all the residents of a particular small town or village.

The genetic information carried by members of a population constitutes that population's **gene pool.** At first glance, it might seem that a population that is well adapted to its environment must be highly homozygous because you assume that the most favorable allele at each locus is present at a high frequency. In addition, a look at most populations of plants and animals reveals many phenotypic similarities among individuals. However, a large body of evidence indicates that, in reality, most populations contain a high degree of heterozygosity. This built-in genetic diversity is often concealed, so to speak, because it is not necessarily apparent phenotypically; hence, detecting it is not a simple task. Nevertheless, the diversity within a population can be revealed by several methods.

Detecting Genetic Variation by Artificial Selection

One way to determine whether genetic variation affects a phenotypic character is to use artificial selection. A phenotype that is not affected by genetic variation will not respond to selection; if genetic variation does have an effect, the phenotype will change over a few generations. A dramatic example of this test is the domestic dog. The broad array of sizes, shapes, colors, and behaviors seen in different breeds of dogs all arose from the effects of selection on the genetic variation present in wild wolves, from which all domestic dogs are descended. Genetic and archeological evidence indicates that the domestication of dogs took place at least 15,000 years ago and possibly much earlier. On a shorter time scale, laboratory selection experiments on the fruit fly *Drosophila melanogaster* have caused significant changes over a few generations in almost every phenotype imaginable, including size, shape, developmental rate, fecundity, and behavior.

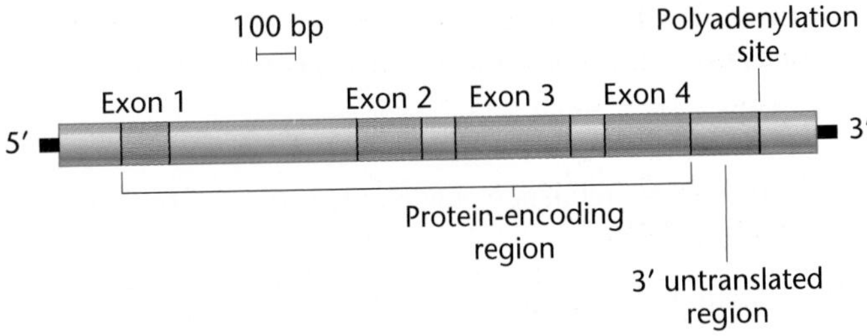

FIGURE 23–1 Organization of the *Adh* locus of *Drosophila melanogaster.*

	Exon 3	Intron 3	Exon 4
Consensus *Adh* sequence:	C C C C	G G A A T	C T C C A* C T A G
Strain			
Wa-S	T T • A	C A • T A	A C • • • • • • • •
Fl1-S	T T • A	C A • T A	A C • • • • • • • •
Ja-S	• • • •	• • • • •	• • • T • T • C A
Fl-F	• • • •	• • • • •	• • G T C T C C •
Ja-F	• • A •	• • G • •	• • G T C T C C •

FIGURE 23–2 DNA sequence variation in parts of the *Drosophila Adh* gene in a sample of the eleven laboratory strains derived from the five natural populations. The dots represent nucleotides that are the same as the consensus sequence; letters represent nucleotide polymorphisms. An A/C polymorphism (A*) in codon 192 creates the two *Adh* alleles (F and S). All other polymorphisms are silent or non-coding.

Variations in Nucleotide Sequence

The most direct way to estimate genetic variation is to compare the nucleotide sequences of genes carried by individuals in a population. In one study, Martin Kreitman examined the *alcohol dehydrogenase* locus (*Adh*) in *Drosophila melanogaster* (Figure 23–1). This locus has two alleles, the *Adh-f* and the *Adh-s* alleles. The encoded proteins differ by only a single amino acid (thr versus lys at codon 192). To determine whether the amount of genetic variation detectable at the protein level (one amino acid difference) corresponds to the variation at the nucleotide level, Kreitman cloned and sequenced *Adh* genes from five natural populations of *Drosophila* (Figure 23–2).

The 11 cloned genes isolated from these five populations contained a total of 43 nucleotide variations in the *Adh* sequence of 2721 base pairs. These variations are distributed throughout the gene: 14 in exon-coding regions, 18 in introns, and 11 in the untranslated flanking regions. Of the 14 variations in coding regions, only one leads to an amino acid replacement—the one in codon 192, producing the two alleles. The other 13 coding-region nucleotide substitutions do not lead to amino acid replacements and are silent variations in this gene.

Among the most intensively studied human genes is the locus encoding the cystic fibrosis transmembrane conductance regulator (CFTR). Recessive loss-of-function mutations in the *CFTR* locus cause **cystic fibrosis,** a disease that affects secretory glands and lungs, leading to susceptibility to bacterial infections. More than 1500 different mutations in the *CFTR* gene have been identified. Among these are missense mutations, amino acid deletions, nonsense mutations, frameshifts, and splice defects.

Figure 23–3 shows a map of the 27 exons in the *CFTR* locus, with most exons identified by function. The histogram above the map shows the locations of some of the disease-causing mutations and the number of copies of each that have been found. One mutation, a 3-bp deletion in exon 10 called *ΔF508* accounts for 67 percent of all mutant cystic fibrosis alleles, but several other mutations were found in at least 100 of the chromosomes surveyed. In populations of European ancestry, between 1 in 44 and 1 in 20 individuals are heterozygous carriers of mutant alleles. Note that Figure 23–3 includes only the sequence variants that alter the function of the CFTR protein. There are undoubtedly many more *CFTR* alleles with silent sequence variants that do not change the amino acid sequence of the protein and do not affect its function.

Studies of other organisms, including the rat, the mouse, and the mustard plant *Arabidopsis thaliana,* have produced similar estimates of nucleotide diversity in various genes. These studies indicate that there is an enormous reservoir of genetic variability within most populations and that, at the

FIGURE 23–3 The locations of disease-causing mutations in the cystic fibrosis gene. The histogram shows the number of copies of each mutation geneticists have found. (The vertical axis is on a logarithmic scale.) The genetic map below the histogram shows the locations and relative sizes of the 27 exons of the *CFTR* locus. The boxes at the bottom indicate the functions of different domains of the CFTR protein.

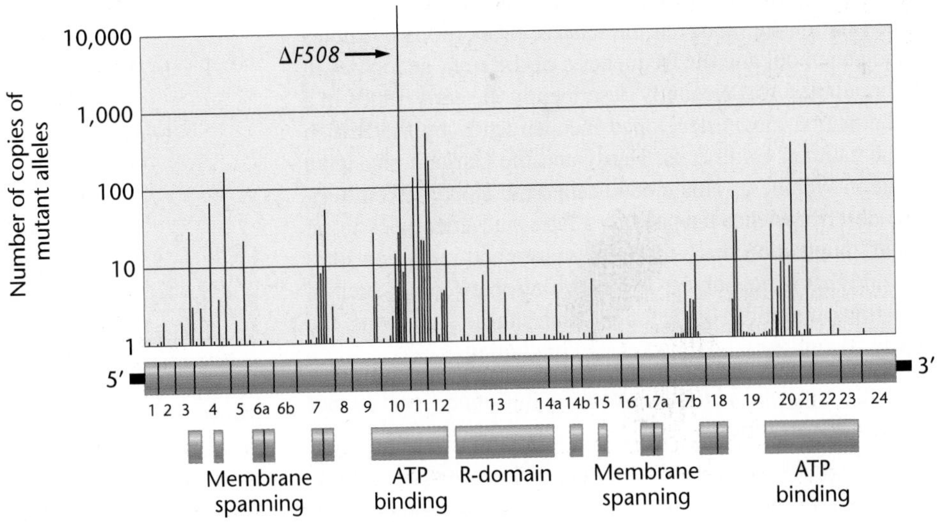

DNA level most, and perhaps all, genes exhibit diversity from individual to individual. Alleles representing these variations are distributed among members of a population.

> **ESSENTIAL POINT**
>
> Genetic variation is widespread in most populations and provides a reservoir of alleles that serve as the basis for evolutionary changes within the population.

23.2 The Hardy–Weinberg Law Describes Allele Frequencies and Genotype Frequencies in Populations

Populations are dynamic; they expand and contract through changes in birth and death rates, migration, or contact with other populations. Often some individuals within a population will produce more offspring than others, contributing a disproportionate fraction of their alleles to the next generation. Thus, differential reproduction in a population can, over time, lead to changes in the allele and genotype frequencies in subsequent generations. Changes in allele frequencies in a population that do not result in reproductive isolation are examples of microevolution. In the following sections, we will discuss microevolutionary changes in population gene pools, and in later sections, we will consider macroevolution and the process of speciation.

Often when we examine a single genetic locus in a population, we find that distribution of the alleles at this locus produces individuals with different genotypes. Key elements of population genetics are the calculation of allele frequencies and genotype frequencies in the population, and the determination of how these frequencies change from one generation to the next. Population geneticists use these calculations to answer questions such as: How much genetic variation is present in a population? Are genotypes randomly distributed in time and space, or do discernible patterns exist? What processes affect the composition of a population's gene pool? Do these processes produce genetic divergence among populations that may lead to the formation of new species?

The relationship between the relative proportions of alleles in the gene pool and the frequencies of different genotypes in the population was elegantly described in the early 1900s in a mathematical model developed independently by the British mathematician Godfrey H. Hardy and the German physician Wilhelm Weinberg. This model, called the **Hardy–Weinberg law,** describes what happens to alleles and genotypes in an "ideal" population that is infinitely large and randomly mating, and that is not subject to any evolutionary forces such as mutation, migration, or selection. Under these conditions, the Hardy–Weinberg model makes two predictions:

1. The frequencies of the alleles in the gene pool do not change over time.
2. If two alleles at a locus, *A* and *a*, are considered, then as we will show later in this chapter, after one generation of random mating, the frequencies of genotypes *AA*:*Aa*:*aa* in the population can be calculated as

$$p^2 + 2pq + q^2 = 1$$

where p = frequency of allele *A* and q = frequency of allele *a*.

A population that meets these criteria, and in which the frequencies p and q of two alleles at a locus do result in the predicted genotypic frequencies, is said to be in Hardy–Weinberg equilibrium. As we will see later in this chapter, it is rare for a real population to conform totally to the Hardy–Weinberg model and for all allele and genotype frequencies to remain unchanged for generation after generation.

The Hardy–Weinberg model uses Mendelian principles of segregation and simple probability to explain the relationship between allele and genotype frequencies in a population. We can demonstrate how this works by considering the example of a single autosomal locus with two alleles, *A* and *a*, in a population where the frequency of *A* is 0.7 and the frequency of *a* is 0.3. Note that $0.7 + 0.3 = 1$, indicating that all the alleles for gene *A* present in the gene pool are accounted for. We assume that individuals mate randomly, following Hardy–Weinberg requirements, so for any one zygote, the probability that the female gamete will contain *A* is 0.7, and the probability that the male gamete will contain *A* is also 0.7. The probability that *both* gametes will contain *A* is $0.7 \times 0.7 = 0.49$. Thus we predict that genotype *AA* will occur 49 percent of the time. The probability that a zygote will be formed from a female gamete carrying *A* and a male gamete carrying *a* is $0.7 \times 0.3 = 0.21$, and the probability of a female gamete carrying *a* being fertilized by a male gamete carrying *A* is $0.3 \times 0.7 = 0.21$, so the frequency of genotype *Aa* is $0.21 + 0.21 = 0.42 = 42$ percent. Finally, the probability that a zygote will be formed from two gametes carrying *a* is $0.3 \times 0.3 = 0.09$, so the frequency of genotype *aa* is 9 percent. As a check on our calculations, note that $0.49 + 0.42 + 0.09 = 1.0$, confirming that we have accounted for all of the zygotes. These calculations are summarized in Figure 23–4.

Eggs \ Sperm	fr(*A*) = 0.7	fr(*a*) = 0.3
fr(*A*) = 0.7	fr(*AA*) = 0.7 × 0.7 = 0.49	fr(*Aa*) = 0.7 × 0.3 = 0.21
fr(*a*) = 0.3	fr(*aA*) = 0.3 × 0.7 = 0.21	fr(*aa*) = 0.3 × 0.3 = 0.09

FIGURE 23–4 Calculating genotype frequencies from allele frequencies. Gametes represent samples drawn from the gene pool to form the genotypes of the next generation. In this population, the frequency of the *A* allele is 0.7, and the frequency of the *a* allele is 0.3. The frequencies of the genotypes in the next generation are calculated as 0.49 for *AA*, 0.42 for *Aa*, and 0.09 for *aa*. Under the Hardy–Weinberg law, the frequencies of *A* and *a* remain constant from generation to generation.

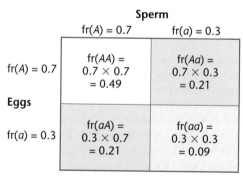

We started with the frequency of a particular allele in a specific gene pool, and we calculated the probability that certain genotypes would be produced from this pool. When the zygotes develop into adults and reproduce, what will be the frequency distribution of alleles in the new gene pool? Under the Hardy–Weinberg law, we assume that all genotypes have equal rates of survival and reproduction. This means that in the next generation, all genotypes contribute equally to the new gene pool. The *AA* individuals constitute 49 percent of the population, and we can predict that the gametes they produce will constitute 49 percent of the gene pool. These gametes all carry allele *A*. Similarly, *Aa* individuals constitute 42 percent of the population, so we predict that their gametes will constitute 42 percent of the new gene pool. Half (0.5) of these gametes will carry allele *A*. Thus, the frequency of allele *A* in the gene pool is $0.49 + (0.5)0.42 = 0.7$. The other half of the gametes produced by *Aa* individuals will carry allele *a*. The *aa* individuals constitute 9 percent of the population, so their gametes will constitute 9 percent of the new gene pool. All these gametes carry allele *a*. Thus, we can predict that the allele *a* in the new gene pool is $(0.5)0.42 + 0.09 = 0.3$. As a check on our calculation, note that $0.7 + 0.3 = 1.0$, accounting for all of the gametes in the gene pool of the new generation.

We have arrived where we began: with a gene pool where the frequency of allele *A* is 0.7 and the frequency of allele *a* is 0.3. For the general case of the Hardy–Weinberg model, we use variables instead of numerical values for the allele frequencies. Imagine a gene pool in which the frequency of allele *A* is p and the frequency of allele *a* is q, such that $p + q = 1$. If we randomly draw male and female gametes from the gene pool and pair them to make a zygote, the probability that both will carry allele *A* is $p \times p$. Thus, the frequency of genotype *AA* among the zygotes is p^2. The probability that the female gamete carries *A* and the male gamete carries *a* is $p \times q$, and the probability that the female gamete carries *a* and the male gamete carries *A* is $q \times p$. Thus, the frequency of genotype *Aa* among the zygotes is $2pq$. Finally, the probability that both gametes carry *a* is $q \times q$, making the frequency of genotype *aa* among the zygotes q^2. Therefore, the distribution of genotypes among the zygotes is

$$p^2 + 2pq + q^2 = 1$$

This is summarized in Figure 23–5.

These calculations demonstrate the two main predictions of the Hardy–Weinberg model. Allele frequencies in our population do not change from one generation to the next, and genotype frequencies after one generation of random mating can be predicted from the allele frequencies. In other words, this population does not change or evolve with respect to the locus we have examined. Remember, however, the assumptions about the theoretical population described by the Hardy–Weinberg model:

1. Individuals of all genotypes have equal rates of survival and equal reproductive success—that is, there is no selection.
2. No new alleles are created or converted from one allele into another by mutation.

		Sperm	
		fr(A) = p	fr(a) = q
Eggs	fr(A) = p	fr(AA) = p^2	fr(Aa) = pq
	fr(a) = q	fr(aA) = qp	fr(aa) = q^2

FIGURE 23–5 The general description of allele and genotype frequencies under Hardy–Weinberg assumptions. The frequency of allele *A* is p, and the frequency of allele *a* is q. After mating, the three genotypes *AA*, *Aa*, and *aa* have the frequencies p^2, $2pq$, and q^2, respectively.

3. Individuals do not migrate into or out of the population.
4. The population is infinitely large, which in practical terms means that the population is large enough that sampling errors and other random effects are negligible.
5. Individuals in the population mate randomly.

These assumptions are what make the Hardy–Weinberg model so useful in population genetics research. By specifying the conditions under which the population cannot evolve, the Hardy–Weinberg model identifies the real-world forces that cause allele frequencies to change. In other words, by holding certain conditions constant, Hardy–Weinberg isolates the forces of evolution and allows them to be quantified. Application of this model can also reveal "neutral genes" in a population gene pool—those not being operated on by the forces of evolution.

The Hardy–Weinberg model has three additional important consequences. First, it shows that dominant traits do not necessarily increase from one generation to the next. Second, it demonstrates that **genetic variability** can be maintained in a population since, once established in an ideal population, allele frequencies remain unchanged. Third, if we invoke Hardy–Weinberg assumptions, then knowing the frequency of just one genotype enables us to calculate the frequencies of all other genotypes at that locus. This is particularly useful in human genetics because we can calculate the frequency of heterozygous carriers for recessive genetic disorders even when all we know is the frequency of affected individuals.

NOW SOLVE THIS

Problem 5 on page 505 asks you to calculate the frequencies of a dominant and a recessive allele in a population where you know the phenotype frequencies.

Hint: Determine which allele frequency (p or q) you must estimate first when homozygous dominant and heterozygous genotypes have the same phenotype.

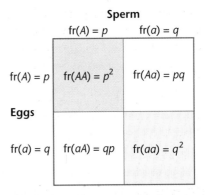

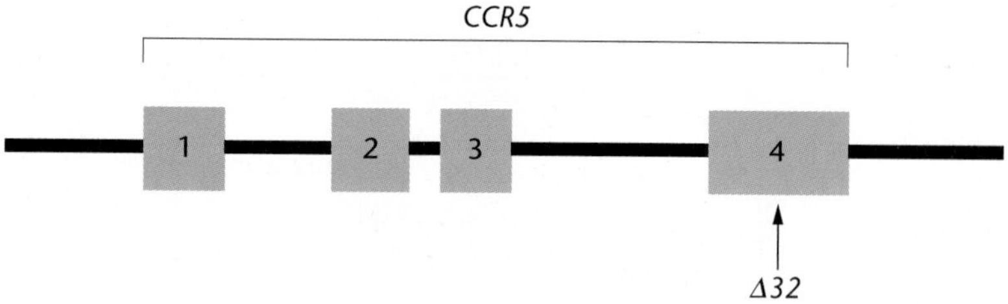

FIGURE 23–6 Organization of the *CCR5* gene in region 3p21.3. The gene contains 4 exons and 2 introns. The arrow shows the location of the 32-bp deletion in exon 4 that confers resistance to HIV-1 infection.

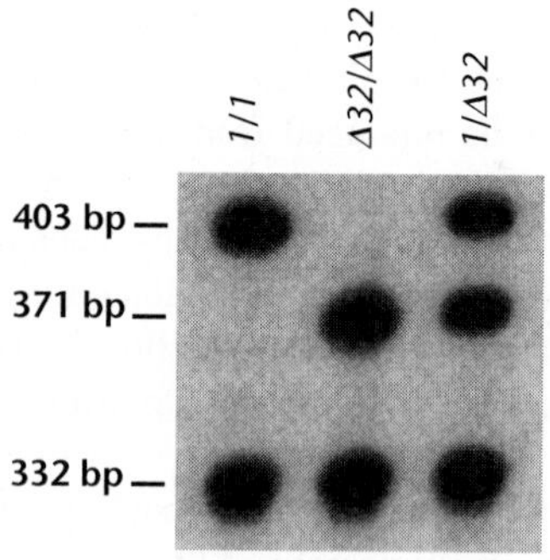

FIGURE 23–7 Allelic variation in the *CCR5* gene. Michel Samson and colleagues used PCR to amplify a part of the *CCR5* gene containing the site of the 32-bp deletion, cut the resulting DNA fragments with a restriction enzyme, and ran the fragments on an electrophoresis gel. Each lane reveals the genotype of a single individual. The *1* allele produces a 332-bp fragment and a 403-bp fragment; the Δ*32* allele produces a 332-bp fragment and a 371-bp fragment. Heterozygotes produce three bands.

23.3 The Hardy–Weinberg Law Can Be Applied to Human Populations

To show how allele frequencies are measured in a real population, let's consider genetic factors that influence an individual's susceptibility to infection by HIV-1, the virus responsible for AIDS (acquired immunodeficiency syndrome). A small number of individuals who make high-risk choices (such as unprotected sex with HIV-positive partners) remain uninfected, some of whom are homozygous for a mutant allele of a gene called *CCR5*.

The *CCR5* gene (Figure 23–6) encodes a protein called the C–C chemokine receptor-5, often abbreviated CCR5. Chemokines are signaling molecules associated with the immune system. The CCR5 protein is also a receptor for strains of HIV-1, allowing it to gain entry to cells. The mutant allele of the *CCR5* gene contains a 32-bp deletion in one of its coding regions, making the encoded protein shorter and nonfunctional. In individuals homozygous for this mutation, HIV-1 cannot enter their cells. The normal allele is called *CCR51* (also called *1*), and the mutant allele is called *CCR5-Δ32* (also called Δ*32*).

Δ*32*/Δ*32* individuals are resistant to HIV-1 infection. Heterozygotes (*1*/Δ*32*) are susceptible to HIV-1 infection but progress more slowly to AIDS. Table 23.1 summarizes the genotypes possible at the *CCR5* locus and the phenotypes associated with each.

The discovery of the *CCR5-Δ32* allele generates two important questions: Which human populations harbor the Δ*32* allele, and how common is it? To address these questions, several teams of researchers surveyed people from a variety of populations. Genotypes were determined by direct analysis of DNA (Figure 23–7). In one population, 79 individuals had genotype *1/1*, 20 were (*1*/Δ*32*), and 1 was Δ*32*/Δ*32*. This population has 158 *1* alleles carried by the *1/1* individuals plus 20 *1* alleles carried by *1*/Δ*32* individuals, for a total of 178. The frequency of the *CCR51* allele in the sample population is thus $178/200 = 0.89 = 89$ percent. Copies of the *CCR5-Δ32* allele were carried by 20 *1*/Δ*32* individuals, plus 2 carried by the Δ*32*/Δ*32* individual, for a total of 22. The frequency of the *CCR5-Δ32* allele is thus $22/200 = 0.11 = 11$ percent. Notice that $p + q = 1$, confirming that we have accounted for the entire gene pool. Table 23.2 shows two methods for computing the frequencies of the *1* and Δ*32* alleles in the population surveyed.

Can we expect the *CCR5-Δ32* allele to increase in human populations in which it is currently rare? From what we now know about the Hardy–Weinberg law, if the population meets the five assumptions, then the frequency of the Δ*32* allele in human populations will not change. However, as we shall see in the last half of the chapter, when the assumptions of the Hardy–Weinberg law are not met—because of natural selection, mutation, migration, or genetic drift—the allele frequencies in a population may change from one generation to the next. The Hardy–Weinberg law tells geneticists where to look to find the causes of evolution in populations.

Testing for Hardy–Weinberg Equilibrium

One way to establish whether one (or more) of the Hardy–Weinberg assumptions does not hold in a given population is to determine whether the population's genotypes are in equilibrium. To do this, we first determine the frequencies of the genotypes, either directly from the phenotypes (if heterozygotes are recognizable) or by analyzing proteins or DNA sequences. We then calculate the allele frequencies from the genotype frequencies, as demonstrated earlier. Finally, we use the allele frequencies in the parental generation to predict the offspring's genotype frequencies. According to the Hardy–Weinberg law, the genotype frequencies are predicted to fit the $p^2 + 2pq + q^2 = 1$ relationship. If they do not, then one or more of the assumptions are invalid for the population in question.

TABLE 23.1 *CCR5* Genotypes and Phenotypes

Genotype	Phenotype
1/1	Susceptible to sexually transmitted strains of HIV-1
1/Δ*32*	Susceptible but may progress to AIDS slowly
Δ*32*/Δ*32*	Resistant to most sexually transmitted strains of HIV-1

TABLE 23.2 Methods of Determining Allele Frequencies from Data on Genotypes

A. Counting Alleles

Genotype	*1/1*	*1/Δ32*	*Δ32/32*	**Total**
Number of individuals	79	20	1	100
Number of *1* alleles	158	20	0	178
Number of *Δ32* alleles	0	20	2	22
Total number of alleles	158	40	2	200

Frequency of *CCR51* in sample: 178/200 = 0.89 = 89%
Frequency of *CCR5-Δ32* in sample: 22/200 = 0.11 = 11%

B. From Genotype Frequencies

Genotype	*1/1*	*1/Δ32*	*Δ32/32*	**Total**
Number of individuals	79	20	1	100
Genotype frequency	79/100 = 0.79	20/100 = 0.20	1/100 = 0.01	1.00

Frequency of *CCR51* in sample: 0.79 + (0.5)0.20 = 0.89 = 89%
Frequency of *CCR5-Δ32* in sample: (0.5)0.20 + 0.01 = 0.11 = 11%

To demonstrate this, let's start with a population that includes 283 individuals, of which 223 have genotype *1/1*; 57 have genotype *1/Δ32*; and 3 have genotype *Δ32/Δ32*. These numbers represent genotype frequencies of 223/283 = 0.788, 57/283 = 0.201, and 3/283 = 0.011, respectively. From the genotype frequencies, we compute the *CCR51* allele frequency as 0.89 and the frequency of the *CCR5-Δ32* allele as 0.11. From these allele frequencies, we can use the Hardy–Weinberg law to determine whether this population is in equilibrium. The allele frequencies predict the genotype frequencies as follows:

Expected frequency of genotype

$$1/1 = p^2 = (0.89)^2 = 0.792$$

Expected frequency of genotype

$$1/\Delta 32 = 2pq = 2(0.89)\,(0.11) = 0.196$$

Expected frequency of genotype

$$\Delta 32/\Delta 32 = q^2 = (0.11)^2 = 0.012$$

These expected frequencies are nearly identical to the observed frequencies. Our test of this population has failed to provide evidence that Hardy–Weinberg assumptions are being violated. The conclusion is confirmed by a χ^2 analysis (see Chapter 3). The χ^2 value in this case is tiny: 0.00023. To reject the null hypothesis at even the most generous, accepted level, $p = 0.05$, the χ^2 value would have to be 3.84. (In a test for Hardy–Weinberg equilibrium, the degrees of freedom are given by $k - 1 - m$, where k is the number of genotypes and m is the number of independent allele frequencies estimated from the data. Here, $k = 3$ and $m = 1$ since calculating only one allele frequency allows us to determine the other by subtraction. Thus, we have $3 - 1 - 1 = 1$ degree of freedom.

ESSENTIAL POINT

Populations that are not in Hardy–Weinberg equilibrium may be undergoing changes in allele frequency owing to selection, drift, migration, or nonrandom mating.

NOW SOLVE THIS

Problem 10 on page 505 asks you to determine whether two populations are in Hardy–Weinberg equilibrium, based on their genotype frequencies.

Hint: Start by determining the allele frequencies based on these data.

Calculating Frequencies for Multiple Alleles in Hardy–Weinberg Populations

We commonly find several alleles of a single gene in a population. The ABO blood group in humans (discussed in Chapter 4) is such an example. The locus *I* (isoagglutinin) has three alleles I^A, I^B, and I^O, yielding six possible genotypic combinations ($I^AI^A, I^BI^B, I^OI^O, I^AI^B, I^AI^O, I^BI^O$). Remember that in this case I^A and I^B are codominant alleles and that both of these are dominant to I^O. The result is that homozygous I^AI^A and heterozygous I^AI^O individuals are phenotypically identical, as are I^BI^B and I^BI^O individuals, so we can distinguish only four phenotypic combinations.

By adding another variable to the Hardy–Weinberg equation, we can calculate both the genotype and allele frequencies for the situation involving three alleles. Let p, q, and r represent the frequencies of alleles I^A, I^B, and I^O, respectively. Note that because there are three alleles

$$p + q + r = 1$$

Under Hardy–Weinberg assumptions, the frequencies of the genotypes are given by

$$(p + q + r)^2 = p^2 + q^2 + r^2 + 2pq + 2pr + 2qr = 1$$

If we know the frequencies of blood types for a population, we can then estimate the frequencies for the three alleles of the ABO system. For example, in one population sampled, the following blood-type frequencies are observed: A = 0.53, B = 0.13, O = 0.26. Because the I^O allele is recessive, the population's frequency of type O blood equals the proportion of the recessive genotype r^2. Thus,

$$r^2 = 0.26$$
$$r = \sqrt{0.26}$$
$$r = 0.51$$

Using *r*, we can calculate the allele frequencies for the I^A and I^B alleles. The I^A allele is present in two genotypes, $I^A I^A$ and $I^A I^O$ alleles. The frequency of the $I^A I^A$ genotype is represented by p^2 and the $I^A I^O$ genotype by $2pr$. Therefore, the combined frequency of type A blood and type O blood is given by

$$p^2 + 2pr + r^2 = 0.53 + 0.26$$

If we factor the left side of the equation and take the sum of the terms on the right, we get

$$(p + r)^2 = 0.79$$
$$p + r = \sqrt{0.79}$$
$$p = 0.89 - r$$
$$p = 0.89 - 0.51 = 0.38$$

Having calculated *p* and *r*, the frequencies of allele I^A and allele I^O, we can now calculate the frequency for the I^B allele:

$$p + q + r = 1$$
$$q = 1 - p - r$$
$$= 1 - 0.38 - 0.51$$
$$= 0.11$$

The phenotypic and genotypic frequencies for this population are summarized in Table 23.3.

Calculating Heterozygote Frequency

In another application, the Hardy–Weinberg law allows us to estimate the frequency of heterozygotes in a population. The frequency of a recessive trait can usually be determined by counting such individuals in a sample of the population. With this information and the Hardy–Weinberg law, we can then calculate the allele and genotype frequencies.

Cystic fibrosis, an autosomal recessive trait, has an incidence of about $1/2500 = 0.0004$ in people of northern European ancestry. Individuals with cystic fibrosis are easily distinguished from the population at large by such symptoms as extra-salty sweat, excess amounts of thick mucus in the lungs, and susceptibility to bacterial infections. Because this is a recessive trait, individuals with cystic fibrosis must be homozygous. Their frequency in a population is represented by q^2, provided that mating has been random in the previous generation. The frequency of the recessive allele is therefore

$$q = \sqrt{q^2} = \sqrt{0.0004} = 0.02$$

Since $p + q = 1$, then the frequency of *p* is

$$p = 1 - q = 1 - 0.02 = 0.98$$

In the Hardy–Weinberg equation, the frequency of heterozygotes is $2pq$. Thus,

$$2pq = 2(0.98)(0.02)$$
$$= 0.04 \text{ or } 4 \text{ percent, or } 1/25$$

Thus, heterozygotes for cystic fibrosis are rather common in the population (about 1/25, or 4 percent), even though the incidence of homozygous recessives is only 1/2500, or 0.04 percent. Calculations such as these are estimates because the population may not meet all Hardy–Weinberg assumptions.

NOW SOLVE THIS

Problem 16 on page 505 asks you to calculate the frequency of heterozygous carriers of the recessive trait albinism in a human population.

Hint: First determine the frequency of the albinism allele in this population.

TABLE 23.3 Calculating Genotype Frequencies for Multiple Alleles in a Hardy–Weinberg Population Where the Frequency of Allele $I^A = 0.38$, Allele $I^B = 0.11$, and Allele $I^O = 0.51$

Genotype	Genotype Frequency	Phenotype	Phenotype Frequency
I^AI^A	$p^2 = (0.38)^2 = 0.14$	A	0.53
I^AI^O	$2pr = 2(0.38)(0.51) = 0.39$		
I^BI^B	$q^2 = (0.11)^2 = 0.01$	B	0.12
I^BI^O	$2qr = 2(0.11)(0.51) = 0.11$		
I^AI^B	$2pq = 2(0.38)(0.11) = 0.084$	AB	0.08
I^OI^O	$r^2 = (0.51)^2 = 0.26$	O	0.26

23.4 Natural Selection Is a Major Force Driving Allele Frequency Change

To understand evolution, we must understand the forces that transform the gene pools of populations and can lead to the formation of new species. Chief among the mechanisms transforming populations is **natural selection,** discovered independently by Darwin and by Alfred Russel Wallace. The Wallace–Darwin concept of natural selection can be summarized as follows:

1. Individuals of a species exhibit variations in phenotype—for example, differences in size, agility, coloration, defenses against enemies, ability to obtain food, courtship behaviors, and flowering times.
2. Many of these variations, even small and seemingly insignificant ones, are heritable and passed on to offspring.
3. Organisms tend to reproduce in an exponential fashion. More offspring are produced than can survive. This causes members of a species to engage in a struggle for survival, competing with other members of the community for scarce resources. Offspring also must avoid predators, and in sexually reproducing species, adults must compete for mates.
4. In the struggle for survival, individuals with particular phenotypes will be more successful than others, allowing the former to survive and reproduce at higher rates.

As a consequence of natural selection, populations and species change. The phenotypes that confer improved ability to survive and reproduce become more common, and the phenotypes that confer poor prospects for survival and reproduction may eventually disappear. Under certain conditions, populations that at one time could interbreed may lose that capability, thus segregating their adaptations into particular niches. If selection continues, it may result in the appearance of new species.

Detecting Natural Selection in Populations

Recall that measuring allele frequencies and genotype frequencies using the Hardy–Weinberg law is based on certain assumptions about an ideal population: large population size, lack of migration, presence of random mating, absence of selection and mutation, and equal survival rates of offspring.

However, if all genotypes do not have equal rates of survival or do not leave equal numbers of offspring, then allele frequencies may change from one generation to the next. To see why, let's imagine a population of 100 individuals in which the frequency of allele *A* is 0.5 and that of allele *a* is 0.5. Assuming the previous generation mated randomly, we find that the genotype frequencies in the present generation are $(0.5)^2 = 0.25$ for *AA*, $2(0.5)(0.5) = 0.5$ for *Aa*, and $(0.5)^2 = 0.25$ for *aa*. Because our population contains 100 individuals, we have 25 *AA* individuals, 50 *Aa* individuals, and 25 *aa* individuals. Now suppose that individuals with different genotypes have different rates of survival: All 25 *AA* individuals survive to reproduce, 90 percent or 45 of the *Aa* individuals survive to reproduce, and 80 percent or 20 of the *aa* individuals survive to reproduce. When the survivors reproduce, each contributes two gametes to the new gene pool, giving us $2(25) + 2(45) + 2(20) = 180$ gametes. What are the frequencies of the two alleles in the surviving population? We have 50 *A* gametes from *AA* individuals, plus 45 *A* gametes from *Aa* individuals, so the frequency of allele *A* is $(50 + 45)/180 = 0.53$. We have 45 *a* gametes from *Aa* individuals, plus 40 *a* gametes from *aa* individuals, so the frequency of allele *a* is $(45 + 40)/180 = 0.47$.

These differ from the frequencies we started with. The frequency of allele *A* has increased, whereas the frequency of allele *a* has declined. A difference among individuals in survival or reproduction rate (or both) is an example of **natural selection.** Natural selection is the principal force that shifts allele frequencies within large populations and is one of the most important factors in evolutionary change.

Fitness and Selection

Selection occurs whenever individuals with a particular genotype enjoy an advantage in survival or reproduction over other genotypes. However, selection may vary from less than 1 to 100 percent. In the previous hypothetical example, selection was strong. Weak selection might involve just a fraction of a percent difference in the survival rates of different genotypes. Advantages in survival and reproduction ultimately translate into increased genetic contribution to future generations. An individual organism's genetic contribution to future generations is called its **fitness.** Genotypes associated with high rates of reproductive success are said to have high fitness, whereas genotypes associated with low reproductive success are said to have low fitness.

Hardy–Weinberg analysis also allows us to examine fitness. By convention, population geneticists use the letter *w* to represent fitness. Thus, w_{AA} represents the relative fitness of genotype *AA*, w_{Aa} the relative fitness of genotype *Aa*, and w_{aa} the relative fitness of genotype *aa*. Assigning the values $w_{AA} = 1$, $w_{Aa} = 0.9$, and $w_{aa} = 0.8$ would mean, for example, that all *AA* individuals survive, 90 percent of the *Aa* individuals survive, and 80 percent of the *aa* individuals survive, as in the previous hypothetical case.

Let's consider selection against deleterious alleles. Fitness values $w_{AA} = 1$, $w_{Aa} = 1$, and $w_{aa} = 0$ describe a situation in which *a* is a homozygous lethal allele. As homozygous recessive individuals die without leaving offspring, the frequency of allele *a* will decline. The decline in the frequency of allele *a* is described by the equation

$$q_g = \frac{q_0}{1 + gq_0}$$

where q_g is the frequency of allele *a* in generation *g*, q_o is the starting frequency of *a* (i.e., the frequency of *a* in generation zero), and *g* is the number of generations that have passed.

Figure 23–8, p. 492, shows what happens to a lethal recessive allele with an initial frequency of 0.5. At first, because of the high percentage of *aa* genotypes, the frequency of allele *a* declines rapidly. The frequency of *a* is halved in only two generations. By the sixth generation, the frequency is halved again. By now, however, the majority of *a* alleles are carried by heterozygotes. Because *a* is recessive, these heterozygotes are

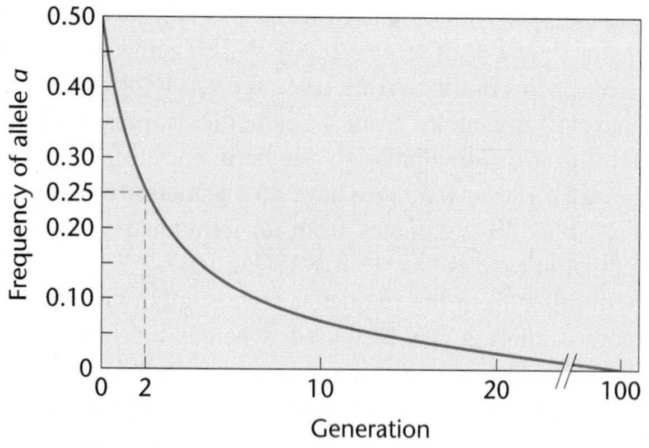

FIGURE 23–8 Change in the frequency of a lethal recessive allele, *a*. The frequency of *a* is halved in two generations and halved again by the sixth generation. Subsequent reductions occur slowly because the majority of *a* alleles are carried by heterozygotes.

not selected against. Consequently, as more time passes, the frequency of allele *a* declines ever more slowly. As long as heterozygotes continue to mate, it is difficult for selection to completely eliminate a recessive allele from a population.

Figure 23–9 shows the outcome of different degrees of selection against a nonlethal recessive allele, *a*. In this case, the intensity of selection varies from strong (red curve) to weak (blue curve), as well as intermediate values (yellow, purple, and green curves). In each example, the frequency of the deleterious allele, *a*, starts at 0.99 and declines over time. However, the rate of decline depends heavily on the strength of selection. When selection is strong and only 90 percent of the heterozygotes and 80 percent of the *aa* homozygotes survive (red curve), the frequency of allele *a* drops from 0.99 to less than 0.01 in about 85 generations. However, when selection is weak, and 99.8 percent of the heterozygotes and 99.6 percent of the *aa* homozygotes survive (blue curve), it takes 1000 generations for the frequency of allele *a* to drop from 0.99 to 0.93. Two important conclusions can be drawn from this example. First, over thousands of generations, even weak selection can cause substantial changes in allele frequencies; because evolution generally occurs over a large number of generations, selection is a powerful force in evolutionary change. Second, for selection to produce rapid changes in allele frequencies, the differences in fitness among genotypes must be large.

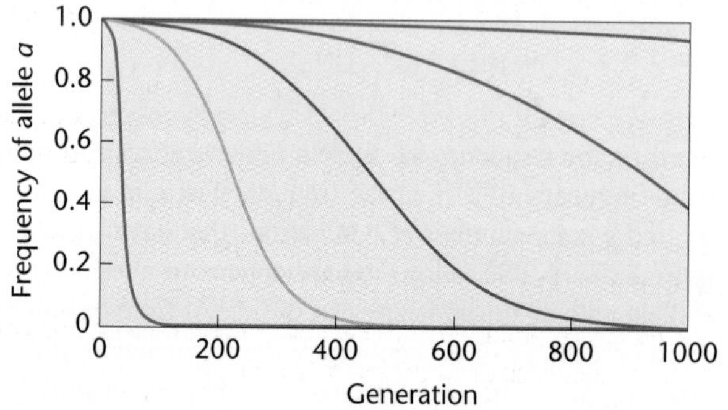

FIGURE 23–9 The effect of selection on allele frequency. The rate at which a deleterious allele is removed from a population depends heavily on the strength of selection.

The manner in which selection affects allele frequencies allows us to make some inferences about the *CCR5*-Δ*32* allele that we discussed earlier. Because individuals with genotype Δ*32*/Δ*32* are resistant to most sexually transmitted strains of HIV-1, while individuals with genotypes *1*/*1* and *1*/Δ*32* are susceptible, we might expect AIDS to act as a selective force causing the frequency of the Δ*32* allele to increase over time. Indeed, it probably will, but the increase in frequency is likely to be slow in human terms. In fact, it will take about 100 generations (about 2000 years) for the frequency of the Δ*32* allele to reach just 0.11. In other words, the frequency of the Δ*32* allele will probably not change much over the next few generations in most populations that currently harbor it.

There Are Several Types of Selection

The phenotype is the result of the combined influence of the individual's genotype at many different loci and the effects of the environment. Selection for traits can be classified as (1) directional, (2) stabilizing, or (3) disruptive.

In **directional selection** (Figure 23–10) phenotypes at one end of the spectrum of phenotypes present in the population become selected for or against, usually as a result of changes in the environment. A carefully documented example comes from research by Peter and Rosemary Grant and their colleagues, who study the medium ground finches (*Geospiza fortis*) of Daphne Major Island in the Galapagos Islands. The beak size of these birds varies enormously. In 1976, for example, some birds in the population had beaks less than 7 mm deep, while others had beaks more than 12 mm deep. In 1977, a severe drought killed some 80 percent of the finches. Big-beaked birds survived at higher rates than small-beaked birds because when food became scarce, the big-beaked birds were able to eat a greater variety of seeds. When the drought ended in 1978, the offspring of the survivors inherited their parents' big beaks. Between 1976 and 1978, the beak depth of the average finch in the Daphne Major population increased by just over 0.5 mm, shifting the average beak size toward one phenotypic extreme.

Stabilizing selection, in contrast, tends to favor intermediate phenotypes, with those at both extremes being selected against. Over time, this will reduce the phenotypic variance in the

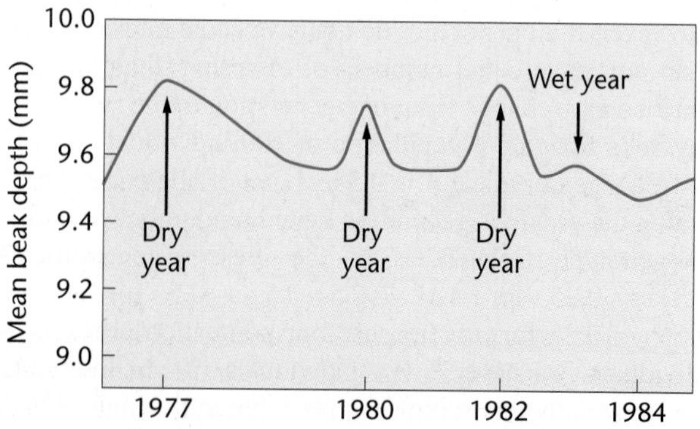

FIGURE 23–10 Beak size in finches during dry years increases because of strong selection. Between droughts, selection for large beak size is not as strong, and birds with smaller beak sizes survive and reproduce, increasing the number of birds with smaller beaks.

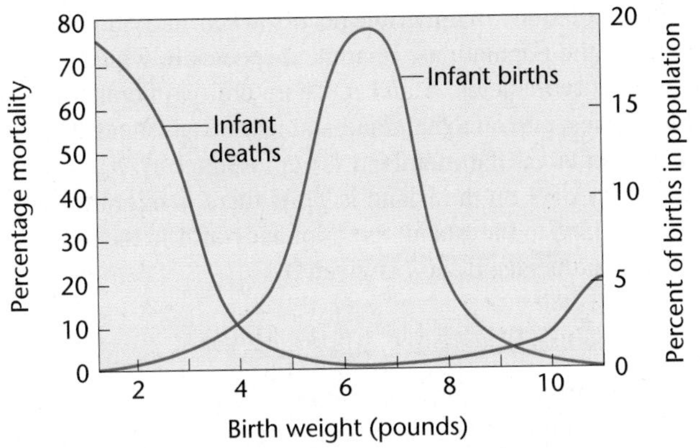

FIGURE 23–11 Relationship between birth weight and mortality in humans.

population but without a significant shift in the mean. One of the clearest demonstrations of stabilizing selection is shown by a study of human birth weight and survival for 13,730 children born over an 11-year period. Figure 23–11 shows the distribution of birth weight and the percentage of mortality at 4 weeks of age. Infant mortality increases on either side of the optimal birth weight of 7.5 pounds. Stabilizing selection acts to keep a population well adapted to its environment.

Disruptive selection is selection against intermediates and for both phenotypic extremes. It can be viewed as the opposite of stabilizing selection because the intermediate types are selected against. This will result in a population with an increasingly bimodal distribution for the trait, as we can see in Figure 23–12. In one set of experiments using *Drosophila,* after several generations of disruptive artificial selection for bristle number, in which only flies with high- or low-bristle numbers were allowed to breed, most flies could be easily placed in a low- or high-bristle category. In natural populations, such a situation might exist for a population in a heterogeneous environment.

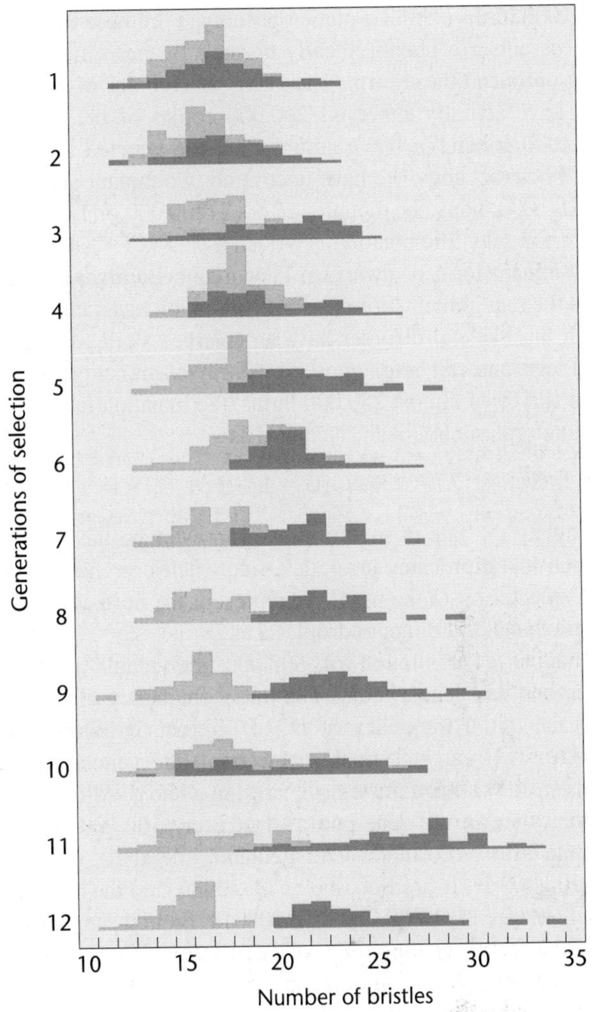

FIGURE 23–12 The effect of disruptive selection on bristle number in *Drosophila.* When individuals with the highest and lowest bristle number were selected, the population showed a nonoverlapping divergence in only 12 generations.

ESSENTIAL POINT

The rate of change under natural selection depends on initial allele frequencies, selection intensity, and the relative fitness of different genotypes.

23.5 Mutation Creates New Alleles in a Gene Pool

Within a population, the gene pool is reshuffled each generation to produce new genotypes in the offspring. The enormous genetic variation present in the gene pool allows assortment and recombination to produce new genotypic combinations continuously. But assortment and recombination do not produce new alleles. **Mutation** alone acts to create new alleles. It is important to keep in mind that mutational events occur at random—that is, without regard for any possible benefit or disadvantage to the organism. In this section, we consider whether mutation, by itself, is a significant factor in changing allele frequencies.

To determine whether mutation is a significant force in changing allele frequencies, we measure the rate at which mutations are produced. As most mutations are recessive, it is difficult to observe mutation rates directly in diploid organisms. Indirect methods use probability and statistics or large-scale screening programs to estimate mutation rates. For certain dominant mutations, however, a direct method of measurement can be used. To ensure accuracy, several conditions must be met:

1. The allele must produce a distinctive phenotype that can be distinguished from similar phenotypes produced by recessive alleles.
2. The trait must be fully expressed or completely penetrant so that mutant individuals can be identified.
3. An identical phenotype must never be produced by nongenetic agents such as drugs or chemicals.

Mutation rates can be stated as the number of new mutant alleles per given number of gametes. Suppose that for a given gene that undergoes mutation to a dominant allele, 2 out of

100,000 births exhibit a mutant phenotype. In these two cases, the parents are phenotypically normal. Because the zygotes that produced these births each carry two copies of the gene, we have actually surveyed 200,000 copies of the gene (or 200,000 gametes). If we assume that the affected births are each heterozygous, we have uncovered two mutant alleles out of 200,000. Thus, the mutation rate is 2/200,000 or 1/100,000, which in scientific notation is written as 1×10^{-5}. In humans, a dominant form of dwarfism known as **achondroplasia** fulfills the requirements for measuring mutation rates. Individuals with this skeletal disorder have an enlarged skull, short arms and legs, and can be diagnosed by X-ray examination at birth. In a survey of almost 250,000 births, the mutation rate (μ) for achondroplasia has been calculated as

$$\mu = 1.4 \times 10^{-5} \pm 0.5 \times 10^{-5}$$

Knowing the rate of mutation, we can estimate the extent to which mutation can cause allele frequencies to change from one generation to the next. We represent the normal allele as *d* and the allele for achondroplasia as *D*.

Imagine a population of 500,000 individuals in which everyone has genotype *dd*. The initial frequency of *d* is 1.0, and the initial frequency of *D* is 0. If each individual contributes two gametes to the gene pool, the gene pool will contain 1,000,000 gametes, all carrying allele *d*. Although the gametes are in the gene pool, 1.4 of every 100,000 *d* alleles mutate into a *D* allele. The frequency of allele *d* is now $(1{,}000{,}000 - 14)/1{,}000{,}000 = 0.999986$, and the frequency of allele *D* is $14/1{,}000{,}000 = 0.000014$. From these numbers, it will clearly be a long time before mutation, by itself, causes any appreciable change in the allele frequencies in this population (Figure 23–13). In other words, mutation generates new alleles but by itself does not alter allele frequencies at an appreciable rate.

23.6 Migration and Gene Flow Can Alter Allele Frequencies

Occasionally, a species becomes divided into populations that are separated geographically. Various evolutionary forces, including selection, can establish different allele frequencies in such populations. **Migration** occurs when individuals move between the populations. Imagine a species in which a given locus has two alleles, *A* and *a*. There are two populations of this species, one on a mainland and one on an island. The frequency of *A* on the mainland is represented by p_m, and the frequency of *A* on the island is p_i. If there is migration from the mainland to the island, the frequency of *A* in the next generation on the island ($p_{i'}$) is given by

$$p_{i'} = (1 - m)p_i + mp_m$$

where *m* represents migrants from the mainland to the island.

As an example of how migration might affect the frequency of *A* in the next generation on the island ($p_{i'}$), assume that $p_i = 0.4$ and $p_m = 0.6$ and that 10 percent of the parents of the next generation are migrants from the mainland ($m = 0.1$). In the next generation, the frequency of allele *A* on the island will therefore be

$$p_{i'} = [(1 - 0.1) \times 0.4] + (0.1 \times 0.6)$$
$$= 0.36 + 0.06$$
$$= 0.42$$

In this case, migration from the mainland has changed the frequency of *A* on the island from 0.40 to 0.42 in a single generation.

These calculations reveal that the change in allele frequency attributable to migration is proportional to the differences in allele frequency between the donor and recipient populations and to the rate of migration. If either *m* is large or p_m is very different from p_i, then a rather large change in the frequency of *A* can occur in a single generation. If migration is the only force acting to change the allele frequency on the island, then equilibrium will be attained only when $p_i = p_m$. These guidelines can often be used to estimate migration in cases where it is difficult to quantify. As *m* can have a wide range of values, the effect of migration can substantially alter allele frequencies in populations, as shown for the I^B allele of the ABO blood group in Figure 23–14.

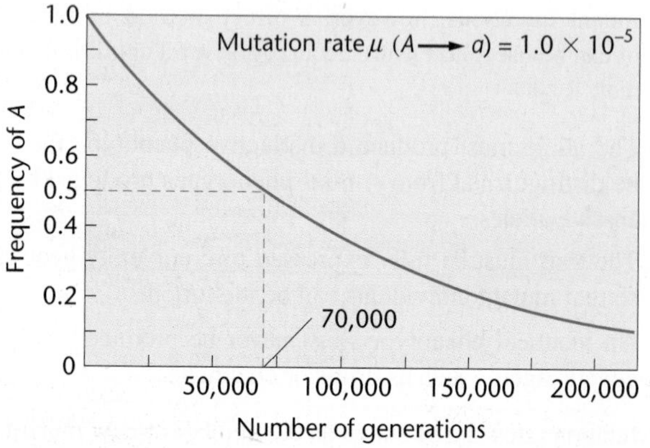

FIGURE 23–13 Replacement rate of an allele by mutation alone, assuming an average mutation rate of 1.0×10^{-5}.

23.7 Genetic Drift Causes Random Changes in Allele Frequency in Small Populations

In small populations, significant random fluctuations in allele frequencies are possible by chance alone. The degree of fluctuation increases as the population size decreases, a situation known as **genetic drift.** In addition to small population size, drift can arise through the **founder effect,** which occurs when a population originates from a small number of individuals. Although the population may later increase to a large size, the genes carried by all members are derived from those of the founders (assuming no mutation, migration, or selection, and the presence of random mating). Drift can also arise via a **genetic bottleneck.** Bottlenecks develop when a large population undergoes a drastic but temporary reduction in

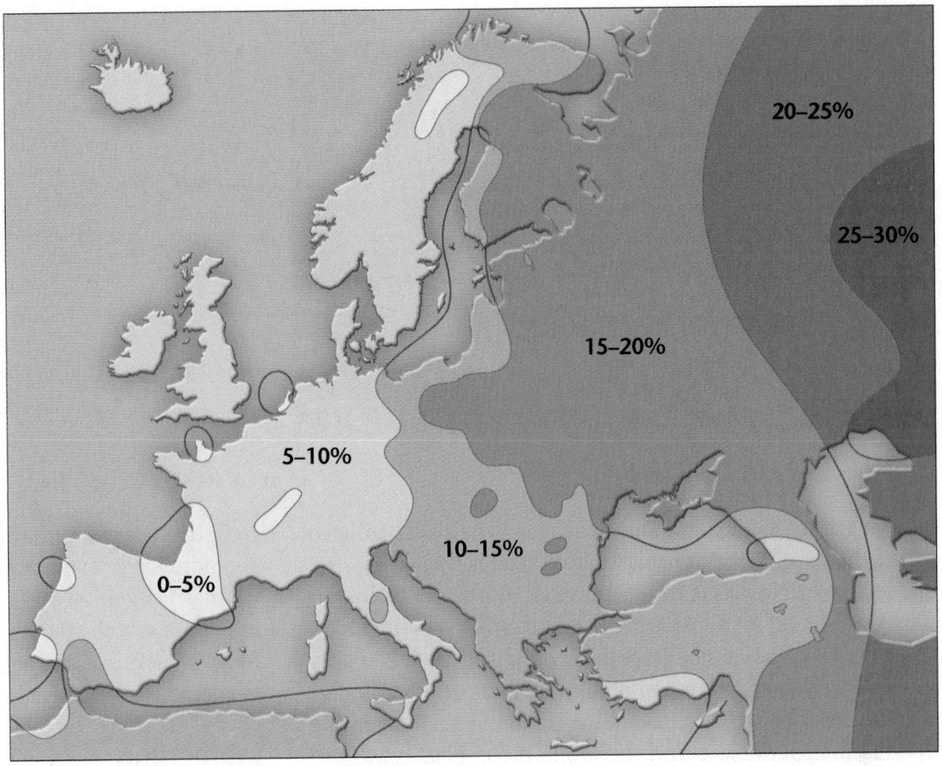

FIGURE 23–14 Affected migration as a force in evolution. The I^B allele of the *ABO* locus is present in a gradient from east to west. This allele shows the highest frequency in central Asia and the lowest in northeast Spain. The gradient parallels the waves of Mongol migration into Europe following the fall of the Roman Empire and is a genetic relic of human history.

numbers. Even though the population recovers, its genetic diversity has been greatly reduced.

Founder Effects in Human Populations

Allele frequencies in certain human populations demonstrate the role of genetic drift in natural populations. Native Americans living in the southwestern United States have a high frequency of oculocutaneous albinism (OCA). In the Navajo, who live primarily in northeast Arizona, albinism occurs with a frequency of 1 in 1500–2000, compared with whites (1 in 36,000) and African-Americans (1 in 10,000). There are four different forms of OCA (OCA1–4), all with varying degrees of melanin deficiency in the skin, eyes, and hair. OCA2 is caused by mutations in the *P* gene, which encodes a plasma membrane protein. To investigate the genetic basis of albinism in the Navajo, researchers screened for mutations in the *P* gene. In their study, all Navajo with albinism were homozygous for a 122.5-kb deletion in the *P* gene, spanning exons 10–20 (Figure 23–15). This deletion allele was not present in 34 individuals belonging to other Native American populations.

Using a set of PCR primers, researchers were able to identify homozygous affected individuals and heterozygous carriers (Figure 23–16, p. 496) and surveyed 134 normally pigmented

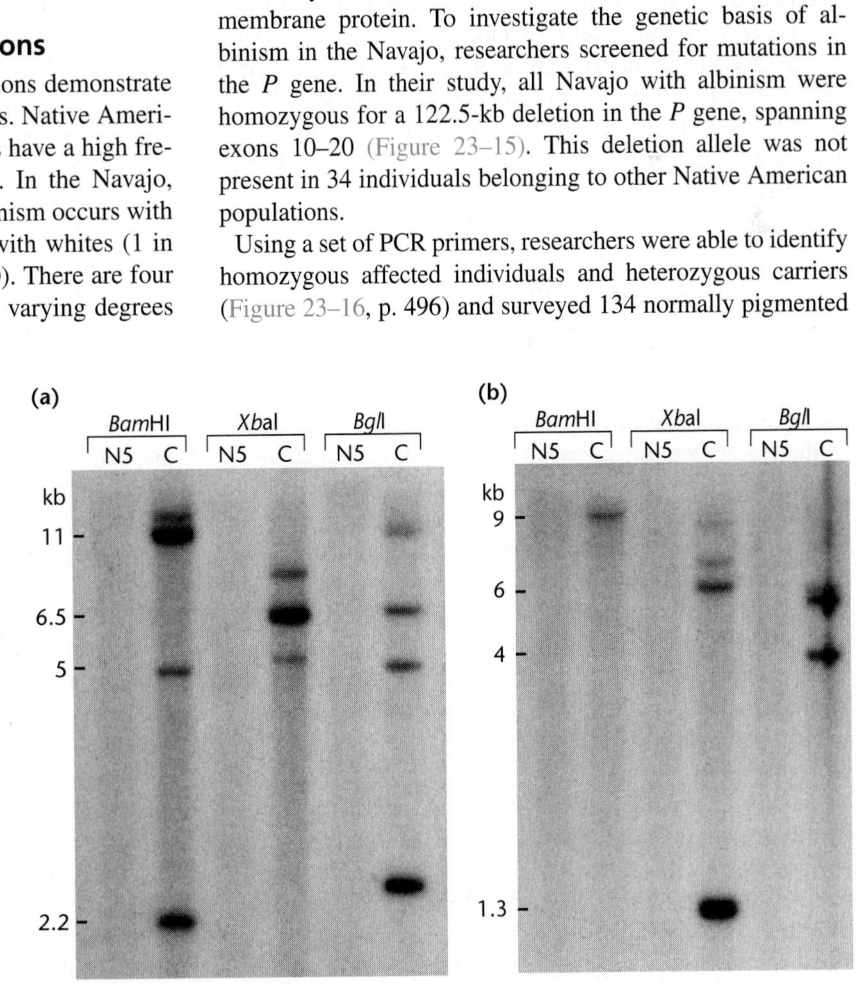

FIGURE 23–15 Genomic DNA digests from a Navajo affected with albinism (N5) and a normally pigmented individual (C). (a) Hybridization with a probe covering exons 11–15 of the *P* gene; there are no hybridizing fragments detected in N5. (b) Hybridization with a probe covering exons 15–20 of the *P* gene; there are no hybridizing fragments detected in N5. This confirms the presence of a deletion in affected individuals. *Courtesy of Murray Brilliant, "A 122.5 kilobase deletion of P gene underlies the high prevalence of oculocutaneous albinism type 2 in the Navajo population." From:* American Journal Human Genetics *72:62–72, Figure. 1, p. 65. Published by University of Chicago Press.*

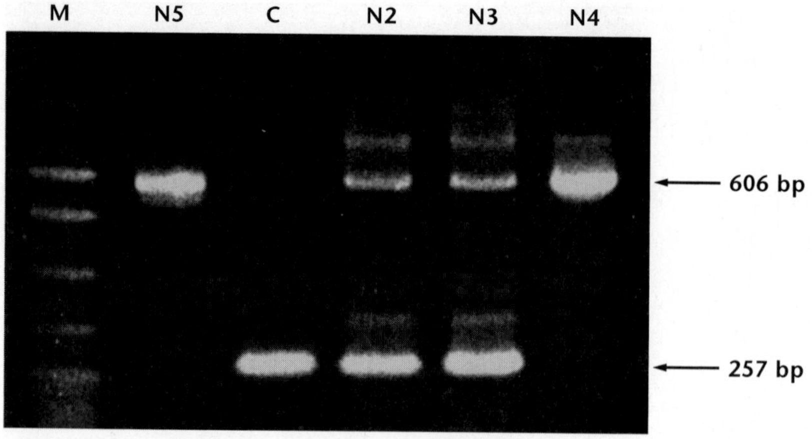

FIGURE 23–16 PCR screens of Navajo affected with albinism (N4 and N5) and the parents of N4 (N2 and N3). Affected individuals (N4 and N5), heterozygous carriers (N2 and N3), and a homozygous normal individual (C) each give a distinctive band pattern, allowing detection of heterozygous carriers in the population. Molecular size markers (M) are in the first lane.
Courtesy of Murray Brilliant, "A 122.5 kilobase deletion of P gene underlies the high prevalence of oculocutaneous albinism type 2 in the Navajo population." From: American Journal Human Genetics *72: 62–72, Figure. 3, p. 67. Published by University of Chicago Press.*

Navajo and 42 members of the Apache, a tribe closely related to the Navajo. Based on this sample, the heterozygote frequency in the Navajo is estimated to be 4.5 percent. No carriers were found in the Apache population that was studied.

The 122.5-kb deletion allele causing OCA2 was found only in the Navajo population and not in members of other Native American tribes in the southwestern United States, suggesting that the mutant allele is specific to the Navajo and may have arisen in a single individual who was one of a small number of founders of the Navajo population. Using other methods, workers estimated the age of the mutation to be between 400 and 11,000 years. To narrow this range, they relied on tribal history. Navajo oral tradition indicates that the Navajo and Apache became separate populations between 600 and 1000 years ago. Because the deletion is not found in the Apaches, it probably arose in the Navajo population after the tribes split. On this basis, the deletion is estimated to be 400 to 1000 years old and probably arose as a founder mutation.

23.8 Nonrandom Mating Changes Genotype Frequency but Not Allele Frequency

We have explored how violations of the first four assumptions of the Hardy–Weinberg law, in the form of selection, mutation, migration, and genetic drift, can cause allele frequencies to change. The fifth assumption is that the members of a population mate at random; in other words, any one genotype has an equal probability of mating with any other genotype in the population. Nonrandom mating can change the frequencies of genotypes in a population. Subsequent selection for or against certain genotypes has the potential to affect the overall frequencies of the alleles they contain, but it is important to note that nonrandom mating *does not itself directly change allele frequencies.*

Nonrandom mating can take one of several forms. In **positive assortive mating** similar genotypes are more likely to mate than dissimilar ones. This often occurs in humans: A number of studies have indicated that many people are more attracted to individuals who physically resemble them (and are therefore more likely to be genetically similar as well). **Negative assortive mating** occurs when dissimilar genotypes are more likely to mate; some plant species have inbuilt pollen/stigma recognition systems that prevent fertilization between individuals with the same alleles at key loci. However, the form of nonrandom mating most commonly found to affect genotype frequencies in population genetics is **inbreeding.**

Inbreeding

Inbreeding occurs when mating individuals are more closely related than any two individuals drawn from the population at random; loosely defined, inbreeding is mating among relatives. For a given allele, inbreeding increases the proportion of homozygotes in the population. A completely inbred population will theoretically consist only of homozygous genotypes.

To describe the intensity of inbreeding in a population, geneticist Sewall Wright devised the **coefficient of inbreeding (*F*).** F quantifies the probability that the two alleles of a given gene in an individual are identical *because they are descended from the same single copy of the allele in an ancestor.* If $F = 1$, all individuals in the population are homozygous, and both alleles in every individual are derived from the same ancestral copy. If $F = 0$, no individual has two alleles derived from a common ancestral copy.

One method of estimating F for an individual is shown in Figure 23–17. The fourth-generation female (shaded pink) is the daughter of first cousins (yellow). Suppose her great-grandmother (green) was a carrier of a recessive lethal allele, *a*. What is the probability that the fourth-generation female will inherit two copies of her great-grandmother's lethal allele? For this to happen, (1) the great-grandmother had to pass a copy of the allele to her son, (2) her son had to pass it to his daughter, and (3) his daughter had to pass it to her daughter (the pink female). Also, (4) the great-grandmother had to pass a copy of the allele to her daughter, (5) her daughter had to pass it to her son, and (6) her son had to pass it to his daughter (the pink female). Each of the six necessary events has an individual probability of $1/2$, and they *all* have to happen, so the probability that the pink female will inherit two copies of her great-grandmother's lethal allele is $(1/2)^6 = 1/64$. To calculate an overall value of F for the pink female as a child of a first-cousin marriage, remember that she could also inherit two copies of any of the other three

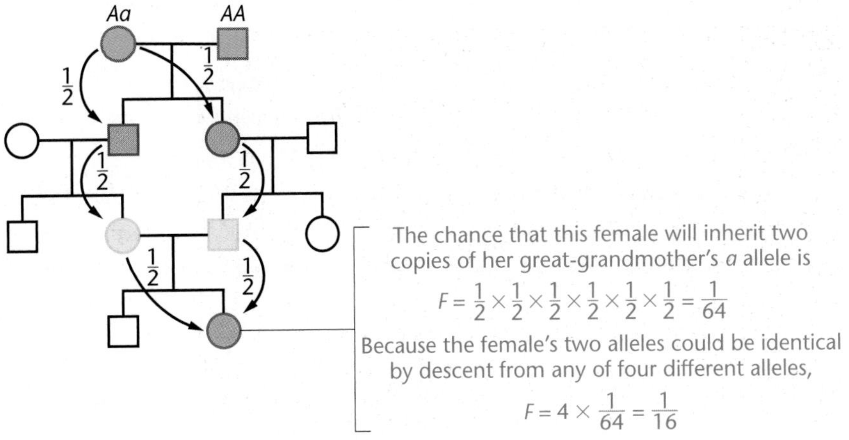

FIGURE 23–17 Calculating the coefficient of inbreeding *(F)* for the offspring of a first-cousin marriage.

The chance that this female will inherit two copies of her great-grandmother's *a* allele is

$$F = \frac{1}{2} \times \frac{1}{2} \times \frac{1}{2} \times \frac{1}{2} \times \frac{1}{2} \times \frac{1}{2} = \frac{1}{64}$$

Because the female's two alleles could be identical by descent from any of four different alleles,

$$F = 4 \times \frac{1}{64} = \frac{1}{16}$$

alleles present in her great-grandparents. Because any of four possibilities would give the pink female two alleles identical by descent from an ancestral copy,

$$F = 4 \times (1/64) = 1/16$$

ESSENTIAL POINT

Nonrandom mating in the form of inbreeding increases the frequency of homozygotes in the population and decreases the frequency of heterozygotes.

NOW SOLVE THIS

Problem 18 on page 505 asks you to calculate the probability that a couple will have a child with cystic fibrosis.

Hint: First, work out the probability that each parent carries the mutant allele.

23.9 Reduced Gene Flow, Selection, and Genetic Drift Can Lead to Speciation

A **species** can be defined as a group of actually or potentially interbreeding organisms that is reproductively isolated in nature from all other such groups. In sexually reproducing organisms, speciation transforms the gene pool of the parental species or divides a single gene pool into two or more separate gene pools. Changes in morphology or physiology and adaptation to an ecological niche may also occur but are not necessary components of the speciation event. Speciation can take place gradually or within a few generations.

Most populations contain considerable genetic variation, and different populations within a species may carry different alleles or allele frequencies at a variety of loci. The genetic divergence of these populations can be caused by natural selection, genetic drift, or both. In an earlier section, we saw that the migration of individuals between populations, together with the gene flow that accompanies that migration, tends to homogenize allele frequencies among populations. In

other words, migration counteracts the tendency of populations to diverge.

When gene flow between populations is reduced or absent, the populations may diverge to the point that members of one population are no longer able to interbreed successfully with members of the other. When populations reach the point where they are reproductively isolated from one another, they have become different species, according to the biological species concept. The genetic changes that result in reproductive isolation between or among populations and that lead to the formation of new species or higher taxonomic groups is an example of macroevolution. We will restrict our discussion to the macroevolutionary changes that lead to the formation of new species.

The biological barriers that prevent or reduce interbreeding between populations are called **reproductive isolating mechanisms,** classified in Table 23.4. These mechanisms may be ecological, behavioral, seasonal, mechanical, or physiological.

Prezygotic isolating mechanisms prevent individuals from mating in the first place. Individuals from different populations may not find each other at the right time, may not recognize each other as suitable mates, or may try to mate but find that they are unable to do so.

Postzygotic isolating mechanisms create reproductive isolation even when the members of two populations are willing and able to mate with each other. For example, genetic divergence may have reached the stage where the viability or fertility of hybrids is reduced. Hybrid zygotes may be formed, but all or most may be inviable. Alternatively, the hybrids may be viable, but be sterile or suffer from reduced fertility. Yet again, the hybrids themselves may be fertile, but their progeny may have lowered viability or fertility. In all these situations, hybrids will not reproduce and are genetic dead-ends. These postzygotic mechanisms act at or beyond the level of the zygote and are generated by genetic divergence.

Postzygotic isolating mechanisms waste gametes and zygotes and lower the reproductive fitness of hybrid survivors. Selection will therefore favor the spread of alleles that reduce the formation of hybrids, leading to the development of prezygotic isolating mechanisms, which in turn prevent interbreeding and the formation of hybrid zygotes and

TABLE 23.4 Reproductive Isolating Mechanisms

Prezygotic Mechanisms
Prevent fertilization and zygote formation

1. **Geographic or ecological:** The populations live in the same regions but occupy different habitats.
2. **Seasonal or temporal:** The populations live in the same regions but are sexually mature at different times.
3. **Behavioral** (only in animals): The populations are isolated by different and incompatible behavior before mating.
4. **Mechanical:** Cross-fertilization is prevented or restricted by differences in reproductive structures (genitalia in animals, flowers in plants).
5. **Physiological:** Gametes fail to survive in alien reproductive tracts.

Postzygotic Mechanisms
Fertilization takes place and hybrid zygotes are formed, but these are nonviable or give rise to weak or sterile hybrids.

1. **Hybrid nonviability or weakness.**
2. **Developmental hybrid sterility:** Hybrids are sterile because gonads develop abnormally or meiosis breaks down before completion.
3. **Segregational hybrid sterility:** Hybrids are sterile because of abnormal segregation into gametes of whole chromosomes, chromosome segments, or combinations of genes.
4. **F_2 breakdown:** F_1 hybrids are normal, vigorous, and fertile, but the F_2 contains many weak or sterile individuals.

Source: From G. Ledyard Stebbins, *Processes of Organic Evolution,* 3rd ed., copyright 1977, p. 143. Reprinted by permission of Prentice Hall, Upper Saddle River, NJ.

offspring. In animal evolution, one of the most effective prezygotic mechanisms is behavioral isolation, involving courtship behavior.

Changes Leading to Speciation

The Isthmus of Panama, which created a land bridge connecting North and South America and simultaneously separated the Caribbean Sea from the Pacific Ocean, formed roughly 3 million years ago. After identifying seven Caribbean species of snapping shrimp (Figure 23–18), researchers matched each one with a similar Pacific species to form a pair. Members of each pair were closer to each other in structure and appearance than either was to any other species in its own ocean. Analysis of allele frequencies and mitochondrial DNA sequences confirmed that the members of each pair were one another's closest genetic relatives.

The interpretation of these data is that, prior to the formation of the isthmus, the ancestors of each pair were members of a single species. When the isthmus closed, each of the seven ancestral species was divided into two separate populations, one in the Caribbean and the other in the Pacific.

FIGURE 23–18 A snapping shrimp (genus *Alpheus*).

Meeting in a dish in a lab for the first time in 3 million years, would Caribbean and Pacific members of a species pair recognize each other as suitable mates? Males and females were paired together, and the relative inclination of Caribbean–Pacific couples to mate versus that of Caribbean–Caribbean or Pacific–Pacific couples was calculated. For three of the seven species pairs, transoceanic couples refused to mate altogether. For the other four species pairs, transoceanic couples were 33, 45, 67, and 86 percent as likely to mate with each other as were same-ocean pairs. Of the same-ocean couples that mated, 60 percent produced viable clutches of eggs. Of the transoceanic couples that mated, only 1 percent produced viable clutches. We can conclude from these results that 3 million years of separation has resulted in complete or nearly complete speciation, involving strong pre- and postzygotic isolating mechanisms for all seven species pairs.

The Rate of Macroevolution and Speciation

How much time is required for speciation? In many cases, divergence of populations and the appearance of new species is a process that occurs over a long period of time. In other cases, however, genetic divergence, reproductive isolation, and speciation can be surprisingly rapid.

The Rift Valley lakes of East Africa support hundreds of species of cichlid fish. Lake Victoria (Figure 23–19), for example, has more than 400 species. Cichlids are highly specialized for different niches. Some eat algae floating on the water's surface, whereas others are bottom feeders, insect feeders, mollusk eaters, and predators on other fish species. Lake Tanganyika has a similar array of species. Genetic analyses indicate that the species in a given lake are all more

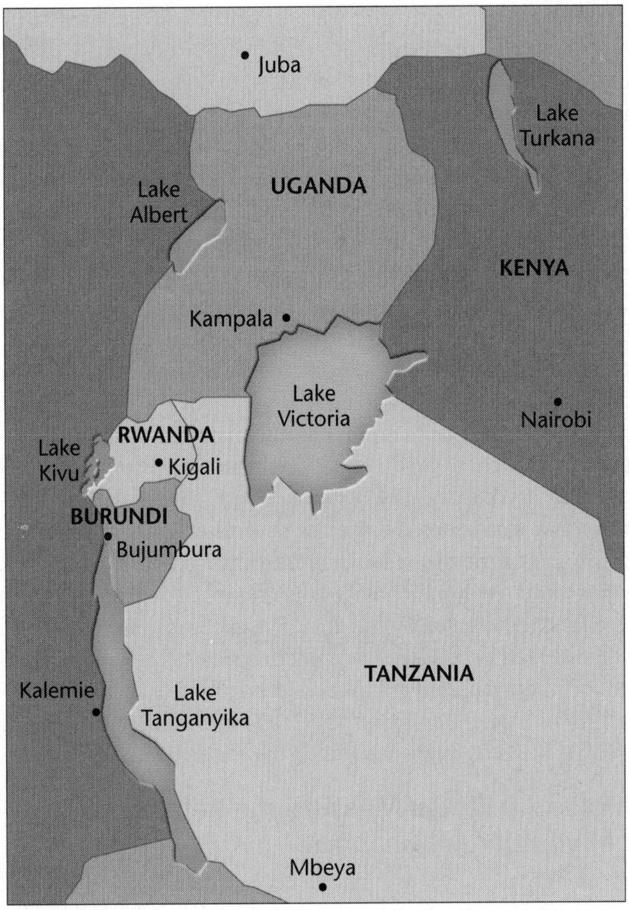

FIGURE 23–19 Lake Victoria in the Rift Valley of east Africa is home to more than 400 species of cichlids.

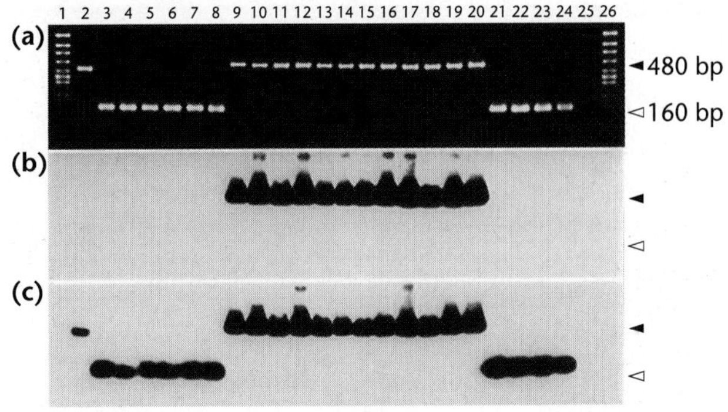

FIGURE 23–20 Use of repetitive DNA elements to trace species relationships in cichlids from Lake Tanganyika. (a) Photograph of an agarose gel showing PCR fragments generated from primers that flank the AFC family of SINES. Large fragments containing this family are present in all samples of genomic DNA from members of the Lamprologini tribe of cichlids (lanes 9–20). DNA from the species in lane 2 has a similar but shorter fragment, which may represent another repetitive sequence. DNA from other species (lanes 3–8 and 21–24) produce short, non-SINE-containing fragments. (b) A Southern blot of the gel from (a) probed with DNA from the AFC family of SINES. AFC is present in DNA from all species in the Lamprologini tribe (lanes 9–20), but not in DNA from other species (lanes 2–8 and 21–24). (c) A second Southern blot of the gel from (a) probed with the genomic sequence at which the AFC SINE inserts. All species examined (lanes 2–24) contain the insertion sequence. The larger fragments in Lamprologini DNA (lanes 9–20) correspond to those in the previous blot, showing that the SINE is inserted at the same site in all tribal species. In sum, these data show that the AFC SINE is present only in members of this tribe, is present in all member species, and is inserted at the same site in all cases. These results are interpreted as showing a common origin for the species of the tribe in question.

closely related to each other than to species from other lakes. The implication is that most or all the species in, say, Lake Tanganyika are descended from a single common ancestor and that they diverged within their home lake.

Lake Victoria is between 250,000 and 750,000 years old, and there is evidence that the lake may have dried out almost completely less than 14,000 years ago. Is it possible that the 400 species in the lake today evolved from a common ancestral species in less than 14,000 years?

In a study of cichlid origins in Lake Tanganyika, Norihiro Okada and colleagues examined the insertion of a novel family of repetitive DNA sequences called short interspersed elements (SINES) (discussed in Chapter 11) into the genomes of cichlid species in Lake Tanganyika. SINES are a type of retroposon, and the random integration of a SINE at a locus is most likely an irreversible event. If a SINE is present at the same locus in the genome of all species examined, this is strong evidence that all of those species descend from a common ancestor. Using a SINE called AFC, Okada's team screened 33 species of cichlids belonging to four groups (called species tribes). In each tribe, the SINE was present at all the sites tested, indicating that the species in each tribe are descended from a single ancestral species (Figure 23–20). If further research using SINES and other molecular markers produces similar results with the Lake Victoria species, it means that the 400 cichlid species in this lake evolved in less than 14,000 years. If confirmed, this finding would represent the fastest evolutionary divergence of species ever documented in vertebrates.

23.10 Genetic Differences Can Be Used to Reconstruct Evolutionary History

Speciation is associated with changes in the genetic structure of populations and with genetic divergence of those populations. Therefore, we should be able to use genetic differences among present-day species to reconstruct their evolutionary histories.

Constructing Phylogenetic Trees from Amino Acid Sequences

In an important early example of phylogeny reconstruction, W. M. Fitch and E. Margoliash assembled data on the amino acid sequence for cytochrome c in a variety of organisms. **Cytochrome c** is a respiratory molecule found in the mitochondria of eukaryotes, and its amino acid sequence has evolved very slowly. For example, its amino acid sequence in humans and chimpanzees is identical, and humans and rhesus monkeys show only one amino acid difference. This similarity is remarkable considering that the fossil record indicates that

TABLE 23.5 Amino Acid Differences and Minimal Mutational Distances between Cytochrome c in Humans and Other Organisms

Organism	(a) Amino Acid Differences	(b) Minimal Mutational Distance
Human	0	0
Chimpanzee	0	0
Rhesus monkey	1	1
Rabbit	9	12
Pig	10	13
Dog	10	13
Horse	12	17
Penguin	11	18
Moth	24	36
Yeast	38	56

Source: From W.M. Fitch and E. Margoliash, Construction of phylogenetic trees, *Science* 155: 279–284, January 20, 1967. Copyright 1967 by the American Association for the Advancement of Science.

the lines leading to humans and monkeys diverged from a common ancestor approximately 20 million years ago.

Column (a) of Table 23.5 shows the number of amino acid differences between cytochrome c in humans and in various other species. The table is broadly consistent with our intuitions about how closely related we are to these other species. For example, we are more closely related to other mammals than we are to insects, and we are more closely related to insects than we are to yeast. Similarly, our cytochrome c differs in 10 amino acids from that of dogs, in 24 amino acids from that of moths, and in 38 amino acids from that of yeast.

More than one nucleotide change, however, may be required to change a given amino acid. When the nucleotide changes necessary for all amino acid differences observed in a protein are totaled, the **minimal mutational distance** between the genes of any two species is established. Column (b) in Table 23.5 shows such an analysis of the genes encoding cytochrome c. As expected, these values are larger than the corresponding number of amino acids separating humans from the other nine organisms listed.

Fitch used data on the minimal mutational distances between the cytochrome c genes of 19 organisms to reconstruct their evolutionary history. The result is an estimate of the phylogenetic tree that unites the species studied (Figure 23–21). The black dots on the tips of the branches represent existing species, whose inferred common ancestors are linked to them by green lines and represented by red dots. The ancestral species evolved and diverged to produce the modern species. The common ancestors are connected to still earlier common ancestors, culminating in a single common ancestor for all the species on the tree, represented by the red dot on the extreme left.

Molecular Clocks Measure the Rate of Evolutionary Change

In many cases, we would like to estimate not only which members of a set of species are most closely related, but also when their common ancestors lived. Sometimes we can do so,

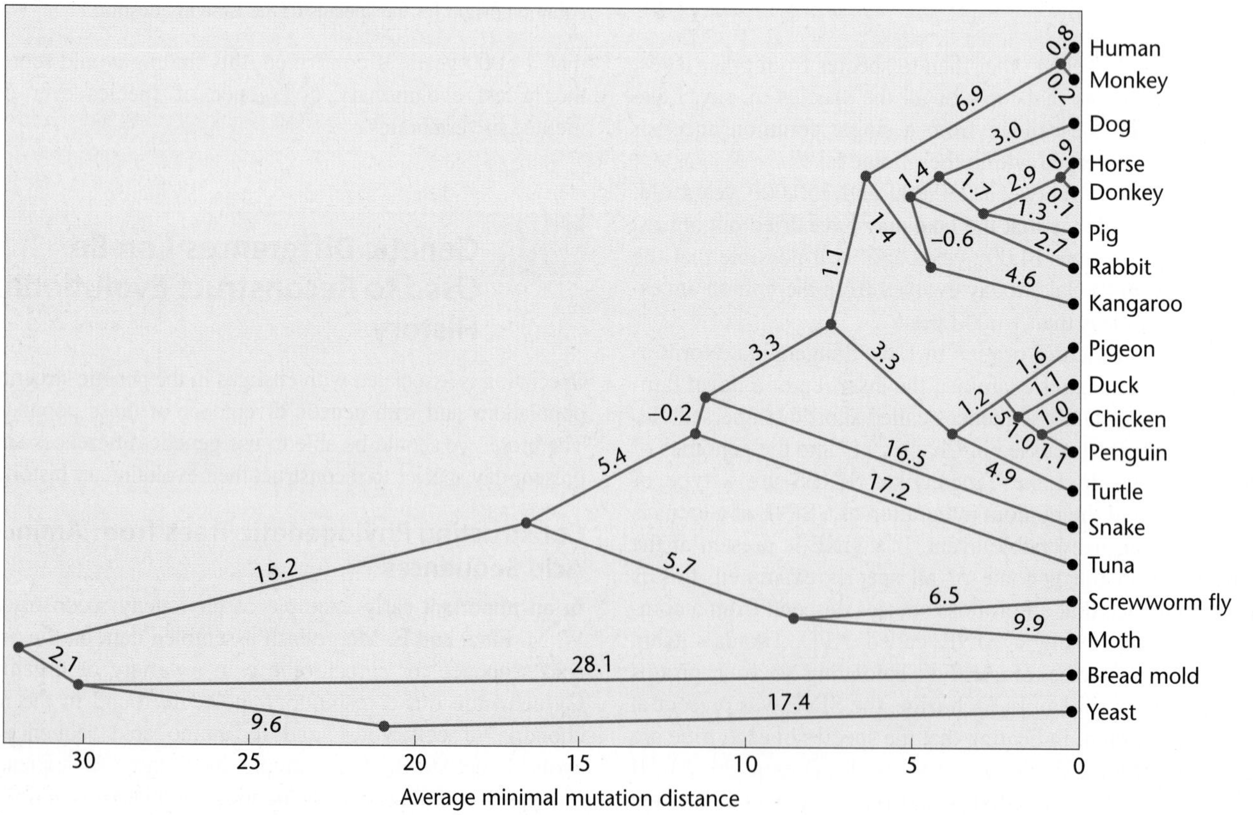

FIGURE 23–21 Phylogenetic tree constructed by comparing homologies in cytochrome c amino acid sequences. *Reprinted with permission from Fitch, W.M. and Margoliash, E. 1967. Construction of phylogenetic trees.* Science *279: 279–284, Figure 2.*

thanks to **molecular clocks**—amino acid sequences or nucleotide sequences in which evolutionary changes accumulate at a constant rate over time.

Research by Walter M. Fitch and colleagues on the influenza A virus shows how molecular clocks are used. Fitch and his associates sequenced part of the hemagglutinin gene from flu viruses that had been isolated at different times over a 20-year period. They then calculated the numbers of nucleotide differences between the various virus samples and constructed an evolutionary tree [Figure 23–22(b)]. Most of the strains that had been collected have gone extinct, leaving no descendants among the more recently isolated viruses. Fitch's group then plotted the number of nucleotide substitutions between the first virus and each subsequent virus against the year in which the virus was isolated [Figure 23–22(a)]. The points all fall very close to a straight line, indicating that nucleotide substitutions in this gene have accumulated at a steady rate. Nucleotide substitutions in the hemagglutinin gene thus serve as a molecular clock. Molecular clocks are used to compare the sequences of new flu viruses as they appear each year and to estimate the time that has passed since each diverged from a common ancestor.

Molecular clocks must be carefully calibrated and used with caution. For example, Fitch's data indicate that strains of influenza A that jump from birds to humans have evolved much more rapidly than strains that have remained in birds. Hence, a molecular clock calibrated from human strains of the virus would be highly misleading if applied to bird strains.

Genetic Divergence between Neanderthals and Modern Humans

Paleontological evidence indicates that the Neanderthals, *Homo neanderthalensis,* lived in Europe and western Asia from some 300,000 to 30,000 years ago. For at least 30,000 years, Neanderthals coexisted with anatomically modern humans (*Homo sapiens*) in several regions. Genetic analysis using DNA sequencing has answered several questions about Neanderthals and modern humans: (1) Were Neanderthals direct ancestors of modern humans? (2) Did Neanderthals and *H. sapiens* interbreed, so that descendants of the Neanderthals are alive today? Or did the Neanderthals die off and become extinct? (3) What can we say about the similarities and differences between our genome and that of the Neanderthals?

These and other questions were answered by researchers who were able to extract and analyze DNA from Neanderthal bones. In 1997, workers extracted mitochondrial DNA (mtDNA) fragments from a Neanderthal skeleton found in Feldhofer cave near Düsseldorf, Germany. As discussed in Chapter 4, mitochondria are maternally inherited organelles, and their genome (see Chapter 11) encodes proteins essential to mitochondrial function.

DNA nucleotide differences between Neanderthal mtDNA and those of more than 2000 present-day humans were used to construct a phylogenetic tree to establish the evolutionary relationships among the ancient Neanderthals and modern human populations [Figure 23–23(a), p. 502]. From the structure of the tree, researchers concluded that Neanderthals are a

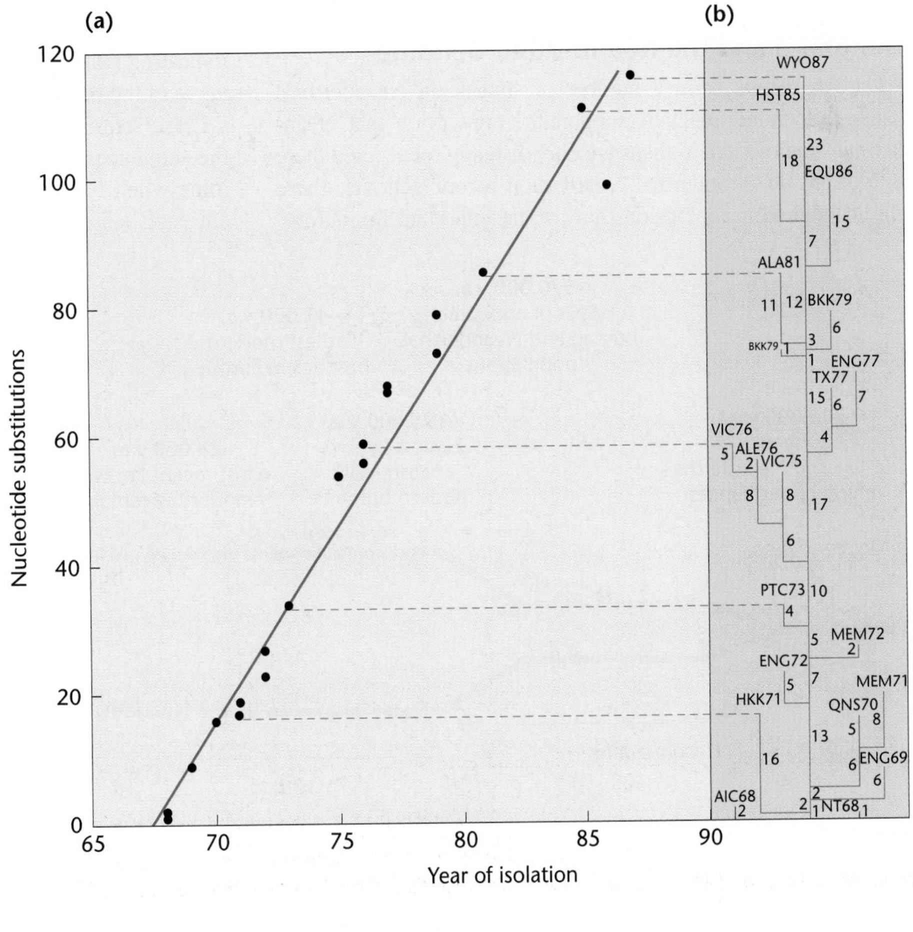

FIGURE 23–22 Phylogenetic molecular clock in the influenza A hemagglutinin gene. (a) Number of nucleotide differences between the first isolate as a function of year of isolation. (b) Estimate of the phylogeny of the isolates.

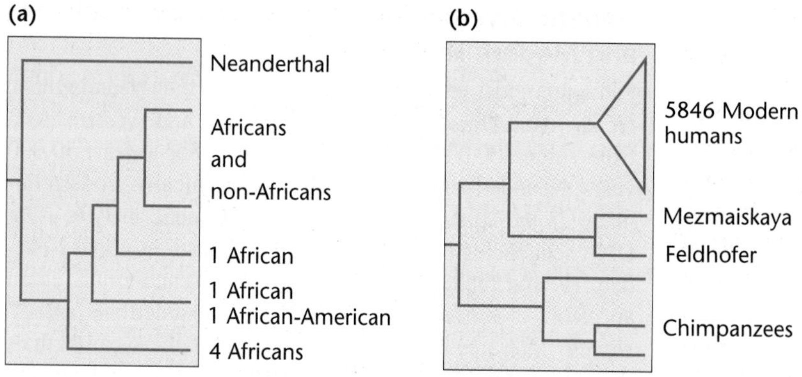

FIGURE 23–23 (a) A phylogenetic tree estimated from mitochondrial DNA sequences of one Neanderthal and over 2000 modern humans. *From Krings, M. et al., 1997.* Cell *90: 19–30. Figure 7A, p.26, copyright 1997 with permission from Elsevier;* (b) A phylogenetic tree estimated from analysis of over 5000 modern humans, the Neanderthal samples from the Feldhofer cave and the Mezmaiskaya cave, and from chimpanzees. *From Ovchinnikov, I.V. et al., 2000. Molecular analysis of Neanderthal DNA from the northern Caucasus.* Nature *404: 490–493 copyright 2000 Macmillan Publishers Ltd.*

distant relative of modern humans. Using a molecular clock calibrated with chimpanzee and human sequences, the researchers calculated that based on mtDNA sequence differences, the last common ancestor of Neanderthals and modern humans lived roughly 600,000 years ago, four times as long ago as the last common ancestor of all modern humans.

Other researchers successfully extracted mitochondrial DNA from Neanderthal bones found in Mezmaiskaya cave in the Caucasus Mountains east of the Black Sea. Although the two Neanderthal sequences (i.e., from Feldhofer and Mezmaiskaya) are from locations more than 1000 miles apart, they vary by only about 3.5 percent. This indicates that the Neanderthal samples derive from a single gene pool. Phylogenetic analysis places the two Neanderthals in a group that is clearly distinct from modern humans [Figure 23–23(b)]. The conclusion from these two studies is that while Neanderthals and humans have a common ancestor, the Neanderthals were a separate hominid line and did not contribute mitochondrial genes to *H. sapiens.*

Genomics and the Neanderthal Genome

Although results from a number of studies on Neanderthals using mitochondrial DNA support the conclusion that Neanderthals and modern humans are only distantly related and there is little or no Neanderthal contribution to our genome, these studies and their conclusions have some important limitations. First, only a few hundred base pairs of mitochondrial DNA were analyzed, and sequencing of more mitochondrial DNA might show interbreeding between the two species. Second, because mitochondria are maternally inherited, they are a record of only female genetic history and exclude any male genetic contributions. Studies of genomic sequences from cell nuclei include both female and male contributions to the genome and are needed to complete the picture and clarify what relationship our species might have with Neanderthals.

Recently, two research groups have successfully recovered, sequenced, and analyzed genomic sequences from Neanderthal remains. Through use of different methods developed independently by each team, about 65,000 base pairs of Neanderthal sequence were obtained by one team, and more than 1 million nucleotides of the Neanderthal genome were sequenced by the other team. Sequence analysis indicates that Neanderthals and modern humans have genomes that are more than 99.5 percent identical. The Neanderthal genomic DNA analyzed to date represents sequences that are present on all human chromosomes, indicating that the DNA sequenced to date is broadly representative of the Neanderthal genome.

These sequence data have been analyzed and compared to the sequences of modern humans and chimpanzees to determine when Neanderthals and humans last shared a common ancestor (Figure 23–24), assuming that chimpanzees and

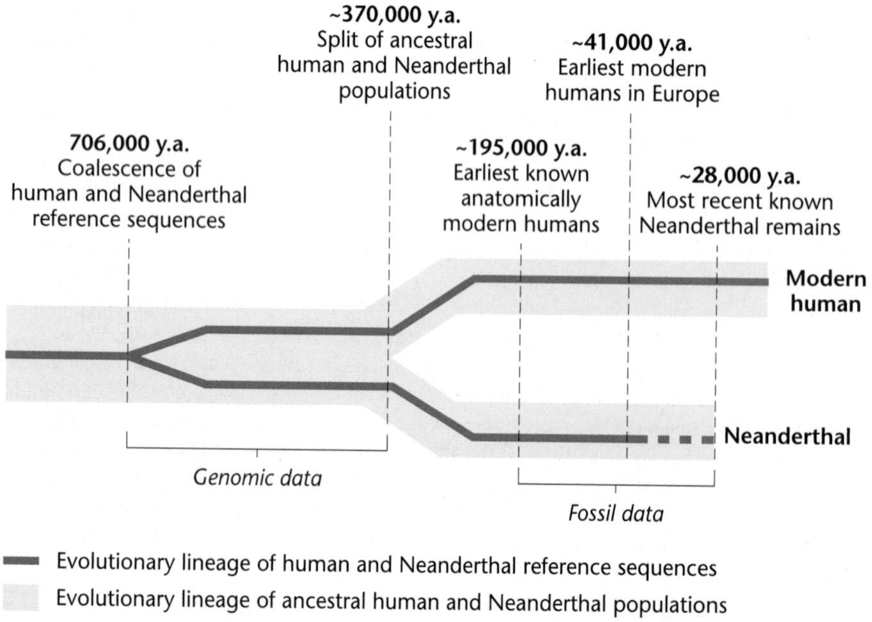

FIGURE 23–24 Estimated times of divergence of human and Neanderthal genomic sequences and, subsequently, of their populations, relative to landmark events in both human and Neanderthal evolution. These estimates are based on sequencing about 65,000 base pairs of Neanderthal DNA (y.a. = years ago). *From Noonan, J.P. et al. 2006. Sequencing and analysis of Neanderthal genomic DNA.* Science *314: 1113–1118.*

humans last shared a common ancestor about 6.5 million years ago. The different methods used by the two teams produced slightly divergent dates for the events in question. One team combined their genomic data with the fossil data to show that Neanderthals and modern humans last shared a common ancestor about 700,000 years ago and that the split between Neanderthals and modern humans occurred about 370,000 years ago. The other team concluded that the human and Neanderthal lines split about 500,000 years ago, a number supported by the analysis of mitochondrial DNA. These studies confirmed the earlier studies and reinforce the conclusions about the relationship between our species and the Neanderthal species. First, Neanderthals are not direct ancestors of our species. Second, Neanderthals and members of our species may have interbred, but from the results available, it appears that Neanderthals did not make major contributions to our genome. As a species, Neanderthals are extinct, but some of their genes may survive as part of our genome. Third, from what little we know about the Neanderthal genome, we share most genes and other sequences with them.

More exciting answers to the question about the similarities and differences between our genome and that of Neanderthals could be derived by sequencing the entire Neanderthal genome. Using several new methods, a number of research groups are now working to isolate, sequence, and analyze the Neanderthal genome, with the goal of completing a draft sequence in a few years. Analysis of the genomes of chimpanzees, Neanderthals, and modern humans would allow us to identify key differences that define our species and, in the process, revolutionize the field of human evolution.

EXPLORING GENOMICS

The Y Chromosome Haplotype Reference Database (YHRD)

In this chapter we examined important principles of population and evolutionary genetics including examples of allele frequency variations in human populations. Among the variations studied by population geneticists are *short tandem repeats (STRs),* which are short sequences of tandem base repeats that are typically 2 to 9 bp in length. Often, STRs are inherited as *haplotypes,* clusters or blocks of sequence variations that are closely linked on a chromosome and typically inherited as a unit.

In this exercise you will use the **Y Chromosome Haplotype Reference Database (YHRD)** to learn about population frequencies for STRs on the Y chromosome.

■ Exercise I: The Y Chromosome Haplotype Reference Database (YHRD)

Y chromosome sequences have been used to track human migration patterns around the world. The **Y Chromosome Haplotype Reference Database (YHRD)** is a great program for examining haplotype frequencies for Y chromosome loci, particularly STRs on the Y chromosome. Y chromosome STR haplotypes have been very useful for tracing paternal lineages in genealogical and kinship testing and for identifying male DNA in crime scene samples.

1. Visit the YHRD site at **http://www.yhrd.org/index.html**. Explore the links at this site to learn more about the purpose of YHRD and features of the Y chromosome that can be used for genealogy and forensic applications.
2. Use the "Haplotype" characteristics link to learn more about STRs on the Y chromosome and the position of specific Y-STR loci; then answer the following questions:
   a. What are minimal haplotypes (minHts)?
   b. Is the locus for DYS390 located on the p or q arm of the Y?
3. Use the "Population analyses" link to learn more about the frequency of minHts in different populations. The "Per population feature" will show you a drop-down menu of different choices. For example, what is the most frequent minHt for people in Verona, Italy? View other populations you are interested in.
4. Use the "Search database" link "GeoSearch" feature to search for a particular haplotype frequency of your choice.
   a. What does it mean if a particular population has a "10" for a particular STR locus such as DYS390?
   b. Explore the "GeoSearch" feature by entering a haplotype combination of your choice. What did you find? Can you find a haplotype combination that does not match any of the haplotypes in the YHRD?
5. Use the "Compare populations" feature of "Population analyses" to compare minHt frequencies for people from "USA[European American]" with frequencies for people from "Birmingham, UK." Do the same to compare "USA[European American]" with "Virginia, USA[Hispanic American]" and to compare "USA[European American]" with "West Africa." What do these results tell us about the ancestry of these populations? Are these data consistent with what you have learned about migration and gene flow? What do these results tell us about the relatedness of these populations based on the Y-minHts evaluated?
6. Explore the site to learn more about Y-STRs in other populations that interest you.

CASE STUDY An unexpected outcome

A newborn screening program identified a baby with a rare autosomal recessive disorder called arginosuccinate aciduria (AGA), which causes high levels of ammonia to accumulate in the blood. Symptoms usually appear in the first week after birth and can progress to include severe liver damage, developmental delay, and mental retardation. AGA occurs with a frequency of about 1 in 70,000 births. There is no history of this disorder in either the father's or mother's family. The above case raises several questions:

1. Since it appears that the unaffected parents are heterozygotes, would it be considered unusual that there would be no family history of the disorder? How would they be counseled about risks to future children?
2. If the disorder is so rare, what is the frequency of heterozygous carriers in the population?
3. What are the chances that two heterozygotes will meet and have an affected child?

INSIGHTS AND SOLUTIONS

1. Tay–Sachs disease is caused by loss-of-function mutations in a gene on chromosome 15 that encodes a lysosomal enzyme. Tay-Sachs is inherited as an autosomal recessive condition. Among Ashkenazi Jews of Central European ancestry, about 1 in 3600 children is born with the disease. What fraction of the individuals in this population are carriers?

Solution: If we let p represent the frequency of the wild-type enzyme allele and q the total frequency of recessive loss-of-function alleles, and if we assume that the population is in Hardy–Weinberg equilibrium, then the frequencies of the genotypes are given by p^2 for homozygous normal, $2pq$ for carriers, and q^2 for individuals with Tay–Sachs. The frequency of Tay–Sachs alleles is thus

$$q = \sqrt{q^2} = \sqrt{\frac{1}{3600}} = 0.017$$

Since $p + q = 1$, we have

$$p = 1 - q = 1 - 0.017 = 0.983$$

Therefore, we can estimate that the frequency of carriers is

$$2pq = 2(0.983)(0.017) = 0.033 \text{ or about 1 in 30}$$

2. A single plant twice the size of others in the same population suddenly appears. Normally, plants of that species reproduce by self-fertilization and by cross-fertilization. Is this new giant plant simply a variant, or could it be a new species? How would you determine which it is?

Solution: One of the most widespread mechanisms of speciation in higher plants is polyploidy, the multiplication of entire sets of chromosomes. The result of polyploidy is usually a larger plant with larger flowers and seeds. There are two ways of testing the new variant to determine whether it is a new species. First, the giant plant should be crossed with a normal-sized plant to see whether the giant plant produces viable, fertile offspring. If it does not, then the two different types of plants would appear to be reproductively isolated. Second, the giant plant should be cytogenetically screened to examine its chromosome complement. If it has twice the number of its normal-sized neighbors, it is a tetraploid that may have arisen spontaneously. If the chromosome number differs by a factor of two and the new plant is reproductively isolated from its normal-sized neighbors, it is a new species.

PROBLEMS AND DISCUSSION QUESTIONS

1. What types of nucleotide substitutions will not be detected by electrophoretic studies of a gene's protein product?
2. Price et al. (1999. *J. Bacteriol.* 181: 2358–2362) conducted a genetic study of the toxin transport protein (PA) of *Bacillus anthracis,* the bacterium that causes anthrax in humans. Within the 2294-nucleotide gene in 26 strains they identified five point mutations—two missense and three synonyms—among different isolates. Necropsy samples from an anthrax outbreak in 1979 revealed a novel missense mutation and five unique nucleotide changes among ten victims. The authors concluded that these data indicate little or no horizontal transfer between different *B. anthracis* strains.
   (a) Which types of nucleotide changes (missense or synonyms) cause amino acid changes?
   (b) What is meant by horizontal transfer?
   (c) On what basis did the authors conclude that evidence of horizontal transfer is absent from their data?
3. The genetic difference between two *Drosophila* species, *D. heteroneura* and *D. sylvestris,* as measured by nucleotide diversity, is about 1.8 percent. The difference between chimpanzees (*P. troglodytes*) and humans (*H. sapiens*) is about the same, yet the latter species are classified in different genera. In your opinion, is this valid? Explain why.
4. The use of nucleotide sequence data to measure genetic variability is complicated by the fact that the genes of higher eukaryotes are complex in organization and contain 5′ and 3′ flanking regions as well as introns. Researchers have compared the nucleotide sequence of two cloned alleles of the *γ-globin* gene from a single individual and found a variation of 1 percent. Those differences include 13 substitutions of one nucleotide for another and 3 short DNA segments that have been inserted in one allele or deleted in the other. None of the changes takes place in the gene's exons (coding regions). Why do you think this is so, and should it change our concept of genetic variation?

5. The ability to taste the compound PTC is controlled by a dominant allele *T*, while individuals homozygous for the recessive allele *t* are unable to taste PTC. In a genetics class of 125 students, 88 can taste PTC and 37 cannot. Calculate the frequency of the *T* and *t* alleles in this population and the frequency of the genotypes.
6. Calculate the frequencies of the *AA*, *Aa*, and *aa* genotypes after one generation if the initial population consists of 0.2 *AA*, 0.6 *Aa*, and 0.2 *aa* genotypes and meets the requirements of the Hardy–Weinberg relationship. What genotype frequencies will occur after a second generation?
7. Consider rare disorders in a population caused by an autosomal recessive mutation. From the frequencies of the disorder in the population given, calculate the percentage of heterozygous carriers:
   (a) 0.0064
   (b) 0.000081
   (c) 0.09
   (d) 0.01
   (e) 0.10
8. What must be assumed in order to validate the answers in Problem 7? See Now Solve This on page 487.
9. In a population where only the total number of individuals with the dominant phenotype is known, how can you calculate the percentage of carriers and homozygous recessives?
10. Determine whether the following two sets of data represent populations that are in Hardy–Weinberg equilibrium (use χ^2 analysis if necessary):
    (a) *CCR5* genotypes: *1/1*, 60 percent; $1/\Delta 32$, 35.1 percent; $\Delta 32/\Delta 32$, 4.9 percent
    (b) Sickle-cell hemoglobin: *AA*, 75.6 percent; *AS*, 24.2 percent; *SS*, 0.2 percent
    See Now Solve This on page 489.
11. If 4 percent of a population in equilibrium expresses a recessive trait, what is the probability that the offspring of two individuals who do not express the trait will express it?
12. Consider a population in which the frequency of allele *A* is $p = 0.7$ and the frequency of allele *a* is $q = 0.3$, and where the alleles are codominant. What will be the allele frequencies after one generation if the following occurs?
    (a) $w_{AA} = 1$, $w_{Aa} = 0.9$, and $w_{aa} = 0.8$
    (b) $w_{AA} = 1$, $w_{Aa} = 0.95$, and $w_{aa} = 0.9$
    (c) $w_{AA} = 1$, $w_{Aa} = 0.99$, $w_{aa} = 0.98$
    (d) $w_{AA} = 0.8$, $w_{Aa} = 1$, $w_{aa} = 0.8$
13. If the initial allele frequencies are $p = 0.5$ and $q = 0.5$ and allele *a* is a lethal recessive, what will be the frequencies after 1, 5, 10, 25, 100, and 1000 generations?
14. Under what circumstances might a lethal dominant allele persist in a population?
15. Assume that a recessive autosomal disorder occurs in 1 of 10,000 individuals (0.0001) in the general population and that in this population about 2 percent (0.02) of the individuals are carriers for the disorder. Estimate the probability of this disorder occurring in the offspring of a marriage between first cousins. Compare this probability to the population at large.
16. If the albino phenotype occurs in 1/10,000 individuals in a population at equilibrium and albinism is caused by an autosomal recessive allele *a*, calculate the frequency of
    (a) the recessive mutant allele
    (b) the normal dominant allele
    (c) heterozygotes in the population
    (d) matings between heterozygotes
    See Now Solve This on page 490.
17. One of the first Mendelian traits identified in humans was a dominant condition known as *brachydactyly.* This gene causes an abnormal shortening of the fingers or toes (or both). At the time, some researchers thought that the dominant trait would spread until 75 percent of the population would be affected (because the phenotypic ratio of dominant to recessive is 3:1). Show that the reasoning was incorrect.
18. A prospective groom, who is normal, has a sister with cystic fibrosis (CF), an autosomal recessive disease. Their parents are normal. The brother plans to marry a woman who has no history of CF in her family. What is the probability that they will produce a CF child? They are both Caucasian, and the overall frequency of CF in the Caucasian population is 1/2500—that is, 1 affected child per 2500. (Assume the population meets the Hardy–Weinberg assumptions.) See Now Solve This on page 497.
19. Describe how populations with substantial genetic differences can form. What is the role of natural selection?
20. Achondroplasia is a dominant trait that causes a characteristic form of dwarfism. In a survey of 50,000 births, five infants with achondroplasia were identified. Three of the affected infants had affected parents, while two had normal parents. Calculate the mutation rate for achondroplasia and express the rate as the number of mutant genes per given number of gametes.
21. A recent study examining the mutation rates of 5669 mammalian genes (17,208 sequences) indicates that, contrary to popular belief, mutation rates among lineages with vastly different generation lengths and physiological attributes are remarkably constant (Kumar, S., and Subramanian S. 2002. *Proc. Natl. Acad. Sci. [USA]* 99: 803–808). The average rate is estimated at 12.2×10^{-9} per bp per year. What is the significance of this finding in terms of mammalian evolution?
22. A form of dwarfism known as Ellis–van Creveld syndrome was first discovered in the late 1930s, when Richard Ellis and Simon van Creveld shared a train compartment on the way to a pediatrics meeting. In the course of conversation, they discovered that they each had a patient with this syndrome. They published a description of the syndrome in 1940. Affected individuals have a short-limbed form of dwarfism and often have defects of the lips and teeth, and polydactyly (extra fingers). The largest pedigree for the condition was reported in an Old Order Amish population in eastern Pennsylvania by Victor McKusick and his colleagues (1964). In that community, about 5 per 1000 births are affected, and in the population of 8000, the observed frequency is 2 per 1000. All affected individuals have unaffected parents, and all affected cases can trace their ancestry to Samuel King and his wife, who arrived in the area in 1774. It is known that neither King nor his wife was affected with the disorder. There are no cases of the disorder in other Amish communities, such as those in Ohio or Indiana.
    (a) From the information provided, derive the most likely mode of inheritance of this disorder. Using the Hardy–Weinberg law, calculate the frequency of the mutant allele in the population and the frequency of heterozygotes, assuming Hardy–Weinberg conditions.
    (b) What is the most likely explanation for the high frequency of the disorder in the Pennsylvania Amish community and its absence in other Amish communities?
23. List the barriers that prevent interbreeding and give an example of each.
24. What are the two groups of reproductive isolating mechanisms? Which of these is regarded as more efficient, and why?
25. In a recent study of cichlid fish inhabiting Lake Victoria in Africa, Nagl et al. (1998. *Proc. Natl. Acad. Sci. [USA]* 95: 14,238–14,243)

examined suspected neutral sequence polymorphisms in noncoding genomic loci in 12 species and their putative river-living ancestors. At all loci, the same polymorphism was found in nearly all of the tested species from Lake Victoria, both lacustrine and riverine. Different polymorphisms at these loci were found in cichlids at other African lakes.

(a) Why would you suspect neutral sequences to be located in noncoding genomic regions?

(b) What conclusions can be drawn from these polymorphism data in terms of cichlid ancestry in these lakes?

26. Given that there are approximately 400 cichlid species in Lake Victoria and that it dried up almost completely about 14,000 years ago, what evidence indicates that extremely rapid evolutionary adaptation rather than extensive immigration occurred?
27. What genetic changes take place during speciation?
28. Some critics have warned that the use of gene therapy to correct genetic disorders will affect the course of human evolution. Evaluate this criticism in light of what you know about population genetics and evolution, distinguishing between somatic gene therapy and germ-line gene therapy.
29. Comparisons of Neanderthal mitochondrial DNA with that of modern humans indicate that they are not related to modern humans and did not contribute to our mitochondrial heritage. However, because Neanderthals and modern humans are separated by at least 25,000 years, this does not rule out some forms of interbreeding causing the modern European gene pool to be derived from both Neanderthals and early humans (called Cro-Magnons). To resolve this question, Caramelli et al. (2003. *Proc. Natl. Acad. Sci. [USA]* 100: 6593–6597) analyzed mitochondrial DNA sequences from 25,000-year-old Cro-Magnon remains and compared them to four Neanderthal specimens and a large data set derived from modern humans. The results are shown in the graph.

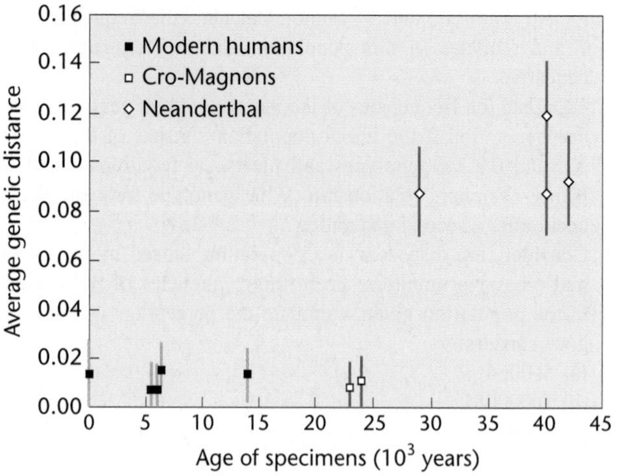

The x-axis represents the age of the specimens in thousands of years; the y-axis represents the average genetic distance. Modern humans are indicated by filled squares; Cro-Magnons, open squares; and Neanderthals, diamonds.

(a) What can you conclude about the relationship between Cro-Magnons and modern Europeans? What about the relationship between Cro-Magnons and Neanderthals?

(b) From these data, does it seem likely that Neanderthals made any contributions to the Cro-Magnon gene pool or the modern European gene pool?

Seeds of rare crop varieties, cryogenically preserved at the U.S. Department of Agriculture National Seed Storage Laboratory.

24

Conservation Genetics

Chapter Concepts

- Species require genetic diversity for long-term survival and adaptation.
- Small, isolated populations are particularly vulnerable to genetic effects.
- Endangered populations that recover in size may not redevelop genetic diversity.
- Conservation and breeding efforts focus on retaining genetic diversity for long-term species survival.

As the twenty-first century progresses, the diversity of life on Earth is under increasing pressure from the direct and indirect effects of explosive human population growth. Approximately 10 million *Homo sapiens* lived on the planet 10,000 years ago. This number grew to 100 million 2000 years ago and to 2.5 billion by 1950. Within the span of a single lifetime, the world's human population more than doubled to 5.5 billion in 1993 and is projected to reach as high as 19 billion by 2100 (Figure 24–1).

The effect of accelerating human population growth on other species has been dramatic. Data from the 2008 World Conservation Union (IUCN) Red List of Threatened Species show that, globally, 21 percent of all mammals, 12 percent of birds, 31 percent of reptiles, 30 percent of amphibians, and 37 percent of fish species are threatened. Plants are not faring much better: the 2008 IUCN Red List also includes over 8400 vascular plant species worldwide. Not only wild species are at risk. Genetic diversity in domesticated plants and animals is also being lost as many traditional crop varieties and livestock breeds disappear. The Food and Agriculture Organization (FAO) estimates that since 1900, 75 percent of the genetic diversity in agricultural crops has been lost. Out of approximately 5000 different breeds of domesticated farm animals worldwide, one-third are at risk.

Why should we be concerned about losing **biodiversity**—that is, the biological variation represented by these different plants and animals? After all, the fossil record shows that many different plants and animals have become extinct in the course of evolution, even before humans existed, and other species have taken their places. Biologists are concerned, however, at the accelerated rate of species extinctions we are witnessing today, all of which can be ascribed directly or indirectly to human impact. Deliberate hunting or harvesting of plants and animals by humans, and habitat destruction through human development activities have greatly reduced the populations of many species. An additional problem in many parts of the world is the deliberate or accidental introduction by humans of invasive nonnative plants or animals that prey on or compete with native species, further jeopardizing their survival. For example, the brown tree snake, *Boiga irregularis* (Figure 24–2), was accidentally transported by ship in the late 1940s to the island of Guam. An aggressive predator, the brown tree snake has so far caused the extinction of 12 bird and 4 lizard species that were previously native to Guam. This introduced species continues to threaten native biodiversity on the island.

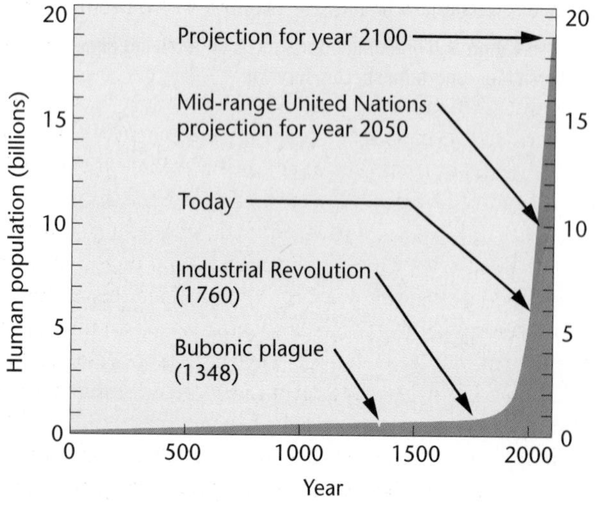

FIGURE 24–1 Growth in human population over the past 2000 years and projected through 2100.

FIGURE 24–2 The brown tree snake (*Boiga irregularis*).

One of the greatest threats to biodiversity, however, may be climate change due to global warming. Scientists are increasingly concerned that this phenomenon will lead to the extinction of many species, especially those adapted to live in colder environments. For example, the IUCN 2008 Red List classifies the 22,000 remaining polar bears (Figure 24–3) as vulnerable due to fears that their habitat will be lost from global warming, and a recent study by scientists from the Woods Hole Oceanographic Institute predicts that loss of sea ice will result in the disappearance of 95 percent of emperor penguin populations by 2100.

As a result of species extinction, biologists fear that **ecosystems**—the complex webs of interdependent but diverse plants and animals found together in the same environment—may collapse if key sustaining species are lost. This portends possible consequences for our own long-term survival. Other scientists have pointed out that we lose unknown economic potential of unexploited plants and animals if we allow them

FIGURE 24–3 The polar bear (*Ursus maritimus*).

to become extinct. Rare species sometimes turn out to be immensely valuable. For example, an obscure bacterium, *Thermus aquaticus,* was first discovered in a thermal pool in Yellowstone National Park and is known to exist in only a few hot springs throughout the world. DNA polymerase isolated from this microorganism functions at very high temperatures and is critical to the polymerase chain reaction (PCR), a process for *in vitro* replication of DNA that we have alluded to throughout this book. This research technique is essential to the multibillion dollar global biotechnology industry. Quite apart from the practical benefits of biodiversity, scientists and nonscientists have made the case that human life will be diminished both spiritually and aesthetically if we do not strive to maintain the fascinating and often beautiful variety of living organisms that share the planet with us.

Conservation biologists work to understand and maintain biodiversity, studying the factors that lead to species decline and the ways species can be preserved. The new field of **conservation genetics** has emerged in the last 25 years as scientists have begun to recognize that genetics will be an important tool in maintaining and restoring population viability. The applications of genetics to conservation biology are multifaceted; in this chapter, we will explore just a few of them. Underlying the increasingly important role of genetics in conservation biology is the recognition that biodiversity depends on genetic diversity and that maintaining biodiversity in the long term is unlikely if genetic diversity is lost.

How Do We Know?

In this chapter, we focus on conservation genetics, emphasizing how geneticists assess genetic diversity and work to maintain species survival. Along the way, we will find many opportunities to consider the methods and reasoning by which several of these practices were developed. From the explanations given in the chapter, what answers would you propose to the following fundamental questions?

1. How do we know the extent of genetic diversity in a species?
2. How do we know that diminished genetic diversity is detrimental to species survival?
3. How do we attempt to "conserve" existing genetic diversity?

24.1 Genetic Diversity Is the Goal of Conservation Genetics

While *biodiversity* refers to the variation represented by all existing species of plants and animals on this planet at any given time, **genetic diversity** is not as easy to define or to study. Genetic diversity can be considered on two levels: *interspecific diversity* and *intraspecific diversity.*

FIGURE 24–4 (a) A tropical rainforest is an ecosystem with high interspecific diversity. (b) A coastal marsh in North Carolina exemplifies an ecosystem with low interspecific diversity.

Diversity between species, or **interspecific diversity,** is the diversity reflected in the number of different plant and animal species present in an ecosystem. Some ecosystems have a very high level of interspecific diversity, such as a tropical rainforest in which hundreds of different plant and animal species may be found within a few square meters [Figure 24–4(a)]. Other ecosystems, especially where plants and animals must adapt to a harsh environment, may have much lower numbers of species [Figure 24–4(b)]. Species inventories—lists of the different plant and animal species found in a particular environment—are useful for identifying diversity hot spots: geographic areas with especially high interspecific diversity, where conservation efforts can be focused. Conservation biologists working at the ecosystem level are interested in preventing species from being lost and in restoring species that were once part of the system but are no longer present. The reestablishment of the gray wolf (*Canis lupus*) in Yellowstone National Park and the release of captive-bred California condors (*Gymnogyps californianus*) into the mountain ranges they previously occupied in the southwestern United States are examples of current attempts by conservation biologists to restore missing species to their ecosystems.

Intraspecific diversity—the diversity within a species—itself has two components: *intrapopulation diversity* is the genetic variation occurring between individuals within a single population of a given species; and *interpopulation diversity* is the variation occurring between different populations of the same species. Genetic variation within populations can be measured as the frequency of individuals in the population that are heterozygous at a given locus or as the number of different alleles at a locus that are present in the population gene pool. When DNA-profiling techniques are used, the percentage of polymorphic loci—represented by bands on the DNA profile that are different in different individuals—can be calculated to indicate the extent of genetic diversity in a population. In outbreeding species, most intraspecific genetic diversity is found at the intrapopulation level.

Significant interpopulation diversity can occur if populations are separated geographically and there is no migration or exchange of gametes between them. Similarly, predominantly inbreeding species (such as self-fertilizing plants) tend to have greater levels of interpopulation than intrapopulation diversity. In these species, a limited number of genotypes dominate an individual population, but there is greater differentiation between populations. Understanding the breeding system and the distribution of genetic variation in an endangered species is important to its conservation, not only to ensure continued production of offspring but also to determine the best strategy for maintaining intraspecific diversity. For example, would it be more effective to preserve a few large populations or many small, distinct ones? This information can then be used to guide conservation or restoration efforts.

Loss of Genetic Diversity

Loss of genetic diversity in nondomesticated species is often associated with a reduction in population size owing in part to excessive hunting or harvesting. For example, biologists blame commercial overfishing for the collapse in the early 1990s of the deep-sea cod populations off the Newfoundland coast. The cod fishery in that area was once one of the most productive in the world, but despite being protected from fishing since 1992, these fish populations have not recovered. Habitat loss is also a major cause of population decline. As the global human population increases, more land is developed for housing and transport systems or is put into agricultural production, reducing or eliminating areas that were once home to wild plants and animals. The shrinking of available habitat reduces populations of wild species and often also isolates them from each other, as individual populations become trapped in pockets of undeveloped land surrounded by areas taken over for agriculture, urban development, or other human uses. This process is known as **population fragmentation.** When populations are no longer in contact with each other, **gene flow,** the gradual exchange of alleles between two populations, brought about through migration or gamete exchange between them, ceases, and an important mechanism for maintaining genetic variation is lost.

In domesticated species, loss of genetic diversity is not usually the result of habitat loss or collapsing population numbers; there is little risk that cows or corn as species will become extinct any time soon. Reduction in diversity within domesticated species can instead be traced to changes in agricultural practice and consumer demand. Modern farming techniques have greatly increased production levels, but they have also led to greater genetic uniformity. As farmers switch to new crop varieties or improved livestock strains on a large scale, they abandon cultivation of many older local types, which may then disappear if efforts are not made to preserve them.

For example, Seed Savers Exchange, an organization working to preserve traditional fruit and vegetable varieties, reports that in 1904 a total of 7098 apple varieties were recorded in North America by U.S. Department of Agriculture scientist W. H. Ragan. Fewer than 14 percent of those varieties can still be found today, with just 15 varieties accounting for over 90 percent of the apples sold in U.S. grocery stores. Modern varieties may be better adapted for modern agricultural production, but many older types contain useful genes that can still play a vital role in survival functions, such as resistance to disease, cold, or drought.

Identifying Genetic Diversity

For many years, population geneticists based estimates of intraspecific diversity on phenotypic differences between individuals, such as different colors of seeds or flowers or variation in markings (Figure 24–5). DNA analysis is a more modern and direct molecular approach for detecting and quantifying genetic differences between individuals. This technique has become an important tool for assessing intra- and interpopulation genetic variation. Nuclear, mitochondrial, and chloroplast DNA can all be analyzed to determine levels of genetic diversity. We have already described applications of DNA analysis using either restriction enzymes or PCR in Chapter 19. Conservation biologists have used similar techniques to examine the levels and distribution of genetic variation both within and between populations and to subsequently guide conservation efforts.

One technique that is widely used to analyze genetic diversity in plant and animal populations takes advantage of the presence within the genome of **STRs (short tandem repeats),** also known as **microsatellites.** STRs consist of short DNA sequences (two to nine bases) repeated a variable number of times and are found throughout the genome. The number of times a repeat is present at a given STR locus often varies between individuals. This variation is detected by sequencing the DNA regions on either side of the STR locus and then creating single-stranded DNA primers complementary to these flanking sequences that will selectively anneal to them during the polymerase chain reaction (PCR), which is explained in more detail in Chapter 17. This allows targeted amplification of the microsatellite regions so that they can be analyzed and the number of repeats determined. Although the STR regions themselves do not contain expressed genes, conservation biologists can use microsatellite variation among individuals within populations as an indirect estimate of the amount of overall genetic variation present.

FIGURE 24–5 Phenotypic variation in seed color and markings in the common bean (*Phaseolus vulgaris*) reveals high levels of intraspecific diversity.

For example, researchers from the University of Arizona recently used microsatellite analysis to compare genetic variation in the three remaining populations of the Mauna Loa silversword, an endangered plant found only in Hawaii. Biologists working to conserve this plant needed to know whether they should try to increase genetic diversity by redistributing seed among the populations. However, STR analysis not only revealed unexpected amounts of diversity still present within populations but also showed that the three populations were genetically quite different from one another. The seed redistribution plan was abandoned in favor of preserving the unique identity of each population.

Another application of DNA markers such as microsatellites in conservation involves *PCR-based DNA fingerprinting*. The use of DNA-profiling techniques in forensic science was discussed in Chapter 19. Similar techniques can be used to uncover illegal trade in endangered plant and animal species that are protected by law. For example, although international trade in ivory is illegal, poaching of African elephants and smuggling of ivory worth millions of dollars continue. To determine the origin of ivory found in a secret compartment in a shipping container in Hong Kong in 2008, an international group of researchers used microsatellite markers to analyze DNA extracted from the seized ivory. They compared this microsatellite profile with more than 600 reference samples of DNA collected from the droppings of wild elephant populations across Africa and found that it matched those of forest elephants from southern Gabon. This finding provided a vital clue about poaching rings that can be used to combat this illicit trade.

Unlawful trade in animal products also occurs in the United States. In a recent study by scientists at the University of Florida, amplification of mtDNA sequences from turtle meat sold in Louisiana and Florida revealed that one quarter of the samples were actually alligator, not turtle. Most of the rest were from small freshwater turtles, indicating that populations of freshwater turtle species were declining through overharvesting and thus were in need of protection. Further refinements to DNA extraction and profiling techniques allowed accurate identification of illegally harvested sea turtle eggs cooked and served in restaurants, and led to prosecution of the poachers.

New methods for analyzing DNA marker data also permit estimation of past changes in populations, based on the rate at which mtDNA is known to accumulate spontaneous point mutations. In a controversial recent study, scientists at Stanford University led by Dr. Stephen Palumbi measured sequence diversity at key points in mtDNA from humpback whales and compared their findings with the amount of diversity expected for a given population size. The Stanford researchers found much more mtDNA sequence diversity than expected and concluded that before being hunted almost to extinction in the nineteenth and twentieth centuries, the global humpback whale population must have numbered over one million, ten times higher than historical estimates based on records from whaling ships. Because current international agreements could permit hunting of whales to resume when populations have recovered to 54 percent of what is thought to be their original carrying capacity, Palumbi's findings are important for the future management of this species.

ESSENTIAL POINT

Conservation genetics applies principles of population genetics to the preservation and restoration of threatened species, with the major goal being the maintenance of both interspecific and intraspecific diversity.

24.2 Population Size Has a Major Impact on Species Survival

Some species have never been numerous, especially those that are adapted to survive in unusual habitats. Biologists refer to such species as *naturally rare. Newly rare* species, on the other hand, are those whose numbers are in decline because of pressures such as habitat loss. Populations of such species may not only be small in number but also fragmented and isolated from other populations. Both decreased population size and increased isolation have important genetic consequences, leading to an increased risk of loss of diversity.

How small must a population be before it is considered endangered? The answer varies somewhat with species, but small populations can quickly become vulnerable to genetic phenomena that increase the risk of extinction. In general, a population of fewer than 100 individuals is considered extremely sensitive to these problems, which include genetic drift, inbreeding, and reduction in gene flow. The effects of such problems on species survival are substantial. Studies of bighorn sheep, for example, have shown that populations of fewer than 50 are highly likely to become extinct within 50 years. Projections based on computer models show that for all species, populations of fewer than 10,000 are likely to be limited in adaptive genetic variation and that at least 100,000 individuals must be present if a population is to show long-term sustainability.

Determining the number of individuals a population must contain in order to have long-term sustainability is complicated by the fact that not all members of a population are equally likely to produce offspring: some will be infertile, too young, or too old. The *effective population size* (N_e) is defined as the number of individuals in a population having an equal probability of contributing gametes to the next generation. N_e is almost always smaller than the *absolute population size* (N). The effective population size can be calculated in different ways, depending on the factors that are preventing individuals in a population from contributing equally to the next generation. In a sexually reproducing population that contains different numbers of males and females, for example, the effective population size is calculated as

$$N_e = \frac{4(N_m N_f)}{N_m + N_f}$$

where N_m is the number of males and N_f the number of females in the population. Hence a population of 100 males and

100 females would have an effective size of $4(100 \times 100)/(100 + 100) = 200$. In contrast, if there were 180 males and only 20 females, the effective population size would be $4(180 \times 20)/(180 + 20) = 72$.

Effective population size is also influenced by fluctuations in absolute population size from one generation to the next. Here, the effective population size is the harmonic mean of the numbers in each generation, so

$$N_e = 1 \Big/ \frac{1}{t}\left(\frac{1}{N_1} + \frac{1}{N_2} \ldots + \frac{1}{N_t}\right)$$

where t is the total number of generations being considered. For example, if a population went through a temporary reduction in size in generation 2, so that $N_1 = 100$, $N_2 = 10$, and $N_3 = 100$, then

$$N_e = 1 \Big/ \frac{1}{3}\left(\frac{1}{100} + \frac{1}{10} + \frac{1}{100}\right) = \frac{1}{0.04} = 25$$

In this case, although the actual mean number of individuals in the population over three generations was 70, the effective population size during that time was only 25. A severe temporary reduction in size such as this is known as a **population bottleneck.** Bottlenecks occur when a population or species is reduced to a few reproducing individuals whose offspring then increase in numbers over subsequent generations to reestablish the population. Although the number of individuals may be restored to healthier levels, genetic diversity in the newly expanded population is often severely reduced because gametes from the handful of surviving individuals functioning as parents only represent a subset of the original gene pool.

NOW SOLVE THIS

Question 13 on page 521 asks you to calculate effective population size N_e for an endangered gorilla population.

Hint: Consider what unusual feature of the social structure of the gorilla population will affect N_e.

Captive-breeding programs, in which a few surviving individuals from an endangered species are removed from the wild to rebuild the population in a protected environment, inevitably create population bottlenecks. Bottlenecks also occur naturally when a small number of individuals from one population migrate to establish a new population elsewhere. When a new population derived from a small subset of individuals has significantly less genetic diversity than the original population, it exhibits the **founder effect** (Chapter 23). Reduced levels of genetic diversity due to a founder effect can persist for many generations, as shown by studies of two species of Antarctic fur seal, *Arctocephalus gazella* and *Arctocephalus tropicalis.* Seal hunters in the eighteenth and nineteenth centuries severely reduced the populations of these species, eliminating them from parts of their natural range in the Southern Ocean.* Although the number of Antarctic fur seals has now rebounded and the two species have recolonized much of their original habitat, mtDNA fingerprinting reveals that a founder effect can still be detected in *A. gazella,* with reduced genetic variation in current populations descended from a handful of surviving individuals.

FIGURE 24–6 The cheetah (*Acinonyx jubatus*).

The cheetah (*Acinonyx jubatus*), shown in Figure 24–6, is another well-studied example of a species with reduced genetic variation resulting from at least one severe population bottleneck that occurred in its recent history. Allozyme studies of South African cheetah populations have shown levels of genetic variation that are less than 10 percent of those found in other mammals. The abnormal spermatozoa and poor reproductive rates commonly observed in cheetahs are thought to be linked to the lack of genetic diversity, although when and how the population bottleneck occurred in this species is still unclear.

24.3 Genetic Effects Are More Pronounced in Small, Isolated Populations

Small isolated populations, such as those found in threatened and endangered species or produced by population fragmentation, are especially vulnerable to genetic drift, inbreeding, and reduction in gene flow. These phenomena act on the gene pool in different ways, but ultimately have similar effects in that they all can further reduce genetic diversity and long-term species viability.

Genetic Drift

If the number of breeding individuals in a population is reduced in size, fewer gametes will form the next generation. The alleles carried by these gametes may not be a representative sample of all those present in the population; purely by chance, some alleles may be underrepresented or not present at all, which will cause changes in allele frequency over time,

*In 2000, this fifth world ocean was delimited by oceanographers, incorporating southern portions of the Atlantic, Indian, and Pacific oceans.

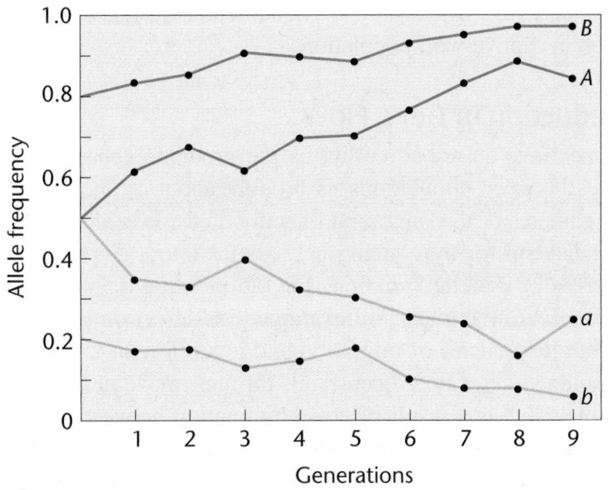

FIGURE 24–7 Change in frequencies over ten generations for two sets of alleles, *A/a* and *B/b*, in a theoretical population subject to genetic drift.

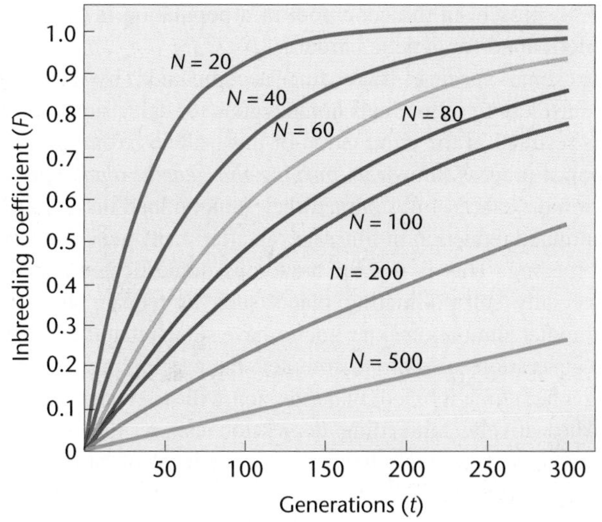

FIGURE 24–8 Increase in inbreeding coefficient (F) in theoretical populations as the population size (N) decreases.

resulting in **genetic drift.** This phenomenon has already been described in Chapter 23. A serious result of genetic drift in populations with a small effective population size is the loss of genetic variation. Genetic drift is a random process, so either deleterious or advantageous alleles can become fixed within a small population. In other words, one allele becomes the only version of that gene present in the gene pool of the population. This means a useful allele can be lost even if it has the potential to increase fitness or long-term adaptability. The probability that an allele will be fixed through drift is the same as its initial frequency. Thus, if a locus has two alleles, *A* and *a,* and if the initial frequency of *A* is 0.8, then the probability of *A* becoming fixed is 0.8, or 80 percent, and the probability of *A* being lost through drift is 0.2 or 20 percent. Figure 24–7 is a graph of the effect of genetic drift on a theoretical population.

Inbreeding

In small populations, the chance of **inbreeding**—meaning mating between closely related individuals—is greater. As described in Chapter 23, inbreeding increases the proportion of homozygotes in a population, thus expanding the possibility that an individual may be homozygous for a deleterious allele. We have already seen in Chapter 23 how the extent of inbreeding in a population can be quantified as the **inbreeding coefficient (*F*),** which measures the probability that two alleles of a given gene are derived from a common ancestral allele. Remember that the inbreeding coefficient is inversely related to the frequency of heterozygotes in the population and can be calculated as

$$F = \frac{2pq - H}{2pq}$$

where $2pq$ is the expected frequency of heterozygotes based on the Hardy–Weinberg law and H is the actual frequency of heterozygotes in the population.

In a declining population that has become small enough for drift to occur, heterozygosity (H) will decrease with each generation. The smaller the effective population size, the more rapid the decrease in H and the resulting increase in F, as can be seen from the equation

$$H_t/H_0 = \left(1 - \frac{1}{2N_e}\right)^t$$

where H_0 is the initial frequency of heterozygotes, H_t is the frequency of heterozygotes after t generations, and N_e is the effective population size. Figure 24–8 compares rates of increase in F for different effective population sizes.

NOW SOLVE THIS

Problem 1 on page 521 asks you to calculate effective population sizes, a heterozygote frequency, and an inbreeding coefficient for a jackal population that crashed and then recovered.

Hint: Calculate N_e for each generation first, and remember that heterozygote frequency and inbreeding coefficient are inversely related.

What effect does inbreeding have on the long-term survival of a species? If numbers of individuals in a species remain high, inbreeding in a single population may not immediately reduce the amount of genetic variation present in the overall gene pool. Self-pollinating plants, for example, often show high levels of homozygosity and relatively little genetic variation within single populations. However, they tend to have considerable variation between *different populations,* each of which has adapted to slightly different local environmental conditions. On the other hand, in most outbreeding species (those where gene flow occurs freely between populations), including all mammals, inbreeding is associated with reduced fitness and lower survival rates among offspring. This **inbreeding depression** can result from increased homozygosity for deleterious alleles. The number of deleterious alleles present in the gene pool of a population is called the **genetic load** (or genetic burden).

In some species, inbreeding accompanied by selection against less fit individuals homozygous for deleterious alleles has resulted in the elimination of these alleles from the gene pool, a process known as *purging the genetic load.* Species that have successfully purged their genetic load do not show continued reduction in fitness, even after many generations of inbreeding. This is true of numerous domesticated species, especially self-pollinating plants such as wheat. However, computer simulation experiments have shown that it may take 50 generations or more to complete the purging process, during which time inbreeding depression will still occur.

Alternatively, inbreeding depression can result from heterozygous individuals having a higher level of fitness than either of the corresponding homozygotes. In this case, the long-term survival of the population requires that inbreeding be avoided and that the levels of all alleles in the gene pool be maintained. Both of these requirements can be difficult to satisfy in a species that has already suffered a significant reduction in population size.

The effects of inbreeding depression and loss of genetic variation in a small, isolated population have been documented in the case of the Isle Royale gray wolves (Figure 24–9). Around 1950, a pair of gray wolves apparently crossed an ice bridge from the Canadian mainland to Isle Royale in Lake Superior. The island had no other wolves and had an abundance of moose, which became the wolves' main food source. By 1980, the Isle Royale wolf population had increased to over 50 individuals. Over the next decade, however, wolf numbers declined to fewer than a dozen with no new litters being born, despite plentiful food and no apparent sign of disease. The genetic variation of the remaining wolves was examined by mtDNA analysis and nuclear DNA fingerprinting. It was found that the Isle Royale wolves had levels of homozygosity that were twice as high as those of wolves in an adjacent mainland population. Furthermore, all the wolves possessed the same mtDNA genotype, consistent with descent from the same female. Hence, the degree of relatedness between individual Isle Royale wolves was equivalent to that of full siblings. This finding suggests that the wolves' reproductive failure was due to inbreeding depression, a phenomenon that has also been seen in captive-wolf populations.

FIGURE 24–9 The Isle Royale gray wolf (*Canis lupus*).

Reduction in Gene Flow

Gene flow, described earlier as the gradual exchange of alleles between populations, is brought about by the dispersal of gametes or the migration of individuals. It is an important mechanism for introducing new alleles into a gene pool and increasing genetic variation. Migration is the main route for gene flow in animals; we examined the effect on population allele frequencies of migration and gene flow in Chapter 23. In plants, gene flow occurs not through movement of individuals but as a result of cross-pollination between different populations and through seed dispersal. Isolation and fragmentation of populations in rare and declining species significantly reduces gene flow and the potential for maintaining genetic diversity. As we have already discussed, habitat loss is a major threat to species survival. It is not unusual for a threatened or endangered species to be restricted to small separate pockets of the remaining habitat. This isolates and fragments the surviving populations so that movement of individuals can no longer occur between them, thus preventing gene flow.

The term **metapopulation** describes a population consisting of spatially separated subpopulations with limited gene flow, especially if local extinctions and replacements of some of the subpopulations occur over time. One well-studied metapopulation is that of the endangered red-cockaded woodpecker (*Picoides borealis*) (Figure 24–10), which was once common in pinewoods throughout the southeastern United States. Habitat loss because of logging has reduced the remaining populations of this bird to small scattered sites isolated from each other, with virtually no migration between them. Studies using allozymes and DNA profiling

FIGURE 24–10 The red-cockaded woodpecker (*Picoides borealis*).

show that the smallest surviving woodpecker populations (where $N = < 100$) have suffered the greatest loss of genetic diversity and are at most risk of inbreeding depression, compared with larger populations where $N = > 100$. Management of the red-cockaded woodpecker now includes efforts to increase the genetic diversity of the smallest populations by introducing birds from different larger populations to artificially re-create the gene flow by migration that would have occurred in the original unfragmented distribution of the species.

Another species in which the effects of gene-flow reduction have been studied is the North American brown bear, *Ursus arctos*. A team of Canadian and U.S. researchers measured heterozygosity in different brown bear populations using amplified microsatellite markers to create DNA profiles for individual bears. The researchers found that levels of heterozygosity in the brown bear population living in and around Yellowstone Park were only two-thirds as high as those in brown bear populations from Canada and mainland Alaska. They concluded that this was due to the isolation and reduced migration of the Yellowstone bears. The habitat of the Canadian and Alaskan bear populations was much less fragmented, allowing migration of individuals and consequent gene flow between populations. The researchers also examined an island population of brown bears on the Kodiak archipelago off the Alaskan coast. Here, heterozygosity was even lower—less than one-half that of the mainland populations—providing further evidence for the effect of small population size, restricted migration, and restricted gene flow on the reduction of genetic diversity.

ESSENTIAL POINT

Major declines in a species' population numbers reduce genetic diversity and contribute to the risk of extinction as a result of genetic drift, inbreeding, or loss of gene flow.

24.4 Genetic Erosion Threatens Species' Survival

The loss of previously existing genetic diversity from a population or species is referred to as **genetic erosion.** Why does it matter if a population loses genetic diversity, especially if the numbers of individuals remain high? Genetic erosion has two important effects on a population. First, it can result in the loss of potentially useful alleles from the gene pool, thus reducing the ability of the population to adapt to changing environmental conditions and increasing its risk of extinction. Several decades before the dawn of modern genetics, Charles Darwin recognized the importance of diversity to long-term species survival and evolutionary success. In *On the Origin of Species* (1859), Darwin wrote:

> *"The more diversified the descendants of any one species . . . by so much will they be better enabled to seize on many widely diversified places in the polity of nature, and so enabled to increase in numbers."*

Although the importance of allelic diversity for population survival under changing environmental conditions can be presumed for an endangered species, it has actually been demonstrated in several weed species whose gene pools contain alleles conferring resistance to chemical herbicide. If the weeds are not sprayed, the plants with these resistance alleles enjoy no particular selective advantage, and the resistance allele often persists at a relatively low frequency. However, if herbicide is applied, the individuals with the resistance allele are much more likely to survive and generate a new resistant population, while populations whose gene pools lack the resistance allele will be eliminated.

The second important effect of genetic erosion is a reduction in levels of heterozygosity. At the population level, reduced heterozygosity will be seen as an increase in the number of individuals homozygous at a given locus. At the individual level, a decrease in the number of heterozygous loci within the genotype of a particular plant or animal will occur. As we have seen, loss of heterozygosity is a common consequence of reduced population size. Alleles may be lost through genetic drift or because individuals carrying them die without reproducing. Smaller populations also increase the likelihood of inbreeding, which inevitably increases homozygosity.

Obviously, once an allele is lost from a gene pool, the potential for heterozygosity is greatly reduced. If there were originally only two alleles at a given locus, loss of one means that heterozygosity at the locus is completely eliminated and the other allele is now fixed. The level of homozygosity that can be tolerated varies with species. Studies of populations showing higher-than-normal levels of homozygosity have documented a range of deleterious effects, including reduced sperm viability and reproductive abnormalities in African lions, increased offspring mortality in elephant seals, and reduced nesting success in woodpeckers.

As yet, no evidence has emerged from field studies that conclusively links the extinction of a wild population to genetic erosion. However, laboratory studies using *Drosophila melanogaster* to model evolutionary events show that loss of genetic variation does reduce the ability of a population to adapt to changing environmental conditions. Fruit flies, with their small size and rapid generation time of only 10 to 14 days, are a useful model organism for studying evolutionary events, especially because large populations can be readily developed and maintained through multiple generations.

In one set of experiments carried out by scientists at Macquarie University in Sydney, Australia, *Drosophila* populations were reduced to a single pair for up to three generations to simulate a population bottleneck and were then allowed to increase in number. The capacity of the bottlenecked populations to tolerate increasing levels of sodium chloride was compared with that of normal outbred populations. Researchers found that bottlenecked populations with low levels of heterozygosity were less tolerant of high salt concentrations (Figure 24–11, p. 516).

Other experiments comparing inbred and outbred *Drosophila* populations exposed to environmental stresses such as high temperatures and ethanol presence have shown that populations with even a low level of inbreeding have a much greater

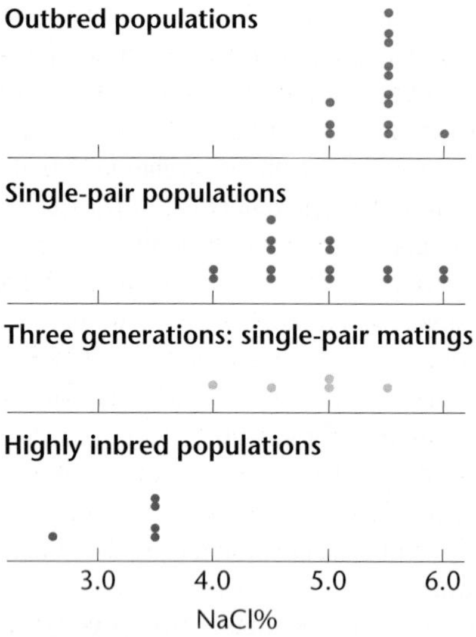

FIGURE 24–11 Effects of bottlenecks in various populations on evolutionary potential in *Drosophila,* as shown by distributions of NaCl concentrations at extinction.

probability of extinction at lower levels of environmental stress than outbred populations. Experimental results such as these indicate that genetic erosion does indeed reduce the long-term viability of a population by reducing its capacity to adapt to changing environmental conditions.

24.5 Conservation of Genetic Diversity Is Essential to Species Survival

Scientists working to maintain biological diversity face several dilemmas. Should they focus on preserving individual populations, or should they take a broader approach by trying to conserve not just one species but all the interdependent plants and animals in an ecosystem? How can genetic diversity be maintained in a species whose numbers are declining? Can genetic diversity lost from a population be restored? Early conservation efforts often focused only on the population size of an endangered species. Biologists now recognize that a complex interplay of factors must be considered during conservation efforts, including the habitat and role of a species within an ecosystem, as well as the importance of genetic variation for long-term survival.

Ex Situ Conservation: Captive Breeding

Ex situ (Latin for off-site) **conservation** involves removing plants or animals from their original habitat to an artificially maintained location such as a zoo or botanic garden. This living collection can then form the basis of a captive-breeding program.

These programs, where a few surviving individuals from an endangered species are removed from the wild to breed, raise their offspring, and rebuild the population in a protected environment, have been instrumental in bringing a number of species back from the brink of extinction. However, such programs can have undesirable genetic consequences that potentially jeopardize the long-term survival of the species even after population numbers have been restored. A captive-breeding program is rarely initiated until very few individuals are left in the wild, when the original genetic diversity of the species is already depleted. The breeding program is frequently based on a small number of captured individuals, so genetic diversity is further reduced by the founder effect. Since captive-breeding facilities often accommodate only a small number of individuals in the breeding group, inbreeding is difficult to avoid, especially for animal species that form harems (breeding groups dominated by a single male), so that N_e is considerably smaller than N. Finally, unintended selection for genotypes more suited to captive-breeding conditions over time can reduce the overall capacity of the recovered population to adapt and survive in the wild.

How can this type of program be managed to minimize these genetic effects? As we have already seen, loss of genetic diversity measured as heterozygosity over t generations can be expressed as

$$H_t/H_0 = \left(1 - \frac{1}{2N_e}\right)^t$$

This equation indicates that the loss of genetic diversity H_t/H_0 will be greater with a smaller effective population size N_e and a larger number of generations t. We can expand this equation for a captive-breeding population as follows:

$$H_t/H_0 = [1 - (1/2\,N_{fo})]\{1 - 1/[2N(N_e/N)]\}^{t-1}$$

where N_{fo} is the effective size of the founding population, N is the mean population size, and N_e is the mean effective population size of the group over t generations. Examination of this equation shows that maximum genetic diversity in a captive-breeding group will be maintained by (1) using the largest possible number of founding individuals to maximize N_{fo}, (2) maximizing N_e/N so that as many individuals as possible produce offspring each generation, and (3) minimizing the number of generations in captivity to reduce t. The importance of maximizing N_{fo} to increase allelic diversity in the founding population can also be seen in Figure 24–12. Successful programs of this sort for endangered species are managed with these three goals in mind. In addition, keeping good pedigree records and exchanging individuals between different breeding programs when possible will reduce inbreeding.

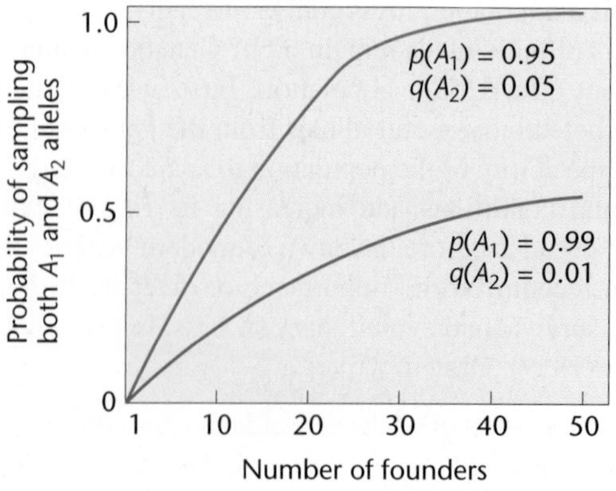

FIGURE 24–12 Effect of captive-population founder number on the probability of maintaining both A_1 and A_2 alleles at a locus.

NOW SOLVE THIS

Problem 14 on page 521 asks you to estimate the potential for genetic erosion in a captive population of rare red pandas.

Hint: If N_e/N is maintained at 0.42 and actual population size at 50, you can calculate N_e. Then use the equation given in the discussion above.

Rescue of the Black-Footed Ferret through Captive Breeding

The black-footed ferret, *Mustela nigripes* (Figure 24–13), is an excellent example of a species that has been successfully subjected to captive breeding. This ferret was once widespread throughout the plains of the western United States, but by the 1970s it was considered to be extinct. Decades of trapping and poisoning of both the ferret and its main prey, the prairie dog, had decimated their populations. However, in 1981 a small surviving colony of black-footed ferrets was discovered on a ranch near Meeteetse, Wyoming. Conservation biologists first tried to conserve this ferret population *in situ,* but in 1985 the colony was infected with canine distemper, which nearly wiped it out. Of the 18 ferrets that were saved and transferred to a captive facility, only 8 were considered to be sufficiently unrelated to found a new population. The breeding program has produced more than 6500 ferrets, and the species is now being reintroduced to parts of its original range. The goal is to establish populations of 1500 ferrets in sustainable wild colonies by 2010. By the end of 2008, more than 2300 captive-bred ferrets had been released at 18 different locations, and reintroduced populations in Arizona, South Dakota, and Wyoming are now actively breeding and considered self-sustaining.

Although the recovery of the black-footed ferret appears to be a success, the small founder group of eight individuals caused a severe bottleneck for this species. Additional loss of genetic diversity occurred in the captive-breeding program because of drift, limited reproduction from several of the founder females, and a high rate of breeding by one founder male. In short, the ferrets still face the risks of inbreeding and further genetic erosion that are characteristic of this type of program.

FIGURE 24–13 The black-footed ferret (*Mustela nigripes*).

Genetic management strategies, such as using DNA markers to identify the most genetically varied individuals, maintaining careful pedigree records to avoid mating closely related animals, and developing techniques for artificial insemination and sperm cryopreservation, have helped to conserve the genetic diversity that remained in the species. Conservation geneticists working on the recovery project estimate that all existing black-footed ferrets share about 12 percent of their genome. This is roughly the equivalent of being full cousins. The effects of this degree of genetic similarity in the expanding ferret population are unclear. Occasional abnormalities such as webbed feet and kinked or short tails have been observed in the captive population, but without a noninbred population available for comparison, it is unclear whether this is evidence that inbreeding is increasing the homozygosity for deleterious alleles.

Scientists disagree as to whether the long-term future of the black-footed ferret is jeopardized by the severe bottleneck and subsequent loss of genetic diversity the species has experienced. Some geneticists suggest that because the one surviving Meeteetse population was so isolated, it may have already become sufficiently inbred to purge any deleterious alleles. Other researchers point out that studies of black-footed ferret DNA extracted from museum specimens show that the ferrets alive today have lost significant genetic diversity compared with earlier prebottleneck populations, and they suggest that loss of fitness because of inbreeding will inevitably be seen over time. This view is supported by a 2008 study using microsatellite analysis to assess levels of genetic diversity in reintroduced ferret populations; the study found that animals in smaller populations with less diversity tended to have shorter limbs and decreased body size. The authors of the study recommended repeated annual introductions of additional ferrets to newly reintroduced colonies to build numbers rapidly and minimize inbreeding effects.

Ex Situ Conservation and Gene Banks

Another form of *ex situ* conservation is provided by establishing **gene banks.** In contrast to housing entire animals or plants, these collections instead provide long-term storage and preservation for reproductive components, such as sperm, ova, and frozen embryos in the case of animals, and seeds, pollen, and cultured tissue in the case of plants. Many more individual genotypes can be preserved for longer periods in a gene bank than in a living collection. Cryopreserved gametes or seeds can be used to reconstitute lost or endangered animals or plants after many years in storage. Because they are expensive to construct and maintain, most gene banks are used to conserve the genetic legacy of domesticated species having economic value.

Gene banks have been established in many countries to help preserve genetic material of agricultural importance, such as traditional crop varieties that are no longer grown or old livestock breeds that are becoming rare. One of the most important *ex situ* collections in the United States is the National Center for Genetic Resources Preservation, a U.S. Department of Agriculture (USDA) facility in Fort Collins, Colorado, which maintains more than 300,000 different crop varieties and related wild species. Some of the accessions are stored as seeds and others as cryogenically preserved tissue from which whole plants can be regenerated. (See this chapter's opening photograph.) Animal genetic resources, including frozen semen and embryos from endangered livestock breeds, are also preserved at this facility.

Ex situ conservation using gene banks, though often vital, has several disadvantages compared to other methods of conservation. A major problem with gene banks is that even large collections cannot contain all the genetic variation that is present in a species. Conservation geneticists attempt to address this problem by identifying a **core collection** for a species. The core collection is a subset of individual genotypes, carefully chosen to contain as much as possible of the species' genetic variation; preserving the core collection takes priority over randomly collecting and preserving large numbers of genotypes. Another disadvantage of *ex situ* conservation is that the artificial conditions under which a species is preserved in a living collection or gene bank often create their own selection pressures. When seeds of a rare plant species are maintained in cold storage, for example, selection may occur for those genotypes better adapted to withstand the low temperatures, and genotypes may be lost that would actually have greater fitness in the plant's natural environment. Yet another problem posed by *ex situ* conservation is that while the greatest biological diversity in both domesticated and nondomesticated species is frequently found in underdeveloped countries, most *ex situ* collections are situated in developed countries that have the resources to establish and maintain them. This leads to conflict over who owns and has access to the potentially valuable genetic resources maintained in such collections.

In Situ Conservation

In situ (Latin for on-site) **conservation** attempts to preserve the population size and biological diversity of a species while it remains in its original habitat. The use of species inventories to identify diversity hot spots is an important tool for determining the best places to establish parks and reserves where plants and animals can be protected from hunting or collecting and where their habitat can be preserved.

For domesticated species, there is increasing interest in "on-farm" preservation in which additional resources and financial incentives encourage farmers to maintain traditional crop varieties and livestock breeds. Nondomesticated species with economic potential have also been targeted for *in situ* preservation. In 1998, the USDA established its first *in situ* conservation sites for a wild plant in order to protect populations of the native rock grape (*Vitis rupestris*) in several eastern states. The rock grape is prized by winegrowers not for its fruit but for its roots; grape vines grafted onto wild rock grape rootstock are resistant to phylloxera, a serious pest of wine grapes.

The advantage of *in situ* conservation for the rock grape, as with other species, is that larger populations with greater genetic diversity can be maintained. Another advantage is that species conserved *in situ* continue to live and reproduce in the environments to which they are adapted, which reduces the likelihood that novel selection pressures will produce undesirable changes in allele frequency. However, the continuing increase in global human population makes setting aside suitable areas for *in situ* conservation an ever greater challenge. For example, even in a large preserve such as Yellowstone National Park, the migration and gene flow of species such as the North American brown bear may eventually be eliminated, with a consequent loss of genetic diversity. This problem is even more acute in smaller, more fragmented areas of protected habitat.

> **ESSENTIAL POINT**
>
> Conservation of genetic diversity depends on *ex situ* methods, such as living collections, captive-breeding programs, and gene banks, as well as *in situ* approaches, such as the establishment of parks and preserves.

Population Augmentation

What genetic considerations should accompany efforts to restore populations or species that are in decline? As we have already seen, populations that experience bottlenecks continue to suffer from low levels of genetic diversity, even after their numbers have recovered. We have also seen that inbreeding and drift contribute to genetic erosion in small populations, and fragmentation interrupts migration and gene flow, further reducing diversity. Captive-breeding programs designed to restore a critically endangered species beginning with only a few surviving individuals risk genetic erosion in the renewed population from founder effect and inbreeding, as in the case of the black-footed ferret.

An alternative strategy used by conservation biologists is **population augmentation**—boosting the numbers of a declining population by transplanting and releasing individuals of the same species captured or collected from more numerous populations elsewhere. Attempts to reestablish gene flow in severely fragmented populations of the endangered red-cockaded woodpecker by augmenting the smallest populations were described earlier. Other population augmentation projects in the United States have involved bighorn sheep and grizzly bears in the Rocky Mountains. This method has also been employed with the Florida panther (Figure 24–14), using an isolated population of less than 30 animals confined to the area around the Big Cypress Swamp and Everglades National Park in south Florida. As we will discuss in the Genetics, Technology, and Society essay at the end of this chapter, DNA-profiling patterns show high levels of inbreeding in the Florida panther, with reduced fitness because of severe reproductive abnormalities and increased susceptibility to parasite infections. In this effort, seven unrelated animals from a captive population of South and North American panthers were released into the Everglades in the 1960s and allowed to interbreed with the Florida population, thus producing more genetically diverse family groups. The case of the Florida panther shows how augmentation can increase population numbers, as well as genetic diversity, if the transplanted individuals are unrelated to those in the population to which they are introduced. However, the Florida panther also demonstrates a potential problem with population augmentation, namely, **genetic swamping.** This occurs when the gene pool of the

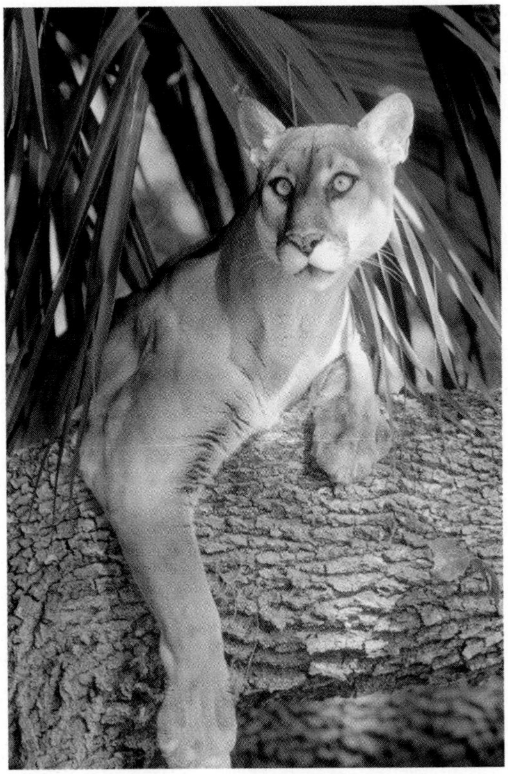

FIGURE 24–14 The Florida panther (*Puma concolor coryi*).

original population is overwhelmed by different genotypes from the transplanted individuals and loses its identity. Further augmentation of the Florida panther population remains controversial, as some biologists argue that the unique features that allow the Florida panther to be classified as a separate subspecies will be lost if breeding takes place with other panthers.

Another difficulty that can be caused by population augmentation is **outbreeding depression,** where reduced fitness occurs in the progeny of matings between genetically diverse individuals. Outbreeding depression occurring in the F_1 generation is thought to be due to the offspring being less well adapted to local environmental conditions than either parent. This phenomenon has been documented in some plant species in which seeds of the same species but from a different location were used to revegetate a damaged area by cross-pollination with the remaining local plants. Outbreeding depression that occurs in the F_2 and later generations is due to the disruption of **coadapted gene complexes**—groups of alleles that have evolved to work together to produce the best level of fitness in an individual. This type of outbreeding depression has been documented in F_2 hybrid offspring from matings between fish from different salmon populations in Alaska. These studies suggest that restoring the most beneficial type and amount of genetic diversity in a population is more complicated than previously thought, reinforcing the argument that the best long-term strategy for species survival is to prevent the loss of diversity in the first place.

GENETICS, TECHNOLOGY, AND SOCIETY

Gene Pools and Endangered Species: The Plight of the Florida Panther

The story of the Florida panther provides a dramatic example of an animal brought to the verge of extinction and the challenges that must be overcome to restore it to a healthy place in the ecosystem. The Florida panther is an endangered population of North American cougars (also known as pumas or mountain lions). These panthers once roamed the southeastern corner of the United States, from South Carolina and Arkansas to the southern tip of Florida; however, they now occupy less than 5 percent of their original habitat, in southern Florida. After human settlement began, panthers were killed by hunting, poisoning, highway collisions, and loss of habitat that supports their prey—primarily wild deer and hogs. By 1967, Florida panthers were listed as endangered, and only about 30 of the animals were left. Population estimates predicted that the Florida panther would be extinct by the year 2055.

Because of geographical isolation and inbreeding, Florida panthers had the lowest levels of genetic heterozygosity of any population of cougar. The loss of genetic diversity manifested itself in the appearance of some severe genetic defects. For example, almost 80 percent of panther males born after 1989 in the Big Cypress region showed an autosomal dominant or X-linked recessive condition known as *cryptorchidism*—failure of one or both testicles to descend. This defect is associated with low testosterone levels and reduced sperm count. Life-threatening congenital heart defects also appeared in the Florida panther population, possibly because of an autosomal dominant gene defect. In addition, some immune deficiencies emerged, making the animals more susceptible to diseases and further contributing to the population's decline. Other less serious genetic features arose, such as a kink in the tail and a whorl of fur on the back.

Over the last two decades, federal and state agencies, as well as private individuals, have implemented a *Florida Panther Recovery Program.* The program's goal is to exceed 500 breeding animals and give the panther a 95 percent probability of survival while retaining up to 90 percent of its genetic diversity. The plan includes strict protection, enlargement and improvement of the panther's habitat, and education of landowners. Wildlife underpasses have been constructed along highways in panther territory, and these have significantly reduced panther highway fatalities (which account for about half of panther deaths). In 2008, the Florida panther population was estimated at around 100 animals. While numbers have been holding steady, this represents little increase over the past five years and is well short of recovery program goals.

To retard the detrimental effects of inbreeding, eight wild female Texas

(*Cont. on the next page*)

panthers were released into Florida panther territory in 1995. Two of the females survived and gave birth to litters of healthy kittens. It is estimated that 40 to 70 of the 100 panthers in south Florida are hybrids of Texas panthers and Florida panthers. None of the hybrids appears to have the detrimental traits of the inbred Florida panthers. Although the population augmentation program has been considered successful, it is of concern to some biologists, who worry that the Texas cougar genes may genetically swamp the distinct Florida panther population.

Paradoxically, the success of the restoration program has become a problem. Now that the population has reached about 100 animals, the Florida panther has almost exceeded the capacity of its existing habitat. Biologists estimate that the population must reach at least 250 individuals in order to be self-sustaining. To reach this number, the panthers will need to expand their territory, putting them in direct competition with human expansion in Florida. Biologists are now evaluating potential new territories for the growing population of Florida panthers—including parts of Louisiana, Arkansas, and South Carolina.

Despite the success of the restoration program, the survival of the Florida panther is far from certain. The panther's comeback will require years of monitoring and frequent intervention. In addition, people must be willing to share their land with wild creatures that are dangerous and do not directly further human interests. Public support for the return of the Florida panther has been strong, however, so there may be hope for this unique, impressive animal.

Your Turn

Take time, individually or in groups, to answer the following questions. Investigate the references and links to help you understand some of the challenges surrounding the Florida Panther Recovery Program.

1. Recent genetic studies suggest that the Florida panther may be too closely related to other North American cougars to merit classification as a subspecies. How were these genetic studies done, and how would you interpret the results of the studies, as they relate to the Florida panther?

To learn about genetic relatedness among American cougars, see Culver, M. et al. 2000. Genetic Ancestry of the American Puma *(Puma concolor). J. Heredity* **91**(3): 186–197.

2. How do you think the data in the Culver et al. study could both support and complicate efforts to restore the endangered Florida panther?

A discussion of the relative merits of genetic and traditional classification methods can be found in the Florida Panther Recovery Plan, 3rd edition 2008, U.S. Fish and Wildlife Service, pages 7–12. *You can access this work through the Florida Panther Net* (**http://www.panther.state.fl.us**).

3. Do you think that genetics or habitat is the most important factor involved in the Florida Panther Recovery Program?

An interesting discussion of this question can be found in the New York Times *article, "A Rare Predator Bounces Back (Now Get It Out Of Here)" at* **http://www.nytimes.com/2006/03/14/science/14pant.html?scp=1sq=florida+panther&st=nyt**.

CASE STUDY The flip side of the green revolution

To increase food production, the Green Revolution (see Chapter 22) fostered the development of inbred strains of high-yield crop plants. This approach was very successful, and it is estimated that over one billion people worldwide now derive all or part of their food supply from these new crops. However, this program has a dark side. In many countries, as they switched to the high-yield strains, farmers abandoned the traditional varieties of crop plants, which contained a high degree of genetic diversity. For example, 50 years ago, farmers in India used to plant more than 30,000 varieties of rice. Now it is estimated that 50 to 75 percent of all rice fields are planted with just 10 varieties. In the United States some 90 percent of all varieties of crop plants grown a century ago have disappeared from seed catalogs and are not preserved in seed banks. This dramatic loss of genetic variation has generated some controversy and raised several questions.

1. Might the Green Revolution represent short-term gains but long-term losses?
2. Why is it important to preserve the genetic diversity of crop plants?
3. Will climate change adversely affect the new and old strains of crop plants and perhaps reduce the food supply?

INSIGHTS AND SOLUTIONS

1. Is a rare species found as several fragmented subpopulations more vulnerable to extinction than an equally rare species found as one larger population? What factors should be considered when managing fragmented populations of a rare species?

Solution: A rare species in which the remaining individuals are divided among smaller, isolated subpopulations can appear to be less vulnerable. If one subpopulation becomes extinct through local causes, such as disease or habitat loss, then the remaining subpopulations may still survive. However, genetic drift will cause smaller populations to experience more rapidly increasing homozygosity over time compared with larger populations. Even with random mating, the change in heterozygosity from one generation to the next because of drift can be calculated as

$$H_1 = H_0(1 - 1/2N)$$

where H_0 is the frequency of heterozygotes in the present generation, H_1 is the frequency of heterozygotes in the next generation, and N is the number of individuals in the population. Thus, in a small population of 50 individuals with an initial heterozygote frequency of 0.5, heterozygosity will decline to $0.5(1 - 1/100) = 0.4995$, a loss of 0.5 percent in just one generation. In a larger population of 500 individuals and the same initial heterozygote frequency, heterozygosity after one generation will be $0.5(1 - 1/100) = 0.4995$, a loss of only 0.05 percent.

Therefore, smaller populations are likely to show the effects of homozygosity for deleterious alleles sooner than larger populations, even with random mating. If populations are fragmented so that movement of individuals or gametes between them is prevented, management options could include transplanting individuals from one subpopulation to another to enable gene flow to occur. Establishment of "wildlife corridors" of undisturbed habitat that connect fragmented populations could be considered. In captive populations, exchange of breeding adults (or their gametes through shipment of preserved semen or pollen) can be undertaken. Promotion of increased population numbers, however, is vital to prevent further genetic erosion through drift.

PROBLEMS AND DISCUSSION QUESTIONS

1. A wildlife biologist studied four generations of a population of rare Ethiopian jackals. When the study began, there were 47 jackals in the population, and analysis of microsatellite loci from these animals showed a heterozygote frequency of 0.55. In the second generation, an outbreak of distemper occurred in the population, and only 17 animals survived to adulthood. These jackals produced 20 surviving offspring, which in turn gave rise to 35 progeny in the fourth generation. See Now Solve This on page 513.
   (a) What was the effective population size for the four generations of this study?
   (b) Based on its effective population size, what is the heterozygote frequency of the jackal population in generation 4?
   (c) What is the inbreeding coefficient in generation 4, assuming an inbreeding coefficient of $F = 0$ at the beginning of the study, no change in microsatellite allele frequencies in the gene pool, and random mating in all generations?
2. Chondrodystrophy, a lethal form of dwarfism, has recently been reported in captive populations of the California condor and has killed embryos in 5 out of 169 fertile eggs. Chondrodystrophy in condors appears to be caused by an autosomal recessive allele with an estimated frequency of 0.09 in the gene pool of this species.
   (a) How do you think California condor populations should be managed in the future to minimize the effect of this lethal allele?
   (b) What are the advantages and disadvantages of attempting to eliminate it from the gene pool?
3. A geneticist is studying three loci, each with one dominant and one recessive allele, in a small population of rare plants. She estimates the frequencies of the alleles at each of these loci as $A = 0.75$, $a = 0.25$; $B = 0.80$, $b = 0.20$; $C = 0.95$, $c = 0.05$. What is the probability that all the recessive alleles will be lost from the population through genetic drift?
4. How are genetic drift and inbreeding similar in their effects on a population? How are they different?
5. You are the manager of a game park in Africa with a native herd of just 16 black rhinos, an endangered species worldwide. Describe how you would manage this herd to establish a viable population of black rhinos in the park. What genetic factors would you take into consideration in your management plan?
6. Compare the causes and effects of inbreeding depression and outbreeding depression.
7. Cloning, using the techniques similar to those pioneered by the Scottish scientists who produced Dolly the sheep, has been proposed as a way to increase the numbers of some highly endangered mammalian species. Discuss the advantages and disadvantages of using such an approach to aid long-term species survival.
8. In a population of wild poppies found in a remote region of the mountains of eastern Mexico, almost all of the members have pale yellow flowers, but breeding experiments show that pale yellow is recessive to deep orange. Using the tools of the conservation geneticists described in this chapter, how could you experimentally determine whether the prevalence of the recessive phenotype among the eastern Mexican poppy population is due to natural selection or simply due to the effects of genetic drift and/or inbreeding?
9. Contrast *ex situ* conservation techniques with *in situ* conservation techniques.
10. Describe how the captive-breeding *ex situ* conservation approach is applied to a severely endangered species.
11. Explain why a low level of genetic diversity in a species is a detriment to the survival of that species.
12. Contrast allozyme analysis with RFLP analysis as measures of genetic diversity.
13. A population of endangered lowland gorillas is studied by conservation biologists in the wild. The biologists count 15 gorillas but observe that the population consists of two harems, each dominated by a different single male. One harem contains eight females and the other has five. What is the effective population size, N_e, of the gorilla population? See Now Solve This on page 512.
14. Twenty endangered red pandas are taken into captivity to found a breeding group. See Now Solve This on page 517.
    (a) What is the probability that at least one of the captured pandas has the genotype B_1/B_2 if $p(B_1) = 0.99$?
    (b) Careful management keeps the N_e/N ratio for the red panda breeding population at 0.42. If the overall size of the captive population is maintained at 50 individuals, what proportion of the heterozygosity present in the founding population will still be present after five generations in captivity?
15. Use your analysis of the red panda in Problem 14 to make suggestions for managing the captive-breeding population of red pandas to maintain as much genetic diversity as possible.
16. *Antechinus agilis,* a small mouse-like marsupial found in Australia, mates for a 10- to 15-day period in August during which females ovulate and after which males die of stress. Individual females rely on stored sperm from multiple male sources to fertilize their eggs. Using DNA profiling, researchers determined that females are capable of releasing a mix of sperm for fertilization—regardless of the time of male access to the female or her time of ovulation—that maintains high genetic heterogeneity in the face of male absenteeism (Shimmin et al., 2000). Does the reproductive strategy of *Antechinus* provide any clues as to the lengths organisms will go to maintain genetic variation?
17. As revealed by microsatellite analysis, habitat loss and fragmentation have led to a significant loss of heterozygosity among nine populations of Florida black bear (*Ursus americanus floridanus*). Researchers determined that although each population was in Hardy–Weinberg equilibrium, mean heterozygosity was

lower in smaller, less interconnected populations, and each population was genetically distinct (Dixon et al., 2007). No relationship was found between the degree of genetic differences and the distance from neighboring populations.

(a) Is their finding of a lower mean heterozygosity in smaller, less interconnected populations expected?

(b) Given that the Florida black bear inhabits relatively populated areas, what would explain the lack of a relationship between genetic differences and interpopulational distances?

(c) Considering that low heterozygosity often contributes to reduced species survival, what steps might be taken to enhance the survival of the Florida black bear?

18. Once nearly extinct, black-footed ferrets, which prey on prairie dogs, have increased in numbers from 18 in 1985 to approximately 700 presently occupying a recovery area of Buffalo Gap National Grassland (O'Neill, 2007). However, because prairie dogs eat grass that feeds cattle, some ranchers want to reduce prairie dog populations in that area, where approximately 30,000 of their cattle graze. A plan for balancing the welfare of the ranchers, the prairie dogs, and the black-footed ferret is under consideration by the U.S. Forest Service.

    (a) What impact would a reduction of prairie dog populations have on black-footed ferret populations?

    (b) Conservationists argue that exposing the black-footed ferret population to a second bottleneck would irreversibly harm ferret survival. Some might argue that since the ferret population has in the past recovered from one bottleneck, it should be able to recover from a second. Do you agree?

    (c) Given that the black-footed ferret is one of America's most endangered mammals, what kind of compromise would you negotiate among the farmer, the prairie dog, and the ferret?

19. Przewalski's horse (*Equus przewalskii*), thought by some biologists to represent the ancestral species from which modern horses were domesticated, is classified as a separate species from the domestic horse, although members of the two species can mate and produce fertile offspring. Przewalski's horse was hunted to extinction in its native habitat in the Asian steppes by the 1920s. The few hundred Przewalski's horses alive today are descended from 12 surviving individuals that had been taken into captivity. The founder breeding group also included a domestic mare. Currently, there is interest in reintroducing Przewalski's horse to its original range. What advice would you give to the conservation biologists managing the reintroduction project?

20. DNA profiles based on different kinds of molecular markers are increasingly used to measure levels of genetic diversity in populations of threatened and endangered species. Do you think these marker-based estimates of genetic diversity are reliable indicators of a population's potential for survival and adaptation in a natural environment? Why, or why not?

21. Seed banks provide managed protection for the conservation of species of economic and noneconomic importance. One study quantified considerable fitness decay in seeds maintained in long-term storage (Schoen et al., 1998. *Proc. Natl. Acad. Sci. [USA]* 95: 394–399). When germination falls below 65–85 percent, regeneration (planting and seed collection) of a finite sample of the stored seed type is recommended. What genetic consequences might you expect to accompany the conservation practice of long-term seed storage and regeneration?

22. According to one writer, "One of the first questions a resource manager asks about threatened and endangered species is: How bad is it?" (Holmes, 2001. *Proc. Natl. Acad. Sci. [USA]* 98: 5072–5077). In other words, the manager seeks a population viability analysis that includes estimates of extinction risk. What factors would you consider significant in providing an estimate of extinction risk for a species?

23. It is believed that most microsatellite DNA sequences are selectively neutral and highly heterozygous in natural populations. Considering that cheetahs underwent a population bottleneck approximately 12,000 years ago, North American pumas 10,000 years ago, and Gir Forest lions 1,000 years ago, in which of these species would you expect to see the highest degree of microsatellite polymorphism? the lowest?

24. For many conservation efforts, scientists lack sufficient data to make definitive conservation decisions. Yet the allocation of resources cannot be delayed until such data are available. In these cases, conservation efforts are often directed toward three classes of species: flagships (high-profile species), umbrellas (species requiring large areas for habitat), and biodiversity indicators (species representing diverse, especially productive habitats) (Andelman and Fagan, 2000. *Proc. Natl. Acad. Sci. [USA]* 97: 5954–5959). In terms of protecting threatened species, what advantages and disadvantages might accompany investing scarce talent and resources in each of these classes?

APPENDIX A

Solutions to Problems and Discussion Questions

Chapter 1 Introduction to Genetics

2. Based on the parallels between Mendel's model of heredity and the behavior of chromosomes, the chromosome theory of inheritance emerged. It states that inherited traits are controlled by genes residing on chromosomes that are transmitted by gametes.

4. A gene variant is called an allele. There can be many such variants in a population, but for a diploid organism, only two such alleles can exist in any given individual.

6. *Genes,* linear sequences of nucleotides, usually exert their influence by producing proteins through the process of transcription and translation. Genes are the functional units of heredity. They often associate with proteins to form *chromosomes*.

8. The central dogma of molecular genetics refers to the relationships among DNA, RNA, and protein.

10. Restriction enzymes (endonucleases) cut double-stranded DNA at particular base sequences. When a vector is cleaved with the same enzyme, complementary ends are created such that ends, regardless of their origin, can be combined and ligated to form intact double-stranded recombinant molecules.

12. Supporters of organismic patenting argue that it is needed to encourage innovation and allow the costs of discovery to be recovered. Others argue that natural substances should not be privately owned and that once owned by a small number of companies, free enterprise will be stifled. Individuals and companies needing vital, but patented products, may have limited access.

14. Model organisms are not only useful, but necessary, for understanding genes that influence human diseases. Given that genetic/molecular systems are highly conserved across broad phylogenetic lines, what is learned in one organism is usually applied to all organisms. Most model organisms have peculiarities, such as ease of growth, genetic understanding, or abundant offspring, that make them especially informative in genetic studies.

16. Safeguards should probably include tests for allergenicity, environmental impact, and likelihood of cross-pollination. In addition, concern would increase if such a crop contained antibiotic-resistant genetic markers and genes conferring toxicity to pests. While not required in the United States, in the interest of the consumer, one might consider labeling such products as genetically modified. On a broader scale, one might reduce vulnerability by using multiple suppliers (if available) and help minimize the domination of the world food supply by a few companies.

Chapter 2 Mitosis and Meiosis

2. Chromosomes that are homologous share many properties including: overall length; position of the centromere (metacentric, submetacentric, acrocentric, telocentric); banding patterns; type and location of genes; and autoradiographic patterns. *Diploidy* is a term often used in conjunction with the symbol $2n$. It means that both members of a homologous pair of chromosomes are present. *Haploidy* specifically refers to the fact that each haploid cell contains *one chromosome of each homologous pair of chromosomes.*

6. Metacentric (a) centromere in the middle, submetacentric (b) centromere off-center, acrocentric (c) centromere toward one end, telocentric (d), centromere at one end. Anaphase configurations are influenced by centromere placement as indicated below.

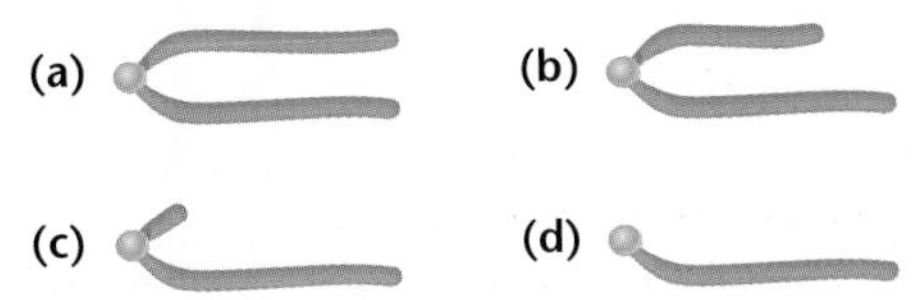

8. Major divisions of the cell cycle include interphase and mitosis. Interphase is composed of four phases: G1, G0, S, and G2. During the S phase, chromosomal DNA doubles. Karyokinesis involves nuclear division, while cytokinesis involves division of the cytoplasm.

10. Compared with mitosis, which maintains a chromosomal constancy, meiosis provides for a reduction in chromosome number and an opportunity for exchange of genetic material between homologous chromosomes.

12. Sister chromatids are genetically identical, except where mutations may have occurred during DNA replication. Nonsister chromatids are genetically similar when on homologous chromosomes or genetically dissimilar when on nonhomologous chromosomes. If crossing over occurs, then chromatids attached to the same centromere will no longer be identical.

14. (a) 8 tetrads.
(b) 8 dyads.
(c) 8 monads migrating to *each* pole.

16. First, through independent assortment of chromosomes at anaphase I of meiosis, daughter cells (secondary spermatocytes and secondary oocytes) may contain different sets of maternally and paternally derived chromosomes. Second, crossing over, which happens at a much higher frequency in meiotic cells as compared to mitotic cells, allows maternally and paternally derived chromosomes to exchange segments, thereby increasing the likelihood that daughter cells (that is, secondary spermatocytes and secondary oocytes) are genetically unique.

18. There would be 16 combinations with the addition of another chromosome pair.

20. $(1/2)^{10}$

22. In angiosperms, meiosis results in the formation of microspores (male) and megaspores (female) that give rise to the haploid male and female gametophyte stage. Micro- and megagametophytes produce the pollen and the ovules, respectively. Following fertilization, the sporophyte is formed.

24. The folded-fiber model is based on each chromatid consisting of a single fiber wound like a skein of yarn. Each fiber consists of DNA and protein. A coiling process occurs during the transition of interphase chromatin to more condensed chromosomes during prophase of mitosis or meiosis. Such condensation leads to a 5000-fold contraction in the length of the DNA within each chromatid. The transition is at the end of interphase and the beginning of prophase when the chromosomes are in the condensation process. This eventually leads to the typically shortened and "fattened" metaphase chromosome.

26. Duplicated chromosomes A^m, A^p, B^m, B^p, C^m, and C^p will align at metaphase, with the centromeres dividing and sister chromatids going to opposite poles at anaphase. Each pole will contain one copy of each of the chromosomes A^m, A^p, B^m, B^p, C^m, and C^p.

28. As long as you have accounted for eight possible combinations in the previous problem, there would be no new ones added in this problem.

30. See the products of nondisjunction of chromosome C at the end of meiosis I, as follows.

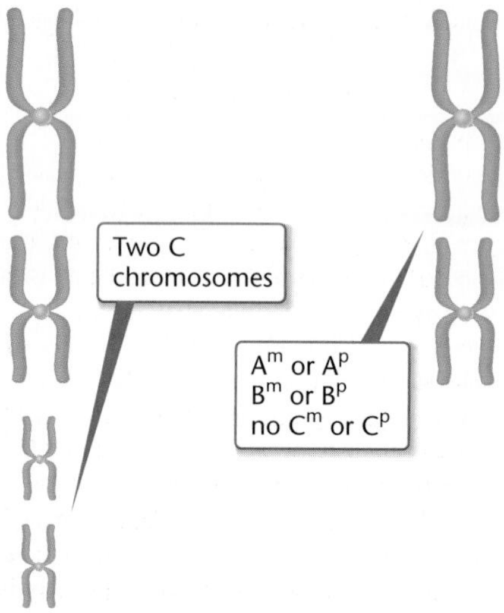

At the end of meiosis II, assuming that, as the problem states, the C chromosomes separate as dyads instead of monads during meiosis II, you would have monads for the A and B chromosomes, and dyads (from the cell on the left) for both C chromosomes as one possibility. However, another possibility exists, as shown below for the products of meiosis II:

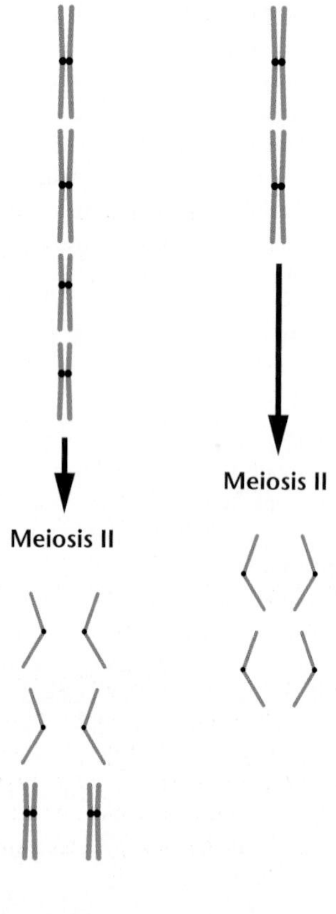

32. (a) 7 + 14 = 21

(b) Given that no homologous chromosome pairing occurs at metaphase I, one would expect 21 univalents and random "1 × 0" separation of chromosomes at anaphase I.

Chapter 3 Mendelian Genetics

2. (a) The fact that they produce an albino child requires that each parent provides an *a* gene to the albino child; thus the parents must both be heterozygous (*Aa*).

(b) The female must be *aa*. Since all the children are normal, one would consider the male to be *AA* instead of *Aa*. However, the male could be *Aa*. Under that circumstance, the likelihood of having six children, all normal, is 1/64.

6. Symbolism:

w = wrinkled seeds g = green cotyledons
W = round seeds G = yellow cotyledons

P_1: *WWGG* × *wwgg*

Gametes produced: One member of each gene pair is "segregated" to each gamete.

WWGG → (*WG*) *wwgg* → (*wg*)

F_1: *WwGg*

$F_1 \times F_1$: *WwGg* × *WwGg*

9/16 *W_G_* round seeds, yellow cotyledons
3/16 *W_gg* round seeds, green cotyledons
3/16 *wwG_* wrinkled seeds, yellow cotyledons
1/16 *wwgg* wrinkled seeds, green cotyledons

Seed Shape	***Cotyledon Color***	***Phenotypes***
3/4 round	3/4 yellow ⇒	9/16 round, yellow
	1/4 green ⇒	3/16 round, green
1/4 wrinkled	3/4 yellow ⇒	3/16 wrinkled, yellow
	1/4 green ⇒	1/16 wrinkled, green

8. In Problem 7, (c) fits this description.

10. 1. Factors (genes) occur in pairs.
2. Some genes have dominant and recessive alleles.
3. Alleles segregate from each other during gamete formation. When homologous chromosomes separate from each other at anaphase I, alleles will go to opposite poles of the meiotic apparatus.
4. One gene pair separates independently from other gene pairs. Different gene pairs on the same homologous pair of chromosomes (if far apart) or on nonhomologous chromosomes will separate independently from each other during meiosis.

12. Homozygosity refers to a condition where both genes of a pair are the same (i.e., *AA* or *GG* or *hh*), whereas heterozygosity refers to the condition where members of a gene pair are different (i.e., *Aa* or *Gg* or *Bb*).

14. (a) 4: *AB, Ab, aB, ab*
(b) 2: *AB, aB*
(c) 8: *ABC, ABc, AbC, Abc, aBC, aBc, abC, abc*
(d) 2: *ABc, aBc*
(e) 4: *ABc, Abc, aBc, abc*
(f) $2^5 = 32$

16. The offspring will occur in a typical 1:1:1:1 as

1/4 *WwGg* (round, yellow)
1/4 *Wwgg* (round, green)
1/4 *wwGg* (wrinkled, yellow)
1/4 *wwgg* (wrinkled, green)

18. (a,b)

Expected ratio	*Observed (o)*	*Expected (e)*
9/16	315	312.75
3/16	108	104.25
3/16	101	104.25
1/16	32	34.75

$$\chi^2 = 0.47$$

χ^2 value is associated with a probability greater than 0.90 for 3 degrees of freedom. The observed and expected values do not deviate significantly. To deal with parts **(b)** and **(c),** it is easier to see the observed values for the monohybrid ratios if the phenotypes are listed:

smooth, yellow	315
smooth, green	108
wrinkled, yellow	101
wrinkled, green	32

For the smooth: wrinkled *monohybrid component,* the smooth types total 423 (315 + 108), while the wrinkled types total 133 (101 + 32).

Expected ratio	*Observed (o)*	*Expected (e)*
3/4	423	417
1/4	133	139

The χ^2 value is 0.35 for 1 degree of freedom, which gives a p value greater than 0.50 and less than 0.90. We fail to reject the null hypothesis and are confident that the observed values do not differ significantly from the expected values.

(c) For the yellow:green portion of the problem, see that there are 416 yellow plants (315 + 101) and 140 (108 + 32) green plants.

Expected ratio	*Observed (o)*	*Expected (e)*
3/4	416	417
1/4	140	139

The χ^2 value is 0.01 for 1 degree of freedom, and the p value is greater than 0.90. We fail to reject the null hypothesis and are confident that the observed values do not differ significantly from the expected values.

20. Use of the $p = 0.10$ as the "critical" value for rejecting or failing to reject the null hypothesis instead of $p = 0.05$ would allow more null hypotheses to be rejected. As the critical p value is increased, it takes a smaller χ^2 value to cause rejection of the null hypothesis. It would take less difference between the expected and observed values to reject the null hypothesis. Therefore, the stringency of failing to reject the null hypothesis is increased.

22. The probability of getting *aabbcc* from the *AaBbCC* × *AABbCc* mating is zero.

24. Genes that are recessive can skip generations and exist in a carrier state in parents. For example, notice that II-4 and II-5 produce a female child (III-4) with the affected phenotype. On these criteria alone, the gene must be viewed as being recessive.

I-1 (*Aa*), I-2 (*aa*), I-3 (*Aa*), I-4 (*Aa*)

II-1 (*aa*), II-2 (*Aa*), II-3 (*aa*), II-4 (*Aa*), II-5 (*Aa*), II-6 (*aa*), II-7 (*AA* or *Aa*), II-8 (*AA* or *Aa*)

III-1 (*AA* or *Aa*), III-2 (*AA* or *Aa*), III-3 (*AA* or *Aa*), III-4 (*aa*), III-5 (probably *AA*), III-6 (*aa*)

IV-1 through IV-7 all *Aa*.

26. The gene is inherited as an autosomal recessive.

I-1 (*aa*), I-2 (*Aa* or *AA*), I-3 (*Aa*), I-4 (*Aa*)

II-1 (*Aa*), II-2 (*Aa*), II-3 (*Aa*), II-4 (*Aa*), II-5 (*aa*), II-6 (*AA* or *Aa*), II-7 (*AA* or *Aa*)

III-1 (*AA* or *Aa*), III-2 (*aa*), III-3 (*AA* or *Aa*)

28. (a) First consider that each parent is homozygous (true-breeding in the question) and since in the F_1 only round, axial, violet, and full phenotypes were expressed, they must each be dominant. Because all genes are on nonhomologous chromosomes, independent assortment will occur.
(b) Round, axial, violet and full would be the most frequent phenotypes: $3/4 \times 3/4 \times 3/4 \times 3/4$
(c) Wrinkled, terminal, white, and constricted would be the least frequent phenotypes: $1/4 \times 1/4 \times 1/4 \times 1/4$
(d) $(3/4 \times 3/4 \times 3/4 \times 3/4) +$
$(1/4 \times 1/4 \times 1/4 \times 1/4) = 82/256$
(e) There would be 16 different phenotypes in the testcross offspring, just as there are 16 different phenotypes in the F_2 generation.

30. (a) Data indicate that *straight* is dominant to *curled* and *short* is dominant to *long*.
Possible symbols would be (using standard *Drosophila* symbolism):

straight wings $= w^+$ curled wings $= w$
short bristles $= b^+$ long bristles $= b$

(b) Cross #1: $w^+/w\,;b^+/b \times w^+/w\,;b^+/b$
Cross #2: $w^+/w\,;b/b \times w^+/w\,;b/b$
Cross #3: $w/w\,;b/b \times w^+/w\,;b^+/b$
Cross #4: $w^+/w^+\,;b^+/b \times w^+/w^+\,;b^+/b$
(one parent could be w^+/w)
Cross #5: $w/w\,;b^+/b \times w^+/w\,;b^+/b$

32. (a) Notice in the first cross that a 3:1 ratio exists for the spiny to smooth phenotypes. Thus, we would predict that the *spiny* allele is dominant to *smooth*. We would also predict that the *purple* allele is dominant to *white*.
(b) One could cross a homozygous purple, spiny plant to a homozygous white, smooth plant. The purple, spiny F_1 would support the hypothesis that *purple* is dominant to *white* and *spiny* is dominant to *smooth*. In the F_2, a 9:3:3:1 ratio would not only support the above hypothesis, it would also indicate the independent inheritance and expression of the two traits.

Chapter 4 Modification of Mendelian Ratios

2. *Incomplete dominance* can be viewed more as a quantitative phenomenon where the heterozygote is intermediate (approximately) between the limits set by the homozygotes.
Codominance can be viewed in a more qualitative manner where both of the alleles in the heterozygote are expressed.

4. $Pp \times Pp$ ⇩

1/4 PP (lethal)
2/4 Pp (platinum)
1/4 pp (silver)

Therefore, the ratio of surviving foxes is 2/3 platinum, 1/3 silver. The P allele behaves as a recessive in terms of lethality (seen only in the homozygote) but as a dominant in terms of coat color (seen in the homozygote).

6. Flower color: RR = red; Rr = pink; rr = white

Flower shape: P = personate; p = peloric

(a) $RRpp \times rrPP \Longrightarrow RrPp$

(b) $RRPP \times rrpp \Longrightarrow RrPp$

(c) $RrPp \times RRpp \Longrightarrow$ { $RRPp$, $RRpp$, $RrPp$, $Rrpp$ }

(d) $RrPp \times rrpp \Longrightarrow$ { $rrPp$, $rrpp$, $RrPp$, $Rrpp$ }

In the cross of the F_1 of (a) to the F_1 of (b), both of which are double heterozygotes, one would expect the following:

$RrPp \times RrPp$

1/4 red
- / 3/4 personate ⟹ 3/16 red, personate
- \ 1/4 peloric ⟹ 1/16 red, peloric

2/4 pink
- / 3/4 personate ⟹ 6/16 pink, personate
- \ 1/4 peloric ⟹ 2/16 pink, peloric

1/4 white
- / 3/4 personate ⟹ 3/16 white, personate
- \ 1/4 peloric ⟹ 1/16 white, peloric

8. This is a case of gene interaction (novel phenotypes) where the yellow and black types (double mutants) interact to give the cream phenotype and epistasis where the cc genotype produces albino.

(a) $AaBbCc \Longrightarrow$ gray (C allows pigment)

(b) $A\_B\_Cc \Longrightarrow$ gray (C allows pigment)

(c) Use the forked line method for this portion.

3/4 $A\_$
- 3/4 $B\_$
  - / 1/2 Cc ⟹ 9/32 gray
  - \ 1/2 cc ⟹ 9/32 albino
- 1/4 bb
  - / 1/2 Cc ⟹ 3/32 yellow
  - \ 1/2 cc ⟹ 3/32 albino

1/4 aa
- 3/4 $B\_$
  - / 1/2 Cc ⟹ 3/32 black
  - \ 1/2 cc ⟹ 3/32 albino
- 1/4 bb
  - / 1/2 Cc ⟹ 1/32 cream
  - \ 1/2 cc ⟹ 1/32 albino

Combining the phenotypes gives (always count the proportions to see that they add up to 1.0):

16/32 albino;
9/32 gray;
3/32 yellow;
3/32 black;
1/32 cream

10. (a) 1/4
(b) 1/2
(c) 1/4
(d) zero

12. Consider two alleles that are autosomal and let

BB = beardless in both sexes
Bb = beardless in females
Bb = bearded in males
bb = bearded in both sexes

P_1: female: bb (bearded) × male: BB (beardless)

F_1: Bb = females beardless; males bearded

F_2:

	1/2 females	1/2 males
1/4 BB	1/8 beardless	1/8 beardless
2/4 Bb	2/8 beardless	2/8 bearded
1/4 bb	1/8 bearded	1/8 bearded

One could test the above model by crossing F_1 (heterozygous) beardless females with bearded (homozygous) males. Comparing these results with the reciprocal cross would support the model if the distributions of sexes with phenotypes were the same in both crosses.

14. Symbolism: Normal wing margins = sd^+; scalloped = sd

(a) P_1: $X^{sd}X^{sd} \times X^+/Y$ ⇩

F_1: 1/2 X^+X^{sd} (female, normal)
1/2 X^{sd}/Y (male, scalloped)

F_2: 1/4 X^+X^{sd} (female, normal)
1/4 $X^{sd}X^{sd}$ (female, scalloped)
1/4 X^+/Y (male, normal)
1/4 X^{sd}/Y (male, scalloped)

(b) P_1: $X^+X^+ \times X^{sd}/Y$ ⇩

F_1: 1/2 X^+X^{sd} (female, normal)
1/2 X^+/Y (male, normal)

F_2: 1/4 X^+X^+ (female, normal)
1/4 X^+X^{sd} (female, normal)
1/4 X^+/Y (male, normal)
1/4 X^{sd}/Y (male, scalloped)

If the *scalloped* gene were not X-linked, then all of the F_1 offspring would be wild (phenotypically) and a 3:1 ratio of normal to scalloped would occur in the F_2.

16. (a) P_1: $X^vX^v; +/+ \times X^+/Y; b^r/b^r$

F_1: 1/2 $X^+X^v; +/b^r$ (female, normal)
1/2 $X^v/Y; +/b^r$ (male, vermilion)

F_2: *Eye color (X)*	*Eye color (autosomal)*	
1/4 females, normal	3/4 normal	3/16
	1/4 brown	1/16
1/4 females, vermilion	3/4 normal	3/16
	1/4 brown	1/16
1/4 males, normal	3/4 normal	3/16
	1/4 brown	1/16
1/4 males, vermilion	3/4 normal	3/16
	1/4 brown	1/16

3/16 = females, normal
1/16 = females, brown eyes
3/16 = females, vermilion eyes
1/16 = females, white eyes
3/16 = males, normal
1/16 = males, brown eyes
3/16 = males, vermilion eyes
1/16 = males, white eyes

(b) P_1: $X^+X^+; b^r/b^r \times X^v/Y; +/+$ ⇨

F_1: $1/2\ X^+X^v; +/b^r$ (female, normal)
$1/2\ X^+/Y; +/b^r$ (male, normal)

F_2: 6/16 = females, normal
2/16 = females, brown eyes
3/16 = males, normal
1/16 = males, brown eyes
3/16 = males, vermilion eyes
1/16 = males, white eyes

(c) P_1: $X^vX^v; b^r/b^r \times X^+/Y; +/+$ ⇨

F_1: $1/2\ X^+X^v; +/b^r$ (female, normal)
$1/2\ X^v/Y; +/b^r$ (male, vermilion)

F_2: 3/16 = females, normal
1/16 = females, brown eyes
3/16 = females, vermilion eyes
1/16 = females, white eyes
3/16 = males, normal
1/16 = males, brown eyes
3/16 = males, vermilion eyes
1/16 = males, white eyes

18. (a) One might hypothesize that two gene pairs are involved in the inheritance of one trait while one gene pair is involved in the other.

(b) Croaking is due to one (dominant/recessive) gene pair, while eye color is due to two gene pairs. Because there is a (9:4:3) ratio regarding eye color, some gene interaction (epistasis) is indicated.

(c) Symbolism: Croaking: *R_* = rib-it; *rr* = knee-deep
Eye color: Since the most frequent phenotype is blue eye, let *A_B_* represent the genotypes. For the purple class, "a 3/16 group," use the *A_bb* genotypes. The "4/16" class (green) would be the *aaB_* and the *aabb* groups.

(d) *AABBrr* × *AAbbRR* which would produce an F_1 of *AABbRr* which would be blue-eyed and rib-it. The F_2 will follow a pattern of a 9:3:3:1 ratio because of homozygosity for the A locus and heterozygosity for both the B and R loci.

9/16 *AAB_R_* = blue-eyed, rib-it

3/16 *AAB_rr* = blue-eyed, knee-deep

3/16 *AAbbR_* = purple-eyed, rib-it

1/16 *AAbbrr* = purple-eyed, knee-deep

20. *A_B_* = solid white (9/16)
*aaB_* = solid white (3/16)
A_bb = black and white spotted (3/16)
aabb = solid black (1/16)

The selection of "*bb*" as giving the spotted phenotype is arbitrary. One could obtain *AAbb* true-breeding black and white-spotted cattle.

22. (a,b) One can envision two pathways leading to the production of green pigment:

⟶ *B* ⟶ Blue pigment ⟶ green pigment
⟶ *Y* ⟶ Yellow pigment ⟶ green pigment

YyBb × *YyBb* (both green)

9/16 *Y_B_* (green)
3/16 *yyB_* (blue)
3/16 *Y-bb* (yellow)
1/16 *yybb* (albino)

P_1: *YYBB* (green) × *yybb* (albino)

or

YYbb × *yyBB*
↓
YyBb(*green*)

Crossing these F_1's gives the observed ratios in the F_2.

24. (a,b) In looking at the pedigrees, one can see that the condition cannot be dominant because it appears in the offspring (II-3 and II-4) and not the parents in the first two cases. The condition is therefore *recessive*. In the second cross, note that the father is not shaded, yet the daughter (II-4) is. If the condition is recessive, then it must also be *autosomal*.

(c) II-1 = *AA* or *Aa*
II-6 = *AA* or *Aa*
II-9 = *Aa*

26. Symbolism:

*A_B_* = black
A_bb = golden
aabb = golden
*aaB_* = brown

The combination of *bb* is epistatic to the *A* locus.

(a) *AAB_* × *aaBB* (other configurations possible but each must give all offspring with *A* and *B* dominant alleles)
(b) *AaB_* × *aaBB* (other configurations are possible but no *bb* types can be produced)
(c) *AABb* × *aaBb*
(d) *AABB* × *aabb*
(e) *AaBb* × *Aabb*
(f) *AaBb* × *aabb*
(g) *aaBb* × *aaBb*
(h) *AaBb* × *AaBb*

Those genotypes that will breed true will be as follows:

black = *AABB*
golden = all genotypes which are *bb*
brown = *aaBB*

28. (a) This is a case of incomplete dominance in which, as shown in the third cross, the heterozygote (palomino) produces a typical 1:2:1 ratio. Therefore one can set the following symbols:

$C^{ch}C^{ch}$ = chestnut
C^cC^c = cremello
$C^{ch}C^c$ = palomino

(b) The F_1 resulting from matings between cremello and chestnut horses would be expected to be all palomino. The F_2 would be expected to fall in a 1:2:1 ratio as in the third cross in part (a) above.

30. Cross #1 = (c)
Cross #2 = (d)
Cross #3 = (b)
Cross #4 = (e)
Cross #5 = (a)

Given that each parental/offspring grouping can only be used once, there are no other combinations.

32. With extranuclear inheritance, the phenotype is determined by the nuclear (maternal effect) or cytoplasmic (organelle or infectious) condition of the parent that contributes the bulk of the cytoplasm to the offspring. Standard Mendelian ratios (3:1) are usually not present. In general, the results of reciprocal crosses differ. In sex-linked inheritance, the pattern is often from grandfather through carrier mother to son. Patterns of extrachromosomal inheritance are often not influenced by the sex of the offspring.

34. (a) The presence of bcd^-/bcd^- males can be explained by the maternal effect: mothers were bcd^+/bcd^-.

(b) The cross

female $bcd^+/bcd^- \times$ male bcd^-/bcd^-

will produce an F_1 with normal embryogenesis because of the maternal effect. In the F_2, any cross having bcd^+/bcd^- mothers will have phenotypically normal embryos. Any cross involving homozygous bcd^-/bcd^- mothers will have problems with embryogenesis.

36. Beatrice, Alice of Hesse, and Alice of Athlone are carriers. There is a 1/2 chance that Princess Irene is a carrier.

Chapter 5 Sex Determination and Sex Chromosomes

4. Sexual differentiation is the response of cells, tissues, and organs to signals provided by the genetic mechanisms of sex determination.

6. In *Drosophila* it is the balance between the number of X chromosomes and the number of haploid sets of autosomes which determines sex. In humans there is a small region on the Y chromosome which determines maleness.

8. Individuals with the 47,XXY complement are males while 45,XO produces females. In *Drosophila* it is the balance between the number of X chromosomes and the number of haploid sets of autosomes which determines sex. In contrast to humans, XO *Drosophila* are males and the XXY complement is female.

10. (a) female $X^{rw}Y \times$ male X^+X^+

F_1:	females:	X^+Y (normal)
	males:	$X^{rw}X^+$ (normal)
F_2:	females:	X^+Y (normal)
		$X^{rw}Y$ (reduced wing)
	males:	$X^{rw}X^+$ (normal)
		X^+X^+ (normal)

(b) female $X^{rw}X^{rw} \times$ male X^+Y

F_1:	females:	$X^{rw}X^+$ (normal)
	males:	$X^{rw}Y$ (reduced wing)
F_2:	females:	$X^{rw}X^+$ (normal)
		$X^{rw}X^{rw}$ (reduced wing)
	males:	X^+Y (normal)
		$X^{rw}Y$ (reduced wing)

12. Males and females share a common placenta and therefore hormonal factors carried in blood. Hormones and other molecular species (transcription factors perhaps) triggered by the presence of a Y chromosome lead to a cascade of developmental events that both suppress female organ development and enhance masculinization.

14. If the male offspring had white eyes and the female offspring were wild type, one might suspect that the attached-X had become unattached.

16. A *Barr body* is a differentially staining chromosome seen in some interphase nuclei of mammals with two X chromosomes. The Barr body is an X chromosome that is considered to be genetically inactive.

18. The *Lyon hypothesis* states that the inactivation of the X chromosome occurs at random early in embryonic development. Such X chromosomes are in some way "marked" such that all clonally related cells have the same X chromosome inactivated.

20. Females may display mosaic retinas with patches of defective color perception. Under these conditions, their color vision may be influenced.

22. Many organisms have evolved over millions of years under the fine balance of numerous gene products. Many genes required for normal cellular and organismic function in *both* males and females are located on the X chromosome. These gene products have nothing to do with sex determination or sex differentiation.

24. In general, information about the primary sex ratio in humans is obtained from abortions and miscarriages. In addition, some studies (Barczyk 2001) using "hamster oocyte–human sperm" have been successful in determining some of the causal factors involved in determining the primary sex ratio.

26. Since there is a region of synapsis close to the *Sry*-containing section on the Y chromosome, crossing over in this region would generate XY translocations, which would lead to the condition described.

28. The presence of the Y chromosome provides a factor (or factors) that leads to the initial specification of maleness. Subsequent expression of secondary sex characteristics must be dependent on the interaction of the normal X-linked *Tfm* allele with testosterone. Without such interaction, differentiation takes the female path. Since the *Tfm* allele is dominant, one would predict that normal male development would only occur if the mutant *Tfm* allele were deleted and a copy of the normal *Tfm* allele were engineered into the genome.

30. Since all haploids are male and half of the eggs are unfertilized, 50% of the offspring would be male at the start; adding the X_a/X_a types gives 25% more male, the remainder X_a/X_b would be female. Overall, 75% of the offspring would be male, while 25% would be female.

32. Because of X-chromosome inactivation in mammals, scientists would be interested in determining whether the nucleus taken from Rainbow (donor) would continue to show such inactivation. Would the inactivated X chromosome retain the property of inactivation?

34. If one assumes that the somatic ovarian cell engaged in X chromosome inactivation, the ovarian somatic cell that Rainbow donated to create CC contained an activated black gene and an inactivated orange gene (from X-inactivation). This would mean that as CC developed, her cells did not change that inactivation pattern.

Chapter 6 Chromosome Mutations: Variation in Number and Arrangement

4. The fact that there is a significant maternal age effect associated with Down syndrome indicates that nondisjunction in older females contributes disproportionately to the number of Down syndrome individuals. In addition, certain genetic and cytogenetic marker data indicate the influence of female nondisjunction.

6. Because an allotetraploid has a possibility of producing bivalents at meiosis I, it would be considered the most fertile of the three. Having an even number of chromosomes to match up at the metaphase I plate, autotetraploids would be considered to be more fertile than autotriploids.

8. It is likely that an interspecific hybridization occurred followed by chromosome doubling. These events probably produced a fertile amphidiploid (allotetraploid).

10. While there is the appearance that crossing over is suppressed in inversion "heterozygotes," the phenomenon extends from the fact that the crossover chromatids end up being abnormal in genetic content. As such, they fail to produce viable (or competitive) gametes or lead to zygotic or embryonic death.

12. In a work entitled *Evolution by Gene Duplication*, Ohno, suggests that gene duplication has been essential in the origin of new genes. If gene products serve essential functions, mutation and therefore evolution would not be possible unless these gene products could be compensated for by products of duplicated, normal genes. The duplicated genes, or the original genes themselves, would be able to undergo mutational "experimentation" without necessarily threatening the survival of the organism.

14. A Turner syndrome female has the sex chromosome composition of XO. If the father had hemophilia, it is likely that the Turner syndrome individual inherited the X chromosome from the father and no sex chromosome from the mother. If nondisjunction occurred in the mother, either during meiosis I or meiosis II, an egg with no X chromosome could be the result.

16. Given the basic chromosome set of nine unique chromosomes (a haploid complement) other forms with the "n multiples" are forms of polyploidy. Organisms with 27 chromosomes ($3n$) are more likely to be sterile because there are trivalents at meiosis I that cause a relatively high number of unbalanced gametes to be formed.

18. The cross would be as follows:

$WWWW \times wwww$

F_1: $WWww$

F_2: $35W$ and $1w$

20. Since two Gl_1 alleles and two ws_3 alleles are present in the triploid, they must have come from the pollen parent. By the wording of the problem, it is implied that the pollen parent contributed an unreduced (2n) gamete; however, another explanation, dispermic fertilization, is possible. In this case two Gl_1ws_3 gametes could have fertilized the ovule.

22. In a paracentric inversion, one recombinant chromatid is dicentric (two centromeres), while the other is acentric (lacks a centromere). In a pericentric inversion, recombinant chromatids have duplications and deletions; however, no acentric or dicentric chromatids are produced.

24. In plants, gametes with aberrant unbalanced genetic complements usually fail to develop normally, leading to aborted pollen or ovules. Therefore, genetic imbalance is revealed prior to fertilization and inviable seeds result. In animals, aberrant or unbalanced gametes tend to function; however, significant embryonic (and subsequent) developmental abnormalities are likely to occur.

26. (a,b,c) It is likely that the translocation described above is the cause of the miscarriages. Segregation of the chromosomal elements will produce approximately half unbalanced gametes. The chance of a normal child is approximately one in two, however; half of the normal children will be translocation carriers. Should she abandon her attempts to have a child of her own? The answer to this question is more one of personal choice than science. It is the task of the scientific community to provide accurate information within the limits of technology. Generally speaking, this information is provided to individuals so that they can make informed decisions. In this case, the woman has been given information that probably fits her circumstance.

28. (a) The father must have contributed the abnormal X-linked gene.

(b) Since the son is XXY and heterozygous for anhidrotic dysplasia, he must have received both the defective gene and the Y chromosome from his father. Thus nondisjunction must have occurred during meiosis I.

(c) This son's mosaic phenotype is caused by X-chromosome inactivation, a form of dosage compensation in mammals.

30. Below is a description of breakage/reunion events which illustrate such a translocation in relatively small, similarly sized, chromosomes 19 (metacentric) and 20 (metacentric/submetacentric). The case described here is shown occurring before S phase duplication. The same phenomenon is shown in the text as occurring after S phase. Since the likelihood of such a translocation is fairly small in a general population, inbreeding probably played a significant role in allowing the translocation to "meet itself."

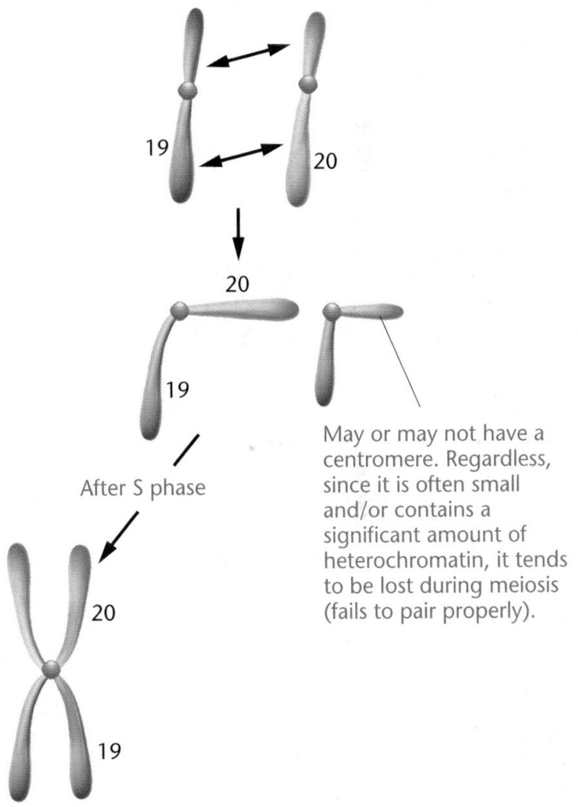

19
20
20
19
May or may not have a centromere. Regardless, since it is often small and/or contains a significant amount of heterochromatin, it tends to be lost during meiosis (fails to pair properly).
After S phase
20
19

32. This female will produce meiotic products of the following types:

normal: 18 + 21
translocated: 18/21
translocated plus 21: 18/21 + 21
deficient: 18 only

Note: The 18/21 + 18 gamete is not formed because it would require separation of primarily homologous chromosomes at anaphase I.

Fertilization with a normal 18 + 21 sperm cell will produce the following offspring:

normal: 46 chromosomes
translocation carrier: 45 chromosomes 18/21 + 18 + 21
trisomy 21: 46 chromosomes 18/21 + 21 + 21
monosomic: 45 chromosomes 18 + 18 + 21, lethal

Chapter 7 Linkage and Chromosome Mapping in Eukaryotes

2. First, in order for chromosomes to engage in crossing over, they must be in proximity. It is likely that the side-by-side pairing that occurs during synapsis is the earliest time during the cell cycle that chromosomes achieve that necessary proximity. Second, chiasmata are visible during prophase I of meiosis, and it is likely that these structures are intimately associated with the genetic event of crossing over.

4. Because crossing over occurs at the four-strand stage of the cell cycle (that is, after S phase), notice that each single crossover involves only two (50%) of the four chromatids.

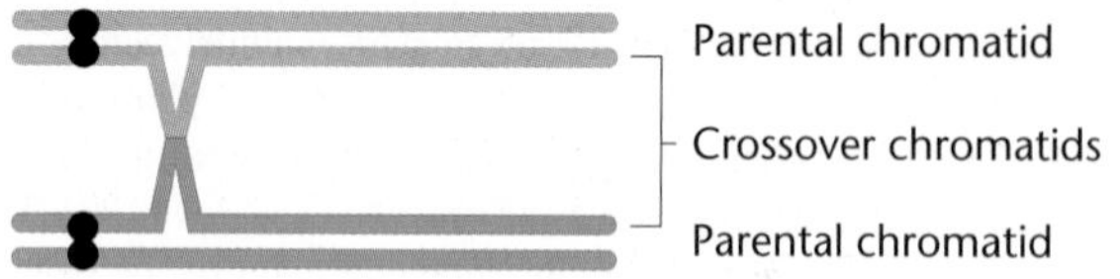

6. Positive interference occurs when a crossover in one region of a chromosome interferes with crossovers in nearby regions. Such interference ranges from zero (no interference) to 1.0 (complete interference). Interference is often explained by a physical rigidity of chromatids such that they are unlikely to make sufficiently sharp bends to allow crossovers to be close together.

8.

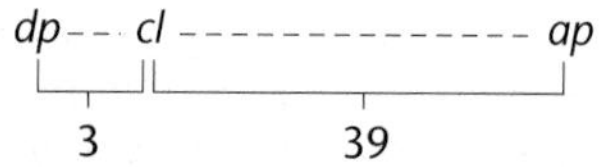

10. The question is whether the arrangement in the parents is *coupled* $RY/ry \times ry/ry$ or *not coupled* $Ry/rY \times ry/ry$.
Notice that the most frequent phenotypes in the offspring, the parentals, are colored, green (88) and colorless, yellow (92). This indicates that the heterozygous parent in the testcross is coupled $RY/ry \times ry/ry$. There would be 10 map units between the loci.

12. $PZ/pz \times pz/pz$
Adding the crossover percentages together (6.9 + 7.1) gives 14 percent, which would be the map distance between the two genes.

14.

	female A	female B	Frequency
NCO	3, 4	7, 8	first
SCO	1, 2	3, 4	second
SCO	7, 8	5, 6	third
DCO	5, 6	1, 2	fourth

16. (a) $y\ w\ +/+\ +ct \times y\ w\ +/Y$

(b)

(c) There were $0.185 \times 0.015 \times 1000 = 2.775$ double crossovers expected.

(d) Because the cross to the F_1 males included the normal (wild-type) gene for *cut wings*, it would not be possible to unequivocally determine the genotypes from the F_2 phenotypes for all classes.

18. The results would change because of no crossing over in males.

20. (a,b)

$$a - b = \frac{32 + 38 + 0 + 0}{1000} \times 100$$
$$= 7 \text{ map units}$$

$$b - c = \frac{11 + 9 + 0 + 0}{1000} \times 100$$
$$= 2 \text{ map units}$$

(c) The progeny phenotypes that are missing are $++c$ and $a\,b+$, which, of 1000 offspring, 1.4 would be expected. Perhaps by chance or some other unknown selective factor, they were not observed.

22. These observations as well as the results of other experiments indicate that the synaptonemal complex is required for crossing over.

24. Since the genetic map is more accurate when relatively small distances are covered and when large numbers of offspring are scored, this map would probably not be too accurate with such a small sample size.

26. $A\_B\_$ 33/64
A_bb 15/64
$aaB\_$ 15/64
$aabb$ 1/64

28. By having microscopically visible markers on the chromosomes, Creighton and McClintock were able to show that homologous chromosomal material physically exchanged segments during crossing over.

30. $a\,b\,c = 168$
$+ + + = 168$
$a + + = 20$
$+ b\,c = 20$
$+ + c = 10$
$a\,b + = 10$
$+ b + = 2$
$a + c = 2$

The map distances would be computed as follows:

$$a - b = \frac{20 + 20 + 2 + 2}{400} \times 100$$
$$= 11 \text{ map units}$$

$$b - c = \frac{10 + 10 + 2 + 2}{400} \times 100$$
$$= 6 \text{ map units}$$

32. B^+ = wild eye shape
B = Bar-eye shape
m^+ = wild wings
m = miniature wings
e^+ = wild body color
e = ebony body color

Mapping the distance between B and m would be as follows:

$$\frac{(57 + 64)}{(226 + 218 + 57 + 64)} \times 100 = \frac{121}{565} \times 100 = 21.4 \text{ map units}$$

We would conclude that the *ebony* locus is either far away from B and m (50 map units or more), or it is on a different chromosome. In fact, *ebony* is on a different chromosome.

Chapter 8 Genetic Analysis and Mapping in Bacteria and Bacteriophages

2. (a) The requirement for physical contact between bacterial cells during conjugation was established by placing a filter in a U-tube such that the medium could be exchanged but the bacteria could not come in contact. Under this condition, conjugation does not occur.

(b) By treating cells with streptomycin, an antibiotic, it was shown that recombination would not occur if one of the two bacterial strains was inactivated. However, if the other was similarly treated, recombination would occur. Thus, directionality was suggested, with one strain being a donor strain and the other being the recipient.

(c) An F^+ bacterium contains a circular, double-stranded, structurally independent, DNA molecule which can direct recombination.

4. Bacteria that are F^+ possess the F factor, while those that are F^- lack the F factor. In Hfr cells, the F factor is integrated into the bacterial chromosome, and in F′ bacteria, the F factor is free of the bacterial chromosome yet possesses a piece of the bacterial chromosome.

6. In an Hfr $\times$ F^- cross, the F factor is directing the transfer of the donor chromosome. It takes at least 90 minutes to transfer the entire chromosome. Because the F factor is the last element to be transferred and the conjugation tube is fragile, the likelihood for complete transfer is low.

8. Transformation requires *competence* on the part of the recipient bacterium, meaning that only under certain conditions are bacterial cells capable of being transformed. Transforming DNA must be *double-stranded* to begin with, yet is converted to a single-stranded structure upon insertion into the host cell. The most efficient length of the transforming DNA is about 1/200 of the size of the host chromosome. Transformation is an energy-requiring process, and the number of sites on the bacterial cell surface is limited.

10. Notice that the incorporation of loci a^+ or b^+ occurs much more frequently than the incorporation of b^+ and c^+ together (210 to 1) and the incorporation of all three genes $a^+b^+c^+$ occurs relatively infrequently. If *a* and *b* loci are close together and both are far from locus *c,* then fewer crossovers would be required to incorporate the two linked loci compared to all three loci. If all three loci were close together, then the frequency of incorporation of all three would be similar to the frequency of incorporation of any two contiguous loci, which is not the case.

14. A single plaque is a clearing of bacteria resulting from the lytic action of millions of bacteriophage.

16. A lytic cycle occurs as bacteriophage enter a bacterial host and form progeny phage after a relatively short period of time. *Lysogeny* is a complex process whereby certain temperate phage can enter a bacterial cell and instead of following a lytic developmental path, integrate their DNA into the bacterial chromosome.

18. In their experiment, a filter was placed between the two auxotrophic strains, which would not allow contact. F-mediated conjugation requires contact, and without that contact, such conjugation cannot occur. The treatment with DNase showed that the filterable agent was not naked DNA.

20. Cotransduction of genes in generalized transduction allows linkage relationships to be determined because the closer two genes are to each other, the higher the likelihood that they will be physically "linked" together in a single DNA strand during transduction.

22. Viral recombination occurs when there is a sufficiently high number of infecting viruses, so that there is a high likelihood that more than one type of phage will infect a given bacterium. Under this condition, phage chromosomes can recombine by crossing over.

24. Assume that 0.1 ml of the plaque suspension is typically used in a plaque assay.

(a) The original concentration is greater than 10^5.

(b) The original concentration is about 1.4×10^7.

(c) Remembering that 0.1 ml is typically used in the plaque assay, the initial concentration of phage is less than 10^7. Coupling this information with the calculations in part (b) above, it would appear that the initial concentration of phage is around 1×10^7 and the failure to obtain plaques in this portion of the experiment is expected and due to sampling error.

26. Because the frequency of double transformants is quite high (compare the trp^+tyr^+ transformants in A and B experiments), one may conclude that the genes are quite closely linked together.

28. (a) Rifampicin eliminates the donor strain which is rif^s.

(b) <u>*b a*   *c*   F</u>

(c) The recombinants must be replated on a rifampicin medium to determine which ones are sensitive.

30. Since *g* cotransforms with *f,* it is likely to be in the *c b f* "linkage group" and would be expected to cotransform with each. One would not expect transformation with *a, d,* or *e.*

Chapter 9 DNA Structure and Analysis

2. Abundant knowledge of the structural and enzymatic properties of proteins generated a bias that worked to favor proteins as the hereditary substance. In addition, proteins were composed of as many as 20 different subunits (amino acids), thereby providing ample structural and functional variation for the multiple tasks that must be accomplished by the genetic material. The tetranucleotide hypothesis (structure) provided insufficient variability to account for the diverse roles of the genetic material.

4. Transformation is dependent on a macromolecule (DNA) that can be extracted and purified from bacteria. During such purification, however, other macromolecular species may contaminate the DNA. Specific degradative enzymes, proteases, RNase, and DNase were used to selectively eliminate components of the extract, and, if transformation is concomitantly eliminated, then the eliminated fraction is the transforming principle. DNase eliminates DNA and transformation. Therefore it must be the transforming principle.

6. Actually, phosphorus is found in approximately equal amounts in DNA and RNA. Therefore, labeling with ^{32}P would "tag" both RNA and DNA. However, the T2 phage, in its mature state, contains very little if any RNA. Therefore DNA would be interpreted as being the genetic material in T2 phage.

8. The early evidence would be considered indirect in that at no time was there an experiment, like transformation in bacteria, in which genetic information in one organism was transferred to another using DNA. Rather, by comparing DNA content in various cell types (sperm and somatic cells) and observing that the *action* and *absorption* spectra of ultraviolet light were correlated, DNA was considered to be the genetic material. This suggestion was supported by the fact that DNA was shown to be the genetic material in bacteria and some phage. Direct evidence for DNA being the genetic material comes from a variety of observations, including gene transfer that has been facilitated by recombinant DNA techniques.

10. Refer to the text for illustrations. Linkages among the three components require the removal of water (H_2O).

12.

Guanine:	2-amino-6-oxypurine
Cytosine:	2-oxy-4-aminopyrimidine
Thymine:	2,4-dioxy-5-methylpyrimidine
Uracil:	2,4-dioxypyrimidine

16. Because in double-stranded DNA, A=T and G≡C (within limits of experimental error), the data presented would have indicated a lack of pairing of these bases in favor of a single-stranded structure or some other nonhydrogen-bonded structure. Alternatively, from the data it would appear that A=C and T=G, which would negate the chance for typical hydrogen bonding since opposite charge relationships do not exist. Therefore, it is quite unlikely that a tight helical structure would form at all.

18. Three main differences between RNA and DNA are the following:
1. Uracil in RNA replaces thymine in DNA.
2. Ribose in RNA replaces deoxyribose in DNA.
3. RNA often occurs as both single- and partially double-stranded forms, whereas DNA most often occurs in a double-stranded form.

20. The nitrogenous bases of nucleic acids (nucleosides, nucleotides, and single- and double-stranded polynucleotides) absorb UV light maximally at wavelengths 254 to 260 nm. Using this phenomenon, one can often determine the presence and concentration of nucleic acids in a mixture. Since proteins absorb UV light maximally at 280 nm, this is a relatively simple way of dealing with mixtures of biologically important molecules. UV absorption is greater in single-stranded molecules (hyperchromic shift) as compared to double-stranded structures. Therefore, one can easily determine, by applying denaturing conditions, whether a nucleic acid is in the single- or double-stranded form. In addition, A-T rich DNA denatures more readily than G-C rich DNA. Therefore, one can estimate base content by denaturation kinetics.

22. *A hyperchromic effect* is the increased absorption of UV light as double-stranded DNA (or RNA for that matter) is converted to single-stranded DNA. If one monitors the UV absorption with a spectrophotometer during the melting process the hyperchromic shift can be observed. The T_m is the point on the profile (temperature) at which half (50%) of the sample is denatured.

24. The chemical basis for molecular hybridization is hydrogen bonding between the bases of each strand: AT and GC pairs.

26.
1. As shown, the extra phosphate is not normally expected.
2. In the adenine ring, a nitrogen is at position 8 rather than position 9.
3. The bond from the C-1' to the sugar should form with the N at position 9 (N-9) of the adenine.
4. The dinucleotide is a "deoxy" form, therefore each C-2' should not have a hydroxyl group. Notice the hydroxyl group at C-2' on the sugar of the adenylic acid.
5. At the C-5 position on the thymine residue, there should be a methyl group.
6. There are too many bonds at the N-3 position on the thymine.
7. There are too few bonds at the C-5 of thymine.

28. Since cytosine pairs with guanine and uracil pairs with adenine, the result would be a base substitution of G:C to A:T after rounds of replication.

30. Without knowing the exact bonding characteristics of hypoxanthine or xanthine, it may be difficult to predict the likelihood of each pairing type. It is likely that both are of the same class (purine or pyrimidine) because the names of the molecules indicate a similarity. In addition, the diameter of the structure is constant which, under the model to follow, would be expected. In fact, hypoxanthine and xanthine are both purines.

Because there are equal amounts of A, T, and H, one could suggest that they are hydrogen bonded to each other; the same may be said for C, G, and X. Given the molar equivalence of erythrose and phosphate, an alternating sugar-phosphate-sugar backbone as in "earth-type" DNA would be acceptable. A model of a triple helix would be acceptable, since the diameter is constant. Given the chemical similarities to "earth-type" DNA, it is probable that the unique creature's DNA follows the same structural plan.

32. In comparing DNA migration to RNA, even though RNA molecules have the same charge-to-mass ratios, they also exist in a variety of shapes. Complementary intrastrand base pairing can make more compact structures compared to the more relaxed, open conformation. For electrophoretic size comparisons, RNA molecules must be denatured to eliminate secondary structural variables.

Chapter 10 DNA Replication and Recombination

2. By labeling the pool of nitrogenous bases of the DNA of *E. coli* with the heavy isotope ^{15}N, it would be possible to "follow" the "old" DNA.

4. (a) Under a conservative scheme all of the newly labeled DNA will go to one sister chromatid, while the other sister chromatid will remain unlabeled.

(b) Under a dispersive scheme, all of the newly labeled DNA will be interspersed with unlabeled DNA. Both sister chromatids would appear as evenly labeled structures.

6. The *in vitro* replication requires a DNA template, a primer to give a double-stranded portion, a divalent cation (Mg^{++}), and all four of the deoxyribonucleoside triphosphates: dATP, dCTP, dTTP, and dGTP. The lower case "d" refers to the deoxyribose sugar.

8. Several analytical approaches showed that the products of DNA polymerase I were probably copies of the template DNA. *Base composition* was used initially to compare both templates and products. Within experimental error, those data strongly suggested that the DNA replicated faithfully.

10. Each precursor (dNTP) to DNA synthesis is added to the 3' end of a growing chain by the removal of the terminal phosphates and the formation of a covalent bond. The 3'-OH provided by the 3' nucleotide directly participates in the formation of that covalent bond.

14.

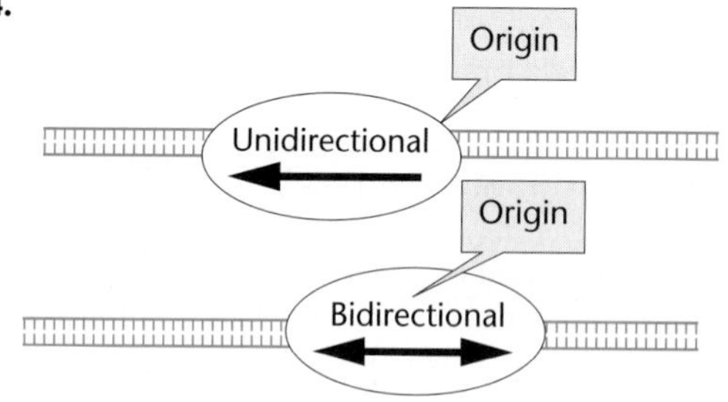

In continuous DNA synthesis, no short fragments are created because the direction of replication follows the replication fork. In discontinuous replication, short DNA segments are created because the direction of synthesis is opposite the replication fork.

16. *Okazaki fragments* are relatively short (1000 to 2000 bases in prokaryotes) DNA fragments that are synthesized in a discontinuous fashion on the lagging strand during DNA replication. DNA *ligase* is required to form phosphodiester linkages in gaps that are generated when DNA polymerase I removes RNA primer and meets newly synthesized DNA ahead of it. *Primer* RNA is formed by RNA primase to serve as an initiation point for the production of DNA strands on a DNA template.

18. Eukaryotic DNA is replicated in a manner that is very similar to that of *E. coli*. Synthesis is bidirectional, continuous on one strand and discontinuous on the other, and the requirements of synthesis (four deoxyribonucleoside triphosphates, divalent cation, template, and primer) are the same. Okazaki fragments of eukaryotes are about one-tenth the size of those in bacteria. Because there is a much greater amount of DNA to be replicated and DNA replication is slower, there are multiple initiation sites for replication in eukaryotes (and increased DNA polymerase per cell) in contrast to the single replication origin in prokaryotes. Replication occurs at different sites during different intervals of the S phase.

20. (a) about 4,000,000 bp.

(b) 1.3 mm.

22. Gene conversion is now considered a result of heteroduplex formation that is accompanied by mismatched bases. When these mismatches are corrected, the "conversion" occurs.

24. Telomerase activity is present in germ-line tissue to maintain telomere length from one generation to the next.

26. If replication is conservative, the first autoradiographs would have label distributed only on one side (chromatid) of the metaphase chromosome as shown below.

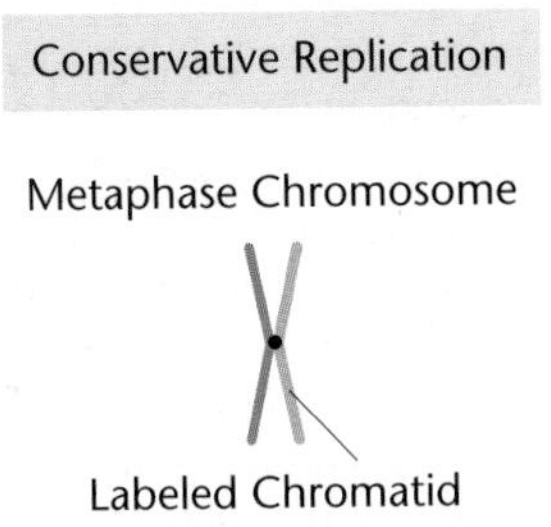

28. If the DNA contained parallel strands in the double helix and the polymerase would be able to accommodate such parallel strands, there would be continuous synthesis and no Okazaki fragments. The telomere problem would only be at one end. Several other possibilities exist. If the DNA strands were replicated as complete single strands, the synthesis could begin at the opposite free ends. In addition, if the DNA existed only as a single strand, the same results would occur.

30. Conservative replication can be eliminated.

Chapter 11 Chromosome Structure and DNA Sequence Organization

2. By having a circular chromosome, no free ends present the problem of linear chromosomes, namely, complete replication of terminal sequences.

4. Mitochondrial DNA varies in size from 16 kb in humans to over 350 kb in some plants and codes for ribosomal, transfer and messenger RNAs. The protein-synthesizing apparatus and many components of cellular respiration are jointly formed from nuclear and mitochondrial genes. Chloroplast DNA codes for ribosomal RNAs, as well as tRNAs and mRNAs for ribosomal proteins and photosynthesis. Nuclear genes also contribute to the pool of functional proteins in chloroplasts.

6. Mitochondria and chloroplasts contain rRNAs more similar to bacterial forms than those found in eukaryotes. This finding supports the endosymbiont hypothesis for the origin of such organelles from bacteria.

8. Since eukaryotic chromosomes are "multirepliconic" in that there are multiple replication forks along their lengths, one would expect to see multiple clusters of radioactivity.

10. Long interspersed elements (LINEs) are repetitive transposable DNA sequences in humans. The most prominent family, designated **L1,** is about 6.4 kb each and is represented about 100,000 times. LINEs are often referred to as retrotransposons because their mechanism of transposition resembles that used by retroviruses.

12. Because of the diverse cell types of multicellular eukaryotes, a variety of gene products is required, which may be related to the increase in DNA content per cell. In addition, the advantage of diploidy automatically increases DNA content per cell. However, when the question is seen in another way, it is likely that a much higher *percentage* of the genome of a prokaryote is actually involved in phenotype production than in a eukaryote. Eukaryotes have evolved the capacity to obtain and maintain what appears to be large amounts of "extra" perhaps "junk" DNA. Prokaryotes, on the other hand, with their relatively short life cycle, are extremely efficient in their accumulation and use of their genome. In addition, the eukaryotic genome is divided into separate entities (chromosomes) to perhaps facilitate the partitioning process in mitosis and meiosis.

14. Nucleosomes are octomeric structures of two molecules of each histone (H2A, H2B, H3, and H4) except H1. Between the nucleosomes and complexed with linker DNA is histone H1. A 146-base-pair sequence of DNA wraps around the nucleosome.

16. *Heterochromatin* is chromosomal material that stains deeply and remains condensed when other parts of chromosomes, euchromatin, are otherwise pale and decondensed. Heterochromatic regions replicate late in S phase and are relatively inactive in a genetic sense because there are few genes present, or if they are present, they are repressed. Telomeres and the areas adjacent to centromeres are composed of heterochromatin.

18. Volume of DNA:

$$3.14 \times 10\ \text{Å} \times 10\ \text{Å} \times (50 \times 10^4\ \text{Å}) = 1.57 \times 10^8\ \text{Å}^3$$

Volume of capsid:

$$4/3\,(3.14 \times 400\ \text{Å} \times 400\ \text{Å} \times 400\ \text{Å}) = 2.67 \times 10^8\ \text{Å}^3$$

Because the capsid head has a greater volume than the volume of DNA, the DNA will fit into the capsid.

20. Volume of the nucleus $= 5.23 \times 10^{11}\text{nm}^3$
Volume of the chromosome $= 1.9 \times 10^{11}\text{nm}^3$
Therefore, the percentage of the volume of the nucleus occupied by the chromatin is

$$\frac{1.9 \times 10^{11}\text{nm}^3}{5.23 \times 10^{11}\text{nm}^3} \times 100 = \text{about } 36.3\%$$

22. The findings that natural chemical modification of nucleosomal components, as indicated in the question, increases gene activity suggests that changes in the binding of nucleosomes to DNA enables genes to be more accessible to factors that promote gene function.

24. Dividing 3×10^9 base pairs by 10^6 gives an average of 3000 base pairs or 3 kb between *Alu* sequences.

Chapter 12 The Genetic Code and Transcription

2. (a) If the code contained six nucleotides (a sextuplet code), then the translation system is "out of phase" until the sixth "+" or "−" is encountered. In this case, the "out-of-phase" region would probably be more extensive and likely cause more amino acid alterations; however, the reading frame would eventually be established.

(b) Given a sextuplet code, restoration of the reading frames would only occur with the addition or loss of six nucleotides.

4. There are two possibilities for establishing the reading frames: ACA, if one starts at the first base, and CAC, if one starts at the second base. These would code for two different amino acids (ACA = threonine; CAC = histidine) and would produce repeating polypeptides that would alternate *thr-his-thr-his*. . . or *his-thr-his-thr*. . .

Given the sequence CUACUACUACUA, notice the different reading frames producing three different sequences, each containing the same amino acid.

Codons:	CUA	CUA	CUA	CUA. . .
Amino Acids:	leu	leu	leu	leu. . .
	UAC	UAC	UAC	UAC. . .
	tyr	tyr	tyr	tyr. . .
	ACU	ACU	ACU	ACU. . .
	thr	thr	thr	thr. . .

If a tetranucleotide is used, such as ACGUACGUACGU. . .

Codons:	ACG	UAC	GUA	CGU	ACG
Amino Acids:	thr	tyr	val	arg	thr
	CGU	ACG	UAC	GUA	CGU
	arg	thr	tyr	val	arg
	GUA	CGU	ACG	UAC	GUA
	val	arg	thr	tyr	val
	UAC	GUA	CGU	ACG	UAC
	tyr	val	arg	thr	tyr

Notice that the sequences are the same except that the starting amino acid changes.

6. From the repeating polymer ACACA. . . one can say that threonine is either CAC or ACA. From the polymer CAACAA. . . with ACACA. . ., ACA is the only codon in common. Therefore, threonine would have the codon ACA.

8. The basis of the technique is that if a trinucleotide contains bases (a codon) that are complementary to the anticodon of a charged tRNA, a relatively large complex is formed containing the ribosome, the tRNA, and the trinucleotide. This complex is trapped in the filter whereas the components by themselves are not trapped. If the amino acid on a charged, trapped tRNA is radioactive, then the filter becomes radioactive.

10. Apply the most conservative pathway of change.

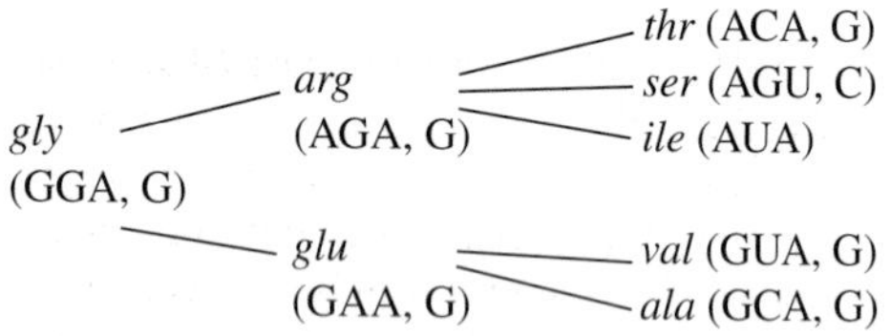

12. Because Poly U is complementary to Poly A, double-stranded structures will be formed. In order for an RNA to serve as a messenger RNA it must be single-stranded, thereby exposing the bases for interaction with ribosomal subunits and tRNAs.

14. (a)

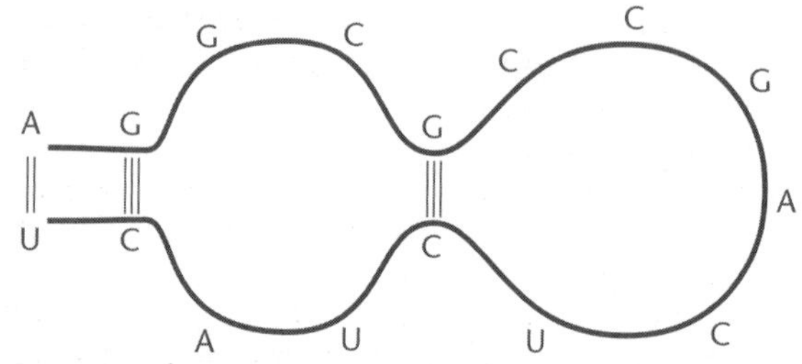

(b) TCCGCGGCTGAGATGA (use complementary bases, substituting T for U)

(c) GCU

(d) Assuming that the AGG. . . is the 5′ end of the mRNA, then the sequence would be

arg-arg-arg-leu-tyr

16. (a) Starting from the 5′ end and locating the AUG triplets, one finds two initiation sites leading to the following two sequences:

met-his-thr-tyr-glu-thr-leu-gly

met-arg-pro-leu-asp (or glu)

(b) In the shorter of the two reading sequences (the one using the internal AUG triplet), a UGA triplet was introduced at the second codon. While not in the reading frames of the longer polypeptide (using the first AUG codon), the UGA triplet eliminates the product starting at the second initiation codon.

18. The central dogma of molecular genetics, and to some extent all of biology, states that DNA produces, through transcription, RNA, which is "decoded" (during translation) to produce proteins.

20. RNA polymerase from *E. coli* is a complex, large (almost 500,000 daltons) molecule composed of subunits ($\alpha,\beta,\beta',\sigma$) in the proportion $\alpha2,\beta,\beta',\sigma$ for the holoenzyme. The β subunit provides catalytic function, while the sigma (σ) subunit is involved in recognition of specific promoters. The core enzyme is the protein without the sigma.

22. While some folding (from complementary base pairing) may occur with mRNA molecules, they generally exist as single-stranded structures that are quite labile. Eukaryotic mRNAs are generally processed such that the 5′ end is "capped" and the 3′ end has a considerable string of adenine bases. It is thought that these features protect the mRNAs from degradation. Such stability of eukaryotic mRNAs probably evolved with the differentiation of nuclear and cytoplasmic functions. Because prokaryotic cells exist in a less stable environment (nutritionally and physically, for example) than many cells of multicellular organisms, rapid genetic response to environmental change is likely to be adaptive. To accomplish such rapid responses, a labile gene product (mRNA) is advantageous. A pancreatic cell, which is developmentally stable and exists in a relatively stable environment, could produce more insulin on stable mRNAs for a given transcriptional rate.

24.

Proline:	C_3, and one of the C_2A triplets
Histidine:	one of the C_2A triplets
Threonine:	one C_2A triplet, and one A_2C triplet
Glutamine:	one of the A_2C triplets
Asparagine:	one of the A_2C triplets
Lysine:	A_3

26. The advantage would be that if sequence homologies can be identified for a variety of HIV isolates, then perhaps a single or a few vaccines could be developed for the multitude of subtypes that infect various parts of the world. In other words, the wider the match of a vaccine to circulating infectives, the more likely the efficacy. On the other hand, the more finely aligned a vaccine is to the target, the more likely it is that new or previously undiscovered variants will escape vaccination attempts.

Chapter 13 Translation and Proteins

2. Transfer RNAs are "adaptor" molecules in that they provide a way for amino acids to interact with sequences of bases in nucleic acids. Amino acids are specifically and individually attached to the 3' end of tRNAs that possess a three-base sequence (the anticodon) to base-pair with three bases of mRNA. Messenger RNA, on the other hand, contains a copy of the triplet codes that are stored in DNA. The sequences of bases in mRNA interact, three at a time, with the anticodons of tRNAs.
Enzymes involved in transcription include the following: RNA polymerase (*E. coli*), and RNA polymerase I, II, III (eukaryotes). Those involved in translation include the following: aminoacyl tRNA synthetases, peptidyl transferase, and GTP-dependent release factors.

4. The sequence of base triplets in mRNA constitutes the sequence of codons. A three-base portion of the tRNA constitutes the anticodon.

6. An amino acid in the presence of ATP, Mg^{++}, and a specific aminoacyl synthetase produces an amino acid-AMP enzyme complex (+ PP_i). This complex interacts with a specific tRNA to produce the aminoacyl tRNA.

8. While too much phenylalanine and its derivatives cause PKU in phenylketonurics, too little will restrict protein synthesis.

10. Tyrosine is a precursor to melanin, skin pigment. Individuals with PKU fail to convert phenylalanine to tyrosine and even though tyrosine is obtained from the diet, at the population level, individuals with PKU have a tendency for less skin pigmentation.

12.

trp-8, *trp*-2, *trp*-3, *trp*-1

precursor - - → AA - - → IGP - - → I - - → TRY

14. The fact that enzymes are a subclass of the general term *protein,* a *one-gene:one-protein* statement might seem to be more appropriate. However, some proteins are made up of subunits, each different type of subunit (polypeptide chain) being under the control of a different gene. Under this circumstance, the *one-gene:one-polypeptide* might be more reasonable. It turns out that many functions of cells and organisms are controlled by stretches of DNA, which either produce no protein product (operator and promoter regions, for example) or have more than one function as in the case of overlapping genes and differential mRNA splicing. A simple statement regarding the relationship of a stretch of DNA to its physical product is difficult to justify.

16. Each chain represents a primary structure of amino acids connected by covalent peptide bonds. Secondary structures are determined by hydrogen bonding between components of the peptide bonds. Alpha helices and pleated sheets result. Tertiary structures are formed from interactions of the amino acid side chains, while the quaternary level results from the associations of chains shown above.

18. In the late 1940s, Pauling demonstrated a difference in the electrophoretic mobility of HbA and HbS (sickle-cell hemoglobin) and concluded that the difference had a chemical basis. Ingram determined that the chemical change occurs in the primary structure of the globin portion of the molecule using the fingerprinting technique. He found a change in the 6th amino acid in the β chain.

20. Dividing 20 by 0.34 gives the number of nucleotides (about 59) occupied by a ribosome. Dividing 59 by three gives the approximate number of triplet codes, approximately 20.

22. Since all higher levels of protein structure are dependent on the sequence of amino acids (primary structure), it is the primary structure that is most influential in determining protein structure and function.

24. Enzymes function to regulate catabolic and anabolic activities of cells. They influence (lower) the *energy of activation* thus allowing chemical reactions to occur under conditions that are compatible with living systems.

26. When an expectant mother returns to consumption of phenylalanine in her diet, she subjects her baby to higher than normal levels of phenylalanine throughout its development. Since increased phenylalanine is toxic, many (approximately 90 percent) newborns are severely and irreversibly retarded at birth.

28. Even though three gene pairs are involved, notice that because of the pattern of mutations, each cross may be treated as monohybrid (a) or dihybrid (b,c).

(a) F_1: *AABbCC* = speckled
F_2: 3 *AAB_CC* = speckled
1 *AAbbCC* = yellow

(b) F_1: *AABbCc* = speckled
F_2: 9 *AAB_C_* = speckled
3 *AAB_cc* = green
3 *AAbbC_* = yellow } 4
1 *AAbbcc* = yellow }

(c) F_1: *AaBBCc* = speckled
F_2: 9 *A_BBC_* = speckled
3 *A_BBcc* = green
3 *aaBBC_* = colorless } 4
1 *aaBBcc* = colorless }

30. The normal glutamic acid is a negatively charged amino acid, whereas valine carries no net charge and lysine is positively charged. Given these significant charge changes, one would predict some, if not considerable, influence on protein structure and function. Such changes could stem from internal changes in folding or interactions with other molecules in the RBC, especially other hemoglobin molecules.

32.

c, *b*, *a*

pink - - → rose - - → orange - - → purple

The above hypothesis could be tested by conducting a backcross as given below:

AaBbCc × *aabbcc*

The cross should give a

4(pink):2(rose):1(orange):1(purple) ratio

34. The antibacterial action of evernimicin is probably at the A site on the ribosome.

Chapter 14 Gene Mutation, Transposition, and DNA Repair

2. When conducting genetic screens, one assumes that all the cells of an organism are genetically identical. Therefore, the organism responds to the screen and enables detection. If a somatic cell of a multicellular organism is mutated, it is highly unlikely that the organism will be sufficiently altered to respond to a screen.

4. Each gene and its product function in an environment that has also evolved, or co-evolved. A coordinated output of each gene product is required for life. Deviations from the norm, caused by mutation, are likely to be disruptive because of the complex and interactive environment in which each gene product must function. However, on occasion a beneficial variation occurs.

6. A *conditional* mutation is one that produces a wild phenotype under one environmental condition and a mutant phenotype under a different condition.

8. All three of the agents are mutagenic because they cause base substitutions. Deaminating agents oxidatively deaminate bases such that cytosine is converted to uracil and adenine is converted to hypoxanthine. Uracil pairs with adenine, and hypoxanthine pairs with cytosine. Alkylating agents donate an alkyl group to the amino or keto groups of nucleotides, thus altering base-pairing affinities. 6-ethyl guanine acts like adenine, thus pairing with thymine. Base analogs such as 5-bromouracil and 2-amino purine are incorporated as thymine and adenine respectively, yet they pair with guanine and cytosine respectively.

10. X-rays are of higher energy and shorter wavelength than UV light. They have greater penetrating ability and can create more disruption of DNA.

12. *Photoreactivation* can lead to repair of UV-induced damage. An enzyme, photoreactivation enzyme, will absorb a photon of light to cleave thymine dimers. *Excision repair* involves the products of several genes, DNA polymerase I, and DNA ligase to clip out the UV-induced dimer, fill in, and join the phosphodiester backbone in the resulting gap. The excision repair process can be activated by damage that distorts the DNA helix. *Recombinational* *repair* is a system that responds to DNA that has escaped other repair mechanisms at the time of replication. If a gap is created on one of the newly synthesized strands, a "rescue operation or SOS response" allows the gap to be filled. Many different gene products are involved in this repair process: *rec*A, *lex*A. In SOS repair, the proofreading by DNA polymerase III is suppressed, and this therefore is called an "error-prone system."

14. Both Duchenne muscular dystrophy (DMD) and Becker muscular dystrophy (BMD) are recessive caused by X-linked mutations. DMD, the more severe of the two, causes rapid progression of muscle degeneration and involvement of the heart and lungs. Males with DMD die in their early 20s. BMD does not involve the heart or lungs and progresses more slowly. The gene responsible for both—the *dystrophin* gene—is large, consisting of about 2.5 million base pairs. More than 70 percent of mutations in the *dystrophin* gene that lead to DMD and BMD are deletions and insertions. Generally, DMD mutations change the reading frame of the *dystrophin* gene, while BMD mutations usually do not. Most DMD frameshift mutations lead to premature termination of translation and a truncated protein. The majority of BMD gene mutations alter the internal sequence of the *dystrophin* transcript and protein, but do not alter the reading frame thus producing a modified but somewhat functional dystrophin protein with less severe consequences to the phenotype.

16. *Xeroderma pigmentosum* is a form of human skin cancer caused by perhaps several rare autosomal genes that interfere with the repair of damaged DNA. Studies with heterokaryons provided evidence for complementation, indicating that as many as seven different genes may be involved. The photoreactivation repair enzyme appears to be involved.

18. It is possible that through the reduction of certain environmental agents that cause mutations, mutation rates might be reduced. On the other hand, certain industrial and medical activities actually concentrate mutagens (radioactive agents and hazardous chemicals). Unless human populations are protected from such agents, mutation rates might increase. If one asks about the accumulation of mutations (not rates) in human populations as a result of improved living conditions and medical care, then it is likely that as the environment becomes less harsh (through improvements), more mutations will be tolerated as selection pressure decreases. However, as individuals live longer and have children at a later age, some studies indicate that older males accumulate more gametic mutations.

20. The *dystrophin* gene is very large and composed of 97 exons and 2.6 Mb in length. Given the size of this gene and the number of exons/introns, many opportunities exist for mutational upset.

22. There are several ways in which an unexpected mutant gene may enter a pedigree. If a gene is incompletely penetrant, it may be present in a population and only express itself under certain conditions. It is unlikely that the gene for hemophilia behaved in this manner. If a gene's expression is suppressed by another mutation in an individual, it is possible that offspring may inherit a given gene and not inherit its suppressor. Such offspring would have hemophilia. Since all genetic variations must arise at some point, it is possible that the mutation in Queen Victoria's family was new, arising in the father.

Lastly, it is possible that the mother was heterozygous and by chance, no other individuals in her family were unlucky enough to receive the mutant gene.

24. Your study should include examination of the following short-term aspects: immediate assessment of radiation amounts distributed in a matrix of the bomb sites as well as a control area not receiving bomb-induced radiation, radiation exposure as measured by radiation sickness and evidence of radiation poisoning from tissue samples, abortion rates, birthing rates, and chromosomal studies. Long-term assessment should include sex-ratio distortion (males being more influenced by X-linked recessive lethals than females), chromosomal studies, birth and abortion rates, cancer frequency and type, and genetic disorders. In each case data should be compared to the control site to see if changes are bomb-related. In addition, to attempt to determine cause-effect, it is often helpful to show a dose response. Thus, by comparing the location of individuals at the time of exposure to the matrix of radiation amounts, one may be able to determine whether those most exposed to radiation suffer the most physiologically and genetically. If a positive correlation is observed, then statistically significant conclusions may be possible.

26. (a) For those organisms that generate energy by aerobic respiration, a process occurs that involves the reduction of molecular oxygen. Partially reduced species are produced as intermediates and by-products of such molecular action: O_2^-, H_2O_2, and OH^-. These species are potent electrophilic oxidants that escape mitochondria and attack numerous cellular components. Collectively, these are called reactive oxygen species (ROS).

(b)

oxoGuanine

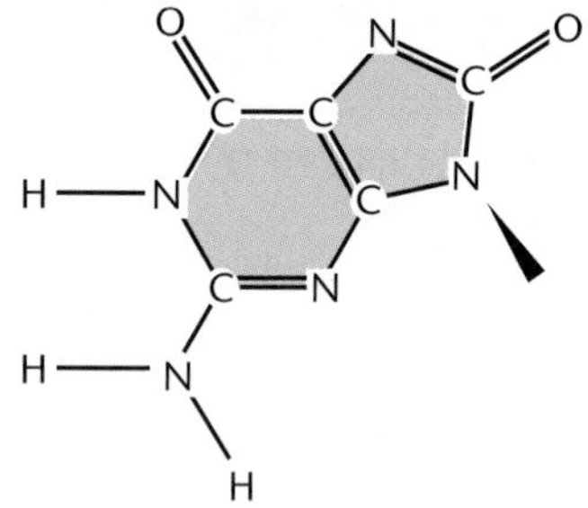

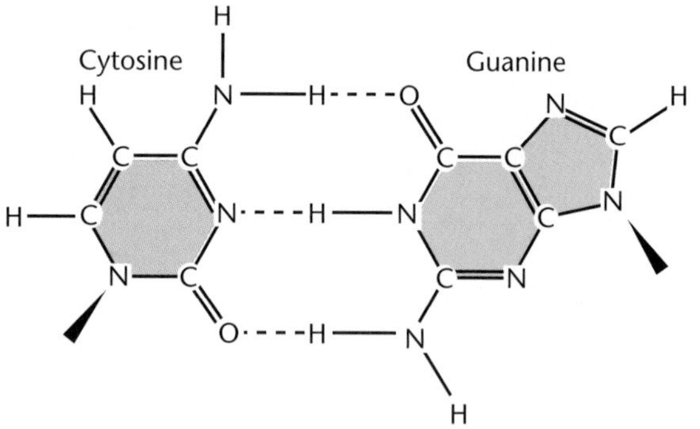

When casually examining the structures in the above diagrams, it is not immediately obvious that oxoG:A pairs should occur. However, hydrogen bonding can occur to any other base, including self pairs. Homopurine (A:A, G:G) and heteropurine (A:G) pairs represent anomalous pairing possibilities even with nonaltered bases. While G:C is undoubtedly the most stable, several mispairs are actually stronger than the A-T pair. Base pairing is complicated by the fact that the purines possess two H-bonding faces: the Watson-Crick face, involving ring positions 1 and 6 for Adenine and, 1, 2, and 6 for Guanine; and the Hoogsteen face involving ring positions 6 and 7. The typical pairing mode is indicated as *wc,* where pairing occurs on the Watson-Crick face in the normal orientation, even for the mispair A:G. Alteration of pairing and favoring of the Hoogsteen face can occur with the alteration generated by oxoGuanine. Indeed, triple helix configurations commonly involve the Hoogsteen face.

(c) If not repaired (see below), the first round of replication involves the pairing of oxoG to Adenine (see above), while in the next round of replication, Adenine pairs with its normal Thymine. Therefore, if one starts with a G:C pair, one ends up with an A:T pair.

(d) It turns out that G:G>T:A transversions are quite commonly found in human cancers and are especially prevalent in the tumor suppressor gene *p53*. Thus, the cellular defense system has been extensively studied. One component is a triphosphatase that cleanses the nucleotide precursor pool by removing the two outermost phosphates from oxo-dGTP. Another involves a DNA glycosylase that initiates repair of misreplicated oxoG:A by hydrolyzing the glycosidic bond linking the adenine base to the sugar. Another is a DNA gycosylase/lyase system that recognizes oxoG opposite cytosine. Of the three systems, the DNA glycosylases are probably the most effective.

28. Individuals with xeroderma pigmentosum (XP) are much more likely to contract skin cancer in youth than non-XP individuals. By age 20, approximately 80 percent of the XP population has skin cancer compared with approximately 4 percent in the non-XP group. XP individuals lack one or more genes involved in DNA repair.

30. First, while less likely, one might suggest that transposons, for one reason or another, are more likely to insert in noncoding regions of the genome. One might also suggest that they are more stable in such regions. Second, and more likely, it is possible that transposons insert rather randomly and that selection eliminates those that have interrupted coding regions of the genome. Since such regions are more likely to influence the phenotype, selection is more likely to influence such regions.

32. (a) 1. When a region of DNA contains repeated segments, it is possible that the alignment of these segments may be offset as pictured below. Should there be a crossover between the elements within a segment, shortening and lengthening of the segments can occur. (Note: both of the figures below were obtained from http://biol.lf1.cuni.cz/ucebnice/en/repetitive_dna.htm)

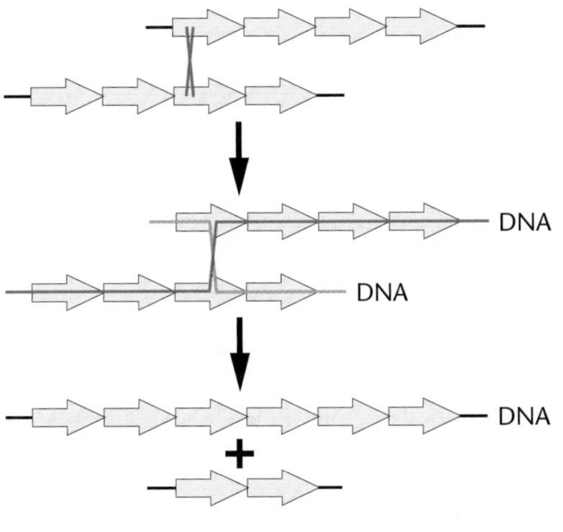

2. When a loop forms because of within-strand base pairing of complementary repeats within a repetitive segment, it provides an opportunity for the polymerase to add repeats by continuing to replicate beyond the loop as shown below.

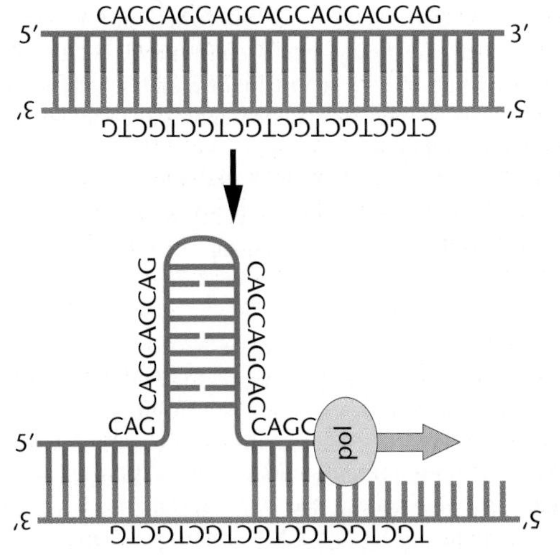

(b) A variety of regulatory signals occur both upstream and downstream from the coding regions of genes. Repeats often upset such regulatory signals. While introns are normally removed from RNAs during maturation, repeats with introns may upset the splicing process or may actually influence regulatory sequences that are contained in introns.

(c) The coding region of a gene specifies a sequence of amino acids in a protein. Having a stretch of repetitive amino acids, even if short, may have a profound influence on protein function. Trinucleotide repeat expansion probably occurs to the same degree within and outside exons, but there is stronger selection against protein sequence changes. Changes in upstream and downstream sequences would also be significantly influenced by repeat expansion and be selected against.

Chapter 15 Regulation of Gene Expression

2. Under *negative* control, the regulatory molecule interferes with transcription, while in *positive* control, the regulatory molecule stimulates transcription. Negative control requires a molecule to be removed from the DNA for transcription to occur. Positive control requires a molecule to be added to the DNA for transcription to occur.

4. (a) Due to the deletion of a base early in the *lac* Z gene, there will be a shift of all the reading frames downstream from the deletion. It is likely that either premature chain termination of translation will occur (from the introduction of a nonsense triplet in a reading frame) or the normal chain termination will be ignored. Regardless, a mutant condition for the Z gene will be likely. If such a cell is placed on a lactose medium, it will be incapable of growth because β-galactosidase is not available.

(b) If the deletion occurs early in the *A* gene, one might expect impaired function of the *A* gene product, but it will not influence the use of lactose as a carbon source.

6. $I^+ O^+ Z^+$ = Because of the function of the active repressor from the I^+ gene, and no lactose to influence its function, there will be **No Enzyme Made**.

$I^+ O^c Z^+$ = There will be a **Functional Enzyme Made** because the constitutive operator is in *cis* with a *Z* gene. The lactose in the medium will have no influence because of the constitutive operator. The repressor cannot bind to the mutant operator.

$I^- O^+ Z^-$ = There will be a **Nonfunctional Enzyme Made** because with I^- the system is constitutive, but the Z gene is mutant. The absence of lactose in the medium will have no influence because of the nonfunctional repressor. The mutant repressor cannot bind to the operator.

$I^- O^+ Z^-$ = There will be a **Nonfunctional Enzyme Made** because with I^- the system is constitutive, but the Z gene is mutant. The lactose in the medium will have no influence because of the nonfunctional repressor. The mutant repressor cannot bind to the operator.

$I^- O^+ Z^+/F' I^+$ = There will be **No Enzyme Made** because in the absence of lactose, the repressor product of the I^+ gene will bind to the operator and inhibit transcription.

$I^+ O^c Z^+/F' O^+$ = Because there is a constitutive operator in *cis* with a normal Z gene, there will be **Functional Enzyme Made.** The lactose in the medium will have no influence because of the mutant operator.

$I^+ O^+ Z^-/F' I^+ O^+ Z^+$ = Because there is lactose in the medium, the repressor protein will not bind to the operator and transcription will occur. The presence of a normal Z gene allows a **Functional and Nonfunctional Enzyme to be Made.** The repressor protein is diffusible, working in *trans*.

$I^- O^+ Z^-/F' I^+ O^+ Z^+$ = Because there is no lactose in the medium, the repressor protein (from I^+) will repress the operators and there will be **No Enzyme Made.**

$I^s O^+ Z^+/F' O^+$ = With the product of I^s there is binding of the repressor to the operator and therefore **No Enzyme Made.** The lack of lactose in the medium is of no consequence because the mutant repressor is insensitive to lactose.

$I^+ O^c Z^+/F' O^+ Z^+$ = The arrangement of the constitutive operator (O^c) with the Z gene will cause a **Functional Enzyme to be Made.**

8. Catabolite repression is a mechanism whereby glucose, a catabolite of lactose, inhibits the synthesis of β-galactosidase. When glucose and lactose are both present, glucose is preferentially used as the energy source. When glucose is exhausted, β-galactosidase synthesis occurs and lactose is metabolized. Thus, catabolite repression balances the use of glucose and lactose through regulation of β-galactosidase synthesis.

10. (a) With no lactose and no glucose, the operon is off because the *lac* repressor is bound to the operator, and although CAP is bound to its binding site, it will not override the action of the repressor.

(b) With lactose added to the medium, the *lac* repressor is inactivated and the operon is transcribing the structural genes. With no glucose, the CAP is bound to its binding site, thus enhancing transcription.

(c) With no lactose present in the medium, the *lac* repressor is bound to the operator region, and since glucose inhibits adenyl cyclase, the CAP protein will not interact with its binding site. The operon is therefore "off."

(d) With lactose present, the *lac* repressor is inactivated; however, since glucose is also present, CAP will not interact with its binding site. Under this condition transcription is severely diminished and the operon can be considered to be "off."

12. Attenuation functions to reduce the synthesis of tryptophan when it is in full supply. It does so by reducing transcription of the *tryptophan* operon. The same phenomenon is observed when tryptophan activates the repressor to shut off transcription of the *tryptophan* operon.

14. First, notice that in the first row of data, the presence of *tm* in the medium causes the production of active enzyme from the wildtype arrangement of genes. From this one would conclude that the system is *inducible*. To determine which gene is the structural gene, look for the *IE* function and see that it is related to C. Therefore, C codes for the **structural gene.** Because when B is mutant, no enzyme is produced, B must be the **promoter.**

Notice that when genes A and D are mutant, constitutive synthesis occurs; therefore, one must be the operator and the other gene codes for the repressor protein. To distinguish these functions, one must remember that the repressor operates as a diffusible substance and can be on the host chromosome or the F factor (functioning in *trans*). However, the operator can only operate in *cis*. In addition, in *cis*, the constitutive operator is dominant to its wildtype allele, while the mutant repressor is recessive to its wildtype allele.

Notice that the mutant A gene is dominant to its wild-type allele, whereas the mutant d allele is recessive (behaving as wild type in the first row). Therefore, the A locus is the **operator,** and the D locus is the **repressor** gene.

16. Because the operon by itself (when mutant as in strain #3) gives constitutive synthesis of the structural genes, the *cis*-acting system is supported. The *cis*-acting element is most likely part of the operon.

20. Transcription factors are proteins that are *necessary* for the initiation of transcription. However, they are not *sufficient* for the initiation of transcription. To be activated, RNA polymerase II requires a number of transcription factors. Transcription factors contain at least two functional domains: one binds to the DNA sequences of promoters and/or enhancers, while the other interacts with RNA polymerase or other transcription factors. Some transcription factors bind to other transcription factors without themselves binding to DNA.

22. In general, one determines the influence of various regulatory elements by removing necessary elements or adding extra elements. Assay systems determine the relative levels of gene expression after such action.

24.

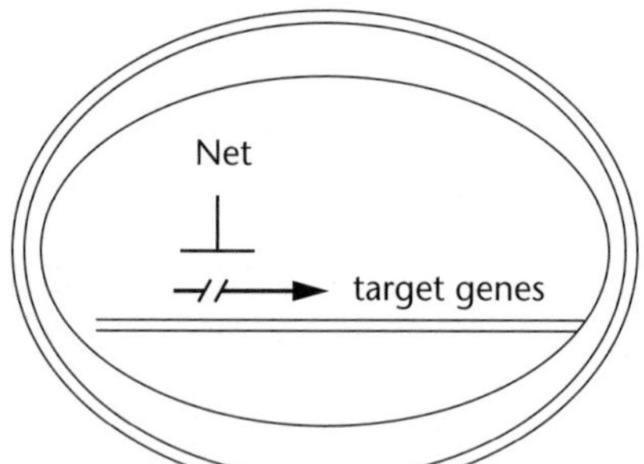

Neutral Conditions

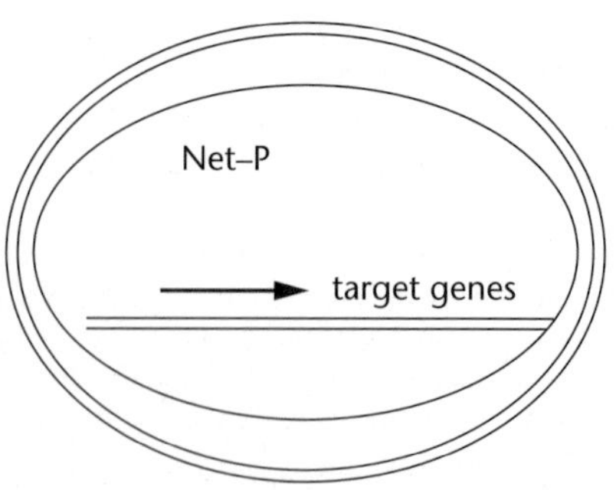

Phosphorylated Net

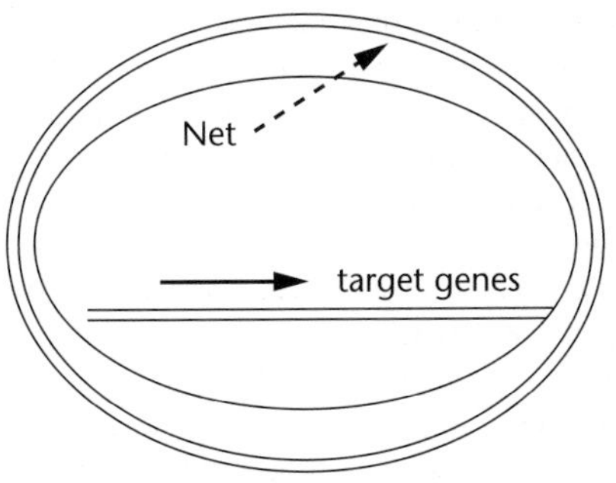

Sketches modified from Ducret et al., *Molecular and Cellular Biology* 1999 19: 7076–7087.

26. Given that DNA methylation plays a role in gene expression in mammals, any change in DNA methylation, plus or minus, can potentially have a negative impact on progeny development. In addition, since $m^5C >>> $ thymine, transitions are likely to cause mutations in coding regions of DNA, when methylation patterns change, new sites for mutation arise. Should mutations occur at a higher rate in previously unmethylated sites (genes), embryonic development is likely to be affected.

28. Methylation of CpGs causes a reduction in luciferase expression, which is somewhat proportional to the amount of methylation and patch size. Methylation within the transcription unit more drastically reduces luciferase expression compared with methylation outside the transcription unit. A high degree of methylation outside the transcription unit (593 CpGs) has as great of an impact on depressing transcription as the same degree of methylation within the transcription unit.

30. When splice specificity is lost, one might observe several classes of altered RNAs: (1) a variety of nonspecific variants producing RNA pools with many lengths and combinations of exons and introns, (2) incomplete splicing where introns and exons are erroneously included or excluded in the mRNA product, and (3) a variety of nonsense products, which result in premature RNA decay or truncated protein products. It is presently unknown as to whether cancer-specific splices initiate or result from tumorigenesis. Given the complexity of cancer induction and the maintenance of the transformed cellular state, gene products that are significant in regulating the cell cycle may certainly be influenced by alternative splicing and thus contribute to cancer.

Chapter 16 Cancer and Regulation of the Cell Cycle

2. The G1 stage begins after mitosis and is involved in the synthesis of many cytoplasmic elements. In the S phase DNA synthesis occurs. G2 is a period of growth and preparation for mitosis. Most cell-cycle time variation is caused by changes in the duration of G1. G0 is the nondividing state.

4. Kinases regulate other proteins by adding phosphate groups. Cyclins bind to the kinases, switching them on and off. CDK4 binds to cyclin D, moving cells from G1 to S. At the G2/mitosis border a CDK1 (cyclin-dependent kinase) combines with another cyclin (cyclin B). Phosphorylation occurs, bringing about a series of changes in the nuclear membrane via caldesmon, cytoskeleton, and histone H1.

6. To say that a particular trait is inherited conveys the assumption that when a particular genetic circumstance is present, it will be revealed in the phenotype. When one discusses an inherited predisposition, one usually refers to situations where a particular phenotype is expressed in families in some consistent pattern. However, the phenotype may not always be expressed or may manifest itself in different ways.

8. Apoptosis, or programmed cell death, is a genetically controlled process that leads to death of a cell. It is a natural process involved in morphogenesis and a protective mechanism against cancer formation. During apoptosis, nuclear DNA becomes fragmented, cellular structures are disrupted, and the cells dissolve.

10. The nonphosphorylated form of pRB binds to transcription factors such as E2F, causing inactivation and suppression of the cell cycle. Phosphorylation of pRB activates the cell cycle by releasing transcription factors (E2F) to advance the cell cycle. With the phosphorylation site inactivated in the PSM-RB form, phosphorylation cannot occur, thereby leaving the cell cycle in a suppressed state.

12. Various kinases can be activated by breaks in DNA. One kinase, called ATM and/or a kinase called Chk2, phosphorylates BRCA1 and p53. The activated p53 arrests replication during the S phase to facilitate DNA repair. The activated BRCA1 protein, in conjunction with BRCA2, mRAD51, and other nuclear proteins, is involved in repairing the DNA.

14. Proto-oncogenes are those that normally function to promote or maintain cell division. In the mutant state (oncogenes), they induce or maintain uncontrolled cell division; that is, there is a gain-of-function. Generally, this gain-of-function takes the form of increased or abnormally continuous gene output. On the other hand, loss-of-function is generally attributed to tumor-suppressor genes that function to halt passage through the cell cycle. When such genes are mutant, they have lost their capacity to halt the cell cycle.

16. Unfortunately, it is common to spend enormous amounts of money dealing with diseases after they occur rather than concentrating on disease prevention. Too often pressure from special interest groups or lack of political stimulus retards advances in education and prevention. Obviously, it is less expensive, both in terms of human suffering and money, to seek preventive measures for as many diseases as possible. However, having gained some understanding of the mechanisms of disease, in this case cancer, it must also be stated that no matter what preventive measures are taken it will be impossible to completely eliminate disease from the human population. It is extremely important, however, that we increase efforts to educate and protect the human population from as many hazardous environmental agents as possible.

18. Normal cells are often capable of withstanding mutational assault because they have checkpoints and DNA repair mechanisms in place. When such mechanisms fail, cancer may be a result. Through mutation, such protective mechanisms are compromised in cancer cells, and as a result they show higher than normal rates of mutation, chromosomal abnormalities, and genomic instability.

20. Certain environmental agents such as chemicals and X-rays cause mutations. Since genes control the cell cycle, mutations in cell-cycle control genes, or those that impact on cell-cycle control, can lead to cancer.

22. No, she will still have the general population risk of about 10 percent. In addition, it is possible that genetic tests won't detect all breast cancer mutations.

24. A benign tumor is a multicellular cell mass that is usually localized to a giving anatomical site. Malignant tumors are those generated by cells that have migrated to one or more secondary sites.

26. As with many forms of cancer, a single gene alteration is not the only requirement. The authors (Bose et al.) state: "but only infrequently do the cells acquire the additional changes necessary to produce leukemia in humans." Some studies indicate that variations (often deletions) in the region of the breakpoints may influence expression of CML.

28. Since there are multiple routes that lead to cancer, one would expect complex regulatory systems to be involved. More specifically, while in some cases, downregulation of a gene, such as an oncogene, may be a reasonable cancer therapy, downregulation of a tumor-suppressor gene would be undesirable in therapy. Various levels of methylation (hypermethylation and hypomethylation) influence gene activity and therefore, can cause cancer.

30. (a,b) Even though there are changes in the *BRCA1* gene, they don't always have physiological consequences. Such neutral polymorphisms make screening difficult in that one can't always be certain that a mutation will cause problems for the patient.
(c) The polymorphism in *PM2* is probably a silent mutation because the third base of the codon is involved.
(d) The polymorphism in *PM3* is probably a neutral missense mutation because the first base is involved.

32. Various alterations in gene activity related to hyper- and hypomethylation have been associated with numerous cancers. Hypermethylation usually leads to a suppression of gene activity. *Hormonal response genes* often synthesize receptors that respond to a variety of hormones such as androgens, retinoic acid, and estrogens. For example, prostate cancer is often associated with suppressed *ESR1* and *ESR2* activity, both of which are estrogen receptors. *Cell-cycle control genes* are involved in both upregulation and downregulation of the cell cycle as they influence cyclins and cyclin-dependent kinases. The cyclin-dependent kinase inhibitor *CDKN2A,* when hypermethylated, fails to inhibit cyclin-dependent kinase and thereby contributes to cancer progression. A number of *tumor cell invasion genes* have been described, some of which interfere with intercellular adhesion. When such genes are suppressed through hypermethylation, cadherin–catenin adherence mechanisms are compromised that can influence cell-to-cell contacts. One of the most common causes of cancer is a breakdown in *DNA repair* mechanisms. Hypermethylation suppresses the genes that normally supply DNA repair components. Proper *signal transduction* is required to maintain cell-cell adherence and response to extracellular signals. For example, when *CD44,* an integral membrane protein that is involved in matrix adhesion and signal transduction, is hypermethylated, cells cannot maintain proper cell-cell contact and communication, without which cell-cycle control is compromised.

Chapter 17 Recombinant DNA Technology and Gene Cloning

2. *Reverse transcriptase* is often used to promote the formation of cDNA (complementary DNA) from a mRNA molecule. Eukaryotic mRNAs typically have a 3′ polyA tail, as indicated in the following diagram. The poly-dT segment provides a double-stranded section that serves to prime the production of the complementary strand.

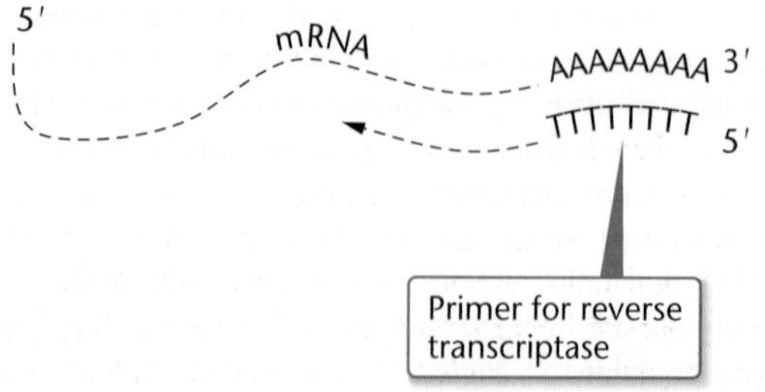

4. It is believed that the protein interacts with the major groove of the DNA helix. This information comes from the structure of the proteins that have been sufficiently well studied to suggest that the DNA major groove and "fingers" or extensions of the protein form the basis of interaction.

6. This segment contains the palindromic sequence of GGATCC, which is recognized by the restriction enzyme *Bam*HI. The double-stranded sequence is the following
CCTAGG
GGATCC

8. Because of their small size, plasmids are relatively easy to separate from the host bacterial chromosome, and they have relatively few restriction sites. They can be engineered fairly easily (i.e., polylinkers and reporter genes added). YACs (yeast artificial chromosomes) contain telomeres, an origin of replication, and a centromere and are extensively used to clone DNA in yeast. With selectable markers (*TRP1* and *URA3*) and a cluster of restriction sites, DNA inserts ranging from 100 kb to 1000 kb can be cloned and inserted into yeast. Since yeast, being eukaryotes, undergo many of the typical RNA and protein processing steps of other, more complex eukaryotes, the advantages are numerous when working with eukaryotic genes.

10. This problem can be solved by the following expressions:
*Not*I 4^8
*Hin*fI $4 \times 4 \times 1 \times 4 \times 4$
*Xho*II $2 \times 4 \times 4 \times 4 \times 4 \times 2$

12. (a) Bacteria that have been transformed with the recombinant plasmid will be resistant to tetracycline and therefore tetracycline should be added to the medium.
(b) Colonies that grow on a tetracycline medium should be tested for growth on an ampicillin medium either by replica plating or some similar controlled transfer method. Those bacteria that do not grow on the ampicillin medium probably contain the *Drosophila* DNA insert.
(c) Resistance to both antibiotics by a transformed bacterium could be explained in several ways. First, if cleavage with the *Bam*H1 was incomplete, then no change in biological properties of the uncut plasmids would be expected. Also, it is possible that the cut ends of the plasmid were ligated together in the original form with no insert.

14. For the antibiotic resistance to be present, the ligation will reform the plasmid into its original form. However, two of the plasmids can join to form a dimer by each rejoining to form a single complex.

16. Because of complementary base pairing, the 3' end of the DNA strand often loops back onto itself, thereby providing a primer for DNA polymerase I.

18. A filter is used to bind the DNA from the colonies containing recombinant plasmids. A labeled probe is constructed from the protein sequence of EF1a. Since it is highly conserved, it should show considerable complementation to the human EF-1a cDNA. It is used to detect, through hybridization, the DNA of interest. Cells with the desired clone are then picked from the original plate, and the plasmid is isolated from the cells.

20.

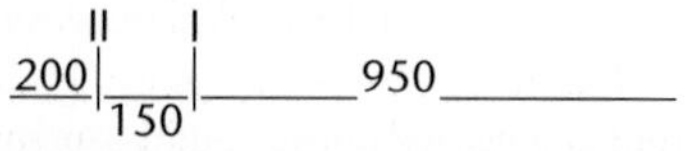

22. Option (b) fits the expectation because the thick band in the offspring probably represents the bands at approximately the same position in both parents. The likelihood of such a match is expected to be low in the general population.

24. Assuming that one has knowledge of the amino acid sequence of the protein, using the genetic code a DNA strand can be made which can be prepared for cloning into an appropriate vector or amplified by PCR. A variety of labeling techniques can then be used to identify complementary base sequences contained in the genomic library. One must know at least a portion of the amino acid sequence of the protein in order for the procedure to be applied.

Some problems can occur through degeneracy in the genetic code (not allowing construction of an appropriate probe), pseudogenes in the library (hybridizations with inappropriate fragments in the library), and variability of DNA sequences in the library due to introns (causing poor or background hybridization).

To overcome some of these problems, one can construct a variety of relatively small probes of different types that take into account the degeneracy in the code. By varying the conditions of hybridization (salt and temperature), one can reduce undesired hybridizations.

26. (a) Heating to 90–95°C denatures the double-stranded DNA so that it dissociates into single strands. It usually takes about five minutes, depending on the length and GC content of the DNA.

(b) Lowering the temperature to 50–70°C allows the primers to bind to the denatured DNA.

(c) Bringing the temperature to 70–75°C allows the heat stable DNA polymerase an opportunity to extend the primers by adding nucleotides to the 3′ ends of each growing strand. Each PCR is designed with specific temperatures (not ranges) based on the characteristics of the DNAs (template and primers).

28. ddNTPs are analogs of the "normal" deoxyribonucleotide triphosphates (dNTPs), but they lack a 3′-hydroxyl group. As DNA synthesis occurs, the DNA polymerase occasionally inserts a ddNTP into a growing DNA strand. Since there is no 3′-hydroxyl group, chain elongation cannot take place and resulting fragments are formed, which can be separated by electrophoresis. Where the ddNTP was incorporated, the length of each strand, and therefore the position of the particular ddNTP, is established and used to eventually provide the base sequence of the DNA.

30. (a) The overall size of the fragment is 12kb. From the A + N digest, sites A and N must be 1 kb apart. N must be 2 kb from an E site. Pattern #5 is the likely choice. Notice that digest A + N breaks up the 6kb E fragment.

(b) By drawing lines through sections that hybridize to the probe, one can see that the only place of consistent overlap to the probe is the 1kb fragment between A and N.

32. (a) Short tandem repeats of the Y chromosome (Y-STRs) vary considerably among individuals and populations. By amplifying Y-STRs by PCR and separating the amplified products by electrophoresis, one can genotypically type an individual as one does with a standard fingerprint. Because tissue samples are often left at the scene of a violent crime, DNA fingerprints are sometimes more available than standard fingerprints. Linking an individual with the time and place of a significant event has multiple forensic applications. Eliminating an individual as a suspect also has important forensic applications.

(b) The nonrecombining region of the Y is maintained strictly in the male population. Of special relevance in forensic applications would be the elimination of half the population (females) from a suspect group.

(c) Because different ethnic groups show different levels of Y-STR polymorphism, different final probabilities occur as products of individual probabilities. Since these probabilities are used to match individuals in forensics, ethnic variations must be taken under consideration.

(d) While there are many potential uses of DNA samples, generally a "match" is determined by multiplying the occurrence probabilites of each haplotype to arrive at the overall probability (product) of a genotype occurring in a population. If an individual's genotype matches that found in DNA at a crime scene, depending on the frequecies of the haplotypes, one might be able to say that the individual was at the crime scene. However, contamination, inappropriate genotyping, and laboratory expertise may give both false positive or negative results. Identical twins will have identical DNA fingerprints and may complicate forensic applications.

Chapter 18 Genomics, Bioinformatics, and Proteomics

2. Whole-genome shotgun sequencing involves randomly cutting the genome into numerous smaller segments. Overlapping sequences are used to identify segments that were once contiguous, eventually producing the entire sequence. This approach often has difficulty with repetitive regions of the genome. Map-based sequencing relies on known landmarks (genes, nucleotide polymorphisms, etc.) to orient the alignment of cloned fragments that have been sequenced. Compared to whole-genome sequencing, the map-based approach is somewhat cumbersome and time consuming. Whole-genome sequencing has become the most common method for assembling genome with map-based cloning being used to resolve the problems often encountered during whole-genome sequencing.

4. The question as to how to define an organism's genome is complicated by a variety of symbiotic relationships that are known to exist in virtually all organisms. Plasmids are capable of carrying both essential and nonessential genes of the host. To complicate the matter, it is likely that all cells contain nonessential genes. In all likelihood, an organism's genome will probably come to encompass all genetic elements that can be shown to be stable cellular inhabitants.

6. The main goals of the Human Genome Project are to establish, categorize, and analyze functions for human genes.

8. One initial approach to annotating a sequence is to compare the newly sequenced genomic DNA to the known sequences already stored in various databases. The National Center for Biotechnology Information (NCBI) provides access to BLAST (Basic Local Alignment Search Tool) software that directs searches through databanks of DNA and protein sequences. A segment of DNA can be compared to sequences in major databases such as GenBank to identify matches that align in whole or in part. One might seek similarities of a sequence on chromosome 11 in a mouse and find that or similar sequences in a number of taxa. BLAST will compute a similarity score or identity value to indicate the degree of similarity of sequences.

10. The human genome is composed of over 3 billion nucleotides in which about 2 percent code for genes. Genes are unevenly distributed over chromosomes with clusters of gene-rich separated by gene-poor ones (deserts). Human genes tend to be larger and contain more and larger introns than in invertebrates such as *Drosophila*. It is estimated that at least half of the genes generate products by alternative splicing. Hundreds of genes have been transferred from bacteria into vertebrates. Duplicated regions are common, which may facilitate chromosomal rearrangement. The human genome appears to contain approximately 20,000 protein-coding genes; however, there is still uncertainty as to the total number.

12. Bacterial genes are densely packed in the chromosome. The protein-coding genes are mostly organized in polycistronic transcription units without introns. Eukaryotic genes are less densely packed in chromosomes, and protein-coding genes are mostly organized as single transcription units with introns.

14. One usually begins to annotate a sequence by comparing it, often using BLAST, to the known sequences already stored in various databases. Similarity to other annotated sequences often provides insight as to a sequence's function. Hallmarks to annotation include the identification of gene regulatory sequences found upstream of genes (promoters, enhancers, and silencers), downstream elements (termination sequences), and triplet nucleotides that are part of the coding region of the gene. In addition 5' and 3' splice sites that are used to distinguish exons from introns as well as polyadenylation sites are also used in annotation. Similar hallmarks are used to annotate prokaryotic genes; however, because prokaryotic genes don't contain introns, their annotation is sometimes less complicated.

16. Metagenomics is a relatively new discipline that examines the genomes from entire communities of microorganisms in environmental samples of water, air, and soil. Virtually every environment on Earth is being sampled in metagenomics projects. A major initiative is a global expedition called the *Sorcerer II* Global Ocean Sampling (GOS) in which researchers travel the globe by yacht, and sample flora and fauna. Metagenomics is teaching us more about millions of species of bacteria and viruses, of which only a few thousand have been well characterized.

18. Aneuploidy in humans occurs for the sex chromosomes (X and Y) and three of the autosomes (13, 18, and 21). Other aneuploids are apparently not compatible with survival. Extra or missing X chromosomes are apparently tolerated because of dosage compensation, while Y chromosome aneuploids are most likely compatible with survival because of general paucity of Y-linked genes. Notice that the number of genes on chromosomes 13, 18, and 21 are the lowest for the autosomes. It is probably not coincidental that chromosomes with the fewest genes and no known mechanism for dosage compensation are the only ones that survive as human aneuploids.

20. While the β-globin gene family is a relatively large (60kb) sequence and restriction analyses show that it is composed of six genes, one is a pseudogene and therefore does not produce a product. The five functional genes each contain two similarly-sized introns, which, when included with noncoding flanking regions (5′ and 3′) and spacer DNA between genes, account for the 95% mentioned in the question.

22. (a) To annotate a gene one identifies gene regulatory sequences found upstream of genes (promoters, enhancers, and silencers), downstream elements (termination sequences) and in-frame triplet nucleotides that are part of the coding region of the gene. In addition 5′ and 3′ splice sites that are used to distinguish exons from introns as well as polyadenylation sites are also used in annotation.

(b) Similarity to other annotated sequences often provides insight as to a sequence's function and may serve to substantiate a particular genetic assignment. Direct sequencing of cDNAs from various tissues and developmental stages aid in verification.

(c) Taking an average of 22,500 for the estimated number of genes in the human genome and computing the percentage represented by 3141 gives 13.96%. It might be safe to say that chromosome 1 is gene rich.

24. Two factors may be significant if causing a similar gene to function one way in one species and another way in a closely related species. Despite the fact that humans and chimps share significant sequence overlap, there are still approximately 35 million single base differences and about 5 million deletion/addition differences. Such changes influence the molecular environment in which a gene is expressed. Second, the external environment, especially in terms of carbohydrate availability and metabolism, has been different during the evolution of these two species. Such environmental differences may engage a different genetic background (therefore proteome) in which a particular gene is expressed. A protein functioning in one molecular environment may function quite differently in a slightly different environment. Such complexities in gene expression must be addressed when therapies are developed using model organisms.

Chapter 19 Applications and Ethics of Genetic Engineering and Biotechnology

2. Because of the recent rise in food sensitivities (allergies and adverse reactions), the public should probably have access to the contents of all foods. GMOs have the potential for possessing suites of gene products that might be atypical for a given food and unless consumers know of that possibility, harm could result. Some argue that consumers have a right to know about GMOs as a matter of principle regardless of potential health risks.

4. Glyphosate (a herbicide) inhibits EPSP, a chloroplast enzyme involved in the synthesis of the amino acids phenylalanine, tyrosine, and tryptophan. To generate glyphosate resistance in crop plants, a fusion gene was created that introduced a viral promoter to control the EPSP synthetase gene. The fusion product was placed into the Ti vector and transferred to *A. tumifaciens,* which was used to infect crop cells. Calluses were selected on the basis of their resistance to glyphosate. Resistant calluses were later developed into transgenic plants. There is a remote possibility that such an "accident" can occur as suggested in the question. However, in retracing the steps to generate the resistant plant in the first place, it seems more likely that the trait will not "escape" from the plant; rather, the engineered *A. tumifaciens* may escape, infect, and transfer glyphosate resistance to pest species.

6. Short tandem repeats are very similar to VNTRs, but the repeat motif is much shorter (2–9 base pairs). STRs have been used to generate a marker panel for DNA profiling. STR typing is less expensive, less labor intensive, and quicker to perform than traditional DNA typing.

8. Kleter and Peijnenburg used the BLAST tool from the http://www.ncbi.nlm.nih.gov/BLAST Web site to conduct a series of alignment comparisons of transgenic sequences with sequences of known allergenic proteins. Of 33 transgenic proteins screened for identities of at least 6 contiguous amino acids found in allergenic proteins, 22 gave positive results.

10. From a purely scientific viewpoint, there will be no added danger to consuming cow's milk from cloned animals. However, some individuals may have an aversion to organismic cloning and supporting such activities through consumption of products of cloned organisms may be viewed negatively on moral grounds. It is likely that public sentiment will pressure for the labeling of "cloned products" on the grounds that consumers should be able to make an informed choice as to the origin of such products.

12. As with all therapies, the cure must be less hazardous than the disease. In the case of viral-mediated gene therapy, the antigenicity of the virus must not interfere with the delivery system; such antigenicity can cause inflammation or more severe immunologic responses. Combating the host immune response may involve the use of immunosuppressive drugs or modification of the vector. The duration of desired gene expression at the diseased site is an issue. Short-period expression may require repeated exposure to the vehicle, which may present undesired responses. For some diseases, local gene therapy through inhalation or injection may produce fewer side effects than systemic exposure. Adenoviruses appear to be particularly useful for gene therapy because they can infect nondividing cells and they can accept relatively large amounts of additional DNA (30 kb or more).

14. (a,b) One of the main problems with gene therapy is delivery of the desired virus to the target tissue in an effective manner. Several of the problems involving the use of retroviral vectors are the following: (1) Integration into the host must be cell specific so as not to damage nontarget cells. (2) Retroviral integration into host-cell genomes only occurs if the host cell is replicating. (3) Insertion of the viral genome might influence nontarget, but essential, genes. (4) Retroviral genomes have a low cloning capacity and cannot carry large inserted sequences as are many human genes. (5) There is a possibility that recombination with host viruses will produce an infectious virus that may do harm.

(c) The question posed here plays on the practical versus the ethical. It would certainly be more efficient (although perhaps more difficult technically) to engineer germ tissue, for once it is done in a family, the disease would be eliminated. However, there are considerable ethical problems associated with germ-plasm therapy. It recalls previous attempts of the eugenics movements of past decades, which involved the use of selective breeding to purify the human stock. Some present-day biologists have said publically that germ-line gene therapy will *not* be conducted.

16. Since both mutations occur in the CF gene, children who possess both alleles will suffer from CF. With both parents heterozygous, each child born will have a 25 percent chance of developing CF.

18. One method is to use the amino acid sequence of the protein to produce the gene synthetically. Alternatively, since the introns are spliced out of the hnRNA in the production of mRNA, if mRNA can be obtained, it can be used to make DNA (cDNA) through the use of reverse transcriptase.

20. At this point there is considerable reluctance to allow the open sharing of genetic information among institutions. In general, the establishment of governmental databases containing our most intimate information is viewed with skepticism. It is likely that considerable time and discussion will elapse before such databases are established.

22. The child in question is a carrier of the deletion in the β-globin gene, just as the parents are carriers. Its genotype is therefore $\beta^A\beta^o$.

Chapter 20 Developmental Genetics

2. Because the egg represents an isolated, "closed" system that can be mechanically, environmentally, and to some extent biochemically manipulated, various conditions may be developed that allow one to study facets of gene regulation.

4. (a-d) Genes that control early development are often dependent on the deposition of their products (mRNA, transcription factors, various structural proteins, etc.) in the egg by the mother. Such maternal-effect genes control early events such as defining anterior-posterior polarity. Such products are placed in eggs during oogenesis and are activated immediately after fertilization. The phenotypes of maternal-effect mutations vary from determination of general body plan to eye pigmentation and direction of shell coiling.

6. (a,b) Zygotic genes are activated or repressed depending on their response to maternal-effect gene products. Three subsets of zygotic genes divide the embryo into segments. These segmentation genes are normally transcribed in the developing embryo, and their mutations have embryonic lethal phenotypes. The maternal genotype contains zygotic genes, and these are passed to the embryo as with any other gene.

8. Because the polar cytoplasm contains information to form germ cells, one would expect such a transplantation procedure to generate germ cells in the anterior region. Work done by Illmensee and Mahowald in 1974 verified this expectation.

10. First, one may determine whether levels of hnRNA are consistent among various cell types of interest. If the hnRNA pools for a given gene are consistent in various cell types, then transcriptional control can be eliminated as a possibility. Support for translational control can be achieved directly by determining, in different cell types, the presence of a variety of mRNA species with common sequences. This can be accomplished only in cases where sufficient knowledge exists for specific mRNA trapping or labeling. Clues as to translational control via alternative splicing can sometimes be achieved by examining the amino acid sequence of proteins. Similarities in certain structural/functional motifs may indicate alternative RNA processing.

12. The "gain-of-function" *Antp* mutation causes the wild-type *Antennapedia* gene to be expressed in the head, and mutant flies have legs on the head in place of antenna. Such mutations are often dominant.

14. Homeotic genes encode DNA binding domains that influence gene expression and any factor that influences gene expression may, under some circumstances, influence cell-cycle control. However attractive this model is, there have been no homeotic transformations noted in mammary glands, so the typical expression of mutant homeotic genes in insects is not revealed in mammary tissue according to Lewis (2000). A substantial number of experiments will be needed to establish a functional link between homeotic gene mutation and cancer induction. Mutagenesis and transgenesis experiments are likely to be most productive in establishing a cause-effect relationship.

16. Two coupled approaches might be used. First, one could make transgenic flies that contain a series of deletions spanning all segments of the *bicoid* mRNA; and the coding region, 5' and 3' untranslated regions. Comparison of stabilities of individual, deleted mRNAs with controls would indicate whether a particular segment of the mRNA contains a degradation signal sequence. If a degradation-sensitive region or signal sequence is located by deletion, that same intact region, when ligated to a noninvolved, nondegraded mRNA (like a ribosomal protein or tubulin mRNA) should foster degradation in a manner similar to the *bicoid* mRNA.

18.

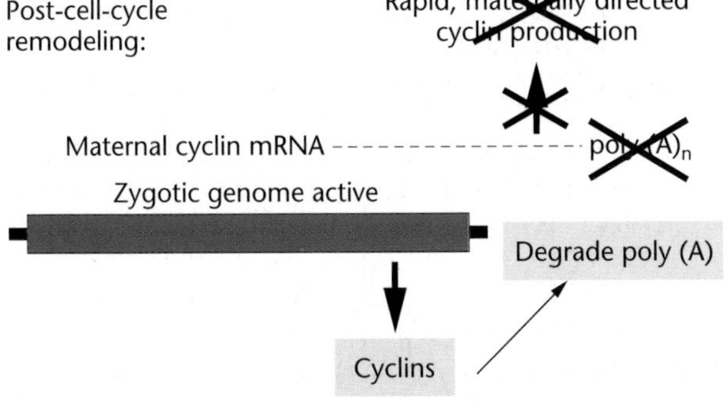

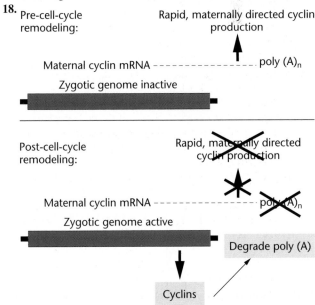

20. (a) The term *rescued* is often used when the introduction of genes from an outside source (within or among species) restores the wild-type phenotype from a mutant organism.

(b) Results such as these, and there are many like them, indicate the extreme conservation of protein structure and function across phylogenetically distant organisms. Such results attest to the conservation from a distant common ancestor of fundamental molecular species during development. Failure to adhere to a common developmental theme is rewarded by death.

22. The *Polycomb* gene family induces changes in chromatin that influence *Hox* gene expression. A gene in *Arabidopsis* has significant homology to the *Polycomb* gene family and works by altering chromatin structure. Such parallel functions indicate that mechanisms of regulation are conserved over vast evolutionary distances.

24. (a) If the *her-1*$^+$ product acts as a negative regulator, then when the gene is mutant, suppression over *tra-1*$^+$ is lost and hermaphroditism would be the result. This hypothesis fits the information provided.

(b) The double mutant should be male because even though there is no suppression from *her-1*$^-$, there is no *tra-1*$^+$ product to support hermaphrodite development.

Chapter 21 Genetics and Behavior

2. If the trait is determined by a dominant gene, that trait should appear in the offspring, probably half of them if the gene was in the heterozygous state. If the gene is recessive and homozygous, then one may not see expression in the offspring of the first cross; however, if one crosses the F_1, the trait might appear in approximately 1/4 of the offspring. Such ratios would not be expected if the trait is polygenic and that generates complexities in any study. Modifications of these patterns would be expected if the mode of inheritance is X-linked or shows other modifications of typical Mendelian ratios.

4. From the information provided, one can conclude that the tasting trait is determined by a dominant gene. Notice that a 3:1 ratio of tasters to nontasters is obtained in the third cross. This result strongly argues for the *TT* or *Tt* condition providing taste of PTC. Information from other crosses does not contradict the dominant nature of PTC tasting.

6. Concordance for traits such as addictive behavior in monozygotic and dizygotic twins has been applied to many types of behaviors. Monozygotic twins are genetically identical, whereas dizygotic twins are genetically like siblings. Since members of each set of twins are likely to be reared under similar environmental conditions, it is possible to get some estimate of the genetic contribution to a given behavior. A higher rate of concordance for monozygotic twins compared with dizygotic twins often indicates a genetic component to a trait. Although such studies suggest that a genetic component exists, they do not reveal the precise genetic basis of pathological gambling.

8. The simplest model would place each component in a linear pathway as shown below:

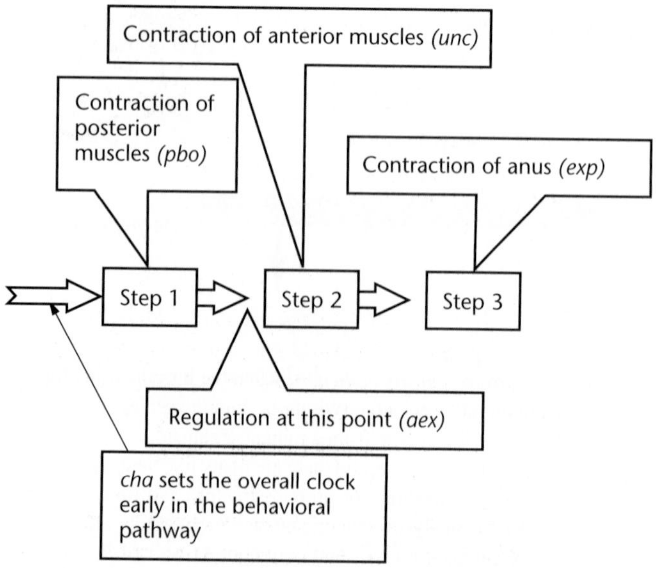

10. Usually, if traits fail to breed true that are considered to be homozygous (the pure breeds of dogs), it is likely that the trait is determined by conditioning or complex genetic factors that each have minor influences.

12. Examine the data and notice that the magnitude of change attributed to the X chromosome is relatively small when compared to the unselected line. Notice that chromosome 2 contributes relatively strongly to negative geotaxis, while chromosome 3 contributes fairly strongly to positive geotaxis. Because this is a whole chromosome comparison, it is possible that there are strong positive geotaxis alleles and strong negative geotaxis alleles on the same chromosome that cancel each other.

Chapter 22 Quantitative Genetics and Multifactorial Traits

2. (a) *Polygenes* are those genes involved in determining continuously varying or multiple-factor traits.

(b) *Additive alleles* are those alleles that account for the hereditary influence on the phenotype in an additive way.

(c) *Correlation* is a statistic that varies from -1 to $+1$ and describes the extent to which variation in one trait is associated with variation in another. It does not imply that a cause-and-effect relationship exists between two traits.

(d) *Monozygotic twins* are derived from a single fertilized egg and are thus genetically identical to each other. *Dizygotic twins* arise from two eggs fertilized by two sperm cells. They have the same genetic relationship as siblings.

(e) *Heritability* is a measure of the degree to which the phenotypic variation of a given trait is due to genetic factors.

(f) QLT stands for Quantitative Trait Loci, which are situations where multiple genes contribute to a quantitative trait.

4. If you add the numbers given for the ratio, you obtain the value of 16, which is indicative of a dihybrid cross. The distribution is that of a dihybrid cross with additive effects.

(a) Because a dihybrid result has been identified, two loci are involved in the production of color. There are two alleles at each locus for a total of four alleles.

(b,c) Because the description of red, medium-red, etc., gives us no indication of a *quantity* of color in any form of units, we would not be able to actually quantify a unit amount for each change in color. We can say that each gene additive allele provides an equal unit amount to the phenotype and the colors differ from each other in multiples of that unit amount. The number of additive alleles needed to produce each phenotype is given below:

1/16	= dark red	= *AABB*
4/16	= medium-dark red	= 2*AABb* 2*AaBB*
6/16	= medium red	= *AAbb* 4*AaBb* *aaBB*
4/16	= light red	= 2*aaBb* 2*Aabb*
1/16	= white	= *aabb*

(d) F_1 = all light red
F_2 = 1/4 medium red
2/4 light red
1/4 white

6. (a, b) There are four gene pairs involved.

(c) 3 cm

(d) A typical F_1 cross that produces a "typical" F_2 distribution would be where all gene pairs are heterozygous (*AaBbCcDd*),

independently assorting, and additive. Many possible sets of parents that would give an F_1 of this type.
An example for the parents:

$$AABBccdd \times aabbCCDD$$

(d) Since the *aabbccdd* genotype gives a height of 12 cm and each upper-case allele adds 3 cm to the height, there are many possibilities for an 18 cm plant:

AAbbccdd,
AaBbccdd,
aaBbCcdd, etc.

Any plant with seven upper-case letters will be 33 cm tall:

AABBCCDd,
AABBCcDD,
AABbCCDD, for examples.

8. Ridge count and height have the highest heritability values.

10. Height, general body structure, skin color, and perhaps most common behavioral traits including intelligence

12. (a) *For back fat:*

Broad-sense heritability $= H^2 = 12.2/30.6 = 0.398$
Narrow-sense heritability $= h^2 = 8.44/30.6 = 0.276$

For body length:

Broad-sense heritability $= H^2 = 26.4/52.4 = 0.504$
Narrow-sense heritability $= h^2 = 11.7/52.4 = 0.223$

(b) Of the two traits, selection for back fat would produce more response.

14. (a) For Vitamin A

$$h^2_A = V_A/V_P = V_A/(V_E + V_A + V_D) = 0.097$$

For Cholesterol

$$h^2_A = 0.223$$

(b) Cholesterol content should be influenced to a greater extent by selection.

16. Given that both narrow-sense heritability values are relatively high, it is likely that a farmer would be able to alter both milk protein content and butterfat by selection. The value of 0.91 for the correlation coefficient between protein content and butterfat suggests that if one selects for butterfat, protein content will increase. However, correlation coefficients describe the extent to which variation in one quantitative trait is associated with variation in another and does not reveal the underlying causes of such variation. Assuming that these dairy cows had been selected for high butterfat in the past and increased protein content followed that selection (for butterfat), it is likely that selection for butterfat would continue to correlate with increased protein content. However, there may well be a point where physiological circumstances change and selection for high butterfat may be at the expense of protein content.

18. Given the realized heritability value of 0.4, it is unlikely that selection experiments would cause a rapid and/or significant response to selection. A minor response might result from intense selection.

20. Since the rice plants are genetically identical, V_G is zero and $H^2 = V_G/V_P =$ zero. Broad-sense heritability is a measure to which the phenotypic variance is due to genetic factors. In this case, with genetically identical plants, H^2 is zero, and the variance observed in grain yield is due to the environment. Selection would not be effective in this strain of rice.

22. (a) The most direct explanation would involve two gene pairs, with each additive gene contributing about 1.2 mm to the phenotype.

(b) The fit to this backcross supports the original hypothesis.

(c) These data do not support the simple hypothesis provided in part (a).

(d, e) With these data, one can see no distinct phenotypic classes suggesting that the environment may play a role in eye development or that there are more genes involved.

24. For traits including blood type, eye color, and mental retardation, there is a fairly significant difference between MZ and DZ groups and therefore a high genetic component. However, for measles, the difference is not as significant, indicating a greater role of the environment. Hair color has a significant genetic component, as do idiopathic epilepsy, schizophrenia, diabetes, allergies, cleft lip, and club foot. The genetic component to mammary cancer is present but minimal according to these data.

26. As with many traits that are caused by numerous loci acting additively, some genes have more influence on expression than others. In addition, environmental factors may play a role in the expression of some polygenic traits. In the case of brachydactyly, numerous modifier genes in the genome which can influence brachydactyly expression. Examination of OMIM (*Online Mendelian Inheritance of Man*) through http://www.ncbi.nlm.nih.gov/ will illustrate this point.

Chapter 23 Population and Evolutionary Genetics

2. (a) Missense mutations cause amino acid changes.

(b) Horizontal transfer refers to the process of passing genetic information from one organism to another without producing offspring. In bacteria, plasmid transfer is an example of horizontal transfer.

(c) The fact that none of the isolates shared identical nucleotide changes indicates that there is little genetic exchange among different strains. Each alteration is unique, most likely originating in an ancestral strain and maintained in descendents of that strain only.

4. Many sections of DNA in a eukaryotic genome are not reflected in a protein product. Indeed, there are many sections of DNA that are not even transcribed and/or have no apparent physiological role. Such regions are more likely to tolerate nucleotide changes compared with those regions with a necessary physiological impact. Introns, for example, show sequence variation, which is not reflected in a protein product. Exons, on the other hand, code for products that are usually involved in production of a phenotype and, as such, are subject to selection.

6. $p = 0.5$

$q = 1 - p = 0.5$

Frequency of AA = 0.25 or 25%

Frequency of Aa = 0.5 or 50%

Frequency of aa = 0.25 or 25%

8. In order for the Hardy–Weinberg equations to apply, the population must be in equilibrium.

10. (a) The equilibrium values will be as follows:

Frequency of $l/l = p^2 = 0.6014$ or 60.14%

Frequency of $l/\Delta32 = 2pq = 0.3482$ or 34.82%

Frequency of $\Delta32/\Delta32 = q^2 = 0.0504$ or 5.04%

Comparing these equilibrium values with the observed values strongly suggests that the observed values are drawn from a population in equilibrium.

(b) The equilibrium values will be as follows:

Frequency of $AA = p^2 = 0.7691$ or 76.91%

Frequency of $AS = 2pq = 0.2157$ or 21.57%

Frequency of $SS = q^2 = 0.0151$ or 1.51%

Comparing these equilibrium values with the observed values suggests that the observed values may be drawn from a population that is not in equilibrium. Notice that there are more heterozygotes than predicted and fewer *SS* types. To test for a Hardy–Weinberg equilibrium, apply the chi-square test as follows.

$$\chi^2 = \frac{\Sigma(o - e)^2}{e}$$
$$= 1.47$$

In calculating degrees of freedom in a test of gene frequencies, the "free variables" are reduced by an additional degree of freedom because one estimated a parameter (p or q) used in determining the expected values. Therefore, there is one degree of freedom, even though there are three classes. Entering the χ^2 table with 1 degree of freedom gives a value of 3.84 at the 0.05 probability level. Since the χ^2 value calculated here is smaller, the null hypothesis (the observed values fluctuate from the equilibrium values by chance and chance alone) should not be rejected. Thus the frequencies of *AA, AS, SS* sampled a population that is in equilibrium.

12. The following formula calculates the frequency of an allele in the next generation for any selection scenario, given the frequencies of a and A in this generation and the fitness of all three genotypes.

$$q_{g+1} = \left[w_{Aa}p_g q_g + w_{aa}q_g^{\,2}\right]/\left[w_{AA}p_g^{\,2} + w_{Aa}2p_g q_g + w_{aa}q_g^{\,2}\right]$$

where q_{g+1} is the frequency of the a allele in the next generation, q_g is the frequency of the a allele in this generation, p_g is the frequency of the A allele in this generation, and each "w" represents the fitness of their respective genotypes.

(a) $q_{g+1} = [.9(.7)(.3) + .8(.3)^2/[1(.7)^2 + .9(2)(.7)(.3) + .8(.3)^2$

$q_{g+1} = 0.278 \quad p_{g+1} = 0.722$

(b) $q_{g+1} = 0.289 \quad p_{g+1} = 0.711$

(c) $q_{g+1} = 0.298 \quad p_{g+1} = 0.702$

(d) $q_{g+1} = 0.319 \quad p_{g+1} = 0.681$

14. Since a dominant lethal gene is highly selected against, it is unlikely that it will exist at too high a frequency, if at all. However, if the gene shows incomplete penetrance or late age of onset (after reproductive age), it may remain in a population.

16. Given that the recessive allele a is present in the homozygous state (q^2) at a frequency of 0.0001, the value of q is 0.01 and $p = 0.99$.

(a) q is 0.01

(b) $p = 1 - q$ or .99

(c) $2pq = 2(.01)(.99)$
$= 0.0198$ (or about $1/50$)

(d) $2pq \times 2pq$
$= 0.0198 \times 0.0198 = 0.000392$ or about $1/255$

18. The overall probability of the couple producing a CF child is $98/2500 \times 2/3 \times 1/4$.

20. Because three of the affected infants had affected parents, only two "new" alleles, from mutation, enter into the problem. The allele is dominant; therefore each new case of achondroplasia arose from a single new mutation. There are 50,000 births; therefore 100,000 gametes (genes) are involved. The frequency of mutation is therefore given as follows: 2/100,000 or 2×10^{-5}.

22. (a) The gene is most likely recessive because all affected individuals have unaffected parents and the condition clearly runs in families. For the population, since $q^2 = .002$, then $q = .045$, $p = .955$, and $2(pq) = 0.086$. For the community, since $q^2 = .005$, then $q = .07$, $p = .93$, and $2(pq) = 0.13$.

(b) The "founder effect" is probably operating here. Relatively small, local populations that are relatively isolated in a reproductive sense tend to show differences in gene frequencies when compared with larger populations. In such small populations, homozygosity is increased as a gene has a higher probability of "meeting itself" due to inbreeding.

24. Reproductive isolating mechanisms are grouped into prezygotic and postzygotic. Prezygotic mechanisms are most efficient because they occur before resources are expended in the processes of mating.

26. A number of studies using SINES, repetitive DNA, and neutral polymorphisms (see above problem) indicate that most, if not all, cichlid species in Lake Victoria evolved from a single ancestral species. If that is the case, then this finding would represent the most rapid evolutionary radiation ever documented for vertebrates.

28. Somatic gene therapy, like any therapy, allows some individuals to live more normal lives than those not receiving therapy. As such, the ability of such individuals to contribute to the gene pool increases the likelihood that less fit alleles will enter and be maintained in the gene pool. This is a normal consequence of therapy, genetic or not, and in the face of disease control and prevention, societies have generally accepted this consequence. Germ-line therapy could, if successful, lead to limited, isolated, and infrequent removal of an allele from a gene lineage. However, given the present state of the science, its impact on the course of human evolution will be diluted and negated by a host of other factors that afflict humankind.

Chapter 24 Conservation Genetics

2. (a,b) First, if detailed records are kept of the breeding partners of the captive birds, then knowledge of heterozygotes should be available. Breeding programs could be established to restrict matings between those carrying the lethal gene. Such "kinship management" is often used in captive populations. If kinship records are not available, it is often possible to establish kinship using genetic markers such as DNA microsatellite polymorphisms. Using such markers, one can often identify mating partners and link them to their offspring. By coupling knowledge of mating partners with the likelihood of producing a lethal genetic combination, selective matings can often be used to minimize the influence of a deleterious gene. In addition, such markers can be used to establish matings that optimize genetic mixing, thus reducing inbreeding depression.

4. Both genetic drift and inbreeding tend to drive populations toward homozygosity. Genetic drift is more common when the effective breeding size of the population is low. When this condition prevails, inbreeding is also much more likely. They are different in that inbreeding can occur when certain population structures or behaviors favor matings between relatives, regardless of the effective size of the population. Inbreeding tends to increase the frequency of both homozygous classes at the expense of the heterozygotes. Genetic drift can lead to fixation of one allele or the other, thus producing a single homozygous class.

6. Inbreeding depression, over time, reduces the level of heterozygosity, usually a selectively advantageous quality of a species. When homozygosity increases (through loss of heterozygosity), deleterious alleles are likely to become more of a load on a population. Outbreeding depression occurs when there is a reduction in fitness of progeny from genetically diverse individuals. It is usually attributed to offspring being less well adapted to the local environmental conditions of the parents. Even though forced outbreeding may be necessary to save a threatened species, where population numbers are low, it significantly and permanently changes the genetic makeup of the species.

8. Often, molecular assays of overall heterozygosity can indicate the degree of inbreeding and/or genetic drift. An allele whose frequency is dictated by inbreeding will not be uniquely influenced. That is, other alleles would be characterized by decreased heterozygosity as well. So, if the genome in general has a relatively high degree of heterozygosity, the gene is probably influenced by selection rather than inbreeding and/or genetic drift.

10. Generally, threatened species are captured and bred in an artificial environment until sufficient population numbers are achieved to ensure species survival. Next, genetic management strategies are applied to breed individuals in such a way as to increase genetic heterozygosity as much as possible. If plants are involved, seed banks are often used to facilitate long-term survival.

12. Allozymes are variants of a given allele often detected by electrophoresis. Such variation may or may not impact on the fitness of an individual. The greater the allozyme variation, the more genetically heterogeneous the individual. It is generally agreed that such genetic diversity is essential for long-term survival. All other factors being equal, allozyme variation is more likely to reflect physiological variation than AFLP variation because AFLP regions are not necessarily found in protein-coding regions of the genome. AFLP allows one to detect very small amounts of genetic diversity in a population and is unlikely to encounter an organism that is not in some way variable compared to other organisms (within and among species).

14. (a) 0.396 (b) 0.886
Therefore, there is a loss of approximately 11.4 percent heterozygosity after five generations.

16. Data from *Antechinus* provide insight as to the significance of genetic diversity to the survival of a species. Because such mechanisms (i.e., sperm mixing) are in place, there must be considerable evolutionary rewards. In this case, maintaining diversity must offset the cost (if any) of evolving such a mechanism.

18. (a) Since prairie dogs are the main food source for black-footed ferrets, a reduction in prairie dogs would likely stress black-footed ferrets. Unless alternate food sources are available and utilized by the ferrets, their numbers would decline.

(b) Because a population survives a population bottleneck does not mean that the population is in a healthy state. Usually bottlenecks reduce genetic variability upon which survival and adaptation depends. A second bottleneck, while perhaps not having immediate ramifications, would likely have a negative impact on the long-term survival of the species. One would expect additional reductions in genetic diversity.

(c) Because extinction of an organism is irreversible, one might consider the fate of the ferret as the highest priority. If the prairie dog population is reduced very slowly, the ferret population may succeed in finding alternative food sources, but this is doubtful. Since the ferret population is the most fragile of the three (ferret, prairie dog, cattle) and represents one of America's most endangered mammals, this case will test the strength of laws designed to protect such species. In some situations, compromise to the point of mutual agreement is not possible.

20. DNA profiles indicate the degree of heterogeneity in DNA sequences and therefore the degree of genetic variation. While non-coding DNA sequences represent the bulk of sequence diversity, such information can be helpful in determining gene flow, ancestry, and overall inter- and intra-population diversity. Since diversity *per se* appears to be essential for long-term species survival in natural environments, one would expect that the assessment of diversity by any tool will be a useful predictor.

22. A census of population size and range would be needed to establish levels of habitat exploitation and probable number of effective breeding pairs. From this information, an estimation of long-term habitat support can be provided, along with the probability of genetic drift eroding genetic variability. Effective breeder estimates will also provide a method for estimating inbreeding depression. It would be helpful to conduct surveys on a season-to-season and year-to-year basis to decrease the possibility of sampling in an atypical season or year. In would be important to determine the age and stage-specific structure of the endangered population. Few young or juveniles might indicate that reproductive capacities are in decline. It would be important to determine the general level biodiversity and carrying capacity of the habitat as well as the genetic diversity of the species in question. Nuclear, mitochondrial, and chloroplast DNA profiles can be used to assess intrapopulation and interpopulation variation as well as to aid in determining migration and breeding patterns.

23. While flagship species (often large mammals) may make it possible to gather considerable public support and funding, they may reduce support for species, which may have a greater impact on a community of species. Primary producers (plants) are a necessary component of a diverse and supportive habitat. If one focuses on a flagship species within an area, it is possible that other areas will suffer more dramatically because foundational species are lost. Using umbrella species to protect a large geographic area in hopes of protecting other species in that area is a reasonable approach. However, the size of an area is not necessarily a primary factor in determining species success. Diversity and productivity of a habitat are major contributors to species success. Since land is at a premium, it may be wiser in the long run to select umbrella species in diverse and productive habitats rather than on the basis of land size. By selecting sets of species that show considerable biodiversity, one increases the likelihood of protecting a sufficiently rich habitat to support many species. Such habitats are often of considerable economic value, thereby making their availability limited.

APPENDIX B

Selected Readings

Chapter 1: Introduction to Genetics

Amman, N. H. 2008. In defense of GM crops. *Science* 322: 1465–1466.

Bilen, J., and Bonini, N. M. 2005. *Drosophila* as a model for human neurodegenerative disease. *Annu. Rev. Genet.* 39: 153–171.

Campbell, A. M., and Heyer, L. J. (2007). *Discovering genomics, proteomics, and bioinformatics,* 2nd ed. San Francisco, CA: Benjamin Cummings.

Dale, P. J., Clarke, B., and Fontes, E. M. G. 2002. Potential for the environmental impact of transgenic crops. *Nature Biotech.* 20: 567–574.

Daya, S., and Berns, K. I. 2008. Gene therapy using adeno-associated virus vectors. *Clin. Microbiol. Rev*. 21: 83–93.

Ehrnhoefer, D/E., Butland, S. L., Pouladi, M. A., and Hayden, M. R. 2009. Mouse models of Huntington disease: Variations on a theme. *Dis. Models Mech.* 2: 123–129.

Lonberg, N. 2005. Human antibodies from transgenic animals. *Nature Biotech.* 23: 1117–1125.

Pearson, H. 2006. What is a gene? *Nature* 441: 399–401.

Pray, C. E., Huang, J., Hu, R., and Rozelle, S. 2002. Five years of Bt cotton in China—The benefits continue. *The Plant Journal* 31: 423–430.

Primrose, S. B., and Twyman, R. M. 2004. *Genomics: Applications in human biology.* Oxford: Blackwell Publishing.

Weinberg, R. A. 1985. The molecules of life. *Sci. Am.* (Oct.) 253: 48–57.

Wisniewski, J-P., Frange, N., Massonneau, A., and Dumas, C. 2002. Between myth and reality: Genetically modified maize, an example of a sizeable scientific controversy. *Biochimie* 84: 1095–1103.

Chapter 2: Mitosis and Meiosis

Alberts, B., et al. 2007. *Molecular biology of the cell,* 5th ed. New York: Garland Publ.

Brachet, J., and Mirsky, A. E. 1961. *The cell: Meiosis and mitosis,* Vol. 3. Orlando, FL: Academic Press.

DuPraw, E. J. 1970. *DNA and chromosomes.* New York: Holt, Rinehart & Winston.

Glotzer, M. 2005. The molecular requirements for cytokinesis. *Science* 307: 1735–1739.

Glover, D. M., Gonzalez, C., and Raff, J. W. 1993. The centrosome. *Sci. Am.* (June) 268: 62–68.

Golomb, H. M., and Bahr, G. F. 1971. Scanning electron microscopic observations of surface structures of isolated human chromosomes. *Science* 171: 1024–1026.

Hartwell, L. H., and Karstan, M. B. 1994. Cell cycle control and cancer. *Science* 266: 1821–1828.

Hartwell, L. H., and Weinert, T. A. 1989. Checkpoint controls that ensure the order of cell cycle events. *Science* 246: 629–634.

Mazia, D. 1961. How cells divide. *Sci. Am.* (Jan.) 205: 101–120.

———. 1974. The cell cycle. *Sci. Am.* (Jan.) 235: 54–64.

McIntosh, J. R., and McDonald, K. L. 1989. The mitotic spindle. *Sci. Am.* (Oct.) 261: 48–56.

Nature Milestones. 2001. *Cell Division*. London: Nature Publishing Group.

Westergaard, M., and von Wettstein, D. 1972. The synaptinemal complex. *Annu. Rev. Genet.* 6: 71–110.

Chapter 3: Mendelian Genetics

Bennett, R. L., et al. 1995. Recommendations for standardized human pedigree nomenclature. *Am. J. Hum. Genet.* 56: 745–752.

Carlson, E. A. 1987. *The gene: A critical history,* 2nd ed. Philadelphia: Saunders.

Dunn, L. C. 1965. *A short history of genetics.* New York: McGraw-Hill.

Henig, R. M. 2001. *The monk in the garden: The lost and found genius of Gregor Mendel, the father of genetics.* New York: Houghton-Mifflin.

Klein, J. 2000. Johann Mendel's field of dreams. *Genetics* 156: 1–6.

Miller, J. A. 1984. Mendel's peas: A matter of genius or of guile? *Sci. News* 125: 108–109.

Olby, R. C. 1985. *Origins of Mendelism,* 2nd ed. London: Constable.

Orel, V. 1996. *Gregor Mendel: The first geneticist.* Oxford: Oxford University Press.

Peters, J., ed. 1959. *Classic papers in genetics.* Englewood Cliffs, NJ: Prentice-Hall.

Sokal, R. R., and Rohlf, F. J. 1995. *Biometry,* 3rd ed. New York: W. H. Freeman.

Stern, C., and Sherwood, E. 1966. *The origins of genetics: A Mendel source book.* San Francisco: W. H. Freeman.

Stubbe, H. 1972. *History of genetics: From prehistoric times to the rediscovery of Mendel's laws.* Cambridge, MA: MIT Press.

Sturtevant, A. H. 1965. *A history of genetics.* New York: Harper & Row.

Tschermak-Seysenegg, E. 1951. The rediscovery of Mendel's work. *J. Hered.* 42: 163–172.

Welling, F. 1991. Historical study: Johann Gregor Mendel 1822–1884. *Am. J. Med. Genet.* 40: 1–25.

Chapter 4: Modification of Mendelian Ratios

Bartolomei, M. S., and Tilghman, S. M. 1997. Genomic imprinting in mammals. *Annu. Rev. Genet.* 31: 493–525.

Brink, R. A., ed. 1967. *Heritage from Mendel.* Madison: University of Wisconsin Press.

Bultman, S. J., Michaud, E. J., and Woychik, R. P. 1992. Molecular characterization of the mouse *agouti* locus. *Cell* 71: 1195–1204.

Carlson, E. A. 1987. *The gene: A critical history,* 2nd ed. Philadelphia: Saunders.

Cattanach, B. M., and Jones, J. 1994. Genetic imprinting in the mouse: Implications for gene regulation. *J. Inherit. Metab. Dis.* 17: 403–420.

Choman, A. 1998. The myoclonic epilepsy and ragged-red fiber mutation provides new insights into human mitochondrial function and genetics. *Am. J. Hum. Genet.* 62: 745–751.

Dunn, L. C. 1966. *A short history of genetics.* New York: McGraw-Hill.

Feil, R., and Khosla, S. 1999. Genomic imprinting in mammals: An interplay between chromatin and DNA methylation. *Trends in Genet.* 15: 431.

Foster, M. 1965. Mammalian pigment genetics. *Adv. Genet.* 13: 311–339.

Grant, V. 1975. *Genetics of flowering plants.* New York: Columbia University Press.

Harper, P. S., et al. 1992. Anticipation in myotonic dystrophy: New light on an old problem. *Am. J. Hum. Genet.* 51: 10–16.

Howeler, C. J., et al. 1989. Anticipation in myotonic dystrophy: Fact or fiction? *Brain* 112: 779–797.

Morgan, T. H. 1910. Sex-limited inheritance in *Drosophila. Science* 32: 120–122.

Peters, J. A., ed. 1959. *Classic papers in genetics.* Englewood Cliffs, NJ: Prentice-Hall.

Phillips, P. C. 1998. The language of gene interaction. *Genetics* 149: 1167–1171.

Race, R. R., and Sanger, R. 1975. *Blood groups in man,* 6th ed. Oxford: Blackwell.

Sapienza, C. 1990. Parental imprinting of genes. *Sci. Am.* (Oct.) 363: 52–60.

Siracusa, L. D. 1994. The *agouti* gene: Turned on to yellow. *Cell* 10: 423–428.

Wallace, D. C. 1997. Mitochondrial DNA in aging and disease. *Sci. Am.* (Aug.) 277: 40–59.

———. 1999. Mitochondrial diseases in man and mouse. *Science* 283: 1482–1488.

Wallace, D. C., et al. 1988. Familial mitochondrial encephalomyopathy (MERRF): Genetic, pathophysiological and biochemical characterization of a mitochondrial DNA disease. *Cell* 55: 601–610.

Waters, D. J., and Wildasin, K. 2006. Cancer clues from pet dogs. *Sci. Am.* (Dec.) 295: 96–101.

Yoshida, A. 1982. Biochemical genetics of the human blood group ABO system. *Am. J. Hum. Genet.* 34: 1–14.

Chapter 5: Sex Determination and Sex Chromosomes

Amory, J. K., et al. 2000. Klinefelter's syndrome. *Lancet* 356: 333–335.

Arnold, A. P., Itoh, Y., and Melamed, E. 2008. A bird's-eye view of sex chromosome dosage compensation. *Annu. Rev. Genomics Hum. Genet.* 9:109–27.

Court-Brown, W. M. 1968. Males with an XYY sex chromosome complement. *J. Med. Genet.* 5: 341–359.

Davidson, R., Nitowski, H., and Childs, B. 1963. Demonstration of two populations of cells in human females heterozygous for glucose-6-phosphate dehydrogenase variants. *Proc. Natl. Acad. Sci. (USA)* 50: 481–485.

Hodgkin, J. 1990. Sex determination compared in *Drosophila* and *Caenorhabditis*. *Nature* 344: 721–728.

Hook, E. B. 1973. Behavioral implications of the humans XYY genotype. *Science* 179: 139–150.

Irish, E. E. 1996. Regulation of sex determination in maize. *BioEssays* 18: 363–369.

Jacobs, P. A., et al. 1974. A cytogenetic survey of 11,680 newborn infants. *Ann. Hum. Genet.* 37: 359–376.

Jegalian, K., and Lahn, B. T. 2001. Why the Y is so weird. *Sci. Am.* (Feb.) 284: 56–61.

Koopman, P., et al. 1991. Male development of chromosomally female mice transgenic for *Sry. Nature* 351: 117–121.

Lahn, B. T., and Page, D. C. 1997. Functional coherence of the human Y chromosome. *Science* 278: 675–680.

Lucchesi, J. 1983. The relationship between gene dosage, gene expression, and sex in *Drosophila. Dev. Genet.* 3: 275–282.

Lyon, M. F. 1972. X-chromosome inactivation and developmental patterns in mammals. *Biol. Rev.* 47: 1–35.

McMillen, M. M. 1979. Differential mortality by sex in fetal and neonatal deaths. *Science* 204: 89–91.

Ng, K., Pullirsch, D., Leeb, M., and Wutz, A. 2007. Xist and the order of silencing. *EMBO Reports* 8: 34–39.

Penny, G. D., et al. 1996. Requirement for Xist in X chromosome inactivation. *Nature* 379: 131–137.

Pieau, C. 1996. Temperature variation and sex determination in reptiles. *BioEssays* 18: 19–26.

Straub, T., and Becker, P. B. 2007. Dosage compensation: the beginning and end of a generalization. *Nat. Rev. Genet.* 8: 47–57.

Westergaard, M. 1958. The mechanism of sex determination in dioecious flowering plants. *Adv. Genet.* 9: 217–281.

Whitelaw, E. 2006. Unravelling the X in sex. *Dev. Cell* 11: 759–762.

Witkin, H. A., et al. 1996. Criminality in XYY and XXY men. *Science* 193: 547–555.

Xu, N., Tsai, C-L., and Lee, J. T. 2006. Transient homologous chromosome pairing marks the onset of X inactivation. *Science* 311: 1149–1152.

Chapter 6: Chromosome Mutations: Variation in Number and Arrangement

Antonarakis, S. E. 1998. Ten years of genomics, chromosome 21, and Down syndrome. *Genomics* 51: 1–16.

Ashley-Koch, A. E., et al. 1997. Examination of factors associated with instability of the *FMR1* CGG repeat. *Am. J. Hum. Genet.* 63: 776–785.

Beasley, J. O. 1942. Meiotic chromosome behavior in species, species hybrids, haploids, and induced polyploids of *Gossypium. Genetics* 27: 25–54.

Blakeslee, A. F. 1934. New jimson weeds from old chromosomes. *J. Hered.* 25: 80–108.

Boue, A. 1985. Cytogenetics of pregnancy wastage. *Adv. Hum. Genet.* 14: 1–58.

Carr, D. H. 1971. Genetic basis of abortion. *Ann. Rev. Genet.* 5: 65–80.

Croce, C. M. 1996. The *FHIT* gene at 3p14.2 is abnormal in lung cancer. *Cell* 85: 17–26.

DeArce, M. A., and Kearns, A. 1984. The fragile X syndrome: The patients and their chromosomes. *J. Med. Genet.* 21: 84–91.

Feldman, M., and Sears, E. R. 1981. The wild gene resources of wheat. *Sci. Am.* (Jan.) 244: 102–112.

Galitski, T., et al. 1999. Ploidy regulation of gene expression. *Science* 285: 251–254.

Gersh, M., et al. 1995. Evidence for a distinct region causing a catlike cry in patients with 5p deletions. *Am. J. Hum. Genet.* 56: 1404–1410.

Hassold, T. J., et al. 1980. Effect of maternal age on autosomal trisomies. *Ann. Hum. Genet.* (London) 44: 29–36.

Hassold, T. J., and Hunt, P. 2001. To err (meiotically) is human: The genesis of human aneuploidy. *Nat. Rev. Gen.* 2: 280–291.

Hassold, T., and Jacobs, P. A. 1984. Trisomy in man. *Annu. Rev. Genet.* 18: 69–98.

Hecht, F. 1988. Enigmatic fragile sites on human chromosomes. *Trends Genet.* 4: 121–122.

Hulse, J. H., and Spurgeon, D. 1974. *Triticale. Sci. Am.* (Aug.) 231: 72–81.

Kaiser, P. 1984. Pericentric inversions: Problems and significance for clinical genetics. *Hum. Genet.* 68: 1–47.

Lewis, E. B. 1950. The phenomenon of position effect. *Adv. Genet.* 3: 73–115.

Lewis, W. H., ed. 1980. *Polyploidy: Biological relevance.* New York: Plenum Press.

Lynch, M., and Conery, J. S. 2000. The evolutionary fate and consequences of duplicate genes. *Science* 290: 1151–1154.

Madan, K. 1995. Paracentric inversions: A review. *Hum. Genet.* 96: 503–515.

Nelson, D. L., and Gibbs, R. A. 2004. The critical region in trisomy 21. *Science* 306: 619–621.

Ohno, S. 1970. *Evolution by gene duplication.* New York: Springer-Verlag.

Oostra, B. A., and Verkerk, A. J. 1992. The fragile X syndrome: Isolation of the *FMR-1* gene and characterization of the fragile X mutation. *Chromosoma* 101: 381–387.

Patterson, D. 1987. The causes of Down syndrome. *Sci. Am.* (Aug.) 257: 52–61.

Patterson, D., and Costa, A. 2005. Down syndrome and genetics—A case of linked histories. *Nature Reviews Genetics* 6: 137–145.

Shepard, J., et al. 1983. Genetic transfer in plants through interspecific protoplast fusion. *Science* 21: 683–688.

Shepard, J. F. 1982. The regeneration of potato plants from protoplasts. *Sci. Am.* (May) 246: 154–166.

Stranger, B. E., et al., 2007. Relative impact of nucleotide and copy number variation on gene expression phenotypes. *Science* 315: 848–854.

Taylor, A. I. 1968. Autosomal trisomy syndromes: A detailed study of 27 cases of Edwards syndrome and 27 cases of Patau syndrome. *J. Med. Genet.* 5: 227–252.

Tjio, J. H., and Levan, A. 1956. The chromosome number of man. *Hereditas* 42: 1–6.

Wilkins, L. E., Brown, J. X., and Wolf, B. 1980. Psychomotor development in 65 home-reared children with cri-du-chat syndrome. *J. Pediatr.* 97: 401–405.

Chapter 7: Linkage and Chromosome Mapping in Eukaryotes

Allen, G. E. 1978. *Thomas Hunt Morgan: The man and his science.* Princeton, NJ: Princeton University Press.

Chaganti, R., Schonberg, S., and German, J. 1974. A manyfold increase in sister chromatid exchange in Bloom syndrome lymphocytes. *Proc. Natl. Acad. Sci.* 71: 4508–4512.

Creighton, H. S., and McClintock, B. 1931. A correlation of cytological and genetical crossing over in *Zea mays. Proc. Natl. Acad. Sci.* 17: 492–497.

Douglas, L., and Novitski, E. 1977. What chance did Mendel's experiments give him of noticing linkage? *Heredity* 38: 253–257.

Ellis, N. A., et al. 1995. The Bloom syndrome gene product is homologous to RecQ helicases. *Cell* 83: 655–666.

Ephrussi, B., and Weiss, M. C. 1969. Hybrid somatic cells. *Sci. Am.* (Apr.) 220: 26–35.

Latt, S. A. 1981. Sister chromatid exchange formation. *Annu. Rev. Genet.* 15: 11–56.

Lindsley, D. L., and Grell, E. H. 1972. *Genetic variations of Drosophila melanogaster.* Washington, DC: Carnegie Institute of Washington.

Morgan, T. H. 1911. An attempt to analyze the constitution of the chromosomes on the basis of sex-linked inheritance in *Drosophila. J. Exp. Zool.* 11: 365–414.

Morton, N. E. 1955. Sequential test for the detection of linkage. *Am. J. Hum. Genet.* 7: 277–318.

———. 1995. LODs—Past and present. *Genetics* 1 40: 7–12.

Neuffer, M. G., Jones, L., and Zober, M. 1968. *The mutants of maize.* Madison, WI: Crop Sci. Soc. of America.

Perkins, D. 1962. Crossing over and interference in a multiply marked chromosome arm of *Neurospora. Genetics* 47: 1253–1274.

Ruddle, F. H., and Kucherlapati, R. S. 1974. Hybrid cells and human genes. *Sci. Am.* (July) 231: 36–49.

Stahl, F. W. 1979. *Genetic recombination.* New York: W. H. Freeman.

Stern, C. 1936. Somatic crossing over and segregation in *Drosophila melanogaster. Genetics* 21: 625–631.

Sturtevant, A. H. 1913. The linear arrangement of six sex-linked factors in *Drosophila,* as shown by their mode of association. *J. Exp. Zool.* 14: 43–59.

———. 1965. *A history of genetics.* New York: Harper & Row.

Voeller, B. R., ed. 1968. *The chromosome theory of inheritance: Classical papers in development and heredity.* New York: Appleton-Century-Croft.

Wellcome Trust Case Control Consortium, 2007. Genome association study of 14,000 cases of seven common diseases and 3,000 shared controls. *Nature* 447: 661–676.

Wolff, S., ed. 1982. *Sister chromatid exchange.* New York: Wiley-Interscience.

Chapter 8: Genetic Analysis and Mapping in Bacteria and Bacteriophages

Adelberg, E. A. 1960. *Papers on bacterial genetics.* Boston: Little, Brown.

Benzer, S. 1962. The fine structure of the gene. *Sci. Am.* (Jan.) 206: 70–86.

Birge, E. A. 1988. *Bacterial and bacteriophage genetics—An introduction.* New York: Springer-Verlag.

Brock, T. 1990. *The emergence of bacterial genetics.* Cold Spring Harbor, NY: Cold Spring Harbor Press.

Bukhari, A. I., Shapiro, J. A., and Adhya, S. L., eds. 1977. *DNA insertion elements, plasmids, and episomes.* Cold Spring Harbor, NY: Cold Spring Harbor Press.

Cairns, J., Stent, G. S., and Watson, J. D., eds. 1966. *Phage and the origins of molecular biology.* Cold Spring Harbor, NY: Cold Spring Harbor Press.

Campbell, A. M. 1976. How viruses insert their DNA into the DNA of the host cell. *Sci. Am.* (Dec.) 235: 102–113.

Hayes, W. 1968. *The genetics of bacteria and their viruses,* 2nd ed. New York: Wiley.

Hershey, A. D., and Rotman, R. 1949. Genetic recombination between host range and plaque-type mutants of bacteriophage in single cells. *Genetics* 34: 44–71.

Hotchkiss, R. D., and Marmur, J. 1954. Double marker transformations as evidence of linked factors in deoxyribonucleate transforming agents. *Proc. Natl. Acad. Sci. (USA)* 40: 55–60.

Jacob, F., and Wollman, E. L. 1961. Viruses and genes. *Sci. Am.* (June) 204: 92–106.

Kohiyama, M., et al. 2003. Bacterial sex: Playing voyeurs 50 years later. *Science* 301: 802–803.

Kruse, H., and Sorum, H. 1994. Transfer of multiple drug resistance plasmids between bacteria of diverse origins in natural microenvironments. *Appl. Environ. Microbiol.* 60: 4015–4021.

Lederberg, J. 1986. Forty years of genetic recombination in bacteria: A fortieth anniversary reminiscence. *Nature* 324: 627–628.

Luria, S. E., and Delbruck, M. 1943. Mutations of bacteria from virus sensitivity to virus resistance. *Genetics* 28: 491–511.

Lwoff, A. 1953. Lysogeny. *Bacteriol. Rev.* 17: 269–337.

Miller, J. H. 1992. *A short course in bacterial genetics.* Cold Spring Harbor, NY: Cold Spring Harbor Press.

Miller, R. V. 1998. Bacterial gene swapping in nature. *Sci. Am.* (Jan.) 278: 66–71.

Morse, M. L., Lederberg, E. M., and Lederberg, J. 1956. Transduction in *Escherichia coli* K12. *Genetics* 41: 141–156.

Novick, R. P. 1980. Plasmids. *Sci. Am.* (Dec.) 243: 102–127.

Smith-Keary, P. F. 1989. *Molecular genetics of Escherichia coli.* New York: Guilford Press.

Stahl, F. W. 1987. Genetic recombination. *Sci. Am.* (Nov.) 256: 91–101.

Stent, G. S. 1966. *Papers on bacterial viruses,* 2nd ed. Boston: Little, Brown.

Wollman, E. L., Jacob, F., and Hayes, W. 1956. Conjugation and genetic recombination in *Escherichia coli* K12. *Cold Spring Harb. Symp. Quant. Biol.* 21: 141–162.

Zinder, N. D. 1958. Transduction in bacteria. *Sci. Am.* (Nov.) 199: 38–46.

Chapter 9: DNA Structure and Analysis

Adleman, L. M. 1998. Computing with DNA. *Sci. Am.* (Aug.) 279: 54–61.

Alloway, J. L. 1933. Further observations on the use of pneumococcus extracts in effecting transformation of type *in vitro. J. Exp. Med.*57: 265–278.

Avery, O. T., MacLeod, C. M., and McCarty, M. 1944. Studies on the chemical nature of the substance inducing transformation of pneumococcal types: Induction of transformation by a desoxyribonucleic acid fraction isolated from pneumococcus type III. *J. Exp. Med.* 79: 137–158. (Reprinted in Taylor, J. H. 1965. *Selected papers in molecular genetics.* Orlando, FL: Academic Press.)

Britten, R. J., and Kohne, D. E. 1968. Repeated sequences in DNA. *Science* 161: 529–540.

Chargaff, E. 1950. Chemical specificity of nucleic acids and mechanism for their enzymatic degradation. *Experientia* 6: 201–209.

Darnell, J. E. 1985. RNA. *Sci. Am.* (Oct.) 253: 68–87.

Dawson, M. H. 1930. The transformation of pneumococcal types: I. The interconvertibility of type-specific *S. pneumococci. J. Exp. Med.* 51: 123–147.

Dickerson, R. E., et al. 1982. The anatomy of A-, B-, and Z-DNA. *Science* 216: 475–485.

Dubos, R. J. 1976. *The professor, the Institute and DNA: Oswald T. Avery, his life and scientific achievements.* New York: Rockefeller University Press.

Felsenfeld, G. 1985. DNA. *Sci. Am.* (Oct.) 253: 58–78.

Franklin, R. E., and Gosling, R. G. 1953. Molecular configuration in sodium thymonucleate. *Nature* 171: 740–741.

Griffith, F. 1928. The significance of pneumococcal types. *J. Hyg.* 27: 113–159.

Guthrie, G. D., and Sinsheimer, R. L. 1960. Infection of protoplasts of *Escherichia coli* by subviral particles. *J. Mol. Biol.* 2: 297–305.

Hershey, A. D., and Chase, M. 1952. Independent functions of viral protein and nucleic acid in growth of bacteriophage. *J. Gen. Phys.* 36: 39–56. (Reprinted in Taylor, J. H. 1965. *Selected papers in molecular genetics.* Orlando, FL: Academic Press.)

Levene, P. A., and Simms, H. S. 1926. Nucleic acid structure as determined by electrometric titration data. *J. Biol. Chem.* 70: 327–341.

McCarty, M. 1985. *The transforming principle: Discovering that genes are made of DNA.* New York: W. W. Norton.

Olby, R. 1974. *The path to the double helix.* Seattle: University of Washington Press.

Pauling, L., and Corey, R. B. 1953. A proposed structure for the nucleic acids. *Proc. Natl. Acad. Sci. (USA)* 39: 84–97.

Rich, A., Nordheim, A., and Wang, A. H.-J. 1984. The chemistry and biology of left-handed Z-DNA. *Annu. Rev. Biochem.* 53: 791–846.

Spizizen, J. 1957. Infection of protoplasts by disrupted T2 viruses. *Proc. Natl. Acad. Sci. (USA)* 43: 694–701.

Varmus, H. 1988. Retroviruses. *Science* 240: 1427–1435.

Watson, J. D. 1968. *The double helix.* New York: Atheneum.

Watson, J. D., and Crick, F. C. 1953a. Molecular structure of nucleic acids: A structure for deoxyribose nucleic acids. *Nature* 171: 737–738.

———. 1953b. Genetic implications of the structure of deoxyribose nucleic acid. *Nature* 171: 964.

Wilkins, M. H. F., Stokes, A. R., and Wilson, H. R. 1953. Molecular structure of desoxypentose nucleic acids. *Nature* 171: 738–740.

Chapter 10: DNA Replication and Recombination

Blackburn, E. H. 1991. Structure and function of telomeres. *Nature* 350: 569–572.

DeLucia, P., and Cairns, J. 1969. Isolation of an *E. coli* strain with a mutation affecting DNA polymerase. *Nature* 224: 1164–1166.

Gilbert, D. M. 2001. Making sense of eukaryotic DNA replication origins. *Science* 294: 96–100.

Greider, C. W. 1996. Telomeres, telomerase, and cancer. *Sci. Am.* (Feb.) 274: 92–97.

_______. 1998. Telomerase activity, cell proliferation, and cancer. *Proc. Natl. Acad. Sci. (USA)* 95: 90–92.

Holliday, R. 1964. A mechanism for gene conversion in fungi. *Genet. Res.* 5: 282–304.

Holmes, F. L. 2001. *Meselson, Stahl, and replication of DNA: A history of the "most beautiful experiment in biology."* New Haven, CT: Yale University Press.

Kim, J., Kaminker, P., and Campisi, J. 2002. Telomeres, aging and cancer: In search of a happy ending. *Oncogene* 21: 503–511.

Kornberg, A. 1960. Biological synthesis of DNA. *Science* 131: 1503–1508.

Kornberg, A., and Baker, T. 1992. *DNA replication,* 2nd ed. New York: W. H. Freeman.

Meselson, M., and Stahl, F. W. 1958. The replication of DNA in *Escherichia coli. Proc. Natl. Acad. Sci. (USA)* 44: 671–682.

Mitchell, M. B. 1955. Aberrant recombination of pyridoxine mutants of *Neurospora. Proc. Natl. Acad. Sci. (USA)* 41: 215–220.

Okazaki, T., et al. 1979. Structure and metabolism of the RNA primer in the discontinuous replication of prokaryotic DNA. *Cold Spring Harbor Symp. Quant. Biol.* 43: 203–222.

Radman, M., and Wagner, R. 1988. The high fidelity of DNA duplication. *Sci. Am.* (Aug.) 259: 40–46.

———. 1987. Genetic recombination. *Sci. Am.* (Feb.) 256: 90–101.

Taylor, J. H., Woods, P. S., and Hughes, W. C. 1957. The organization and duplication of chromosomes revealed by autoradiographic studies using tritium-labeled thymidine. *Proc. Natl. Acad. Sci. (USA)* 48: 122–128.

Wang, J. C. 1987. Recent studies of DNA topoisomerases. *Biochim. Biophys. Acta* 909: 1–9.

Whitehouse, H. L. K. 1982. *Genetic recombination: Understanding the mechanisms.* New York: Wiley.

Chapter 11: Chromosome Structure and DNA Sequence Organization

Angelier, N., et al. 1984. Scanning electron microscopy of amphibian lampbrush chromosomes. *Chromosoma* 89: 243–253.

Beerman, W., and Clever, U. 1964. Chromosome puffs. *Sci. Am.* (Apr.) 210: 50–58.

Carbon, J. 1984. Yeast centromeres: Structure and function. *Cell* 37: 352–353.

Chen, T. R., and Ruddle, F. H. 1971. Karyotype analysis utilizing differential stained constitutive heterochromatin of human and murine chromosomes. *Chromosoma* 34: 51–72.

DuPraw, E. J. 1970. *DNA and chromosomes.* New York: Holt, Rinehart & Winston.

Gall, J. G. 1981. Chromosome structure and the *C*-value paradox. *J. Cell Biol.* 91: 3s–14s.

Hewish, D. R., and Burgoyne, L. 1973. Chromatin substructure. The digestion of chromatin DNA at regularly spaced sites by a nuclear deoxyribonuclease. *Biochem. Biophys. Res. Comm.* 52: 504–510.

Korenberg, J. R., and Rykowski, M. C. 1988. Human genome organization: *Alu,* LINES, and the molecular organization of metaphase chromosome bands. *Cell* 53: 391–400.

Kornberg, R. D. 1975. Chromatin structure: A repeating unit of histones and DNA. *Science* 184: 868–871.

Kornberg, R. D., and Klug, A. 1981. The nucleosome. *Sci. Am.* (Feb.) 244: 52–64.

Luger, K., et al. 1997. Crystal structure of the nucleosome core particle at 2.8 resolution. *Nature* 389: 251–256.

Moyzis, R. K. 1991. The human telomere. *Sci. Am.* (Aug.) 265: 48–55.

Olins, A. L., and Olins, D. E. 1974. Spheroid chromatin units (ν bodies). *Science* 183: 330–332.

———. 1978. Nucleosomes: The structural quantum in chromosomes. *Am. Sci.* 66: 704–711.

Singer, M. F. 1982. SINES and LINES: Highly repeated short and long interspersed sequences in mammalian genomes. *Cell* 28: 433–434.

Chapter 12: The Genetic Code and Transcription

Barrell, B. G., Air, G., and Hutchinson, C. 1976. Overlapping genes in bacteriophage ϕX174 *Nature* 264: 34–40.

Barrell, B. G., Banker, A. T., and Drouin, J. 1979. A different genetic code in human mitochondria. *Nature* 282: 189–194.

Bass, B. L., ed. 2000. *RNA editing.* Oxford: Oxford University Press.

Brenner, S., Jacob, F., and Meselson, M. 1961. An unstable intermediate carrying information from genes to ribosomes for protein synthesis. *Nature* 190: 575–580.

Brenner, S., Stretton, A. O. W., and Kaplan, D. 1965. Genetic code: The nonsense triplets for chain termination and their suppression. *Nature* 206: 994–998.

Cattaneo, R. 1991. Different types of messenger RNA editing. *Annu. Rev. Genet.* 25: 71–88.

Cech, T. R. 1986. RNA as an enzyme. *Sci. Am.* (Nov.) 255(5): 64–75.

_______. 1987. The chemistry of self-splicing RNA and RNA enzymes. *Science* 236: 1532–1539.

Chambon, P. 1981. Split genes. *Sci. Am.* (May) 244: 60–71.

Cramer, P., et al. 2000. Architecture of RNA polymerase II and implications for the transcription mechanism. *Science* 288: 640–649.

Crick, F. H. C. 1962. The genetic code. *Sci. Am.* (Oct.) 207: 66–77.

_______. 1966a. The genetic code: III. *Sci. Am.* (Oct.) 215: 55–63.

_______. 1966b. Codon–anticodon pairing: The wobble hypothesis. *J. Mol. Biol.* 19: 548–555.

Crick, F. H. C., Barnett, L., Brenner, S., and Watts-Tobin, R. J. 1961. General nature of the genetic code for proteins. *Nature* 192: 1227–1232.

Darnell, J. E. 1983. The processing of RNA. *Sci. Am.* (Oct.) 249: 90–100.

Dickerson, R. E. 1983. The DNA helix and how it is read. *Sci. Am.* (Dec.) 249: 94–111.

Dugaiczk, A., et al. 1978. The natural ovalbumin gene contains seven intervening sequences. *Nature* 274: 328–333.

Fiers, W., et al. 1976. Complete nucleotide sequence of bacteriophage MS2 RNA: Primary and secondary structure of the replicase gene. *Nature* 260: 500–507.

Gamow, G. 1954. Possible relation between DNA and protein structures. *Nature* 173: 318.

Hall, B. D., and Spiegelman, S. 1961. Sequence complementarity of T2-DNA and T2-specific RNA. *Proc. Natl. Acad. Sci. (USA)* 47: 137–146.

Hamkalo, B. 1985. Visualizing transcription in chromosomes. *Trends Genet.* 1: 255–260.

Khorana, H. G. 1967. Polynucleotide synthesis and the genetic code. *Harvey Lectures* 62: 79–105.

Miller, O. L., Hamkalo, B., and Thomas, C. 1970. Visualization of bacterial genes in action. *Science* 169: 392–395.

Nirenberg, M. W. 1963. The genetic code: II. *Sci. Am.* (Mar.) 190: 80–94.

O'Malley, B., et al. 1979. A comparison of the sequence organization of the chicken ovalbumin and ovomucoid genes. In *Eucaryotic gene regulation,* R. Axel et al., eds., pp. 281–299. Orlando, FL: Academic Press.

Reed, R., and Maniatis, T. 1985. Intron sequences involved in lariat formation during pre-mRNA splicing. *Cell* 41: 95–105.

Ridley, M., 2006. *Francis Crick: Discoverer of the genetic code.* New York: HarperCollins.

Sharp, P. A. 1994. Nobel lecture: Split genes and RNA splicing. *Cell* 77: 805–815.

Steitz, J. A. 1988. Snurps. *Sci. Am.* (June) 258(6): 56–63.

Volkin, E., and Astrachan, L. 1956. Phosphorus incorporation in *E. coli* ribonucleic acids after infection with bacteriophage T2. *Virology* 2: 149–161.

Watson, J. D. 1963. Involvement of RNA in the synthesis of proteins. *Science* 140: 17–26.

Woychik, N. A., and Jampsey, M. 2002. The RNA polymerase II machinery: Structure illuminates function. *Cell* 108: 453–464.

Chapter 13: Translation and Proteins

Anfinsen, C. B. 1973. Principles that govern the folding of protein chains. *Science* 181: 223–230.

Bartholome, K. 1979. Genetics and biochemistry of phenylketonuria—Present state. *Hum. Genet.* 51: 241–245.

Beadle, G. W., and Tatum, E. L. 1941. Genetic control of biochemical reactions in *Neurospora. Proc. Natl. Acad. Sci. (USA)* 27: 499–506.

Beet, E. A. 1949. The genetics of the sickle-cell trait in a Bantu tribe. *Ann. Eugenics* 14: 279–284.

Brenner, S. 1955. Tryptophan biosynthesis in *Salmonella typhimurium. Proc. Natl. Acad. Sci. (USA)* 41: 862–863.

Doolittle, R. F. 1985. Proteins. *Sci. Am.* (Oct.) 253: 88–99.

Frank, J. 1998. How the ribosome works. *Amer. Scient.* 86: 428–439.

Garrod, A. E. 1902. The incidence of alkaptonuria: A study in chemical individuality. *Lancet* 2: 1616–1620.

_______. 1909. *Inborn errors of metabolism.* London: Oxford University Press. (Reprinted 1963, Oxford University Press, London.)

Garrod, S. C. 1989. Family influences on A. E. Garrod's thinking. *J. Inher. Metab. Dis.* 12: 2–8.

Ingram, V. M. 1957. Gene mutations in human hemoglobin: The chemical difference between normal and sickle-cell hemoglobin. *Nature* 180: 326–328.

Koshland, D. E. 1973. Protein shape and control. *Sci. Am.* (Oct.) 229: 52–64.

Lake, J. A. 1981. The ribosome. *Sci. Am.* (Aug.) 245: 84–97.

Maniatis, T., et al. 1980. The molecular genetics of human hemoglobins. *Annu. Rev. Genet.* 14: 145–178.

Neel, J. V. 1949. The inheritance of sickle-cell anemia. *Science* 110: 64–66.

Nirenberg, M. W., and Leder, P. 1964. RNA codewords and protein synthesis. *Science* 145: 1399–1407.

Nomura, M. 1984. The control of ribosome synthesis. *Sci. Am.* (Jan.) 250: 102–114.

Pauling, L., Itano, H. A., Singer, S. J., and Wells, I. C. 1949. Sickle-cell anemia: A molecular disease. *Science* 110: 543–548.

Ramakrishnan, V. 2002. Ribosome structure and the mechanism of translation. *Cell* 108: 557–572.

Rich, A., and Houkim, S. 1978. The three-dimensional structure of transfer RNA. *Sci. Am.* (Jan.) 238: 52–62.

Rich, A., Warner, J. R., and Goodman, H. M. 1963. The structure and function of polyribosomes. *Cold Spring Harbor Symp. Quant. Biol.* 28: 269–285.

Richards, F. M. 1991. The protein-folding problem. *Sci. Am.* (Jan.) 264: 54–63.

Rould, M. A., et al. 1989. Structure of *E. coli* glutaminyl-tRNA synthetase complexed with $tRNA^{gln}$ and ATP at 2.8 resolution. *Science* 246: 1135–1142.

Srb, A. M., and Horowitz, N. H. 1944. The ornithine cycle in *Neurospora* and its genetic control. *J. Biol. Chem.* 154: 129–139.

Warner, J., and Rich, A. 1964. The number of soluble RNA molecules on reticulocyte polyribosomes. *Proc. Natl. Acad. Sci. (USA)* 51: 1134–1141.

Wimberly, B. T., et al. 2000. Structure of the 30*S* ribosomal subunit. *Nature* 407: 327–333.

Yusupov, M. M., et al. 2001. Crystal structure of the ribosome at 5.5A resolution. *Science* 292: 883–896.

Chapter 14: Gene Mutation, Transposition, and DNA Repair

Aartsma-Rus, A. et al. 2006. Entries in the Leiden Duchenne muscular dystrophy database: an overview of mutations types and paradoxical cases that confirm the reading-frame rule. *Muscle Nerve* 34: 135–144.

Berneburg, M., and Lehmann, A. R. 2001. Xeroderma pigmentosum and related disorders: Defects in DNA repair and transcription. *Adv. Genet.* 43:71–102.

Cairns, J., Overbaugh, J., and Miller, S. 1988. The origin of mutants. *Nature* 335: 142–145.

Cleaver, J. E. 2005. Cancer in xeroderma pigmentosum and related disorders of DNA repair. *Nature Reviews Cancer* 5: 564–571.

Comfort, N. C. 2001. *The tangled field: Barbara McClintock's search for the patterns of genetic control.* Cambridge, MA: Harvard University Press.

Friedberg, E. C., Walker, G. C., and Siede, W. 1995. *DNA repair and mutagenesis.* Washington, DC: ASM Press.

Jiricny, J. 1998. Eukaryotic mismatch repair: An update. *Mutation Research* 409: 107–121.

Miki, Y. 1998. Retrotransposal integration of mobile genetic elements in human disease. *J. Human Genet.* 43: 77–84.

Mirkin, S. M. 2007. Expandable DNA repeats and human disease. *Nature* 447: 932–940.

Mortelmans, K., and Zeigler, E. 2000. The Ames *Salmonella*/microsome mutagenicity assay. *Mutation Research* 455: 29–60.

O'Driscoll, M., and Jeggo, P. A. 2006. The role of double-strand break repair – insights from human genetics. *Nature Reviews Genetics* 7: 45–51.

O'Hare, K. 1985. The mechanism and control of *P* element transposition in *Drosophila. Trends Genet.* 1: 250–254.

Radman, M., and Wagner, R. 1988. The high fidelity of DNA duplication. *Sci. Am.* (Aug.) 259: 40–46.

Chapter 15: Regulation of Gene Expression

Beckwith, J. R., and Zipser, D., eds. 1970. *The lactose operon.* Cold Spring Harbor, NY: Cold Spring Harbor Laboratory Press.

Bertrand, K., et al. 1975. New features of the regulation of the tryptophan operon. *Science* 189: 22–26.

Black, D. L. 2000. Protein diversity from alternative splicing: a challenge for bioinformatics and postgenomic biology. *Cell* 103: 367–370.

Black, D. L. 2003. Mechanisms of alternative pre-mRNA splicing. *Annu. Rev. Biochem.* 72: 291–336.

Bumcrot, D., et al. 2006. RNAi therapeutics: a potential new class of pharmaceutical drugs. *Nature Chemical Biology* 2: 711–718.

Butler, J. E. F., and Kadonaga, J. T. 2002. The RNA polymerase II core promoter: A key component in the regulation of gene expression. *Genes Dev.* 16: 2583–2592.

Dillon, N. 2006. Gene regulation and large-scale chromatin organization in the nucleus. *Chromosome Res.* 14: 117–126.

Gilbert, W., and Müller-Hill, B. 1966. Isolation of the *lac* repressor. *Proc. Natl. Acad. Sci.* (USA) 56: 1891–1898.

_______. 1967. The *lac* operator is DNA. *Proc. Natl. Acad. Sci. (USA)* 58: 2415–2421.

Henikoff, S. 2008. Nucleosome destabilization in the epigenetic regulation of gene expression. *Nature Reviews Genetics* 9: 15–20.

Jackson, D. A. 2003. The anatomy of transcription sites. *Curr. Opin. Cell Biol.* 15: 311–317.

Jacob, F., and Monod, J. 1961. Genetic regulatory mechanisms in the synthesis of proteins. *J. Mol. Biol.* 3: 318–356.

Mello, C. C., and Conte, D., Jr. 2004. Revealing the world of RNA interference. *Nature* 431: 338–342.

Newbury, S. F. 2006. Control of mRNA stability in eukaryotes. *Biochem. Soc. Trans.* 34: 30–34.

Rana, T. M. 2007. Illuminating the silence: understanding the structure and function of small RNAs. *Nature Mol. Cell Biol.* 8: 23–34.

Wheeler, T. M. and Thornton, C. A. 2007. Myotonic dystrophy: RNA-mediated muscle disease. *Curr. Opin. Neurol.* 20: 572–576.

Yanofsky, C. 1981. Attenuation in the control of expression of bacterial operons. *Nature* 289: 751–758.

Yanofsky, C., and Kolter, R. 1982. Attenuation in amino acid biosynthetic operons. *Annu. Rev. Genet.* 16: 113–134.

Chapter 16: Cancer and Regulation of the Cell Cycle

Bernards, R., and Weinberg, R. A. 2002. A progression puzzle. *Nature* 418: 823.

Brown, M. A. 1997. Tumor suppressor genes and human cancer. *Adv. Genet.* 36: 45–135.

Compagni, A., and Christofori, G. 2000. Recent advances in research on multistage tumorigenesis. *Brit. J. Cancer* 83: 1–5.

Esteller, M. 2007. Cancer epigenomics: DNA methylomes and histone-modification maps. *Nature Reviews Cancer* 8: 286–297.

Futreal, P. A., et al. 2004. A census of human cancer genes. *Nature Reviews Cancer* 4: 177–183.

Hartwell, L. H., and Kastan, M. B. 1994. Cell cycle control and cancer. *Science* 266: 1821–1827.

Lengauer, C., Kinzler, K. W., and Vogelstein, B. 1997. Genetic instability in colorectal cancer. *Nature* 386: 623–627.

Nurse, P. 1997. Checkpoint pathways come of age. *Cell* 91: 865–867.

Raff, M. 1998. Cell suicide for beginners. *Nature* 396: 119–122.

Sherr, C. J. 1996. Cancer cell cycles. *Science* 274: 1672–1677.

Varmus, H. 2006. The new era in cancer research. *Science* 312: 1162–1164.

Vazquez, J. et al. 2008. The genetics of the p53 pathway, apoptosis and cancer therapy. *Nature Reviews Drug Discov.* 7: 979–987.

Vogelstein, B., and Kinzler, K. W. 1993. The multistep nature of cancer. *Trends in Genetics* 9: 138–141.

Weinberg, R. A. 1995. The retinoblastoma protein and cell cycle control. *Cell* 81: 323–330.

Yokota, J. 2000. Tumor progression and metastasis. *Carcinogenesis* 21: 497–503.

Chapter 17: Recombinant DNA Technology and Gene Cloning

Brownlee, C. 2005. Danna and Nathans: Restriction enzymes and the boon to modern molecular biology. *Proc. Nat. Acad. Sci.* 103: 5909.

Cohen, S. N. 1975. The manipulation of genes. *Sci. Am.* (Jan.) 233: 24–33.

Flotte, T. R. 2005. Adeno-associated virus-based gene therapy for inherited disorders. *Pediatr. Res.* 58: 1143–1147.

Mardis, E. R. 2008. Next-generation DNA sequencing methods. *Annu. Rev. Genomics Hum. Genet.* 9: 387–402.

Metzker, M. L. 2005. Emerging technologies in DNA sequencing. *Genome Res.* 15: 1767–1776.

Mullis, K. B. 1990. The unusual origin of the polymerase chain reaction. *Sci. Am.* (Apr.) 262: 56–65.

Pingoud, A., et al. 2005. Type II restriction endonucleases: structure and mechanism. *Cell Mol. Life Sci.* 62: 685–707.

Rothberg, J. M., and Leamon, J. H. 2008. The development and impact of 454 sequencing. *Nat. Biotech.* 26: 1117–1124.

Sanger, F., et al. 1977. DNA sequencing with chain-terminating inhibitors. *Proc. Natl. Acad. Sci. (USA)* 74: 5463–5467.

Schambach, A., and Baum, C. 2008. Clinical application of lentiviral vectors—concepts and practice. *Curr. Gene Ther.* 8: 474–482.

Southern, E. 1975. Detection of specific sequences among DNA fragments separated by gel electrophoresis. *J. Mol. Biol.* 98: 503–507.

Chapter 18: Genomics, Bioinformatics, and Proteomics

Andersen, J., et al. 2002. Directed proteomic analysis of the human nucleolus. *Curr. Biol.* 12: 1–11.

Asara, J. M., et al. 2007. Protein sequences from Mastodon and *Tyrannosaurus Rex* revealed by mass spectrometry. *Science* 316: 280–285.

Butler, J. M. 2006. Genetics and genomics of core short tandem repeat loci used in human identity testing. *J. Forensic Sci.* 51: 253–265.

Chandonia, J. M. 2006. The impact of structural genomics: expectations and outcomes. *Science* 311: 347–351.

Church, G. M. 2006. Genomes for all. *Sci. Am.* (Jan.) 294: 47–54.

Domon, B., and Aebersold, R. 2006. Mass spectrometry and protein analysis. *Science* 312: 212–217.

Fleming, K., et al. 2006. The proteome: structure, function and evolution. *Philos. Trans. R. Soc. Lond. B Biol. Sci.* 361: 441–451.

Hood, L., Heath, J. R., Phelps, M. E., and Lin, B. 2004. Systems biology and new technologies enable predictive and preventative medicine. *Science* 306: 640–643.

Hoskins, R. A., et al., 2007. Sequence finishing and mapping of *Drosophila melanogaster* heterochromatin. *Science* 316: 1625–1628.

International Human Genome Sequencing Consortium 2004. Finishing the euchromatic sequence of the human genome. *Nature* 431: 931–945.

International Human Genome Sequencing Consortium. 2001. Initial sequencing and analysis of the human genome. *Nature* 409: 860–921.

Joyce, A. R., and Palsson, B. O. 2006. The model organism as a system integrating "omics" data sets. *Nat. Rev. Mol. Cell Biol.* 7: 901–909.

Nei, M., and Rooney, A. P. 2005. Concerted and birth-and-death evolution of multigene families. *Annu. Rev. Genet.* 39: 197–218.

Noonan, J. P., et al. 2006. Sequencing and analysis of Neanderthal DNA. *Science* 314: 1113–1118.

Ochman, H., and Davilos, L. M. 2006. The nature and dynamics of bacterial genomes. *Science* 311: 1730–1733.

Pennisi, E. 2006. The dawn of Stone Age genomics. *Science* 314: 1068–1071.

Rhesus Macque Genome Sequencing and Analysis Consortium. 2007. Evolutionary and biomedical insights from the Rhesus Macaque genome. *Science* 316: 222–234.

Schweitzer, M. H., et al., 2007. Analyses of soft tissue from *Tyrannosaurus rex* suggest the presence of protein. *Science* 316: 277–280.

Sea Urchin Genome Sequencing Consortium. 2006. The genome of the sea urchin *Stronglyocentrotus purpuratus*. *Science* 314: 941–956.

Switnoski, M., Szczerbal, I., and Nowacka, J. 2004. The dog genome map and its use in mammalian comparative genomics. *J. Appl. Physiol.* 45: 195–214.

Thieman, W. J. and Palladino, M. A. 2009. *Introduction to biotechnology,* 2nd ed. San Francisco, CA: Benjamin Cummings.

Venter, J. C., et al. 2001. The sequence of the human genome. *Science* 291: 1304–1351.

Venter, J. C., et al. 2004. Environmental genome shotgun sequencing of the Sargasso Sea. *Science* 304: 6–74.

Yooseph, S., Sutton, G., Rusch, D. B., Halpern, A. L., Williamson, S .J., et al. 2007. *The Sorcerer II* global ocean sampling expedition: Expanding the universe of protein families. *PLoS Biology* 5(3): 0432–0466.

Chapter 19: Applications and Ethics of Genetic Engineering and Biotechnology

Cavanna-Calvo, M., et al. 2000. Gene therapy of severe combined immunodeficiency (SCID)-XI disease. *Science* 288: 669–672.

Chrispeels, M. J., and Sadava, D. E. 2003. *Plants, Genes, and Crop Biotechnology,* 2nd ed. Sudbury, MA: Jones and Barlett Publishers.

Dale, P. J., Clarke, B., and Fontes, E. M. G. 2002. Potential for the environmental impact of transgenic crops. *Nat. Biotechnol.* 20: 567–574.

Dykxhoorn, D. M., and Liberman, J. 2006. Knocking down disease with siRNAs. *Cell* 126: 231–235.

Engler, O. B., et al. 2001. Peptide vaccines against hepatitis B virus: From animal model to human studies. *Mol. Immunol.* 38: 457–465.

Friend, S. H., and Stoughton, R. B. 2002. The magic of microarrays. *Sci. Am.* (Feb.) 286: 44–50.

Hunter, C. V., Tiley, L. S., and Sang, H. M. 2005. Developments in transgenic technology: applications for medicine. *Trends Mol. Med.* 11: 293–300.

Knoppers, B. M., Bordet, S., and Isasi, R. M. 2006. Preimplantation genetic diagnosis: an overview of socio-ethical and legal considerations. *Annu. Rev. Genom. Hum. Genet.* 7: 201–221.

O'Connor, T. P., and Crystal, R. G. 2006. Genetic medicines: treatment strategies for hereditary disorders. *Nature Reviews Genetics* 7: 261–276.

Pray, C. E., Huang, J., and Rozelle, S. 2002. Five years of Bt cotton in China: The benefits continue. *The Plant J.* 31: 423–430.

Schillberg, S., Fischer, T., and Emans, N. 2003. Molecular farming of recombinant antibodies in plants. Cell *Mol. Life Sci.* 60: 433–445.

Shastry, B. S. 2006. Pharmacogenetics and the concept of individualized medicine. *Pharmacogenetics Journal* 6: 16–21.

Stix, G. 2006. Owning the stuff of life. *Sci. Am.* (Feb.) 294: 76–83.

Wang, D. G., et al. 1998. Large-scale identification, mapping, and genotyping of single-nucleotide polymorphisms in the human genome. *Science* 280: 1077–1082.

Whitelaw, C. B. A. 2004. Transgenic livestock made easy. *Trends in Biotechnology* 22(4): 257–259.

Varsha, M. S. 2006. DNA fingerprinting in the criminal justice system: An overview. *DNA Cell Biol.* 25: 181–188.

Chapter 20: Developmental Genetics

Davidson, E. H., and Levine, M. S. 2008. Properties of developmental gene regulatory networks. *Proc. Nat. Acad. Sci.* 115: 20063–20066.

De-Leon, S. B-T., and Davidson, E. H. 2007. Gene regulation: Gene control networks in development. *Annu. Rev. Biophys. Biomol. Struct.* 36: 191–212.

Goodman, F. 2002. Limb malformations and the human *HOX* genes. *Am. J. Med. Genet.* 112: 256–265.

Gridley, T. 2003. Notch signaling and inherited human diseases. *Hum. Mol. Genet.* 12: R9–R13.

Inoue, T., et al. 2005. Gene regulatory special feature: Transcriptional network underlying *Caenorhabditis elegans* vulval development. *Proc. Nat. Acad. Sci.* 102: 4972–4977.

Kestler, H. A., Wawra, C., Kracher, B., and Kuhl, M. 2008. Network modeling of signal transduction: establishing the global view. *Bioessays* 30: 1110–1125.

Krizek, B. A., and Fletcher, J. C. 2006. Molecular mechanisms of flower development: An armchair guide. *Nat. Rev. Genet.* 6: 688–698.

Moens, C. B., and Selleri, L. 2006. Hox cofactors in vertebrate development. *Dev. Biol.* 291: 193–206.

Nüsslein-Volhard, C., and Weischaus, E. 1980. Mutations affecting segment number and polarity in *Drosophila. Nature* 287: 795–801.

Poellinger, L., and Lendahl, U. 2008. Modulating Notch signaling by pathway-intrinsic and pathway-extrinsic mechanisms. *Curr. Opin. Genet. Dev.* 18: 449–454.

Reyes, J. C. 2006. Chromatin modifiers that control plant development. *Curr. Opin. Plant Biol.* 9: 21–27.

Schroeder, M. D., et al., 2004. Transcriptional control in the segmentation gene network of *Drosophila. PLOS Biol.* 2: 1396–1410.

Schvartsman, S. Y., Coppey, M., and Berezhkovskii, A. 2008. Dynamics of maternal morphogen gradients in *Drosophila. Curr. Opin. Genet. Dev.* 18: 342–347.

Verakasa, A., Del Campo, M., and McGinnis, W. 2000. Developmental patterning genes and their conserved functions: From model organisms to humans. *Mol. Genet. Metabol.* 69: 85–100.

Wang, M., and Sternberg, P. W. 2001. Pattern formation during *C. elegans* vulval induction. *Curr. Top. Dev. Biol.* 51: 189–220.

Chapter 21: Genetics and Behavior

Davis, R. L. 2005. Olfactory memory formation in *Drosophila*: From molecular to systems neuroscience. *Annu. Rev. Neurosci.* 28: 275–302.

Desbonnet, L., Waddington, J. L., O'Tuathaigh, C. M. 2009. Mutant models for genes associated with schizophrenia. *Biochem. Soc. Trans.* 37: 308–312.

Dickson, B. J. 2008. Wired for sex: the neurobiology of *Drosophila* mating decisions. *Science* 322: 904–909.

Flint, J., and Shifman, S. 2008. Animal models of psychiatric diseases. *Curr. Opin. Genet. Dev.* 18: 235–240.

Ganetzky, B. 1994. Cysteine strings, calcium channels and synaptic transmission. *BioEssays* 16: 461–463.

Hakak, Y., et al., 2001. Genome-wide expression analysis reveals dysregulation of myelination-related genes in chronic schizophrenia. *Proc. Nat. Acad. Sci. (USA)* 98: 4746–4751.

Hovatta, I., and Barlow, C. 2008. Molecular genetics of anxiety in mice and men. *Ann. Med.* 40: 92–109.

Mackay, T. F. C. and Anholt, R. H. 2006. Of flies and man: *Drosophila* as a model for human complex traits. *Annu. Rev. Hum. Genet.* 7: 39–367.

Reddy, P. H., et al. 1999. Transgenic mice expressing mutated full-length *HD* cDNA: A paradigm for locomotor changes and selective neuronal loss in Huntington's disease. *Phil. Trans. R. Soc. Lond.* B: 354: 1035–1045.

Ricker, J. P., and Hirsch, J. 1988. Genetic changes occurring over 500 generations in lines of *Drosophila melanogaster* selected divergently for geotaxis. *Behav. Genet.* 18: 13–24.

Rockenstein, E., Crews, L., and Masliah, E. 2007. Transgenic animal models of neurodegenerative diseases and their application to treatment development. *Advanced Drug Del. Rev.* 59: 1093–1102.

Siegel, R. W., Hall, J. C., Gailey, D. A., and Kyriacou, C. P. 1984. Genetic elements of courtship in *Drosophila*: Mosaics and learning mutants. *Behav. Genet.* 14: 383–410.

Stoltenberg, S. F., Hirsch, J., and Berlocher, S. H. 1995. Analyzing correlations of three types in selected lines of *Drosophila melanogaster* that have evolved stable extreme geotactic performance. *J. Comp. Psychol.* 105: 85–94.

Toma, D. P., White, K. P., Hirsch, J., and Greenspan, R. J. 2002. Identification of genes involved in *Drosophila melanogaster* geotaxis, a complex behavioral trait. *Nat. Genet.* 31: 349–353.

Turri, M. G., et al. 2001. QTL analysis identifies multiple behavioral dimensions in ethological tests of anxiety in laboratory mice. *Curr. Biol.* 11: 725–734.

Tully, T. 1996. Discovery of genes involved with learning and memory: An experimental synthesis of Hirschian and Benzerian perspectives. *Proc. Nat. Acad. Sci. (USA)* 93: 13460–13467.

Willis-Owen, S. A. G., and Flint, J. 2006. The genetic basis of emotional behavior in mice. *Eur. J. Hum. Genet.* 14: 721–728.

Chapter 22: Quantitative Genetics and Multifactorial Traits

Browman, K. W. 2001. Review of statistical methods of QTL mapping in experimental crosses. *Lab Animal* 30: 44–52.

Cong, B., Liu, J., and Tanksley, S. 2002. Natural alleles at a tomato fruit size quantitative trait locus differ by heterochronic regulatory mutations. *Proc. Nat. Acad. Sci.* 99: 13606–13611.

Cong, B., Barrero, L. S., and Tanksley, S. D. 2008. Regulatory changes in a YABBY-like transcription factor led to evolution of extreme fruit size during tomato domestication. *Nat. Genet.* 40: 800–804.

Crow, J. F. 1993. Francis Galton: Count and measure, measure and count. *Genetics* 135: 1.

Falconer, D. S., and Mackay, F. C. 1996. *Introduction to quantitative genetics,* 4th ed. Essex, England: Longman.

Farber, S. 1980. *Identical twins reared apart.* New York: Basic Books.

Feldman, M. W., and Lewontin, R. C. 1975. The heritability hangup. *Science* 190: 1163–1166.

Frary, A., et al. 2000. *fw2.2*: A quantitative trait locus key to the evolution of tomato fruit size. *Science* 289: 85–88.

Haley, C. 1991. Use of DNA fingerprints for the detection of major genes for quantitative traits in domestic species. *Anim. Genet.* 22: 259–277.

———. 1996. Livestock QTLs: Bringing home the bacon. *Trends Genet.* 11: 488–490.

Kaminsky, A. A., et al., 2009. DNA methylation profiles in monozygotic and dizygotic twins. *Nat. Genet.* 41: 240–245.

Lander, E., and Botstein, D. 1989. Mapping Mendelian factors underlying quantitative traits using RFLP linkage maps. *Genetics* 121: 185–199.

Lander, E., and Schork, N. 1994. Genetic dissection of complex traits. *Science* 265: 2037–2048.

Lynch, M., and Walsh, B. 1998. *Genetics and analysis of quantitative traits.* Sunderland, MA: Sinauer Associates.

Macy, T. F. C. 2001. Quantitative trait loci in *Drosophila. Nature Reviews Genetics* 2: 11–19.

Newman, H. H., Freeman, F. N., and Holzinger, K. T. 1937. *Twins: A study of heredity and environment.* Chicago: University of Chicago Press.

Paterson, A., Deverna, J., Lanini, B., and Tanksley, S. 1990. Fine mapping of quantitative traits loci using selected overlapping recombinant chromosomes in an interspecific cross of tomato. *Genetics* 124: 735–742.

Tanksley, S. D. 2004. The genetic, developmental and molecular bases of fruit size and shape variation in tomato. *Plant Cell* 16: S181–S189.

Zar, J. H. 1999. *Biostatistical analysis,* 4th ed. Upper Saddle River, NJ: Prentice Hall.

Chapter 23: Population and Evolutionary Genetics

Abzanhov, A., et al., 2004. Bmp4 and morphological variation of beaks in Darwin's finches. *Science* 305: 1462–1465.

Ansari-Lari, M. A., et al. 1997. The extent of genetic variation in the *CCR5* gene. *Nature Genetics* 16: 221–222.

Carrington, M., Kissner, T., et al. 1997. Novel alleles of the chemokine-receptor gene *CCR5*. *Am. J. Hum. Genet.* 61: 1261–1267.

Green, R. E., et al. 2006. Analysis of one million base pairs of Neanderthal DNA. *Nature* 444: 330–336.

Karn, M. N., and Penrose, L. S. 1951. Birth weight and gestation time in relation to maternal age, parity and infant survival. *Ann. Eugen.* 16: 147–164.

Kerr, W. E., and Wright, S. 1954. Experimental studies of the distribution of gene frequencies in very small populations of *Drosophila melanogaster. Evolution* 8: 172–177.

Knowlton, N., et al., 1993. Divergence in proteins, mitochondrial DNA, and reproductive compatibility across the Isthmus of Panama. *Science* 260: 1629–1632.

Kreitman, M. 1983. Nucleotide polymorphism at the alcohol dehydrogenase locus of *Drosophila melanogaster. Nature* 304: 412–417.

Lamb, R. S., and Irish, V. F. 2003. Functional divergence within the *APETALA3/PISTILLATA* floral homeotic gene lineages. *Proc. Natl. Acad. Sci.* 100: 6558–6563.

Leibert, F., et al. 1998. The *DCCR5* mutation conferring protection against HIV-1 in Caucasian populations has a single and recent origin in northeastern Europe. *Hum. Molec. Genet.* 7: 399–406.

Markow, T., et al. 1993. HLA polymorphism in the Havasupai: Evidence for balancing selection. *Am. J. Hum. Genet.* 53: 943–952.

Nedelkov, D., et al., 2005. Investigating diversity in human plasma proteins. *Proc. Nat. Acad. Sci.* 102: 10852–10857.

Noonan, J. P., et al. 2006. Sequencing and analysis of Neanderthal genomic DNA. *Science* 314: 111–1118.

Reznick, D. N., and Ricklefs, R. E., 2009. Darwin's bridge between microevolution and macroevolution. *Nature* 457: 837–842.

Riesenberg, L. H., and Willis, J. H. 2007. Plant speciation. *Science* 317: 910–914.

Salzburger, W. 2009. The interaction of sexually and naturally selected traits in the adaptive radiations of cichlid fishes. *Mol. Evol.* 18: 169–185.

Schulter, D. 2009. Evidence for ecological speciation and its alternative. *Science* 323: 737–741.

Stiassny, M. L. J., and Meyer, A. 1999. Cichlids of the Rift Lakes. *Sci. Am.* (Feb.) 280: 64–69.

Takahashi, K., 1998. A novel family of short interspersed repetitive elements (SINES) from cichlids: The pattern of insertion of SINES at orthologous loci support the proposed monophyly of four major groups of cichlid fishes in Lake Tanganyika. *Mol. Biol. Evol.* 15: 391–407.

Yi, Z., et al., 2003. A 122.5 kilobase deletion of the *P* gene underlies the high prevalence of oculocutaneous albinism type 2 in the Navajo population. *Am. J. Hum. Genet.* 72: 62–72.

Chapter 24: Conservation Genetics

Baker, C. S., and Palumbi, S. R. 1994. Which whales are hunted? A molecular genetic approach to monitoring whaling. *Science* 265: 1538–1539.

Daniels, S. J., and Walters, J. R. 2000. Inbreeding depression and its effects on natal dispersal in red-cockaded woodpeckers. *Condor* 102: 482–491.

Dobson, A., and Lyles, A. 2000. Black-footed ferret recovery. *Science* 288: 985.

Frankham, R. 1995. Conservation genetics. *Ann. Rev. Genet.* 29: 305–327.

Friar, E. A., et al. 2001. Population structure in the endangered Mauna Loa silversword, *Argyroxiphium kauense,* and its bearing on reintroduction. *Molecular Ecology* 10: 1657–1663.

Gharrett, A. J., and Smoker, W. W. 1991. Two generations of hybrids between even-year and odd-year pink salmon (*Oncorhynchus gorbuscha*): A test for outbreeding depression? *Canadian J. Fish. Aquat. Sci.* 48: 426–438.

Hedrick, P. W. 2001. Conservation genetics: Where are we now? *Trends Ecol. Evol.* 16: 629–636.

Lacy, R. C. 1997. Importance of genetic variation to the viability of mammalian populations. *J. Mammalogy* 78: 320–335.

Moore, M. K., et al. 2003. Use of restriction fragment length polymorphisms to identify sea turtle eggs and cooked meats to species. *Conserv. Genet.* 4: 95–103.

Paetkau, D., et al. 1998. Variation in genetic diversity across the range of North American brown bears. *Conserv. Biol.* 12: 418–429.

Ralls, K., et al. 2000. Genetic management of chondrodystrophy in California condors. *Animal Conserv.* 3: 145–153.

Roman, J., and Bowen, B. W. 2000. The mock turtle syndrome: Genetic identification of turtle meat purchased in the southeastern United States of America. *Animal Conserv.* 3: 61–65.

Roman, J., and Palumbi, S. R. 2003. Whales before whaling in the North Atlantic. *Science* 301: 508–510.

Wayne, R. K., et al. 1991. Conservation genetics of the endangered Isle Royale gray wolf. *Conserv. Biol.* 5: 41–51.

Wilson, E. O., ed. 1988. *Biodiversity.* Washington, DC: National Academy of Sciences.

Wynen, L. P., et al. 2000. Postsealing genetic variation and population structure of two species of fur seal. *Mol. Ecol.* 9: 299–314.

Glossary

accession number An identifying number or code assigned to a nucleotide or amino acid sequence for entry and cataloging in a database.

acrocentric chromosome Chromosome with the centromere located very close to one end. *Human* chromosomes 13, 14, 15, 21, and 22 are acrocentric.

active site The substrate-binding site of an enzyme; in non-enzymatic molecules, the site directly imparting function to the molecule.

adaptation A heritable component of the phenotype that confers an advantage in survival and reproductive success. The process by which organisms adapt to current environmental conditions.

additive genes See *polygenic inheritance*.

albinism A condition caused by the lack of melanin production in the iris, hair, and skin. In humans, it is most often inherited as an autosomal recessive trait.

allele Any one of alternate forms of the same gene, often distinguished by alternate phenotypic expressions.

allele-specific oligonucleotide (ASO) Synthetic nucleotides, usually 15–20 bp in length, that under carefully controlled conditions will hybridize only to a perfectly matching complementary sequence.

allelic exclusion In a plasma cell heterozygous for an immunoglobulin gene, the selective action of only one allele.

allelism test See *complementation test*.

allopatric speciation Process of speciation associated with geographic isolation.

allopolyploid Polyploid condition formed by the union of two or more distinct chromosome sets with a subsequent doubling of chromosome number.

allosteric effect Conformational change in the active site of a protein brought about by interaction with an effector molecule.

allotetraploid An allopolyploid containing two genomes derived from different species.

allozyme An allelic form of a protein that can be distinguished from other forms by electrophoresis.

alternative splicing Generation of different protein molecules from the same pre-mRNA by incorporation of a different set and order of exons into the mRNA product.

***Alu* sequence** A DNA sequence of approximately 300 bp found interspersed within the genomes of primates that is cleaved by the restriction enzyme *Alu*I. In humans, 300,000–600,000 copies are dispersed throughout the genome and constitute some 3–6 percent of the genome. See *short interspersed elements*.

Ames test A bacterial assay developed by Bruce Ames to detect mutagenic compounds; it assesses reversion to histidine independence in the bacterium *Salmonella typhimurium*.

amino acids Aminocarboxylic acids comprising the subunits that are covalently linked to form proteins.

aminoacyl tRNA Covalently linked combination of an amino acid and a tRNA molecule. Also referred to as a charged tRNA.

amniocentesis A procedure in which fluid and fetal cells are withdrawn from the amniotic layer surrounding the fetus; used to test for fetal defects.

amphidiploid Same as *allotetraploid*.

analog A chemical compound that differs structurally from a similar compound and whose chemical behavior is the same. Used experimentally to provide improved detection during analysis. See *base analog*.

anaphase Stage of mitosis or meiosis in which chromosomes begin moving to opposite poles of the cell.

aneuploidy A condition in which the chromosome number is not an exact multiple of the haploid set.

annotation Analysis of genomic nucleotide sequence data to identify the protein-coding genes, the nonprotein-coding genes, and the regulatory sequences and function(s) of each gene.

antibody Protein (immunoglobulin) produced in response to an antigenic stimulus with the capacity to bind specifically to an antigen.

anticodon In a tRNA molecule, the nucleotide triplet that binds to its complementary codon triplet in an mRNA molecule.

antigen A molecule, often a cell-surface protein, that is capable of eliciting the formation of antibodies.

antisense RNA An RNA molecule (synthesized *in vivo* or synthetic) with a ribonucleotide sequence that is complementary to part of an mRNA molecule.

apoptosis A genetically controlled program of cell death, activated as part of normal development or as a result of cell damage.

artificial selection See *selection*.

ascospore A meiotic spore produced in certain fungi, formed within an ascus.

assortative mating Nonrandom mating between males and females of a species. Positive assortative mating selects mates with the same genotype; negative selects mates with opposite genotypes.

attached-X chromosome Two conjoined X chromosomes that share a single centromere and thus migrate together during cell division.

attenuator A nucleotide sequence between the promoter and the structural gene of some bacterial operons that regulates the transit of RNA polymerase, reducing transcription of the related structural gene.

autogamy A process of self-fertilization resulting in homozygosis.

autonomously replicating sequences (ARS) Origins of replication, about 100 nucleotides in length, found in yeast chromosomes. ARS elements are also present in organelle DNA.

autopolyploidy Polyploid condition resulting from the duplication of one diploid set of chromosomes.

autoradiography Production of a photographic image by radioactive decay. Used to localize radioactively labeled compounds within cells and tissues or to identify radioactive probes in various blotting techniques. See *Southern blotting*.

autosomes Chromosomes other than the sex chromosomes. In humans, there are 22 pairs of autosomes.

auxotroph A mutant microorganism or cell line that requires a nutritional substance for growth that can be synthesized, and is not required by the wild-type strain.

backcross A cross between an F_1 heterozygote and one of the P_1 parents (or an organism with a genotype identical to one of the parents).

bacteriophage A virus that infects bacteria, using it as the host for reproduction (also, *phage*).

balanced lethals Recessive, nonallelic lethal genes, each carried on different homologous chromosomes. When organisms carrying balanced lethal genes are interbred, only organisms with genotypes identical to the parents (heterozygotes) survive.

balanced polymorphism Genetic polymorphism maintained in a population by natural selection.

Barr body Densely staining DNA-positive mass seen in the somatic nuclei of mammalian females. Discovered by Murray Barr, this body represents an inactivated X chromosome.

base analog A purine or pyrimidine base that differs structurally from one normally used in biological systems whose chemical behavior is the same.

base pair See *nucleotide pair*.

base substitution A single base change in a DNA molecule that produces a mutation. There are two types of substitutions: *transitions,* in which a purine is substituted for a purine, or a pyrimidine for a pyrimidine; and *transversions,* in which a purine is substituted for a pyrimidine or vice versa.

B-DNA The conformation of DNA which is most often found in cells and serves as the basis of the Watson-Crick double-helical model.

β-galactosidase A bacterial enzyme, encoded by the *lacZ* gene, that converts lactose into galactose and glucose.

bidirectional replication A mechanism of DNA replication in which two replication forks move in opposite directions from a common origin.

bioinformatics The design and application of software and computational methods for the storage, analysis, and management of biological information such as nucleotide or amino acid sequences.

biotechnology Commercial and/or industrial processes that utilize biological organisms or products.

bivalents Synapsed homologous chromosomes in the first prophase of meiosis.

BLAST (Basic Local Alignment Search Tool) Any of a family of search engines designed to compare or query nucleotide or amino acid sequences with sequences in databases. BLAST also calculates the statistical significance of the matches.

Bombay phenotype A rare variant of the ABO antigen system in which affected individuals do not have A or B antigens and thus appear to have blood type O, even though their genotype may carry unexpressed alleles for the A and/or B antigens.

bottleneck See *population bottleneck*.

bovine spongiform encephalopathy (BSE) A fatal, degenerative brain disease of cattle (transmissible to humans and other animals) caused by prion infection. Also known as mad cow disease.

broad heritability That proportion of total phenotypic variance in a population that can be attributed to genotypic variance.

CAAT box A highly conserved DNA sequence found in the untranslated promoter region of eukaryotic genes. This sequence is recognized by transcription factors.

CAP Catabolite activator protein; a protein that binds cAMP and regulates the activation of inducible operons.

carcinogen A physical or chemical agent that causes cancer.

carrier An individual heterozygous for a recessive trait.

catabolite activator protein See *CAP*.

catabolite repression The selective inactivation of an operon by a metabolic product of the enzymes encoded by the operon.

***cdc* mutations** A class of cell division cycle mutations in yeasts that affect the timing and progression through the cell cycle.

cDNA (complementary DNA) DNA synthesized from an RNA template by the enzyme reverse transcriptase.

cDNA library A collection of cloned cDNA sequences.

cell cycle The sequence of growth phases of an individual cell; divided into G1 (gap 1), S (DNA synthesis), G2 (gap 2), and M (mitosis). A cell withdrawn from the cell cycle is said to enter the G0 stage.

CEN The DNA region of centromeres critical to their function. In yeasts, fragments of chromosomal DNA, about 120 bp in length, that when inserted into plasmids confer the ability to segregate during mitosis.

centimorgan (cM) A unit of distance between genes on chromosomes representing 1 percent crossing over between two genes. Equivalent to 1 map unit (mu).

central dogma The concept that genetic information flow progresses from DNA to RNA to proteins. Although exceptions are known, this idea is central to an understanding of gene function.

centric fusion See *Robertsonian translocation*.

centriole A cytoplasmic organelle composed of nine groups of microtubules, generally arranged in triplets. Centrioles function in the generation of cilia and flagella and serve as foci for the spindles in cell division.

centromere The specialized heterochromatic chromosomal region at which sister chromatids remain attached after replication, and the site to which spindle fibers attach to the chromosome during cell division. Also known as the primary constriction.

centrosome Region of the cytoplasm containing a pair of centrioles.

chaperone A protein that facilitates the correct folding of a polypeptide into a functional three-dimensional shape.

character An observable phenotypic attribute of an organism, such as seed color in plants.

chemotaxis Movement of a cell or an organism in response to a chemical gradient.

chiasma (pl., chiasmata) The crossed strands of nonsister chromatids seen in diplotene of the first meiotic division. Regarded as the cytological evidence for exchange of chromosomal material, or crossing over.

ChIP-on-chip A technique that combines chromatin immunoprecipitation (ChIP) with microarrays (chips) to identify and localize DNA sites that bind DNA-binding proteins of interest. These sites may be enhancers, promoters, transcription initiation sites, or other functional DNA regions.

chi-square (χ^2) analysis Statistical test to test the null hypothesis that an observed set of data is equivalent to a theoretical expectation.

chloroplast A self-replicating cytoplasmic organelle containing chlorophyll. The site of photosynthesis.

chorionic villus sampling (CVS) A technique of prenatal diagnosis in which chorionic fetal cells are retrieved intravaginally and used to detect cytogenetic and biochemical defects in the embryo.

chromatid One of the longitudinal subunits of a replicated chromosome.

chromatin The complex of DNA, RNA, histones, and nonhistone proteins that make up uncoiled chromosomes, characteristic of the eukaryotic interphase nucleus.

chromatin remodeling A process in which the structure of chromatin is altered by a protein complex, resulting in changes in the transcriptional state of genes in the altered region.

chromatography Technique for the separation of a mixture of solubilized molecules by their differential migration over a substrate.

chromocenter In polytene chromosomes, an aggregation of centromeres and heterochromatic elements where the chromosomes appear to be attached together.

chromomere A coiled, beadlike region of a chromosome, most easily visualized during cell division. The aligned chromomeres of polytene chromosomes are responsible for their distinctive banding pattern.

chromosomal aberration Any duplication, deletion, or rearrangement of the otherwise diploid chromosomal content of an organism, also called a chromosomal mutation.

chromosomal polymorphism Possession of alternative structures or arrangements of a chromosome among members of a population.

chromosome In prokaryotes, a DNA molecule containing the organism's genome; in eukaryotes, a DNA molecule complexed with RNA and proteins to form a threadlike structure containing genetic information arranged in a linear sequence and visible during mitosis and meiosis.

chromosome banding Technique for the differential staining of mitotic or meiotic chromosomes to produce a characteristic banding pattern; or selective staining of certain chromosomal regions such as centromeres, the nucleolus organizer regions, and GC- or AT-rich regions.

chromosome map A diagram showing the location of genes on chromosomes.

chromosome puff A localized uncoiling and swelling in a polytene chromosome, usually regarded as a sign of active transcription.

chromosome theory of inheritance The idea put forward independently by Walter Sutton and Theodor Boveri that chromosomes are the carriers of genes and the basis for the Mendelian mechanisms of segregation and independent assortment.

chromosome walking A method for analyzing long stretches of DNA. The end of a cloned segment of DNA is subcloned and used as a probe to identify other clones that overlap the first clone.

***cis*-acting element** A DNA sequence that regulates the expression of a gene located on the same chromosome. This contrasts with a *trans*-acting element where regulation is under the control of a sequence on the same homolog.

***cis* configuration** The arrangement of two wild-type or mutant genes (or two mutational sites within a single gene) present on the same homolog, such as

$$\frac{a^1 \quad a^2}{+ \quad +}$$

This contrasts with the *trans* configuration, where the two genes or alleles are located on opposite homologs.

***cis–trans* test** A genetic test to determine whether two mutations are located within the same cistron (or gene).

cistron That portion of a DNA molecule, often referring to a gene, coding for a single-polypeptide chain.

cline A gradient of genotype or phenotype distributed over a geographic range.

clone Identical molecules, cells, or organisms derived from a single ancestor by asexual or parasexual methods; for example, a DNA segment that has been inserted into a plasmid or chromosome of a phage or a bacterium and replicated to produce many copies, or an organism of identical genetic composition to that used in its production.

codominance Condition in which the phenotypic effects of a gene's alleles are fully and simultaneously expressed in the heterozygote.

codon A triplet of nucleotides that specifies a particular amino acid or a start or stop signal in the genetic code. Sixty-one codons specify the amino acids used in proteins, and three codons, called stop codons, signal termination of growth of the polypeptide chain.

coefficient of coincidence A ratio of the observed number of double crossovers divided by the expected number of such crossovers.

coefficient of inbreeding The probability that two alleles present in a zygote are descended from a common ancestor.

coefficient of selection (*s*) A measurement of the reproductive disadvantage of a given genotype in a population. For example, if for genotype *aa* only 99 of 100 individuals reproduce, then the selection coefficient is 0.01.

colinearity The linear relationship between the nucleotide sequence in a gene (or the RNA transcribed from it) and the order of amino acids in the polypeptide chain specified by the gene.

competence In bacteria, the transient state or condition during which the cell can bind and internalize exogenous DNA molecules, making transformation possible.

complementation test A genetic test to determine whether or not two mutations occur within the same gene (or cistron).

complete linkage A condition in which two genes are located so close to each other that no recombination occurs between them.

complex trait A trait whose phenotype is determined by the interaction of multiple genes and environmental factors.

concordance Pairs or groups of individuals identical in their phenotype. In twin studies, a condition in which both twins exhibit or fail to exhibit a trait under investigation.

conditional mutation A mutation that is expressed only under a certain condition; that is, a wild-type phenotype is expressed under certain (permissive) conditions and a mutant phenotype under other (restrictive) conditions.

conjugation Temporary fusion of two single-celled organisms for the sexual transfer of genetic material.

consanguineous Related by a common ancestor within the previous few generations.

consensus sequence The sequence of nucleotides in DNA or amino acids in proteins most often present in a particular gene or protein under study in a group of organisms.

contig A continuous DNA sequence reconstructed from overlapping DNA sequences derived by cloning or sequence analysis.

continuous variation Phenotype variation in which quantitative traits range from one phenotypic extreme to another in an overlapping or continuous fashion.

cosmid A vector designed to allow cloning of large segments of foreign DNA. Cosmids are composed of the *cos* sites of phage λ inserted into a plasmid. In cloning, the recombinant DNA molecules are packaged into phage protein coats, and after infection of bacterial cells, the recombinant molecule replicates and can be maintained as a plasmid.

covalent bond A nonionic chemical bond, formed by the sharing of electrons.

CpG island A short region of regulatory DNA found upstream of genes and which contain unmethylated stretches of sequence with a high frequency of C and G nucleotides.

Creutzfeldt–Jakob disease (CJD). See *prion.*

cri du chat syndrome A clinical syndrome produced by a deletion of a portion of the short arm of chromosome 5 in humans. Afflicted infants have a distinctive cry that sounds like a cat.

crossing over The exchange of chromosomal material (parts of chromosomal arms) between homologous chromosomes by breakage and reunion.

C-terminal amino acid In a peptide chain, the terminal amino acid that carries a free carboxyl group. It constitutes the C terminus end of the polypeptide chain.

***C* value** The haploid amount of DNA present in a genome.

***C* value paradox** The paradox that there is no apparent relationship between the increased size of the genome and the evolutionary complexity of species.

cyclic adenosine monophosphate (cAMP) An important regulatory molecule in both prokaryotic and eukaryotic organisms.

cyclins In eukaryotic cells, a class of proteins that are synthesized and degraded in synchrony with the cell cycle and regulate passage through stages of the cycle.

cytogenetics A branch of biology in which the techniques of both cytology and genetics are used in genetic investigations.

cytokinesis The division or separation of the cytoplasm during mitosis or meiosis.

cytoplasmic inheritance See *extranuclear inheritance.*

cytoskeleton An internal array of microtubules, microfilaments, and intermediate filaments that confers shape and the ability to move to a eukaryotic cell.

dalton (Da) A unit of mass equal to that of the hydrogen atom, which is 1.67×10^{-24} gram. A unit used in designating molecular weights.

degenerate code The fact that a given amino acid may be represented by more than one codon.

deletion A chromosomal mutation, also referred to as a deficiency, involving the loss of chromosomal material.

de novo Newly arising; synthesized, or created from less complex precursors.

deoxyribonuclease (DNase) A class of enzyme that breaks down DNA into oligonucleotide fragments by introducing single-stranded or double-stranded breaks into the double helix.

deoxyribonucleic acid (DNA) A double-helical molecule consisting of linear polymers of deoxyribonucleotides held together by hydrogen bonds. The primary carrier of genetic information. See also *double helix.*

deoxyribose The five-carbon sugar associated with the deoxyribonucleotides in DNA, characterized by the absence of a 2'-hydroxyl group.

dermatoglyphics The study of the surface ridges of the skin, especially of the hands and feet.

determination Establishment of a specific pattern of future gene activity and developmental fate for a given cell, usually prior to any manifestation of the cell's future phenotype.

diakinesis The final stage of meiotic prophase I, in which the chromosomes become tightly coiled and compacted and move toward the periphery of the nucleus.

dicentric chromosome A chromosome having two centromeres, which is pulled in opposite directions during anaphase of cell division.

dicer An enzyme (a ribonuclease) that cleaves double-stranded RNA (dsRNA) and pre-micro RNA (miRNA) to form small interfering RNA (siRNA) molecules about 20–25 nucleotides long that serve as guide molecules for the degradation of mRNA molecules with sequences complementary to the siRNA.

dideoxynucleotide A nucleotide containing a deoxyribose sugar lacking a 3′ hydroxyl group. It stops further chain elongation when incorporated into a growing polynucleotide and is used in the Sanger method of DNA sequencing.

differentiation The complex process of change by which cells and tissues attain their adult structure and functional capacity, characterized by differential gene activity.

dihybrid cross A genetic cross involving two characters in which the parents possess different forms (traits) of each character (e.g., yellow, round × green, wrinkled peas).

diploid (2*n*) A condition in which each chromosome exists in pairs; having two of each chromosome.

diplotene The stage of meiotic prophase I immediately following pachytene. In diplotene, the sister chromatids begin to separate, and chiasmata become visible. These cross-like overlaps move toward the ends of the chromatids (terminalization).

directed mutagenesis See *gene targeting.*

directional selection A selective force that changes the frequency of an allele in a given direction, either toward fixation or toward elimination.

discontinuous variation Pattern of variation for a trait whose phenotypes fall into two or more distinct classes.

discordance In twin studies, a situation where one twin expresses a trait but the other does not.

disjunction The separation of chromosomes during the anaphase stage of cell division.

disruptive selection Simultaneous selection for phenotypic extremes in a population, usually resulting in the production of two phenotypically discontinuous strains.

dizygotic twins Twins produced from separate fertilization events; two ova fertilized independently. Also known as fraternal twins.

DNA See *deoxyribonucleic acid.*

DNA fingerprinting A molecular method for identifying an individual member of a population or species. A unique pattern of DNA fragments is obtained by restriction enzyme digestion followed by Southern blot hybridization using minisatellite probes. See also *DNA profiling; STR sequences.*

DNA gyrase One of a class of enzymes known as topoisomerases, which during DNA replication reduces molecular tension caused by supercoiling.

DNA helicase An enzyme that participates in DNA replication by unwinding the double helix near the replication fork.

DNA ligase An enzyme that forms a covalent bond between the 5′ end of one polynucleotide chain and the 3′ end of another polynucleotide chain. It is also called polynucleotide-joining enzyme.

DNA microarray An ordered arrangement of DNA sequences or oligonucleotides on a substrate (often glass). Microarrays are used in quantitative assays of DNA-DNA or DNA-RNA binding to measure profiles of gene expression (for example, during development or to compare the differences in gene expression between normal and cancer cells).

DNA polymerase An enzyme that catalyzes the synthesis of DNA from deoxyribonucleotides utilizing a template DNA molecule.

DNA profiling A method for identification of individuals that uses variations in the length of short tandem repeating DNA sequences (STRs), which are widely distributed in the genome.

DNase See also *deoxyribonuclease, endonuclease, exonuclease, restriction endonuclease.*

dominance The expression of a trait in the heterozygous condition.

dosage compensation A genetic mechanism that equalizes the levels of expression of genes at loci on the X chromosome. In mammals, this is accomplished by random inactivation of one X chromosome, leading to Barr body formation.

double helix The model for DNA structure proposed by Watson and Crick, in which two antiparallel hydrogen-bonded polynucleotide chains are wound into a right-handed helical configuration 2 nm in diameter, with 10 base pairs per full turn.

Duchenne muscular dystrophy An X-linked recessive genetic disorder caused by a mutation in the gene for dystrophin, a protein found in muscle cells. Affected males show a progressive weakness and wasting of muscle tissue. Death ensues by about age 20 caused by respiratory infection or cardiac failure.

duplication A chromosomal aberration in which a segment of the chromosome is repeated.

dyad The products of tetrad separation or disjunction at meiotic prophase I. Each dyad consists of two sister chromatids joined at the centromere.

electrophoresis A technique that separates a mixture of molecules by their differential migration through a stationary medium (such as a gel) under the influence of an electrical field.

ELSI (Ethical, Legal, Social Implications) A program established by the National Human Genome Research Institute in 1990 as part of the Human Genome Project to sponsor research on the ethical, legal, and social implications of genomic research and its impact on individuals and social institutions.

embryonic stem cells Cells derived from the inner cell mass of early blastocyst mammalian embryos. These cells are pluripotent, meaning they can differentiate into any of the embryonic or adult cell types characteristic of the organism.

endonuclease An enzyme that hydrolyzes internal phosphodiester bonds in a single- or double-stranded polynucleotide chain.

endoplasmic reticulum (ER) A membranous organelle system in the cytoplasm of eukaryotic cells. In rough ER, the outer surface of the membranes is ribosome-studded; in smooth ER, it is not.

endopolyploidy The increase in chromosome sets within somatic nuclei that results from endomitotic replication.

endosymbiotic theory The proposal that self-replicating cellular organelles such as mitochondria and chloroplasts were originally free-living organisms that entered into a symbiotic relationship with nucleated cells.

enhancer A DNA sequence that enhances the transcription of nearby structural genes. Enhancers can act over a distance of thousands of base pairs and can be located upstream, downstream, or internal to the gene they affect, differentiating them from promoters.

enzyme A protein or complex of proteins that catalyzes a specific biochemical reaction by lowering the energy of activation that is normally required to initiate the reaction.

epigenesis The idea that an organism or organ arises through the sequential appearance and development of new structures, in contrast to preformationism, which holds that development is the result of the assembly of structures already present in the egg.

epigenetics The study of modifications in an organism's gene function or phenotypic expression that are not attributable to alterations in the nucleotide sequence (gene mutation) of the organism's DNA.

episome In bacterial cells, a circular genetic element that can replicate independently of the bacterial chromosome or integrate and replicate as part of the chromosome.

epistasis Nonreciprocal interaction between nonallelic genes such that one gene influences or interferes with the expression of another gene, leading to a specific phenotype.

equational division A division stage where the number of centromeres is not reduced by half but where each chromosome is split into longitudinal halves that are distributed into two daughter nuclei. Chromosome division in mitosis and the second meiotic division are examples of equational divisions. See also *reductional division.*

equatorial plate See *metaphase plate.*

euchromatin True chromatin, or chromosomal regions that are relatively uncoiled during the interphase portion of the cell cycle. Euchromatic regions contain most of the structural genes.

eugenics A movement advocating the improvement of the human species by selective breeding. Positive eugenics refers to the promotion of breeding between people thought to possess favorable genes, and negative eugenics refers to the discouragement of breeding among those thought to have undesirable traits.

eukaryotes Organisms having true nuclei and membranous organelles and whose cells demonstrate mitosis and meiosis.

euphenics Medical or genetic intervention to reduce the impact of defective genotypes.

euploid Polyploid with a chromosome number that is an exact multiple of a basic chromosome set.

evolution Descent with modification. The emergence of new kinds of plants and animals from preexisting types.

excision repair Removal of damaged DNA segments followed by repair. Excision can include the removal of individual bases (base repair) or of a stretch of damaged nucleotides (nucleotide repair).

exon The DNA segments of a gene that contain the sequences that, through transcription and translation, are eventually represented in the final polypeptide product.

exonuclease An enzyme that breaks down nucleic acid molecules by breaking the phosphodiester bonds at the 3′- or 5′-terminal nucleotides.

expressed sequence tag (EST) All or part of the nucleotide sequence of a cDNA clone. ESTs are used as markers in the construction of genetic maps.

expression vector Plasmids or phages carrying promoter regions designed to cause expression of inserted DNA sequences.

expressivity The degree or range in which a phenotype for a given trait is expressed.

extranuclear inheritance Transmission of traits by genetic information contained in cytoplasmic organelles such as mitochondria and chloroplasts.

F^- cell A bacterial cell that does not contain a fertility factor and that acts as a recipient in bacterial conjugation.

F^+ cell A bacterial cell that contains a fertility factor and that acts as a donor in bacterial conjugation.

F factor An episomal plasmid in bacterial cells that confers the ability to act as a donor in conjugation (also called *fertility factor*).

F′ factor A fertility factor that contains a portion of the bacterial chromosome.

F_1 generation First filial generation; the progeny resulting from the first cross in a series.

F_2 generation Second filial generation; the progeny resulting from a cross of the F_1 generation.

F pilus On bacterial cells possessing an F factor, a filament-like projection that plays a role in conjugation.

fate map A diagram of an embryo showing the location of cells whose developmental fate is known.

fertility factor See *F factor.*

filial generations See *F_1, F_2 generations.*

fingerprint The unique pattern of ridges and whorls on the tip of a human finger. Also, the pattern obtained by enzymatically cleaving a protein or nucleic acid and subjecting the digest to two-dimensional chromatography or electrophoresis. See also *DNA fingerprinting.*

FISH See *fluorescence* in situ *hybridization.*

fitness A measure of the relative survival and reproductive success of a given individual or genotype.

fluctuation test A statistical test developed by Salvadore Luria and Max Delbrück that demonstrated that bacterial mutations arise spontaneously, in contrast to being induced by selective agents.

fluorescence *in situ* hybridization (FISH) A method of *in situ hybridization* that utilizes probes labeled with a fluorescent tag, causing the site of hybridization to fluoresce when viewed using ultraviolet light.

f-met See *formylmethionine.*

folded-fiber model A model of eukaryotic chromosome organization in which each sister chromatid consists of a single chromatin fiber composed of double-stranded DNA and proteins wound like a tightly coiled skein of yarn.

formylmethionine (f-met) A molecule derived from the amino acid methionine by attachment of a formyl group to its terminal amino group. This is the first monomer used in the synthesis of all bacterial polypeptides. Also known as *N*-formyl methionine.

forward genetics The classical approach used to identify a gene controlling a phenotypic trait in the absence of knowledge of the gene's location in the genome or its DNA sequence. Accomplished by isolating mutant alleles and mapping the gene's location, most traditionally using recombination analysis. Once mapped, the gene may be cloned and further studied at the molecular level. An approach contrasted with *reverse genetics.*

founder effect A form of genetic drift. The establishment of a population by a small number of individuals whose genotypes carry only a fraction of the different kinds of alleles in the parental population.

fragile site A heritable gap, or nonstaining region, of a chromosome that can be induced to generate chromosome breaks.

fragile X syndrome A human genetic disorder caused by the expansion of a CGG trinucleotide repeat and a fragile site at Xq27.3 within the *FMR-1* gene. Fragile X syndrome is the most common form of mental retardation.

frameshift mutation A mutational event leading to the insertion of one or more base pairs in a gene, shifting the codon reading frame in all codons that follow the mutational site.

fraternal twins Same as *dizygotic twins*.

gain-of-function mutation A mutation that produces a phenotype different from that of the normal allele and from any loss of function alleles.

gamete A specialized reproductive cell with a haploid number of chromosomes.

gap genes Genes expressed in contiguous domains along the anterior–posterior axis of the *Drosophila* embryo that regulate the process of segmentation in each domain.

gene The fundamental physical unit of heredity, whose existence can be confirmed by allelic variants and which occupies a specific chromosomal locus. A DNA sequence coding for a polypeptide.

gene amplification The process by which gene sequences are selected and differentially replicated either extrachromosomally or intrachromosomally.

gene chip See *DNA microarray.*

gene conversion The process of nonreciprocal recombination by which one allele in a heterozygote is converted into the corresponding allele.

gene family A number of closely related genes derived from a common ancestral gene by duplication and divergence over evolutionary time.

gene flow The gradual exchange of genes between two populations; brought about by the dispersal of gametes or the migration of individuals.

gene interaction Production of novel phenotypes by the interaction of products of alleles of different genes.

gene knockout The introduction of a *null mutation* into a gene that is subsequently introduced into an organism using transgenic techniques, whereby the organism loses the function of the gene. Often used in mice. See also *gene targeting*.

gene mutation See *point mutation*.

gene pool The total of all alleles possessed by all reproductive members of a population.

gene targeting A transgenic technique used to create and introduce a specifically altered gene into an organism.

generalized transduction The process of transduction whereby any gene in the bacterial genome is recombined in a process mediated by a bacteriophage.

genetically modified organism (GMO) A plant or animal whose genome carries a gene transferred from another species by recombinant DNA technology and expressed to produce a gene product.

genetic anticipation The phenomenon in which the severity of symptoms in genetic disorders increases and the age of onset decreases from generation to generation. It is caused by the expansion of trinucleotide repeats within or near a gene and was first observed in myotonic dystrophy.

genetic background The collective genome of an organism, as it impacts on the expression of a gene under investigation.

genetic code The deoxynucleotide triplets that encode the 20 amino acids or specify termination of translation.

genetic counseling Analysis of risk for genetic defects in a family and the presentation of options available to avoid or ameliorate possible risks.

genetic drift Random variation in allele frequency from generation to generation, most often observed in small populations.

genetic engineering The technique of altering the genetic constitution of cells or individuals by the selective removal, insertion, or modification of individual genes or gene sets.

genetic equilibrium A condition relevant to population genetic studies in which allele frequencies are neither increasing nor decreasing.

genetic load Average number of recessive lethal genes carried in the heterozygous condition by an individual in a population.

genetic polymorphism The stable coexistence of two or more distinct genotypes for a given trait in a population. When the frequencies of two alleles for such a trait are in equilibrium, the condition is called balanced polymorphism.

genetics The branch of biology concerned with study of inherited variation. More specifically, the study of the origin, transmission, and expression of genetic information.

genome The haploid set of hereditary information encoded in the DNA of an organism, including both the protein-coding and non-protein-coding sequences.

genomic imprinting The process by which the expression of a gene depends on whether it has been inherited from a male or a female parent. Also referred to as parental imprinting.

genomic library A collection of clones that contains all the DNA sequences of an organism's genome.

genomics A subdiscipline of the field of genetics generated by the union of classical and molecular biology with the goal of sequencing and understanding genes, gene interaction, genetic elements, and the structure of genomes.

genotype The allelic or genetic constitution of an organism; often, the allelic composition of one or a limited number of genes under investigation.

germ line An embryonic cell lineage that forms the reproductive cells (eggs and sperm).

GMO See *genetically modified organism.*

Goldberg–Hogness box A short nucleotide sequence 20–30 bp upstream from the initiation site of eukaryotic genes to which RNA polymerase II binds. The consensus sequence is TATAAAA. Also known as a TATA box.

gyrase See *DNA gyrase*.

haploid (*n*) A cell or organism having one member of each pair of homologous chromosomes. Also, the gametic chromosome number.

haploinsufficiency In a diploid organism, a condition in which an individual possesses only one functional copy of a gene and the other is inactivated by mutation. The protein produced by the single copy is insufficient to produce a normal phenotype, leading to an abnormal phenotype. In humans, this condition is present in many autosomal dominant disorders.

haplotype A set of alleles from closely linked loci carried by an individual inherited as a unit.

HapMap Project An international effort by geneticists to identify haplotypes (closely linked genetic markers on a single chromosome) shared by certain individuals as a way of facilitating efforts to identify, map, and isolate genes associated with disease or disease susceptibility.

Hardy–Weinberg law The principle that genotype frequencies will remain in equilibrium in an infinitely large, randomly mating population in the absence of mutation, migration, and selection.

helicase See *DNA helicase*.

helix–turn–helix (HTH) motif In DNA-binding proteins, the structure of a region in which a turn of four amino acids holds two α helices at right angles to each other.

hemizygous Having a gene present in a single dose in an otherwise diploid cell. Usually applied to genes on the X chromosome in heterogametic males.

heredity Transmission of traits from one generation to another.

heritability A relative measure of the degree to which observed phenotypic differences for a trait are genetic.

heterochromatin The heavily staining, late-replicating regions of chromosomes that are prematurely condensed in interphase. Thought to be devoid of structural genes.

heteroduplex A double-stranded nucleic acid molecule in which each polynucleotide chain has a different origin. It may be produced as an intermediate in a recombinational event or by the *in vitro* reannealing of single-stranded, complementary molecules.

heterogametic sex The sex that produces gametes containing unlike sex chromosomes. In mammals, the male is the heterogametic sex.

heterokaryon A somatic cell containing nuclei from two different sources.

heterozygote An individual with different alleles at one or more loci. Such individuals will produce unlike gametes and therefore will not breed true.

Hfr Strains of bacteria exhibiting a high frequency of recombination. These strains have a chromosomally integrated F factor that is able to mobilize and transfer part of the chromosome to a recipient F^- cell.

histocompatibility antigens See *HLA*.

histone code Various chemical modifications applied to histone tails (the free ends of histone molecules, projecting from nucleosomes). These modifications influence DNA–histone interactions and promote or repress transcription.

histones Positively charged proteins complexed with DNA in the nucleus. They are rich in the basic amino acids arginine and lysine, and function in coiling DNA to form nucleosomes.

HLA Cell-surface proteins produced by histocompatibility loci and involved in the acceptance or rejection of tissue and organ grafts and transplants.

Holliday structure In DNA recombination, an intermediate seen in transmission electron microscope images as an X-shaped structure showing four single-stranded DNA regions.

homeobox A sequence of about 180 nucleotides that encodes a sequence of 60 amino acids called a *homeodomain,* which is part of a DNA-binding protein that acts as a transcription factor.

homeotic mutation A mutation that causes a tissue normally determined to form a specific organ or body part to alter its differentiation and form another structure.

homogametic sex The sex that produces gametes that do not differ with respect to sex chromosome content; in mammals, the female is homogametic.

homologous chromosomes Chromosomes that synapse or pair during meiosis and that are identical with respect to their genetic loci and centromere placement.

homozygote An individual with identical alleles for a gene or genes of interest. These individuals will produce identical gametes (with respect to the gene or genes in question) and will therefore breed true.

human immunodeficiency virus (HIV) An RNA-containing human retrovirus associated with the onset and progression of AIDS.

hybrid vigor The general superiority of a hybrid over a purebred for the traits' health and resilience.

hydrogen bond A weak electrostatic attraction between a hydrogen atom covalently bonded to an oxygen or nitrogen atom and an atom that contains an unshared electron pair.

hypervariable regions The regions of antibody molecules that attach to antigens. These regions have a high degree of diversity in amino acid content.

identical twins Same as *monozygotic twins*.

immunoglobulin (Ig) The class of serum proteins having the properties of antibodies.

imprinting See *genomic imprinting*.

inborn error of metabolism A genetically controlled biochemical disorder; usually an enzyme defect that produces a clinical syndrome.

inbreeding Mating between closely related organisms.

inbreeding depression A decrease in viability, vigor, or growth in progeny after several generations of inbreeding.

incomplete dominance Expressing a heterozygous phenotype that is distinct from the phenotype of either homozygous parent. Also called *partial dominance*.

independent assortment The independent behavior of each pair of homologous chromosomes during their segregation in meiosis I. The random distribution of maternal and paternal homologs into gametes.

inducer An effector molecule that activates transcription.

inducible enzyme system An enzyme system under the control of an inducer, a regulatory molecule that acts to block a repressor and allow transcription.

initiation codon The nucleotide triplet AUG that in an mRNA molecule codes for incorporation of the amino acid methionine as the first amino acid in a polypeptide chain.

insertion sequence See *IS element*.

***in situ* hybridization** A cytological molecular hybridization technique for pinpointing the chromosomal location of DNA sequences complementary to a given nucleic acid or polynucleotide.

interference (*I*) A measure of the degree to which one crossover affects the incidence of another crossover in an adjacent region of the same chromatid. Negative interference increases the chance of another crossover; positive interference reduces the probability of a second crossover event.

interphase In the cell cycle, the interval between divisions.

intron An intervening sequence of DNA that lies between coding regions in a gene. Introns are transcribed but are spliced out of the RNA product and are not represented in the polypeptide encoded by the gene.

inversion A chromosomal aberration in which a chromosomal segment has been reversed.

inversion loop The chromosomal configuration resulting from the synapsis of homologous chromosomes, one of which carries an inversion.

in vitro Literally, *in glass*; outside the living organism; occurring in an artificial environment.

in vivo Literally, *in the living*; occurring within the living body of an organism.

IS element A mobile DNA segment that is transposable to any of a number of sites in the genome.

isoagglutinogen An antigenic factor or substance present on the surface of cells that is capable of inducing the formation of an antibody (e.g., the A and B antigens on the surface of human red blood cells).

isolating mechanism Any barrier to the exchange of genes between different populations of a group of organisms. In general, isolation can be classified as spatial, environmental, or reproductive.

isozyme Any of two or more distinct forms of an enzyme that have identical or nearly identical chemical properties but differ in some property such as net electrical charge, pH optima, number and type of subunits, or substrate concentration.

κ particles DNA-containing cytoplasmic particles found in certain strains of *Paramecium aurelia* capable of releasing a toxin, paramecin, that kills other sensitive strains.

karyotype The chromosome complement of a cell or an individual. Often used to refer to the arrangement of metaphase chromosomes in a sequence according to length and centromere position.

kilobase (kb) A unit of length in DNA or RNA consisting of 1000 nucleotides.

kinetochore A fibrous structure with a size of about 400 nm, located within the centromere. It appears to be the site of microtubule attachment during division.

Klinefelter syndrome A genetic disorder in human males caused by the presence of one or more extra X chromosomes. Klinefelter males are usually XXY instead of XY. This syndrome includes enlarged breasts, small testes, sterility, and mild mental retardation.

knockout mice Mice created by a process in which a normal gene is cloned, inactivated by the insertion of a marker (such as an antibiotic resistance gene), and transferred to embryonic stem cells, where the altered gene will replace the normal gene (in some cells). These cells are injected into a blastomere embryo, producing a mouse that is then bred to yield mice homozygous for the mutated gene. See also *gene targeting*.

Kozak sequence A short nucleotide sequence adjacent to the initiation codon that is recognized as the translational start site in eukaryotic mRNA.

***lac* repressor protein** A protein that binds to the operator in the *lac* operon and blocks transcription.

lagging strand During DNA replication, the strand synthesized in a discontinuous fashion, in the direction opposite of the replication fork. See also *Okazaki fragment*.

lampbrush chromosomes Meiotic chromosomes characterized by extended lateral loops that reach maximum extension during diplotene. Although most intensively studied in amphibians, these structures also occur in vertebrate oocytes.

lariat structure A structure formed by an intron by means of a 5′ to 3′ bond during processing and removal of that intron from an mRNA molecule.

leader sequence That portion of an mRNA molecule from the 5′ end to the initiating codon, often containing regulatory or ribosome binding sites.

leading strand During DNA replication, the strand synthesized continuously in the direction of the replication fork.

leptotene The initial stage of meiotic prophase I, during which the chromosomes become visible and are often arranged with one or both ends gathered at one spot on the inner nuclear membrane (the so-called bouquet configuration).

lethal gene A gene whose expression results in death of the organism at some stage of its life cycle.

leucine zipper In DNA-binding proteins, a structural motif characterized by a stretch in which every seventh amino acid residue is leucine, with adjacent regions containing positively charged amino acids. Leucine zippers on two polypeptides may interact to form a dimer that binds to DNA.

linkage The condition in which genes have their loci present on the same chromosome, causing them to be inherited as a unit, provided that they are not separated by crossing over during meiosis. Also used to designate bacterial genes that are contiguous along the chromosome characterizing the species.

locus (pl., loci) The site or place on a chromosome where a particular gene is located.

lod score (LOD) A statistical method used to determine whether two loci are linked or unlinked. A lod (*l*og *o*f the o*d*ds) score of 4 indicates that linkage is 10,000 times more likely than nonlinkage. By convention, lod scores of 3–4 are interpreted as linkage.

long interspersed elements (LINES) Repetitive DNA sequences several thousand nucleotides long that are interspersed in the genomes of higher organisms.

long terminal repeat (LTR) A sequence of several hundred base pairs found at both ends of a retroviral DNA.

loss-of-function mutation Mutations that produce alleles with reduced or no function.

Lyon hypothesis The proposal describing the random inactivation of the maternal or paternal X chromosome in somatic cells of mammalian females early in development. All daughter cells will have the same X chromosome inactivated as in the cell they descended from, producing a mosaic pattern of expression of X chromosome genes.

lysis The disintegration of a cell brought about by the rupture of its membrane.

lysogenic bacterium A bacterial cell carrying the DNA of a temperate bacteriophage integrated into its chromosome.

lysogeny The process by which the DNA of an infecting phage becomes repressed and integrated into the chromosome of the bacterial cell it infects.

lytic phase The condition in which a bacteriophage invades, reproduces, and lyses the bacterial cell. Lysogenic bacteria may be induced to enter the lytic phase.

major histocompatibility (MHC) loci Loci encoding antigenic determinants responsible for tissue specificity. In humans, the HLA complex; and in mice, the H2 complex. See also *HLA*.

map unit A measure of the genetic distance between two genes, corresponding to a recombination frequency of 1 percent. See also *centimorgan*.

maternal effect Phenotypic effects in offspring attributable to genetic information transmitted through the oocyte derived from the maternal genome.

maternal inheritance The transmission of traits strictly through the maternal parent, usually due to DNA found in the cytoplasmic organelles, the mitochondria, or chloroplasts.

mean The arithmetic average.

meiosis The process of cell division in gametogenesis or sporogenesis during which the diploid number of chromosomes is reduced to the haploid number.

melting profile (T_m) The temperature at which a population of double-stranded nucleic acid molecules is one-half dissociated into single strands.

merozygote A partially diploid bacterial cell containing, in addition to its own chromosome, a chromosome fragment introduced into the cell by transformation, transduction, or conjugation.

messenger RNA (mRNA) An RNA molecule transcribed from DNA and translated into the amino acid sequence of a polypeptide.

metabolism The collective chemical processes by which living cells are created and maintained. In particular, the chemical basis of generating and utilizing energy in living organisms.

metacentric chromosome A chromosome with a centrally located centromere and therefore has chromosome arms of equal length.

metafemale In *Drosophila,* a poorly developed female of low viability with a ratio of X chromosomes to sets of autosomes exceeding 1.0. Previously called a *superfemale*.

metagenomics The study of DNA recovered from organisms obtained from the environment as opposed to laboratory cultures. Often used for estimating the diversity of organisms in an environmental sample.

metamale In *Drosophila,* a poorly developed male of low viability with a ratio of X chromosomes to sets of autosomes below 0.5. Previously called a *supermale*.

metaphase The stage of cell division in which condensed chromosomes lie in a central plane between the two poles of the cell and during which the chromosomes become attached to the spindle fibers.

metaphase plate The plane in which mitotic or meiotic chromosomes collect at the equator of the cell during metaphase.

metastasis The process by which cancer cells spread from the primary tumor to other parts of the body.

methylation Enzymatic transfer of methyl groups from S-adenosylmethionine to biological molecules, including phospholipids, proteins, RNA, and DNA. Methylation of DNA is associated with the regulation of gene expression and with epigenetic phenomena such as imprinting.

MHC See *major histocompatibility loci.*

Microarray See *DNA microarray.*

microRNA (miRNA) Single-stranded RNA molecules approximately 20–23 nucleotides in length that regulate gene expression by participating in the degradation of mRNA.

microsatellite A short, highly polymorphic DNA sequence of 1–4 base pairs, widely distributed in the genome, that are used as molecular markers in a variety of methods. Also called *simple sequence repeats* (*SSRs*).

minimal medium A medium containing only the essential nutrients needed to support the growth and reproduction of wild-type strains of an organism. Usually comprised of inorganic components that include a carbon and nitrogen source.

minisatellite Series of short tandem repeat sequences (STRs) 10–100 nucleotides in length that occur frequently throughout the genome of eukaryotes. Because the number of repeats at each locus is variable, the loci are known as variable number tandem repeats (VNTRs). Used in DNA fingerprinting. See also *VNTRs* and *STR sequences*.

mismatch repair A form of excision repair of DNA in which the repair mechanism is able to distinguish between the strand with the error and the strand that is correct.

missense mutation A mutation that alters a codon to that of another amino acid and thus results in an alteration in the translation product.

mitochondrial DNA (mtDNA) Double-stranded, self-replicating circular DNA found in mitochondria that encodes mitochondrial ribosomal RNAs, transfer RNAs, and proteins used in oxidative respiratory functions of the organelle.

mitogen A substance that stimulates mitosis in nondividing cells.

mitosis A form of cellular reproduction producing two progeny cells, identical genetically to the progenitor cell, that is, the production of two cells from one, each with the same chromosome complement as the parent cell.

mode In a set of data, the value occurring with the greatest frequency.

model genetic organism An experimental organism conducive to efficiently conducted research whose genetics is intensively studied on the premise that the findings can be applied to other organisms.

molecular clock In evolutionary studies, a method that counts the number of differences in DNA or protein sequences in order to measure the time elapsed since two species diverged from a common ancestor.

monohybrid cross A genetic cross involving only one character (e.g., $AA \times aa$).

monosomic An aneuploid condition in which one member of a chromosome pair is missing; having a chromosome number of $2n - 1$.

monozygotic twins Twins produced from a single fertilization event; the first division of the zygote produces two cells, each of which develops into an embryo. Also known as *identical twins*.

mRNA See *messenger RNA*.

mtDNA See *mitochondrial DNA*.

multigene family A gene set descended from a common ancestor exhibiting duplication and subsequent divergence. The globin genes are an example of a multigene family.

multiple alleles In a population of organisms, three or more alleles of the same gene.

mutagen Any agent that causes an increase in the spontaneous rate of mutation.

mutant A cell or organism carrying an altered or mutant gene.

mutation The process that produces an alteration in DNA or chromosome structure; in genes, the source of new alleles.

mutation rate The frequency with which mutations take place at a given locus or in a population.

natural selection Differential reproduction among members of a species owing to variable fitness conferred by genotypic differences.

neutral mutation A mutation with no immediate adaptive significance or phenotypic effect.

noncrossover gamete A gamete whose chromosomes have undergone no genetic recombination.

nondisjunction A cell division error in which homologous chromosomes (in meiosis) or the sister chromatids (in mitosis) fail to separate and migrate to opposite poles; responsible for defects such as monosomy and trisomy.

nonsense codons The nucleotide triplets (UGA, UAG, and UAA) in an mRNA molecule that signals the termination of translation.

nonsense mutation A mutation that creates a nonsense codon.

NOR See *nucleolar organizer region.*

normal distribution A probability function that approximates the distribution of random variables. The normal curve, also known as a Gaussian or bell-shaped curve, is the graphic display of a normal distribution.

northern blot An analytic technique in which RNA molecules are separated by electrophoresis and transferred by capillary action to a nylon or nitrocellulose membrane. Specific RNA molecules can then be identified by molecular hybridization to a labeled nucleic acid probe.

N terminus The free amino group of the first amino acid in a polypeptide. By convention, the structural formula of polypeptides is written with the N terminus at the left.

nuclease An enzyme that breaks bonds in nucleic acid molecules.

nucleolar organizer region (NOR) A chromosomal region containing the genes for rRNA; most often found in physical association with the *nucleolus*.

nucleolus The nuclear site of ribosome biosynthesis and assembly; usually associated with or formed in association with the DNA comprising the *nucleolar organizer region*.

nucleoside In nucleic acid chemical nomenclature, a purine or pyrimidine base covalently linked to a ribose or deoxyribose sugar molecule.

nucleosome In eukaryotes, a nuclear complex consisting of four pairs of histone molecules wrapped by two turns of a DNA molecule. The major structure associated with the organization of chromatin in the nucleus.

nucleotide In nucleic acid chemical nomenclature, a nucleoside covalently linked to one or more phosphate groups. Nucleotides containing a single phosphate linked to the 5′ carbon of the ribose or deoxyribose sugar are the building blocks of nucleic acids.

nucleotide pair The base-paired nucleotides (A with T or G with C) on opposite strands of a DNA molecule that are hydrogen-bonded to each other.

nucleus The membrane-bound cytoplasmic organelle of eukaryotic cells that contains the chromosomes and nucleolus.

null allele A mutant allele that produces no functional gene product. Usually inherited as a recessive trait.

null hypothesis (H_0) Used in statistical tests, the hypothesis that there is no real difference between the observed and expected data sets, whereby the observed deviation is attributed to chance.

Okazaki fragment The small, discontinuous strands of DNA produced on the lagging strand during DNA synthesis.

oligonucleotide A linear sequence of about 10–20 nucleotides connected by 5′–3′ phosphodiester bonds.

oncogene A gene whose activity promotes uncontrolled proliferation in eukaryotic cells. Usually a mutant gene derived from a *proto-oncogene*.

open reading frame (ORF) A nucleotide sequence organized as triplets that encodes the amino acid sequence of a polypeptide, including an initiation codon and a termination codon.

operator region In bacterial operons, a DNA region that interacts with a specific repressor protein to regulate the expression of an adjacent gene or gene set.

operon A genetic unit consisting of one or more structural genes encoding polypeptides, and operator region that regulates the transcriptional activity of the adjacent structural genes.

origin of replication (ori) A site where DNA replication begins along the length of a chromosome.

orthologs Genes with sequence similarity found in two or more related species that arose from a single gene in a common ancestor.

pachytene The stage in meiotic prophase I when the synapsed homologous chromosomes split longitudinally (except at the centromere), producing a group of four chromatids called a tetrad.

pair-rule genes Genes expressed as stripes around the blastoderm embryo during development of the *Drosophila* embryo.

palindrome In genetics, a sequence of DNA base pairs that reads the same backward or forward. Since strands run antiparallel to one another in DNA, the base sequences on the two strands read the same backward and forward. For example:

5′-GAATTC-3′
3′-CTTAAG-5′

Palindromic sequences are noteworthy as recognition sites along DNA, often providing the substrates for restriction endonucleases.

paracentric inversion A chromosomal inversion that does not include the region containing the centromere.

paralogs Two or more genes in the same species derived by duplication and subsequent divergence from a single ancestral gene.

parthenogenesis Development of an egg without fertilization.

partial diploids See *merozygote*.

partial dominance See *incomplete dominance*.

pedigree In human genetics, a diagram showing the ancestral relationships and transmission of genetic traits over several generations in a family.

P element In *Drosophila,* a transposable DNA element responsible for hybrid dysgenesis.

penetrance The frequency, expressed as a percentage, with which individuals of a given genotype manifest at least some degree of a specific mutant phenotype associated with a trait.

peptide bond The covalent bond between the amino group of one amino acid and the carboxyl group of another amino acid, creating a dipeptide.

pericentric inversion A chromosomal inversion that involves both arms of the chromosome and thus the centromere.

phage See *bacteriophage*.

pharmacogenomics The study of how genetic variation influences the efficacy of pharmaceutical drugs in individuals.

phenotype The overt appearance of a genetically controlled trait.

phenylketonuria (PKU) A hereditary condition in humans that is associated with the inability to metabolize the amino acid phenylalanine due to the loss of activity of the enzyme phenylalanine hydroxylase.

Philadelphia chromosome The product of a reciprocal translocation in humans that contains the short arm of chromosome 9, carrying the *C-ABL* oncogene, and the long arm of chromosome 22, carrying the *BCR* gene. This translocation is associated with chronic myelogenous leukemia (CML).

phosphodiester bond In nucleic acids, the system of covalent bonds by which a phosphate group links adjacent nucleotides, extending from the 5′ carbon of one pentose sugar (ribose or deoxyribose) to the 3′ carbon of another pentose sugar in the neighboring nucleotide. Phosphodiester bonds create the backbone of nucleic acid molecules.

photoreactivation enzyme (PRE) An exonuclease that catalyzes the light-activated excision of ultraviolet-induced thymine dimers from DNA.

photoreactivation repair Light-induced repair of damage caused by exposure to ultraviolet light. Associated with an intracellular enzyme system.

phyletic evolution The gradual transformation of one species into another over time; so-called vertical evolution.

pilus A filament-like projection from the surface of a bacterial cell. Often associated with cells possessing F factors.

plaque On an otherwise opaque bacterial lawn, a clear area caused by the growth and reproduction of a single bacteriophage.

plasmid An extrachromosomal, circular DNA molecule that replicates independently of the host chromosome.

pleiotropy Condition in which a single mutation causes multiple phenotypic effects.

ploidy A term referring to the basic chromosome set or to multiples of that set.

pluripotent The capacity of a cell or an embryo part to differentiate into all types characteristic of an adult, a capacity progressively restricted during development. Used interchangeably with *totipotent*.

point mutation A mutation that can be mapped to a single locus. At the molecular level, a mutation that results in the substitution of one nucleotide for another. Also called a *gene mutation*.

polar body Produced in females at either the first or second meiotic division of gametogenesis, a discarded cell that contains one of the nuclei of the division process, but almost no cytoplasm as a result of an unequal cytokinesis.

polycistronic mRNA A messenger RNA molecule that encodes the amino acid sequence of two or more polypeptide chains in adjacent structural genes.

polygenic inheritance The transmission of a phenotypic trait whose expression depends on the additive effect of a number of genes.

polylinker A segment of DNA that has been engineered to contain multiple sites for restriction enzyme digestion. Polylinkers are usually found in engineered vectors such as plasmids.

polymerase chain reaction (PCR) A method for amplifying DNA segments that depends on repeated cycles of denaturation, primers, and DNA polymerase–directed DNA synthesis.

polymerases Enzymes that catalyze the formation of DNA and RNA from deoxynucleotides and ribonucleotides, respectively.

polymorphism The existence of two or more discontinuous, segregating phenotypes in a population.

polynucleotide A linear sequence of 20 or more nucleotides, joined by 5′–3′ phosphodiester bonds. See also *oligonucleotide*.

polypeptide A molecule made up of amino acids joined by covalent peptide bonds. This term is used to denote the amino acid chain before it assumes its functional three-dimensional configuration.

polyploid A cell or individual having more than two haploid sets of chromosomes.

polyribosome A structure composed of two or more ribosomes associated with mRNA and engaged in translation. Also called a *polysome*.

polytene chromosome Literally, a many-stranded chromosome, one that has undergone numerous rounds of DNA replication without separation of the replicated strands, which remain in exact parallel register. The result is a giant chromosome with aligned chromomeres displaying a characteristic banding pattern, most often studied in *Drosophila* larval salivary gland cells.

population A local group of individuals belonging to the same species that are actually or potentially interbreeding.

population bottleneck A drastic reduction in population size and consequent loss of genetic diversity, followed by an increase in population size. The rebuilt population has an altered gene pool with reduced variability caused by genetic drift.

position effect Change in expression of a gene associated with a change in the gene's location within the genome.

positional cloning The identification and subsequent cloning of a gene in the absence of knowledge of its polypeptide product or function. Based on the determination of its precise chromosomal location, the process begins with knowledge of an approximate location of the gene within the genome that is then refined, often by examining the cosegregation of mutant phenotypes with known genetic map locations or DNA markers (e.g., *RFLP polymorphisms*).

posttranscriptional modification Changes made to pre-mRNA molecules during conversion to mature mRNA. These include the addition of a methylated cap at the 5′ end and a poly-A tail at the 3′ end, excision of introns, and exon splicing.

posttranslational modification The processing or modification of the translated polypeptide chain by enzymatic cleavage, or the addition of phosphate groups, carbohydrate chains, or lipids.

postzygotic isolation mechanism A factor that prevents or reduces inbreeding by acting after fertilization to produce nonviable, sterile hybrids or hybrids of lowered fitness.

preadaptive mutation A mutational event that later becomes of adaptive significance.

preformationism See *epigenesis.*

preimplantation genetic diagnosis (PGD) The removal and genetic analysis of unfertilized oocytes, polar bodies, or single cells from an early embryo (3–5 days old) following *in vitro* fertilization.

prezygotic isolation mechanism A factor that reduces inbreeding by preventing courtship, mating, or fertilization.

Pribnow box In prokaryotic genes, a 6-bp sequence to which the sigma (σ) subunit of RNA polymerase binds, upstream from the beginning of transcription. The consensus sequence for this box is TATAAT.

primary protein structure The sequence of amino acids in a polypeptide chain.

primary sex ratio Ratio of males to females at fertilization, often expressed in decimal form (e.g., 1.40).

primer In nucleic acids, a short length of RNA or single-stranded DNA required for the initiating synthesis directed by polymerases.

prion An infectious pathogenic agent devoid of nucleic acid and composed of a protein, PrP, known to cause scrapie, a degenerative neurological disease in sheep; bovine spongiform encephalopathy (BSE, or mad cow disease) in cattle; and similar diseases in humans, including kuru and Creutzfeldt–Jakob disease.

probability Ratio of the frequency of a given outcome to the frequency of all possible outcomes.

proband An individual who is the focus of, or who called attention to, the genetic study leading to the production of a pedigree tracking the inheritance of a genetically determined trait of interest. Formerly known as a *propositus.*

probe A macromolecule such as DNA or RNA that has been labeled and can be detected by an assay such as autoradiography or fluorescence microscopy. Probes are used to identify target molecules, genes, or gene products.

prokaryotes Organisms lacking nuclear membranes and true chromosomes. Bacteria and blue–green algae are examples of prokaryotic organisms.

promoter An upstream region of a gene serving a regulatory function and to which RNA polymerase binds prior to the initiation of transcription.

prophage A bacteriophage genome integrated into a bacterial chromosome that is replicated along with the bacterial chromosome. Bacterial cells carrying prophages are said to be *lysogenic* and to be capable of entering the *lytic cycle,* whereby the phage is reproduced.

propositus (female, proposita) See *proband.*

protein A molecule composed of one or more polypeptides, each composed of amino acids covalently linked together. Proteins demonstrate *primary, secondary, tertiary,* and *quaternary structure.*

protein domain Amino acid sequences with specific conformations and functions that are structurally and functionally distinct from other regions on the same protein.

proteome The entire set of proteins expressed by a cell, a tissue, or an organism at a given time.

proteomics The study of the expressed proteins present in a cell at a given time.

proto-oncogene A gene that functions to initiate, facilitate, or maintain cell growth and division. Proto-oncogenes can be converted to *oncogenes* by mutation.

protoplast A bacterial or plant cell with the cell wall removed. Sometimes called a *spheroplast.*

prototroph A strain (usually of a microorganism) that is capable of growth on a defined, minimal medium. Wild-type strains are usually regarded as prototrophs and contrasted with *auxotrophs.*

pseudoalleles Genes that behave as alleles to one another by complementation but can be separated from one another by recombination.

pseudoautosomal region A region present on the human Y chromosome that is also represented on the X chromosome. Genes found in this region of the Y chromosome have a pattern of inheritance that is indistinguishable from genes on autosomes.

pseudodominance The expression of a recessive allele on one homolog owing to the deletion of the dominant allele on the other homolog.

pseudogene A nonfunctional gene with sequence homology to a known structural gene present elsewhere in the genome. It differs from the functional version by insertions or deletions and by the presence of flanking direct-repeat sequences of 10–20 nucleotides.

puff Same as *chromosome puff.*

punctuated equilibrium A pattern in the fossil record of long periods of species stability, punctuated with brief periods of species divergence.

quantitative inheritance Same as *polygenic inheritance.*

quantitative trait loci (QTLs) Two or more genes that act on a single polygenic trait.

quantum speciation Formation of a new species within a single or a few generations by a combination of selection and drift.

quaternary protein structure Types and modes of interaction between two or more polypeptide chains within a protein molecule.

race A genotypically or geographically distinct subgroup within a species.

random mating Mating between individuals without regard to genotype.

reading frame A linear sequence of codons in a nucleic acid.

reannealing Formation of double-stranded DNA molecules from denatured single strands.

recessive An allele whose potential genetic expression is overridden in the heterozygous condition by a dominant allele.

reciprocal cross A pair of crosses in which the genotype of the female in one is present as the genotype of the male in the other, and vice versa.

reciprocal translocation A chromosomal aberration in which nonhomologous chromosomes exchange parts.

recombinant DNA A DNA molecule formed by joining two heterologous molecules. A term also applied to the technology associated with the use of DNA molecules produced by *in vitro* ligation of DNA from two different organisms.

recombinant gamete A gamete containing a new combination of genes produced by crossing over during meiosis.

recombination The process that leads to the formation of new gene combinations on chromosomes.

reductional division The chromosome division that halves the number of centromeres and thus reduces the chromosome number by half. The first division of meiosis is a reductional division. See also *equational division.*

redundant genes Gene sequences present in more than one copy per haploid genome (e.g., ribosomal genes).

regulatory site A DNA sequence that functions in the control of expression of other genes, usually involving an interaction with another molecule.

repetitive DNA sequence A DNA sequence present in many copies in the haploid genome.

replication The process whereby DNA is duplicated.

replication fork The Y-shaped region of a chromosome associated with the site of replication.

replicon The unit of DNA replication, originating with DNA sequences necessary for the initiation of DNA replication. In bacteria, the entire chromosome is a replicon.

replisome The complex of proteins, including DNA polymerase, that assembles at the bacterial replication fork to synthesize DNA.

repressible enzyme system An enzyme or group of enzymes whose synthesis is regulated by the intracellular concentration of certain metabolites.

repressor A protein that binds to a regulatory sequence adjacent to a gene and blocks transcription of the gene.

reproductive isolation Absence of interbreeding between populations, subspecies, or species. Reproductive isolation can be brought about by extrinsic factors, such as behavior, or intrinsic barriers, such as hybrid inviability.

resistance transfer factor (RTF) A component of R plasmids that confers the ability to transfer the R plasmid between cells by conjugation.

restriction endonuclease A bacterial nuclease that recognizes specific nucleotide sequences in a DNA molecule, often a *palindrome,* and cleaves or nicks the DNA at those sites. Provides defense for invading DNA segments in bacteria and is useful in the construction of *recombinant DNA molecules.*

restriction fragment length polymorphism (RFLP) Variation in the length of DNA fragments generated by restriction endonucleases. These variations are caused by mutations that create or abolish cutting sites for restriction enzymes. RFLPs are inherited in a codominant fashion and are extremely useful as genetic markers.

restrictive transduction See *specialized transduction.*

retrotransposon Major components of many eukaryotic genomes, mobile genetic elements that can be copied by means of an RNA intermediate and inserted at a distant locus.

retrovirus A type of virus that uses RNA as its genetic material and employs the enzyme reverse transcriptase during its life cycle.

reverse genetics An experimental approach used to discover the function of a gene, after the gene has been identified, cloned, and sequenced. The cloned gene may be knocked out (e.g., by *gene targeting*) or have its expression altered (e.g., by *RNA interference* or *transgenic overexpression*) and the resulting phenotype studied. An approach contrasted with *forward genetics.*

reverse transcriptase A polymerase that uses RNA as a template to transcribe a single-stranded DNA molecule as a product.

reversion A mutation that restores the wild-type phenotype.

R factor (R plasmid) A bacterial plasmid that carries antibiotic resistance genes. Most R plasmids have two components: an r-determinant, which carries the antibiotic resistance genes, and the resistance transfer factor (RTF).

RFLP See *restriction fragment length polymorphism.*

Rh factor An antigenic system first described in the rhesus monkey, which in humans is the genetic basis of the immunological incompatibility disease called erythroblastosis fetalis (hemolytic disease of the newborn).

ribonucleic acid (RNA) A nucleic acid similar to DNA but characterized by the pentose sugar ribose, the pyrimidine uracil, and the single-stranded nature of the polynucleotide chain. Several forms are recognized, including ribosomal RNA, messenger RNA, transfer RNA, and a variety of small regulatory RNA molecules.

ribose The five-carbon sugar associated with ribonucleosides and ribonucleotides associated with RNA.

ribosomal RNA (rRNA) The RNA molecules that are the structural components of the ribosomal subunits. In prokaryotes, these are the 16*S*, 23*S*, and 5*S* molecules; in eukaryotes, they are the 18*S*, 28*S*, and 5*S* molecules. See also *Svedberg coefficient (S).*

ribosome A ribonucleoprotein organelle consisting of two subunits, each containing RNA and protein molecules. Ribosomes are the site of translation of mRNA codons into the amino acid sequence of a polypeptide chain.

RNA See *ribonucleic acid.*

RNA editing Alteration of the nucleotide sequence of an mRNA molecule after transcription and before translation. There are two main types of editing: substitution editing, which changes individual nucleotides, and insertion/deletion editing, in which individual nucleotides are added or deleted.

RNA interference (RNAi) Inhibition of gene expression in which a protein complex (RNA-induced silencing complex, or RISC), containing a partially complementary RNA strand binds to an mRNA, leading to degradation or reduced translation of the mRNA.

RNA polymerase An enzyme that catalyzes the formation of an RNA polynucleotide strand using the base sequence of a DNA molecule as a template.

RNase A class of enzymes that hydrolyzes RNA.

Robertsonian translocation A form of chromosomal aberration in which breaks occur in the short arms of two acrocentric chromosomes and the long arms of these chromosomes fuse at the centromere. Also called *centric fusion.*

rRNA See *ribosomal RNA.*

RTF See *resistance transfer factor.*

S_1 nuclease A deoxyribonuclease that cuts and degrades single-stranded molecules of DNA.

satellite DNA DNA that forms a minor band when genomic DNA is centrifuged in a cesium salt gradient. This DNA usually consists of short sequences repeated many times in the genome.

SCE See *sister chromatid exchange.*

secondary protein structure The α-helical or β-pleated-sheet formations in a polypeptide, dependent on hydrogen bonding between certain amino acids.

secondary sex ratio The ratio of males to females at birth, usually expressed in decimal form (e.g., 1.06).

secretor An individual who has soluble forms of the blood group antigens A and/or B present in saliva and other body fluids. This condition is caused by a dominant autosomal gene independent of the *ABO* locus (*I* locus).

segment polarity genes Genes that regulate the spatial pattern of differentiation within each segment of the developing *Drosophila* embryo.

segregation The separation of maternal and paternal homologs of each homologous chromosome pair into gametes during meiosis.

selection The changes that occur in the frequency of alleles and genotypes in populations as a result of differential reproduction.

selection coefficient (*s*) A quantitative measure of the relative fitness of one genotype compared with another. Same as *coefficient of selection.*

semiconservative replication A mode of DNA replication in which a double-stranded molecule replicates in such a way that each daughter molecule is composed of one parental (old) and one newly synthesized strand.

semisterility A condition in which a percentage of all zygotes are inviable.

sex chromatin body See *Barr body.*

sex chromosome A chromosome, such as the X or Y in humans, or Z or W in birds, which is involved in sex determination.

sexduction Transmission of chromosomal genes from a donor bacterium to a recipient cell by means of the F′ factor.

sex-influenced inheritance Phenotypic expression that is conditioned by the sex of the individual. A heterozygote may express one phenotype in one sex and the alternate phenotype in the other sex (e.g., pattern baldness in humans).

sex-limited inheritance A trait that is expressed in only one sex even though the trait may not be X-linked.

sex ratio See *primary sex ratio* and *secondary sex ratio.*

sexual reproduction Reproduction through the fusion of gametes, which are the haploid products of meiosis.

Shine–Dalgarno sequence The nucleotides AGGAGG that serve as a ribosome-binding site in the leader sequence of prokaryotic genes. The 16*S* RNA of the small ribosomal subunit contains a complementary sequence to which the mRNA binds.

short interspersed elements (SINES) Repetitive sequences found in the genomes of higher organisms, such as the 300-bp *Alu* sequence.

short tandem repeats See *STR sequences.*

shotgun experiment The cloning of random fragments of genomic DNA into a vehicle such as a plasmid or phage, usually to produce a library from which clones of specific interest can be selected.

sibling species Species that are morphologically almost identical but that are reproductively isolated from one another.

sickle-cell anemia A genetic disease in humans caused by an autosomal recessive gene, often fatal in the homozygous condition if untreated. Caused by a mutation leading to an alteration in the amino acid sequence of the β chain of globin.

sickle-cell trait The phenotype exhibited by individuals heterozygous for the sickle-cell gene.

sigma (σ) factor In bacterial RNA polymerase, a polypeptide subunit that recognizes the DNA binding site for the initiation of transcription.

single nucleotide polymorphism (SNP) A variation in a single nucleotide pair in DNA, as detected during genomic analysis. Usually present in at least 1% of a population, SNPs are useful as genetic markers.

single-stranded binding proteins (SSBs) In DNA replication, proteins that bind to and stabilize the single-stranded regions of DNA that result from the action of unwinding proteins.

sister chromatid exchange (SCE) A crossing over event that can occur in meiotic and mitotic cells involving the reciprocal exchange of chromosomal material between sister chromatids joined by a common centromere. Such exchanges can be detected cytologically after BrdU is incorporated into the replicating chromosomes.

site-directed mutagenesis A process that uses a synthetic oligonucleotide containing a mutant base or sequence as a primer for inducing a mutation at a specific site in a cloned gene.

small nuclear RNA (snRNA) Abundant species of small RNA molecules ranging in size from 90 to 400 nucleotides that in association with proteins form RNP particles known as snRNPs or *snurps.* Located in the nucleoplasm, snRNAs have been implicated in the processing of pre-mRNA and may have a range of cleavage and ligation functions.

SNP See *single nucleotide polymorphism.*

snurps See *small nuclear RNA.*

solenoid structure A feature of eukaryotic chromatin conformation that is generated by the supercoiling of nucleosomes.

somatic cells All cells other than the germ cells or gametes in an organism.

somatic mutation A nonheritable mutation occurring in a somatic cell.

somatic pairing The pairing of homologous chromosomes in somatic cells.

SOS DNA repair The induction of enzymes for repairing damaged DNA in *Escherichia coli.* The response involves activation of an enzyme that cleaves a repressor, activating a series of genes involved in DNA repair.

Southern blotting Developed by Edwin Southern, a technique in which DNA fragments produced by restriction enzyme digestion are separated by electrophoresis and transferred by capillary action to a nylon or nitrocellulose membrane. Specific DNA fragments can be identified by hybridization to a complementary radioactively labeled nucleic acid probe using the technique of *autoradiography.*

spacer DNA DNA sequences found between genes. Usually, the genes are repetitive DNA segments.

specialized transduction Genetic transfer of only specific host genes by transducing phages.

speciation The process by which new species of plants and animals arise.

species A group of actually or potentially interbreeding individuals that is reproductively isolated from other such groups.

spheroplast See *protoplast.*

spindle fibers Cytoplasmic fibrils formed during cell division that are involved with the separation of chromatids at the anaphase stage of mitosis and meiosis as well as their movement toward opposite poles in the cell.

spliceosome The nuclear macromolecule complex within which splicing reactions occur to remove introns from pre-mRNAs.

spontaneous mutation A mutation that is not induced by a mutagenic agent.

spore Produced by some bacteria, plants, and invertebrates, a unicellular body or cell encased in a protective coat. It is capable of surviving in unfavorable environmental conditions and gives rise to a new individual upon germination. In plants, spores are the haploid products of meiosis.

SRY The sex-determining region of the Y chromosome, found near the chromosome's pseudoautosomal boundary. Accumulated evidence indicates that this gene's product is the testis-determining factor (TDF).

stabilizing selection Preferential reproduction of those individuals having genotypes close to the mean for the population. A selective elimination of genotypes at both extremes.

standard deviation (s) A quantitative measure of the amount of variation in a sample of measurements from a population.

standard error A quantitative measure of the variation of the means in a sample of measurements from a population.

strain A group of organisms with common ancestry that has physiological or morphological characteristics of interest for genetic study or domestication.

STR sequences Short tandem repeats 2–9 base pairs long that are found within minisatellites. These sequences are used to prepare DNA profiles in forensics, paternity identification, and other applications.

structural gene A gene that encodes the amino acid sequence of a polypeptide chain.

submetacentric chromosome A chromosome with the centromere placed so that one arm of the chromosome is slightly longer than the other.

subspecies A morphologically or geographically distinct interbreeding population of a species.

sum law The law that holds that the probability of one of two mutually exclusive events occurring is the sum of their individual probabilities.

supercoiled DNA A DNA configuration in which the helix is coiled upon itself. Supercoils can exist in stable forms only when the ends of the DNA are not free, as in a covalently closed circular DNA molecule.

superfemale See *metafemale*.

supermale See *metamale*.

suppressor mutation A mutation that acts to completely or partially restore the function lost by a mutation at another site.

Svedberg coefficient unit (*S*) A unit of measure for the rate at which particles (molecules) sediment in a centrifugal field. This rate is a function of several physicochemical properties, including size and shape. A rate of 1×10^{-13} sec is defined as one Svedberg coefficient unit.

sympatric speciation Speciation occurring in populations that inhabit, at least in part, the same geographic range.

synapsis The pairing of homologous chromosomes at meiosis.

synaptonemal complex (SC) An organelle consisting of a tripartite nucleoprotein ribbon that forms between the paired homologous chromosomes in the pachytene stage of the first meiotic division. May be essential to crossing over.

syndrome A group of characteristics or symptoms associated with a disease or an abnormality. An affected individual may express a number of these characteristics but not necessarily all of them.

syntenic test In somatic cell genetics, a method for determining whether two genes are on the same chromosome.

systems biology A field that identifies and analyzes gene and protein networks to gain an understanding of intracellular regulation of metabolism, intra- and intercellular communication, and complex interactions within, between, and among cells.

TATA box See *Goldberg–Hogness box*.

tautomeric shift A reversible isomerization, brought about by a rare proton shift, altering the location of a hydrogen atom. In DNA, tautomeric shifts alter base-pairing properties that can lead to mutations during replication.

TDF (testis-determining factor) The product of the *SRY* gene on the Y chromosome; it controls the developmental switch point for the development of the indifferent gonad into a testis.

telocentric chromosome A chromosome in which the centromere is located at its very end.

telomerase The enzyme that adds short, tandemly repeated DNA sequences to the ends of eukaryotic chromosomes.

telomere The heterochromatic terminal region of a chromosome.

telophase The stage of cell division in which the daughter chromosomes have reached the opposite poles of the cell and reverse the stages characteristic of prophase, re-forming the nuclear envelopes and uncoiling the chromosomes. Telophase ends with cytokinesis, which divides the cytoplasm and splits cell into two.

telophase I The stage in the first meiotic division when duplicated chromosomes reach the poles of the dividing cell.

temperature-sensitive mutation A conditional mutation that produces a mutant phenotype at one temperature range and a wild-type phenotype at another.

terminalization The movement of chiasmata toward the ends of chromosomes during the diplotene stage of the first meiotic division.

tertiary protein structure The three-dimensional conformation of a polypeptide chain in space, achieving maximum thermodynamic stability and functional capacity.

testcross A cross between an individual whose genotype at one or more loci may be unknown and an individual who is homozygous recessive for the gene or genes in question.

tetrad The four chromatids that make up paired homologs in the prophase of the first meiotic division. In algae and fungi, tetrad denotes the four haploid cells produced by a single meiotic division.

tetrad analysis A method that analyzes gene linkage and recombination in organisms with a predominant haploid phase in their life cycle. See *tetrad*.

thymine dimer In a polynucleotide strand, a lesion consisting of two adjacent thymine bases that become joined by a covalent bond. Usually caused by exposure to ultraviolet light, this lesion inhibits DNA replication.

T_m See *melting profile*.

topoisomerase A class of enzymes that convert DNA from one topological form to another, e.g., DNA gyrase that reduces supercoiling upstream from the *replication fork*.

totipotent See *pluripotent*.

trait Any detectable phenotypic variation of a particular inherited character.

***trans*-acting element** See *cis*-acting element.

***trans* configuration** An arrangement in which two mutations are located on opposite homologs, such as:

$$\frac{a^1 \quad +}{+ \quad a^2}$$

in contrast to a *cis arrangement,* in which the sites are located on the same homolog.

transcription Transfer of genetic information from DNA by the synthesis of a complementary RNA molecule under the direction of RNA polymerase.

transcriptome The set of mRNA molecules present in a cell at any given time.

transdetermination Change in developmental fate of a cell or group of cells.

transduction Virally mediated bacterial recombination. Also used to describe the transfer of eukaryotic genes mediated by a retrovirus.

transfer RNA (tRNA) A small ribonucleic acid molecule playing an essential role in *translation,* which adapts a triplet codon to its prescribed amino acid.

transformation Heritable change in a cell or an organism brought about by exogenous DNA. Known to occur naturally and also used in *recombinant DNA* studies.

transgenic organism An organism whose genome has been modified by the introduction of external DNA sequences into the germ cells.

translation The derivation of the amino acid sequence of a polypeptide from the base sequence of an mRNA molecule in association with a ribosome and dependent on *tRNAs*.

translocation A chromosomal mutation associated with the reciprocal or nonreciprocal transfer of a chromosomal segment from one chromosome to another. Also denotes the movement of mRNA through the ribosome during translation.

transmission genetics The field of genetics concerned with heredity and the mechanisms by which genes are transferred from parent to offspring.

transposable element A DNA segment capable of translocation to other sites in the genome. Usually flanked at each end by short inverted repeats of 20–40 base pairs, insertion into a structural gene can produce a mutant phenotype.

trinucleotide repeat A tandemly repeated cluster of three nucleotides (such as CTG) within or near a gene. Certain diseases (myotonic dystrophy, Huntington disease) are caused by expansion in copy number of such repeats.

triploidy The condition in which a cell or organism possesses three haploid sets of chromosomes.

trisomy The condition in which a cell or organism possesses two copies of each chromosome except for one, which is present in three copies (designated $2n + 1$).

tRNA See *transfer RNA*.

tumor-suppressor gene A gene whose product functions to suppress unrestricted cell division, particularly of tumorous growth.

unequal crossing over A crossover between two improperly aligned homologs, producing one homolog with three copies of a region and the other with one copy of that region.

unwinding proteins Nuclear proteins that act during DNA replication to destabilize and unwind the DNA helix ahead of the replicating fork.

variable number tandem repeats (VNTRs) Short, repeated DNA sequences (of 2–20 nucleotides) present as tandem repeats between two restriction enzyme recognition sites. Variation in the number of repeats creates DNA fragments of differing lengths following restriction enzyme digestion. Used in early versions of *DNA fingerprinting*.

variable region Portion of an immunoglobulin molecule that exhibits many amino acid sequence differences between antibodies of differing specificities.

variance (s^2) A statistical measure of the variation of values from a central value, calculated as the square of the standard deviation.

vector In recombinant DNA, an agent such as a phage or plasmid into which a foreign DNA segment will be inserted and utilized to transform host cells.

VNTRs See *variable number tandem repeats*.

western blot An analytical technique in which proteins are separated by gel electrophoresis and identified through hybridization to a labeled antibody.

wild type The most commonly observed phenotype or genotype, designated as the norm or standard.

wobble hypothesis An idea proposed by Francis Crick, stating that the third base in an anticodon can align in several ways to allow it to recognize more than one base in the codons of mRNA.

X chromosome The sex chromosome present in species where females are the homogametic sex (XX).

X chromosome inactivation In mammalian females, the random cessation early in development of transcriptional activity of either the maternally or paternally derived X chromosome, providing a mechanism of dosage compensation. All progeny cells inactivate the same X chromosome. See also *Barr body, Lyon hypothesis, XIST*.

XIST A locus in the X-chromosome inactivation center that controls inactivation of the X chromosome in mammalian females.

X-linkage The pattern of inheritance resulting from genes located on the X chromosome.

X-ray crystallography A technique for determining the three-dimensional structure of molecules by analyzing X-ray diffraction patterns produced by crystals of the molecule under study.

YAC A cloning vector in the form of a yeast artificial chromosome, constructed using chromosomal elements from yeast, including telomeres (from a ciliate), centromeres, origin of replication, and marker genes.

Y chromosome The sex chromosome in species where the male is heterogametic (XY).

Y-linkage Mode of inheritance shown by genes located on the Y chromosome.

Z-DNA An alternative "zig-zag" structure of DNA in which the two antiparallel polynucleotide chains form a left-handed double helix.

zinc finger A class of DNA-binding domains found in proteins containing a characteristic pattern of cysteine and histidine residues that complex with zinc ions, creating finger-like structures.

zygote The diploid cell produced by the fusion of haploid gametic nuclei.

zygotene A stage of meiotic prophase I in which the homologous chromosomes synapse and pair along their entire length, forming bivalents. The synaptonemal complex forms at this stage.

Credits

PHOTOS

Chapter 1 Introduction to Genetics

CO1 (top) Sinclair Stammers/Photo Researchers **CO1 (left)** Mark Smith/Photo Researchers **CO1 (right)** Alberto Salguero/Wikemedia Commons **F1.2** Biophoto Associates/Science Source/Photo Researchers **F1.3** Sovereign/Phototake NYC **F1.4** ISM/Phototake NYC **F1.6 (top)** Carolina Biological Supply Company/Phototake NYC **F1.6 (bottom)** Carolina Biological Supply Company/Phototake NYC **F1.7** Biozentrum, University of Basel/Science Photo Library/Photo Researchers **F1.10** Manuel C. Peitsch/CORBIS- NY **F1.12** Oliver Meckes & Nicole Ottawa/Photo Researchers **F1.14** Roslin Institute **F1.15** SSPL/The Image Works **F1.17** Guillaume Paumier/Wikemedia Commons **F1.18(a)** John Paul Endress/Pearson Learning Photo Studio **F1.18(b)** David M. Phillips/Visuals Unlimited/Getty Images **F1.19(a)** Jeremy Burgess/Photo Researchers **F1.19(b)** David M. Phillips/Visuals Unlimited/Getty Images/Getty Images

Chapter 2 Mitosis and Meiosis

CO2 Andrew S. Bajer, University of Oregon **F2.2** CNRI/Science Photo Library/Photo Researchers **F2.4** David Ward, Yale University **F2.7(a-g)** Andrew S. Bajer, University of Oregon **F2.13(a)** Biophoto Associates/Photo Researchers **F2.13(b)** Andrew Syred/SPL/Photo Researchers

Chapter 3 Mendelian Genetics

CO3 Archiv/Photo Researchers **p43 (left and right)** Mike Dunton/The Victory Seed Company **p57** Saim Nadir/Shutterstock

Chapter 4 Modification of Mendelian Ratios

CO4 Dale C. Spartas/DCS Photo **F4.1** John Kaprielian/Photo Researchers **F4.3 (left)** The Jackson Laboratory, Bar Harbor, Maine **F4.3 (right)** Fred Habegger/Grant Heilman Photography **F4.8** Irene Vandermolen/Animals Animals/Earth Scenes **F4.10 (left and right)** Carolina Biological Supply Company/Phototake NYC **F4.12** Ishihara, "Test for Color Deficiency." Courtesy Kanehara & Co., Ltd. Offered Exclusively in The USA by Graham-Field Health Products, Atlanta, GA. **F4.13** Hans Reinhard/Photoshot Holdings/Bruce Coleman **F4.14** Debra P. Hershkowitz/Photoshot Holdings/Bruce Coleman **F4.15 (top)** Tanya Wolff, Washington University School of Medicine **F4.15 (center)** Joel C. Eissenberg, Dept. of Biochemistry and Molecular Biology, St. Louis University Medical Center. **F4.15 (bottom)** Tanya Wolff, Washington University School of Medicine **F4.16(a)** Jane Burton/Photoshot Holdings/Bruce Coleman **F4.16(b)** Steve Klug/William S. Klug **F4.18** Pam Collins/Photo Researchers **F4.19 (left and right)** Ronald A. Butow, Department of Molecular Biology and Oncology, University of Texas Southwestern Medical Center **F4.20(a-b)** Alan Pestronk, Dept. of Neurology, Washington University School of Medicine, St. Louis **p85 (top left)** Ralph Somes **p85 (top right)** J. James Bitgood/University of Wisconsin, Animal Sciences Dept. **p85 (bottom left)** Ralph Somes **p85 (bottom right)** J. James Bitgood/University of Wisconsin, Animal Sciences Dept. **p90 (top left)** CORBIS NY **p90 (top right and bottom)** Ardea/Retna **p91 (top)** Nigel J. H. Smith/Animals Animals/Earth Scenes **p91 (bottom)** Francois Gohier/Photo Researchers

Chapter 5 Sex Determination and Sex Chromosomes

CO5 Wellcome Trust Medical Photographic Library **F5.1** Biophoto Assoc./Photo Researchers **F5.2** Artem Solovev/Fotolia, LLC Royalty Free **F5.3** Maria Gallegos, University of California, San Francisco **F5.5(a-b)** Catherine G. Palmer, Indiana University **F5.7 (left and right)** Michael Abbey/Photo Researchers **F5.9(a)** W. Layer/Okapia/Photo Researchers **F5.9(b)** William S. Klug **p110** Associated Press/Courtesy of Texas A & M University

Chapter 6 Chromosome Mutations: Variation in Number and Arrangement

CO6 Evelin Schrock, Stan du Manoir and Tom Reid, NIH **F6.2 (left)** Courtesy of the Greenwood Genetic Center, Greenwood, SC **F6.2 (right)** Kristy-Anne Glubish/Design Pics/Alamy Images Royalty Free **F6.4** David D. Weaver, Indiana University **F6.8** Courtesy of National Cotton Council of America **F6.11 (left)** Courtesy of University of Washington Medical Center Pathology **F6.11 (right)** Ray Clarke, Cri du chat Syndrome Support Group, UK **F6.13 (left, center, right)** Mary Lilly/Carnegie Institution of Washington **F6.17** Jorge Yunis. From Yunis and Chandler, 1979. **F6.18** Custom Medical Stock Photo.

Chapter 7 Linkage and Chromosome Mapping in Eukaryotes

CO7 B. John Cabisco/Visuals Unlimited **F7.14** Namboori B. Raju, Stanford University. Fig 3C pg 29 from Raju, N.B. 2008. Six decades of Neurospora ascus biology at Stanford. *Fung. Biol. Rev.* 22: 26–35. **F7.16** Sheldon Wolff & Judy Bodycote/Laboratory of Radiobiology and Environmental Health, University of California, San Francisco.

Chapter 8 Genetic Analysis and Mapping in Bacteria and Bacteriophages

CO8 Dennis Kunkel/Phototake NYC **F8.2** Michael G. Gabridge/Visuals Unlimited **F8.11(a)** Gopal Murti/Science Photo Library/Photo Researchers **F8.13** M. Wurtz/Biozentrum, University of Basel/Science Photo Library/Photo Researchers **F8.15** Bruce Iverson/Bruce Iverson, Photomicrography **F8.18** William S. Klug

Chapter 9 DNA Structure and Analysis

CO9 Ken Eward/Science Source/Photo Researchers **F9.2 (top and bottom)** Bruce Iverson/Bruce Iverson, Photomicrography **F9.4** Oliver Meckes/Max-Planck-Institut-Tubingen/Photo Researchers **F9.11** Photo by M. H. F. Wilkins. Courtesy of King's College London, England. **F9.15** Ventana Medical Systems **F9.16** William S. Klug

Chapter 10 DNA Replication and Recombination

CO10 Gopal Murti/Science Photo Library/Photo Researchers **F10.5 (top)** Walter H. Hodge/Peter Arnold **F10.5 (center and bottom)** Figure from "Molecular Genetics," Pt. 1 pp. 74–75, J. H. Taylor (ed). ©1963 and renewed 1991, reproduced with permission from Elsevier Science **F10.6** Reprinted from *CELL*, Vol. 25, 1981, pp 659, Sundin and Varshavsky, (1 figure), with permission from Elsevier Science. A. Varshavsky. **F10.14** Reproduced by permission from H. J. Kreigstein and D. S. Hogness, *Proceedings of the National Academy of Sciences* 71:136 (1974), p. 137, Fig. 2. **F10.15** Harold Weintraub, Howard Hughes Medical Institute, Fred Hutchinson Cancer Center/"Essential Molecular Biology" 2e, Freifelder & Malachinski, Jones & Bartlett, Fig. 7-24, pp. 141. **F10.18** David Dressler, Oxford University, England

Chapter 11 Chromosome Structure and DNA Sequence Organization

CO11 Science VU/BMRL/Visuals Unlimited **F11.1(a)** M. Wurtz/Biozentrum, University of Basel/Science Photo Library/Photo Researchers **F11.1(b)** William S. Klug **F11.2** Science Source/Photo Researchers **F11.3** Gopal Murti/Science Photo Library/Photo Researchers **F11.4** Don W. Fawcett/Kahri/Dawid/Science Source/Photo Researchers **F11.5** Richard D. Kolodnar/Dana-Farber Cancer Institute **F11.6** Brian Harmon and John Sedat, University of California, San Francisco **F11.7** Science Source/Photo Researchers **F11.8(a)** William S. Klug **F11.8(b)** Omikron/Photo Researchers **F11.9** From Olins and Olins, 1978, Fig. 1 and 4 **F11.13** From Pardue & Gall, 1972. Keter.

Chapter 12 The Genetic Code and Transcription

CO12 Oscar L. Miller/Science Photo Library/Photo Researchers **F12.10** Bert W. O'Malley, Baylor College of Medicine

Chapter 13 Translation and Proteins

CO13 Reprinted from the front cover of *Science,* Vol. 292, May 4, 2001 "Crystal structure of a Thermus thermophilus 70S ribosome containing three bound transfer RNAs(top) and exploded views showing its different molecular components (middle and bottom). Image provided by Albion Baucom (baucom@biology.ucsc.edu). American Association for the Advancement of Science. **F13.9(a)** "The Structure and Function of Polyribosomes." Alexander Rich, Jonathan R. Warner and Howard M. Goodman, 1963. Reproduced by permission of the Cold Spring Harbor Laboratory Press. Cold Spring Harbor Symp. *Quant. Biol.* 28 (1963) fig. 4C (top), p. 273. 1964. **F13.9(b)** E.V. Kiseleva **F13.13(a)** Dennis Kunkel/Phototake NYC **F13.13(b)** Francis Leroy/Biocosmos/Science Photo Library/Photo Researchers **F13.18** Kenneth Eward/BioGrafx/Science Source/Photo Researchers

Chapter 14 Gene Mutation, Transposition, and DNA Repair

CO14 MaizeGDB/Courtesy M. G. Neuffer **F14.14 (left and right)** W. Clark Lambert/University of Medicine & Dentistry of New Jersey

Chapter 15 Regulation of Gene Expression

CO15 T. Cremer/I. Solovei/Dr. F. Habermann/Biozentrum (LMU)

Chapter 16 Cancer and Regulation of the Cell Cycle

CO16 SPL/Photo Researchers **F16.1(a-b)** Hesed M. Padilla-Nash, Antonio Fargiano, and Thomas Ried. Affiliation is Section of Cancer Genomics, Genetics Branch, Center for Research, National Cancer Institute, National

Institutes of Health, Bethesda, MD 20892 **F16.2(a-b)** Manfred Schwab, DKFZ, Heidelberg, Germany.

Chapter 17 Recombinant DNA Technology and Gene Cloning

CO17 D. VoTrung/Phototake **F17.4** Gopal Murti/Science Photo Library/Photo Researchers **F17.6** Custom Medical Stock Photo **F17.15** Laurie Ann Achenbach, Southern Illinois University at Carbondale **F17.18** Dr. Suzanne McCutcheon

Chapter 18 Genomics, Bioinformatics, and Proteomics

F18.7(a) Holt Studios International/Photo Researchers **F18.10** David McCarthy/SPL/Photo Researchers **F18.11** Deanne Fitzmaurice/CORBIS All Rights Reserved **F18.16 (top)** Volker Steger/SPL/Photo Researchers **F18.16 (bottom)** Cancer Genetics Branch/National Human Genome Research Institute/NIH **F18.17** Jeff Rotman/Nature Picture Library **F18.19** Swiss Institute of Bioinformatics

Chapter 19 Applications and Ethics of Genetic Engineering and Biotechnology

CO19 From The Roslin Institute - roslin.design@bbsrc.ac.uk http://www.roslin.ac.uk/imagelibrary/index.php *Trends in Biotechnology* Volume 22, Issue 4, April 2004, Pages 157–159 doi:10.1016/j.tibtech.2004.02.005. **F19.3** From Doebley, *J. Plant Cell,* 2005 Nov; 17(11): 2859–72. John Doebley/University of Wisconsin. **F19.5(a)** Michael A. Palladino **F19.5(b)** Scotts Miracle-Gro Company **F19.7** Reprinted with permission from "New Genes Boost Rice Nutrients" by I. Potrykus and P. Beyer, *Science,* August 13, 1999, Vol. 285, p. 994. **F19.8** Choy Hew, National University of Singapore, and Garth Fletcher, Memorial University of Newfoundland, St. John's, Newfoundland. **F19.13** GeneChip(R) Human Genome U133 Plus 2.0 Array. Affymetrix **F19.14** Cancer Genetics Branch/National Human Genome Research Institute/NIH **F19.15** Reprinted with permission from Ash Alizadeh, *Nature Magazine* 2000: 403, pgs. 503–511, figure 4 left panel. © 2000 Macmillian Magazines **F19.16** Whole genome expression profiling. From Lorence, M. C. Application of High-Density Microarrays for Expression and Genotype Analysis in Infectious Disease, American Biotechnology Laboratory, pp. 10–12, January 2006. Figure, p.10 Matthew Lorence **F19.18(a)** Van De Silva

Chapter 20 Developmental Genetics

CO20 Edward B. Lewis, California Institute of Technology **F20.5** Jim Langeland, Stephen Paddock, and Sean Carroll, University of Wisconsin at Madison **F20.7(a)** Peter A. Lawrence "The Making of a Fly", Blackwells Scientific, 1992. **F20.7(b)** Peter A. Lawrence and P. Johnston, "Development", 105, 761–767 (1989). **F20.8** Jim Langeland, Steve Paddock and Sean Carroll. HHMI, Dept. Molecular Biology, Univ. Wisconsin **F20.9** Reprinted by permission from Macmillan Publishers Ltd.: Fig 1 on p. 244 from: *British Dental Journal* 195:243–48 2003. Greenwood, M. and Meechan, J. G. "General medicine and surgery for dental practitioners. Macmillan Magazines Limited. **F20.10** Reproduced by permission from Fig. 2 on p. 766 in *Cell* 89: 765–71, May 30, 1997. Elsevier Science Mike J. Owen. **F20.11(a-b)** Reproduced by permission from T. Kaufmann, et al., *Advanced Genetics* 27:309–362, 1990. F. Rudolf Turner, Indiana University. **F20.15** P. Barber/Custom Medical Stock Photo **F20.16** Elliot M. Meyerowitz, California Institute of Technology, Division of Biology **F20.17(a-b)** Max-Planck-Institut fur Entwicklungsbiologie **F20.19** Jose Luis Riechmann, Division of Biology, California Institute of Technology. From *Science* 2002: 295, pp. 1482–85.

Chapter 21 Genetics and Behavior

CO21 Eve. G. Stringham and E. Peter M. Candido, *Journal of Experimental Zoology* 266:227–233 (1993). Publication of Wiley-Liss, University of British Columbia. **F21.1** Photo provided by MED Associates, St. Albans, VT 05478, www.med-associates.com. **F21.2** Jonathan Flint **F21.13** ISM/CMEABG/UCBL1/Phototake NYC

Chapter 22 Quantitative Genetics and Multifactorial Traits

CO22 Joao Virissimo/Shutterstock **F22.8** Steven D. Tanksley, Cornell University

Chapter 23 Population and Evolutionary Genetics

CO23 Edward S. Ross, California Academy of Sciences **F23.7** Michel Samson, Frederick Libert, et al., "Resistance to HIV-1 infection in Caucasian individuals bearing mutant alleles of the CCR-5 chemokine receptor gene." Reprinted with permission from *Nature* [vol. 382, 22 August 1996, p. 725, Fig. 3]. © 1996 Macmillan Magazines Limited. **F23.15(a)** Murray Brilliant, "A 122.5 kilobase deletion of P gene underlies the high prevalence of oculocutaneous albinism type 2 in the Navajo population: From *American Journal Human Genetics* 72: 62–72, fig. 1 p. 65. Published by University of Chicago Press. **F23.15(b)** Murray Brilliant, "A 122.5 kilobase deletion of P gene underlies the high prevalence of oculocutaneous albinism type 2 in the Navajo population: From *American Journal Human Genetics* 72: 62–72, fig. 2 p. 66. Published by University of Chicago Press. **F23.16** Murray Brilliant, "A 122.5 kilobase deletion of P gene underlies the high prevalence of oculocutaneous albinism type 2 in the Navajo population: From *American Journal Human Genetics* 72: 62–72, fig. 3 p. 67. Published by University of Chicago Press. **F23.18** Carl C. Hansen/Nancy Knowlton/Smithsonian Institution Photo Services **F23.20** Fig. 5 in *Molecular Biology Evolution* 15 (4) pages 391–407, Kazuhiko Takahasi/Norihiro Okada. The Society for Molecular Biology and Evolution.

Chapter 24 Conservation Genetics

CO24 Scott Bauer/USDA/ARS/Agricultural Research Service **F24.2** John Mitchell/Photo Researchers **F24.3** Thomas Pickard/istockphoto **F24.4(a)** Tui De Roy/Minden Pictures **F24.4(b)** Sarah Ward/Department of Soil and Crop Services/Colorado State University **F24.5** Sarah Ward/Department of Soil and Crop Services/Colorado State University **F24.6** Alison Wilson/istockphoto **F24.9** Art Wolfe/Getty Images/Stone Allstock **F24.10** Tim Thompson/CORBIS- NY **F24.13** Jim Brandenburg/Minden Pictures **F24.14** Lynn Stone/Animals Animals/Earth Scenes.

FIGURES AND TEXT

Chapter 7 Linkage and Chromosome Mapping in Eukaryotes

p152 Stig Blixt, "Why didn't Gregor Mendel find linkage?" *Nature*, Vol. 256, p. 206, 1975. © 1975 Nature Publishing Group. Used with permission.

Chapter 10 DNA Replication and Recombination

F10.5 J. H. Taylor, *Molecular Genetics,* 9/E, Pt.1, pp. 74–75, 1963. © 1963 and renewed 1991 by Elsevier. Used with permission. **F10.6** From Sundin & Varshavsky, "Arrested segregation leads to accumulation of highly intertwined catenated dimmers," *Cell*, Vol. 25, 1981, p. 659. © 1981, Elsevier. Used with permission.

Chapter 15 Regulation of Gene Expression

F15.20 D. L. Black, "Protein diversity from alternative splicing" *Cell*, Vol. 103, fig. 1, p. 368, 2000. © 2000, with permission from Elsevier.

Chapter 18 Genomics, Bioinformatics, and Proteomics

F18.7 Analysis of the genome sequence of the flowering plant *Arabidopsis Thaliana, Nature* 408:796–815, 2000. Nature Publishing Group. Used with permission. **F18.15** S. Yooseph et al., "The Sorcerer II Global ocean sampling expedition, *PloS Biology* 5(3): 0432–0466, fig. 1, p. 0436, 2007. Used with permission. **F18.17(b)** M. P. Samanatga et al., "The t transcriptome of the sea urchin embryo," *Science* 314:96–962, f. 2, p. 961, 2006. Used with permission from AAAS.

Chapter 19 Applications and Ethics of Genetic Engineering and Biotechnology

F19.2 Reprinted with permission from Biodesign Institute, Arizona State University, Box 5001, Tempe, AZ 85287-5001. **F19.4** I. Potrykus and P. Beyer, "New Genes boost rice nutrients" *Science*, August 13, 1999, Vol. 285, p. 994 **F19.9** P. Rainard, "Tackling mastitis in dairy cows," *Nature Biotechnology* 23 (4): 430–432, 2005. Nature Publishing Group. Used with permission.

Chapter 21 Genetics and Behavior

F21.3 Adapted from S.A.G. Willis-Owen and J. Flint, "The genetic basis of emotional behavior in mice," *European Journal of Human Genetics* 14:721–728, plot 1, p. 723. 2006 Nature Publishing Group. Used with permission. **T21.1** Adapted from table 1, p. 882 in M. B. Sokolowski, "*Drosophilia*: Genetics meets behavior," *Nature Reviews Genetics* 2: 879–890, 2001. Nature Publishing Group. Used with permission. **F21.11** Adapted from B. Riley and K. S. Kendler, "Molecular genetics studies of schizophrenia," *European Journal of Human Genetics* 14:669–680, fig. 1, p. 670, 2006. Nature Publishing Group. Used with permission.

Chapter 23 Population and Evolutionary Genetics

F23.2 M. Kreitman, "Nucleotide polymorphism at the alcohol dehydrogenase locus of *Drosophilia melanogaster*," *Nature* 304(4): 412–417, table, 1, p. 413. © 1983 Nature Publishing Group. Used with permission. **F23.3** Reprinted from L. Tsui, "The spectrum of cystic fibrosis mutations," *Trends in Genetics* 8:392–398, fig. 1, 1992. © 1992 Elsevier Used with permission. **F23.22** From W. M. Fitch et al, "Positive Darwinian revolution in human influenza A viruses," *PNAS* 88:4270–73, fig. 1, 1991. Walter M. Fitch. Used with permission. **F23.23** M. Krings, et al., *Cell* vol. 90, pp. 19–30, f 7a, p. 29, 1997. © 1997, with kind permission of Elsevier. **F23.24** J. Noonan et al., "Sequencing and analysis of Neanderthal genomic DNA," *Science* 314:1113–1118, fig. 3a, p. 1115 and fig. 6, p. 1117. © 2006. Used with permission from AAAS. **p506** Caramelli et al., "Evidence for a genetic discontinuity between Neanderthals and 24,000-year-old anatomically modern Europeans," *PNAS* 100:6593–6597, fig. 2, p. 6596, 2003j. © National Academy of Sciences. Used with permission.

Index

Note: Pages in **bold** indicate figures; pages in *italic* indicate tables; *n* indicates note.

D

H

J

K

L

S

T